Prof. Dr. E. Grubert
Institut für Organische Chemie
und Makromolekulare Chemie
der Friedrich-Schiller-Universität
Humboldtstraße 10
07743 JENA

PROGRESS IN HETEROCYCLIC CHEMISTRY

VOLUME 22

Related Titles of Interest Published by Elsevier

Books

CARRUTHERS: Cycloaddition Reactions in Organic Synthesis
CLARIDGE: High-Resolution NMR Techniques in Organic Chemistry, 2nd edition
FINET: Ligand Coupling Reactions with Heteroatomic Compounds
GAWLEY & AUBÉ: Principles of Asymmetric Synthesis
HASSNER & STUMER: Organic Syntheses Based on Name Reactions, 2nd edition
KATRITZKY: Advances in Heterocyclic Chemistry
KATRITZKY, RAMSDEN, JOULE & ZHDANKIN: Handbook of Heterocyclic
    Chemistry, 3rd Edition
KURTI & CZAKO: Strategic Applications of Named Reactions in Organic Synthesis
LEVY & TANG: The Chemistry of C-Glycosides
LI & GRIBBLE: Palladium in Heterocyclic Chemistry: A Guide for the Synthetic Chemist,
    2nd Edition
MATHEY: Phosphorus-Carbon Heterocyclic Chemistry: The Rise of a New Domain
McKILLOP: Advanced Problems in Organic Reaction Mechanisms
OBRECHT: Solid Supported Combinatorial and Parallel Synthesis of
    Small-Molecular-Weight Compound Libraries
WONG & WHITESIDES: Enzymes in Synthetic Organic Chemistry

Major Reference Works

LIU & MANDER: Comprehensive Natural Products II: Chemistry and Biology
KATRITZKY & REES: Comprehensive Heterocyclic Chemistry I (CD-Rom)
KATRITZKY, REES & SCRIVEN: Comprehensive Heterocyclic Chemistry II
KATRITZKY, RAMSDEN, SCRIVEN & TAYLOR: Comprehensive Heterocyclic Chemistry III
KATRITZKY, METH-COHN & REES: Comprehensive Organic Functional Group
    Transformations
KATRITZKY, TAYLOR: Comprehensive Organic Functional Group Transformations II
TROST & FLEMING: Comprehensive Organic Synthesis

Our reference works are also available online via www.sciencedirect.com

Journals

BIOORGANIC & MEDICINAL CHEMISTRY
BIOORGANIC & MEDICINAL CHEMISTRY LETTERS
CARBOHYDRATE RESEARCH
HETEROCYCLES (distributed by Elsevier)
PHYTOCHEMISTRY
PHYTOCHEMISTRY LETTERS
TETRAHEDRON
TETRAHEDRON: ASYMMETRY
TETRAHEDRON LETTERS

Full details of all Elsevier publications: see www.elsevier.com

PROGRESS IN HETEROCYCLIC CHEMISTRY

VOLUME 22

Edited by

GORDON W. GRIBBLE
*Department of Chemistry, Dartmouth College,
Hanover, New Hampshire, USA*

JOHN A. JOULE
*The School of Chemistry, The University of Manchester,
Manchester, UK*

Amsterdam • Boston • Heidelberg • London • New York • Oxford
ELSEVIER Paris • San Diego • San Francisco • Singapore • Sydney • Tokyo

Elsevier
The Boulevard, Langford Lane, Kidlington, Oxford OX5 1GB, UK
Radarweg 29, PO Box 211, 1000 AE Amsterdam, The Netherlands

First edition 2011

Copyright © 2011 Elsevier Ltd. All Rights Reserved.

No part of this publication may be reproduced, stored in a retrieval system or transmitted in any form or by any means electronic, mechanical, photocopying, recording or otherwise without the prior written permission of the publisher.

Permissions may be sought directly from Elsevier's Science & Technology Rights Department in Oxford, UK: phone (+44) (0) 1865 843830; fax (+44) (0) 1865 853333; email: permissions@elsevier.com. Alternatively you can submit your request online by visiting the Elsevier web site at http://elsevier.com/locate/permissions, and selecting Obtaining permission to use Elsevier material.

British Library of Cataloguing in Publication Data
A catalogue record for this book is available from the British Library.

Library of Congress Cataloging-in-Publication Data
A catalog record for this book is available from the Library of Congress.

For information on all Elsevier publications
visit our web site at elsevierdirect.com

Printed and bound in Great Britain
11 12 10 9 8 7 6 5 4 3 2 1

ISBN: 978-0-08-096685-4

Working together to grow
libraries in developing countries

www.elsevier.com | www.bookaid.org | www.sabre.org

ELSEVIER   BOOK AID International   Sabre Foundation

CONTENTS

Foreword xi
Editorial Advisory Board Members xiii

1. **Recent Advances in the C-2 Regioselective Direct Arylation of Indoles**   1
   Tanya C. Boorman, Igor Larrosa

   1.1. Introduction   1
   1.2. Cross-Coupling Reactions   2
   1.3. Direct Arylation Reactions   5
   1.4. Oxidative Coupling Reactions   17
   1.5. Conclusions   19
   References   19

2. **Heterocyclic Dyes: Preparation, Properties, and Applications**   21
   S. Shaun Murphree

   2.1. Introduction   21
   2.2. Five-Membered Rings Containing Oxygen: Furanoids   22
   2.3. Six-Membered Rings Containing Oxygen: Pyranoids   24
   2.4. Five-Membered Rings Containing Nitrogen   31
   2.5. Six-Membered Rings Containing Nitrogen   42
   2.6. Five-Membered Rings Containing Sulfur   49
   References   53

3. **Three-Membered Ring Systems**   59
   Stephen C. Bergmeier, David J. Lapinsky

   3.1. Introduction   59
   3.2. Epoxides   59
   3.3. Aziridines   70
   References   79

4. **Four-Membered Ring Systems**   85
   Benito Alcaide, Pedro Almendros

   4.1. Introduction   85
   4.2. Azetidines, Azetines, and Related Systems   85

4.3. Monocyclic 2-Azetidinones (β-Lactams)	88
4.4. Fused and Spirocyclic β-Lactams	92
4.5. Oxetanes, Dioxetanes, Dioxetanones and 2-Oxetanones (β-Lactones)	94
4.6. Thietanes, β-Sultams, and Related Systems	98
4.7. Silicon and Phosphorus Heterocycles: Miscellaneous	99
References	101

5.1. Five-Membered Ring Systems: Thiophenes and Se/Te Derivatives 109

Edward R. Biehl

5.1.1. Introduction	109
5.1.2. Reviews, Accounts and Books on Thiophene Chemistry	109
5.1.3. Synthesis of Thiophenes	110
5.1.4. Elaboration of Thiophenes and Benzothiophenes	117
5.1.5. Synthesis of Thiophenes for Use in Material Science	127
5.1.6. Thiophenes Derivatives in Medicinal Chemistry	131
5.1.7. Selenophenes and Tellurophenes	134
References	136

5.2. Five-Membered Ring Systems: Pyrroles and Benzo Analogs 143

Jonathon S. Russel, Erin T. Pelkey, Sarah J. P. Yoon-Miller

5.2.1. Introduction	143
5.2.2. Synthesis of Pyrroles	143
5.2.3. Reactions of Pyrroles	151
5.2.4. Synthesis of Indoles	158
5.2.5. Reactions of Indoles	165
5.2.6. Oxindoles and Spirooxindoles	170
5.2.7. Carbazoles	171
5.2.8. Azaindoles and Carboline Analogs	172
5.2.9. Indole Natural Products	173
References	174

5.3. Five-Membered Ring Systems: Furans and Benzofurans 181

Kap-Sun Yeung, Zhen Yang, Xiao-Shui Peng, Xue-Long Hou

5.3.1. Introduction	181
5.3.2. Reactions	182
5.3.3. Synthesis	188
References	208

5.4. Five Membered Ring Systems: With More than One N Atom — 217
Larry Yet

- 5.4.1. Introduction — 217
- 5.4.2. Pyrazoles and Ring-Fused Derivatives — 217
- 5.4.3. Imidazoles and Ring-Fused Derivatives — 226
- 5.4.4. 1,2,3-Triazoles and Ring-Fused Derivatives — 235
- 5.4.5. 1,2,4- Triazoles and Ring-Fused Derivatives — 241
- 5.4.6. Tetrazoles and Ring-Fused Derivatives — 245
- References — 248

5.5. Five-Membered Ring Systems: With N and S (Se) Atoms — 259
Y.-J. Wu, Bingwei V. Yang

- 5.5.1. Introduction — 259
- 5.5.2. Thiazoles — 259
- 5.5.3. Isothiazoles — 285
- 5.5.4. Thiadiazoles and Selenodiazoles — 297
- 5.5.5. Selenazoles, 1,3-Selenadolidines and Telenazoles — 302
- References — 304

5.6. Five-Membered Ring Systems: With O & S (Se, Te) Atoms — 309
R. Alan Aitken, Lynn A. Power

- 5.6.1. 1,3-Dioxoles and Dioxolanes — 309
- 5.6.2. 1,3-Dithioles and Dithiolanes — 311
- 5.6.3. 1,3-Oxathioles and Oxathiolanes — 314
- 5.6.4. 1,2-Dioxolanes — 315
- 5.6.5. 1,2-Dithioles and Dithiolanes — 315
- 5.6.6. 1,2-Oxathioles and Oxathiolanes — 316
- 5.6.7. Three Hetero Atoms — 316
- References — 317

5.7. Five-Membered Ring Systems with O & N Atoms — 321
Stefano Cicchi, Franca M. Cordero, Donatella Giomi

- 5.7.1. Isoxazoles — 321
- 5.7.2. Isoxazolines — 325
- 5.7.3. Isoxazolidines — 327
- 5.7.4. Oxazoles — 331
- 5.7.5. Oxazolines — 335
- 5.7.6. Oxazolidines — 339
- 5.7.7. Oxadiazoles — 342
- References — 343

6.1. Six-Membered Ring Systems: Pyridines and Benzo Derivatives 349
Philip E. Alford

- 6.1.1. Introduction — 349
- 6.1.2. Pyridines — 350
- 6.1.3. Quinolines — 370
- 6.1.4. Isoquinolines — 377
- 6.1.5. Special Topic: Supramolecular Chemistry — 382
- References — 384

6.2. Six-Membered Ring Systems: Diazines and Benzo Derivatives 393
Michael M. Miller, Albert J. DelMonte

- 6.2.1. Introduction — 393
- 6.2.2. Pyridazines and Benzo Derivatives — 394
- 6.2.3. Pyrimidines and Benzo Derivatives — 401
- 6.2.4. Pyrazines and Benzo Derivatives — 413
- References — 419

6.3. Triazines, Tetrazines and Fused Ring Polyaza Systems 427
Dmitry N. Kozhevnikov, Anton M. Prokhorov

- 6.3.1. Triazines — 427
- 6.3.2. Tetrazines — 436
- 6.3.3. Fused [6]+[5] Polyaza Systems — 439
- 6.3.4. Fused [6]+[6] Polyaza Systems — 443
- References — 444

6.4. Six-Membered Ring Systems: With O and/or S Atoms 449
John D. Hepworth, B. Mark Heron

- 6.4.1. Introduction — 449
- 6.4.2. Heterocycles Containing One Oxygen Atom — 450
- 6.4.3. Heterocycles Containing One Sulfur Atom — 478
- 6.4.4. Heterocycles Containing Two or More Oxygen Atoms — 480
- 6.4.5. Heterocycles Containing Two or More Sulfur Atoms — 482
- 6.4.6. Heterocycles Containing Both Oxygen and Sulfur in the Same Ring — 483
- References — 484

7. Seven-Membered Rings — 491

Jason A. Smith, Peter P. Molesworth, Christopher J. T. Hyland, John H. Ryan

7.1. Introduction	491
7.2. Seven-Membered Systems Containing One Heteroatom	491
7.3. Seven-Membered Systems Containing Two Heteroatoms	503
7.4. Seven-Membered Systems Containing Three or More Heteroatoms	517
7.5. Seven-Membered Systems of Pharmacological Significance	518
7.6. Future Directions	526
References	527

8. Eight-Membered and Larger Rings — 537

George R. Newkome

8.1. Introduction	537
8.2. Carbon–Oxygen Rings	538
8.3. Carbon–Nitrogen Rings	541
8.4. Carbon–Sulfur Rings	545
8.5. Carbon–Selenium Rings	546
8.6. Carbon–Nitrogen–Oxygen Rings	546
8.7. Carbon–Nitrogen–Sulfur Rings	550
8.8. Carbon–Sulfur–Oxygen Rings	550
8.9. Carbon–Oxygen–Silicon Rings	551
8.10. Carbon–Oxygen–Phosphorus Rings	551
8.11. Carbon–Nitrogen–Phosphorus Rings	552
8.12. Carbon–Nitrogen–Selenium Rings	552
8.13. Carbon–Nitrogen–Sulfur–Oxygen Rings	552
8.14. Carbon–Nitrogen–Phosphorus–Sulfur Rings	553
8.15. Carbon–Nitrogen–Metal Rings	553
8.16. Carbon–Nitrogen–Oxygen–Metal Rings	555
8.17. Carbon–Nitrogen–Sulfur/Phosphorus-Metal Rings	556
References	556

Index — 563

FOREWORD

This is the 22nd annual volume of *Progress in Heterocyclic Chemistry*, and covers the literature published during 2009 on most of the important heterocyclic ring systems. References are incorporated into the text using the journal codes adopted by *Comprehensive Heterocyclic Chemistry*, and are listed in full at the end of each chapter. This volume opens with two specialized reviews. The first, by Tanya Boorman and Igor Larrosa explores 'Recent advances in the C-2 regioselective direct arylation of indoles', a topic of enormous current interest, and the second, by Shaun Murphree reviews the recent resurgence of 'Heterocyclic dyes: preparation, properties, and applications'. The remaining chapters examine the 2009 literature on the common heterocycles in order of increasing ring size and the heteroatoms present.

The Index is not fully comprehensive; however, again this year the Contents pages list all the subheadings of the chapters that we hope will considerably improve accessibility for readers. We are delighted to welcome some new contributors to this volume and we continue to be indebted to the veteran nucleus of authors for their expert and conscientious coverage. We are also grateful to our colleagues at Elsevier Science for supervising the publication of this volume, which is typeset for the first time.

We hope that our readers find this series to be a useful guide to modern heterocyclic chemistry. As always, we encourage both suggestions for improvements and ideas for review topics.

Gordon W. Gribble
John A. Joule

Editorial Advisory Board Members Progress in Heterocyclic Chemistry

2008 - 2009

PROFESSOR M. BRIMBLE (CHAIRMAN)
University of Auckland, New Zealand

PROFESSOR D. ST CLAIR BLACK
University of New South Wales
Australia

PROFESSOR M.A. CIUFOLINI
University of British Columbia
Canada

PROFESSOR T. FUKUYAMA
University of Tokyo
Japan

PROFESSOR A. FÜRSTNER
Max Planck Institut
Germany

PROFESSOR R. GRIGG
University of Leeds
UK

PROFESSOR H. HIEMSTRA
University of Amsterdam
The Netherlands

PROFESSOR D.W.C. MACMILLAN
Princeton University
USA

PROFESSOR M. SHIBASAKI
University of Tokyo
Japan

PROFESSOR L. TIETZE
University of Göttingen
Germany

PROFESSOR P. WIPF
University of Pittsburgh
USA

Information about membership and activities of the International Society of Heterocyclic Chemistry (ISCH) can be found on the World Wide Web at http://webdb.unigraz.at/~kappeco/ISHC/index.html

CHAPTER 1

Recent Advances in the C-2 Regioselective Direct Arylation of Indoles

Tanya C. Boorman, Igor Larrosa

Queen Mary University of London, School of Biological and Chemical Sciences, Joseph Priestley Building, Mile End Road, London E1 4NS
i.larrosa@qmul.ac.uk

1.1. INTRODUCTION

Indole, or benzo[b]pyrrole (Figure 1), is a privileged motif that enjoys widespread inclusion in molecules, both naturally occurring and designed, which find applications in pharmaceutical, agrochemical and materials industries.[1] Consequently, the ability to synthesise substituted indoles as easily, economically and efficiently as possible is a high-priority aim, common to a range of disciplines.[2]

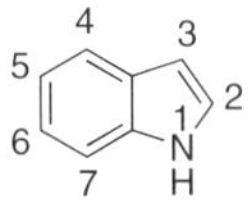

Figure 1 Numbering of the indole ring. A functionalisation can be described in terms of the position at which substitution occurs, e.g. N-1 or C-3.

Arylindoles are particularly ubiquitous substructures that can be found in many biologically active compounds.[3] This review will focus on recent methods for the synthesis of 2-arylindoles *via* arylation at C-2; excellent reviews are available that also cover arylation at C-3.[4]

Three main strategies have been adopted to date for the C-2 arylation of indoles (Scheme 1). All these approaches rely on the use of a transition metal as a catalyst, and vary depending on the level of pre-functionalization on the substrate starting materials. Firstly, there are processes that involve a traditional cross-coupling reaction between two pre-functionalized substrates, usually a metallated indole and a haloarene. Secondly, there are direct arylation methodologies where only the arene coupling partner is pre-functionalized, either as a haloarene or as an organometallic reagent. Thirdly, oxidative couplings can be used, where neither the indole nor the arene coupling partner is pre-functionalized, representing the most atom- and step-efficient methodologies.

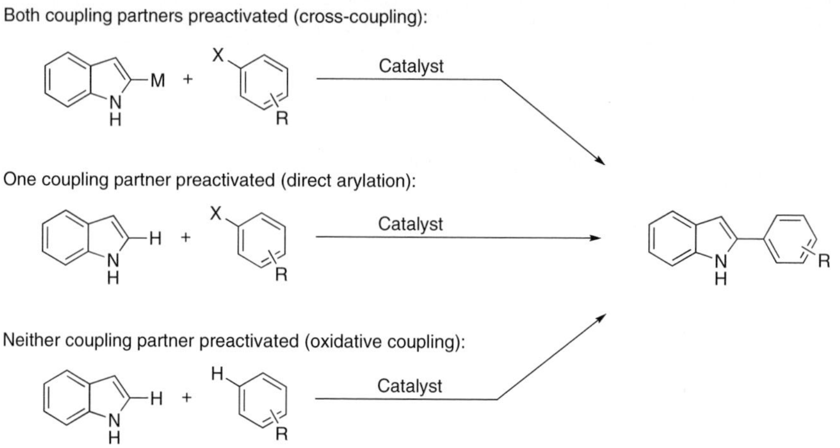

Scheme 1 Three approaches to C-2 indole arylation.

In the following sections, recent examples for these three strategies for the C-2 arylation of indoles will be presented and compared.

1.2. CROSS-COUPLING REACTIONS

1.2.1 Stille Coupling

In 1994 Labadie et al. described the Stille cross-coupling of a number of N-protected indolyltin species with aryl halides, producing a range of 2-arylindoles (Scheme 2).[5] Their method entailed installation of a tributyltin group at the C-2 position, to give an indolyltin compound, which was then employed in the Pd-catalysed cross-coupling reaction to yield the C-2 arylated indole in good yield. A further deprotection step gave the arylated free indole in excellent yield.

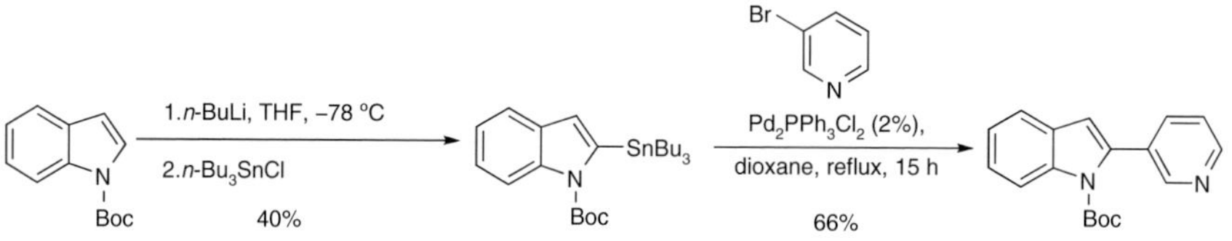

Scheme 2 Stille cross-coupling of N-protected indolyltin with 3-bromopyridine. Boc = tert-butyloxycarbonyl.

Hudkins et al. also reported the use of a Stille cross-coupling reaction to prepare a number of 2-arylindoles, but in this case using carbon dioxide as an effective yet easily removable protecting group (Scheme 3).[6] Thus, starting from indole, they prepared the N-protected indolyltin coupling partner over five steps in near quantitative yield. This indole was then coupled to an aryl halide in a Pd-catalysed process, giving the products in good to excellent yields. Conveniently, the product was readily deprotected during the work-up, avoiding further deprotection steps.

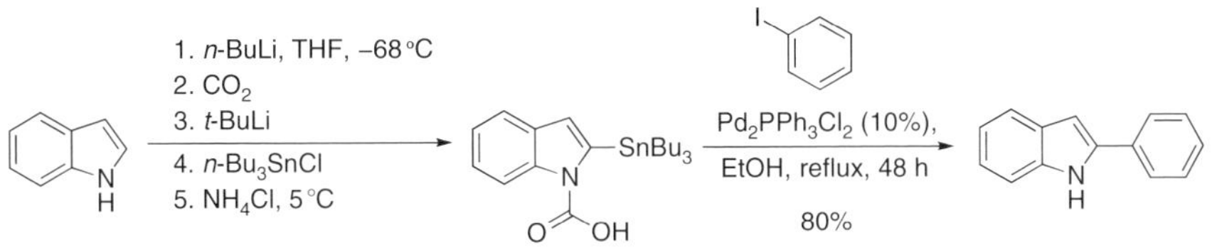

Scheme 3 Preparation of an N-protected indolyltin species and its subsequent cross-coupling.

In general, the indolyltin compound intermediates were reported to be relatively unstable, especially in the presence of traces of acid or base, requiring storage at low temperatures and to be used shortly after preparation.

1.2.2 Suzuki Coupling

In 2006, Fu et al. reported the Suzuki cross-coupling of a number of heteroarylboronic acids and (hetero)aryl halides,[7] and among them was the coupling of an N-protected indol-2-ylboronic acid and 3-bromopyridine (Scheme 4). Whereas other heteroarylboronic acids gave arylated products in good to excellent yields, the C-2 arylation of indole unfortunately proceeded with low yield and removal of the Boc protecting group, greatly hindering its general applicability.

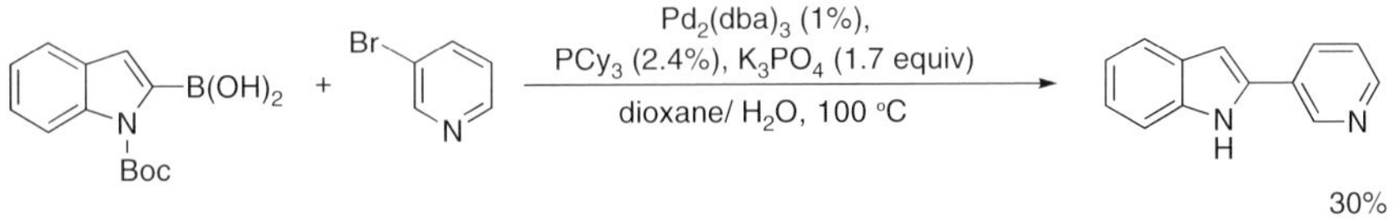

Scheme 4 Suzuki cross-coupling of an N-protected indol-2-ylboronic acid and 3-bromopyridine.

Molander et al. utilised a range of heteroaryltrifluoroborates in Suzuki cross-coupling reactions with (hetero)aryl halides (Scheme 5).[8] The use of these organoboron compounds prevents typical problems encountered with the traditionally used boronic acids, such as their limited shelf life and tendency to undergo protodeboronation during reaction. As a result, these reagents are easier to handle and large excesses of the aryl boron coupling partners are not required to achieve good yields. However, the use of indol-2-ylboronic acid is still necessary, as it is the starting material for the aryltrifluoroborate salt (prepared in 58% yield). The coupling of

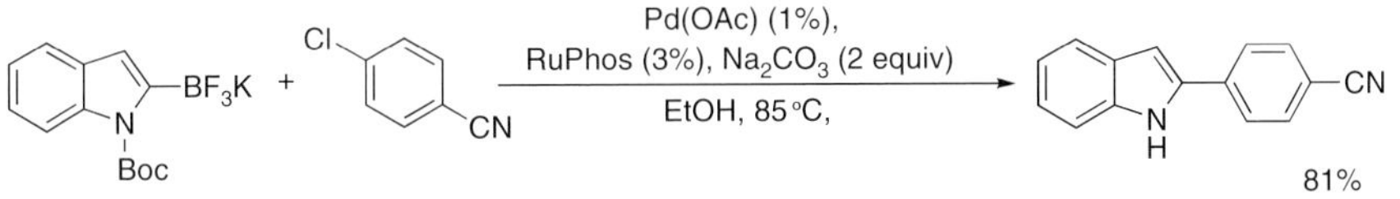

Scheme 5 Suzuki cross-coupling of potassium indolyltrifluoroborate and aryl chloride. RuPhos = 2-dicyclohexylphosphino-2′,6′-diisopropoxybiphenyl.

N-protected potassium indolyltrifluoroborate with (hetero)aryl halides was achieved, although with loss of the Boc protecting group, to give the C-2 arylated product in moderate to good yields.

1.2.3 Other Couplings

In 2000, Wang et al. reported the use of a Negishi coupling in the synthesis of 2-arylindoles for use as bidentate ligands in 4-coordinate boron complexes (Scheme 6).[9] This process, adapted from one reported by the same authors a few years earlier,[10,11] involves N-protection, lithiation and transmetallation to provide the indolylzinc chloride. This C-2 metallated indole was then cross-coupled with 8-bromoquinoline, in a Pd-catalysed process, to achieve C-2 indole arylation, which after deprotection yielded the final arylated free indole in excellent yield. This procedure has also been used by Amat et al. while exploring new routes to indolo[2,3-a]quinolizines.[12]

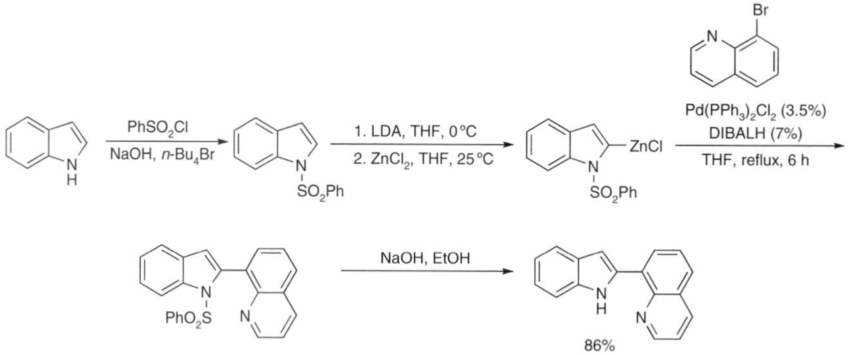

Scheme 6 Four-step procedure for the C-2 arylation of indole. LDA = lithium diisopropylamide; DIBALH = diisobutylaluminium hydride

Indium has also been used to pre-functionalise indoles at the C-2 position in a methodology reported by Sarandeses et al. (Scheme 7).[13] In this case, lithiation of the indole followed by transmetallation with $InCl_3$ generated an indolylindium species, which then participated in a Pd-catalysed cross-coupling to provide the 2-arylindole. This method was applied to a range of aryl and heteroaryl iodides, giving products in good to excellent yields and with preservation of the N-protection. The authors compared this to Stille and Suzuki couplings, as described above, which commonly lead to poor yields and N-deprotection. Another advantage of this

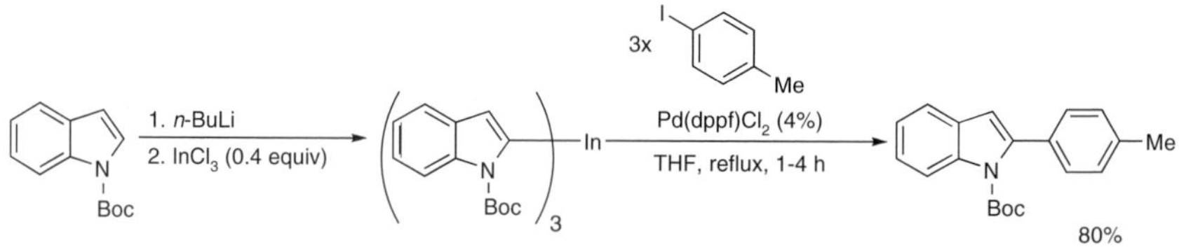

Scheme 7 Metallation, transmetallation and cross-coupling of N-Boc protected indole.

method is that the indium centre is 3-coordinate and so only a third of an equivalent of this metal is required.

All of the methods discussed so far have required the use of two preactivated coupling partners. This burdens them with inherently poor step- and atom-economy, placing them at odds with the ever-increasing trend towards 'green chemistry' as a result of the amount of waste material, time and energy that they consume. Furthermore, the metallated indole coupling partners, as many other heterocycles, tend to suffer from low stability *via* protodemetallation.

1.3. DIRECT ARYLATION REACTIONS

A far more attractive approach to the arylation of indoles involves the direct arylation of indole at the C-2 position. This removes the requirement for its pre-functionalisation thus saving synthetic steps and the stoichiometric metallic waste, leading to increased step- and atom-economy.[14] In this context, a range of methodologies have been developed in the last few decades that involve only the pre-functionalisation of the aryl coupling partner, either as an (pseudo)haloarene or as an organometallic reagent. In most of these approaches, the generally inert C–H bond at the C-2 position of indole is activated by the transition metal catalyst. This is a challenging process, as C–H bonds are strong and ubiquitous, and therefore special attention has to be paid to the regioselectivity. In general, the direct arylation of indoles can proceed either at the C-2 or the C-3 positions, and this distinction is defined by a combination of many factors, including the nature of the catalyst, solvent, ligands, additives and base used. This review will focus on the C-2 arylation of indoles but when a methodology is able to produce selectively both regioisomers it will also be mentioned.

1.3.1 Palladium-Catalysed Couplings

The first Pd-catalysed direct C-2 indole arylation was reported by Ohta *et al.* in 1985,[15] and involved the reaction of indole with chloropyrazines. This reaction was revisited and expanded two decades later by Sames *et al.*, who reported a Pd-catalysed regioselective C-2 coupling of N-substituted indoles in 2004 (Scheme 8).[16] This methodology benefits from very low catalyst loadings (0.5 mol%), fairly broad substituent compatibility and moderate to very good yields. However it suffers from competitive homocoupling of the aryl halide, a problem countered by low catalytic

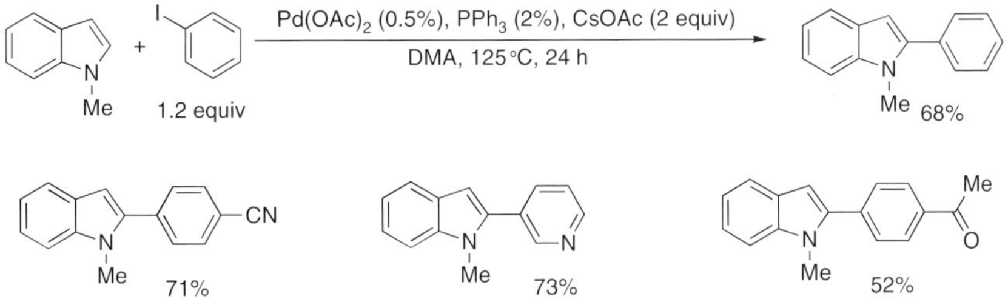

Scheme 8 Selective C-2 arylation of N-methylindole.

loading at the expense of indole scope. This method, therefore, was only suitable for indoles with alkyl or aryl N-substituents, precluding the use of N-protecting groups as a route to arylated free indoles. Interestingly, though indoles are normally nucleophilic at C-3, the 2-arylindole was obtained as the major product in all cases, although when using *ortho*-substituted iodoarenes a side reaction leading to the 3-arylindole starts to become significant.

Detailed mechanistic studies for this Pd-catalysed direct indole arylation were reported in 2005 (Scheme 9).[17] Following kinetic isotope effect (KIE) studies, the authors proposed that both C-2 and C-3 arylation pathways proceed *via* electrophilic addition of a Pd(II) species at the more electron-rich C-3 position. This C-3 palladated indole may then either undergo deprotonation and reductive elimination to give the 3-arylindole, or experience a C-3 to C-2 Pd migration, which then would lead to the C-2 arylated product. An alternative pathway to C-2 arylation may involve non-electrophilic addition of Pd at C-2, however the authors noted that this route is not supported by the observed KIE and that this pathway generally requires the presence of an *ortho*-directing group. A final possibility is a Heck-type carbometallation at C-3 leading to arylation at C-2, which the authors noted cannot be excluded.

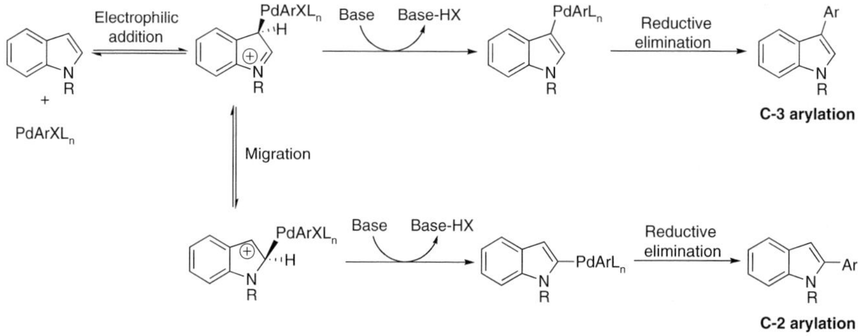

Scheme 9 Proposed mechanisms for C-2 and C-3 arylation of indoles.

The same authors later developed a method for the C-2 arylation of various N-SEM-protected azoles, including indoles, a less reactive substrate, not suited to their previous method (Scheme 10).[18] They reasoned that use of an N-heterocyclic

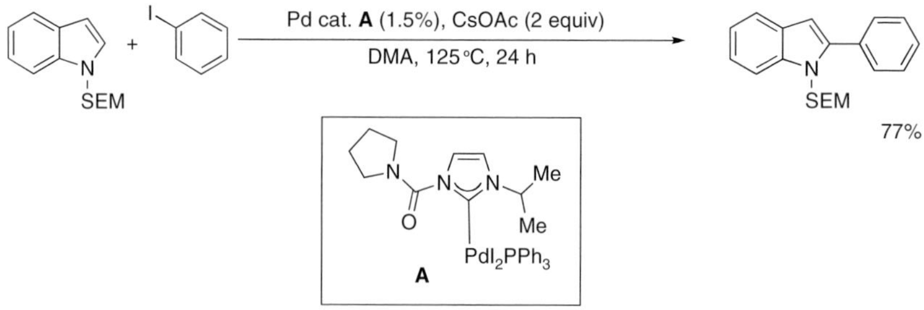

Scheme 10 C-2 arylation of SEM protected indole. SEM = 2-(trimethylsilyl)ethoxymethyl.

carbene (NHC) spectator ligand on Pd would disfavour the previously problematic homocoupling. Subsequent optimisation highlighted the importance of steric bulk rather than electronic properties for the NHC ligand in this role. Application of the optimised conditions allowed the arylation of a range of functionalised SEM-protected indoles, showing much improved selectivity over their previous system, with biphenyl production no longer an issue.

More recently, the same authors published a method for the arylation of free indoles,[19] noting the beneficial effect of a phosphine-free system (Scheme 11). They successfully arylated a range of functionalised free indoles, observing good C-2/C-3 regioselectivity and no N-arylation, with moderate to good yields. They highlighted the ability of this system to cope with sterically demanding substrates by performing the coupling of 3-methylindole and 2-iodotoluene, obtaining the arylindole in 28% isolated yield.

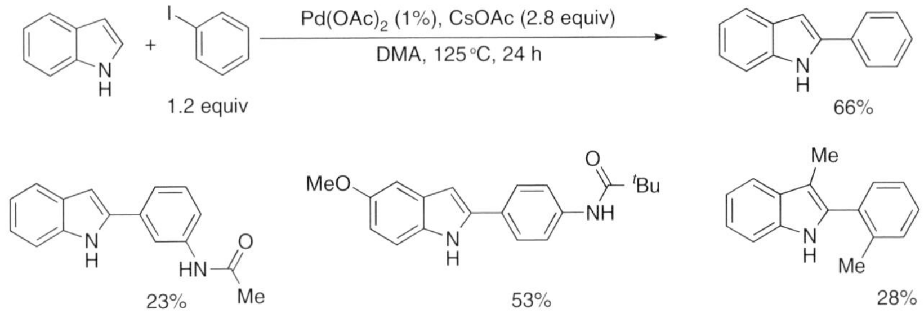

Scheme 11 Phosphine-free Pd-catalysed free indole arylation.

Miura *et al.* have also reported a methodology for the C-2 arylation of indoles that requires the presence of a C-3 carboxymethyl substituent (Scheme 12).[20] Interestingly, after the direct C-2 arylation is performed *ortho* to the ester, the resulting substrate allows for a subsequent C-3 arylation. Indeed, ester hydrolysis, followed by a decarboxylative C-3 arylation is used to afford the C-2 and C-3 bisarylated indoles with good overall yields despite the high temperatures required.

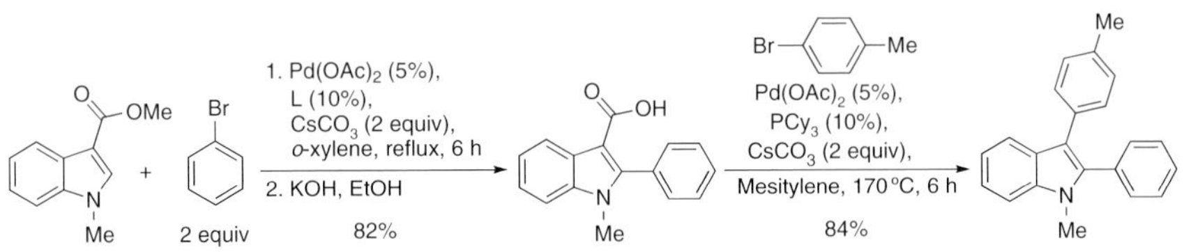

Scheme 12 Direct C-2 and decarboxylative C-3 diarylation. L = P(biphenyl-2-yl)(*tert*-butyl)$_2$.

Bellina, Rossi *et al.* have reported a base-free direct arylation of unprotected imidazoles,[21] achieving excellent yields, and eliminating the need to pre-protect the base-sensitive N-H functionalities (Scheme 13). When these conditions were applied to the C-2 arylation of free (NH-) indole, a 35% of the C-2 arylindole was obtained.

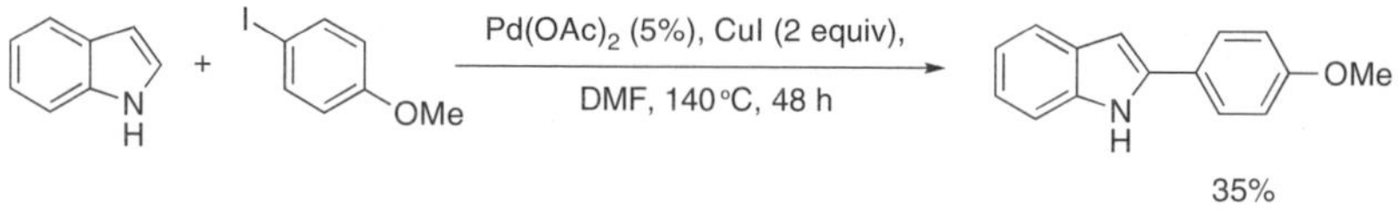

Scheme 13 Direct C-2 arylation under base-free conditions.

Bhanage et al. sought to develop a general Pd-catalysed method for the selective C-2 arylation of a range of heterocycles (Scheme 14).[22] The use of Pd(TMHD)$_2$ as catalyst provided products in good to excellent yields, furthermore the authors noted that this catalyst is very stable and not air-sensitive, in this respect, offering advantages over the phosphine-based systems typically used. Two examples were reported of the application of this method to N-methylindole, generating N-methyl-2-phenylindole and N-methyl-2-(4-nitrophenyl)indole in 96% and 67% yields, respectively.

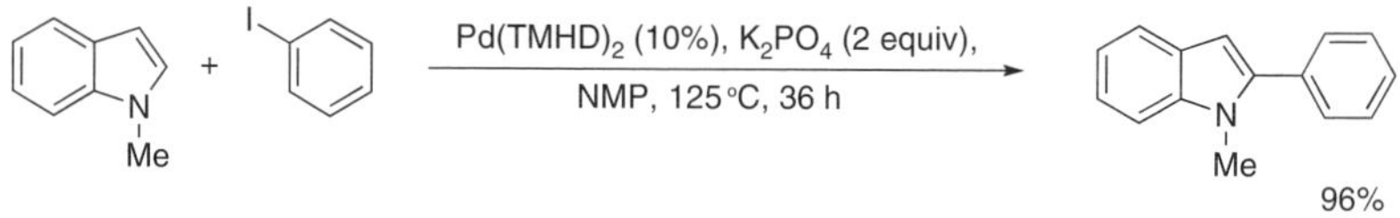

Scheme 14 Direct C-2 arylation of N-methylindole. TMHD = 2,2,6,6-tetramethyl-3,5-heptanedioate; NMP = N-methyl-2-pyrrolidine.

As part of a larger study into the use of chlorine substituents to induce regioselectivity in the direct arylation of heteroarenes, a method for obtaining C-2 arylated indole products as single regioisomers was recently described by Fagnou's group (Scheme 15).[23] Chlorination of N-methylindole at C-3 with N-chlorosuccinimide (NCS) and application of their arylation conditions provided products in good yields. Alternatively, chlorination at C-2 allowed regioselective C-3 arylation to proceed with similarly good yields. This method benefits from its simplicity. Indeed, the authors noted the ease with which the indole can be chlorinated and subsequently dechlorinated or used as a handle for further transformations.

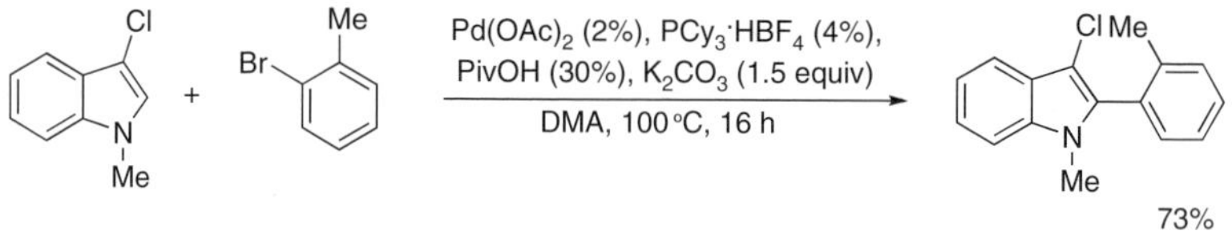

Scheme 15 The use of 3-Cl to induce regioselectivity.

Shibahara and Murai recently reported the direct arylation of a number of heteroarenes with aryl iodides, catalysed by the cationic Pd complex Pd(phen)$_2$(PF$_6$)$_2$ (Scheme 16).[24] Among the heteroarenes tested, N-methylindole gave the C-2 arylation product in excellent yield. Although the mechanism is yet to be elucidated, the authors noted that the observed KIE (1.0) indicates that the C–H bond breaking is not involved in the rate-limiting step.

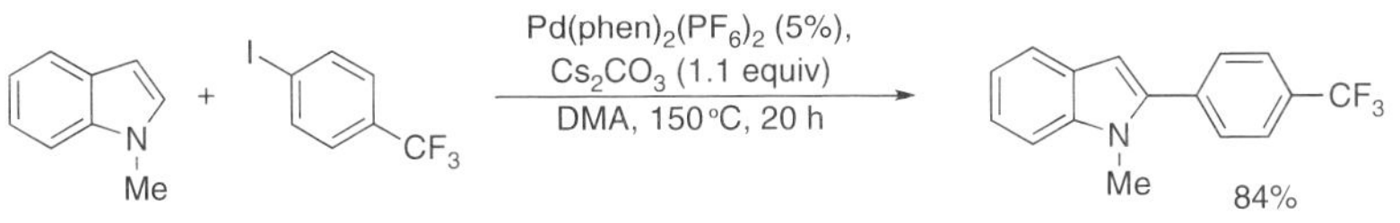

Scheme 16 The use of a cationic Pd complex. Phen = 1,10-phenanthroline.

In a methodology that takes advantage of intramolecular delivery of the arene to the C-2 position of the indole, Lautens *et al.* reported the annulation of N-bromoalkylindoles to generate six- and seven-membered ring annulated indoles (Scheme 17).[25] They applied the procedure to a number of substituted indoles and aryl iodides, with electron donating and withdrawing groups both well accommodated, producing consistently good to excellent yields.

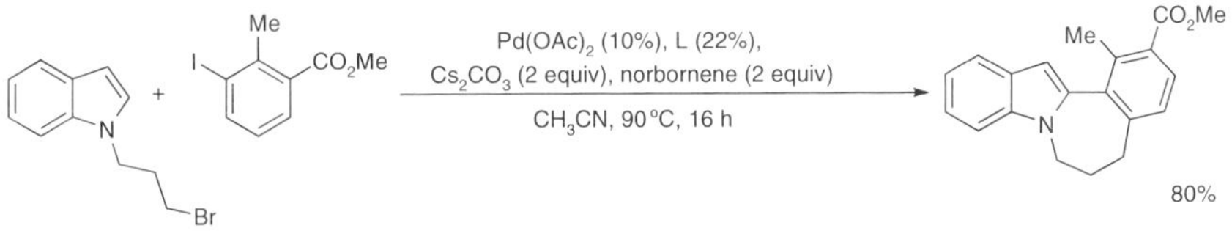

Scheme 17 Indole annulation *via* direct C-2 arylation. L = tri-2-furylphosphine.

The authors proposed an interesting tandem *ortho*-alkylation/direct C-2 arylation process (Scheme 18), where the initial Pd(II)-aryl species undergoes norbornene-assisted

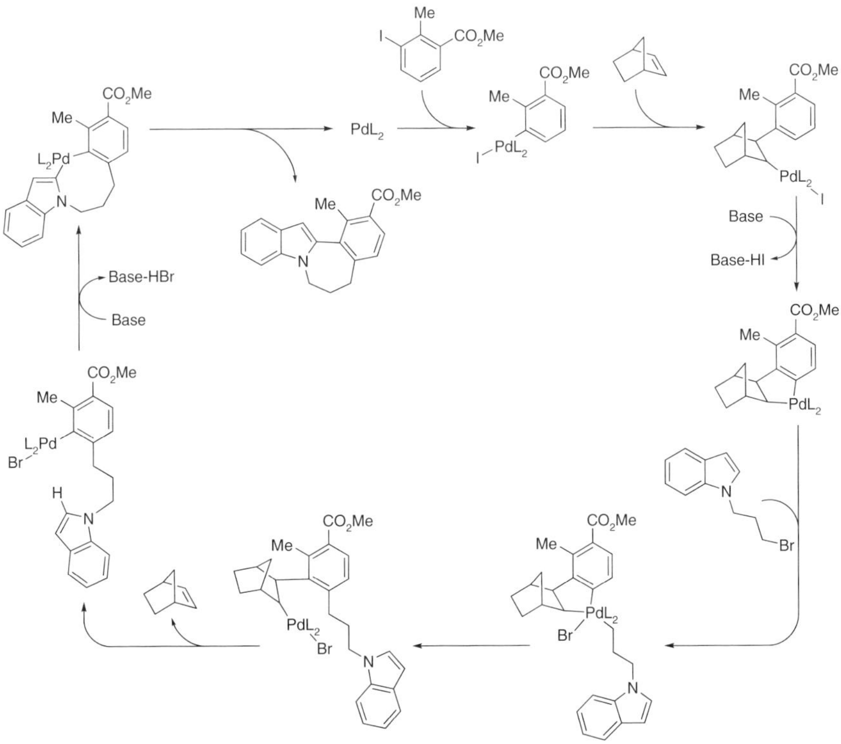

Scheme 18 Proposed reaction mechanism for the formation of annulated indoles.

ortho-alkylation *via* a Pd(IV) intermediate. The alkylated Pd(II)-aryl species thus generated performs an intramolecular C-2 arylation on the indole. The authors suggest that this occurs *via* C-3 electrophilic addition and C-3 to C-2 Pd migration, as proposed by Sames *et al.*,[17] although the possibility of a Heck-type carbopalladation cannot be excluded.

The same authors subsequently extended this method to the functionalisation of more heteroarenes and also to eight-membered ring-annulated indoles, albeit with lower yields than their six- and seven-membered counterparts.[26]

All the methodologies for C–H arylations presented so far require high temperatures and long reaction times to proceed. This in turn, limits the substrate scope and functional group compatibility of these approaches to 2-arylindoles.

In 2006, Sanford *et al.* reported the first indole arylation to proceed at room temperature (Scheme 19).[27] In their approach, both *N*-methylindole and indole itself were shown to react at temperatures in the range 25-60 °C in good to excellent yields and with high C-2 regioselectivity. This method is compatible with substituted indoles bearing a range of electron-withdrawing and -donating substituents, including bromide, a potentially useful substituent for further manipulation. However, instead of a haloarene, the more oxidising and less commercially available Ar_2IBF_4 species are required as coupling partners in this process.

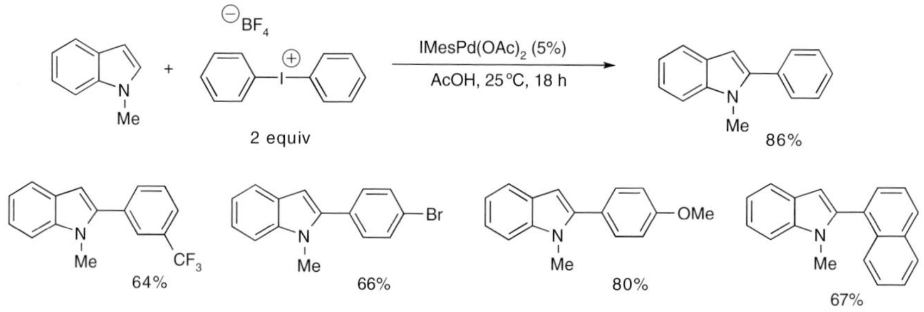

Scheme 19 Room temperature C-2 arylation of *N*-methylindole with Ar_2IBF_4. IMes = 1,3-bis(2,4,6-trimethylphenyl)imidazol-2-ylidene.

To rationalise the higher reactivity of this system, Sanford *et al.* propose a $Pd^{II/IV}$ catalytic cycle, bypassing the turnover–limiting electrophilic palladation step involved in a $Pd^{0/II}$ cycle, a process hindered by an already electron-rich Pd(II) σ-aryl species (Scheme 20).[17] Use of a comparatively electron-deficient Pd(II) catalyst provides faster electrophilic indole palladation, giving a Pd(II) σ-indole species. Oxidative addition with the diaryliodonium species yields a Pd(IV) complex, which then undergoes reductive elimination to produce the C-2 arylated indole.

In 2008, Larrosa *et al.* reported a room temperature C–H arylation of indoles that allows the use of iodoarenes as the coupling partners (Scheme 21).[28] This methodology provides C-2 arylated indoles with complete regioselectivity and good to excellent yields under remarkably mild conditions, and allows the use

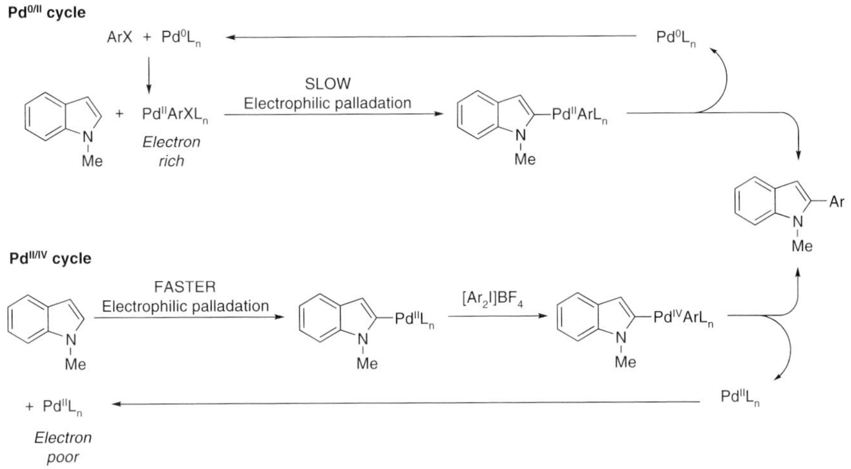

Scheme 20 Traditional Pd$^{0/II}$ catalytic cycle and proposed Pd$^{II/IV}$ cycle.

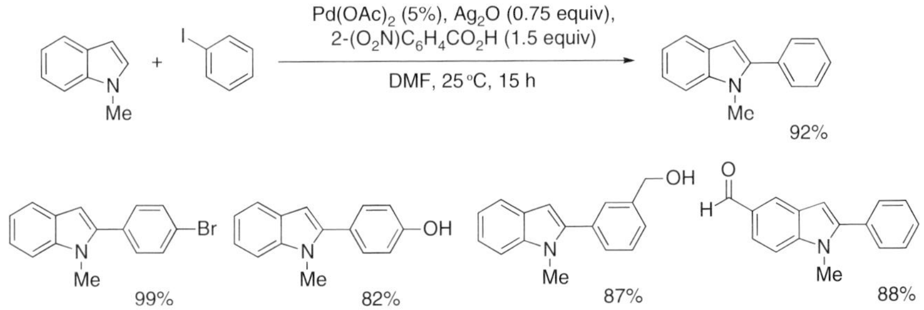

Scheme 21 Pd-catalysed direct C-2 indole arylation with ArI at room temperature.

of sensitive functional groups such as unprotected alcohols, phenols, or aldehydes. The authors also noted that the reactions are not sensitive to air or moisture.

The authors suggested that the Pd(II) species, resulting from oxidative addition with the iodoarene, can be rendered more electrophilic (Scheme 22), thus enhancing

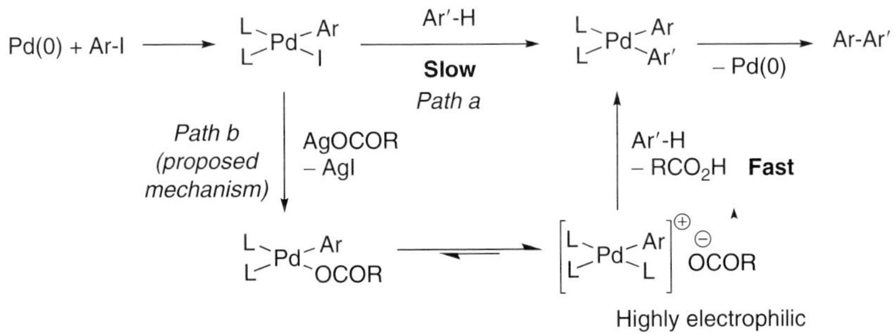

Scheme 22 Proposed mechanism of Pd-catalysed room temperature C-2 arylation of indole.

the rate of the C–H activation step, by replacing the iodine substituent with a carboxylate ligand (*path a* versus *path b*). This is accomplished by the use of an Ag-carboxylate, generated *in situ*, as the halide extractor and carboxylate donor. The Pd(II) carboxylate thus generated facilitates the electrophilic palladation of the indole, allowing the reaction to proceed under much milder reaction conditions.

As recently reported by Albericio, Lavilla *et al.*,[29] a modification of the method developed by Larrosa *et al.*[28] allowed the direct C-2 arylation of *N*-acetyl-tryptophan methyl ester with a range of substituted aryl iodides, in excellent yields and selectivity (Scheme 23). For this methodology mild reaction conditions are not simply an attractive 'selling point' but essential, with the amino acid stereocentre susceptible to racemisation. However, higher temperatures were required, presumably due to the increase in the number of coordination points in tryptophan compared to indole. The use of microwave irradiation as the heat source provided the best results, permitting excellent preservation of the amino acid stereochemistry. Subsequently, this method was applied to tryptophan-containing peptide chains, which required the adjustment of the conditions to allow for the increased fragility of these substrates. The same catalyst and additives in a buffered aqueous media, irradiated at 80 °C for 10 minutes, allowed the arylation of a range of tryptophan peptides in good to excellent yield.

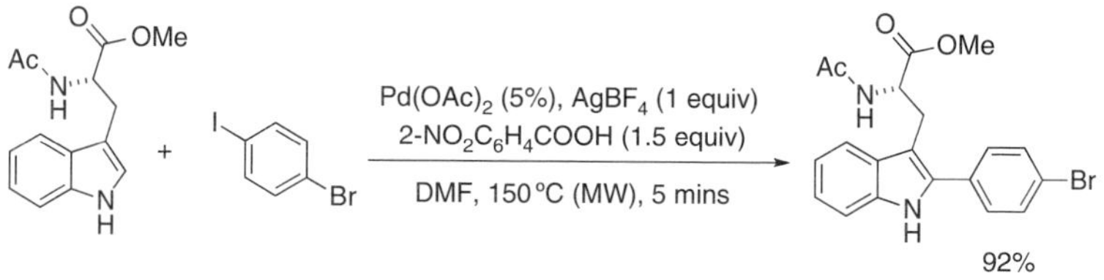

Scheme 23 Pd-catalysed direct arylation of *N*-acetyl-tryptophan methyl ester.

Azaindoles bear an obvious resemblance to indoles, sharing the same indene framework, but with an extra nitrogen. Therefore when Fagnou *et al.* turned their attention to the direct C-2 arylation of azaindoles[30] (while working towards direct arylation at the azine ring) they turned to published methods for direct C-2 indole arylation. Having surveyed a number of methods, the authors reported that, with slight temperature modification, application of the conditions developed by Larrosa *et al.*[28] generated the C-2 arylated products, with high selectivity and good yields, over a range of substituted azaindoles and aryl iodides (Scheme 24).

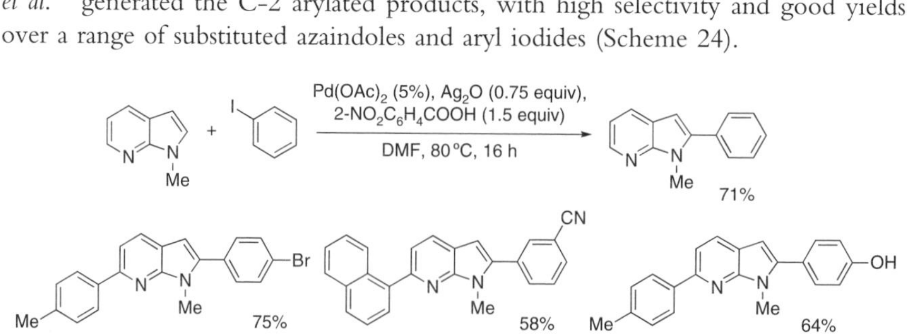

Scheme 24 Direct C-2 arylation of *N*-methyl-7-azaindoles.

A common feature of all these methodologies is that a (pseudo)haloarene is used as the coupling partner. On the contrary, Shi et al.,[31] explored the use of arylboronic acids as the coupling partners in oxidative couplings (Scheme 25): very mild conditions were reported for the arylation of a range of substituted indoles with high selectivity and good to excellent yields. Since both coupling partners are nucleophilic, an oxidant is required to complete the Pd catalytic cycle. Remarkably, oxygen was found to perform this role. Although the exclusion of aryl halides means that the reaction does not generate any halogenated waste, which is environmentally advantageous, the authors conceded that most arylboronic acids are synthesised from halide precursors.

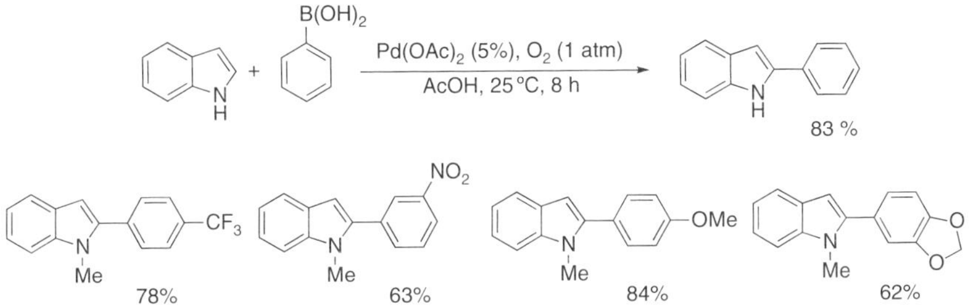

Scheme 25 Pd-catalysed direct C-2 arylation of free (N-H) and N-methylindole with phenylboronic acid.

Having previously investigated the use of 2,2,6,6-tetramethylpiperidine N-oxyl radical (TEMPO), as a mild oxidant, in the direct arylation of (hetero)arenes with arylboronic acids,[32] Studer et al. sought to extend their substrate scope to include indoles (Scheme 26).[33] They did so successfully, achieving high C-2 selectivity and good yields with both N-methylindole and indole itself. Interestingly, when the same conditions were applied to indoles N-protected with benzoyl or acetyl, a trans-2,3-disubstituted 2,3-dihydroindole was obtained. The authors propose that the reaction proceeds through a Pd$^{0/II}$ catalytic cycle, involving electrophilic indole palladation at the C-3 position and migration to C-2, as suggested by others.[17] At this point, the N-methyl or free indole intermediates are deprotonated/rearomatised and reductive elimination yields the C-2 arylated product. However, the N-protected indoles are slower to deprotonate, as a result of increased stability arising from the protecting group, leaving the cationic intermediate highly susceptible to trans-trapping by TEMPO$^-$. Subsequent reductive elimination gives an arylcarboaminoxylation product in good to excellent yields. Furthermore, easy removal of TEMPO yields a 3-hydroxydihydroindole product.

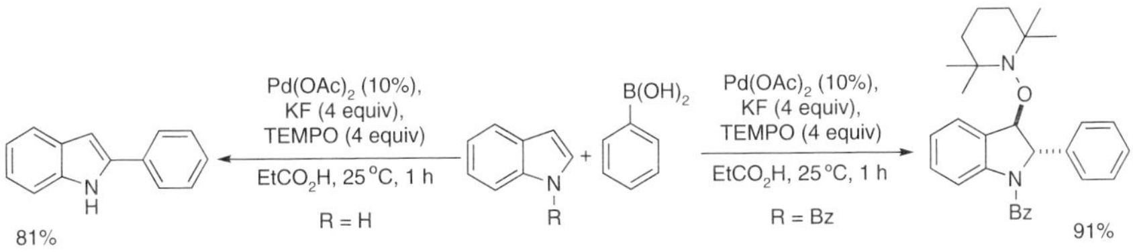

Scheme 26 C-2 arylation proceeding via divergent pathways. R = H or benzoyl

Another methodology to utilize organometallic reagents as the arylating coupling partner was reported by Zhang et al.[34] This method allows the selective C-2 arylation of free and N-methyl indoles using organotrifluoroborate salts, regarded as a more robust alternative to aryl boronic acids (Scheme 27).[35] The authors note that the reaction is not sensitive to air or moisture, in fact air is used as the oxidant, along with 10% Cu(OAc)$_2$. A wide range of substituted aryltrifluoroborate potassium salts were surveyed, giving products in moderate to excellent yields. Arenes possessing an *ortho-* or electron-withdrawing substituent produced the poorer yields, an effect far more pronounced for free indole compared to N-methylindole, which coupled to potassium 2-methylphenyltrifluoroborate in 81% yield cf. 36% for the free indole. A number of substituted indoles were also coupled, again with moderate to excellent yields with electron-withdrawing substituents leading to the lower yields.

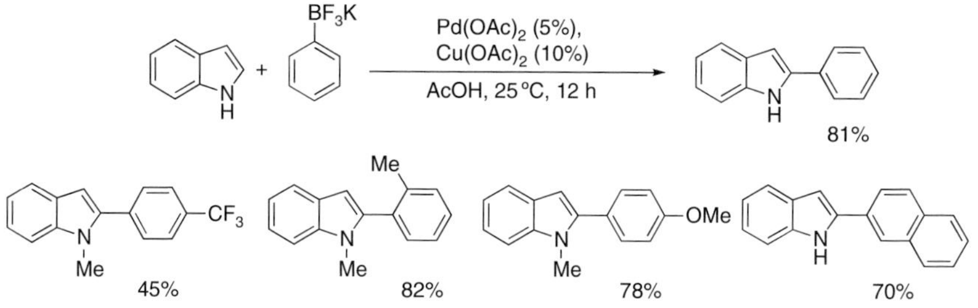

Scheme 27 Direct C-2 arylation of free (N-H) and N-methylindole with phenyltrifluoroborate salt.

1.3.2 Copper-Catalysed Couplings

Copper has also been shown to mediate the intermolecular direct arylation of indoles in a recent report by Gaunt et al. (Scheme 28).[36] Oxidative addition of the diaryliodonium coupling partner to the Cu(I) catalyst generates a highly electrophilic Cu(III)-aryl species, allowing easy electrophilic metallation of the indole. Reductive elimination of the resulting Cu(III) complex provides the arylindole. The ease of this metallation step is reflected in the mild reaction conditions employed in this coupling

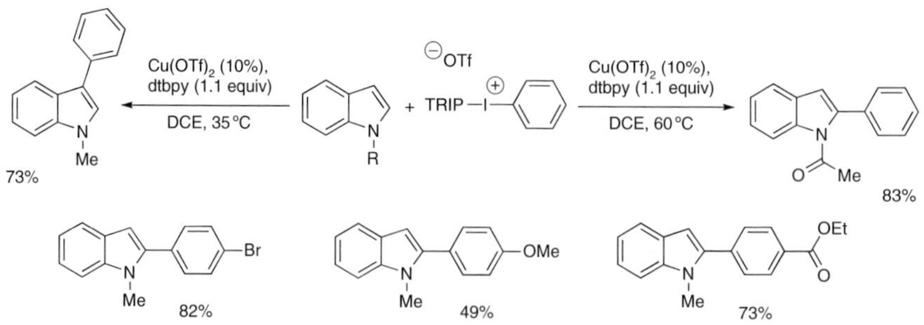

Scheme 28 Selective C-2 or C-3 arylation. TRIP = 2,4,6-tri-*iso*propylphenyl, dtbpy = 2,6-di-*tert*-butylpyridine.

(35 °C), generating the products with high selectivity and in good yields. Remarkably, this Cu-catalysed system is capable of accessing both C-2 and C-3 arylated indoles by simply changing the N-protecting group. Indeed, when N-methyl indoles are used as starting materials, C-3 arylation is observed exclusively, whereas the use of an acetyl N-protecting group leads to C-2 arylation. This is proposed to be the result of the acetyl group promoting migration of the Cu(III)-aryl species from the initial C-3 position to C-2 *via* coordination. Although this method requires the use of expensive or commercially unavailable diaryliodonium coupling partners, it provides high selectivity, good yields and also demonstrates the control available through careful design.

More recently, Joseph *et al.* reported a Cu-catalysed methodology for the C-2 arylation of N-pyridinylindoles with iodoarenes (Scheme 29).[37] Despite this process requiring the presence of an aldehyde or nitrile group at the C-3 position of the

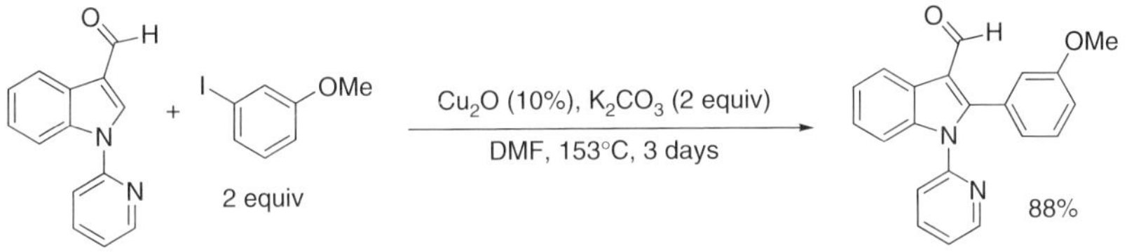

Scheme 29 Cu-mediated C-2 arylation of indole.

indole, high temperatures and long reaction times, good yields were generally obtained, and no protection was required for the aldehyde functionality.

Domínguez, SanMartin *et al.* have employed a Cu-catalysed indole arylation in their route to isoindolo[2,1-*a*]indoles.[38] The reaction proceeds *via* initial N-arylation, after which the resulting *N*-(2-halobenzyl)indole undergoes an intramolecular Cu-catalysed C-2 arylation to give isoindolo[2,1-*a*]indole (Scheme 30). This multi-step

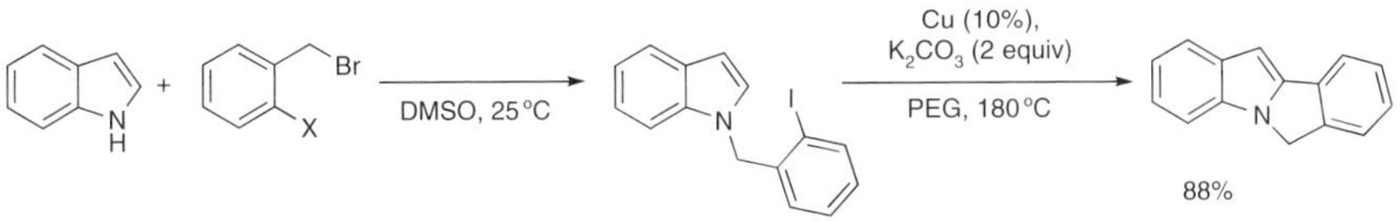

Scheme 30 Route to isoindolo[2,1-*a*]indole proceeding *via* direct C-2 arylation. PEG = poly(ethylene glycol).

procedure requires very high temperatures (180 °C) and its scope and generality were not explored.

The use of a Cu catalyst in a tandem process involving an indole arylation has recently been disclosed by Larock *et al.* (Scheme 31).[39] This impressive process involves consecutive Cu-mediated N-1–C and C-2–C bond formation. The authors

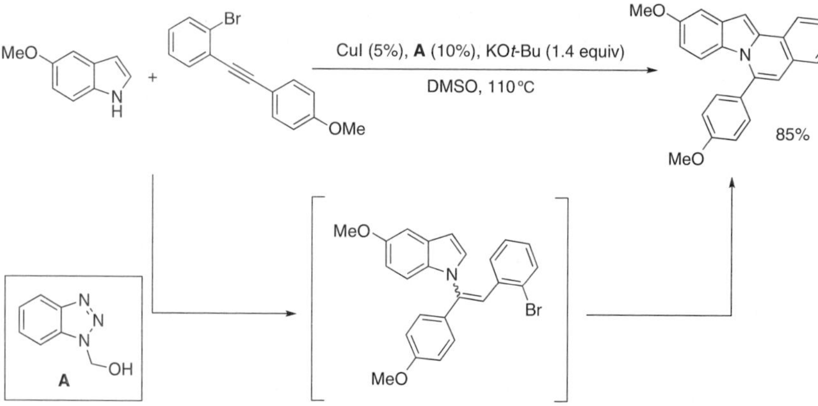

Scheme 31 Cu-catalysed tandem N and C-2 functionalisation of indole.

propose the initial generation of an enamine intermediate, which then undergoes intramolecular C-2 arylation to give the indolo[2,1-a]isoquinoline product. A wide range of functionalised indoles and haloarylalkynes were coupled with good to excellent yields. Interestingly, bromoarenes are used in this case as the aryl donor coupling partners.

1.3.3 Rhodium-Catalysed Couplings

Another transition metal, rhodium, has also been used for the C-2 arylation of indoles. Sames et al. described the highly selective (>50:1) C-2 arylation of free indoles using a Rh(I) catalyst and aryl iodide coupling partners (Scheme 32).[40] The authors proposed a concerted metallation deprotonation of the indole, assisted by a pivaloate ligand at the Rh(III) species, with subsequent reductive elimination to yield the product. This method is compatible with a range of indole substituents, including acidic ones, retaining high C-2 selectivity with moderate to good yields. The authors

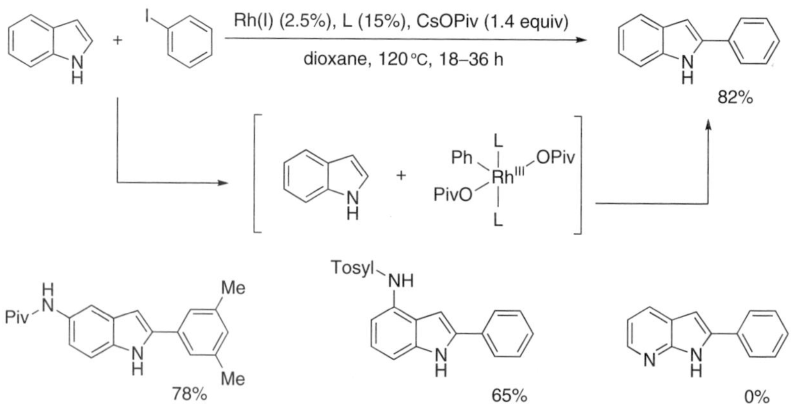

Scheme 32 Rh(I)-catalysed free indole arylation. The reaction is proposed to proceed via a 5-coordinate Rh(III) complex. Rh(I) = [Rh(cyclooctene)$_2$Cl$_2$]$_2$, L = [p-(CF$_3$)C$_6$H$_4$]$_3$P.

note, however, that the reaction did not proceed with sterically demanding substrates or in the presence of a basic nitrogen substituent, with 7-azaindole proving inert to arylation.

1.4. OXIDATIVE COUPLING REACTIONS

The use of C–H activation in previous methods for direct C-2 arylation of indole resulted in the need for only one pre-functionalised coupling partner. Continuing along this path, the ultimate application of C–H activation would be the coupling of two unactivated coupling partners, requiring no pre-functionalisation at all. Such oxidative coupling procedures benefit from greatly increased step- and atom-economy as no activating functional groups are involved. Furthermore, if oxygen were used as the terminal oxidant then the waste produced would be water. However, along with the thermodynamic and regioselectivity problems common to all C–H activation methodologies, a further challenge is present: the need for the catalyst to perform the C–H bond activation of both substrates sequentially, while avoiding

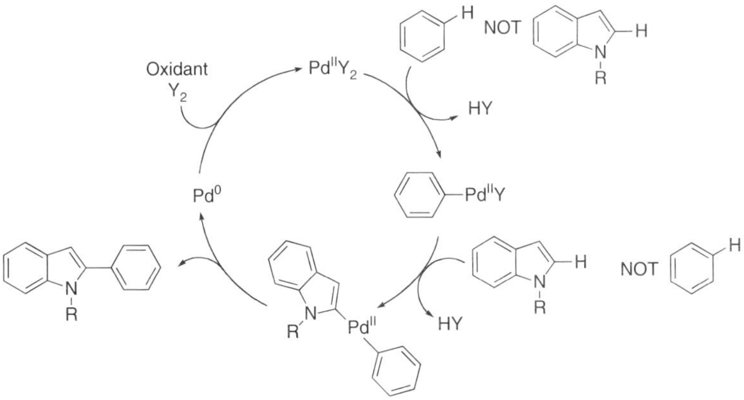

Scheme 33 Pd-catalysed oxidative coupling of benzene and indole illustrating the challenges posed by catalyst selectivity.

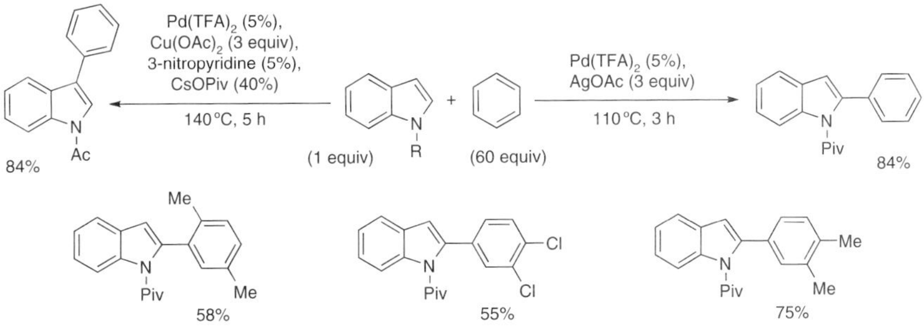

Scheme 34 Direct C-2 arylation of N-pivaloylindole with an Ag oxidant and C-3 arylation of N-acetylindole with a Cu oxidant. TFA = trifluoroacetic acid.

the homocoupling of each of them. A general mechanism for this type of transformation highlighting this problem is shown in Scheme 33.

The first example of one of these oxidative coupling reactions with indoles was reported by Fagnou et al. in 2007, achieving the selective cross-coupling of N-substituted indoles and benzenes (Scheme 34).[41] The homocoupling of the indole substrate is avoided by using an excess (30-60 equiv) of the benzene coupling partner. Remarkably, it is possible to obtain with high regioselectivity both the C-2 and the C-3 arylated indoles, just by changing the oxidant (and presumably the N-protecting group on the indole) in this cross-coupling: while the 2-arylindoles are obtained from N-pivaloyl protected indoles when using AgOAc as the oxidant, 3-arylindoles are obtained from N-acetyl indoles when using $Cu(OAc)_2$ as the oxidant. A range of substituents were examined in both the indole and the benzene coupling partners, observing generally good yields and selectivities. Notably, despite the huge excesses used for the benzene coupling partner, homocoupling was not observed for either arene, meaning that a complete reversal in catalyst selectivity is achieved.

In the same year, DeBoef also reported a double C–H activation based oxidative

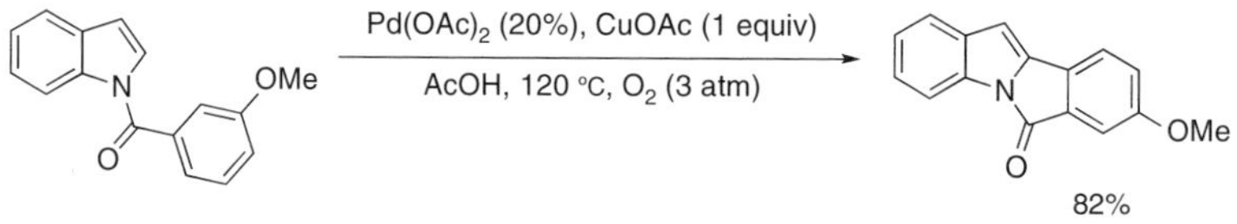

Scheme 35 Intramolecular Pd-catalysed double C–H activation.

coupling that can afford C-2 arylated indoles in an inter- or an intramolecular reaction (Scheme 35).[42] This process uses molecular oxygen as the terminal oxidant, an important step towards 'greener' chemistry. However it suffers from requirements of relatively high catalyst loading, stoichiometric Cu additive and high temperatures.

The same authors later reported more general conditions for the C-2 arylation of indoles, requiring the use of AgOAc as the oxidant (Scheme 36).[43] As before, this method benefits from involving a double C–H activation, thus avoiding the prefunctionalisation of either coupling partner. However, it suffers from the need to

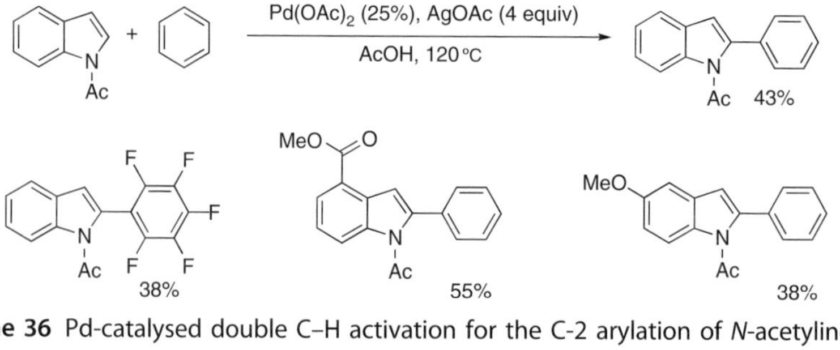

Scheme 36 Pd-catalysed double C–H activation for the C-2 arylation of N-acetylindole.

employ comparatively high catalyst loadings, excess silver oxidant and high temperatures in order to generate moderate product yields. It is, nonetheless, a significant step towards the development of these highly challenging oxidative coupling methodologies.

1.5. CONCLUSIONS

The last decade has seen truly remarkable advances in the C-2 arylation of indoles. This area has benefited from being the 'training ground' for the development of novel C–H activation based cross-coupling methodologies, which has provided an arsenal of methods with which to perform C-2 indole arylations. However, there is no room for complacency: more efficient methodologies, using lower catalyst loadings, milder conditions, avoiding pre-functionalisation of the substrates, compatible with unprotected functional groups, are still necessary and their development will undoubtedly occupy the minds of many synthetic chemists in the next decade.

REFERENCES

1. a) R.J. Sundberg, in *The Chemistry of Indoles*, Academic Press, 1970; b) R.K. Brown, in *Indoles* (Ed.: W.J. Houlihan), Wiley-Interscience, New York, 1972; c) R.J. Sundberg, in *Indoles (Best Synthetic Methods)*, Academic Press, New York, 1996; d) F.-R. Chen and J. Huang, *Chem. Rev.*, **2005**, *105*, 4671.
2. J.A. Joule, 'Indole and its Derivatives'. in *Science of Synthesis (Houben-Weyl Methods of Molecular Transformations)* (Ed.: E.J. Thomas), Vol 10. Thieme, Stuttgart, 2000; chap. 10.13.
3. S. Cacchi and G. Fabrizi, *Chem. Rev.*, **2005**, *105*, 2873.
4. a) M. Bandini and A. Eichholzer, *Angew. Chem. Int. Ed.*, **2009**, *48,* 9608; b) L. Joucla and L. Djakovitch, *Adv. Synth. Catal.*, **2009**, *351,* 673.
5. S.S. Labadie and E. Teng, *J. Org. Chem.*, **1994**, *59,* 4250.
6. R.L. Hudkins, J.L. Diebold and F.D. Marsh, *J. Org. Chem.*, **1995**, *60,* 6218.
7. N. Kudo, M. Perseghini and G.C. Fu, *Angew. Chem. Int. Ed.*, **2006**, *45,* 1282.
8. G.A. Molander, B. Canturk and L.E. Kennedy, *J. Org. Chem.*, **2009**, *74,* 973.
9. Q.-D. Liu, M.S. Mudadu, R. Thummel, Y. Tao and S. Wang, *Adv. Funct. Mater.*, **2005**, *15,* 143.
10. S.-F. Liu, Q. Wu, H.L. Schmider, H. Aziz, N.-X. Hu, Z. Popovic and S. Wang, *J. Am. Chem. Soc.*, **2000**, *122,* 3671.
11. Q.-D. Liu, M.S. Mudadu, H.L. Schmider, R. Thummel, Y. Tao and S. Wang, *Organometallics*, **2002**, *21,* 4743.
12. M. Amat, N. Llor, G. Pshenichnyi and J. Bosch, *Arkivoc.*, **2002**, *73*.
13. M.A. Pena, J.P. Sestelo and L.A. Sarandeses, *J. Org. Chem.*, **2007**, *v, 72,* 1271.
14. a) For general reviews on C–H arylation see: D. Alberico, M.E. Scott and M. Lautens, *Chem. Rev.*, **2007**, *107,* 174; (b) L.-C. Campeau, D.R. Stuart and K. Fagnou, *Aldrichimica Acta*, **2007**, *20,* 35; (c) I.V. Seregin and V. Gevorgyan, *Chem. Soc. Rev.*, **2007**, *36,* 1173; (d) L. Ackermann, R. Vicente and A.R. Kapdi, *Angew. Chem. Int. Ed.*, **2009**, *48,* 9792.
15. a) Y. Akita, A. Inoue, K. Yamamonto, A. Ohta, T. Kurihara and M. Shimizu, *Heterocycles*, **1985**, *23,* 2327; b) Y. Akita, Y. Itagaki, S. Takizawa and A. Ohta, *Chem. Pharm. Bull.*, **1989**, *37,* 1477.
16. B.S. Lane and D. Sames, *Org. Lett.*, **2004**, *6,* 2897.
17. B.S. Lane, M.A. Brown and D. Sames, *J. Am. Chem. Soc.*, **2005**, *127,* 8050.
18. B.B. Touré, B.S. Lane and D. Sames, *Org. Lett.*, **2006**, *8,* 1979.
19. X. Wang, D.V. Gribkov and D. Sames, *J. Org. Chem.*, **2007**, *72,* 1476.
20. M. Miyasaka, A. Fukushima, T. Satoh, K. Hirano and M. Miura, *Chem. Eur. J.*, **2009**, *15,* 3674.
21. a) F. Bellina, S. Cauteruccio and R. Rossi, *Eur. J. Org. Chem.*, **2006**, *63,* 1379; b) F. Bellina, C. Calandri, S. Cauteruccio and R. Rossi, *Tetrahedron*, **2007**, 1970.

22 N.S. Nandurkar, M.J. Bhanushali, M.D. Bhor and B.M. Bhanage, *Tetrahedron Lett.*, **2008**, *49*, 1045.
23 B. Liégault, I. Petrov, S.I. Gorelsky and K. Fagnou, *J. Org. Chem.*, **2010**, *75*, 1047.
24 F. Shibahara, E. Yamaguchi and T. Murai, *Chem. Commun.*, **2010**, 2471.
25 C. Bressy, D. Alberico and M. Lautens, *J. Am. Chem. Soc.*, **2005**, *127*, 13148.
26 C. Blaszykowski, E. Aktoudianakis, D. Alberico, C. Bressy, D.G. Hulcoop, F. Jafarpour, A. Joushaghani, B. Laleu and M. Lautens, *J. Org. Chem.*, **2008**, *73*, 1888.
27 N.R. Deprez, D. Kalyani, A. Krause and M.S. Sanford, *J. Am. Chem. Soc.*, **2006**, *128*, 4972.
28 N. Lebrasseur and I. Larrosa, *J. Am. Chem. Soc.*, **2008**, *130*, 2926.
29 J. Ruiz-Rodríguez, F. Albericio and R. Lavilla, *Chem. Eur. J.*, **2010**, *16*, 1124.
30 M.P. Huestis and K. Fagnou, *Org. Lett.*, **2009**, *11*, 1357.
31 S.-D. Yang, C.-L. Sun, Z. Fang, B.-J. Li, Y.-Z. Li and Z.-J. Shi, *Angew. Chem. Int. Ed.*, **2008**, *47*, 1473.
32 S. Kirchber, T. Vogler and A. Studer, *Synlett.*, **2008**, 2841.
33 S. Kirchberg, R. Fröhlich and A. Studer, *Angew. Chem. Int. Ed.*, **2009**, *48*, 4235.
34 J. Zhao, Y. Zhang and K. Cheng, *J. Org. Chem.*, **2008**, *73*, 7428.
35 G.A. Molander and N. Ellis, *Acc. Chem. Res.*, **2007**, *40*, 275.
36 R.J. Phipps, N.P. Grimster and M.J. Gaunt, *J. Am. Chem. Soc.*, **2008**, *130*, 8172.
37 C. Sagnes, G. Fournet and B. Joseph, *Synlett.*, **2009**, 433.
38 N. Barbero, R. SanMartin and E. Domínguez, *Tetrahedron Lett.*, **2009**, *50*, 2129.
39 A.K. Verma, T. Kesharwani, J. Singh, V. Tandon and R.C. Larock, *Angew. Chem. Int. Ed.*, **2009**, *48*, 1138.
40 X. Wang, B.S. Lane and D. Sames, *J. Am. Chem. Soc.*, **2005**, *127*, 4996.
41 a) D.R. Stuart and K. Fagnou, *Science*, **2007**, *316*, 1172; b) D.R. Stuart, E. Villemure and K. Fagnou, *J. Am. Chem. Soc.*, **2007**, *129*, 12072.
42 T.A. Dwight, N.R. Rue, D. Charyk, R. Josselyn and B. DeBoef, *Org. Lett.*, **2007**, *9*, 3137.
43 S. Potavathri, A.S. Dumas, T.A. Dwight, G.R. Naumiec, J.M. Hammann and B. DeBoef, *Tetrahedron Lett.*, **2008**, *49*, 4050.

CHAPTER 2

Heterocyclic Dyes: Preparation, Properties, and Applications

S. Shaun Murphree
Allegheny College, Meadville, PA 16335
smurphre@allegheny.edu

2.1. INTRODUCTION

Dyestuff chemistry is often neglected among synthetic chemists, yet the case for the scientific relevance of colorants would be difficult to overstate. Indeed, the synthesis of dyes is intertwined with the birth of modern industrial organic chemistry <06CLT235>, and some excellent treatises focus on the modern manufacture and applications of dyes <B-03MI001; B-01MI001>. Moreover, a recent resurgence in dye chemistry has been fueled by exciting applications emerging from a variety of novel and interdisciplinary research topics.

Heterocyclic compounds are used as naturally occurring <B-08M1101> and synthetic <03RPC15> food dyes; they constitute a large portion of colorants used for inkjet printing <05RPC1>, photography <02ISJ187>, and xerography <01JTA245>; and they are well represented among the commercially significant cellulosic and textile dyes <02RPC80>. Freeman and Mock have written a fine review of the large-scale synthesis of these industrially important colorants, including many heterocyclic classes <B-07MI499>.

Among smaller scale products, the so-called functional dyes show promise in potentially paradigm-shifting technologies, such as the direct splitting of water with visible light <09ACR1966>, and they are important in optoelectronic applications, such as solar cell technology <09AG(I)2474> and digital optical storage, as well as in biomedical applications, such as photodynamic therapy (PTD) and photoinduced electron transfer (PET) imaging <09IJP1>. Heterocyclic dyes are still used in traditional histochemical protocols, but they are also key components in more advanced fluorescent probes <06CLT1; 09CRV190; 10COCB64>. On a more theoretical level, solvatochromic dyes are employed as reference compounds to determine empirical solvent polarity parameters <94CRV2319>.

The realm of dyestuff chemistry is vast, and the field of heterocyclic dyes is but marginally smaller. Thus, a comprehensive review of any kind is impossible within the sweep of this chapter. Rather, the present work seeks to provide a sampling of the many interesting and timely investigations into heterocyclic dyes, focusing primarily on the literature of the past year from the perspective of a synthetic chemist. The material is organized according to ring systems using an unconventional but useful classification scheme, in which indocyanines are tagged as "pyrroloids", for example. For structures containing more than one type of ring system, an attempt is made to categorize according to some defining characteristic, although this may at times appear arbitrary. Finally, a strict delineation between dyes and pigments has been not been drawn, so as to allow for a more fluid discussion of synthetic protocols.

2.2. FIVE-MEMBERED RINGS CONTAINING OXYGEN: FURANOIDS

Benzodifuranones, exemplified by C.I. Disperse Red 356 (**1**), are used as disperse dyes in the textile industry for red shades. Introduced nearly thirty years ago by ICI <80DP(1)103>, this chromophore is still the subject of innovation. For example, a series of unsymmetrical benzodifuranones have been prepared, which are appended with stronger electron-donating groups on one side, leading to a significant bathochromic shift. Thus, p-hydroxymandelic acid (**2**) is converted to the arylbenzofuranone **4**, which is condensed with an aryltartronate (**8**) in the presence of ammonium persulfate to give the blue dye **9** <01DP(48)107>. Another strategy to shift the absorbance to longer wavelengths is to increase conjugation in the core of the chromophore (i.e., naphthodifuranones) <01DP(48)121>. These products exhibit very good dyeing properties, although hues tend to be dull <02CLT125>.

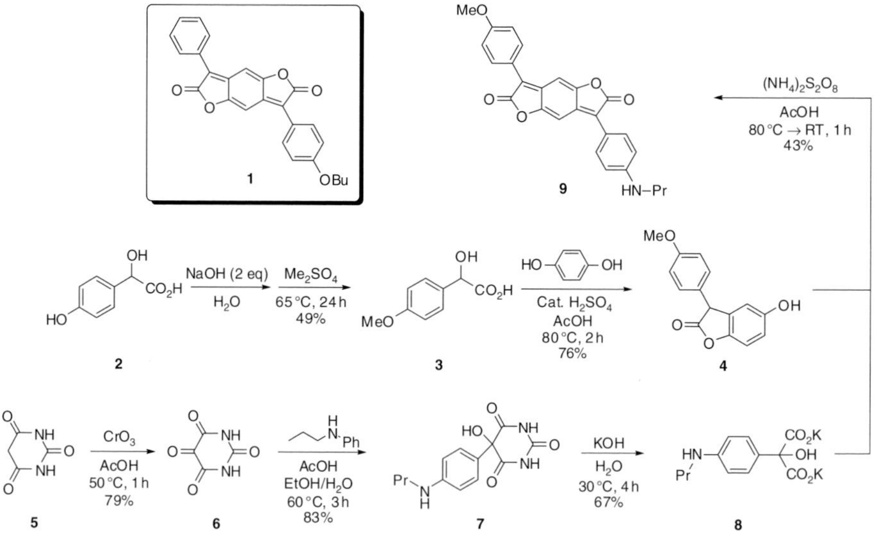

The structurally related benzodifurans (BDFs) are of interest in optoelectronic applications, and a high-yielding and flexible synthesis of tetra-aryl BDFs (**12**) has been achieved through a clever sequence of Sonogashira coupling, base-promoted double cyclization, and Negishi cross-coupling. These products have been shown to function as efficient hole-transport materials <07JA11902>.

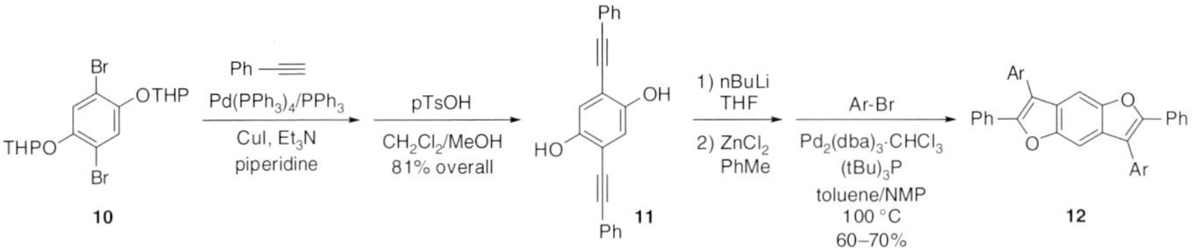

Aminobenzodifurans (**14**) can be obtained through a novel protocol by the basic treatment of aryl benzoquinones (**13**) with cyanoacetic esters <02PJC1419>. Subsequent treatment with sebacoyl chloride yields polymers (**15**) with tunable chemo-physical properties and high thermal stability <09CRC622>.

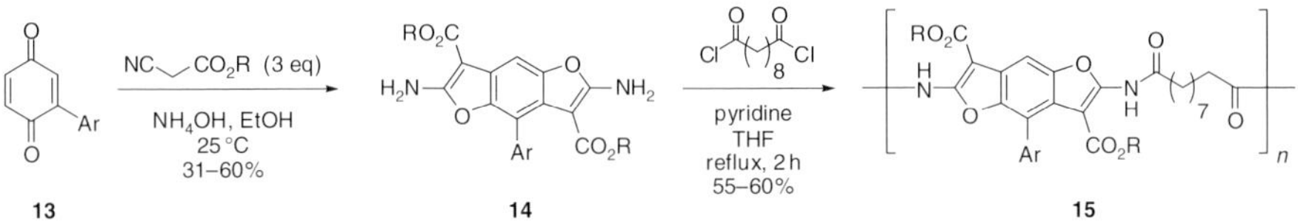

Furanoid dyes have shown promise in dye-sensitized solar cell (DSSC) applications. For example, the chromophoric compound **19** containing a benzofuran donor and cyanoacrylate acceptor exhibits a solar-to-electric conversion efficiency of 4.7% in a nanocrystalline TiO_2 solar cell <07JOC3652>. The structurally similar but more compact dye **23** can achieve efficiencies of 7.2% in an acetonitrile-based electrolyte <09JPC(C)7469>. Suzuki-type couplings have also been used to attach the furan moiety to the arylamine donor <09OL97>.

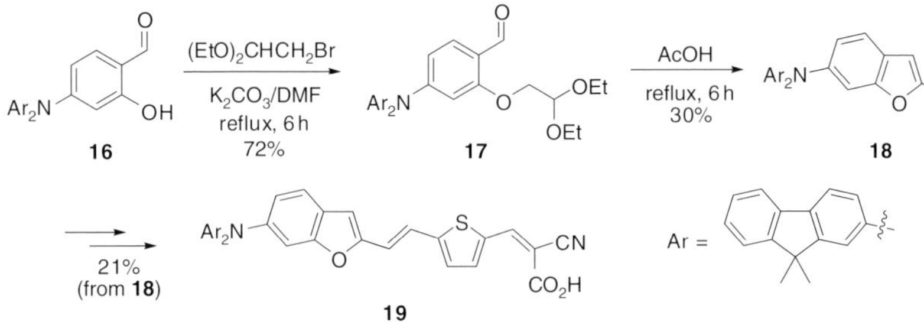

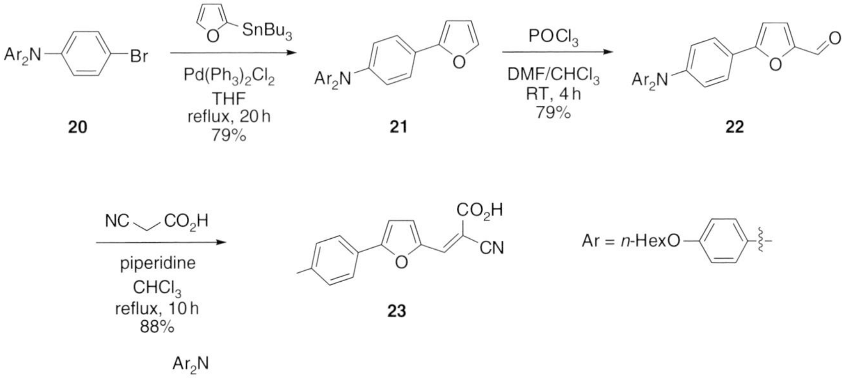

Tetracyclic naphthofuranones **25**, easily obtained from the condensation of gem-dialkyl naphthalenones **24** with m-(dibutylamino)phenol, exhibit strong blue and green fluorescence emission properties in the solid state. The steric bulk of the gem-dialkyl substituents prevent the otherwise planar chromophores from adopting a closely packed crystalline architecture. Instead, X-ray crystal structures reveal a herringbone pattern, which avoids fluorescence quenching in the solid phase <07TL5791>.

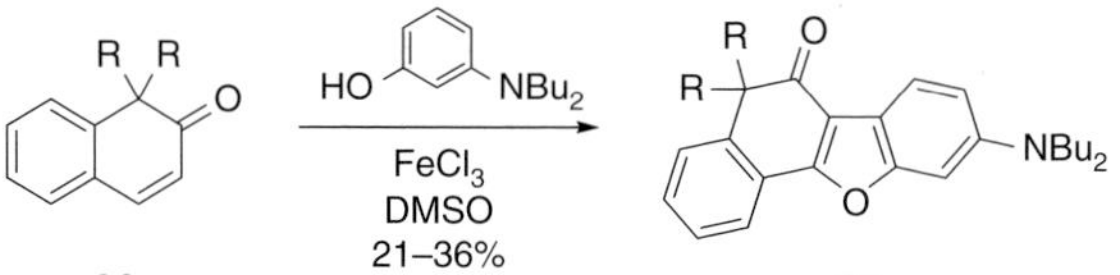

2.3. SIX-MEMBERED RINGS CONTAINING OXYGEN: PYRANOIDS

Among the more common pyranoid chromophores are the xanthene dyes, and they find broad application. For example, Fluorol Yellow 88 (**26**, C.I. Solvent Green 4) has been used as a voltage stabilizer in polyolefin electrical cable insulation <80USP001> and as tracer for leak detection in refrigerant systems <88USP001>. Rhodamine B (**27**, C.I. Basic Violet 10) has been used for decades in conventional dyeing and printing applications. Due to its fluorescent properties, it provides a very bright magenta shade. At higher pH, the cationic species is converted to Rhodamine B Base (**28**), although this equilibrium can be shifted by supramolecular hosts, such as aminomethylated resorcinarenes <09NJC1148>.

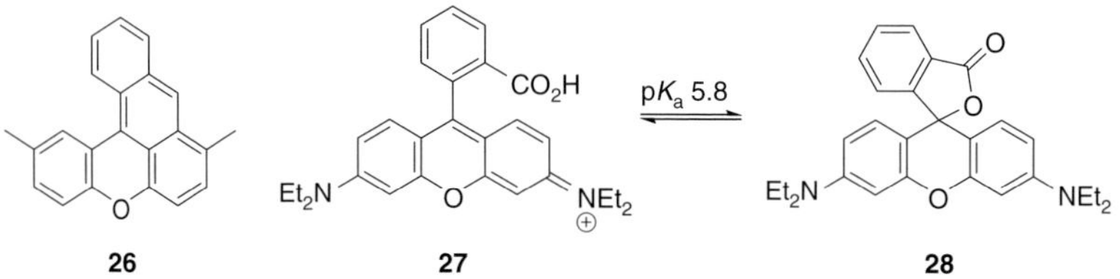

Interest in the xanthene dyes has exploded in recent years, and one piece of evidence can be found in the number of reports surrounding the construction of the basic xanthene-type skeletons in the past few months. For example, a number of catalytic systems have been developed for assembling 14-aryl dibenzoxanthenes (e.g. **30**) from aldehydes and β-naphthol, including tantalum(V) chloride <09BMCL5590>, the Brønsted acidic ionic liquid 1-methyl-3-propane sulfonic-imidazolium hydrosulfate <09DP(80)30>, tetrafluoroboric acid on a silica gel support <09JCCS381; 09SC580>, a scandium bis(perfluorooctanesulfonyl)imide complex in perfluorodecalin <09JFC(130)989>, silica supported ammonium dihydrogen phosphate under ultrasonic irradiation <09US7>, phosphorus pentoxide on alumina <10DP(85) 133> or neat <10TL422>, indium trichloride <10TL422>, and cellulosic sulfuric acid <09JMOC(A304)85>. All of these catalysts are amenable to a variety of aryl aldehydes, and additional methods are summarized in the report by Kumar and co-workers <10TL442>.

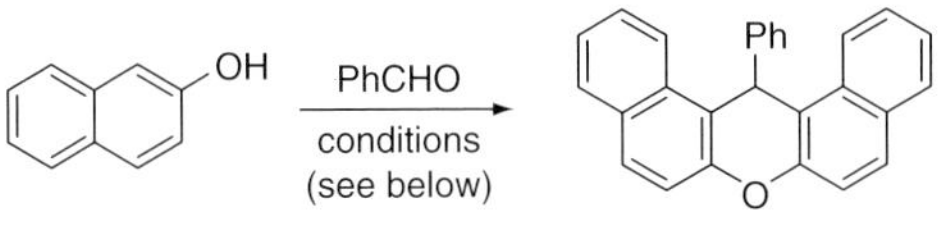

Entry	Conditions	Yield	No. of other aldehydes	Reference
1	TaCl$_5$ (10 mol%), neat, reflux, 1 h	95%	14	<09BMCL5590>
2	[MIMPS]HSO$_4$ (25 mol%), neat, 100 °C, 7 min	93%	13	<09DP(80)30>
3	HBF$_4$·SiO$_2$ (3 mol%), neat, 125 °C, 30 min	94%	11	<09JCCS381>
4	HBF$_4$·SiO$_2$ (2 mol%), neat, 120 °C, 1 h	92%	24	<09SC580>
5	Sc[N(SO$_2$C$_8$F$_{17}$)$_2$]$_3$ (1 mol%), C$_{10}$F$_{18}$, 110 °C, 5 h	93%	10	<09JFC(130)989>
6	NH$_4$H$_2$PO$_4$·SiO$_2$, H$_2$O, ultrasound, 40 °C, 40 min	88%	15	<09US7>
7	P$_2$O$_5$·Al$_2$O$_3$ (18 mol%), MW, 12 min	85%	19	<09DP(85)133>
8	P$_2$O$_5$ (20 mol%), neat, 80 °C, 45 min	88%	20	<10TL442>
9	InCl$_3$ (30 mol%), neat, 80 °C, 55 min	78%	20	<10TL442>
10	Cellulose sulfuric acid, neat, 110 °C, 1.5 h	95%	13	<09JMOC(A304)85>

Access to the xanthene system itself can be gained via the reaction of benzyne with salicylaldehyde derivatives to give xanthols (e.g. **32**) <09OL169> or with salicylideneacetone derivatives to give 9-alkylxanthenes (e.g. **33**) <10JOC506>. A tandem alkylation-cyclization protocol has also been reported, in which bromobenzylacetates (e.g., **34**) and substituted phenols are treated sequentially with iron(III) chloride and cesium carbonate under microwave conditions to give variously substituted xanthenes <10OL100>.

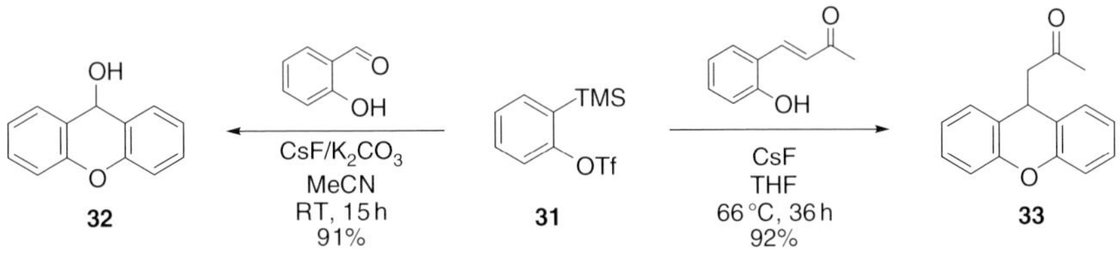

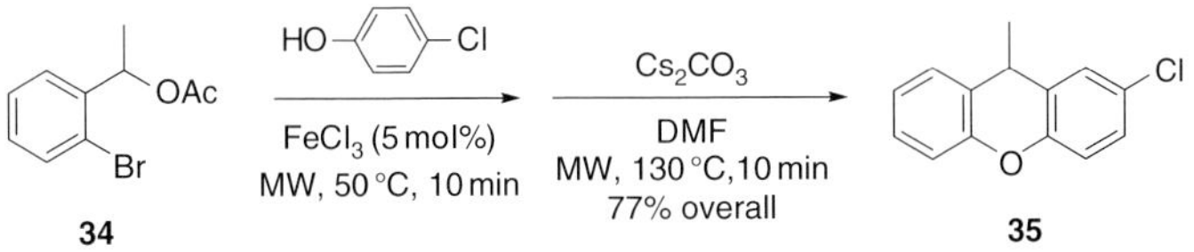

The rhodamine dyes have become highly prized templates for constructing tailor-made fluorescent probes <B-99MI001; 09CSR2410>. Just to name a few applications reported in the last year, Rhodamine B hydrazide (**36**, RBH) is easily prepared from Rhodamine B by refluxing in methanolic hydrazine. This colorless, non-fluorescent spiro compound rapidly forms the corresponding hydrazone **37** in the presence of diacetyl (or benzaldehyde), consequently restoring the fluorescent chromophore <09JF601>. The free amine functionality in RBH also allows for conjugation with other reactive or coordinating moieties to provide selective probes. For example, condensation with 5-chlorosalicylaldehyde yields a dual chemosensor (**38**) that can allow for naked eye detection of copper(II) ion and fluorescence emission detection of vanadium <10DP(86)50>.

Reaction of RBH with potassium thiocyanate in acidic medium converts the free amine functionality into a thiourea moiety to give spiro compound **39**, which can quantitatively detect copper(II) ion through a reversible coordinative mechanism <09SNA(B135)625>. Thiourea **39** can also be used to detect mercury(II) ion via an irreversible mercury-mediated desulfurization, which converts the hydrazide thiourea array into a 1,3,4-oxadiazole ring <09CHP261>. Reversible mercury detection can be achieved in the 100 nM to 10 μM range using the thiorhodamine hydrazone **40**, prepared by reaction of RBH with Lawesson's reagent followed by condensation with an aminocoumarin aldehyde. The mode of detection is believed to be through tetradentate coordination of thioamide-hydrazone sequence and the coumarin carbonyl <10ANCA(663)85>. Along similar lines of fusing two chromophoric species, a novel rhodamine-naphthalimide dyad has been synthesized by condensing RBH with a functionalized 1,8-naphthalic anhydride. The design element in this case is to use the naphthalimide moiety as a receptor, which then engages in fluorescence resonance energy transfer (FRET) with the rhodamine chromophore <09SNA(B143)42>.

The linker motif can be extended by the use of other diamines. For example, the ethylene analog of RBH (i.e. **41**) is obtained in good yield under similar conditions using ethylenediamine instead of hydrazine. Condensation of the free amine with 2,4-pentanedione provides β-ketoimine **42**, which appears to be a promising chemosensor for iron(III) ion. Detection limits are in the 0.1 μM range using fluorescence spectroscopy and the 2 μM range using UV-Vis spectroscopy. Presumably, initial binding of the iron to the β-ketoimine moiety induces opening of the spirolactam to complete the chelation, thereby creating the fluorescent chromophore <10SNA(B145)433>.

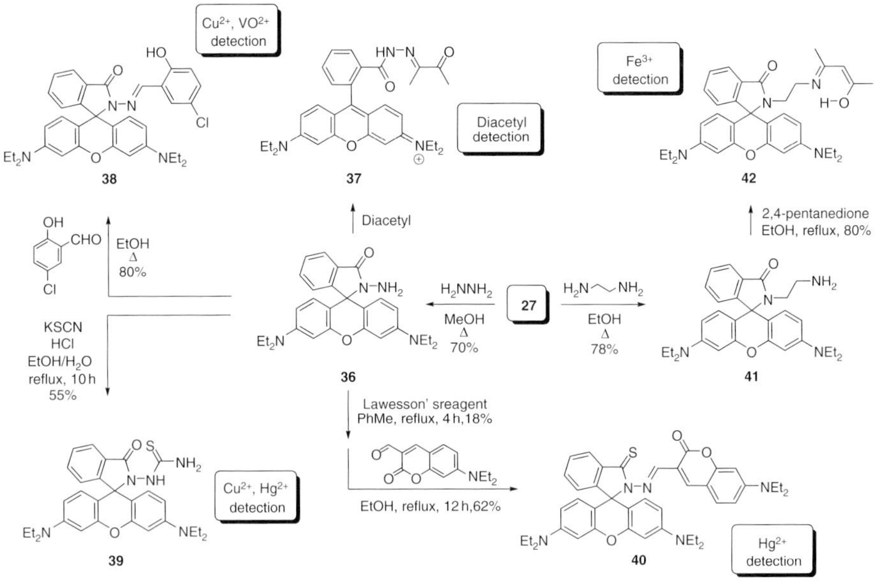

Yatzeck and co-workers <08BMCL5864> used a very clever design in the construction of a fluorescent probe (**43**) for cytochrome P450 activity in vivo. Specifically, they were interested in measuring the activity of the cytochrome isozyme CYP1A1 in lung tissue as an assay for carcinogenicity. Toward this end, they employed the "trimethyl lock" approach to temporarily cloak the fluorescence of rhodamine derivatives, whereby a sterically encumbered *o*-hydroxycinnamic acid side chain rapidly hydrolyzes upon the formation of a phenolate <05JA1652>. Since ethyl ethers are very effective substrates for CYP1A1, the phenolic oxygen was capped with an ethyl group; thus, cytochromic activity leads to rapid generation of the fluorescent species **45**.

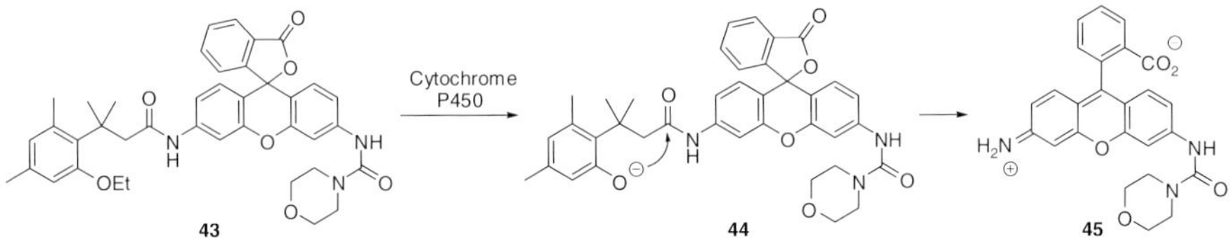

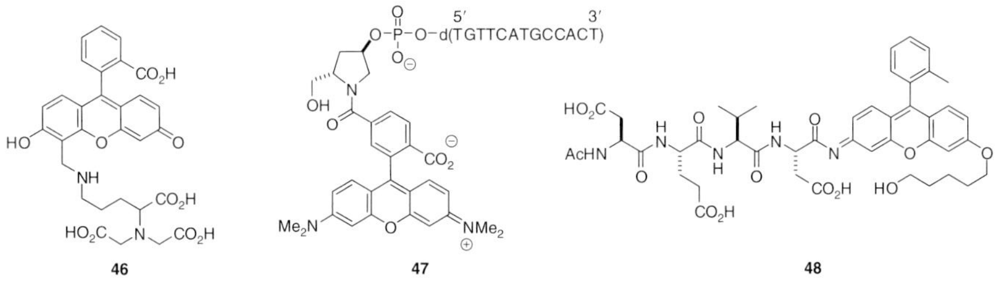

A wide variety of innovative approaches have been reported for the introduction of selective receptors onto the rhodamine-type fluorophores, including a nitrilotriacetic acid-modified fluorescein derivative (**46**) used in the labeling of a hexahistidine tagged protein <09BMCL2285>, a 5-carboxytetramethylrhodamine (or TAMRA) with an oligonucleotide appendage (i.e., **47**) investigated as a DNA probe <09BCC1673>, a dichlorofluorescein equipped with crown ether binding units for the selective detection of copper(II) ion <09DP(82)341>, and a latent chromophore (i.e. **48**) embellished with an oligopeptide for the purposes of assaying protease activity. In this latter example, a further innovation is accomplished in the design of the fluorophore itself, which is a structural hybrid of Tokyo Green and Rhodamine 110; the eastern phenolic residue (a feature of Tokyo Green) provides a handle for potential immobilization, while the fluorophorically responsive amino residue on the western end is useful for reporting purposes <09OL405>. Rhodamine derivatives have also been developed for multi-color single-molecule switching fluorescent microscopy and nanoscopy <09CEJ10762; 10CEJ158>, and fluorine-18 labeled Rhodamine B has been evaluated as a potential PET myocardial perfusion imaging agent <10ARI(68)96>.

The coumarins are another important class of pyranoid dyes. For example, C.I. Disperse Yellow 232 (**49**) <90DP(13)301> is used for dyeing polyester a bright yellow shade. Kodak's Coumarin 152 (**50**) finds application in dye lasers <85JPC294>, and the benzocoumarin splitomycin (**51**) exhibits an array of biological activity, including sirtuin inhibition <09BMC7031>.

The novel iminocoumarin dye **54**, possessing the anchoring functionality of a carboxylic acid, has been evaluated as a photosensitizer in a nanocrystalline TiO_2 solar cell <09SE(83)574>. Other chemical modifications of the coumarin system allow for the introduction of receptors for chemosensing applications, as is exemplified by the benzothiazolyl coumarin **55**, which is equipped with a dipicolylamino moiety for binding to zinc(II) ions. The extension of conjugation provided by the benzothiazole system allows for detection in the visible region. This probe was introduced into living RAW264 cells for the observation of intracellular Zn^{2+} concentrations via ratiometric fluorescence microscopy <09IC7630>. The coumarin nucleus is also central to the elaborately designed turn-on fluorescent biosensor **56**. In this system, the fluorophore is covalently attached to a protein ligand and an azoic fluorescence quencher via a labile sulfonate linkage. When the ligand binds to a target protein surface, the sulfonate undergoes nucleophilic cleavage, separating the coumarin fluorophore from the quencher. This system has shown high target specificity towards carbonic anhydrase II (CAII) inhibitors <09JA9046>. Finally, a very simple probe for gold(III) has been suggested by using the Au(III)-mediated cyclization of aryl alkynyl esters to coumarins <10OL932>.

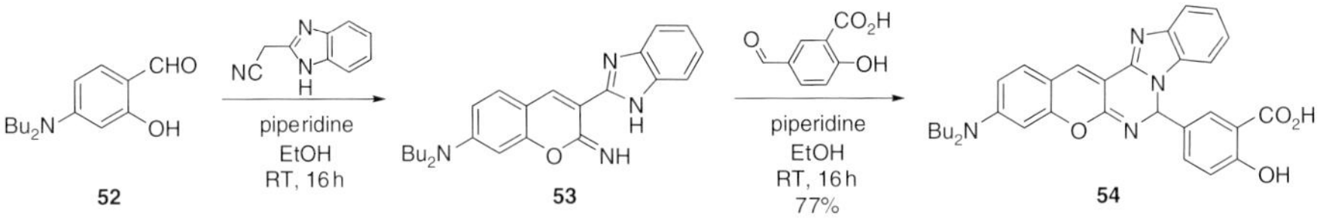

There are some important merocyanine-type dyes that incorporate the pyran moiety. For example, 4-(dicyanomethylene)-2-methyl-6-(4-dimethylaminostyryl)-4*H*-pyran (**57**, DCM1) is a laser dye which has found can be used to advantage as the red dopant in organic multilayer white LEDs <96JA1213>. The structurally similar 4-(dicyanomethylene)-2-*tert*-butyl-6-(1,1,7,7-tetramethyljulolidyl-9-enyl)-4H-pyran (**58**, DCJTB) is the subject of a Kodak patent on electroluminescent devices <99USP001> and constitutes one of the most efficient red dopants for Alq3-based OLEDs <04CM4389>. The symmetrical D-π-A-π-D dye **59** exhibits a power conversion efficiency of 1.5% when incorporated into a solution-processable organic solar cell <09JPC(C)12911>.

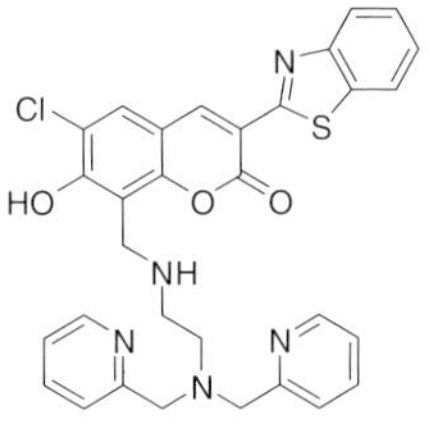

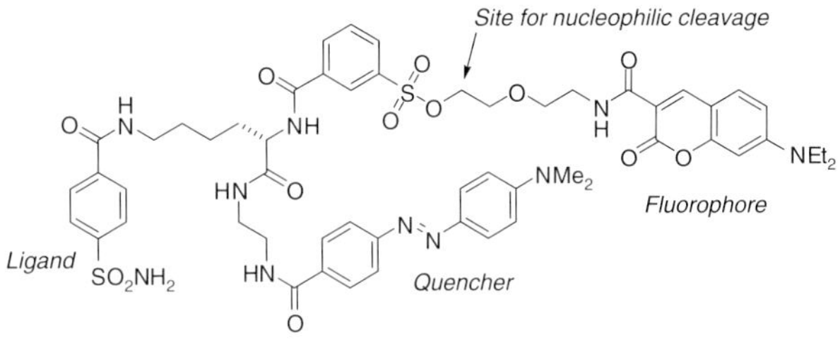

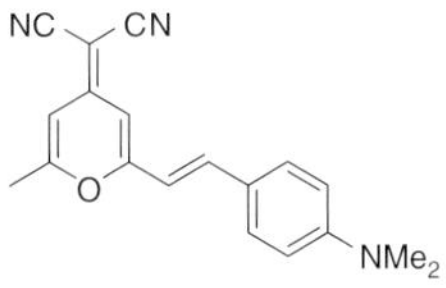

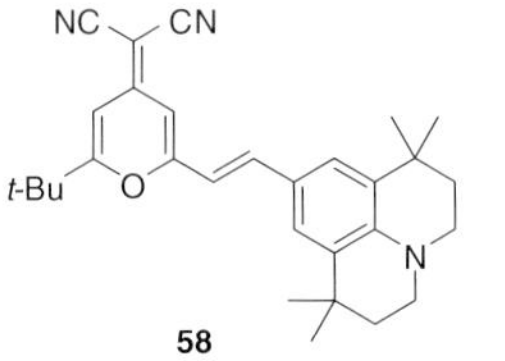

In an effort to identify a less expensive alternative to DCJTB, Zhao and co-workers <09DP(82)316> have described the synthesis of indoline analog **62**, which obviates the need for the costly tetramethyljulolidine moiety in DCJTB. Interestingly, dye **62** exhibited higher luminescent efficiency than DCJTB.

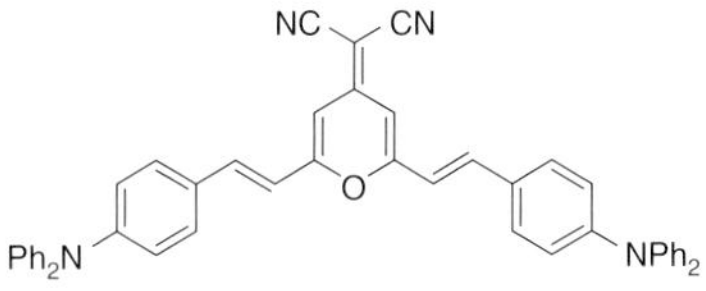

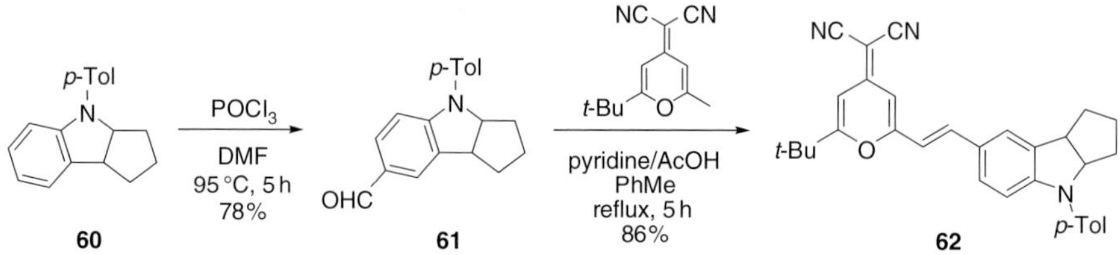

The V-shaped nonlinear optical chromophore **66** was synthesized starting with the readily available 2,6-di-*tert*-butyl-4-methylpyrylium triflate (**63**), which was first elaborated using Vilsmeier-type chemistry and then condensed onto the commercially available (2,6-dimethyl-4*H*-pyran-4-ylidene)malononitrile, providing for a more robust access to these substrates <10JOC1684>. Evidence in other symmetrical systems of this type suggests that the central pyran unit does not function as a donor itself, but rather as a spacer <09TL2920>.

A DCM-based inorganic-organic hybrid material was prepared by first synthesizing a hydroxyethyl analog (**69**), which was then covalently bound to an alkoxysilane moiety to form compound **70**. This organic component was connected to an inorganic silica network via a sol-gel process. The hybrid films thus formed exhibited photoluminescent emission around 650 nm, making them candidates for LED applications <09JSGS(52)362>. An analogous chromophore (**73**) incorporating a barbituric acid acceptor domain exhibits useful solvatochromic properties <09DP(80)314>.

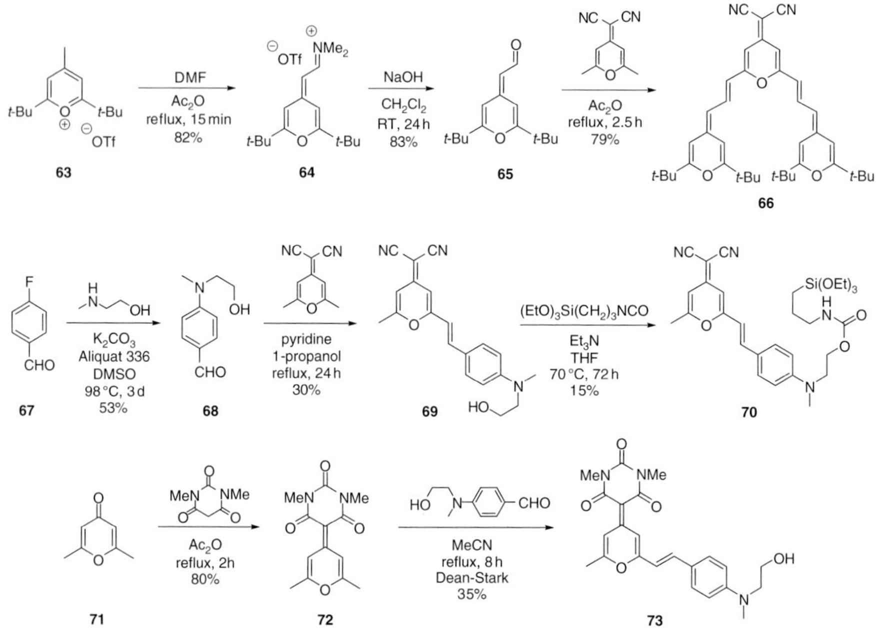

2.4. FIVE-MEMBERED RINGS CONTAINING NITROGEN

2.4.1 Rings with One Nitrogen Atom: Pyrroloids

The indolenes are a class of merocyanine dyes which terminate in an indole ring. The cyanines are among the oldest dyes, having been studied for over a century, and their utility spans the old and the new applications <00CRV1973>. For example, C.I. Basic Yellow 11 (**74**) is employed in textile dyeing and printing, paper dyeing, inkjet printing, ball-point pen applications, and more recently in optical data storage <09USP001>. The synthesis of indocyanine dyes has been recently reviewed <08THC1>; however, progress in the field continues apace. Most preparative routes construct the indole moiety using traditional methodologies for accessing indoles, typically the Fischer indole synthesis <00JCS(P1)1045>.

Kiyose and co-workers <09CEJ9191> designed a near-infrared ratiometric fluorescent FRET probe for pH detection based on two merocyanine units connected by a rigid spacer. The synthesis launches from 2,3,3-trimethyl-3*H*-indole (**76**), a common precursor available from the reaction of phenylhydrazine (**75**) with isopropyl methyl ketone under conventional conditions <02JHC391> or in ionic Brønsted acidic ionic liquid <07EJO1007>. After quaternization with 3-iodopropionic acid, the resulting indolenium species **77** undergoes condensation onto malonaldehyde dianilide to give the symmetrical indocyanine **79**, the FRET donor component. Further elaboration via one of the carboxylate functionalities yields the dyad **80**, which exhibits efficient FRET at neutral pH (85% at pH 6.8) with a marked reduction of efficiency in acidic medium (8.1% at pH 2.1). The observed pK_a of 4.9 could be advantageous for in vivo measurements. Similar architectural approaches have been used in designing NIR probes for human serum albumin <09JHC925> and for measuring oxidative stress within cells <10JA2795>.

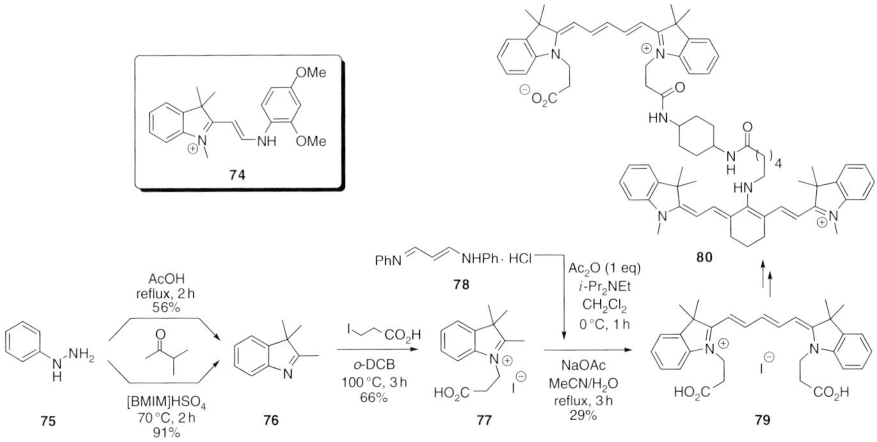

Malonaldehyde dianilide is the usual 3-carbon polyene chain precursor for the pentamethine (or Cy5 type) indocyanines; however, there have been some innovations with regard to polymer-bound reagents. Thus, for the synthesis of water-soluble dyes, a PEG-supported polyene precursor (**85**) has been developed and applied to the synthesis of both symmetrical and non-symmetrical dyes. In the latter case, the supported reagent **85** offers distinct advantage with respect to purification of the intermediate singly condensed species <09T5257>. An alternative approach has been described using a "catch and release" protocol, in which the singly condensed dye is captured by a Merrifield resin-bound sulfonyl chloride (PS-SO$_2$Cl) and then released by reaction with the second indolene substrate <05JOC2939>.

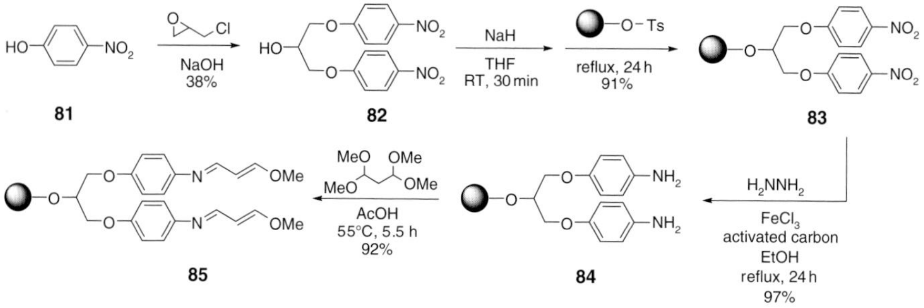

The benzo analogs of the indolenes have recently been examined with respect to the impact of dye structural elements on fluorescent properties. Thus, commercially available benz[*cd*]indol-2(1*H*)-one (**86**) was converted to the 2-(methylthio)benzo[*cd*]indolium derivative **88** through treatment with Lawesson's reagent and subsequent methylation according to an existing procedure <60JCS1537>. Addition of the 2-methyl substituent was accomplished by reaction with the anion of Meldrum's acid followed by quaternization of the nitrogen and hydrolysis. Condensation onto formaldehyde then provided the trimethine dye **92**. The fluorescent lifetimes of these benzoindolium derivatives proved to be about 30% lower than their indolium counterparts <08JPCP(A200)438>. An alternative route to the initial 2-methylation, which avoids the use of Lawesson's reagent, is available via Grignard chemistry. For example, benz[*cd*]indol-2(1*H*)-one (**86**) was treated with 1,4-butanesultone to give the potassium sulfonate **93**, which was converted to the tetrabutylammonium salt (**94**) by the addition of tetrabutylammonium chloride. Reaction with methyl Grignard and subsequent elimination of hydroxide provided the benzoindolium substrate **95** in very good yield, which was used to generate a variety of visible and NIR dyes <09JHC84>.

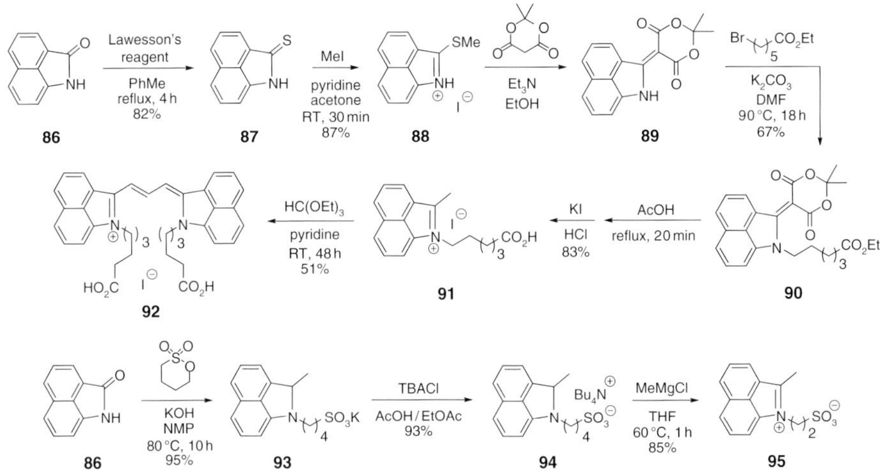

Having been studied for well over a century, symmetrical indocyanines have a long and august heritage. Although relative newcomers, unsymmetrical dyes incorporating two different heterocycles are of increasing interest as chemosensors, due in part to their heightened sensitivity to environmental conditions <07JA5710>. The architectural flexibility gained by exiting the symmetrical paradigm allows for fine-tuning of various physical and chemical properties, including solubility and fluorescence parameters. The general synthetic strategy typically involves the stepwise addition of each indolium salt onto a suitable polyene precursor, as demonstrated by the preparation of the pentamethine dye **98** containing a carboxylated indole and a benzo[*e*]indole moiety. In combination with other cyanine dyes, this chromophore exhibited a power conversion efficiency of 3.0% in a nanocrystalline TiO_2 dye-sensitized solar cell <09SM1028>.

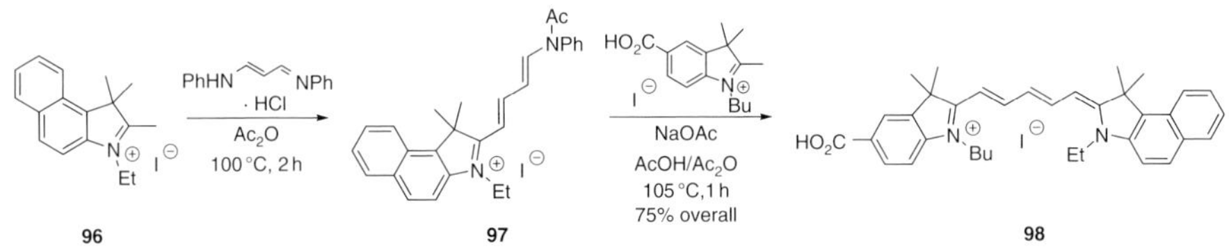

The polymethine bridge can also be used to graft indolenes onto other heterocyclic subunits to provide unique properties. For example, the pyrimidine-fused benzoindolene **99**, derived from the reaction of a symmetrical cyanine with barbituric acid, is useful pH sensitive NIR fluorescent probe with a pK_a of 3.5 <09OL29>. The coumarin-hemicyanine hybrid **100** has been developed as a far-red emitting fluorogenic probe for penicillin G acylase (PGA) with high stability towards biological thiols <08OL4175>.

Even symmetrical cyanines can be modified to extend their functionality. For example, the 5-carboxylate groups on a Cy5 dye were converted to TEMPO-type residues bearing nitroxide radical centers. The resulting hybrid dye (**101**) exhibited a twofold increase increase in fluorescence intensity when treated with sodium ascorbate, making it an intriguing candidate for probing reactive oxygen species in vivo <09SA(A71)2030>.

99-H **99**

100

101

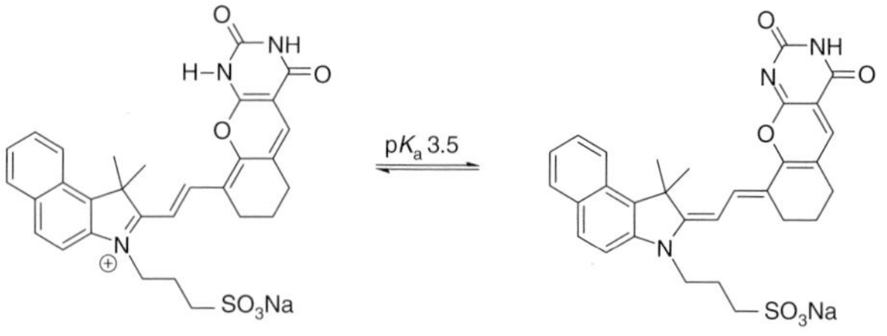

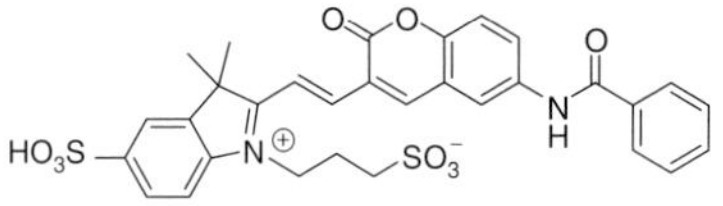

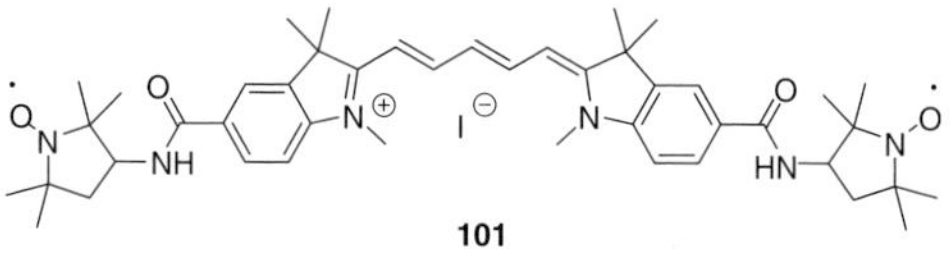

The remarkably stable squarylium moiety confers some interesting and useful properties to chromophores, and the squarylium dyes often incorporate heterocyclic components as donors and/or acceptors <01CLT61; 05ACR449; 08THC133>. For example, Song and co-workers <09DP(82)396> prepared novel water-soluble symmetrical squarylium dyes (**104**) by condensing indolium compounds onto squaric acid, which becomes part of the polymethine backbone. In studies of these dyes, it was found that water-soluble dyes with benzyl *N*-substituents (i.e. **104a**) exhibit greater photostability than their methylated cousins, and changing the *N*-substituent in less polar dyes to a bulkier alkylcarbamoylmethyl group (i.e. **104b**) inhibits aggregation and thereby enhances fluorescence and photostability in water <10DP(85)43>. Panchromatic squaraines for dye-sensitized solar cell applications have been constructed by equipping structures of type **104** with a *N,N*-diethylthiobarbituric acid residue condensed onto the squaric carbonyl functionality <09CHSC621>.

102a $R^1 = SO_3K$ **103a** $R^2 = Bn$ **104a** $R^1 = SO_3K$; $R^2 = Bn$ (20% overall)

102b $R^1 = H$ **103b** $R^2 = CH_2CONEt_2$ **104b** $R^1 = H$; $R^2 = CH_2CONEt_2$ (21% overall)

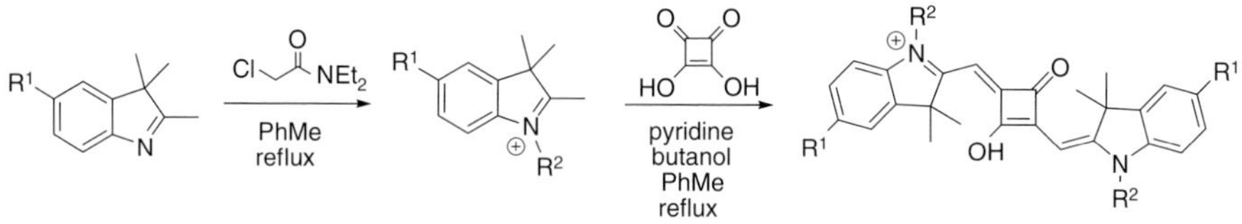

The indolospirobenzofurans constitute a fascinating group of latent indolene fluorophores that are compact and easily accessible. For example, Roxburgh and co-workers <09DP(83)31; 09DP(82)226> prepared a series of these compounds with various substitution patterns to study the equilibrium between the leuco spirohemiaminal form (**109**) and the open-chain zwitterionic chromogenic (or merocyanine) form (**110**). In the case of the unsubstituted variant (**109a**), the leuco form predominates at room temperature, with ring opening becoming more prevalent as temperature increases. However, by introducing steric crowding via the pyran ring (**109b**) or the indole ring (**109c**) more merocyanine form is observed at room temperature, presumably because the latter form relieves the steric tension. These results suggest that photodynamic control over these chromophores can be exerted by careful consideration of predictable non-bonding interactions.

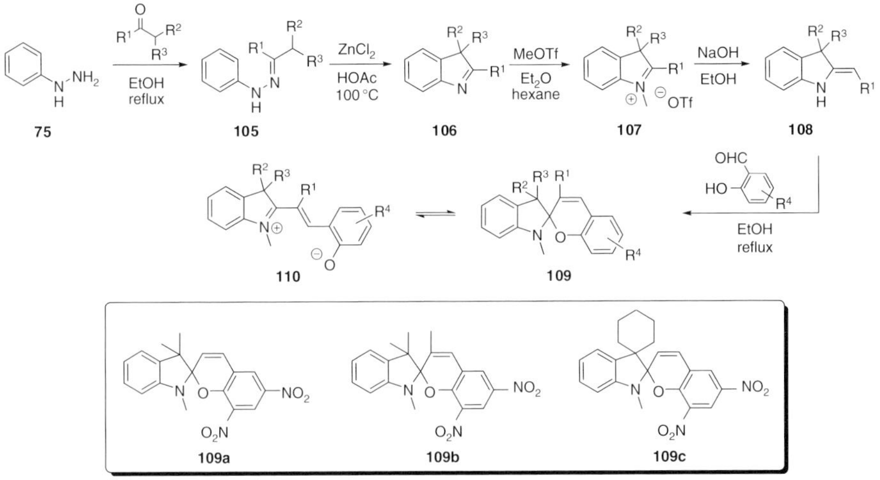

The leuco-merocyanine equilibrium can also be influenced by other structural features. For example, the bis(spiropyran) podand **113** was rapidly and efficiently constructed using the previously reported oxybis(ethanediyloxy)bis(salicylaldehyde) **112** <91JOC225>. Upon complexation with alkaline earth metals, podand **113** isomerized from the spiropyran form to the merocyanine form, a phenomenon not observed in the presence of alkali metals. Furthermore, naked eye observation allowed for discrimination between the Ca^{2+} complex from the Mg^{2+} complex <09DP(80)98>.

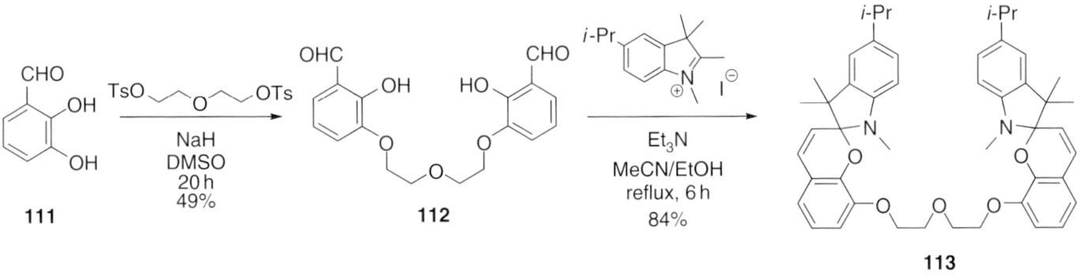

Another important family of pyranoid dyes is constituted by the carbazoles. As one example, Indanthren Olive R (**114**, C.I. Vat Black 27) is a long established textile dye with a dull shade but very good lightfastness. It is currently of interest in applications for military camouflage fabric, as it can be used to color synthetic materials to match the reflectance profile of leaves in the NIR region <01USP001>.

Carbazoles have also joined the ranks of designer dyes. For example, carbazole **119** was synthesized and evaluated for use in dye-sensitized solar cells. The design elements include a carboxylic acid to serve as both an anchoring group for the TiO_2 and an electron density acceptor, and (for these systems) a somewhat uncommon cyclohexadiene unit to fill the role of a nonaromatic but conformationally rigid π spacer. The energy conversion efficiency for DSSCs incorporating this dye can reach 4.4% <09OL377>. The structurally similar chromophore **125**, incorporating a thiophene spacer and a cyanoacrylic acid acceptor, exhibits an open-circuit voltage of 0.939 V and an energy conversion efficiency of 5.2% in a DSSC with a Br^-/Br_3^--containing electrolyte <09OL5542>.

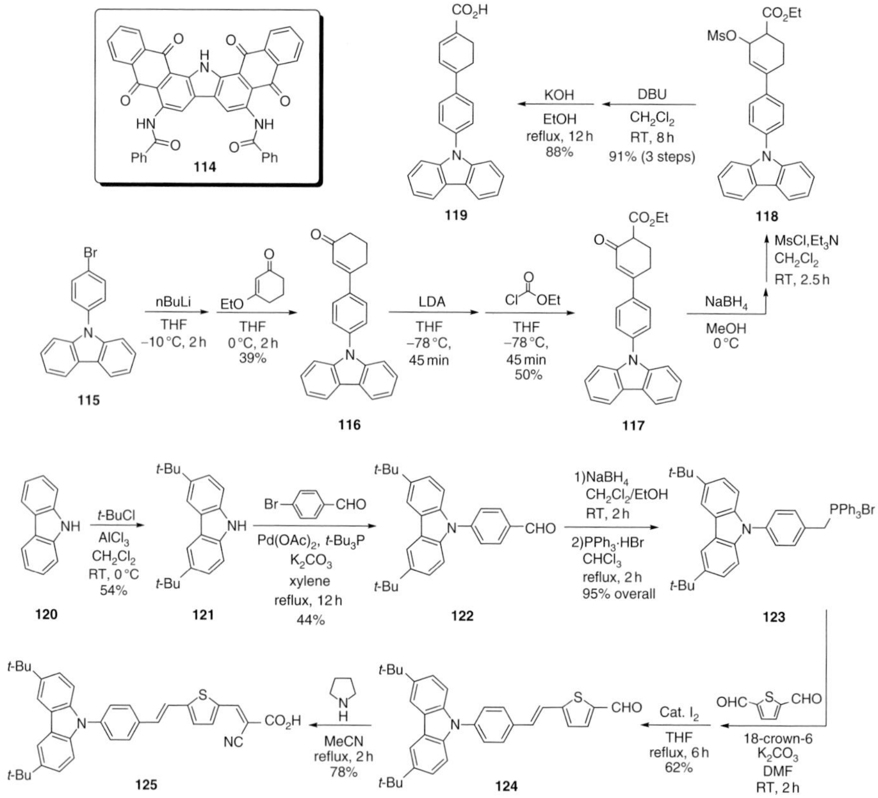

While the previous examples feature elaboration of the carbazole at the N-position, there are also preparative routes that involve chemistry at the carbazole 3-position. For instance, the two-photon absorption chromophore **129** was assembled via a Wadsworth-Horner-Emmons olefination of 9-butylcarbazole-3-carbaldehyde using phosphonate **128** as a precursor <10JL654>. The analogous 9-ethylcarbazole-3-carbaldehyde **130** was incorporated into a pyridyl structure (**131**) as the aldehyde component of a Chichibabin reaction, the ketone component being *p*-iodoacetophenone. The diiodopyridine **131** was then polymerized using Suzuki chemistry to form an optically active conjugated polymer containing fluorine, pyridine, and carbazole moieties. When subjected to an applied bias voltage of 2.5 V, the emission wavelength of the polymer exhibited a bathochromic shift from 439 nm (blue) to 551 nm (yellow) <09DP(82)109>.

Substituents on the carbazole need not be connected by conjugation. For example, the C_2-symmetric 3,6-disubstituted carbazole-based bisboronic acid **135** was shown to be an effective enantioselective fluorescent sensor for tartaric acid. Instead of direct influence on the chromophore through conjugation, the remote binding sites report to the fluorophore by way of photoinduced electron transfer (PET). The binding of tartaric acid with **135** gives rise to an enantiospecific response, with L-tartaric acid leading to an enhancement of fluorescence and D-tartaric acid diminishing the intensity. Interestingly, the fluorescence in general is more intense at higher pH, a phenomenon the authors rationalize by a reverse PET effect involving the protonated amine <09JOC1333>.

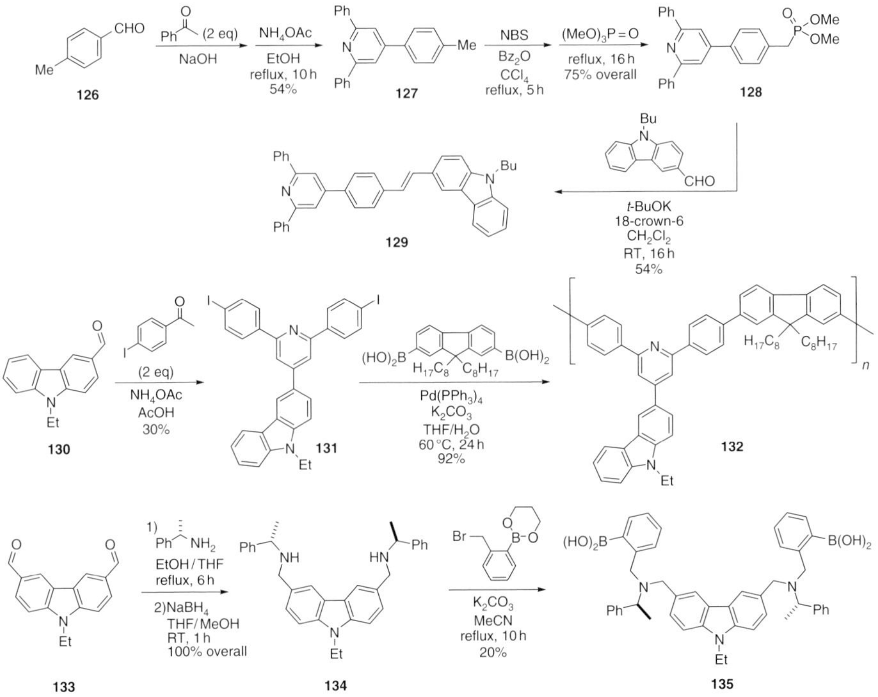

The 3,6-disubstitution motif has also been applied to the construction of interesting polymeric arrays. For example, the wide band-gap polymer **137**, prepared from a Suzuki coupling between a bis(*p*-bromophenyl)silane and the carbazole-3,6-bis(dioxaborolane) derivative **136**, has been used as an effective host for ternary phosphorescent iridium complexes in a white polymer LED <09JPS(A)4784; 10OE498>. The indolocarbazole **138**, equipped with reactive oxetane moieties, was employed in the preparation of cross-linked hole transporting structures in multilayer OLEDs <10DP(85)183>, and the hexylthiophene-functionalized indolocarbazole **139** showed promise as a sensitizer for TiO_2-based DSSC applications <09JPC(C)13409>.

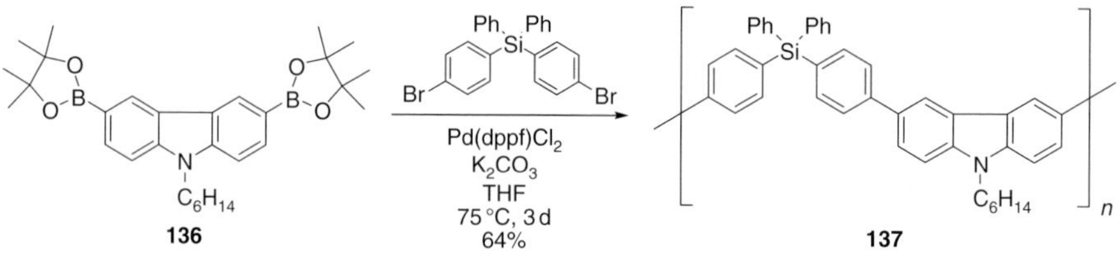

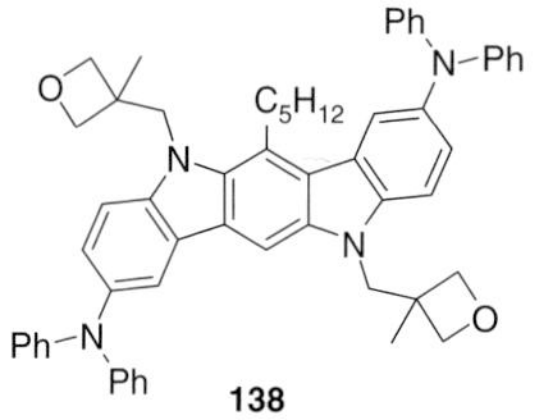

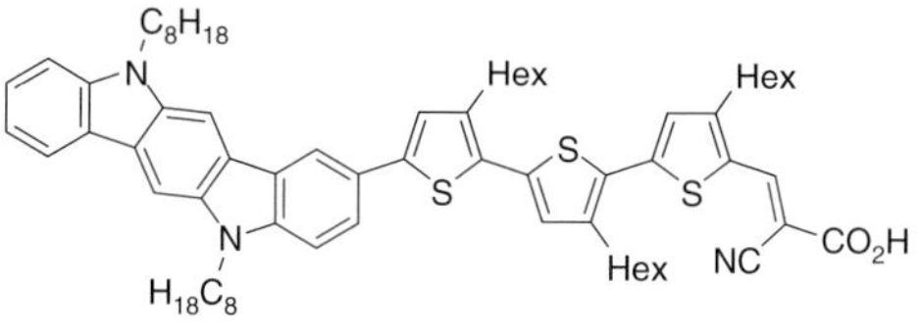

The boron dipyrromethene (BODIPY) dyes are the subject of a flurry of research activity in recent years. First synthesized serendipitously in the late 1960s <68LA(718)208>, the BODIPY dyes have come into their own as bioanalytically relevant fluorophores, thanks to their high extinction coefficients and fluorescence quantum yields. An excellent review has summarized the synthesis and spectroscopic properties of this intriguing dye class <07CRV4891>, so the present work will touch on just a few of the more recent highlights.

Last year two groups reported independently the first synthesis of the unsubstituted core BODIPY compound (**145**), which turns out to have high photostability even though it is thermally labile. Tram and co-workers <09DP(82)392> accessed the target molecule through a DDQ oxidation of dipyrromethane (**143**), which is ultimately derived in moderate yield from pyrrole (**140**) <02T2405>, followed by treatment with boron trifluoride etherate. On the other hand, Schmitt and co-workers <09JF755> reported a procedure starting with pyrrole, which is condensed with pyrrole-2-carboxaldehyde and treated with boron trifluoride etherate in one pot. Either way, the yields from pyrrole are quite modest.

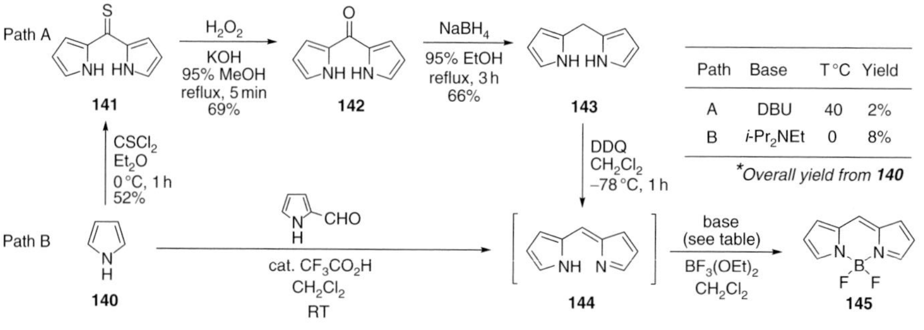

Path	Base	T °C	Yield
A	DBU	40	2%
B	i-Pr$_2$NEt	0	8%

*Overall yield from **140**

The novel benzoBODIPY dye **152** exhibited interesting properties, with a solution-phase absorption at 750 nm and extinction coefficient of 83,000/M·cm. After spin-coating onto a glass plate, the absorption shifted dramatically to 922 nm. The synthesis of the dye started with methyl *o*-anisate (**146**), which was converted in two steps to the acyl hydrazone **148**. In the presence of lead acetate, this compound underwent an oxidative rearrangement <07TL7181> to form the 1,2-diacetylbenzene derivative **150**, which was converted to the dipyrrole **151** via a Paal-Knorr synthesis, and in two subsequent steps to the target compound <10TL1600>.

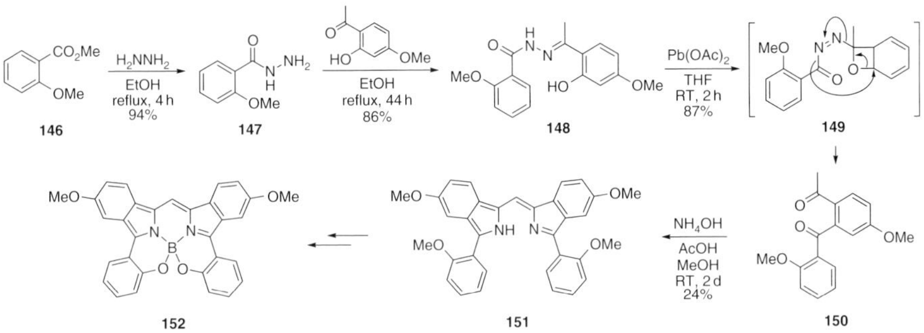

In the matched set department, a range of fourteen rationally designed BODIPY dyes of type **153**, dubbed the "Keio Fluors", span the visible and near-IR spectrum from 547 nm to 743 nm. These dyes exhibit extinction coefficients from 140,000 to 316,000/M·cm and quantum yields between 0.56 and 0.98, and they are relatively insensitive to solvent polarity <08JA1550; 09CEJ1096>. Other more specialized innovations have also been reported. For example, one general drawback of the BODIPYs is that their hydrophobicity limits aqueous applications. In one answer to this problem, water solubility has been conferred by equipping the chromophore with dimethylpropargylamine residues, which are further quaternized by treatment with propanesultone to give zwitterionic dye **154** <09OL2049>. Moving in the opposite direction in solubility, the diiodo BODIPY monomer **155** was synthesized for incorporation into polymeric dyes <09MM6529>.

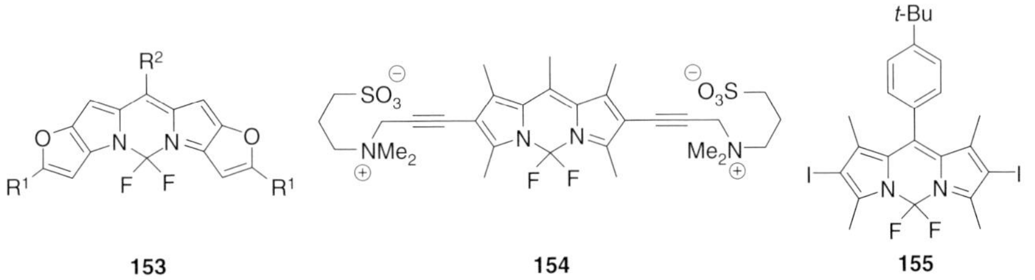

The diketopyrrolopyrroles (DPPs) are high-performance pigments used in plastics and paint formulations <B-09MI165>. One of the most commercially important (and thus most readily available) DPPs is C.I. Pigment Red 256 (**156**) <97CSR203>. Recently, Yamagata and co-workers < 10TL1596> converted

156 to the corresponding morpholino derivative (157) and demonstrated its utility as a gas-phase acid sensor. DPP dyes used for dye-sensitized solar cells <10JPC(C)1343>. Elaborated DPPs have also been developed for DSSC applications <10JPC(C)1343>, and DPP-cyanine hybrids have been constructed for use as near-IR dyes and fluorophores <09CEJ4857>.

Among other interesting developments in pyrroloid dyes, the structurally simple phenolic dipyrrole 158 can serve as a chromogenic probe for the ratiometric determination of water in organic solvents <09BKC197>, and the unsymmetrical bipyrrole 159 shows promise as an electrochromic dye, since anodically induced color changes can be triggered at low potentials <09JOC9497>. The dipyrroloazo dye 160 displays the interesting ability to differentiate colorimetrically among fluoride, dihydrogenphosphate, and acetate anions in solution <09SNA(B141)116>. Finally, a very clever Ca^{2+} probe (i.e. 161) incorporating an indole subunit has been grafted onto a guanine residue for use with the engineered DNA alkyltransferase known as SNAP-tag, which selectively reacts with O-benzylguanines. Once docked to a protein of interest, SNAP-tag facilitates nucleophilic attack at the benzylic position, covalently binding the fluorophore to the protein to report local Ca^{2+} concentrations <09ACB179>.

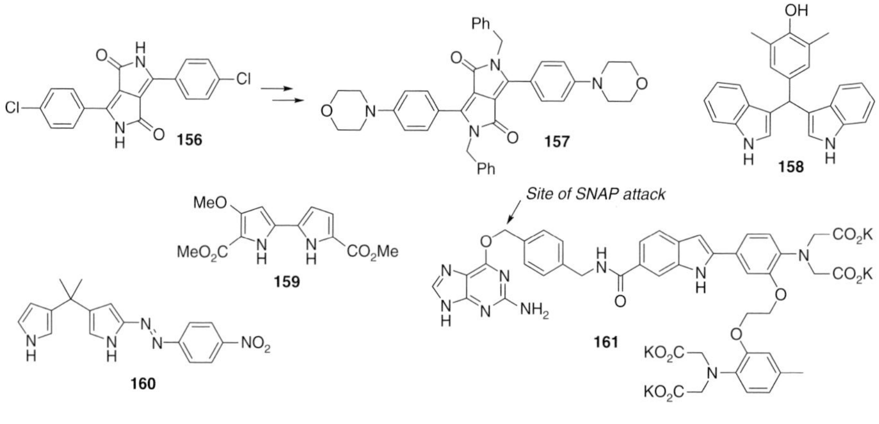

2.4.2 Rings with Two Nitrogen Atoms: Imidazoloids

The imidazole ring system is occasionally found in traditional dyes <05COL105> and some fluorescent whitening agents (FWAs) <07USP001>. In addition to these conventional applications, some recent innovations involving imidazole derivatives have been reported. For example, a cyanuric chloride-based reactive UV-absorber incorporating a benzimidazole (i.e. 166) was developed for creating UV-opaque textile materials. The sulfonated benzimidazole component (163) was synthesized by the condensation of o-phenylenediamine (162) with p-aminobenzoic acid in hot concentrated sulfuric acid. The remainder of the synthesis proceeded via standard reactive dye chemistry <09JAPS(112)3605>.

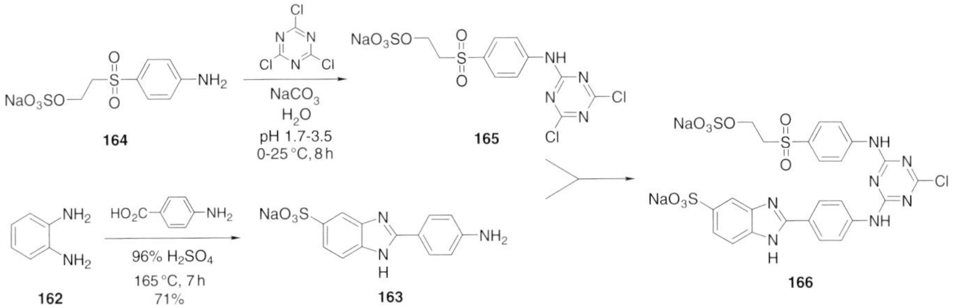

An acceptor-donor-acceptor (A-D-A) sensor (**169**), in which the imidazole ring serves as the donor, was designed as a turn-on fluorescent probe for cyanide. Similar to the previous example, the imidazole ring was constructed through the condensation of an aldehyde with 2,3-diaminonaphthoquinone (**167**). When incorporated into a cetyltrimethylammonium bromide (CTAB) micellar system in aqueous medium, compound **169** exhibited a strong increase in emission at 460 nm upon addition of 50 µM CN⁻ <09JOC3919>.

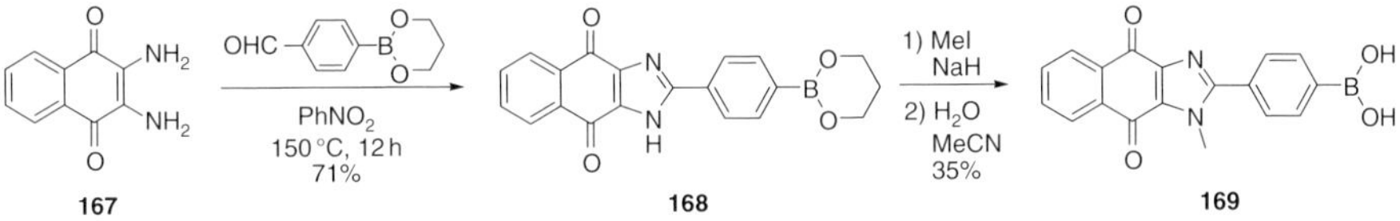

The Debus synthesis can also be used for the preparation of the imidazole nucleus. Thus, 1,10-phenanthroline-5,6-dione (**170**) was refluxed in acetic acid with a thienaldehyde derivative in the presence of ammonium acetate to give the fused thiophenylimidazole **171**, which exhibited interesting solvatochromic behavior <09DP(80)329>. Similar methodology was used for the synthesis of the novel Y-type two-photon absorbing fluorophore **172**, which could be used as a sensor for cysteine and homocysteine <07TL2329>.

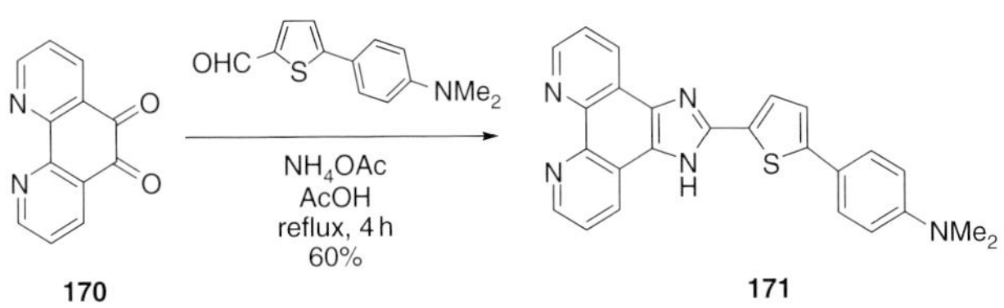

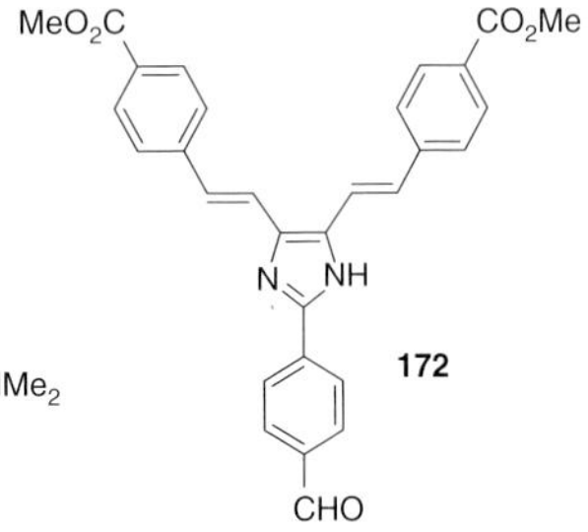

Finally, imidazoles feature in some very interesting photochromic molecules. For example, hexaarylbiimidazole (**173**, HABI), which is itself colorless, is known to undergo photolytic cleavage into the highly colored triphenylimidazolyl radical **174** with a time constant of 80 fs; however, the reversion (or thermal bleaching) occurs too slowly for the compound to be of practical photochromic utility.

However, Kishimoto and Abe have designed a paracyclophane analog **175**, in which the imidazolyl radicals are constrained so that recombination is facilitated. In fact, thermal bleaching for the diradical **176** has a half-life of 10 ms at 40°C, making it applicable to real-time image processing at video frame rates <09JA4227>. Photochromism is also observed in 1,3-diazabicyclo[3.1.0]hexenes (**177**), which serve as latent colorigenic azomethine ylides (**178**) <09JPO559>.

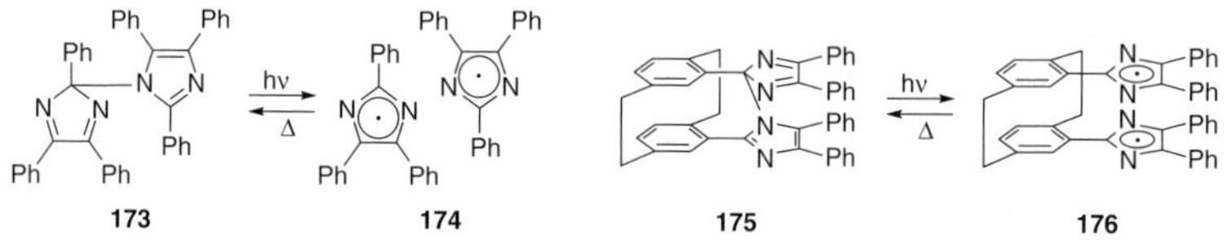

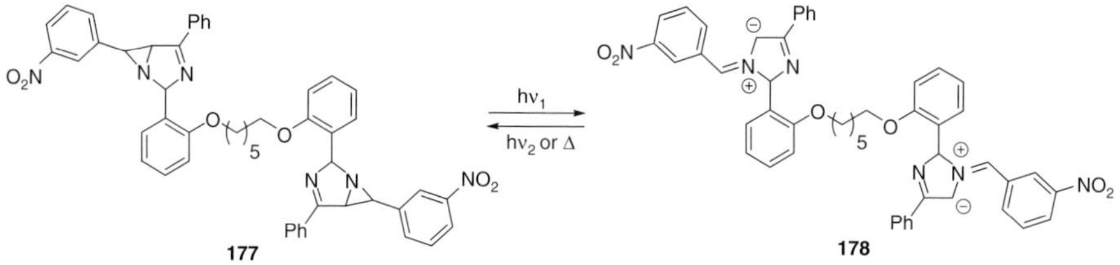

2.5. SIX-MEMBERED RINGS CONTAINING NITROGEN

2.5.1 Rings with One Nitrogen Atom: Pyridinoids

One of the more compact pyridinoid chromophores is contained in the 1,8-naphthalimide dyes. Representing this family is C.I. Disperse Yellow 11 (**179**), a fluorescent dye used in textiles, "neon" art materials, and optical storage <03USP001>. This substructure provides a handy template for functional elaboration. For example, a blue-emitting napthalimide moiety was incorporated into a larger compound (**185**) containing a triazine UV-absorber and a hindered amine radical scavenger (HALS) <10JPCP(A210)89>.

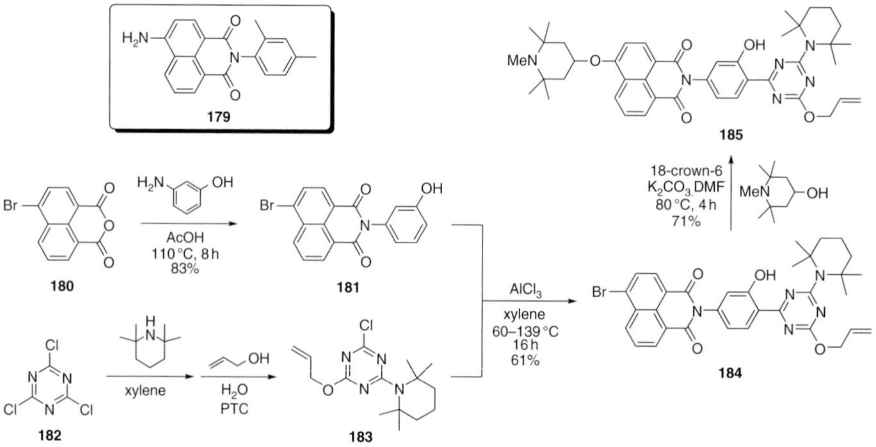

Elaboration of the naphthalimides can be undertaken in two general ways: by attaching functionality to the imide nitrogen (usually done via condensation) and through substitution on the naphthalene ring. Thus, en route to the blue-light emitting fluorophore **187**, 6-bromonaphthalic anhydride (**180**) is first condensed with ethanolamine to provide the *N*-hydroxyethyl group, and then subjected to nucleophilic aromatic substitution to append the 6-phenoxy substituent <10DP(86)190>. A similar approach is seen in the preparation of dye **190**, a very photostable "on-off" pH switcher and copper(II)-selective metal ion fluorescent probe <09JF127>.

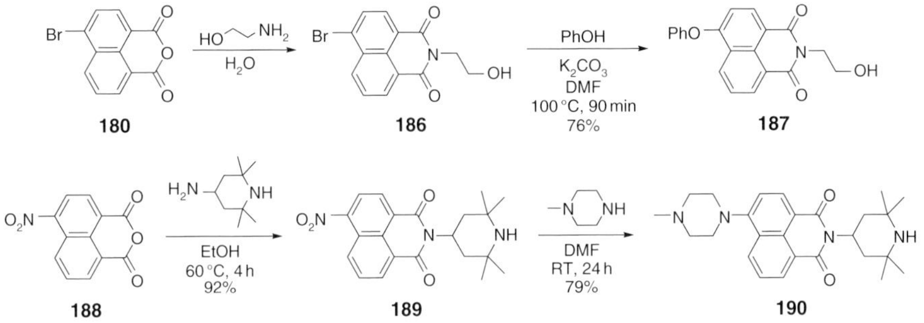

Naphthalene bisimide dyes can provide interesting symmetrically functionalized species. For example, refluxing 1,4,5,8-naphthalenetetracarboxylic dianhydride (**191**, NTCDA) with *N,N*-bis(2-pyridynmethyl)ethane-1,2-diamine in toluene yields the naphthalene bisimide **192**, which was treated with zinc(II) nitrate to give the zinc complex **193**. In the presence of pyrophosphate, compound **193** forms a sandwich-type duplex structure leading to excimer emission. This behavior was highly selective for pyrophosphate even in the presence of ATP and phosphate, an application of increasing research interest <07JA3828; 09ACR23>.

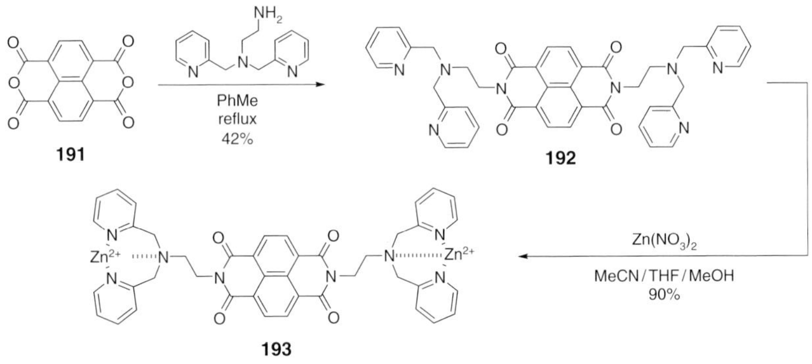

Moving to the next level of aromatic real estate, the perylenes, typified by C.I. Pigment Red 179 (**194**), are high-performance colorants of broad and substantial commercial significance <B-09MI261>. In addition to classical applications, the perylenes are also the subject of continued innovation. For example, introducing sterically bulky aliphatic substituents at the imide nitrogen, as seen in the symmetrical derivative **196**, confers solubility in a wide range of organic solvents <09DP(83) 297>. These swallow-tail perylenes have also been engineered for use in liquid crystal applications, as in the oligoethylene oxide derivative **197**. The unsymmetrical substitution pattern allows for tuning of the mesophase width <09JA14442>.

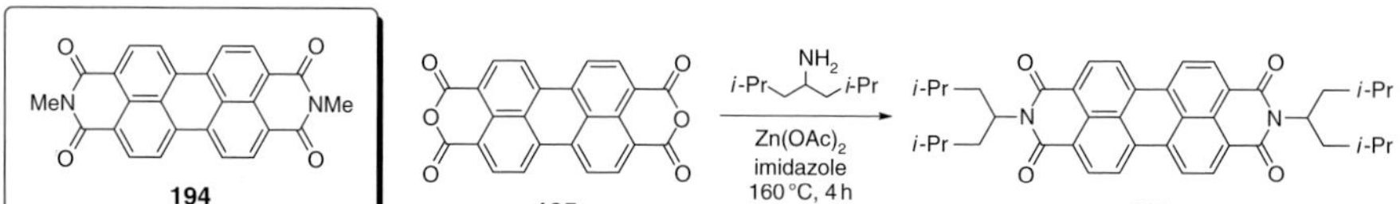

The perylenes can also be elaborated at the angular (or bay) positions. Thus, perylene-3,4,9,10-tetracarboxylic dianhydride (**195**, PTCDA) was brominated in the presence of iodine to give the 1,7-dibromo derivative **198**, which was then refluxed with 2,6-diisopropylaniline in propionic acid to provide the dibromoperylene diimide **199**. Subsequent treatment with hydroquinone under basic conditions led to a dual nucleophilic aromatic substitution reaction. The adduct (**201**) thus formed was converted to the target compound **202** by coupling to pyrenebutanoic acid. This macromolecular perylene diimide exhibited fast photoinduced energy transfer and efficient electron transfer from the pyrene quencher moiety to the perylene chromophore. Cyclic voltammetry suggests that the angular pyrene units also slightly narrow the band gap energy. The absorption maximum for the dye was centered around 539 nm in organic solvents, shifting to 582 nm in aqueous medium <10DP(86)32>.

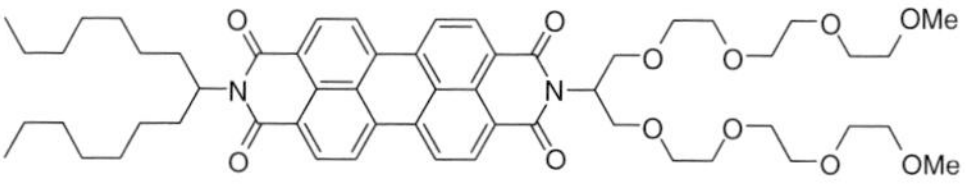

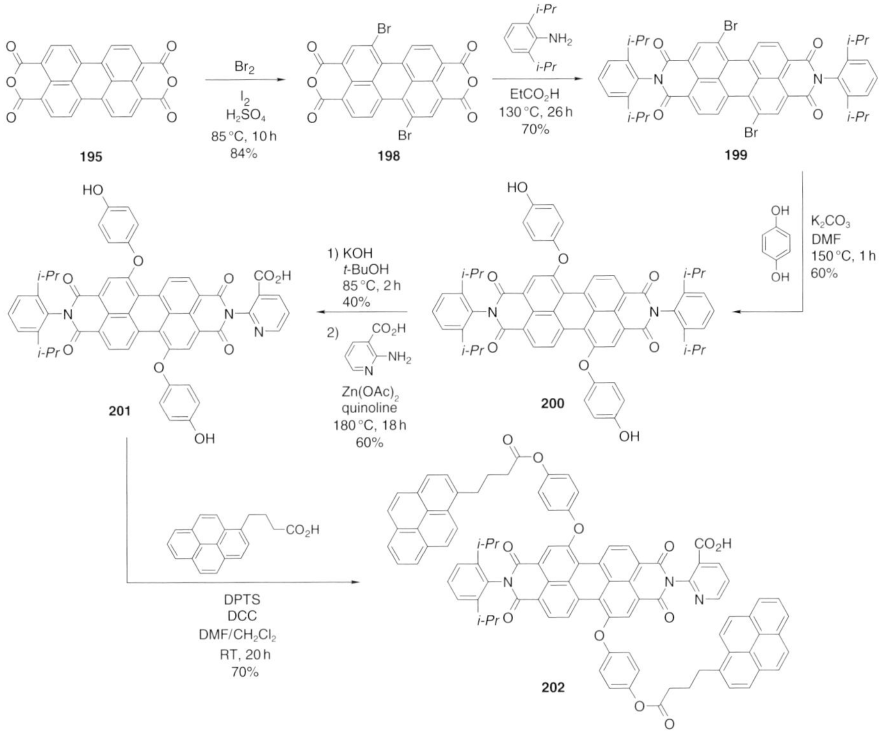

The angular substitution motif was also incorporated into the construction of the novel 1,7-bis(perfluorophenyl) substituted perylene dye **204**, which was evaluated as a potential building block in organic photovoltaic devices. In this case, the substituents were attached by means of a copper-mediated Ullmann coupling <09HCA2525>. The angularly disulfonated perylene dye **205** was designed for water solubility, so that it could be applied in inkjet printed liquid crystal display color filters <09DP(81)45>.

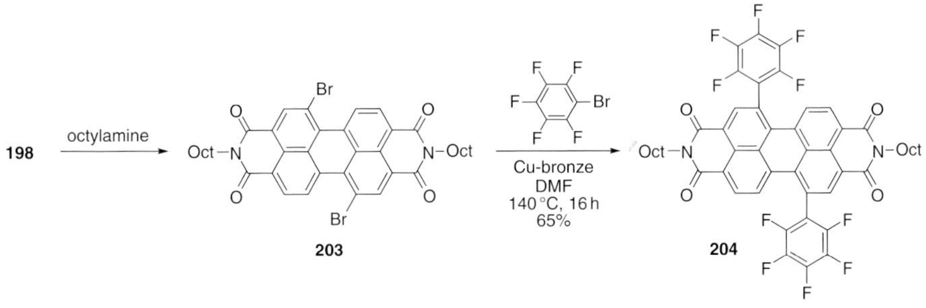

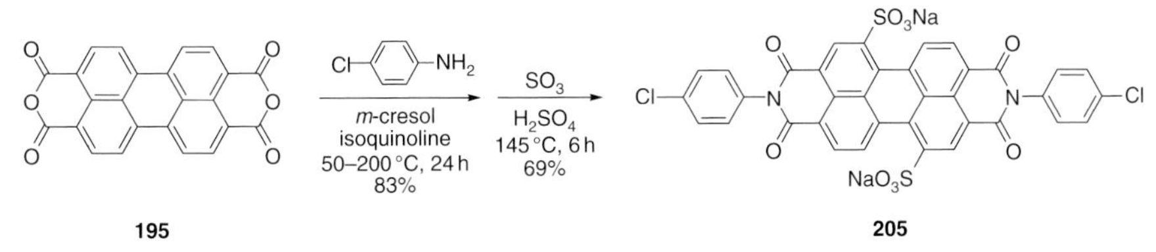

In an interesting architectural twist, the first Z-shaped perylene bisimide (i.e. **209**) was recently synthesized <06JA702>. This fluorescent dye, which can be dissolved in polar organic solvents or incorporated into polystyrene films, exhibits a solution absorption maximum of 491 nm with an extinction coefficient of 29,000/ M·cm. The fluorescence emission is centered about 517 nm with a quantum yield of 0.67. The synthesis features a Diels-Alder reaction between *N*-octylmaleimide and photoenol **207**.

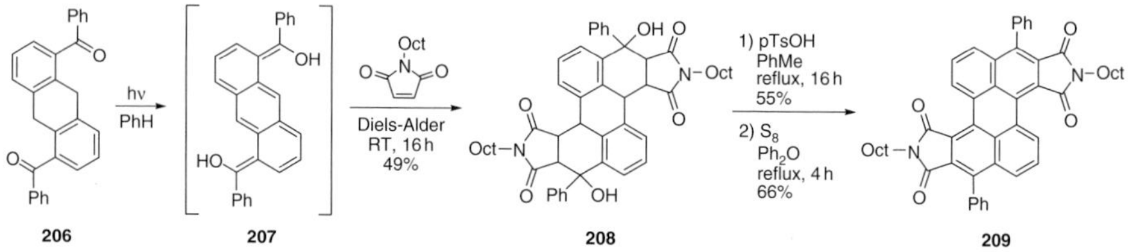

Perylene 3,4-dicarboxyanhydride (or PMA) derivatives offer templates for designing unsymmetrical perylenes with interesting optical properties. For example, compound **211** is one of a series of so-called rainbow perylene monoimides whose absorption properties and electrochemical potentials could be fine-tuned by modifying the substituent in the 9-position, which is conveniently attached via a Pd-mediated amination of 9-bromo-PMA <09CEJ878>. Some of these dyes find potential application in solid-state DSSCs <09JPC(C)14595>.

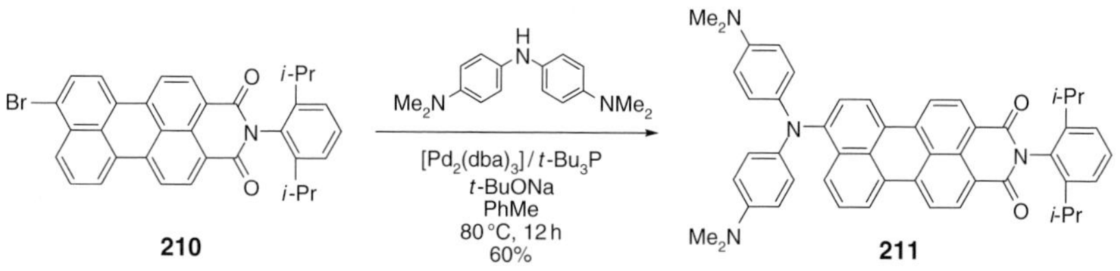

The pyridine ring finds itself in other chromogenically interesting architectures. For example, the tetrasubstituted anthrazoline **216** was prepared by the Friedländer condensation of *p*-phenylbenzaldehyde onto 2,5-diamino-1,4-dibenzoylbenzene (**215**), which was ultimately derived from pyromellitic dianhydride (**212**). The product exhibited high thermal stability ($T_d > 450°C$) and melt transition temperature ($T_m = 440°C$), and a relatively low band gap (2.64 eV) <09DP(81)218>. Even the azaperylene system offers some intriguing possibilities, and Gryko and co-workers <10JOC1297> have reported an improved synthesis of the excited-state intramolecular proton-transfer (ESIPT) chromophore 12-hydroxy-1-azaperylene (**220**).

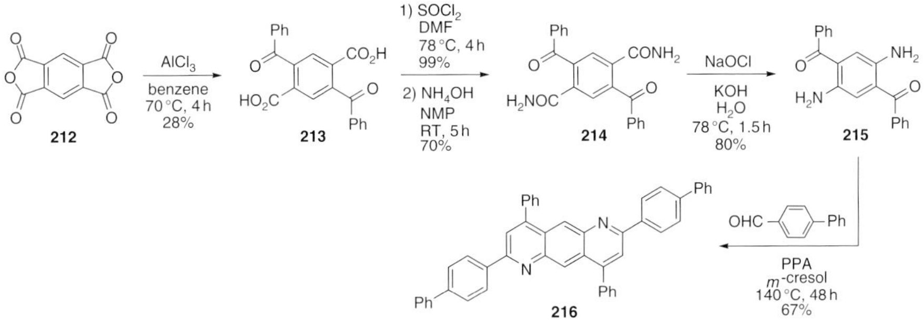

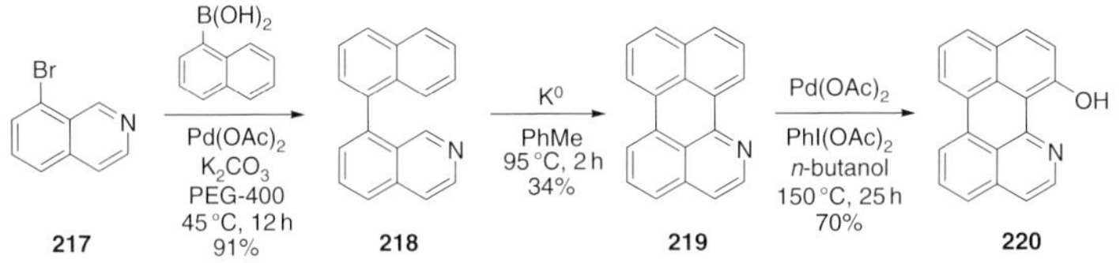

2.5.2 Rings with Two Nitrogen Atoms: Pyrazinoids

The pyrazine ring is a very convenient and easily accessible heteroaromatic template for constructing complex and functionally interesting chromophores. For example, the novel blue-green emitting quinoxaline **226** was prepared by condensation of *p*-dibromobenzil (**224**) with 4-bromo-1,2-phenylenediamine, followed by Suzuki coupling using the triphenylamine-derived boronic acid **223**. With a HOMO energy level of -5.38 (eV), compound **226** might be a useful model for designing hole-transporting and electron-transporting materials for OLED devices <09DP(83)269>.

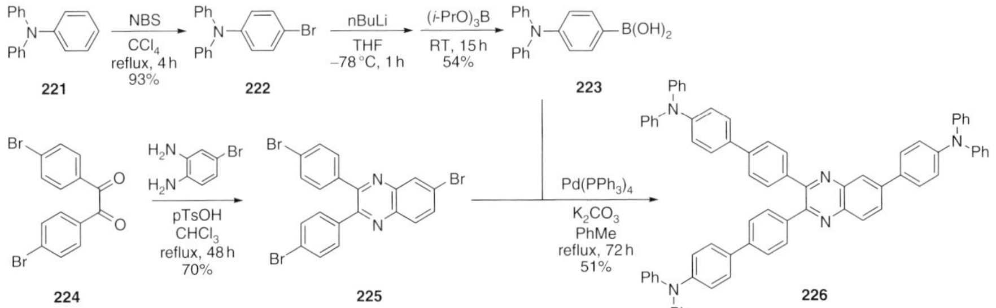

Similarly, the condensation of cinnamil (**227**) with phenylenediamine provided rapid and very high-yielding access to the bis(phenylvinyl)quinoxaline **228**, which exhibited a fluorescence emission maximum at 449 nm <09DP(81)245>. A panchromatic series of fluorophores were prepared about the quinoxaline nucleus by modifying electronic features at various positions on the heterocycle. For example, the dimethylaminophenyl derivative **231**, with an absorption maximum of 437 nm and photoluminescent maximum at 570 nm, was synthesized by condensing benzil (**229**) with 1,4-

dibromo-2,3-phenylenediamine to give the dibromoquinoxaline **230**, to which the auxochromic moieties were appended using Suzuki coupling <09JOC3175>.

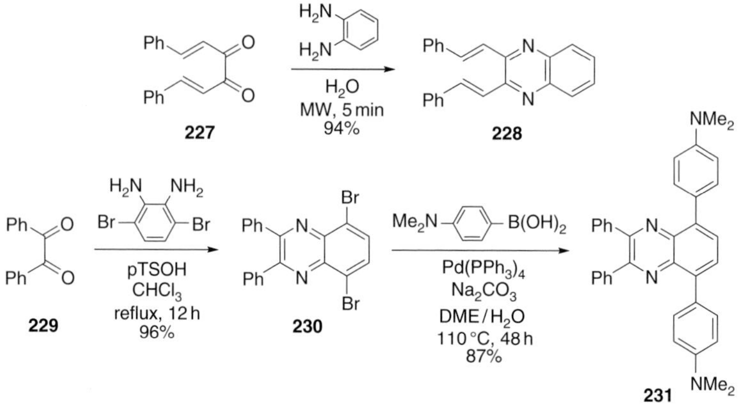

A series of film-forming, low-bandgap chromophores, exemplified by structure **236**, were rationally designed through molecular modeling to be near-IR absorbing fluorophores. Thus, compound **236** exhibits an electrochemically derived bandgap of 1.27 eV, an absorption maximum in toluene of 746 nm and a photoluminescent maximum at 1035 nm <09CEJ8902>. Quinoxaline derivatives have also been developed for use as visible-wavelength, oxidizable polymerization sensitizers <09DP(82)365; 09JPCP(A202)115>.

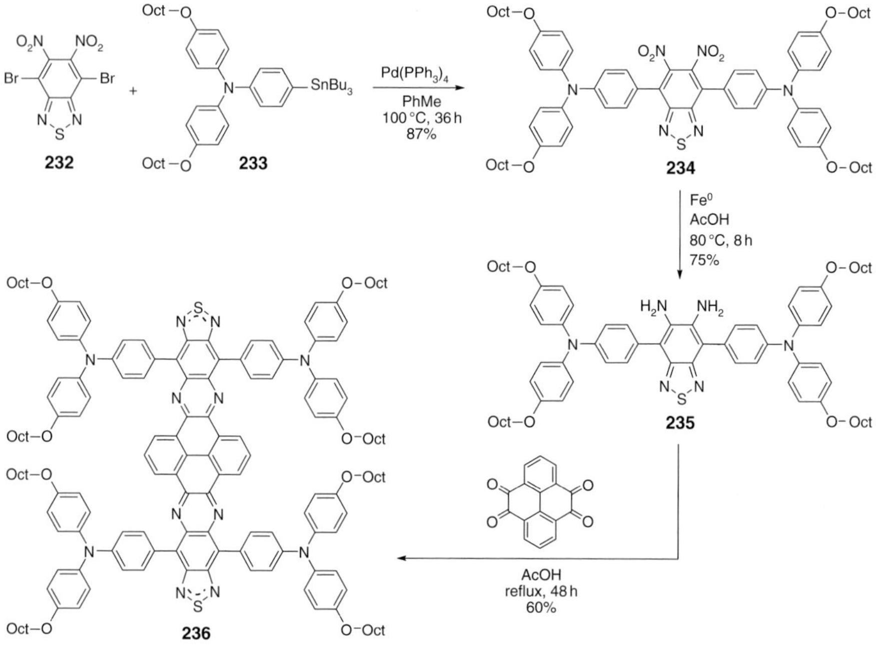

2.6. FIVE-MEMBERED RINGS CONTAINING SULFUR

2.6.1 Rings with One Sulfur Atom: Thiophenoids

The thiophene moiety is frequently used as a photostable π linker between donor and acceptor groups in dyes for solar cell applications. For example, the merocyanine dye **241**, which tethers a diphenylamino donor to a cyanoacrylic acid acceptor via a terthiophene linker, is prepared in good yield by monobromination of diethyl-terthiophene (**237**), followed by palladium-mediated amination, formylation, and condensation. Photovoltaic devices doped with dye **241** attained a maximum monochromatic incident photon to current efficiency (IPCE) of 80% and an overall efficiency of about 7% in full sunlight, retaining 73% of its initial efficiency after 1,000 hours of irradiation. Introducing a branched architecture in the thiophene linker region increases this value to 96% <10CM1836>.

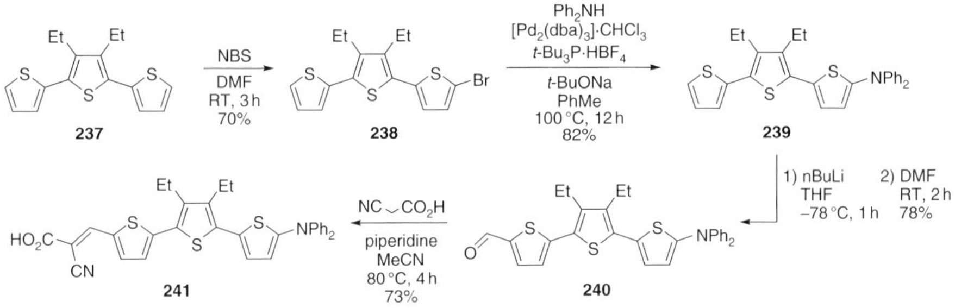

Thiophene serves double duty in the merocyanine dye **244**, which incorporates both a terthiophene linker region and a dithienothiophene donor group. This compound exhibits an absorption maximum of 552 nm and a bandgap of 2.25 eV. When incorporated into a TiO_2-based DSSC, dye **244** provided a conversion efficiency of 5.02% <10OL16>.

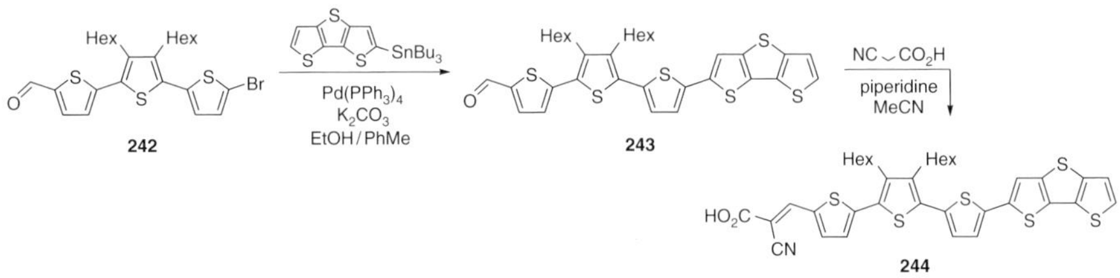

Some topographically intriguing octopoles with a tunable core have been designed around a "star-shaped" motif. One example involves the assembly of three thienylacetylene arms onto a central benzene ring. The chromophore thus formed (**246**) exhibits an absorption maximum at 290 nm, a fluorescence maximum at 365 nm, a phosphorescence maximum at 480 nm, and a static hyperpolarizability value of 21 in methylene chloride. The depolarization ratio derived from NLO optical measurements, reflects the octopolar configuration of the molecule <09CEJ8223>. The triazinyl ring is another convenient scaffold for elaborating the star-shaped theme, as demonstrated by the two photon-absorbing dye **251**. Synthesis of this compound began with a triflic acid-mediated cyclotrimerization of 2-cyanothiophene (**247**). Subsequent bromination with NBS provided the tribromothienyl-triazine **249**, which was elaborated using the Suzuki coupling. This dye exhibits two-photon absorption activity in the range of 720–880 nm, with a 2PA cross section of 1508 at 850 nm <09EJO5587>.

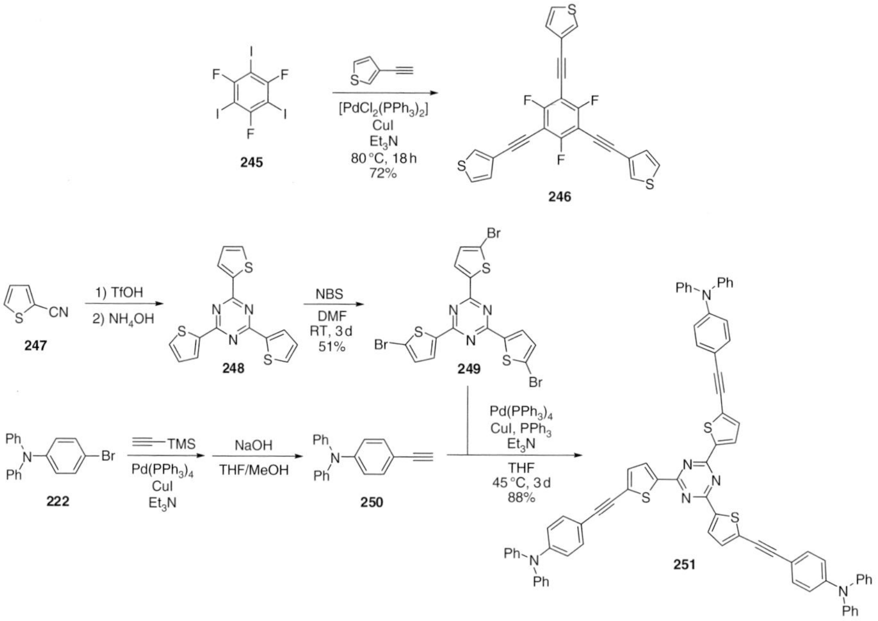

The thiophene ring can also be incorporated into a coupling component for the preparation of heterocyclic azo dyes. For example, dye **253** was produced from the coupling of *p*-nitroaniline diazo onto the dimethylaminobithiophene **252**. A series of such dyes was prepared by varying the diazo component to provide thermally stable solvatochromatic probes <09DP(83)59>.

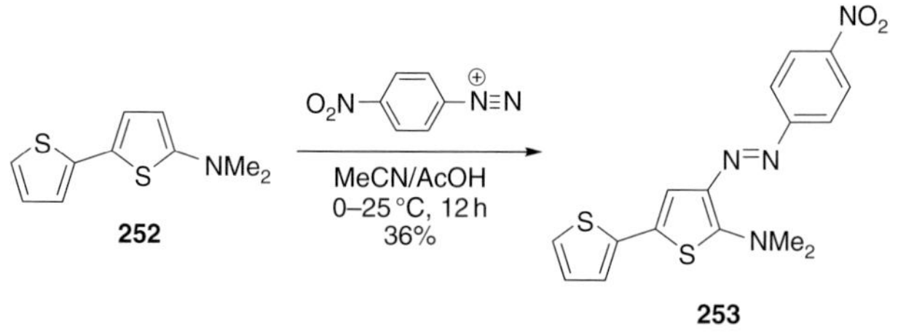

Diarylethenes, and in particular dithienylethenes (DTEs), have been the subject of intense research in recent years because of their potential application in optoelectronics and optical data storage <00CRV1685>. Their photoaddressability derives from the dual capability of UV-mediated electrocyclization (i.e. **254** → **255**) and visible light-promoted cycloreversion (or photobleaching). Hermes and co-workers

<09TL1614> have reported a rapid synthesis of perfluoropentenyl bridged DTEs (e.g. **259**) involving a Dixon reaction with octafluorocyclopentene and subsequent Suzuki coupling. The methodology is amenable to the preparation of both symmetrical and unsymmetrical products.

Other recently reported methodologies allow for access to variously functionalized DTEs. For example, thiophene (**260**) can be elaborated to the diformyl DTE **265**, the carbonyl groups providing convenient handles for further construction. The 2-alkyl groups can also be easily modified by using different bromoalkanes in the first step <09JMST(921)89>. Similar protocols have been disclosed for obtaining DTEs with 2-methoxy substituents <09JPO954> and 2-fluoro substituents <09JMST(936)29>.

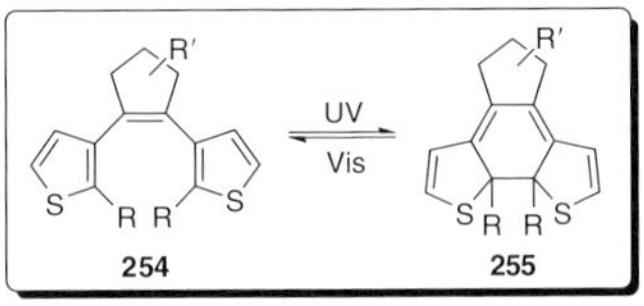

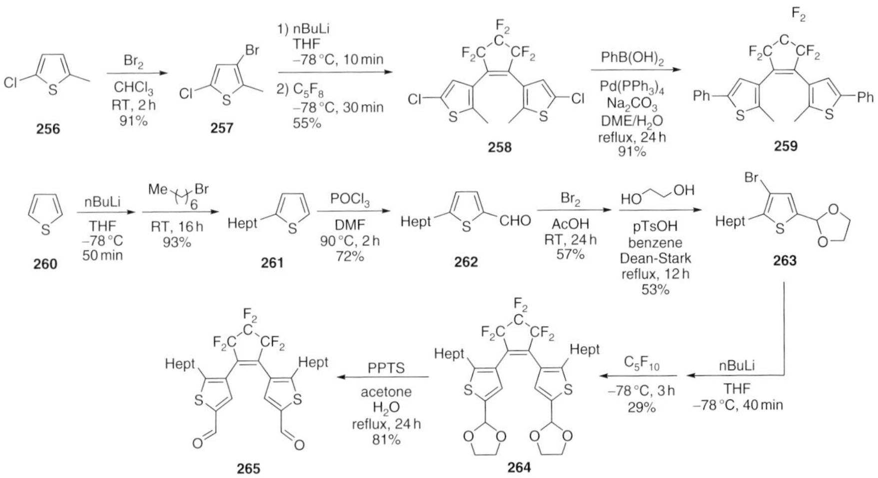

2.6.2 Rings with Two Sulfur Atoms: Dithioloids

The dithiole subunit can be used as a donor group in merocyanine dyes, as demonstrated in dye **267**, which uses a thiobarbituric acceptor moiety <03OL3143>. The first molecular hyperpolarizabilities of these dyes can be fine-tuned by the incorporation of a proaromatic spacer group (e.g. **268**), which serves to stabilize the zwitterionic excited state (e.g. **269**) by contributing aromaticity. Interestingly, these compounds exhibit negative $\mu\beta$ values in spite of the fact that crystallographic data and modeling studies support a quinoid ground state <05JA8835>. The non-linear optical properties of these dyes can also be improved by using 1,1,3-tricyano-2-phenylpropene as a very strong acceptor, as illustrated by the dithiafulvene compound **271** <07JOC6440>.

266 267

268 269

270 271

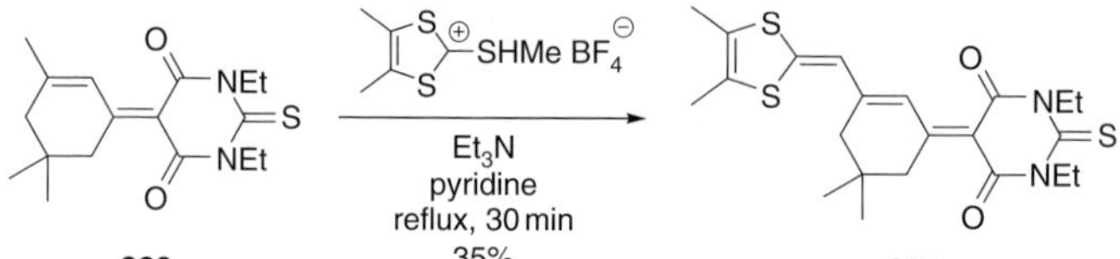

The tetrathiafulvalene (TTF) array is also a useful thienyl component in dyes. For example, García and co-workers <09OL5398> have reported a rapid and flexible synthesis of redox active chromophores of type **275** featuring the Wadsworth-Horner-Emmons condensation of tetrathiafulvalene carboxaldehyde with the anion of tris(phosphonate) **274**. Another innovative use of this moiety is found in the novel fructose sensor **276**, comprised of an anthracene-tetrathiafulvalene PET dyad connected to an arylboronic acid saccharide docking region <05JOC5729>.

272 273 274 275 276

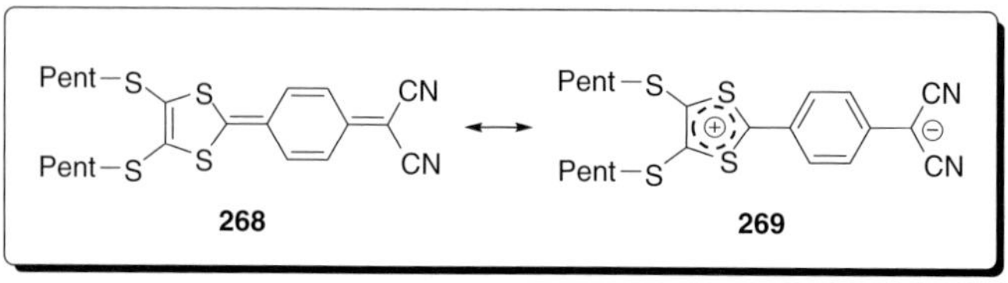

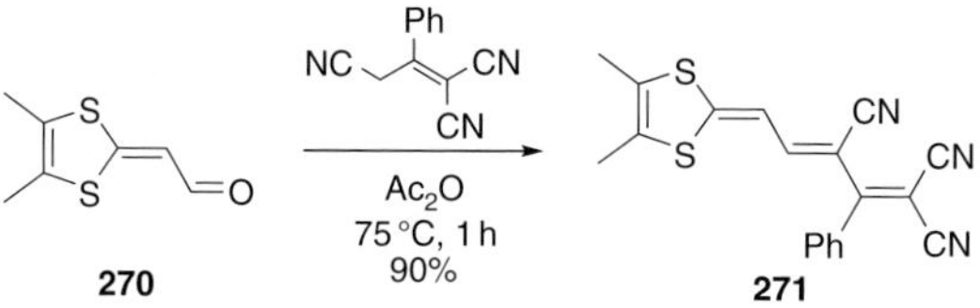

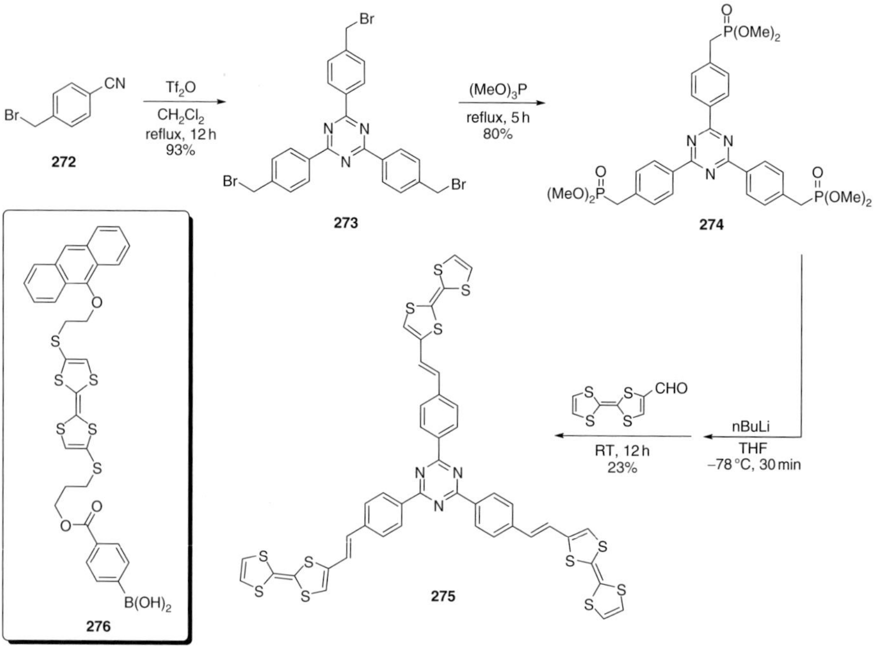

REFERENCES

B-99MI001	W.T. Mason, Fluorescent and Luminescent Probes for Biological Activity. 2nd ed. Academic Press, San Diego (1999).
B-01MI001	R.M. Christie, *Colour Chemistry* Royal Society of Chemistry, Cambridge (2001).
B-03MI001	H. Zollinger, Color Chemistry: Syntheses, Properties, and Applications of Organic Dyes and Pigments. 3rd ed. Wiley-VCH, Weinheim (2003).
B-07MI499	H.S. Freeman and G.N. Mock, (J.A. Kent, ed.)**1**Springer, New York (2007).
B-08M1101	A.M. Pintea, "Food Colorants. Chemical and Functional Properties," (C. Socaciu, ed.). CRC Press, Baton Rouge (2008).
B-09MI165	O. Wallquist and R. Lenz, "High Performance Pigments," (E.B. Faulkner and R.J. Schwartz, eds.), 2nd ed. Wiley-VCH, Weinheim (2009).
B-09MI261	B. Thompson, "High Performance Pigments," (E.B. Faulkner and R.J. Schwartz, eds.), 2nd ed. Wiley-VCH, Weinheim (2009).
60JCS1537	G.E. Ficken and J.D. Kendall, *J. Chem. Soc.*, 1537 (1960).
68LA(718)208	A. Treibs and F.H. Kreuzer, *Liebigs. Ann. Chem.*, **718**, 208 (1968).
80DP(1)103	C.W. Greenhalgh, J.C. Carey and D.F. Newton, *Dyes Pigm.*, **1**, 103 (1980).
80USP001	H.J. Davis, *U.S. Patent 4,216,101* (1980).
85JPC294	G. Jones, W.R. Jackson, C.Y. Choi and W.R. Bergmark, *J. Phys. Chem.*, **89**, 294 (1985).
88USP001	P. Manher, *U.S. Patent Application 4,758,366 A* (1988).
90DP(13)301	N.R. Ayyangar, K.V. Srinivasan and T. Daniel, *Dyes Pigm.*, **13**, 301 (1990).
91JOC225	F.C.J.M. van Veggel, M. Bos, S. Harkema, H. van de Bovenkamp, W. Verboom, J. Reedijk and D.N. Reinhoudt, *J. Org. Chem.*, **56**, 225 (1991).
94CRV2319	C. Reichardt, *Chem. Rev.*, **94**, 2319 (1994).
96JA1213	M. Strukelj, R.H. Jordan and A. Dodabalapur, *J. Am. Chem. Soc.*, **118**, 1213 (1996).
97CSR203	Z. Hao and A. Iqbal, *Chem. Soc. Rev.*, **26**, 203 (1997).
99USP001	C.H. Chen, K.P. Klubek and J. Shi, *U.S. Patent 5,908,581* (1999).
00CRV1685	M. Irie, *Chem. Rev.*, **100**, 1685 (2000).
00CRV1973	A. Mishra, R.K. Behera, P.K. Behera, B.K. Mishra and G.B. Behera, *Chem. Rev.*, **100**, 1973 (2000).
00JCS(P1)1045	G.W. Gribble, *J. Chem. Soc., Perkin Trans.*, **1**, 2000, 1045.
01CLT61	S.H. Kim and S.K. Han, *Color. Technol.*, **117**, 61 (2001).
01DP(48)107	G. Hallas and C. Yoon, *Dyes Pigm.*, **48**, 107 (2001).
01DP(48)121	G. Hallas and C. Yoon, *Dyes Pigm.*, **48**, 121 (2001).
01JTA245	V.R. Kanetkar, P. Bineesh, P.S.R. Kumar and G.G. Pawar, *J. Textil. Assoc.*, 245 (2001).
01USP001	H.R. Mach and G. Krabbe, *U.S. Patent Application 2001/004780 A1* (2001).
02CLT125	C. Yoon and G. Hallas, *Color. Technol.*, **118**, 125 (2002).
02ISJ187	P. Bergthaller, *Imag. Sci. J.*, **50**, 187 (2002).
02JHC391	T. Kappe, F. Frühwirth, P. Roschger, B. Jocham, J. Kremsner and W. Stadlbauer, *J. Heterocycl. Chem.*, **39**, 391 (2002).
02PJC1419	M.D. Obushak, R.L. Martyak and V.S. Matiychuk, *Pol. J. Chem.*, **76**, 1419 (2002).
02RPC80	G.N. Mock, *Rev. Prog. Color,* **32**, 80 (2002).
02T2405	M.J. Plater, S. Aiken and G. Bourhill, *Tetrahedron,* **58**, 2405 (2002).
03OL3143	R. Andreu, J. Garín, J. Orduna, R. Alcalá and B. Villacampa, *Org. Lett.*, **5**, 3143 (2003).
03RPC15	D. Frick, *Rev. Prog. Color,* **33**, 15 (2003).
03USP001	H. Berneth, U. Claussen, S. Kostromine, R. Neigl, J. Rübner and R. Ruhmann, *U.S. Patent 6,620,920 B1* (2003).
04CM4389	C.T. Chen, *Chem. Mater.,* **16**, 4389 (2004).
05ACR449	A. Ajayaghosh, *Acc. Chem. Res.,* **38**, 449 (2005).
05COL105	N. Sekar and S.S. Mahajan, *Colourage,* 105 (2005).
05JA1652	S.S. Chandran, K.A. Dickson and R.T. Raines, *J. Am. Chem. Soc.*, **127**, 1652 (2005).

05JA8835	R. Andreu, M.J. Blesa, L. Carrasquer, J. Garín, J. Orduna, B. Villacampa, R. Alcalá, J. Casado, M.C.R. Delgado, J.T.L. Navarrete and M. Allain, *J. Am. Chem. Soc.*, **127**, 8835 (2005).
05JOC2939	S.J. Mason, J.L. Hake, J. Nairne, W.J. Cummins and S. Balasubramanian, *J. Org. Chem.*, **70**, 2939 (2005).
05JOC5729	Z. Wang, D. Zhang and D. Zhu, *J. Org. Chem.*, **70**, 5729 (2005).
05RPC1	M. Fryberg, *Rev. Prog. Color*, **35**, 1 (2005).
06CLT1	J.A. Kiernan, *Color. Technol.*, **122**, 1 (2006).
06CLT235	I. Holme, *Color. Technol*, **122**, 235 (2006).
06JA702	F. Ilhan, D.S. Tyson, D.J. Stasko, K. Kirschbaum and M.A. Meador, *J. Am. Chem. Soc.*, **128**, 702 (2006).
07CRV4891	A. Loudet and K. Burgess, *Chem. Rev.*, **107**, 4891 (2007).
07EJO1007	D.-Q. Xu, W.-L. Yang, S.-P. Luo, B.-T. Wang, J. Wu and Z.-Y. Xu, *Eur. J. Org. Chem.*, 1007 (2007).
07JA3828	H.N. Lee, Z. Xu, S.K. Kim, K.M.K. Swamy, Y. Kim, S.J. Kim and J. Yoon, *J. Am. Chem. Soc.*, **129**, 3828 (2007).
07JA5710	G.L. Silva, V. Ediz, D. Yaron and B.A. Armitage, *J. Am. Chem. Soc.*, **129**, 5710 (2007).
07JA11902	H. Tsuji, C. Mitsui, L. Ilies, Y. Sato and E. Nakamura, *J. Am. Chem. Soc.*, **129**, 11902 (2007).
07JOC3652	I. Jung, J.K. Lee, K.H. Song, K. Song, S.O. Kang and J. Ko, *J. Org. Chem.*, **72**, 3652 (2007).
07JOC6440	S. Alías, R. Andreu, M.J. Blesa, S. Franco, J. Garín, A. Gragera, J. Orduna, P. Romero, B. Villacampa and M. Allain, *J. Org. Chem.*, **72**, 6440 (2007).
07TL2329	M. Zhang, M. li, Q. Zhao, F. Li, D. Zhang, J. Zhang, T. Yi and C. Huang, *Tetrahedron Lett.*, **48**, 2329 (2007).
07TL5791	Y. Ooyama, T. Mamura and K. Yoshida, *Tetrahedron Lett.*, **48**, 5791 (2007).
07TL7181	A. Kotali, I.S. Lafazanis and P.A. Harris, *Tetrahedron Lett.*, **48**, 7181 (2007).
07USP001	V.P. Eliu and J. Hauser, *U.S. Patent Application 2007/0249840 A1* (2007).
08BMCL5864	M.M. Yatzeck, L.D. Lavis, T.-Y. Chao, S.S. Chandran and R.T. Raines, *Bioorg. Med. Chem. Lett.*, **18**, 5864 (2008).
08JA1550	K. Umezawa, Y. Nakamura, H. Makino, D. Citterio and K. Suzuki, *J. Am. Chem. Soc.*, **130**, 1550 (2008).
08JPCP(A200)438	H. Lee, M.Y. Berezin, M. Henary, L. Strekowski and S. Achilefu, *J. Photochem. Photobiol. Chem.*, **200**, 438 (2008).
08OL4175	J.-A. Richard, M. Massonneau, P.-Y. Renard and A. Romieu, *Org. Lett.*, **10**, 4175 (2008).
08THC1	M. Mojzych and M. Henary, *Top. Heterocycl. Chem.*, **14**, 1 (2008).
08THC133	S. Yagi and H. Nakazumi, *Top. Heterocycl. Chem.*, **14**, 133 (2008).
09ACB179	M. Bannwarth, I.R. CorrêaJr., M. Sztretye, S. Pouvreau, C. Fellay, A. Aebischer, L. Royer, E. Ríos and K. Johnsson, *ACS. Chem. Biol.*, **4**, 179 (2009).
09ACR23	S.K. Kim, D.H. Lee, J.-I. Hong and J. Yoon, *Acc. Chem. Res.*, **42**, 23 (2009).
09ACR1966	W.J. Youngblood, S.-H.A. Lee, K. Maeda and T.E. Mallouk, *Acc. Chem. Res.*, **42**, 1966 (2009).
09AG(I)2474	A. Mishra, M.K.R. Fischer and P. Bäuerle, *Angew. Chem. Int. Ed.*, **48**, 2474 (2009).
09BCC1673	M.V. Kvach, I.A. Stepanova, I.A. Prokhorenko, A.P. Stupak, D.A. Bolibrukh, V.A. Korshun and V.V. Shmanai, *Bioconjugate Chem.*, **20**, 1673 (2009).
09BKC197	K.N. Kim, K.C. Song, J.H. Noh and S.K. Chang, *Bull. Korean Chem. Soc.*, **30**, 197 (2009).
09BMC7031	B.D. Sanders, B. Jackson, M. Brent, A.M. Taylor, W. Dang, S.L. Berger, S.L. Schreiber, K. Howitz and R. Marmorstein, *Bioorg. Med. Chem.*, **17**, 7031 (2009).
09BMCL2285	M. Kamato, N. Umezawa, N. Kato and T. Higuchi, *Bioorg. Med. Chem. Lett.*, **19**, 2285 (2009).
09BMCL5590	A.K. Bhattacharya, K.C. Rana, M. Mujahid, I. Sehar and A.K. Saxena, *Bioorg. Med. Chem. Lett.*, **19**, 5590 (2009).

09CEJ878	C. Li, J. Schöneboom, Z. Liu, N.G. Pschirer, P. Erk, A. Herrmann and K. Müllen, *Chem. Eur. J.*, **15**, 878 (2009).
09CEJ1096	K. Umezawa, A. Matsui, Y. Nakamura, D. Citterio and K. Suzuki, *Chem. Eur. J.*, **15**, 1096 (2009).
09CEJ4857	G.M. Fischer, M. Isomäki-Krondahl, I. Göttker-Schnetmann, E. Daltrozzo and A. Zumbusch, *Chem. Eur. J.*, **15**, 4857 (2009).
09CEJ8223	M.M. Oliva, J. Casado, J.T.L. Navarrete, G. Hennrich, S. van Cleuvenbergen, I. Asselberghs, K. Clays, M.C.R. Delgado, J.-L. Brédas, J.S.S. de Melo and L. De Cola, *Chem. Eur. J.*, **15**, 8223 (2009).
09CEJ8902	M. Luo, H. Shadnia, G. Qian, X. Du, D. Yu, D. Ma, J.S. Wright and Z.Y. Wang, *Chem. Eur. J.*, **15**, 8902 (2009).
09CEJ9191	K. Kiyose, S. Aizawa, E. Sasaki, H. Kojima, K. Hanaoka, T. Terai, Y. Urano and T. Nagano, *Chem. Eur. J.*, **15**, 9191 (2009).
09CEJ10762	V.N. Belov, M.L. Bossi, J. Fölling, V.P. Boyarskiy and S.W. Hell, *Chem. Eur. J.*, **15**, 10762 (2009).
09CHP261	M. Zhao, X.-F. Yang, S. He and L. Wang, *Chem. Paper*, **63**, 261 (2009).
09CHSC621	L. Beverina, R. Ruffo, C.M. Mari, G.A. Pagani, M. Sassi, F. DeAngelis, S. Fantacci, J.-H. Yum, M. Grätzel and M.K. Nazeeruddin, *ChemSusChem*, **2**, 621 (2009).
09CRC622	U. Caruso, B. Panunzi, G.N. Roviello, G. Roviello, M. Tingoli and A. Tuzi, *C.R. Chimie*, **12**, 622 (2009).
09CRV190	M. Sameiro and T. Gonçalves, *Chem. Rev.*, **109**, 190 (2009).
09CSR2410	M. Beija, C.A.M. Afonso and J.M.G. Martinbo, *Chem. Soc. Rev.*, **38**, 2410 (2009).
09DP(80)30	K. Gong, D. Fang, H.-L. Wang, X.-L. Zhou and Z.-L. Liu, *Dyes Pigm.*, **80**, 30 (2009).
09DP(80)98	S. Yagi, S. Nakamura, D. Watanabe and H. Nakazumi, *Dyes Pigm.*, **80**, 98 (2009).
09DP(80)314	S. Wang and S.-H. Kim, *Dyes. Pigm.*, **80**, 314 (2009).
09DP(80)329	R.M.F. Batista, S.P.G. Costa, M. Belsley and M.M.M. Raposo, *Dyes Pigm.*, **80**, 329 (2009).
09DP(81)45	Y.D. Kim, J.P. Kim, O.S. Kwon and I.H. Cho, *Dyes Pigm.*, **81**, 45 (2009).
09DP(81)218	S. Liu, P. Jiang, G. Song, R. Liu and H. Zhu, *Dyes Pigm.*, **81**, 218 (2009).
09DP(81)245	P. Thirumurugan, D. Muralidharan and P.T. Perumal, *Dyes Pigm.*, **81**, 245 (2009).
09DP(82)226	C.J. Roxburgh, P.G. Sammes and A. Abdullah, *Dyes Pigm.*, **82**, 226 (2009).
09DP(82)316	P. Zhao, H. Tang, Q. Zhang, Y. Pi, M. Xu, R. Sun and W. Zhu, *Dyes Pigm.*, **82**, 316 (2009).
09DP(82)341	M.H. Kim, J.H. Noh, S. Kim, S. Ahn and S.K. Chang, *Dyes Pigm.*, **82**, 341 (2009).
09DP(82)365	R. Podsiadły, A.M. Szymczak and K. Podemska, *Dyes Pigm.*, **82**, 365 (2009).
09DP(82)392	K. Tram, H. Yan, H.A. Jenkins, S. Vassiliev and D. Bruce, *Dyes Pigm.*, **82**, 392 (2009).
09DP(82)396	B. Song, Q. Zhang, W.-H. Ma, X.-J. Peng, X.-M. Fu and B.-S. Wang, *Dyes Pigm.*, **82**, 396 (2009).
09DP(83)31	C.J. Roxburgh, P.G. Sammes and A. Abdullah, *Dyes Pigm.*, **83**, 31 (2009).
09DP(83)59	M.M.M. Raposo, A.M.F.P. Ferreira, M. Amaro, M. Belsley and J.C.V.P. Moura, *Dyes Pigm.*, **83**, 59 (2009).
09DP(83)269	H. Wang, G. Chen, Y. Liu, L. Hu, X. Xu and S. Ji, *Dyes Pigm.*, **83**, 269 (2009).
09DP(83)297	G. Türkmen, S. Erten-Ela and S. Icli, *Dyes Pigm.*, **83**, 297 (2009).
09EJO5587	L. Zou, Z. Liu, X. Yan, Y. Liu, Y. Fu, J. Liu, Z. Huang, X. Chen and J. Qin, *Eur. J. Org. Chem.*, 5587 (2009).
09HCA2525	T. Schnitzler, C. Li and K. Müllen, *Helv. Chim. Acta*, **92**, 2525 (2009).
09IC7630	S. Mizukami, S. Okada, S. Kimura and K. Kikuchi, *Inorg. Chem.*, **48**, 7630 (2009).
09IJP1	R.M. El-Shishtawy, *Int. J. Photoen.*, **1** (2009).
09JA4227	Y. Kishimoto and J. Abe, *J. Am. Chem. Soc.*, **131**, 4227 (2009).
09JA9046	S. Tsukiji, H. Wang, M. Miyagawa, T. Tamura, Y. Takaoka and I. Hamachi, *J. Am. Chem. Soc.*, **131**, 9046 (2009).

09JA14442	A. Wicklein, A. Lang, M. Muth and M. Thelakkat, *J. Am. Chem. Soc.*, **131**, 14442 (2009).
09JAPS(112)3605	L. Kubáč, J. Akrman, L. Burgert, D. Dvorský and P. Grüner, *J. Appl. Polymer. Sci.*, **112**, 3605 (2009).
09JCCS381	G.-Y. Fu, Y.-X. Huang, X.-G. Chen and X.-L. Liu, *J. Chin. Chem. Soc.*, **56**, 381 (2009).
09JF127	V.B. Bojinov, N.I. Georgiev and P. Bosch, *J. Fluoresc.*, **19**, 127 (2009).
09JF601	X. Li, A. Duerkop and O.S. Wolfbeis, *J. Fluoresc.*, **19**, 601 (2009).
09JF755	A. Schmitt, B. Hinkeldey, M. Wild and G. Jung, *J. Fluoresc.*, **19**, 755 (2009).
09JFC(130)989	M. Hong and C. Cai, *J. Fluorine Chem.*, **130**, 989 (2009).
09JHC84	M. Henary, M. Mojzych, M. Say and L. Strekowski, *J. Heterocyclic Chem.*, **46**, 84 (2009).
09JHC925	E. Wolinska, M. Henary, E. Paliakov and L. Strekowski, *J. Heterocyclic Chem.*, **46**, 925 (2009).
09JMOC(A304)85	J.V. Madhav, Y.T. Reddy, P.N. Reddy, M.N. Reddy, S. Kuarm, P.A. Crooks and B. Rajitha, *J. Mol. Catal. Chem.*, **304**, 85 (2009).
09JMST(921)89	S. Pu, S. Zhu, Y. Rao, G. Liu and H. Wei, *J. Mol. Struct.*, **921**, 89 (2009).
09JMST(936)29	W. Liu, S. Pu and G. Liu, *J. Mol. Struct.*, **936**, 29 (2009).
09JOC1333	F. Han, L. Chi, X. Liang, S. Ji, S. Liu, F. Zhou, Y. Wu, K. Han, J. Zhao and T.D. James, *J. Org. Chem.*, **74**, 1333 (2009).
09JOC3175	H.-J. Son, W.-S. Han, D.-H. Yoo, K.-T. Min, S.-N. Kwon, J. Ko and S.O. Kang, *J. Org. Chem.*, **74**, 3175 (2009).
09JOC3919	M. Jamkratoke, V. Ruangpornvisuti, G. Tumcharern, T. Tuntulani and B. Tomapatanaget, *J. Org. Chem.*, **74**, 3919 (2009).
09JOC6592	S.G. Robinson, V.A. Sauro and R.H. Mitchell, *J. Org. Chem.*, **74**, 6592 (2009).
09JOC9497	T. Tshibaka, I.U. Roche, S. Dufresne, W.D. Lubell and W.G. Skene, *J. Org. Chem.*, **74**, 9497 (2009).
09JPC(C)7469	R. Li, X. Lv, D. Shi, D. Zhou, Y. Cheng, G. Zhang and P. Wang, *J. Phys. Chem. C*, **113**, 7469 (2009).
09JPC(C)12911	L. Xue, J. He, X. Gu, Z. Yang, B. Xu and W. Tian, *J. Phys. Chem. C*, **113**, 12911 (2009).
09JPC(C)13409	X.-H. Zhang, Z.-S. Wang, Y. Cui, N. Koumura, A. Furube and K. Hara, *J. Phys. Chem. C*, **113**, 13409 (2009).
09JPC(C)14595	U.B. Cappel, M.H. Karlsson, N.G. Pschirer, F. Eickemeyer, J. Schöneboom, P. Erk, G. Boschloo and A. Hagfeldt, *J. Phys. Chem. C*, **113**, 14595 (2009).
09JPCP(A202)115	R. Podsiadły, *J. Photochem. Photobiol. Chem.*, **202**, 115 (2009).
09JPO559	H. Kiyani, N.O. Mahmoodi, K. Tabatabaeian and M. Zanjanchi, *J. Phys. Org. Chem.*, **22**, 559 (2009).
09JPO954	S. Pu, W. Liu and W. Miao, *J. Phys. Org. Chem.*, **22**, 954 (2009).
09JPS(A)4784	T. Fei, G. Cheng, D. Hu, P. Lu and Y. Ma, *J. Polym. Sci. Polym. Chem., Part A*, **47**, 4784 (2009).
09JSGS(52)362	Y. Cui, J. Yu, J. Gao, Z. Wang and G. Qian, *J. Sol-Gel. Sci. Technol.*, **52**, 362 (2009).
09MM6529	F.E. Alemdaroglu, S.C. Alexander, D. Ji, D.K. Prusty, M. Börsch and A. Herrmann, *Macromolecules*, **42**, 6529 (2009).
09NJC1148	K. Helttunen, P. Prus, M. Luostarinen and M. Nissinen, *New J. Chem.*, **33**, 1148 (2009).
09OL29	H. Lee, M.Y. Berezin, K. Guo, J. Kao and S. Achilefu, *Org. Lett.*, **11**, 29 (2009).
09OL97	J.T. Lin, P.-C. Chen, Y.-S. Yen, Y.-C. Hsu, H.-H. Chou and M.-C. Yeh, *Org. Lett.*, **11**, 97 (2009).
09OL169	K. Okuma, A. Nojima, N. Matsunaga and K. Shioji, *Org. Lett.*, **11**, 169 (2009).
09OL377	K.-F. Chen, Y.-C. Hsu, Q. Wu, M.-C.P. Yeh and S.-S. Sun, *Org. Lett.*, **11**, 377 (2009).
09OL405	J. Li and S.Q. Yao, *Org. Lett.*, **11**, 405 (2009).
09OL2049	S.L. Niu, G. Ulrich, R. Ziessel, A. Kiss, P.-Y. Renard and A. Romieu, *Org. Lett.*, **11**, 2049 (2009).
09OL5398	A. García, B. Insuasty, M.Á. Herranz, R. Martínez-Álvarez and N. Martín, *Org. Lett.*, **11**, 5398 (2009).

09OL5542	C. Teng, X. Yang, C. Yuan, C. Li, R. Chen, H. Tian, S. Li, A. Hagfeldt and L. Sun, *Org. Lett.*, **11**, 5542 (2009).
09SA(A71)2030	S. Sato, M. Tsunoda, M. Suzuki, M. Kutsuna, K. Takido-uchi, M. Shindo, H. Mizuguchi, H. Obara and H. Ohya, *Spectrochim. Acta, Part A*, **71**, 2030 (2009).
09SC580	Y.-H. Liu, S.-Y. Tao, L.-Q. Lei and Z.-H. Zhang, *Synth. Commun.*, **39**, 580 (2009).
09SE(83)574	V. Kandavelu, H.-S. Huang, J.-L. Jian, T.C.K. Yang, K.-L. Wang and S.-T. Huang, *Sol. Energ.*, **83**, 574 (2009).
09SM1028	W. Wu, F. Meng, J. Li, X. Teng and J. Hua, *Synth. Met.*, **159**, 1028 (2009).
09SNA(B135)625	M. Zhao, X.-F. Yang, S. He and L. Wang, *Sensor. Actuator. B. Chem.*, **135**, 625 (2009).
09SNA(B141)116	S.M.S. Chauhan, T. Bisht and B. Garg, *Sensor. Actuator. B. Chem.*, **141**, 116 (2009).
09SNA(B143)42	V.B. Bojinov, A.I. Venkova and N.I. Georgiev, *Sensor. Actuator. B. Chem.*, **143**, 42 (2009).
09T5257	L.-L. Jiang, B.-L. Li, F.-T. Lv, L.-F. Dou and L.-C. Wang, *Tetrahedron*, **65**, 5257 (2009).
09TL1614	S. Hermes, G. Dassa, G. Toso, A. Bianco, C. Bertarelli and G. Zerbi, *Tetrahedron Lett.*, **50**, 1614 (2009).
09TL2920	R. Andreu, L. Carrasquer, J. Garín, M.J. Modrego, J. Orduna, R. Alicanta, B. Villacampa and M. Allain, *Tetrahedron Lett.*, **50**, 2920 (2009).
09US7	G.H. Mahdavinia, S. Rostamizadeh, A.M. Amani and Z. Emdadi, *Ultrason. Sonochem.*, **16**, 7 (2009).
09USP001	P. Steffanut, J.-C. Graciet, L. Lucke and M.A. Winter, *U.S. Patent Application 2009/0018318 A1* (2009).
10ANCA(663)85	Q.-J. Ma, X.-B. Zhang, X.-H. Zhao, Z. Jin, G.-J. Mao, G.-L. Shen and R.-Q. Yu, *Anal. Chim. Acta*, **663**, 85 (2010).
10ARI(68)96	T.K. Heinrich, V. Gottumukkala, E. Snay, P. Dunning, F.H. Fahey, S.T. Treves and A.B. Packard, *Appl. Radiat. Isot.*, **68**, 96 (2010).
10CEJ158	K. Kolmakov, V.N. Belov, J. Bierwagen, C. Ringemann, V. Müller, C. Eggeling and S.W. Hell, *Chem. Eur. J.*, **16**, 158 (2010).
10CM1836	M.K.R. Fischer, S. Wenger, M. Wang, A. Mishra, S.M. Zakeeruddin, M. Grätzel and P. Bäuerle, *Chem. Mater.*, **22**, 1836 (2010).
10COCB64	J.O. Escobedo, O. Rusin, S. Lim and R.M. Strongin, *Curr. Opin. Chem. Biol.*, **14**, 64 (2010).
10DP(85)43	B. Wang, J. Fan, S. Sun, L. Wang, B. Song and X. Peng, *Dyes Pigm.*, **85**, 43 (2010).
10DP(85)133	A. Zarei, A.R. Hajipour and L. Khadooz, *Dyes Pigm.*, **85**, 133 (2010).
10DP(85)183	S. Lengvinaite, J.V. Grazulevicius, R. Gu, W. Dehaen, V. Jankauskas, B. Zhang and Z. Xie, *Dyes Pigm.*, **85**, 183 (2010).
10DP(86)32	H. Dinçalp, S. Kizilok and S. İçli, *Dyes Pigm.*, **86**, 32 (2010).
10DP(86)50	F.-J. Huo, J. Su, Y.-Q. Sun, C.-X. Yin, H.-B. Tong and Z.-X. Nie, *Dyes Pigm.*, **86**, 50 (2010).
10DP(86)190	Y. Wang, X. Zhang, B. Han, J. Peng, S. Hou, Y. Huang, H. Sun, M. Xie and Z. Lu, *Dyes Pigm.*, **86**, 190 (2010).
10JA2795	D. Oushiki, H. Kojima, T. Terai, M. Arita, K. Hanaoka, Y. Urano and T. Nagano, *J. Am. Chem. Soc.*, **132**, 2795 (2010).
10JL654	J.-X. Yang, L. Li, C.-X. Wang, Y.P. Tian, F. Wu, X.-J. Xing, C.-K. Wang, H.-H. Tang, W.-H. Huang, X.-T. Tao and M.-H. Jiang, *J. Lumin.*, **130**, 654 (2010).
10JOC506	X. Huang and T. Zhang, *J. Org. Chem.*, **75**, 506 (2010).
10JOC1297	D.T. Gryko, J. Piechowska and M. Gałęzowski, *J. Org. Chem.*, **75**, 1297 (2010).
10JOC1684	R. Andreu, E. Galán, J. Garín, V. Herrero, E. Lacarra, J. Orduna, R. Alicante and B. Villacampa, *J. Org. Chem.*, **75**, 1684 (2010).
10JPC(C)1343	S. Qu, W. Wu, J. Hua, C. Kong, Y. Long and H. Tian, *J. Phys. Chem. C*, **114**, 1343 (2010).
10JPCP(A210)89	V.B. Bojinov, I.P. Panova, D.B. Simeonov and N.I. Georgiev, *J. Photochem. Photobiol. Chem.*, **210**, 89 (2010).

10OE498	G. Cheng, T. Fei, Y. Zhao, Y. Ma and S. Liu, *Org. Electron.*, **11**, 498 (2010).
10OL16	H.-Y. Yang, Y.-S. Yen, Y.-C. Hsu, H.H. Chou and J.T. Lin, *Org. Lett.*, **12**, 16 (2010).
10OL100	X. Xu, X. Xu, H. Li, X. Xie and Y. Li, *Org. Lett.*, **12**, 100 (2010).
10OL932	J.H. Do, H.N. Kim, J. Yoon, J.S. Kim and H.J. Kim, *Org. Lett.*, **12**, 932 (2010).
10SNA(B145)433	L. Dong, C. Wu, X. Zeng, L. Mu, S.F. Xue, Z. Tao and J.X. Zhang, *Sensor. Actuator. B. Chem.*, **145**, 433 (2010).
10TL422	R. Meesala, R. Nagarajan, *Tetrahedron Lett.*, **51**, 422 (2010).
10TL442	R. Kumar, G.C. Nandi, R.K. Verma and M.S. Singh, *Tetrahedron Lett.*, **51**, 442 (2010).
10TL1596	T. Yamagata, J. Kuwabara and T. Kanbara, *Tetrahedron Lett.*, **51**, 1596 (2010).
10TL1600	Y. Kubo, Y. Minowa, T. Shoda and K. Takeshita, *Tetrahedron Lett.*, **51**, 1600 (2010).

CHAPTER 3
Three-Membered Ring Systems

Stephen C. Bergmeier*, David J. Lapinsky**
*Department of Chemistry & Biochemistry, Ohio University, Athens, OH, USA
bergmeis@ohio.edu
**Division of Pharmaceutical Sciences, Duquesne University, Pittsburgh, PA, USA
lapinskyd@duq.edu

3.1. INTRODUCTION

This review covers the chemical literature on epoxides and aziridines for the year 2009. As in previous years, this review is not comprehensive, but rather covers a selection of synthetically useful and interesting reactions. Three-membered ring systems, epoxides and aziridines in particular, are excellent synthetic intermediates. This is largely due to their ability to be converted into other functional groups such as diols, diamines, and amino alcohols to name but a few. The chapter has been divided into two sections, one covering epoxides and the other covering aziridines. Each of these sections has been further divided into two additional sections, one on the synthesis of the aziridine or epoxide and one on the reactions of the heterocycle. In addition to the usual coverage of epoxides and aziridines, a number of reports on the synthesis and reactions of azirines and oxaziridines are included as well. As is typical, reactions of epoxides often are reported along with a similar reaction of an aziridine and we will note such reactions as they occur.

3.2. EPOXIDES
3.2.1. Preparation of Epoxides

The epoxide ring system continues to be one of the most studied heterocyclic ring systems due to its utility as a synthetic intermediate and potential biological activity. A SciFinder search on the term epoxides provides over 600 answers for the year 2009. Metal-catalyzed epoxidations continue to be an attractive method for the synthesis of epoxides. Most of the work in the past year has focused on the identification of new catalysis systems as well as improving the general utility through attachment of the catalyst systems to insoluble supports. A review on the organocatalyzed enantioselective epoxidation of α,β-enones has been communicated this past year <09S1551>. In addition, a review on the hydrolytic kinetic resolution of racemic epoxides has been reported <09SL1367>.

A very interesting use of an oxo-version of an RCM catalyst for epoxidation has been reported <09TL5344>. The Mo-catalyst **1** readily epoxidizes a variety of olefins with good chemoselectivity. Electron-rich olefins are epoxidized selectively over electron-poor olefins.

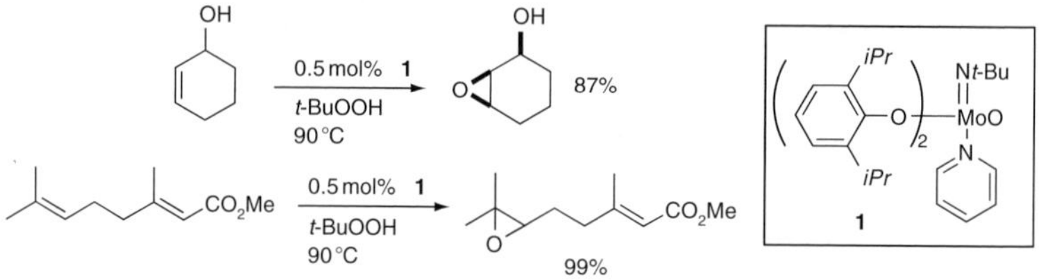

A Mo-based catalyst linked to a solid support (MCM-41,**2**) provides excellent yields of the corresponding epoxides <09MI853; 09JMOC110>. A similar Mo-catalyst with a chiral proline-derived ligand has been shown to provide moderate levels of enantioselectivity in the epoxidation of styrene <09TL2509>. Additionally, a number of manganese-derived catalysts have been reported in the past year <09JOC1135, 09OL3622, 09MI348>.

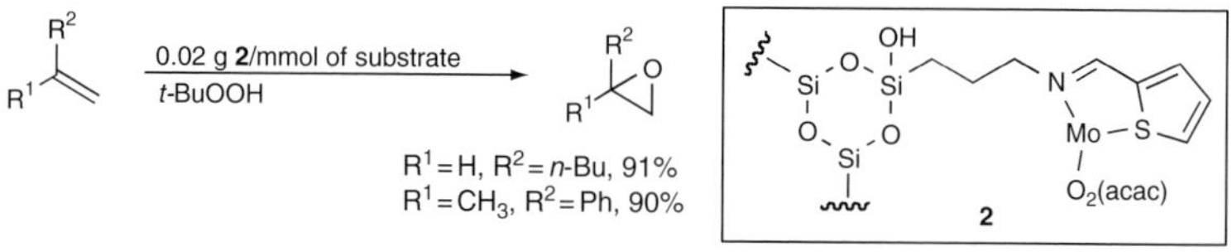

A rather unique tungsten/selenium catalyst has been developed for the epoxidation of allylic and homoallylic alcohols <09JA6997>. Treatment of allylic and homoallylic alcohols with $[SeO_4\{WO(O_2)\}_2]^{2-}$ and H_2O_2 provided the corresponding epoxides in generally good yields albeit with no diastereoselectivity.

Titanium and vanadium have long been used in epoxidation reactions. Two new ligand/oxidant combinations have been developed for epoxidations. In a somewhat typical reaction, $VOSO_4$ and ligand **3** provides good yields of the hydroxyepoxides with good to moderate levels of enantioselectivity <09JOC3350>. A rather interesting epoxidation using a catalyst derived from $Ti(OiPr)_4$ and ligand **4** provides a moderate yield of styrene oxide with good enantioselectivity <09AG(I)7432>.

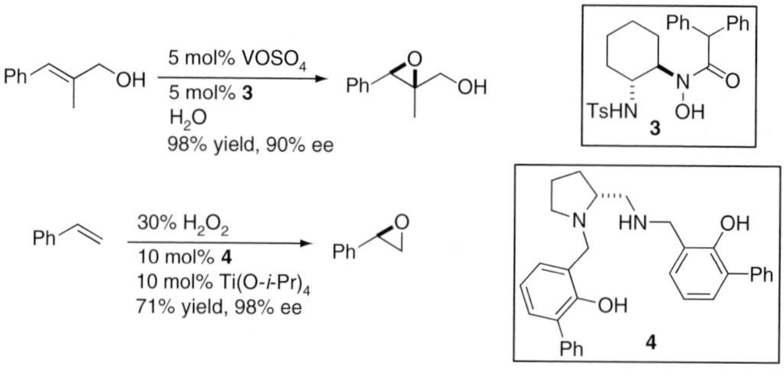

An interesting self-supported catalyst has been developed for the epoxidation of chalcone and chalcone derivatives. Treatment of **5** with Et_2Zn provided an insoluble catalyst <09TL2200>. Combination of 10 mol% of the resulting catalyst with cumene hydroperoxide yielded the corresponding epoxides with good yield and high levels of enantioselectivity. A useful aspect of this catalyst system is reusability of the catalyst and lack of need for a separate solid support.

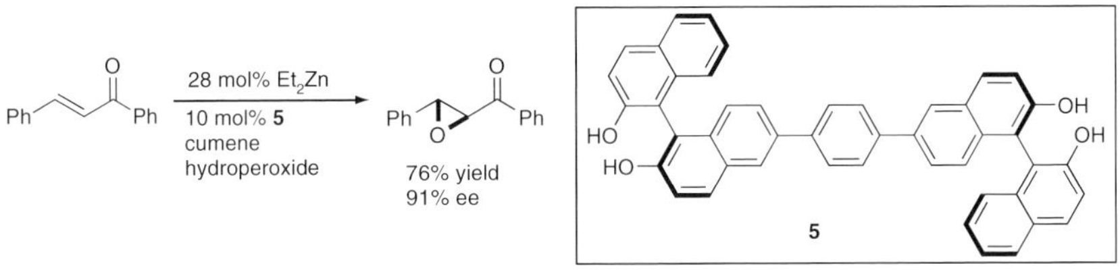

The directed epoxidation of cyclic allylic alcohols has long been a useful method for controlling the relative stereochemistry of epoxidation. A recent publication has examined the directed epoxidation of allylic amines <09JOC6735>. These authors found that an initial protonation of the amine, followed by epoxidation with a carefully controlled quantity of mCPBA provided excellent stereoselectivity in some cases.

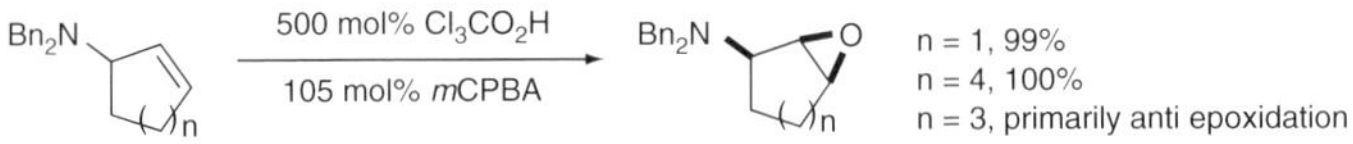

A unique and potentially very useful method of epoxidation has been reported this past year. Photolytic epoxidation using O_2, benzaldehyde, and a simple fluorescent lamp provided very good yields of a wide variety of epoxides <09SL3024>.

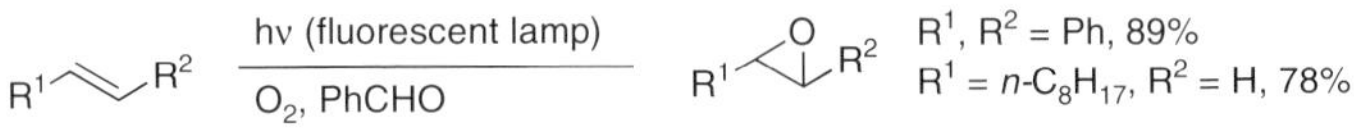

A very interesting palladium-catalyzed epoxidation/arylation method has been reported <09JA2052>. Reaction of a tertiary allylic alcohol with a palladium catalyst and an aryl- or alkenylhalide provides excellent yields of the aryl or alkenylated epoxides. The primary limitation is the necessity of using a tertiary alcohol in the reaction. This reaction has been shown to be effective with aziridines as well as epoxides <09AG(I)7224>.

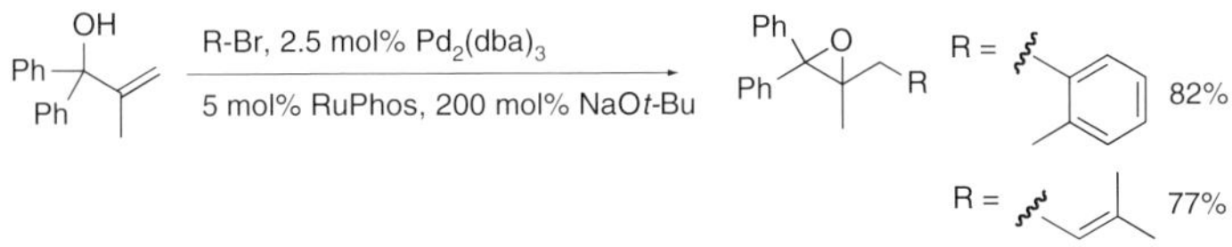

Research into the optimization and understanding of chiral dioxiranes for epoxidation has continued. New developments include the synthesis of fluoro-epoxides <09JOC8377>, optimization of the synthesis of *cis*-epoxides <09OL5150>, and new sugar-derived ketone dioxirane precursors <09EJO6009; 09JOC6335; 09TA883>. A potentially useful dioxirane method using dehydro cholic acid as the dioxirane precursor has been reported <09S1123>. Treatment of dehydro cholic acid with sodium perborate ($NaBO_3 \cdot 4H_2O$) followed by reaction with the olefin provides the epoxides in very good yield with good chemo- and stereoselectivity.

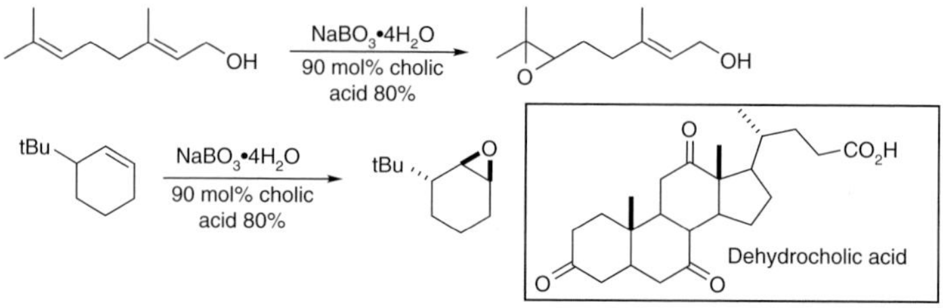

Another dioxo species that has been useful in epoxidations are the gem-dihydroperoxides. It has recently been reported that the gem-dihydroperoxide derived from verbenone can be used to enantioselectively epoxidize naphthoquinones <09TL4629>. Treatment of a naphthoquinone with a verbenone-derived gem-hydroperoxide provides the corresponding epoxides in good to excellent yield and moderate enantioselectivity.

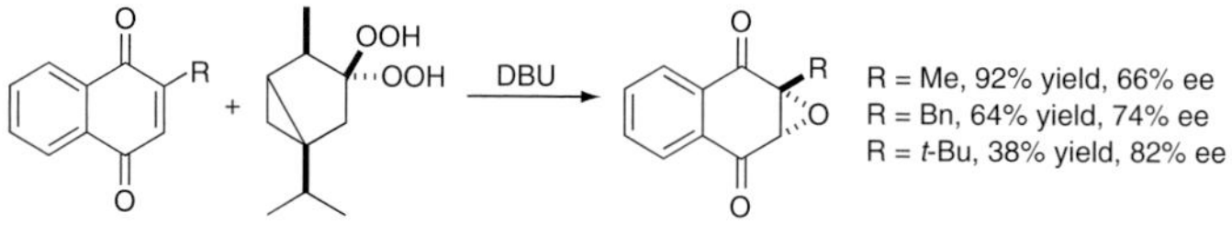

Binaphthyl-derived iminium salts continue to be examined as chiral epoxidation reagents <09T2910, 09EJO3413>. A recent study has shown that the counterion can have a significant effect on the level of enantioselectivity <09MI596>. The use of the common Ph_4B anion can have deleterious effects on prolonged epoxidation conditions as this counterion undergoes decomposition under the oxidative conditions to give biphenyl. It was suggested that the SbF_6 anion is a better counterion.

Julia-Colonna-type epoxidation methods continue to be developed. A poly(leucine) derivative linked to an imidazolium salt has provided an operationally simple method for the asymmetric epoxidation of chalcone derivatives in excellent yield and enantioselectivity <09TL5225>. A guanidine-linked bis-urea has been used as the catalyst for the epoxidation of chalcones in excellent yields and with good enantioselectivity <09SL667>.

Enzymes continue to receive attention as valuable reagents for the synthesis of chiral epoxides. An excellent review on different enzymes and reaction conditions has been published in the past year <09MI239>. The use of enzymatic epoxidation in a continuous flow reactor has also been reported <09MI27>. An interesting study on the use of several different enzymes (YcnD and YhdA from *Bacillus subtilis* and Lot6p from *Saccharomyces cerevisiae*) for epoxidation has been reported <09OBC1115>. These FMN-dependent oxidoreductases were examined for their ability to epoxidize α,β-unsaturated ketones and aldehydes.

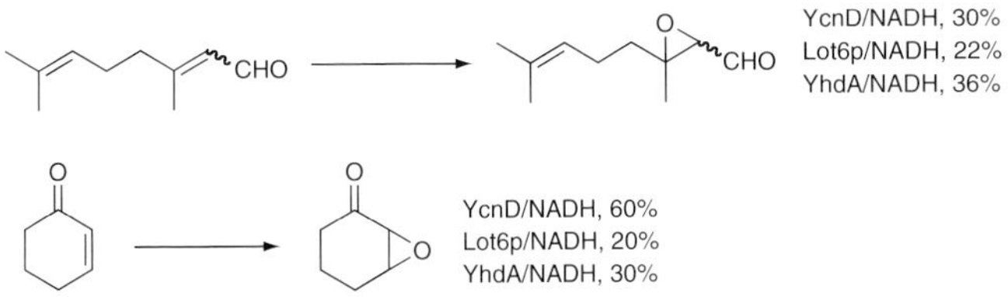

The addition of a nucleophile containing an extant leaving group to an aldehyde, e.g., Darzens condensation, continues to be a fertile ground for the development of new methods for epoxide synthesis. For example, the addition of allylzinc to α-haloketones provides ready access to allyl-substituted epoxides via a Darzens-type approach <09MI10732>. A sulfur ylide approach for the synthesis of hydroxyepoxides provides a useful method <09CC5763>. A sulfur ylide derived from methionine provides the corresponding epoxy amide in good yields and excellent enantioselectivity. It is exciting to see that the use of an aliphatic aldehyde provided similar yields and enantioselectivity as aryl aldehydes. A two-step reduction of the amide provided the epoxy alcohols in generally good yields.

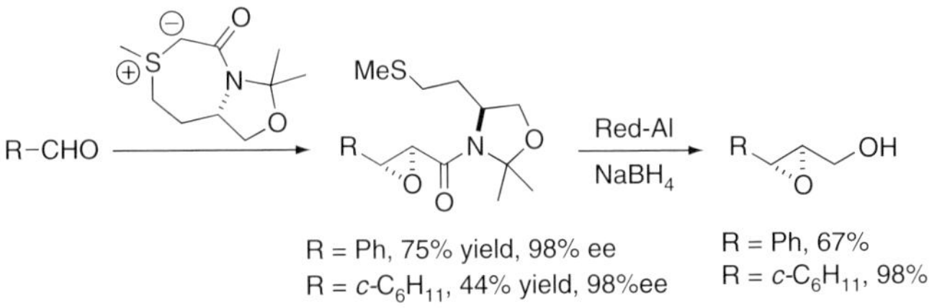

A similar type of reaction is the reaction of an aldehyde with some type of diazocarbonyl derivative. The reaction of diazoamides with aldehydes provides another nice route for the synthesis of epoxy amides <09AG(I)6503>.

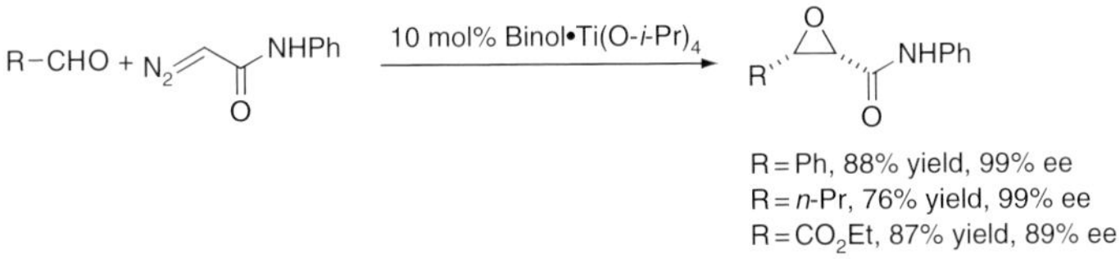

The formation of epoxides from more readily obtainable epoxides continues to be an exciting area of chemistry. An interesting example involving the decarbonylation of an epoxy aldehyde has been reported in the past year <09SL2076>. Reaction of an epoxy aldehyde with a rhodium catalyst at high temperatures provides decarboxylated epoxides in moderate yields. The corresponding ketones are also obtained as a byproduct of this reaction.

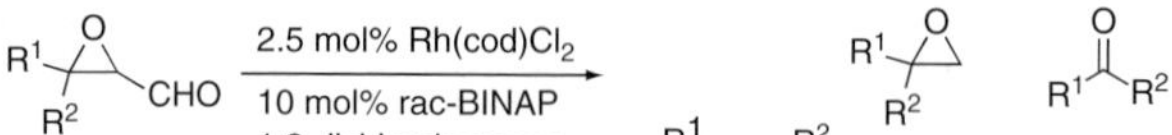

R¹	R²		
Ph	H	63%	26%
n-heptyl	H	64%	23%
Ph	Me	40%	20%

Reaction conditions: 2.5 mol% Rh(cod)Cl₂, 10 mol% rac-BINAP, 1,2-dichlorobenzene, 135–145 °C, 14h.

An additional example of the synthesis of epoxides from more readily obtainable epoxides is the asymmetric allylic oxidation of a meso epoxide <09OL3314>. The reaction of an allylic epoxide with t-Bu-perbenzoate, a Cu-catalyst, and a chiral ligand (**6**) provided a moderate yield of a benzoate-substituted epoxide in 85% ee. The enantiomeric purity can be dramatically improved via hydrolysis and conversion to the p-nitrobenzoate ester which can be recrystallized to >99% ee.

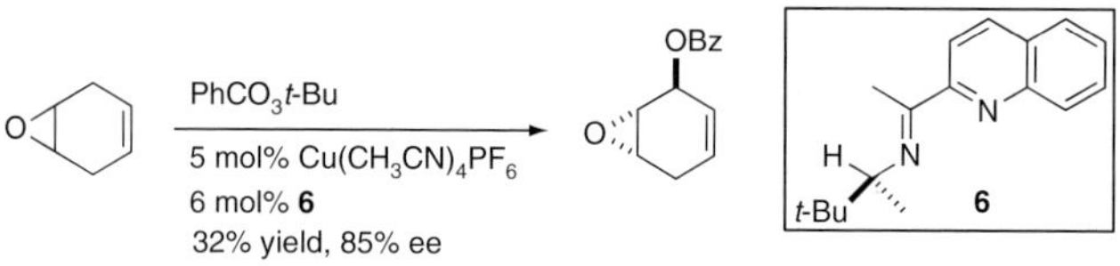

Reaction: PhCO₃t-Bu, 5 mol% Cu(CH₃CN)₄PF₆, 6 mol% **6**, 32% yield, 85% ee.

3.2.2. Reactions of Epoxides

The primary type of epoxide reaction remains the nucleophilic ring-opening reaction. Many of these studies focus on improving regioselectivity and determining milder reaction conditions. A review covering the development of endo-selective epoxide opening reactions in water has been published in the past year <09CSR3175>. The ring-opening reaction of an epoxide with an alcohol in the absence of any additional solvent has been found to be catalyzed by Er(OTf)₃ <09S3433>. This reaction can be carried out with a variety of alcohols including t-BuOH and phenol. Mesoporous aluminosilicate has also been identified as a useful method for the ring-opening reaction of an epoxide with an alcohol in the absence of additional solvent <09OBC2559>. When this reaction is carried out in an aprotic solvent (CH₂Cl₂) in the absence of any alcohol, only the product of a Meinwald rearrangement is obtained. NaHSO₄ can effectively open epoxides to provide either the sulfonate ester (CH₂Cl₂ as solvent) or the diol (H₂O added to the reaction) <09T985>.

Conditions	Result
0.1 mol% Er(OTf)₃, i-PrOH, no solvent	100%, R = i-Pr
50 mg/mmol mesoporous aluminosilicate, i-PrOH	90%, R = i-Pr
NaHSO₄, CH₂Cl₂	70%, R = SO₃H
NaHSO₄, H₂O, CH₂Cl₂	75%, R = H

Intramolecular ring-opening reactions of epoxides with oxygen nucleophiles provide ready access to a variety of heterocyclic ring systems. As shown below, hydrogenolysis of a benzyl ether, followed by treatment with base, provided a dihydrobenzofuran in good yield <09S1886>. The presence of a high concentration of sodium anion was crucial to the success of the cyclization reaction.

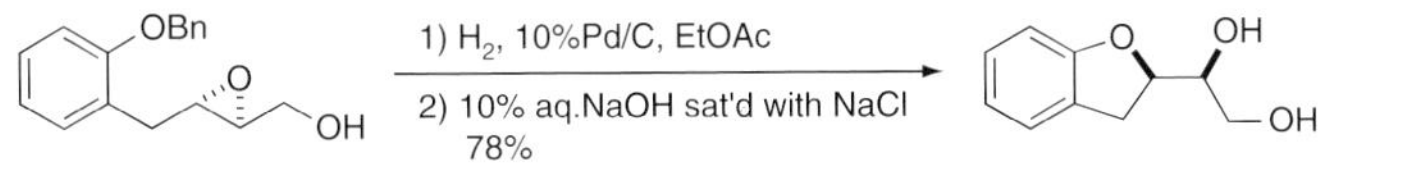

While carbonyls have been used as nucleophiles to open epoxides, the following example is interesting in that the epoxide is never isolated. Treatment of an unsaturated aldol with typical directed epoxidation conditions directly provides the lactone as the product <09OL2896>. In this rearrangement, the carbonyl opens the epoxide to form an intermediate acyliminium ion, which is hydrolyzed by the addition of water to provide the lactone. A similar reaction between an isolated epoxy ester provides the corresponding lactone in excellent yields <09TL2536>.

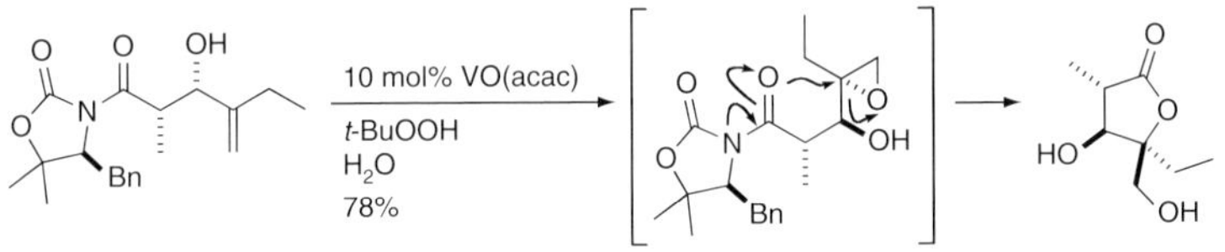

Two interesting and almost identical examples of ring opening of epoxides with a thiol or selenol using a disulfide or diselenide have been reported. Treatment of an epoxide with diphenyl disulfide, K_2CO_3, and Rongalite® ($NaHSO_2 \cdot CH_2O \cdot 2H_2O$) provides the hydroxysulfide in excellent yield <09S3267; 09T5240>. The reaction is also reported to work using diphenyl diselenide with aziridines <09S3267>. Treatment of epoxides with sodium sulfinate has been shown to provide β-hydroxy sulfones in good yield <09TL5009>. As exemplified by the reaction of styrene oxide and methods A, B, or C, this reaction is not an acid-catalyzed process. β-Hydroxy phosphonates can be prepared by treatment of cyclohexene oxide with triethyl phosphite and Al(OTf)$_3$ <09T7691>. Other examples of thiol opening of epoxides include the use of a chiral heterobimetallic Ti-Ga-Salen complex, which provides the hydroxysulfides with up to 92% ee <09TL548; 09MI920>. The reaction of silylated thiols and selenols with epoxides has been reported to be catalyzed by PhONBu$_4$ <09TL2808>.

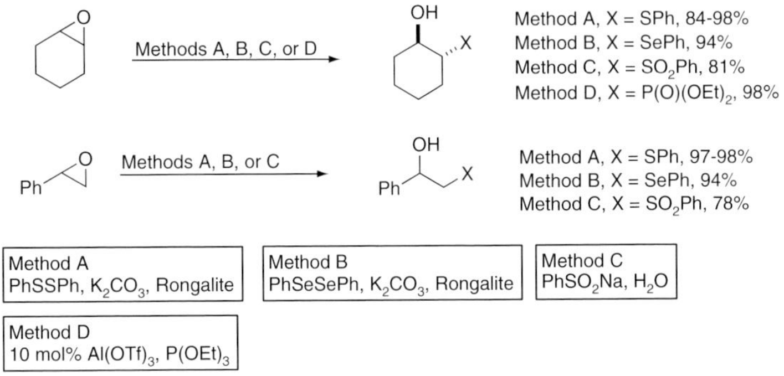

Several useful examples of ring opening reactions of epoxides with amine nucleophiles have been reported this past year. The reaction of an α-amino ester with epoxides in trifluoroethanol leads to excellent yields of the corresponding aminoalcohols <09OBC2026>. This provides an excellent and mild method to attach an alkyl group to the N-terminus of an amino acid or even a peptide.

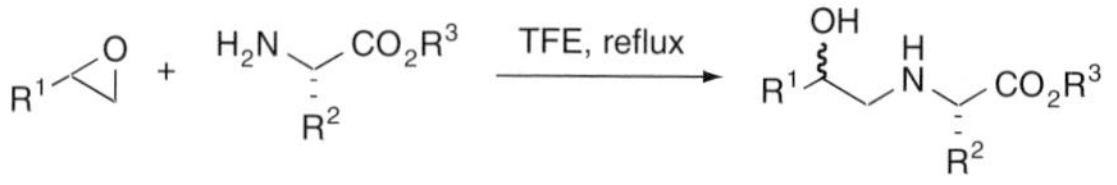

R^1	R^2	R^3	%Yield
PhOCH$_2$	i-Pr	Bn	92
	Ph	Me	85
	H	Et	54
PhthNCH$_2$	i-Pr	Me	96
n-C$_{10}$H$_{21}$	i-Pr	Bn	78

Treatment of a racemic alkynyl epoxide with an aniline derivative, a copper catalyst, and (R)-DTBM-MeO-BIPHEP as a chiral ligand, provided an excellent yield of the ring-opened product with moderate enantioselectivity <09JOC7603>. This reaction is not some type of resolution, but rather involves the formation of an allenylidene intermediate, which then reacts with the amine to provide the product.

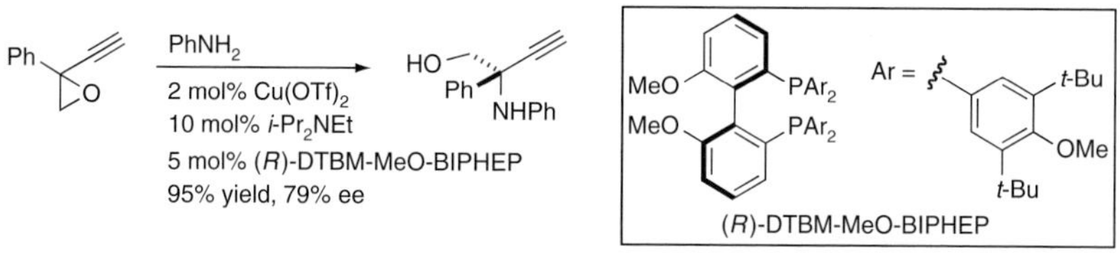

A very interesting example of a kinetic resolution of a vinyl epoxide with a variety of amine derivatives has been reported <09OL3258>. Treatment of butadiene monoepoxide with a hydroxylamine (or hydrazine) derivative, a Pd-catalyst, and chiral ligand **7** provides the amino alcohol in good yields with moderate enantioselectivity. This is a very useful example of a kinetic resolution of vinyl epoxides in that it provides an amino alcohol product.

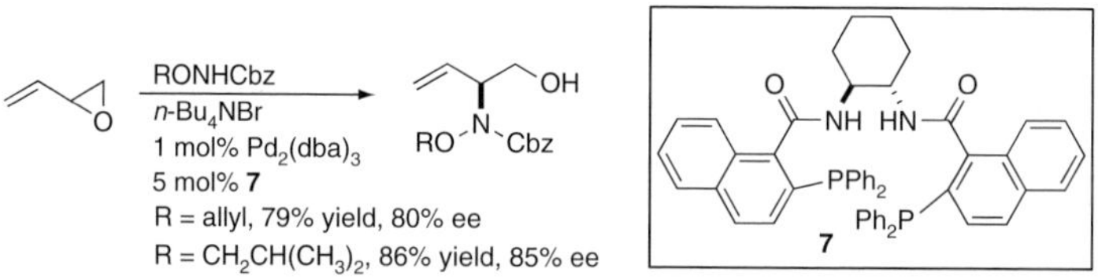

An interesting indole synthesis makes use of an epoxide opening by an amine as the key initial step <09TL1529>. Treatment of the epoxide with an aliphatic amine followed by heating leads to an amino alcohol, which then undergoes an S$_N$Ar/elimination pathway to the isolated indole.

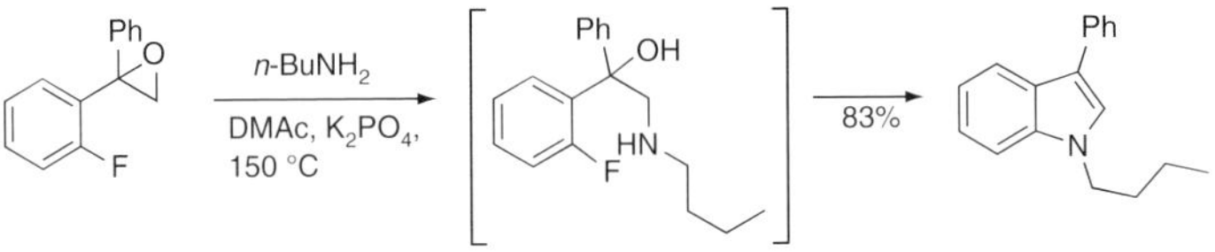

Radical reactions of epoxides provide useful methods to invert the "normal" reactivity of epoxides. In one example, the reaction of a substituted cyclohexene oxide with Cp_2TiCl and a nitrile provided good yields of the ring-opened and acylated products with generally good stereocontrol <09JOC3913>. An intramolecular variant of this reaction has been reported in which the epoxide is opened to form a radical, which then undergoes an intramolecular cyclization to provide a 2:1 mixture of diastereomeric bicyclo[3.1.1]heptenes <09JOC1798>. The Ti-catalyzed reduction/elimination of an epoxide has been shown to provide excellent yields of an allylic alcohol <09T10837>. Titanium-induced radical formation of epoxides has also been used to efficiently reduce terminal epoxides to primary alcohols <09T4984>.

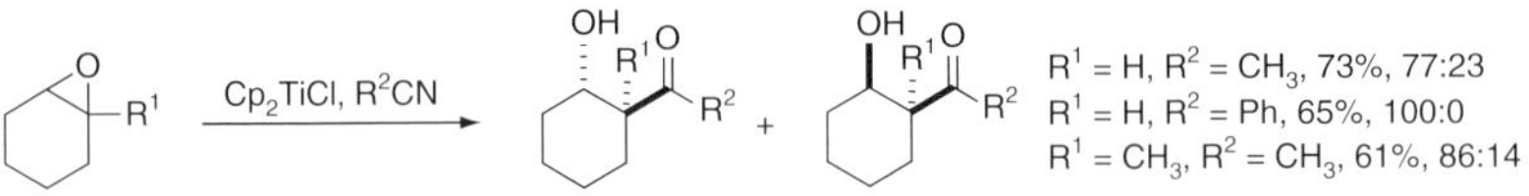

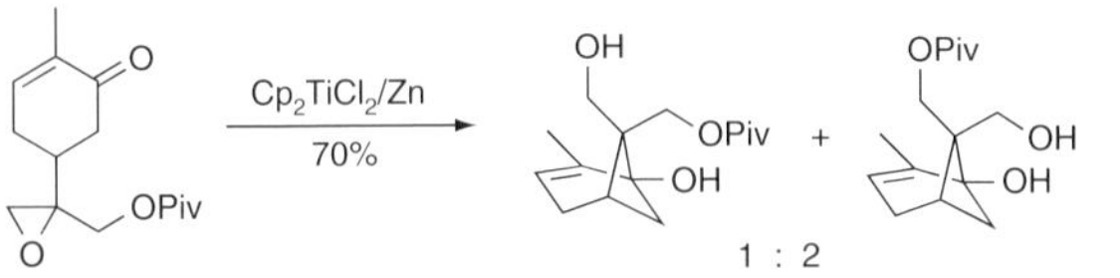

Another example of the formal reversal of reactivity of an epoxide is the recently reported Ni-catalyzed borylation of allylic epoxides <09OL3762>. Reaction of an allylic epoxide with bispinacolatoborane under Ni-catalysis provided an intermediate allylborate. Reaction with an aldehyde then provides the typical allylborate addition product with excellent diastereocontrol. This reaction also works well with aziridines.

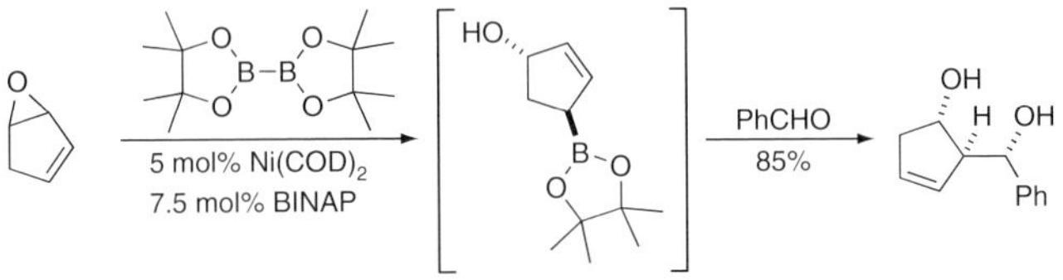

Carbanions are regularly used in ring-opening reactions with epoxides. For example, the Cu-catalyzed reaction of a pyridylmethyl anion provides the ring-opened product in good yields <09JOC1374>. A study on methods to provide both stereoisomers of the ring-opening product of a sulfone substituted allylic epoxide has been reported <09JA9150>. The use of Kochi's reagent (Li_2CuCl_4) in combination with Me$_2$Zn provides the syn ring-opened product, while Me$_3$Al/MeCu combination gave the expected anti ring-opened product.

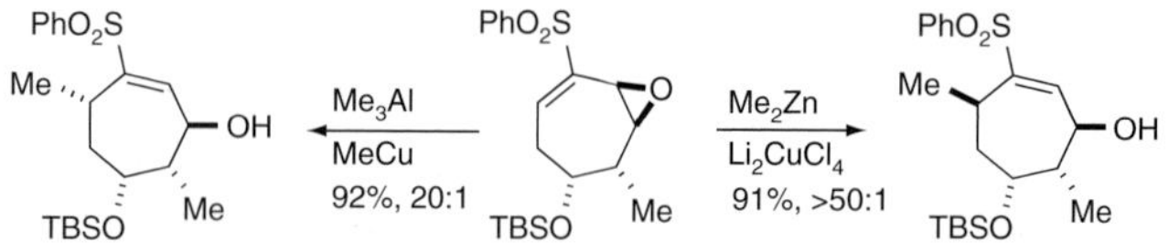

An interesting intramolecular reaction of the anion derived from a sulfone and an epoxide can provide either the 5-*exo* or the 6-*endo* cyclization product depending upon the reaction conditions <09T8668>. The simple anion derived by the deprotonation of the allylsulfone provided an ~2:1 mixture of the 6-*endo*:5-*exo* cyclization products. The addition of a Lewis acid to the reaction provided exclusively the 5-*exo* cyclization product as a mixture of diastereomers.

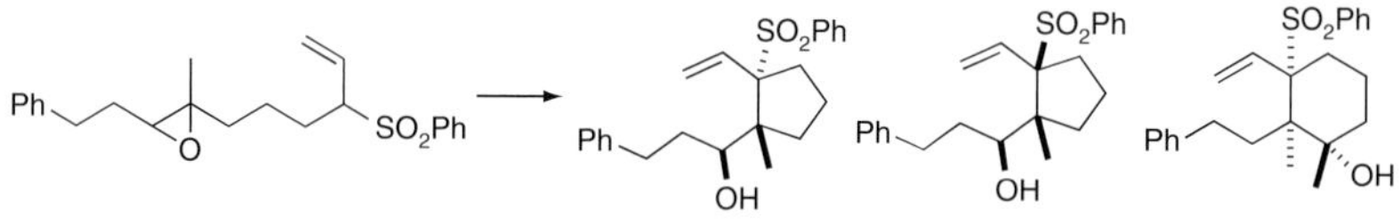

n-BuLi		16	35
n-BuLi/Me$_3$Al		66	12

An interesting intramolecular cyclization reaction involving a conjugate addition followed by cyclization has been used to generate highly substituted cyclobutane rings <09SL1157>. The sulfone-stabilized anion generated by the conjugate addition of lithium trimethylsilylacetylide cleanly undergoes a 4-*exo* cyclization to provide a highly functionalized cyclobutane.

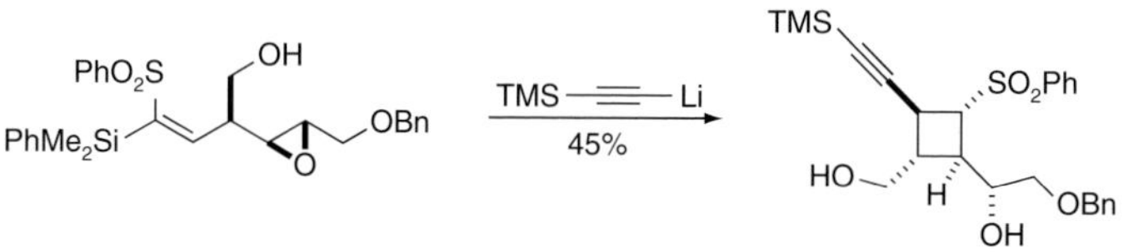

The reaction of epoxides with π-nucleophiles such as aromatic rings provides a useful method for synthesizing functionalized alcohols. Both inter- and intramolecular reactions have been reported. A solvent free method for the reaction of indole or pyrrole with styrene oxide using SbCl$_3$ on montmorillonite K-10 provided the ring-opened products in excellent yields <09TL916>.

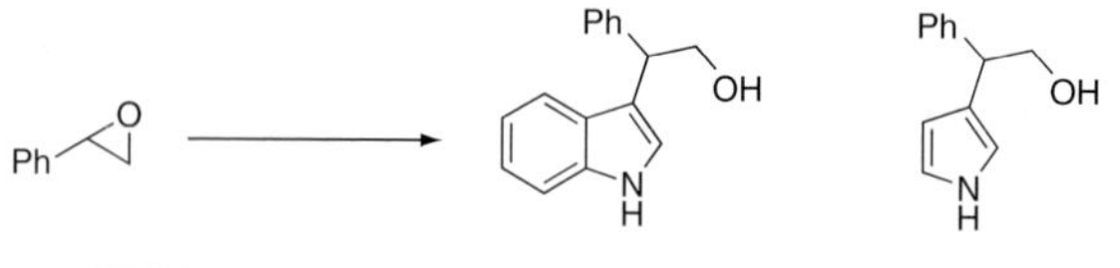

Indole, 10 mol% SbCl$_3$/K-10	90%
Pyrrole, 10 mol% SbCl$_3$/K-10	88%

In addition to the intermolecular variants, several examples of intramolecular cyclizations of epoxides with π-nucleophiles have been reported. An approach to a hexahydroxanthene system using a polyene-type cyclization provides the cyclized product in 25% yield <09TL3881>. A cyclization to form a 7-membered ring points out the importance of an electron-rich aromatic π-nucleophile as the 4-methoxy-substituted aromatic ring provides the cyclized product in 97% yield, while a simple benzene ring provided cyclized product in only 25% yield <09T9884>. An example of an 8-*endo* cyclization of a Co-complexed acetylenic vinyl oxirane has been reported providing the benzofused 8-membered ring in excellent yield <09TL26>. An interesting example highlighting the selectivity of this reaction is shown below <09OL3750>. When R = alkyl or another non-reactive group, only the product of the *exo*-epoxide cyclization, a tetrahydroquinoline, is obtained. However, when R = Bn, an *endo*-epoxide rearrangement takes place providing the vicinal aminoalcohol as the sole product.

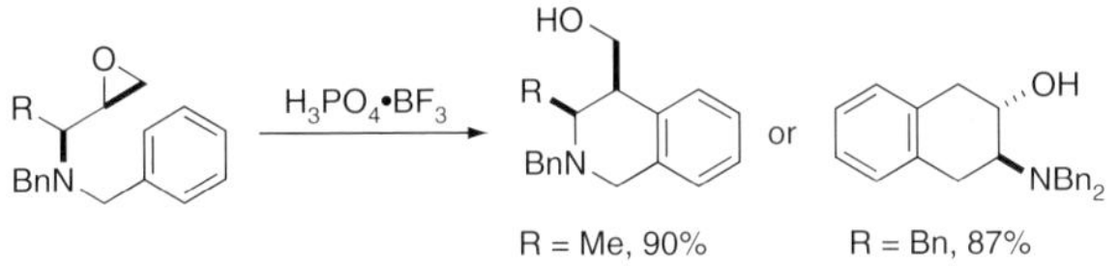

In addition to more typical carbon nucleophiles, silyl derivatives can also act as nucleophiles under appropriate reaction conditions. For example, the reaction of butadiene monoepoxide with a triethoxysilane under Cu-catalysis provides good yields of the S_N2' products as a mixture of the *E:Z* isomers <09MI8713>. An interesting 3-component coupling reaction makes use of a similar epoxide opening reaction in tandem with a Sonagashira coupling <09AG(I)391>.

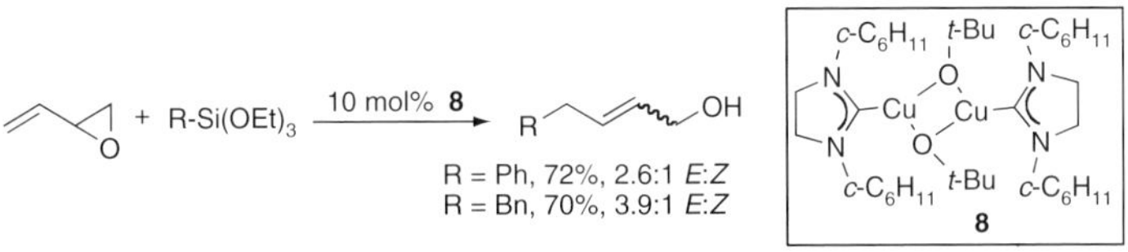

A very interesting reaction of a silacyclopropane with butadiene monoepoxide has been used to prepare a silaoxetane <09JA14182>. These highly reactive allylsilanes readily react with electrophiles such as aldehydes to provide a *trans*-dioxasilacyclooctene in good yields. This reaction works with a variety of vinyl epoxides to provide more highly substituted dioxasilacyclooctenes.

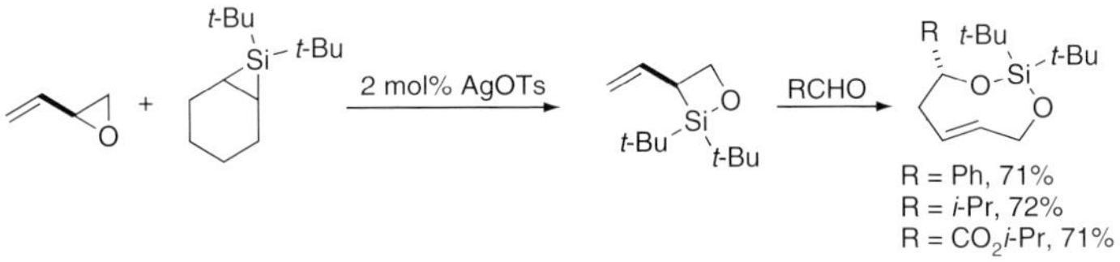

Rearrangement reactions of alkynyl epoxides continue to garner attention due to the wide range of structures available. For example, the Rh-catalyzed reaction of an alkynyl, epoxide with phenylboronic acid provides the allenol in good yield and excellent diastereoselectivity <09JOC6050>. The rhodium complexes with the alkyne, which leads to a syn-selective ring opening. A subsequent coupling with the phenylboronic acid provides the allenol in excellent yield. A similar reaction has also been reported using slightly altered reaction conditions <09H193>. A very similar alkynyl epoxide was readily converted to the fused-ring furan derivative upon treatment with AgOTf <09JOC4360>. Fused-ring epoxides as well as monocyclic epoxides work well in the reaction; however, the reaction with cyclopentene oxide provided no product. Terminal alkynes do not work in the reaction and only provide unidentified byproducts.

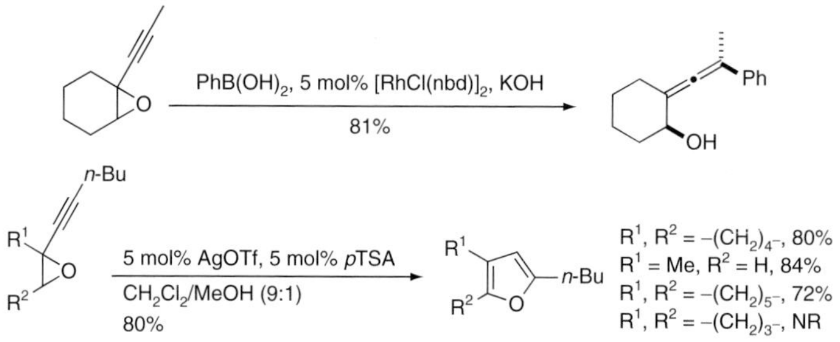

The same reaction can also be catalyzed by $PtCl_2$, which provided the fused-ring furans in similar yields <09S2454>. Other examples of the rearrangement of alkynyl epoxides have also been reported, including an example in which the resulting furan product undergoes a subsequent Friedel-Crafts reaction with furan <09OBC2501>. Other metal-catalyzed examples include intramolecular cyclizations of alkynes with epoxides <09T2643> and the cyclization of epoxides with allenes <09OL3490>.

3.3. AZIRIDINES
3.3.1. Preparation of Aziridines

The catalytic aziridination of alkenes with transition metal species in combination with suitable oxidants and coordinating ligands continues to attract significant attention. With respect to copper catalysis, $Cu(acac)_2$ proved an effective and selective catalyst in the aziridination of fatty acid methyl esters by (N-tosylimino)phenyliodinane <09S3769>. Likewise, an efficient copper iodide-catalyzed aziridination of alkenes with sulfamate esters and sulfonamides as the nitrogen source and iodosylbenzene as the oxidant was reported <09TL161>. This reaction provided aziridines in good to excellent yields under mild conditions utilizing a variety of alkenes containing electron-withdrawing, electron-donating, and sterically encumbered substrate combinations.

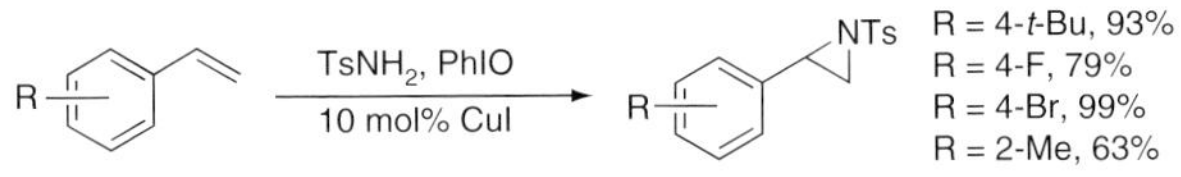

R—⟨Ar⟩—CH=CH2 →(TsNH2, PhIO, 10 mol% CuI)→ R—⟨Ar⟩—CH—CH2 with NTs
R = 4-*t*-Bu, 93%
R = 4-F, 79%
R = 4-Br, 99%
R = 2-Me, 63%

A highly efficient aziridination system based on trichloroethoxysulfonyl azide (TcesN$_3$) as a nitrene source and cobalt-based catalyst [Co(**P6**)] was developed <09CC4266>. This represented the first highly enantioselective and effective catalytic system for asymmetric aziridination of a broad range of simple olefins without the need of additional functionalities in the substrate for secondary binding interactions. The simple protocol established for this system allowed for recycling of the catalyst and proved complementary to previously reported Cu/ArI=NTs-based systems.

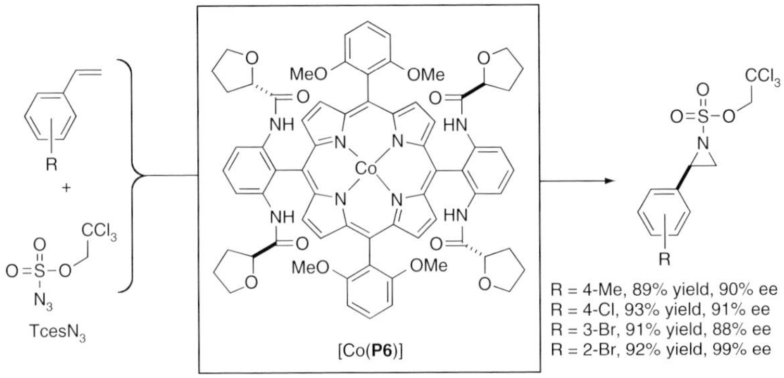

[Co(**P6**)]

R = 4-Me, 89% yield, 90% ee
R = 4-Cl, 93% yield, 91% ee
R = 3-Br, 91% yield, 88% ee
R = 2-Br, 92% yield, 99% ee

Diastereoselective control through hydrogen bonding was reported for the aziridination of chiral allylic alcohols by acetoxyaminoquinazolinone with lead tetraacetate <09JOC9452>. In a similar light, oxidation of primary amines with lead tetraacetate effected efficient intramolecular aziridinations to form corresponding 1-azatricyclo[2.2.1.0^{2,6}]heptanes and novel monoterpene analogues <09JA11998>. A mechanism was put forth where the amine displaces an acetate ligand from lead tetraacetate to generate lead(IV) amide intermediate RNHPb(OAc)$_3$ as the proposed aziridinating species.

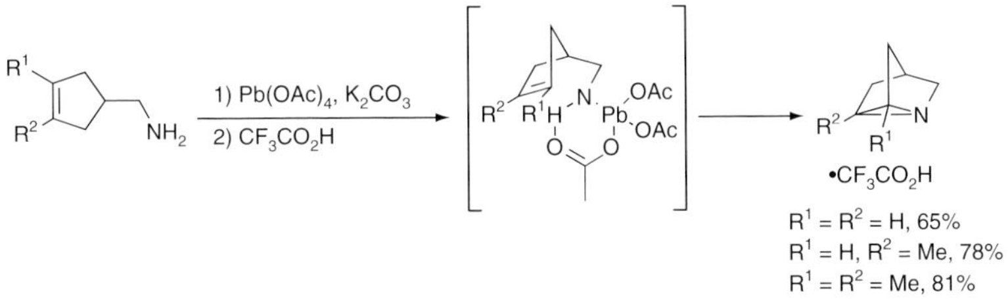

$R^1 = R^2 = H$, 65%
$R^1 = H, R^2 = Me$, 78%
$R^1 = R^2 = Me$, 81%

A highly stereoselective synthesis of aminoglycosides was achieved via an intramolecular rhodium-catalyzed and substrate-controlled aziridination of olefinic glycals <09OBC1284>. The resulting unstable aziridine glycal intermediates were treated in one pot with a host of nucleophiles (O-, S-, and N-based) towards stereoselective generation of polyfunctionalized α- or β-aminosaccharides.

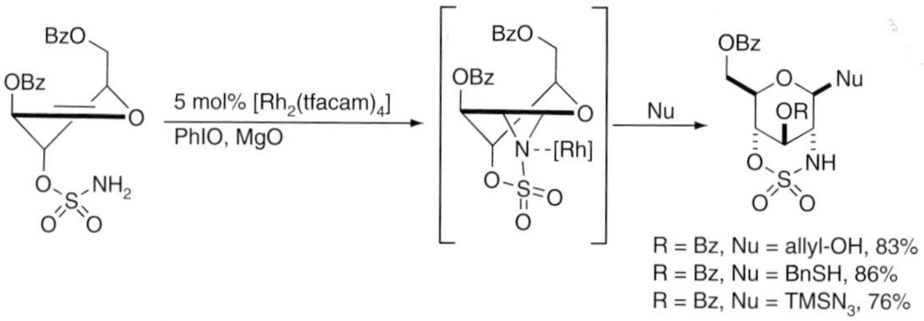

[Scheme: Rh-catalyzed aziridination with BzO-protected substrate]

R = Bz, Nu = allyl-OH, 83%
R = Bz, Nu = BnSH, 86%
R = Bz, Nu = TMSN$_3$, 76%

A new synthetic method for the preparation of aziridines by palladium-catalyzed reactions of allylamines with alkenyl or aryl halides has been reported <09AG(I)7224>. Studies utilizing deuterium-labeled substrates showed that this reaction proceeds via *syn* aminopalladation to produce a carbon-nitrogen bond concomitant to carbon-carbon bond formation. This reaction also proceeds with epoxides <09JA2052>.

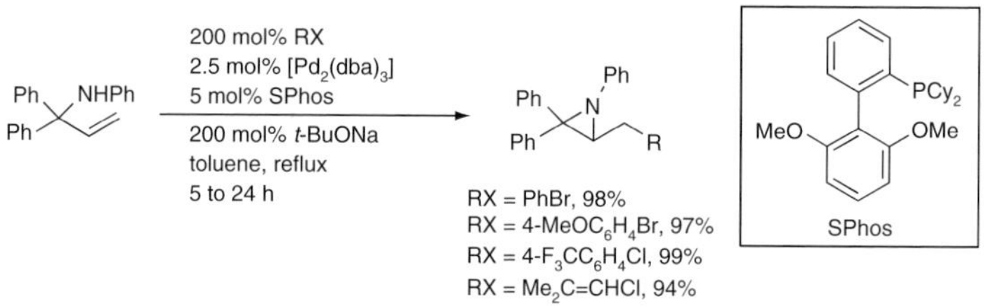

[Scheme: Pd-catalyzed aziridination of allylamine with aryl halides]

200 mol% RX
2.5 mol% [Pd$_2$(dba)$_3$]
5 mol% SPhos
200 mol% *t*-BuONa
toluene, reflux
5 to 24 h

RX = PhBr, 98%
RX = 4-MeOC$_6$H$_4$Br, 97%
RX = 4-F$_3$CC$_6$H$_4$Cl, 99%
RX = Me$_2$C=CHCl, 94%

SPhos

The aza-MIRC (Michael-Initiated Ring Closure) reaction constitutes an intriguing approach towards obtaining chiral aziridines in one synthetic step. This strategy was used to prepare multifunctionalized aziridines in high overall yields (82 – 92%) with a diastereomeric ratio up to 99:1 <09T484>. Both the chiral aldehyde employed and the electron-withdrawing groups (EWG) present on the generated Michael acceptor strongly controlled diastereomeric induction.

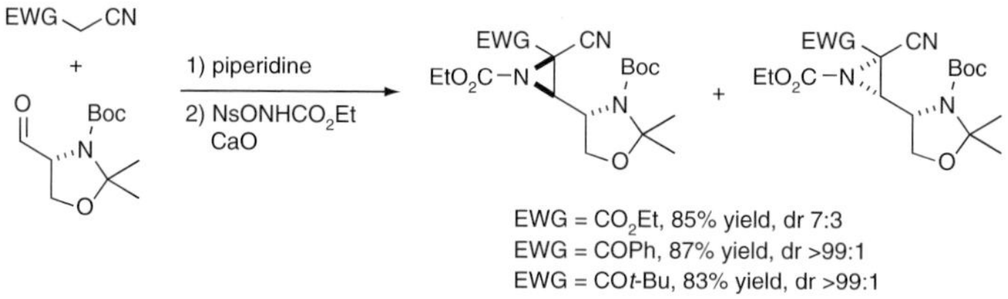

[Scheme: aza-MIRC reaction]

1) piperidine
2) NsONHCO$_2$Et
   CaO

EWG = CO$_2$Et, 85% yield, dr 7:3
EWG = COPh, 87% yield, dr >99:1
EWG = CO*t*-Bu, 83% yield, dr >99:1

Several examples of aziridines prepared via a 1,2-amino leaving group motif, either generated *in situ* or as part of the starting material, have been reported <09TL476; 09TA1969; 09JOC7548; 09AG(I)5760>. This aziridination strategy traditionally features an amine lone pair or an amide anion facilitating an intramolecular S$_N$2 reaction. A particularly intriguing example of this reaction is the highly regio- and stereoselective electrophilic addition of (*Z*)-alk-2-en-4-ynoates and TsNBr$_2$, resulting in a novel class of bromine-substituted aziridines containing three contiguous stereocenters <09OL5698>.

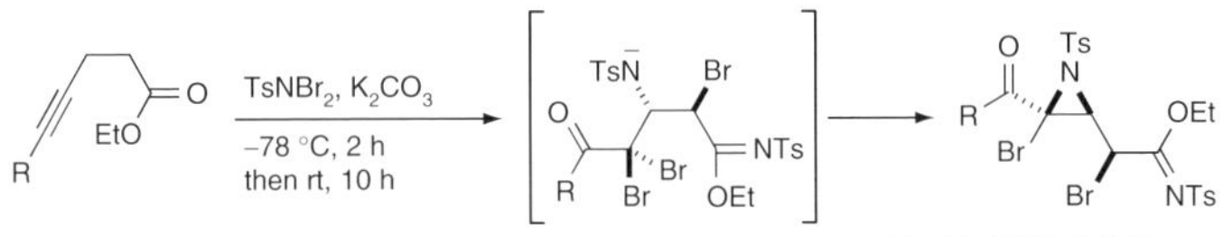

[Scheme: alkyne ester + TsNBr₂, K₂CO₃, –78 °C, 2 h then rt, 10 h → dibromo intermediate → aziridine with Ts, Br, CO-R, OEt, NTs substituents]

R = Ph, 71%, dr 91:9
R = *m*-Me-C₆H₄, 66%, dr 93:7
R = *p*-Cl-C₆H₄, 70%, dr 90:10
R = cyclopropyl, 68%, dr 87:13

In addition to alkenes, imines are tremendously popular aziridine precursors via an aza-Darzens or Darzens-like approach. Such an approach for aziridine preparation has been recently reviewed <09EJO4911>. Chiral aziridines were obtained by treating chiral sulfinyl imines with sulfur ylides generated either via decarboxylation of carboxymethylsulfonium betaines <09T70; 09SC2405> or deprotonation of (*S*)-dimethyl-[2-(*p*-toluenesulfinyl)phenyl]sulfonium salt <09JOC4217>. In a similar context, the first organocatalytic aziridination of *N*-tosyl imines with phenacyl bromide derivatives was reported via *in situ* generation of ammonium ylides <09SL3123>. This one-pot process provided aziridines in high yields with *trans* diastereoselectivity and could be made enantioselective by utilizing a chiral tertiary amine as the catalyst.

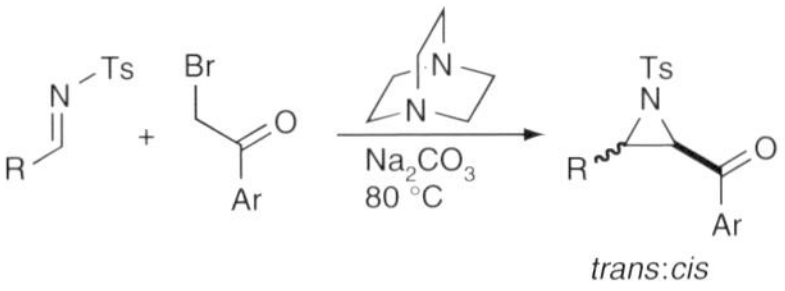

[Scheme: R-CH=N-Ts + Br-CH₂-C(O)-Ar, DABCO catalyst, Na₂CO₃, 80 °C → N-Ts aziridine with Ar and C(O) substituents, trans:cis]

R = Ar = Ph, 85%, dr 94:6
R = 4-OMeC₆H₄, Ar = 4-ClC₆H₄, 87%, dr 95:5
R = 4-NO₂C₆H₄, Ar = 4-ClC₆H₄, 84%, dr 94:6
R = *n*-Bu, Ar = 4-NO₂C₆H₄, 81%, dr 91:9

Additional reports of aziridine synthesis strategies employing imines include facile iodine(III)-induced oxidative cycloaddition with methylene compounds <09TL3857> and catalyst-free diastereoselective nitroaziridination with 1-bromonitroalkanes <09TL5420>. With respect to the synthesis of terminal aziridines, chiral (R_s,*R*)- and (R_s,*S*)-*N*-(*tert*-butylsulfinyl)-2-aryl-aziridines generated from imines were transformed into (*R*)- and (*S*)-2-aryl-1-benzylaziridines in good to excellent yield (74 – 94%) and enantiomeric excess (83 – 99% ee) via a short three-step procedure <09SL1265>. Additionally, good yields and diastereomeric ratios of terminal *N*-*tert*-butylsulfinyl azirdines were obtained upon Grignard or organocerium addition to *N*-(2-chloroethylidene)-*tert*-butylsulfinamide <09S1923>. In a similar approach as shown below, halogenated anions generated *in situ* by treatment of diiodo- or chloroiodomethane with methyllithium were allowed to react with chiral 2-aminoaldimines to yield enantiopure aziridines <09JOC2452>.

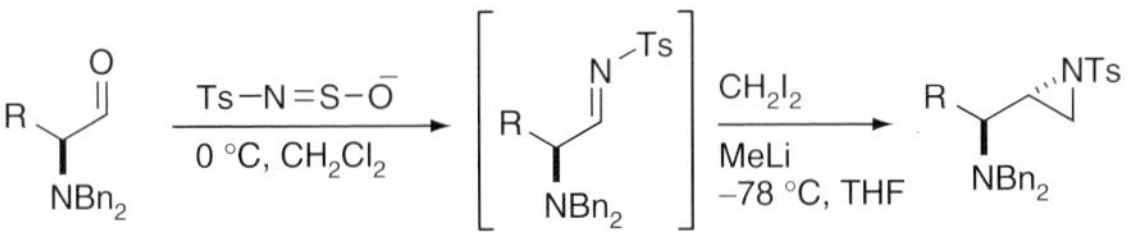

[Scheme: R-C(O)-CH(NBn₂) + Ts–N=S=O, 0 °C, CH₂Cl₂ → R-CH=N-Ts with NBn₂; then CH₂I₂, MeLi, –78 °C, THF → aziridine with NTs, NBn₂]

R = Me, 54%, dr 7/1
R = *i*-Bu, 58%, dr 5/1
R = Bn, 61%, dr 9/1

Chiral and achiral aziridination of imines with carbenes generated from diazo compounds continues to attract significant attention. Transition metal complexes

<09OBC187; 09OM3611>, Lewis or Brønsted acids <09JA15615; 09OL3036; 09OL2445; 09SL2715>, and organocatalysts <09MI2579> have been reported to promote this reaction. One unique report featured application of an F^+ organocatalyst in a mild, efficient, and straightforward protocol to generate structurally diverse racemic N-aryl and N-H aziridines in good yields and with good stereoselectivity <09OL4552>. Using this methodology, the racemic cis-N-H aziridine highlighted below was subsequently transformed in two steps to the antimicrobial chloramphenicol.

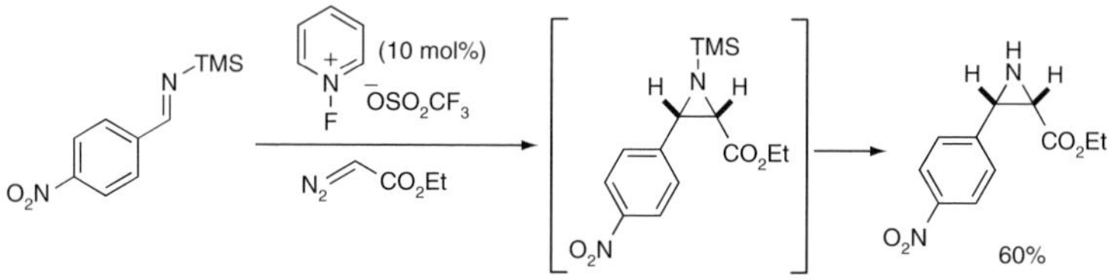

Given the aziridine moiety holds promise to serve as a versatile entity in the construction and design of bioactive agents, understanding the mechanism of its formation within natural products may lead to the re-engineering of biosynthetic pathways towards preparing new aziridine-containing secondary metabolites <09EJM3373>. In this light, isotope-tracer experiments and cloning of the biosynthetic gene cluster were reported towards elucidating the pathway and mechanism of aziridine formation in the polyketide antimicrobial azicemicin <09JA18066>.

3.3.2. Reactions of Aziridines

The reactions of aziridines (like epoxides) are largely dominated by nucleophilic ring-opening reactions. Recently, the desymmetrization of meso-aziridines with a variety of nucleophiles has attracted significant attention, including metal-based <09MI147; 09CC5722> and chiral Brønsted acid-catalyzed <09OL5186; 09OL3330> versions. A review on the ring opening of meso-aziridines with bimetallic catalysts has been reported <09AG(I)2082>. An in situ prepared chiral catalyst from commercially available Ti(O-i-Pr)$_4$ and (R)-binol was reported to catalyze the highly enantioselective ring-opening of meso-aziridines with anilines affording valuable chiral 1,2-diamines in high yields and up to 99% ee <09AG(I)4849>. Additionally as shown below, dimeric yttrium-salen complexes were reported to catalyze the highly enantioselective ring-opening of meso-aziridines by TMSCN and TMSN$_3$ <09AG(I)1126; 09SCI1662>. This work featured proposal of a dimetallic mechanism based on the solid state structure of the dimer in order to explain the dramatic differences in selectivity between mono- and dimeric catalysts.

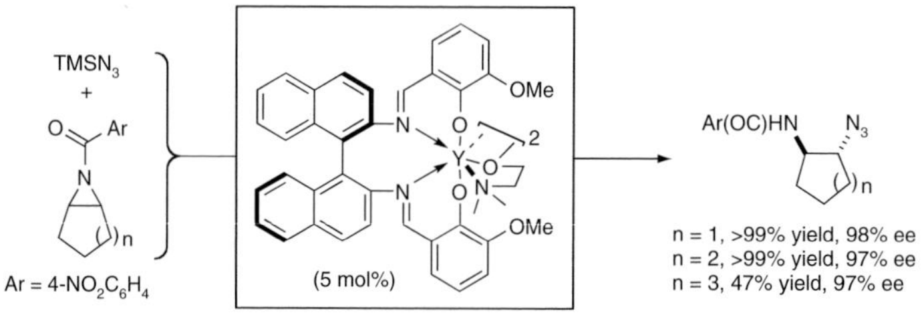

A number of interesting reactions involving oxygen-based nucleophiles, including alcohols for diversity-oriented syntheses of alkaloids <09TL3230> and carbohydrates <09TA902; 09OL2675>, have been reported. Tandem aziridine ring-opening/closing cascade reactions continue to attract significant attention via utilization of halogenated <09JOC7013> or propargyl <09OL1119; 09JOC8814> alcohols. In a similar light as exemplified below, *trans*-3,4-2*H*-1,4-benzoxazines were synthesized in good to excellent yields by domino aziridine ring opening with *o*-iodophenols followed by copper-catalyzed Goldberg coupling cyclization (intramolecular C(aryl)-N(amide) bond formation) <09OL1923>.

R = H, 99%
R = Cl, 94%
R = *t*-Bu, 85%

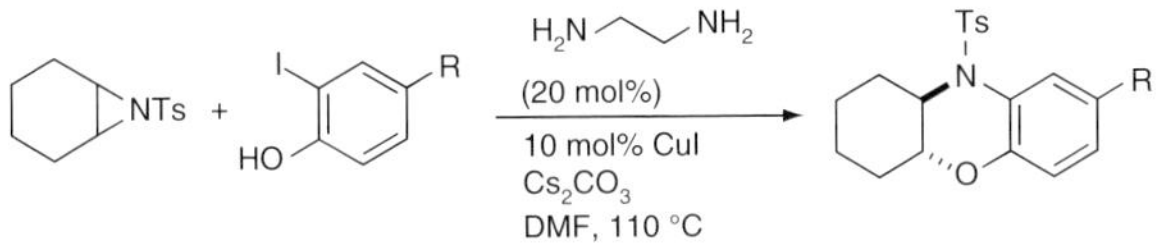

The oxygen within carbonyls can also be used to facilitate aziridine ring opening. Zirconyl chloride was found to be an efficient catalyst for the cycloaddition reaction of aziridines with carbon dioxide leading to preferential formation of 5-aryl-2-oxazolidinones under solvent-free conditions <09T6204>. Additionally, $Sc(OTf)_3$ was found to effectively catalyze the synthesis of 5-alkyl-1,3-oxazolidines in good yields and excellent regioselectivity via condensation of 2-alkyl-*N*-tosylaziridine with a host of ketones and aldehydes <09CC3928>.

Ar = 4-$NO_2C_6H_4$, isomer ratio 19:1, 86%
Ar = 2-furanyl, isomer ratio >50:1, 82%
Ar = 3-OHC_6H_4, isomer ratio 20:1, 41%

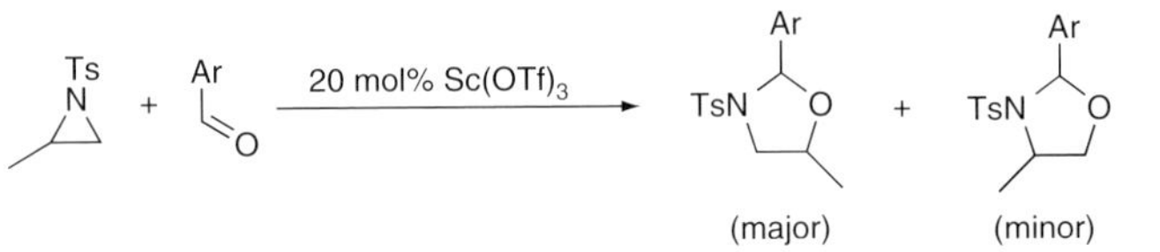

With respect to sulfur-based nucleophiles, quantum chemical computations of appropriate model systems were reported for the rational design of aziridine-based inhibitors of cysteine proteases <09JPC5282> and structurally diverse taurines were obtained from thiazolidine-2-thiones generated via reaction of aziridines with carbon disulfide <09EJO5841>. As exemplified below, optimized conditions were reported for the synthesis of functionalized unsymmetrical disulfides by treating aziridines with symmetrical disulfides in the presence of benzyltriethylammonium tetrathiomolybdate as a sulfur transfer reagent <09JOC7958>.

R = Et, 71%
R = *i*-Pr, 72%
R = *t*-Bu, 70%
R = CH_2CH_2OH, 69%

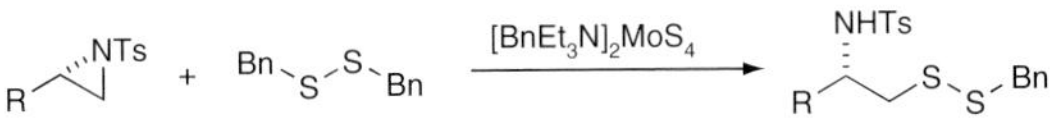

Nitrogen-based nucleophiles continue to remain popular in ring-opening reactions of aziridines. A facile and practical synthesis of a cannabinoid-1 antagonist was reported via regio- and stereoselective ring-opening of an aziridinium ion by an aniline nucleophile <09T9067>. Stable, water-soluble aminosugar K-252a, staurosporine, and rebeccamycin analogs were prepared by nucleophilic opening of C_2-symmetric N-activated bis-aziridines with bis-indolylmaleimides <09OBC706>. Functionalized tetrahydrotriazines were obtained under aerobic conditions via a novel and efficient copper-catalyzed tandem reaction of N-tosylaziridines and hydrazones <09OL5678>. As shown below, a novel base-mediated intramolecular aziridine ring opening reaction via carbon-carbon bond cleavage to afford ring expanded imidazolidin-4-ones was reported <09CC422>.

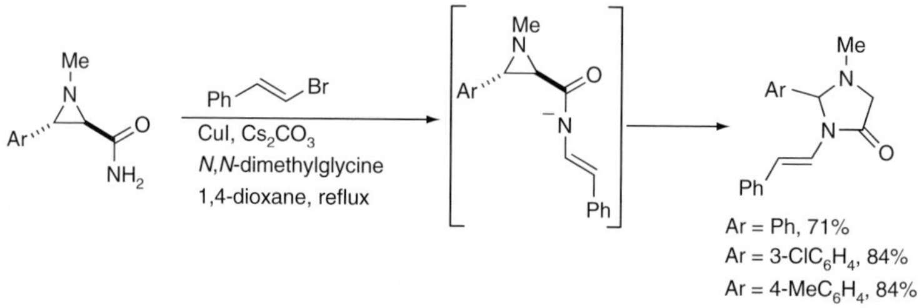

Carbanion nucleophiles continue to be examined in ring opening reactions of aziridines. The scope and limitation of ring-opening reactions of sulfonyl-activated aziridines with lithiated dithianes was reported <09OBC502>. Grignard reagents were utilized to open aziridine rings as part of a four-component reaction for the preparation of α-amino phosphonates <09JOC3573> and the asymmetric total syntheses of (−)-renieramycins M, G, and (−)-jorumycin <09OL5558>. As shown below, palladium-catalyzed decarboxylative ring expansion of a γ-methylidene-δ-valerolactone in the presence of an aziridine led to synthesis of an azepane <09OL5642>.

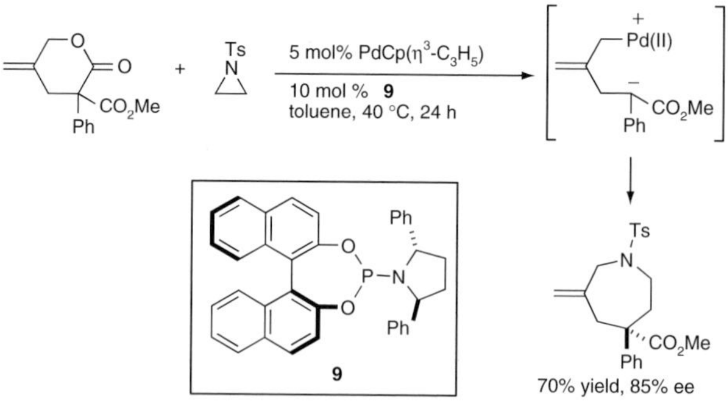

With respect to π-nucleophiles, a chiral terpenoid bearing a 3-amino group on the A ring was synthesized via an $InBr_3$ catalyzed aziridinyl bio-inspired polyene cyclization <09CC3738>. Additionally, it was found that tether-substituted aziridine-allylsilanes cyclize differently compared to their unsubstituted counterparts upon treatment with $BF_3 \cdot OEt_2$ <09T741>. The tether-substituted substrates provided a 6-endo-type product as the major product and it was shown that steric interactions between the *N*-substituent of the aziridine and substitution on the tether could control the distribution of products.

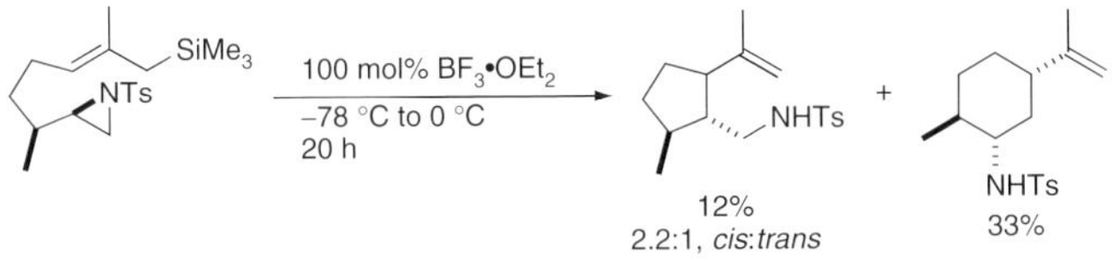

Additional studies have been pursued employing aziridines as masked 1,3-dipoles or three-atom components in both intra- and intermolecular annulation reactions <09AG(I)9026>. Highly functionalized tetrahydropyridines were accessed by phosphine-promoted [3+3] annulation of aziridines with allenoates <09JA6318>. Likewise, aziridines have also been utilized in a [3+2] cycloaddition fashion <09TL6180; 09OL2615>. As shown below, 2,3-dihydropyrroles were synthesized by formal [3+2] cycloaddition between aziridines and nonactivated alkynes upon catalysis by Lewis or Brønsted acids <09JA7528>.

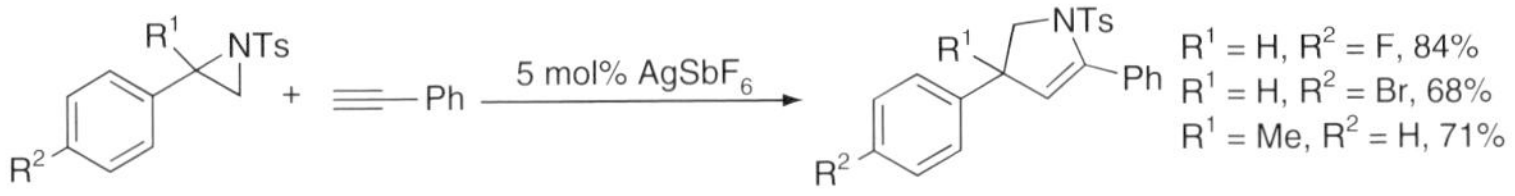

Propargylic aziridines continue to garner significant attention as precursors to substituted pyrroles. Reports included iodine- <09TL6268>, platinum- <09S2454>, and gold-mediated <09OL2293; 09TL6944> versions of this reaction. In this light, an unusual tandem cyclization/ring-opening/Wagner-Meerwein process provided trisubstituted and cycloalkene-fused pyrroles upon PPh$_3$AuCl/AgOTf-catalyzed rearrangement of propargylic aziridines <09OL4002>.

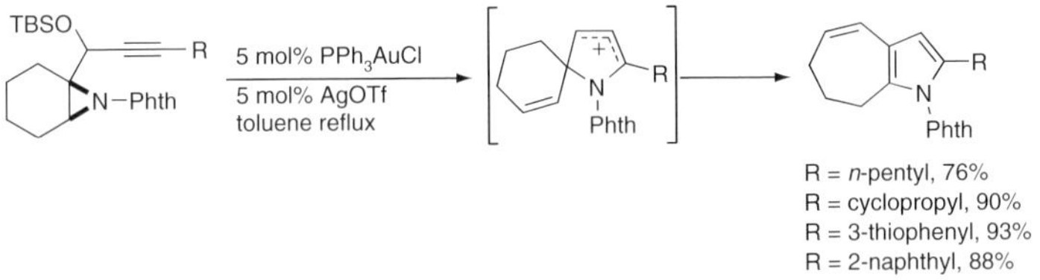

Lithiated aziridines were featured in a number of reports including a detailed study of the lithiation-electrophilic trapping of *N*-sulfonyl ethylene aziridines <09OBC335>, the synthesis of optically active arylaziridines via regio- and stereospecific lithiation of *N*-Bus-phenylaziridine <09OL325>, and stereocontrolled synthesis of β-amino alcohols from lithiated aziridines and boronic esters <09AG(I)1149>. A particularly interesting report featured a new and convenient method for preparing ortho-functionalized aziridines based on benzylic lithiation of simple and easily available *o*-tolylaziridines <09JOC6319>. As exemplified below, these lithiated aziridines were trapped with ketones and aldehydes then subsequently converted to isochromans upon acid-catalyzed cyclization.

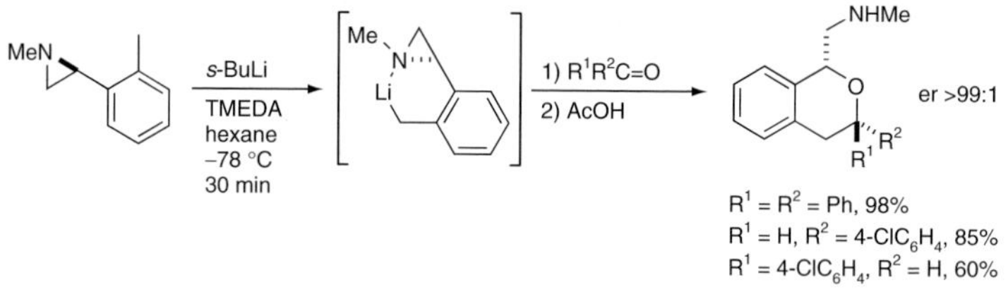

With respect to aziridines functioning as catalysts, bridging metallaaziridines were shown to facilitate intramolecular α-alkylation reactions via selective C-H activation α to primary amines <09JA2116>. Aziridine-functionalized tridentate sulfonyl catalysts proved highly efficient in the enantioselective addition of diethyl zinc to alkyl and aryl aldehydes <09TA2311>. Likewise as exemplified below, (2S)-1-ferrocenyl-methylaziridin-2-yl(diphenol)methanol catalyzed the addition of dimethylzinc to arylaldehydes providing chiral alcohol products with enantioselectivity up to 97.5% ee <09TA288>.

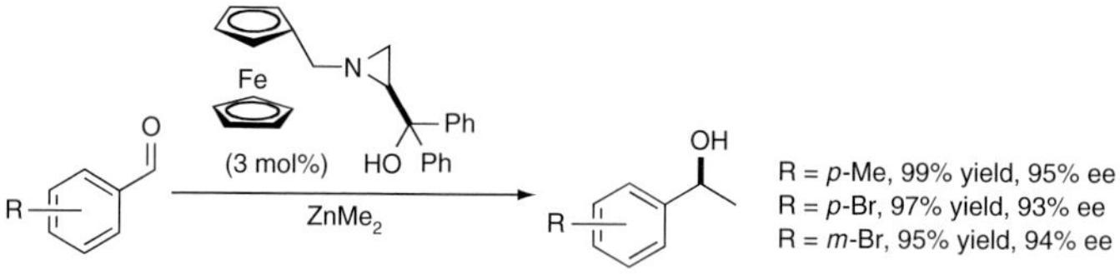

Reactions involving diaziridinones and oxaziridines as three-membered ring heterocycles continue to be investigated. Copper-catalyzed diamination of a disubstituted terminal olefin using di-*tert*-butylaziridinone as a nitrogen source provided access to a potent NK_1 antagonist <09OL2365>. A sulfonyloxaziridine was found to be a convenient reagent for solid-phase synthesis of N-terminal peptide hydroxylamines for chemoselective ligations <09JA3864>. Anionic halocuprate(II) complexes were investigated as catalysts for oxaziridine-mediated aminohydroxylation of olefins <09JOC5545>. As exemplified below, structurally diverse heterocyclic compounds could be obtained upon intramolecular amination of sp³-hybridized C-H bonds with dihydrostilbene-derived oxaziridines <09JA12560>.

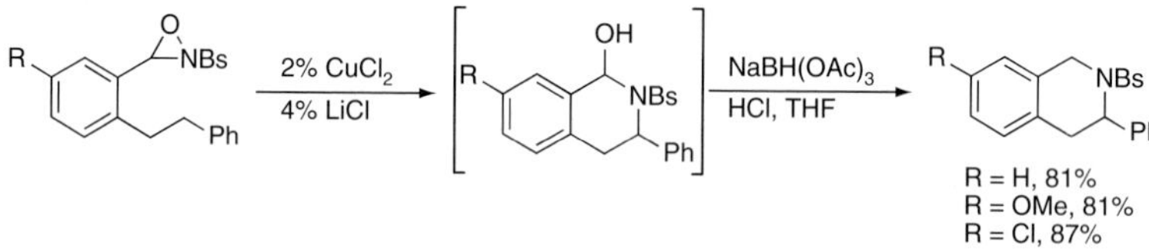

Azirines are structurally interesting compounds which represent unsaturated derivatives of N-H aziridines. Nucleophilic 1,4-disubstituted-1,3-dienes were reported to undergo diastereoselective Diels-Alder cycloaddition to [(1R)-10-(N,N-diethylsulfamoyl)isobornyl]-2H-azirine <09TA1378>. Novel photochromic oxazines were obtained upon Rh(II)-catalyzed condensation of ethyl 2-acyl-2-diazoacetates with 2H-azirines <09TL6509>. Additionally, phenyliodine (III) diacetate (PIDA) was found to mediate the synthesis of 2*H*-azirines from their corresponding substituted enamines <09OL2643>. In turn, indole-3-carbonitriles or isoxazoles could be generated from the 2-aryl-2*H*-azirines via thermal rearrangement.

R = H, 72%
R = F, 76%
R = Cl, 72%

R = H, 85%
R = F, 90%
R = Cl, 94%

REFERENCES

09AG(I)391	M. Jeganmohan, S. Bhuvaneswari and C.-H. Cheng, *Angew. Chem. Int. Ed.*, **48**, 391 (2009).
09AG(I)1126	B. Wu, J.C. Gallucci, J.R. Parquette and T.V. RajanBabu, *Angew. Chem. Int. Ed.*, **48**, 1126 (2009).
09AG(I)1149	F. Schmidt, F. Keller, E. Vedrenne and V.K. Aggarwal, *Angew. Chem. Int. Ed.*, **48**, 1149 (2009).
09AG(I)2082	C. Schneider, *Angew. Chem. Int. Ed.*, **48**, 2082 (2009).
09AG(I)4849	S. Peruncheralathan, H. Teller and C. Schneider, *Angew. Chem. Int. Ed.*, **48**, 4849 (2009).
09AG(I)5760	M. Karpf and R. Trussardi, *Angew. Chem. Int. Ed.*, **48**, 5760 (2009).
09AG(I)6503	W.-J. Liu, B.-D. Lv and L.-Z. Gong, *Angew. Chem. Int. Ed.*, **48**, 6503 (2009).
09AG(I)7224	S. Hayashi, H. Yorimitsu and K. Oshima, *Angew. Chem. Int. Ed.*, **48**, 7224 (2009).
09AG(I)7432	K. Matsumoto, T. Oguma and T. Katsuki, *Angew. Chem. Int. Ed.*, **48**, 7432 (2009).
09AG(I)9026	P. Dauban and G. Malik, *Angew. Chem. Int. Ed.*, **48**, 9026 (2009).
09CC422	J.Y. Wang, Y. Hu, D.X. Wang, J. Pan, Z.T. Huang and M.X. Wang, *Chem. Commun.*, **28**, 422 (2009).
09CC3738	Y.J. Zhao, L.J.S. Tan, B. Li, S.M. Li and T.P. Loh, *Chem. Commun.*, **28**, 3738 (2009).
09CC3928	B. Kang, A.W. Miller, S. Goyal and S.T. Nguyen, *Chem. Commun.*, **28**, 3928 (2009).
09CC4266	V. Subbarayan, J.V. Ruppel, S. Zhu, J.A. Perman and X.P. Zhang, *Chem. Commun.*, **28**, 4266 (2009).
09CC5722	K. Seki, R. Yu, Y. Yamazaki, Y. Yamashita and S. Kobayashi, *Chem. Commun.*, **28**, 5722 (2009).
09CC5763	F. Sarabia, S. Chammaa, M. Garcia-Castro and F. Martin-Galvez, *Chem. Commun.*, **28**, 5763 (2009).
09CSR3175	C.J. Morten, J.A. Byers, A.R. Van Dyke, I. Vilotijevic and T.F. Jamison, *Chem. Soc. Rev.*, **38**, 3175 (2009).
09EJM3373	F.M.D. Ismail, D.O. Levitsky and V.M. Dembitsky, *Eur. J. Med. Chem.*, **44**, 3373 (2009).
09EJO3413	P.C.B. Page, B.R. Buckley, M.M. Farah and J.A. Blacker, *Eur. J. Org. Chem.*, **33**, 3413 (2009).
09EJO4911	J. Sweeney, *Eur. J. Org. Chem.*, **33**, 4911 (2009).
09EJO5841	N. Chen, W. Jia and J. Xu, *Eur. J. Org. Chem.*, **33**, 5841 (2009).
09EJO6009	J.M. Vega-Perez, M. Vega Holm, M.L. Martinez, E. Blanco and F. Iglesias-Guerra, *Eur. J. Org. Chem.*, **33**, 6009 (2009).
09H193	M. Yoshida, M. Hayashi, K. Matsuda and K. Shishido, *Heterocycles*, **77**, 193 (2009).
09JA2052	S. Hayashi, H. Yorimitsu and K. Oshima, *J. Am. Chem. Soc.*, **131**, 2052 (2009).
09JA2116	J.A. Bexrud, P. Eisenberger, D.C. Leitch, P.R. Payne and L.L. Schafer, *J. Am. Chem. Soc.*, **131**, 2116 (2009).
09JA3864	T. Fukuzumi and J.W. Bode, *J. Am. Chem. Soc.*, **131**, 3864 (2009).
09JA6318	H. Guo, Q. Xu and O. Kwon, *J. Am. Chem. Soc.*, **131**, 6318 (2009).

09JA6997	K. Kamata, T. Hirano, S. Kuzuya and N. Mizuno, *J. Am. Chem. Soc.*, **131**, 6997 (2009).
09JA7528	P.A. Wender and D. Strand, *J. Am. Chem. Soc.*, **131**, 7528 (2009).
09JA9150	W.P. Hong, A. El-Awa and P.L. Fuchs, *J. Am. Chem. Soc.*, **131**, 9150 (2009).
09JA11998	H. Hu, J.A. Faraldos and R.M. Coates, *J. Am. Chem. Soc.*, **131**, 11998 (2009).
09JA12560	C.P. Allen, T. Benkovics, A.K. Turek and T.P. Yoon, *J. Am. Chem. Soc.*, **131**, 12560 (2009).
09JA14182	M. Prevost and K.A. Woerpel, *J. Am. Chem. Soc.*, **131**, 14182 (2009).
09JA15615	G. Hu, L. Huang, R.H. Huang and W.D. Wulff, *J. Am. Chem. Soc.*, **131**, 15615 (2009).
09JA18066	Y. Ogasawara and H. Liu, *J. Am. Chem. Soc.*, **131**, 18066 (2009).
09JMOC110	B. Monteiro, S.S. Balula, S. Gago, C. Grosso, S. Figueiredo, A.D. Lopes, A.A. Valente, M. Pillinger, J.P. Lourenco and I.S. Goncalves, *J. Mol. Catal. A*, **297**, 110 (2009).
09JOC1135	H. Kilic, W. Adam and P.L. Alsters, *J. Org. Chem.*, **74**, 1135 (2009).
09JOC1374	J.R. Vyvyan, R.C. Brown and B.P. Woods, *J. Org. Chem.*, **74**, 1374 (2009).
09JOC1798	M. Martin-Rodriguez, R. Galan-Fernandez, A. Marcos-Escribano and F.A. Bermejo, *J. Org. Chem.*, **74**, 1798 (2009).
09JOC2452	J.M. Concellon, H. Rodriguez-Solla, P.L. Bernad and C. Simal, *J. Org. Chem.*, **74**, 2452 (2009).
09JOC3350	A.V. Malkov, L. Czemerys and D.A. Malyshev, *J. Org. Chem.*, **74**, 3350 (2009).
09JOC3573	P.M. Mumford, G.J. Tarver and M. Shipman, *J. Org. Chem.*, **74**, 3573 (2009).
09JOC3913	A. Fernandez-Mateos, S.E. Madrazo, P.H. Teijon and R.R. Gonzalez, *J. Org. Chem.*, **74**, 3913 (2009).
09JOC4217	Y. Arroyo, A. Meana, J.F. Rodriguez, M.A. Sanz-Tejedor, I. Alonso and J.L.G. Ruano, *J. Org. Chem.*, **74**, 4217 (2009).
09JOC4360	A. Blanc, K. Tenbrink, J.-M. Weibel and P. Pale, *J. Org. Chem.*, **74**, 4360 (2009).
09JOC5545	T. Benkovics, J. Du, I.A. Guzei and T.P. Yoon, *J. Org. Chem.*, **74**, 5545 (2009).
09JOC6050	T. Miura, M. Shimada, P. de Mendoza, C. Deutsch, N. Krause and M. Murakami, *J. Org. Chem.*, **74**, 6050 (2009).
09JOC6319	M. Dammacco, L. Degennaro, S. Florio, R. Luisi, B. Musio and A. Altomare, *J. Org. Chem.*, **74**, 6319 (2009).
09JOC6335	O.A. Wong, O.B. Wang, M.-X. Zhao and Y. Shi, *J. Org. Chem.*, **74**, 6335 (2009).
09JOC6735	C.W. Bond, A.J. Cresswell, S.G. Davies, A.M. Fletcher, W. Kurosawa, J.A. Lee, P.M. Roberts, A.J. Russell, A.D. Smith and J.E. Thomson, *J. Org. Chem.*, **74**, 6735 (2009).
09JOC7013	M.K. Ghorai, D. Shukla and K. Das, *J. Org. Chem.*, **74**, 7013 (2009).
09JOC7548	B. Ritzen, M.C.M. vanOers, F.L. vanDelft and F.P.J.T. Rutjes, *J. Org. Chem.*, **74**, 7548 (2009).
09JOC7603	G. Hattori, A. Yoshida, Y. Miyake and Y. Nishibayashi, *J. Org. Chem.*, **74**, 7603 (2009).
09JOC7958	D. Sureshkumar, V. Ganesh, R.S. Vidyarini and S. Chandrasekaran, *J. Org. Chem.*, **74**, 7958 (2009).
09JOC8377	O.A. Wong and Y. Shi, *J. Org. Chem.*, **74**, 8377 (2009).
09JOC8814	M. Bera and S. Roy, *J. Org. Chem.*, **74**, 8814 (2009).
09JOC9452	M. Cakici, S. Karabuga, H. Kilic, S. Ulukanli, E. Sahin and F. Sevin, *J. Org. Chem.*, **74**, 9452 (2009).
09JPC5282	V. Buback, M. Mladenovic, B. Engels and T. Schirmeister, *J. Phys. Chem. B*, **113**, 5282 (2009).
09MI27	C. Wiles, M.J. Hammond and P. Watts, *Beilstein J. Org. Chem.*, **5**, 27 (2009).
09MI147	R. Yu, Y. Yamashita and S. Kobayashi, *Adv. Synth. Catal.*, **351**, 147 (2009).
09MI239	W.J. Choi, *Appl. Microbiol. Biotech.*, **84**, 239 (2009).
09MI348	I. Garcia-Bosch, X. Ribas and M. Costas, *Adv. Syn. Catal.*, **351**, 348 (2009).
09MI596	R. Novikov, G. Bernardinelli and J. Lacour, *Adv. Syn. Catal.*, **351**, 596 (2009).
09MI853	S. Tangestaninejad, M. Moghadam, C. Mirkhani, I. Mohammadpoor-Baltork and K. Ghani, *Cat. Commun.*, **10**, 853 (2009).
09MI920	J. Sun, M. Yang, F. Yuan, X. Jia, X. Yang, Y. Pan and C. Zhu, *Adv. Synth. Catal.*, **351**, 920 (2009).

09MI2579	S.P. Bew, R. Carrington, D.L. Hughes, J. Liddle and P. Pesce, *Adv. Synth. Catal.*, **351**, 2579 (2009).
09MI8713	J.R. Herron, V. Russo, E.J. Valente and Z.T. Ball, *Chem. Eur. J.*, **15**, 8713 (2009).
09MI10732	M. Zhang, Y. Hu and S. Zhang, *Chem. Eur. J.*, **15**, 10732 (2009).
09OBC187	X. Zhang, M. Yan and D. Huang, *Org. Biomol. Chem.*, **7**, 187 (2009).
09OBC335	J. Huang, S.P. Moore, P. O'Brien, A.C. Whitwood and J. Gilday, *Org. Biomol. Chem.*, **7**, 335 (2009).
09OBC502	K. Sakakibara and K. Nozaki, *Org. Biomol. Chem.*, **7**, 502 (2009).
09OBC706	S. Delarue-Cochin and I. McCort-Tranchepain, *Org. Biomol. Chem.*, **7**, 706 (2009).
09OBC1115	N.J. Mueller, C. Stueckler, M. Hall, P. Macheroux and K. Faber, *Org. Biomol. Chem.*, **7**, 1115 (2009).
09OBC1284	R. Lorpitthaya, K.B. Sophy, J.-L. Kuo and X.-W. Liu, *Org. Biomol. Chem.*, **7**, 1284 (2009).
09OBC2026	C. Philippe, T. Milcent, B. Crousse and D. Bonnet-Delpon, *Org. Biomol. Chem.*, **7**, 2026 (2009).
09OBC2501	K.-G. Ji, X.-Z. Shu, J. Chen, S.-C. Zhao, Z.-J. Zheng, X.-Y. Liu and Y.-M. Liang, *Org. Biomol. Chem.*, **7**, 2501 (2009).
09OBC2559	M.W.C. Robinson, A.M. Davies, R. Buckle, I. Mabbett, S.H. Taylor and A.E. Graham, *Org. Biomol. Chem.*, **7**, 2559 (2009).
09OL325	B. Musio, G.J. Clarkson, M. Shipman, S. Florio and R. Luisi, *Org. Lett.*, **11**, 325 (2009).
09OL1119	L. Wang, Q.B. Liu, D.S. Wang, X. Li, X.W. Han, W.J. Xiao and Y.G. Zhou, *Org. Lett.*, **11**, 119 (2009).
09OL1923	R.K. Rao, A.B. Naidu and G. Sekar, *Org. Lett.*, **11**, 1923 (2009).
09OL2293	P.W. Davies and N. Martin, *Org. Lett.*, **11**, 2293 (2009).
09OL2365	Y. Wen, B. Zhao and Y. Shi, *Org. Lett.*, **11**, 2365 (2009).
09OL2445	T. Akiyama, T. Suzuki and K. Mori, *Org. Lett.*, **11**, 2445 (2009).
09OL2615	S. Wang, Y. Zhu, Y. Wang and P. Lu, *Org. Lett.*, **11**, 2615 (2009).
09OL2643	X. Li, Y. Du, Z. Liang, X. Li, Y. Pan and K. Zhao, *Org. Lett.*, **11**, 2643 (2009).
09OL2675	V. Di Bussolo, A. Fiasella, L. Favero, F. Bertolini and P. Crotti, *Org. Lett.*, **11**, 2675 (2009).
09OL2896	I.R. Davies, M. Cheeseman, R. Green, M.F. Mahon, A. Merritt and S.D. Bull, *Org. Lett.*, **11**, 2896 (2009).
09OL3036	X. Zeng, X. Zeng, Z. Xu, M. Lu and G. Zhong, *Org. Lett.*, **11**, 3036 (2009).
09OL3258	I. Mangion, N. Strotman, M. Drahl, J. Imbriglio and E. Guidr, *Org. Lett.*, **11**, 3258 (2009).
09OL3314	Q. Tan and M. Hayashi, *Org. Lett.*, **11**, 3314 (2009).
09OL3330	G.D. Sala and A. Lattanzi, *Org. Lett.*, **11**, 3330 (2009).
09OL3490	M.A. Tarselli, J.L. Zuccarello, S.J. Lee and M.R. Gagne, *Org. Lett.*, **11**, 3490 (2009).
09OL3622	M. Wu, B. Wang, S. Wang, C. Xia and W. Sun, *Org. Lett.*, **11**, 3622 (2009).
09OL3750	J.M. Concellon, P. Tuya, V. del Solar, S. Garcia-Granda and M.R. Diaz, *Org. Lett.*, **11**, 3750 (2009).
09OL3762	S. Crotti, F. Bertolini, F. Macchia and M. Pineschi, *Org. Lett.*, **11**, 3762 (2009).
09OL4002	X. Zhao, E. Zhang, Y.Q. Tu, Y.Q. Zhang, D.Y. Yuan, K. Cao, C.A. Fan and F.M. Zhang, *Org. Lett.*, **11**, 4002 (2009).
09OL4552	S.P. Bew, S.A. Fairhurst, D.L. Hughes, L. Legentil, J. Liddle, P. Pesce, S. Nigudkar and M.A. Wilson, *Org. Lett.*, **11**, 4552 (2009).
09OL5150	C.P. Burke and Y. Shi, *Org. Lett.*, **11**, 5150 (2009).
09OL5186	S.E. Larson, J.C. Baso, G. Li and J.C. Antilla, *Org. Lett.*, **11**, 5186 (2009).
09OL5558	Y.C. Wu and J. Zhu, *Org. Lett.*, **11**, 5558 (2009).
09OL5642	R. Shintani, M. Murakami, T. Tsuji, H. Tanno and T. Hayashi, *Org. Lett.*, **11**, 5642 (2009).
09OL5678	D. Hong, X. Lin, Y. Zhu, M. Lei and Y. Wang, *Org. Lett.*, **11**, 5678 (2009).
09OL5698	R. Shen and X. Huang, *Org. Lett.*, **11**, 5698 (2009).
09OM3611	M. Ranocchiari and A. Mezzetti, *Organometallics*, **28**, 3611 (2009).
09S1123	O. Bortolini, G. Fantin and M. Fogagnolo, *Synthesis*, 1123 (2009).
09S1551	A. Russo and A. Lattanzi, *Synthesis*, 1551 (2009).
09S1886	S.K. Dinda, S.K. Das and G. Panda, *Synthesis*, 1886 (2009).
09S1923	D.M. Hodgson, J. Kloesges and B. Evans, *Synthesis*, 1923 (2009).

09S2454	M. Yoshida, M. Al-Amin and K. Shishido, *Synthesis*, 2454 (2009).
09S3267	V. Ganesh and S. Chandrasekaran, *Synthesis*, 3267 (2009).
09S3433	R. Dalpozzo, M. Nardi, M. Oliverio, R. Paonessa and A. Procopio, *Synthesis*, 3433 (2009).
09S3769	D.P. Bottega, M. Martinelli and M. Koetz, *Synthesis*, 3769 (2009).
09SC2405	D.C. Forbes, S.V. Bettigeri, S.R. Amin, C.J. Bean, A.M. Law and R.A. Stockman, *Synth. Commun.*, **39**, 2405 (2009).
09SCI1662	B. Wu, J.R. Parquette and T.V. RajanBabu, *Science*, **326**, 1662 (2009).
09SL667	S. Tanaka and K. Nagasawa, *Synlett*, 667 (2009).
09SL1157	M. Adachi, E. Yamauchi, T. Komada and M. Isobe, *Synlett*, 1157 (2009).
09SL1265	E. Leemans, S. Mangelinckx and N. De Kimpe, *Synlett*, 1265 (2009).
09SL1367	P. Kumar and P. Gupta, *Synlett*, 1367 (2009).
09SL2076	B. Morandi and E.M. Carreira, *Synlett*, 2076 (2009).
09SL2715	Y. Zhang, Z. Lu and W.D. Wulff, *Synlett*, 2715 (2009).
09SL3024	N. Tada, H. Okubo, T. Miura and A. Itoh, *Synlett*, 3024 (2009).
09SL3123	L.D.S. Yadav and R.K. Garima, *Synlett*, 3123 (2009).
09T70	D.C. Forbes, S.V. Bettigeri, S.A. Patrawala, S.C. Pischek and M.C. Standen, *Tetrahedron*, **65**, 70 (2009).
09T484	S. Fioravanti, S. Morea, A. Morreale, L. Pellacani and P.A. Tardella, *Tetrahedron*, **65**, 484 (2009).
09T741	D.J. Lapinsky, A.B. Pulipaka and S.C. Bergmeier, *Tetrahedron*, **65**, 741 (2009).
09T985	H. Cavdar and N. Saracoglu, *Tetrahedron*, **65**, 985 (2009).
09T2643	Z. Wang, X. Lin, R.L. Luck, G. Gibbons and S. Fang, *Tetrahedron*, **65**, 2643 (2009).
09T2910	P.C.B. Page, P. Parker, B.R. Buckley, G.A. Rassias and D. Bethell, *Tetrahedron*, **65**, 2910 (2009).
09T4984	A. Gansaeuer, M. Otte, F. Piestert and C.-A. Fan, *Tetrahedron*, **65**, 4984 (2009).
09T5240	W. Guo, J. Chen, D. Wu, J. Ding, F. Chen and H. Wu, *Tetrahedron*, **65**, 5240 (2009).
09T6204	Y. Wu, L.N. He, Y. Du, J.Q. Wang, C.X. Miao and W. Li, *Tetrahedron*, **65**, 6204 (2009).
09T7691	S. Sobhani and A. Vafaee, *Tetrahedron*, **65**, 7691 (2009).
09T8668	K. Ota, T. Kurokawa, E. Kawashima and H. Miyaoka, *Tetrahedron*, **65**, 8668 (2009).
09T9067	E.B. Villhauer, W.C. Shieh, Z. Du, K. Vargas, L. Ciszewski, Y. Lu, M. Girgis, M. Lin and M. Prashad, *Tetrahedron*, **65**, 9067 (2009).
09T9884	S. Nagumo, T. Miura, M. Mizukami, I. Miyoshi, M. Imai, N. Kawahara, H. Akita, *Tetrahedron*, **65**, 9884 (2009).
09T10837	J. Justicia, T. Jimenez, S.P. Morcillo, J.M. Cuerva and J.E. Oltra, *Tetrahedron*, **65**, 10837 (2009).
09TA288	M.C. Wang, Q.J. Zhang, G.W. Li and Z.K. Liu, *Tetrahedron Asymmetr.*, **20**, 288 (2009).
09TA902	A. Schafer, D. Henkensmeier, L. Kroger and J. Thiem, *Tetrahedron Asymmetr.*, **20**, 902 (2009).
09TA883	T.K.M. Shing and T. Luk, *Tetrahedron Asymmetr.*, **20**, 883 (2009).
09TA1378	M.J. Alves, C. Costa and M.M. Duraes, *Tetrahedron Asymmetr.*, **20**, 1378 (2009).
09TA1969	J.A. Groeper, J.B. Eagles and S.R. Hitchcock, *Tetrahedron Asymmetr.*, **20**, 1969 (2009).
09TA2311	S. Lesniak, M. Rachwalski, E. Sznajder and P. Kielbasinski, *Tetrahedron Asymmetr.*, **20**, 2311 (2009).
09TL26	S. Nagumo, Y. Ishii, G. Sato, M. Mizukami, M. Imai, N. Kawahara and H. Akita, *Tetrahedron Lett.*, **50**, 26 (2009).
09TL161	J.W.W. Chang, T.M.U. Ton, Z. Zhang, Y. Xu and P.W.H. Chan, *Tetrahedron Lett.*, **50**, 161 (2009).
09TL476	M.K. Ghorai, K. Ghosh and A.K. Yadav, *Tetrahedron Lett.*, **50**, 476 (2009).
09TL548	J. Sun, F. Yuan, M. Yang, Y. Pan and C. Zhu, *Tetrahedron Lett.*, **50**, 548 (2009).
09TL916	Y.-H. Liu, Q.-S. Liu and Z.-H. Zhang, *Tetrahedron Lett.*, **50**, 916 (2009).
09TL1529	C. Chen and R.A. Reamer, *Tetrahedron Lett.*, **50**, 1529 (2009).
09TL2200	H. Wang, Z. Wang and K. Ding, *Tetrahedron Lett.*, **50**, 2200 (2009).
09TL2509	Y. Wang, Z. Wu, Z. Li and X.-G. Zhou, *Tetrahedron Lett.*, **50**, 2509 (2009).
09TL2536	S. Antoniotti and E. Dunach, *Tetrahedron Lett.*, **50**, 2536 (2009).

09TL2808	A. Capperucci, C. Tiberi, S. Pollicino and A. Degl'Innocenti, *Tetrahedron Lett.*, **50**, 2808 (2009).
09TL3230	A.M. Taylor and S.L. Schreiber, *Tetrahedron Lett.*, **50**, 3230 (2009).
09TL3857	R. Fan, L. Wang, Y. Ye and J. Zhang, *Tetrahedron Lett.*, **50**, 3857 (2009).
09TL3881	J.D. Neighbors, J.J. Topczewski, D.C. Swenson and D.F. Wiemer, *Tetrahedron Lett.*, **50**, 3881 (2009).
09TL4629	A. Bunge, H.-J. Hamann, E. McCalmont and J. Liebscher, *Tetrahedron Lett.*, **50**, 4629 (2009).
09TL5009	S. Narayana Murthy, B. Madhav, V. Prakash Reddy, K. Rama Rao and Y.V.D. Nageswar, *Tetrahedron Lett.*, **50**, 5009 (2009).
09TL5225	W. Qiu, L. He, Q. Chen, W. Luo, Z. Yu, F. Yang and J. Tang, *Tetrahedron Lett.*, **50**, 5225 (2009).
09TL5344	J.C. Anderson, N.M. Smith, M. Robertson and M.S. Scott, *Tetrahedron Lett.*, **50**, 5344 (2009).
09TL5420	L.D.S. Yadav and G.R. Kapoor, *Tetrahedron Lett.*, **50**, 5420 (2009).
09TL6180	F.M. Ribeiro Laia and T.M.V.D. Pinho e Melo, *Tetrahedron Lett.*, **50**, 6180 (2009).
09TL6268	M. Yoshida, M. Al-Amin and K. Shishido, *Tetrahedron Lett.*, **50**, 6268 (2009).
09TL6509	V.A. Khlebnikov, M.S. Novikov, A.F. Khlebnikov and N.V. Rostovskii, *Tetrahedron Lett.*, **50**, 6509 (2009).
09TL6944	D.D. Chen, X.L. Hou and L.X. Dai, *Tetrahedron Lett.*, **50**, 6944 (2009).

CHAPTER 4

Four-Membered Ring Systems

Benito Alcaide*, Pedro Almendros**
*Grupo de Lactamas y Heterociclos Bioactivos, Departamento de Química Orgánica I, Unidad Asociada al CSIC, Facultad de Química, Universidad Complutense de Madrid, 28040-Madrid, Spain
alcaideb@quim.ucm.es
**Instituto de Química Orgánica General, CSIC, Juan de la Cierva 3, 28006-Madrid, Spain
Palmendros@iqog.csic.es

4.1. INTRODUCTION

Research in the field of four-membered heterocycles, where a non-carbon atom is part of the ring, has been actively pursued in 2009 within Organic Chemistry, Inorganic Chemistry, Medicinal Chemistry, and Material Science. In particular, the synthesis and chemistry of oxygen- and nitrogen-containing heterocycles dominate the field in terms of the number of publications. It should be clearly stated that a comprehensive description of all aspects of the vast research area of four-membered heterocyclic chemistry can not be fully covered in this chapter. Instead, the present overview presents a personal selection of the topics which we believe are the most relevant.

4.2. AZETIDINES, AZETINES, AND RELATED SYSTEMS

The synthesis and reactivity of 3-haloazetidines, 3-sulfonyloxyazetidines, and 2-methylenezetidines have been reviewed in several contributions <09COR827; 09COR852>. A review on azetidines as new tools for the synthesis of higher nitrogen heterocycles has been published <09SL3053>. A novel series of azetidinyl ketolides **1** for the treatment of susceptible and multidrug resistant community-acquired respiratory tract infections has been discovered <09JMC7446>. Theoretical and experimental results have provided evidence that the 2-alkyl-2-carboxyazetidine scaffold **2** is able to efficiently induce γ-turns when incorporated into short peptides, irrespective of their localization in the peptide chain <09JOC8203>. Chiral azetidino amino alcohol ligands **3** bearing an additional stereogenic center were readily prepared and used as catalysts for the asymmetric addition of alkynylzinc to aromatic aldehydes <09TA2616>. The phosphine-free Sonogashira coupling reaction of aryl halides catalysed by palladium(II) complexes of azetidine-derived polyamines **4** under mild conditions has been accomplished <09T1630>. The enantioselective epoxidation of chalcone has been catalyzed by an azetidin-2-ylmethanol derivative <09S1551>. Prepared by mixing 3,3-dinitroazetidine (DNAZ) and 3-nitro-1,2,4-triazol-5-one (NTO) in ethanol solution, non-isothermal decomposition kinetics, heat capacity and adiabatic time-to-explosion of NTO•DNAZ have been described <09MI1068>. Crystal structure determinations on Cu(II) and Zn(II) complexes of tridentate and quadridentate azetidine derivatives show the ligands to facilitate square-pyramidal and trigonal bipyramidal coordination geometries <09MI65>. DFT calculations have been performed in order to simulate the ring enlargement of azetidine to the corresponding 3-chloropyrrolidine <09MI26>. The chemistry and pharmacology of nicotinic ligands based on 6-[5-(azetidin-2-ylmethoxy)pyridin-3-yl]hex-5-yn-1-ol for possible use in treatment of depression have been reported <09MI1279>.

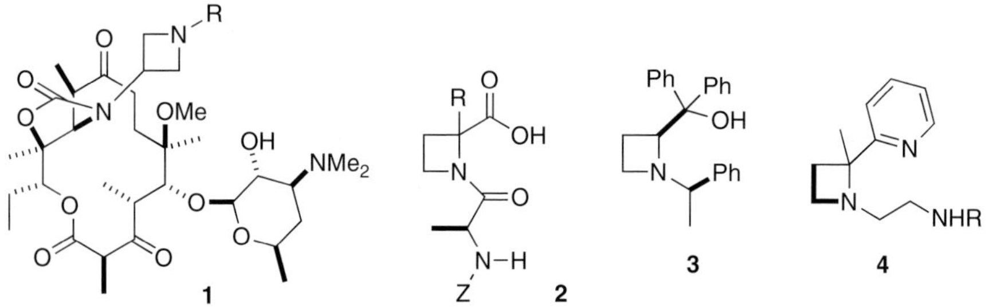

The development of two synthetic routes to the azetidine-based drug CE-178,253-26, a CB_1 antagonist for the treatment of obesity, has been documented <09T3292>. Base-induced cyclization of enantiopure (2-aminoalkyl)oxiranes allowed the stereospecific formation of 2-(hydroxymethyl)azetidines **5** <09JOC7859>. Simple azetidines have been synthesized in good yields via cyclization of 3-aminopropyl sulfates in water under the influence of microwave-assisted heating <09TL6590>. A new route to two 2-azetidinylglycine derivatives has been developed from Garner's aldehyde <09TA1213>. A convenient and rapid synthesis of L-azetidine-2-carboxylic acid **6** has been described starting from commercially available L-aspartic acid via double activation with the SES group, with conservation of the chiral center <09EJO2729>. An efficient copper-mediated synthesis of ynamides, including azetidinyl derivatives, from 1,1-dibromo-1-alkenes has been reported <09AGE4381>.

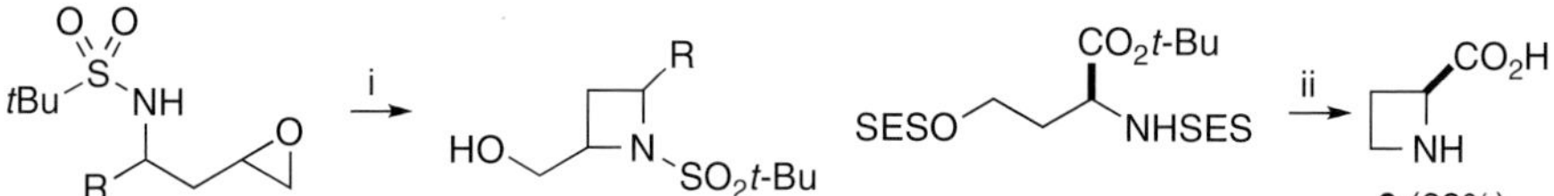

Key: i) K_2CO_3, 80 °C. ii) (a) Cs_2CO_3, MeCN, MW; (b) HF, 0 °C.

The first synthesis of 3-fluoroazetidine-3-carboxylic acid **7**, a cyclic fluorinated β-amino acid derivative with high potential as a building block in medicinal chemistry, has been disclosed <09JOC2250>. The synthesis of 3-aryl-3-azetidinyl acetic acid esters, including a novel spiroazetidine ring system, by rhodium(I)-catalysed conjugate addition of organoboron reagents has been reported <09TL3909>. Treatment of N-tosylaldimines with acetophenone at room temperature in the presence of $BF_3 \cdot OEt_2$ as a catalyst furnished N-tosyl β-amino ketones, which, after subsequent reduction and cyclization, afforded 2,4-disubstituted N-tosylazetidines **8** <09JOC9505>.

Enantiomerically pure *N*-allyl azetidinium ions undergo a stereoselective [2,3]-sigmatropic shift to give azetidines with an α-quaternary center <09SL767>. Two pairs of *cis* and *trans* 1-benzyl- and 1-allyl-2-cyano-3-phenylazetidines have been synthesized and subjected to flash vacuum thermolysis <09T9322>. Highly efficient and enantioselective biotransformations of racemic azetidine-2-carbonitriles and their synthetic applications have been carried out <09JOC6077>. Indoles undergo smooth alkylation with *N*-tosylazetidines in the presence of indium(III) bromide to produce the corresponding C-3-substituted indoles <09SL727>.

Key: i) (a) NBS, Et$_3$N·3HF; (b) NaCNBH$_3$. ii) (a) H$_2$, Pd/C, Boc$_2$O; (b) CAN, MeCN–H$_2$O, −10 °C. iii) 5 mol% RuCl$_3$, NaIO$_4$, MeCN–H$_2$O. iv) (a) NaBH$_4$, MeOH; (b) TsCl, KOH, THF.

Beginning with a 1,3-dibromopropane derivative, readily available from inexpensive epibromohydrin, a scaleable synthesis of the novel 6-oxo-2-azaspiro[3.3]heptane ring system exemplified by compound **9** has been described <09OL3523>. The transformation of *anti*-Mannich adducts into azetidin-2-amides has been described <09AGE3353>. Azetidin-2-imines have been isolated as minor components of the copper-catalyzed multicomponent reaction of sulfonyl azides, terminal alkynes and α,β-unsaturated imines <09ASC1768>. A bicyclic azetidine bearing a CF$_3$ substituent has been formed as major component during the iodocyclization of a chiral allylmorpholinone <09OL209>. The synthesis of tricyclic 3-aminopyridines, including fused azetidine **10**, through intramolecular Co(I)-catalyzed [2+2+2] cycloaddition between ynamides, nitriles, and alkynes has been accomplished <09CEJ2129>. The first synthesis of 2,6-diazabicyclo[3.2.0]heptanes **11**, a fused azetidine with considerable potential as a building block in medicinal chemistry, has been developed from *trans*-3-hydroxy-L-proline <09TL7280>. A polyhydroxylated 1-azabicyclo[5.2.0]nonane, a fused azetidine isomeric with the indolizidine castanospermine, has been obtained as a minor component *en route* to the alkaloid <09JOC8886>.

Key: i) (a) EtO$_2$CCH$_2$CN, K$_2$CO$_3$; (b) NaBH$_4$; (c) TsCl, Et$_3$N. ii) (a) LiAlH$_4$; (b) Boc$_2$O. iii) (a) Pd(OH)$_2$, MeOH, Δ; (b) Dess–Martin peridodinane. iv) (a) SOCl$_2$; (b) Boc$_2$O. v) (a) NaN$_3$, DMF; (b) PPh$_3$, THF–H$_2$O. vi) (a) CbzCl, NaOH; (b) BF$_3$·Et$_2$O, DIBAL-H. vii) (a) *t*-BuOK, THF; (b) H$_2$, Pd/C, MeOH.

A strategy for copper-mediated S_N2 type nucleophilic ring-opening followed by [4+2] cycloaddition reactions of 2-aryl-N-tosylazetidines **12** with nitriles to afford substituted tetrahydropyrimidines **13** has been reported <09TL1105>. The [6+2] cycloaddition of 2-vinylazetidines with tosyl isocyanate proceeded smoothly in the absence of a catalyst to yield 1,3-diazocinones **14** <09OL5438>. The ring expansion of enantiomerically pure 2-alkenyl azetidines provides 1,2,3,6-azocines upon reaction with activated alkynes <09SL3182>. Simple N-substituted azetidines heated with diazocarbonyl compounds in the presence of catalytic $Cu(acac)_2$, furnish substituted pyrrolidines via [1,2]-shift <09JOC2832>. A synthesis of substituted nonracemic homomorpholines via an S_N2-type ring opening of activated azetidines by suitable halogenated alcohols in the presence of Lewis acid, followed by base-mediated intramolecular ring closure of the resulting haloalkoxy amine, has been described <09JOC7013>. An approach for the preparation of oxazocines through a Ag(I)-catalyzed cascade reaction of azetidines with propargyl alcohols has been presented <09JOC8813>. Azetines have been proposed as intermediates of the gold-catalyzed reaction between 1,3-dienes and aldimines yielding cyclopent-2-enimines <09OL13>. N-Heterocyclic carbenes have been found to be efficient catalysts for the formal [2+2] cycloaddition of aryl(alkyl)ketenes and diazenedicarboxylates to give the corresponding aza-β-lactams **15** in good yields with up to 91% ee <09JOC7585>. The catalytic asymmetric cycloaddition of ketenes with nitroso compounds leads to the synthesis of 1,2-oxazetidin-3-ones **16**, which in addition to serving as potentially bioactive target molecules, can be transformed into other important classes of compounds, such as α-hydroxycarboxylic acids <09AGE2391>. The rearrangement of β-chloro N-oxides to hydroxylamines is stereospecific in accord with the presence of a cyclic oxazetidinium intermediate, which is opened by nucleophiles <09JOC2254>.

Key: i) RCN, $Cu(OTf)_2$, 80 °C.

4.3. MONOCYCLIC 2-AZETIDINONES (β-LACTAMS)

A compendium on the generation of the β-lactam ring covering different methods has been assembled <09CSY325>. The stereoselectivity in the synthesis of 2-azetidinones from ketenes and imines via the Staudinger reaction has been reviewed <09ARK21>. The preparation of β-lactams by catalytic asymmetric reactions of ketenes and ketene enolates has been reviewed <09T6771>. An overview presenting recent progress on the metal-catalyzed one-step synthesis of heterocycles, including the β-lactam nucleus, has been published <09CEJ302>. A statin therapy in combination with ezetimibe **17** in patients with a high risk of atherosclerotic disease has been used successfully <09MI2180; 09MI798>. The effects of ezetimibe add-on therapy for high-risk patients with dyslipidemia have been documented <09MI41>. The new biocatalyst *Burkholderia cenocepacia* has proved efficient for the bioreduction of ezetimibe <09MI1369>. A simple high-performance liquid chromatography method has been developed and validated for the simultaneous determination of ezetimibe and simvastatin from their combination drug products <09MI527>. *In vitro* inhibition assays against human histone deacetylase (HDAC) isoforms have shown an interesting isoform-selectivity of β-lactams towards HDAC6 and HDAC8 <09MI1991>. 3-Substituted-3-hydroxy-β-lactams **18**, with two new adjacent stereogenic centers, have been prepared in a single step by a rhodium-catalyzed, three-component reaction between azetidine-2,3-diones, ethyl diazoacetate and alcohols <09JOC8421>. Pd(0)/InI-mediated allylic addition to 4-acetoxy-2-azetidinone has provided derivatized cyclopentenes **19** with high regio- and diastereoselectivity <09OL1293; 09JOC5730>. A series of chiral *trans*-β-lactams has been obtained via Staudinger cycloaddition with low diastereoselectivity (up to 54% *de*) induced by a chiral amine component of the imine <09T10339>. A stereoselective synthesis of novel 3-methylthio-β-lactams and their Lewis acid mediated functionalization has been described <09T10060>. The cycloaddition reactions of 1,3-diazabuta-1,3-dienes with conjugated alkynyl ketenes have provided interesting azetidinones bearing an alkynyl moiety <09T4664>. The halodecarboxylation reaction of 4-alkylidene-β-lactams has been described <09EJO4541>.

Key: i) Rh$_2$(OAc)$_4$, R^{3}OH, CH$_2$Cl$_2$, Δ.

A convenient method for the synthesis of azetidin-2-ones **20** using electrochemical oxidation has been exploited <09T9742>. The one-pot synthesis of *N-tert*-butyl-*trans*-α-ethoxycarbonyl-β-phenyl-β-lactam by the octacarbonyldicobalt-catalyzed carbonylation of ethyl diazoacetate in the presence of *N-tert*-butylbenzaldimine has been described <09EJO1994>. Mannich adducts have been transformed into *trans*-azetidin-2-ones and α-fluoro-β-lactams <09AGE1838; 09AGE7604>. The copper-catalyzed synthesis of β-lactams by intramolecular C—H insertion of diazocompounds has been achieved <09OBC4777>. Asymmetric imidazolium-dithiocarboxylates have been found to be highly selective catalysts for the Staudinger reaction <09CC1040>. 3,4-Diaryl β-lactams **21** have been prepared with high stereoselectivity by a palladium-catalyzed [2+2] carbonylative cycloaddition of benzyl halides with heteroarylidene amines <09TA368>. It has been reported that α-oxoamides are stable axially chiral atropisomers, undergoing enantiospecific photochemical γ-hydrogen abstraction to yield β-lactams <09JA11314>. A β-lactam derivative has been synthesized in an application of the catalytic asymmetric alkylation of α-cyanocarboxylates and acetoacetates with an alkyl halide under phase-transfer conditions <09TA2530>. α-Alkylidene-β-lactams **22** have been prepared in good yields by olefin cross metathesis of electron-poor α-methylene-β-lactams with 1,1-disubstituted alkenes <09TL1020>. The [2+2]-cycloaddition reaction of glycosidic enol ethers has provided glycosylated β-lactams with *trans*-only stereoselectivity <09TA1646>. Ultrafast time-resolved infrared spectroscopy study of the photochemistry of *N,N*-diethyldiazoacetamide has shown that in chloroform, β-lactam is formed immediately after the laser pulse <09JA9646>.

Key: i) −2e, 4F/mol, NaI (0.5 equiv), MeCN. ii) CO (400 psi), Pd(OAc)$_2$, Et$_3$N, PPh$_3$, THF.

The copper-catalyzed skeletal rearrangement of *O*-propargyl arylaldoximes has produced 4-arylidene-2-azetidinones **23** in good to excellent yields <09JA2804>. Access to poly-β-peptides with functionalized side-chains and end-groups *via* controlled ring-opening polymerization of β-lactams has been described <09JA1589>. A series of mono- and bimetallic Pd and Pt macrocyclic β-lactam molecules has been synthesized <09CEJ6940>. A study using a combination of experiments and DFT calculations has been conducted to understand the torquoselectivity and the electronic effects of the Staudinger reaction <09JA1542>. A stereoselective synthesis of (1′*S*,3*R*,4*R*)-4-acetoxy-3-(2′-fluoro-1′-trimethylsilyloxyethyl)-2-azetidinone as a new fluorine-containing intermediate towards β-lactams, has been described <09TL2676>. A number of 2-azetidinones **24** have been synthesized in good yields by a novel reaction between Schiff bases, substituted acetic acids and alkoxymethylene-*N,N*-dimethyliminium salts <09TL1568>. The diastereo- and enantioselective synthesis of functionalized β-lactams from oxiranecarbaldimines and lithium ester enolates has been carried out <09EJO5653>. The stereodivergent synthesis of both *cis*- and *trans*-β-lactams using thermal rearrangement of aminocyclobutenones has been documented <09OL3266>. The enantioselective Kinugasa reaction of nitrones with terminal alkynes in the presence of 20 mol% of IndaBox–Cu(OTf)$_2$ and di-*sec*-butylamine (1.5 equiv) produced β-lactams with a high level of enantiomeric excess <09TL4969>. Micelle-promoted, copper-catalyzed multicomponent Kinugasa reactions have been studied in aqueous media <09TL1893>. The preparation of β-lactams by Mannich-type addition of ethyl trimethyl∼silylacetate to *N*-(2-hydroxyphenyl)aldimine sodium salts has been achieved <09SL2437>. It has been shown that sequence-random nylon-3 copolymers containing β-lactam nuclei can mimic favorable properties of host-defense peptides <09JA16779>, and structure-activity relationships in this polymer family have also been determined <09JA9735>.

23 (61–92%) **24** (79–94%)

Key: i) 10 mol % CuBr, toluene, 100 °C. ii) R³CH₂COOH, Et₃N, CH₂Cl₂.

The diastereoselective synthesis of *trans*-β-lactams using a phosphonium fluoride multifunctional catalyst has been achieved <09SL1651>. Cyclization of diazoacetamides catalyzed by *N*-heterocyclic carbene dirhodium(II) complexes has yielded β- and γ-lactams <09S3519>. Antiaromaticity plays the prime role in suppressing the β-stabilizing effect of silicon in 3-silylated monocyclic β-lactams <09OL5722>. The Staudinger ketene–imine cycloaddition reaction has been applied to bis-*o*-allyloxyarylideneamines affording the corresponding bisallyloxyazetidinones as the *cis,cis* diastereomers, exclusively obtained as a mixture of *cis,syn,cis* and *cis,anti,cis* <09JOC4305>. Mechanistic details of the Mg^{2+} ion-activated enantioselective reduction of methyl benzoylformate have been investigated at a B3LYP/6-31G* theory level, using peptide NADH models rigidified with a β-lactam ring <09JOC6691>. The diastereoselective synthesis of bicyclic γ-lactams **25** by ring expansion of monocyclic chloro-β-lactams via *N*-acyliminium intermediates has been accomplished <09JOC1644>. Aminophosphonic acids, potential inhibitors of penicillin-binding proteins, have been obtained starting from 4-acetoxyazetidinone <09EJO85>. An enantioselective formal synthesis of (*S*)-dapoxetine **26**, a selective serotonin re-uptake inhibitor against depression and a potential cure of premature ejaculation in men, has been achieved from a substituted 3-hydroxy β-lactam in 17% overall yield <09T2605>. The synthesis of (*E*)-arylimino-acetonitriles has been described *via* thermal fragmentation of 1-aryl-4-cyano-β-lactams <09T10581>. The mechanism of the *N*-heterocyclic carbene-catalyzed ring-expansion of 4-formyl-β-lactams to succinimides has been studied using DFT methods at the B3LYP/6-31G** level <09T3432>. A highly enantioselective synthesis of oseltamivir **27** has been achieved starting from L-methionine, in which Staudinger -reaction is utilized for the alignment of three contiguous chiral centers of oseltamivir <09SL787>. The enantioselective total synthesis of (−)-himandrine benefits from the hydrolysis of an *N*-vinyl β-lactam <09JA9648>. A first generation process for the synthesis of sitagliptin, a dipeptidyl peptidase inhibitor for the treatment of type 2 diabetes mellitus, has been documented <09JA8798>. β-Lactam-based preparation of novel 2-alkoxy-3-amino-3-arylpropan-1-ols and 5-alkoxy-4-aryl-1,3-oxazinanes with antimalarial activity has been published <09JMC4058>. The preparation of fucose–saccharosamine disaccharide glycals using β-lactam chemistry has been developed <09OL4850>. Isoxazoline-fused cispentacins have been prepared by the 1,3-dipolar cycloaddition of nitrile oxides to β-amino esters derived from β-lactams <09TL2605>. An enantioselective synthesis of (2*S*,3*R*,4*R*)-D-*xylo*-phytosphingosine has been achieved from a β-lactam derived from D-mannitol triacetonide <09TL3296>. The efficient synthesis of 3,4- and 4,5-dihydroxy-2-amino-cyclohexanecarboxylic acid enantiomers has been reported <09TA2220>. Fused [4.2.0]aminocyclobutane-containing δ-lactams have been accessed from N-vinyl-β-lactams <09OL1281>. The enzymatic synthesis of carnosine derivatives from a β-lactam and a protected α-amino acid has been performed <09TA1641>.

25 (38–57%)   **26**   **27**

Key: i) AgBF$_4$, pyridine, toluene.

4.4. FUSED AND SPIROCYCLIC β-LACTAMS

The principal characteristics of carbapenems and their clinical implications have been reviewed <09CME564>. A review article on recent advances in the chemistry and biology of naturally occurring antibiotics has appeared <09AGE660>. A fully automated method for the detection of β-lactam antibiotics, including six penicillins (amoxicillin, ampicillin, cloxacillin, dicloxacillin, oxacillin, and penicillin G) and four cephalosporins (cefazolin, ceftiofur, cefoperazone, and cefalexin) in bovine milk samples based on online solid-phase extraction-liquid chromatography/electrospraytandem mass spectrometry has been developed <09ANC4285>. It has been discovered that 30 β-lactams derived from (3R,4R)-3-[(R)-1'-(t-butyldimethylsilyloxy)-ethyl]-4-acetoxy-2-azetidinone are selective inhibitors of human fatty acid amide hydrolase versus human monoacylglycerol lipase <09JMC7054>. The reactions between the only β-lactamase inhibitors in clinical use, tazobactam, sulbactam, and clavulanic acid, with a class A β-lactamase have been examined in single crystals using a Raman microscope <09JA2338>. The structural basis of the inhibition of class A β-lactamases and penicillin-binding proteins by 6-β-iodopenicillanate have been published <09JA15262>. Structures and labelling mechanisms of the fluorescent probes on β-lactam cleavage by class A β-lactamases have been described <09JA5016>. The structural characterization of a true dizinc metallo-β-lactamase has been performed <09JA11642>. 6-Aminopenicillanic acid **28** and two of its derivatives have been evaluated as catalysts for use in direct crossed-aldol reactions for the first time <09EJO3155>. The synthesis of 2-azetidinones incorporating carbenechromium(0) moieties and their use in the preparation of penicillin- and cephalosporin-containing peptides has been achieved <09EJO2998>. An efficient biocatalyst for the esterification of 7-aminocephalosporanic acid has been designed <09EJO1384>. The synthesis of regio- and stereoselectively deuterium-labelled derivatives of L-glutamate semialdehyde for studies on carbapenem biosynthesis has been accomplished <09OBC2770>. An enantiocontrolled entry to the spiro-β-lactam core of chartellines has been developed by oxidative nitrogen atom transfer methodology based on chiral Rh–nitrenoid species <09CC6265>. The synthesis of spirocyclic β-lactams **29** by palladium-catalyzed domino cycloisomerization/cross-coupling of α-allenols and Baylis–Hillman acetates has been performed <09CEJ3344>. Spirocyclic β-lactams have been synthesized under mild conditions by using (chloromethylene)dimethylammonium chloride as a versatile acid activator reagent for the direct [2+2] ketene–imine cycloaddition <09T2927>.

28

29 (44–71%)

Key: i) 5 mol% Pd(OAc)$_2$, K$_2$CO$_3$, TDMPP, DMSO, RT.

A novel Pd(II)-catalyzed C–H lactamization reaction, including the formation of spiro-β-lactams has been achieved <08JA14058>. New oxidative dearomatization procedures leading to spiro β-lactams have been developed <09OL2820>. Regiospecific synthesis of 6α-(1R-hydroxyoctyl)penicillanic acid **30** and 6β-(1R-hydroxyoctyl) penicillanic acid as mechanistic probes of class D β-lactamases have been achieved <09OL2515>. Efficient synthesis of epithienamycin A in its readily deprotected form **31** has been reported where three contiguous stereocenters are established in a single catalytic asymmetric azetidinone-forming reaction <09OL3606>. An entry to 6-ruthenocenyl-substituted penicillins has been accomplished <09CEJ593>. 3-Arylthiomethyl-Δ^3-cephems **32**, possessing various substituents on the arylthio moiety, undergo chemoselective and product-selective electrooxidation to give various products <09S3449>. By using a ring-closing metathesis strategy, non-traditional bicyclic β-lactams featuring high conformational adaptability have been prepared with the aim of developing novel inhibitors of penicillin-binding proteins <09EJO1757>. A non-traditional approach for designing reactive β-lactams and possibly new antibacterial agents has been explored, based on large, flexible 1,3-bridged 2-azetidinones featuring a "planar amide" instead of the traditional "twisted amide" found in the penicillin family <09EJM2071>.

30   **31**   **32**

A systematic investigation of the metal (Au, Ag, Pt, Pd, and La)-catalyzed heterocyclization reaction of 2-azetidinone-tethered γ-allenols establishes a regiocontrolled versatile route to a variety of enantiopure fused tetrahydrofuran-, dihydropyran-, and tetrahydrooxepine-β-lactams **33–36** <09CEJ1901>. Additionally, the mechanisms of these metal-catalyzed cycloetherification reactions were investigated theoretically <09CEJ1909>. The stereoselective insertion of allyl-seleno moieties at the C-4 position of azetidinones, and further ring-closing metathesis, afforded novel selenium-containing bicyclic β-lactams **37** <09OBC2591>. A short, enantioselective Lewis acid-catalyzed synthesis of 3,4-benzo-5-oxacephams has been reported <09EJO338>. The asymmetric Kinugasa reaction of cyclic nitrones and nonracemic acetylenes allows construction of the carbapenam skeleton <09JOC3094>. A cinchona alkaloid can catalyze the enantioselective synthesis of 4-aryloxyazetidinones and 3,4-benzo-5-oxacephams <09JOC5687>. An entry to 4-aryl-azetidinones via alkylation of nucleophilic arenes using four-membered acyliminium cations has been accomplished <09T4440>. N-Heterocyclic carbene-catalyzed reactions of α,β-unsaturated aldehydes and a variety of electrophiles allow the efficient preparation of a diverse array of annulation products, including bicyclic β-lactams <09JA8714>. Chiral induction during the photoelectrocyclization of pyridones included within an achiral hydrophobic capsule yields bicyclic β-lactams <09T7277>. Enantiopure 1,3-dioxolanyl-substituted 1,2-oxazines rearrange under Lewis acidic conditions to provide bicyclic products, including β-lactams <09CEJ11632>. A diastereocontrolled Lewis acid-catalyzed preparation of enantiopure carbacepham derivatives startis from 2-azetidinone-tethered enals <09CAJ1604>. Chemo- and regioselectivity control in the palladium-catalyzed O–C cyclization of γ,δ-allendiols provides access to bicyclic β-lactams <09CEJ2496>. An efficient Cu-promoted preparation of bis(β-lactam) fused cyclobutenes in a totally controlled fashion using alkyne homocoupling as well as double [2+2] cyclization in a cascade sequence has been accomplished <09CEJ9987>.

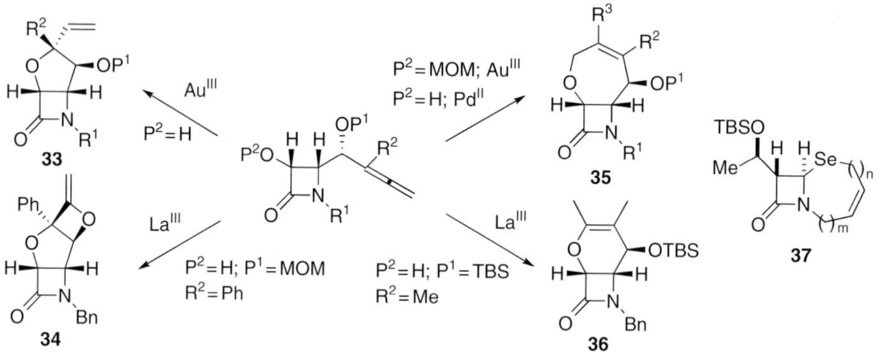

4.5. OXETANES, DIOXETANES, DIOXETANONES AND 2-OXETANONES (β-LACTONES)

A review aiming to provide an overview of the synthesis and reactivity of small strained spiroheterocycles, including spirooxetanes, and to illustrate their applications in synthetic endeavors has appeared <09T5879>. Hydrolytic kinetic resolution as an emerging tool in the synthesis of bioactive molecules, such as tetrahydrolipstatin, has been reviewed <09SL1367>. The total syntheses of sesquiterpenes from *Illicium* species, including the highly oxygenated cage architecture of anisatin which contains a spiro β-lactone has been reviewed <09T6271>. The reactivity of oxazolones and their application in the synthesis of natural products, including β-lactones, has been reviewed <09S2825>. An overview on future perspectives in the total synthesis of natural products such as paclitaxel **38** has appeared <09JOC951>. It has been reported that an N-aroyltransferase has broad aroyl CoA specificity *in vitro* with analogues of N-dearoylpaclitaxel <09JA5994>. The synthesis of 4-deacetyl-1-dimethylsilyl-7-triethylsilylbaccatin III has been carried out <09JOC2186>. A simple route to the paclitaxel side-chain and its analogues based on the (R)-proline-catalyzed addition of aldehydes to N-(phenylmethylene)benzamides, followed by oxidation of the resulting protected α-hydroxy-β-benzoylaminoaldehydes has been presented <09CEJ4044>. A water-soluble derivative of the chemotherapeutic agent taxol has been synthesized with bioconjugation functionality and attached to capsids of the bacteriophage <09AGE9493>. An overview on the basic concepts of rotational-echo double nuclear magnetic resonance and its application to the structural study of natural products such as taxol in biological matrices has been reported <09CC5664>. The biomimetic transannular oxa-conjugate addition approach to the 2,6-disubstituted dihydropyran of laulimalide has yielded an unprecedented transannular oxetane **39** <09JOC1454>. 3-(Silyloxy)oxetanes and 3-aminooxetanes have been obtained through the Paternò-Büchi reactions of aldehydes with silyl enol ethers and enamides <09S4268>. *o*-Naphthoquinone methides have been generated by photolysis of 3-ethoxymethyl- and 1-(ethoxymethyl)-2-naphthols, 2*H*-naphthoxetes **40** and **41** being the precursors <09JA11892>.

Using quantum-chemical calculations it was found that the main DNA lesions induced by solar UV radiation, in addition to being repaired *via* a mechanism involving an oxetane intermediate, a non-oxetane pathway is also probable <09JA17793>. Introduction of oxetan-3-yl groups into heteroaromatic systems that have found important uses in the drug discovery industry, such as the preparation of heteroaryloxetane **42** derived from the marketed inhibitor gefitinib, has been achieved using a radical addition method (Minisci reaction) <09JOC6354>. Enantioenriched tetrahydrofurans have been accessed by enantioselective intramolecular openings of oxetanes **43** catalyzed by (salen)Co(III) complexes <09JA2786>. 2,2-Disubstituted oxetanes have been synthesized using a one-pot double methylene transfer, catalyzed by a chiral heterobimetallic La/Li complex <09AGE1677>. A fused oxetane has been proposed as intermediate for the rearrangement of a propargylic aziridine into a pyrrole <09OL4002>. Polycyclic aldehydes have been prepared by protolytic oxa-metathesis of fused strained oxetane systems <09OL3886>. The wavelength-dependent diastereodifferentiating Paternò–Büchi reaction of chiral cyanobenzoates with diphenylethene has provided chiral oxetanes <09JA17076>. A library of 1,4-disubstituted 1,2,3-triazoles including oxetan-3-yl substituents has been synthesized using a copper flow reactor <09ASC849>. A solvent-controlled oxidative cyclization for the divergent synthesis of highly functionalized oxetanes has been published <09OL3156>. The strong substituent effect of alkynyl groups on the highly stereoselective olefination of alkynyl ketones with ynolates through oxetene intermediates has been described <09JA2092>. The effects of C–S and C–Se bonds on torquoselectivity for the stereoselective olefination of α-thio- and α-selenoketones with ynolates through oxetene intermediates have been reported <09T8832>. Decomposition pathways for the thermolysis of 1,2-dioxetanedione **44** have been postulated <09JA2770>. The unimolecular chemiluminescent decomposition of unsubstituted dioxetanone **45** has been studied at the complete active space self-consistent field level of theory combined with the multistate second-order multiconfigurational perturbation theory energy correction <09JA6181>. The proposed mechanism for the gold-catalyzed oxidative cleavage of aryl-substituted alkynyl ethers using molecular oxygen involves a dioxetane intermediate <09CC4046>. The chemiluminescence of bicyclic dioxetanes bearing a hydroxyaryl group is enhanced when these dioxetanes are decomposed with alkaline metal ions <09TL2340>.

Key: i) 10 mol% (salen)Co(III) complex, MeCN, RT.

Enantiomerically pure cyclopropylboronic esters have been utilized in the synthesis of a cyclopropylamine, a key building block for the total synthesis of the proteasome inhibitor belactosin A **46** <09EJO5998>. A three-dimensional structure-activity relationship study of belactosin A and its stereo- and regioisomers has been documented <09OBC1868>. The first asymmetric total synthesis of the β-lactone-containing natural product vittatalactone featured the divergent synthesis of two diastereomers to assign the absolute configuration of the natural product <09OL4767>. The utility of the asymmetric synthesis of *anti*-aldol segments *via* a nonaldol route has been demonstrated by the synthesis of the lipase inhibitor (−)-tetrahydrolipstatin **47** <09JOC4508>. An asymmetric synthesis of (−)-

tetrahydrolipstatin from a β-hydroxy-δ-oxo sulfoxide has been described <09SL1285>. A palladium-catalyzed Wacker-type reaction to convert an alkene into a ketone, highly diastereoselective reduction of a β-hydroxy ketone, selective oxidation of a diol, and modular synthesis were the key features of the successful approach to the asymmetric synthesis of (−)-tetrahydrolipstatin <09T10083>. The formal synthesis of salinosporamide A **48**, a potent 20S proteasome inhibitor and anti-cancer therapeutic, has been achieved *via* N-heterocyclic carbene-catalyzed intramolecular lactonization from enals <09T4957>. A new series of coenzyme A-tethered polyketide synthase extender units have been discovered related to the biosynthesis of the anticancer agents of the salinosporamide A family, from the marine bacterium *Salinispora tropica* <09JA10376>. The synthesis of a propellane derivative of salinosporamide A having increased stability under physiological-like conditions, took advantage of a substrate-controlled stereoselective Ugi 4-center 3-component reaction to construct the required *syn*-bicyclic pyroglutamic acid framework <09T5899>. Short, practical, and scalable syntheses of (±)-7-methylomuralide and (−)-7-methylomuralide **49** have been developed from N-trichloroethoxycarbonyl glycine <09JA5746>. A formal synthesis of salinosporamide A starting from D-glucose has been accomplished <09S2983>.

The highly diastereo- and enantioselective synthesis of β-trifluoromethyl-β-lactones **50** bearing two contiguous stereocenters has been realized by chiral N-heterocyclic carbene-catalyzed formal cycloaddition of alkyl(aryl)ketenes and trifluoromethyl ketones <09OL4029>. A general method for the organocatalytic dimerization of ketoketenes by tri-*n*-butylphosphine to yield ketoketene dimers **51** has been described <09JOC1777>. KHMDS and KO*t*-Bu have been applied as highly active Lewis base catalysts for the formal [2+2] cycloaddition of ketenes with aldehydes to afford β-lactones <09OBC4009>. A tethered Lewis acid-Lewis base bifunctional catalyst promotes the rapid asymmetric [2+2] cycloaddition of ketene and aldehydes <09SL1675>. Chiral triazolium-derived N-heterocyclic carbene catalysts promote the direct annulation of α,β-unsaturated aldehydes and achiral α-hydroxy enones to afford cyclopentane-fused β-lactones with high enantioselectivity <09OL677>. The asymmetric synthesis, structure, and reactivity of unexpectedly stable spiroepoxy-β-lactones, including facile conversion to tetronic acids, have been reported <09JOC4772>. The desymmetrization of 1,3-diketones using N-heterocyclic carbenes has resulted in the formation of enantioenriched cyclopentenes through the expulsion of carbon dioxide in a bicyclic β-lactone intermediate <09S687>. The one-pot synthesis of 2,3-dihydro-1,5-benzodiazepin-2-ones **52** bearing a phosphono-succinate substituent, from diketene, has been described

<09T2684>. α-Stabilized phosphorus ylides have been obtained from the reaction between primary amines, diketene, and a dialkyl acetylenedicarboxylate in the presence of triphenylphosphine <09S464>. The solid-phase generation of acetylketene through sequential treatment of ketene dimer with hydrogen chloride and sodium carbonate has been reported <09TL1295>. The ring-opening polymerization of a mixture of enantiomerically pure, but different, β-lactone monomers, using an yttrium complex as initiator, proceeds readily at room temperature to give the corresponding highly alternating polyester <09JA16042>. The enantioselective syntheses of the glucosidase inhibitors schulzeines B and C have been achieved using a Pictet–Spengler reaction of a β-lactone-derived, masked bishomoserine aldehyde <09OL1143>. Aromatic aldehydes and ketones react with ketene under Lewis acid catalysis to produce β-lactones, which react *in situ* with another molecule of ketene to produce 3-arylglutaric anhydrides <09TL2334>. The opening of a β-lactone has been used during the stereocontrolled synthesis of iriomoteolide, a smaller but equally cytotoxic congener of amphidinolides <09AGE8780>. The synthesis of highly substituted pyrazolo[3,4-*b*]pyridine-5-carboxamide and 1,4-dihydro-1,8-naphthyridine-3-carboxamide derivatives starting from diketene has been developed via two different one-pot four-component reactions <09TL2911; 09TL6355>. The reaction of ketoketene dimer β-lactones with organolithium reagents affords 1,3-diketones <09TL6919>.

Key: i) 12 mol% NHC derived from L-pyroglutamic acid, 10 mol% Cs_2CO_3, toluene, −40 °C. ii) CH_2Cl_2, RT.

4.6. THIETANES, β-SULTAMS, AND RELATED SYSTEMS

The intramolecular *S*-vinylation of thiols with bromides has been implemented without the help of an additional ligand for the preparation of thietanes **53** <09JOC459>. The synthesis of the highly constrained adenosine derivative **54** featuring at spirothietane at C-4′, which may be considered as a rigid analogue of methylthioadenosine, has been described <09TL463>. A full account of the intramolecular vinylic substitution of bromoalkenes having an acetylthio moiety, which give 2-alkylidenethietane derivatives **55**, has been presented <09T6888>. A synthetic route to functionalized thietanes has been developed by employing diastereoselective three-component coupling reaction of *O*,*O*-diethyl hydrogen phosphorodithioate, aromatic aldehydes, and activated olefins <09SL1055>. A computational study at the MP2(Full)/6-311++G(d,p)//MP2(Full)/6-31+G(d) level of the ammonolysis of halogen-substituted thietanes has been performed in the gas phase and in acetonitrile <09OBC4496>.

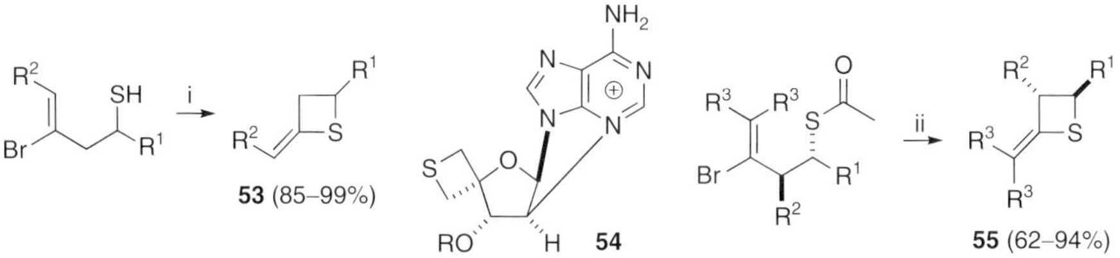

Key: i) 10 mol% CuI, K$_3$PO$_4$·3H$_2$O, dioxane, 90 °C. ii) K$_2$CO$_3$, MeOH, N,N-dimethylimidazolidinone, 120 °C.

β-Thiolactones **56** monosubstituted in the 3-position by alkyl and carbamoyl groups undergo nucleophilic ring opening with arenethiolates through a process involving an S$_N$2-type attack at the 4-position leading to 3-arylthiopropionates substituted at the 2-position <09JOC3389>. 1-Thietane 1,1-dioxides **57** have been prepared by oxidation of 2-alkylidenethietanes **55** with m-chloroperbenzoic acid <09T6888>. β-Sultams **58**, biologically interesting sulfonyl analogues of β-lactams, have been prepared by an organocatalytic asymmetric formal [2+2]-cycloaddition approach of non-nucleophilic imines with alkyl sulfonyl chlorides <09CEJ8204>. 1,3-Disubstituted thioureas react with bis(trichloromethyl) carbonate in the presence of a base such as NaHCO$_3$ to form 4-(arylimino)-1,3-thiazetidin-2-ones **59** almost quantitatively <09SL607>.

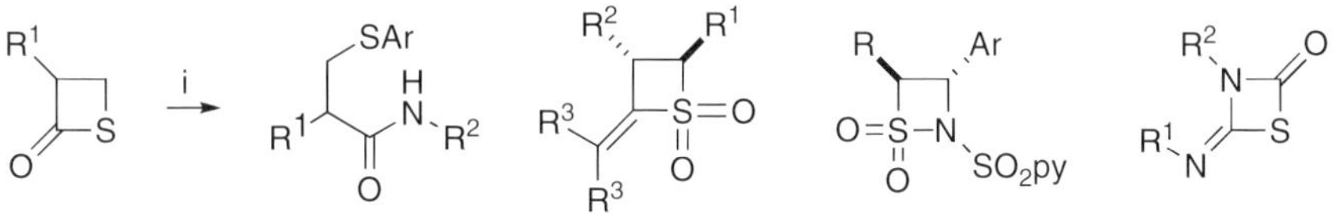

Key: i) (a) ArSH, Cs$_2$CO$_3$; (b) R^{2}NH$_2$.

4.7. SILICON AND PHOSPHORUS HETEROCYCLES: MISCELLANEOUS

The development, mechanistic investigations, and applications in natural product total synthesis of palladium-catalyzed cross-coupling reactions of silicon derivatives, including siletanes have been discussed <09JOC2915>. The synthesis, reactivity, and indirect oxidative cleavage of *para*-siletanylbenzyl ethers **60** have been reported <09JOC1876>. The synthesis, structure, photoluminescence and photoreactivity of 2,3-diphenyl-4-neopentyl-1-silacyclobut-2-enes have been studied <09CEJ8625>. A remarkable base-stabilized bis(silylene) **61** with a silicon(I)-silicon(I) bond has been prepared by the reduction of amidinatotrichlorosilane with potassium graphite <09AGE8536>. A bridging square-planar *cyclo*-P$_4$ unit has been formed by reduction of a zirconium diamidodiphosphine macrocycle in the presence of white phosphorus <09AGE115>. Triphosphacyclobutadiene intermediates are involved in [2+4]-cycloaddition reactions with an organic diene and a phosphaalkyne to yield tetraphosphabenzenes <09AGE934>. A new catalytic system for the C–N bond formation between aryl bromides/chlorides with amines, using Pd_2dba_3 and an inexpensive cyclodiphosphazane ligand **62** has been presented <09TL6004>. Chlorination of 1,2,3,4-tetracyclohexyl-*cyclo*-tetraphosphine by $PhICl_2$ or PCl_5 in the presence of Me_3SiOTf or $GaCl_3$ has provided a stepwise approach to salts of the first *cyclo*-phosphino-chlorophosphonium cations $[Cy_4P_4Cl]^+$**63** and $[Cy_4P_4Cl_2]^{2+}$**64** <09JA17943>.

An efficient two-step method for the preparation of a series of novel 2,5-disubstituted 1,3,4-selenadiazoles by selenating with 2,4-bis(phenyl)-1,3-diselenadiphosphetane-2,4-diselenide (Woollins' reagent) **65** has been reported <09EJO1612>. Woollins' reagent 65 reacts with cyanamides in refluxing toluene to afford a series of novel selenazadiphospholaminediselenides <09T6074>. 2-Imidazolines and imidazoles have been accessed by an aza-Wittig sequence involving the azaoxaphosphatane intermediate **66** <09CC1900>. The organocatalytic α-fluorination of aldehydes and trapping of the intermediate providing optically active propargylic fluorides proceeds *via* an oxaphosphatane intermediate <09JA7153>. It has been reported that diselenadisiletane **67**, formed from direct reaction of a racemic silylene with elemental selenium, gives the first bis(silaselenone) with two donor-stabilized Si=Se bonds upon hydrolysis with water <09AGE4069>. A one-step protocol for the conversion of carboxylic acids to thioesters, uses Lawesson's reagent **68** <09TL6684>. A mild one-pot protocol for the synthesis of 1,3,4-oxadiazoles from carboxylic acids and acylhydrazides using the Burgess reagent has been developed <09TL6435>. The convenient oxidative synthesis of a 16-electron organophosphorus iron sandwich complex analogue of bis(η^4-cyclobutadiene)iron(0) has been accomplished <09AGE3104>. Unusual structural rearrangements that involve reversible Si–C(sp^3) and Si–C(sp^2) bond activation in four-membered nickel and palladium silyl pincer complexes have been reported <09AGE8568>. The rhodium-catalyzed [2+2+2] cycloaddition of terminal alkyl alkynes and alkenyl isocyanates proceeds through a CO migration pathway involving a rhodaazetine intermediate <09AGE2379>. The reaction of diazo compounds with alkenes affords metathesis or cyclopropanation products depending on the steric and electronic properties of the substituents in the ruthenacyclobutane intermediate <09CEJ1516>. A ruthenacyclobutane is involved in the proposed mechanism of the triple-cascade catalysis to generate the stereochemical core of aromadendranediol <09AGE4349>. The synthesis of a four-membered thorium tuck-in complex has been developed <09CEJ12204>.

A ruthenium amido pincer complex catalyzes the dehydrocoupling of dimethylamine-borane to its four-membered cyclic dimer (Me$_2$NBH$_2$)$_2$ <09CEJ10339>. Typical zeolite prenucleating solutions containing Ga and Ge have been studied by mass spectroscopy, pointing to a Ga and Ge incorporation into the oligomers in different ways, such as the heteroatom of a four-membered ring <09CEJ5920>. The proposed mechanism for the conversion of aryl azides into amines using Al(OTf)$_3$/NaI based on a mass spectrometric analysis involves an aluminatriazete <09CEJ7215>. The formation of a 2-platinaoxetane from an oxo complex and norbornene is mediated by a hydroxo complex, which itself may react with norbornene to give a protonated 2-platinaoxetane <09JA8736>. Dumbbell-like Au−Fe$_3$O$_4$ nanoparticles have been made and coupled with Herceptin and a platinadioxetane complex <09JA4216>. N-(1′-Alkoxy)cyclopropyl-2-haloanilines have been transformed into 3,4-dihydro-2(1H)-quinolinones via palladium-catalyzed cyclopropane ring expansion involving a four-membered aza-palladacycle intermediate <09OL1043>. Conjugated dienes have been produced with complete regio- and stereoselectivity by the titanocene(II)-promoted alkylation of propargyl carbonates via the formation of 2,3,4-trisubstituted titanacyclobutenes <09CC3375>. A double cubane structure has been prepared in organoplatinum(IV) chemistry <09CC1487>. Treatment of vinylidene ruthenium complexes with methyl propiolate in the presence of a catalytic amount of HBF$_4$ results in the corresponding cycloaddition products, an unusual ruthenacyclobutene species <09CC2911>. The synthesis and characterization of a novel four-membered germanium bismethanediide complex has been accomplished <09CC6816>.

REFERENCES

09AGE115	W.W. Seidel, O.T. Summerscales, B.O. Patrick and M.D. Fryzuk, *Angew. Chem. Int. Ed.*, **48**, 115 (2009).
09AGE660	K.C. Nicolaou, J.S. Chen, D.J. Edmonds and A.A. Estrada, *Angew. Chem. Int. Ed.*, **48**, 660 (2009).
09AGE934	N.A. Piro and C.C. Cummins, *Angew. Chem. Int. Ed.*, **48**, 934 (2009).
09AGE1677	T. Sone, G. Lu, S. Matsunaga and M. Shibasaki, *Angew. Chem. Int. Ed.*, **48**, 1677 (2009).
09AGE1838	T. Kano, Y. Yamaguchi and K. Maruoka, *Angew. Chem. Int. Ed.*, **48**, 660 (2009).
09AGE2379	R.T. Yu, E.E. Lee, G. Malik and T. Rovis, *Angew. Chem. Int. Ed.*, **48**, 2379 (2009).
09AGE2391	M. Dochnahl and G.C. Fu, *Angew. Chem. Int. Ed.*, **48**, 2391 (2009).
09AGE3104	R. Wolf, J.C. Slootweg, A.W. Ehlers, F. Hartl, B. de, M. Bruin, Lutz, A.L. Spek and K. Lammertsma, *Angew. Chem. Int. Ed.*, **48**, 3104 (2009).
09AGE3353	Y. Xu, G. Lu, S. Matsunaga and M. Shibasaki, *Angew. Chem. Int. Ed.*, **48**, 3353 (2009).
09AGE4069	A. Mitra, J.P. Wojcik, D. Lecoanet, T. Müller and R. West, *Angew. Chem. Int. Ed.*, **48**, 4069 (2009).

09AGE4349	B. Simmons, A.M. Walji and D.W.C. MacMillan, *Angew. Chem. Int. Ed.*, **48**, 4349 (2009).
09AGE4381	A. Coste, G. Karthikeyan, F. Couty and G. Evano, *Angew. Chem. Int. Ed.*, **48**, 4381 (2009).
09AGE7604	X. Han, J. Kwiatkowski, F. Xue, K.-W. Huang and Y. Lu, *Angew. Chem. Int. Ed.*, **48**, 7604 (2009).
09AGE8536	S.S. Sen, A. Jana, H.W. Roesky and C. Schulzke, *Angew. Chem. Int. Ed.*, **48**, 8536 (2009).
09AGE8568	S.J. Mitton, R. McDonald and L. Turculet, *Angew. Chem. Int. Ed.*, **48**, 8568 (2009).
09AGE8780	R. Cribiú, C. Jäger and C. Nevado, *Angew. Chem. Int. Ed.*, **48**, 8780 (2009).
09AGE9493	W. Wu, S.C. Hsiao, Z.M. Carrico and M.B. Francis, *Angew. Chem. Int. Ed.*, **48**, 9493 (2009).
09ANC4285	L. Kantiani, M. Farré, M. Sibum, C. Postigo, M. López de Alda and D. Barceló, *Anal. Chem.*, **81**, 4285 (2009).
09ARK21	J. Xu, *Arkivoc*, **ix**, 21 (2009).
09ASC849	A.R. Bogdan and N.W. Sach, *Adv. Synth. Catal.*, **351**, 849 (2009).
09ASC1768	W. Lu, W. Song, D. Hong, P. Lu and Y. Wang, *Adv. Synth. Catal.*, **351**, 1768 (2009).
09CAJ1604	B. Alcaide, P. Almendros, C. Pardo, C. Rodríguez-Ranera and A. Rodríguez-Vicente, *Chem. Asian. J.*, **4**, 1604 (2009).
09CC1040	O. Sereda, A. Blanrue and R. Wilhelm, *Chem. Commun.*, 1040 (2009).
09CC1487	M.S. Safa, M.C. Jennings and R.J. Puddephatt, *Chem. Commun.*, 1487 (2009).
09CC1900	P. Loos, M. Riedrich and H.-D. Arndt, *Chem. Commun.*, 1900 (2009).
09CC2911	M. Yamaguchi, Y. Arikawa, Y. Nishimura, K. Umakoshi and M. Onishi, *Chem. Commun.*, 2911 (2009).
09CC3375	Y. Yatsumonji, Y. Atake, A. Tsubouchi and T. Takeda, *Chem. Commun.*, 3375 (2009).
09CC4046	A. Das, R. Chaudhuri and R.-S. Liu, *Chem. Commun.*, 4046 (2009).
09CC5664	S. Matsuoka and M. Inoue, *Chem. Commun.*, 5664 (2009).
09CC6265	S. Sato, M. Shibuya, N. Kanoh and Y. Iwabuchi, *Chem. Commun.*, 6265 (2009).
09CC6816	C. Foo, K.-C. Lau, Y.-F. Yang and C.-W. So, *Chem. Commun.*, 6816 (2009).
09CEJ302	B.A. Arndtsen, *Chem. Eur. J.*, **15**, 302 (2009).
09CEJ593	M.L. Lage, I. Fernández, M.J. Mancheño, M. Gómez-Gallego and M.A. Sierra, *Chem. Eur. J.*, **15**, 593 (2009).
09CEJ1516	M. Basato, C. Tubaro, A. Biffis, M. Bonato, G. Buscemi, F. Lighezzolo, P. Lunardi, C. Vianini, F. Benetollo, A. Del and Zotto, *Chem. Eur. J.*, **15**, 1516 (2009).
09CEJ1901	B. Alcaide, P. Almendros, T. Martínez del Campo, E. Soriano and J.L. Marco-Contelles, *Chem. Eur. J.*, **15**, 1901 (2009).
09CEJ1909	B. Alcaide, P. Almendros, T. Martínez del Campo, E. Soriano and J.L. Marco-Contelles, *Chem. Eur. J.*, **15**, 1909 (2009).
09CEJ2129	P. García, S. Moulin, Y. Miclo, D. Leboeuf, V. Gandon, C. Aubert and M. Malacria, *Chem. Eur. J.*, **15**, 2129 (2009).
09CEJ2496	B. Alcaide, P. Almendros, R. Carrascosa and T. Martínez del Campo, *Chem. Eur. J.*, **15**, 2496 (2009).
09CEJ3344	B. Alcaide, P. Almendros, T. Martínez del Campo and M.T. Quirós, *Chem. Eur. J.*, **15**, 3344 (2009).
09CEJ4044	P. Dziedzic, P. Schyman, M. Kullberg and A. Córdova, *Chem. Eur. J.*, **15**, 4044 (2009).
09CEJ5920	B.B. Schaack, W. Schrader and F. Schüth, *Chem. Eur. J.*, **15**, 5920 (2009).
09CEJ6940	D. Pellico, M. Gómez-Gallego, P. Ramírez-López, M.J. Mancheño, M.A. Sierra and M.R. Torres, *Chem. Eur. J.*, **15**, 6940 (2009).
09CEJ7215	A. Kamal, N. Markandeya, N. Shankaraiah, C.R. Reddy, S. Prabhakar, C.S. Reddy, M.N. Eberlin and L.S. Santos, *Chem. Eur. J.*, **15**, 7215 (2009).
09CEJ8204	M. Zajac and R. Peters, *Chem. Eur. J.*, **15**, 8204 (2009).
09CEJ8625	D. Yan, M.D. Thomson, M. Backer, M. Bolte, R. Hahn, R. Berger, W. Fann, H.G. Roskos and N. Auner, *Chem. Eur. J.*, **15**, 8625 (2009).
09CEJ9987	B. Alcaide, P. Almendros and C. Aragoncillo, *Chem. Eur. J.*, **15**, 9987 (2009).
09CEJ10339	A. Friedrich, M. Drees and S. Schneider, *Chem. Eur. J.*, **15**, 10339 (2009).
09CEJ11632	A. Al-Harrasi, F. Pfrengle, V. Prisyazhnyuk, S. Yekta, P. Koóš and H.-U. Reissig, *Chem. Eur. J.*, **15**, 11632 (2009).

09CEJ12204	W.J. Evans, J.R. Walensky and J.W. Ziller, *Chem. Eur. J.*, **15**, 12204 (2009).
09CME564	M. Bassetti, L. Nicolini, S. Esposito, E. Righi and C. Viscoli, *Curr. Med. Chem.*, **16**, 564 (2009).
09COR827	W. Van Brabandt, S. Mangelinckx, M. D'hooghe, B. Van Driessche and N. De Kimpe, *Curr. Org. Chem.*, **13**, 827 (2009).
09COR852	K.A. Tehrani and N. De Kimpe, *Curr. Org. Chem.*, **13**, 852 (2009).
09CSY325	M.T. Aranda, P. Pérez-Faginas and R. González-Muñiz, *Curr. Org. Synth.*, **6**, 325 (2009).
09EJM2071	A. Urbach, G. Dive, Be. Tinant, V. Duval and J. Marchand-Brynaert, *Eur. J. Med. Chem.*, 2071 (2009).
09EJO85	J. Beck, S. Gharbi, A. Herteg-Fernea, L. Vercheval, C. Bebrone, P. Lassaux, A. Zervosen and J. Marchand-Brynaert, *Eur. J. Org. Chem.*, 85 (2009).
09EJO338	A. Kozioł, J. Frelek, M. Woźnica, B. Furman and M. Chmielewski, *Eur. J. Org. Chem.*, 338 (2009).
09EJO1384	I. Serra, D.A. Cecchini, D. Ubiali, E.M. Manazza, A.M. Albertini and M. Terreni, *Eur. J. Org. Chem.*, 1384 (2009).
09EJO1612	G. Hua, Y. Li, A.L. Fuller, A.M.Z. Slawin and J.D. Woollins, *Eur. J. Org. Chem.*, 1612 (2009).
09EJO1757	A. Urbach, G. Dive and J. Marchand-Brynaert, *Eur. J. Org. Chem.*, 1757 (2009).
09EJO1994	E. Fördõs, R. Tuba, L. Párkányi, T. Kégl and F. Ungváry, *Eur. J. Org. Chem.*, 1994 (2009).
09EJO2729	M. Bouazaoui, J. Martinez and F. Cavelier, *Eur. J. Org. Chem.*, 2729 (2009).
09EJO2998	M.A. Sierra, M. Rodríguez-Fernández, L. Casarrubios, M. Gómez-Gallego and M.J. Mancheño, *Eur. J. Org. Chem.*, 2998 (2009).
09EJO3155	E. Emer, P. Galletti and D. Giacomini, *Eur. J. Org. Chem.*, 3155 (2009).
09EJO4541	P. Galletti, A. Quintavalla, C. Ventrici and D. Giacomini, *Eur. J. Org. Chem.*, 4541 (2009).
09EJO5653	K. Michel, R. Fröhlich and E.-U. Würthwein, *Eur. J. Org. Chem.*, 5653 (2009).
09EJO5998	J. Pietruszka and G. Solduga, *Eur. J. Org. Chem.*, 5998 (2009).
09JA1542	J. Zhang, D.A. Kissounko, S.E. Lee, S.H. Gellman and S.S. Stahl, *J. Am. Chem. Soc.*, **131**, 1542 (2009).
09JA1589	J. Zhang, D.A. Kissounko, S.E. Lee, S.H. Gellman and S.S. Stahl, *J. Am. Chem. Soc.*, **131**, 1589 (2009).
09JA2092	T. Yoshikawa, S. Mori and M. Shindo, *J. Am. Chem. Soc.*, **131**, 2092 (2009).
09JA2338	M. Kalp, M.A. Totir, J.D. Buynak and P.R. Carey, *J. Am. Chem. Soc.*, **131**, 2338 (2009).
09JA2770	R. Bos, S.A. Tonkin, G.R. Hanson, C.M. Hindson, K.F. Lim and N.W. Barnett, *J. Am. Chem. Soc.*, **131**, 2770 (2009).
09JA2786	R.N. Loy and E.N. Jacobsen, *J. Am. Chem. Soc.*, **131**, 2786 (2009).
09JA2804	I. Nakamura, T. Araki and M. Terada, *J. Am. Chem. Soc.*, **131**, 2804 (2009).
09JA4216	C. Xu, B. Wang and S. Sun, *J. Am. Chem. Soc.*, **131**, 4216 (2009).
09JA5016	S. Mizukami, S. Watanabe, Y. Hori and K. Kikuchi, *J. Am. Chem. Soc.*, **131**, 5016 (2009).
09JA5746	R.A. Shenvi and E.J. Corey, *J. Am. Chem. Soc.*, **131**, 5746 (2009).
09JA5994	D.M. Nevarez, Y.A. Mengistu, I.N. Nawarathne and K.D. Walker, *J. Am. Chem. Soc.*, **131**, 5994 (2009).
09JA6181	F. Liu, Y. Liu, L. De Vico and R. Lindh, *J. Am. Chem. Soc.*, **131**, 6181 (2009).
09JA7153	H. Jiang, A. Falcicchio, K.L. Jensen, M.W. Paixão, S. Bertelsen and K.A. Jørgensen, *J. Am. Chem. Soc.*, **131**, 7153 (2009).
09JA8714	P.-C. Chiang, M. Rommel and J.W. Bode, *J. Am. Chem. Soc.*, **131**, 8714 (2009).
09JA8736	N.M. Weliange, E. Szuromi and P.R. Sharp, *J. Am. Chem. Soc.*, **131**, 8736 (2009).
09JA8798	K.B. Hansen, Y. Hsiao, F. Xu, N. Rivera, A. Clausen, M. Kubryk, S. Krska, T. Rosner, B. Simmons, J. Balsells, N. Ikemoto, Y. Sun, F. Spindler, C. Malan, E.J.J. Grabowski and J.D. Armstrong,III, *J. Am. Chem. Soc.*, **131**, 8798 (2009).
09JA9646	Y. Zhang, G. Burdziński, J. Kubicki and M.S. Platz, *J. Am. Chem. Soc.*, **131**, 9646 (2009).
09JA9648	M. Movassaghi, M. Tjandra and J. Qi, *J. Am. Chem. Soc.*, **131**, 9648 (2009).
09JA9735	B.P. Mowery, A.H. Lindner, B. Weisblum, S.S. Stahl and S.H. Gellman, *J. Am. Chem. Soc.*, **131**, 9735 (2009).

09JA10376	Y. Liu, C. Hazzard, A.S. Eustáquio, K.A. Reynolds and B.S. Moore, *J. Am. Chem. Soc.*, **131**, 10376 (2009).
09JA11314	A.J.-L. Ayitou, J.L. Jesuraj, N. Barooah, A. Ugrinov and J. Sivag, *J. Am. Chem. Soc.*, **131**, 11314 (2009).
09JA11642	R.M. Breece, Z. Hu, B. Bennett, M.W. Crowder and D.L. Tierney, *J. Am. Chem. Soc.*, **131**, 11642 (2009).
09JA11892	S. Arumugam and V.V. Popik, *J. Am. Chem. Soc.*, **131**, 11892 (2009).
09JA15262	E. Sauvage, A. Zervosen, G. Dive, R. Herman, A. Amoroso, B. Joris, E. Fonzé, R.F. Pratt, A. Luxen, P. Charlier and F. Kerff, *J. Am. Chem. Soc.*, **131**, 15262 (2009).
09JA16042	J.W. Kramer, D.S. Treitler, E.W. Dunn, P.M. Castro, T. Roisnel, C.M. Thomas and G.W. Coates, *J. Am. Chem. Soc.*, **131**, 16042 (2009).
09JA16779	M.-R. Lee, S.S. Stahl, S.H. Gellman and K.S. Masters, *J. Am. Chem. Soc.*, **131**, 16779 (2009).
09JA17076	K. Matsumura, T. Mori and Y. Inoue, *J. Am. Chem. Soc.*, **131**, 17076 (2009).
09JA17793	T. Domratcheva and I. Schlichting, *J. Am. Chem. Soc.*, **131**, 17793 (2009).
09JA17943	J.J. Weigand, N. Burford, R.J. Davidson, T.S. Cameron and P. Seelheim, *J. Am. Chem. Soc.*, **131**, 17943 (2009).
09JMC4058	M. D'hooghe, S. Dekeukeleire, K. Mollet, C. Lategan, P.J. Smith, K. Chibale and N. De Kimpe, *J. Med. Chem.*, **52**, 4058 (2009).
09JMC7054	M. Feledziak, C. Michaux, A. Urbach, G. Labar, G.G. Muccioli, D.M. Lambert and J. Marchand-Brynaert, *J. Med. Chem.*, **52**, 7054 (2009).
09JMC7446	T.V. Magee, S.L. Ripp, B. Li, R.A. Buzon, L. Chupak, T.J. Dougherty, S.M. Finegan, D. Girard, A.E. Hagen, M.J. Falcone, K.A. Farley, K. Granskog, J.R. Hardink, M.D. Huband, B.J. Kamicker, T. Kaneko, M.J. Knickerbocker, J.L. Liras, A. Marra, I. Medina, T.-T. Nguyen, M.C. Noe, R.S. Obach, J.P. O'Donnell, J.B. Penzien, U.D. Reilly, J.R. Schafer, Y. Shen, G.G. Stone, T.J. Strelevitz, J. Sun, A. Tait-Kamradt, A.D.N. Vaz, D.A. Whipple, D.W. Widlicka, D.G. Wishka, J.P. Wolkowski and M.E. Flanagan, *J. Med. Chem.*, **52**, 7446 (2009).
09JOC459	Q. Zhao, L. Li, Y. Fang, D. Sun and C. Li, *J. Org. Chem.*, **74**, 459 (2009).
09JOC951	K.C. Nicolau, *J. Org. Chem.*, **74**, 951 (2009).
09JOC1454	S.R. Houghton, L. Furst and C.N. Boddy, *J. Org. Chem.*, **74**, 1454 (2009).
09JOC1644	S. Dekeukeleire, M. D'hooghe and N. De Kimpe, *J. Org. Chem.*, **74**, 1644 (2009).
09JOC1777	A.A. Ibrahim, G.D. Harzmann and N.J. Kerrigan, *J. Org. Chem.*, **74**, 1777 (2009).
09JOC1876	S.F. Tlais, H. Lam, S.E. House and G.B. Dudley, *J. Org. Chem.*, **74**, 1876 (2009).
09JOC2186	M.E. Ondari and K.D. Walker, *J. Org. Chem.*, **74**, 2186 (2009).
09JOC2250	E. Van Hende, G. Verniest, F. Deroose, J.-W. Thuring, G. Macdonald and N. De Kimpe, *J. Org. Chem.*, **74**, 2250 (2009).
09JOC2254	U.K. Wefelscheid and S. Woodward, *J. Org. Chem.*, **74**, 2254 (2009).
09JOC2832	T.M. Bott, J.A. Vanecko and F.G. West, *J. Org. Chem.*, **74**, 2832 (2009).
09JOC2915	S.E. Denmark, *J. Org. Chem.*, **74**, 2915 (2009).
09JOC3094	S. Stecko, A. Mames, B. Furman and M. Chmielewski, *J. Org. Chem.*, **74**, 3094 (2009).
09JOC3389	D. Crich and K. Sana, *J. Org. Chem.*, **74**, 3389 (2009).
09JOC4305	Y.A. Ibrahim, T.F. Al-Azemi, M.D.A. El-Halim and E. John, *J. Org. Chem.*, **74**, 4305 (2009).
09JOC4508	A.K. Ghosh, K. Shurrush and S. Kulkarni, *J. Org. Chem.*, **74**, 4508 (2009).
09JOC4772	R.J. Duffy, K.A. Morris, R. Vallakati, W. Zhang and D. Romo, *J. Org. Chem.*, **74**, 4772 (2009).
09JOC5687	A. Kozioł, B. Furman, J. Frelek, M. Woźnica, E. Altieri and M. Chmielewski, *J. Org. Chem.*, **74**, 5687 (2009).
09JOC5730	C. Cesario and M.J. Miller, *J. Org. Chem.*, **74**, 5730 (2009).
09JOC6077	D.-H. Leng, D.-X. Wang, J. Pan, Z.-T. Huang and M.-X. Wang, *J. Org. Chem.*, **74**, 6077 (2009).
09JOC6354	M.A.J. Duncton, M.A. Estiarte, R.J. Johnson, M. Cox, D.J.R. O'Mahony, W.T. Edwards and M.G. Kelly, *J. Org. Chem.*, **74**, 6354 (2009).
09JOC6691	J.M. Aizpurua, C. Palomo, R.M. Fratila, P. Ferrón, A. Benito, E. Gómez-Bengoa, J.I. Miranda and J.I. Santos, *J. Org. Chem.*, **74**, 6691 (2009).
09JOC7013	M.K. Ghorai, D. Shukla and K. Das, *J. Org. Chem.*, **74**, 7013 (2009).

09JOC7585	X.-L. Huang, X.-Y. Chen and S. Ye, *J. Org. Chem.*, **74**, 7585 (2009).
09JOC7859	M. Medjahdi, J.C. González-Gómez, F. Foubelo and M. Yus, *J. Org. Chem.*, **74**, 7859 (2009).
09JOC8203	J.L. Baeza, G. Gerona-Navarro, K. Thompson, M.J. Pérez de Vega, L. Infantes, M.T. García-López, R. González-Muñiz and M. Martín-Martínez, *J. Org. Chem.*, **74**, 8203 (2009).
09JOC8421	B. Alcaide, P. Almendros, C. Aragoncillo, R. Callejo, M.P. Ruiz and M.R. Torres, *J. Org. Chem.*, **74**, 8421 (2009).
09JOC8813	M. Bera and S. Roy, *J. Org. Chem.*, **74**, 8813 (2009).
09JOC8886	T. Jensen, M. Mikkelsen, A. Lauritsen, T.L. Andresen, C.H. Gotfredsen and R. Madsen, *J. Org. Chem.*, **74**, 8886 (2009).
09JOC9505	B. Das, P. Balasubramanyam, B. Veeranjaneyulu and G.C. Reddy, *J. Org. Chem.*, **74**, 9505 (2009).
09MI26	F. Couty and M. Kletskii, *J. Mol. Struc-Theochem.*, **908**, 26 (2009).
09MI41	Mi. Yamaoka-Tojo, T. Tojo, R. Kosugi, Y. Hatakeyama, Y. Yoshida, Y. Machida, N. Aoyama, T. Masuda and T. Izumi, *Lipids Health Dis.*, **8**, 41 (2009).
09MI65	Y.H. Lee, H.H. Kim, P. Thuery, J.M Harrowfield, W.T. Lim, B.J. Kim and Y. Kim, *J. Incl. Phenom. Macro.*, **65**, 65 (2009).
09MI527	M. Hefnawy, M. Al-Omar and S. Julkhuf, *J. Pharm. Biomed. Anal.*, **50**, 527 (2009).
09MI798	H. Teoh, A.A. Mendelsohn, S.G. Goodman, S. Jaffer, R.Y.Y. Chen, S. Tjia, L. Theriault, A. Langer and L.A. Leiter, *Am. J. Cardiol.*, **104**, 798 (2009).
09MI1068	H. Ma, B. Yan, Z. Li, Y. Guan, J. Song, K. Xu and R. Hu, *J. Hazard. Mater.*, **169**, 1068 (2009).
09MI1279	A.P. Kozikowski, J.B. Eaton, K.M. Bajjuri, S.K. Chellappan, Y. Chen, S. Karadi, R. He, B. Caldarone, M. Manzano, P.-W. Yuen and R.J. Lukas, *Chem. Med. Chem.*, **4**, 1279 (2009).
09MI1369	A. Singh, A. Basit and U.C. Banerjee, *J. Ind. Microbiol. Biotechnol.*, **36**, 1369 (2009).
09MI1991	P. Galletti, A. Quintavalla, C. Ventrici, G. Giannini, W. Cabri, S. Penco, G. Gallo, S. Vincenti and D. Giacomini, *Chem. Med. Chem*, **4**, 1991 (2009).
09MI2180	J.J.P. Kastelein and M.L. Bots, *New. Engl. J. Med.*, **361**, 2180 (2009).
09OBC1868	K. Yoshida, K. Yamaguchi, A. Mizuno, Y. Unno, A. Asai, T. Sone, H. Yokosawa, A. Matsuda, M. Arisawa and S. Shuto, *Org. Biomol. Chem.*, **7**, 1868 (2009).
09OBC2591	D.R. Garud, D.D. Garud and M. Koketsu, *Org. Biomol. Chem.*, **7**, 2591 (2009).
09OBC2770	C. Ducho, R.B. Hamed, E.T. Batchelar, J.L. Sorensen, B. Odell and C.J. Schofield, *Org. Biomol. Chem.*, **7**, 2770 (2009).
09OBC4009	S. Tabassum, O. Sereda, P.V.G. Reddy and R. Wilhelm, *Org. Biomol. Chem.*, **7**, 4009 (2009).
09OBC4496	H.D. Banks, *Org. Biomol. Chem.*, **7**, 4496 (2009).
09OBC4777	C. Martín, T.R. Belderraín and P.J. Pérez, *Org. Biomol. Chem.*, **7**, 4777 (2009).
09OL13	S. Suárez-Pantiga, E. Rubio, C. Álvarez-Rúa and J.M. González, *Org. Lett.*, **11**, 13 (2009).
09OL209	C. Caupéne, G. Chaume, L. Ricard and T. Brigaud, *Org. Lett.*, **11**, 209 (2009).
09OL677	J. Kaeobamrung and J.W. Bode, *Org. Lett.*, **11**, 677 (2009).
09OL1043	T. Tsuritani, Y. Yamamoto, M. Kawasaki and T. Mase, *Org. Lett.*, **11**, 1043 (2009).
09OL1281	L.L.W. Cheung and A.K. Yudin, *Org. Lett.*, **11**, 1281 (2009).
09OL1293	C. Cesario and M.J. Miller, *Org. Lett.*, **11**, 1293 (2009).
09OL1143	G. Liu and D. Romo, *Org. Lett.*, **11**, 1143 (2009).
09OL2515	S.A. Testero, P.I. O'Daniel, Q. Shi, M. Lee, D. Hesek, A. Ishiwata, B.C. Noll and S. Mobashery, *Org. Lett.*, **11**, 2515 (2009).
09OL2820	J. Liang, J. Chen, F. Du, X. Zeng, L. Li and H. Zhang, *Org. Lett.*, **11**, 2820 (2009).
09OL3156	Y. Ye, C. Zheng and R. Fan, *Org. Lett.*, **11**, 3156 (2009).
09OL3266	I. Hachiya, T. Yoshitomi, Y. Yamaguchi and M. Shimizu, *Org. Lett.*, **11**, 3266 (2009).
09OL3523	M.J. Meyers, I. Muizebelt, J. van Wiltenburg, D.L. Brown and A. Thorarensen, *Org. Lett.*, **11**, 3523 (2009).
09OL3606	M.J. Bodner, R.M. Phelan and C.A. Townsend, *Org. Lett.*, **11**, 3606 (2009).
09OL3886	R.A. Valiulin and A.G. Kutateladze, *Org. Lett.*, **11**, 3886 (2009).
09OL4002	X. Zhao, E. Zhang, Y.-Q. Tu, Y.-Q. Zhang, D.-Y. Yuan, K. Cao, C.-A. Fan and F.-M. Zhang, *Org. Lett.*, **11**, 4002 (2009).

09OL4029	X.-N. Wang, P.-L. Shao, H. Lv and S. Ye, *Org. Lett.*, **11**, 4029 (2009).
09OL4767	Y. Schmidt and B. Breit, *Org. Lett.*, **11**, 4767 (2009).
09OL4850	B.R. Balthaser and F.E. McDonald, *Org. Lett.*, **11**, 4850 (2009).
09OL5438	S. Koya, K. Yamanoi, R. Yamasaki, I. Azumaya, H. Masu and S. Saito, *Org. Lett.*, **11**, 5438 (2009).
09OL5722	S.S. Bag, R. Kundu, A. Basak and Z. Slanina, *Org. Lett.*, **11**, 5722 (2009).
09S464	A. Alizadeh, N. Zohreh and L.-G. Zhu, *Synthesis*, 464 (2009).
09S687	E.M. Phillips, M. Wadamoto and K.A. Scheidt, *Synthesis*, 687 (2009).
09S1551	A. Russo and A. Lattanzi, *Synthesis*, 1551 (2009).
09S2825	N.M. Hewlett, C.D. Hupp and J.J. Tepe, *Synthesis*, 2825 (2009).
09S2983	T. Momose, Y. Kaiya, J. Hasegawa, T. Sato and N. Chida, *Synthesis*, 2983 (2009).
09S3449	H. Tanaka, Y. Tokumaru, K.-i. Fukui, M. Kuroboshi, S. Torii, A. Jutand and C. Amatore, *Synthesis*, 3449 (2009).
09S3519	L.F.R. Gomes, A.F. Trindade, N.R. Candeias, L.F. Veiros, P.M.P. Gois and C.A.M. Afonso, *Synthesis*, 3519 (2009).
09S4268	F. Vogt, K. Jödicke, J. Schröder and T. Bach, *Synthesis*, 4268 (2009).
09SL607	C. Jin, C. Liu and W. Su, *Synlett*, 607 (2009).
09SL727	J.S. Yadav, B.V.S. Reddy, G. Narasimhulu and G. Satheesh, *Synlett*, 727 (2009).
09SL767	F. Couty and G. Evano, *Synlett*, 767 (2009).
09SL787	T. Oshitari and T. Mandai, *Synlett*, 787 (2009).
09SL1055	L.D.S. Yadav, R. Kapoor and Garima, *Synlett*, 1055 (2009).
09SL1285	S. Raghavan and K. Rathore, *Synlett*, 1285 (2009).
09SL1367	P. Kumar and P. Gupta, *Synlett*, 1367 (2009).
09SL1675	S. Chidara and Y.-M. Lin, *Synlett*, 1675 (2009).
09SL1651	C.J. Abraham, D.H. Paull, C. Dogo-Isonagie and T. Lectka, *Synlett*, 1651 (2009).
09SL2437	V. Gembus, T. Poisson, S. Oudeyer, F. Marsais and V. Levacher, *Synlett*, 2437 (2009).
09SL3053	B. Drouillat, F. Couty and J. Marrot, *Synlett*, 3053 (2009).
09SL3182	B. Drouillat, F. Couty and V. Razafimahaléo, *Synlett*, 3182 (2009).
09T1630	D.-H. Lee, Y.H. Lee, J.M. Harrowfield, I.-M. Lee, H.I. Lee, W.T. Lim, Y. Kim and M.-J. Jin, *Tetrahedron*, **65**, 1630 (2009).
09T2605	P.M. Chincholkar, A.S. Kale, V.K. Gumaste and A.R.A.S. Deshmukh, *Tetrahedron*, **65**, 2605 (2009).
09T2684	A. Alizadeh, N. Zohreh and L.-G. Zhu, *Tetrahedron*, **65**, 2684 (2009).
09T2927	A. Jarrahpour and M. Zarei, *Tetrahedron*, **65**, 2927 (2009).
09T3292	T.A. Brandt, S. Caron, D.B. Damon, J. DiBrino, A. Ghosh, D.A. Griffith, S. Kedia, J.A. Ragan, P.R. Rose, B.C. Vanderplas and L. Wei, *Tetrahedron*, **65**, 3292 (2009).
09T3432	L.R. Domingo, M.J. Aurrell and M. Arnó, *Tetrahedron*, **65**, 3432 (2009).
09T4440	B. Zambroń, M. Masnyk, B. Furman and M. Chmielewski, *Tetrahedron*, **65**, 4440 (2009).
09T4664	G. Abbiati, A. Contini, D. Nava and E. Rossi, *Tetrahedron*, **65**, 4664 (2009).
09T5879	R.J. Duffy, K.A. Morris and D. Romo, *Tetrahedron*, **65**, 5879 (2009).
09T5899	M. Vamos and Y. Kobayashi, *Tetrahedron*, **65**, 5899 (2009).
09T6074	G. Hua, Q. Zhang, Y. Li, A.M.Z. Slawin and J.D. Woollins, *Tetrahedron*, **65**, 6074 (2009).
09T6771	D.H. Paull, A. Weatherwax and T. Lectka, *Tetrahedron*, **65**, 6771 (2009).
09T6888	M.-Y. Lei, Y.-J. Xiao, W.-M. Liu, K. Fukamizu, S. Chiba, K. Ando and K. Narasaka, *Tetrahedron*, **65**, 6888 (2009).
09T7277	A.K. Sundaresan, C.L.D. Gibb, B.C. Gibb and V. Ramamurthy, *Tetrahedron*, **65**, 7277 (2009).
09T8832	T. Yoshikawa, S. Mori and M. Shindo, *Tetrahedron*, **65**, 8832 (2009).
09T9322	A. Chrostowska, A. Dargelos, A. Graciaa, S. Khayar, S. Leśniak, R.B. Nazarski, T.X.M. Nguyen, M. Maciejczyk and M. Rachwalski, *Tetrahedron*, **65**, 9322 (2009).
09T4957	J.R. Struble and J.W. Bode, *Tetrahedron*, **65**, 4957 (2009).
09T6271	D. Urabe and M. Inoue, *Tetrahedron*, **65**, 6271 (2009).
09T9742	D. Minato, S. Mizuta, M. Kuriyama, Y. Matsumura and O. Onomura, *Tetrahedron*, **65**, 9742 (2009).
09T10060	S.S. Bari, A. Reshma, Bhalla and G. Hundal, *Tetrahedron*, **65**, 10060 (2009).

09T10083	S. Raghavan and K. Rathore, *Tetrahedron*, **65**, 10083 (2009).
09T10339	A.R. Todorov, V.B. Kurteva, R.P. Bontchev and N.G. Vassilev, *Tetrahedron*, **65**, 10339 (2009).
09T10581	S. Leśniak, A. Chrostowska, D. Kuc, M. Maciejczyk, S. Khayar, R.B. Nazarski and Ł. Urbaniak, *Tetrahedron*, **65**, 10581 (2009).
09TA368	L. Troisi, E. Pindinelli, V. Strusi and P. Trinchera, *Tetrahedron: Asymmetry*, **20**, 368 (2009).
09TA1213	L. Thander, K. Sarkar and S.K. Chattopadhyay, *Tetrahedron: Asymmetry*, **20**, 1213 (2009).
09TA1641	P. D'Arrigo, L.T. Kanerva, X.-G. Li, C. Saraceno, S. Servi and D. Tessaro, *Tetrahedron: Asymmetry*, **20**, 1641 (2009).
09TA1646	I. Pérez-Sánchez and E. Turos, *Tetrahedron: Asymmetry*, **20**, 1646 (2009).
09TA2220	G. Benedek, M. Palkó, E. Wéber, T.A. Martinek, E. Forró and F. Fülöp, *Tetrahedron: Asymmetry*, **20**, 2220 (2009).
09TA2530	K. Nagata, D. Sano, Y. Shimizu, M. Miyazaki, T. Kanemitsu and T. Itoh, *Tetrahedron: Asymmetry*, **20**, 2530 (2009).
09TA2616	J.-L. Niu, M.-C. Wang, L.-j. Lu, G.-L. Ding, H.-J. Lu, Q.-T. Chen and M.-P. Song, *Tetrahedron: Asymmetry*, **20**, 2616 (2009).
09TL463	G.S.G. De Carvalho, J.-L. Fourrey, R.H. Dodd and A.D. Da Silva, *Tetrahedron Lett.*, **50**, 463 (2009).
09TL1020	Y. Liang, R. Raju, T. Le, C.D. Taylor and A.R. Howell, *Tetrahedron Lett.*, **50**, 1020 (2009).
09TL1105	M.K. Ghorai, K. Das and A. Kumar, *Tetrahedron Lett.*, **50**, 1105 (2009).
09TL1295	K. Bell, D.V. Sadasivam, I.R. Gudipati, H. Ji and D. Birney, *Tetrahedron Lett.*, **50**, 1295 (2009).
09TL1568	A. Jarrahpour and M. Zarei, *Tetrahedron Lett.*, **50**, 1568 (2009).
09TL1893	C.S. McKay, D.C. Kennedy and J.P. Pezacki, *Tetrahedron Lett.*, **50**, 1893 (2009).
09TL2334	H. Matsunaga, K. Ikeda, K. Iwamoto, Y. Suzuki and M. Sato, *Tetrahedron Lett.*, **50**, 2334 (2009).
09TL2340	M. Matsumoto, F. Kakuno, A. Kikkawa, N. Hoshiya, N. Watanabe and H.K. Ijuin, *Tetrahedron Lett.*, **50**, 2340 (2009).
09TL2605	L. Kiss, M. Nonn, E. Forró, R. Sillanpää and F. Fülöp, *Tetrahedron Lett.*, **50**, 2605 (2009).
09TL2676	I. Plantan, M. Stephan, U. Urleb and B. Mohar, *Tetrahedron Lett.*, **50**, 2676 (2009).
09TL2911	A. Shaabani, M. Seyyedhamzeh, A. Maleki, M. Behnam and F. Rezazadeh, *Tetrahedron Lett.*, **50**, 2911 (2009).
09TL3296	G. Pandey and D.K. Tiwari, *Tetrahedron Lett.*, **50**, 3296 (2009).
09TL3909	P.N. Collier, *Tetrahedron Lett.*, **50**, 3909 (2009).
09TL4969	T. Saito, T. Kikuchi, H. Tanabe, J. Yahiro and T. Otani, *Tetrahedron Lett.*, **50**, 4969 (2009).
09TL6355	A. Shaabani, M. Seyyedhamzeh, A. Maleki and M. Behnam, *Tetrahedron Lett.*, **50**, 6355 (2009).
09TL6004	R.R. Suresh and K.C.K. Swamy, *Tetrahedron Lett.*, **50**, 6004 (2009).
09TL6435	C. Li and H.D. Dickson, *Tetrahedron Lett.*, **50**, 6435 (2009).
09TL6590	B.A. Burkett, S.Z. Ting, G.C.S. Gan and C.L.L. Chai, *Tetrahedron Lett.*, **50**, 6590 (2009).
09TL6684	Y. Rao, X. Li, P. Nagorny, J. Hayashida and S.J. Danishefsky, *Tetrahedron Lett.*, **50**, 6684 (2009).
09TL6919	A.A. Ibrahim, S.M. Smith, S. Henson and N.J. Kerrigan, *Tetrahedron Lett.*, **50**, 6919 (2009).
09TL7280	C. Napolitano, M. Borriello, F. Cardullo, D. Donati, A. Paio and S. Manfredini, *Tetrahedron Lett.*, **50**, 7280 (2009).

CHAPTER 5.1

Five-Membered Ring Systems: Thiophenes and Se/Te Derivatives

Edward R. Biehl
Southern Methodist University, Dallas, TX 75275 USA
ebiehl@smu.edu

5.1.1. INTRODUCTION

A tremendous amount of synthetic effort has been expended in preparing a wide variety of thiophenes and benzo[b]thiophenes with important drug activities or for use as valuable precursors. Additionally, the properly design and synthesis of thiophenes for polymeric, electronic, superconducting, and non-linear optical materials continued unabated this year. Due to space limitations, we will focus mainly on the synthetic aspects of thiophene and Se/Te chemistry. Reports with a common flavor have been grouped together whenever possible. We should mention that many thiophene derivatives used as important synthetic intermediates, conducting materials, polymeric semiconductors, etc., are now available from Sigma Aldrich at matsci@sial.com.

5.1.2. REVIEWS, ACCOUNTS AND BOOKS ON THIOPHENE CHEMISTRY

The chemistry of thiophenes has appeared in several review articles. For example, a critical survey of the diverse and creative ways of synthesizing 2,5-dihydrothiophene as well as other 2,5-dihydrofurans and pyrroles was published <09OBC1761>. A special issue review of synthesis, reactivity and electronic structure of multifarious five-membered heteroaryl and heteroaryl azides with main focus on the azido transfer reaction appeared this year <09ARK(i)97>. A review on new directions in medical biosensors using poly(3,4-ethylenedioxy thiophene) derivative-based electrodes summarizes newly developed methods associated with the application of REDOT to diagnostic sensing <09ABC637>. Tetrahedron Report 893 discusses recent advances in the synthesis of (hetero)aryl-substituted thiophenes, benzothiophene and thienyl[2,3-b]heteroarenes via transition metal-catalyzed direct hetero (arylation) reactions <09T10269> and Tetrahedron report 882 is devoted to recent applications of Sonogashira cross coupling reactions in heterocyclic synthesis <09T7761>. A review on organic photovoltaics with special interest in developing organic solar cells appeared in Accounts of Chemical Research in which various thiophene derivatives such as poly(3-hexylthiophene (P3HT), fullerene/thiophene, etc., are major players <09ACR(11)1691>. A handbook of thiophene-based materials and their applications to organic electronics and photonics was published this year <B-09M1001> and functional oligothiophenes as molecular design for multidimensional nanoarchitectures and their applications were reviewed by Bauerle *et al.* <09CRV1141>.

5.1.3. SYNTHESIS OF THIOPHENES

5.1.3.1 Thiophene Rings

Several reports of novel and convenient syntheses of thiophene rings have appeared this year. Many of these preparations gave multi-substituted thiophene analogs. For example, a novel Cu(1)-catalyzed tandem addition/cycloisomerization reaction of sulfonylalkylidenethiiranes **1** with terminal alkynes **2** gave trifunctionalized thiophenes **3** <09CC4729>. The reaction tolerated aromatic rings containing electron-rich (*e.g.* amino group) and electron-poor groups (*e.g.* nitro), as well as thiophene, furan and pyridine, and alkyl chains with alcohol or ester functionality.

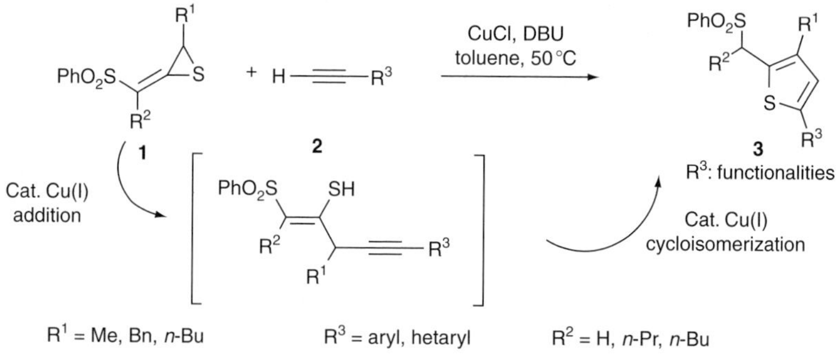

A convenient one-pot procedure for the preparation of benzo[*b*]thiophenes and selenophenes from the reaction of *o*-halo-ethynylbenzene precursors with appropriate chalcogenides was reported <09OL2473>. By treating the appropriate mono- **4**, bis- **5**, and tris-substituted *o*-chloroethynylbenzenes **6** with NaX where X = or Se, the corresponding mono- **7**, bis- **8** and tris-chalcogenophene-annulated benzenes **9** were obtained in good to high yields.

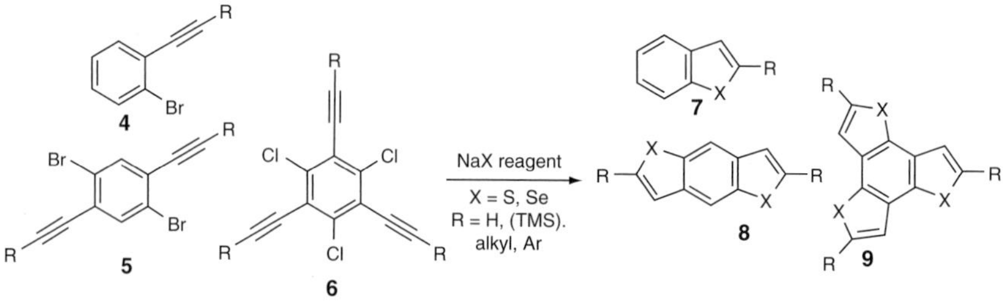

A novel direct one-step route to potentially biologically active 3-substituted 3-cyanobenzo[b]thiophenes **11** from polarized ketene dithioacetals derived from o-bromoarylacetonitriles **10** was reported <09JOC5496>. The synthesis of **11** probably proceeds via an intramolecular radical cyclization protocol, which involves tandem attack of a carbon-centered radical on sulfur and subsequent CS bond cleavage of the resulting adduct.

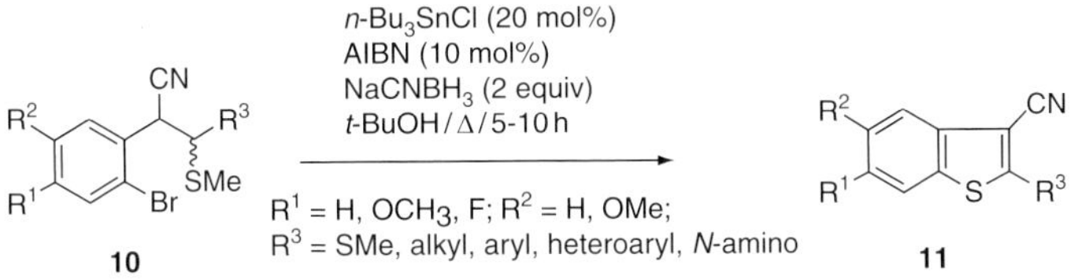

2,2-Disubstituted benzo[b]thiophen-3(2H)-ones **14** and 2-substituted 3-hydroxybenzo[b]thiophenes **15** were easily prepared by the tandem sulfanylation-acylation of (chlorocarbonyl)phenyl hypochlorothioite **12** and C–H acids **13**, such as β–diketones, β–ketoesters, β–cyano esters, diethyl malonate, and malonitrile <09PS1115>.

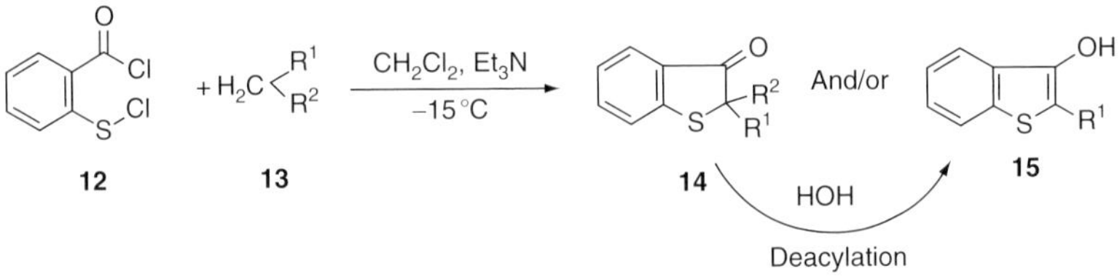

Kobayashi et al. has described a new synthesis of 3-arylbenzo[b]thiophenes **17** using an interrupted Pummerer reaction of 2-(1-arylvinyl)phenyl ethyl sulfoxides **16** <09T2430>. However, when the Pummerer reaction was not interrupted, complicated mixtures of products were obtained. A mechanism similar to that reported by Bates et al. was proposed <92JOC3094>.

A one-pot, four-step synthesis of new tetra-substituted thiophenes **22**, shown below, was developed <09T10453>.

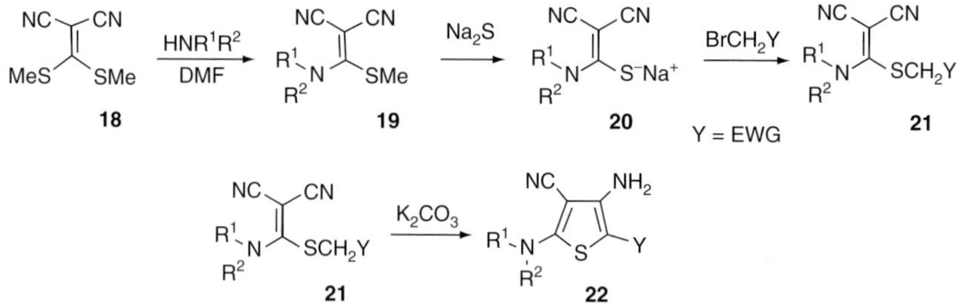

A one-pot three-step synthesis of 2,3-disubstituted thieno[2,3-*b*]pyridines **24** from halopyridyl ketones **23** has been described <09H2993>. The three steps involved successive treatment of **22** with Na_2S, $BrCH_2EWG$ and NaH. Similarly, thieno[2,3-*c*]- and thieno[3,2-*c*]pyridines were synthesized from appropriate ketones.

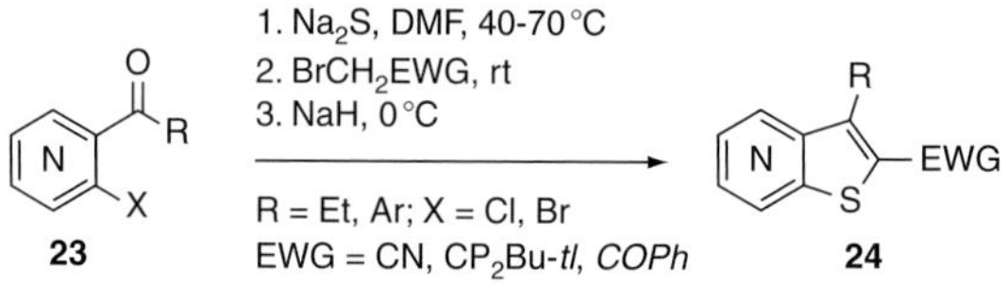

The first regioselective synthesis of poly-substituted thiophenes from Baylis-Hillman adducts appeared <09TL6480>. The synthetic pathway involves the reaction of appropriate Baylis-Hillman vinyl acetate **25** with ethyl mercaptoacetate to give 2,3,4-trisubstituted tetrahydrothiophenes **27** via sequential S_N2 to **26** and Michael reaction. The tetrahydro derivatives **27** are then aromatized to 2,3,4-trisubstituted thiophenes **28** by DDQ oxidation.

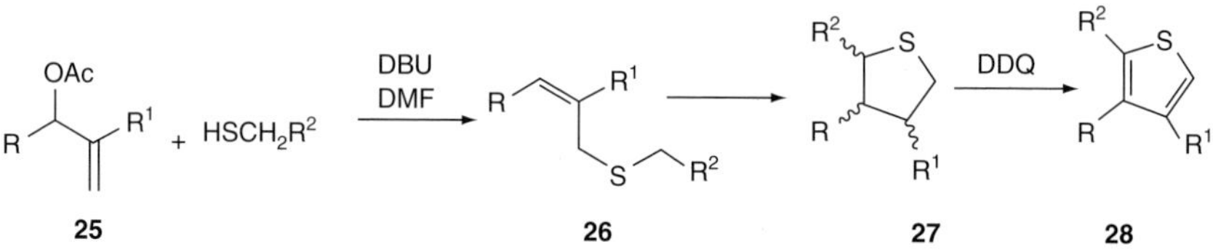

The Gewald procedure <66CB99> continues to find use in the synthesis of thiophenes. For example, this procedure was used for the preparation of certain 2-amino-7-oxotetrabenzo[*b*]thiophenes (**30**) <09PS2078> by treating cyclohexan-1,3-dione (**29**) with cyanomethylenes. These products were subsequently transformed into potential antifungal annulated heterocycles. A short Gewald synthesis of 3,6-disubstituted *N*-2- thienyl derivatives of indole was also carried out in which 2-amino-3,6-disubstituted thiophenes were produced <09TL6562>. In another report, 4-aryl-substituted 2-amino-3-benzoylthiophenes were prepared by a $TiCl_4$ promoted Knoevenagel condensation of a phenacyl chloride with an α,β−unsaturated nitrile followed by Gewald step using Na_2S <09JMC4543>.

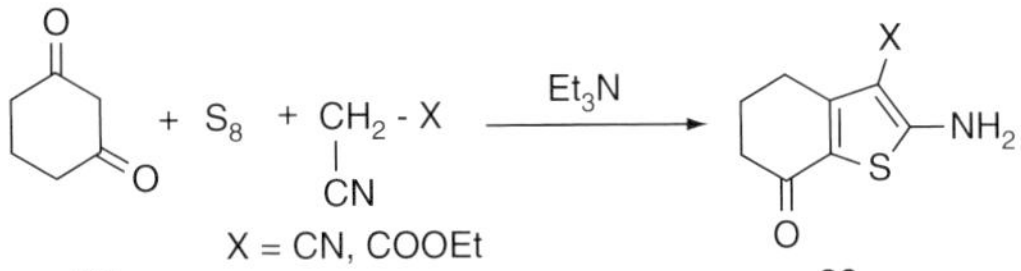

29 **30**

Several microwave-assisted thiophene ring syntheses have been reported this year. For example, several 2-amino-5-nitrobenzo[*b*]thiophenes **32** were prepared in good to excellent yields by subjecting 1-(2-chloro-5-nitrophenyl)ethanone **31** to microwave heating for 10 min at 90 °C in the presence of a mixture of elemental sulfur, a primary amine, NaOAc and DMF <09JHC599>. In addition, minor amounts of 2-amino-5-nitrophenylethanones **33** were formed.

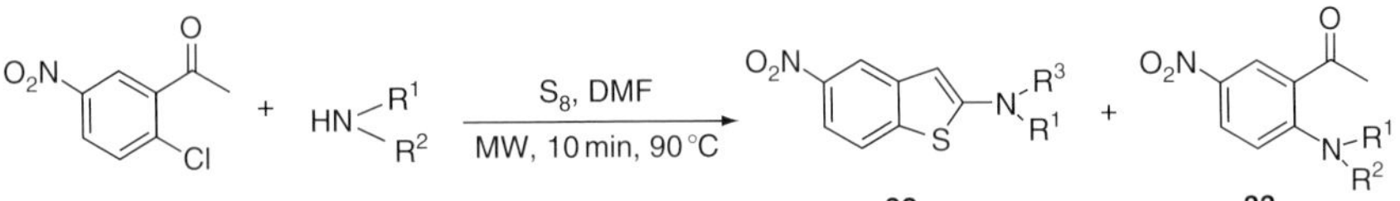

31 **32** **33**

The key step in the synthesis of **32** probably involves intramolecular addition of a thioamine anion to the 2-chloro atom of the benzo ring to give a Meisenheimer complex, whereas **33** most likely is formed by the usual S_NAr pathway. Indeed, when these reactions are conducted in the absence of S_8, compounds **33** were obtained in excellent yields. Interestingly, treatment of **31** with either S_8 or Se in the presence of NH_4Cl and DMF, the corresponding 3-methylbenzo[*d*]isothiazole **34** and benzo[*d*]selenazole **35** were obtained in yields of 90% and 70%, respectively <09THC17>.

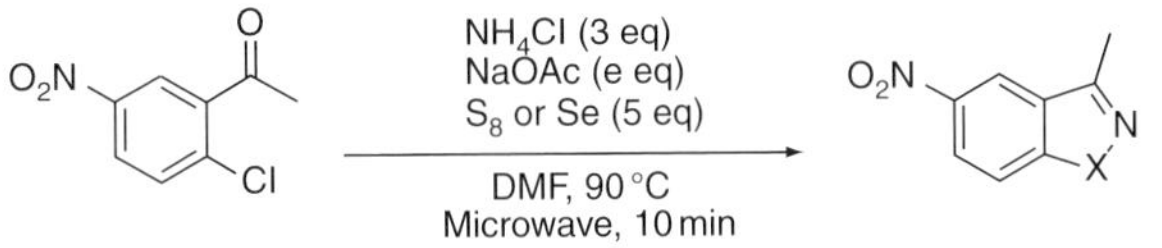

31

34 X = S = 90%
35 X = Se = 70%

The proposed mechanism for formation of **34** (used as typical example) is shown below.

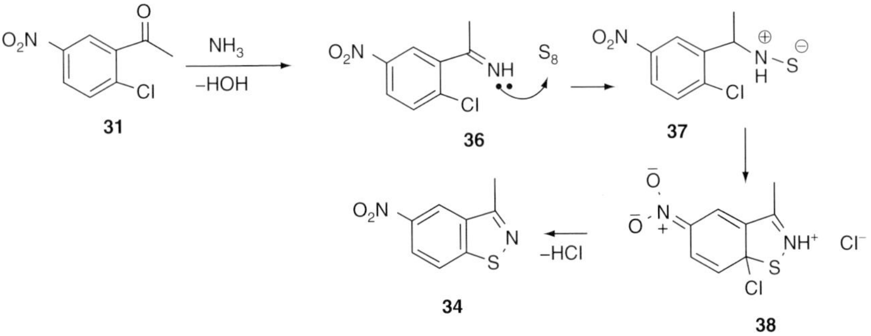

31 **36** **37** **34** **38**

A novel iodine-mediated cyclization reaction yielding 6-iodomethyl 2-benzothiophene-1(3H-thione) thiophenes **42** was reported <09H169>. The synthesis involves treating α-methylstyrene derivatives **39** with n-BuLi and CS_2 to generate lithium 2-(vinyl)dithiobenzoate **40**, which then reacts with I_2 to give iodonium bridged cation **41**. Intramolecular addition of sulfide ion onto the alkene leads to concomitant ring opening to give **42**. Interestingly, the reaction involving other α−substituents gives a mixture of **42** and isothiochromene-1-thiones.

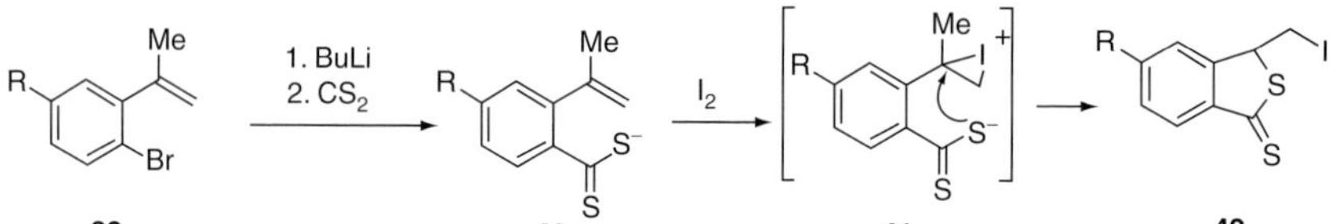

Larock et al. <09JCO900> have generated a library of multi-substituted benzo-[b]thiophenes using solution-phase parallel synthetic methodology. The requisite 3-iodobenzo[b]thiophenes **44** were prepared by electrocyclic cyclization of electron-rich alkynes **43** in the presence of iodine. Elaboration of the 3-iodo compound **44** was carried out using Pd-catalyzed Suzuki-Miyaura, Sonogashira, Heck, and carboalkoxyation chemistry. Subsequently, Larock et al. <09JOC1141> studied alkyne cyclization competing reactions between several 2-substituted alkynes and found that SeMe and SMe 3-substitutents were the most reactive.

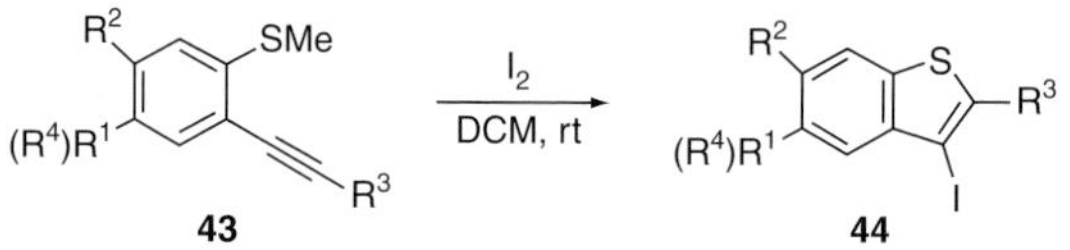

Benzothiophene-based palladacycles **46**, which are resistance to air and moisture, were made by a unique intramolecular thiopalladation of thioanisole substituted propargyl imines **46** <09OM3966>.

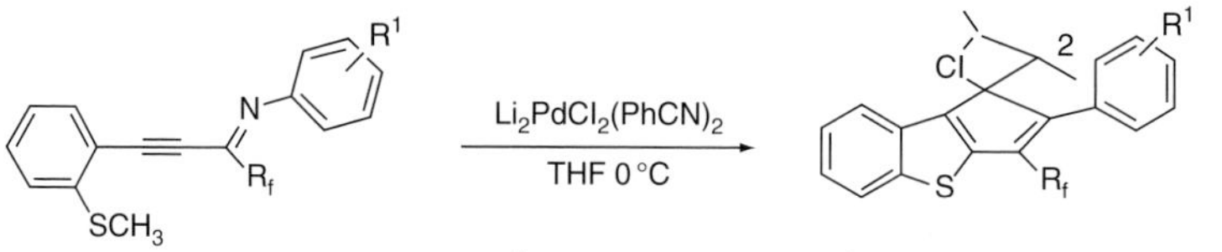

In addition, 3-amino-5-chlorobenzo[b]thiophene-2-carboxylate **48**, which was prepared by the condensation of fluorobenzonitrile **47** with methyl 2-mercaptoacetate, reacted with formamide to give the 2-chloroacetimidoyl derivative **49** which is a lead compound in the synthesis of 3H-benzo[4,5]thieno[3,2-d]pyrimidin-4-ones <09JMC6621>.

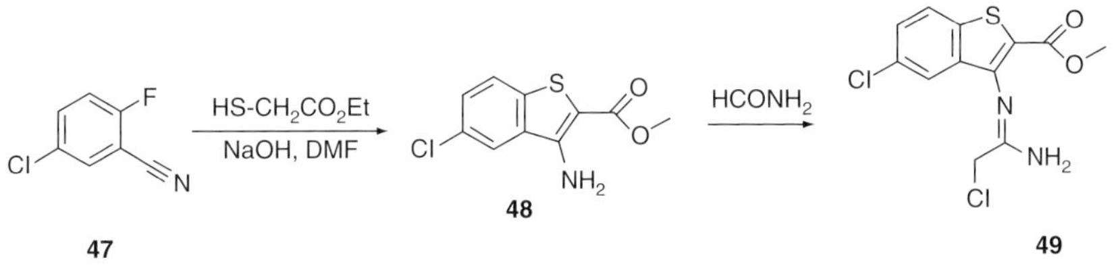

Thiophenes as well as pyrroles and furans, were readily prepared by the dehydrative cyclization of propargyl alcohols using Au[t-Bu)$_2$(o-biphenyl)]/Cl/AgAgOTf or AuCl as catalyst <09OL4624>. Finally, axial 1,9-dithiaalkane-bridged thianthrene 10-oxides were synthesized from by the dilithiation of thianthrene derivatives followed by addition of sulfur, followed by a dibromoalkane and were found to exhibit unusual photochemical activities <09TL1381>.

5.1.3.2 Fused Thiophene Rings

Several methods for constructing fused thiophenes appeared in the 2009 literature. For example, thieno[3,2-*b*]thiophene **53a** was prepared in an incredible yield of 94% by the reaction of **50a** with *t*-BuLi with 1,2-bis(2,2-diethoxyethyl)sulfide **51a** to give the open chain **52a** which underwent ring closure in the presence Amberlyst to **53a**. This yield is significantly higher than those prepared by previous methods. This methodology also produced the first-reported thieno[3,2-*b*]furan **53b** <09OL3144>.

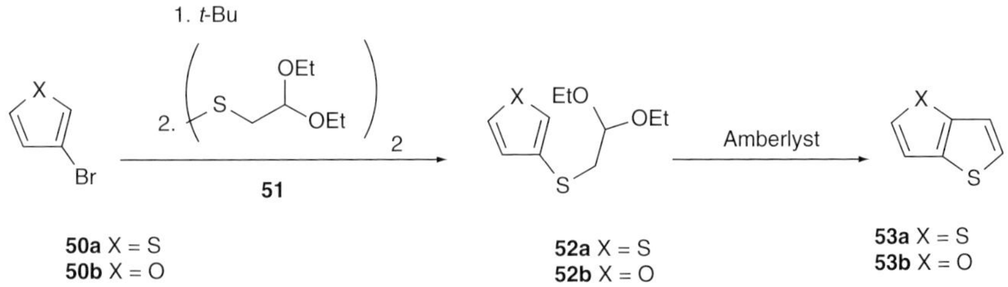

The synthesis of dithienyl[3,4-*b*:3′,4′-*d*]thiophene **56** was accomplished by introducing TMS at the 3, and 4 positions of 2,5-dibromothiophene **54**. The resulting tetrasubstituted thiophenes **55** were then converted to **56** by successive sulfidation, intramolecular cyclization and TMS removal <09JOC4747>. The presence of the TMS group in **55** protects against possible side reactions of the α–hydrogens during the initial Br/Li exchange resulting in much higher yields than those previously obtained using **54** as starting material <71JOC1645>.

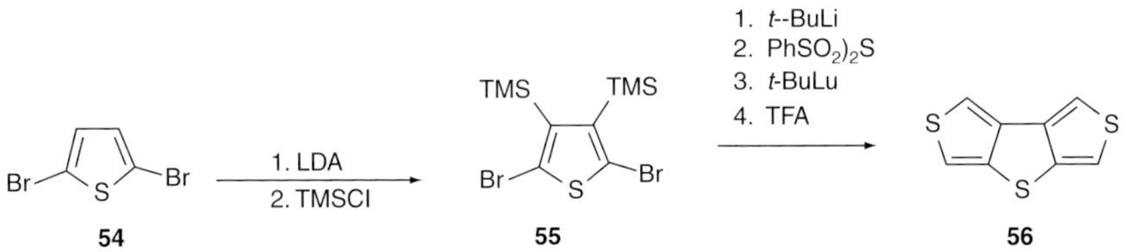

The synthesis of two new synthons for organic semiconductors, namely *syn*-2,6-bis(tributyltin) and *anti*-benzo[1,2-*b*:5,4-*b*]dithiophene, were reported using a Stille coupling protocol <09ARK(v)90>. The synthesis of *syn* isomer **60**, given below as typical example, involves an interesting construction of the two thiophene rings onto **59** via an intramolecular cyclization of dithiol intermediate **58** (prepared from **57**).

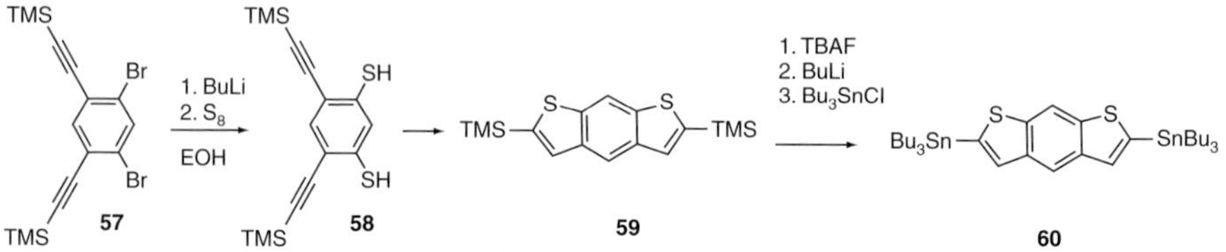

The first study of the reaction of ninhydrin with benzo[*b*]thiophenes was reported this year and was found to produce very interesting and unexpected benzo[*b*]thiophene derivatives including a novel fluorenone compound **61** fused to benzo[*b*]thiophene rings, a phthalide conjoined in a spiro configuration **62** along with an isocoumarin **63** and an isomeric isocoumarin **64** <09H2467>.

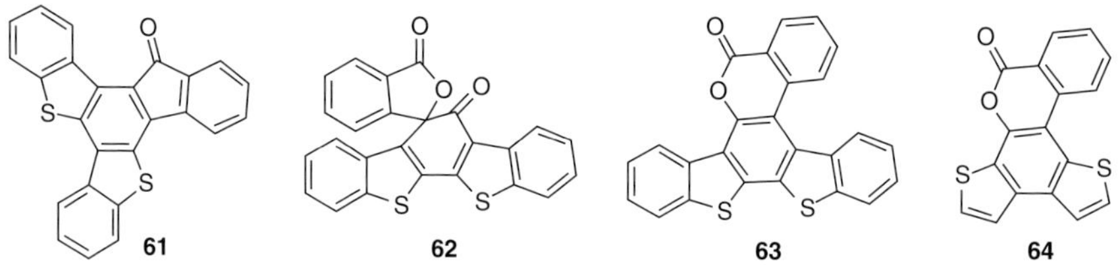

An interesting way to prepare sodium thiophenethiolate salts **66** from **65** is shown below <09SC1781>. These salts readily form novel thiophene thio-glycosides.

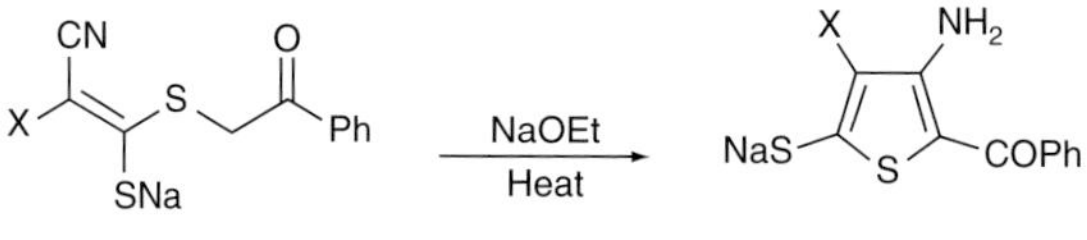

Finally, a neat method for preparing the first parent heteroindoxyl, 4,5-dihydrothieno[3,2-*b*]pyrrole-6-one was described. The method involved flash vacuum pyrolysis (FCP) at 350 °C of 2-acetyl-3-azidothiophene to give 3-methylthieno[3,2-*c*]isoxazole, which upon further FCP at 550 °C affords the parent compound <09JOC4278>. The chemistry of 4,5-dihydrothieno[3,2-*b*]pyrrole-6-one was found to be very exciting. For example, its reactions with diazonium salts, isatin and dimethyl acetylenedicarboxylate occurred at the methylene carbon to give a hydrazone, an indirubin analog and a succinate derivative, respectively, whereas oxidation gives a heteroindigotin, a major ingredient in indigo.

5.1.4. ELABORATION OF THIOPHENES AND BENZOTHIOPHENES

Crossed-coupling reactions continued to be extensively used in the elaboration of thiophene rings. Larock *et al.* <09JCO900> generated a library using solution-phase parallel synthesis of multi-substituted benzo[*b*]thiophenes, a real *tour de force*. The synthetic procedure involved preparing 3-iodobenzo[*b*]thiophene **43** by electrocyclic cyclization of electron-rich alkynes using iodine (this was discussed in Section 5.1.3.1) which were further elaborated by palladium-catalyzed Suzuki-Miyaura, Sonogashira and Heck chemistry to afford multi-substituted benzo[*b*]thiophenes **67** <09JCO900>.

Several new cross-coupling reactions have been developed that eliminate the use of organic metallic derivatives and thus are more economically and environmentally attractive than traditional cross-coupling procedures that involve organometallic derivatives. A ligand-less palladium-catalyzed method with low catalyst loadings has recently been reported <09GC425>, which directs addition of aryl bromide **69** to the 5-position of thiophene-2-carbaldehydes **68** yielding **70**. This is a significant synthetic advance since several 5-aryl-thiophene-2-carbaldehydes **70** are important reagents in the synthesis of lead compounds for biologically active compounds <03BMC4729>.

A metal-free oxidative cross-coupling of unfunctionalized aromatic compounds with thiophenes using PhI(OH)OTs (PIFA) in fluoro alcohol media to give 2-arylated thiophenes has also been developed by Kita *et al.* <09JA1668>. This work is particularly important in that the first successful example of a cross-coupling between two molecules of heteroaromatics, *i.e.*, thiophene and pyrrole was reported. An excellent summary of Kita's extensive hypervalent iodine (III) work has appeared this year <09T10797>. The method shows a broad scope of functional group compatibilities, and crossed-coupled products are formed at room temperature in high yields with no evidence of any oligomer formation. A mechanism has been suggested which involves addition PIFA (AKA Koser's reagent) to **71** to give a CT-complex **72** that undergoes a one-electron transfer (SET) to the cation-radical **73**. Species **73** then reacts with thiophene to give radical cation **74**, which is aromatized by the PIFA radical anion formed in the collapse of **75**.

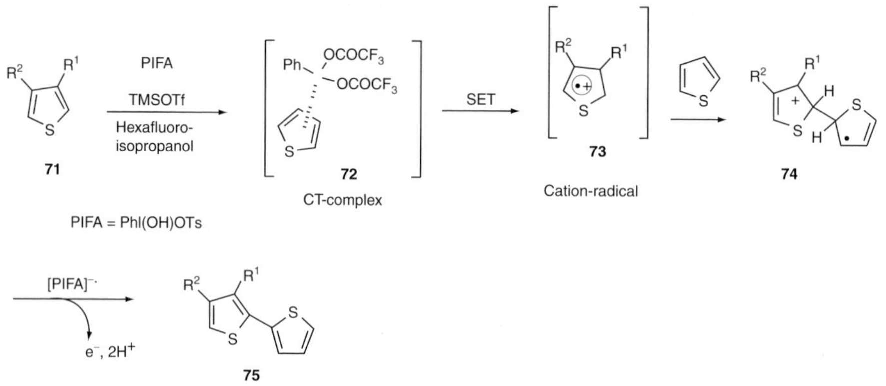

Kita has also developed a clean and direct synthesis of α,α-bithiophenes by metal-free oxidation coupling of recyclable hypervalent iodine (III) <09CPB710>. Interestingly, Fagnou et al. in a featured article in the Journal of Organic Chemistry established broadly applicable reaction conditions for the palladium-catalyzed direct 2-arylation of heteroatom-containing aromatic compounds with aryl bromides <09JOC1826>. The use of pivalic acid in accelerating direct 2-arylation of a wide variety of heterocycles was particularly noteworthy.

A facile synthesis of α-substituted thiophenes from functionalized 2-aminothiophenes by homo- and cross-coupling reactions has been developed <09TL4670>. Since 2-aminothiophenes can exhibit properties similar to the 2-imino tautomers, diazotization reactions are unsuitable for introducing iodine at the 2-position. However, if a strong electron-attracting group, such as cyano, carboxylate, etc., is introduced at the 5-position, the diazotization/iodination reaction sequence can occur thus facilitating Ullmann homo-coupling and Sonogashira cross-coupling reactions.

Suzuki coupling reaction has also been used to prepare terarylenes by the coupling of dibutyl(2-methyl-benzo[b]thiophen-3-yl)borane with 2,3-dibromothiophene derivatives <09TL5288>. Functionalized benzothiophenes **78** were synthesized by a two-fold Heck reaction with **76** that yielded diene **77** <09TL4962>. The diene then underwent a domino 6π-electrocyclization to **78** which was subsequently reduced by Pd/C to **79**. (S)-t-Butyl-4-(4-chlorophenylthiophen-2-yl)methyl carbamate was prepared by a rhodium-catalyzed enantioselective addition of arylboronic acids to t-butyl phenylsulfonyl(thiophene-2-ylcarbamate <09OS360>. A domino Pd-catalyzed Heck intermolecular direct arylation reaction has also been reported <09OL4560>.

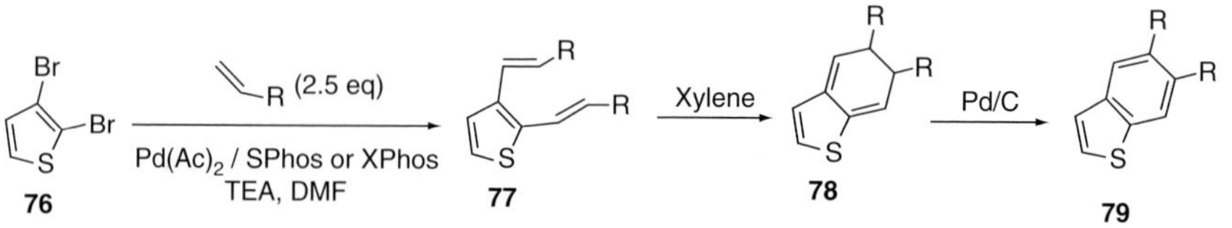

Prior to 2009, direct 3- and 4-arylation of thiophenes was rarely studied and the selective bimolecular palladium-catalyzed direct 3- and 4- arylation of

2,5-disubstituted thiophenes had not been reported. However, recently a palladium-catalyzed (Pd(OAc)$_2$ direct 3- or 4-arylation of 2,5-disubstituted thiophenes using aryl bromides was described <09TL2778>. 3-Arylthiophenes were obtained using 2,5-dimethylthiophene in fair to moderate yields. In addition, the 4-arylation of unsymmetrical 2-acetyl-5-methylthiophene was observed. This chemistry should spur the study of these novel compounds,

A wide variety of potassium heteroaryltrifluoroborates were prepared and general conditions were developed for Suzuki-Miyaura cross-coupling reactions with aryl and hetaryl halides which gave heterobiaryls <09JOC973>. Prior to this study, attempts to use heterocyclic boronic acids usually failed due to their instability.

Ar (HetAr)-X + HetAr-BF$_3$K $\xrightarrow[\text{Ethanol. 85 °C}]{\text{Pd(OAc)}_2\text{/ RuPhos} \quad \text{Na}_2\text{CO}_3}$ Ar(Het)-HetAr

X = Cl, Br, I, OTf

Problems associated using unstable heterocyclic boronic acids in Suzuki-Miyaura reactions has been alleviated by carrying out a slow-release cross-coupling with air-stable MIDA (*N*-methyliminodiacetic acid) heterocyclic boronates **80** <09JA6961>. This method is superior to the boronic acid surrogate method such as trifluoroborate salts <09JOC973>, trialkoxy or trihydroxy salts, and bulky boronic acid classes, since none of these surrogates provide air-stable substitutes for all three boronic acid classes. The trick here is that MIDA boronic acids have the ability to release the corresponding boronic acids from the MIDA boronic acid *in situ*. It is particularly noteworthy that aryl chlorides react readily with MIDA heterocyclic boronates under these conditions.

Mechanistic studies on palladium-catalyzed CH arylation of 2,3-dibromothiophenes in the presence of AgNO$_3$/KF were carried out <09BCJ555>. The results strongly suggest that palladium-CH arylations proceed through an electrophilic substitution pathway.

Expanding upon the pioneering work of Fujiwara *et al.*, <81JOC851> and the recent finding that lithium salts have a special effect on the metal-catalyzed alkenations of thiophenes <08T5982>, a palladium-catalyzed alkenation of thiophenes by regioselective C-H bond functionalization was developed <09TL2758>. The process involved treating thiophenes **81** and alkenes **82** with Pd(OAc)$_2$ AgOAc and pyridine to give predominantly *E*-alkenes.

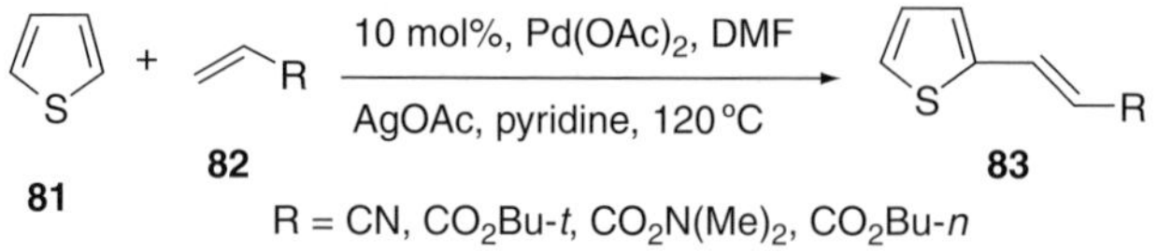

An improved method for the Suzuki-Miyaura cross-coupling for the preparation of 3-vinylthiophene and other heterocyclic bromides (monomers for the synthesis of SERS-active polymers) was developed <09TL5467> in which a thiophene-3-yl boronic acid was used. Previous methods for carrying out Suzuki-Miyaura reaction using 3-bromothiophene with vinyl trifluoroborate or vinyl silanes gave 3-vinylthiophene; however, the yields were low and long reaction times were required <06JOC9681>.

TMPZnCl·LiCl has been found to be an active selective base for the directed zincation of sensitive aromatic and heteroaromatics <09OL1837>. For example, zincation of benzo[*b*]thiophene-3-carbaldehyde **84** using TMPZnCl·LiCl and subsequent trapping of zinc salt **85** with various electrophiles gave 2-aryl derivatives of benzo[*b*]thiophene-3-carbaldehyde **86**. Since zincation of aromatic rings with sensitive groups such as nitro and aldehyde are tolerated, this new method should serve for the preparation of heterocycles possessing sensitive groups.

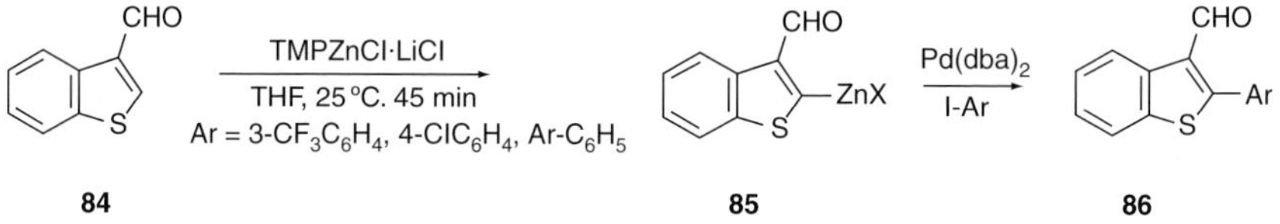

Denmark *et al.* published an exhaustive study of cross-coupling reactions on aromatic and heteroaromatic silanolates with aromatic and heteroaromatic halides <09JA3104>. They found that cross-coupling of potassium (dibenzothienyl)-dimethylsilanolate with aryl bromides using (t-Bu$_3$P)$_2$Pd gave aryl-heteroaryl coupled products in good to excellent yields. Although the reactions of potassium (dibenzothienyl)-dimethylsilanolate with aryl chlorides were not reported, sodium (difuranyl)-dimethylsilanolate reacted with aryl chlorides smoothly to give the coupled product in good yields.

A practical synthesis of 2-bromo-3-formylthiophene **90** was developed in which 3-bromothiophene **87** was converted to 3-iodothiophene **88** by selective lithium/iodine exchange which was then treated with NBS to give 2-bromo-3-iodothiophene **89**. Treatment of **89** with EtMgCl followed by quenching with DMF gave **90** <09SC3315>.

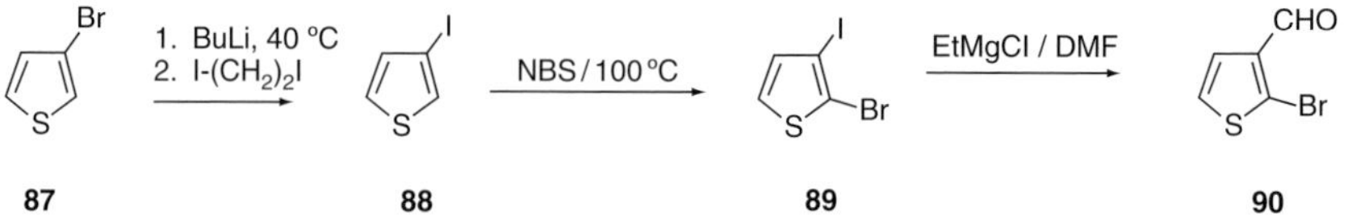

Bromothiophenes have also served as precursors in the preparation for brominated **94** and arylated anthraquinones **95** <09MO1013>. The reaction involves the *in situ* treatment of brominated thiophenes **91** with *m*-CPBA to give thiophene S-oxides **92** which cycloadds to naphthoquinone **93**. The resulting bromoanthraquinones **94** are then subjected to Suzuki-Miyaura cross-coupling to yield arylated anthraquinones **95**. The anthraquinone structures were determined by NMR and UV spectroscopy which revealed at least three major bands corresponding to $\pi-\pi$ transitions.

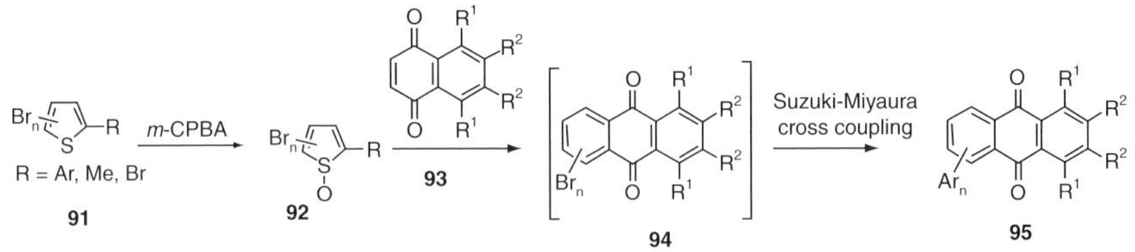

Dibenzothiophene was converted to its sulfone by the tantalum (V) catalyzed oxidation with 30% H_2O_2 <09TL1180> and to its sulfone and sulfoxide by oxidation with sodium periodate catalyzed by reusable supported $Mn(Br_8TPPCl)$ catalysts under various reaction conditions <09ACA61> and by Ceric ammonium nitrate (CAN)/Bronsted Acidic Ionic Liquid <09PS705>. Dicarbomethoxycarbene can be generated by photolysis of S-C sulfonium ylides **96** derived from thiophenes <09OL955>. Depending on the substitution of the thiophene the initial singlet **97** and triplet spin states of carbene **98** can be manipulated. These carbenes were trapped with (*Z*)-4-octene and the distribution of three-membered ring products allowed determination of the chemistry of the various spins states. These results, although preliminary, strongly indicate that their approach should apply to other carbenes.

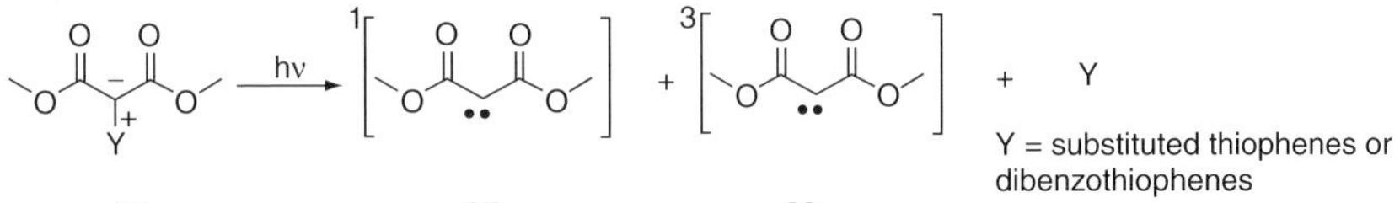

Condensation approaches to the synthesis of thiophenes were once again popular. For example, further functionalization of the thiophene ring was effected by condensation of dibenzothiophene carboxaldehydes **99** with certain aryl acetic acids which gave the 2,3-diaryl acrylates **100**. Subsequent esterification followed by reduction of the ester so formed gave allylic alcohols **101** that underwent Eschenmoser-Claisen [3,3]-sigmatropic rearrangement by the reaction of *N*,*N*-dimethylacetamide in the presence of TBNF as catalyst to give 3,4-diaryl α,β-unsaturated amides **103** <09SC332>.

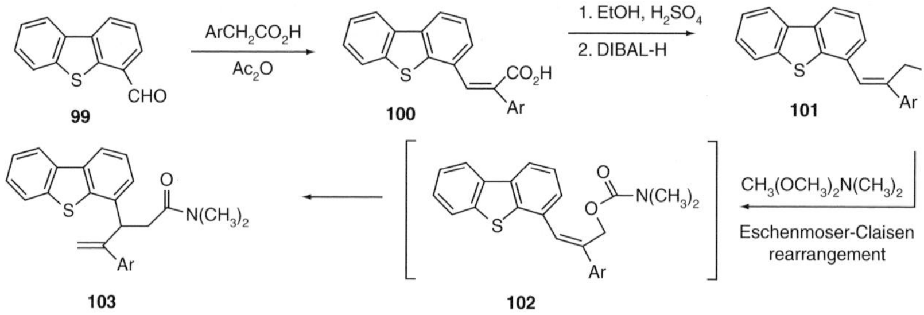

A one-pot, three-component synthesis of 2-(β-acylaminoketonyl)thiophene **106** involving the condensation of 2-acetylthiophene **104** with an aromatic aldehyde **105** and nitrile in the presence of SiCl$_4$/ZnCl$_2$ was described <09PS220>.

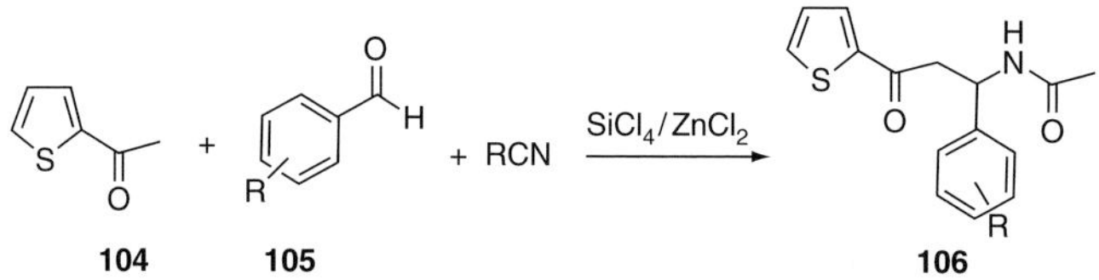

The major problems in preparing β-aminoketones by the classical Mannich reaction are the drastic conditions and long reaction times (∼7 h) required. However, Pessoa-Mahana et al. prepared 2-aminoketo thiophene derivatives in 4-9 min by a solvent-free microwave-promoted Michael addition of various azanucleophiles to benzo[b]thiophen-2-yl-2-propenones in the presence of Si/MnO$_2$ <09ARK(xi) 316>. In another study, aminoalkylation reactions were applied to two annulated derivatives to prepare mono- and bis-substituted Mannich bases <09TL3750>.

Several new highly substituted dihydrothiophene-S,S-dioxides **109** were prepared by base-mediated condensation of the corresponding sulfone **107** and bis-imidoyl chloride <09PS1161>. The vicinal amino-imino substructures were found to be excellent chelating ligands with metals such as palladium. Various lipophilic sulfones were also synthesized; they should find use in cyclization reactions.

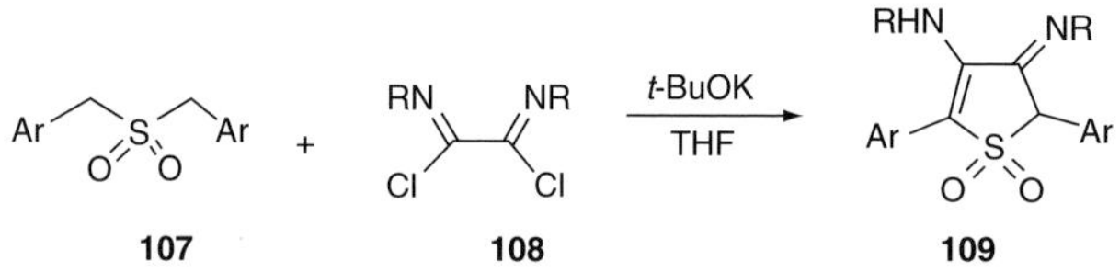

Interestingly, thiophene S-oxides, unlike thiophene S,S-dioxides, have been little studied because of synthetic difficulties in stopping at the mono oxidation stage. This problem has been solved and the resulting exciting chemistry of thiophene S-oxides and related compounds with special emphasis on Diels-Alder cycloadditions as well as photochemistry etc. Photochemistry and electrochemistry are discussed <09ARK(ix)96>.

2-Cyanobenzo[b]thiophene was prepared by the catalytic (TBAF) dehydration of the corresponding 2-carboxamide <09OL2461>. The replacement of the isobutenyl moiety in the acyclic monoterpenes linalool, geraniol, nerol, and citronellol by a thiophene substituent was accomplished <09JFA2088> by treating readily available 2- and 3-methylthiophenes and 2- and 3-thiophenecarbaldehydes by literature procedures <94BSF900, 91JOC6720>, respectively. The sensory and microbial activities of the resulting thienyl analogs indicate that they may be suitable for perfumery and cosmetology. Thiophenecarbaldehydes also found use for the preparation of 2-(thiophen-3-yl)vinylphosphoric acid <09SC1511>. The main steps of the synthesis involved adding the thiophenecarboxaldehyde to lithium diethyl methylenephosphonate and dehydrating and hydrolyzing the resulting intermediate.

The elaborations of the thiophenes using gold and lead were reported this year. For example, lead (IV) tetra(2-thiophenecarboxylate) **111**, which is easily prepared by the metathesis of lead tetraacetate with 4 equiv of 2-thiophenecarboxylic acid and 4-penten-1-ols **110** with alkyl groups at positions 2-4, reacted to give mixtures of tetrahydropyrans **112** and tetrahydrofurans **113**. These mixtures were separated by column chromatography <09TL1097>.

The reaction between acetylenes and (2-phenylsulfinyl)thiophene under gold-catalyzed intermolecular oxyarylation conditions yielded, unexpectedly, 1-phenyl-2-(2-(phenylthio)thiophen-3-yl)ethanone <09OL4906>. Theoretical studies indicated that the novel gold-catalyzed reaction proceeded *via* a concerted mechanism.

Many studies concerning functionalization of thiophenes appeared in 2009. For example, starting from 2,5-dichlorothiophene **114** a full functionalization of the thiophene ring was achieved <09OL445>. By using TMPMgCl-LiCl positions 3 and 4 can be successively magnesiated and trapped with different electrophiles to give 3,4-difunctionalized dichlorothiophenes **115**. Fully functionalized thiophenes **116** were subsequently prepared by sequential magnesiation with TMPMgCl/LiCl and appropriate electrophilic trapping agents. The synthesis of a thiophene analog of atorvastatin was reported.

A type II anion relay chemistry (ARC) has been applied in a one-pot synthesis of 2,3-disubstituted thiophenes **121** <09OL1861>. The linchpin in this reaction is the formylation of **117** to give **118**. Treatment of **118** gives with RLi leads to a 1,4 Brook migration yielding the anion **120** which is trapped with suitable electrophiles. The electrophilic step works best when the electrophilic reagent is added to a 1:2 mixture of HMPA and ether at -78 °C.

Functionalization at the 2-position of thiophene is well established; however, functionalization at the 3-position is less studied. A new method involving the use of TMPMgCl/LiCl and iPrMgCl/LiCl gave 3-substituted thiophenes *via* magnesiation and sulfoxide-magnesium exchange <09CC3536>. The kinetics of bromine-magnesium exchange reactions in heteroaryl bromides using *i*PrMgCl/LiCl has been studied <09OL3502>. The relative rates ($k_{rel}[t_{1/2}]$) of bromothiophenes toward *i*PrMgCl/LiCl were found to increase in order of 3-bromothiophene (5.93 x 10^1) < 3-bromobenzo[*b*]thiophene (5.53 x 10^2) << 2-bromothiophene (1.50 x 10^5). Interestingly, the relative rate of 2-bromothiophene was comparable to 3,5-dibromopyridine (1.54 x 10^5) and somewhat higher (5.28 x 10^4) than 3,4-dibromofuran.

A tandem decarboxylation of 2- and 3-aminothiophenecarboxylic acids **122** to **123**, and an inverse electron-demand Diels-Alder reaction between **123** and 1,3,5-triazines **124** gave the respective thieno[2,3-*d*]- and thieno[3,2-*d*]pyrimidines **126** via intermediate **125** <09TL2874>. Of the several solvent systems evaluated, DMF-AcOH gave best yields. This reaction takes advantage of the latent dienophilic nature of aminothiophenecarboxylic acids <01TL8419>.

X = H, Ph, CF$_2$Cl, CF$_3$, CO$_2$Et R^1, R^2 = H, Ph, Me

The stereochemical and regioselective syntheses of thiophenes were again actively studied. A practical catalytic asymmetric synthesis of enantioenriched diaryl-, aryl-, heteroaryl-, and diheteroarylmethanols was described <09JA12483>. For example, 3-bromothiophene **127** is converted into **128** by the reaction of *n*-BuLi and EtZnCl. Subsequent treatment of **128** with the appropriate aryl or heteroaryl aldehyde leads to the corresponding enriched methanol **129** with >90% ee. The reaction of 3-benzo[*b*]thiophene is given below as a typical example. The metalation reaction requires EtZnCl since the 3-lithio derivatives are unstable. Also, to ensure that the methanols are obtained in high ee, it is crucial that TEEDA (tetraethylene diamine) be present in the reaction.

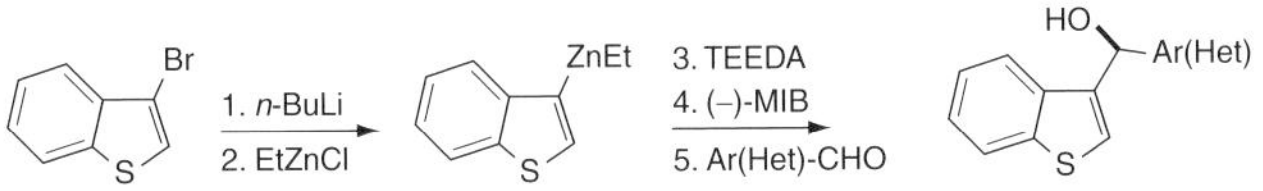

127 **128** **129**

Ar = Ph; Het = 2-thienyl, 2-furanyl, benzo[*b*]-3-furanyl

The enantioselective conjugate addition of 2-heteroaryl titanates and zinc reagents **130**, including 2-thienyl reagents, to conjugated cyclic Michael acceptors such as cyclic enones, unsaturated lactones, and unsaturated lactams **131** in the presence of the chiral catalyst system, [Rh(cod)acac] and (*R*)–MeO-BIPHEP, was developed <09OL4200>. This gives 3-(2-thienyl) adducts **132** in good yields and with high diastereroselectivities (up to 96%). This method precludes the use of 2-heteroarylboronic acids, which are unstable during trans-tin metal catalyzed processes.

130 **131** **132**

n = 0.1

ML_n = ZhCl, $Ti(OiPr)_4Li$

Z = CH_2, O, NMe

The enantioselective conjugate addition of 2-heteroaryl titanates and zinc reagents, including 2-thienyl reagents, to conjugated cyclic Michael acceptors such as cyclic enones, unsaturated lactones and unsaturated lactams in the presence of the chiral catalyst system, [Rh(cod)acac] and (*R*)–MeO-BIPHEP, was developed <09OL4200>. The corresponding 3-(2-thienyl) adducts were obtained in good yields and with high diastereroselectivities (up to 96%). This method precludes the use of 2-heteroarylboronic acids, which are unstable during trans-tin metal catalyzed processes.

New chiral thiophene-Salen chromium complexes were prepared for use in asymmetric Henry reaction of aldehydes <09JOC2242>. Enantiomeric excesses up to 72% were obtained. A rhodium-catalyzed addition/cyclization reaction was used to prepare several polycyclic heteroaromatics **135** by treating thiophene carboxylic acids **133** with norbornene to give Rh complex **134** which is reduced to **135** <09JOC1809>. This method affords an efficient route to heterocyclic aromatic compounds.

133 **134** **135**

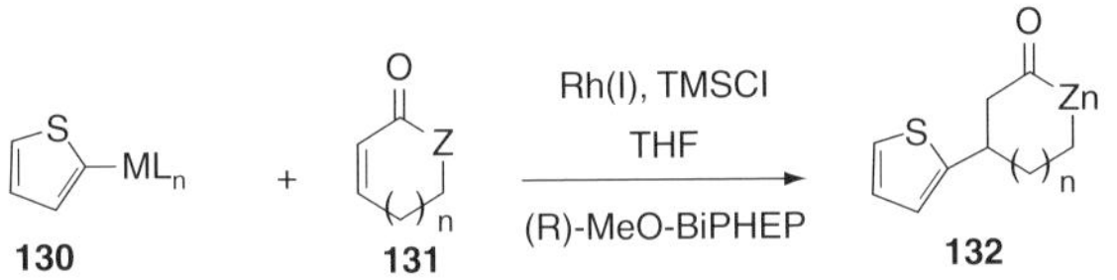

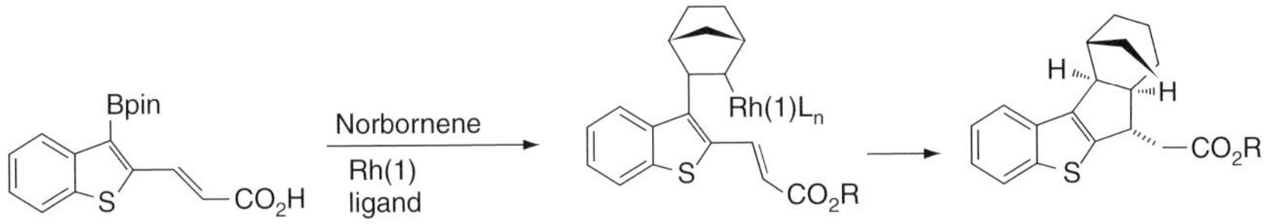

Diastereoselective domino reactions of chiral-2-subsituted 1-(2′,2′3′,3′-tetramethylcyclopropyl)-alkan-1-ols **136** were obtained by carbonyl addition of tetramethylcyclopropyllithium to the respective aldehyde <09JOC4747>. Under Bronsted acid conditions, **136** serve as substrates for Friedel-Crafts alkylation with 2-methylthiophene **137**. The rearranged products **138** (*syn*) and **139** (*anti*) were formed in good yields. The diastereoselectivity of this reaction is excellent (*syn*-preference for *t*-Bu (*syn:anti* = 97:3) and good (*anti* preference for Ph, CN, PO(OEt) with *anti:syn* = 75:25). Unexpectedly, when the same reactions were carried out using the carbethoxy derivative of **136**, cis and trans isomers **140** were obtained with diastereomeric ratio (dr) = 64:36 for R = Me.

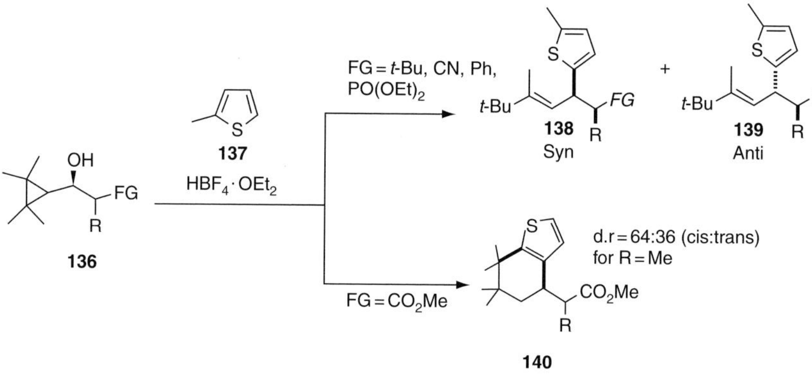

Several additional studies involving Friedel-Crafts chemistry were reported this year. Thus, certain thiophenes possessing electron attracting groups (EAG) **141** were readily converted to **143** by treatment with benzyl chloride **142** in the presence of the Pd(OPiv)$_2$ catalyst <09OL4160>. This is particularly important since Friedel-Crafts addition of electrophiles to thiophenes possessing EAG groups were rare before this study.

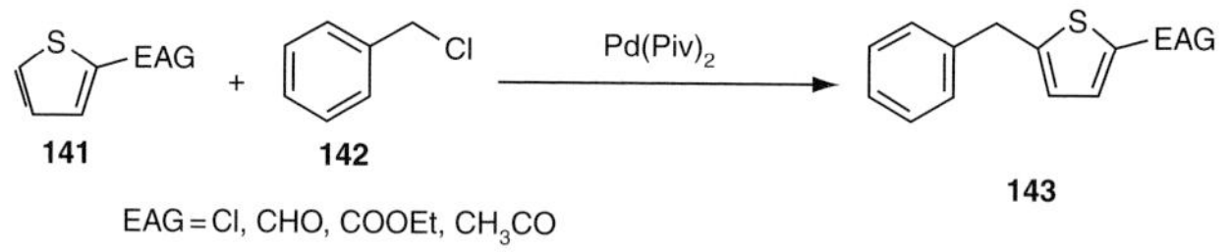

Friedel-Crafts acylations of thiophene with various acyl chlorides were performed in the presence of MoO$_2$Cl$_2$ <09TL1407>. In all cases, only 2-acylthiophenes were formed in moderate yields (61–66%). In other studies, pyrrolidinyl ketones added to the 5-position of 2-phenylthiophene by a tandem Pd(II)/indium(III)-catalyzed aminochloro-carbonylation/Friedel-Crafts acylation reaction <09JA3124> and 2- and 3-*N*-hydroxy-2-phenylacetimidoyl chloride derivatives of 2-thiophene and 3-benzo[*b*]thiophene, respectively, were synthesized by a Friedel-Crafts acylation reaction involving β-nitrostyrene <09T2436>.

Functionalized hydroxy thiophene motifs were prepared as potential bioisosteres of phenolic amides and phenolic sulfonamides <09TL5005>. The hydroxy thiophenes **148** were prepared by a multi-step reaction in which the 3-methoxy ester **144** was demethylated to the 4-hydroxy ester **145**. This ensured that bromination would occur at the 3-position. Subsequent hydroxy methylation of **145** gave **146** which was subjected to Buchwald palladium catalyzed amination using $Ph_2C=NH$ to give **147** which was hydrolyzed to 3-amino derivative **148**.

5.1.5. SYNTHESIS OF THIOPHENES FOR USE IN MATERIAL SCIENCE

Interest in thiophene dendrimers continued unabated during the year. Thiophene dendrimers as entangled photo sensor materials were investigated and found to have good potential for application in quantum optical devices <09JA973>. Functional oligothiophenes including fused thiophenes, thienothiophenes, macrocyclic thiophenes, and dendritic oligothiophenes, have been reviewed with respect to their molecular design for multidimensional nanoarchitectures and their applications <09CRV1141>. *Syn* and *anti* cyclophanes having oligothiophene units have been prepared for the first time <09TL4509>. Self-assembling carbohydrate-functionalized oligothiophenes have been synthesizing using mild Sonogashira cross-coupling conditions <09OL5098>. Acenes, such as pentacene, are π-conjugated ring-fused systems that have been extensively studied because of their potential for use in optical electronic devices. Due to certain issues with this system, *i.e.*, poor solubility in common organic solvents and poor stability at high temperatures, heteroacenes have attracted increasing interest. Most heteroacenes studied are thiophene base symmetrical systems. Recent research efforts have been directed toward heteroacenes possessing two or more different heterocycles. 2,2′-Indolo[3,2-*b*;4,5-*b*′]thiophenes (DITs) were the first in the thiophene/pyrrole series to be published <08JOC4638>. Preliminary results looked encouraging; however, the unsymmetrical DIT isomers, which would be expected to have higher dipole moments and increased solubility in polar solvents, were not reported, probably due to synthetic issues. In a later study, both symmetrical and unsymmetrical heteroacene containing thiophene and pyrrole rings, *i.e.*, dibenzo[*b*]thienopyrroles, were synthesized in two-step reactions <09OL3358>. For example, unsymmetric dibenzothienopyrroles were prepared in a two-step reaction in which 3-bromobenzo[*b*]thiophene was aminated with a primary amine by a palladium-catalyzed process to give a *N*-functionalized 3-aminobenzo[*b*]thiophene. This amine presumably dimerizes to an unsymmetric product upon treatment with n-BuLi and $CuCl_2$. The symmetric thienopyrroles were prepared by a palladium cross-catalyzed amination reaction using alkyl or aryl amines and also by a copper catalyzed amination reaction. All synthesized heteroacenes were highly soluble in organic solvents and stable at ambient temperatures. The products were characterized by NMR spectroscopy (^{1}H and ^{13}C) and X-ray crystallography.

Several other acene thiophenes were synthesized and studied, such as [7]helicene **154** and a double helicene, a D_2-symmetric dimer of 3,3'-bis(dithieno-[2,3-b:3',2'-d]thiophene <09JOC408>. The key intermediate in this synthesis (dimer) **152** is readily prepared from 2,5-bis-trimethylsilanyldithieno[2,3-b:3',2'-d]thiophene **149** in three steps; namely, bromination to **150**, LDA-catalyzed halogen dance to **151**, followed by $CuCl_2$-catalyzed cyclization. Intermediate **152** is then converted to the dianion which can either be converted to the annelated helicene **154** either by treatment with bis(phenylsulfonyl)sulfide **153** followed by removal of the TMS group by TFA or oxidized to a double helicene (not shown here). Structures of these products were ascertained by NMR (^{1}H and ^{13}C), UV/vis and X-ray analysis. The crystal structures of the products showed strong $\pi-\pi$ and S—S interactions and UV/vis spectra showed some π delocalization.

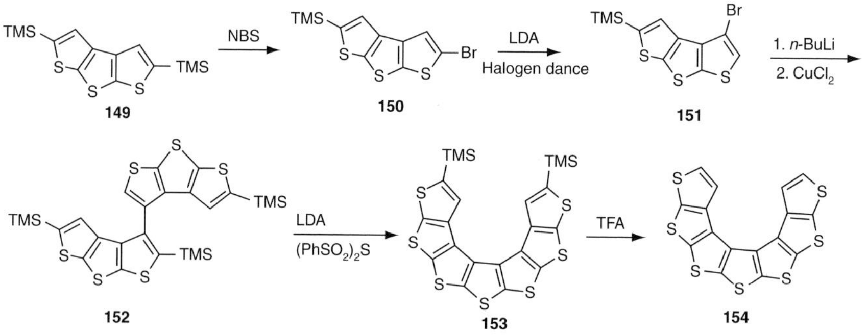

Also, novel organometallic Ru(II) and Fe(II) complexes with tetrathia-[7]-helicene derivative ligands were synthesized <09POL621>. A typical example is shown below. The thiophene fused 3,7-diphenyl[1]benzothieno[3,2-b]benzothiophene with modified with diarylamino groups **156** was prepared by palladium-catalyzed Suzuki-Miyaura cross-coupling between 2,7-diiodo[1]benzothieno[3,2-b]benzothiophene and 4-(N,N-diarylamino)phenylboronic acid **155** <09CL420>. This modification resulted in vastly improved electronic properties.

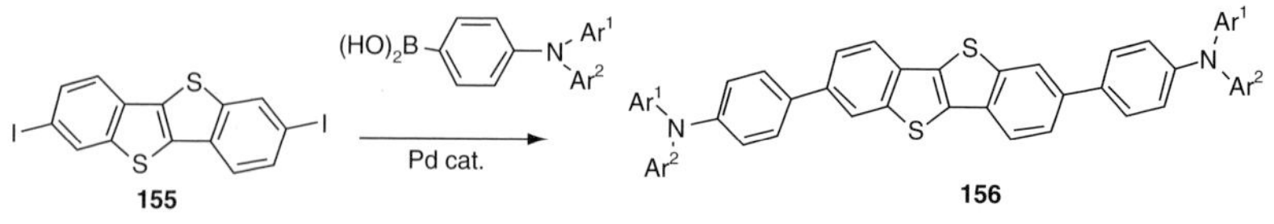

Zirconium-mediated coupling reaction involving 2,3-diiodo- and 2,3,4,5-tetraiodothiophenes was used to prepare substituted thiophene-fused acenes <09OL3702>. Also, new oligothiophene-pentacene hybrids were prepared by a Suzuki cross-coupling reaction <09OL2563>. These compounds have significantly higher solubility and improved thermal and photooxidation stabilities. Moderate anisotropic field-effect mobility was observed in dinaphtho[2,3-b:2′,3′-f]thiopheno[3,2-b]thiophenes **157**, which are single-crystals transistors <09APL223308>.

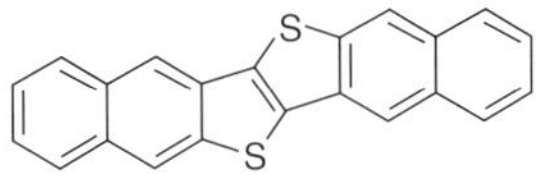

157

Two molecular ensembles of thiophene-extended tetrathiafulvalene-thiophene were prepared by Stille coupling and Horner-Wittig reaction in the key steps <09TL6897>. Phototropic compounds based on 1,2-dithienylethenes **161** for possible application to photo and electronic devices have also been reported. A facile three-step synthetic process shown below for the synthesis of the symmetric isomer <09TL1614> has three important aspects that deserve comment: (1) it takes advantage of the selective 4-bromination of **158** and subsequent debromination **159** by BuLi; 2) it uses Pd(PPh$_3$)$_4$ in the Suzuki-Miyaura cross coupling of **160** which is known to cross-couple hetaryl chlorides <06TL1993>; and 3) it allows for the introduction of either one substituent or two of the same or different substituents.

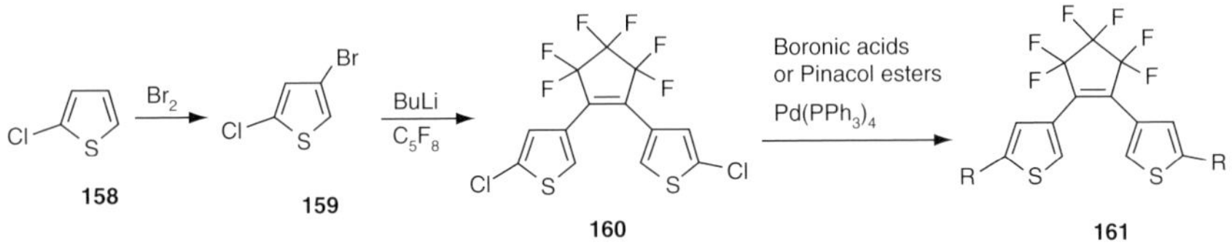

158 **159** **160** **161**

In addition, the optic and proton dual-control of fluorescence of rhodamine based on photochromic diarylethenes has been accomplished <09TL1588> and the synthesis and optoelectronic properties of diarylethene derivatives having benzothiophene and n-alkylthiophene units has been reported <09JST100>. Synthesis of monomers and their conversion to a wide variety of polymers for use in the material science were prepared. Thus, the synthesis and characterization of benzo[d]thiophene tethered with dibenzo-heterocycles were prepared and found to have potential use as OLEDs <09T822>. In addition, a series of 2,7-disubstituted hexafluoro-2-thienylfluorenes **163** were prepared as potential building blocks for electron transporting materials <09JOC820>. The synthesis involves a nucleophilic addition of phenyllithium and thienyllithium onto octafluoro-2-thienylfluorene **162**. The structures of the building blocks **163** were confirmed by X-ray analysis.

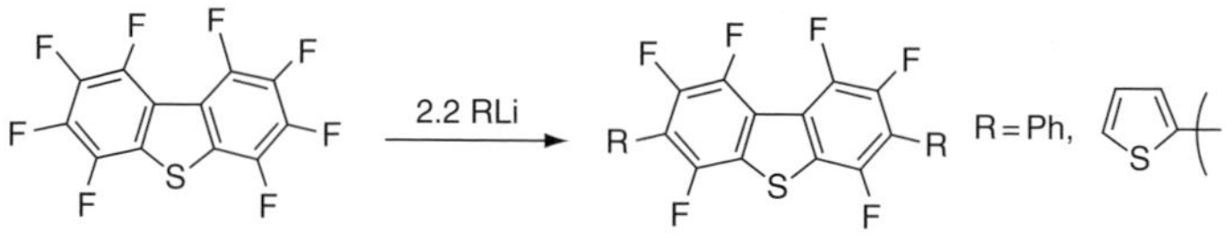

[Scheme: compound **162** (octafluorodibenzothiophene) + 2.2 RLi → compound **163** (R-substituted), R = Ph, thienyl]

Copolymers of alkyl- and azothiophenes were also synthesized <09JAPS680>. For example, a variety of low band gap copolymers based on indenofluorene and thiophene derivates were prepared by Suzuki cross-coupling reaction of appropriate 2,5-dibromothiophenes and 11,11,12,12,-tetrahexylindeno-fluorene-2,7-bis(trimethylene) boronate <09JPS(A)5044>. In addition, a novel photoresponsive terarylene having a leaving methoxy group was prepared that showed a relatively high fluorescence quantum yield after UV-light irradiation <09OL1475>.

Another class of mixed thiophene-octahomotetraoxacalixarenes was prepared and their complexing properties with C_{60} and C_{70} fullerenes were studied <09TL4289>. The ^{1}H NMR and ^{13}C NMR spectra were in agreement with proposed structures. The simplicity of the spectra indicated the calixarenes were symmetric; e.g., ^{13}C NMR revealed only six signals appearing at low fields (aryl and thiophenyl carbons) and five signals at high fields (the remaining carbons). The structures were further confirmed by X-ray analysis. Unfortunately, solution complexation studies with C_{60} and C_{70} fullerenes were inconclusive.

A new series of semiconducting polymers containing 3,6-dimethylthieno[3,2-b] thiophene were described <09CM2650>. The key step in this process is a Suzuki-Miyaura cross-coupling reaction between readily available 2,5-dibromo-3,6-dimethylthieno[3,2-b]thiophene and 4,4,5,5-tetramethyl-1,3,2-dioxaborolane using Pd(PPh$_3$)$_4$ as catalyst in toluene. The polymers were prepared by using FeCl$_3$-mediated oxidative coupling polymerizations in refluxing chlorobenzene. In addition, a wide variety of 5-aryl-5′-formyl-2,2′-bithiophenes were synthesized by Vilsmeier-Haack-Arnold reaction (VHA) or via a Suzuki-Miyaura cross-coupling reaction <09T2079>. The one-step Suzuki cross-coupling involving **164** and **165** shown below gave desired 2,2′-bithiophenes **166** in good yields whereas the VHA reactions (not shown) gave lower yields, and required four-steps. These bithiophenes have been shown to be new precursors for nonlinear optical (NLO) materials.

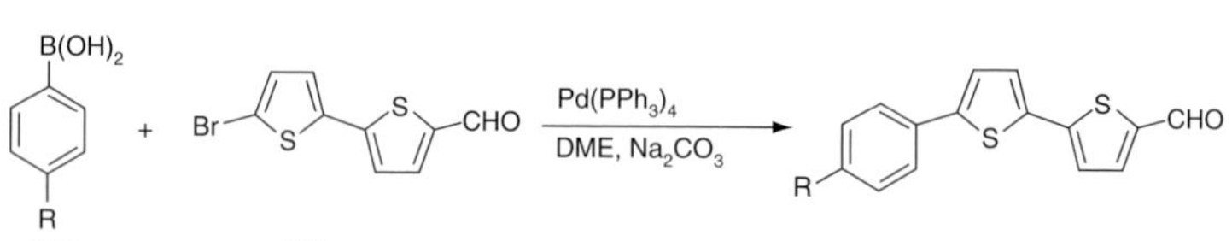

[Scheme: **164** (R-C$_6$H$_4$-B(OH)$_2$) + **165** (Br-thiophene-thiophene-CHO) → Pd(PPh$_3$)$_4$, DME, Na$_2$CO$_3$ → **166** (R-C$_6$H$_4$-thiophene-thiophene-CHO)]

Other interesting syntheses of oligomers containing thiophene rings reported this year include: intramolecular cyclization of thiophene-base [7] helicenes to quasi-[8] circulenes <09JOC9105>, thiophene/thieno[3,2-b]thiophene co-oligomers, which are fused-ring analogues of sexithiophene, by a combination of Stille cross-coupling and oxidative homocoupling reactions <09JOC9112>, a series of conjugated thiophene-based trithienylvinylene compounds bearing dithiocarbonate by successive Wittig and trialkylphosphite-mediated cross coupling reactions <09JOC9188>, and novel star-shaped thiophene oligomers were synthesized by first preparing terthienobenzene (TTB) by a ring closure reaction with a 1,3,5-trialdehyde and ethyl mercaptoacetate to give the TBB-triester. The triesters were saponified and decarboxylated to give TTB which was then treated with NBS to give tribromo-TTB. Subsequent Stille couplings with tributyltinthiophenes gave star-shaped thiophene oligomers. The oligomers were then electropolymerized to give polymers with unusual properties, which are discussed <09OL3230>. Scanning tunneling microscopy studies were carried on supramolecular ordering in the prochiral triacid derivative TTBTA-fullerene monolayers <09JA16844>. Finally, dithieno-1,2-dihydro-1,2-diborin and its dianion were synthesized. Interestingly, whereas the neutral 1,2-borin has a twisted ring the dianion is planar and has characteristic π-conjugation through the B-B bond, which gives rise to an aromatic 14-π electron system <09JA10850>. We conclude this section by noting the giant cyclo[n]thiophenes with extended π-conjugation up to 35 thiophene rings were prepared by Bauerle *et al.* <09AG6632>.

5.1.6. THIOPHENES DERIVATIVES IN MEDICINAL CHEMISTRY

A new approach for the stereoselective synthesis of chiral 2,3,4-trisubstituted tetrahydrothiophenes (a class of thiophenes with a wide range of biological activities from commercially available tri-O-acetyl-D-glucal) was demonstrated <09TL6941>. An expeditious synthesis of 5-nitrobenzo[*b*]thiophene-2-carboxaldehyde **172**, an important intermediate in the preparation of lipoxygenase inhibitors, anti-tumor agents, and psychotropic antagonist, was developed <09JCR(S)359>. Unlike the nine-step synthesis previously reported <60FES396), the expeditious procedure requires a single step in which a solution of triethylamine (TEA) in DMF is added slowly to a mixture of 1,4-dithiane-2,5-diol **167** (dimer of mercaptoacetaldehyde) and 2-chloro-5-nitrobenzaldehyde **169**. The addition of TEA then liberates mercaptoacetaldehyde **168** whose sulfur atom adds to the 2-chloro carbon by a S$_N$Ar reaction to give complex **170**. Concomitant elimination of chloride ion and aromatization of **170** gives rise to the dialdehyde **171** which undergoes concomitant cyclization and dehydration to **172**.

A facile Gewald domino protocol for the regioselective synthesis of novel 2-amino-5-aryl[2,3-*b*]thiophenes **175** was reported in which 5-aryldihydro-3(2*H*)-thiophenenones **173**, NCCH$_2$X **174**, and sulfur are subjected to microwave heating <09TL6191>. *In vitro* activity studies showed that several derivatives of **175** were powerful antimycobacterial agents.

The synthesis of 3,4-di(substituted)oxy-N^2,N^3-bis(substituted)thiophene-2,5-dicarbohydrazides, a new class of anticonvulsants, also appeared in the literature <09EJM4376>. The titled compounds were prepared by a condensation reaction between ethyl thiogycolate with diethyl oxalate yielding the 3,4-dihydroxythiophene-2,5-diester that was derivatized using different alkyl azides.

An interesting synthesis of a series of novel 3-benzoyl-3-ferrocenylbenzo[b]thiophenes involves the acid catalyzed intramolecular cyclization of 1-ferrocenyl-2-[3-(methoxyphenyl)thio]ethanone <09OM5412>. Several of these compounds exhibited strong antitumor properties.

Several 3-(aryl)benzothieno[2,3-c]pyran-1-ones were prepared by Sonogashira coupling of either 3-bromobenzo[b]thiophene-2-carboxylic acid, and methyl 3-bromo- or methyl-6-methoxybenzo[b]thiophene-2-carboxylate with aryl acetylenes. The resulting adducts were then subjected to electrophilic intramolecular cyclization using iodine in TFA. The majority of titled compounds showed high antitumor activity <09EJM1893>. Several novel derivatives of pyridylbenz[b]thiophene-2-carboxamides and benzo[b]thieno[2,3-e]naphthridin-2-ones were prepared by amidation of 3-chlorobenzo[b]thiophene-2-carbonyl chloride with several aromatic amines <09JMC2482>. Interestingly, the starting 3-chloro compound was prepared by intramolecular cyclization of cinnamic acid brought about by treatment with thionyl chloride in pyridine. A study of the mechanism of their antitumor action indicated that minor structural changes resulted in major difference in action provoked major differences of antitumor action mechanisms. The first thiophene thioglycosides were prepared by treating sodium thiophenethiolate salts with 2,3,5,6-tetra-O-acetyl-α-D-glucopyranosyl bromides <09JCC161>. The methyl ester derivatives serve as a valuable precursor to 3H-benzo[4,5]thieno[3,2-d]pyrimidin-4-ones, which are important lead compounds in medicinal synthesis.

A variety of substituted bis-(hydroxyphenyl) thiophenes were prepared and their ability to inhibit of 17β–HSD1 was evaluated <09JMC6724>. 2-Fluoro-4-(5-(3-hydroxyphenyl)thiophen-2-yl)phenol was found to be the most active and selective inhibitor for treatment of estrogen-dependent diseases. A new family of aromatic 2-(thio)ureidocarboxylic acids **176** and **177** were prepared and found to be particularly potent inhibitors of multidrug resistance-associated protein (MRP1) <09JMC4586>.

The reaction of benzo[b]-5,6-dicarboxaldehyde with dibenzyl ketone, thiodiacetic acid, dimethyl ester, hydrazine, *p*-toluidine, *p*-aminoacetophenone and nitromethane gave the corresponding seven-, six- and five-membered rings fused to benzo[b]thiophenes <09JCR326>. These condensed products were found to be good reagents for fluorescence analysis of amino acids.

A new series of 5-aryl-2-(trifluoroacetyl) thiophenes were synthesized and found to be novel, selective and stable inhibitors of Class II histone deacetylases HDAC4 <09JMC6782>. Additional studies of this series produced 3-(5-(2,2,2-trifluoroacetyl)thiophen-2-yl)benzoic acid that displayed a 40-fold increase in selectivity over HDAC4.

A plethora of novel quaternary derivatives of (3R)-quinuclidinol esters were prepared and one compound, aclidinium bromide, was found to be most effective for once-daily maintenance treatment of COPD (chronic obstructive pulmonary disease) <09JMC5076>.

A highly enantioselective Friedel-Crafts reaction of a series of 2-thiophenes with glyoxylates gave hydroxy(thiophene-2-yl) acetates <09OL4636>. A formal synthesis of duloxetine, a serotonin norepinephrine reuptake inhibitor, was accomplished starting with thiophene and *n*-butyl glyoxalate. Additionally, a new class of fully functionalized thiophene was found to be protein farnesyltransferase inhibitors, of which the 3-(4-chlorophenyl)-4-cyano-5-thioalkylthiophene-2-carboxylic acids **178** and **179** are particularly strong inhibitors <09JMC6205).

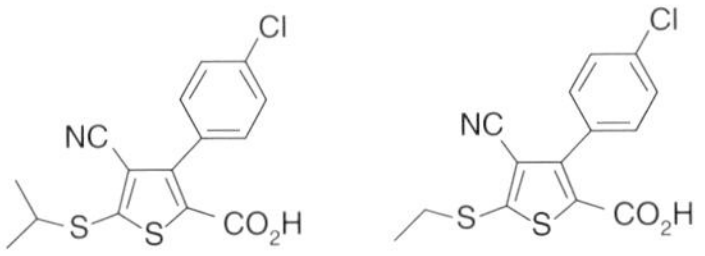

Tertiary thienyl alcohols exhibit well-known biological activity; however, the first synthesis involving direct asymmetric catalytic addition of trithiophenylaluminum **181** to ketones **180** to give chiral alcohols **182** was just reported recently <09OL3386>. The synthesis of (S)-tiemonium iodide, sold as an anticholinergic/spasmolytic drug, was demonstrated.

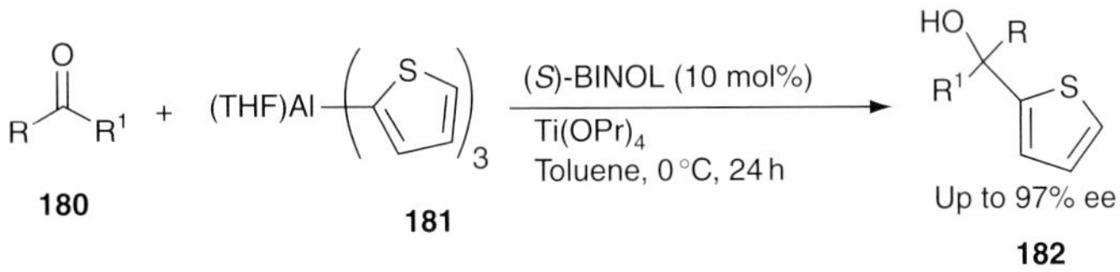

A photocontrolled molecular switch, shown below, that uses two different wavelengths of light to toggle between its two structural forms **183** (colorless) and **184** (blue) has been demonstrated <09JA15966>. Using light to moderate biochemical agents in living organisms offers great potential for advancing biomedical technologies.

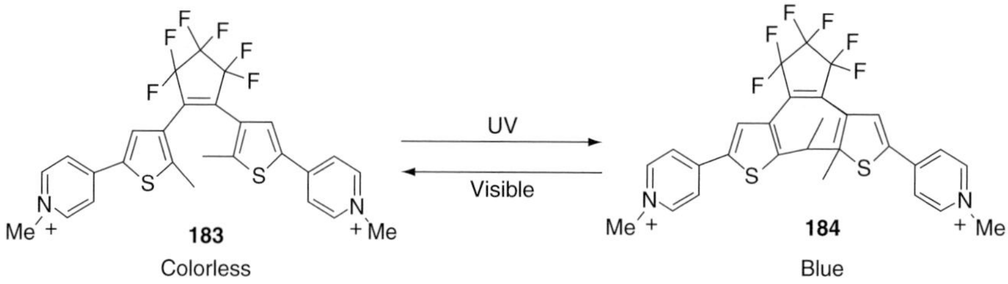

5.1.7. SELENOPHENES AND TELLUROPHENES

Although not as actively studied as thiophenes, the uses of selenophenes continued to increase in intensity, especially in the material science area. For example, a synthetic selenophene containing a multi-ring polymer was prepared by a straightforward synthesis involving Stille coupling to give a 1,4-bis-selenophene-2,5-dialkoxybenzene derivative, which was subjected to electrochemical polymerization <09SM361>.

An improved, mild platinum-catalyzed synthesis of multisubstituted benz[*b*]-selenophenes **186** developed involving a C-Se bond addition to an alkyne **185** <09EJO5509>. Previous syntheses of 2,3-disubstituted selenophenes failed due to the air-sensitivity of the Se-H group and poisoning of the metal catalyst by selenium metal <09T10453>.

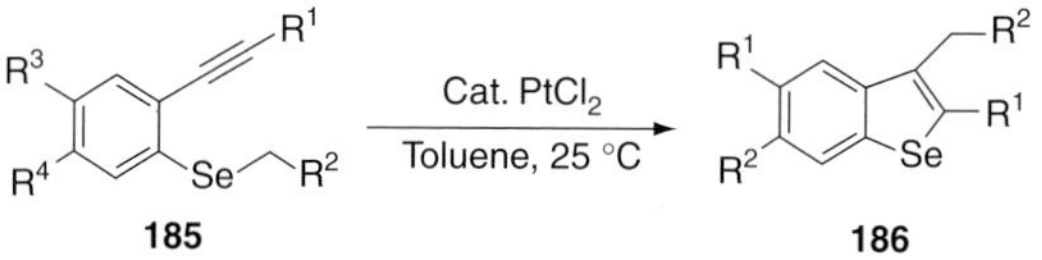

Multifunctional dihydroselenophenes **191** were prepared presumably by a [3+2] cycloaddition of thiazole carbene-derived C-C-Se 1,3-dipoles **187** to allene **188** to form adduct **189**, which collapses to imine **190** and then hydrolyzed to **191** <09OBC3264>.

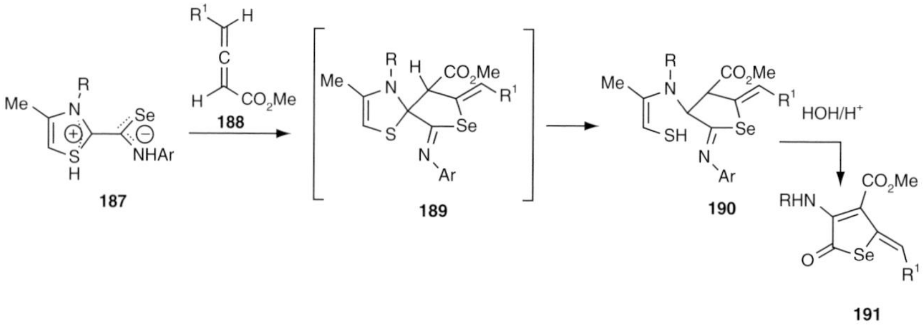

Five-membered selenium and sulfur heterocycles were accomplished by a one-pot Dzhemilev reaction <09CHE317>. The method involves cyclometalation of alkenes, allenes, and acetylenes using alkyl and haloalkyl derivatives of Al and Mg in presence of catalytic amounts of Ti and Zr complexes. The resulting alumina- and magnesa-carbocycles are treated with selenium or sulfur to give titled compounds.

Novel selenophene-based semiconducting copolymers were prepared by a microwave-assisted Stille coupling reaction or by oxidative coupling <09JMAC3490>. In addition, new semiconducting polymers containing selenophene[3,2-*b*]selenophene were prepared in similar manner as that described for 3,6-dimethyl(thieno[3,2-*b*]thiophene) for use in organic thin-film transistors <09CM2650>.

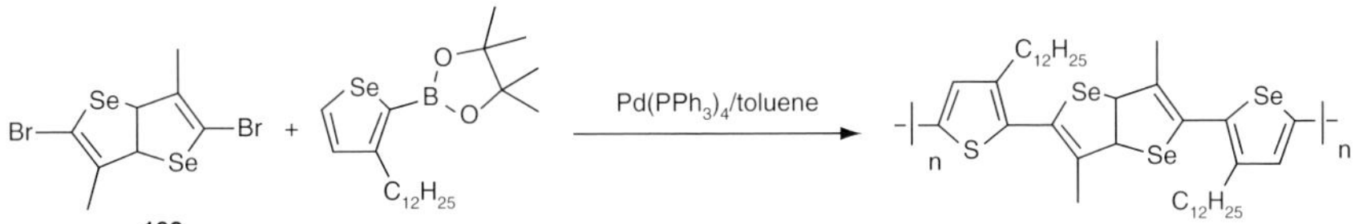

For the first time, a selenophene, compound **195**, can be used as a stable unit in dye-sensitive solar cells <09JPC7469>. The key steps included a Stille cross coupling reaction.

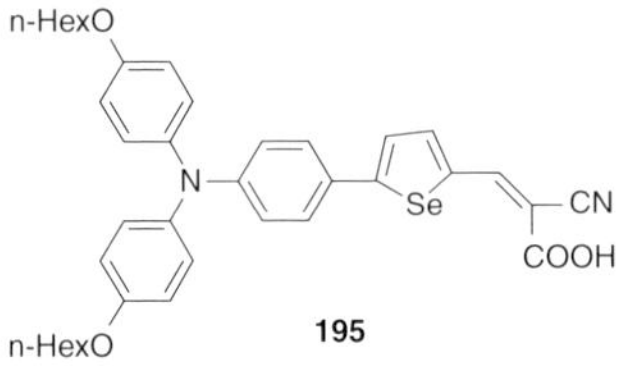

A variety of 2-(selenophene-2-yl) pyrroles were prepared from 2-acylselenophene *via* their oximes and acetylene and were electropolymerized to electronchromic nanofilms <09CH6435>. These new systems should aid the study of the interface between selenophene and pyrrole chemistry.

Several theoretical studies on 2-selenophenes were carried out this year. Frustrated rotations in single-molecule junctions in oligo- and polyselenophenes were determined by comparing the conductance of 1,4-bis-(methyseleno)benzene with that of 2,3,6,7-tetrahydrobenzo[1,2-*b*:4,5-*b*']diselenophene <09JA10820>. Since the electron lone pair is rigidly locked in the latter and not in the former, a relationship between the conductance and orientation of the lone pair was obtained. Also, a study on how the non-covalent Se—Se=O interaction stabilize selenoxides at naphthalene 1,8-position was discussed on structural and theoretical grounds <09NJC196>.

2-Selenophen-2-pyrroles were prepared in a one-pot procedure involving the reaction of 2-acylselenophenes and acetylenes and were subsequently electropolymerized to electronchromic nanofilms <09CH6435>. The facile synthesis of these exotic heterocycles should lead to further studies in medicine, biochemistry, and advanced materials.

We have learned more about the physicochemical and organic molecular electronic properties of selenophenes through some detailed investigations. For example, 3-alkynyl selenophene (ASP) was prepared and found to be especially potent at a dose range of 5-50 mg/kg, and produced systemic anti-hyperalgesic and antinociceptive actions in mice <09PBB419>. In addition, ASP was found to have a hepatoprotective effect on acute liver injury induced by D-galactosamine and lipopolysaccharide in rats <09EMP20>.

Single crystals of perylo[1,12-b,c,d]selenophene [PESE] were synthesized and microribbons of PESE were grown by usual techniques <09APL153306>. Single crystal X-ray analysis showed that the structural features of PESO (short S—S, extraordinary solid-state packing) make PESE an excellent candidate for use in organic/molecular electronics. Reddy and Anand have synthesized and characterized unprecedented aromatic 30π expanded isophlorins composed of thiophene, selenophene, and furans <09JA15433>.

As usual, tellurophenes were little studied this year. However, some exciting chemistry was reported. For example, 3,4-dimethoxytellurophene was synthesized by the reaction of 2,3-dimethoxy-1,3-butadiene and $TeCl_2$ (prepared by the reaction of $TeCl_4$ and $Me_3SiSiMe_3$) in the presence of sodium acetate and hexane <09OL1487>. Electropolymerization of 3,4-dimethoxytellurophene gave a mixture of poly-4 and possibly poly-5. Due to the instability of poly-5, a definite assignment could not be made although UV/VIS revealed bands that matched the calculated band gap of poly-5.

Having developed a convenient method for preparing multi-gram quantities of tellurophene <08JOM2463>, Stephens and Sweat were able to prepare 2-(tributylstannyl)- and 2,5-bis(trimethylstanyl)-tellurophene and carry out the first reported Stille cross-coupling using a mixed catalyst system consisting of tetrakis(triphenylphosphine)palladium(0) and copper(I) iodide, together with cesium fluoride as additive, in N,N-dimethylformamide <09S3214>. Several 2-aryl- and 2,5-diaryltellurophenes were prepared in good yields. This seminal study should lead to several significant advances in the material and medical sciences. A somewhat related advance was reported by Shimad et al. <09TL6651> who carried out a regioselective synthesis of polysubstituted pyridines by a hetero-Diels Alder reaction of isotellurazoles with acetylenic dienophiles. Tellurium is extruded during the Diels-Alder reaction.

Cycloaddition/cycloreversion reactions between TCNQ and mono or bis-[2-(azulen-1-yl]thiophene and other benzene derivatives were carried out <09EJO4316>. The adducts were fully characterized by spectroscopic means. CV and DPV measurements showed the TCNQ adducts behave as redox-active chromophores.

REFERENCES

60FES396	S. Rossi and R. Trave, *Farm. Ediz. Scient.*, **15**, 396 (1960).
66CB99	K. Gewald, E. Schinke and H. Bottcher, *Chem. Ber.*, **94**, 99 (1966).
71JOC1645	F. De Jong and M.J. Janssen, *J. Org. Chem.*, **36**, 1645 (1971).

81JOC851	Y. Fujiwara, O. Maruyama, M. Yoshidomi and H. Taniguchi, *J. Org. Chem.*, **46**, 851 (1981).
91JOC6720	Y. Nishiyama, M. Yoshida, S. Ohkawa and S. Hamanaka, *J. Org. Chem.*, **56**, 6720 (1991).
92JOC3094	D.K. Bates, R.T. Winters and J.A. Picard, *J. Org. Chem.*, **57**, 3094 (1992).
94BSF900	G. Dujardin and J.M. Poirier, *Bull. Soc. Chim. Fr*, **131**, 900 (1994).
01TL8419	Q. Dang, Y. Liu and Z. Sun, *Tetrahedron Lett.*, **42**, 8419 (2001).
03BMC4729	T. Noguchi, M. Hasegawa, K. Tomisawa and M. Mitsukuchi, *Biorg. Med. Chem.*, **11**, 4729 (2003).
06JOC9681	G.A. Molander, *J. Org. Chem.*, **71**, 9681 (2006).
06TL1993	W. Sun, B. Yu and F.E. Kuhn, *Tetrahedron Lett.*, **46**, 1993 (2006).
08JOC4638	T. Qi, W. Qiu, Y. Liu, H. Zhang, X. Gao, Y. Liu, K. Lu, C. Du, G. Yu and D. Zhu, *J. Org. Chem.*, **73**, 4638 (2008).
08JOM2463	D.P. Sweat and C.E. Stephens, *J. Organomet.*, **14**, 2463 (2008).
08T5982	A. Maehara, T. Saton and M. Miura, *Tetrahedron*, **64**, 5982 (2008).
09ABC637	N. Rozlosnik, *Anal. Bioanal. Chem.*, **395**, 637 (2009).
09ACA61	V. Mirkhani, M. Moghadam, S. Tangestaninejad, I. Mohammdpoor-Baltork, H. Kargar and M. Araghi, *App. Catal. A*, **353**, 61 (2009).
09ACR(11)1691	*Acc. Chem. Res.*, **11**, 1691 (2009).
09AG6632	F. Zhang, G. Goetz, H.D.F. Winkler, C.A. Schalley and P. Bauerle, *Angew. Chem. Int. Ed.*, **48**, 6632 (2009).
B-09M1001	I.F. Perepichka, and D.F. Perepichka (eds.), (2009). Handbook of Thiophene-Based Materials: Applications in Organic Electronics and Photonics, Wiley, New York (2009).
09APL153306	L. Tan, W. Jung, L. Jiang, S. Jiang, Z. Wang, S. Yan and W. Hu, *Appl. Phys. Lett.*, **94**, 153306 (2009).
09APL223308	M. Uno, Y. Tominari, M. Yamagishi, I. Doi, E. Miyazaki, K. Takiniya and J. Takeya, *Appl. Phys. Lett.*, **94**, 223308 (2009).
09ARK(i)97	P. Zanirato, *ARKIVOC*, (i), 97 (2009).
09ARK(v)90	M.P. Boone, Y. Dienes and T. Baumgartner, *ARKIVOC*, (v), 90 (2009).
09ARK(ix)96	T. Thiemann, D.J. Watson, A.O. Brett, J. Iniesta, F. Marken and Y.-q. Li, *ARKIVOC*, (ix), 96 (2009).
09ARK(xi)316	H. Pessoa-Mahana, M. Gonzalez, M. Gonzalez, D. Pessoa-Mahana, R. Araya-Maturana, N. Ron and C. Saitz, *ARKIVOC*, (xi), 316 (2009).
09BCJ555	A. Sugie, H. Furukawa, Y. Suzaki, K. Osakada, M. Akita, D. Monguchi and A. Mori, *Bull. Chem. Soc. Jpn.*, **82**, 555 (2009).
09CC3536	L. Melzig, C.B. Rauhut and P. Knochel, *Chem. Commun.*, 3536 (2009).
09CC4729	Y. Zhang, M. Bian, W. Yao, J. Gu and C. Ma, *Chem. Commun.*, 4729 (2009).
09CH6435	B.A. Trofimov, E.Y. Schmidt, A.I. Mikhaleva, C. Pozo-Gonzalo, J.A. Pomposo, M. Salsamendi, N.I. Portzuk, N.V Zorina, A.V. Afonin, A.V. Vashchenko, E.P. Levanova and G.G. Levkovskaya, *Chemistry*, **15**, 6435 (2009).
09CHE317	V.A. D'yakonov, A.G. Iragimov, A.A. Makarovv, R.K. Timerkhanov, R.A. Tuktarova, O.A. Trapeznikova and L.F. Galimova, *Chem. Heterocycl. Compd. (English Trans.)*, **45**, 317 (2009).
09CL420	T. Izawa, H. Mori, Y. Shinmura, M. Iwatani, E. Miyazaki, K. Takimiya, H.-W. Hung, M. Yahiro and C. Adachi, *Chem. Lett.*, **38**, 420 (2009).
09CM2650	H. Kong, Y.K. Jung, N.S. Cho, I.-N. Kang, J.-H. Park, S. Cho and H.-K. Shim, *Chem. Mater.*, **21**, 2650 (2009).
09CPB710	T. Dohi, K. Morimoto, C. Ogawa, H. Furoka and Y. Kita, *Chem. Pharm. Bull.*, **57**, 710 (2009).
09CRV1141	A. Mishra, C.-Q. Ma and P. Bauerie, *Chem. Rev.*, 1141 (2009).
09EJM1893	M.-J. Queiroz, R.C. Calhelha, L.A. Vale-Silva, E. Pinto and M.S-J. Nascimeto, *Eur. J. Med. Chem.*, **44**, 1893 (2009).
09EJM4376	R. Kulandasamy, A.V. Adhikari and J.P. Stables, *Eur. J. Med. Chem.*, **44**, 4376 (2009).
09EJO4316	T. Shoji, S. Ito, K. Toyota, T. Iwamoto, M. Yasunami and N. Morita, *Eur. J. Org. Chem.*, 4316 (2009).
09EJO5509	T. Sato, I. Nakamura and M. Terada, *Eur. J. Org. Chem.*, 5509 (2009).

09EMP20	E.A. Wilhelm, C.R. Jesse, S.S. Roman, C.W. Nogueira and L. Savegnago, *Exp. Mol. Pathol.*, **87**, 20 (2009).
09GC425	J. Robert, F. Pozgan and H. Doucet, *Green. Chem.*, **1**, 425 (2009).
09H169	S. Fukamachi, H. Konishi and K. Kobayashi, *Heterocycles*, **78**, 169 (2009).
09H2467	N. Suzue, R. Ishie, H. Watabayashi and K. Kobayashi, *Heterocycles*, **78**, 2467 (2009).
09H2993	K. Kobayashi, T. Kozuki and H. Konishi, *Heterocycles*, **78**, 2993 (2009).
09JA973	M.R. Harpham, O. uzer, C.-Q. Ma, P. Bauerle and T. Goodson, *J. Am. Chem. Soc.*, **131**, 973 (2009).
09JA1668	Y. Kita, K. Morimoto, M. Ito, C. Ogawa, A. Goto and T. Dohi, *J. Am. Chem. Soc.*, **131**, 1668 (2009).
09JA3104	S.E. Denmark, R.C. Smith, W-T.T. Chang and J.M. Muhuhi, *J. Am. Chem. Soc.*, **131**, 3104 (2009).
09JA3124	T.A. Cernak and T.H. Kambert, *J. Am. Chem. Soc.*, **131**, 3124 (2009).
09JA6961	D.M. Knapp, E.P. Gillis and M.D. Burke, *J. Am. Chem. Soc.*, **131**, 6961 (2009).
09JA10820	Y.S. Park, J.R. Widawky, M. Kamenetska, M.L. Steigerwald, M.S. Hybertsen, C. Nuckolls and L. Venkataraman, *J. Am. Chem. Soc.*, **131**, 10820 (2009).
09JA10850	A. Wakamiya, K. Mori, Y. Araki and S. Yamaguchi, *J. Am. Chem. Soc.*, **131**, 10850 (2009).
09JA12483	L. Salvi, J.G. Kim and P.J. Walsh, *J. Am. Chem. Soc.*, **131**, 12483 (2009).
09JA15433	J.S. Reddy and V.G. Anand, *J. Am. Chem. Soc.*, **131**, 15433 (2009).
09JA15966	U. Al-Atar, R. Fernandes, B. Johnson, D. Baillie and N.R. Branda, *J. Am. Chem. Soc.*, **131**, 15966 (2009).
09JA16844	J.M. MacLeod, O. Ivasenko, C. Fu, T. Taerum, F. Rosel and D.F. Perepichka, *J. Am. Chem. Soc.*, **131**, 16844 (2009).
09JAPS680	V.C. Concalves, L.M.M. Costa, M.R. Cardoso, C.R. Mendonca and D.T. Balough, *J. App. Poly. Sci.*, **114**, 680 (2009).
09JCC161	G.H. Elgemeie, W.A. Zaghard, K.M. Amin and K.M. Nasr, *J. Carbohydr. Chem.*, **28**, 161 (2009).
09JCO900	C.-H. Cho, B. Neuenswander, G.H. Lushington and R.C. Larock, *J. Comb. Chem.*, **11**, 900 (2009).
09JCR326	M.A. El-Boral and H.F. Rick, *J. Chem. Res.*, 326 (2009).
09JCR(S)359	H. Wei, M. Sun and M. Ji, *J. Chem. Res.*, 359 (2009).
09JFA2088	R. Bonikowski, M. Sikora, J. Kula and A. Kunicka, *J. Sci. Food Agric.*, **89**, 2088 (2009).
09JHC599	A. Rais, H. Ankati and E.R. Biehl, *J. Heterocycl. Chem.*, **46**, 599 (2009).
09JMC2482	K. Ester, M. Hranjec, I. Pianatanida, I. Caleta, I. Jarak, K. Pavelic, M. Kralj and G. Karminski-Zamola, *J. Med. Chem.*, **52**, 2482 (2009).
09JMC4543	L. Aurelio, C. Valant, B.L. Flynn, P.M. Sexton, A. Christopoulos and P.J. Seammells, *J. Med. Chem.*, **52**, 4543 (2009).
09JMC4586	H.-G. Hacker, S. Leyers, J. Wiendlocha, M. Gutschow and M. Wiese, *J. Med. Chem.*, **52**, 4586 (2009).
09JMC5076	M. Prat, D. Fernandez, M.A. Buil, M. Crespo, G. Casals, M. Ferrer, L. Tort, J. Castro, J.M. Monleon, A. Gavalda, M. Miralpeix, I. Ramos, T. Dome-nech, D. Vilella, F. Anton, J.M. Huerta, S. Espinosa, M. Lopez, S. Sentellas, M. Gonzales, J. Albert, V. Segarra, A. Cardenas, J. Beleta and H. Ryder, *J. Med. Chem.*, **52**, 5076 (2009).
09JMC6205	S. Lethu, M. Ginisty, D. Bose and J. Dubois, *J. Med. Chem.*, **52**, 6205 (2009).
09JMC6621	Z.-F. Tao, L.A. Hasvold, J.D. Leverson, E.K. Han, R. Guan, E.F. Johnson, V.S. Stoll, K.D. Stewart, G. Stamper, N. Son, J.J. Bouska, Y. Luo, T.J. Sowin, N.-H. Lin, V.S. Giranda, S.H. Rosenberg and T.D. Penning, *J. Med. Chem.*, **52**, 6621 (2009).
09JMC6724	E. Bey, S. Marchais-Oberwinkler, M. Negri, P. Kruchten, A. Oster, T. Klein, A. Spadaro, R. Werth, M. Frotscher, B. Birk and R.W. Hatmann, *J. Med. Chem.*, **52**, 6724 (2009).
09JMC6782	J.M. Ontoria, S. Altamura, A.D. Marco, F. Ferrerigno, R. Laufer, E. Miraglia, M.C. Palumbi, M. Scarpelli, C. Schultz-Faderecht, S. Serafini, C. Steinkuehler and P. Jones, *J. Med. Chem.*, **52**, 6782 (2009).

09NJC196	S. Hayashi, W. Nakanishi, A. Furuta, J. Drabowicz, T. Sasamon and N. Tokitch, *New. J. Chem.*, **333**, 196 (2009).
09JOC408	C. Li, J. Shi, L. Xu, Y. Wang, Y. Cheng and H. Wang, *J. Org. Chem.*, **74**, 408 (2009).
09JOC820	K. Geramita, J. McBee and T.D. Tilley, *J. Org. Chem.*, **74**, 820 (2009).
09JOC973	G.A. Molander, B. Canturk and L.E. Kennedy, *J. Org. Chem.*, **74**, 973 (2009).
09JOC1141	S. Mehta, J.P. Waldo and R.C. Larock, *J. Org. Chem.*, **74**, 1141 (2009).
09JOC1809	N.-W. Tseng and M. Lautens, *J. Org. Chem.*, **74**, 1809 (2009).
09JOC1826	B. Liegault, D. Lapointe, L. Caron, A. Vlassova and K. Fagnou, *J. Org. Chem.*, **74**, 1826 (2009).
09JOC2242	A. Zulauf, M. Mellah and E. Schultz, *J. Org. Chem.*, **74**, 2242 (2009).
09JOC4278	A.P. Gaywood and H. McNab, *J. Org. Chem.*, **74**, 4278 (2009).
09JOC4747	D. Stadler and T. Bach, *J. Org. Chem.*, **74**, 4747 (2009).
09JOC5496	P.P. Singh, A.K. Yadav, H. Ila and H. Junjappa, *J. Org. Chem.*, **74**, 5496 (2009).
09JOC9105	A. Rajea, M. Miyasaka, S. Ziao, P.J. Boratynski, M. Pink and S. Rajea, *J. Org. Chem.*, **74**, 9105 (2009).
09JOC9112	J.T. Henssler, Z. Zhang and A.J. Matzger, *J. Org. Chem.*, **74**, 9112 (2009).
09JOC9188	Z. Chen, N.R. de Tacconi and R.L. Elsenbaumer, *J. Org. Chem.*, **74**, 9188 (2009).
09JPC7469	R. Li, X. Lu, D. Shi, D. Zhou, Y. Cheng, G. Zhang and P. Wang, *J. Phys. Chem.*, **113**, 7469 (2009).
09JPS(A)5044	W.-C. Yen, B. Pal, I.-S. Yang, Y.C. Hung, S.T. Lin, C.Y. Chao and W.-F. Su, *J. Polym. Sci.*, **47**, 5044 (2009).
09JST100	S. Pu, M. Li, C. Fan and H.L.L. Shen, *J. Mol. Struct.*, **919**, 100 (2009).
09MO1013	T. Thiemann, Y. Tanaka and J. Iniesta, *Molecules*, **14**, 1013 (2009).
09OBC1761	M. Brichacek and J.T. Njardarson, *Org. Biomol. Chem.*, **7**, 1761 (2009).
09OBC3264	J.-H. Zhang and V. Cheng, *Org. Biomol. Chem.*, **7**, 3264 (2009).
09OL445	F.M. Piller and P. Knochel, *Org. Lett.*, **11**, 445 (2009).
09OL955	W.S. Jenks, M.J. Heying and E.M. Rockafellow, *Org. Lett.*, **11**, 955 (2009).
09OL1475	H. Nakagawa, S. Kawai, T. Nakashima and T. Kawai, *Org. Lett.*, **11**, 1475 (2009).
09OL1487	A. Patra, Y.H. Wijsboom, G. Leitus and M. Bendikov, *Org. Lett.*, **11**, 1487 (2009).
09OL1837	M. Mosrin and P. Knochel, *Org. Lett.*, **11**, 1837 (2009).
09OL1861	N.O. Devarie-Baez, W.-S. Kim, A.B. Smith, III and M. Xian, *Org. Lett.*, **11**, 1861 (2009).
09OL2461	S. Zhou, K. Junge, D. Addis, S. Das and M. Beller, *Org. Lett.*, **11**, 2461 (2009).
09OL2473	T. Kashiki, S. Shinamura, M. Kohara, E. Miyazaki, K. Takimiya, M. Ikeda and H. Kuwabara, *Org. Lett.*, **11**, 2473 (2009).
09OL2563	J. Wang, K. Liu, Y.-Y. Liu, C.-L. Song, Z.-F. Shi, J.-B. Peng, H.-L. Zhang and X.P. Cao, *Org. Lett.*, **11**, 2563 (2009).
09OL3144	J.T. Henssler and A.J. Matzger, *Org. Lett.*, **11**, 3144 (2009).
09OL3230	T. Taerum, O. Lukoyanoa, R.G. Wylie and D.F. Perepichka, *Org. Lett.*, **11**, 3230 (2009).
09OL3358	G. Balaji and S. Valiyaveettil, *Org. Lett.*, **11**, 3358 (2009).
09OL3386	D.B. Biradar, S. Zhou and H.-M. Gau, *Org. Lett.*, **11**, 3386 (2009).
09OL3502	L. Shi, Y. Chu, P. Knochel and H. Mayr, *Org. Lett.*, **11**, 3502 (2009).
09OL3702	Y. Ni, K. Nakajima and K.-I.T. Takahashi, *Org. Lett.*, **11**, 3702 (2009).
09OL4160	D. Lapointe and K. Fagnou, *Org. Lett.*, **11**, 4160 (2009).
09OL4200	A.J. Smith, L.K. Abbott and S.F. Martin, *Org. Lett.*, **11**, 4200 (2009).
09OL4636	J. Majer, P. Kwiatkowski and J. Jurczak, *Org. Lett.*, **11**, 4636 (2009).
09OL4560	O. Rene, D. Lapointe and K. Fagnou, *Org. Lett.*, **11**, 4560 (2009).
09OL4624	A. Aponick, C.-Y. Li, J. Malinge and E.F. Marques, *Org. Lett.*, **11**, 4624 (2009).
09OL4906	A.B. Cuenca, S. Montserrat, K.M. Hossain, G. Mancha, A. Lied, M. Medio-Simon, G. Ujaque and G. Asensio, *Org. Lett.*, **11**, 4906 (2009).
09OL5098	S. Schmid, E. Mena-Osteritz, A. Kopyshev and P. Bauerle, *Org. Lett.*, **11**, 5098 (2009).
09OM3966	P.R. Likhar, S.M. Salian, S. Roy, M.L. Kantam, B. Sridhar, K.V. Mohan and B. Jagadeesh, *Organometallics*, **28**, 3966 (2009).

09OM5412	A.P. Ferreira, J.J. Ferreira da Silva, M.T. Duarte, M. Fatima Minas da Piedade, M.P. Robalo, S.G. Harjivan, C. Marzano, V. Gandin and M.M. Marques, *Organometallics*, **28**, 5412 (2009).
09OS360	M. Storgaard and J.A. Ellman, *Org. Synth.*, **86**, 360 (2009).
09PBB419	E.A. Wilhelm, C.R. Jess, C. Bortolatta, C.W. Nogueira and L. Savengnago, *Pharmacol. Biochem. Behav.*, **93**, 419 (2009).
09POL621	M.H. Garcia, P. Florindo, M. de Fatima, M. Piiedade and S. Maiorana, *Polyhedron*, **28**, 621 (2009).
09PS220	D.S. Badaway, E. Abdel-Galil, E.M. Kandeel, W.M. Basyouni, K.A.M. El-Bayouki and T.K. Khatab, *Phosphorus, Sulfur, Silicon, Relat. Elem.*, **184**, 220 (2009).
09PS705	A.R. Hajipour, L. Khazdooz and A.E. Ruoho, *Phosphorus, Sulfur, Silicon, Relat. Elem.*, **184**, 705 (2009).
09PS1115	J.M. Lochowski, P. Potaczek, F.F. Abdel-Latif and M.A. Ameen, *Phosphorus, Sulfur, Silicon, Relat. Elem.*, **184**, 1115 (2009).
09PS1161	G. Buehrdel, E. Petrlikova, P. Herzigova, R. Beckert and H. Goerls, *Phosphorus, Sulfur, Silicon, Relat. Elem.*, **184**, 1161 (2009).
09PS2078	R.M. Mohareb, S.M. Sherif and H.E.-D. Moustafa, *Phosphorus, Sulfur, Silicon, Relat. Elem.*, **184**, 2078 (2009).
09S3214	D.P. Sweat and C.E. Stephens, *Synthesis*, **19**, 3214 (2009).
09SC332	V.S.P.R. Lingham, R. Vinodkumar, K. Mukkanti, A. Thomas and B. Gopalan, *Synth. Commun.*, **39**, 332 (2009).
09SC1511	I. Linzaga, J. Escalanta, M.W. Nicho and M. Guizado-Rodriquez, *Synth. Commun.*, **39**, 1511 (2009).
09SC1781	G.H. Elgemeie, S.H. Elsayed and A.S. Hassan, *Synth. Commun.*, **39**, 1781 (2009).
09SC3315	M. Sonoda, S. Kinoshita, T. Luu, H. Fukuda, K. Miki, R. Umeda and Y. Tobe, *Synth. Commun.*, **39**, 3315 (2009).
09SM361	Y.A. Udum, S. Tarkuc and L. Toppare, *Synth. Met.*, **159**, 361 (2009).
09T822	N.S. Kumar, J.A. Clement and A.K. Mohanakrishnan, *Tetrahedron*, **65**, 822 (2009).
09T2079	C. Herbivo, A. Comel, G. Kirsch and M.M.M. Raposo, *Tetrahedron*, **65**, 2079 (2009).
09T2430	K. Kobayashi, M. Horiuchi, S. Fukamachi and H. Konishi, *Tetrahedron*, **65**, 2430 (2009).
09T2436	Z. Tu, R. Raju, T.-R. Liou, V. Kavala, C.-W. Kuo and Y. Jang, *Tetrahedron*, **65**, 2436 (2009).
09T7761	M.M. Heravi and S. Sadjadi, *Tetrahedron*, **65**, 7761 (2009).
09T10269	F. Bellina and R. Rossi, *Tetrahedron*, **65**, 10269 (2009).
09T10453	D. Thomae, E. Perspicace, D. Henryon, Z. Xu, S. Schneider, S. Hesse, G. Kirsch and P. Seck, *Tetrahedron*, **65**, 10453 (2009).
09T10797	T. Dohi, M. Iko, N. Yamaoka, K. Morimoto and H. Fujioka, *Tetrahedron*, **65**, 10797 (2009).
09THC17	A. Rais, H. Ankati and E.R. Biehl, *Trends Heterocycl. Chem.*, **14**, 17 (2009).
09TL1097	I.F. Cottrell, M.G. Moloney and K. Smithies, *Tetrahedron Lett.*, **50**, 1097 (2009).
09TL1180	M. Kinhara, J. Yamamoto, T. Naguchi and Y. Hirai, *Tetrahedron Lett.*, **50**, 1180 (2009).
09TL1381	S. Suwabe, A. Okuhara, T. Sugahara, K. Suzuki, K. Kunimasa, T. Nakajima, Y. Kumafuji, Y. Osawa, T. Yoshimura and H. Morita, *Tetrahedron Lett.*, **50**, 1381 (2009).
09TL1407	R.G. de Moronha, A.C. Fernandes and C.C. Romao, *Tetrahedron Lett.*, **50**, 1407 (2009).
09TL1588	H. Zheng, W. Zhou, M. Yuan, X. Yin, Z. Zuo, C. Ouyang, H. Liu, Y. Li and D. Zhu, *Tetrahedron Lett.*, **50**, 1588 (2009).
09TL1614	S. Hermes, G. Dassa, G. Toso, C. Bianco, G. Bertarelli and Zerbi, *Tetrahedron Lett.*, **50**, 1614 (2009).
09TL2758	J. Zhao, L. Huang, K. Cheng and Y. Zhang, *Tetrahedron Lett.*, **50**, 2758 (2009).
09TL2874	Q. Dang, E. Carruli, F. Tian, F.W. Dang, T. Gibson, W. Li, H. Bai, M. Chunc and S.J. Hecker, *Tetrahedron Lett.*, **50**, 2874 (2009).
09TL2778	J.J. Dong, J. Roger and H. Doucet, *Tetrahedron Lett.*, **50**, 2778 (2009).

09TL3750	X. Chen and R.L. Elsenbaumer, *Tetrahedron Lett.*, **50**, 3750 (2009).
09TL4289	H. Al-Saralerh, L.N. Dawe and P.E. Georghiou, *Tetrahedron Lett.*, **50**, 4289 (2009).
09TL4509	A. Tsuge, T. Hara and T. Moriguchi, *Tetrahedron Lett.*, **50**, 4509 (2009).
09TL4670	Z. Puterova, A. Andicsova, J. Moncol, C. Rabong and D. Vegh, *Tetrahedron Lett.*, **50**, 4670 (2009).
09TL4962	S.-M.T. Toguem, M. Hussain, I. Malik and A. Villinger, *Tetrahedron Lett.*, **50**, 4962 (2009).
09TL5005	J. Chao, A.G. Taveras and C.J. Aki, *Tetrahedron Lett.*, **50**, 5005 (2009).
09TL5288	Y.-C. Jeong, C.G. Gao, I.S. Lee, S.I. Yang and K.-H. Ahn, *Tetrahedron Lett.*, **50**, 5288 (2009).
09TL5467	R. Perez-Pineiro, S. Dai, R. Alvarez-Puebla, J. Wigginton, B.J. Al-Hourani and H. Fenniri, *Tetrahedron Lett.*, **50**, 5467 (2009).
09TL6191	K. Balamurugan, S. Perumal, A.S.K. Reddy, P. Yogeeswari and D. Sriram, *Tetrahedron Lett.*, **50**, 6191 (2009).
09TL6480	H.S. Lee, S.H. Kim and J.N. Kim, *Tetrahedron Lett.*, **50**, 6480 (2009).
09TL6562	H.B. Borate, S.P. Sawargave and S.R. Maujan, *Tetrahedron Lett.*, **50**, 6562 (2009).
09TL6651	K. Shimad, Y. Takata, Y. Osaki, A. More-Oka, H. Kogawa, M. Sakuraba, S. Aoyagi, Y. Takichi and S. Ogawa, *Tetrahedron Lett.*, **50**, 6651 (2009).
09TL6897	M. Shao and Y. Zhao, *Tetrahedron Lett.*, **50**, 6897 (2009).
09TL6941	P. Besada, M. Perez, G. Gomez and Y. Fall, *Tetrahedron Lett.*, **50**, 6941 (2009).

CHAPTER 5.2

Five-Membered Ring Systems: Pyrroles and Benzo Analogs

Jonathon S. Russel*, Erin T. Pelkey**, Sarah J.P. Yoon-Miller**
*St. Norbert College, De Pere, WI, 54115, USA
Jonathon.Russel@snc.edu
**Hobart and William Smith Colleges, Geneva, NY, 14456, USA
pelkey@hws.edu

5.2.1. INTRODUCTION

The synthesis and chemistry of pyrroles, indoles, and fused ring systems reported during 2009 are included in this monograph. Pyrroles and especially indoles continue to draw a lot of attention from the scientific community due to their prevalence in natural products and wide range of biological and materials science applications. Pyrroles (by *ETP* and *SY-M*) and indoles (by *JSR*) are treated in separate sections. New this year, coverage of the synthesis of pyrrole and indole natural products are limited to examples that involve the development or improvement of synthetic methodology related to ring synthesis and/or substitution. Review articles and monographs from 2009 will be mentioned in the relevant sections.

5.2.2. SYNTHESIS OF PYRROLES

De novo pyrrole syntheses have been organized systematically into intramolecular and intermolecular approaches as well as by the location of the new bonds that describe the pyrrole ring-forming step (two examples illustrated below). Multi-component reactions leading to pyrroles and pyrrole syntheses by transformation of other heterocycles appear at the end of this section. Njardarson has written a review article that details synthetic approaches to 2,5-dihydropyrroles (and structurally related heterocycles) <09OBC1761>.

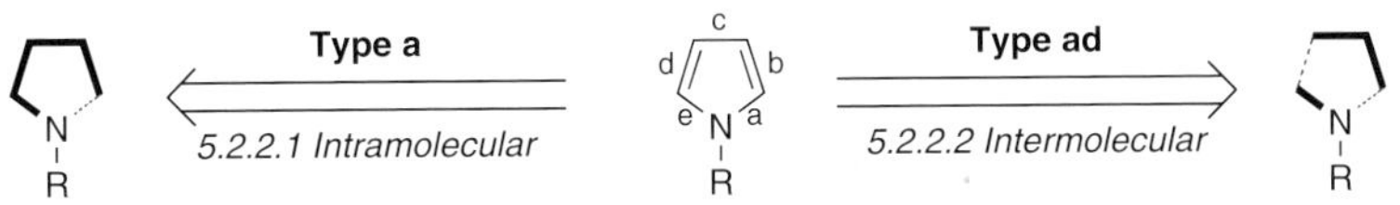

5.2.2.1 Intramolecular Approaches to Pyrroles

Intramolecular Type a

'Type a' approaches to pyrroles mainly fall into three general categories: (1) cyclo-condensation reactions of γ-aminoketones; (2) metal-mediated 5-*endo* cyclizations of homopropargyl amines; and (3) metal-mediated 5-*exo* cyclizations of γ-aminoalkynes.

de Kimpe reported a gold-catalyzed synthesis of β-fluoropyrroles <09OL2920>. Treatment of sulfonamides **1** with AuCl$_3$ led to fluoropyrroles **2** via a 5-*endo*-cyclization and subsequent dehydrofluorination. Neither silver(I) salts nor DBU promoted this cyclization. Similar gold-catalyzed syntheses of 2-arylpyrroles were reported later by Aponick <09OL4624> and Akai <09OL5002>. They each employed dehydrative 5-*endo* cyclizations of sulfonamide-substituted propargylic alcohols. The Knölker group observed successful 5-*endo*-cyclizations of *N*-sulfon-amide-protected homopropargylamines mediated by a silver(I) catalyst <09AG(I)8042>. The products of these cyclizations, 2-aryl-2,3-dihydropyrroles, were then converted into 2-arylpyrroles by elimination of the sulfonamide group induced by *t*-BuOK. These 2-arylpyrroles served as building blocks for the preparation of pentahalopseudilins, inhibitors of myosin ATPase. Dembinski and co-workers developed a novel route to 2,5-disubstituted pyrroles via a zinc chloride-catalyzed 5-*endo* cyclization of homopropargyl azides <09T1268>.

Prandi and co-workers reported the synthesis of β-alkoxypyrroles via a palladium-catalyzed cyclization of (*E*)-alkoxydienylamines <09OL3914>. Davies reported a gold-catalyzed ring expansion of alkynyl aziridines **3** that gave pyrroles <09OL2293>. The regiochemical outcome of the cyclization was directed by the counterion. Treatment of **3** with PPh$_3$AuOTs catalyst system gave 2,5-disubstituted pyrrole **4** while the same reaction with the corresponding triflate system gave the rearranged 2,4-disubstituted pyrrole **5**. Additional studies involving similar ring expansions were reported by the research groups of Tu <09OL4002>, Hou <09TL6944>, and Yoshida <09S2454, 09TL6268>.

Xu investigated a novel approach to coumarin-fused pyrroles (*e.g.*, **6**) that employed a palladium-catalyzed 5-*endo*-cyclization of 4-alkynyl-3-aminocoumarins

<09ASC2005>. The pyrrolocoumarin ring system is found in a number of the lamellarin natural products. Tehrani and co-workers prepared unsymmetrical benzo [*f*]isoindole-4,9-diones (*e.g.*, **7**) via an intramoleculear cyclization (and oxidation) of 3-(aminomethyl)-2-(bromomethyl)-1,4-naphthoquinones <09EJO4882>. The latter were prepared using a Kochi-Anderson oxidative decarboxylation method involving 1,4-naphthoquinones and *N*-trifluoroacetyl-α-amino acids.

Shin and co-workers prepared isoindoles (*e.g.*, **8**) via a gold-catalyzed 5-*endo*-cyclization of *o*-alkynylaryl (*Z*)-ketoximes <09OBC4744>. Interestingly, the stereochemistry of the oxime changed the outcome of the reaction, as exposing the corresponding (*E*)-ketoximes to the same reaction conditions produced isoquinoline-*N*-oxides.

Intramolecular Type b

Solé prepared isoindole-1-carboxylates (*e.g.*, **9**) via an intramolecular α-arylation of amino esters <09OBC3382>. Due to the relative instability of these isoindoles, the crude reaction mixtures were treated directly with dienophiles, which gave stable, bridged cycloadducts. McNab and co-workers produced *N*-unsubstituted 3-hydroxypyrroles by the flash vacuum pyrolysis of amino ester derivatives that incorporated Meldrum's acid <09S2531>. They were able to generate the parent, 3-hydroxypyrrole (**10**), utilizing this method <09S2535>.

Intramolecular Type c

Two established methods for preparing pyrroles via 'type c' processes include the intramolecular cyclization of enaminones and the ring-closing metathesis (RCM) of 4-aza-1,6-dienes (and subsequent oxidation). Langer prepared suitably substituted enaminones via a regioselective Staudinger reaction between 1,3-diketones and azides <09S227>. Chattopadhyay and co-workers prepared pyrrole-substituted derivatives of amino acids using an RCM approach <09TA1719>.

5.2.2.2 Intermolecular Approaches to Pyrroles

Intermolecular Type ab

Three different approaches to pyrroles from α,β-unsaturated imines **11** have appeared in the last year. First, Iwasawa reported a synthesis of 1,2-disubstituted pyrroles **12** by treatment of **11** with terminal alkynes in the presence of a rhodium(I) catalyst <09AG(I)8318>. A mechanism was proposed that included a formal [4+1] cycloaddition of a rhodium vinylidene complex with **11** followed by desilylation. Second, Zhu reported a conceptually similar [4+1] cycloaddition approach to pyrroles involving isocyanides in the place of alkynes <09OL1555>. Treatment of **11** with isocyanides in the presence of AlCl$_3$ gave highly functionalized 2-amino-5-cyanopyrroles **13** in good yields. Finally, Arndtsen disclosed a novel phosphine-mediated approach to complex pyrroles <09OL1369>. Mixing **11** with acid chlorides and PPh$_3$ in the presence of DBU led to pyrroles **14**. The proposed mechanism includes an intramolecular Wittig-like reaction involving an amide carbonyl.

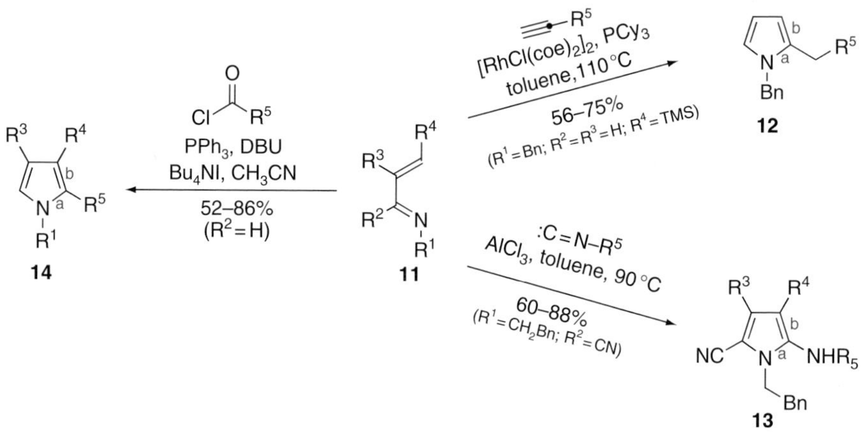

Müller reported a palladium/copper-mediated synthesis of β-iodopyrroles via a coupling-addition-cyclocondensation sequence involving propargyl amines and acid chlorides and sodium iodide <09OL2269>.

Intermolecular Type ac

Several established pyrrole name reactions involve 'type ac' cyclizations including: Knorr (α-aminocarbonyls + 1,3-dicarbonyls), Trofimov (oximes + alkynes) <09CEJ5823>, and Piloty-Robinson (hydrazines + aldehydes/ketones). Komeyama and co-workers developed a variation on the Knorr pyrrole synthesis that incorporated propargylamines instead of α-aminocarbonyls <09CL224>. Treatment of propargylamines with 1,3-dicarbonyls in the presence of a bismuth(III) catalyst gave highly substituted pyrroles.

Several variations on the Piloty-Robinson pyrrole synthesis have appeared during the past year leading to symmetrically substituted pyrroles **17**. Takamura heated hydrazones **18** in the presence of SOCl$_2$ and methanol to produce **17** via intermediate **15** <09TL4762>. This methodology proved useful for completing formal syntheses of polycitones and storniamide. Ciez generated diimine intermediates **16** by the titanium-mediated homocoupling of azido esters **19** <09OL4282>. Yu and co-workers generated intermediates **16** via the homocoupling of enamine esters **20** <09ASC2063, 09SL2529>.

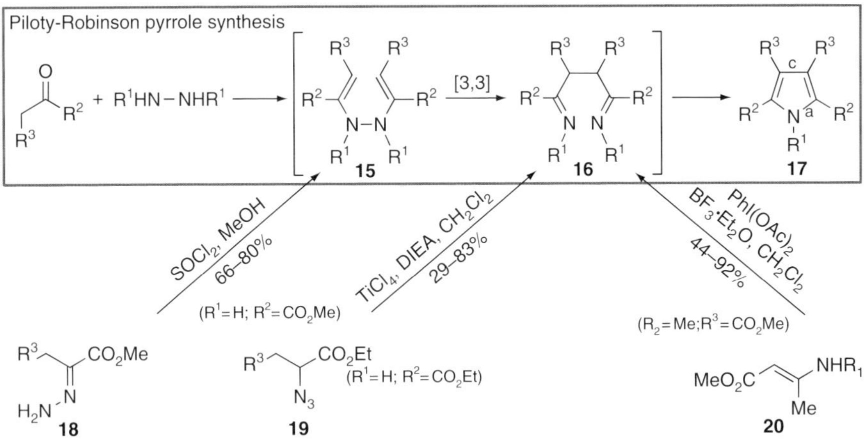

Zhong and co-workers reported an approach to N-hydroxypyrroles via a reaction sequence related to the Knorr pyrrole synthesis <09AG(I)758>. Their methodology involved a domino reaction of α-carbonyl oximes with α,β-unsaturated aldehydes. Mantellini and co-workers produced pyrrole-2,3-dicarboxylates via the cyclocondensation of α-aminohydrazones with dimethyl acetylenedicarboxylate <09ASC715>.

Intermolecular Type ad

Opatz and co-workers developed a novel route to 2,3,5-triarylpyrroles **23** (and 2,3,5-substituted pyrroles with alkyl and aryl substituents) utilizing a cyclocondensation that took advantage of the chemistry of the nitrile functional group <09JOC8243>. Treatment of α-aminonitriles **21** and enones **22** with $TiCl_4$ gave **23**. The mechanism included imine formation, deprotonation, electrocyclization, and loss of the nitrile group.

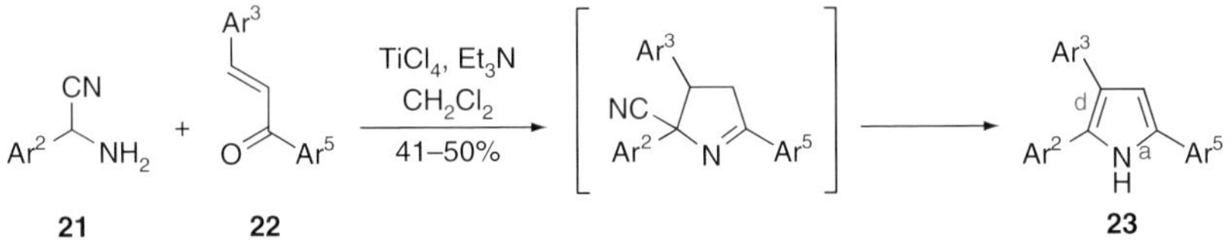

Intermolecular Type ae

Cyclocondensation reactions between primary amines and 1,4-dicarbonyl compounds (Paal-Knorr) or 2,5-dialkoxytetrahydrofurans (Clauson-Kaas) are the most utilized de novo pyrrole syntheses. Selected recent examples that demonstrate the continued versatility of the Paal-Knorr reaction include syntheses of the lamellarins <09T4283>, prodigiosin R1 <09JA14579>, agelastin A <09AG(I)3802>, anti-Alzheimer drug candidates <09JMC3377>, and pyrrole-modified proteins <09BMC7548>. In a Paal-Knorr approach to 2,5-dimethyl-N-tosylpyrroles, Gryko found that the bis-acetal derived from 2,5-hexanedione was necessary for the reaction to proceed <09S1147>. Williams and co-workers employed an Ru-catalyzed isomerization of alkyne-1,4-diols **24** to 1,4-diketones for the synthesis of furans, pyridazines, and pyrroles <09T8981>. For pyrroles, they treated **24** with a Ru catalyst system followed by primary amines giving 1,2,5-trisubstituted pyrroles **25**. Recent advances involving the Clauson-Kaas reaction include: Fe catalyst <09SL2245>, Bi catalyst under microwave irradiation <09TL5445>, and microwave irradiation with no catalyst <09TL4807>.

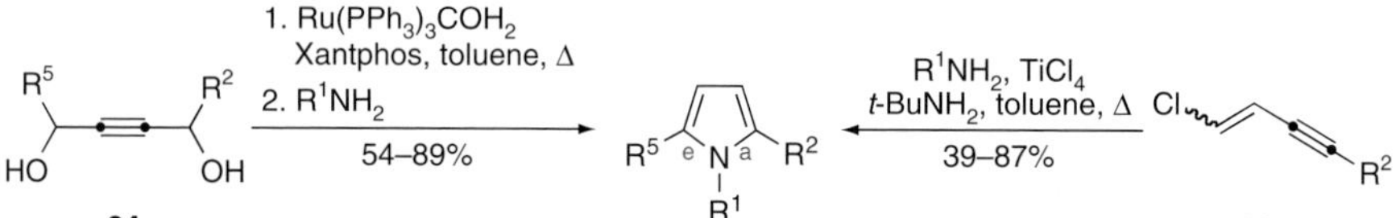

Several metal-mediated 'type ae' cyclization strategies that incorporate hydroaminations and/or cross-coupling reactions have been reported. Ackermann reported the synthesis of 1,2-disubstituted pyrroles **27** via a titanium-catalyzed hydroamination/cyclization of chloroenynes **26** with primary amines <09OL2031>. Chloro-substituted propargyl alcohols could also be utilized as coupling partners. On a similar note, Li and co-workers reported further progress in the double copper-mediated cross-coupling of 1,4-diiodo-1,3-butadienes with primary amides, which gave N-acylated pyrroles <09T8961>.

Liu and co-workers utilized a domino reaction that combined an amination and intramolecular hydroamination to convert enynols **28** into highly substituted pyrroles **29** <09ASC129>. The reaction sequence employed a gold/silver catalyst system. Similarly, Liu and co-workers developed a synthesis of **29** utilizing a palladium-catalyzed domino reaction starting with enynyl acetates **30** <09T1424>. Tejedor and co-workers reported the synthesis of chain-substituted tetra-substituted pyrroles (similar to **29**) that involved the cyclocondensation of 1,4-diynes with primary amines <09CEJ838>.

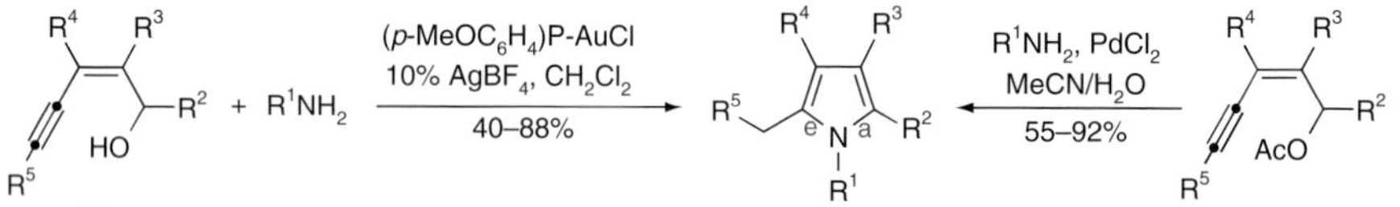

Shi reported an extension of the "methylenecyclopropane fragmentation" approach to pyrroles <JOC5983>. Heating alkylidenecyclopropylcarboxaldehydes in the presence of primary amines produced N-aminopyrroles.

Intermolecular Type bd

Several established pyrrole name reactions involve 'type bd' cyclizations including the Hinsberg (iminodiacetates + oxalates), Barton-Zard (activated isocyanides + nitroalkenes), Montforts (activated isocyanides + vinyl sulfones), van Leusen (Tos-Mic + electron-deficient alkenes), and Huisgen (1,3-dipoles + alkynes). Iwao reported an extension of his Hinsberg approach to 3,4-diarylpyrrole-2,5-dicarboxylates that included an N-benzyl protecting group <09H(77)1105>. The Montforts pyrrole synthesis was utilized by Pathak to prepare pyrrole-substituted carbohydrate derivatives <09JOC669>. The versatility of the van Leusen pyrrole synthesis was further demonstrated by the preparation of 2,2′-bipyrroles <09OL77>, 3-indolylpyrroles <09SC531>, and β-nitropyrroles <09S1485>. The de Meijere group published a complete account of their new approach to highly substituted pyrroles via the cyclocondensation of activated isocyanides and electron-deficient alkynes <09CEJ227>.

Dolbier attempted to utilize the van Leusen pyrrole synthesis to prepare SF_5-substituted pyrroles <09JOC5626>. In the event, treatment of acrylate **31** with TosMic in the presence of NaH failed to give **32**; by-product **33** was obtained instead via elimination of the SF_5 group rather than the tosyl group. The problem was solved by employing a 1,3-dipolar cycloaddition approach. Heating aziridine **35** (1,3-dipole precursor) in the presence of alkyne **34** led to 3-pyrroline **36**. Treatment of **36** with DDQ followed by triflic acid gave pyrrole **37**.

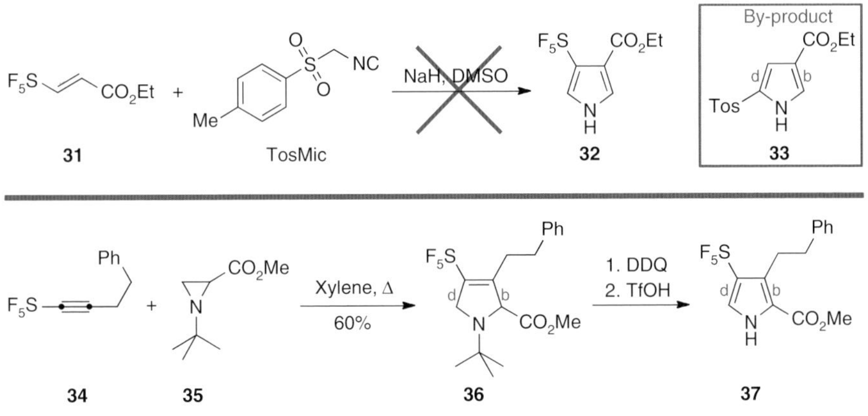

The generation of nitrogen ylides has been reported in syntheses of various fused pyrrole analogs via 1,3-dipolar cycloadditions including pyrrolo[1,2-a]quinolines <09H(78)177> and indolizines <09TL6981>. Caira and co-workers attempted to prepare pyrrolo[1,2-a]quinazolines utilizing a 1,3-dipolar cycloaddition but obtained simple N-arylpyrroles instead due to an expected ring fragmentation <09SL3336>. Finally, Kauhaluoma employed 1,3-dipolar cycloadditions of Münchnones (1,3-oxazolium-5-oxides) under microwave irradiation for the preparation of pyridyl-substituted pyrroles <09T9702>.

Intermolecular Type ace

Huang and co-workers reported an interesting palladium-catalyzed cross-coupling reaction leading to fused tricyclic and tetracyclic pyrroles that formed three new bonds ('type ace' cyclization) in the process <09ASC3118>. For example, treatment of enone **38** and azide **39** with a palladium catalyst gave tetracyclic pyrrole **40**. The authors proposed a complex, multi-step mechanism to explain this novel transformation.

Multi-component Reactions

Several three-component reaction sequences have been reported in the last year for the preparation of highly functionalized pyrroles. These reactions were classified as three-component by counting the number of components that are incorporated into the pyrrole product. Here are descriptions of the three-component reaction sequences: α-bromoacetophenone, 2,4-pentanedione, primary amines catalyzed by β-cyclodextrins gave 3-acetylpyrroles <09HCA2118>; arylglyoxals, activated acetylenes, and anilines mediated by triphenylphosphine gave pyrrole-2,3-dicarboxylates <09SL1115>; enones, aldehydes, primary amines catalyzed by thiazolium salts gave heteroaryl-substituted pyrroles <09SC3833>; 3-phenylacylideneoxindoles, β-ketoesters, ammonium acetate catalyzed by $InCl_3$ gave oxindole-substituted pyrroles <09TL3959>; and benzoylnitromethanes, activated acetylenes, and isoquinoline gave pyrrolo[2,1-*a*]isoquinolines <09T2067>.

5.2.2.3 Transformations of Heterocycles to Pyrroles

This section discusses synthetic routes to pyrroles from non-pyrrole heterocycles. As mentioned in section 5.2.2.1, the ring expansion of alkynyl-substituted aziridines to pyrroles has been well studied <09OL2293, 09OL4002, 09S2454, 09TL6268, 09TL6944>.

The Tunge group has explored a novel redox amination approach to *N*-substituted pyrroles from simple 3-pyrroline <09JA16626>. The best conditions found involved heating 3-pyrroline and aldehydes in the presence of benzoic acid. de Kimpe reported a synthesis of 3-fluoropyrroles by treating 1-pyrrolines with Selectfluor <09JOC1377>. Ohmura and co-workers developed an approach to boryl-substituted isoindoles starting from isoindolines <09JA6070>. They heated isoindolines and pinacolborane in the presence of a Pd catalyst and Me_2S to give the boryl-substituted isoindoles.

Murakami reported a novel approach to highly substituted pyrroles utilizing a nickel-catalyzed alkyne insertion reactions involving 1,2,3-triazoles <09CC1470>.

Heating N-sulfonyl-1,2,3-triazoles **41** and 5-decyne in the presence of a nickel catalyst system gave pyrroles **42** in moderate yields.

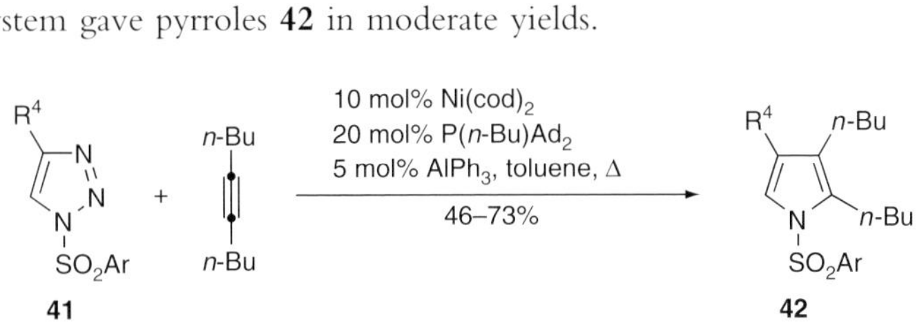

The Padwa group has developed a novel multi-step sequence that transforms furans **43** into 2,4-disubstituted pyrroles **45** via a 3-pyrrolin-2-one intermediate **44** <09OL1233, 09T6720>. The key step is the oxidative rearrangement of furanyl carbamate **43** into the 5-methoxy-3-pyrrolin-2-one **44**.

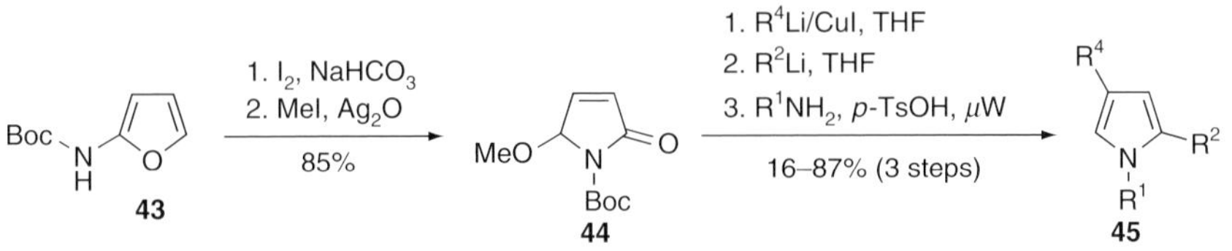

Zanatta and co-workers reported a novel approach that transformed dihydrofurans into 3-trifluoroacetylpyrroles via a ring-opening, oxidation, cyclization sequence <09SL755>.

5.2.3. REACTIONS OF PYRROLES

5.2.3.1 Substitutions at Pyrrole Nitrogen

N-substitution reactions of pyrroles have been extensively investigated. In an attempt to utilize green chemistry, Staszak and colleagues employed the less toxic dimethyl carbonate for N-methylation of pyrroles <09OPRD1199>. Le and colleagues also explored green chemistry through the use of ionic liquids as both a catalyst and solvent in the N-substitution of pyrroles <09H(78)2013>.

Another example of N-alkylation was demonstrated by Wang and colleagues in the synthesis of indolizines through a tandem S_N2 reaction/intramolecular nucleophilic addition of pyrrole-2-carboxaldehydes and ethyl γ-bromocrotonates <09H(78)725>. They found that mild conditions (strong polar aprotic solvents and carbonate bases) gave the indolizines in good yields (62-86%). N-alkylation was also studied by Trost in a palladium-catalyzed asymmetric allylic alkylation using pyrroles as nucleophiles <09CEJ6910>. The combination of Cs_2CO_3, CH_2Cl_2, and [Pd(π-C_3H_5)Cl]$_2$ gave pyrrole **49** in the highest yield (83%). This methodology was used in a synthesis of agelastatin A, a marine natural product with anticancer activity.

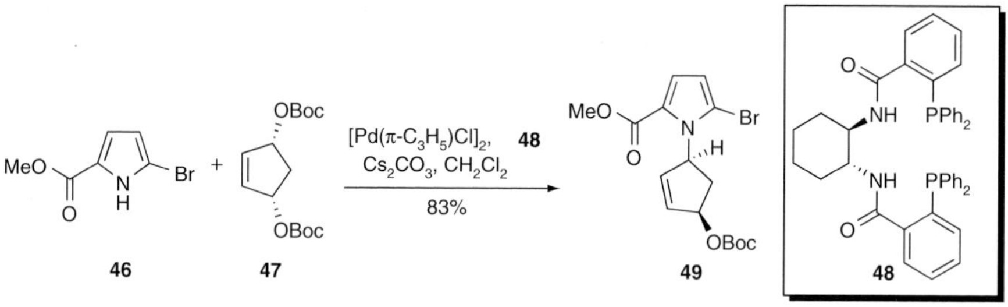

In developing an enantioselective total synthesis of the myrmicarins, complex alkaloids with bi- and tricyclic pyrroloindolizine cores, Ondrus and Movassaghi explored *N*-vinylation of substituted pyrroles with vinyl triflates <09CC4151>. A similar *N*-vinylation was studied by Xi and colleagues using a combination of CuI and ethylenediamine as a catalyst in the cross-coupling of indoles and pyrroles <09JOC6371>.

In a synthesis of agelastatin A, Davis and colleagues optimized an intramolecular Michael addition with a pyrrole substrate to form the tricyclic core in 81% yield <09SC1914>. Feldman and colleagues also explored the synthesis of marine alkaloids, specifically dibromophakellstatin, dibromophakellin, and dibromoagelaspongin <09S3162>. To form the tetracyclic skeleton of these compounds, an oxidative cyclization was employed.

5.2.3.2 Substitutions at Pyrrole Carbon
Electrophilic
Dallemagne and colleagues studied electrophilic halogenation on a tripentone with an electron-withdrawing group on the 2-position of the pyrrole <09BMC7783>. They found bromination was best achieved at C4 using NBS (80% yield), while both mono- and di-bromination occurred when Br_2 was used. Pratt and colleagues also studied regioselective electrophilic halogenation in pyrroles in their synthesis of the antifungal antibiotic, pyrrolnitrin <09OL1051>. In order to obtain 3-chloropyrrole, a regioselective bromination was first utilized, followed by lithium-halogen exchange and quenching with an electrophilic chlorine source.

Electrophilic sulfenylation has also been studied as the sulfenyl group has been shown to effectively mask the 2-position of pyrroles; it also has high utility in reactions such as acylation, nitration, and condensation with aldehydes. Thompson and colleagues optimized a reaction and workup utilizing $MgBr_2$ and NaOH, respectively <09SL112>. In the reaction, $MgBr_2$ acts to activate phthalimide **51**, while NaOH serves to remove any unreacted **51**, thus increasing the yield and allowing for purification and improved stability of the product **52**.

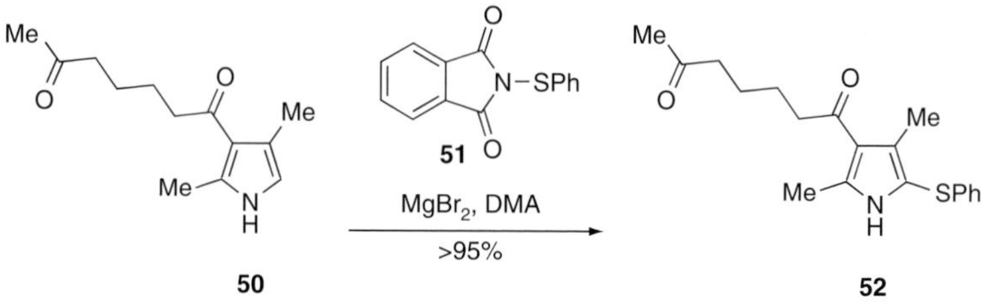

Another example of electrophilic substitution in pyrroles can be seen through acylations, specifically, the Vilsmeier-Haack reaction. In a synthesis of BODIPY dyes, Dehaen and colleagues utilized halogenation followed by an *in situ* acylation (Vilsmeier-Haack or trifluoroacylation) in THF to give acylpyrrole **54** in good yield <09CC4515>.

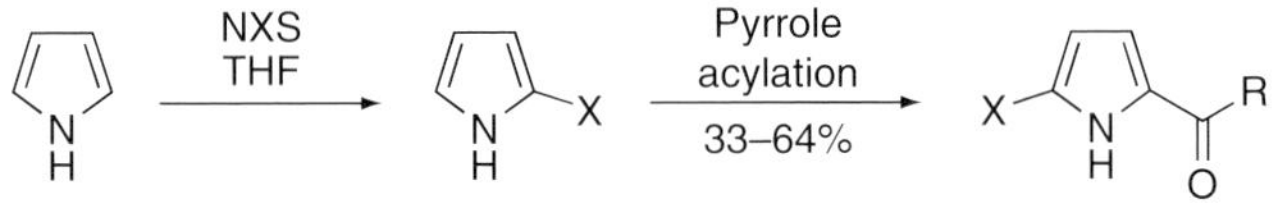

Trofimov and colleagues also studied the Vilsmeier-Haack reaction as pyrrole-2-carbaldehydes are very useful building blocks for the synthesis of pharmaceuticals and analogs of natural compounds <09S587>. They found that when formylating 1-vinylpyrroles, the substitution of oxalyl chloride for phosphorus oxychloride gave the desired products in higher yields and milder reaction conditions were possible (rt instead of -78 °C). A separate study of acylations, at both the two and three positions, was carried out by Gryko and colleagues in their synthesis of tripyrranes, building blocks for porphyrinoids <09JOC5610>. Acylation was accomplished by creating an anhydride *in situ* using 4-methylbenzoic acid and trifluoroacetic acid anhydride to form 2-acylpyrroles in 46% yield, or 4-methylbenzoyl chloride and aluminum chloride to form 3-acylpyrroles in 59% yield. Additionally, Bitar and Frontier utilized a Vilsmeier-Haack formylation to form a 2-acylpyrrole in a 92% yield in a synthesis of (±)-rosephilin <09OL49>.

To obtain 3-acylpyrrole, a Friedel-Crafts acylation was employed by Nguyen and colleagues in a regioselective acylation with *t*-butylpyrrole utilizing trifluoroacetic acid anhydride and tin chloride in toluene to give the desired product <09TL6807>. A Friedel-Crafts acylation was also used by Oshima and colleagues in a synthesis of celastramycin A <09OL1693>. Pyrrole **55** underwent acylation with **56** to give β-benzoylpyrrole **57** in a 43% yield, subsequent demethylation of **57** with BBr$_3$ completed the synthesis of the reported structure of celastramycin A **58** (the correct structure is substituted at C2).

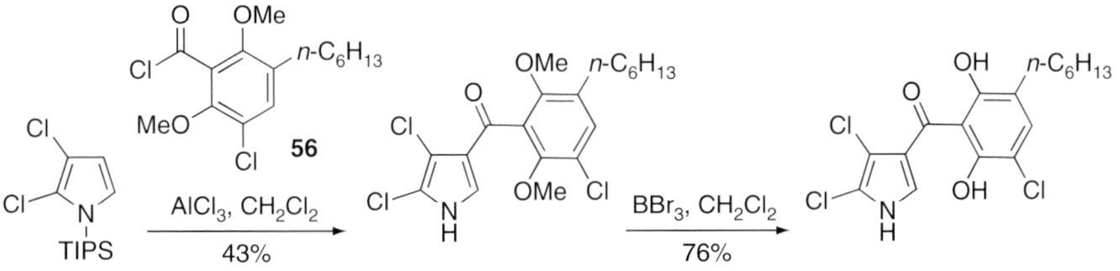

Electrophilic alkylation, specifically trifluoromethylation, has been examined in pyrroles by Nenajdenko and colleagues <09S3905>. Another study of electrophilic alkylation by Meshram and colleagues found a Friedel-Crafts alkylation using electron-deficient alkenes and zinc acetate gave the desired pyrrole products in excellent yields <09HCA1002>.

Electrophilic, Stereoselective

Catalytic asymmetric Friedel-Crafts alkylation has been widely studied, as it is one of the most direct methods for introducing a new stereogenic center on an *N*-heteroaromatic compound. Yokoyama and Arai studied this reaction utilizing pyrrole and nitroalkenes and found the highest yields to occur with a newly developed imidazoline-aminophenol ligand-CuOTf complex <09CC3285>. You and colleagues also studied asymmetric Friedel-Crafts alkylations with pyrrole and nitroalkenes **59**; however, they employed chiral phosphoric acids as catalysts, finding **60** to be the most effective in providing chiral product **61** in a 94% yield with 89% ee <09JOC6899>. In addition, catalytic asymmetric Friedel-Crafts alkylation of pyrroles and trifluoromethyl ketones has been used by Pedro and colleagues to provide pyrroles with a trifluoromethyl substituted tertiary alcohol moiety in excellent yields <09OL441>.

Organometallics

In addition to electrophilic aromatic substitution, metal-mediated cross-coupling reactions also play an important role in pyrrole transformations. Iwao and colleagues investigated Suzuki-Miyaura cross-coupling of 3,4-bis(triflates) and found that reactions with *N*-benzyl protected substrates gave the desired products in much higher yields (96-100%) than unprotected substrates (3%) <09H(77)1105>. The Suzuki-Miyaura cross-coupling reaction was also utilized by Álvarez and colleagues in a synthesis of the marine alkaloids, the dictyodendrins <09OBC860>. The dictyodendrins consist of a pyrrole fragment and an indole fragment; to synthesize the pyrrole fragment, a pyrrole boronic ester was coupled with *p*-bromoanisole to give the 3-arylpyrrole in 81% yield. The final steps of the synthesis include a Suzuki-Miyaura cross-coupling of the pyrrole and the indole fragment, followed by a 6π-electrocyclization to accomplish the final ring closure. 3-Arylpyrroles were also used in the synthesis of pyrrolnitrin by Pratt and colleagues <09OL1051>. TIPS protected pyrrole **62** was coupled with **63** in a Suzuki-Miyaura cross-coupling reaction to give 3-arylpyrrole **64** in an 89% yield, this was then deprotected to give 3-arylpyrrole **65** in a 74% yield.

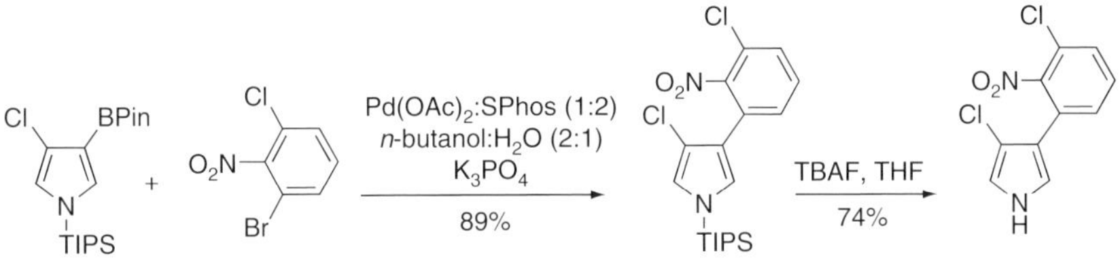

[Scheme: compound **62** (Cl, BPin pyrrole with N-TIPS) + **63** (O₂N, Cl, Br arene) → Pd(OAc)₂:SPhos (1:2), n-butanol:H₂O (2:1), K₃PO₄, 89% → **64** → TBAF, THF, 74% → **65**]

Metal-mediated cross-coupling reactions of pyrroles have also been used in the synthesis of natural products. Banwell and colleagues utilized a Suzuki-Miyaura cross-coupling reaction in their synthesis of ningalin B, a marine alkaloid isolated from *Didemnum* in Western Australia <09AJC683>. In order to obtain ningalin B, the pyrrole was coupled with the boronic acid, followed by a spontaneous lactonization with the ester residue originally present in the pyrrole. A similar synthesis was used by Iwao and colleagues in their synthesis of the marine alkaloid, lamellarin D <09JOC8143>. Syntheses for these natural products were also studied by Gupton and colleagues <09T4283>.

In addition to using palladium in cross-coupling reactions, indium has been used as a catalyst by Tsuchimoto and colleagues to introduce alkyl groups onto a β-position of pyrroles in a regioselective manner <09OL2129>. Titanium has also been used in cross-coupling reactions of pyrroles by Martin and colleagues <09OL4200>. 2-Heteroaryl titanates, including pyrrole, and a Michael acceptor were combined in the presence of a rhodium catalyst to provide the desired products in good yields.

Alkynylation has also been widely studied in pyrroles. Langer and colleagues demonstrated this in their synthesis of tetra(1-alkynyl)pyrroles using Sonogashira cross-coupling reactions <09SL838>. Tetrabromopyrrole **66** was coupled with phenylacetylene **67** in the presence of palladium and copper iodide to give pyrrole **68** in a 25% yield. Interestingly, a copper-free Sonogashira cross-coupling was studied by Pellet-Rostaing and colleagues using 3-haloindolizinones as substrates <09SL2617>.

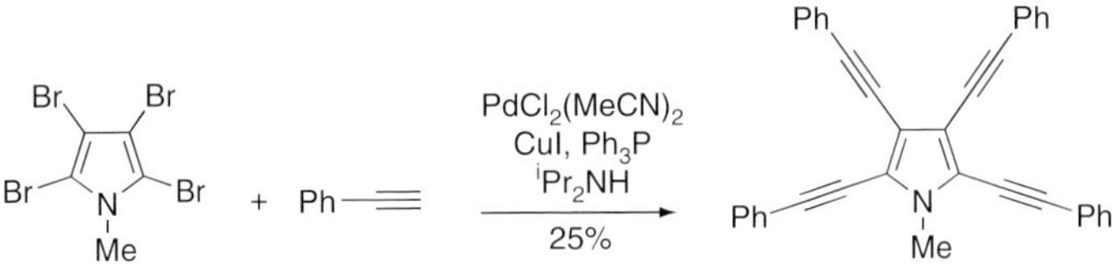

[Scheme: **66** (tetrabromo-N-methylpyrrole) + **67** (PhC≡CH) → PdCl₂(MeCN)₂, CuI, Ph₃P, ⁱPr₂NH, 25% → **68** (tetra(phenylethynyl)-N-methylpyrrole)]

C-H Activation

Arylation of pyrroles often occurs with the help of C-H activation. Gryko and colleagues demonstrated this in their study of the direct arylation of pyrroles with iodoarenes to form 2-arylpyrroles <09JOC9517>. Another similar study was carried out by Doucet and Roger in their study of regioselective C-2 or C-5 arylation using a palladium acetate catalyst <09ASC1977>. Palladium catalyzed arylation was also studied by Fagnou and colleagues <09JOC1826>. Pyrrole **69** was treated with aryl bromide **70** in the presence of palladium acetate and pivalic acid to give 2-arylpyrrole **71**; it was found that pivalic acid (*in situ* generated potassium pivalate) accelerates arylation.

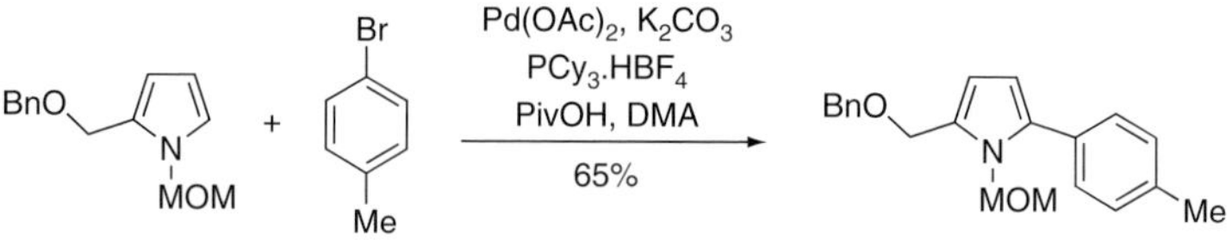

2-Heteroarylation has been studied in pyrroles by Kita and colleagues in a metal-free cross-coupling reaction; compounds such as **72** were synthesized using PhI(OH)OTs and TMSBr in HFIP in a 51% yield <09JA1668>. Pellet-Rostaing and colleagues also studied 2-heteroarylation; however, they utilized an indolizine substrate and a palladium catalyst to obtain **73** in 78 % yield <09JOC3160>. A palladium/norbornene system was employed by Catellani and colleagues in a synthesis of heteroatom-containing *o*-tetraryls **74** that involved intermolecular aryl-aryl and aryl-heteroaryl bond formation in sequence through direct C-H functionalization <09CEJ7850>.

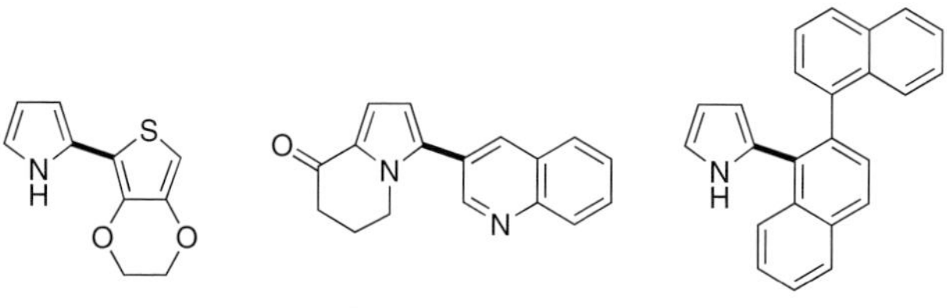

An intramolecular tandem Suzuki-Miyaura cross-coupling/direct arylation reaction of dihalovinyl pyrroles to synthesize *N*-fused heterocycles was used by Lautens and Chai <09JOC3054>. They found that the addition of water has a positive effect on this reaction. Zhang and colleagues studied palladium-catalyzed vinylation of pyrroles; they found this reaction to be highly regioselective with the use of bidentate nitrogen ligands producing only the branched α-product <09OL5606>. C-H bond functionalization was also studied by Kitamura and Oyamada in the hydroarylation of alkynes to give vinylpyrroles <09T3842>. They found platinum catalysts, such as K_2PtCl_4/AgOTf, to be more effective than palladium catalysts. Interestingly, gold catalysts were used by Waser and colleagues in the direct alkynylation of pyrroles at the two-position using a benziodoxolone-based hypervalent iodine reagent <09AG(I)9346>.

Ring Annelation

Pyrroles also undergo ring annelation. This was demonstrated by Settambolo in her synthesis of (-)-indolizidine 167B utilizing an intramolecular cyclodehydration of a

4-pyrrolylbutanal <09H(79)219>. In a synthesis of (±)-rhazinal, Trauner and Bowie also employed ring annelation through a palladium catalyzed Heck cyclization <09JOC1581>. During the key oxidative cyclization, pyrrole **75** was transformed to tetrahydroindolizine **76**, which contains the requisite quaternary stereocenter, in 69% yield.

An environmentally and economically friendly iron-catalyzed Nazarov cyclization was used by Itoh and colleagues to form fused carbocycles at the pyrrole carbon <09ASC123>. A Nazarov cyclization was also employed in a formal synthesis of (−)-rosephilin <09OL49>. Satoh, Miura, and colleagues exhibited an indole synthesis via palladium-catalyzed oxidative coupling reactions of pyrroles with alkynes <09JOC7481>. In addition to the palladium catalyst, these reactions utilized Cu $(OAc)_2$ as an oxidant, and LiOAc as a base to improve the efficiency of the reaction. Gold and platinum catalysts were used by Beller and colleagues in the formation of pyrroloazepinone derivatives through a 7-*endo*-dig cyclization <09AG(I)7212>. Other fused heterocycles, such as pyrrolopyridines, were synthesized by Groth and colleagues in a regioselective cyclocondensation utilizing AcOH <09SL456>. Pyrrolopyridine derivatives were also synthesized by Arai and Yokoyama in a highly diastereoselective Pictet-Spengler cyclization <09CC3285>.

Ring annelation can also be accomplished through a Diels-Alder reaction. This was demonstrated by Noland and colleagues in their one-pot synthesis of tetrahydroindoles from 2-alkylpyrroles, cyclic ketones, maleimides, and an acid catalyst <09JHC503>. Noland and colleagues also explored cycloadditions with 2- and 3-vinylpyrroles <09JHC1154, 09JHC1285>. Solé and Serrano reported a palladium-catalyzed intramolecular arylation of α-amino acid esters to obtain isoindoles followed by a sequential Diels-Alder reaction with alkynes to acquire the corresponding bicyclic cycloadducts <09OBC3382>.

Miscellaneous

Other reactions involving the pyrrole ring include a study by Kluger and Mundle on the mechanism of an acid catalyzed decarboxylation <09JA11674>. In addition, Le Bideau and colleagues reported an acid catalyzed rearrangement of 2,5-bis(silylated)-*N*-Boc pyrroles to their corresponding 2,4-species <09JOC8890>.

5.2.3.3 Functionalization of the Side-Chain

In addition to reactions at a pyrrole ring carbon, reactions utilizing the side-chain have also been explored. Tsuchimoto, Shirakawa, and colleagues reported an indium-catalyzed cleavage of carbon-pyrrolyl bonds to synthesize methanes with four different carbon substituents from *gem*-dipyrrolylalkanes <09EJO2437>. In these reactions, a pyrrolyl group acted as a leaving group and was replaced with a certain range of carbon nucleophiles.

Functionalization of the side-chain was also demonstrated by Dolphin and Li in their synthesis of divinylpyrroles <09EJO3562>. Diacylpyrrole **77** was regioselectively reduced at the four-position (due to the steric effect of the sulfonyl protecting group) to carbinol **78** in a 63% yield, this was then converted to 4-vinylpyrrole **79** in an 82% yield. The same steps were also utilized to obtain the divinylpyrrole in good yield. Another example of side-chain reduction was reported by Hou and Hsu in the formation of hemiaminals from 1-pyrrolyl carbamates utilizing lithium aluminum hydride <09TL7169>.

Gryko and colleagues demonstrated side-chain oxidation with 2,5-dialkylpyrroles and cerium(V)ammonium nitrate <09S1147>. They found the yield of the reaction strongly depends on the functional groups present and the details of the reaction conditions. In a synthesis of urea and carbonylurea derivatives, Ong and Liu explored pyrrole acyl azide chemistry <09T8389>. Reactions of the pyrrole acyl azide were used to introduce the urea functionality.

5.2.3.4 Transformations of Pyrroles to other Heterocycles

Lastly, pyrroles can also be transformed into other heterocycles. Pelloux-Léon and Minassian and colleagues reported an eleven-step total synthesis of penmacric acid from N-triisopropylsilylpyrrole <09OBC4512>. Penmarcric acid, a pyroglutamic acid derivative, was isolated from *Pentaclethra macrophylla*, a leguminous tree from Western Africa, and has been used as an anti-inflammatory in the traditional medicines.

5.2.4. SYNTHESIS OF INDOLES

From the modest ornamental frameworks of *Psilocybe* extracts, to the elaborate polycyclic arrays of the *Aspidosperma* alkaloids, the diverse molecular architectures that characterize the indole family of natural products continue to inspire the development of innovative methods for ring synthesis and functionalization. As an opening illustration, vinblastine **82**, an isolate of the Madagascan periwinkle (*Catharanthus roseus* (L.) G. Don) has been synthesized by Boger and co-workers <09JA4904> through tactical extension of a coupling protocol first described by the Kutney group <88T325>. Accordingly, FeCl$_3$-promoted coupling of catharanthine **80** and vindoline **81** provided the requisite all-carbon quaternary juncture of vinblastine with complete selectivity for the natural diastereomeric form <09JA4904>. Sequential treatment of the bis-indole adduct with $Fe_2(ox)_3$-$NaBH_4$ and oxygen (air) afforded vinblastine as a 2:1 mixture of epimeric alcohols with preferential formation of the desired β-alcohol.

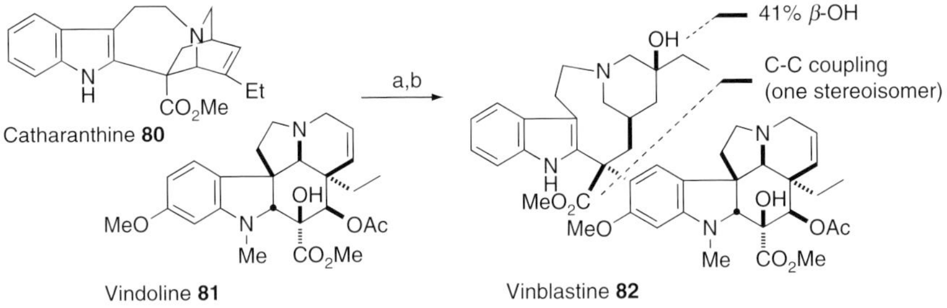

Catharanthine **80**
Vindoline **81**
Vinblastine **82**

a) $FeCl_3$, 25 °C, 2 h; b) $Fe_2(ox)_3$-$NaBH_4$, air, 0 °C

The following sections of this chapter will highlight recent activity in the general areas of indole ring construction and methodology for elaboration of the indole core. Following a listing of recent reviews, the remaining indole syntheses will be categorized utilizing a systematic approach. Intramolecular approaches (type I) and intermolecular approaches (type II) are classified by the number and location of the new bonds that describe the indole forming step (2 examples shown below). In addition, oxindoles, carbazoles, azaindoles, and carbolines will be treated separately.

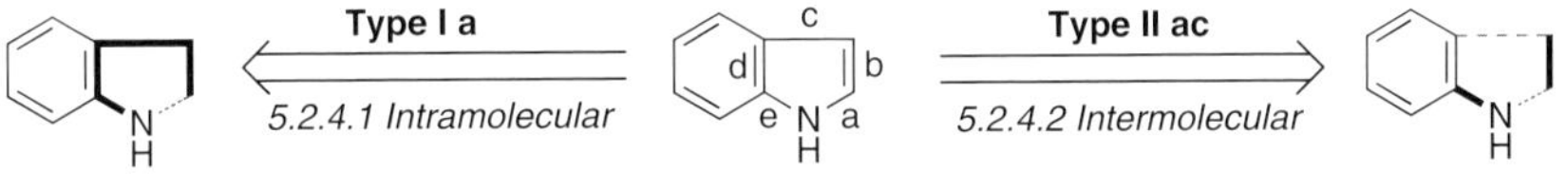

Type I a — 5.2.4.1 Intramolecular
Type II ac — 5.2.4.2 Intermolecular

A few general reviews on natural product synthesis have appeared that highlight select advances in the area of indole chemistry. In one example, The Baran group has laid out a series of guidelines for building elements of chemoselectivity into strategic plans for natural product synthesis <09ACR530>. The same group has outlined a collection of routes to complex natural products that avoid protection/deprotection sequences <09NC193>. Kim and Movassaghi have published a broad account of biomimetic strategies for alkaloid natural product synthesis <09CSR3035>.

Review articles on indole chemistry have covered a range of topics including applications of transition metal-catalysis for the arylation of indole at N, C2, or C3 <09ASC673> as well as applications of transition metals for the synthesis of biologically active carbazoles <09CL8>. Cascade reactions for indole and quinoline ring construction have been reviewed <09CAJ1037>, as have methods for the functionalization of the indole core <09AG(I)9608>. A few monographs have appeared detailing the chemistry of ethynylindoles <09CHC501>, indoles with simple and non-rearranged monoterpenoid scaffolds <09NPR803>, enantiopure 2-substituted indolines <09TA2193>, and spirooxindoles <09CSR3160, 09S3003>.

5.2.4.1 Intramolecular Approaches

Intramolecular Type Ia

Intramolecular strategies for Ia-type bond installation have relied heavily on utilization of *o*-alkynyl aniline cyclization precursors. In one study, Larock and co-workers have developed a procedure for the synthesis of 3-sulfenyl- or 3-selenylindoles <09JOC6802, 09OL173>. Phenylsulfenyl indole **83** was prepared in 87% yield via an *n*-Bu$_4$NI-promoted electrophilic cyclization of the corresponding *N,N*-dimethyl-alkynylaniline and benzenesulfenyl chloride. An alternate route to 3-sulfenylindoles has been reported by Li, Zhang, and co-workers that involved the Pd-catalyzed ring annulation of *o*-allyl anilines with disulfides <09ASC2615>. Numerous other transition-metal catalysts have been explored for promoting Ia bond installation including copper <09ASC3107, 09TL4878, 09JOC7052>, gold <09SL763, 09S311, 09OL5162>, indium <09JOC1418>, mercury <09AG(I)1244> platinum <09TL2075>, palladium <09SL1817, 09T8916>, rhodium <09OL1329>, and zinc <09TL2943>.

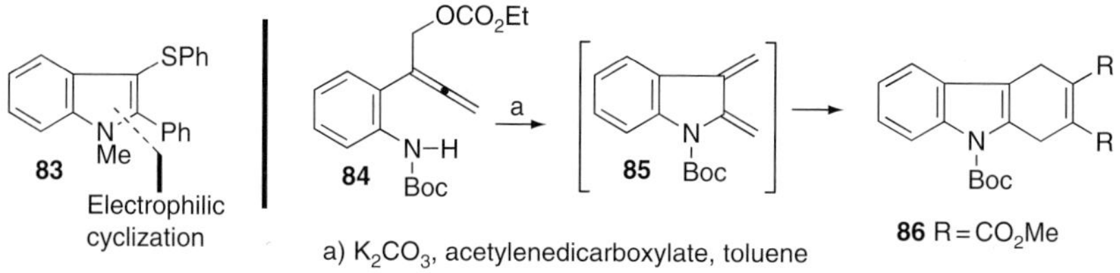

The Mukai group has employed *o*-allenyl anilines for the generation of indole 2,3-quinodimethane intermediates that were transformed to dihydro- or tetrahydrocarbazoles <09JOC6402>. A sequence involving quinodimethane formation followed by [4+2] cycloaddition onto dimethyl acetylenedicarboxylate provided the tricyclic scaffold of **86** in 93% yield. Feldman, López, and co-workers have observed the formation of 1,2 or 2,3-annulated indoles, *e.g.*, **88** and **89**, through thermal, photochemical, or photochemical/CuI-promoted cascade cyclizations of *ortho*-allenyl azide starting materials <09JOC4958>. In one example, irradiation of azide **87** in the presence of CuI afforded a 10:1 mixture of the indole tetracycles **88** and **89** in 70% overall yield <09JOC4958>.

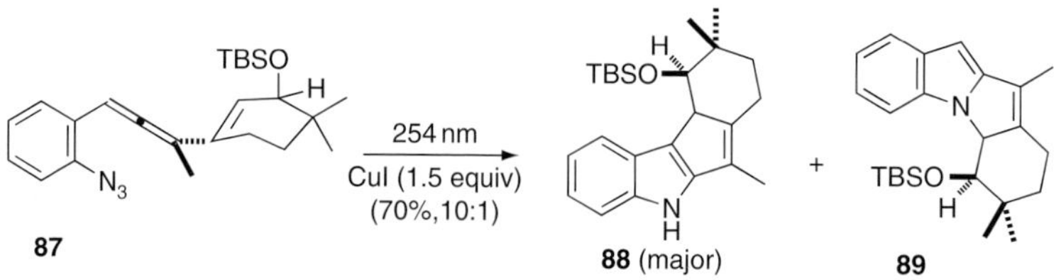

A small selection of other approaches to Ia bond installation include a gold-catalyzed cyclization of transient aryl isourea intermediates <09JOC6874>; a microwave-promoted cycloisomerization of *o*-alkynyl anilines in water that was free of acid, base, or catalyst <09TL6877>; and a Pd-catalyzed synthesis of 2-aroylindoles from *o*-gem-dibromoanilines and boronic acids <09OL4608>.

Intramolecular Type Ib

A few examples of indole Ib bond construction have appeared that involve cyclization of isothiocyanides <09S1786, 09TL3853>. In one account, Otani, Saito, and co-workers have disclosed a methodology for the preparation of thieno[2,3-*b*] indoles, e.g., **93**, by means of a triflic acid-promoted cyclization of isothiocyanide **90** with **91** <09TL3853>. Isocyanides have also been employed for Ib-type ring closure and have been applied to the synthesis of 2-alkyl- or 2-aryl-indoles <09T7523> and 1-arylindole-3-carboxylates <09H(78)161>.

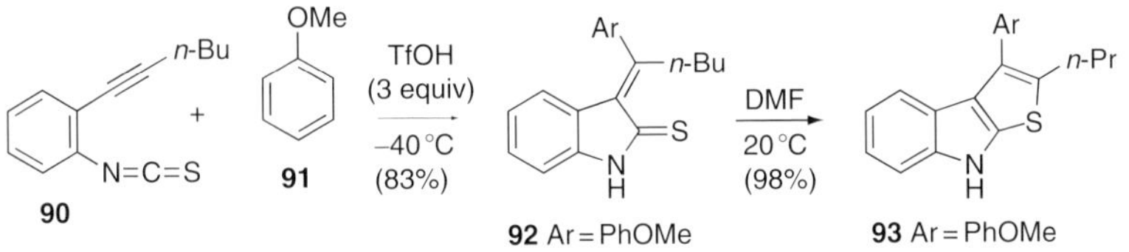

A pair of reports has appeared that detail the synthesis of indoles from aniline derived imines. In an investigation by Zhou and Doyle, Lewis acid-catalyzed nucleophilic addition of the diazoacetate side-chain of **94** to the pendant imine functionality afforded the 2,3-disubstituted indole **95** in 80% yield <09JOC9222>. Alternatively, the quinazolinone ring system was fused onto indole **97** via ring-closure of the triphenylphosphonium intermediate **96**, which, in turn, was available from condensation of the corresponding aniline/aldehyde partners <09JOC5337>.

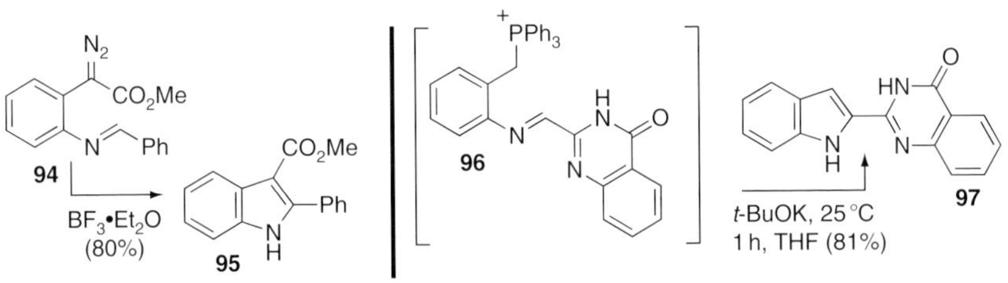

Intramolecular Type Ie

An eclectic group of tactics has been applied for promoting indole Ie-type ring closure. In one report, Laufer and co-workers have employed an intramolecular S_NAr route to *N*,3-disubstituted indoles or 3-hydroxyl indolines through displacement of fluoride by an *ortho*-tethered amine functionality <09TL1529>. A Hemetsberger-Knittel indole synthesis, a transformation involving the condensation of an aldehyde and azidoacetate, has been described by Laufer and coworkers for the synthesis of indole-2-carboxylates under microwave conditions <09TL1708>. In a final example of Ie ring closure, Hsieh and Dong have prepared 3-arylindols from aryl-tethered nitroalkenes via a sequence of nitro-group reduction/C-H bond amination using Pd-catalysis with carbon monoxide as the reducing agent <09T3062>.

Intramolecular Type Ic

A series of 2-arylindoles have been prepared by Fuwa and Sasaki by means of a Pd-catalyzed Ic-type cyclization of aniline derived *o*-bromo enecarbamates **98** <09JOC212>; the requisite enecarbamates **98** were derived from a Suzuki–Miyaura coupling of enolphosphates with boronic acids. Installation of the indole Ic bond has also been achieved by Saito and Hanzawa through an amino-Claisen rearrangement of *N*-propargylaniline starting materials <09JOC1517>. In the event, Rh(I)-catalyzed sigmatropic rearrangement of **100** afforded the *o*-allenyl aniline intermediate **101** that underwent Rh-promoted cyclization to afford dimethylindole **102** in 97% yield.

5.2.4.2 Intermolecular Approaches

Intermolecular Type IIae

The intermolecular fusion (type IIae bond construction) of *o*-iodoanilines with terminal alkynes (Sonogashira-type couplings) or internal alkynes (variations of the Larock heteroannulation) remain among the most highly investigated methodologies for the construction of indole ring systems. While analogous to many of the Ia type cyclizations of *o*-alkynyl anilines described above, the examples in this section typically involve a spontaneous cyclization step or have been developed as a one-pot protocol for annulation. In one example, Dooleweerdt, Ruhland, and Skrydstrup have described a protocol for the synthesis of 2-aminoindoles, *e.g.*, indole **103**, via Pd-catalyzed coupling of *o*-iodoanilines with ynamides <09OL221>. In the event, the alkyne coupling was followed by a spontaneous intramolecular hydroamination to close the indole ring. The Sanz group has exploited an interesting solvent effect for the preparation of either 5,7-dinitroindoles or 7-amino-5-nitroindoles through implementation of Sonogashira coupling reactions carried out in DMA (*N,N*-dimethylacetamide) or DMF, respectively <09TL4423>. For example, dinitroindole **104** was prepared in 75% yield via Pd-catalyzed coupling of 2-bromo-4,6-dinitroaniline with 1-hexyne in DMA, while formation of amino-nitroindole **105** was observed in 40% yield using DMF as the solvent; DMF derived ammonium formate has been postulated as the source of hydrogen for the nitro group reduction. In one further example, a sequence of heteroannulation and silicon-based cross-coupling has been employed by Denmark and Bair for the synthesis of 2,3-disubstituted indoles <09T3120>. Accordingly, indole-2-silanol derivative **106** was stitched together using the Larock indole synthesis in 71% yield; the silyl functionality was further manipulated to access 3-cyclopropyl-2-phenylindole in 86% yield.

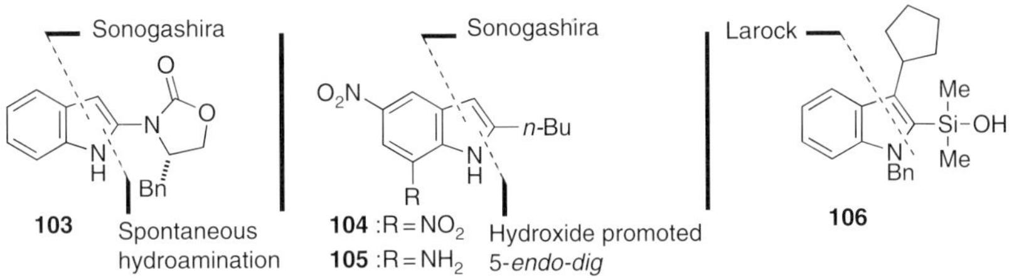

Outside the realm of alkyne coupling reactions, the Jia group has prepared a collection of tryptamine derivatives through the condensation/Pd-catalyzed cyclization of *o*-iodoanilines with *N*-protected 4-aminobutanal, *e.g.*, **107-108** <09T9075>; the methodology has been extended to the synthesis of psilocin **109** as well as 6,7-dimethoxy-1-methyltryptamine, an intermediate employed in Corey's aspidophytine synthesis <99JA6771>.

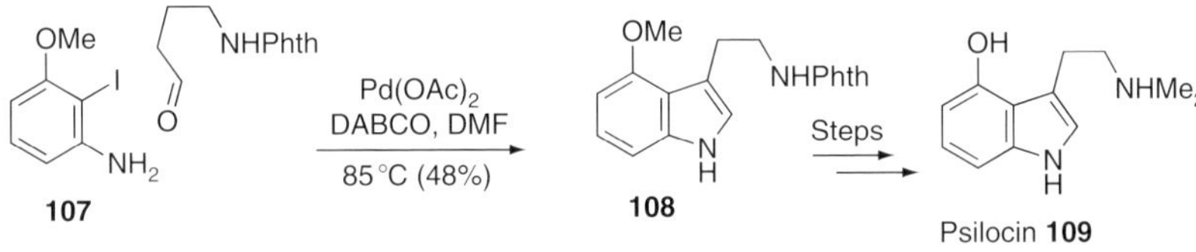

Annulation of aniline derivates through direct C-H activation continues to provide clean access to indole ring systems; an inherent virtue of this route to indole scaffolds is that it does not require an *ortho*-halogen on aniline progenitors. In one report, Jiao and co-workers have prepared a vast array of functionalized indole scaffolds, *e.g.*, **110**, **111**, through the Pd-catalyzed fusion of anilines with alkynes using molecular oxygen as the oxidizing agent <09AG(I)4572>. An alternate strategy for ring installation on *ortho*-unsubstituted anilines has been described by Moody and co-workers who have devised a modified Bischler reaction that involved coupling of anilines with α-diazo-β-ketoesters <09T8995>. Thus, Cu(II)-promoted insertion of a diazo-derived carbene onto aniline N-H afforded **112** that underwent in the same pot an acid-catalyzed intramolecular condensation to generate indole **113**.

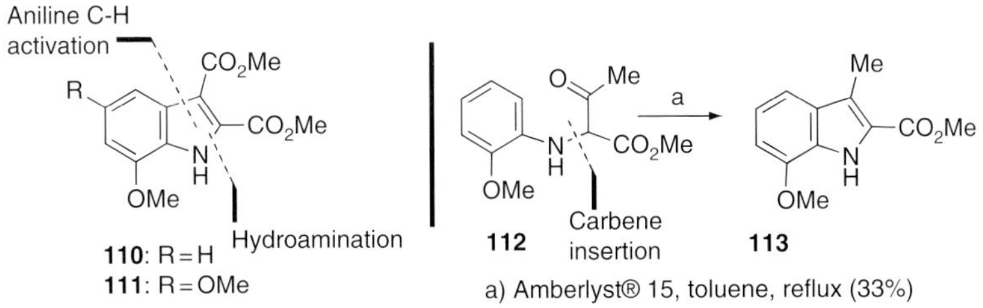

In two additional examples of type IIae bond construction, the Sorensen group has choreographed a three-component interrupted Ugi reaction for the synthesis of 3-aminoindoles from anilines <09T3096>, while Nicholas and co-workers have prepared 3-arylindoles via Fe(II)-promoted cyclocondensation of arylhydroxylamines with alkynes <09T3829>.

A diverse array of alkaloid scaffolds have been accessed through the judicious tailoring of the classic Fischer indole synthesis, an archetypical strategy for IIae ring formation. The Wood group has employed a Fischer indolization, under slightly modified conditions, in their work on the core structure of (±)-actinophyllic acid. Accordingly, the phenylhydrazone **114** was prepared from the corresponding β-keto amide via treatment with phenylhydrazine and scandium(III) triflate; microwave irradiation of **114** in the presence of anhydrous $ZnCl_2$ afforded the polycyclic indole **115** in 42% yield <09OL4532>. In an unrelated report on Fischer methodology, the Garg group has prepared pyrrolidino- and piperidinoindoline scaffolds, *e.g.*, **116**, through strategic implementation of an interrupted Fischer cyclization sequence <09OL3458>.

A key mechanistic feature of the Fischer indole synthesis involves the forging of a new carbon-carbon bond (the indole c-bond) through the [3,3]-sigmatropic rearrangement of enehydrazine intermediates. Accordingly, Murphy and co-workers have explored a Wittig approach for the direct preparation of enehydrazines en route to indole ring systems <09TL3290>. Interestingly, while treatment of acetylhydrazine **118** with methylenetriphenylphosphorane afforded, unexpectedly, the oxindole **117** in 76% yield, alternative use of the Petasis reagent (a titanium-based source of a methylene unit) afforded the desired indole **119** in 53% yield. Numerous other reports on Fischer indolization chemistry have appeared <09OL5454, 09SL3016, 09SC2506, 09T4212, 09T461, 09AJC1027>.

Indole ring synthesis has also been carried out though type IIae intramolecular bond construction. In one example, the Borate group has prepared *N*-thienyl substituted indoles through the coupling of styrene-derived epoxides and 2-formylaminothiophene condensation partners <09TL6562>. In one example, the 3-hydroxyindoline **120** was prepared in 85% yield, then dehydrated in a separate step to afford the corresponding indole nucleus (83%). Ackermann and co-workers have employed a Pd-catalyzed coupling of primary amines with *o*-alkynyl chlorobenzenes to afford indoles, *e.g.*, **121**, bearing sterically demanding *N*-substituents <09T8930, 09SL1219>. An alternate strategy involving the Cu(I)-catalyzed fusion of *t*-butyl carbamates with *o*-haloalkenyl aryl halides has been reported by the Willis group; Boc-protected indole **122** was prepared in 62% yield <09OBC432>.

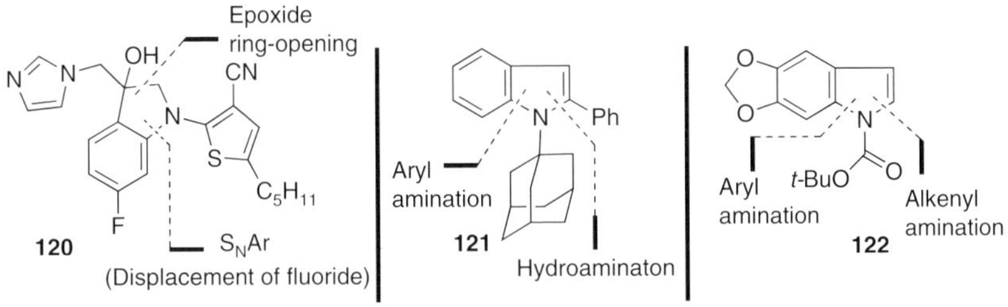

5.2.5. REACTIONS OF INDOLES

5.2.5.1 Pericyclic Transformations

Cycloadditions across the indole 2,3-π system or the benzenoid ring (indolyne chemistry) have been utilized for the rapid elucidation of indole-bound polycyclic arrays. The Vanderwal group has reported on an intramolecular cycloaddition strategy to set four of the five requisite ring systems common to the *Strychnos*, *Aspidosperma*, and *Iboga* mono-terpene alkaloids <09JA3472>. As illustrated below, *t*-butoxide-mediated anionic cyclization of the tryptamine-derived Zincke aldehyde **123** afforded the tetracycle **124** as a single stereoisomer in 84% yield. Buszek and co-worker have observed good regioselectivity for the cycloaddition of 2-*t*-butylfuran across 6,7-indolynes with preferential formation of the more sterically congested adducts, *e.g.*, **125**; similar cycloadditions across 4,5- or 5,6-indolynes gave 1:1 mixtures of Diels-Alder products <09TL63>. The indolyne methodology has been applied to the total synthesis of (±)-trikentrin A and (±)-herbindole <09TL7113, 09OL201>.

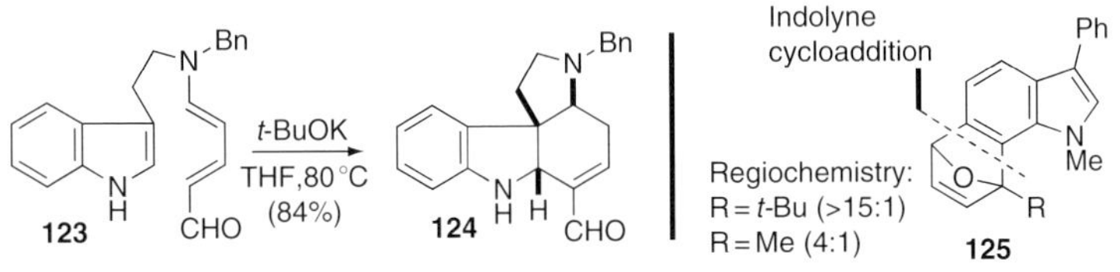

The Majumdar group has described a pair of strategies, a Knoevenagel/hetero-Diels-Alder route <09TL3889>, as well as a *thio*-Claisen rearrangement <09JHC62>, for the synthesis of thiopyrano[2,3-*b*]indole motifs. In one example, Diels-Alder cyclization of the Knoevenagel condensation adduct **128** afforded the corresponding pyrano-fused thiopyranoindole **129** in 96% yield <09TL3889>.

5.2.5.2 Substitution at C3/C2

A great deal of effort has been put forth toward the development of methodology for the installation of substituents at indole C3 with a high degree of stereocontrol. In one study, Ferienga and Roelfes have observed good enantioselectivity for the conjugate addition of indoles to α,β-unsaturated 2-acylimidazoles with asymmetric induction imparted by a Cu-DNA complex <09AG(I)3346>. In an unrelated strategy for C3 functionalization, the Kobayashi group has observed an interesting inversion in the stereochemical outcome of epoxide ring-opening by indole depending on the choice of metal-**133** catalyst system. While the use of Cu(II)-**133** afforded predominately **131**, the enantiomeric partner **132** was obtained when the catalyst **133** was paired with Sc(III) <09CL904>.

a) $Cu(O_3SC_{11}H_{23})_2$, H_2O. b) $Sc(O_3SC_{11}H_{23})_2$, H_2O.

The Carbery group has observed seven-membered ring annulation across indole C3-C2 to set a tricyclic core and two new stereocenters via a double Friedel–Crafts alkylation of indoles with divinyl ketones <09OL1175>. A sequential Friedel–Crafts alkylation has also been observed by the You group who established the C3-fluorenyl linkage of **134** with good enantiocontrol through treatment of indole with a biphenyl-2-carbaldehyde electrophile and an axially chiral phosphoric acid catalyst <09CEJ8709>. A broad spectrum of other approaches to C3-functionalizaton through the development of asymmetric Friedel–Crafts chemistry has been described <09JOC6878, 09CEJ3351, 09EJO4833, 09SL2115, 09S3994, 09TL5602, 09ASC2433, 09ASC772, 09ASC1517>.

A few additional reports in the area of indole C3 functionalization have included a study by Ackermann and Barfüßer on the Pd-catalyzed synthesis of 3-arylindoles from arylhalides <09SL808>, as well as an investigation by the Baran group on the synthesis of 3,3-disubstituted indolines and indolenines using bisaryl λ^3-iodanes for setting the C3 quaternary centers <09T3149>. A Pd-catalyzed route to 3-alkynylindoles has been reported by Gu and Wang <09TL763>. Outside of C-C bond installation, a pair of reports on the C3 sulfenylation of indole have appeared <09S1520, 09S4183>, as has a free radical approach to indole C3 thiocyanides <09TL347>.

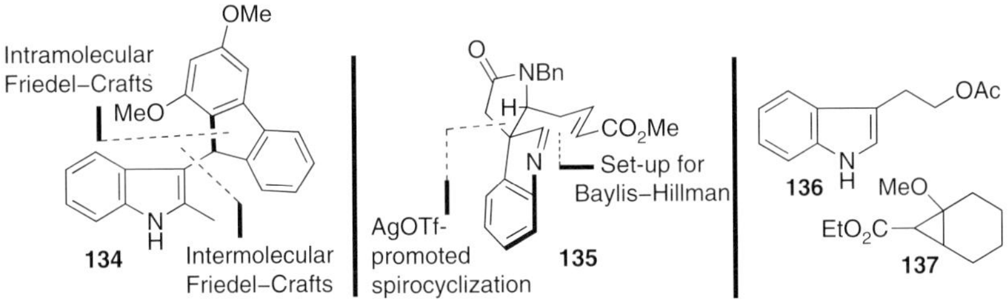

Various tactics have been explored for precision installation of indole C2 functionality. In an approach to the tetracyclic A-E ring core of the *Strychnos* alkaloids, Sirasani and Andrade first set the C3 spirocycle of **135** (the C-ring) using a AgOTf-promoted displacement of bromide to set up an intramolecular aza-Baylis-Hillman reaction to establish the C2 linkage and close the E ring <09OL2085>. En route to the pentacyclic frameworks of tetrahydroisoquinocarbazoles, Bajtos and Pagenkopf have installed the requisite C2 linkage via alkylation of indole **136** at C2 with the 2-alkoxycyclopropanoate ester **137** <09OL2780>. Methodology for arylation at indole C2 has also been described <09S3617, 09S2447>.

5.2.5.3 Substitution at Nitrogen

New methodologies for the selective functionalization of indole nitrogen have frequently been pursued within the context of establishing N- to C2 ring junctures. In a report from Rogness and Larock, indole N- to C2 annulation, as depicted for **138,** has been achieved by means of an addition of indole N- to arynes followed by subsequent ring closure of the aromatic anion onto a C2 pendant methyl ester <09TL4003>. An alternate N-C2 ring-forming strategy has been described by Enders and co-workers who prepared **139** and related analogs by means of an organocatalytic aza-Michael addition of indole onto chiral α,β-unsaturated iminium ions, followed by spontaneous intramolecular aldol-type ring closure onto indole-2-carbaldehyde functionality <09S4119>. In a final example, an intriguing route to annulated indoles has been developed by the Wang group who have described an N-heterocyclic carbene-promoted ring opening of formylcyclopropanes to generate bi-functional intermediates **141** for electrophilic trapping of indole nitrogen and nucleophilic addition to C2 <09JOC4379>.

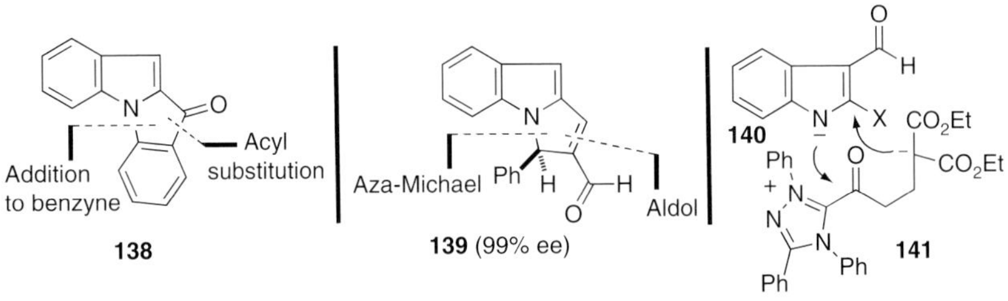

A variety of other reports on indole *N*-substitution have appeared that include methods for the synthesis of *N*-acylindoles <09OL1651>, *N*-*tert*-prenylindoles

<09AG(I)7025>, N-arylindoles <09T10459, 09CPB321, 09T4619>, N-vinylindoles <09JOC5603>, and enantioenriched N-allylindoles <09AG(I)7841, 09AG(I)5737>.

5.2.5.4 Functionalization of the Benzene Ring

While the indolization of substituted anilines remains one of the most reliable strategies for establishing functionality on the indole benzenoid ring, there continues to be significant interest in the development of alternative entry points to benzenoid substitution. For example, the Wipf group has devised an intramolecular Diels-Alder sequence for the synthesis of 4-substituted indoles, e.g., **144** <09CC104>. In the event, microwave-promoted [4+2] cycloaddition of the alkenyl-tethered furan afforded the indole **144** following dehydration to the core aromatic. The C4 linkage of the seven-membered azepinoindole **145** has been installed by Ishikura and co-workers using a Pictet-Spengler transformation <09EJO5752, 09H(77)825>, while a Pd-catalyzed vinylation of a 4-chlorotryptophan has been employed by Jia for the construction of the analogous ring-system **146** <09JOC6859>. In a final example, the Garg group has prepared a number of 4-, 5-, or 6-substituted indoles, e.g., 5-phenylaminoindole, through the trapping of 4,5- or 5,6-indolynes <09OL1007>.

145: R = H
146: R = CO_2H

a) Microwave, o-dichlorobenzene

5.2.5.5 Elaboration of the Indole Side-chains

Numerous well conceived tactics for discrete bond installation have been tucked away within reports of multi-step natural product syntheses. In a series of independent studies, a diverse range of synthetic approaches have been exploited for elaboration of the iminoethanocarbazole scaffold common to a variety of alkaloids. The MacMillan group has forged the pentacyclic framework of *Strychnos* alkaloid (+)-minfiensine **150** using an organocatalytic asymmetric [4+2] cycloaddition strategy; the stereoselectivity of the Diels-Alder transformation was imparted via an iminium ion chiral auxiliary pendant to the alkyne dienophile <09JA13606>. As outlined below, cyclization of a 2-vinylindole diene **147** with chiral alkyne **148** afforded the all-carbon quaternary center at C3 of **149** and set up an intramolecular cyclization to set the hexahydropyrroloindole ring juncture at C2 (87%, 96% ee). A 6-*exo*-dig radical cyclization was employed to stitch together the piperidine E ring and reveal the pentacyclic skeleton of the natural product **149**.

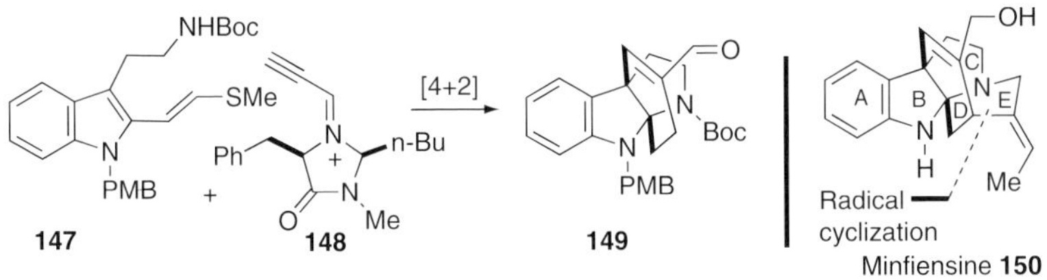

An alternate approach to the C-D ring system of minfiensine **150** has been investigated by Padwa and co-workers; the strategy involved implementation of an intramolecular [4+2] cycloaddition of alkene-tethered imidofurans with subsequent ring-opening of the bridged hemiaminal cycloadduct to afford the characteristic 6-5 fused skeletal system of the C-D ring juncture <09TL3145>. Efforts to reveal the indole moiety via Ia type intramolecular ring closure of a pendant aryl amine were unsuccessful.

As an extension of their own work on (±)-minfiensine <08AG(I)3618>, Qin and co-workers have constructed the pentacyclic skeleton of the *akuammiline* alkaloid vincorine that contains a seven-membered E ring <09JA6013>. As depicted below, implementation of a Cu(I)-catalyzed intramolecular cyclopropanation across indole C2-C3, followed by sequential cyclopropane ring opening/pyrrolidine ring closure at the anomeric indole C2 position set the C3 quaternary center housed within the framework of **151**. The pentacyclic core of **151** was completed with an intramolecular allylation (displacement of chloride) to close the seven-membered E ring.

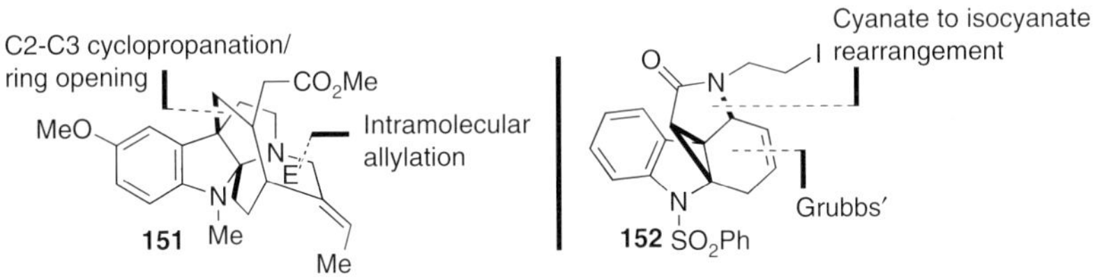

A cyclopropanation/ring-opening across indole C2-C3 also has been employed by Gagnon and Spino to set a key all-carbon stereocenter at indole C3 in their synthesis of (+)-aspidofractinine (the enantiomer of the natural alkaloid). As outlined for **152**, the diastereofacial selectivity of cyclopropanation (also a chemoselective process) was transmitted via the neighboring C-N stereocenter of the lactam ring of **152** that, in turn, was set using an asymmetric cyanate to isocyanate rearrangement. The chiral auxiliary employed for the cyanate rearrangement was cleaved using Grubbs' second-generation catalyst, which, in the same transformation, fused the 6-membered E-ring of the core hexahydrocarbazole.

The structurally intriguing dihydrotryptamine-dihydropyrimidene ring juncture that defines the indoline C2 quaternary center of hinckdentine A **153** has been installed by the Kawasaki group through implementation of a Mannich-type strategy <09OL197>. In the event, the key bond was set in racemic fashion through the acid-catalyzed coupling of a 2-hydroxyindolin-3-one derived iminium ion electrophile and a nucleophilic TMS-enol ether.

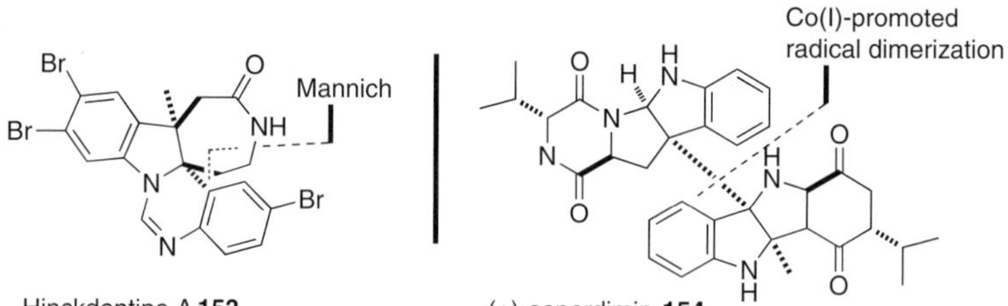

[Structures: Hinckdentine A **153**; (+)-asperdimin **154** (Co(I)-promoted radical dimerization; Mannich)]

de Lera and co-workers have prepared a collection of bispyrrolidinoindoline diketopiperazine alkaloids, including (+)-asperdimin **154** (a revised structure), through a radical dimerization process <09CEJ9928>. While radical coupling of a C3-phenylselenide to access the C3-C3′ bis-quaternary ring juncture proceeded in low yield (30%), adoption of a Co(I)-mediated dimerization of a C3-bromide, a strategy previously disclosed by Movassaghi and Schmidt <07AG(I)3725, 08SL313>, afforded the desired bis-hexahydropyrroloindole scaffold in an improved 54% yield <09CEJ9928>.

Finally, Delgado and Blakey have described a cascade annulation strategy for accessing the core subunit **156** of the malagashanine type alkaloids, members of the *Strychnos* family possessing unique stereochemistry at carbon α to indole C3 (i.e. *trans*-fusion rather than *cis*-fusion of the pyrrolidine C ring) <09EJO1506>.

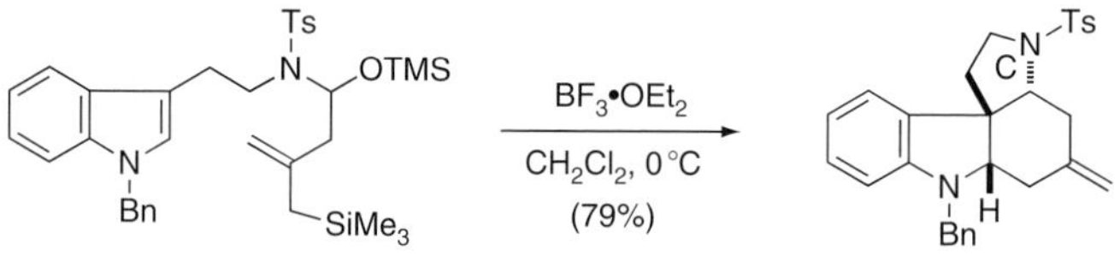

[Scheme: **155** → **156**, BF$_3$·OEt$_2$, CH$_2$Cl$_2$, 0 °C (79%)]

5.2.6. OXINDOLES AND SPIROOXINDOLES

A tremendous amount of effort continues to be directed toward the selective installation of the C3 quaternary centers that characterize the distinctive family of oxindole alkaloids. Numerous asymmetric approaches to all-carbon <09JA14, 09JA9900, 09CEJ7846, 09AG(I)8037>, oxygen-bound <09JOC283, 09JA6946; 09AG(I)6313>, or nitrogen-bound <09JOC4537, 09JOC8935, 09CC6264> oxindole C3 quaternary centers have been described.

Oxindoles bearing a 3-methylene functionality have been popular synthetic targets as well as precursors for further elaboration of the oxindole skeleton. The Murakami group has prepared 3-(amidoalkylidene)oxindoles, *e.g.*, **157**, with excellent selectivity for the Z-stereoisomer via a sequence of Pd-catalyzed intramolecular oxidative cyclization of an *o*-alkynyl-isocyanate, followed by an intermolecular amination <09OL2141>. Building from a variety of 3-methyleneoxindole scaffolds, Gong and co-workers have orchestrated a series of three-component organocatalytic spirocyclization sequences involving azomethine ylide addition across oxindole C3 exocyclic π-systems <09JA13819>. Thus, spirooxindole **161** was prepared from oxindole **158**, aldehyde **159**, and amine **160**, by means of a regio- and enantioselective 1,3-dipolar cycloaddition directed by a chiral phosphoric acid.

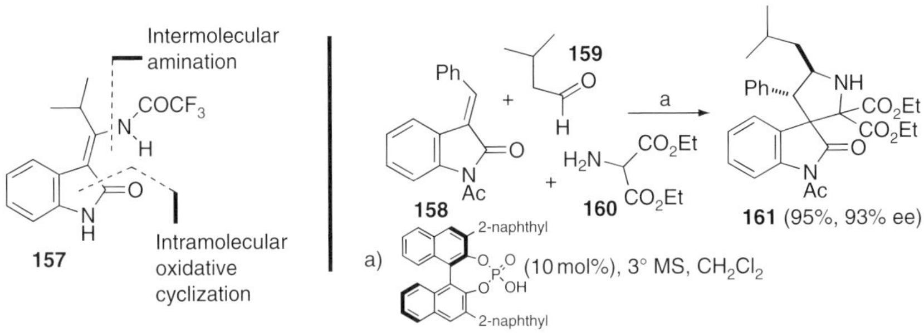

In a final example, a novel approach to the core framework of the welwitindolinone natural products has been reported by Simpkins and co-workers who prepared an α,β-unsaturated oxindole subunit through oxidative transformation of an indole ring-system using singlet oxygen <09TL3283>.

5.2.7. CARBAZOLES

The tricyclic carbazole framework has served as a popular platform for the development of transition metal-catalyzed coupling reactions. For example, the Miura group has reported the synthesis of tetrasubstituted carbazoles via the coupling of indole-3-carboxylic acids with an excess of an internal alkyne partner using Pd-catalysis in the presence of a copper additive <09OL2337, 09JOC7481>. A Rh-catalyzed oxidative coupling of internal alkynes with *o*-aminobenzoic acids has been described by the same group <09JOC3478>. Ackermann and co-workers have developed a Pd-catalyzed route to carbazole scaffolds through the fusion of anilines with 1,2-dihalobenzenes <09S3493>. Alternatively, an intramolecular route to the carbazole framework of **163** has been implemented by the Ma group by means of a Pt-promoted ring closure of indole-2-allenol **162** <09CC4572>.

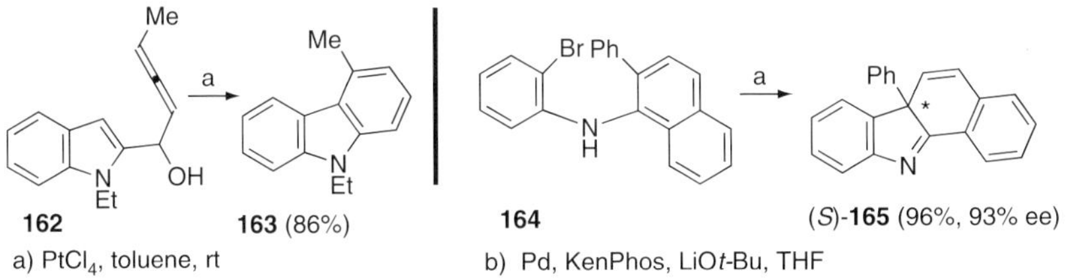

A number of approaches to carbazole ring synthesis have involved the fusion of biaryl amine substrates. In one intriguing report, Buchwald and co-workers who have established the all-carbon quaternary center of (*S*)-**165** in 99% ee through an intramolecular dearomatization of the biaryl precursor **164** using KenPhos as the chiral ligand on a Pd catalyst system <09JA6676>. A fully aromatized carbazole core has been constructed by the Menedez group who stitched together a biaryl amine via a microwave-assisted Pd-catalyzed double C-H activation sequence <09EJO4614>. Rhodium-catalyzed coupling of biaryl azides has also been described <09JOC3225>.

Finally, outside the realm of metal-promoted coupling reactions, a pair of reports have appeared that detail intramolecular electrocyclic routes to carbazole ring systems via annulation across the indole 2,3-bond <09H(79)955, 09T3582>.

5.2.8. AZAINDOLES AND CARBOLINE ANALOGS

As prominent features within the substructure of numerous naturally occurring alkaloids, the synthesis of azaindole and carboline heterocycles continues to be of significant interest. In one example, Suzenet and co-workers have prepared 4- and 6-azaindoles by means of a Fischer indole synthesis using pyridylhydrazine precursors that contained an electron donation group, e.g., **166** <09OL5142>. Alternatively, the Song group observed the formation of diazaindole **169** upon treatment of the dianion **167** with ethyl benzoate **168** <09TL3952>. A series of substituted carbolines, e.g., **170**, have been prepared by Cuny and co-workers; the transformations involved the Pd-catalyzed coupling of aryl amines and 2,3-dibromopyridines <09JOC3152>.

Methodology for the synthesis of tetracyclic indole[2,3-b]quinolines has been described that involved the fusion of aryl amines with indole-3-carboxaldehyde in the presence of catalytic iodine; the reactions required two full equivalents of aryl amine partners with one equivalent serving as a leaving group during the aromatization of the tetracyclic core <09JOC8369>. The Takemoto group has reported an alternate route to the indole[2,3-b]quinoline ring systems via a Pd-catalyzed annulation of unsaturated isothioureas using a copper additive <09CL772>.

Padwa and co-workers have established the tetrahydro-β-carboline subunit found nestled within the broader indole[a]quinolizine framework of (±)-yohimbinone through the fusion of an indole-2-oxime **171** with 2,3-bis(phenylsulfonyl)-1,3-butadiene **172** <09JOC3491>. A sequence of conjugate addition of oxime **171** to **172** followed by a [3+2]-dipolar cycloaddition afforded the oxa-bridged intermediate **173** that was transformed to the functionalized quinolizine **174** upon reduction of the N-O bridge system.

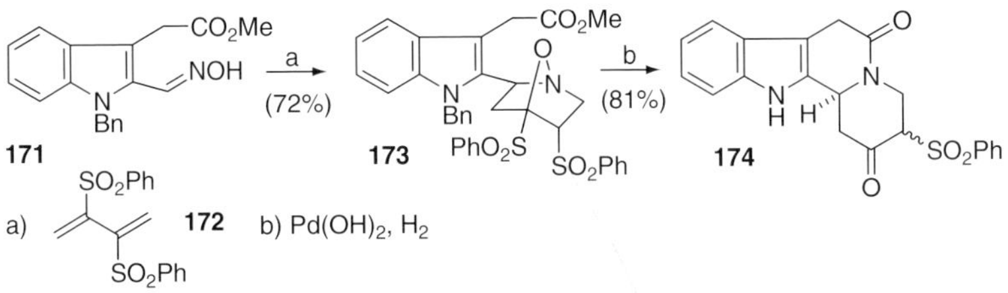

A few other strategies for preparation of indoloquinolizidine have been described. In one study by the Hunter group, a 6-*exo*-trig radical cyclization was used to set the 6,6-ring system of the quinolizidine **175** <09TL6342>. Alternatively, the Cook group has employed a Pictet–Spengler reaction for preparation of the THβ-carboline subunit housed within *Corynanthe* type alkaloids, *e.g.*, mitragynine **177** <09JOC264>. Upon treatment of tryptamine and aldehyde coupling partners with acetic acid, a 1:3 ratio of *cis* to *trans* isomers of the THβ-carboline **176** was obtained; however, the *cis*-adduct was readily epimerized to afford the desired *trans*-adduct in 90% yield. The quinolizine ring system of **177** was stitched together by means of a Ni(0)-mediated cyclization.

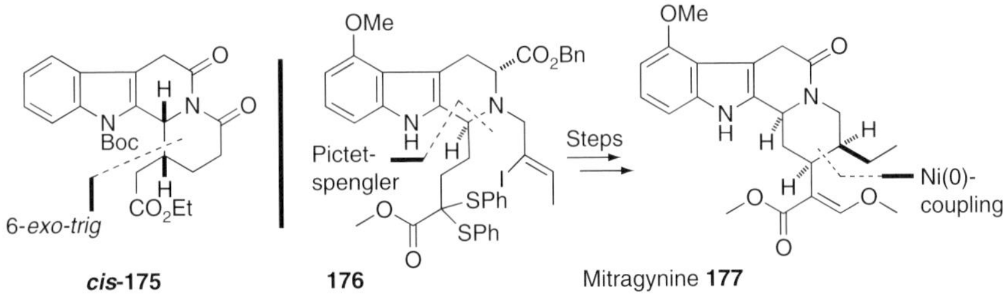

Finally, as a mainstay for construction of the THβ-carboline ring system, the Pictet–Spengler reaction has found application in the development of various enantioselective organocatalytic methodologies <09OL887, 09OL2579, 09JA10796>.

5.2.9. INDOLE NATURAL PRODUCTS

The laborious extraction of alkaloids from natural sources continues to reveal unprecedented structural types as well as new variations on familiar molecular themes. A small collection of recently characterized isolates include the alkaloid conolutinine **178** from the stem-bark of *Malayan Tabernaemontana corymbosa* <09TL752>, melohenine A **179**, an extract of the Chinese cane *Melodinus henryi* <09OL4834>, and finally, pseudocerosine **180**, a blue pigment of the marine flatworm *Pseudoceros indicus* <09OL1111>.

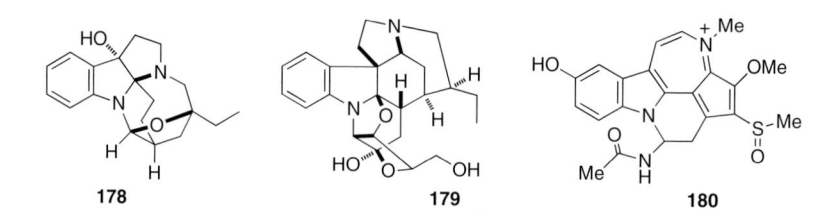

REFERENCES

88T325	J. Vukovic, A.E. Goodbody, J.P. Kutney and M. Misawa, *Tetrahedron*, **44**, 325 (1988).
99JA6771	F. He, Y. Bo, J.D. Altom and E.J. Corey, *J. Am. Chem. Soc.*, **121**, 6771 (1999).
07AG(I)3725	M. Movassaghi and M.A. Schmidt, *Angew. Chem. Int. Ed.*, **46**, 3725 (2007).
08AG(I)3618	L. Shen, M. Zhang, Y. Wu and Y. Qin, *Angew. Chem. Int. Ed.*, **47**, 3618 (2008).
08SL313	M.A. Schmidt and M. Movassaghi, *Synlett*, 313 (2008).
09ACR530	R.A. Shenvi, D.P. O'Malley and P.S. Baran, *Acc. Chem. Res.*, **42**, 530 (2009).
09AG(I)758	B. Tan, Z. Shi, P.J. Chua, Y. Li and G. Zhong, *Angew. Chem. Int. Ed.*, **48**, 758 (2009).
09AG(I)1244	H. Yamamoto, I. Sasaki, Y. Hirai, K. Namba, H. Imagawa and M. Nishizawa, *Angew. Chem. Int. Ed.*, **48**, 1244 (2009).
09AG(I)3346	A.J. Boersma, B.L. Feringa and G. Roelfes, *Angew. Chem. Int. Ed.*, **48**, 3346 (2009).
09AG(I)3802	P.M. When and J. Du Bois, *Angew. Chem. Int. Ed.*, **48**, 3802 (2009).
09AG(I)4572	Z. Shi, C. Zhang, S. Li, D. Pan, S. Ding, Y. Cui and N. Jiao, *Angew. Chem. Int. Ed.*, **48**, 4572 (2009).
09AG(I)5737	H.-L. Cui, X. Feng, J. Peng, J. Lei, K. Jiang and Y.-C. Chen, *Angew. Chem. Int. Ed.*, **48**, 5737 (2009).
09AG(I)6313	J. Itoh, S.B. Han and M.J. Krische, *Angew. Chem. Int. Ed.*, **48**, 6313 (2009).
09AG(I)7025	M.R. Luzung, C.A. Lewis and P.S. Baran, *Angew. Chem. Int. Ed.*, **48**, 7025 (2009).
09AG(I)7212	M. Gruit, D. Michalik, A. Tillack and M. Beller, *Angew. Chem. Int. Ed.*, **48**, 7212 (2009).
09AG(I)7841	L.M. Stanley and J.F. Hartwig, *Angew. Chem. Int. Ed.*, **48**, 7841 (2009).
09AG(I)8037	S. Ma, X. Han, S. Krishnan, S.C. Virgil and B.M. Stoltz, *Angew. Chem. Int. Ed.*, **48**, 8037 (2009).
09AG(I)8042	R. Martin, A. Jäger, M. Böh, S. Richter, R. Federov, D.J. Manstein, H.O. Gutzeit and H.-J. Knölker, *Angew. Chem. Int. Ed.*, **48**, 8042 (2009).
09AG(I)8318	A. Mizuno, H. Kusama and N. Iwasawa, *Angew. Chem. Int. Ed.*, **48**, 8318 (2009).
09AG(I)9346	J.P. Brand, J. Charpentier and J. Waser, *Angew. Chem. Int. Ed.*, **48**, 9346 (2009).
09AG(I)9608	M. Bandini and A. Eichholzer, *Angew. Chem. Int. Ed.*, **48**, 9608 (2009).
09AJC683	K. Hasse, A.C. Willis and M.G. Banwell, *Aust. J. Chem.*, **62**, 683 (2009).
09AJC1027	L. Zhong and G.-K. Chuah, *Aust. J. Chem.*, **62**, 1027 (2009).
09ASC123	M. Fujiwara, M. Kawatsura, S. Hayase, M. Nanjo and T. Itoh, *Adv. Synth. Catal.*, **351**, 123 (2009).
09ASC129	Y. Lu, X. Fu, H. Chen, X. Du, X. Jia and Y. Liu, *Adv. Synth. Catal.*, **351**, 129 (2009).
09ASC673	L. Joucla and L. Djakovitch, *Adv. Synth. Catal.*, **351**, 673 (2009).
09ASC715	O.A. Attanasi, S. Berretta, L. De Crescentini, G. Favi, G. Giorgi and F. Mantellini, *Adv. Synth. Catal.*, **351**, 715 (2009).
09ASC772	L. Hong, L. Wang, C. Chen, B. Zhang and R. Wang, *Adv. Synth. Catal.*, **351**, 772 (2009).
09ASC1517	Y. Lu, X. Du, X. Jia and Y. Liu, *Adv. Synth. Catal.*, **351**, 1517 (2009).
09ASC1977	J. Roger and H. Doucet, *Adv. Synth. Catal.*, **351**, 1977 (2009).
09ASC2005	L. Chen and M.-H. Xu, *Adv. Synth. Catal.*, **351**, 2005 (2009).
09ASC2063	J.-Y. Wang, X.-P. Wang, Z.-S. Yu and W. Yu, *Adv. Synth. Catal.*, **351**, 2063 (2009).

09ASC2433	G. Blay, I. Fernández, A. Monleón, M.C. Muñoz, J.R. Pedro and C. Vila, *Adv. Synth. Catal.*, **351**, 2433 (2009).
09ASC2615	Y.-J. Guo, R.-Y. Tang, J.-H. Li, P. Zhong and X.-G. Zhang, *Adv. Synth. Catal.*, **351**, 2615 (2009).
09ASC3107	Z. Shen and X. Lu, *Adv. Synth. Catal.*, **351**, 3107 (2009).
09ASC3118	X. Huang, S. Zhu and R. Shen, *Adv. Synth. Catal.*, **351**, 3118 (2009).
09BMC7548	L. Lu, X. Gu, L. Hong, J. Laird, K. Jaffe, J. Choi, J. Crabb and R.G. Salomon, *Bioorg. Med. Chem.*, **17**, 7548 (2009).
09BMC7783	V. Perri, C. Rochais, T. Cresteil, P. Dallemagne and S. Rault, *Bioorg. Med. Chem.*, **17**, 7783 (2009).
09CAJ1037	J. Barluenga, F. Rodríguez and F.J. Fañanás, *Chem. Asian J.*, **351**, 1037 (2009).
09CC104	F. Petronijevic, C. Timmons, A. Cuzzupe and P. Wipf, *Chem. Commun.*, 104 (2009).
09CC1470	T. Miura, M. Yamauchi and M. Murakami, *Chem. Commun.*, 1470 (2009).
09CC3285	N. Yakoyama and Y. Arai, *Chem. Commun.*, 3285 (2009).
09CC4151	A.E. Ondrus and M. Movassaghi, *Chem. Commun.*, 4151 (2009).
09CC4515	V. Leen, E. Braeken, K. Luckermans, C. Jackers, M. Van der Auweraer, N. Boens and W. Dehaen, *Chem. Commun.*, 4515 (2009).
09CC4572	W. Kong, C. Fu and S. Ma, *Chem. Commun.*, 4572 (2009).
09CC6264	S. Sato, M. Shibuya, N. Kanoh and Y. Iwabuchi, *Chem. Commun.*, 6264 (2009).
09CEJ227	A.V. Lygin, O.V. Larionov, V.S. Korotkov and A. de Meijere, *Chem. Eur. J.*, **15**, 227 (2009).
09CEJ838	D. Tejedor, S. López-Tosco, J. González-Platas and F. García-Tellado, *Chem. Eur. J.*, **15**, 838 (2009).
09CEJ3351	Y.-F. Sheng, G.-Q. Li, Q. Kang, A.-J. Zhang and S.-L. You, *Chem. Eur. J.*, **15**, 3351 (2009).
09CEJ5823	E.Y. Schmidt, B.A. Trofimov, A.I. Mikhaleva, N.V. Zorina, N.I. Protzuk, K.B. Petrushenko, I.A. Ushakov, M.Y. Dvorko, R. Méallet-Renault, G. Clavier, T.T. Vu, H.T.T. Tran and R.B. Pansu, *Chem. Eur. J.*, **15**, 5823 (2009).
09CEJ6910	B.M. Trost and G. Dong, *Chem. Eur. J.*, **15**, 6910 (2009).
09CEJ7846	P. Galzerano, G. Bencivenni, F. Pesciaioli, A. Mazzanti, B. Giannichi, L. Sambri, G. Bartoli and P. Melchiorre, *Chem. Eur. J.*, **15**, 7846 (2009).
09CEJ7850	Della Ca´, G. Maestri and M. Catellani, *Chem. Eur. J.*, **15**, 7850 (2009).
09CEJ8709	F.-L. Sun, M. Zeng, Q. Gu and S.-L. You, *Chem. Eur. J.*, **15**, 8709 (2009).
09CEJ9928	C. Pérez-Balado, P. Rodríguez-Graña and A.R. de Lera, *Chem. Eur. J.*, **15**, 9928 (2009).
09CHC501	D.L. Tarshits, N.M. Przhiyalgovskaya, V.N. Buyanov and S.Y. Tarasov, *Chem. Heterocycl. Comp.*, **45**, 501 (2009).
09CL8	H.-J. Knölker, *Chem. Lett.*, **38**, 8 (2009).
09CL224	K. Komeyama, M. Miyagi and K. Takaki, *Chem. Lett.*, **38**, 224 (2009).
09CL772	H. Takeda, T. Ishida and Y. Takemoto, *Chem. Lett.*, **38**, 772 (2009).
09CL904	M. Kokubo, T. Naito and S. Kobayashi, *Chem. Lett.*, **38**, 904 (2009).
09CPB321	H. Xu and L.-I. Fan, *Chem. Pharm. Bull.*, **57**, 321 (2009).
09CSR3035	J. Kim and M. Movassaghi, *Chem. Soc. Rev.*, **38**, 3035 (2009).
09CSR3160	K.A. Miller and R.M. Williams, *Chem. Soc. Rev.*, **38**, 3160 (2009).
09EJO1506	R. Delgado and S.B. Blakely, *Eur. J. Org. Chem.*, 1506 (2009).
09EJO2437	T. Tsuchimoto, T. Ainoya, K. Aoki, T. Wagatsuma and E. Shirakawa, *Eur. J. Org. Chem.*, 2437 (2009).
09EJO3562	Y. Li and D. Dolphin, *Eur. J. Org. Chem.*, 3562 (2009).
09EJO4614	V. Sridharan, M.A. Martín and J.C. Menéndez, *Eur. J. Org. Chem.*, 4614 (2009).
09EJO4833	S.C. McKeon, H. Müller-Bunz and P.J. Guiry, *Eur. J. Org. Chem.*, 4833 (2009).
09EJO4882	J. Deblander, S. Van Aeken, J. Jacobs, N. De Kimpe and K.A. Tehrani, *Eur. J. Org. Chem.*, 4882–4892 (2009).
09EJO5752	K. Yamada, Y. Namerikawa, T. Haruyama, Y. Miwa, R. Yanada and M. Ishikura, *Eur. J. Org. Chem.*, 5752 (2009).
09H(77)825	K. Yamada, Y. Namerikawa, T. Abe and M. Ishikura, *Heterocycles*, **77**, 825 (2009).
09H(77)1105	T. Fukuda, Y. Hayashida and M. Iwao, *Heterocycles*, **77**, 1105 (2009).
09H(78)161	S. Fukamachi, H. Konishi and K. Kobayashi, *Heterocycles*, **78**, 161 (2009).

09H(78)177	N.A. Kheder, E.S. Darwish and K.M. Dawood, *Heterocycles*, **78**, 177 (2009).
09H(78)2013	Z.-G. Le, T. Zhong, Z.-B. Xie and J.-P. Xu, *Heterocycles*, **79**, 2013 (2009).
09H(78)725	Y.-Q. Ge, J. Jia, H. Yang, G.-L. Zhao, F.-X. Zhan and J. Wang, *Heterocycles*, **79**, 725 (2009).
09H(79)219	R. Settambolo, *Heterocycles*, **79**, 219 (2009).
09H(79)955	S. Tohyama, T. Choshi, S. Azuma, H. Fujioka and S. Hibino, *Heterocycles*, **79**, 955 (2009).
09HCA1002	H.M. Meshram, D.A. Kumar and B.C. Reddy, *Helv. Chim. Acta*, **92**, 1002 (2009).
09HCA2118	S.N. Murthy, B. Madhav, A.V. Kumar, K.R. Rao and Y.V.D. Nageswar, *Helv. Chim. Acta*, **92**, 2118 (2009).
09JA14	T.A. Duffey, S.A. Shaw and E. Vedejs, *J. Am. Chem. Soc.*, **131**, 14 (2009).
09JA1668	Y. Kita, K. Morimoto, M. Ito, C. Ogawa, A. Goto and T. Dohi, *J. Am. Chem. Soc.*, **131**, 1668 (2009).
09JA3472	D.B.C. Martin and C.D. Vanderwal, *J. Am. Chem. Soc.*, **131**, 3472 (2009).
09JA4904	H. Ishikawa, D.A. Colby, S. Seto, P. Va, A. Tam, H. Kakei, T.J. Rayl, I. Hwang and D.L. Boger, *J. Am. Chem. Soc.*, **131**, 4904 (2009).
09JA6013	M. Zhang, X. Huang, L. Shen and Y. Qin, *J. Am. Chem. Soc.*, **131**, 6013 (2009).
09JA6070	T. Ohmura, A. Kijima and M. Suginome, *J. Am. Chem. Soc.*, **131**, 6070 (2009).
09JA6676	J. Garca-Fortanet, F. Kessler and S.L. Buchwald, *J. Am. Chem. Soc.*, **131**, 6676 (2009).
09JA6946	D. Tomita, K. Yamatsugu, M. Kanai and M. Shibasaki, *J. Am. Chem. Soc.*, **131**, 6946 (2009).
09JA9900	A.M. Taylor, R.A. Altman and S.L. Buchwald, *J. Am. Chem. Soc.*, **131**, 9900 (2009).
09JA10796	M.E. Muratore, C.A. Holloway, A.W. Pilling, R.I. Storer, G. Trevitt and D.J. Dixon, *J. Am. Chem. Soc.*, **131**, 10796 (2009).
09JA11674	S.O.C. Mundle and R. Kluger, *J. Am. Chem. Soc.*, **131**, 11674 (2009).
09JA13606	S.B. Jones, B. Simmons and D.W.C. MacMillan, *J. Am. Chem. Soc.*, **131**, 13606 (2009).
09JA13819	X.-H. Chen, Q. Wei, S.-W. Luo, H. Xiao and L.-Z. Gong, *J. Am. Chem. Soc.*, **131**, 13819 (2009).
09JA14579	M.D. Clift and R.J. Thomson, *J. Am. Chem. Soc.*, **131**, 14579 (2009).
09JA16626	N.K. Pahadi, M. Paley, R. Jana, S.R. Waetzig and J.A. Tunge, *J. Am. Chem. Soc.*, **131**, 16626 (2009).
09JHC62	K.C. Majumdar, S. Alam and B. Chattopadhyay, *J. Heterocycl. Chem.*, **46**, 62 (2009).
09JHC503	W.E. Noland, N.P. Lanzatella, E.P. Sizova, L. Venkatraman and O.V. Afanasyev, *J. Heterocycl. Chem.*, **46**, 503 (2009).
09JHC1154	W.E. Noland, N.P. Lanzatella, L. Venkatraman, N.F. Anderson and G.C. Gullickson, *J. Heterocycl. Chem.*, **46**, 503 (2009).
09JHC1285	W.E. Noland and N.P. Lanzatella, *J. Heterocycl. Chem.*, **46**, 1285 (2009).
09JMC3377	R. Faghih, S.M. Gopalakrishnan, J.H. Gronlien, J. Malysz, C.A. Briggs, C. Wetterstrand, H. Ween, M.P. Curtis, K.A. Sarris, G.A. Gfesser, R. El-Kouhen, H.M. Robb, R.J. Radek, K.C. Marsh, W.H. Bunnelle and M. Gopalakrishnan, *J. Med. Chem.*, **53**, 3377 (2009).
09JOC212	H. Fuwa and M. Sasaki, *J. Org. Chem.*, **74**, 212 (2009).
09JOC264	J. Ma, W. Yin, H. Zhou, X. Liao and J.M. Cook, *J. Org. Chem.*, **74**, 264 (2009).
09JOC283	H. Lai, Z. Huang, Q. Wu and Y. Qin, *J. Org. Chem.*, **74**, 283 (2009).
09JOC669	R. Bhattacharya, A.K. Atta, D. Dey and T. Pathak, *J. Org. Chem.*, **74**, 669 (2009).
09JOC1377	R. Surmont, G. Verniest, F. Colpaert, G. Macdonald, J.W. Thuring, F. Deroose and N. De Kimpe, *J. Org. Chem.*, **74**, 1377 (2009).
09JOC1418	K. Murai, S. Hayashi, N. Takaichi, Y. Kita and H. Fujioka, *J. Org. Chem.*, **74**, 1418 (2009).
09JOC1517	A. Saito, S. Oda, H. Fukaya and Y. Hanzawa, *J. Org. Chem.*, **74**, 1517 (2009).
09JOC1581	A.L. Bowie, Jr. and D. Trauner, *J. Org. Chem.*, **74**, 1581 (2009).
09JOC1826	B. Liégault, D. Lapointe, L. Caron, A. Vlassova and K. Fagnou, *J. Org. Chem.*, **74**, 1826 (2009).
09JOC3054	D.I. Chai and M. Lautens, *J. Org. Chem.*, **74**, 3054 (2009).
09JOC3152	J.K. Laha, P. Petrou and G.D. Cuny, *J. Org. Chem.*, **74**, 3152 (2009).

09JOC3160	S. Gracia, C. Cazorla, E. Métay, S. Pellet-Rostaing and M. Lemaire, *J. Org. Chem.*, **74**, 3160 (2009).
09JOC3225	B.J. Stokes, B. Jovanovic, H. Dong, K.J. Richert, R.D. Riell and T.G. Driver, *J. Org. Chem.*, **74**, 3225 (2009).
09JOC3478	M. Shimizu, K. Hirano, T. Satoh and M. Miura, *J. Org. Chem.*, **74**, 3478 (2009).
09JOC3491	C.J. Stearman, M. Wilson and A. Padwa, *J. Org. Chem.*, **74**, 3491 (2009).
09JOC4379	D. Du, L. Li and Z. Wang, *J. Org. Chem.*, **74**, 4379 (2009).
09JOC4537	G. Lesma, N. Landoni, T. Pilati, A. Sacchetti and A. Silvani, *J. Org. Chem.*, **74**, 4537 (2009).
09JOC4958	K.S. Feldman, D.K. Hester, II, M.R. Iyer, P.J. Munson, C.S. López and O.N. Faza, *J. Org. Chem.*, **74**, 4958 (2009).
09JOC5337	G.A. Kraus and H. Guo, *J. Org. Chem.*, **74**, 5337 (2009).
09JOC5603	G. Fridkin, N. Boutard and W.D. Lubell, *J. Org. Chem.*, **74**, 5603 (2009).
09JOC5610	M. Galezowski, J. Jazwinski, J.P. Lewtak and D.T. Gryko, *J. Org. Chem.*, **74**, 5610 (2009).
09JOC5626	W.R. Dolbier, Jr. and Z. Zheng, *J. Org. Chem.*, **74**, 5626 (2009).
09JOC6371	Q. Liao, Y. Wang, L. Zhang and C. Xi, *J. Org. Chem.*, **74**, 6371 (2009).
09JOC6402	F. Inagaki, M. Mizutani, N. Kuroda and C. Mukai, *J. Org. Chem.*, **74**, 6402 (2009).
09JOC6802	Y. Chen, C.-H. Cho, F. Shi and R.C. Larock, *J. Org. Chem.*, **74**, 6802 (2009).
09JOC6859	Z. Xu, Q. Li, L. Zhang and Y. Jia, *J. Org. Chem.*, **74**, 6859 (2009).
09JOC6874	N.-Y. Huang, M.-G. Liu and M.-W. Ding, *J. Org. Chem.*, **74**, 6874 (2009).
09JOC6878	Y. Hui, Q. Zhang, J. Jiang, L. Lin, X. Liu and X. Feng, *J. Org. Chem.*, **74**, 6878 (2009).
09JOC6899	Y.-F. Sheng, Q. Gu, A.-J. Zhang and S.-L. You, *J. Org. Chem.*, **74**, 6899 (2009).
09JOC7052	Y. Ohta, H. Chiba, S. Oishi, N. Fujii and H. Ohno, *J. Org. Chem.*, **74**, 7052 (2009).
09JOC7481	M. Yamashita, H. Horiguchi, K. Hirano, T. Satho and M. Miura, *J. Org. Chem.*, **74**, 7481 (2009).
09JOC8143	T. Ohta, T. Fukuda, F. Ishibashi and M. Iwao, *J. Org. Chem.*, **74**, 8143 (2009).
09JOC8243	I. Bergner, C. Wiebe, N. Meyer and T. Opatz, *J. Org. Chem.*, **74**, 8243 (2009).
09JOC8369	P.T. Parvatkar, P.S. Parameswaran and S.G. Tilve, *J. Org. Chem.*, **74**, 8369 (2009).
09JOC8890	J.-H. Mirebeau, M. Haddad, M. Henry-Ellinger, G. Jaouen, J. Louvel and F. Le Bideau, *J. Org. Chem.*, **74**, 8890 (2009).
09JOC8935	T. Bui, M. Borregan and C.F. Barbas, III, *J. Org. Chem.*, **74**, 8935 (2009).
09JOC9222	L. Zhou and M.P. Doyle, *J. Org. Chem.*, **74**, 9222 (2009).
09JOC9517	D.T. Gryko, O. Vakuliuk, D. Gryko and B. Koszarna, *J. Org. Chem.*, **74**, 9517 (2009).
09NC193	I.S. Young and P.S. Baran, *Nat. Chem.*, **1**, 193 (2009).
09NPR803	M. Ishikura and K. Yamada, *Nat. Prod. Rep.*, **26**, 803 (2009).
09OBC432	R.C. Hodgkinson, J. Schulz and M.C. Willis, *Org. Biomol. Chem.*, **7**, 432 (2009).
09OBC860	C. Ayats, R. Soley, F. Albericio and M. Álvarez, *Org. Biomol. Chem.*, **7**, 860 (2009).
09OBC1761	M. Brichacek and J.T. Njardarson, *Org. Biomol. Chem.*, **7**, 1761 (2009).
09OBC3382	D. Solé and O. Serrano, *Org. Biomol. Chem.*, **7**, 3382 (2009).
09OBC4512	C. Berini, N. Pelloux-Léon, F. Minassian and J.-N. Denis, *Org. Biomol. Chem.*, **7**, 4512 (2009).
09OBC4744	H.-S. Yeom, Y. Lee, J.-E. Lee and S. Shin, *Org. Biomol. Chem.*, **7**, 4744 (2009).
09OL49	A.Y. Bitar and A.J. Frontier, *Org. Lett.*, **11**, 49 (2009).
09OL77	D. Sánchez-García, J.I. Borrell and S. Nonell, *Org. Lett.*, **11**, 77 (2009).
09OL173	Y. Chen, C.-H. Cho and R.C. Larock, *Org. Lett.*, **11**, 173 (2009).
09OL197	K. Higuchi, Y. Sato, M. Tsuchimochi, K. Sugiura, M. Hatori and T. Kawasaki, *Org. Lett.*, **11**, 197 (2009).
09OL201	K.R. Buszek, N. Brown and K. Luo, *Org. Lett.*, **11**, 201 (2009).
09OL221	K. Dooleweerdt, T. Ruhland and T. Skrydstrup, *Org. Lett.*, **11**, 221 (2009).
09OL441	G. Blay, I. Fernández, A. Monleón, R. Pedro and C. Vila, *Org. Lett.*, **11**, 441 (2009).
09OL887	R.S. Klausen and E.N. Jacobsen, *Org. Lett.*, **11**, 887 (2009).
09OL1007	S.M. Bronner, K.B. Bahnck and N.K. Garg, *Org. Lett.*, **11**, 1007 (2009).
09OL1051	M.D. Morrison, J.J. Hanthorn and D.A. Pratt, *Org. Lett.*, **11**, 1051 (2009).

09OL1111	P.J. Schupp, C. Kohlert-Schupp, W.Y. Yoshida and T.K. Hemscheidt, *Org. Lett.*, **11**, 1111 (2009).
09OL1175	A.C. Silvanus, S.J. Heffernan, D.J. Liptrot, G. Kociok-Kühn, B.I. Andrews and D.R. Carbery, *Org. Lett.*, **11**, 1175 (2009).
09OL1233	S. Kiren, X. Hong, C.A. Leverett and A. Padwa, *Org. Lett.*, **11**, 1233 (2009).
09OL1329	N. Isono and M. Lautens, *Org. Lett.*, **11**, 1329 (2009).
09OL1369	Y. Lu and B.A. Arndtsen, *Org. Lett.*, **11**, 1369 (2009).
09OL1555	P. Fontaine, G. Masson and J. Zhu, *Org. Lett.*, **11**, 1555 (2009).
09OL1651	B.E. Maki and K.A. Scheidt, *Org. Lett.*, **11**, 1651 (2009).
09OL1693	H. Kikuchi, M. Sekiya, Y. Katou, K. Ueda, T. Kabeya, S. Kurata and Y. Oshima, *Org. Lett.*, **11**, 1693 (2009).
09OL2031	L. Ackermann, R. Sandmann and L.T. Kaspar, *Org. Lett.*, **11**, 2031 (2009).
09OL2085	G. Sirasani and R.B. Andrade, *Org. Lett.*, **11**, 2085 (2009).
09OL2129	T. Tsuchimoto, T. Wagatsuma, K. Aoki and J. Shimotori, *Org. Lett.*, **11**, 2129 (2009).
09OL2141	T. Miura, T. Toyoshima, Y. Takahashi and M. Murakami, *Org. Lett.*, **11**, 2141 (2009).
09OL2269	E. Merkul, C. Boersch, W. Frank and T.J.J. Müller, *Org. Lett.*, **11**, 2269 (2009).
09OL2293	P.W. Davies and N. Martin, *Org. Lett.*, **11**, 2293 (2009).
09OL2337	M. Yamashita, K. Hirano, T. Satoh and M. Miura, *Org. Lett.*, **11**, 2337 (2009).
09OL2579	M.J. Wanner, R.N.A. Boots, B. Eradus, R. de Gelder, J.H. van Maarseveen and H. Hiemstra, *Org. Lett.*, **11**, 2579 (2009).
09OL2780	B. Bajtos and B.L. Pagenkopf, *Org. Lett.*, **11**, 2780 (2009).
09OL2920	R. Surmont, G. Verniest and N. De Kimpe, *Org. Lett.*, **11**, 2920 (2009).
09OL3458	B.W. Boal, A.W Schammel and N.K. Garg, *Org. Lett.*, **11**, 3458 (2009).
09OL3914	M. Blangetti, A. Deagostino, C. Prandi, S. Tabasso and P. Venturello, *Org. Lett.*, **11**, 3914 (2009).
09OL4002	X. Zhao, E. Zhang, Y.-Q. Tu, Y.-Q. Zhang, D.-Y. Yuan, K. Cao, C.-A. Fan and F.-M. Zhang, *Org. Lett.*, **11**, 4002 (2009).
09OL4200	A.J. Smith, L.K. Abott and S.F. Martin, *Org. Lett.*, **11**, 4200 (2009).
09OL4282	D. Ciez, *Org. Lett.*, **11**, 4282 (2009).
09OL4532	R.G. Vaswani, J.J. Day and J.L. Wood, *Org. Lett.*, **11**, 4532 (2009).
09OL4608	M. Arthuis, R. Pontikis and J.-C. Florent, *Org. Lett.*, **11**, 4608 (2009).
09OL4834	T. Feng, X.-H. Cai, Y. Li, Y.-Y. Wang, Y.-P. Liu, M.-J. Xie and X.-D. Luo, *Org. Lett.*, **11**, 4834 (2009).
09OL5002	M. Egi, K. Azechi and S. Akai, *Org. Lett.*, **11**, 5002 (2009).
09OL5142	M. Jeanty, J. Blu, F. Suzenet and G. Guillaumet, *Org. Lett.*, **11**, 5142 (2009).
09OL5162	Y. Yamane, X. Liu, A. Hamasaki, T. Ishida, M. Haruta, T. Yokoyama and M. Tokunaga, *Org. Lett.*, **11**, 5162 (2009).
09OL5454	I.-K. Park, S.-E. Suh, B.-Y. Lim and C.-G. Cho, *Org. Lett.*, **11**, 5454 (2009).
09OL5606	Y. Yang, K. Cheng and Y. Zhang, *Org. Lett.*, **11**, 5606 (2009).
09OPRD1199	M.L. Laurila, N.A. Magnus and M.A. Staszak, *Org. Proc. Res. Dev.*, **13**, 1199 (2009).
09S227	E. Bellur, M.A. Yawer, A. Riahi, O. Fatunsin, C. Fischer and P. Langer, *Synthesis*, 227 (2009).
09S311	K.C. Majumdar, B. Chattopadhyay and S. Samanta, *Synthesis*, 311 (2009).
09S587	A.I. MIkhaleva, A.V. Ivanov, E.V. Skital'steva, I.A. Ushakov, A.M. Vasil'tsov and B.A. Trofimov, *Synthesis*, 587 (2009).
09S1147	R. Voloshchuk, M. Galezowski and D.T. Gryko, *Synthesis*, 1147 (2009).
09S1485	I.R. Baxendale, C.D. Buckle, S.V. Ley and L. Tamborini, *Synthesis*, 1485 (2009).
09S1520	J.S. Yadav, B.V.S. Reddy, Y.J. Reddy and K. Praneeth, *Synthesis*, 1520 (2009).
09S1786	S. Fukamachi, H. Konishi and K. Kobayashi, *Synthesis*, 1786 (2009).
09S2447	P. Kassis, V. Bénéteau, J.-Y. Mérour and S. Routier, *Synthesis*, 2447 (2009).
09S2454	M. Yoshida, M. Al-Amin and K. Shishido, *Synthesis*, 2454 (2009).
09S2531	L. Hill, G.A. Hunter, S.H. Imam, H. McNab and W.J. O'Neill, *Synthesis*, 2531 (2009).
09S2535	L. Hill, S.H. Imam, H. McNab and W.J. O'Neill, *Synthesis*, 2535 (2009).
09S3003	B.M. Trost and M.K. Brennan, *Synthesis*, 3003 (2009).
09S3162	K.S. Feldman, M.D. Fodor and A.P. Skoumbourdis, *Synthesis*, 3162 (2009).
09S3493	L. Ackermann, A. Althammer and P. Mayer, *Synthesis*, 3493 (2009).

09S3617	A. Seggio, G. Priem, F. Chevallier and F. Mongin, *Synthesis*, 3617 (2009).
09S3905	V.M. Muzalevskiy, A.V. Shastin, E.S. Balenkova, G. Haufe and V.G. Nenajdenko, *Synthesis*, 3905 (2009).
09S3994	Z.-J. Wang, J. Jian-Guo, X. Lv and W. Bao, *Synthesis*, 3994 (2009).
09S4119	D. Enders, C. Wang and G. Raabe, *Synthesis*, 4119 (2009).
09S4183	X.-L. Fang, R.-Y. Tang, P. Zhong and J.-H. Li, *Synthesis*, 4183 (2009).
09SC531	R. Balamurugan, R. Sureshbabu, G.G. Rajeshwaran and A.K. Mohanakrishnan, *Synth. Commun.*, **39**, 531 (2009).
09SC1914	F.A. Davis, J. Zhang, Y. Zhang and H. Qiu, *Synth. Commun.*, **39**, 1914 (2009).
09SC2506	A. Sudhakara, H. Jayadevappa, K.M. Mahadevan and V. Hulikal, *Synth. Commun.*, **39**, 2506 (2009).
09SC3833	X. Jing, X. Pan, Z. Li, X. Bi, C. Yan and H. Zhu, *Synth. Commun.*, **38**, 3833 (2009).
09SL112	H.M. Gillis, L. Greene and A. Thompson, *Synlett*, 112 (2009).
09SL456	U. Groth, V.O. Iarochenko, Y. Wang and T. Wesch, *Synlett*, 456 (2009).
09SL755	N. Zanetta, A.D. Wouters, L. Fantinel, F.M. da Silva, R. Barichello, P.E.A. da Silva, D.F. Ramos, H.G. Bonacorso and A.P. Martins, *Synlett*, 755 (2009).
09SL763	W. Fu, C. Xu, G. Zou, D. Hong, D. Deng, Z. Wang and B. Ji, *Synlett*, 763 (2009).
09SL808	L. Ackermann and S. Barfüber, *Synlett*, 808 (2009).
09SL838	F. Ullah, T.T. Dang, J. Heinicke, A. Villinger and P. Langer, *Synlett*, 838 (2009).
09SL1115	M. Anary-Abbasinejad, K. Charkhati and H. Anaraki-Ardakani, *Synlett*, 1115 (2009).
09SL1219	L. Ackermann, R. Sandmann and M.V. Kondrashov, *Synlett*, 1219 (2009).
09SL1817	S. Cacchi, G. Fabrizi and E. Filisti, *Synlett*, 1817 (2009).
09SL2115	T. Tian, B.-J. Pei, Q.-H. Li, H. He, L.-Y. Chen, X. Zhou, W.-H. Chan and A.W.M. Lee, *Synlett*, 2115 (2009).
09SL2245	N. Azizi, A. Khajeh-Amiri, H. Ghafuri, M. Bolourtchian and M.R. Saidi, *Synlett*, 2245 (2009).
09SL2529	J.-Y. Wang, S.-P. Liu and W. Yu, *Synlett*, 2529 (2009).
09SL2617	S. Gracia, E. Métay, S. Pellet-Rostaing and M. Lemaire, *Synlett*, 2617 (2009).
09SL3016	N. Wache and J. Christoffers, *Synlett*, 3016 (2009).
09SL3336	F. Dumitrascu, E. Georgescu, M.R. Caira, F. Georgescu, M. Popa, B. Draghici and D.G. Dumitrescu, *Synlett*, 3336 (2009).
09T461	E. Yasui, M. Wada and N. Takamura, *Tetrahedron*, **65**, 461 (2009).
09T1268	P. Wyrebek, A. Sniady, N. Bewick, Y. Li, A. Mikus, K.A. Wheeler and R. Dembinski, *Tetrahedron*, **65**, 1268 (2009).
09T1424	Y.-J. Bian, X.-Y. Liu, K.-G. Ji, X.-Z. Shu, L.-N. Guo and Y.-M. Liang, *Tetrahedron*, **65**, 1424 (2009).
09T2067	I. Yavari, M. Piltan and L. Moradi, *Tetrahedron*, **65**, 2067 (2009).
09T3062	T.H.H. Hsieh and V.M. Dong, *Tetrahedron*, **65**, 3062 (2009).
09T3096	J.S. Schneekloth, Jr., J. Kim and E.J. Sorensen, *Tetrahedron*, **65**, 3096 (2009).
09T3120	S.E. Denmark and J.D. Baird, *Tetrahedron*, **65**, 3120 (2009).
09T3149	K. Eastman and P.S. Baran, *Tetrahedron*, **65**, 3149 (2009).
09T3582	R. Sureshbabu, R. Balamurugan and A.K. Mohanakrishnan, *Tetrahedron*, **65**, 3582 (2009).
09T3829	A.A. Lamar and K.M. Nicholas, *Tetrahedron*, **65**, 3829 (2009).
09T3842	J. Oyamada and T. Kitamura, *Tetrahedron*, **65**, 3842 (2009).
09T4212	W. Wan, J. Hou, H. Jiang, Y. Wang, S. Zhu, H. Deng and J. Hao, *Tetrahedron*, **65**, 4212 (2009).
09T4283	J.T. Gupton, B.C. Giglio, J.E. Eaton, E.A. Rieck, K.L. Smith, M.J. Keough, P.J. Barelli, L.T. Firich, J.E. Hempel, T.M. Smith and R.P.F. Kanters, *Tetrahedron*, **65**, 4283 (2009).
09T4619	R.K. Rao, A.B. Naidu, E.A. Jaseer and G. Sekar, *Tetrahedron*, **65**, 4619 (2009).
09T6720	S. Kiren, X. Hong, C.A. Leverett and A. Padwa, *Tetrahedron*, **65**, 6720 (2009).
09T7523	K. Kobayashi, K. Iitsuka, S. Fukamachi and H. Konishi, *Tetrahedron*, **65**, 7523 (2009).
09T8389	M.-C. Liu and C.W. Ong, *Tetrahedron*, **65**, 8389 (2009).
09T8916	I. Ambrogio, S. Cacchi, G. Fabrizi and A. Prastaro, *Tetrahedron*, **65**, 8916 (2009).
09T8930	L. Ackermann, R. Sandmann, M. Schinkel and M.V. Kondrashov, *Tetrahedron*, **65**, 8930 (2009).

09T8961	E. Li, X. Xu, H. Li, H. Zhang, X. Xu, X. Yuan and Y. Li, *Tetrahedron*, **65**, 8961 (2009).
09T8981	S.J. Pridmore, P.A. Slatford, J.E. Taylor, M.K. Whittlesey and J.M.J. Williams, *Tetrahedron*, **65**, 8981 (2009).
09T8995	M.A. Honey, A.J. Blake, I.B. Campbell, B.D. Judkins and C.J. Moody, *Tetrahedron*, **65**, 8995 (2009).
09T9075	C. Hu, H. Qin, Y. Cui and Y. Jia, *Tetrahedron*, **65**, 9075 (2009).
09T9702	K. Harju, N. Manevski and J. Yli-Kauhaluoma, *Tetrahedron*, **65**, 9702 (2009).
09T10459	S. Haneda, Y. Adachi and M. Hayashi, *Tetrahedron*, **65**, 10459 (2009).
09TA1719	K. Sarkar, S.K. Singha and S.K. Chattopadhyay, *Tetrahedron. Asymmetry*, **20**, 1719 (2009).
09TA2193	S. Anas and H.B. Kagan, *Tetrahedron. Asymmetry*, **20**, 2193 (2009).
09TL63	N. Brown, D. Luo, D. Vander Velde, S. Yang, A. Brassfield and K.R. Buszek, *Tetrahedron Lett.*, **50**, 63 (2009).
09TL347	X.-Q. Pan, M.-Y. Lei, J.-L. Zou and W. Zhang, *Tetrahedron Lett.*, **50**, 347 (2009).
09TL752	K.-H. Lim, T. Etoh, M. Hayashi, K. Komiyama and T.-S. Kam, *Tetrahedron Lett.*, **50**, 752 (2009).
09TL763	Y. Gu and X.-m. Wang, *Tetrahedron Lett.*, **50**, 763 (2009).
09TL1529	C.-y. Chen and R.A. Reamer, *Tetrahedron Lett.*, **50**, 1529 (2009).
09TL1708	F. Lehmann, M. Holm and S. Laufer, *Tetrahedron Lett.*, **50**, 1708 (2009).
09TL2075	I. Nakamura, Y. Sato, S. Konta and M. Terada, *Tetrahedron Lett.*, **50**, 2075 (2009).
09TL2943	K. Okuma, J.-i. Seto, K.-i. Sakaguchi, S. Ozaki, N. Nagahora and K. Shioji, *Tetrahedron Lett.*, **50**, 2943 (2009).
09TL3145	D.R. Bobeck, S. France, C.A. Leverett, F. Sánchez-Cantalejo and A. Padwa, *Tetrahedron Lett.*, **50**, 3145 (2009).
09TL3283	V. Boissel, N.S. Simpkins and G. Bhalay, *Tetrahedron Lett.*, **50**, 3283 (2009).
09TL3290	K. Hisler, A.G.J. Commeureuc, S.-z. Zhou and J.A. Murphy, *Tetrahedron Lett.*, **50**, 3290 (2009).
09TL3853	T. Otani, S. Kunimatsu, T. Takahashi, H. Nihei and T. Saito, *Tetrahedron Lett.*, **50**, 3853 (2009).
09TL3889	K.C. Majumdar, A. Taher and K. Ray, *Tetrahedron Lett.*, **50**, 3889 (2009).
09TL3952	J.J. Song, Z. Tan, J.T. Reeves, D.R. Fandrick, H. Lee, N.K. Yee and C.H. Senanayake, *Tetrahedron Lett.*, **50**, 3952 (2009).
09TL3959	G. Shanthi and P.T. Perumal, *Tetrahedron Lett.*, **50**, 3959 (2009).
09TL4003	D.C. Rogness and R.C. Larock, *Tetrahedron Lett.*, **50**, 4003 (2009).
09TL4423	R. Sanz, V. Guilarete and A. Pérez, *Tetrahedron Lett.*, **50**, 4423 (2009).
09TL4762	E. Yasui, M. Wada and N. Takamura, *Tetrahedron Lett.*, **50**, 4762 (2009).
09TL4807	M.A. Wilson, G. Filzen and G.S. Welmaker, *Tetrahedron Lett.*, **50**, 4807 (2009).
09TL4878	M. Layek, A.V.D. Rao, V. Gajare, D. Kalita, D.K. Barange, A. Islam, K. Mukkanti and M. Pal, *Tetrahedron Lett.*, **50**, 4878 (2009).
09TL5445	S. Rivera, D. Bandyopadhyay and B.K. Banik, *Tetrahedron Lett.*, **50**, 5445 (2009).
09TL5602	K. Akagawa, T. Yamashita, S. Sakamoto and K. Kudo, *Tetrahedron Lett.*, **50**, 5602 (2009).
09TL6268	M. Yoshida, M. Al-Amin and K. Shishido, *Tetrahedron Lett.*, **50**, 6268 (2009).
09TL6342	M.W. Smith, R. Hunter, D.J. Patten and W. Hinz, *Tetrahedron Lett.*, **50**, 6342 (2009).
09TL6562	H.B. Borate, S.P. Sawargave and S.R. Maujan, *Tetrahedron Lett.*, **50**, 6562 (2009).
09TL6807	D.V. Nguyen, R.A. Schiksnis and E.L. Michelotti, *Tetrahedron Lett.*, **50**, 6807 (2009).
09TL6877	A. Carpita and A. Ribecai, *Tetrahedron Lett.*, **50**, 6877 (2009).
09TL6944	D.-D. Chen, X.-L. Hou and L.-X. Dai, *Tetrahedron Lett.*, **50**, 6944 (2009).
09TL6981	Y. Shang, M. Zhang, S. Yu, K. Ju, C. Wang and X. He, *Tetrahedron Lett.*, **50**, 6981 (2009).
09TL7113	N. Brown, D. Luo, J.A. Decapo and K.R. Buszek, *Tetrahedron Lett.*, **50**, 7113 (2009).
09TL7169	H.-C. Hsu and D.-R. Hou, *Tetrahedron Lett.*, **50**, 7169 (2009).

CHAPTER 5.3

Five-Membered Ring Systems: Furans and Benzofurans*

Kap-Sun Yeung*, Zhen Yang**, Xiao-Shui Peng[†], Xue-Long Hou[‡]

*Bristol-Myers Squibb Research and Development, 5 Research Parkway, P.O.Box 5100, Wallingford, Connecticut 06492, USA
kapsun.yeung@bms.com
**Key Laboratory of Bioorganic Chemistry and Molecular Engineering of Ministry of Education, Department of Chemical Biology, College of Chemistry, Peking University, Beijing 100871, China
zyang@pku.edu.cn
[†]Department of Chemistry, The Chinese University of Hong Kong, Shatin, New Territories, Hong Kong SAR, China
xspeng@cuhk.edu.hk
[‡]Shanghai-Hong Kong Joint Laboratory in Chemical Synthesis and State Key Laboratory of Organometallic Chemistry, Shanghai Institute of Organic Chemistry, The Chinese Academy of Sciences, 354 Feng Lin Road, Shanghai 200032, China
xlhou@mail.sioc.ac.cn

5.3.1. INTRODUCTION

This article aims to review papers that were published in 2009 on reactions and syntheses of furans, benzofurans and their derivatives. Reviews published in 2009 covered the recent syntheses of furans and 2,5-dihydrofurans <09H(78)1109>, synthesis of furans <09S3353>, synthesis of 2,5-dihydro- and tetrahydrofurans <09T10745>, synthesis of tetrahydrofurans <09P1930; 09TA2537> and synthesis of 2,5-dihydrofurans <09CSR3222>; 09OBC1761>.

Reviews on the synthesis of naturally-occurring polyketide-derived or polycyclic ether-derived tetrahydrofurans were also published <09NPR266; 09S2651; 09SL1525>.

Many new naturally-occurring molecules containing tetrahydrofuran and dihydrofuran rings were identified in 2009. References on compounds whose biological activities were not mentioned are: <09H(78)2565; 09HCA165; 09HCA370; 09HCA409; 09HCA1873; 09HCA2361; 09JNP2005; 09OL3012; 09TL467; 09TL4747>.

Articles on those naturally-occurring compounds containing tetrahydrofuran or dihydrofuran skeletons whose biological activities were assessed are: <09HCA1341; 09HCA2746; 09JNP190; 09JNP912; 09JNP921; 09JNP994; 09JNP1678; 09JNP2040; 09JOC5502; 09OL57; 09OL1353; 09OL2153; 09P1277; 09T164; 09T7016; 09TA381; 09TL5182>.

References on those furan-containing compounds whose biological activities were not mentioned are: <09HCA121; 09HCA1191; 09HCA2071; 09JNP1305;

*Dedicated to Professor Henry N. C. Wong on the occasion of his 60th birthday.

09JNP1361; 09JNP1657; 09JNP857; 09JNP2084; 09OL617; 09OL2281; 09P635; 09P1305>; 09T3425; 09TL2132>.

Naturally-occurring compounds containing furan skeletons whose biological activities were assessed were mentioned in the following papers: <09HCA139; 09HCA1184; 09HCA2101; 08HCA2338; 09JNP685; 09JNP714; 09JNP976; 09JNP1213; 09JNP1331; 09JNP1870; 09JNP2110; 09P256; 09P1233; 09P2047; 09T1708; 09T6029; 09T7408>.

References of those benzo[b]furan- or dihydrobenzo[b]furan-containing compounds whose biological activities were not mentioned are: <09H(77)793; 09H(78)1557; 09H(78)1581; 09HCA195; 09HCA928; 09HCA1203; 09JNP966; 09P1474; 09TL2516>.

References on those naturally occurring compounds containing benzo[b]furan or dihydrobenzo[b]furan skeletons whose biological activities were assessed are: <09HCA1260; 09JNP63; 09JNP194; 09JNP573; 09JNP621; 09JNP852; 09JNP976; 09JNP1379; 09JNP1529; 09JNP1702; 09JNP1980; 09P216; 09P403; 09P1309; 09P2053>.

5.3.2. REACTIONS

5.3.2.1 Furans

Cycloadditions of furans continued to be developed in 2009. The reactivity of 5-aminofuraldehydes in Diels–Alder reactions can be increased by conversion into 2-furylimine derivatives <09OL1817>. The intramolecular [4+2] cycloaddition of a complex amidofuran was applied to the synthesis of the core of welwitindolinone, as shown below <09OL3782>.

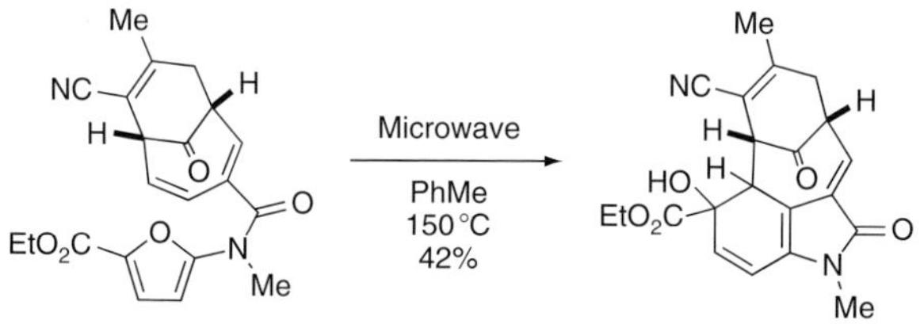

The intramolecular amidofuran Diels–Alder reaction approach was extended to the synthesis of 4-substituted indoles from β-amino-alcohol substrates, which underwent cycloaddition and subsequent double dehydrative aromatization under microwave-promoted conditions, as exemplified below <09CC104>.

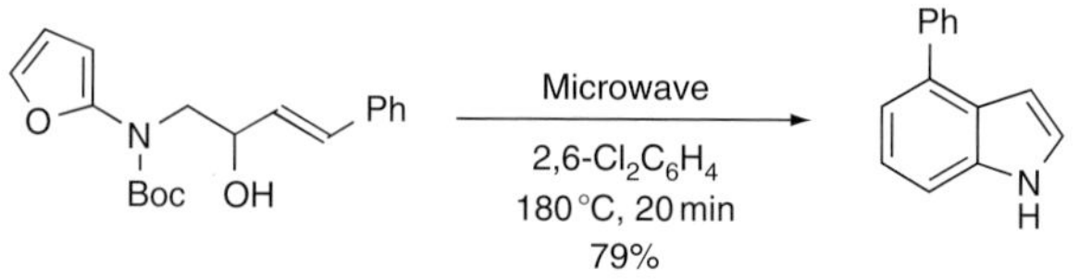

Another variant of intramolecular amidofuran Diels−Alder reaction provided, after subsequent elimination of MeOH, indolines or tetrahydroquinolines depending on the tether length <09OL4462>. An interesting example is shown below.

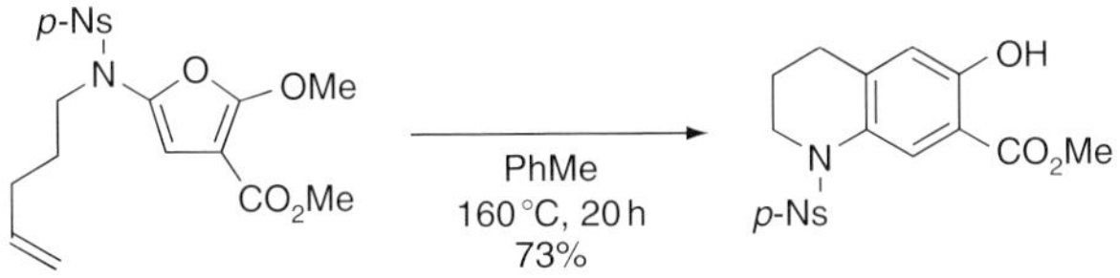

In a synthesis of the BCD rings of cortistatin A, a [2+4] cyclopropene-furan cycloaddition occurred after addition of cyclopropenyllithium to the aldehyde shown below to provide the *exo* product <09OL3938>. Subsequent cyclopropylcarbinol rearrangement transformed the oxatricycle into the 7-membered B-ring.

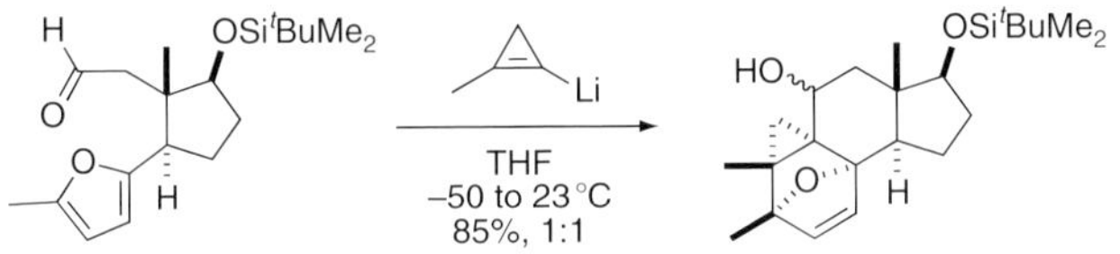

The diastereoselectivity of a Diels−Alder cycloaddition of chiral furanyl sultams with benzynes was influenced by substitution on the furan ring. For example, the 3-methylfuran substrate shown below provided a substantial increase in selectivity relative to the unsubstituted parent <09OL4688>.

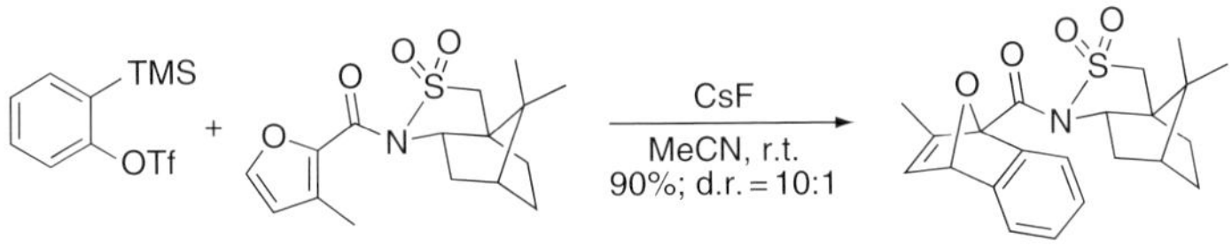

Furan participated in gold(I)-catalyzed intramolecular [4+2] and [4+3] cycloadditions with allenes at room temperature, depending on the ligands used. For example, Au(I) complexed with an arylphosphite π-accepting ligand promoted the [4+2] reaction <09JA6348>. Whilst employing a σ-donating N-imidazolium carbene (IPr) ligand, the [4+3] cycloaddition was the dominant pathway, as shown below <09CEJ3336>. This reaction was shown in 2008 to proceed under $PtCl_2$ catalysis and an atmosphere of CO at high temperature.

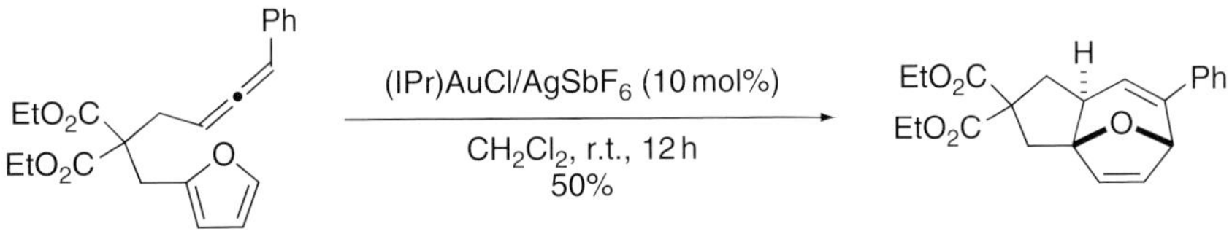

Another interesting example of the Pt-catalyzed intramolecular [4+3] cycloaddition of an allene furanophane followed by rearrangement is shown below <09TL2685>.

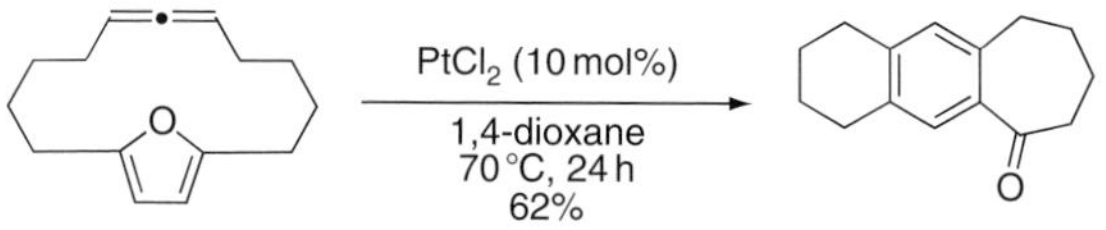

The intramolecular [4+2] cycloaddition of furans with terminal unsubstituted allenamides, however, proceeded under thermal uncatalyzed conditions, as illustrated below <09OL3430>.

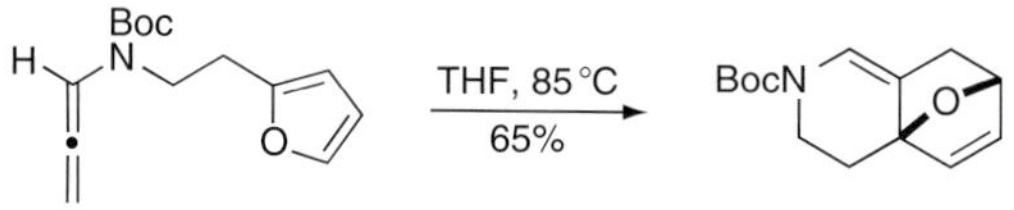

Furan reacted with dioxins, as oxyallyl cation equivalents in a Au/Ag-catalyzed [4+3] cycloaddition <09TL5701>. The [4+3] cycloaddition of furan with an epoxy enol silane-derived oxyallyl cation was optimized by tuning the reactivity of the enol silane with a more bulky triethylsilyl group and conducting the reaction at low temperature, as illustrated below <09JA4556>.

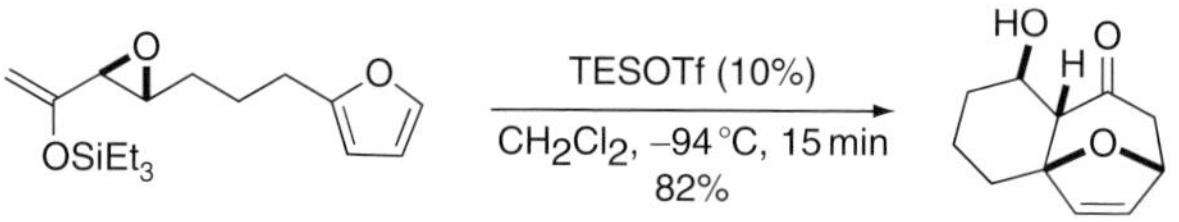

A respectable enantiomeric excess of 55% was achieved in the challenging enantiotopic selective cycloisomerization of furyldialkynes catalyzed by a chiral gold complex <09CEJ13318>. A new gold-catalyzed intramolecular *6-endo-dig* cyclization of a furan-phenoxyalkyne was shown to proceed rapidly at room temperature to furnish tetracyclic structures, as shown below. This reaction is more efficient for mono-substituted furan substrates <09AGE5848>.

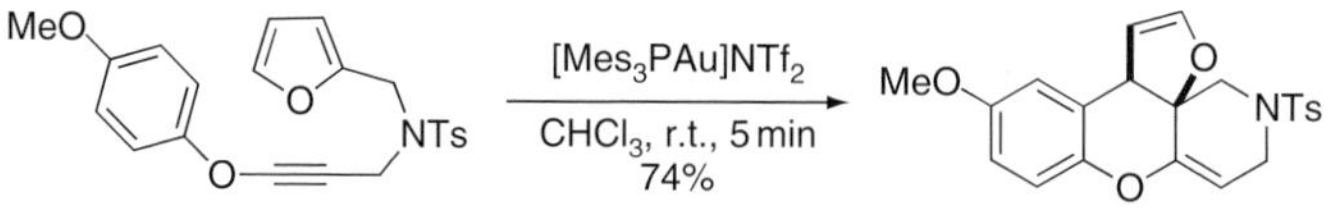

A related gold-catalyzed reaction of furan, as illustrated below, is a one-pot tandem Friedel–Crafts/furan-alkyne cycloisomerization with a pent-2-en-4-yn-1-ol <09OL3838>.

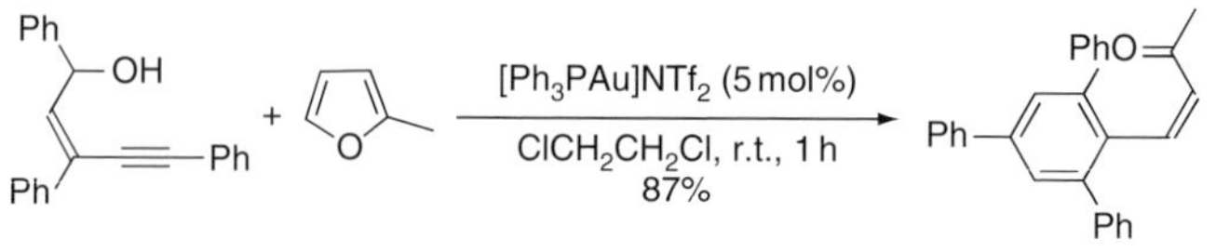

As shown in the representative example below, a one-pot four-component reaction of 2-furaldehyde furnished isoindolinone derivatives after a sequence of Ugi condensation, intramolecular Diels—Alder reaction and aromatization <09JOC8859>.

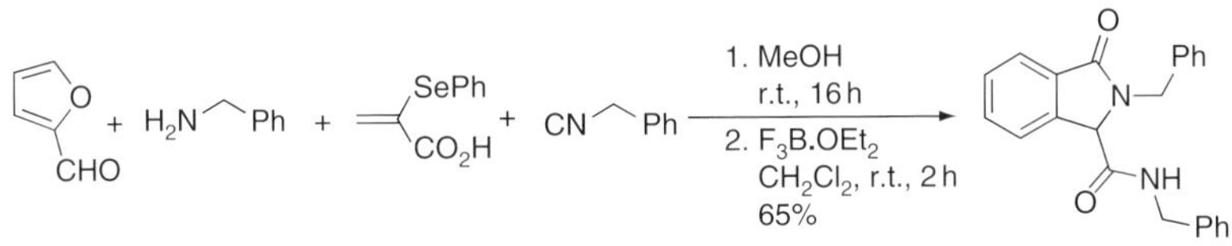

Rh- or Cu-catalyzed reaction of furans with α-diazomethanesulfonate and phosphonate provided a mixture of cyclopropane and ring—opened 1,4-diene products <08T7092>. Furans reacted with 9-ethynyfluoren-9-yl acetate in a Pt/Pd mixed metal-catalyzed one-pot reaction to generate 9-fluorenylidenes regio- and stereoselectively <09SL1937>. The initial mixture of (Z,Z)- and (Z,E)-products, formed from the regioselective reaction of furans with a platinum carbenoid, was isomerized to the all (E,E)-isomer under palladium catalysis. As represented below, reaction of 2-lithio-5-methoxyfuran with magnesium alkylidene carbenoids, generated from vinyl sulfoxides, provided interesting allenes conjugated with α,β-unsaturated methyl esters <08T3509>.

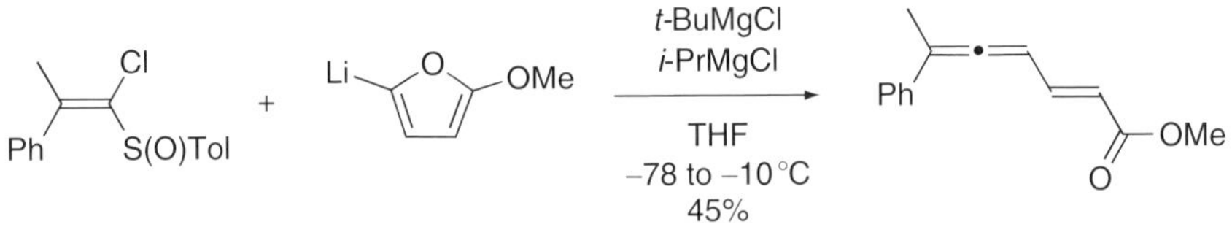

Pd-catalyzed heterogeneous hydrogenation of a chiral furan-2-carboxamide in the presence of titanium tetraethoxide in NMP furnished the corresponding tetrahydrofuran in 95% de <09SL461>. An iodine-mediated oxidative rearrangement of a 2-furanyl carbamate provided 5-methoxypyrrol-2(5H)-one, which was used as a versatile intermediate for the preparation of 2,4-disubstituted pyrroles <09OL1233>. A one-pot asymmetric synthesis of pyranones from furfurals through the enantioselective addition of dialkylzinc to the aldehyde moiety and subsequent NBS-promoted Achmatowicz rearrangement of the furfuryl alcohol, was developed <09OL2703>. Achmatowicz rearrangements of two chiral furfuryl alcohols were used to construct two tetrahydropyran units during a total synthesis of the complex marine polyether norhalichondrin B <09AGE2346>. As illustrated below, a "homologous" Achmatowicz oxidation of 2-(β-hydroxyalkyl)furans by singlet oxygen produced 3-keto-tetrahydrofurans, presumably via a Michael addition to the 1,4-enedione intermediate that is derived from the decomposition of the furan endo-peroxide <09OL313>. However, 2-substituted-5-unsubstituted furans did not furnish the expected products.

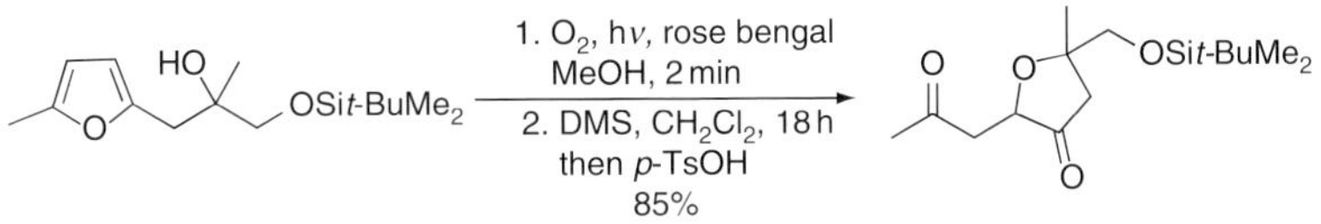

Interestingly, in the presence of an additional γ-hydroxyl group, the initially formed furan *endo*-peroxide intermediate could be trapped by the γ-hydroxyl, leading to a γ-spiroketal, the major isomer being shown below <09OL4556, 09JOC9546>.

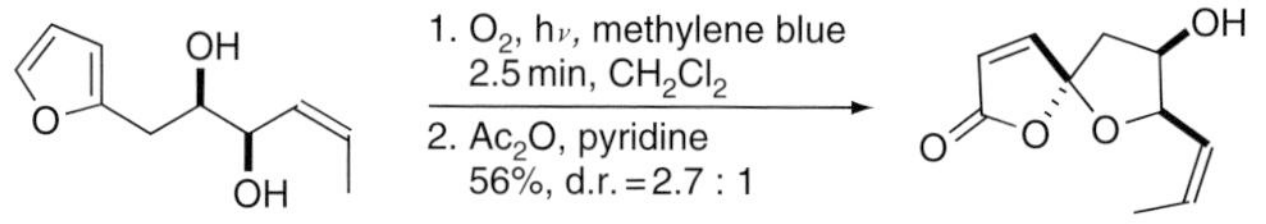

A 2-trimethylsilylfuran was oxidized regioselectively to an α,β-butenolide using singlet oxygen during a biomimetic total synthesis of (±)-pallavicibolide A, as shown below <09AGE2351>. Notably, the two pendant alkene moieties remained intact during the photooxidation. The desired diastereomer was obtained in 45% yield after deprotection of the acetonide.

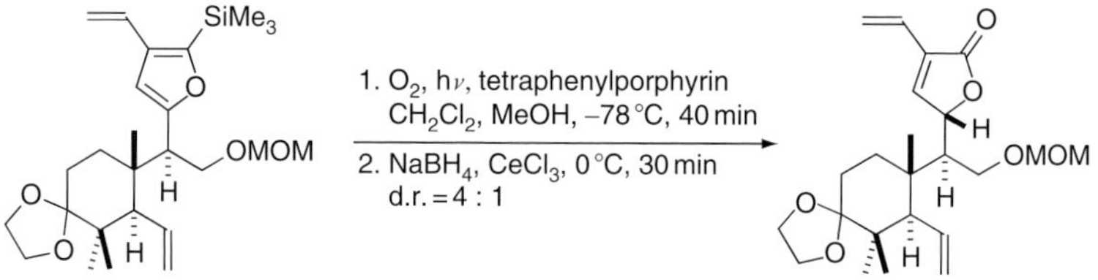

Recent developments in various asymmetric addition reactions of silyloxyfurans were reviewed <09SL1525>. The diastereo- and enantioselective addition of trimethylsilyloxyfuran to α-keto-imine esters as catalyzed by a chiral Ag/imine catalyst <09JA570>, as well as to methyl pyruvate, catalyzed by a Cu/sulfoximine catalyst <09CEJ1566>, were also reported. Boron trifluoride-promoted Michael addition of furan and 2-trimethylsilyloxyfuran to dienyl sulfonamides followed by aromatization provided 3-substituted *N*-phenylsulfonamides <09EJO3871>. Addition of 2-trimethylsilyloxyfuran to nitroalkenes was catalyzed by a BINOL-based phosphoric acid <09SL2629>. As shown in the following prototypical example, vinylogous Mannich addition of silyloxyfurans to *N*-acyl- as well as -sulfonylisoquinolinium ions is feasible, providing adducts with high diastereoselectivity <09OL4044>. A Diels–Alder like transition state was proposed for this reaction.

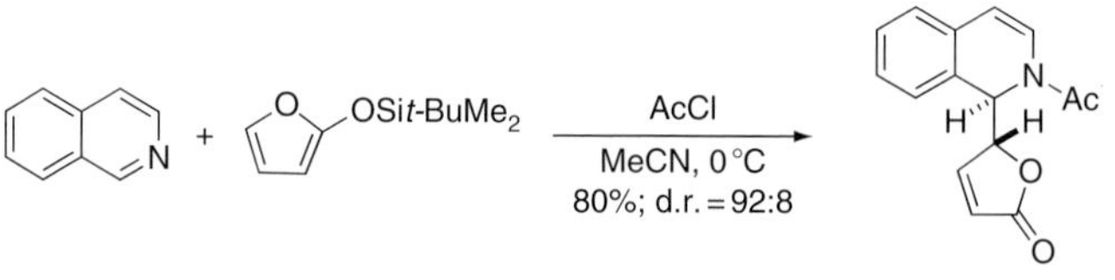

5.3.2.2 Di- and Tetrahydrofurans

A one-pot synthesis of N-substituted 3-trifluoroacetylpyrroles, proceeded *via* a ring opening 1,4-addition of primary amines to 3-trifluoroacetyl-4,5-dihydrofuran followed by oxidation using PCC and cyclization <09SL755>. 2,3-Dihydrofuran effectively trapped a phenyl cation generated by photo-induced heterolytic cleavage of a chlorobenzene. Addition of methanol to the intermediate oxonium ion provided a 2,3-*cis*-disubstituted product, as shown below <09OL349>.

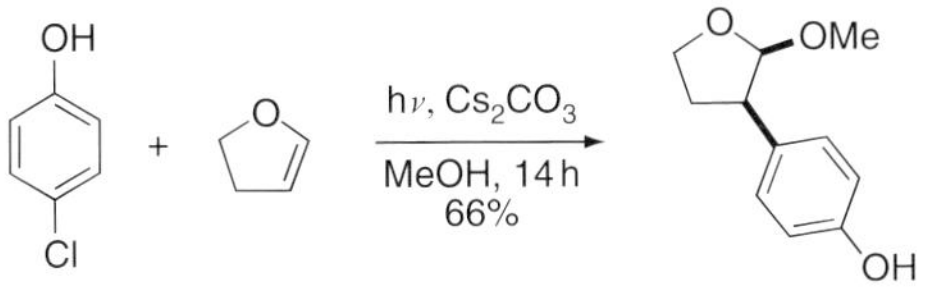

Isomerization of 7-oxabenzonorbornadienes to naphthols was catalyzed by [RuCl$_2$(CO)$_3$]$_2$. The regioselectivity of this reaction was determined by the electronic nature of the C–O bond <09JOC7570>. Ring-opening of 7-oxabenzonorbornadienes by alkylaluminum reagents, catalyzed by copper thiophene-2-carboxylate and using SimplePhos as a chiral ligand, provided 1,2-*trans* alcohol products with up to 94% ee <09TL3474>. An unusual regiodivergent resolution of racemic unsymmetrical 7-oxabenzonorbornadienes under cationic Rh-catalyzed conditions was discovered, as shown in the following example. The reagent-controlled pathway provided the tertiary alcohol with high enantioselectivity <09JA444>.

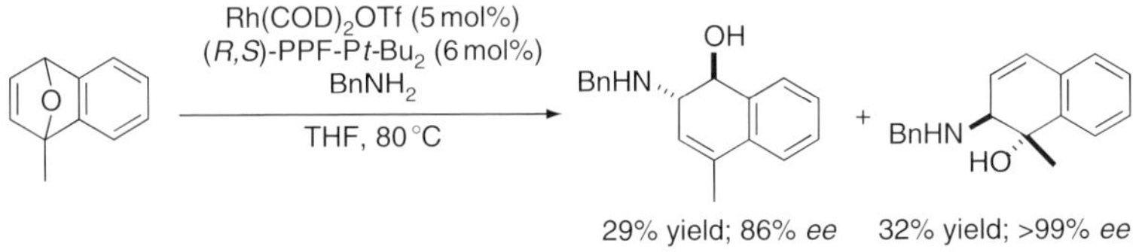

Regioselective ring-opening allylation and azidation of 2-aryl-2,5-dihydrofurans were promoted by a gold catalyst <09OL5304>. While allylation occurred at the 2-position, as illustrated below, azidation with trimethylsilylazide occurred at the 4-position. The observed epimerization of the C-2 stereogenic center was probably due to the involvement of a benzyl cation intermediate.

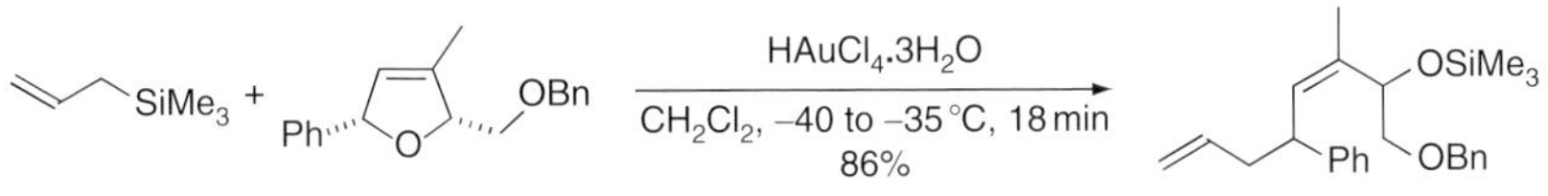

A highly Z- and enantioselective ring opening/cross metathesis reaction of oxabicyclo[3.2.1]octenes with styrene derivatives, catalyzed by chiral stereogenic-at-Mo/adamantylimido complexes, was developed <09JA3844>. Intramolecular nucleophilic ring opening of oxabicyclo[2.2.1]heptenes by amines and amides provided [6,6] and [5,6]-fused bicycles using a Lewis acid. Remarkably, ring opening of the less reactive oxabicyclo[2.2.1]heptanes also occurred under the same conditions, as demonstrated below <09AGE6296>.

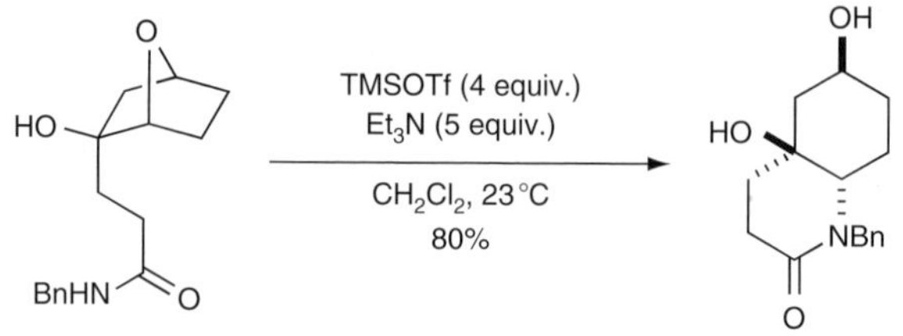

Tetrahydrofuran reacts with indoles in a process of iron-catalyzed C-2–H bond oxidation and C–O ring cleavage to provide symmetrical and unsymmetrical 1,1-bisindolylmethane products. An example is shown below <09JOC8848>.

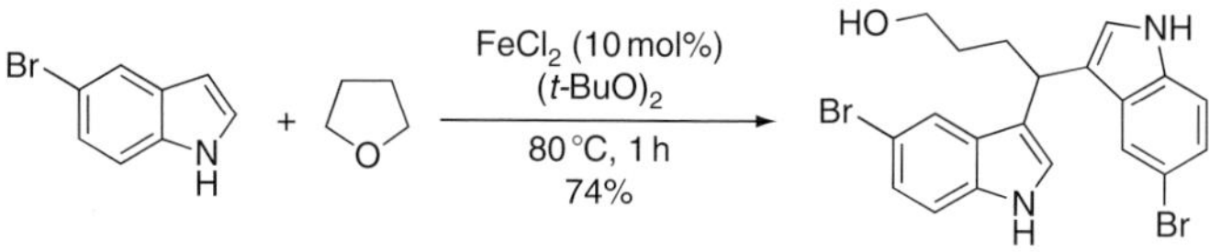

An interesting imidazolinium carbene-catalyzed ring expansion of tetrahydrofurfurals to lactones *via* the formation of a Breslow-type intermediate is shown below <09OL891>.

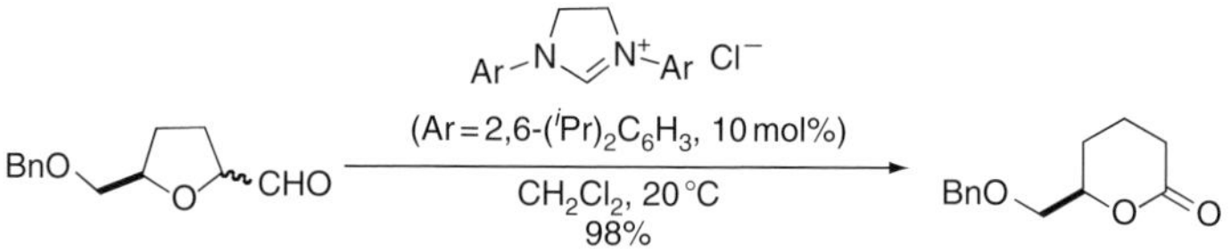

5.3.3. SYNTHESIS

5.3.3.1 Furans

A method for the synthesis of furo[3,4-c]furanones involves DDQ treatment of the corresponding furo[3,4-c]pyranones <09SL1951>. A synthesis of salvinorin A containing 3-substituted furan ring was reported <09T4820>. A pentacyclic furanosteroid containing trisubstituted furan ring was synthesized *via* a multi-step sequence in acceptable total yield <09JOC5429>. Lewis acid-catalyzed reaction of 2-methylfuran with substituted benzaldehydes afforded aryl difuranyl methanes in high yields <09EJO1132>. 2,5-Dimethyl-3,4-di-*tert*-butylfuran was obtained in 34% yield as an unstable oil from the reaction of 3,4-di-*tert*-butylhex-3-en-2,5-dione during a study of the synthesis of tetra-*tert*-butylethylene <09EJO2141>. A reaction of *o*-substituted aryl iodides with 2-methylfuran provided the corresponding furans in high yields by using a Pd/norbornene catalyst <09CEJ7850>. A highly efficient catalyst system using FeCl$_3$ for the Garcia-Gonzalez reaction between aldose sugars with β-ketoesters produced the corresponding polyhydroxyalkyl- and C-glycosyl furans <09S2278>. A variant of the transformation of alkynyl-oxiranes into trisubstituted furans using a Ag-catalyst was developed <09JOC4360>; a mechanistic study showed that the reaction proceeded through a cascade pathway <09JOC5342>. Fluorine-containing tetrasubstituted furans were prepared in good yields by direct arylation of 3-fluorofurans in the presence of Pd(PPh$_3$)$_2$Cl$_2$ as a catalyst <09T1673>. A new procedure for the transformation of (Z)-2-en-4-yn-1-ols to fully substituted furans utilised a gold catalyst <09T1839>. A multi-component reaction of isocyanides, dialkyl acetylenedicarboxylates and 2-hydroxy-1-aryl-2-(arylamino)ethanone provided tetrasubstituted furans in high yields <09SL2676>.

Disubstituted furans were synthesized through Pd-catalyzed carbonylative cyclization of γ-propynyl 1,3-diketones with aryl iodides and CO in moderate to high yields <09JOC8904>.

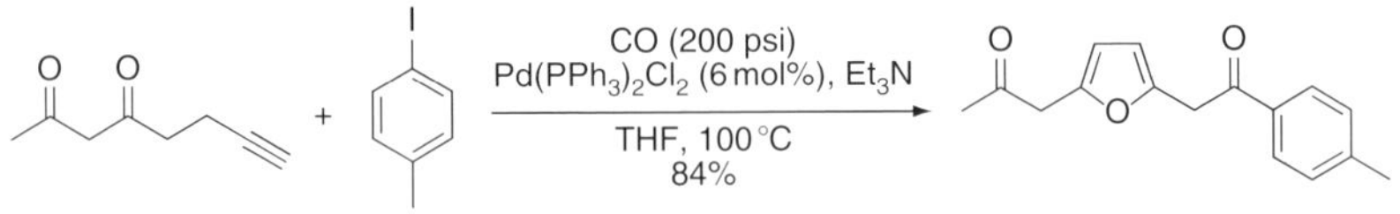

A direct alkenylation of furans using a Pd-catalyst was reported. Reaction of furan and 2-methylfuran with acrylates in the presence of Pd(OAc)$_2$ and AgOAc provided the corresponding α-substituted furans in good yields <09TL2758>.

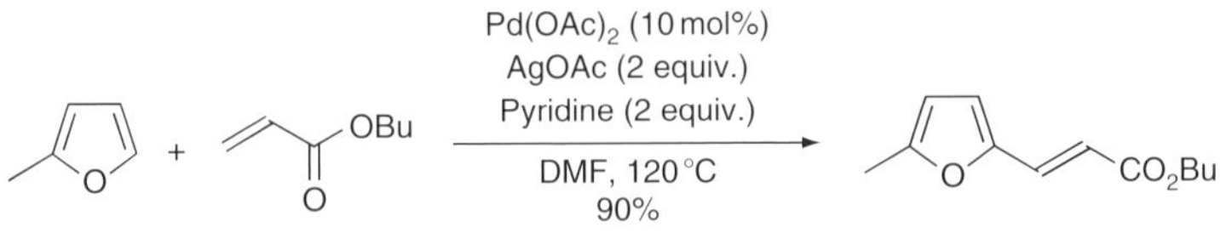

2,5-Disubstituted furans were synthesized from terminal alkynes through 1,3-dienylalkyl ether intermediates in a one-pot reaction catalyzed by Ru and Cu complexes <09AGE1681>.

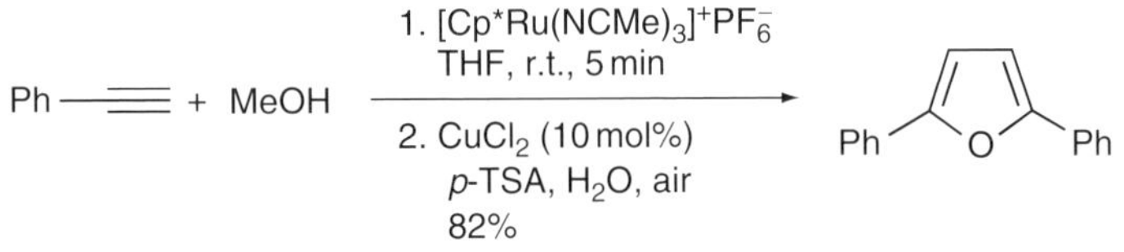

Isomerization followed by intramolecular cyclization of alkyne-1,4-diols using a Ru-catalyst afforded 2,5-disubstituted furans in good yields together with a little 1,4-diketone as by-product <09T8981>.

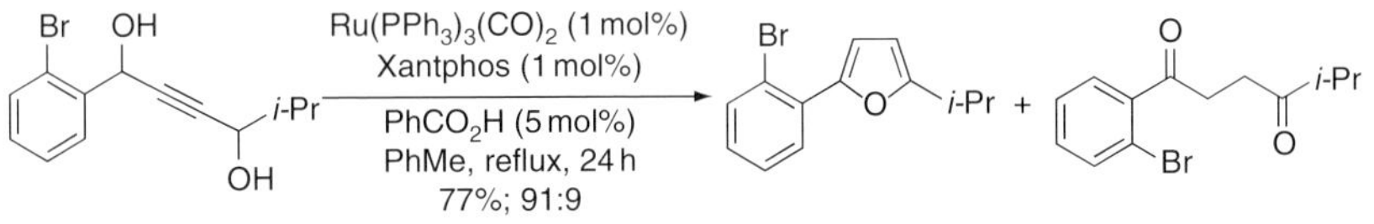

Hydroarylation of alkynes with furans under mild conditions using a Pt-catalyst gave rise to the double-hydroarylation products in good yields. Mono-hydroarylation adducts could also be obtained if the reactivity of reagents or the reaction conditions were changed <09T3842>.

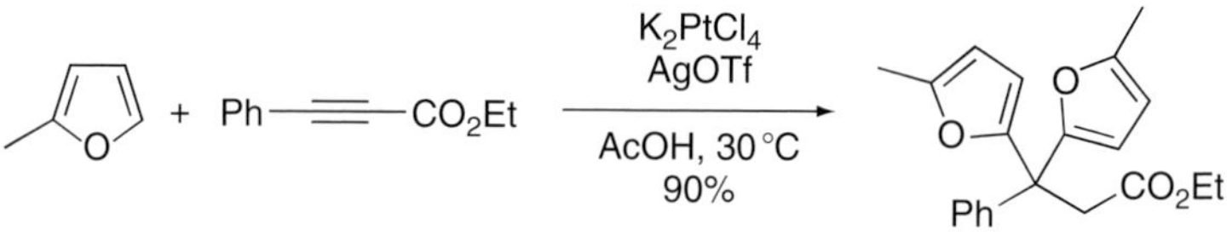

Reaction of electrophiles with 2-ethylidene-3-tosylmethyl-2,3-dihydrofuran, obtained from decarboxylative Claisen rearrangement of 1-(furan-2-yl)ethyl 2-tosylacetate, provided 3,4-disubstituted furans in good yields and stereoselectivity <09TL3503>.

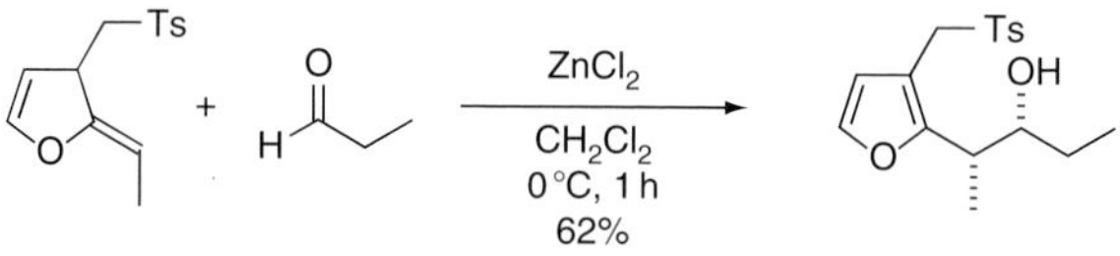

2,3-Disubstituted furans were effectively synthesized from 2-*tert*-butyldimethylsilyl-3-formylfurans in one-pot by multi-component anion relay chemistry <09OL1861>.

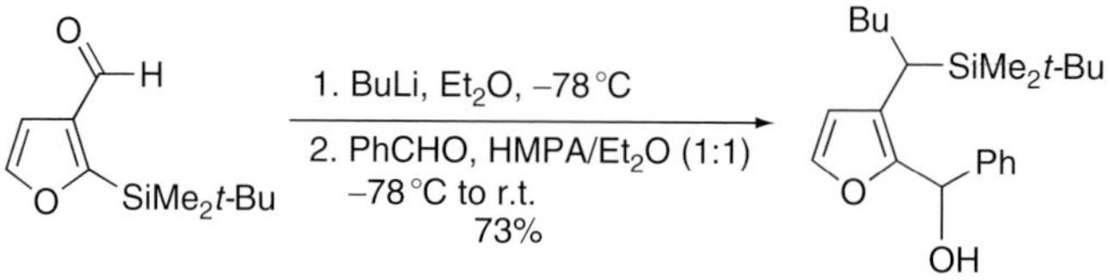

Reaction of 3-furylmethyl ether with BuLi or LDA afforded 2,3-disubstituted furans *via* Wittig rearrangement <09H(77)433>.

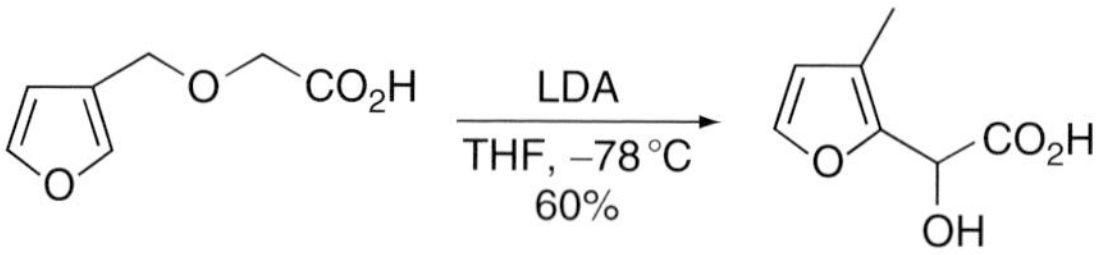

Enantiomerically pure 2-mono- and 2,3-disubstituted furans were synthesized in high yields by the mixed Lewis acids-catalyzed reaction of hex-2-enopyranosides that were derived from commercially available 3,4,6-tri-O-acetyl-D-glucal <09CEJ6041>.

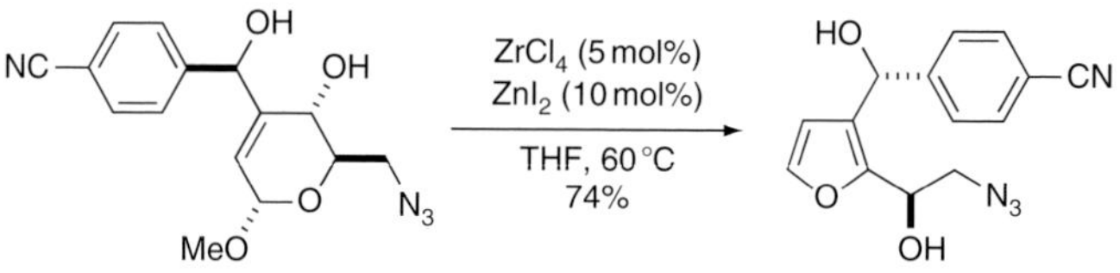

3-Bromo-2,5-disubstituted furans were synthesized by the reaction of a γ,δ-epoxy-bromoacrylate in the present of Lewis acid. 2-Methylated product was obtained using Me$_3$Al–H$_2$O, while 2-ethoxy furan was produced when Me$_3$Al-(CF$_3$)$_2$CHOH was used <09H(77)201>.

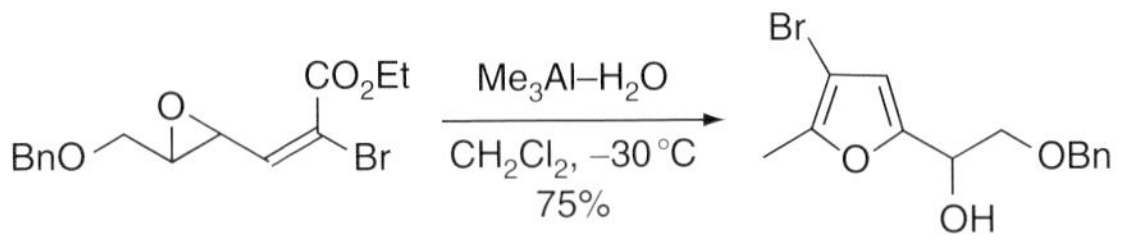

Difuranylketones were obtained in moderate to good yields through Pd-catalyzed carbonylation of allenyl ketones in the presence of p-benzoquinone <09TL4744>.

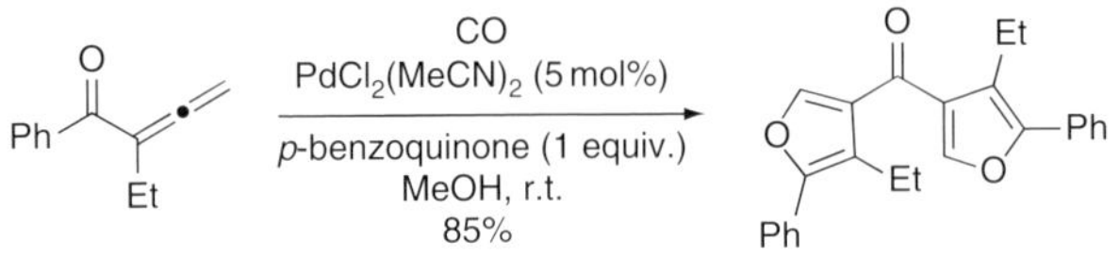

Cu-Catalyzed reaction of diazo compounds with bis(propargylic) esters delivered cyclobutene-substituted furans in good yields <09AGE7569>.

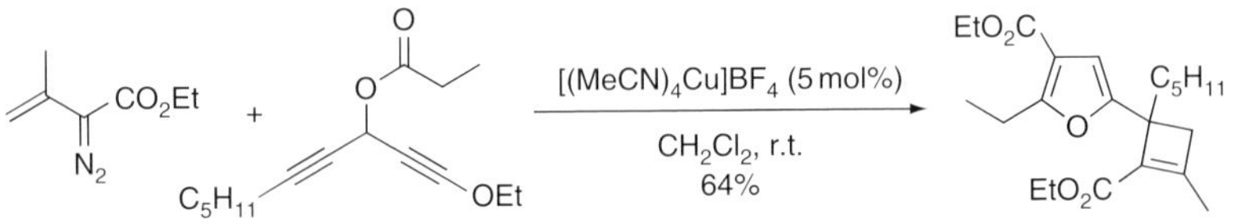

A scandium triflate-catalyzed cascade reaction of β-ketoesters and dihydroxyacetone acetal afforded methylenomycin furans, which belong to a newly discovered class of bacterial signaling molecules <09OL2984>.

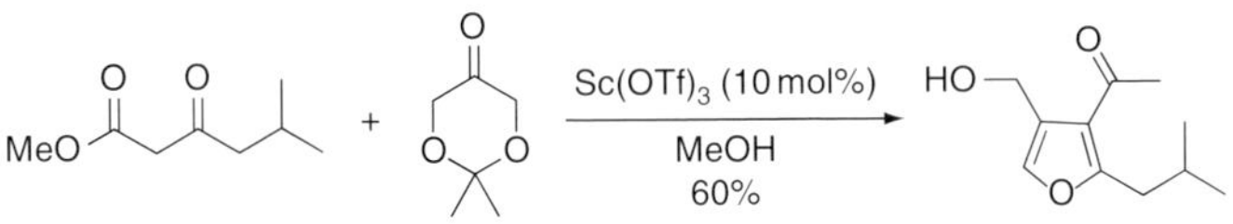

Furo[2,3-d]pyrimidines were prepared via an inverse electron demand Diels–Alder reaction of 2-amino-5-substituted furans as the dienophilic component and 1,3,5-triazines <09TL6758>.

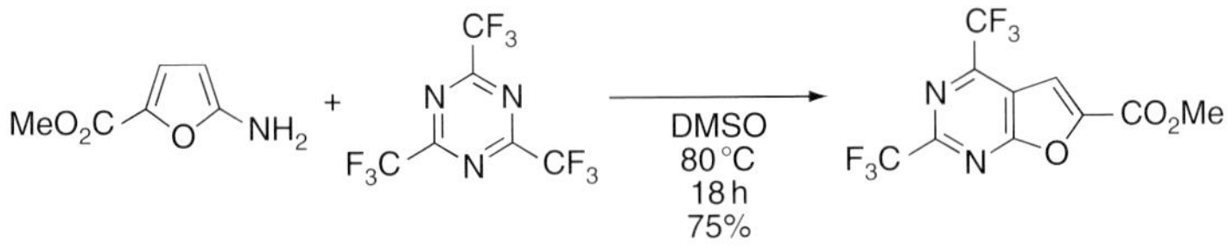

Fused trisubstituted furans were synthesized in good yields via K_2CO_3-mediated intramolecular Michael addition followed by acid-catalyzed elimination under mild conditions <09TL6751>.

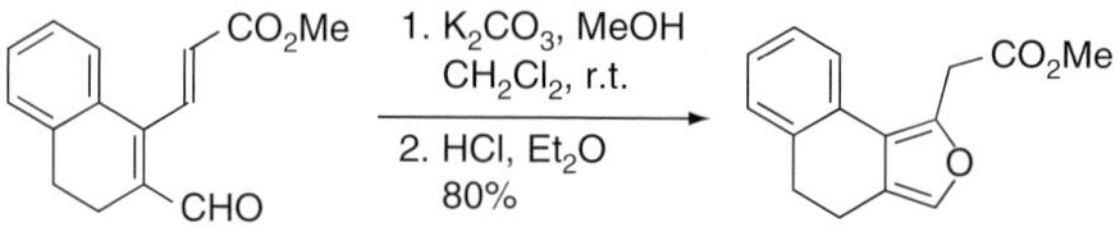

Treatment of a propargylic dithioacetal with BuLi followed by TFA provided trisubstituted furans, which served as the precursors in the synthesis of furanylene-*meta*-phenylene oligomers. Their photophysics was studied <09T9749>.

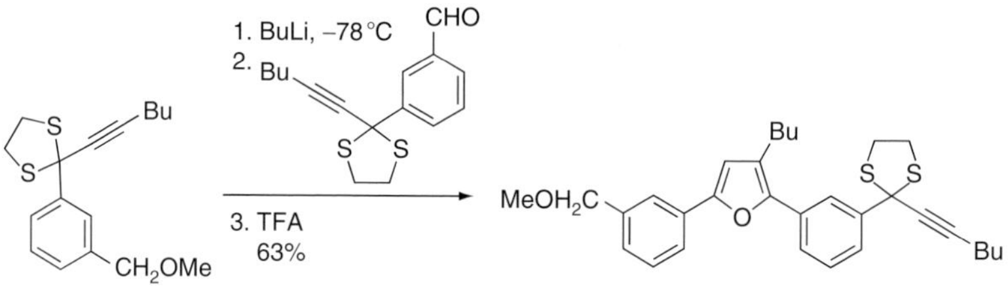

Highly efficient Au-catalyzed intramolecular cyclization of hydroxymethyl propargyl alcohols was independently reported by two groups. Various gold salts and complexes can be used and high yields of substituted furans were obtained with even as low as 0.05 mol% of catalyst <09OL4624; 09OL5002>. Similar cycloalkane fused trisubstituted furans were also synthesized in high yields by Au-catalyzed cycloisomerization of 2-alkynylcycloalk-2-enols <09SL1990>.

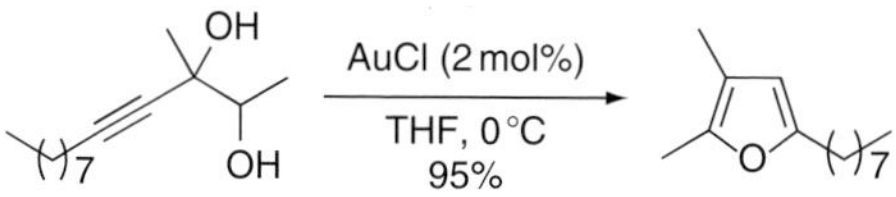

A general procedure for the synthesis of 2-aminofurans was reported, thus reaction of ynamides with diazo malonate or phenyl iodonium malonate ylide, in the presence of Rh-catalyst, provided aminofurans in moderate to good yields <09OL4462>.

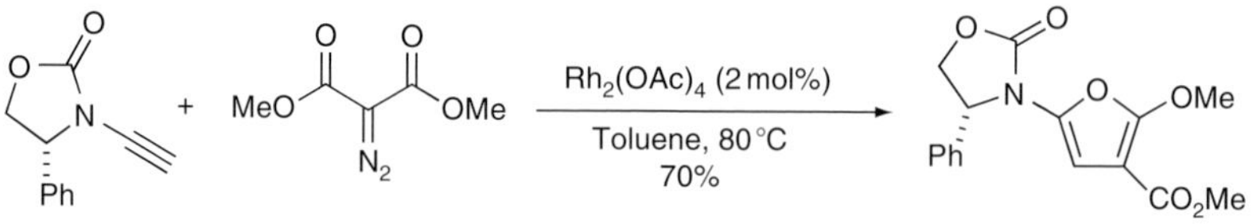

Dehydrogenative coupling reaction of furans with styrene was realized using $Pd(OAc)_2$ as a catalyst under oxidative conditions, providing *trans*-alkenylfurans regio- and stereoselectively in moderate to good yields <09OL4096>.

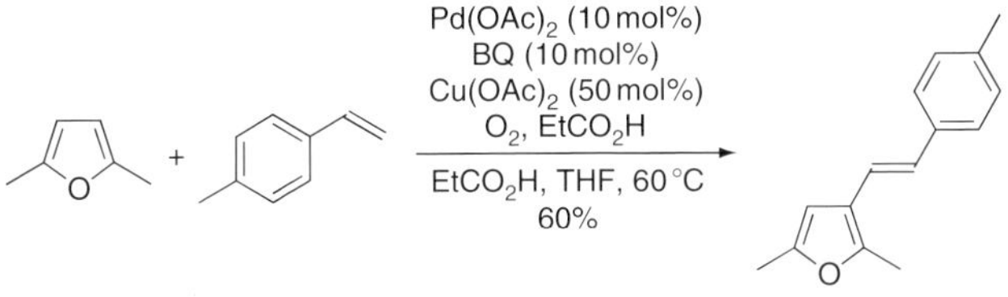

Optically active furans with chiral substituents were prepared in high enantiomeric excess by Michael addition of furans to α,β-unsaturated enones in the presence of a oxazaborolidinone as catalyst, as shown below <09OL5206>. Chiral 2,5-disubstituted furans were also obtained by asymmetric Diels–Alder reactions of 2-methylfuran with chiral (E)-nitro alkenes derived from D-mannose and D-galactose followed by treatment of the adducts with silica gel or CF_3COOH <09TA1999>.

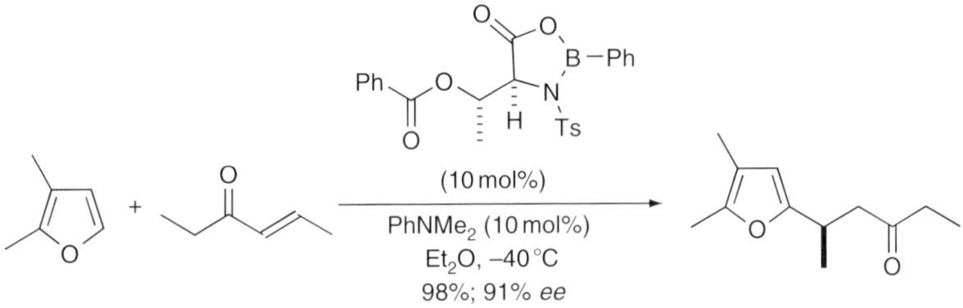

A trisubstituted furan was formed by an unprecedented reduction/iminium ion formation/diastereoselective C–C bond-forming cyclization cascade via a single hydride delivery with DIBAL followed by treatment with chilled HCl and subsequently heating. This protocol was used in a synthesis of (−)-nakadomarin <09JA16632>.

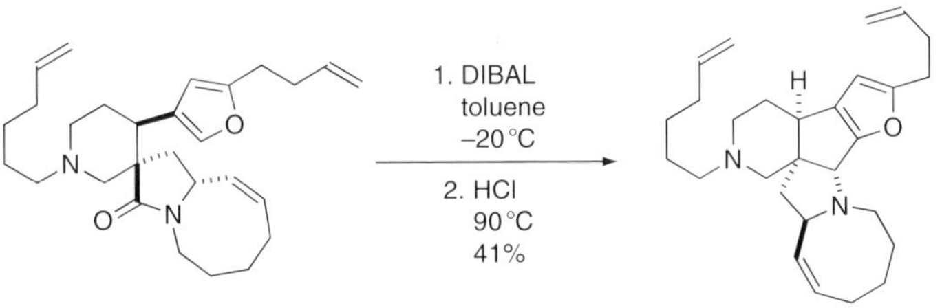

A three-component domino reaction using a Pd(II) catalyst affords tetrasubstituted furans in good to excellent yields, as shown below. A variety of O-, N- and C-based nucleophiles, including olefin-tethered ones, were suitable for the reaction <09ASC617>. A similar reaction using vinyl ketones or acrolein instead of allyl chloride was also described <09CEJ9303>.

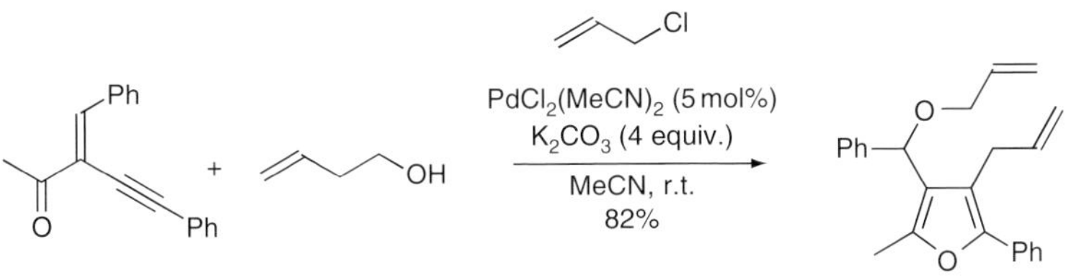

A procedure for the synthesis of tetrasubstituted furans in high regioselectivity via Rh-catalyzed carbonylation/cyclization of 1-(1-alkynyl)cyclopropyl ketones was developed <09CEJ5208>.

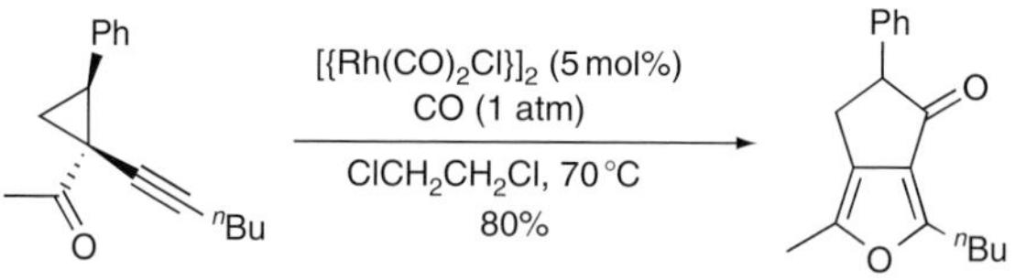

Substituted furans were obtained by Cu- and Au-catalyzed formal [4+3] cycloaddition of 1-(1-alkynyl)cyclopropyl ketones with nitrones <09CEJ8975>. Tetrasubstituted furans were prepared from the reaction of 2-(1-alkynyl)-2-alken-1-ones with nitrones via 1,3-dipolar cycloadditions <09AGE5505>.

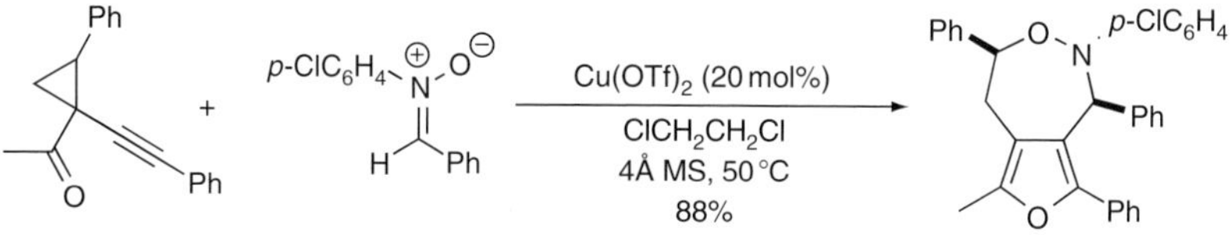

Reaction of β-acyloxy-α,β-unsaturated acetylenic ketones with organocuprates gave rise to the addition products, which underwent cyclization followed by dehydration to afford substituted furans in moderate to high yields <09JOC3430>.

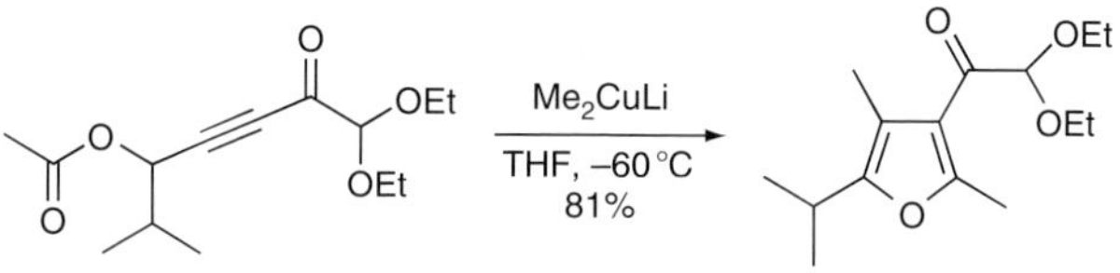

Copper-catalyzed domino reaction of 1,5-enynes, produced in situ from alkynols and diethyl butynedioate, through rearrangement/dehydration/oxidation/carbene oxidation sequence delivered fully substituted furans regioselectively in moderate to good yields <09OL1931>.

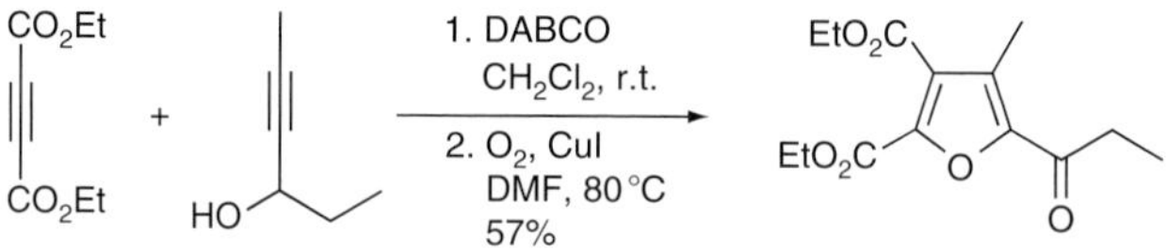

Annulation of homopropargylic alcohols produced mercurio-substituted furans in the presence of mercury acetate, which were converted to the corresponding iodofurans in high yields after treatment with iodine <09TL3263>.

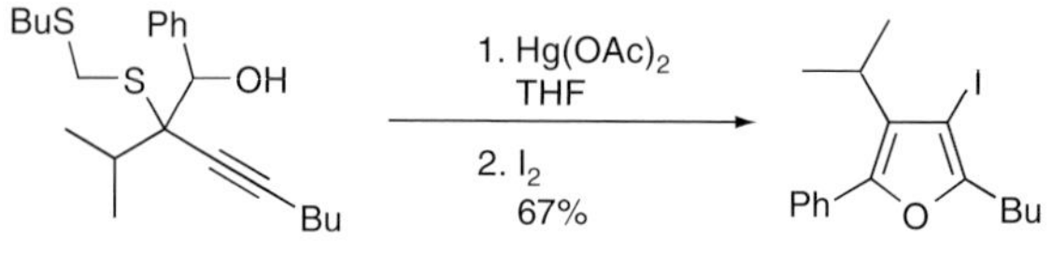

A sequential 2,3-difunctionalization of 2-arylsulfinyl furans was realized by using TMPMgCl·LiCl and i-PrMgCl·LiCl. Treatment of the products with the same reagent allowed selective functionalization of all four positions on the furan ring <09CC3536>.

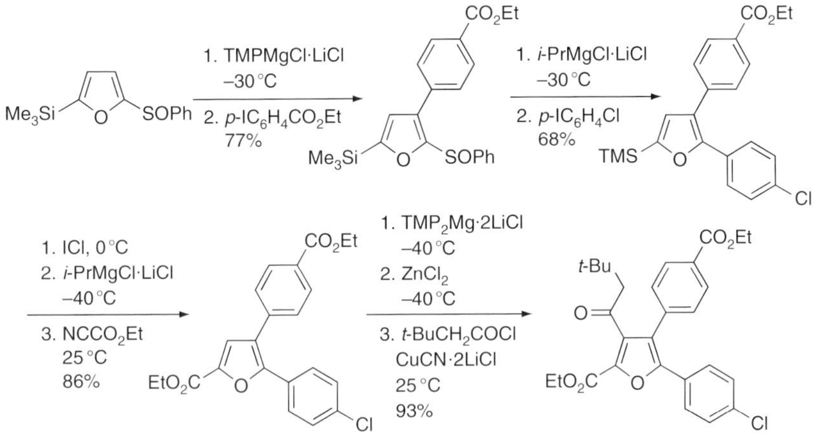

Pd-Catalyzed reaction of 3-alkynyl-4-methoxycoumarins with aryl iodides provided 2,3-disubstituted furo[3,2-c]coumarins in good yields <09OL5254>.

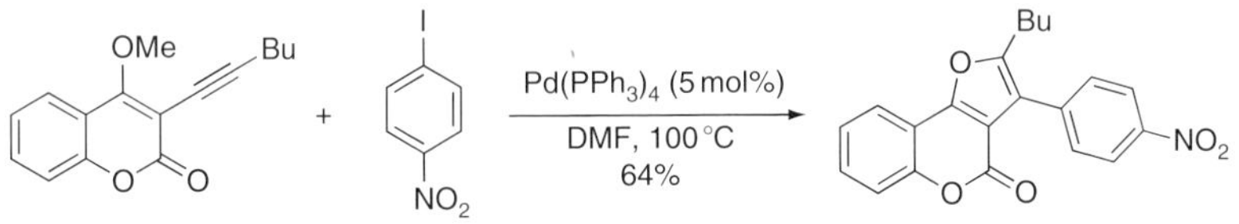

5.3.3.2 Di- and Tetrahydrofurans

Many methods and approaches for the synthesis of di- and tetrahydrofurans were reported in 2009, with transition metal-catalyzed ring-forming cyclizations and cycloisomerizations as the major reaction types. Notable and novel transformations are highlighted below. An interesting sequential reaction involving Pd-catalyzed coupling, propargyl-allenyl isomerization and Alder-ene cycloaddition provided a facile synthesis of 2,3-dihydrofuran derivatives, not otherwise readily available, as depicted below <09JOC4118>.

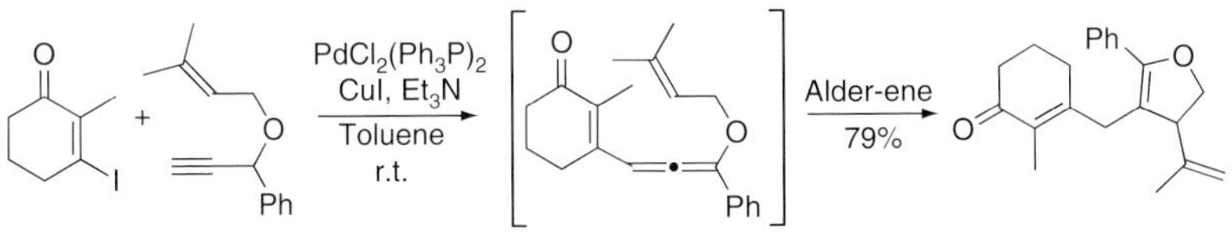

Cu-Catalyzed aerobic annulative demetalation of molybdenum complexes provided dihydrofurans in high yields and stereoselectivity, as illustrated below <09JA12546>.

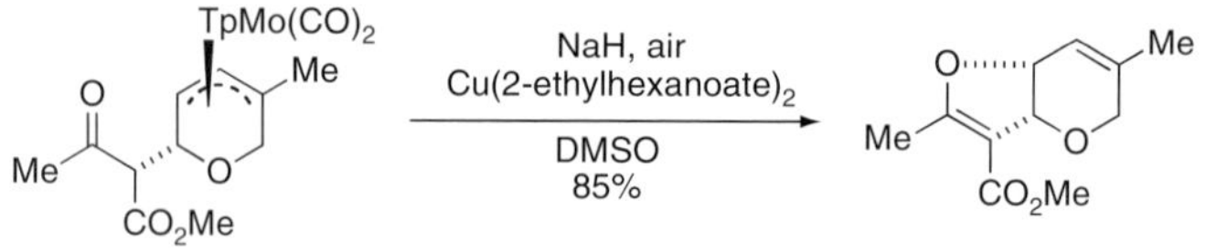

As shown below, oxacycloisomerization of cyclic 1-hydroxy-2-propargyl derivatives using Grubbs I catalyst furnished bicyclic dihydrofurans <09SL1597>.

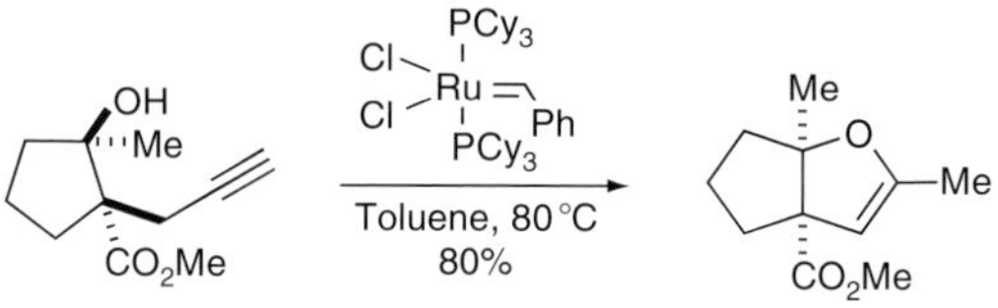

Oxidation of tetrahydrofuran-fused cyclopropyl alcohols by IBX allowed the direct formation of 2,3-dihydrofurans <09OL2317>. 2,3-Dihydrofuran-containing [5,6]-spiroketals were obtained by ring expansion of donor-acceptor-substituted cyclopropanols in the presence of IBX and Yb(OTf)$_3$, as shown below <09JOC8779>.

A series of 2,5-dihydrofurans were obtained from a mild and direct C–C bond formation reaction between propargylic alcohols and olefins in the presence of a silver catalyst <09OL3206>. Functionalized tetrasubstituted dihydrofurans were prepared from a cobalt-mediated [3+2]-annulation reaction of alkenes with enones and imines <09OL3698>. 2,5-Dihydrofurans were also synthesized through [2+2+2] cycloaddition of functionalized 1,6-alkynes in the presence of a nickel catalyst <09AGE7687>. An efficient formation of highly functionalized dihydro- and tetrahydrofuran products was achieved via Rh-catalyzed three-component coupling between aldehydes, α-alkyl-α-diazoesters and dipolarophiles with selectivity over dioxolane formation, as shown below <09JA1101>.

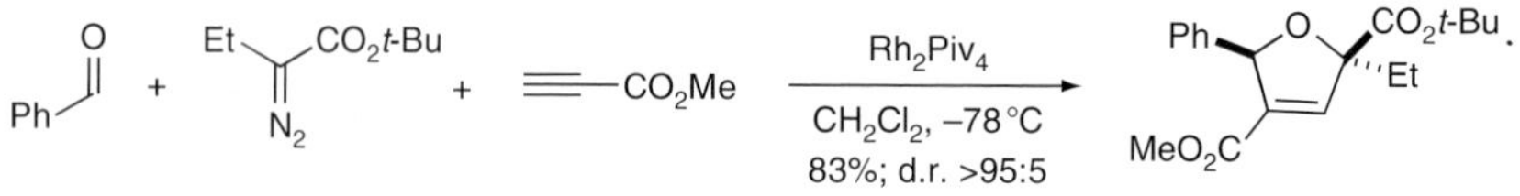

As demonstrated below, a Pd(II)-catalyzed tandem-cyclization reaction of α,ω-bisallenols furnished 2,5-dihydrofuran-fused bicyclic skeletons <09OL1205>.

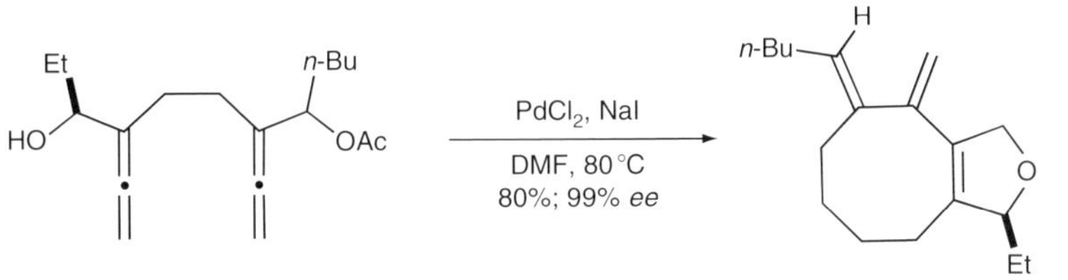

The synthesis of 3-bromo-2,5-dihydrofurans was realized by electrophilic cyclization of 4-substituted tertiary 2,3-allenols using NBS in water <09JOC8733>.

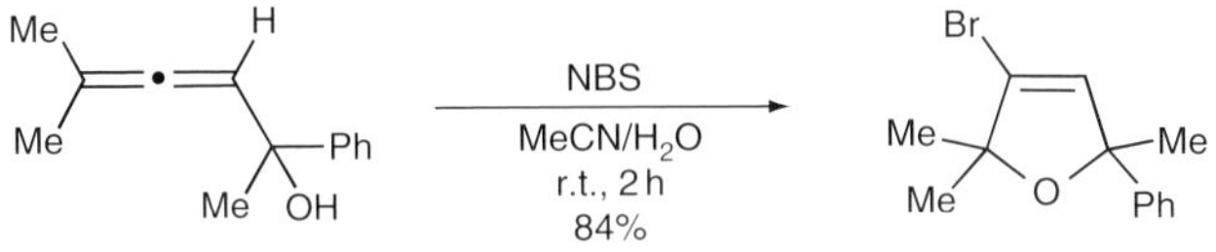

Tetrahydrofurans containing an exo-methylene group were prepared by a variety of methods. Intramolecular hydroalkoxylation/cyclization of terminal alkynyl alcohols using lanthanide catalysts furnished 2-methylenetetrahydrofuran derivatives <09JA263>. Chiral methylenetetrahydrofuran derivatives were prepared from enynes by Rh-catalyzed asymmetric Pauson-Khand reaction <09TL6068>. Fused tetracyclic skeletons containing tetrahydrofuran rings were obtained through Rh-catalyzed [2+2+2+1] cycloaddition reaction of cyclohexene-diynes and carbon monoxide <09CC4569>. A cationic rhodium(I)/H8-BINAP complex catalyzed [2+2+2] cycloaddition of a variety of 1,6- and 1,7-diynes with both electron-deficient and electron-rich carbonyl compounds to afford 3-methylenetetrahydrofurans under mild reaction conditions <09EJO2737>. As shown below, the phosphane-catalyzed [3+2] annulation of γ-methyl allenoates with aldehydes provided an efficient synthetic approach to 2-(ethoxycarbonylmethylene)tetrahydrofurans <09CEJ8698>.

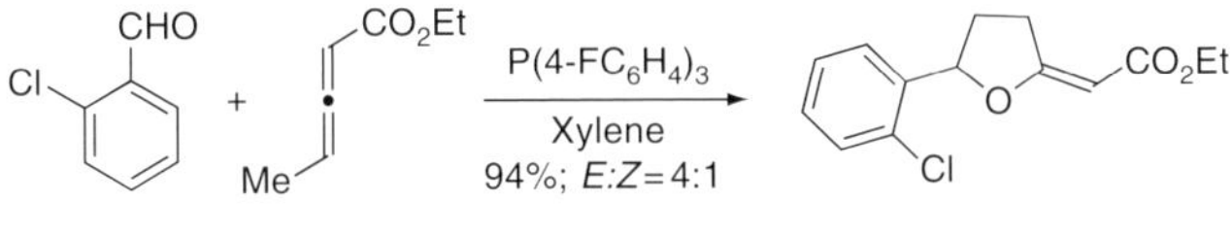

3-Methylenetetrahydrofuryl species were obtained *via* Rh-catalyzed cycloaddition between diazoketoesters and allene esters in good yields, as represented below <09CEJ12926>.

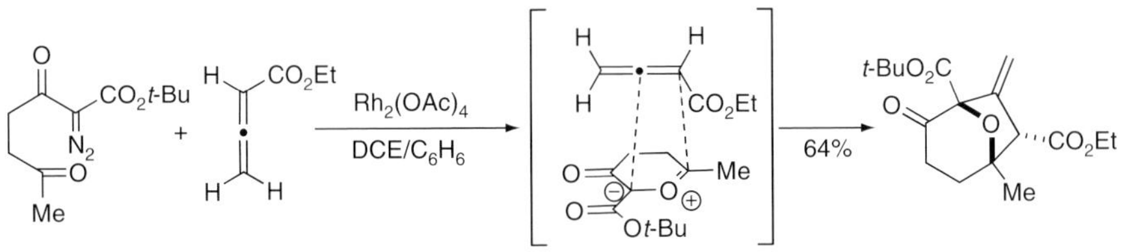

A gold-catalyzed cycloisomerization of O-tethered 1,6-enynes in the presence of an external nucleophile, such as electron-rich aromatic ring, also resulted in the formation of 3-methylenetetrahydrofuran species <09T1911>.

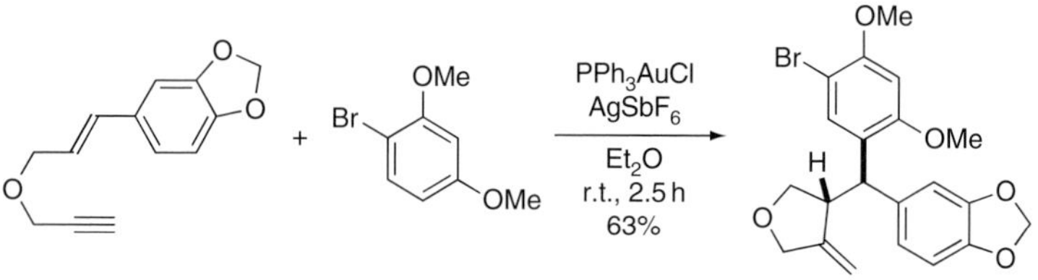

FeCl$_3$- and FeBr$_3$-promoted cyclization/halogenations of alkynyl diethyl acetals was applied to the preparation of the corresponding tetrahydrofuran derivatives <09OL2113>.

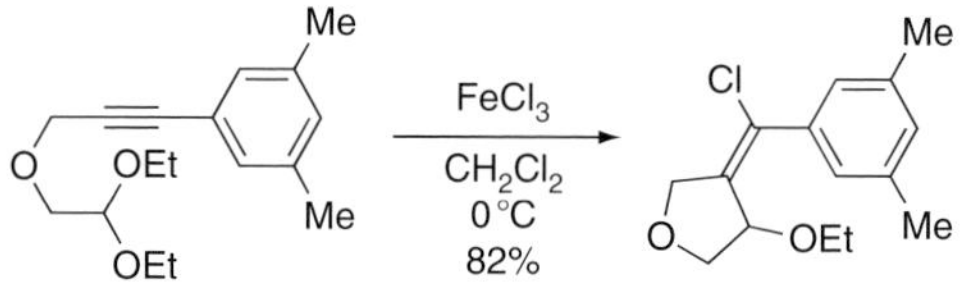

Radical cyclization of an O-tethered enyne in the presence of diphenyl disulfide and tripropylamine led to 5-methylene substituted tetrahydrofuran rings <09OL3298>. Exo-methylene tetrahydrofuran derivatives were also obtained by an efficient alkoxy radical relay cyclization from linear substrates, which was employed in a total synthesis of (−)-amphidinolide K, as shown below <09OL2019>.

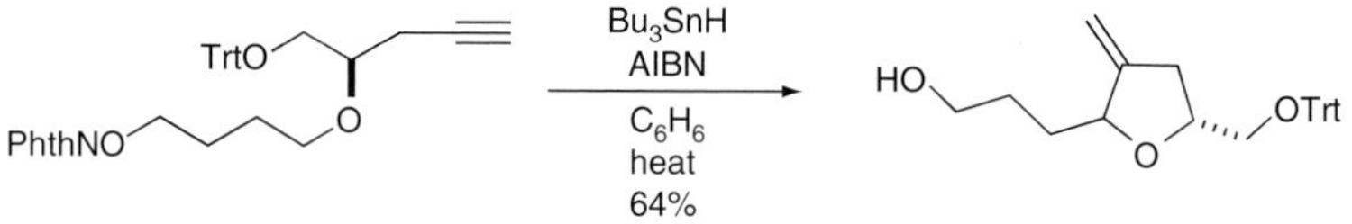

Oxygen-tethered yne-enones were transformed to tetrahydrofuran derivatives by a platinum-catalyzed hydrogenative cyclization <09EJO6091>. An iron-catalyzed, hydrogen-mediated method for the reductive cyclization of O-tethered enynes and diynes was developed to prepare the corresponding tetrahydrofuran derivatives <09JA8772>. As illustrated below, a synthesis of novel bicyclic tetrahydrofuran derivatives by a Mo(CO)$_6$-mediated intramolecular Pauson-Khand reaction of substituted diethyl 3-allyloxy-1-propynylphosphonates was also demonstrated <09JOC1029>.

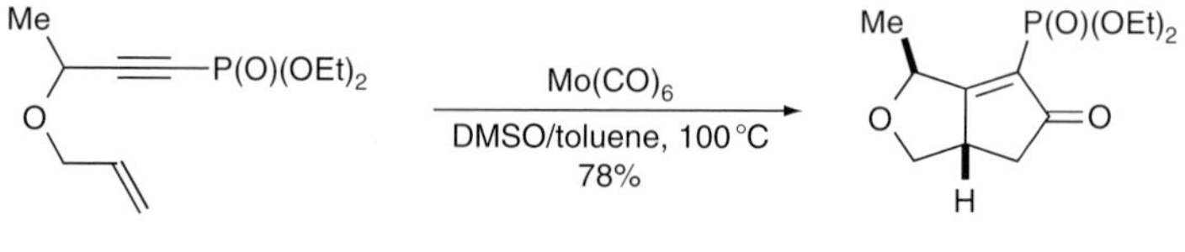

A stereoselective gold-catalyzed [2+2+2] cycloaddition of ketoenynes substituted at the propargylic position was applied to the synthesis of the fused tetrahydrofuran cores of (+)-orientalol F and pubinernoid B <09CC7327>. Complex tetrahydrofurans were also synthesized *via* a stereocontrolled platinum-catalyzed [4+2]-cycloaddition/annulation of enynals with allylic alcohols <09JA2090>. A total synthesis of (+)-virgatusin was completed through a key $AlCl_3$-catalyzed [3+2] cycloaddition of cyclopropane and piperonal in a selective, high-yielding fashion <09CC5135>. In 2009, this process of tetrahydrofuran synthesis was rendered catalytic enantioselective *via* a dynamic kinetic asymmetric [3+2] cycloaddition of racemic cyclopropanes and aldehydes, as shown below <09JA3122>.

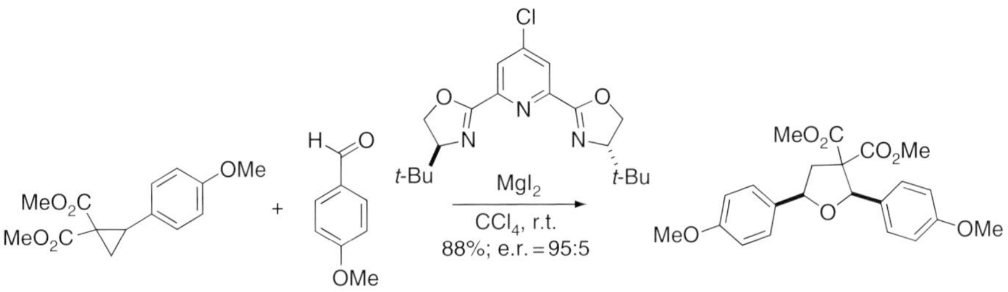

In an asymmetric synthesis of (+)-polyanthellin A, the crucial tetrahydrofuran intermediate was assembled *via* a related $MADNTf_2$-catalyzed formal [3+2] cycloaddition of a donor-acceptor cyclopropane and a β-silyloxy aldehyde <09JA10370>.

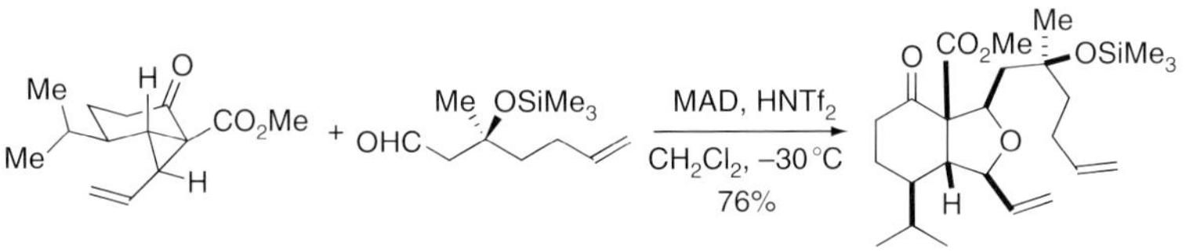

Tetrahydrofuran derivatives were obtained by an intramolecular hydroalkoxylation/cyclization of unactivated alkenols in the presence of lanthanide triflate-derived ionic liquids <09OL1523>. Intramolecular cyclization of γ- and δ-alkenols could be achieved to afford tetrahydrofurans in the presence of $Cu(OTf)_2$ as a catalyst <09T10334>. Tetrahydrofurans were also prepared *via* a Co-catalyzed aerobic oxidation of alkenols <09TL960>, as well as from a Ag-catalyzed intramolecular cyclization of γ-allenols <09CC7125>. As shown below, substituted tetrahydrofurans were synthesized through *in situ* reductive and brominative alkenol cyclization in an aerobic cobalt-catalyzed reaction <09JA12918>.

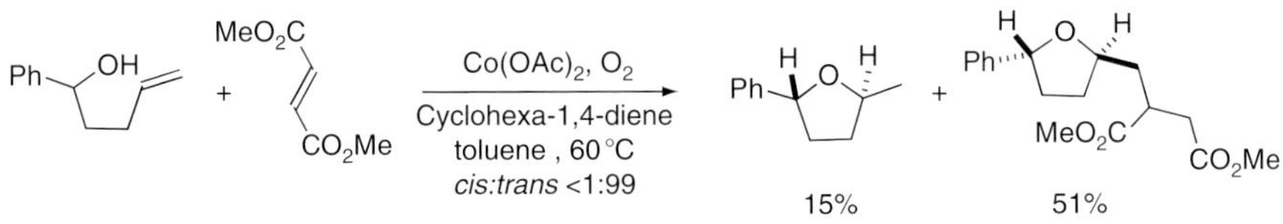

A stereocontrolled synthesis of 3-substituted 2-oxaspiro[4.5]decanes and [4.4] nonanes involving tetrahydrofuran rings was developed by using a Prins-pinacol annulation approach <09OL3834>.

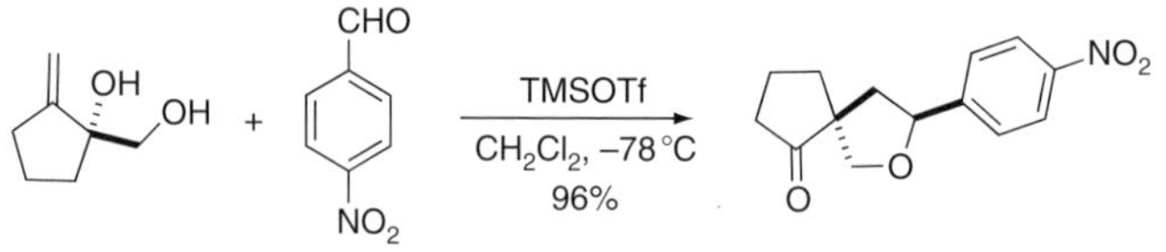

Access to enantio-enriched tetrahydrofurans was realized *via* Co-catalyzed enantioselective intramolecular opening of oxetanes <09JA2786>. As shown below, the strained bicyclic acetal, 2,7-dioxabicyclo[2.2.1]heptane, was readily converted into *trans*-2,5-disubstituted tetrahydrofurans <09OL3958>.

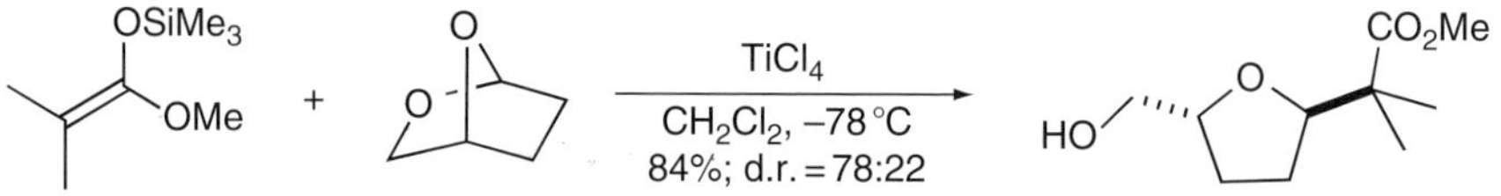

The tetrahydrofuran-containing central core of anthecularin was constructed *via* a [5+2] 1,3-dipolar cycloaddition of a pyrylium-3-olate intermediate, as shown below <09OBC639>.

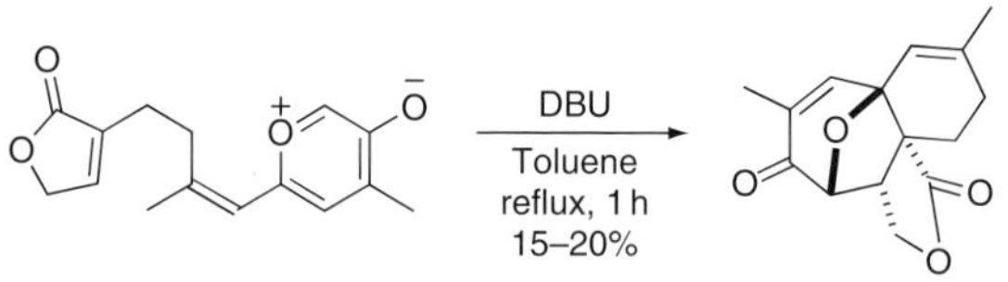

The tetrahydrofuran-containing pentacyclic core of the cortistatins was obtained *via* a hypervalent iodine-induced tandem intramolecular oxidative dearomatization and nitrile oxide cycloaddition <09OL5394>.

Another novel iodine-catalyzed tandem cyclization-cycloaddition reaction of *ortho*-alkynyl-substituted benzaldehydes was applied to the synthesis of oxabicyclo[3.2.1]octane skeletons found in a variety of natural products, as demonstrated below <09CC5451>.

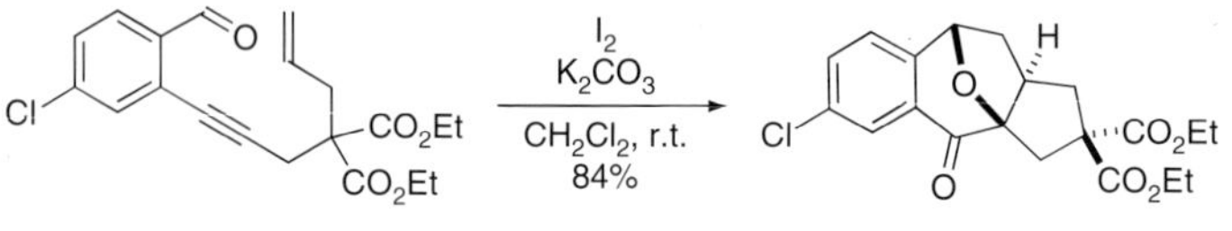

5.3.3.3 Benzo[b]furans and Related Compounds

The Pd-catalyzed coupling of 3-iodobenzofuran with benzofuran-3-stanane provided a dibenzofuran, an intermediate in the total synthesis of kynapcin-24, as shown below <09S1175>. A 3-substituted benzofuran was utilized for the synthesis of viniferifurans, melibatol A and shoreaphenol <09OBC2788>. The Pd-catalyzed C-2–H and C-2–Si activation was applied to the homo-coupling of benzofurans <09SL1941>. Benzofuran-based bisphosphines were prepared via Pd-catalyzed Suzuki-Miyaura reaction of aryl chlorides with arylboronic acids, and Hartwig-Buchwald amination of aryl bromides with aniline derivatives <09TL2239>.

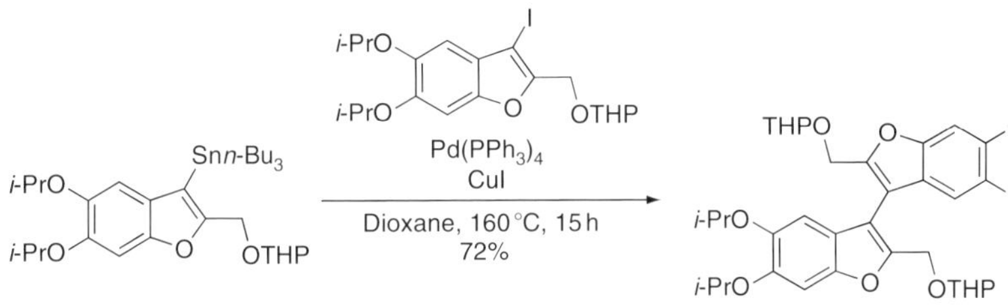

As depicted below, an interesting benzofuran-based photo-Fries rearrangement was utilized in a synthesis of kendomycin <09AGE6032>. Substituted benzofurans were used as substrates in the construction of the scaffolds of morphine, codeine and thebaine <09JA11402>. 3-Acyl-substituted benzofurans were employed as efficient dienophiles in normal electron demand [4 + 2] cycloadditions <09JOC1237>.

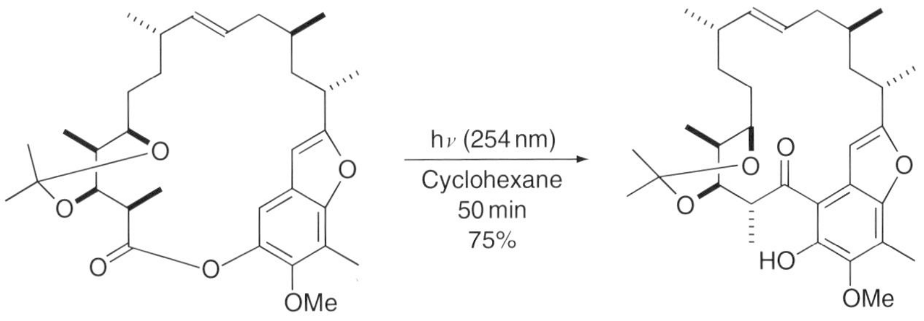

Furo[2,3-b]pyridines bearing ester, amide and ketone groups at the 2-position were prepared from substituted (2-fluoropyridin-3-yl)(phenyl)methanones, as exemplified below <09TL781>. Brønsted acid-promoted cyclization of substituted 3-alkynyl-2-pyridones was also found to be useful in the synthesis of furo[2,3-b]pyridin-4(7H)-ones and related quinolinones <09TL614>. The synthesis of structurally diverse furo[2,3-d]pyrimidines was achieved by base-mediated coupling of substituted pyrimidin-4(3H)-one with an α-bromoketone <09OBC1829>. Furo[3,2-c]pyridin-6(5H)-ones were obtained from the reaction of 5-iodo-4-methoxy-2-pyridones with terminal alkynes under microwave-enhanced Sonogashira conditions <09TL3299>. Benzo[4,5]furo[3,2-c]pyridines could also be made via Pd-catalyzed intramolecular Heck reactions <09TL4492>. A cyclopamine-derived furo[4,5-b]pyridine was made as a potent inhibitor of Hedgehog signaling pathway <09OL2824>.

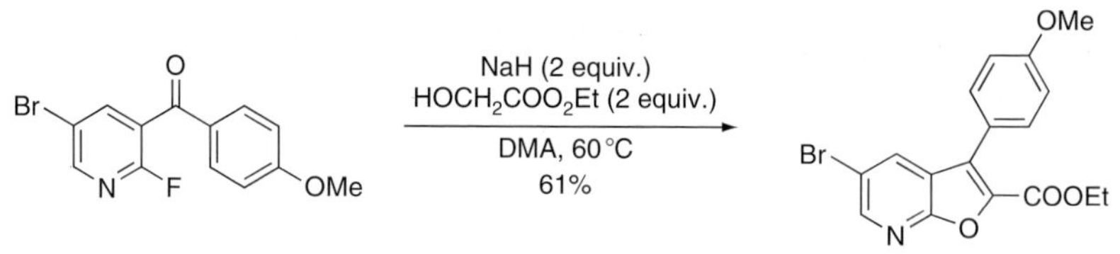

2,3-Dihydrobenzofuran-based tri-, tetra-, and pentacyclic adducts were obtained *via* conjugate addition of δ-ketoesters to 1,4-naphthoquinones in the presence of Yb(OTf)$_2$ <09T1716>. Benzofuran-fused benzopyranones were prepared by LDA-induced migration of biaryl O-carbamates <09JOC4094>.

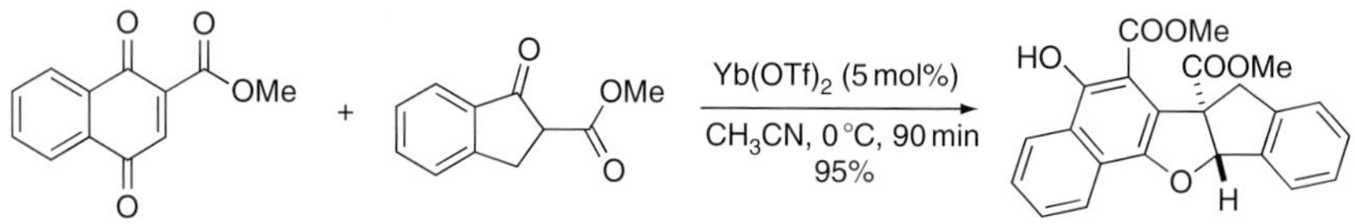

As illustrated below, a 1,2-Brook rearrangement generated a silyloxy carbene, which reacted with benzylic C–H bond to form 2,3-dihydrobenzofurans <09AGE784>. Reaction of 2-hydroxyaryl-α,β-unsaturated ketones with dimethylsulfonium carbonylmethylides also furnished 2,3-dihydrobenzofurans <09T3918>. The Rh(II)-catalyzed C–H insertion was applied to the asymmetric synthesis of neolignans, (-)-epi-conocarpan and (+)-conocarpan <09JOC4418>. Carbene-initiated O–H insertion was found to be effective in the synthesis of benzofurans <09T8995>. Naphthol[1,2-*b*]furan was obtained *via* cycloisomerization of allenyl ketones, followed by 6π-electrocyclization <09JOC2224>. A diazo(trimethylsilyl)methylmagnesium bromide-derived carbene was utilized in the synthesis of benzofuran-3-carboxylates <09T8708>, and the synthesis of benzo[*b*]naphtho[2,1-*d*]furans *via* Dötz intramolecular benzannulation was also reported <09TL4128>.

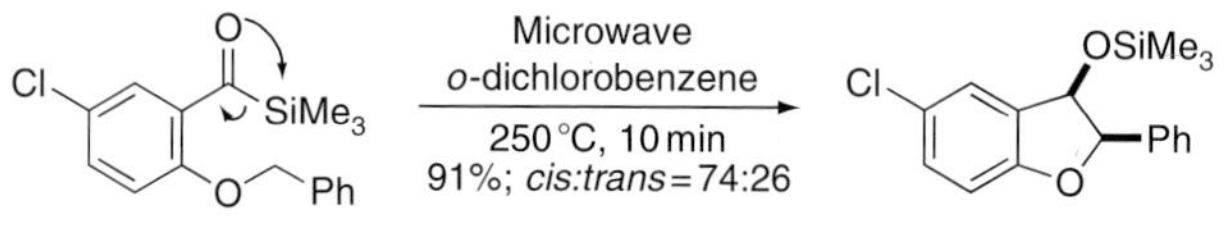

A phosphine-catalyzed domino reaction was applied to the synthesis of both *cis*-2,3-dihydrobenzofurans <09OL137>, as shown below, and 2,3-disubstituted benzo[*b*]furans <09TL2353>.

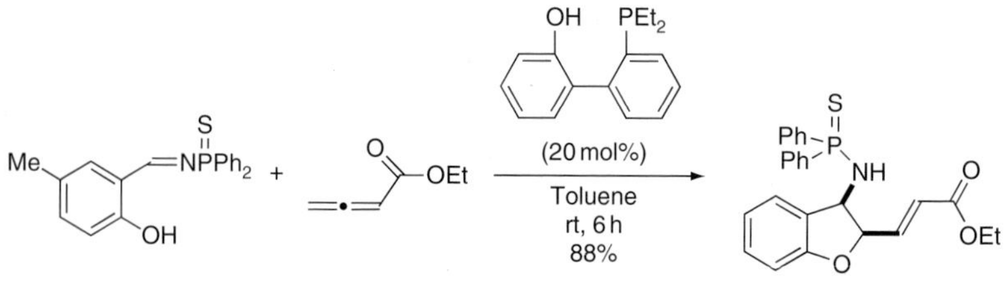

As shown below, structurally diverse benzofuropyrazoles were generated *via* an intramolecular 1,3-dipolar cycloaddition, followed by dehydrohalogenation <09JOC891>. A Pd-catalyzed Suzuki-Miyaura coupling/intramolecular direct arylation of dichlorovinyl ether and organoboronic acids was utilized to construct 2-substituted benzofurans <09OL5478>. An optimized synthetic methodology, which allows for facile and scalable synthesis of thieno[3,2-*b*]furans, was reported <09OL3144>.

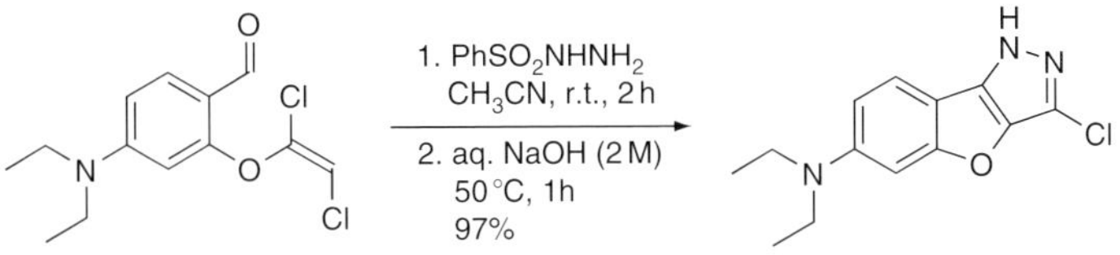

4-Keto-4,5,6,7-tetrahydrobenzofurans were efficiently generated from triketone substrates under microwave irradiation <09TL274>, as depicted below. Functionalized benzotrifurans were prepared from corresponding benzotrifuranones <09OL4314>.

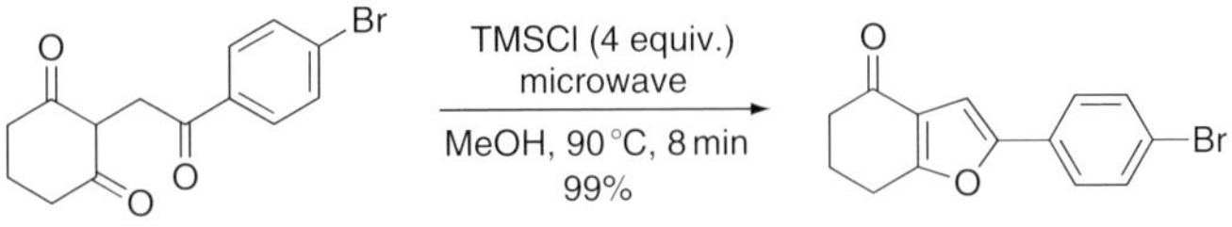

Electrophilic cyclization of 2-chalcogenealkylanisoles was applied to the synthesis of 2-chalcogen-benzo[*b*]furans, as shown below <09JOC2153>. The reactivity of diarylalkyne substrates towards benzofuran formation using different electrophiles (I_2, ICl, NBS and PhSeCl) was studied <09JOC1141>. *p*-Toluenesulfonic acid-mediated cyclization of *o*-(1-alkynyl)anisoles was also explored in the synthesis of 2-aryl substituted benzofurans <09TL3588>.

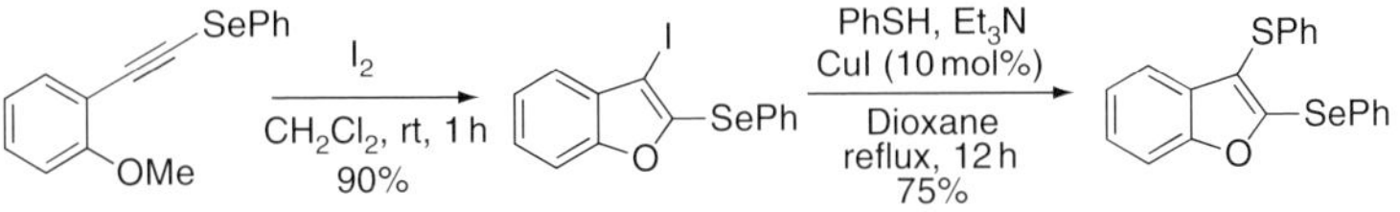

In a synthesis of psoralidin, the 6*H*-benzofuro[3,2-*c*]chromen-6-one core was constructed by condensation of a phenyl acetate with an acid chloride followed by intramolecular cyclization, as shown below <09JOC2750>. 6*H*-Benzofuro[3,2-*c*] chromen-6-one was also prepared by intramolecular Heck reaction <09TL3753>. Natural products moracins O and P were synthesized using a Sonogashira reaction followed by *in situ* cyclization, as key steps <09CC1879>. The 2,3-dihydrobenzofuran core of chafurosides A was formed by a Mitsunobu reaction <09OL2233>. In the enantiodivergent synthesis of codeine, its core structure was constructed by an intramolecular Diels–Alder and a Pd-catalyzed intramolecular Heck reaction <09T4569; 09T9862>.

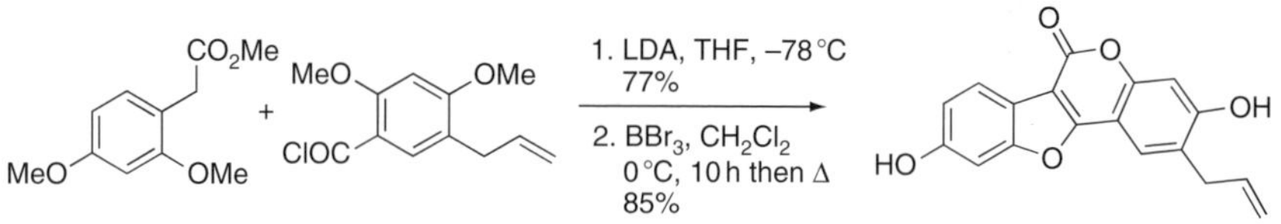

Phosphazene base-catalyzed carbon–carbon bond formation was utilized in the synthesis of 2,3-disubstituted benzofurans, as shown below <09CC5248>. A similar type of base was also employed in synthesis of 2,3-disubstituted benzofurans from o-benzyloxybenzophenones <09TL7180>.

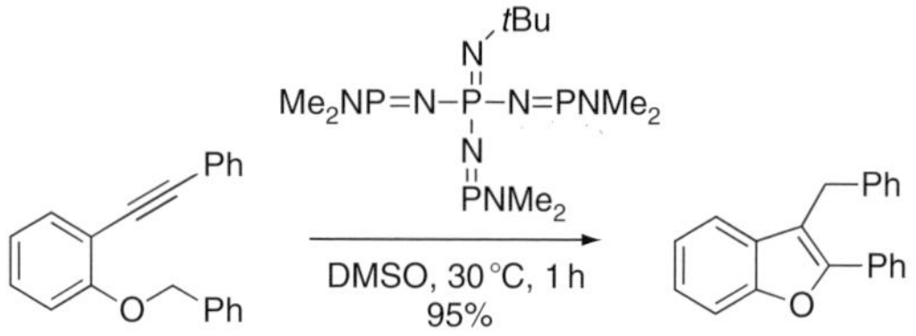

Benzofuran-based spirocyclic compounds were generated from an indole-based enol ether by treatment with a solution of H_2SO_4 in $MeOH/H_2O$, as depicted below <09JHC62>. The benzofuran scaffold could also be derived from O-arylhydroxylamine hydrochlorides and ketones in the presence of methanesulfonic acid <09SL3003>. 7-Hydroxybenzofuran-4-carboxylates were prepared via addition of silyl enol ethers to o-benzoquinone esters <09OL2165>.

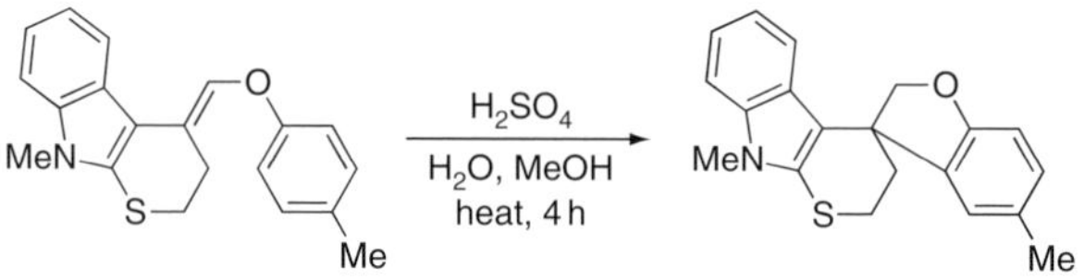

3-Acyl-2-aminobenzofurans were prepared by Pd-catalyzed cycloisomerization of 2-(cyanomethyl)phenyl esters, as illustrated below <09CC3466>. Benzofuran-2,3-diamines were assembled via a three-component reaction of a secondary amine, a 2-hydroxybenzaldehyde and an isocyanide in the presence of silica gel <09TL5625>. A three-component reaction was applied for the synthesis of naphtho[1,2-b]furans, naphtho[2,1-b]furans, and furo[3,2-c]chromenes <09SL2542>.

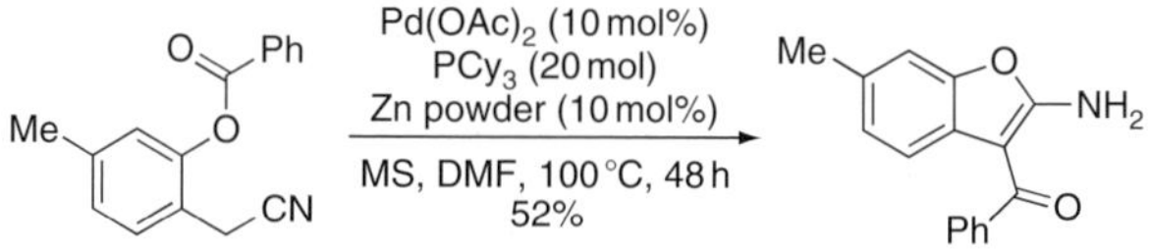

5-Halo-6-substituted benzo[b]naphtho[2,1-d]furans were generated from a sequential halopalladation/decarboxylation/C–C bond formation reactions as shown below <09OL1139>, and 3-alkenylbenzofurans could be constructed by

Pd-catalyzed tandem intramolecular oxypalladation/Heck type coupling of 2-alkynylphenols and alkenes <09OL1083>. The Rh-catalyzed cyclization of o-alkynyl phenols was employed for the synthesis of 2,3-disubstituted benzofurans <09OL1329>. The dibenzofuran scaffold was synthesized either via Pd-catalyzed oxidative coupling of benzofuran carboxylic acids with alkynes <09OL2337>, or Rh-catalyzed [2+2+2] cycloaddition with phenol-linked 1,6-diynes <09OL2361>.

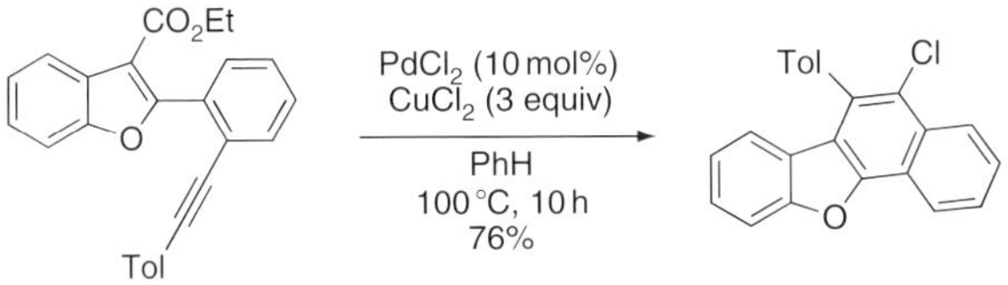

As exemplified below, Pd-catalyzed tandem ring opening/ring closing reaction of diazabicyclic alkenes was applied to generation of dihydrobenzofuran-based heterocycles <09JA5042>, and similar type of scaffold were also constructed by Pd-catalyzed direct arylation of benzoic acids via tandem decarboxylation/C–H activation <09JA4194>. Dibenzofurans could also be obtained by domino 'twofold Heck/6π-electrocyclization' reactions of 2,3-di- and 2,3,5-tribromobenzofurans <09TL3929>. 2-Bromobenzofurans were prepared via CuI-catalyzed cross-coupling of gem-dibromo olefins <09CC5236>.

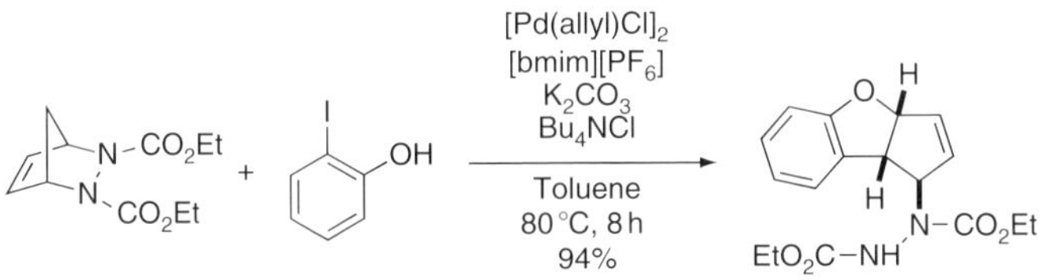

A series of sequential reactions was proposed to account for a one-step transformation of coatline B to matlamine, as shown below <09OL3020>.

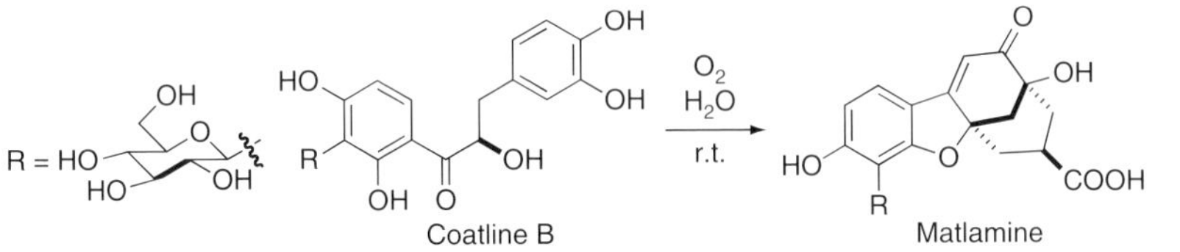

In the total synthesis of (±)-rocaglamide, the key step was a peracid-mediated oxidative Nazarov cyclization for the formation of its dihydrobenzofuran core <09JA7560>, and a similar type of dihydrobenzofuran core was also derived from sequential photochemical [3+2] cycloaddition, rearrangement and reduction <09JA1607>. FeCl$_3$-mediated ring closure of α-aryl ketones was found to be effective for the formation of 3-functionalized benzofurans via oxidative aromatic C–O bond formation <09OL4978>.

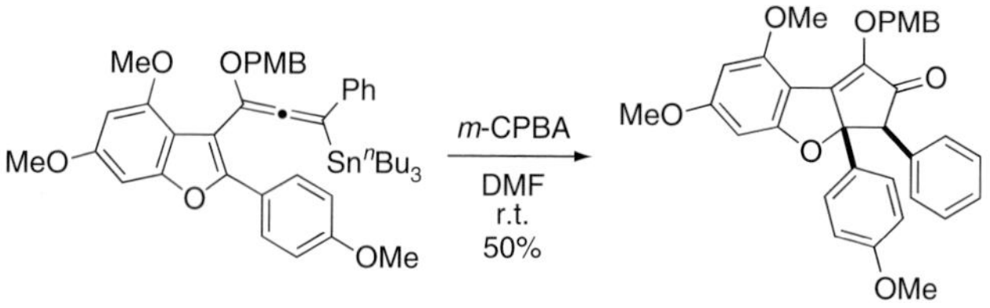

The synthesis of 4,6,7-trisubstituted benzofurans was achieved *via* the reaction of furfural imines with stabilized alkynylcarbene complexes as shown below <09JA14628>.

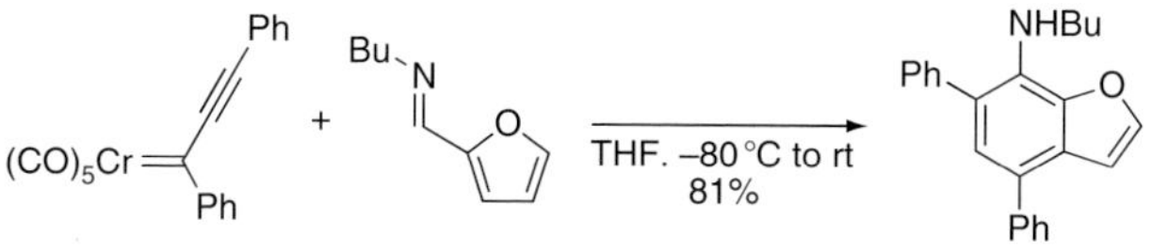

2-Substituted 5-hydroxybenzofurans were synthesized from alkyne-containing quinols *via* a sequence of Pt-catalyzed domino dienone-phenol rearrangement/intramolecular 5-*endo-dig* cyclization reactions, as represented below <09JOC8492>. Substituted benzofurans were also prepared by Au-catalyzed homogeneous reaction of furan-yn-ols <09T9021>, and Au-catalyzed intramolecular hydroarylation of terminal alkynes <09JOC8901>.

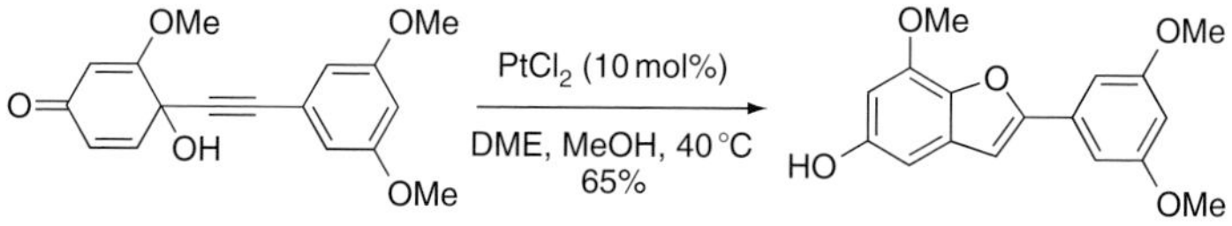

2,3-Dihydrobenzofuran-3-ols were formed *via* sequential alkynylation-cyclization of terminal alkynes with salicylaldehydes by using a Cy_3P-silver chloride complex as a catalyst in water, as shown below <09JOC3378>. An intramolecular carbonickelation of iodoaryl propargylic ethers was found to be effective for the formation of structurally diverse benzofurans <09CC4753>.

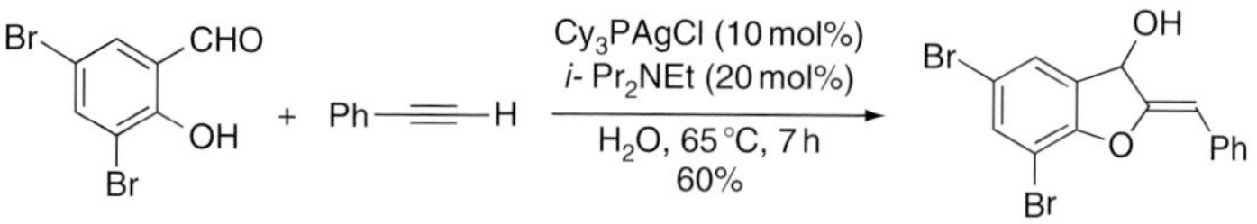

2,3-Dihydrobenzofurans with all-carbon quaternary centers were created by a alkene carboacylation reaction *via* quinoline-directed, rhodium-catalyzed C–C bond activation, as shown below <09JA412>, and application of Rh-catalyzed addition/cyclization to the synthesis of benzofuran-based polycyclic heteroaromatics was also reported <09JOC1809>.

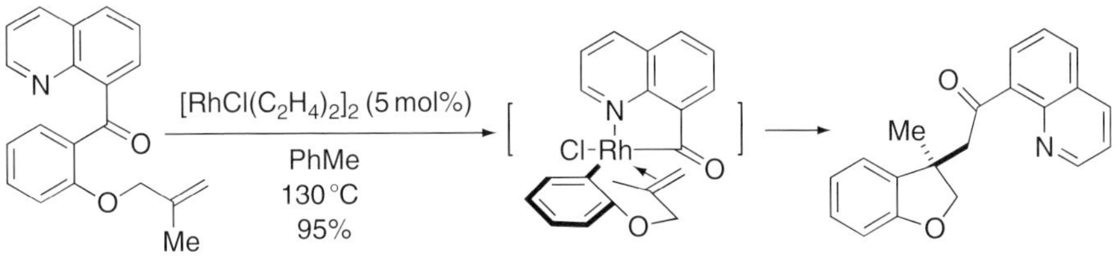

5.3.3.4 Benzo[c]furans and Related Compounds

Various 8-oxabicyclo[3.2.1]octane derivatives could be synthesized from propargylic esters through platinum-catalyzed cyclization followed by 1,2-benzene migration <09EJO117>.

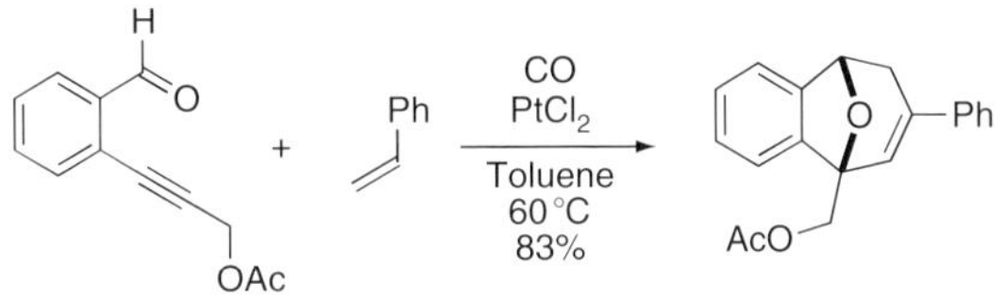

Substituted 1,3-dihydrobenzo[c]furans were obtained by a platinum-catalyzed cycloisomerization of a 1,6-enyne coupled with rearrangement of a propargylic ester. 1,3-Acyloxy migration followed by Diels–Alder type reaction are the key steps of the transformation <09JOC474>.

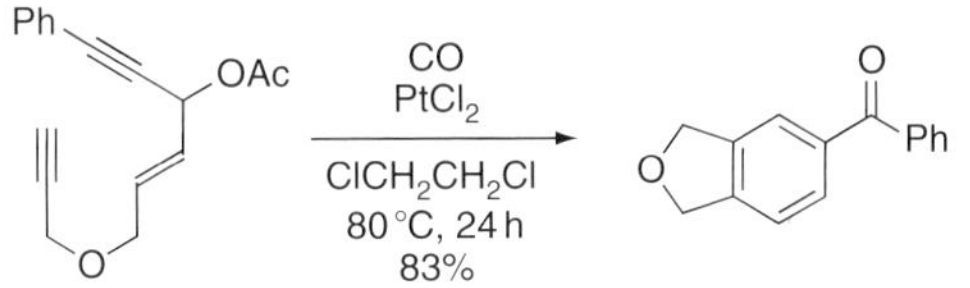

A rhodium(I)-xylyl-BINAP catalyzed asymmetric [2+2+2] cycloaddition of achiral conjugated aryl ynamides with various diynes was reported to lead to the corresponding optically enriched N,O-biaryls, featuring 1,3-dihydrobenzo[c]furans <09T5001>. Related benzo[c]furans were synthesized by Rh(I)-catalyzed inter- and intramolecular [2+2+2] cyclization of diynes with α,β-unsaturated enones, and this was applied to the synthesis of (−)-alcyopterosin I <09JOC2907>. Biologically interesting 2,5-dihydrofuran-fused quinones were synthesized *via* a sequence of ruthenium-catalyzed [2+2+2] cycloaddition of an ether-tethered diiododiyne with alkyne, copper-catalyzed Ullmann coupling followed by oxidation <09JOC4324>. Substituted 1,3-dihydrobenzo[c]furans was synthesized by cationic rhodium(I)/BINAP complex-catalyzed decarboxylative [2+2+2] cycloaddition of 1,6- and 1,7-diynes with vinylene carbonate, as shown below <09OL1337>.

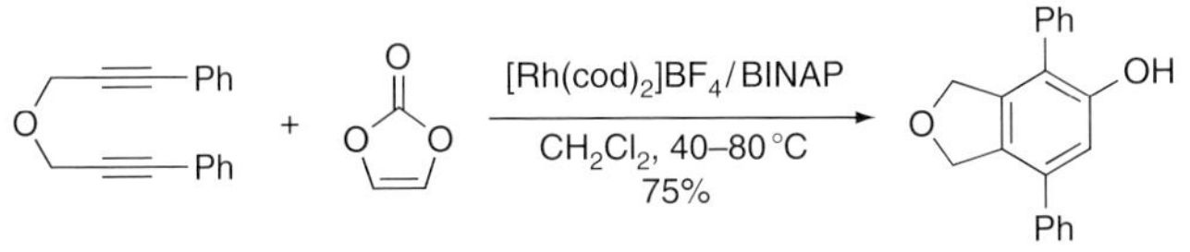

A reductive opening of 4- and 5-halophthalans proceeded through halogen-lithium exchange followed by a second reductive lithiation, leading to polyfunctionalized benzylic alcohols after subsequent reaction with electrophiles <09H(77)991>.

The synthesis of (E)-3-(isobenzofuran-3(1H)-ylidene)indolin-2-ones were achieved by Pd-catalyzed oxidative intramolecular C–H functionalization of 3-(2-(hydroxymethyl)aryl)-N-methyl-N-arylpropiolamides <09JOC3569>. A phosphazene base-catalyzed tandem addition-cyclization reaction between o-alkynylbenzadehyde and a nucleophile, such as i-PrOH, was applied to the synthesis of isobenzofuran derivatives <09SL638>. Related 2,3-dihydrobenzo[c]furans were also obtained through an intramolecular defluorinative cyclization reaction <09JOC2850>. Isobenzofurans were also synthesized by the intramolecular cyclization of dihydroisoindoles with methyl propiolate, dimethyl acetylenedicarboxylate or acetyl acetylene <09TL4851>.

ACKNOWLEDGEMENTS

The authors thank Prof. Henry N. C. Wong for advice and assistance. XLH acknowledges with thanks supports from the National Natural Science Foundation of China, National Outstanding Youth Fund, the Chinese Academy of Sciences, and Shanghai Committee of Science and Technology. KSY thanks Dr Nicholas A. Meanwell for support.

REFERENCES

09AGE784	Z. Shen and V.M. Dong, *Angew. Chem. Int. Ed.*, **48**, 784 (2009).
09AGE1681	M. Zhang, H.F. Jiang, H. Neumann, M. Beller and P.H. Dixneuf, *Angew. Chem. Int. Ed.*, **48**, 1681 (2009).
09AGE2346	K.L. Jackson, J.A. Henderson, H. Motoyoshi and A.J. Phillips, *Angew. Chem. Int. Ed.*, **48**, 2346 (2009).
09AGE2351	J.-Q. Dong and H.N.C. Wong, *Angew. Chem. Int. Ed.*, **48**, 2351 (2009).
09AGE5505	F. Liu, Y. Yu and J. Zhang, *Angew. Chem. Int. Ed.*, **48**, 5505 (2009).
09AGE5848	A.S.K. Hashmi, M. Rudolph, J. Huck, W. Frey, J.W. Bats and M. Hamzic, *Angew. Chem. Int. Ed.*, **48**, 5848 (2009).
09AGE6032	T. Magauer, H.J. Martin and J. Mulzer, *Angew. Chem. Int. Ed.*, **48**, 6032 (2009).
09AGE6296	C.S. Schindler, S. Diethelm and E.M. Carreira, *Angew. Chem. Int. Ed.*, **48**, 6296 (2009).
09AGE7569	J. Barluenga, L. Riesgo, L.A. López, E. Rubio and M. Tomás, *Angew. Chem. Int. Ed.*, **48**, 7569 (2009).
09AGE7687	P.A. Wender, J.P. Christy, A.B. Lesser and M.T. Gieseler, *Angew. Chem. Int. Ed.*, **48**, 7687 (2009).
09ASC617	Y. Xiao and J. Zhang, *Adv. Synth. Catal.*, **351**, 617 (2009).
09CC104	F. Petronijevic, C. Timmons, A. Cuzzupe and P. Wipf, *Chem. Commun.*, 104 (2009).

09CC1879	N. Kaur, Y. Xia, Y. Jin, N.T. Dat, K. Gajulapati, Y. Choi, Y.-S. Hong, J.J. Lee and K. Lee, *Chem. Commun.*, 1879 (2009).
09CC3466	M. Murai, K. Miki and K. Ohe, *Chem. Commun.*, 3466 (2009).
09CC3536	L. Melzig, C.B. Rauhut and P. Knochel, *Chem. Commun.*, 3536 (2009).
09CC4569	J.J. Kaloko, Y.H.G. Teng and I. Ojima, *Chem. Commun.*, 4569 (2009).
09CC4753	M. Durandetti, L. Hardou, M. Clément and J. Maddaluno, *Chem. Commun.*, 4753 (2009).
09CC5135	S.D. Sanders, A. Ruiz-Olalla and J.S. Johnson, *Chem. Commun.*, 5135 (2009).
09CC5236	S.G. Newman, V. Aureggi, C.S. Bryan and M. Lautens, *Chem. Commun.*, 5236 (2009).
09CC5248	C. Kanazawa, K. Goto and M. Terada, *Chem. Commun.*, 5248 (2009).
09CC5451	Y.-X. Xie, Z.-Y. Yan, B. Qian, W.-Y. Deng, D.-Z. Wang, L.-Y. Wu, X.-Y. Liu and Y.-M. Liang, *Chem. Commun.*, 5451 (2009).
09CC7125	J.L. Arbour, H.S. Rzepa, A.J.P. White and K.K. Hii, *Chem. Commun.*, 7125 (2009).
09CC7327	E. Jimenez-Nunez, K. Molawi and A.M. Echavcrren, *Chem. Commun.*, 7327 (2009).
09CEJ1566	M. Frings, I. Atodiresei, J. Runsink, G. Raabe and C. Bolm, *Chem. Eur. J.*, **15**, 1566 (2009).
09CEJ3336	B. Trillo, F. López, S. Montserrat, G. Ujaque, L. Castedo, A. Lledós and J.L. Mascareñas, *Chem. Eur. J.*, **15**, 3336 (2009).
09CEJ5208	Y. Zhang, Z. Chen, Y. Xiao and J. Zhang, *Chem. Eur. J.*, **15**, 5208 (2009).
09CEJ6041	M. Saquib, I. Husain, B. Kumar and A.K. Shaw, *Chem. Eur. J.*, **15**, 6041 (2009).
09CEJ7850	N.D. Ca', G. Maestri and M. Catellani, *Chem. Eur. J.*, **15**, 7850 (2009).
09CEJ8698	S. Xu, L. Zhou, R. Ma, H. Song and Z. He, *Chem. Eur. J.*, **15**, 8698 (2009).
09CEJ8975	Y. Bai, J. Fang, J. Ren and Z. Wang, *Chem. Eur. J.*, **15**, 8975 (2009).
09CEJ9303	R. Liu and J. Zhang, *Chem. Eur. J.*, **15**, 9303 (2009).
09CEJ12926	L. Rout and A.M. Harned, *Chem. Eur. J.*, **15**, 12926 (2009).
09CEJ13318	A.S.K. Hashmi, M. Hamzic, F. Rominger and J.W. Bats, *Chem. Eur. J.*, **15**, 13318 (2009).
09CSR3222	C.S. Schindler and E.M. Carreira, *Chem. Soc. Rev.*, **38**, 3222 (2009).
09EJO117	X.-Z. Shu, S.-C. Zhao, K.-G. Ji, Z.-J. Zheng, X.-Y. Liu and Y.-M. Liang, *Eur. J. Org. Chem.*, 117 (2009).
09EJO1132	S. Genovese, F. Epifano, C. Peluccini and M. Curini, *Eur. J. Org. Chem.*, 1132 (2009).
09EJO2141	O. Klein, H. Hopf and J. Grunenberg, *Eur. J. Org. Chem.*, 2141 (2009).
09EJO2737	Y. Otake, R. Tanaka and K. Tanaka, *Eur. J. Org. Chem.*, 2737 (2009).
09EJO3871	M.-A. Giroux, K.C. Guérard, M.-A. Beaulieu, C. Sabot and S. Canesi, *Eur. J. Org. Chem.*, 3871 (2009).
09EJO6091	M.P. Shinde, X. Wang, E.J. Kang and H.Y. Jang, *Eur. J. Org. Chem.*, 6091 (2009).
09H(77)201	F. Yoshimura, M. Takahashi, K. Tanino and M. Miyashita, *Heterocycles*, **77**, 201 (2009).
09H(77)433	M. Tsubuki, H. Okita, K. Kaneko, A. Shigihara and T. Honda, *Hetereocles*, **77**, 433 (2009).
09H(77)793	N. Yoshikado, S. Taniguchi, N. Kasajima, F. Ohashi, K.-I. Doi, T. Shibata, T. Yoshida and T. Hatano, *Heterocycles*, **77**, 793 (2009).
09H(77)991	D. Garcia, F. Foubelo and M. Yus, *Heterocycles*, **77**, 991 (2009).
09H(78)1109	K.C. Majumdar, S. Muhuri, R.U. Islam and B. Chattopadhyay, *Heterocycles*, **78**, 1109 (2009).
09H(78)1557	S.-H. Tao, S.-H. Qi, S. Zhang, Z.-H. Xiao and Q.-X. Li, *Heterocycles*, **78**, 1557 (2009).
09H(78)1581	N. Iwasaki, M. Baba, H. Aishan, Y. Okada and T. Okuyama, *Heterocycles*, **78**, 1581 (2009).
09H(78)2565	R. Liu and J.-K. Liu, *Heterocycles*, **78**, 2565 (2009).
09HCA121	Z.-Y. Yang, Y.-H. Yin and L.-H. Hu, *Helv. Chim. Acta*, **92**, 121 (2009).
09HCA139	J. Cui, Z. Deng, M. Xu, P. Proksch, Q. Li and W. Lin, *Helv. Chim. Acta*, **92**, 139 (2009).
09HCA165	Y.-Q. Su, Y.-H. Shen, S. Lin, J. Tang, J.-M. Tian, X.-H. Liu and W.-D. Zhang, *Helv. Chim. Acta*, **92**, 165 (2009).
09HCA195	T. Ito, N. Abe, Y. Masuda, M. Nasu, M. Oyama, R. Sawa, Y. Takahashi and M. Iinuma, *Helv. Chim. Acta*, **92**, 195 (2009).

09HCA370	H.-H. Wei, H.-H. Xu, H.-H. Xie, L.-X. Xu and X.-Y. Wei, *Helv. Chim. Acta*, **92**, 370 (2009).
09HCA409	G. Xu, L.-Y. Peng, L. Tu, X.-L. Li, Y. Zhao, P.-T. Zhang and Q.-S. Zhao, *Helv. Chim. Acta*, **92**, 409 (2009).
09HCA928	Y.-K. Li, Q.-J. Zhao, J. Hu, Z. Zou, X.-Y. He, H.-B. Yuan and X.-Y. Shi, *Helv. Chim. Acta*, **92**, 928 (2009).
09HCA1184	L. Yuan, X. Lin, P.-J. Zhao, J. Ma, Y.-J. Huang and Y.-M. Shen, *Helv. Chim. Acta*, **92**, 1184 (2009).
09HCA1191	C.-Y. Zhou, H.-Y. Tang, X. Li and Y.-S. Sang, *Helv. Chim. Acta*, **92**, 1191 (2009).
09HCA1203	T. Ito, N. Abe, M. Oyama, T. Tanaka, J. Murata, D. Darnaedi and M. Iinuma, *Helv. Chim. Acta*, **92**, 1203 (2009).
09HCA1260	S. He, L. Jiang, B. Wu, J. Zhou and Y.-J. Pan, *Helv. Chim. Acta*, **92**, 1260 (2009).
09HCA1341	Y. Li, A.-H. Gao, H. Huang, J. Li, E. Mollo, M. Gavagnin, G. Cimino, Y.-C. Gu and Y.-W. Guo, *Helv. Chim. Acta*, **92**, 1341 (2009).
09HCA1873	N.-Y. Ji, X.-M. Li, K. Li and B.-G. Wang, *Helv. Chim. Acta*, **92**, 1873 (2009).
09HCA2071	H. Nagano, A. Torihata, M. Matsushima, R. Hanai, Y. Saito, M. Baba, Y. Tanio, Y. Okamoto, Y. Takashima, M. Ichihara, X. Gong, C. Kuroda and M. Tori, *Helv. Chim. Acta*, **92**, 2071 (2009).
09HCA2101	Y.-C. Shen, P.-S. Shih, Y.-S. Lin, Y.-C. Lin, Y.-H. Kuo, Y.-C. Kuo and A.T. Khalil, *Helv. Chim. Acta*, **92**, 2101 (2009).
08HCA2338	C.-R. Zhang, S.-P. Yang, X.-Q. Chen, Y. Wu, X.-C. Zhen and J.-M. Yue, *Helv. Chim. Acta*, **92**, 2338 (2009).
09HCA2361	H.-J. Lee, S.-M. Seo, O.-K. Lee, H.-J. Jo, H.-Y. Kang, D.-H. Choi, K.-H. Paik and M. Khan, *Helv. Chim. Acta*, **92**, 2361 (2009).
09HCA2746	J.-L. Sun, A.-J. Deng, Y. Li, Z.-H. Li, H. Chen and H.-L. Qin, *Helv. Chim. Acta*, **92**, 2746 (2009).
09JA263	S.Y. Seo, X. Yu and T.J. Marks, *J. Am. Chem. Soc.*, **131**, 263 (2009).
09JA412	A.M. Dreis and C.J. Douglas, *J. Am. Chem. Soc.*, **131**, 412 (2009).
09JA444	R. Webster, C. Böing and M. Lautens, *J. Am. Chem. Soc.*, **131**, 444 (2009).
09JA570	L.C. Wieland, E.M. Vieira, M.L. Snapper and A.H. Hoveyda, *J. Am. Chem. Soc.*, **131**, 570 (2009).
09JA1101	A. DeAngelis, M.T. Taylor and J.M. Fox, *J. Am. Chem. Soc.*, **131**, 1101 (2009).
09JA1607	T.E. Adams, M.E. Sous, B.C. Hawkins, S. Himer, G. Holloway, M.L. Khoo, D.J. Owen, G.P. Savage, P.J. Scammells and M.A. Rizzacasa, *J. Am. Chem. Soc.*, **131**, 1607 (2009).
09JA2090	Y.-C. Hsu, C.-M. Ting and R.-S. Liu, *J. Am. Chem. Soc.*, **131**, 2090 (2009).
09JA2786	R.N. Loy and E.N. Jacobsen, *J. Am. Chem. Soc.*, **131**, 2786 (2009).
09JA3122	A.T. Parsons and J.S. Johnson, *J. Am. Chem. Soc.*, **131**, 3122 (2009).
09JA3844	I. Ibrahem, M. Yu, R.R. Schrock and A.H. Hoveyda, *J. Am. Chem. Soc.*, **131**, 3844 (2009).
09JA4194	C. Wang, I. Piel and F. Glorius, *J. Am. Chem. Soc.*, **131**, 4194 (2009).
09JA4556	W.K. Chung, S.K. Lam, B. Lo, L.L. Liu, W.-T. Wong and P. Chiu, *J. Am. Chem. Soc.*, **131**, 4556 (2009).
09JA5042	J. John, E. Suresh and K.V. Radhakrishnan, *J. Am. Chem. Soc.*, **131**, 5042 (2009).
09JA6348	P. Mauleón, R.M. Zeldin, A.Z. Gonzàlez and D. Toste, *J. Am. Chem. Soc.*, **131**, 6348 (2009).
09JA7560	J.A. Malona, K. Cariou and A.J. Frontier, *J. Am. Chem. Soc.*, **131**, 7560 (2009).
09JA8772	K.T. Sylvester and P.J. Chirik, *J. Am. Chem. Soc.*, **131**, 8772 (2009).
09JA10370	M.J. Campbell and J.S. Johnson, *J. Am. Chem. Soc.*, **131**, 10370 (2009).
09JA11402	G. Stork, A. Yamashita, J. Adams, G.R. Schulte, R. Chesworth, Y. Miyazaki and J.J. Farmer, *J. Am. Chem. Soc.*, **131**, 11402 (2009).
09JA12546	W. Chen and L.S. Liebeskind, *J. Am. Chem. Soc.*, **131**, 12546 (2009).
09JA12918	D. Schuch, P. Fries, M. Donges, B. Menendez and J. Hartung, *J. Am. Chem. Soc.*, **131**, 12918 (2009).
09JA14628	J. Barluenga, A. Gómez, J. Santamaría and M. Tomás, *J. Am. Chem. Soc.*, **131**, 14628 (2009).
09JA16632	P. Jakubec, D.M. Cokfield and D.J. Dixon, *J. Am. Chem. Soc.*, **131**, 16632 (2009).
09JHC62	K.C. Majumdar, S. Alam and B. Chattopadhyay, *J. Heterocycl. Chem.*, **46**, 62 (2009).

09JNP63	M. Mori-Hongo, H. Yamaguchi, T. Warashina and T. Miyase, *J. Nat. Prod.*, **72**, 63 (2009).
09JNP190	M. Kladi, C. Vagias, P. Papazafiri, S. Brogi, A. Tafi and V. Roussis, *J. Nat. Prod.*, **72**, 190 (2009).
09JNP194	M. Mori-Hongo, H. Takimoto, T. Katagiri, M. Kimura, Y. Ikeda and T. Miyase, *J. Nat. Prod.*, **72**, 194 (2009).
09JNP573	Q. Mi, J.M. Pezzuto, N.R. Farnsworth, M.C. Wani, A.D. Kinghorn and S.M. Swanson, *J. Nat. Prod.*, **72**, 573 (2009).
09JNP621	R.-Q. Mei, Y.-H. Wang, G.-H. Du, G.-M. Liu, L. Zhang and Y.-X. Cheng, *J. Nat. Prod.*, **72**, 621 (2009).
09JNP685	H.-D. Chen, S.-P. Yang, Y. Wu, L. Dong and J.-M. Yue, *J. Nat. Prod.*, **72**, 685 (2009).
09JNP714	X. Fang, Y.-T. Di, C.-S. Li, Z.-L. Geng, Z. Zhang, Y. Zhang, Y. Lu, Q.-T. Zheng, S.-Y. Ynag and X.-J. Hao, *J. Nat. Prod.*, **72**, 714 (2009).
09JNP852	F. Martin, A.-E. Hay, V.R.Q. Condoretty, D. Cressend, M. Reist, M.P. Gupta, P.-A. Carrupt and K. Hostettmann, *J. Nat. Prod.*, **72**, 852 (2009).
09JNP857	D.A. Barancelli, A.C. Mantovani, C. Jesse, C.W. Nogueira and G. Zeni, *J. Nat. Prod.*, **72**, 857 (2009).
09JNP912	Y. Li, D. Ye, X. Chen, X. Lu, Z. Shao, H. Zhang and Y. Che, *J. Nat. Prod.*, **72**, 912 (2009).
09JNP921	Y.-M. Yu, J.-S. Yang, C.-Z. Peng, V. Caer, P.-Z. Cong, Z.-W. Zou, Y. Lu, S.-Y. Yang and Y.-C. Gu, *J. Nat. Prod.*, **72**, 921 (2009).
09JNP966	G. Ni, Q.-J. Zhang, Z.-F. Zheng, R.-Y. Chen and D.-Q. Yu, *J. Nat. Prod.*, **72**, 966 (2009).
09JNP976	Y. Matsuno, J. Deguchi, T. Hosoya, Y. Hirasawa, C. Hirobe, M. Shiro and H. Morita, *J. Nat. Prod.*, **72**, 976 (2009).
09JNP994	S.-L. Wu, J.-H. Su, Z.-H. Wen, C.-H. Hsu, B.-W. Chen, C.-F. Dai, Y.-H. Kuo and J.-H. Sheu, *J. Nat. Prod.*, **72**, 994 (2009).
09JNP1213	Q. Zhang, A.J. Peoples, M.T. Rothfeder, W.P. Millett, B.C. Pescatore, L.L. Ling and C.M. Moore, *J. Nat. Prod.*, **72**, 1213 (2009).
09JNP1305	B.-D. Lin, C.-R. Zhang, S.-P. Yang, S. Zhang, Y. Wu and J.-M. Yue, *J. Nat. Prod.*, **72**, 1305 (2009).
09JNP1331	A.S. Kate, J.K. Pearson, B. Ramanathan, K. Richard and R.G. Kerr, *J. Nat. Prod.*, **72**, 1331 (2009).
09JNP1361	L.M. Kutrzeba, D. Ferreira and J.K. Zjawiony, *J. Nat. Prod.*, **72**, 1361 (2009).
09JNP1379	T. Murata, K. Sasaki, K. Sato, F. Yoshizaki, H. Yamada, H. Mutoh, K. Umehara, T. Miyase, T. Warashina, H. Aoshima, H. Tabata and K. Matsubara, *J. Nat. Prod.*, **72**, 1379 (2009).
09JNP1529	L.F.L. Barros, A. Barison, M.J. Salvador, R. Mello-Silva, E.C. Cabral, M.N. Eberlin and M.E.A. Stefanello, *J. Nat. Prod.*, **72**, 1529 (2009).
09JNP1657	M.-Y. Li, S.-X. Yang, J.-Y. Pan, Q. Xiao, T. Satyanandamurty and J. Wu, *J. Nat. Prod.*, **72**, 1657 (2009).
09JNP1678	W.-L. Xiao, Y.-Q. Gong, R.-R. Wang, Z.-Y. Weng, X. Luo, X.-N. Li, G.-Y. Yang, F. He, J.-X. Pu, L.-M. Yang, Y.-T. Zheng, Y. Lu and H.-D. Sun, *J. Nat. Prod.*, **72**, 1678 (2009).
09JNP1702	T.B. Cesar, J.A. Manthey and K. Myung, *J. Nat. Prod.*, **72**, 1702 (2009).
09JNP1870	C. Mahidol, H. Prawat, S. Sangpetsiripan and S. Ruchirawat, *J. Nat. Prod.*, **72**, 1870 (2009).
09JNP1980	M.J. Lear, O. Simon, T.L. Foley, M.D. Burkart, T.J. Baiga, J.P. Noel, A.G. Dipasquale, A.L. Rheingold and J.J.L. Clair, *J. Nat. Prod.*, **72**, 1980 (2009).
09JNP2005	F. Wang, X.-L. Cheng, Y.-J. Li, S. Shi and J.-K. Liu, *J. Nat. Prod.*, **72**, 2005 (2009).
09JNP2040	P.-T. Thuong, T.-T. Dao, T.-H.-M. Pham, P.-H. Nguyen, T.-V.-T. Le, K.-Y. Lee and W.-K. Oh, *J. Nat. Prod.*, **72**, 2040 (2009).
09JNP2084	B.-D. Lin, T. Yuan, C.-R. Zhang, L. Dong, B. Zhang, Y. Wu and J.-M. Yue, *J. Nat. Prod.*, **72**, 2084 (2009).
09JNP2110	M.-Y. Li, X.-B. Yang, J.-Y. Pan, G. Feng, Q. Xiao, J. Sinkkonen, T. Satyanandamurty and J. Wu, *J. Nat. Prod.*, **72**, 2110 (2009).
09JOC474	L. Lu, X.-Y Liu, X.-Z. Shu, K. Yang, K.-G. Ji and Y.-M. Liang, *J. Org. Chem.*, **74**, 474 (2009).

09JOC891	Z.S. Sales and N.S. Mani, *J. Org. Chem.*, **74**, 891 (2009).
09JOC1029	D. Moradov, A.A.A. Al Quntar, M. Youssef, R. Smoum, A. Rubinstein and M. Srebnik, *J. Org. Chem.*, **74**, 1029 (2009).
09JOC1141	S. Mehta, J.P. Waldo and R.C. Larock, *J. Org. Chem.*, **74**, 1141 (2009).
09JOC1237	N. Chopin, H. Gérard, I. Chataigner and S.R. Piettre, *J. Org. Chem.*, **74**, 1237 (2009).
09JOC1809	N.-W. Tseng and M. Lautens, *J. Org. Chem.*, **74**, 1809 (2009).
09JOC2153	F. Manarin, J.A. Roehrs, R.M. Gay, R. Brandão, P.H. Menezes, C.W. Nogueira and G. Zeni, *J. Org. Chem.*, **74**, 2153 (2009).
09JOC2224	H. Wei, H. Zhai and P.-F. Xu, *J. Org. Chem.*, **74**, 2224 (2009).
09JOC2750	P. Pahari and J. Rohr, *J. Org. Chem.*, **74**, 2750 (2009).
09JOC2850	L. Zhang, W. Zhang, J. Liu and J. Hu, *J. Org. Chem.*, **74**, 2850 (2009).
09JOC2907	A.L. Jones and J.K. Snyder, *J. Org. Chem.*, **74**, 2907 (2009).
09JOC3378	M. Yu, R. Skouta, L. Zhou, H.-F Jiang, X. Yao and C.-J. Li, *J. Org. Chem.*, **74**, 3378 (2009).
09JOC3430	L.K. Sydnes, B. Holmelid, M. Sengee and M. Hanstein, *J. Org. Chem.*, **74**, 3430 (2009).
09JOC3569	P. Peng, B.-X. Tang, S.-F. Pi, Y. Liang and J.-H. Li, *J. Org. Chem.*, **74**, 3569 (2009).
09JOC4094	C.A. James, A.L. Coelho, M. Gevaert, P. Forgione and V. Snieckus, *J. Org. Chem.*, **74**, 4094 (2009).
09JOC4118	R. Shen, S. Zhu and X. Huang, *J. Org. Chem.*, **74**, 4118 (2009).
09JOC4324	Y. Yamamoto, R. Takuma, T. Hotta and K. Yamashita, *J. Org. Chem.*, **74**, 4324 (2009).
09JOC4360	A. Blanc, K. Tembrink, J.-M. Weibel and P. Pale, *J. Org. Chem.*, **74**, 4360 (2009).
09JOC4418	Y. Natori, H. Tsutsui, N. Sato, S. Nakamura, H. Nambu, M. Shiro and S. Hashimoto, *J. Org. Chem.*, **74**, 4418 (2009).
09JOC5342	A. Blanc, K. Tembrink, J.-M. Weibel and P. Pale, *J. Org. Chem.*, **74**, 5342 (2009).
09JOC5429	Y. Lang, F.E.S. Souza, X. Xu, N.J. Taylor, A. Assoud and R. Rodrigo, *J. Org. Chem.*, **74**, 5429 (2009).
09JOC5502	S. Sato, F. Iwata, T. Mukai, S. Yamada, J. Takeo, A. Abe and H. Kawahara, *J. Org. Chem.*, **74**, 5502 (2009).
09JOC7570	M. Ballantine, M.L. Menard and W. Tam, *J. Org. Chem.*, **74**, 7570 (2009).
09JOC8492	I. Kim, K. Kim and J. Choi, *J. Org. Chem.*, **74**, 8492 (2009).
09JOC8733	J. Li, W. Kong, Y. Yu, C. Fu and S. Ma, *J. Org. Chem.*, **74**, 8733 (2009).
09JOC8779	C. Brand, G. Rauch, M. Zanoni, B. Dittrich and D.B. Werz, *J. Org. Chem.*, **74**, 8779 (2009).
09JOC8848	X. Guo, S. Pan, J. Liu and Z. Li, *J. Org. Chem.*, **74**, 8848 (2009).
09JOC8859	Z. Yang, P. Tang, J.F. Gauuan and B.F. Molino, *J. Org. Chem.*, **74**, 8859 (2009).
09JOC8901	R.V. Menon, A.D. Findlay, A.C. Bissember and M.G. Banwell, *J. Org. Chem.*, **74**, 8901 (2009).
09JOC8904	Y. Li and Z. Yu, *J. Org. Chem.*, **74**, 8904 (2009).
09JOC9546	X. Huang and J. Xu, *J. Org. Chem.*, **74**, 9546 (2009).
09NPR266	A.R. Gallimore, *Nat. Prod. Rep.*, **26**, 266 (2009).
09OBC639	Y. Li, C.C. Nawrat, G. Pattenden and J.M. Winne, *Org. Biomol. Chem.*, **7**, 639 (2009).
09OBC1761	M. Brichacek and J.T. Njardason, *Org. Biomol. Chem.*, **7**, 1761 (2009).
09OBC1829	C.L. Gibson, J.K. Huggan, A. Kennedy, L. Kiefer, J.W. Lee, C.J. Suckling, C. Clements, A.L. Harvey, W.N. Hunter and L.B. Tulloch, *Org. Biomol. Chem.*, **7**, 1829 (2009).
09OBC2788	I. Kim and J. Choi, *Org. Biomol. Chem.*, **7**, 2788 (2009).
09OL57	R. Ratnayake, D. Covell, T.T. Ransom, K.R. Gustafson and J.A. Beutler, *Org. Lett.*, **11**, 57 (2009).
09OL137	X. Meng, Y. Huang and R. Chen, *Org. Lett.*, **11**, 137 (2009).
09OL313	M. Tofi, K. Koltsida and G. Vassilikogiannakis, *Org. Lett.*, **11**, 313 (2009).
09OL349	S. Lazzaroni, S. Protti, M. Fagnoni and A. Albini, *Org. Lett.*, **11**, 349 (2009).
09OL617	T. Yuan, S.-P. Yang, C.-R. Zhang, S. Zhang and J.-M. Yue, *Org. Lett.*, **11**, 617 (2009).
09OL891	L. Wang, K. Thai and M. Gravel, *Org. Lett.*, **11**, 891 (2009).
09OL1083	C. Martinez, R. Álvarez and J.M. Aurrecoechea, *Org. Lett.*, **11**, 1083 (2009).

09OL1139	X.-C. Huang, F. Wang, Y. Liang and J.-H. Li, *Org. Lett.*, **11**, 1139 (2009).
09OL1205	Y. Deng, Y. Shi and S. Ma, *Org. Lett.*, **11**, 1205 (2009).
09OL1233	S. Kiren, X. Hong, C.A. Leverett and A. Padwa, *Org. Lett.*, **11**, 1233 (2009).
09OL1329	N. Isono and M. Lautens, *Org. Lett.*, **11**, 1329 (2009).
09OL1337	H. Hara, M. Hirano and K. Tanaka, *Org. Lett.*, **11**, 1337 (2009).
09OL1353	S.-X. Huang, L.-X. Zhao, S.-K. Tang, C.-L. Jiang, Y. Duan and B. Shen, *Org. Lett.*, **11**, 1353 (2009).
09OL1523	A. Dzudza and T.J. Marks, *Org. Lett.*, **11**, 1523 (2009).
09OL1817	R. Medimagh, S. Marque, D. Prim, J. Marrot and S. Chatti, *Org. Lett.*, **11**, 1817 (2009).
09OL1861	N.O. Devarie-Baez, W.S. Kim, A.B. Smith, III and M. Xian, *Org. Lett.*, **11**, 1861 (2009).
09OL1931	H. Gao, H. Jiang, W. Yao and X. Liu, *Org. Lett.*, **11**, 1931 (2009).
09OL2019	H. Zhu, J.G. Wickenden, N.E. Campbell, J.C.T. Leung, K.M. Johnson and G.M. Sammis, *Org. Lett.*, **11**, 2019 (2009).
09OL2113	T. Xu, Z. Yu and L. Wang, *Org. Lett.*, **11**, 2113 (2009).
09OL2153	Y. Tsunematsu, O. Ohno, K. Konishi, K. Yamada, M. Suganuma and D. Uemura, *Org. Lett.*, **11**, 2153 (2009).
09OL2165	M.E. Jung and F. Perez, *Org. Lett.*, **11**, 2165 (2009).
09OL2233	T. Furuta, M. Nakayama, H. Suzuki, H. Tajimi, M. Inai, H. Nukaya, T. Wakimoto and T. Kan, *Org. Lett.*, **11**, 2233 (2009).
09OL2281	J. Luo, J.-S. Wang, J.-G. Luo, X.-B. Wang and L.-Y. Kong, *Org. Lett.*, **11**, 2281 (2009).
09OL2317	T.F. Schneider, J. Kaschel, B. Dittrich and D.B. Werz, *Org. Lett.*, **11**, 2317 (2009).
09OL2337	M. Yamashita, K. Hirano, T. Satoh and M. Miura, *Org. Lett.*, **11**, 2337 (2009).
09OL2361	Y. Komine, A. Kamisawa and K. Tanaka, *Org. Lett.*, **11**, 2361 (2009).
09OL2703	K. Cheng, A.R. Kelly, R.A. Kohn, J.F. Dweck and P.J. Walsh, *Org. Lett.*, **11**, 2703 (2009).
09OL2824	J.D. Winkler, A. Isaacs, L. Holderbaum, V. Tatart and N. Dahmane, *Org. Lett.*, **11**, 2824 (2009).
09OL2984	J.B. Davis, J.D. Bailey and J.K. Sello, *Org. Lett.*, **11**, 2984 (2009).
09OL3012	S.-T. Lin, S.-K. Wang, S.-Y. Cheng and C.-Y. Duh, *Org. Lett.*, **11**, 3012 (2009).
09OL3020	A.U. Acuña, F. Amat-Guerri, P. Morcillo, M. Liras and B. Rodríguez, *Org. Lett.*, **11**, 3020 (2009).
09OL3144	J.T. Henssier and A.J. Matzger, *Org. Lett.*, **11**, 3144 (2009).
09OL3206	K.-G. Ji, X.-Z. Shu, S.-C. Zhao, H.-T. Zhu, Y.-N. Niu, X.-Y. Liu and Y.-M. Liang, *Org. Lett.*, **11**, 3206 (2009).
09OL3298	T. Taniguchi, T. Fujii, A. Idota and H. Ishibashi, *Org. Lett.*, **11**, 3298 (2009).
09OL3430	A.G. Lohse and R.P. Hsung, *Org. Lett.*, **11**, 3430 (2009).
09OL3698	J.M. Schomaker, F.D. Toste and R.G. Bergman, *Org. Lett.*, **11**, 3698 (2009).
09OL3782	B.M. Trost and P.J. McDougall, *Org. Lett.*, **11**, 3782 (2009).
09OL3834	S.N. Chavre, P.R. Ullapu, S.-J. Min, J.K. Lee, A.N. Pae, Y. Kim and Y.S. Cho, *Org. Lett.*, **11**, 3834 (2009).
09OL3838	Y. Chen, Y. Lu, G. Li and Y. Liu, *Org. Lett.*, **11**, 3838 (2009).
09OL3938	P. Magnus and R. Littich, *Org. Lett.*, **11**, 3938 (2009).
09OL3958	GK. Friestad and H.J. Lee, *Org. Lett.*, **11**, 3958 (2009).
09OL4044	P. Hermange, M. Elise, T.H. Dau, P. Retailleau and R.H. Dodd, *Org. Lett.*, **11**, 4044 (2009).
09OL4096	C. Aout, E. Thiery, J. Le Bras and J. Muzart, *Org. Lett.*, **11**, 4096 (2009).
09OL4314	Y. Li, A.J. Lampkins, M.B. Baker, B.G. Sumpter, J. Huang, K.A. Abboud and R.K. Castellano, *Org. Lett.*, **11**, 4314 (2009).
09OL4462	H. Li and R.P. Hsung, *Org. Lett.*, **11**, 4462 (2009).
09OL4556	E. Pavlakos, T. Georgiou, M. Tofi, T. Montagnon and G. Vassilikogiannakis, *Org. Lett.*, **11**, 4556 (2009).
09OL4624	A. Aponick, C.Y. Li, J. Malinge and E.F. Marques, *Org. Lett.*, **11**, 4624 (2009).
09OL4688	R. Webster and M. Lautens, *Org. Lett.*, **11**, 4688 (2009).
09OL4978	Z. Liang, W. Hou, Y. Du, Y. Zhang, Y. Pan, D. Mao and K. Zhao, *Org. Lett.*, **11**, 4978 (2009).
09OL5002	M. Egi, K. Azechi and S. Akai, *Org. Lett.*, **11**, 5002 (2009).

09OL5206	S. Adachi, F. Tanaka, K. Watanabe and T. Harada, *Org. Lett.*, **11**, 5206 (2009).
09OL5254	G. Raffa, M. Rusch, G. Balme and N. Monteiro, *Org. Lett.*, **11**, 5254 (2009).
09OL5304	Y. Sawama, Y. Sawama and N. Krause, *Org. Lett.*, **11**, 5304 (2009).
09OL5394	J.L. Frie, C.S. Jeffrey and E.J. Sorensen, *Org. Lett.*, **11**, 5394 (2009).
09OL5478	L.M. Geary and P.G. Hultin, *Org. Lett.*, **11**, 5478 (2009).
09P216	G.D.W.F. Kapche, C.D. Fozing, J.H. Donfack, G.W. Fotso, D. Amadou, A.N. Tchana, M. Bezabih, P.F. Moundipa, B.T. Ngadiui and B.M. Abegaz, *Phytochemistry*, **70**, 216 (2009).
09P256	P.P. Yadav, R. Maurya, J. Sarkar, A. Arora, S. Kanojiya, S. Sinha, M.N. Srivastava and R. Raghubir, *Phytochemistry*, **70**, 256 (2009).
09P403	W.K.P. Shiu and S. Gibbons, *Phytochemistry*, **70**, 403 (2009).
09P635	C. Argyropoulou, A. Kartioti and H. Skaltsa, *Phytochemistry*, **70**, 635 (2009).
09P1233	C. Kihampa, M.H.H. Nkunya, C.C. Joseph, S.M. Magesa, A. Hassanali, M. Heydenreich and E. Kleinpeter, *Phytochemistry*, **70**, 1233 (2009).
09P1277	K. Matsunami, H. Otsuka, K. Kondo, T. Shinzato, M. Kawahata, K. Yamaguchi and Y. Takeda, *Phytochemistry*, **70**, 1277 (2009).
09P1305	B. Zhang, S.-P. Yang, S. Yin, C.-R. Zhang, Y. Wu and J.-M. Yue, *Phytochemistry*, **70**, 1305 (2009).
09P1309	E.D. Coy-Barrera, L.E. Cuca-Suarez and M. Sefkow, *Phytochemistry*, **70**, 1309 (2009).
09P1474	M.-E.F. Hegazy, M.H.A. Ei-Razek, F. Nagashima, Y. Asakawa and P.W. Pare, *Phytochemistry*, **70**, 1474 (2009).
09P1930	N.G. Kesinger and J.F. Stevens, *Phytochemistry*, **70**, 1930 (2009).
09P2047	J.O. Odalo, C.C. Joseph, M.H.H. Nkunya, I. Sattler, C. Lange, H.-M. Dahse and U. Mollman, *Phytochemistry*, **70**, 2047 (2009).
09P2053	T.T. Dao, P.H. Nguyen, P.T. Thuong, K.W. Kang, M. Na, D.T. Ndinteh, J.T. Mbafor and W.K. Oh, *Phytochemistry*, **70**, 2053 (2009).
09S1175	L.Y. Yang, C.-F. Chang, Y.-C. Huang, Y.-J. Lee, C.-C. Hu and T.-H. Tseng, *Synthesis*, 1175 (2009).
09S2278	L. Nagarapu, M.V. Chary, A. Satyender, B. Supriya and R. Bantu, *Synthesis*, 2278 (2009).
09S2651	S. Rosenberg and R. Leino, *Synthesis*, 2651 (2009).
09S3353	F.D. Simone and J. Waser, *Synthesis*, 3353 (2009).
09SL461	M. Sebek, J. Holz, A. Börner and K. Jähnisch, *Synlett*, 461 (2009).
09SL638	C. Kanazawa, A. Ito and M. Terada, *Synlett*, 638 (2009).
09SL755	N. Zanatta, A.D. Wouters, L. Fantinel, F.M. da Silva, R. Barichello, P.E.A. da Silva, D.F. Ramos, H.G. Bonacorso and M.A.P. Martins, *Synlett*, 755 (2009).
09SL1525	G. Casiraghi, F. Zanardi, L. Battistini and G. Rassu, *Synlett*, 1525 (2009).
09SL1597	A. Taleb, M. Lahrech, S. Hacini, J. Thibonnet and J. Parrain, *Synlett*, 1597 (2009).
09SL1937	K. Miki, Y. Senda, T. Kowada and K. Ohe, *Synlett*, 1937 (2009).
09SL1941	S. Matsuda, M. Takahashi, D. Monguchi and A. Mori, *Synlett*, 1941 (2009).
09SL1951	T. Meng, C.A. Fuhrer and R. Häner, *Synlett*, 1951 (2009).
09SL1990	C. Praveen, P. Kiruthiga and P.T. Pertumal, *Synlett*, 1990 (2009).
09SL2542	M. Adib, M. Mahdavi, S. Bagherazadeh and H.R. Bijianzadeh, *Synlett*, 2542 (2009).
09SL2629	A. Scettri, V. De Sio, R. Villano and M.R. Acocella, *Synlett*, 2629 (2009).
09SL2676	M.H. Mosslemin, M. Anary-Abbasinejad and H. Anaraki-Ardakani, *Synlett*, 2676 (2009).
09SL3003	F. Contiero, K.M. Jones, E.A. Matts, A. Porzelle and N.C.O. Tomkinson, *Synlett*, 3003 (2009).
09T164	C. Lei, J.-X. Pu, S.-X. Huang, J.-J. Chen, J.-P. Liu, L.-B. Yang, Y.-B. Ma, W.-L. Xiao, X.-N. Li and H.-D. Sun, *Tetrahedron*, **65**, 164 (2009).
09T1673	P. Li, Z. Chai, G. Zhao and S.-Z. Zhu, *Tetrahedron*, **65**, 1673 (2009).
09T1708	G. Fontana, G. Savona, B. Rodriguez, C.M. Dersch, R.B. Rothman and T.E. Prisinzano, *Tetrahedron*, **65**, 1708 (2009).
09T1716	N. Venkata, S. Mudiganti, S. Claessens and N. De Kimpe, *Tetrahedron*, **65**, 1716 (2009).
09T1839	X. Du, F. Song, Y. Lu, H. Chen and Y. Liu, *Tetrahedron*, **65**, 1839 (2009).
09T1911	L. Leseurre, C.M. Chao, T. Seki, E. Genin, P.Y. Toullec, J.-P. Genet and V. Michelet, *Tetrahedron*, **65**, 1911 (2009).

09T3425	J. Luo, J.-S. Wang, X.-B. Wang, X.-F. Huang, J.-G. Luo and L.-Y. Kong, *Tetrahedron*, **65**, 3425 (2009).
09T3509	N. Mori, K. Obuchi, T. Katae, J. Sakurada and T. Satoh, *Tetrahedron*, **65**, 3509 (2009).
09T3842	J. Oyamada and T. Kitamura, *Tetrahedron*, **65**, 3842 (2009).
09T3918	S. Malik, U.K. Nadir and P.S. Pandey, *Tetrahedron*, **65**, 3918 (2009).
09T4569	L.B. Nielsen, R. Slamet and D. Wege, *Tetrahedron*, **65**, 4569 (2009).
09T4820	H. Hagiwara, Y. Suka, T. Nojima, T. Hoshi and T. Suzuki, *Tetrahedron*, **65**, 4820 (2009).
09T5001	J. Oppenheimer, W.L. Johnson, R. Figueroa, R. Hayashi and R.P. Hsung, *Tetrahedron*, **65**, 5001 (2009).
09T6029	A.R. Diaz-Marrero, G. Porras, M. Cueto, L. Dcroz, M. Lorenzo and A.S.J. Darias, *Tetrahedron*, **65**, 6029 (2009).
09T7016	B.-W. Chen, Y.-C. Wu, M.Y. Chiang, J.-H. Su, W.-H. Wang, T.-Y. Fan and J.-H. Sheu, *Tetrahedron*, **65**, 7016 (2009).
09T7092	M. Marinozzi, M.C. Fulco, L. Amori, M. Fiumi and R. Pellicciari, *Tetrahedron*, **65**, 7092 (2009).
09T7408	X. Fang, Q. Zhang, C.-J. Tan, S.-Z. Mu, Y. Lu, Y.-B. Lu, Q.-T. Zheng, Y.-T. Di and X.-J. Hao, *Tetarhedron*, **65**, 7408 (2009).
09T8708	Y. Hari, R. Kondo, K. Date and T. Aoyama, *Tetrahedron*, **65**, 8708 (2009).
09T8981	S.J. Pidmore, P.A. Slatford, J.E. Taylor, M.K. Whittlesey and J.M.J. Williams, *Tetrahedron*, **65**, 8981 (2009).
09T8995	M.A. Honey, A.J. Blake, I.B. Campbell, B.D. Judkins and C.J. Moody, *Tetrahedron*, **65**, 8995 (2009).
09T9021	A.S.K. Hashmi and M. Wölfle, *Tetrahedron*, **65**, 9021 (2009).
09T9749	C.M. Chou, C.H. Chen, C.L. Lin, K.W. Yang, T.S. Lim and T.Y. Luh, *Tetrahedron*, **65**, 9749 (2009).
09T9862	H. Leisch, A.T. Omori, K.J. Finn, J. Gilmet and T. Bissett, *Tetrahedron*, **65**, 9862 (2009).
09T10334	L.A. Adrio, L.S. Quek, J.G. Taylor and K.K. Hii, *Tetrahedron*, **65**, 10334 (2009).
09T10745	V. Nair and A. Deepthi, *Tetrahedron*, **65**, 10745 (2009).
09TA381	H. Maeda, T. Okumura, Y. Yoshimi and K. Mizuno, *Tetrahedron: Asymmetry*, **20**, 381 (2009).
09TA1999	N. Araújo, M.V. Gil, E. Román and J.A. Serrano, *Tetrahedron: Asymmetry*, **20**, 1999 (2009).
09TA2537	G. Jalce, X. Franck and B. Figadere, *Tetrahedron: Asymmetry*, **20**, 2537 (2009).
09TL274	S. Goncalves, A. Wagner, C. Mioskowski and R. Baati, *Tetrahedron Lett.*, **50**, 274 (2009).
09TL467	O. Hofer, S. Pointinger, L. Brecker, K. Peter and H. Greger, *Tetrahedron Lett.*, **50**, 467 (2009).
09TL614	E. Bossharth, P. Desbordes, N. Monteiro and G. Balme, *Tetrahedron Lett.*, **50**, 614 (2009).
09TL781	G.L. Beutner, J.T. Kuethe and N. Yasuda, *Tetrahedron Lett.*, **50**, 781 (2009).
09TL960	B.M. Perez and J. Hartung, *Tetrahedron Lett.*, **50**, 960 (2009).
09TL2132	Z.-L. Geng, X. Fang, Y.-T. Di, Q. Zhang, Y. Zeng, Y.-M. Shen and X.-J. Hao, *Tetrahedron Lett.*, **50**, 2132 (2009).
09TL2239	T. Mino, Y. Naruse, S. Kobayashi, S. Oishi, M. Sakamoto and T. Fijita, *Tetrahedron Lett.*, **50**, 2239 (2009).
09TL2353	H. Li, J. Liu, B. Yan and Y. Li, *Tetrahedron Lett.*, **50**, 2353 (2009).
09TL2516	T. Ito, N. Abe, M. Oyama and M. Iinuma, *Tetrahedron Lett.*, **50**, 2516 (2009).
09TL2685	B.W. Gung and D.T. Craft, *Tetrahedron Lett.*, **50**, 2685 (2009).
09TL2758	J. Zhao, L. Huang, K. Cheng and Y. Zhang, *Tetrahedron Lett.*, **50**, 2758 (2009).
09TL3263	C.W. Chen and T.Y. Luh, *Tetrahedron Lett.*, **50**, 3263 (2009).
09TL3299	D. Conreaux, T. Delaunay, P. Desbordes, N. Monteiro and G. Balme, *Tetrahedron Lett.*, **50**, 3299 (2009).
09TL3474	R. Millet, T. Bernardez, L. Palais and A. Alexakis, *Tetrahedron Lett.*, **50**, 3474 (2009).
09TL3503	J.E. Camp and D. Craig, *Tetrahedron Lett.*, **50**, 3503 (2009).
09TL3588	M. Jacubert, A. Hamze, O. Provot, J.-F. Peyrat, J.-D. Brion and M. Alami, *Tetrahedron Lett.*, **50**, 3588 (2009).

09TL3753	D.P. Sant'Ana, V.D. Pinho, M.C.L.S. Maior and P.R.R. Costa, *Tetrahedron Lett.*, **50**, 3753 (2009).
09TL3929	M. Hussain, N.T. Hung and P. Langer, *Tetrahedron Lett.*, **50**, 3929 (2009).
09TL4128	S. Sen, P. Kulkarmi, K. Borate and N.R. Pai, *Tetrahedron Lett.*, **50**, 4128 (2009).
09TL4492	W.S. Yoon, S.J. Lee, S.K. Kang, D.-C. Ha and J.D. Ha, *Tetrahedron Lett.*, **50**, 4492 (2009).
09TL4744	K. Kato, T. Mochida, H. Takayama, M. Kimura, H. Moriyama, A. Takeshita, Y. Kanno, Y. Inouye and H. Akita, *Tetrahedron Lett.*, **50**, 4744 (2009).
09TL4747	N. Tanaka, Y. Kakuguchi, H. Ishiyama, T. Kubota and J. Kobayashi, *Tetrahedron Lett.*, **50**, 4747 (2009).
09TL4851	L.G. Voskressensky, L.N. Kulikova, A. Kleimenov, N. Guranova, T.N. Borisova and A.V. Varlamov, *Tetrahedron Lett.*, **50**, 4851 (2009).
09TL5182	S.B. Singh, H. Jayasuriya, K.B. Herath, C. Zhang, J.G. Ondeyka, D.L. Zink, S. Ha, G. Parthasarathy, J.W. Becker, J. Wang and S.M. Soisson, *Tetrahedron Lett.*, **50**, 5182 (2009).
09TL5625	A. Ramazani, A.T. Mahyari, M. Rouhani and A. Rezaei, *Tetrahedron Lett.*, **50**, 5625 (2009).
09TL5701	M. Harmata and C. Huang, *Tetrahedron Lett.*, **50**, 5701 (2009).
09TL6068	Y.H. Choi, J. Kwak and N. Jeong, *Tetrahedron Lett.*, **50**, 6068 (2009).
09TL6751	S. Samanta, R. Jana and J.K. Ray, *Tetrahedron Lett.*, **50**, 6751 (2009).
09TL6758	Q. Dang and Y. Liu, *Tetrahedron Lett.*, **50**, 6758 (2009).
09TL7180	G.A. Kraus and V. Gupta, *Tetrahedron Lett.*, **50**, 7180 (2009).

CHAPTER 5.4

Five Membered Ring Systems: With More than One N Atom

Larry Yet
AMRI
Larry.Yet@amriglobal.com

5.4.1. INTRODUCTION

The synthesis and chemistry of pyrazoles, imidazoles, 1,2,3-triazoles, 1,2,4-triazoles, and tetrazoles were actively pursued in 2009. No attempt has been made to incorporate all the exciting chemistry and biological applications that were published this year.

5.4.2. PYRAZOLES AND RING-FUSED DERIVATIVES

A review on the recent developments on aminopyrazole chemistry was published <09ARK198>.

Hydrazine addition to 1,3-difunctional groups is the most common method for the preparation of pyrazoles. Baylis-Hillman adducts **1** were used in a regioselective synthesis of 1,5-diarylpyrazoles **2** with phenylhydrazine hydrochloride under ultrasound irradiation <09ARK168>. The reaction of α,β-chalcone ditosylates **3** with phenylhydrazine hydrochloride led to a 1,2-aryl shift, providing a novel route for the synthesis of 1,4,5-trisubstituted pyrazoles **4** <09T10175>. The one-pot and regioselective synthesis of a novel series of 3-aryl(heteroaryl)-5-trifluoromethyl-5-hydroxy-4,5-dihydro-1H-pyrazolyl-1-carbohydrazides and bis-(3-aryl-5-trifluoromethyl-5-hydroxy-4,5-dihydro-1H-pyrazol-1-yl)methanones from the condensation reactions of 4-alkoxy-4-aryl(heteroaryl)-1,1,1-trifluoroalk-3-en-2-ones with carbohydrazide was reported <09ARK174>. Reaction of α,β-unsaturated ketones **5** with hydrazine monohydrate provided intermediate **6**, which was dehydrogenated with DDQ or p-chloranil in dioxane at reflux to give indazoles **7** <09T2298>. A solvent-free route to β-enamino dichloromethyl ketones and its application to the synthesis of 5-dichloromethyl-1H-pyrazoles was reported <09JHC1247>. Treatment of 1,3-diketones and allyltrimethylsilane with cerium ammonium nitrate followed by cerium-catalyzed addition of substituted hydrazines afforded tetrasubstituted pyrazoles <09SL1490>. A series of 4,6-disubstituted 2-(1H-pyrazolyl)-1,3,5-triazines were obtained from the reaction of methylenedicarbonyl compounds with 4,6-disubstituted-2-hydrazinyl-1,3,5-triazines <09TL2505>. Two convenient regioselective syntheses of 1-N-alkyl-3-aryl-4-(pyrimidin-4-yl)pyrazoles were reported <09TL1377>. A seven-step synthesis of 1-substituted

5-(2-acylaminoethyl)-1H-pyrazole-4-carboxamides as the pyrazole analogues of histamine was developed <09T7151>. A one-pot three-step process for the regioselective synthesis of 1,3,4-trisubstituted-1H-pyrazoles from β-keto esters was reported <09TL696>.

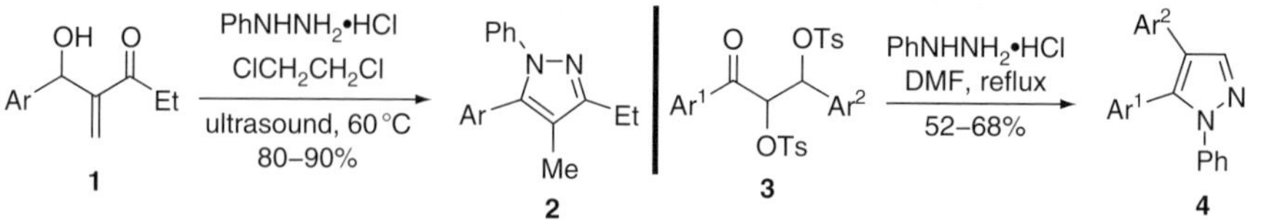

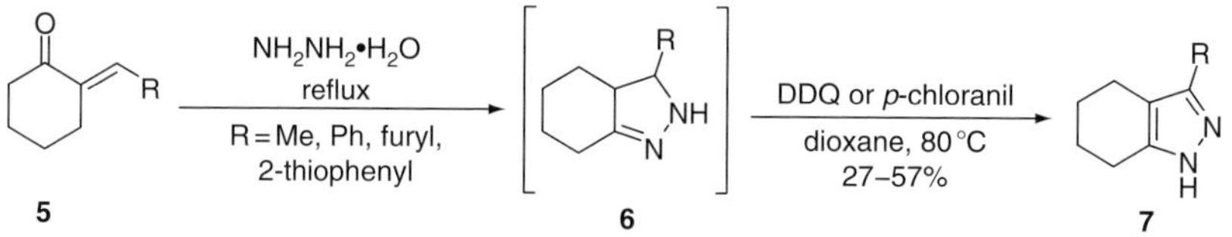

Hydrazones are useful intermediates in the preparation of pyrazoles. One-pot synthesis of semicarbazone, thiosemicarbazone, and hydrazone derivatives of 1-phenyl-3-arylpyrazole-4-carboxaldehyde from acetophenone phenylhydrazones using Vilsmeier–Haack reagent was reported <09SC316>. Iodobenzene diacetate was utilized efficiently for the oxidation of N-substituted hydrazones of chalcones **8** to 1,3,5-trisubstituted pyrazoles **9** under mild conditions <09SC2169>. A spiro-pyrazoline intermediate via a tandem 1,3-dipolar cycloaddition/elimination with α-chlorohydrazones provided entry to a novel synthesis of 1,3,5-trisubstituted pyrazoles <09TL291>.

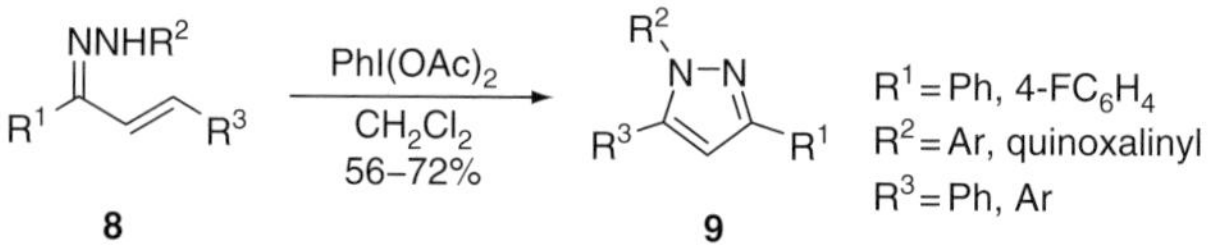

Alkyne precursors reacted with different reagents in the preparation of pyrazole and indazole derivatives. A domino reaction sequence consisting of a regioselective coupling of monosubstituted hydrazines with 2-halophenylacetylenes **10**, followed by an intramolecular hydroamination through a 5-exo-dig cyclization and subsequent isomerization of the exocyclic double bond to give the aromatic 2H-indazole **11** was reported <09AG(E)6879>. Scandium(III) triflate 1,3-dipolar cycloaddition of nitrile imines with functionalized acetylenes provided a regiocontrolled synthesis of 4- and 5-substituted pyrazoles <09SL2328>.

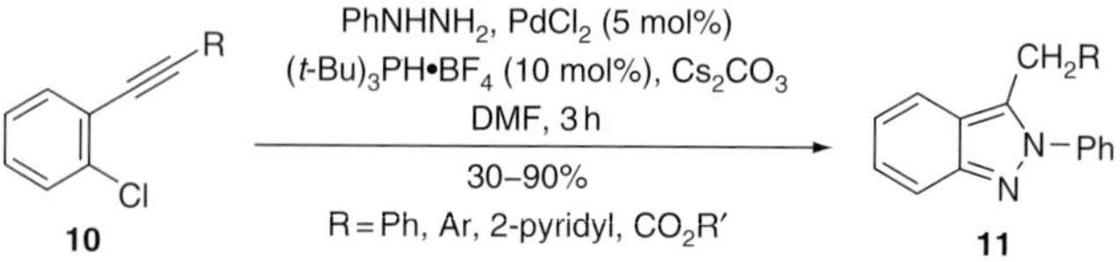

Diazo, diazonium and azo compounds can be employed as precursors in the preparation of pyrazoles and indazoles. A series of 3-acyl-4-amino-1-aryl-1*H*-pyrazoles **13** was prepared from the reaction of β-enaminones **12** with aryldiazonium tetrafluoroborates <09JHC650>. A series of pyrazoles **15** were prepared in good yields via 1,3-dipolar cycloaddition of diazoacetate compounds to terminal alkynes **14** promoted by Zn(OTf)$_2$ under mild conditions <09TL2443>. Readily available 2-chloromethylarylzinc reagents **16** reacted with aryldiazonium tetrafluoroborates to provide polyfunctional indazoles **17**, which underwent selective metalations to new polycyclic aromatics <09OL4270>. A facile one-pot procedure for the regioselective synthesis of pyrazole-5-carboxylates **19** by 1,3-dipolar cycloaddition of ethyl diazoacetate with DBU in acetonitrile with α-methylene carbonyl compounds **18** was described <09TL5978>. The one-pot synthesis of multisubstituted pyrazole derivatives was achieved *via* catalyst-free 1,3-dipolar cycloaddition of ethyl diazoacetate and nitroalkenes as the key step and elimination of a nitro or bromide leaving group followed by intramolecular proton transfer with satisfactory yields <09OBC4352>. A simple one-pot method for the synthesis of diversely functionalized pyrazoles **22** from aryl nucleophiles **20**, di-*tert*-butyloxycarbonylazodicarboxylate, and 1,3-dicarbonyl compounds **21** was presented <09OL2097>.

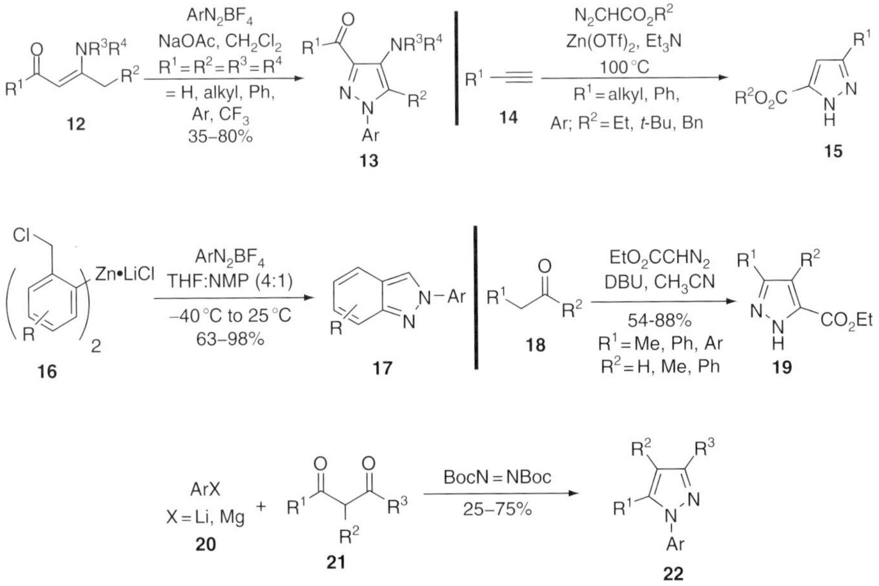

Other heterocycles can be utilized in the preparation of pyrazoles. Isothiazoles were converted to pyrazoles in neat hydrazine <09T7023>. The application of a Suzuki cross coupling approach to a range of *C*-4 substituted sydnones **24** from a 4-bromosydnone **23** followed by reaction with alkynyl boronate ester **25** afforded highly-functionalized pyrazoles **26** <09JOC396>. This strategy was applied to the synthesis of the withasomnines <09OBC4052>. Solid-supported version of mesoionic sydnones **27** reacted *in situ* in 1,3-dipolar cycloaddition reactions with dimethyl acetylenedicarboxylate (DMAD) and traceless cleavage of the products gave *N*-unsubstituted pyrazoles **28** <09OL2219>. 6-(Trifluoromethyl)comanic acid reacted regioselectively with phenylhydrazine in water to give 5-[3,3,3-trifluoro-2-(phenylhydrazono)propyl]-1-phenyl-1*H*-pyrazole-3-carboxylic acid while similar reaction in dioxane led to 3-[3,3,3-trifluoro-2-(phenylhydrazono)propyl]-1-phenyl-1*H*-pyrazole-5-carboxylic acid <09TL4446>.

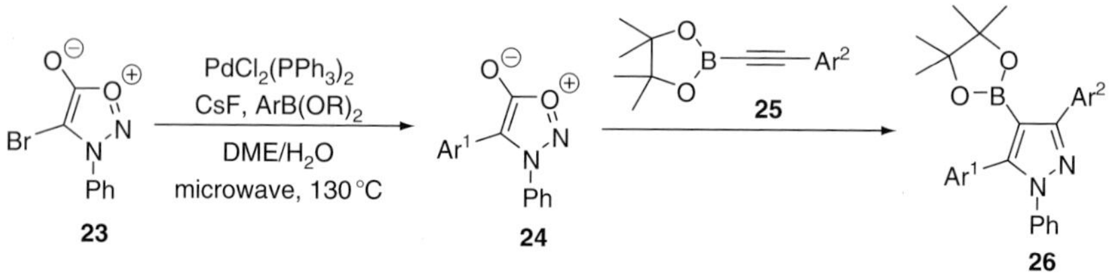

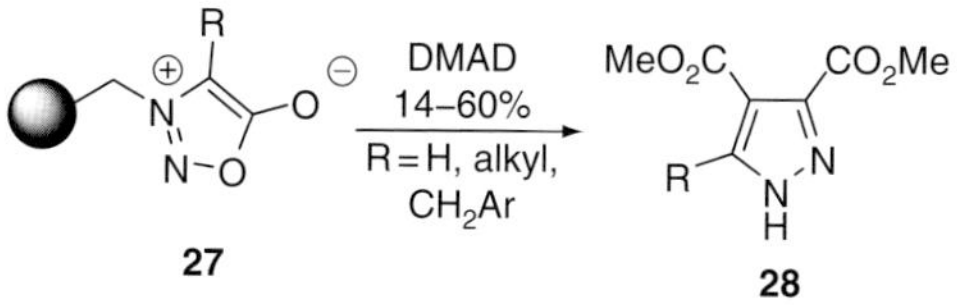

2,3-Dihydro-1*H*-pyrazoles **30** were highly selectively synthesized via the Pd(0)-catalyzed cyclization reaction of 2-substituted 2,3-allenyl hydrazines **29** with aryl iodides in moderate to good yields <09CC4263>.

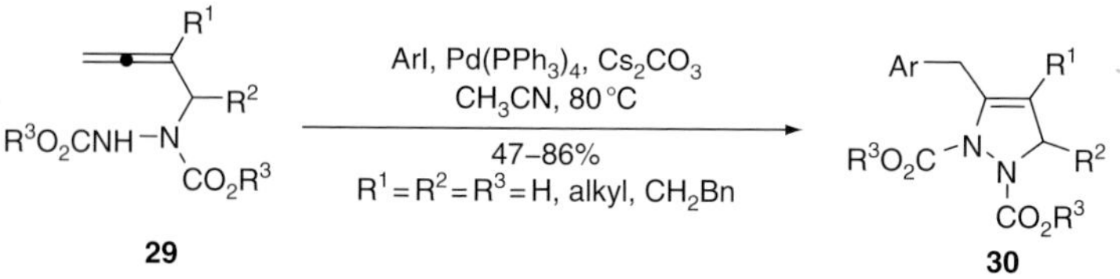

Several approaches were investigated in the preparation of indazoles. Treatment of 2-nitrobenzylamines **31** with methanolic sodium hydroxide furnished 1*N*-hydroxyindazoles **32** regioselectively and in high yield <09OL5078>. Synthesis, reactivity, and NMR spectroscopy of 4,6- and 6,7-difluoro-3-methyl-1*H*-indazoles from 2-fluoroacetophenones was reported <09JHC1408>. Diazo(trimethylsilyl)methylmagnesium bromide readily reacted with various ketones and aldehydes to give the corresponding 2-diazo-(2-trimethylsilyl)ethanols, which were efficiently converted to indazoles bearing hydroxymethyl units at the 3-position by intermolecular [3 + 2] cycloaddition with benzynes <09OBC2804>. A rapid and efficient procedure for the preparation of 5-arylmethyl-3-aminoindazoles was reported from 5-arylmethyl-2-fluorobenzonitriles which were obtained by two different synthetic approaches <09TL3098>. 2-Aryl-2*H*-indazoles **34** were prepared from *N*-(2-nitrobenzylidene)anilines **33** with triethyl phosphite under microwave irradiation <09JHC1309>.

There are few reports of *N*-alkylation and *N*-acylation of pyrazoles. An efficient method for the selective alkylation of indazoles **35** with α-halo esters, lactones, ketones, amides, and bromoacetonitrile provided good to excellent yield of the desired *N*1 products **36** <09OL5054>. *N*-Alkylation of pyrazoles with 2-iodoacetanilides afforded pharmacologically active 2-(1*H*-pyrazol-1-yl)acetamides <09ARK308>. Reactions of 4-nitro-3(5)-pyrazolecarboxylic acid dipotassium salt with different methylating agents in various solvents was investigated to improve the synthesis of isomeric 1-methyl-4-nitro-3- and -5-pyrazolecarboxylic acids <09TL2624>. The 3-amino group of methyl 3-amino-1*H*-pyrazole-5-carboxylate was successfully acylated over *N*1 of the pyrazole with acid anhydrides: acetic anhydride, *tert*-butyl pyrocarbonate, and 2-(2-methoxyethoxy)ethoxyacetic acid/ dicyclohexylcarbodiimide <09SC4122>.

The first catalytic intermolecular C-H arylation of SEM-protected pyrazoles and *N*-alkylpyrazoles laid the foundation for a new approach to the synthesis of complex arylated pyrazoles via a palladium pivalate system <09JA3042>. A chelation-assisted palladium-catalyzed cascade bromination/cyanation reaction of 1-arylpyrazole C-H bonds was developed <09JOC9470>.

There were several reports of cross-coupling reactions of pyrazoles. Copper-catalyzed oxidative *S*-arylation of 1,2-bis(*o*-amino-1*H*-pyrazolyl) disulfides **37** with arylboronic acids for the synthesis of (*o*-amino-1*H*-pyrazolyl)aryl sulfides **38** was developed in the presence of CuI, 1,10-phenanthroline, and oxygen <09S921>. A variety of 1-benzyl-3-heterocyclic pyrazoles **40** were rapidly assembled by a two-step one-pot *N*-benzylation/Suzuki coupling sequence of pyrazoloboronate ester **39** <09TL5479>. An improved synthesis of 1-methyl-1*H*-pyrazole-4-boronic acid pinacol ester via isolation of the corresponding lithium hydroxy ate complex and its application in Suzuki cross-coupling reactions was described <09TL6783>. An efficient and rapid protocol for synthesis of various *N*-protected indazole boronic acid pinacolylesters in good to high yields was developed providing a promising access to new aryl and hydroxyindazole libraries after subsequent Suzuki–Miyaura cross-coupling or hydroxydeboronation reactions <09SL615>. Protected 5-bromoindazoles participated in Buchwald reactions with a range of amines to generate novel derivatives <09JOC6331>.

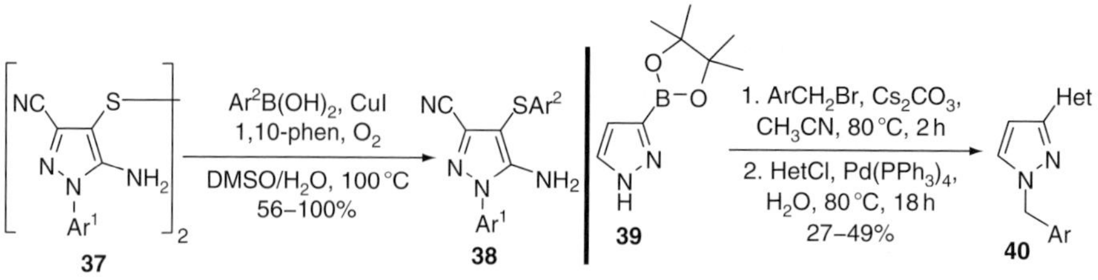

N-Arylation of pyrazoles appeared in several reports. Commercially available FeCl$_3$•6H$_2$O with conformationally rigid diamine ligand **42** was a highly effective catalyst for N-arylation of pyrazoles using aryl and heteroaryl iodides **41** to give pyrazoles **43** <09TL5868>. Solvent-free coupling reactions between pyrazole and aryl bromides **44** were achieved with microwave irradiation using copper(I) iodide and L-lysine or L-glutamine as catalysts to give pyrazoles **43** <09TL1286>. The N-arylation <09T2660> and 4-arylation <09T3529> of 3-alkoxypyrazoles were investigated. N-Arylation reactions of pyrazole and 3-(trifluoromethyl)pyrazole with 2,4-difluoroiodobenzene in the absence and presence of copper catalysis were described <09T855>.

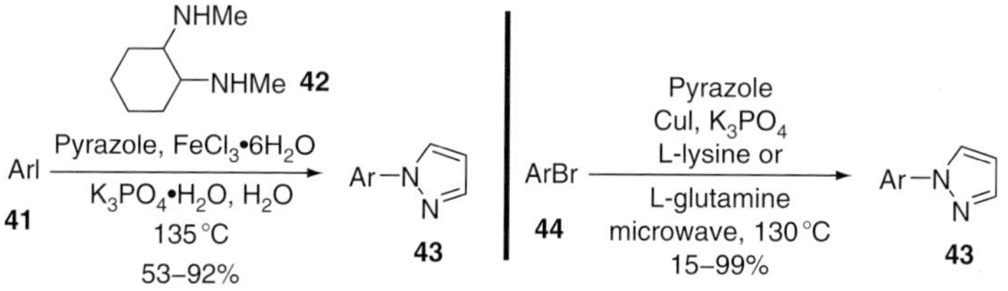

The full functionalization of the pyrazole ring was achieved by successive regioselective metalations using TMPMgCl·LiCl and TMP$_2$Mg·2LiCl and trapping with various electrophiles led to trisubstituted pyrazoles <09OL3326>. 4,5-Dihydrobenzo[g]indazoles were efficiently metallated using hindered Mg- and Zn-TMP amides, which reacted with various electrophiles to furnish novel C3-substituted 4,5-dihydrobenzo[g]indazoles <09S3661>. An efficient and convenient method for the bromination of pyrazolones and 5-hydroxypyrazoles was developed by using N-bromobenzamide in THF at room temperature <09TL9592>. Reaction of indazoles with aldehydes in the presence of p-toluenesulfinic acid afforded the corresponding sulfonyl indazoles which were desulfonylated with sodium amalgam leading to 3-alkylated indazoles <09EJO3184>. Regioselective O-alkylation of indazolinones using (cyanomethylene)triphenthylphosphorane (CMPP) as a Mitsunobu type reagent was described with a variety of aliphatic alcohols <09SL2676>. 1-Phenyl-1H-pyrazol-3-ol was used as a versatile synthon for the preparation of various 1-phenyl-1H-pyrazole derivatives substituted at C-3 and C-4 of the pyrazole nucleus and at the phenyl ring para position <09T7817>. A regio- and stereoselective preparation of substituted 3-pyrazolyl-2-pyrazolines and pyrazoles were prepared by reaction of 4-cyanopyrazole-5-nitrilimine with dipolarophiles <09H(78)911>.

Pyrazole-containing reagents found useful applications in some synthetic transformations. Bis-pyrazole phosphine ligand BippyPhos **45** was effective in the palladium-catalyzed cross-coupling of hydroxylamines with aryl bromides, chlorides, and iodides <09OL233> and both aryl bromides and chlorides were coupled to aryl, benzyl, and aliphatic ureas <09OL947>. Novel bulky pyrazolylphosphine ligands **46** were employed in the Suzuki cross-coupling of aryl chlorides <09TL7217>. A series of polykis(pyrazol-1-yl)benzenes, potential chelating ligands for transition metals, was prepared by nucleophilic substitution of fluorine in 1,2-difluoro-, 1,2,3,4-tetrafluoro-, 1,2,4,5-tetrafluoro-, and hexafluorobenzenes with sodium pyrazolide <09T4652>.

Many methods for the preparation of pyrazole-fused ring systems were published. Several reports revealed the syntheses of pyranopyrazoles. A four-component synthesis of substituted and spiro-conjugated 6-amino-2*H*,4*H*-pyrano[2,3-*c*]pyrazol-5-carbonitriles **47** directly from aromatic aldehydes or heterocyclic ketones, malononitrile, β-ketoesters, and hydrazine hydrate was reported <09JCO914>. An approach to pyrano[2,3-*c*]pyrazoles **48** starting from spirocyclopropanepyrazoles via a ring-opening/cyanomethylation and intramolecular cyclization was described <09JHC782>. A rapid protocol for the multicomponent microwave-assisted organocatalytic domino Knoevenagel/hetero Diels–Alder reaction was developed for the synthesis of 2,3-dihydropyran[2,3-*c*]pyrazoles <09TL6572>. A simple and efficient synthesis of pyranopyrazoles in good yields via a four-component reaction between an aromatic aldehyde, hydrazine hydrate, ethyl acetoacetate and malononitrile under neat conditions was published <09SL2002>. The three-component reaction of the zwitterions generated from dialkyl acetylenedicarboxylate and isocyanides with 3-methyl-1-phenyl-1*H*-pyrazol 5(1H) one afforded pyrano[2,3-*c*]pyrazole derivatives <09T3492>. A new and efficient CuI-catalyzed domino Knoevenagel hetero-Diels-Alder reaction of 1-oxa-1,3-butadienes with terminal activated acetylenes as building blocks afforded pyrano[2,3-*c*]pyrazoles <09SL55>.

Pyrazolopyridines are popular structures. A practical, inexpensive, and rapid method for the stereoselective synthesis of pyrazolo[4,3-*c*]pyridine derivatives **49** via microwave-assisted reactions of 3,5-diarylidenepiperidin-4-ones with phenylhydrazine in ethylene glycol was reported <09JHC849>. A convenient route for the synthesis of pyrazolo[3,4-b]pyridine-3-ones **50** via condensation of 3-amino-1-phenylpyrazolin-5-one with 4-hydroxy-6-methylpyran-2-one was described <09JHC1177>. An expedient synthesis of 3-alkoxymethyl- and 3-aminomethyl-pyrazolo[3,4-b]pyridines **51** was reported <09JOC789>. Reaction of N-substituted or unsubstituted-pyrazol-5(4H)-one, aryl-oxoketene dithioacetals and alkyl amide provided a facile synthesis of pyrazolo[3,4-b]pyridines **52** <09TL3088>. A one-pot four-component reaction of an aliphatic or aromatic amine, diketene, an aromatic aldehyde and 1,3-diphenyl-1H-pyrazol-5-amine in the presence of p-toluenesulfonic acid as a catalyst was developed for the synthesis of pyrazolo[3,4-b]pyridine-5-carboxamide derivatives <09TL2911>. An efficient access to 3,6-disubstituted 1H-pyrazolo[3,4-b]pyridines via a one-pot double S$_N$Ar reaction and pyrazole formation was reported <09TL2293>. Methods for the facile high-yielding synthesis of substituted pyrazolo[3,4-c]pyridines **53** from 2-bromo-5-fluoropyridine were described <09TL383>. A single-step synthesis of pyrazolopyridines and hydropyrazolopyridines with condensation of 3-carbonitrile-5-aminopyrazole with seven substituted α,β-unsaturated aldehydes in acid medium was reported <09ARK258>. A mild one-step synthetic method to access privileged heterobiaryl pyrazolo[3,4-b]pyridines from indole-3-carboxaldehyde derivatives and a variety of aminopyrazoles was developed <09OL5214>. Intramolecular condensation of α,β-unsaturated esters with 3-pyrazolo aldehydes gave pyrazolo[1,5-a]pyridine derivatives <09H(78)197>.

Pyrazolopyrimidines were also desirable structures. A novel approach to regioselective synthesis of new 5-amino-6-arylamino-1H-pyrazolo[3,4-d]pyrimidin-4(5H)-one derivatives via a tandem aza-Wittig and annulation reaction of iminophosphorane, aromatic isocyanates and hydrazine was reported <09JHC256>. 4-(4-Chlorophenylazo)-1H-pyrazole-3,5-diamine reacted with ethyl acetoacetate, benzylidenemalononitrile, ethyl propiolate and malononitrile under microwave irradiation to afford pyrazolo[1,5-a]pyrimidine derivatives <09H(78)2003>. AlCl$_3$-induced C–C bond forming reaction was investigated through the reaction of 7-chloro-5-phenyl-pyrazolo[1,5-a]pyrimidine with arenes and heteroarenes to furnish a novel and highly selective methodology for the preparation of 7-(hetero)aryl substituted-pyrazolopyrimidines in good to excellent yields under mild reaction conditions <09TL354>. A series of pyrazolopyrimidine and pyrazoloquinozoline derivatives were synthesized in one step by multicomponent reactions using 5-aminopyrazole, p-substituted benzoylacetonitrile or dimedone and triethyl orthoesters <09JHC708>. Novel 2,3-substituted-2,4-dihydropyrazolo[4,3-d]pyrimidine-5,7-diones were successfully synthesized in moderate to good yields <09TL6223>.

Some tricyclic fused pyrazole ring systems were published. An efficient two-step synthesis of novel thiazolo[2,3-*b*]pyrazolo[3,4-*f*][1,3,5]triazepines **54** was published <09JHC756>. Benzofuropyrazoles **55** were prepared in three steps from salicaldehydes <09JOC891>. Diversity-oriented synthesis of functionalized *H*-pyrazolo[5,1-*a*]isoquinolines **56** *via* sequential reactions of *N'*-(2-alkynylbenzylidene)hydrazide was described <09OBC4641>. A practical synthesis of pyrazolo[4,3-*e*][1,2,4]triazolo[1,5-*c*]pyrimidines **57** was developed <09TL5617>. Simple procedures for the synthesis of unknown 1,3-dihydropyrazolo[3,4,5-*de*]phthalazines **58** was developed <09T1574>. 1,2-Dihydropyrazolo[1,2-*a*]indazol-3(9*H*)-ones were prepared by the 1,3-dipolar cycloaddition of arynes with azomethine imines <09TL4067>. A series of aryloxymethylquinoxaline oximes, hitherto unknown and synthesized from the corresponding aldehydes, afforded in only one step pyrazolo[1,5-*a*]quinoxalines in the presence of acetic anhydride at high temperatures <09JOC1282>. Efficient synthetic routes to pyrazolo[3,4-*b*]indoles and pyrazolo[1,5-*a*]benzimidazoles via intramolecular palladium- and copper-catalyzed cyclization of 1-aryl/1-unsubstituted 5-(2-bromoanilino)-pyrazole precursors via intramolecular C-C and C-N bond formation was reported <09JOC7046>. A simple synthetic strategy was described for synthesis of the hitherto unreported 6-arylhydrazono-1,3-diphenyl-7-methyl-1*H*-pyrazolo[3',4':4,5]pyrimido[1,6-*b*][1,2,4]triazepin-5(6*H*)-ones <09T644>.

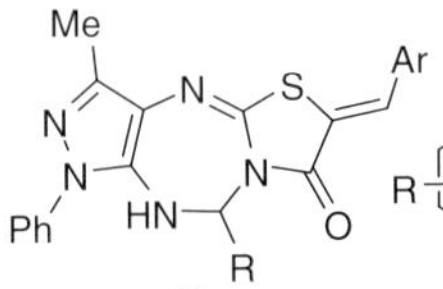

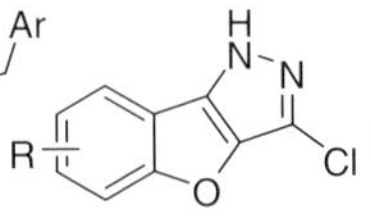

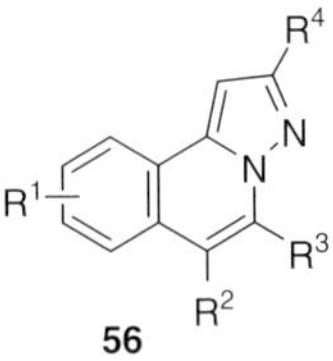

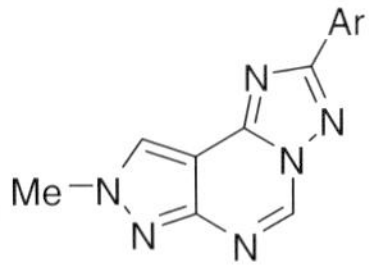

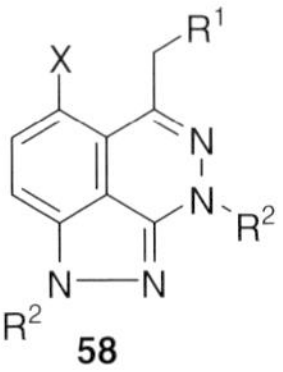

Larger fused pyrazoles were also published. A simple, efficient, and cost-effective method for the synthesis of 2*H*-indazolo[2,1-*b*]phthalazine-1,6,11(13*H*)-trione derivatives **59** by a one-pot, three-component condensation reaction of phthalazide, dimedone, or 1,3-cyclohexanedione and aromatic aldehydes in the presence of trimethylsilyl chloride was described <09JHC728>. 2-Methyl-2,5,6,11,12,13-hexahydro-4*H*-indazolo[5,4-*a*]pyrrolo[3,4-*c*]carbazole-4-one **60** was synthesized utilizing a regioselective Diels-Alder reaction with 5-(1*H*-indol-2-yl)-2-methyl-6,7-dihydro-2*H*-indazole and ethyl *cis*-β-cyanoacrylate <09JHC1185>.

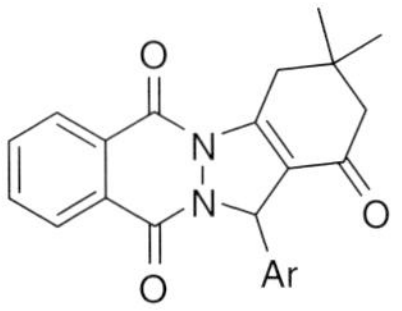

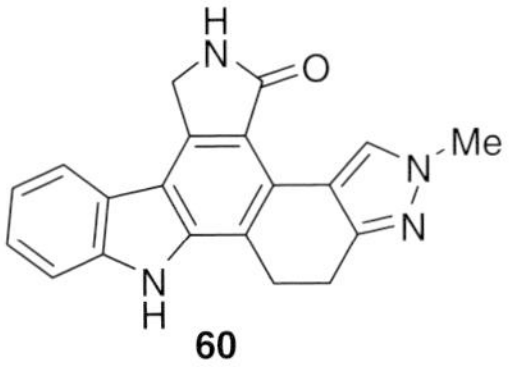

The novel heterocycle 2,3-dihydrooxazolo[3,2-b]indazole was utilized in the synthesis of 1H-indazolones <09OL2760>. Tert-butyl amides resulting from Ugi multicomponent reaction of tert-butyl isocyanide and 5-substituted-1H-pyrazole-3-carboxylic acids with various aldehydes and amines underwent cyclization to 5,6-dihydropyrazolo[1,5-a]pyrazine-4,7-diones in glacial acetic acid under microwave irradiation <09SL260>.

5.4.3. IMIDAZOLES AND RING-FUSED DERIVATIVES

The mechanistic and synthetic chemistry of imidazole-based superelectrophiles was studied <09JOC2502>.

Various methods were reported for the synthesis of imidazoles. A simple highly versatile and efficient synthesis of 2,4,5-trisubstituted imidazoles **62** was achieved by three component cyclocondensation of 1,2-diketo compounds **61**, aldehydes and ammonium acetate using L-proline as a catalyst in methanol at moderate temperatures <09T10155>. Microwave-assisted solvent-free synthesis of tetraaryl imidazoles was achieved from the reaction of benzyl, Schiff base, and ammonium acetate <09JHC278>. A simple and green route to the synthesis of (4 or 5)-aryl-2-aryloyl-(1H)-imidazoles was described from arylglyoxals with ammonium acetate in water; and the two tautomers were studied by NMR techniques <09T6882>. A one-pot, four-component synthesis of 1,2,4-trisubstituted-1H-imidazoles **64** was obtained by heating a mixture of a 2-bromoacetophenones **63**, aldehydes, primary amines, and ammonium acetate under solvent-free conditions <09SL3263>. An efficient method for synthesis of various tetrasubstituted imidazoles, using trifluoroacetic acid as a catalytic support, by four-component condensation of benzil, aldehydes, amines, and ammonium acetate under microwave-irradiation and solvent-free conditions was described <09SC3232>. 2,4,5-Trisubstituted imidazoles **66** were prepared by the three-component condensation of α-hydroxy ketones **65**, aldehydes and ammonium acetate by ceric ammonium nitrate (CAN) catalyzed aerobic oxidation in ethanol at reflux <09SC102>. A microwave-assisted synthesis of 4,5-disubstituted imidazoles from 1,2-diketones and urotropine in the presence of ammonium acetate in acetic acid was reported <09S2319>. An efficient and practical method for the preparation of α-imidazol-1-yl esters from 1,2-diaza-1,3-dienes, α-amino esters, and aldehydes was described <09OL2840>. Four-component microwave-assisted synthesis of 2,4,5-triarylsubstituted imidazoles from aryl cyanides, arylbenzaldehydes and ammonium acetate was reported <09JOC9486>. Addition of benzyl azide to 2-amidoacrylates **67** provided a synthesis of imidazole-4-carboxylates **68** <09JHC1235>. Tetrasubstituted imidazoles were accessible in one or two steps from α-acylaminoimines obtained by the 1,2-addition of deprotonated N-monosubstituted α-aminonitriles to N-acylimines and subsequent elimination of HCN from the primary addition products <09CEJ843>. 2-Imidazolines and imidazoles were accessed by an aza-Wittig sequence featuring novel N-acylation methodology for sulfonamides and optimized conditions for ring closure <09CC1900>.

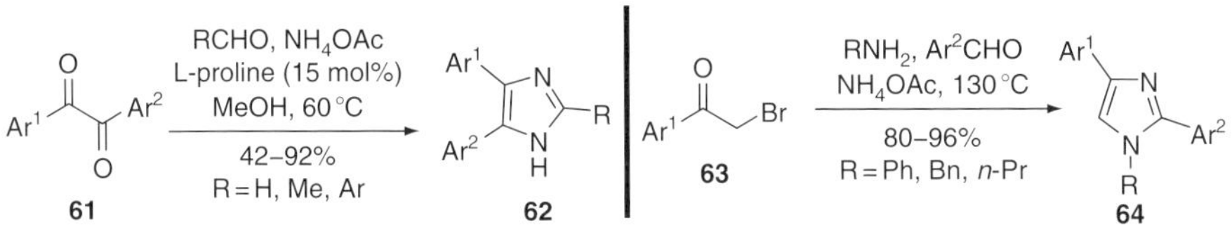

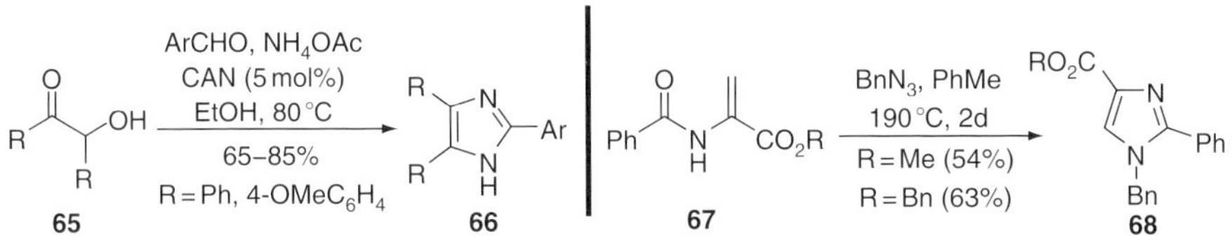

2-Aminoimidazoles were synthesized by various methods. 2-Aminoimidazole and imidazoline derivatives were obtained in three steps through the reduction of *N*-pyridinium imidates into 1,2-dihydropyridine imidates and oxidative addition of guanidine derivatives <09TL6826>. An efficient method for the synthesis of mono- and disubstituted 2-amino-1*H*-imidazoles **70** via microwave-assisted hydrazinolysis of substituted imidazo[1,2-*a*]pyrimidines **69** was reported <09TL5218>. Convenient synthesis of 2-amino-1,5-disubstituted and 2-amino-1,4,5-trisubstituted imidazoles from 1-amidino-3-trityl-thioureas under mild conditions was reported <09TL3955>. Addition–hydroamination reactions of propargyl cyanamides with lanthanide catalyst gave rapid access to highly substituted 2-aminoimidazoles <09AG(E)3116>.

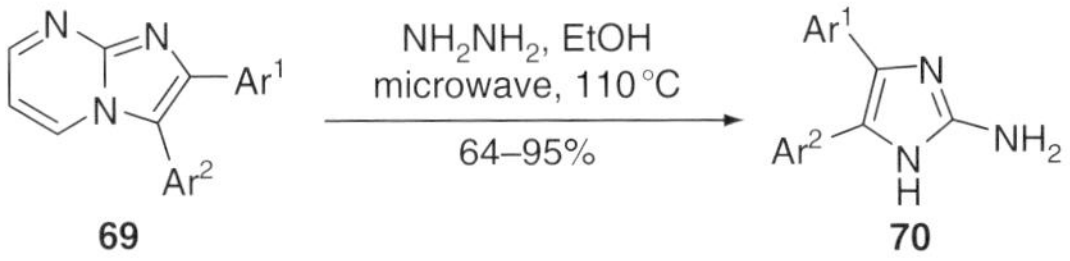

Benzimidazoles were prepared from different precursors. Reaction of benzothiadiazole-4-sulfonyl chloride with a variety of amines followed by highly chemoselective reductive desulfurization gave the intermediate 1,2-phenylenediamines, which were treated with aryl, heteroaryl, and alkyl aldehydes to provide substituted benzimidazole-4-sulfonamides <09TL1219>. A mild and efficient method for the synthesis of *N*-substituted 2-fluoromethylbenzimidazoles via [bis(trifluoroacetoxy)-iodo]benzene-mediated intramolecular cyclization of *N*,*N*'-disubstituted bromodifluoro (or trifluoro) ethanimidamides was described <09SL3299>. α-Chloroaldoxime *O*-methanesulfonates were prepared and employed in the synthesis of functionalized benzimidazoles <09JOC1394>. A rapid and efficient method for the synthesis of novel 2-[6-(arylethynyl)pyridin-3-yl]-1*H*-benzimidazole derivatives under microwave-assisted Sonogashira coupling conditions was developed <09ARK105>.

Alcohols **71** were treated with *o*-aminoaniline in the presence of ruthenium catalyst and crotononitrile leading to the formation of benzimidazoles **72** <09OL2039>. A one-pot synthesis of functionalized benzimidazoles via the cascade reactions of *o*-aminoanilines with terminal alkynes and *p*-tolylsulfonyl azide was reported <09SL2023>.

Reaction of N-methyl-1,2-phenylenediamine **73** with carbonitriles and sodium hydride in toluene provided N-methylbenzimidazoles **74** <09SL63>.

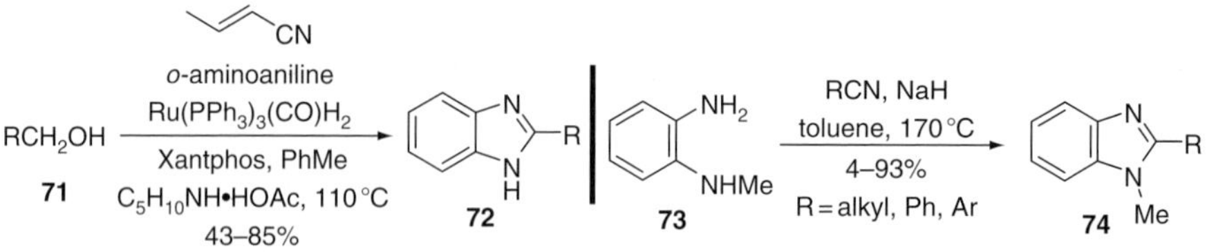

Many similar methods were published for the synthesis of 2-substituted-benzimidazoles **76** from o-phenylenediamines **75** and they are shown in the table below.

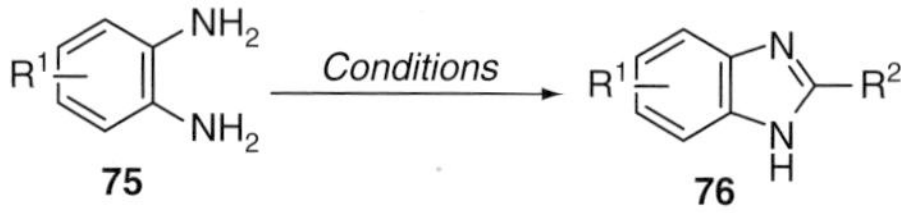

Conditions	Reference
Thioamidinium salts, HTAB, H_2O, 80 °C	09EJO4926
ArCHO, [Pro]NO_2, 25 °C	09JHC74
ArCHO, CA-SiO_2, EtOH, 80 °C	09H(78)2337
2-pyrrol-2-carbaldehydes, TFA, DMSO, 70 °C	09S3603
RCHO, NH_4Br, MeOH, 25 °C	09SC175
RCHO, Amberlite IR-120, EtOH, H_2O	09SC980
RCHO, SiO_2, $ZnCl_2$, 25 °C or MW	09TL1495
RCHO, $CoCl_2 \cdot 6H_2O$, CH_3CN, 25 °C	09SC2339
RCHO, $NaHSO_3$, DMA	09SC2982
ArCHO, H_2O_2, $Fe(NO_3)_3$, 50 °C	09SL569

Copper-catalyzed methods were very popular as a methodology in the preparation of 2-substituted benzimidazoles. A highly practical method for the synthesis of fluorinated benzimidazoles **78** was developed by the CuI/TMEDA-catalyzed cross-coupling reaction of N-(2-haloaryl)trifluoroacetimidoyl chlorides **77** with primary amines <09S1431>. An efficient copper-catalyzed intramolecular arylation of formamidines **79** forming 2-unsubstituted benzimidazoles **80** in excellent yields was reported <09JOC9570>. The synthesis of substituted benzimidazoles and 2-aminobenzimidazoles was described via intramolecular cyclization of o-bromoaryl formamidines using copper(II) oxide nanoparticles in DMSO under air <09JOC8719>. CuI/L-proline catalyzed coupling of aqueous ammonia with 2-iodoacetanilides and 2-iodophenylcarbamates afforded the aryl amination products at room temperature, which underwent *in situ* cyclization under acidic conditions or heating to give substituted 1H-benzimidazoles and 1,3-dihydrobenzimidazol-2-ones, respectively <09JOC7974>. Copper(I)-iodide catalyzed amination of aryl halides with guanidines or amidines provided a facile synthesis of 1H-2-substituted benzimidazoles

<09JOC5742>. S-Alkylation of thioureas **81** followed by Cu-catalyzed intramolecular N-arylation furnished substituted 2-mercaptobenzimidazoles **82** in high yields <09JOC2217>. o-Bromophenyl isocyanide **83** reacted with various primary amines under copper catalysis to afford 1-substituted benzimidazoles **84** in moderate to good yields <09EJO5138>. Preparation of 2-fluoroalkylbenzimidazoles from N-aryltrifluoroacetimidoyl (or bromodifluoroacetimidoyl) chlorides and primary amines was achieved via copper(I)-catalyzed tandem reactions <09CC2338>. N-Substituted S-2-heterobenzimidazoles was synthesized through a Cu(I)-catalyzed cascade intermolecular addition/intramolecular C-N coupling process from o-haloarylcarbodiimides and N- or O-nucleophiles <09JOC5618>.

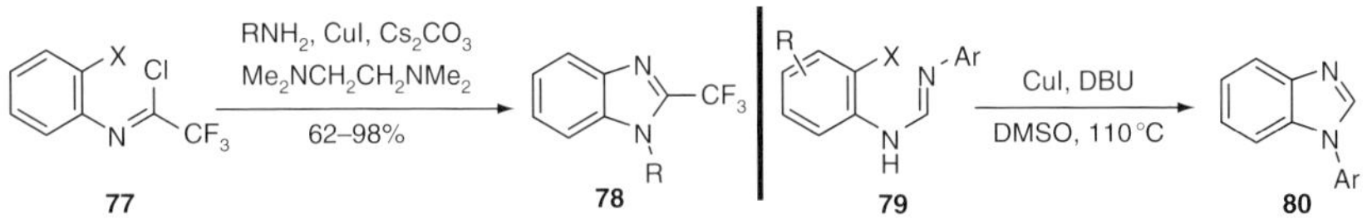

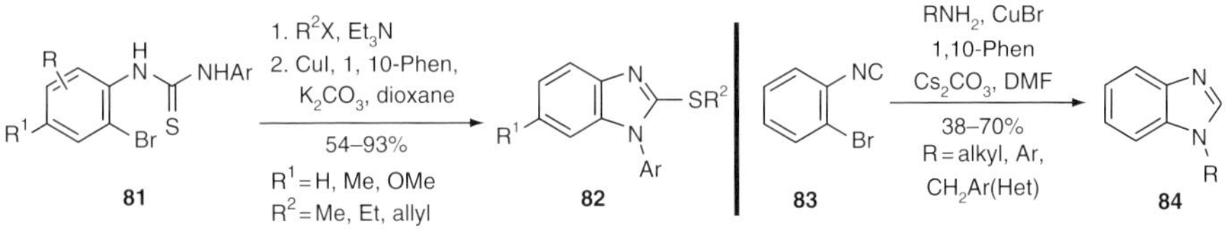

Nucleophilic substitution reactions on 1-methyl-2,4,5-trinitroimidazole **85** to give 5-substituted imidazoles **86** was described <09SC4282>. A facile and efficient method for one-pot N-alkylation of imidazole derivatives from alcohols using triphenylphosphine and tetrabutylammonium iodide in refluxing N,N-dimethylformamide was described <09S3067>. Novel syntheses of 4-(1H-benzo[d]imidazol-2-yl)isoxazol-5-amine and 4-(1H-benzo[d]thiazol-2-yl)isoxazol-5-amine scaffolds was disclosed <09TL1571>. Highly regio- and enantioselective iridium-catalyzed N-allylations of benzimidazoles and imidazoles were developed <09JA8971>. Functionalization reactions of 1-alkynylimidazoles involved the formation of their 2-lithio derivatives followed by addition of various electrophiles was presented and the use of an aldehyde or sulfonimine electrophiles allowed the direct formation of bicyclic ring systems <09TL5194>. Imidazo[1,5-a]pyrazines underwent regioselective C3-metalation and C5/C3-dimetalation to afford a range of functionalized derivatives <09OL5118>. Thermal rearrangement of an N-hydroxyimidazole thiocarbamoyl derivative provided a simple entry into the 4-thioimidazole motif <09ARK17>.

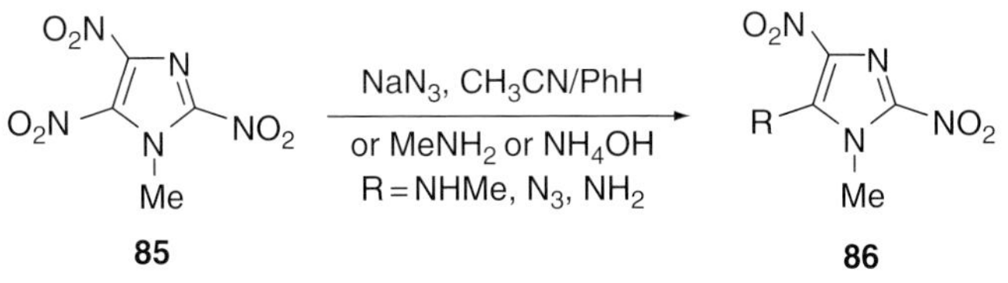

There were numerous published examples of Kumada and Suzuki cross-coupling reactions of imidazoles, benzimidazoles and fused imidazole ring systems. *N*-Substituted 5-(4,4,5,5-tetramethyl-1,3,2-dioxaborolan-2-yl)-1*H*-benzimidazoles were conveniently accessed via microwave-assisted synthesis and subsequent Suzuki–Miyaura cross-coupling with heteroaryl halides proceeded to give a wide variety of heteroaryl-substituted benzimidazoles <09TL1399>. Kumada–Tamao–Corriu cross-coupling of 1-iodo-3-arylimidazo[1,5-*a*]pyridines **87** and aryl Grignard reagents led to 1,3-diarylated imidazo[1,5-*a*]pyridines **88** in good to excellent yields <09T5062>. Palladium-catalyzed Suzuki–Miyaura cross-coupling reactions of 4-bromoimidazole **89** afforded 4-substituted-5-nitro-1*H*-imidazoles **90** under microwave irradiation <09S3150>. Optically active imidazole derivatives featuring an α-amino acid motif substituted at the 2-position were prepared in moderate to good yields by Negishi as well as Suzuki–Miyaura cross-couplings as the key synthetic steps <09S325>. Palladium-catalyzed cross-coupling reactions of 2-iodoimidazole was studied to synthesize imidazole-containing protein farnesyltransferase inhibitors <09OBC2214>.

Direct C–H amination and arylation of imidazoles and benzimidazoles were disclosed. A copper(II) acetate mediated aerobic coupling reaction enabled direct amidation of heterocycles such as imidazoles or benzimidazoles **91** having weakly acidic C–H bonds with a variety of nitrogen nucleophiles to give biologically revelant C-2 aminated heterocycles **92** was published <09OL5178>. The regioselective palladium-catalyzed 5-arylation of 1,2-dimethylimidazole **93** with aryl bromides to give imidazole **94** was best achieved with potassium acetate in DMAC <09T9772>. A novel direct and regioselective Pd/Cu-catalyzed intermolecular oxidative coupling of imidazo[1,2-*a*]pyridines **95** with alkenes to give 3-alkenylimidazo[1,2-*a*]pyridine derivatives **96** in high yields was reported <09S271>. Ni(COD)/AlMe$_2$ binary catalysis was found effective in the direct regioselective alkenylation of imidazoles through C–H bond activation and stereoselective insertion of alkynes with the use of P(*t*-Bu)$_3$ as a ligand to allow exclusive regioselective C(2)-alkenylation, while PCy$_3$ was found effective in the C(5)-alkenylation of C(2)-substituted imidazoles <09TL3463>. The enantioselective intramolecular alkylation of substituted imidazoles with enantiomeric excesses up to 98% was accomplished by rhodium catalyzed C–H bond functionalization with (*S*,*S'*,*R*,*R'*)TangPhos as the chiral ligand <09CC3910>.

Several reports of N-arylation of imidazoles were published. A simple, highly efficient, and environmentally friendly protocol of sulfonato–Cu-(salen)-catalyzed N-arylation of imidazoles in water with a low catalyst loading (2%) and cheap base was reported <09CEJ8971>. N-Arylation of imidazoles with aryl halides catalyzed by a combination of copper(II) sulfate and 1,2-bis(2-pyridyl)-ethane-N,N'-dioxide in water afforded up to 95% yields <09OL3294>. TBAF-assisted N-arylation and benzylation of 1H-benzimidazole with aryl and benzyl halides was demonstrated in the presence of $CuBr_2$ in moderate to good yields <09JOC5675>. A copper-catalyzed methodology for the regioselective N1 coupling of vinyl bromides with benzimidazoles and imidazoles was developed using a combination of copper iodide/ethylenediamine as catalyst <09JOC6371>. Trans-N,N'-bis(1-ethyl-2-benzimidazolylmethylene)cyclohexane-1,2-diimine was the choice ligand for the copper-catalyzed N-arylation of imidazoles <09CEJ10585>.

Imidazole-containing compounds were utilized as reagents for various synthetic transformations. Pyrazine-bridged diimidazolium salts **97** were synthesized and their application in palladium-catalyzed Heck-type coupling reactions were reported <09T909>. A simple and convenient one-pot procedure is reported for the synthesis of 1,2-diketones from the corresponding benzoin-type condensation reaction of aromatic aldehydes in water with N,N-dialkylbenzimidazolium salts **98** as a catalyst and ferric chloride as an oxidizing reagent was reported <09SC492>. Symmetrical ketones were synthesized from Grignard reagents and 1,1'-carbonyldiimidazole <09S2316>. The rate of 1,1'-carbonyldiimidazole mediated amidation of ten aromatic amines was significantly enhanced upon introduction of imidazole hydrochloride as a proton source for acid catalysis <09OPRD106>. An unprecedented transfer of a thiocyanate (-SCN) group from aroyl/acyl isothiocyanate to alkyl or benzylic bromide was observed in the presence of 1-methylimidazole **99** <09OL3382>. A new imidazole-based organocatalyst **100** for asymmetric reactions was developed and was shown to be a very effective catalyst for the Michael reaction involving various nitroolefins and aldehydes in water <09OL3354>. Tosylation of primary alcohols with tosyl chloride was performed effectively with an N-hexadecylimidazole catalyst **101** in water containing potassium carbonate <09OL1757>. Readily available hybrid NH_2/benzimidazole ligands **102** dramatically influenced the outcome of established Ru-based catalysts during asymmetric hydrogenation of aryl ketones <09OL907>. The N-heterocyclic carbene-catalysed oxidative carboxylation of aryl aldehydes with water successfully proceeded when a sulfoxylalkyl-substituted imidazolium salt **103** was used as a catalyst <09OBC4062>. A series of sulfuryl imidazolium salts **104** were prepared and examined as reagents for incorporating trichloroethyl-protected sulfate esters into carbohydrates <09JOC6479>. Chiral *trans*-cyclohexanediamine-benzimidazole organocatalysts **105** promoted the conjugate addition of a wide variety of 1,3-dicarbonyl compounds such as malonates, ketoesters, and 1,3-diketones to nitroolefins in the presence of trifluoroacetic acid as cocatalyst in toluene <09JOC6163>. Imidazole ligand **106** was utilized in a convenient and general palladium-catalyzed coupling reaction of aryl bromides and chlorides with phenols <09CC7330> and in the palladium-catalyzed synthesis of phenols from aryl halides with potassium hydroxide in refluxing water/dioxane mixture <09AG(E)918>. Imidazolium salt **107** promoted efficient C-B bond formation in the synthesis of tertiary and quaternary β-substituted carbons through metal-free catalytic boron conjugate additions to cyclic and acyclic α,β-unsaturated carbonyls <09JA7253>. Asymmetric imidazolinium-dithiocarboxylate **108** was a highly selective catalyst in the Staudinger reaction <09CC1040>.

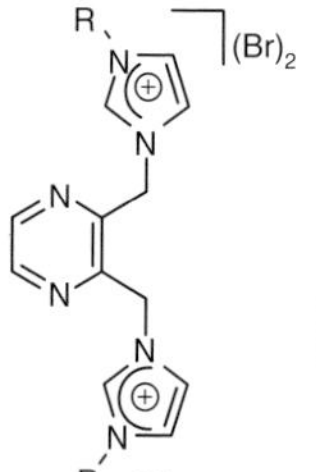

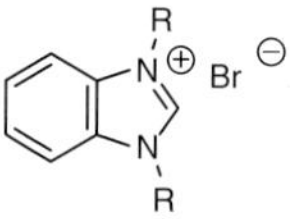

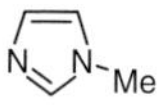

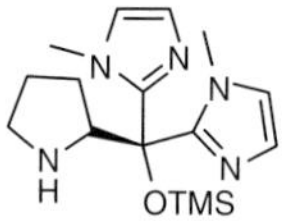

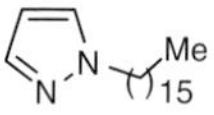

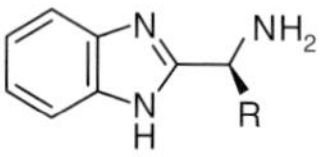

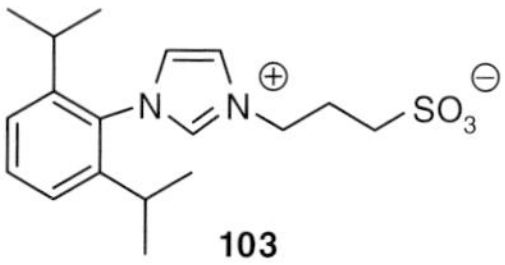

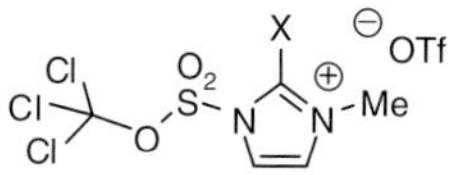

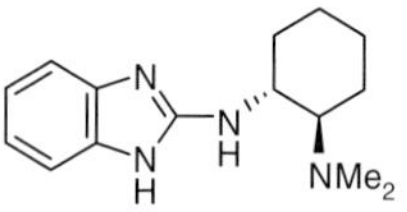

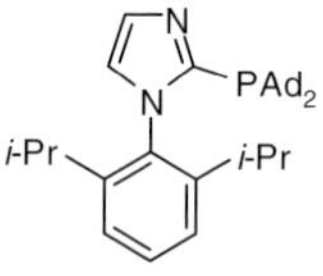

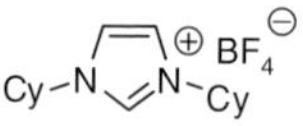

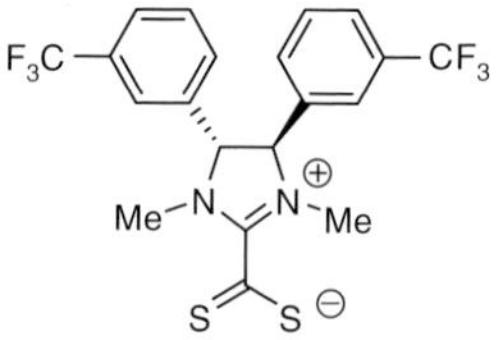

Many methods were developed for the synthesis of imidazole fused-ring systems. Some interesting fused 5,5-rings were reported. 1-Alkynyl-2-(hydroxymethyl)imidazoles underwent 6-endo-dig or 5-exo-dig cyclization under AuCl₃- or base-catalyzed conditions to yield imidazo[1,2-*c*]oxazoles and imidazo[2,1-*c*][1,4]oxazine heterocycles <09JOC9229>. The reaction of iodobenzenes with 2-amino-3-(2-propynyl)thiazolium bromide and 2-imino-3-(2-propynyl)-1,3-benzothiazole, catalyzed by Pd/Cu, led to the formation of 6-substituted-imidazo[2,1-*b*][1,3]thiazoles **109** and 2-substituted imidazo[2,1-*b*][1,3]benzothiazoles **110**, respectively <09TL5459, 09SL2601>.

109   **110** CH₂Ar

Imidazolopyridines were a popular fused bicyclic system. The reaction of 2-amino-1-(2-propynyl)pyridinium bromide with various iodobenzenes, catalyzed by Pd/Cu, in the presence of sodium lauryl sulfate as the surfactant and cesium carbonate as the base, in water afforded 2-substituted imidazo[1,2-*a*]pyridines <09SC1002>. A one-pot synthesis of imidazo[1,5-*a*]pyridines **111** from a carboxylic acids and 2-methylaminopyridines with introduction of various substituents at the 1- and 3-positions was achieved using propane phosphoric acid anhydride in ethyl or *n*-butyl acetate at reflux <09TL4916>. Three-component condensation with trimethylsilyl cyanide or cyanohydrins as the source of cyanide ions and silica sulfuric acid as the catalyst afforded imidazo[1,2-*a*]pyridines **112** <09TL4389>. Reaction of 2,3-diaminopyridine with substituted aryl aldehydes in water under thermal conditions without the use of any oxidative reagent yielded 1*H*-imidazo[4,5-*b*]pyridines **113** by an air oxidative cyclocondensation in one step in excellent yields <09TL1780>. Regioselective nucleophilic substitution of 2-chloro-3-nitropyridine with heterocyclic amides with Buchwald palladium-catalyzed reactions yielded imidazo[4,5-*b*]pyridine-containing polycyclics as novel scaffolds <09TL3798>. Oxidative condensation-cyclization of aldehydes and aryl-2-pyridylmethylamines proceeded in the presence of a stoichiometric amount of elemental sulfur as an oxidant in the absence of catalyst to give 1,3-diarylated imidazo[1,5-*a*]pyridines **114** in good to high yields <09JOC3566>. An efficient synthesis of substituted 1-pyridylimidazo[1,5-*a*]pyridines **115** was accomplished using scandium(III) triflate as catalyst under mild conditions in excellent yields <09SC3546>. Baylis-Hillman substituted allyl isonitriles adducts were utilized in a robust isonitrile based multicomponent reaction in the presence of ammonium chloride to afford substituted imidazo[1,2-*a*]pyridines <09S431>. An Ugi-type multicomponent reaction of heterocyclic amidines with aldehydes and isocyanides catalyzed by zirconium (IV) chloride in PEG-400 was developed for the synthesis of imidazo[1,2-*a*]pyridines <09SL628>. A microwave-assisted Ugi-type multicomponent reaction of heterocyclic amidines with aldehydes and isocyanides catalyzed by zirconium(IV) chloride provided versatile *N*-fused aminoimidazoles was developed <09S3293>. The polystyrene-supported palladium(II) ethylenediamine complex was a highly active and recyclable catalyst for the synthesis of 2-benzylimidazo[2,1-*b*]pyridines through heteroannulation of acetylenic compounds <09JHC100>. 8-Hydroxy-6-methyl-5,6,7,8-tetrahydroimidazo[1,2-*a*]pyridine was formed selectively in high yields from *N*-(β-methallyl)imidazole by a tandem hydroformylation–cyclization sequence

<09CC4944>. Imidazopyridines were prepared from aryl(bromo)methyltrifluoromethyl ketones and 2-aminopyridine <09S2249>.

111 **112** **113** **114** **115**

A palladium-catalyzed/norbornene-mediated one-step synthesis of highly functionalized imidazoles via a sequential alkyl-aryl and aryl-heteroaryl bond formation provided an efficient route to a wide variety of substituted imidazo [5,1-*a*]isoquinolines from readily accessible *N*-bromoalkyl imidazoles and aryl iodides <09JOC1364>. One-pot three-component reaction between isoquinoline, phenacyl bromide derivatives and thiocyanates gave imidazo[2,1-*a*]isoquinolines <09H(78)415>. An efficient, convenient, one-pot synthesis of imidazo [1,2-*c*]quinazolines was accomplished in good yields via the novel reductive cyclization of 2-(2-nitrophenyl)-1*H*-imidazole with isothiocyanates mediated by zinc dust <09JHC971>.

Tricyclic and tetracyclic fused imidazole ring systems were also reported. A facile method was developed for the one-step synthesis of 5-chloro-imidazo[1,5-*a*]quinazolines **116** by cyclization of *N*-acylanthranilic acid with 2-amino acetamide or 2-amino-acetonitrile in the presence of POCl$_3$ under microwave irradiation <09TL6048>. Direct, efficient synthesis of the benzimidazo[2,1-*a*]isoquinoline ring system **117** was achieved with 2-bromoarylaldehydes, terminal alkynes, and 1,2-phenylenediamines by a microwave-accelerated tandem process in which a Sonogashira coupling, 5-endo cyclization, oxidative aromatization, and 6-endo cyclization were performed in a single synthetic operation <09TL4167>. A general and efficient synthesis of imidazo[2,1-*a*]phthalazines **118** by palladium- and copper-mediated coupling of *N*-aminoimidazoles with 2-haloaryl aldehydes with concurrent imine formation was described <09S1715>. Imidazo[1,2-*a*]quinoxalines were accessed via two sequential isocyanide-based multicomponent reactions <09JOC2627>. Polysubstituted pyrido[1,2-*a*]benzimidazoles **119** were efficiently produced in moderate yields in a novel one-pot, four-component reaction from pyridine or 3-picoline, chloroacetonitrile, malononitrile, and aromatic aldehyde in refluxing acetonitrile <09JOC710>. An efficient and versatile method for the synthesis of pyrido- and pyrimido-imidazopyrazines rings using the modified Pictet-Spengler strategy was reported <09JCO720>. The solid-phase synthesis of benzofused benzimidazole tricycles **120** based from resin-bound 3-(2-aminophenylamino)-2-selenoester was disclosed <09JCO515>. Hypervalent iodine mediated oxidation of 1,2-diaminobenzimidazole and its Schiff bases produced 2-aryl-1,2,4-triazolo[1,5-*a*]benzimidazoles <09S1663>.

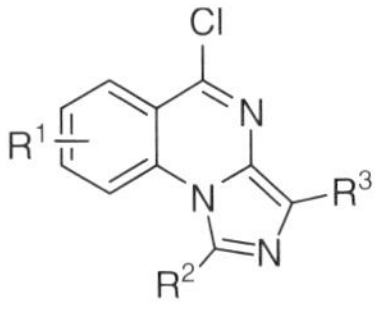

A combinatorial synthesis of 2-substituted benzimidazoles with a polymer-supported hypervalent iodine reagent poly[4-diacetoxyiodo]styrene (PDAIS) was reported <09JCO198>.

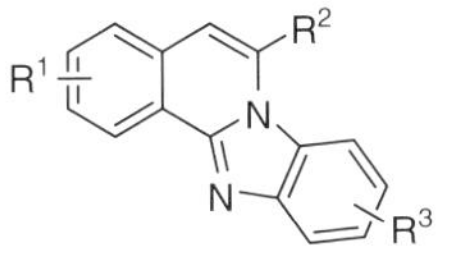

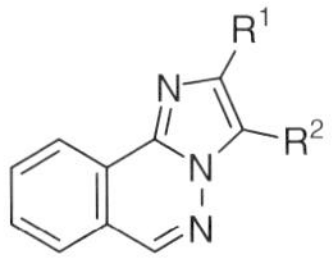

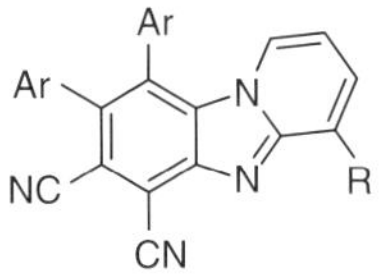

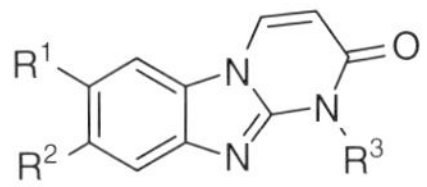

5.4.4. 1,2,3-TRIAZOLES AND RING-FUSED DERIVATIVES

A microreview on click chemistry beyond metal-catalyzed cycloaddition was published <09AG(E)4900>. A review titled, "Cu(I)-Catalyzed Huisgen Azide-Alkyne 1,3-Dipolar Cycloaddition Reaction in Nucleoside, Nucleotide, and Oligonucleotide Chemistry" was published <09CR4207>.

Click chemistry includes a range of reactions that proceed in high yield under ambient conditions, preferably in water, with regioselectivity and a broad tolerance of functional groups. The copper-catalyzed 1,3-dipolar cycloaddition reaction of azides and acetylenes to give 1,2,3-triazoles is known as the "cream of the crop" of all click reactions. A general, rapid, and operationally simple method for the chemo- and regioselective synthesis of 5-iodo-1,4,5-trisubstituted-1,2,3-triazoles **122** from organic azides and iodoalkynes **121** effected by copper(I) iodide in the presence of tris[(1-*tert*-butyl-1*H*-1,2,3-triazolyl)methyl]amine (TTTA) was reported <09AG(E)8018>. Alkynylboronates **123** were useful substrates for the direct synthesis of triazole boronic esters **124** by their thermal cycloaddition with azides <09TL5539>. The [3 + 2] cycloaddition of alkynylboronates and azides provided a direct route to novel triazole boronic esters which can be converted to its bromo intermediate and underwent cross-coupling reactions <09CC436>. Copper-catalyzed click chemistry using acetylene gas was successfully explored under mild conditions in the presence of aliphatic and aryl azides <09SL1453>. The copper-catalyzed 1,3-dipolar cycloaddition of 1-monosubstituted-aryl-1,2,3-triazoles **126** from aryl azides **125** was achieved using calcium carbide as a source of acetylene in an acetonitrile-water mixture <09SL3163>. Aryl halides can be easily transformed in a one-pot procedure into 4-aryl-1,2,3-triazoles with palladium/copper-catalyzed Sonogashira–click reaction sequence, using trimethylsilylacetylene as acetylene surrogate <09S3527>. A fast one-pot microwave-assisted solvent free synthesis of simple alkyl 1,2,3-triazole-4-carboxylate derivatives by 1,3-dipolar cycloaddition reactions with trimethylsilyl azide on the alkyl propiolates and DMAD in high yields was described <09JHC131>. Monodentate phosphoramidite ligands are used to accelerate the copper(I)-catalyzed 1,3-dipolar cycloaddition of azides and alkynes rapidly yielding a wide variety of functionalized 1,4-disubstituted-1,2,3-triazoles <09CC2139>. A convenient synthesis of 4-alkynyl-1,2,3-triazoles and novel unsymmetrically substituted 4,4′-bi-1,2,3-triazole derivatives was devised starting from easily available 1-trimethylsilyl-1,3-butadiyne <09T10573>. One-pot four-component reaction of aldehyde, alcohol, azide, and alkyne **127** underwent smooth coupling by means of acetal formation, azidation, and a subsequent 'click reaction' in the presence of copper(II) triflate and copper metal in acetonitrile to furnish α-alkoxy-1,2,3-triazoles **128** in good yields <09TL6029>.

New copper catalysts were reported for some click reactions. Readily prepared copper nanoparticles were found to effectively catalyze the 1,3-dipolar cycloaddition of a variety of azides and alkynes to give the corresponding 1,2,3-triazoles in excellent yields <09TL2358>. A (2-aminoarenethiolato)copper(I) complex was used as an efficient catalyst for the copper-catalysed Huisgen reaction of azides and terminal alkynes <09EJO5423>. An easily prepared supported copper hydroxide on titanium oxide ($Cu(OH)_x/TiO_2$) showed high catalytic performance for the 1,3-dipolar cycloaddition of organic azides to terminal alkynes in non-polar solvents under anaerobic conditions <09CEJ10464>. Cu(II)–Hydrotalcite was reported to be an efficient heterogeneous catalyst for Huisgen [3+2] cycloaddition of azides and alkynes to give 1,4-disubstituted-1,2,3-triazoles <09CEJ2755>. Cuprous bromide was immobilized as a copper-supported ionic liquid catalyst (Cu-SILC) in the pores of amorphous mercaptopropyl silica gel with the aid of an ionic liquid, [bmim]PF_6 effectively and regioselectively catalyzed Huisgen [3+2] cycloadditions at room temperature in aqueous ethanol to give 1,4-disubstituted-1,2,3-triazoles <09SL643>. A polymer-supported copper catalyst-CuHP20- was easily prepared in water and effectively catalyzed the Huisgen cycloaddition between azides and terminal alkynes <09H(77)521>. A highly active catalyst for Huisgen 1,3-cycloadditions based on the tris(triazolyl)methanol-Cu(I) structure was reported <09OL4680>.

A convenient synthesis of benzofuran- and indole-substituted 1,2,3-triazoles was devised starting from easily available 2-silylalkynyl-substituted benzofuran and indole derivatives <09S3853>. 7-Azido-tetrahydroindazolones underwent efficient copper-catalyzed Huisgen 1,3-dipolar cycloaddition reactions with various alkynes leading to a straightforward synthesis of triazole-functionalized tetrahydroindazolones <09TL3046>. Copper-catalyzed cycloaddition of N-(*tert*-butoxycarbonyl)-3-(1-tosyl-3-butynyl)-1H-indole with various azides readily provided the corresponding (triazolylethyl)indoles <09S3143>. A highly robust and efficient strategy for high-throughput synthesis of a 325-member azide library was achieved using click chemistry <09OBC1821>. A library of 1,4-benzoquinone containing 1,2,3-triazoles was prepared by the application of a highly regioselective copper(I)-catalyzed process <09S1341>. Intramolecular azide-alkyne [3 + 2] cycloaddition was investigated as a versatile route to new heterocyclic structural scaffolds under non-copper thermal reaction conditions <09OBC1921>.

Organic azides can also been generated *in situ* and treated with alkynes in one-pot reactions. The reaction of α-tosyloxy ketones **129**, sodium azide, and terminal alkynes in presence of copper(I) iodide in aqueous polyethylene glycol afforded regioselectively 1,4-disubstituted 1,2,3-triazoles **130** in good yield at ambient temperature via *in situ* formation of the azide <09TL2065>. A one-pot, three-step synthesis of 1,4-disubstituted-1,2,3-triazoles from aldehyde and amine was developed by *in situ* transformation of aldehyde into alkyne, followed by diazo transfer of amine into azide and subsequent cycloaddition <09SL2977>. A one-pot two-step sequence involving a nucleophilic aromatic substitution (S_NAr) of activated fluorobenzenes **131** with azide nucleophile and in situ Huisgen cycloaddition of the resulting aryl azides with alkynes was developed for a rapid access to 1,4-substituted-1,2,3-triazoles **132** <09T3974>. One-pot reaction of α-bromoketones, sodium azide, and terminal acetylenes catalyzed by copper(I) sulfate in aqueous PEG-400 at room temperature to give 1,4-disubstituted-1,2,3-triazoles was published <09SL399>. A novel and efficient way of synthesizing 4,5-disubstituted-1,2,3-(NH)-triazoles **134** through palladium-catalyzed and ultrasonic promoted Sonogashira coupling/1,3-dipolar cycloaddition of acid chlorides, terminal acetylenes **133**, and sodium azide in one pot was developed <09OL3024>. An efficient synthesis of substituted benzotriazoles **137** using an one-pot azide-alkyne 1,3-dipolar cycloaddition "click reaction" with *in situ* generation of the reactive aromatic azide **136** from anilines **135** and benzyne reaction partners used readily available and inexpensive substituted anthranilic acid or *o*-(trimethylsilyl)phenyl triflate as benzyne precursor <09OL1587>. One-pot procedure with primary amines, terminal acetylenes, and diazo transfer reagent imidazole-1-sulfonyl azide hydrochloride delivered a regioselective synthesis of 1,4-disubstituted-1,2,3-triazoles <09SL1391>.

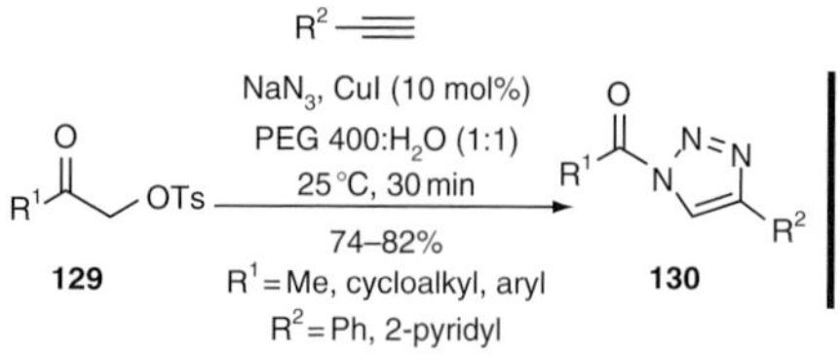

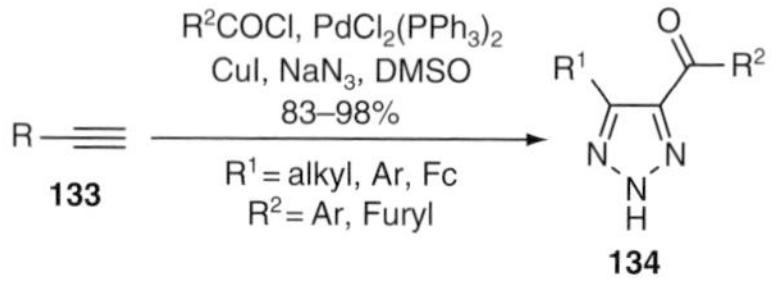

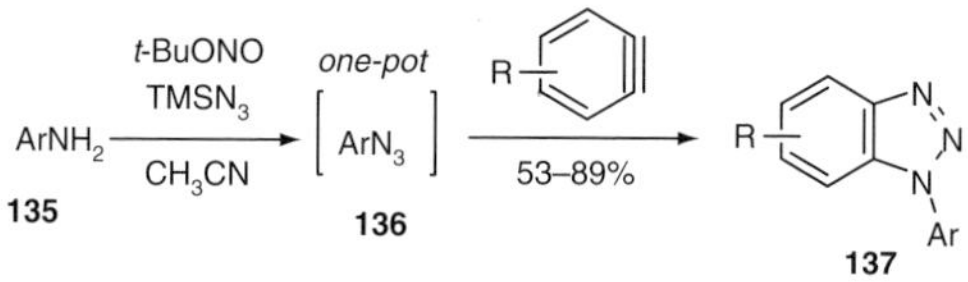

Other methods of 1,2,3-triazole synthesis were also published. Benzotriazoles **140** were prepared by the three-component and two-component microwave-assisted [3+2] cycloadditions of sodium azide to benzyne, generated from the reaction of an *o*-(trimethylsilylaryl) triflate **139** in the presence of arylmethyl halides **138** with either CsF or KF/18-Crown-6 <09TL4677>. *N*-Linked 1,2,3-triazoles were prepared by reaction of azides with vinyl acetate under microwave irradiation <09Sl3275>. Novel 5-aminotriazoles were prepared by the cycloaddition reactions of aryl azides with cyanoacetyl pyrroles and indoles <09S1297>. A convenient synthetic protocol was elaborated for creation of combinatorial libraries of 1-(R^1-phenyl)-5-methyl-*N*-R^2-1*H*-1,2,3-triazole-4-carboxamides **143** from azides **141**, amines, and diketene **142** were selected for the reaction which has proceeded in a one-pot system with high yields and in short time <09JCO481>. *P*-Toluenesulfonyl alkyne underwent copper-free cyclization at room temperature with diverse azido compounds to give 1,4-disubstituted-1,2,3-triazoles with high regioselectivity <09SL1409>. β-Keto sulfones and β-nitrile sulfones were new activated methylenic blocks for the synthesis of 1,2,3-triazoles <09S2321>.

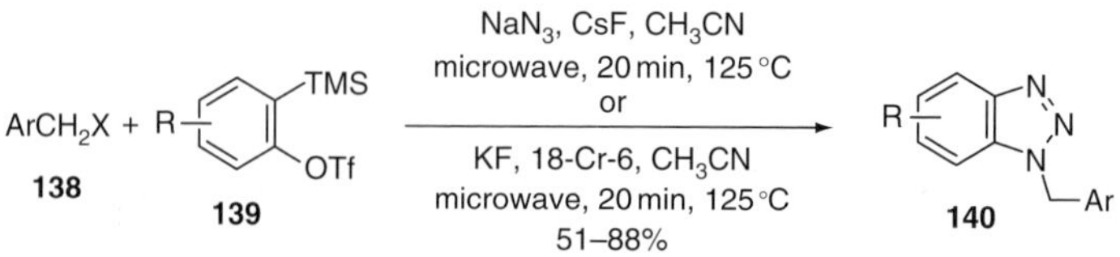

ArN₃ + [142: diketene-like structure] → [143: pyrazole product with RHN-C(O), Me, N=N-Ar]

Reagents: RNH₂, Et₃N, CH₃CN, 25 °C, 53–91%

141 **142** **143**

1,2,3-Triazoles were converted to other structures. The products of the alkylation of sodium 4-nitro-1,2,3-triazolate with ethyl bromide were investigated using ^{1}H, ^{13}C, and ^{15}N NMR spectroscopy <09TL2577>. Treatment of 1,4-disubstituted-1,2,3-triazoles **144** with aryl iodides in the presence of catalytic amount of copper chloride and lithium *tert*-butoxide led to 5-arylated-1,2,3-triazoles **145** <09H(78)645>. Palladium-catalyzed direct arylations from **146** to **145** in polyethylene glycol (PEG) were devised, which set the stage for the development of user-friendly palladium(0)-catalyzed C-H bond functionalizations in the presence of air with a recyclable phosphine ligand-free palladium complex <09OL4922>. Reaction of 1-nitro-4-(1,2,3-triazolyl/tetrazolyl)benzenes with arylacetonitriles in an alcoholic medium in the presence of excess alkali gave novel 2,1-benzisoxazoles <09S2741>. The application of 2-(azidomethyl)arylboronic acids in the synthesis of coumarin-type compounds was reported <09S1673>. Reaction of 4-bromo-*NH*-1,2,3-triazoles with alkyl halides in the presence of potassium carbonate in DMF produced the corresponding 2-substituted 4-bromo-1,2,3-triazoles in a regioselective process and subsequent Suzuki cross-coupling reaction of these bromides provided an efficient synthesis of 2,4,5-trisubstituted triazoles <09OL5490>. Reaction of 4,5-dibromo-1,2,3-triazole **147** with electron-deficient aromatic halides in the presence of potassium carbonate in DMF produced the corresponding 2-aryl-4,5-dibromo-triazoles **148** with high regioselectivity and subsequent debromination of these triazoles by hydrogenation furnished 2-aryltriazoles **149** in excellent yields <09OL5026>. Palladium-catalyzed direct arylation with aryl tosylates in the C5-position of 1,4-disubstituted-1,2,3-triazoles in the presence of S-Phos ligand was disclosed <09AG(E)201>.

144 → **145** ← **146**

Conditions 144→145: ArI, CuCl (10 mol%), LiO*t*-Bu, DMF, 140 °C, 23–99%, R¹ = Tol, *n*-oct, R² = H, alkyl, Ph, Ar

Conditions 146→145: ArBr, Pd(OAc)₂, MesCO₂H, PEG 20,000, K₂CO₃, 120 °C, 52–85%, R¹ = Ph, Ar, Bn, R² = H, Ph, *n*-Bu

147 → **148** → **149**

Conditions 147→148: ArX, K₂CO₃, DMF, 85–97%

Conditions 148→149: HCO₂H, Et₃N, Pd/C, MeOH, 83–95%

Ar needs EWG = NO₂, CF₃, CN, CO₂Me

Several applications of benzotriazole mediated methodology to different synthetic transformations were reported. Pyrido(benzotriazol-1-yl)oxypyrimidine reacted with arylboronic acids under palladium-free, cesium carbonate, hydrogen peroxide, and DME conditions to produce heteroaryl ethers in good yields <09TL5733>. A Dy(OTf)$_3$-mediated selective substitution reaction of N-(α-benzotriazolyl-alkyl)amides with active methylene compounds was reported <09TL5536>. Various β-amido-β-diketones were first synthesized with N-(α-amidoalkyl)benzotriazole-mediated amidoalkylation of 1,3-diketones which underwent rapid condensation with hydrazines to give the corresponding N-[β-(3,5-di and 1,3,5-trisubstituted pyrazol-4-yl)alkyl]amides <09T328>. N-Aryl and N-alkenyl carbamoyl benzotriazoles were prepared in good to excellent yields from acyl azides and benzotriazole via Curtius rearrangement <09SL2461>. Easily synthesized aldoximes were converted to the corresponding nitriles under very mild conditions by a simple reaction with 1H-benzotriazol-1-yloxytris(dimethylamino)phosphonium hexafluorophosphate (BOP) and DBU <09JOC3079>.

Triazole-containing reagents found some applications. Triazole-Au(I) complex **150** was a new class of catalyst with improved thermal stability and reactivity for intermolecular alkyne hydroamination <09JA12100>. Enantiopure cyclohexane fused tetrahydropyrans were synthesized using domino Michael–ketalization with organocatalyst **151** <09CC4985>. 1,2,3-Triazolium-tagged prolines found applications in asymmetric aldol and Michael reactions <09S3975>. Suzuki cross-coupling of heteroaromatic chlorides with various boronic acids were carried out in high yields using highly electron-rich monophosphine ligands **152** <09S3094>. N-Sulfonyl 1,2,3-triazoles **153** participated in rhodium-catalyzed enantioselective cyclopropanation of olefins <09JA18034>.

150 **151** **152** **153**

"Click" chemistry was very active in many fields this year and these applications are reflected in the table below.

Table: Application of Click Chemistry in Different Fields

Click Chemistry Field	References
Amino Acids, Peptides and Peptidomimetics	09AG(E)5042, 09EJO2120, 09EJO4593, 09JOC7165, 09OL5270, 09S133, 09T6156, 09T7935
Biological Systems	09CC7315, 09EJO2611, 09JOC8669, 09OBC3421, 09S3579, 09TL4107,
Carbohydrates	09AG(E)8896, 09CL122, 09JOC7588, 09OBC1097, 09SC830, 09SL2162,

Electronics/Electrochemical Systems	09CC2953, 09CC5573, 09CEJ710
Fluorescent Probes	09AG(E)344, 09OL3008, 09TL7032
Nucleotides and Nucleosides	09ARK152, 09JOC4318, 09JOC6837, 09OBC1374, 09OBC2933, 09OBC4481, 09SL2123, 09TL4101
Polymers	09CC815, 09CC2305
Supramolecular Systems	09CC1712, 09CC3017, 09CEJ7306

Some fused-1,2,3-triazole systems were reported. A wide variety of [1,2,3]triazolo[5,1-c][1,4]benzoxazines **154** were synthesized through palladium-copper catalyzed reactions of 1-azido-2-(prop-2-ynyloxy)benzene with aryl/vinyl iodides <09TL2678>. A one-pot three-component [2+3] cycloaddition for the synthesis of 1-alkyl-1H-naphtho[2,3-d][1,2,3]triazole-4,9-dione and 2-alkyl-2H-naphtho[2,3-d][1,2,3]triazole-4,9-diones **155** was developed <09JOC4414>. A one-pot approach using palladium-copper as catalyst was developed for the synthesis of 1,2,3-triazolo[5,1-c]morpholines **156** <09JOC3612>. The 1,3-dipolar cycloaddition of linear azido alkynes derived from protected β-amino esters proceeded via diastereomeric differentiation to provide *trans*-disubstituted triazolodiazepines **157** in good yields <09JOC2004>. New heterocyclic azides, ethyl 2-azido-4-R^1-5-R^2-3-thiophenecarboxylates, synthesized by diazotization of 2-aminothiophenes and subsequent treatment with sodium azide, were employed in the synthesis of thieno[3,2-e][1,2,3]triazolo[1,5-a]pyrimidin-5(4H)-ones **158** <09T2678>. The deprotonative magnesation and cadmation of [1,2,3]triazolo[1,5-a]pyridines were investigated <09JOC163>. A short and efficient regioselective synthesis of a number of 1,4-disubstituted-1,2,3-triazole derivatives of oxindoles from N-terminal alkyne and alkynyl ether derivatives of Morita-Baylis-Hillman adducts of isatin with *in situ* generated alkyl azide and copper(I) iodide as a catalyst in a 1:1 mixture of *tert*-butanol/water as a solvent system was achieved <09JHC919>. The synthesis of nitro tricycles with a central 1,2,3-triazole ring was obtained by a nitrene-mediated reaction of azidonitrobis(hetaryl) derivatives under thermolysis <09SL1318>.

5.4.5. 1,2,4- TRIAZOLES AND RING-FUSED DERIVATIVES

Various synthetic protocols are available for the preparation of 1,2,4-triazoles and derivatives thereof. Cyclocondensation of amidoguanidines **159** with conventional heating or microwave irradiation in aqueous medium gave 3(5)-amino-1,2,4-triazoles

160 <09TL2124>. Rapid microwave-assisted reaction of acylthioureas **161** with hydrazines afforded N1-substituted 3-amino-1,2,4-triazoles **162** <09TL1667>. The reaction of 4-amino-5-phenyl-3,5-thiaaza-4-pentenoic acid **164** with various nitrilimines **163** led to the formation of substituted carboxymethylthio-1,2,4-triazoles **165** <09SC1847>. Thermal rearrangement of N-1,2,4-oxadiazol-3-yl-hydrazones **166** gave 3-carboxyamido-1,2,4-triazoles **167** under solvent-free conditions <09OL4018>. Hydrazinecarboximidamides **168**, prepared either from thioureas or hydrazinecarbothioamides, reacted with trimethyl orthoformate at 140 °C to give 3-amino-1,2,4-triazoles **169** <09JOC7595>. A series of perfluoroalkyl-1,2,4-triazole-carboxamides was obtained through an ANRORC-like rearrangement (*A*ddition of *N*ucleophile, *R*ing-*O*pening and *R*ing-*C*losure) of 5-perfluoroalkyl-1,2,4-oxadiazole-3-carboxamides with methylhydrazine or hydrazine <09ARK325>. Efficient copper-catalyzed oxidative synthesis of 1,2,4-triazole derivatives **171** from amidines **170** and nitriles as well as triazolo heterocycles was reported <09JA15080>. A facile two-step synthesis of 2,4-dihydro-5-amino-[1,2,4]triazol-3-ones was described <09SL607>.

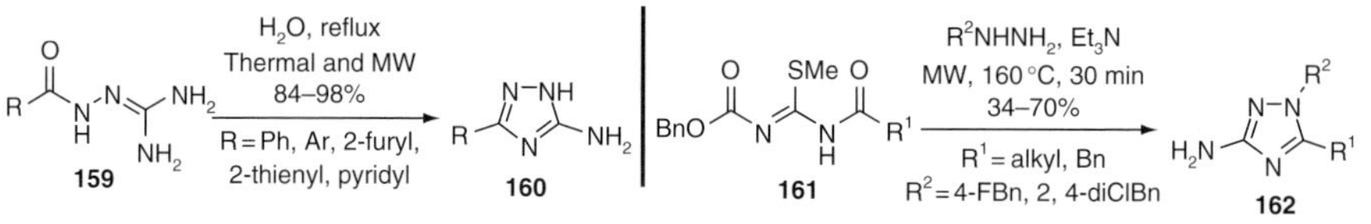

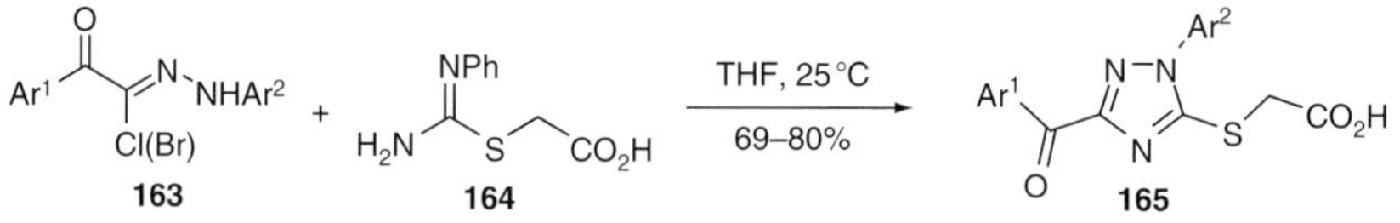

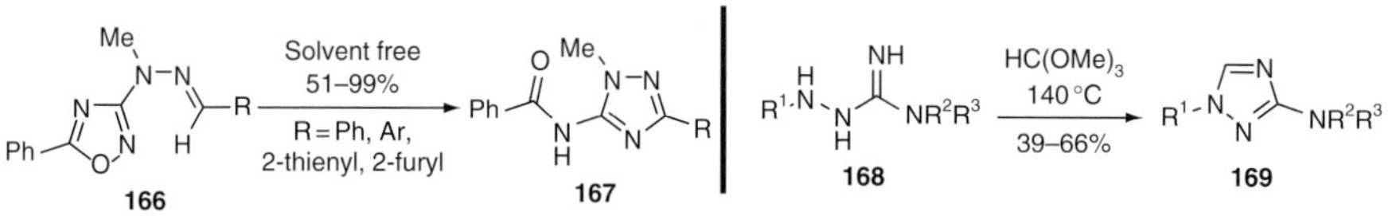

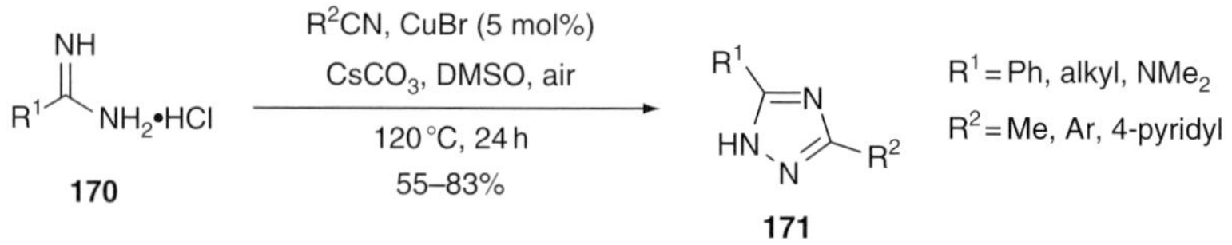

Iodogen was employed as an efficient oxidizing agent for the conversion of urazoles and bis-urazoles to the corresponding 1,2,4-triazoles in good to excellent yields under mild heterogeneous conditions at room temperature <09S2729>. Oxidation of urazoles and bis-urazoles to the corresponding triazolinediones by supported nitric acid on silica gel and polyvinyl pyrrolidone was described <09SC4264>.

There were some literature reports on the reactions of 1,2,4-triazoles. 1,2,4-Triazoles **172** underwent copper-mediated direct arylation with aryl iodides to give 3,5-diaryl-1,2,4-triazoles **173** in the presence of sodium carbonate in warm DMSO <09OL3072>. Trichloromethyl- and tribromomethyltriazoles **174** were displaced with various amines to the corresponding carboxamide derivatives **175** <09SC2585>. α-Dibromides of Boc-protected-4-amino-1,2,4-triazoles were prepared and their molecular structure was confirmed by X-ray crystallography <09H(78)117>.

The use of 1,2,4-triazole reagents in synthetic operations were described. N-Heterocyclic carbene **176** catalyzed the ring expansion of various 2-acyl-1-formylcyclopropanes to 3,4-dihydro-χ-pyrones in good to excellent yields <09OL1623>. An unprecedented catalytic enantioselective [4+2] cycloaddition of alkylarylketenes with N-aryl-N'-benzoyldiazenes or N,N'-dibenzoyldiazenes to give 1,3,4-oxadiazin-6-ones was developed by employing NHC catalysts **177** <09AG(E)192>. Chiral N-heterocyclic carbenes **177** was found to be an efficient catalyst for the formal [4+2] cycloaddition reaction of alkyl(aryl)ketenes and o-quinone methides to give the corresponding 3,3,4-trisubstituted 3,4-dihydrocoumarins in good yields with good diastereoselectivities and excellent enantioselectivities <09ASC2822>. N-Heterocyclic **178** was employed in an operationally simple single step multicatalytic cascade process for the preparation of α-hydroxycyclopentanones containing three contiguous stereocenters <09JA13628>. NHC-catalyst **178** allowed conjugate hydroacylation of 1,4-naphthoquinones for the synthesis of monoacylated 1,4-dihydroxynaphthalene derivatives <09JOC9573>. A triazolinylidene carbene **179** catalyzed the intermolecular Stetter reaction of glyoxamide and alkylidene ketoamides to give 1,4-dicarbonyl products in good to excellent yields, enantioselectivities, and diastereoselectivities <09OL2856>. Highly enantioselective benzoin condensation reactions were achieved using bifunctional protic pentafluorophenyl-substituted triazolium precatalyst **180** <09JOC9214>. The highly diastereo- and enantioselective synthesis of β-trifluoromethyl-β-lactones bearing two contiguous stereocenters was realized by chiral N-heterocyclic carbene-catalyzed formal cycloaddition reaction of alkyl(aryl)ketenes and trifluoromethyl ketones with **181** <09OL4029>. N-heterocyclic **182** carbene-catalyzed reactions of α,β-unsaturated aldehydes and a variety of electrophiles allowed the facile preparation of a diverse array of annulation products including trisubstituted cyclopentenes, γ-lactams, and bicyclic β-lactams <09JA8714>. α-Hydroxyenones underwent clean, catalytic amidations with amines promoted by the combination of an N-heterocyclic carbene **182** and 1,2,4-triazole <09CC4566>. A new NHC catalyst **183** participated in the intermolecular Stetter reaction of nitroalkenes and heteroarylaldehydes in a highly efficient and enantioselective reaction <09JA10872>. A new class of triazolium ion precatalysts incorporating protic substituents was described for the enantioselective benzoin condensation of a range of aromatic aldehydes in 1–62% ee <09OBC3584>. Organocatalytic enantioselective cycloadditions with **184** to give nitrogen-substituted dihydropyran-2-ones were developed <09OL3934>. 1-Phenylsulfanyl[1,2,4]triazole **185**, a novel sulfur transfer reagent, gave excellent product yields in the organocatalytic α-sulfenylation of substituted piperazine-2,5-diones under mild conditions <09TL4310>. Various chiral N-heterocyclic carbenes mediated the enantioselective addition of 2-phenylphenol to unsymmetrical alkylarylketenes, delivering α-alkyl-α-arylacetic acid derivatives with good levels of enantiocontrol up to 84% ee <09ASC3001>. N-Heterocyclic carbenes catalyzed the oxidation of allylic and benzylic alcohols as well as saturated aldehydes to esters with manganese(IV) oxide in excellent yields <09T3102>.

176 R^1 = H, R^2 = Mes
177 R^1 = CPh$_2$OTBS, R^2 = Ph
178 R^1 = H, R^2 = C$_6$F$_5$
179 R^1 = Bn, R^2 = C$_6$F$_5$
180 R^1 = CPh$_2$OTBS, R^2 = C$_6$F$_5$
181 R^1 = CPh$_2$OTBS, R^2 = 2-i-PrC$_6$H$_4$

182

183 **184** **185**

Structurally unique 1,2,4-triazole fused-ring systems were reported. The design and chemical synthesis of [1,2,4]triazol[1,5-c]pyrimidin-5-yl amines **186**, a novel class of VEGFR-2 kinase inhibitors, was disclosed <09TL3809>. General, high-yielding microwave-assisted organic synthesis protocols for the expedient synthesis of functionalized 3,6-disubstituted-[1,2,4]triazolo[4,3-b]pyridazines **187** was described <09TL212>. One-pot reaction of hydrazinopyridine with isothiocyanates followed by desulfurization *in situ* with polymer-supported Mukaiyama's reagent led to the synthesis of 3-amino-[1,2,4]triazolo[4,3-a]pyridines **188** <09JOC5553>. The three-component condensation of 3-amino-5-alkylthio-1,2,4-triazoles with aromatic aldehydes and β-ketoester gave compounds **189** via a regioselective Biginelli-like reaction <09JHC139>. The reaction of 3-amino-1,2,4-triazole with arylidene-5-

acetyl barbituric acid or dehydroacetic acid in refluxing *n*-butanol led to the formation of dihydro-1,2,4-triazolo[1,5-*a*]pyrimidines **190** <09JHC285>. Diethyl maleate reacted with *N*-substituted-hydrazino-carbothioamides to form ethyl [1,2,4]triazolo[3,4-*b*][1,3]thiazine-5-carboxylates **191** <09JHC687>. Efficient synthesis of benzothieno[3,2-*d*]-1,2,4-triazolo[1,5-*a*]pyrimidin-5(1*H*)-ones **192** via a tandem aza-Wittig/heterocumulene-mediated annulation was reported <09JHC903>. Addition of hydrazine to 4,7-dihydro[1,2,4]triazolo[1,5-*a*]pyrimidines gave hydrazine derivatives of 4,5,6,7-tetrahydro[1,2,4]triazolo[1,5-*a*]pyrimidines **193** <09JHC1413>. A novel approach to the synthesis of triazolo[4,3-*b*][1,2,4,5]tetrazines **194** was developed *via* reactions of 4-amino-5-methyl-1,2,4-triazole-3(2*H*)-thione with hydrazonoyl halides using chitosan as a basic catalyst under microwave irradiation <09ARK58>. Cyclization of ethyl 2-ethoxymethylidene-3-polyfluoroalkyl-3-oxopropionates with 3-amino-1*H*-[1,2,4]triazole gave polyfluoroalkylated dihydro[1,2,4-triazolo][1,5-*a*]pyrimidines <09H(78)435>. The reactivity and catalytic performance of 2-ethylpyrido[1,2-*c*][1,2,4]triazol-3-ylidene was comprehensively investigated <09OBC4241>. An expeditious novel approach to 3,6-disubstituted [1,2,4]triazolo[3,4-*b*][1,3,4]thiadiazoles using silica-supported dichlorophosphate as a recoverable cyclodehydrant, carboxylic acids and thiocarbohydrazide as starting materials, and microwave irradiation as thermal source was described <09SC3816>.

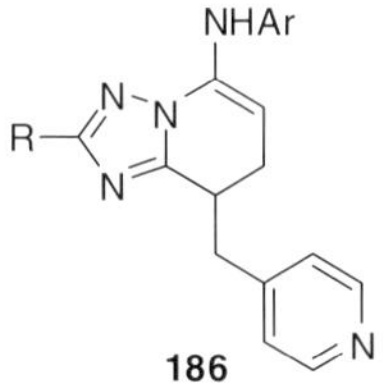

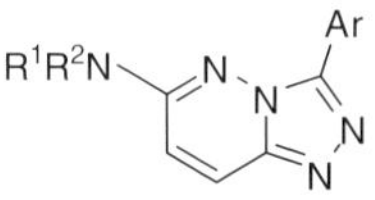

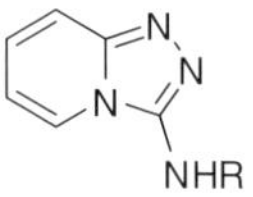

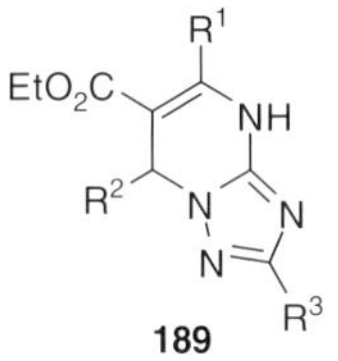

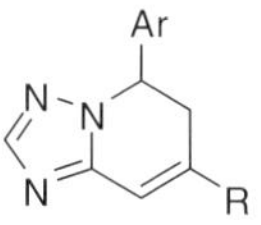

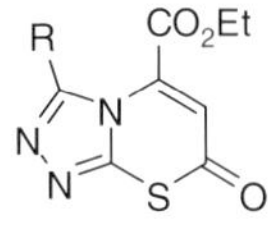

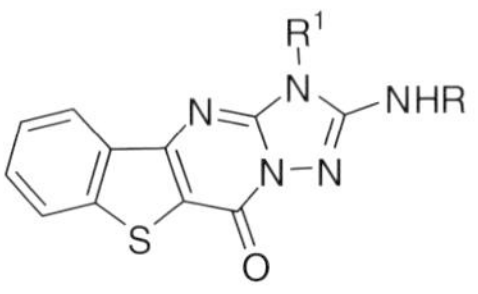

5.4.6. TETRAZOLES AND RING-FUSED DERIVATIVES

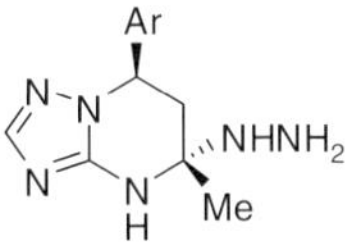

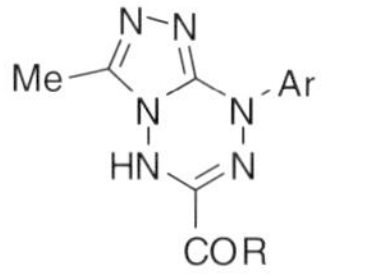

The most common preparation of tetrazoles is the reaction of nitriles with azides. 5-Substituted-1*H*-tetrazoles **196** were obtained from the iron-catalyzed reactions of nitriles **195** with trimethylsilyl azide <09CEJ4543>. 5-Substituted tetrazoles **198** were prepared by treatment of nitriles **197** with sodium azide with triethylammonium chloride in nitrobenzene under microwave conditions <09S2175>. Antimony trioxide and cadmium chloride were found to be efficient Lewis acid catalysts for the synthesis of 5-aryl substituted-1*H*-tetrazoles **200** from aryl nitriles **199** <09SC426, 09SC4479>. Treatment of 2-alkynylbenzonitriles **201** with excess sodium azide and zinc bromide in DMSO at 100 °C under microwave conditions gave tetrazolo[1,5-*a*]pyridines **202** in good yields <09T8367>. An efficient method for the preparation of 5-substituted 1*H*-tetrazole derivatives was reported using $FeCl_3$–SiO_2 as an effective heterogeneous catalyst <09TL4435>. The [2+3] cycloaddition of Boc-α-amino nitrile and sodium azide in the presence of a catalytic amount of zinc bromide yielded the desired Boc-protected tetrazole analogs of amino acids in good yields and purity <09SC395>. Secondary amines **203** reacted with cyanogen azide at ambient temperature in water/acetonitrile to provide tetrazole derivatives **205** directly via intermediate **204** <09EJO3573>. An efficient method for preparation of arylaminotetrazoles from arylcyanamides carrying electron-withdrawing substituents on the aryl ring was reported using natrolite zeolite as a natural catalyst <09T10715>.

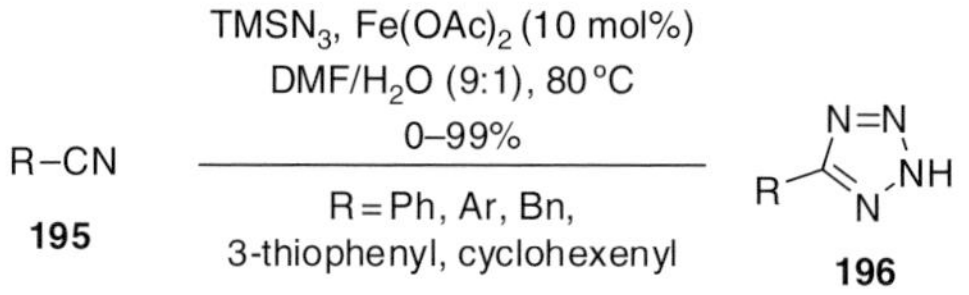

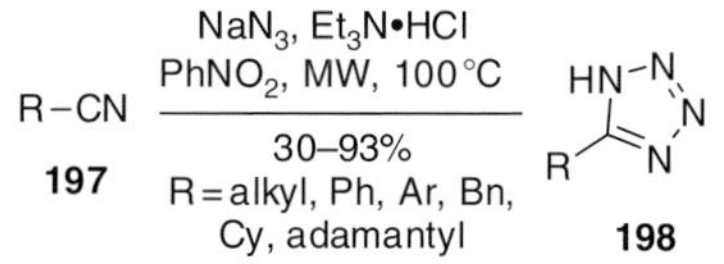

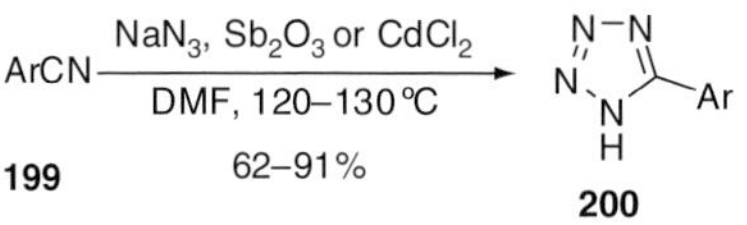

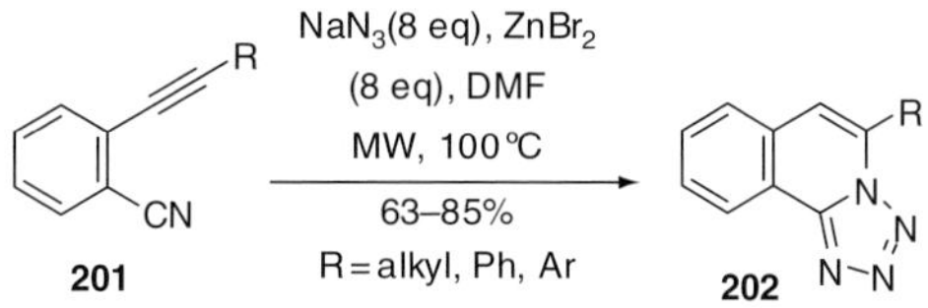

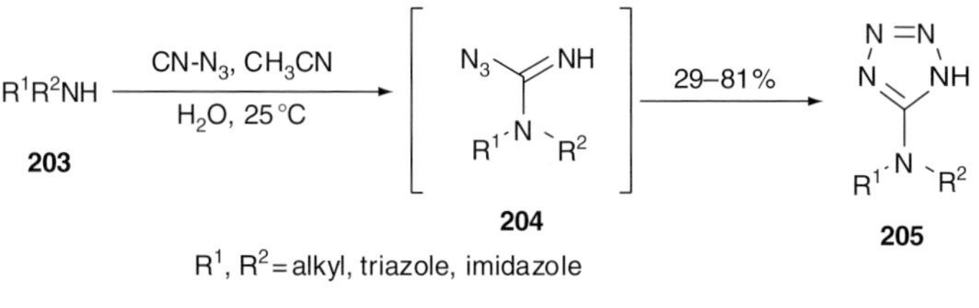

A simple, efficient, and general method was developed for the synthesis of 1-substituted-1*H*-1,2,3,4-tetrazoles via a three-component condensation of amine, trimethyl orthoformate, and sodium azide in the presence of a catalytic amount of indium triflate under solvent-free conditions <09TL2668>.

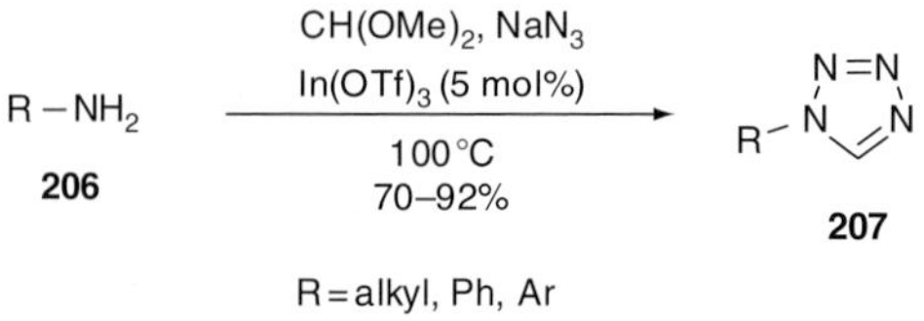

Azido and nitro-containing tetrazoles are highly energetic materials. Treatment of aminotetrazole compounds, which were synthesized from primary amines and cyanogen azide, with excess 100% nitric acid led to energetic mono-, di-, and trisubstituted nitroiminotetrazoles **208** <09AG(E)564>. The synthesis and detonation properties of high energy density materials with ethylene- and propylene bis(nitroiminotetrazolate) as the anions was reported <09CEJ3198>. Disubstituted azidotetrazoles **209** were synthesized by the base-catalyzed activation of the C–F bond in the trifluoromethylazo-substituted cyclic and acyclic alkanes <09CEJ4102>. 5-(1,2,3-Triazol-1-yl)tetrazole derivatives were prepared from an azidotetrazole via click chemistry <09CEJ9897>. The synthesis of functionalized tetrazolyltetrazenes **210** as energetic compounds was reported <09JOC2460>.

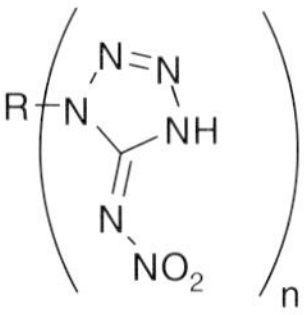

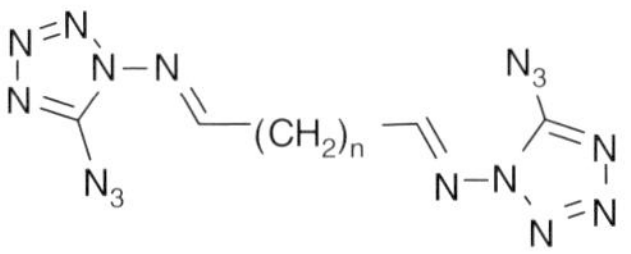

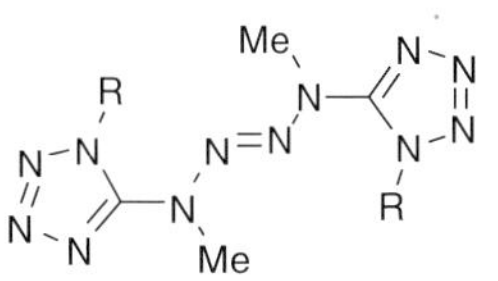

A nitrogen-rich polymer was formed from the reaction of hexamethylene diisocyanate and N-[1-(2-hydroxyethyl)-1H-tetrazol-5-yl]-N-methylhydrazine monomers <09EJO275>. A tetrakis(tetrazolyl) analog of EDTA was reported <09EJO1495>. Six photoreactive tetrazole amino acids were efficiently synthesized either by the de novo Kakehi tetrazole synthesis method or by alkylation of a glycine Schiff base with tetrazole-containing alkyl halides <09OL3570>.

Tetrazole-containing reagents were utilized in several reported reactions. (*S*)-5-Pyrrolidin-2-yltetrazole **211** was employed as an organocatalyst in enantio- and diastereoselective sequential *O*-nitrosoaldol and Grignard addition reactions to give chiral 1,2-diols <09AG(E)3333>. Silver triflate activation of 5-(*p*-methoxybenzylthio)-1-phenyl-1*H*-tetrazole(PMB-ST, **212**) was employed as a *p*-methoxybenzyl protecting group of alcohols in the presence of 2,6-di-*tert*-butyl-4-methylpyridine <09TL5267>. Fluorinated 1-phenyl-1*H*-tetrazol-5-yl sulfone derivatives **213** were employed as general reagents in reactions with ketones and aldehydes for the synthesis of fluoroalkylidenes **214** <09JOC8531>.

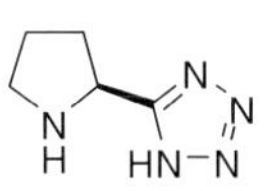

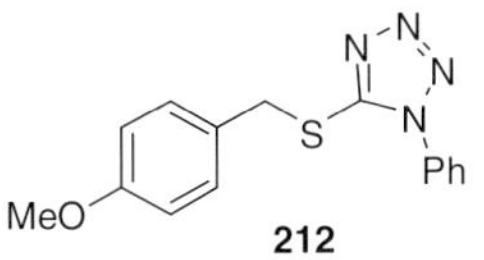

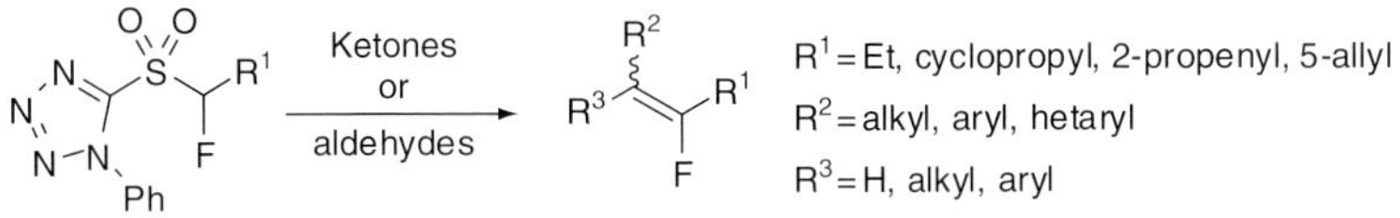

New chiral 5,6,7,8-tetrahydrotetrazolo[1,5-*a*]pyrazines were synthesized from α-amino acid derivatives following click chemistry <09H(77)865>.

REFERENCES

09AG(E)192	X.-L. Huang, L. He, P.-L. Shao and S. Ye, *Angew. Chem., Int. Ed.*, **48,** 192 (2009).
09AG(E)201	L. Ackermann, A. Althammer and S. Fenner, *Angew. Chem., Int. Ed.*, **48,** 201 (2009).
09AG(E)344	P. Kele, G. Mezo, D. Achatz and O.S. Wolfbeis, *Angew. Chem., Int. Ed.*, **48,** 344 (2009).
09AG(E)564	Y.-H. Joo and J.M. Shreeve, *Angew. Chem., Int. Ed.*, **48,** 564 (2009).
09AG(E)918	T. Schulz, C. Torborg, B. Schaffner, J. Huang, A. Zapf, R. Kadyrov, A. Borner and M. Beller, *Angew. Chem., Int. Ed.*, **48,** 918 (2009).
09AG(E)3116	R.L. Giles, J.D. Sullivan, A.M. Steiner and R.E. Looper, *Angew. Chem., Int. Ed.*, **48,** 3116 (2009).
09AG(E)3333	P. Jiao, M. Kawasaki and H. Yamamoto, *Angew. Chem., Int. Ed.*, **48,** 3333 (2009).
09AG(E)4900	C.R. Becer, R. Hoogenboom and U.S. Schubert, *Angew. Chem., Int. Ed.*, **48,** 4900 (2009).
09AG(E)5042	M.P. Ahsanullah, P. Schmieder, R. Kuhne and J. Rademann, *Angew. Chem., Int. Ed.*, **48,** 5042 (2009).
09AG(E)6879	N. Halland, M. Nazar, O. R'kyek, J. Alonso, M. Urmann and A. Lindenschmidt, *Angew. Chem., Int. Ed.*, **48,** 6879 (2009).
09AG(E)8018	J.E. Hein, J.C. Tripp, L.B. Krasnova, K.B. Sharpless and V.V. Fokin, *Angew. Chem., Int. Ed.*, **48,** 8018 (2009).
09AG(E)8896	D. Crich and F. Yang, *Angew. Chem., Int. Ed.*, **48,** 8896 (2009).
09ARK17	L.F.V. Pinto, G.C. Justino, A.J.S.C. Vieira, S. Prabhakar and A.M. Lobo, *Arkivoc*, **v,** 17 (2009).
09ARK58	S.M. Gomha and S.M. Riyadh, *Arkivoc*, **xi,** 58 (2009).
09ARK105	C.N. Raut, R.B. Mane, S.M. Bagul, R.A. Janrao and P.P. Mahulikar, *Arkivoc*, **xi,** 105 (2009).
09ARK152	I. Pérez-Castro, O. Caamaño, F. Fernández, M.D. García, C. López and E. de Clercq, *Arkivoc*, **iii,** 152 (2009).
09ARK168	M. Mamaghani and S. Dastmard, *Arkivoc*, **ii,** 168 (2009).
09ARK174	H.G. Bonacorso, C.A. Cechinel, E.D. Deon, R.C. Sehnem, F.M. Luz, M.A.P.- Martins and N. Zanatta, *Arkivoc*, **ii,** 174 (2009).
09ARK198	H.F. Anwara and M.H. Elnagdi, *Arkivoc*, **i,** 198 (2009).
09ARK258	C. Liu, Z. Li, L. Zhao and L. Shen, *Arkivoc*, **ii,** 258 (2009).
09ARK308	C. Zalaru, F. Dumitrascu, C. Draghici, E. Cristea and I. Tarcomnicu, *Arkivoc*, **ii,** 308 (2009).
09ARK325	A.P. Piccionello, A. Pace, S. Buscemi and N. Vivona, *Arkivoc*, **vi,** 325 (2009).
09ASC2822	H. Lv, L. You and S. Yea, *Adv. Synth. Catal.*, **351,** 2822 (2009).
09ASC3001	C. Concell, N. Duguet and A.D. Smitha, *Adv. Synth. Catal.*, **351,** 3001 (2009).
09CC436	J. Huang, S.J.F. Macdonald and J.P.A. Harrity, *J. Chem. Soc., Chem. Commun.*, 436 (2009).
09CC815	B.M. Cooper, D. Chan-Seng, D. Samanta, X. Zhang, S. Parelkar and T. Emrick, *J. Chem. Soc., Chem. Commun.*, 815 (2009).
09CC1040	O. Sereda, A. Blanrue and R. Wilhelm, *J. Chem. Soc., Chem. Commun.*, 1040 (2009).
09CC1712	G. London, G.T. Carroll, T.F. Landaluce, M.M. Pollard, P. Rudolf and B.L. Feringa, *J. Chem. Soc., Chem. Commun.*, 1712 (2009).
09CC1900	P. Loos, M. Riedrichab and H.D. Arndt, *J. Chem. Soc., Chem. Commun.*, 1900 (2009).
09CC2139	L.S. Campbell-Verduyn, L. Mirfeizi, R.A. Dierckx, P.H. Elsinga and B.L. Feringa, *J. Chem. Soc., Chem. Commun.*, 2139 (2009).
09CC2305	C.E. Evans and P.A. Lovell, *J. Chem. Soc., Chem. Commun.*, 2305 (2009).
09CC2338	J. Zhu, H. Xie, Z. Chen, S. Lia and Y. Wu, *J. Chem. Soc., Chem. Commun.*, 2338 (2009).
09CC2953	M.R. Das, M. Wang, S. Szunerits, L. Gengembrec and R. Boukherroub, *J. Chem. Soc., Chem. Commun.*, 2953 (2009).
09CC3017	M.G. Fisher, P.A. Gale, J.R. Hiscock, M.B. Hursthouse, M.E. Light, F.P. Schmidtchenb and C.C. Tonga, *J. Chem. Soc., Chem. Commun.*, 3017 (2009).

09CC3910	A.S. Tsai, R.M. Wilson, H. Harada, R.G. Bergman and J.A. Ellman, *J. Chem. Soc., Chem. Commun.*, 3910 (2009).
09CC4263	X. Cheng and S. Ma, *J. Chem. Soc., Chem. Commun.*, 4263 (2009).
09CC4566	P.-C. Chiang, Y. Kim and J.W. Bode, *J. Chem. Soc., Chem. Commun.*, 4566 (2009).
09CC4944	P.S. Bauerlein, I.A. Gonzalez, J.J.M. Weemers, M. Lutz, A.L. Spek, D. Vogt and C. Muller, *J. Chem. Soc., Chem. Commun.*, 4944 (2009).
09CC4985	S. Chandrasekhar, K. Mallikarjun, G. Pavankumarreddy, K.V. Raob and B. Jagadeesh, *J. Chem. Soc., Chem. Commun.*, 4985 (2009).
09CC5573	J.L. Bartels, P. Lu, A. Walker, K. Maurerb and K.D. Moeller, *J. Chem. Soc., Chem. Commun.*, 5573 (2009).
09CC7315	J.S. Foot, F.E. Lui and R. Kluger, *J. Chem. Soc., Chem. Commun.*, 7315 (2009).
09CC7330	T. Hu, T. Schulz, C. Torborg, X. Chen, J. Wang, M. Beller and J. Huang, *J. Chem. Soc., Chem. Commun.*, 7330 (2009).
09CEJ710	J.M. Casas-Solvas, E. Ortiz-Salmeron, J.J. Gimenez-Martinez, L. Garcia-Fuentes, L.F. Capitan-Vallvey, F. Santoyo-Gonzalez and A. Vargas-Berenguel, *Chem. Eur. J.*, **15,** 710 (2009).
09CEJ843	C. Kison and T. Opatz, *Chem. Eur. J.*, **15,** 843 (2009).
09CEJ2755	K. Namitharan, M. Kumarraja and K. Pitchumani, *Chem. Eur. J.*, **15,** 2755 (2009).
09CEJ3198	Y.-H. Joo and J.M. Shreeve, *Chem. Eur. J.*, **15,** 3198 (2009).
09CEJ4102	T. Abe, Y.-H. Joo, G.-H. Tao, B. Twamley and J.M. Shreeve, *Chem. Eur. J.*, **15,** 4102 (2009).
09CEJ4543	J. Bonnamour and C. Bolm, *Chem. Eur. J.*, **15,** 4543 (2009).
09CEJ7306	J. Iehl and J.-F. Nierengarten, *Chem. Eur. J.*, **15,** 7306 (2009).
09CEJ8971	Y. Wang, Z. Wu, L. Wang, Z. Li and X. Zhou, *Chem. Eur. J.*, **15,** 8971 (2009).
09CEJ9897	T. Abe, G.-H. Tao, Y.-H. Joo, R.W. Winter, G.L. Gard and J.M. Shreeve, *Chem. Eur. J.*, **15,** 9897 (2009).
09CEJ10464	K. Yamaguchi, T. Oishi, T. Katayama and N. Mizuno, *Chem. Eur. J.*, **15,** 10464 (2009).
09CEJ10585	F. Li and T.S.A. Hor, *Chem. Eur. J.*, **15,** 10585 (2009).
09CL122	E. Yamashita, K. Okubo, K. Negishi and T. Hasegawa, *Chem. Lett.*, **38,** 122 (2009).
09CR4207	F. Amblard, J.H. Cho and R.F. Schinazi, *Chem. Rev.*, **109,** 4207 (2009).
09EJO275	K. Banert, T.M. Klapötke and S.M. Sproll, *Eur. J. Org. Chem.*, 275 (2009).
09EJO1495	F. Touti, P. Maurin and J. Hasserodt, *Eur. J. Org. Chem.*, 1495 (2009).
09EJO2120	V.S. Sudhir, C. Venkateswarlu, O.T.M. Musthafa, S. Sampath and S. Chandrasekaran, *Eur. J. Org. Chem.*, 2120 (2009).
09EJO2611	M. Weïwer, C.-C. Chen, M.M. Kemp and R.J. Linhardt, *Eur. J. Org. Chem.*, 2611 (2009).
09EJO3184	S. Campetella, A. Palmieri and M. Petrini, *Eur. J. Org. Chem.*, 3184 (2009).
09EJO3573	Y.-H. Joo and J.M. Shreeve, *Eur. J. Org. Chem.*, 3573 (2009).
09EJO4593	A. Nadler, C. Hain and U. Diederichsen, *Eur. J. Org. Chem.*, 4593 (2009).
09EJO4926	H.Z. Boeini and K.H. Najafabadi, *Eur. J. Org. Chem.*, 4926 (2009).
09EJO5138	A.V. Lygin and A. de Meijere, *Eur. J. Org. Chem.*, 5138 (2009).
09EJO5423	P. Fabbrizzi, S. Cicchi, A. Brandi, E. Sperotto and G. van Koten, *Eur. J. Org. Chem.*, 5423 (2009).
09H(77)521	Y. Kitamura, K. Taniguchi, T. Maegawa, Y. Monguchi, Y. Kitade and H. Sajiki, *Heterocycles*, **77,** 521 (2009).
09H(77)865	D.K. Mohaptra, P.K. Malty, R.V. Ghorpade and M.K. Gurjar, *Heterocycles*, **77,** 865 (2009).
09H(78)117	T. Yamamoto, G. Tanaka, H. Fukumoto and T.-a. Koizumi, *Heterocycles*, **78,** 117 (2009).
09H(78)197	Y.-Q. Ge, J. Jia, Y. Li, L. Yin and J. Wang, *Heterocycles*, **78,** 197 (2009).
09H(78)415	E. Kianmehr, R. Faramarzi and H. Estiri, *Heterocycles*, **78,** 415 (2009).
09H(78)435	M.V. Goryaeva, Y.V. Burgart, V.I. Saloutin, E.V. Sadchikova and E.N. Ulomskii, *Heterocycles*, **78,** 435 (2009).
09H(78)645	S.-i. Fukuzawa, E. Shimizu and K. Ogata, *Heterocycles*, **78,** 645 (2009).
09H(78)911	A. Mukherjee and K.K. Mahalanabis, *Heterocycles*, **78,** 911 (2009).
09H(78)2003	K.M. Al-Zaydi, *Heterocycles*, **78,** 2003 (2009).
09H(78)2337	A.A. Mohammadi, J. Azizian and N. Karimi, *Heterocycles*, **78,** 2337 (2009).
09JA3042	R. Goikhman, T.L. Jacques and D. Sames, *J. Am. Chem. Soc.*, **131,** 3042 (2009).

09JA7253	K.-s. Lee, A.R. Zhugralin and A.H. Hoveyda, *J. Am. Chem. Soc.*, **131**, 7253 (2009).
09JA8714	P.-C. Chiang, M. Rommel and J.W. Bode, *J. Am. Chem. Soc.*, **131**, 8714 (2009).
09JA8971	L.M. Stanley and J.F. Hartwig, *J. Am. Chem. Soc.*, **131**, 8971 (2009).
09JA10872	D.A. DiRocco, K.M. Oberg, D.M. Dalton and T. Rovis, *J. Am. Chem. Soc.*, **131**, 10872 (2009).
09JA12100	H. Duan, S. Sengupta, J.L. Petersen, N.G. Akhmedov and X. Shi, *J. Am. Chem. Soc.*, **131**, 12100 (2009).
09JA13628	S.P. Lathrop and T. Rovis, *J. Am. Chem. Soc.*, **131**, 13628 (2009).
09JA15080	S. Ueda and H. Nagasawa, *J. Am. Chem. Soc.*, **131**, 15080 (2009).
09JA18034	S. Chuprakov, S.W. Kwok, L. Zhang, L. Lercher and V.V. Fokin, *J. Am. Chem. Soc.*, **131**, 18034 (2009).
09JCO198	A. Kumar, R.A. Maurya and P. Ahmad, *J. Comb. Chem.*, **11**, 198 (2009).
09JCO481	N.T. Pokhodylo, V.S. Matiychuk and M.D. Obushak, *J. Comb. Chem.*, **11**, 481 (2009).
09JCO515	X. Huang, J. Cao and J. Huang, *J. Comb. Chem.*, **11**, 515 (2009).
09JCO720	S. Sharma and B. Kundu, *J. Comb. Chem.*, **11**, 720 (2009).
09JCO914	Y.M. Litvinov, A.A. Shestopalov, L.A. Rodinovskaya and A.M. Shestopalov, *J. Comb. Chem.*, **11**, 914 (2009).
09JHC74	S. Rostamizadeh, R. Aryan, H.R. Ghaieni and A.M. Amani, *J. Heterocycl. Chem.*, **46**, 74 (2009).
09JHC100	M. Bakherad, B. Bahramian, H. Nasr-Isfahani, A. Keivanloo and N. Doostmohammadi, *J. Heterocycl. Chem.*, **46**, 100 (2009).
09JHC131	A.A. Taherpoura and K. Kheradmand, *J. Heterocycl. Chem.*, **46**, 131 (2009).
09JHC139	Q. Chen, L.-L. Jiang, C.-N. Chen and G.-F. Yang, *J. Heterocycl. Chem.*, **46**, 139 (2009).
09JHC256	H.-Q. Wang, W.-P. Zhou, Y.-Y. Wang, C.-R. Lin and Z.-J. Liu, *J. Heterocycl. Chem.*, **46**, 256 (2009).
09JHC278	B.R.P. Kumar, G.K. Sharma, S. Srinath, M.N.B. Suresh and B.R. Srinivasab, *J. Heterocycl. Chem.*, **46**, 278 (2009).
09JHC285	R.V. Rudenko, S.A. Komykhov, V.I. Musatov and S.M. Desenko, *J. Heterocycl. Chem.*, **46**, 285 (2009).
09JHC650	P. Simunek, M. Svobodova and V. Machacek, *J. Heterocycl. Chem.*, **46**, 650 (2009).
09JHC687	A.A. Aly, A.A. Hassan, Y.R. Ibrahim and M. Abdel-Aziz, *J. Heterocycl. Chem.*, **46**, 687 (2009).
09JHC708	B.K. Ghotekar, M.N. Jachak and R.B. Toche, *J. Heterocycl. Chem.*, **46**, 708 (2009).
09JHC728	L. Nagarapu, R. Bantu and H.B. Mereyala, *J. Heterocycl. Chem.*, **46**, 728 (2009).
09JHC756	B. Insuasty, A. Tigreros, H. Martınez, J. Quiroga, R. Abonia, A. Gutierrez, M. Nogueras and J. Cobo, *J. Heterocycl. Chem.*, **46**, 756 (2009).
09JHC782	H. Maruoka, E. Masumoto, T. Eishima, F. Okabe, S. Nishida, Y. Yoshimura, T. Fujioka and K. Yamagata, *J. Heterocycl. Chem.*, **46**, 782 (2009).
09JHC849	J.-H. Peng, W.-J. Hao and S.-J. Tu, *J. Heterocycl. Chem.*, **46**, 849 (2009).
09JHC903	S.-Z. Xu, M.-H. Cao, C.-S. Chen and M.-W. Ding, *J. Heterocycl. Chem.*, **46**, 903 (2009).
09JHC919	P. Shanmugam, M. Damodiran, K. Selvakumar and P.T. Perumal, *J. Heterocycl. Chem.*, **46**, 919 (2009).
09JHC971	D.-Q. Shi, S.-F. Rong, G.-L. Dou and M.-M. Wang, *J. Heterocycl. Chem.*, **46**, 971 (2009).
09JHC1177	S. Fadel, F. Suzenet, A. Hafid, E.M. Rakib, M. Khouili, M.D. Pujol and G. Guillaumet, *J. Heterocycl. Chem.*, **46**, 1177 (2009).
09JHC1185	M. Tao, C.H. Park, K. Josef and R.L. Hudkins, *J. Heterocycl. Chem.*, **46**, 1185 (2009).
09JHC1235	Y.-A. Chang and H. Chang, *J. Heterocycl. Chem.*, **46**, 1235 (2009).
09JHC1247	M.A.P. Martins, R.L. Peres, C.P. Frizzo, E. Scapin, D.N. Moreira, G.F. Fiss, N. Zanatta and H.G. Bonacorso, *J. Heterocycl. Chem.*, **46**, 1247 (2009).
09JHC1309	E.C. Creencia, M. Kosaka, T. Muramatsu, M. Kobayashi, T. Iizuka and T. Horaguchi, *J. Heterocycl. Chem.*, **46**, 1309 (2009).
09JHC1408	C.P. Medina, C. Lopez, R.M. Claramunt and J. Elguero, *J. Heterocycl. Chem.*, **46**, 1408 (2009).

09JHC1413	S.A. Komykhov, K.S. Ostras, K.M. Kobzar, V.I. Musatov and S.M. Desenko, *J. Heterocycl. Chem.*, **46**, 1413 (2009).
09JOC163	G. Bentabed-Ababsa, F. Blanco, A. Derdour, F. Mongin, F. Trecourt, G. Queguiner, R. Ballesteros and B. Abarca, *J. Org. Chem.*, **74**, 163 (2009).
09JOC396	D.L. Browne, J.B. Taylor, A. Plant and J.P.A. Harrity, *J. Org. Chem.*, **74**, 396 (2009).
09JOC710	C.G. Yan, Q.F. Wang, X.K. Song and J. Sun, *J. Org. Chem.*, **74**, 710 (2009).
09JOC789	G.L. Beutner, J.T. Kuethe, M.M. Kim and N. Yasuda, *J. Org. Chem.*, **74**, 789 (2009).
09JOC891	Z.S. Sales and N.S. Mani, *J. Org. Chem.*, **74**, 891 (2009).
09JOC1282	G. Sarodnick, T. Linker, M. Heydenreich, A. Koch, I. Starke, S. Furstenberg and E. Kleinpeter, *J. Org. Chem.*, **74**, 1282 (2009).
09JOC1364	F. Jafarpour and P.T. Ashtiani, *J. Org. Chem.*, **74**, 1364 (2009).
09JOC1394	Y. Yamamoto, H. Mizuno, T. Tsuritani and T. Mase, *J. Org. Chem.*, **74**, 1394 (2009).
09JOC2004	V. Declerck, L. Toupet, J. Martinez and F. Lamaty, *J. Org. Chem.*, **74**, 2004 (2009).
09JOC2217	S. Murru, B.K. Patel, J. Le Bras and J. Muzart, *J. Org. Chem.*, **74**, 2217 (2009).
09JOC2460	J. Heppekausen, T.M. Klapotke and S.M. Sproll, *J. Org. Chem.*, **74**, 2460 (2009).
09JOC2502	M.R. Sheets, A. Li, E.A. Bower, A.R. Weigel, M.P. Abbott, R.M. Gallo, A.A. Mitton and D.A. Klumpp, *J. Org. Chem.*, **74**, 2502 (2009).
09JOC2627	M. Krasavin, S. Shkavrov, V. Parchinsky and K. Bukhryakov, *J. Org. Chem.*, **74**, 2627 (2009).
09JOC3079	M.K. Singh and M.K. Lakshman, *J. Org. Chem.*, **74**, 3079 (2009).
09JOC3566	F. Shibahara, R. Sugiura, E. Yamaguchi, A. Kitagawa and T. Murai, *J. Org. Chem.*, **74**, 3566 (2009).
09JOC3612	C. Chowdhury, S. Mukherjee, B. Das and B. Achari, *J. Org. Chem.*, **74**, 3612 (2009).
09JOC4318	R.A. Youcef, M.D. Santos, S. Roussel, J.-P. Baltaze, N. Lubin-Germain and J. Uziel, *J. Org. Chem.*, **74**, 4318 (2009).
09JOC4414	J. Zhang and C.-W.T. Chang, *J. Org. Chem.*, **74**, 4414 (2009).
09JOC5553	H. Comas, G. Bernardinelli and D. Swinnen, *J. Org. Chem.*, **74**, 5553 (2009).
09JOC5618	X. Lv and W. Bao, *J. Org. Chem.*, **74**, 5618 (2009).
09JOC5675	H.-G. Lee, J.-E. Won, M.-J. Kim, S.-E. Park, K.-J. Jung, B.R. Kim, S.-G. Lee and Y.-J. Yoon, *J. Org. Chem.*, **74**, 5675 (2009).
09JOC5742	X. Deng, H. McAllister and N.S. Mani, *J. Org. Chem.*, **74**, 5742 (2009).
09JOC6163	D. Almas, D.A. Alonso, E. Gomez-Bengoa and C. Najera, *J. Org. Chem.*, **74**, 6163 (2009).
09JOC6331	D.J. Slade, N.F. Pelz, W. Bodnar, J.W. Lampe and P.S. Watson, *J. Org. Chem.*, **74**, 6331 (2009).
09JOC6371	Q. Liao, Y. Wang, L. Zhang and C. Xi, *J. Org. Chem.*, **74**, 6371 (2009).
09JOC6479	L.J. Ingram, A. Desoky, A.M. Ali and S.D. Taylor, *J. Org. Chem.*, **74**, 6479 (2009).
09JOC6837	G. Pourceau, A. Meyer, J.-J. Vasseur and F. Morvan, *J. Org. Chem.*, **74**, 6837 (2009).
09JOC7046	S. Kumar, H. Ila and H. Junjappa, *J. Org. Chem.*, **74**, 7046 (2009).
09JOC7165	A.R. Katritzky, S.R. Tala, N.E. Abo-Dya, K. Gyanda, B.E.-D.M. El-Gendy, Z.K. Abdel-Samii and P.J. Steel, *J. Org. Chem.*, **74**, 7165 (2009).
09JOC7588	V.S. Sudhir, N.Y.P. Kumar, R.B. Baig and S. Chandrasekaran, *J. Org. Chem.*, **74**, 7588 (2009).
09JOC7595	R. Noel, X. Song, R. Jiang, M.J. Chalmers, P.R. Griffin and T.M. Kamenecka, *J. Org. Chem.*, **74**, 7595 (2009).
09JOC7974	X. Diao, Y. Wang, Y. Jiang and D. Ma, *J. Org. Chem.*, **74**, 7974 (2009).
09JOC8531	A.K. Ghosh and B. Zajc, *J. Org. Chem.*, **74**, 8531 (2009).
09JOC8669	M.S. Sandbhor, J.A. Key, I.S. Strelkov and C.W. Cairo, *J. Org. Chem.*, **74**, 8669 (2009).
09JOC8719	P. Saha, T. Ramana, N. Purkait, M.A. Ali, R. Paul and T. Punniyamurthy, *J. Org. Chem.*, **74**, 8719 (2009).
09JOC9214	L. Baragwanath, C.A. Rose, K. Zeitler and S.J. Connon, *J. Org. Chem.*, **74**, 9214 (2009).
09JOC9229	C. Laroche and S.M. Kerwin, *J. Org. Chem.*, **74**, 9229 (2009).

09JOC9470	X. Jia, D. Yang, W. Wang, F. Luo and J. Cheng, *J. Org. Chem.*, **74,** 9470 (2009).
09JOC9486	B. Jiang, X. Wang, F. Shi, S.-J. Tu, T. Ai, A. Ballew and G. Li, *J. Org. Chem.*, **74,** 9486 (2009).
09JOC9570	K. Hirano, A.T. Biju and F. Glorius, *J. Org. Chem.*, **74,** 9570 (2009).
09JOC9573	M.T. Molina, C. Navarro, A. Moreno and A.G. Csaky, *J. Org. Chem.*, **74,** 9573 (2009).
09OBC1097	E. Dijkum, R. Danac, D.J. Hughes, R. Wood, A. Rees, B.L. Wilkinson and A.J. Fairbanks, *Org. Biomol. Chem.*, **7,** 1097 (2009).
09OBC1374	F. Seela, H. Xiong, P. Leonard and S. Budow, *Org. Biomol. Chem.*, **7,** 1374 (2009).
09OBC1821	R. Srinivasan, L.P. Tan, H. Wu, P.-Y. Yang, K.A. Kalesh and S.Q. Yao, *Org. Biomol. Chem.*, **7,** 1821 (2009).
09OBC1921	R. Li, D.J. Jansen and A. Datta, *Org. Biomol. Chem.*, **7,** 1921 (2009).
09OBC2214	J. Kerherve, C. Botuh and J. Dubois, *Org. Biomol. Chem.*, **7,** 2214 (2009).
09OBC2804	Y. Hari, R. Sone and T. Aoyama, *Org. Biomol. Chem.*, **7,** 2804 (2009).
09OBC2933	M.K. Lakshman and J. Frank, *Org. Biomol. Chem.*, **7,** 2933 (2009).
09OBC3421	M. Klein, P. Diner, D. Dorin-Semblat, C. Doerig and M. Grøtli, *Org. Biomol. Chem.*, **7,** 3421 (2009).
09OBC3584	S.E. O'Toole and S.J. Connon, *Org. Biomol. Chem.*, **7,** 3584 (2009).
09OBC4052	R.S. Foster, J. Huang, J.F. Vivat, D.L. Browne and J.P.A. Harrity, *Org. Biomol. Chem.*, **7,** 4052 (2009).
09OBC4062	M. Yoshida, Y. Katagiri, W.-B. Zhu and K. Shishido, *Org. Biomol. Chem.*, **7,** 4062 (2009).
09OBC4241	S. Wei, B. Liu, D. Zhao, Z. Wang, J. Wu, J. Lan and J. You, *Org. Biomol. Chem.*, **7,** 4241 (2009).
09OBC4352	J.-W. Xie, Z. Wang, W.-J. Yang, L.-C. Kong and D.-C. Xu, *Org. Biomol. Chem.*, **7,** 4352 (2009).
09OBC4481	S.A. Diab, A. Hienzch, C. Lebargy, S. Guillarme, E. Pfund and T. Lequeux, *Org. Biomol. Chem.*, **7,** 4481 (2009).
09OBC4641	Z. Chen, M. Su, X. Yu and J. Wu, *Org. Biomol. Chem.*, **7,** 4641 (2009).
09OL233	A. Porzelle, M.D. Woodrow and N.C.O. Tomkinson, *Org. Lett.*, **11,** 233 (2009).
09OL907	Y. Li, K. Ding and C.A. Sandoval, *Org. Lett.*, **11,** 907 (2009).
09OL947	B.J. Kotecki, D.P. Fernando, A.R. Haight and K.A. Lukin, *Org. Lett.*, **11,** 947 (2009).
09OL1587	F. Zhang and J.E. Moses, *Org. Lett.*, **11,** 1587 (2009).
09OL1623	G.-Q. Li, L.-X. Dai and S.-L. You, *Org. Lett.*, **11,** 1623 (2009).
09OL1757	K. Asano and S. Matsubara, *Org. Lett.*, **11,** 1757 (2009).
09OL2039	A.J. Blacker, M.M. Farah, M.I. Hall, S.P. Marsden, O. Saidi and J.M.J. Williams, *Org. Lett.*, **11,** 2039 (2009).
09OL2097	B.S. Gerstenberger, M.R. Rauckhorst and J.T. Starr, *Org. Lett.*, **11,** 2097 (2009).
09OL2219	K. Harju, J. Vesterinen and J. Yli-Kauhaluoma, *Org. Lett.*, **11,** 2219 (2009).
09OL2760	J.S. Oakdale, D.M. Solano, J.C. Fettinger, M.J. Haddadin and M.J. Kurth, *Org. Lett.*, **11,** 2760 (2009).
09OL2840	O.A. Attanasi, E. Caselli, P. Davoli, G. Favi, F. Mantellini, C. Ori and F. Prati, *Org. Lett.*, **11,** 2840 (2009).
09OL2856	Q. Liu and T. Rovis, *Org. Lett.*, **11,** 2856 (2009).
09OL3008	J. Li, M. Hu and S.Q. Yao, *Org. Lett.*, **11,** 3008 (2009).
09OL3024	J. Li, D. Wang, Y. Zhang, J. Li and B. Chen, *Org. Lett.*, **11,** 3024 (2009).
09OL3072	T. Kawano, T. Yoshizumi, K. Hirano, T. Satoh and M. Miura, *Org. Lett.*, **11,** 3072 (2009).
09OL3294	L. Liang, Z. Li and X. Zhou, *Org. Lett.*, **11,** 3294 (2009).
09OL3326	C. Despotopoulou, L. Klier and P. Knochel, *Org. Lett.*, **11,** 3326 (2009).
09OL3354	J. Wu, B. Ni and A.D. Headley, *Org. Lett.*, **11,** 3354 (2009).
09OL3382	C.C. Palsuledesai, S. Murru, S.K. Sahoo and B.K. Patel, *Org. Lett.*, **11,** 3382 (2009).
09OL3570	Y. Wang and Q. Lin, *Org. Lett.*, **11,** 3570 (2009).
09OL3934	S. Kobayashi, T. Kinoshita, H. Uehara, T. Sudo and I. Ryu, *Org. Lett.*, **11,** 3934 (2009).
09OL4018	A.P. Piccionello, A. Pace, S. Buscemi and N. Vivona, *Org. Lett.*, **11,** 4018 (2009).
09OL4029	X.-N. Wang, P.-L. Shao, H. Lv and S. Ye, *Org. Lett.*, **11,** 4029 (2009).
09OL4270	B. Haag, Z. Peng and P. Knochel, *Org. Lett.*, **11,** 4270 (2009).

09OL4680	S. Ozcubukcu, E. Ozkai, C. Jimeno and M.A. Pericas, *Org. Lett.*, **11**, 4680 (2009).
09OL4922	L. Ackermann and R. Vicente, *Org. Lett.*, **11**, 4922 (2009).
09OL5026	X.-j. Wang, L. Zhang, H. Lee, N. Haddad, D. Krishnamurthy and C.H. Senanayake, *Org. Lett.*, **11**, 5026 (2009).
09OL5054	K.W. Hunt, D.A. Moreno, N. Suiter, C.T. Clark and G. Kim, *Org. Lett.*, **11**, 5054 (2009).
09OL5078	F. Lehmann, T. Koolmeister, L.R. Odell and M. Scobie, *Org. Lett.*, **11**, 5078 (2009).
09OL5118	J. Board, J.-X. Wang, A.P. Crew, M. Jin, K. Foreman, M.J. Mulvihill and V. Snieckus, *Org. Lett.*, **11**, 5118 (2009).
09OL5178	Q. Wang and S.L. Schreiber, *Org. Lett.*, **11**, 5178 (2009).
09OL5214	S. Lee and S.B. Park, *Org. Lett.*, **11**, 5214 (2009).
09OL5270	D.J. Lee, K. Mandal, P.W.R. Harris, M.A. Brimble and S.B.H. Kent, *Org. Lett.*, **11**, 5270 (2009).
09OL5490	X.-j. Wang, K. Sidhu, L. Zhang, S. Campbell, N. Haddad, D.C. Reeves, D. Krishnamurthy and C.H. Senanayake, *Org. Lett.*, **11**, 5490 (2009).
09OPRD106	E.K. Woodman, J.G.K. Chaffey, P.A. Hopes, D.R.J. Hose and J.P. Gilday, *Org. Proc. Res. Devel.*, **13**, 106 (2009).
09S133	C.E. Jamookeeah, C.D. Beadle and J.P.A. Harrity, *Synthesis*, 133 (2009).
09S271	J. Koubachi, S. Berteina-Raboin, A. Mouaddib and G. Guillaumet, *Synthesis*, 271 (2009).
09S325	A. Marek, J. Kulhánek and F. Bures, *Synthesis*, 325 (2009).
09S431	M. Nayak, S. Kanojiya and S. Batra, *Synthesis*, 431 (2009).
09S921	P.S. Luo, F. Wang, J.-H. Li, R.-Y. Tang and P. Zhong, *Synthesis*, 921 (2009).
09S1297	N.T. Pokhodylo, V.S. Matiychuk and M.D. Obushak, *Synthesis*, 1297 (2009).
09S1341	F. Algi and M. Balci, *Synthesis*, 1341 (2009).
09S1431	M.-W. Chen, X.-G. Zhang, P. Zhong and M.-L. Hu, *Synthesis*, 1431 (2009).
09S1663	A. Kumar, M. Parshad, R.K. Gupta and D. Kumar, *Synthesis*, 1663 (2009).
09S1673	M.I. Naumov, A.V. Nuchev, N.S. Sitnikov, Y.B. Malysheva, A.S. Shavyrin, I.P. Beletskay, A.E. Gavryushin, S. Combes and A.Y. Fedorov, *Synthesis*, 1673 (2009).
09S1715	A. Heim-Riether and K.R. Gipson, *Synthesis*, 1715 (2009).
09S2175	J. Roh, T.V. Artamonova, K. Vavrova, G.I. Koldobskii and A. Hrabalek, *Synthesis*, 2175 (2009).
09S2249	V.M. Muzalevskiy, V.G. Nenajdenko, A.V. Shastin, E.S. Balenkova and G. Haufe, *Synthesis*, 2249 (2009).
09S2316	D. Bottalico, V. Fiandanese, G. Marchese and A. Punzi, *Synthesis*, 2316 (2009).
09S2319	G. Bratulescu, *Synthesis*, 2319 (2009).
09S2321	N.T. Pokhodylo, V.S. Matiychuk and M.D. Obushak, *Synthesis*, 2321 (2009).
09S2729	A. Khoramabadi-zad, A. Shiri, M.A. Zolfigol and S. Mallakpourb, *Synthesis*, 2729 (2009).
09S2741	N.T. Pokhodylo, Y.O. Teslenko, V.S. Matiychuk and M.D. Obushak, *Synthesis*, 2741 (2009).
09S3067	M.N.S. Rad, A. Khalafi-Nezhad, S. Behrouz, Z. Asrari, M. Behrouz and Z. Amini, *Synthesis*, 3067 (2009).
09S3094	S.M. Spinella, Z.-H. Guan, J. Chen and X. Zhang, *Synthesis*, 3094 (2009).
09S3143	M. Petrini and R.R. Shaikh, *Synthesis*, 3143 (2009).
09S3150	M.D. Crozet, L. Zink, V. Remusat, C. Curti and P. Vanelle, *Synthesis*, 3150 (2009).
09S3293	S.K. Guchhait, C. Madaan and B.S. Thakkar, *Synthesis*, 3293 (2009).
09S3527	K. Lőrincz, P. Kele and Z. Novák, *Synthesis*, 3527 (2009).
09S3579	O. Artyushin, S.N. Osipov, G.-V. Röschenthaler and I.L. Odinets, *Synthesis*, 3579 (2009).
09S3603	B.A. Trofimov, A.V. Ivanov, E.V. Skital'tseva, A.M. Vasil'tsov, I.A. Ushakov, K.B. Petrushenko and A.I. Mikhaleva, *Synthesis*, 3603 (2009).
09S3661	C. Despotopoulou, C. Gignoux, D. McConnell and P. Knochel, *Synthesis*, 3661 (2009).
09S3853	V. Fiandanese, D. Bottalico, G. Marchese, A. Punzi, M.R. Quarta and M. Fittipaldi, *Synthesis*, 3853 (2009).
09S3975	J. Shah, S.S. Khan, H. Blumenthal and J. Liebscher, *Synthesis*, 3975 (2009).
09SC102	A. Shaabani, A. Maleki and M. Behnam, *Synth. Commun.*, **39**, 102 (2009).

09SC175	B.C. Raju, N.D. Theja and J.A. Kumar, *Synth. Commun.*, **39**, 175 (2009).
09SC316	R. Pundeer, P. Ranjan, K. Pannu and O. Prakash, *Synth. Commun.*, **39**, 316 (2009).
09SC395	V.V. Sureshbabu, S.A. Naik and G. Nagendra, *Synth. Commun.*, **39**, 395 (2009).
09SC426	G. Venkateshwarlu, K.C. Rajanna and P.K. Saiprakash, *Synth. Commun.*, **39**, 426 (2009).
09SC492	X. Jing, X. Pan, Z. Li, Y. Shi and C. Yan, *Synth. Commun.*, **39**, 492 (2009).
09SC830	X. Zhang, X. Yang and S. Zhang, *Synth. Commun.*, **39**, 830 (2009).
09SC980	S.D. Sharma and D. Konwar, *Synth. Commun.*, **39**, 980 (2009).
09SC1002	M. Bakherad, A. Keivanloo and M. Hashemi, *Synth. Commun.*, **39**, 1002 (2009).
09SC1847	H.M. Dalloul, *Synth. Commun.*, **39**, 1847 (2009).
09SC2169	R. Aggarwal and R. Kumar, *Synth. Commun.*, **39**, 2169 (2009).
09SC2339	A.T. Khan, T. Parvin and L.H. Choudhury, *Synth. Commun.*, **39**, 2339 (2009).
09SC2585	D. Ellis, *Synth. Commun.*, **39**, 2585 (2009).
09SC2982	T. Yamashita, S. Yamada, Y. Yamazaki and H. Tanaka, *Synth. Commun.*, **39**, 2982 (2009).
09SC3232	M.R. Mohammadizadeh, A. Hasaninejad and M. Bahramzadeh, *Synth. Commun.*, **39**, 3232 (2009).
09SC3546	S.S. Kottawar, S.A. Siddiqui, V.P. Chavan, W.N. Jadhav and S.R. Bhusare, *Synth. Commun.*, **39**, 3546 (2009).
09SC3816	Z. Li and Y. Zhao, *Synth. Commun.*, **39**, 3816 (2009).
09SC4122	A. Kusakiewicz-Dawid, L. Gorecki, E. Masiukiewicz and B. Rzeszotarska, *Synth. Commun.*, **39**, 4122 (2009).
09SC4264	A. Ghorbani-Choghamarani, Z. Chenani and S. Mallakpour, *Synth. Commun.*, **39**, 4264 (2009).
09SC4282	R. Duddu, P.R. Dave, R. Damavarapu, R. Surapaneni and D. Parrish, *Synth. Commun.*, **39**, 4282 (2009).
09SC4479	G. Venkateshwarlu, A. Premalatha, K.C. Rajanna and P.K. Saiprakash, *Synth. Commun.*, **39**, 4479 (2009).
09SL55	M.J. Khoshkholgh, S. Balalaie, H.R. Bijanzadeh and J.H. Gross, *Synlett*, 55 (2009).
09SL63	J. Sluiter and J. Christoffers, *Synlett*, 63 (2009).
09SL260	M. Nikulnikov, S. Tsirulnikov, V. Kysil, A. Ivachtchenko and M. Krasavin, *Synlett*, 260 (2009).
09SL399	K. Kumar, G. Patel and V.B. Reddy, *Synlett*, 399 (2009).
09SL569	K. Bahrami, M.M. Khodaei and F. Naali, *Synlett*, 569 (2009).
09SL607	C. Jin, C. Liu and W. Su, *Synlett*, 607 (2009).
09SL615	F. Crestey, E. Lohou, S. Stiebing, V. Collot and S. Rault, *Synlett*, 615 (2009).
09SL628	S. Guchhait and C. Madaan, *Synlett*, 628 (2009).
09SL643	H. Hagiwara, H. Sasaki, T. Hoshi and T. Suzuki, *Synlett*, 643 (2009).
09SL1318	C. Nyffenegger, E. Pasquinet, F. Suzener, D. Poullain and G. Guillaumet, *Synlett*, 1318 (2009).
09SL1391	N.M. Smith, M.J. Gleaves, R. Jewell, M.W.D. Perry, M.J. Stocks and J.P. Stonehouse, *Synlett*, 1391 (2009).
09SL1409	S.G. Gouin and J. Kovensky, *Synlett*, 1409 (2009).
09SL1453	L.-Y. Wu, Y.-X. Xie, Z.-S. Chen, Y.-N. Niu and Y.-M. Liang, *Synlett*, 1453 (2009).
09SL1490	J.J. Devery, III, P.K. Mohanta, B.M. Casey and R.A. Flowers, II, *Synlett*, 1490 (2009).
09SL2002	A.S. Nagarajan and B.S.R. Reddy, *Synlett*, 2002 (2009).
09SL2023	J. She, Z. Jiang and Y. Wang, *Synlett*, 2023 (2009).
09SL2123	V. Malnuit, M. Duca, A. Manout, K. Bougrin and R. Benhida, *Synlett*, 2123 (2009).
09SL2162	K.P. Kaliappan, P. Kalanidhi and S. Mahapatra, *Synlett*, 2162 (2009).
09SL2328	B.F. Bonini, M.C. Franchini, D. Gentili, E. Locatelli and A. Ricci, *Synlett*, 2328 (2009).
09SL2461	Z. Zhong, X. Wang, L. Kong and X. Zhu, *Synlett*, 2461 (2009).
09SL2601	T.A. Kamali, D. Habibi and M. Nasrollahzadeh, *Synlett*, 2601 (2009).
09SL2676	A. Randall and F. Duval, *Synlett*, 2676 (2009).
09SL2977	S. Maisonneuve and J. Xie, *Synlett*, 2977 (2009).
09SL3163	Y. Jiang, C. Kuang and Q. Yang, *Synlett*, 3163 (2009).
09SL3263	M. Adib, S. Ansari, S. Feizi, J.A. Damavandi and P. Mirzaei, *Synlett*, 3263 (2009).

09SL3275	S.G. Hansen and H.H. Jensen, *Synlett*, 3275 (2009).
09SL3299	J. Zhu, H. Xie, Z. Chen, S. Li and Y. Wu, *Synlett*, 3299 (2009).
09T328	I. Çelik, N. Kanıskan and S. Kokten, *Tetrahedron*, **65**, 328 (2009).
09T644	A.S. Shawali and T.A. Farghaly, *Tetrahedron*, **65**, 644 (2009).
09T855	R. Jitchati, A.S. Batsanov and M.R. Bryce, *Tetrahedron*, **65**, 855 (2009).
09T909	M.C. Jahnke, M. Hussain, F. Hupka, T. Pape, S. Ali, F.E. Hahn and K.J. Cavell, *Tetrahedron*, **65**, 909 (2009).
09T1574	D. Vina, E. del Olmo, J.L. Lopez-Perez and A.S. Feliciano, *Tetrahedron*, **65**, 1574 (2009).
09T2660	S. Guillou, F.J. Bonhomme and Y.L. Janin, *Tetrahedron*, **65**, 2660 (2009).
09T2678	N.T. Pokhodylo, V.S. Matiychuk and M.D. Obushak, *Tetrahedron*, **65**, 2678 (2009).
09T3102	B.E. Maki, A. Chan, E.M. Phillips and K.A. Scheidt, *Tetrahedron*, **65**, 3102 (2009).
09T3492	A. Shaabani, A. Sarvary, A.H. Rezayan and S. Keshipour, *Tetrahedron*, **65**, 3492 (2009).
09T3529	S. Guillou, O. Nesmes, M.S. Ermolenko and Y.L. Janin, *Tetrahedron*, **65**, 3529 (2009).
09T3974	K.A. Dururgkar, R.G. Gonnade and C.V. Ramana, *Tetrahedron*, **65**, 3974 (2009).
09T4652	O. Ivashchuk and V.I. Sorokin, *Tetrahedron*, **65**, 4652 (2009).
09T5062	F. Shibahara, E. Yamaguchi, A. Kitagawa, A. Imai and T. Murai, *Tetrahedron*, **65**, 5062 (2009).
09T6156	A. Paul, J. Einsiedel, R. Waibel, F.W. Heinemann, K. Meyer and P. Gmeine, *Tetrahedron*, **65**, 6156 (2009).
09T6882	B. Khalili, T. Tondro and M.M. Hashemi, *Tetrahedron*, **65**, 6882 (2009).
09T7023	H.A. Ioannidou and P.A. Koutentis, *Tetrahedron*, **65**, 7023 (2009).
09T7151	D. Kralj, M. Friedrich, U. Groselj, S. Kiraly-Potpara, A. Meden, J. Wagger, G. Dahmann, B. Stanovnik and J. Svete, *Tetrahedron*, **65**, 7151 (2009).
09T7817	E.˙ Arbaciauskiene, G. Vilkauskaite, G.A. Eller, W. Holzer and A. Sackus, *Tetrahedron*, **65**, 7817 (2009).
09T7935	C. Li, J. Tang and J. Xie, *Tetrahedron*, **65**, 7935 (2009).
09T8367	C.-W. Tsai, S.-C. Yang, Y.-M. Liu and M.-J. Wu, *Tetrahedron*, **65**, 8367 (2009).
09T9772	J. Roger and H. Doucet, *Tetrahedron*, **65**, 9722 (2009).
09T10155	S. Samai, G.C. Nandi, P. Singh and M.S. Singh, *Tetrahedron*, **65**, 10155 (2009).
09T10175	O. Prakash, D. Sharma, R. Kamal, R. Kumar and R.R. Nair, *Tetrahedron*, **65**, 10175 (2009).
09T10573	V. Fiandanese, D. Bottalico, G. Marchese, A. Punzi and F. Capuzzolo, *Tetrahedron*, **65**, 10573 (2009).
09T10715	M. Nasrollahzadeh, D. Habibi, Z. Shahkarami and Y. Bayat, *Tetrahedron*, **65**, 10715 (2009).
09TL212	L.N. Aldrich, E.P. Lebois, L.M. Lewis, N.T. Nalywajko, C.M. Niswender, C.D. Weaver, P.J. Conn and C.W. Lindsley, *Tetrahedron Lett.*, **50**, 212 (2009).
09TL291	S. Dadiboyena, E.J. Valente and A.T. Hamme, II, *Tetrahedron Lett.*, **50**, 291 (2009).
09TL354	A. Kodimuthali, T.C. Nishad, P.L. Prasunamba and M. Pal, *Tetrahedron Lett.*, **50**, 354 (2009).
09TL383	S.K. Verma and L.V. LaFrance, *Tetrahedron Lett.*, **50**, 383 (2009).
09TL696	S.A. Raw and A.T. Turner, *Tetrahedron Lett.*, **50**, 696 (2009).
09TL1219	M.D. Rosen, Z.M. Simon, K.T. Tarantino, L.X. Zhao and M.H. Rabinowitz, *Tetrahedron Lett.*, **50**, 1219 (2009).
09TL1286	W.S. Chow and T.H. Chan, *Tetrahedron Lett.*, **50**, 1286 (2009).
09TL1377	J.M. Ralph, T.H. Faitg, D.J. Silva, Y. Feng, C.W. Blackledge and J.L. Adams, *Tetrahedron Lett.*, **50**, 1377 (2009).
09TL1399	T.R. Rheault, K.H. Donaldson and M. Cheung, *Tetrahedron Lett.*, **50**, 1399 (2009).
09TL1495	R.G. Jacob, L.G. Dutra, C.S. Radatz, S.R. Mendes, G. Perin and E.J. Lenardão, *Tetrahedron Lett.*, **50**, 1495 (2009).
09TL1571	K. Pattabiraman, R. El-Khouri, K. Modi, L.R. McGee and D. Chow, *Tetrahedron Lett.*, **50**, 1571 (2009).
09TL1667	J. Meng and P.-P. Kung, *Tetrahedron Lett.*, **50**, 1667 (2009).
09TL1780	R.P. Kale, M.U. Shaikh, G.R. Jadhav and C.H. Gill, *Tetrahedron Lett.*, **50**, 1780 (2009).
09TL2065	D. Kumar, V.B. Reddy and R.S. Varma, *Tetrahedron Lett.*, **50**, 2065 (2009).

09TL2124	A.V. Dolzhenko, G. Pastorin, A.V. Dolzhenko and W.K. Chui, *Tetrahedron Lett.*, **50**, 2124 (2009).
09TL2293	Y.-L. Zhong, M.G. Lindale and N. Yasuda, *Tetrahedron Lett.*, **50**, 2293 (2009).
09TL2358	F. Alonso, Y. Moglie, G. Radivoy and M. Yus, *Tetrahedron Lett.*, **50**, 2358 (2009).
09TL2443	S. He, L. Chen, Y.-N. Niu, L.-Y. Wu and Y.-M. Liang, *Tetrahedron Lett.*, **50**, 2443 (2009).
09TL2505	S.N. Mikhaylichenko, S.M. Patel, S. Dalili, A.A. Chesnyuk and V.N. Zaplishny, *Tetrahedron Lett.*, **50**, 2505 (2009).
09TL2577	S.V. Voitekhovich, P.N. Gaponik, A.S. Lyakhov, J.V. Filipova, A.G. Sukhanova, G.T. Sukhanov and O.A. Ivashkevich, *Tetrahedron Lett.*, **50**, 2577 (2009).
09TL2624	A. Regiec, H. Mastalarz, A. Mastalarz and A. Kochel, *Tetrahedron Lett.*, **50**, 2624 (2009).
09TL2668	D. Kundu, A. Majee and A. Hajra, *Tetrahedron Lett.*, **50**, 2668 (2009).
09TL2678	C. Chowdhury, A.K. Sasmal and P.K. Dutta, *Tetrahedron Lett.*, **50**, 2678 (2009).
09TL2911	A. Shaabani, M. Seyyedhamzeh, A. Maleki, M. Behnam and F. Rezazadeh, *Tetrahedron Lett.*, **50**, 2911 (2009).
09TL3046	I. Strakova, M. Turks and A. Strakovs, *Tetrahedron Lett.*, **50**, 3046 (2009).
09TL3088	P. Mizar and B. Myrboh, *Tetrahedron Lett.*, **50**, 3088 (2009).
09TL3098	P. Orsini, M. Menichincheri, E. Vanotti and A. Panzeri, *Tetrahedron Lett.*, **50**, 3098 (2009).
09TL3463	K.S. Kanyiva, F. Löbermann, Y. Nakao and T. Hiyama, *Tetrahedron Lett.*, **50**, 3463 (2009).
09TL3798	C. Salomé, M. Schmitt and J.-J. Bourguignon, *Tetrahedron Lett.*, **50**, 3798 (2009).
09TL3809	A.S. Kiselyov, E.L.P. Chekler, N.B. Chernisheva, L.K. Salamandra and V.V. Semenov, *Tetrahedron Lett.*, **50**, 3809 (2009).
09TL3955	J.C. Kaila, A.B. Baraiya, A.N. Pandya, H.B. Jalani, K.K. Vasu and V. Sudarsanam, *Tetrahedron Lett.*, **50**, 3955 (2009).
09TL4067	F. Shi, R. Mancuso and R.C. Larock, *Tetrahedron Lett.*, **50**, 4067 (2009).
09TL4101	T. Fujino, N. Yamazaki and H. Isobe, *Tetrahedron Lett.*, **50**, 4101 (2009).
09TL4107	M.N. Sakac, A.R. Gakovic, J.J. Csanadi, E.A. Djurendic, O. Klisuric, V. Kojic, G. Bogdanovic and K.M.P. Gaši, *Tetrahedron Lett.*, **50**, 4107 (2009).
09TL4167	N. Okamoto, K. Sakurai, M. Ishikura, K. Takeda and R. Yanada, *Tetrahedron Lett.*, **50**, 4167 (2009).
09TL4310	R. Dubey, N.W. Polaske, G.S. Nichol and B. Olenyuk, *Tetrahedron Lett.*, **50**, 4310 (2009).
09TL4389	A.I. Polyakov, V.A. Eryomina, L.A. Medvedeva, N.I. Tihonova, A.V. Listratova and L.G. Voskressensky, *Tetrahedron Lett.*, **50**, 4389 (2009).
09TL4435	M. Nasrollahzadeh, Y. Bayat, D. Habibi and S. Moshaee, *Tetrahedron Lett.*, **50**, 4435 (2009).
09TL4446	B.I. Usachev, D.L. Obydennov, M.I. Kodess and V.Y. Sosnovskikh, *Tetrahedron Lett.*, **50**, 4446 (2009).
09TL4677	H. Ankati and E. Biehl, *Tetrahedron Lett.*, **50**, 4677 (2009).
09TL4916	J.M. Crawforth and M. Paoletti, *Tetrahedron Lett.*, **50**, 4916 (2009).
09TL5194	C. Laroche and S.M. Kerwin, *Tetrahedron Lett.*, **50**, 5194 (2009).
09TL5218	D.S. Ermolat'ev, E.P. Svidritsky, E.V. Babaev and E. Van der Eycken, *Tetrahedron Lett.*, **50**, 5218 (2009).
09TL5267	S.R. Kotturi, J.S. Tan and M.J. Lear, *Tetrahedron Lett.*, **50**, 5267 (2009).
09TL5459	T.A. Kamali, M. Bakherad, M. Nasrollahzadeh, S. Farhangi and D. Habibi, *Tetrahedron Lett.*, **50**, 5459 (2009).
09TL5479	M.P. Curtis, M.F. Sammons and D.W. Piotrowski, *Tetrahedron Lett.*, **50**, 5479 (2009).
09TL5536	W. Li, Y. Ye, J. Zhang and R. Fan, *Tetrahedron Lett.*, **50**, 5536 (2009).
09TL5539	J. Huang, S.J.F. Macdonald, A.W.J. Cooper, G. Fisher and J.P.A. Harrity, *Tetrahedron Lett.*, **50**, 5539 (2009).
09TL5617	A.V. Dolzhenko, G. Pastorin, A.V. Dolzhenko and W.K. Chui, *Tetrahedron Lett.*, **50**, 5617 (2009).
09TL5733	S. Bardhan, K. Tabei, Z.-K. Wanb and T.S. Mansour, *Tetrahedron Lett.*, **50**, 5733 (2009).
09TL5868	H.W. Lee, A.S.C. Chan and F.Y. Kwong, *Tetrahedron Lett.*, **50**, 5868 (2009).

09TL5978	A. Gioiello, A. Khamidullina, M.C. Fulco, F. Venturoni, S. Zlotsky and R. Pellicciari, *Tetrahedron Lett.*, **50**, 5978 (2009).
09TL6029	J.S. Yadav, B.V.S. Reddy, G.M. Reddy and S.R. Anjum, *Tetrahedron Lett.*, **50**, 6029 (2009).
09TL6048	G. Li, R. Kakarla, S.W. Gerritz, A. Pendri and B. Ma, *Tetrahedron Lett.*, **50**, 6048 (2009).
09TL6223	T. Brady, K. Vu, J.R. Barber, S.C. Ng and Y. Zhou, *Tetrahedron Lett.*, **50**, 6223 (2009).
09TL6572	M. Radi, V. Bernardo, B. Bechi a, D. Castagnolo, M. Pagano and M. Botta, *Tetrahedron Lett.*, **50**, 6572 (2009).
09TL6783	P.R. Mullens, *Tetrahedron Lett.*, **50**, 6783 (2009).
09TL6826	S. Picon, A. Zaparucha and A. Al-Mourabit, *Tetrahedron Lett.*, **50**, 6826 (2009).
09TL7032	K. Varazo, C. Le Droumaguet, K. Fullard and Q. Wang, *Tetrahedron Lett.*, **50**, 7032 (2009).
09TL7217	J. Jackson and A. Xia, *Tetrahedron Lett.*, **50**, 7217 (2009).
09TL9592	Y.-Y. Huang, H.-C. Lin, K.-M. Cheng, W.-N. Su, K.-C. Sung, T.-P. Lin, J.-J. Huang, S.-K. Lin and F.F. Wong, *Tetrahedron Lett.*, **50**, 9592 (2009).

CHAPTER 5.5

Five-Membered Ring Systems: With N and S (Se) Atoms

Y.-J. Wu*, Bingwei V. Yang**
*Bristol Myers Squibb Company, 5 Research Parkway, Wallingford, CT 06492-7660, USA
yong-jin.wu@bms.com
**Bristol Myers Squibb Company, PO Box 4000, Princeton, NJ 08543-4000, USA
bingwei.yang@bms.com

5.5.1. INTRODUCTION

This review chapter focuses on the syntheses and reactions of five-membered heterocyclic ring systems containing nitrogen and sulfur (or selenium) (reported during 2009). The importance of these π-rich heterocycles in medicinal chemistry and natural products is also covered.

5.5.2. THIAZOLES

5.5.2.1 Synthesis of Thiazoles

The Hantzsch reaction discovered in 1889 remains one of the most reliable routes to thiazoles. In a representative example, reaction of α-bromomethylketone **2** with thioamide **1** in a solution of THF and ethanol provides tri(thiazole) **3** in good yield. However, this reaction generates one equivalent (eq.) of hydrogen bromide, which can cause significant loss of optical purity with substrates prone to epimerization under original Hantzsch conditions (refluxing ethanol). As a case in point, reaction of an amino acid derived thioamide **4** with an α-bromocarbonyl compound **5** in refluxing ethanol results in epimerization at the α-stereogenic center. The racemization issue can be overcome by carrying out the Hantzsch thiazole synthesis using the two- or three-step procedure, also called Holzapfel-Meyers-Nicolaou modification <07S3535; 07SL954>. This modification has become a standard protocol in the Hantzsch reactions of chiral substrates vulnerable to racemisation. Thus, cyclocondensation of thioamide **8** with bromide **5** provides the hydroxythiazoline intermediate **9**, which is then dehydrated by treatment with trifluoroacetic anhydride and pyridine to give thiazole **10** <09TA2027>. Exposure of bromide **5** to thiamide **11** leads to thiazoline **12** with no epimerization at the chiral center adjacent to the thiazole ring, and dehydration of **12** to thiazole **13** proceeds smoothly with trifluoroacetic anhydride and pyridine <09AG(E)1>.

Other applications of Hantzsch reaction include the trifluoromethylated thiazoles <09S2249>, and 4-substituted 2-aminothiazolo[4,5-d]pyridazinones <09TL5660>.

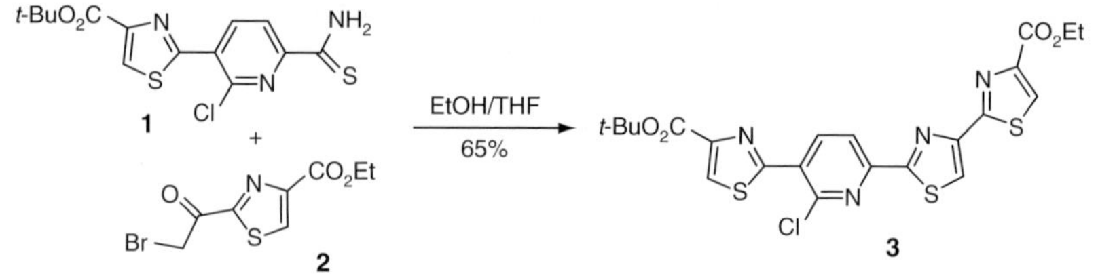

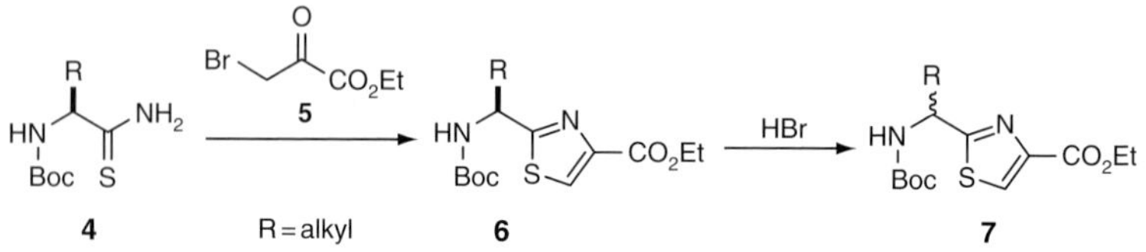

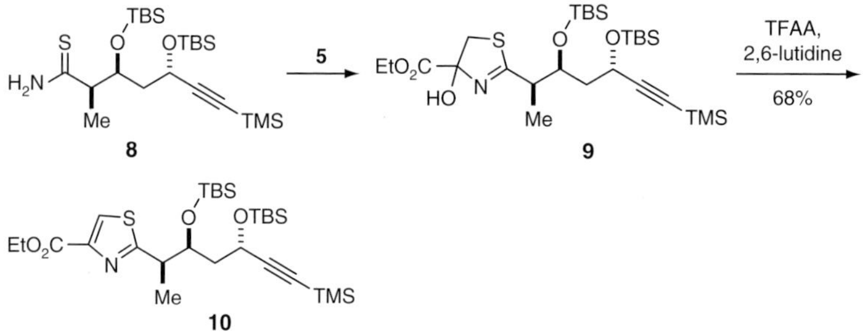

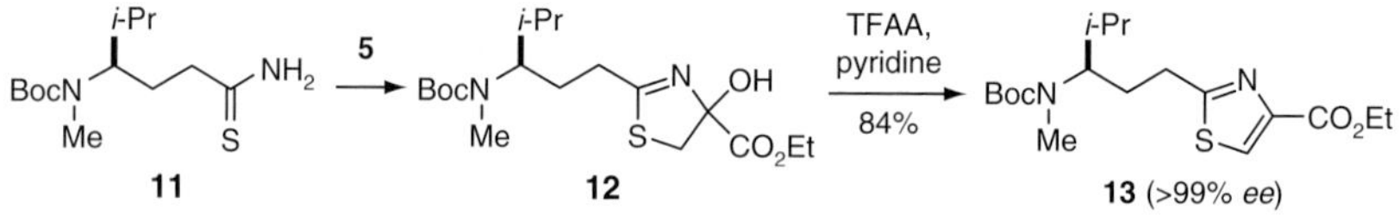

A novel synthesis of thiazoles employs 1,4-dioxaspiro[2.2]pentanes instead of α-halomethyl ketones as in typical Hantzsch reactions. Treatment of the 1,4-dioxaspiro[2.2]pentane derivative **15**, available from allene **14** via epoxidation, with thiobenzamide in chloroform gives thiazoline **16b** <09OL4672>. Under these conditions, the silyl group migrates to the adjacent oxygen. When the same reaction is performed in methanol, the nonsilyl thiazole **16a** is obtained. Dehydrative aromatization of both **16a** and **16b** affords a single thiazole **17**.

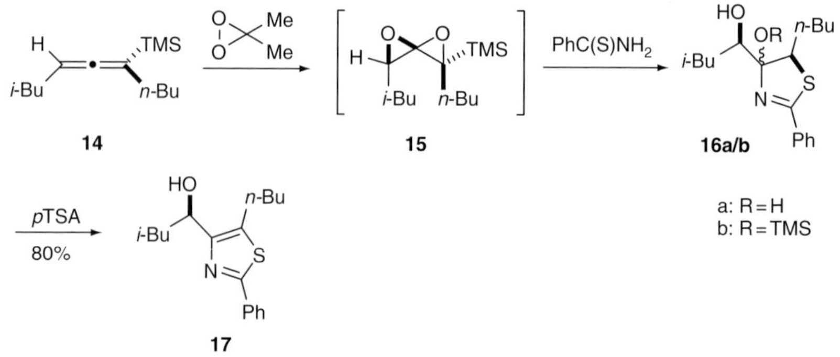

A series of trisubstituted thiazole derivatives **23** has been prepared via a one-pot four-step sequential procedure <09T2982>. Reaction of secondary amines with dimethyl cyanodithioimidocarbonate **18** generates **19**, which, upon treatment with sodium disulfide, provides thiolate **20**. This thiolate undergoes nucleophilic substitution with halide **21** to furnish intermediate **22**, and addition of potassium carbonate brings about cyclization to give thiazole **23**.

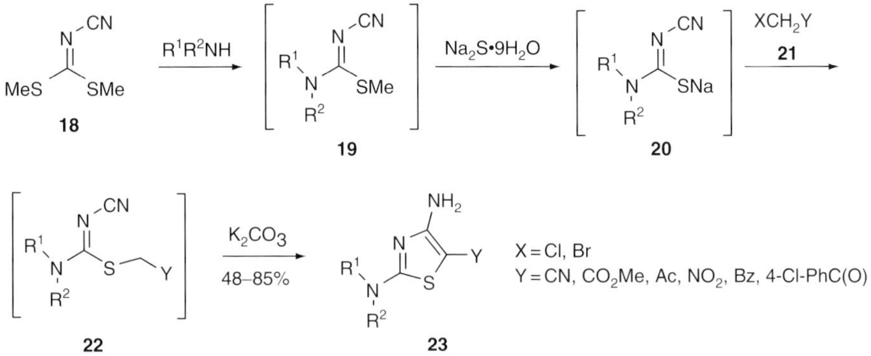

Thiazoles can also be derived from 1,4-dicarbonyl compounds, which are available through one-step four-component Ugi reactions <09JOC7084>. These reactions are conducted using concentrated aqueous ammonia as the amine component and trifluoroethanol as a nonnucleophilic solvent in order to suppress some known side reactions. The Ugi adducts, α-acylaminoketones **24**, are elaborated into a variety of N-substituted 5-aminothiazoles **25** by treatment with Lawesson's reagent. When 1-(isocyanomethyl)-2,4-dimethoxybenzene is used in the Ugi reaction, the resulting thiazole **27** can easily be hydrolyzed to provide the free 5-aminothiazole **28**. Thus, this methodology provides easy access to both N-substituted and free 5-aminothiazoles.

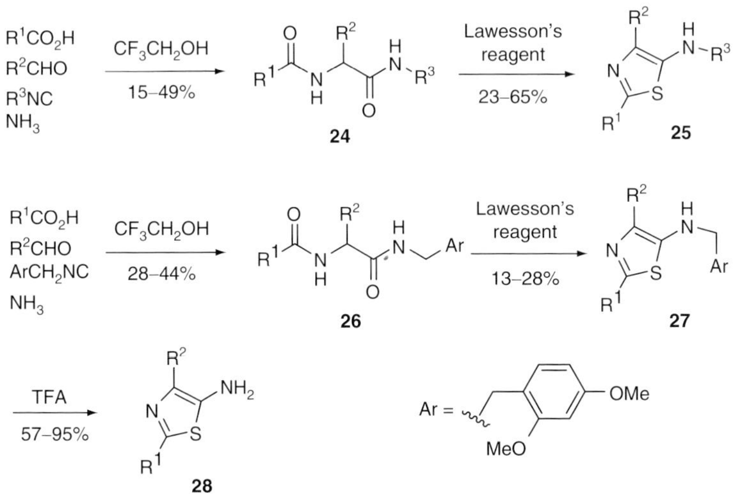

A solution versus fluorous versus solid-phase synthesis of 2,5-disubstituted thiazoles **31** also uses α-amido-β-ketoesters **29a/b/c** <09JOC8988>. The cyclization of fluorous α-amido-β-ketoesters **29c** with Lawesson's reagent gives fluorous thiazoles **30c** in 54-82% yields, which are slightly lower than those obtained with the analogous non-fluorous thiazoles **30a/b** (67-97%). Despite the relatively low yields, the fluorous synthesis offers several advantages, including convenient and fast purification for the intermediate compounds using the fluorous solid-phase extraction cartridges.

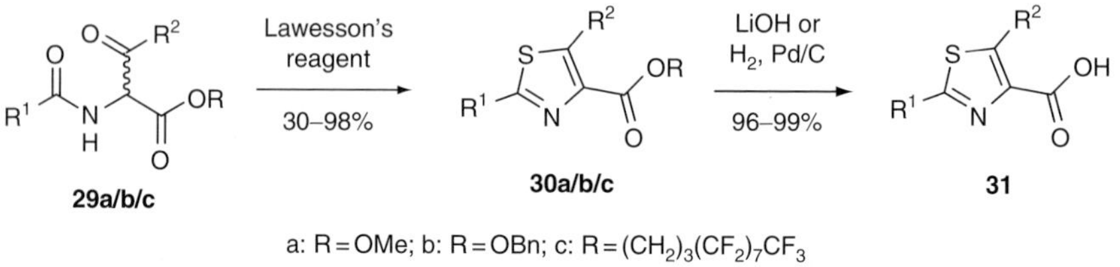

Lawesson's reagent
30–98%
LiOH or H_2, Pd/C
96–99%
29a/b/c
30a/b/c
31
a: R = OMe; b: R = OBn; c: R = $(CH_2)_3(CF_2)_7CF_3$

A new synthesis of substituted thiazoles **36** involves an intramolecular nucleophilic substitution of vinylic bromides **32** with the sulfur from the thioamide moiety at a sp^2 carbon followed by isomerisation <09TL3161>.

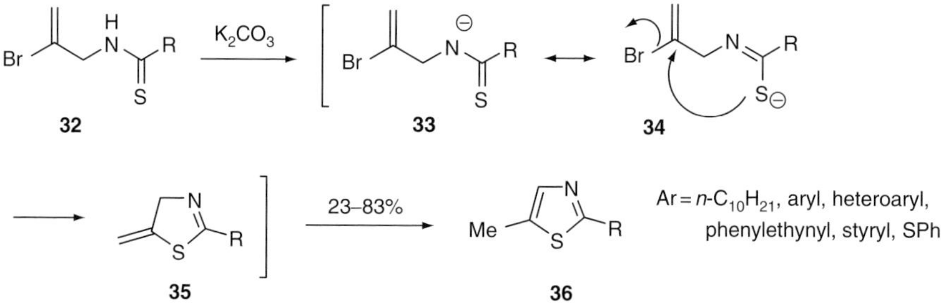

K_2CO_3
32
33
34
35
23–83%
36
Ar = n-$C_{10}H_{21}$, aryl, heteroaryl, phenylethynyl, styryl, SPh

Thiazoles can be prepared from thiazolines by oxidation with activated manganese dioxide as demonstrated in the formation of ketothiazole **38** <09SL1341>. Reaction of L-cysteine ethyl ester hydrochloride with methyl glyoxal affords thiazolidine **37**, which is converted into **38** upon treatment with manganese dioxide.

Methyl glyoxal, $NaHCO_3$
MnO_2
37
38

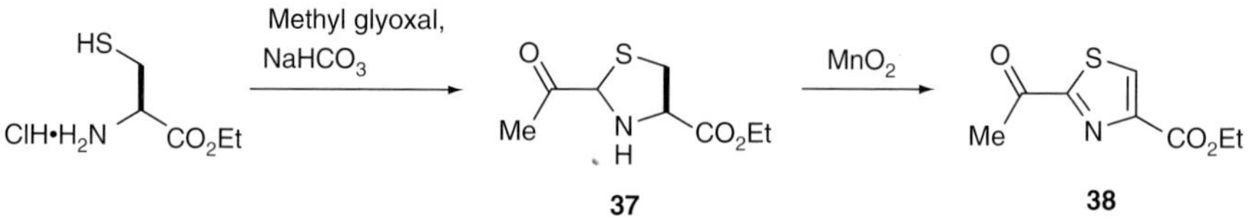

A new cycloaddition reaction of the 3-sulphanyl and 3-selanylpropargyl alcohols **39** with thioamides has been developed for the synthesis of multifunctionalized thiazoles **43** <09OL2952>. The reaction presumably proceeds through the propargyl cation **40**, which isomerizes to the propardienyl cation **41** which is stabilized by either the adjacent sulfur or selenium. Reaction of the propardienyl cation **41** with thiobenzamide gives the propardienylimine intermediate **42**, which undergoes a

5-*exo-trig* ring closure to form thiazole **43**. In this reaction, tetrabutylammonium hydrogensulfate serves as a scavenger to remove the eliminated hydroxyl group. Both the sulfanyl and selanyl are versatile functional groups for further chemical manipulations.

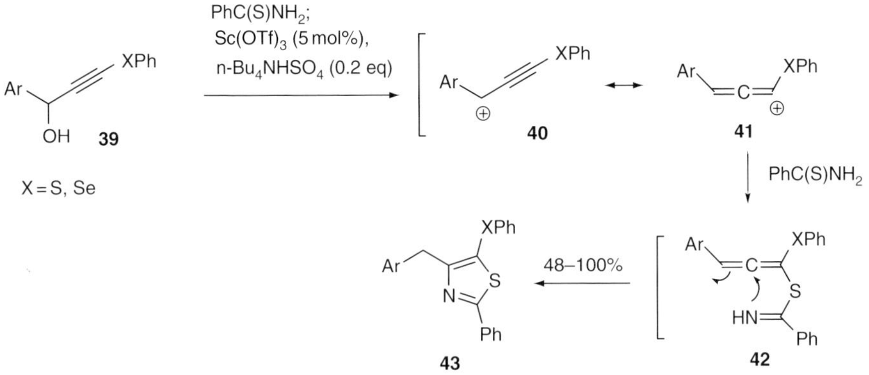

5.5.2.2 Synthesis of Fused Thiazoles

A practical and inexpensive synthesis of substituted benzothiazoles makes use of a copper-catalyzed coupling of 2-haloanilides with metal sulfides <09AGE4222>. The optimized conditions involve 3 eq. of sodium sulfide nonhydrate, 10% mol of copper(I) iodide and 80 °C as reaction temperature. Under these conditions, 2-substituted benzothiazoles are obtained in good yields. According to the proposed reaction mechanism, oxidative addition of copper(I) iodide to aryl halide **44** forms copper complex **45**, a ligand exchange with sodium sulfide leads to complex **46**, which undergoes reductive elimination to give the intermediate sodium benzenethiolate **47**. Ring closure of **47** affords benzothiazole **48**. This methodology has been extended to 2-bromoanilides **49** as substrates, but the reactions conditions are different (potassium sulfide, 140 °C).

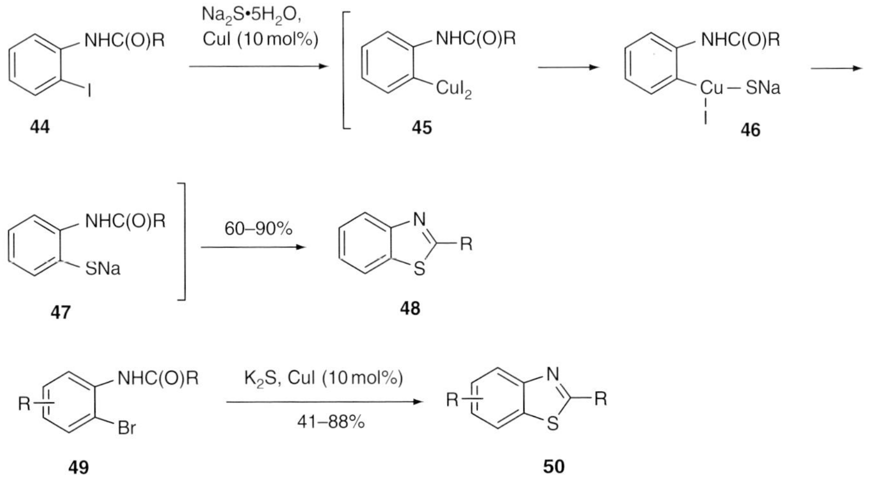

5.5.2.3 Synthesis of Thiazolines

Kelly's biomimetic methodology, first reported in 2003 <03AG(E)83>, has become one of the most effective routes to thiazolines. For example, treatment of β-S-trityl amide **51** with triflic anhydride and triphenylphosphine oxide provides thiazoline **53** in excellent yield <09CAJ111>. In this reaction, the phosphorus-activated amide carbonyl group undergoes nucleophilic attack by the cysteine thiol group to provide the thiazoline moiety (see **52**). Thiazoline **53** has been converted into the highly toxic marine natural product apratoxin.

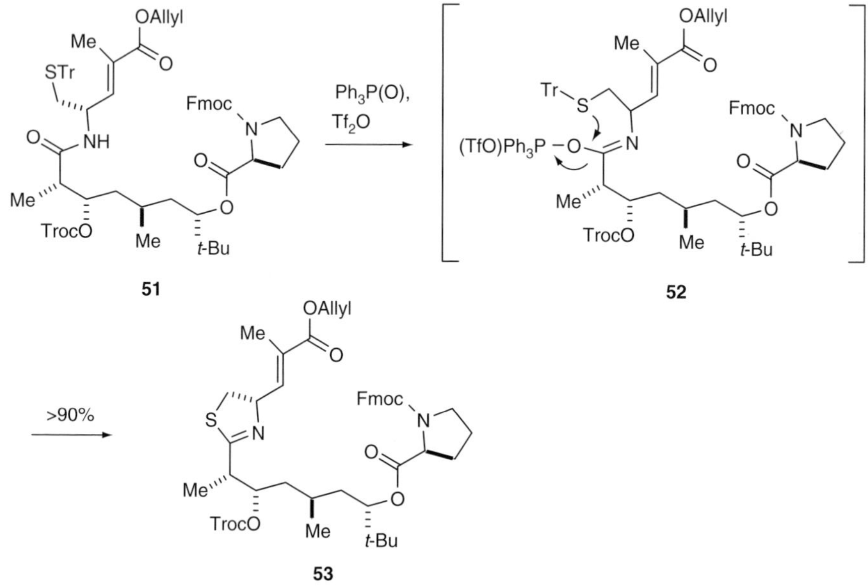

In addition to triflic anhydride and triphenylphosphine as used in Kelly's method, phosphonium anhydride **57** has also been developed as a mild dehydrating agent to convert β-S-trityl amides into thiazolines <09OBC739>. Exposure of trityl derivative **55a/b** to **57** brings about clean cyclization to give thiazolines **56a/b**. Reaction of β-thio amide **54a** with **57** provides thiazoline **56a** directly, but the yield (57%) is lower than that obtained with the two-step process.

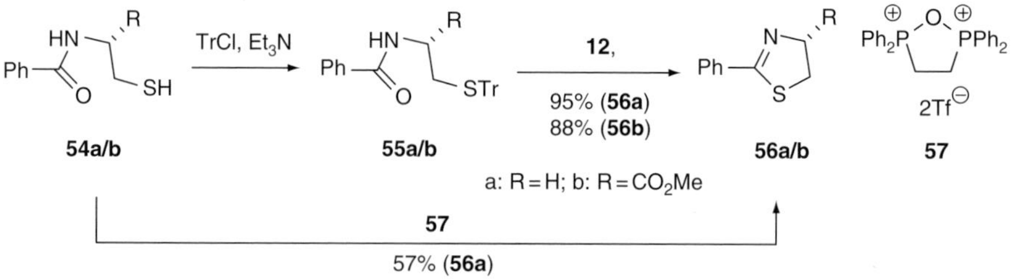

An unexpected thiazoline formation is observed in the synthetic studies on apratoxin <09CAJ111>. Reaction of amide **58** with Lawesson's reagent provides thioamide **59**, which undergoes intramolecular Michael addition to give thiazoline **60**. The yield of this reaction was not disclosed.

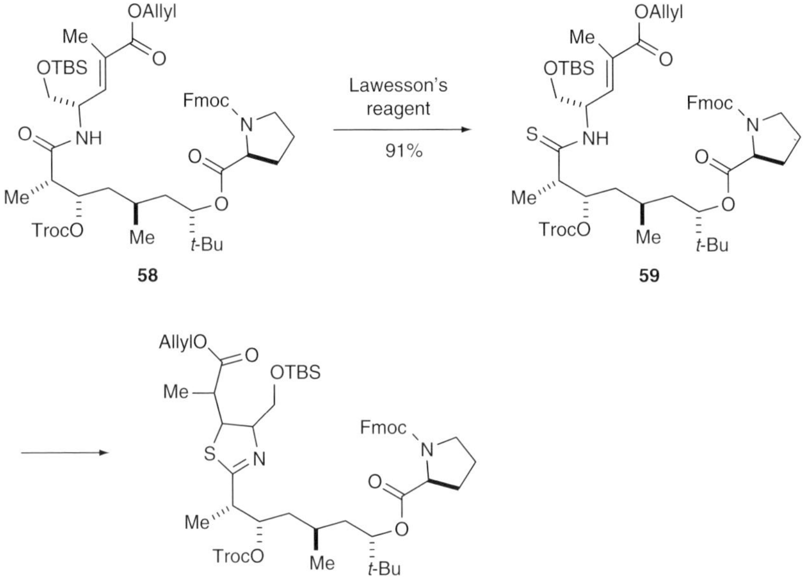

A full account of thiazoline formation via molybdenum(VI) oxides-catalyzed dehydrative cyclization of the cysteine derivatives has been published <09T2102>. The L-cysteine methyl ester derivative **61** is converted to thiazoline **62** in excellent yield using either $(NH_4)_2MoO_4$ or $(NH_4)_6Mo_7O_{24} \cdot 4H_2O$ or $MoO_2(acac)_2$. When the dipeptide substrate Cbz-L-Ala-L-Cys-OMe **63** is used as the substrate, both $(NH_4)_2MoO_4$ and $(NH_4)_6Mo_7O_{24} \cdot 4H_2O$ give poor yields. In contrast, a good yield is obtained with $MoO_2(acac)$, but the resulting product is an 82 : 18 mixture of the desired thiazoline **64** and its epimer **65**. Conceivably, the epimerization at α carbon can be attributed to the relatively high acidity of the catalyst, $MoO_2(acac)$. One of the strategies to reduce the acidity of $MoO_2(acac)_2$ is to introduce appropriate ligands onto molybdenum(VI) oxide. To this end, 2-ethyl-8-quinolinolato complex **68** is developed. This catalyst brings about clean conversion of cysteine-containing dipeptides **66** to the corresponding thiazolines **67** without epimerization.

$(NH_4)_2MoO_4$, 99%; $(NH_4)_6Mo_7O_{24}$•$4H_2O$, 96%; $MnO_2(acac)_2$, 81%

$(NH_4)_2MoO_4$, 26%; $(NH_4)_6Mo_7O_{24}$•$4H_2O$, 85%; $MnO_2(acac)_2$, 81%

R = Bn, Me; PG = Cbz, Boc, Fmoc *dr*: 94/6 to 98/2 **68**

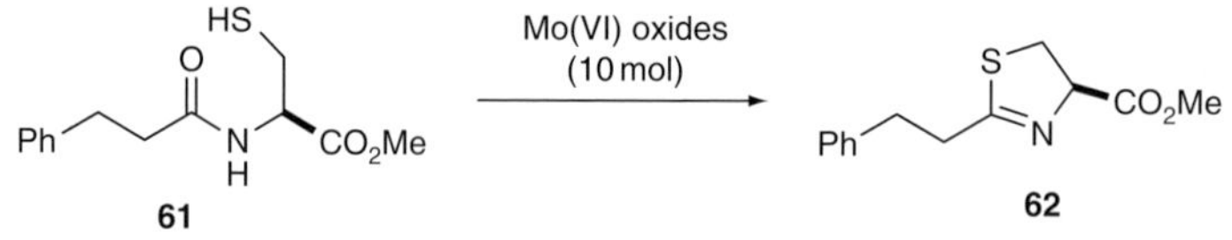

Pattenden's approach to thiazolines, first reported in 1993 <93T5359; 95T7321>, has been applied to the synthesis of largazole analogs <09OL1301>. Cyclocondensation of L-cysteine **69** with nitrile **70** in the presence of triethylamine results in thiazoline-thiazole acid **71** in quantitative yield. Chloro-nitrile **73** also undergoes condensation with α-methyl cysteine **72** to provide thiazoline acid **74** in moderate yield.

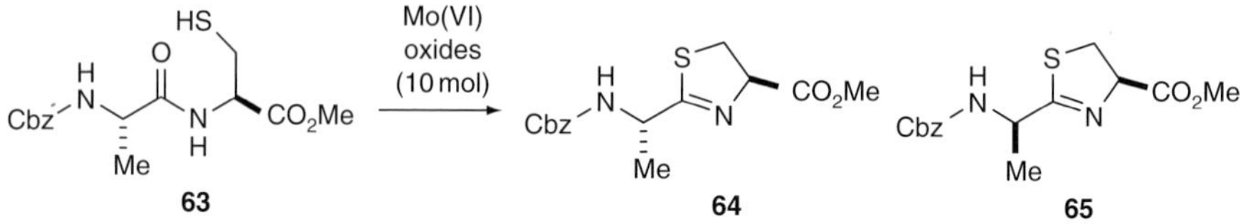

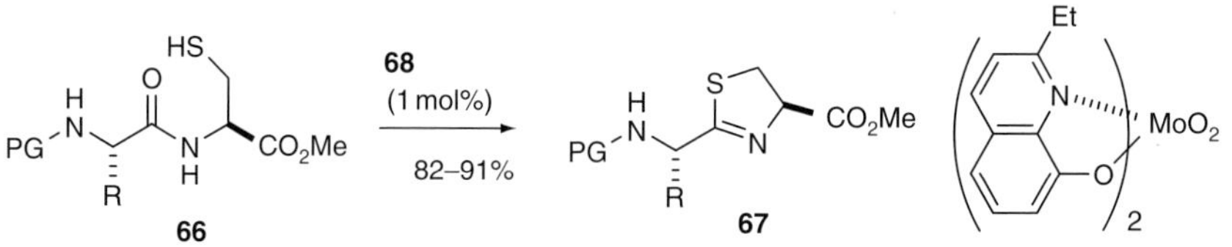

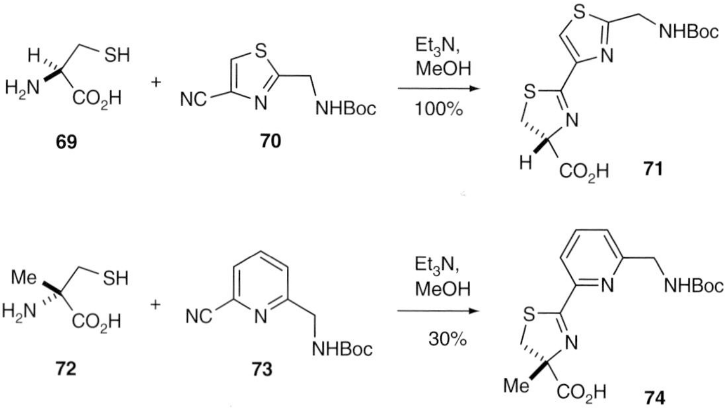

Cyclocondensation of hydroxythioamides provides another efficient approach to thiazolines as demonstrated by the formation of thiazoline **77**. The carbonyl group of the hydroxyamides **75** is first sulfinated using phosphorus pentasulfide, and the resulting hydroxythioamides **76** subsequently undergo cyclocondensation in refluxing pyridine to form thiazolines **77** in excellent yields <09EJOC4833>.

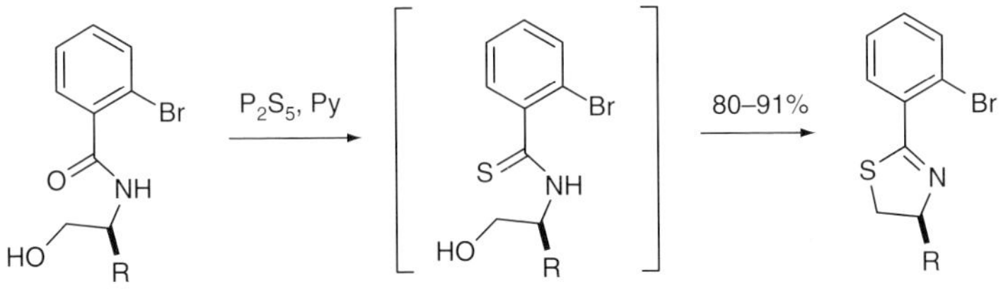

R = *i*-Pr, *t*-Bu, Ph, Bn

A comprehensive review on the thiazoline chemistry has been published <09CR1371>.

5.5.2.4 Reactions of Thiazoles and Fused Derivatives

A facile synthetic route to the thiazol-5-ylboronic acid pinacol ester **79** has been established <09T5739>. 2-Bromothiazole undergoes halogen lithium exchange to give 2-trimethylsilylthiazole **78**. Lithiation at C-5 with *n*-butyllithium, boronation with triisopropyl borate and transesterification with pinacol in the presence of acetic acid generates the stable pinacol ester **79** directly. The trimethylsilyl group is cleaved during the work-up process. Compound **79** is coupled with various aryl halides under typical Suzuki conditions to give the 5-arylthiazoles **80**. Alternatively, thiazoles **80** can be prepared from thiazole and aryl halides using direct C-H arylation reactions <09T5739>.

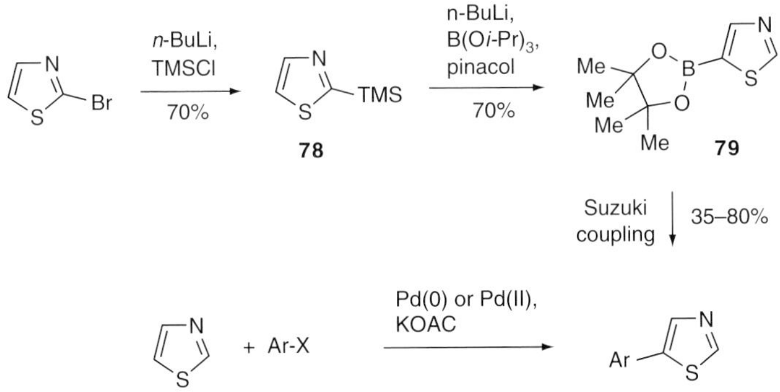

The halogen dance reaction is applied to the synthesis of the 5-substituted 4-bromo-2-chlorothiazole derivatives **84** <09EJOC3228>. Treatment of 5-bromothiazole **81** with 1.2 eq. of LDA at -80 °C brings about full conversion to the halogen dance product **83**, which is trapped with various electrophiles to give the 5-substituted thiazoles **84**. For example, 4-bromo-5-trimethylsilylthiazole **84a** is obtained in 97% yield when 5-bromothiazole **81** is exposed to LDA followed by addition of trimethylsilyl chloride (TMSCl) as the electrophile. A review on the halogen dance reaction has appeared <09EJOC3228>.

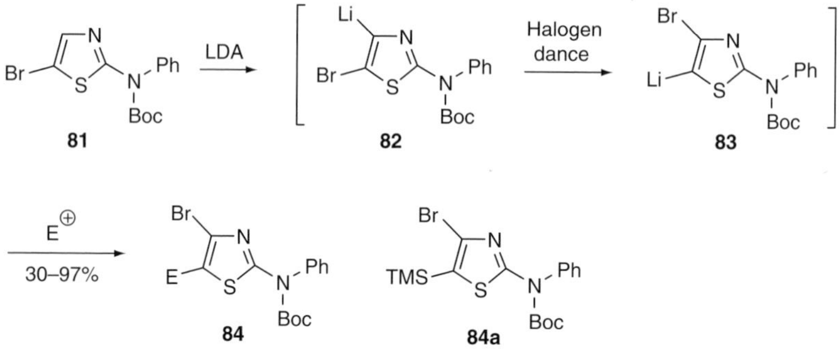

Regioselective halogenation of 2-aminothiazoles provides facile synthesis of mono- and di-halo thiazole derivatives <09JOC2578>. Treatment of **85** with copper(I) halide under modified Sandmeyer conditions (*n*-butyl nitrite, MeCN, 60 °C) gives 2-halothiazoles **86a/b/c** in 33%, 46% and 50% yields, respectively. However, when copper(II) halides are used in the same reactions, 2,5-dihalothiazoles **88a/b** are obtained in 35% and 79% yields, respectively. Reaction of **85** with copper(II) halides in acetonitrile at room temperature affords 2-amino-4-halothiazoles **87a/b** in 51% and 94% yield, respectively. Both thiazoles are converted into 2,4-dihalothiazoles **88a/b** using copper(I) halides under modified Sandmeyer conditions.

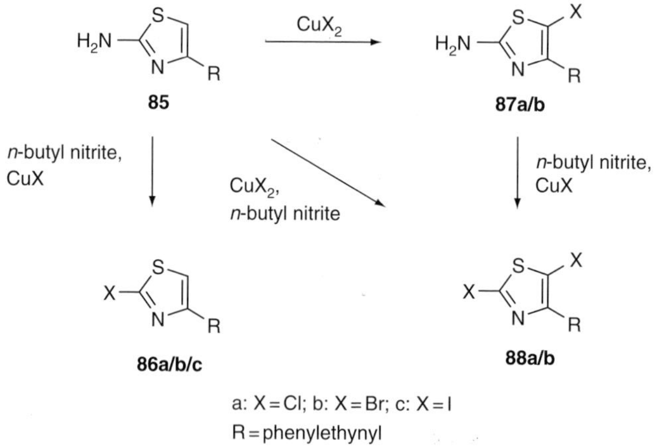

2-Iodobenzothiazole **89a** is prepared from benzothiazole by deprotonation using a mixture of MCl$_2$•TMEDA and lithium 2,2,6,6-tetramethylpiperidide (LiTMP) followed by trapping with iodine <09CEJ10280>. Among the metal chlorides evaluated, calcium provides the highest yield (97%). 2-Chloro and 2-bromobenzothiazoles **89b/c**

are obtained by means of *in situ* generation and trapping of benzothiazol-2-yllithium with carbon tetrachloride and tetrabromide, respectively <09JOC8309>.

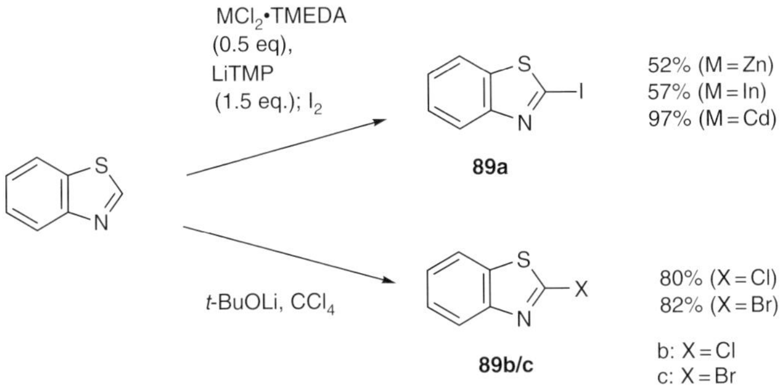

A Glaser–Hay reaction has been performed on the unsubstituted thiazole <09JA17052>. Treatment of thiazole with *iso*-propylmagnesium chloride/lithium chloride/tetramethylpiperidine (1:1:1) and a catalytic amount of copper(I) chloride in the presence of oxygen affords bis(thiazole) **91** in good yield. The reaction proceeds by way of an organocopper intermediate **90**, which dimerizes to forms biaryl **91** under an oxygen atmosphere.

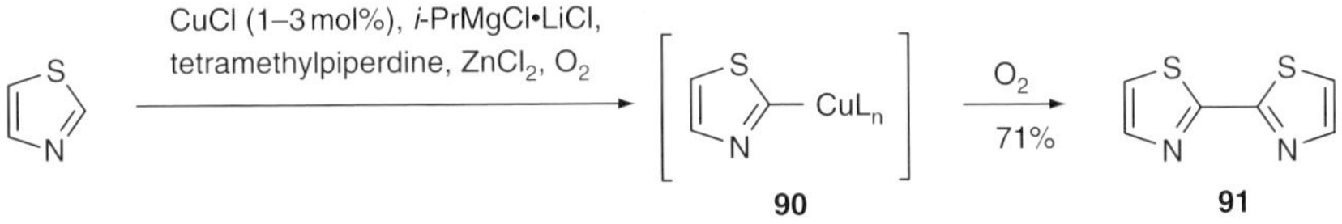

Direct amination of thiazoles and benzothiazoles at the C2 position has been carried out <09OL1607>. The optimal conditions require copper(II) acetate (20 mol %) and triphenylphosphine (40 mol%) under oxygen atmosphere in xylene at 140 °C. A variety of 2-aminothiazoles **93** and benzothiazoles **92** have been prepared using this methodology.

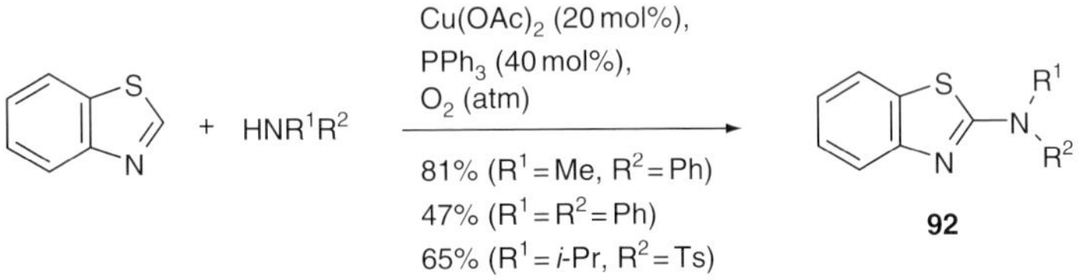

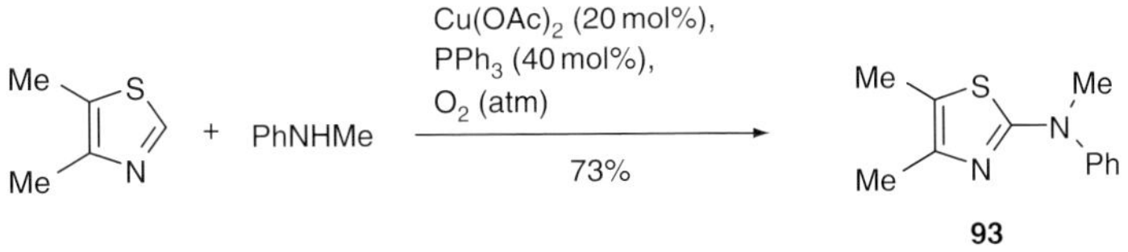

The palladium-catalyzed direct C-5 heteroarylation of chloropyridines and chloroquinolines has been investigated <09JOMC455>. Thus, thiazole **94** and benzothiazole undergo coupling via C-H bond activation/functionalization reactions with chloropyridines and chloroquinolines in low to high yields using air-stable PdCl(dppb)(C_3H_5) as the catalyst. A related direct C-5 arylation involves the palladium-catalyzed direct coupling of 4-thiazole-carboxylate **97** with picolinate **98** to afford pyridylthiazole **99** <09OL3690>.

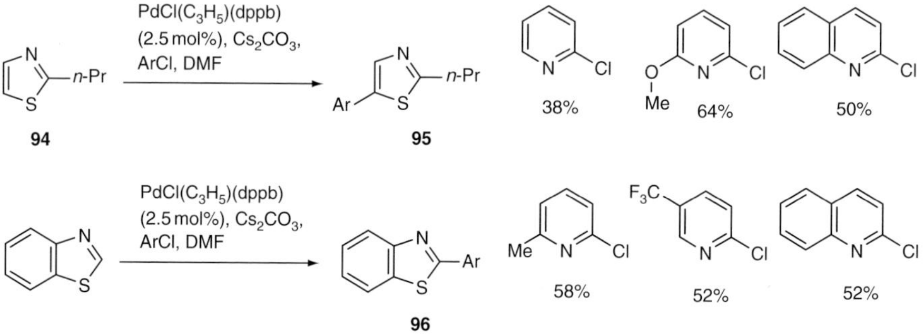

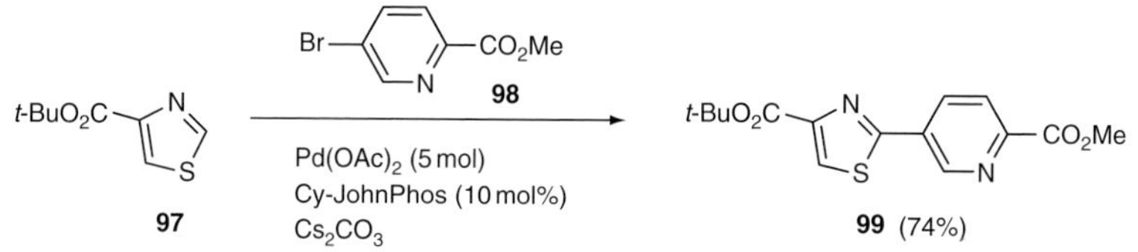

Ligand-free palladium(II) acetate has been shown to catalyze the direct C-5 arylation of thiazole derivatives under very low catalyst concentration <09JOC1179>. This methodology works well with activated aryl bromides but can be problematic with some strongly deactivated or highly sterically hindered aryl bromides.

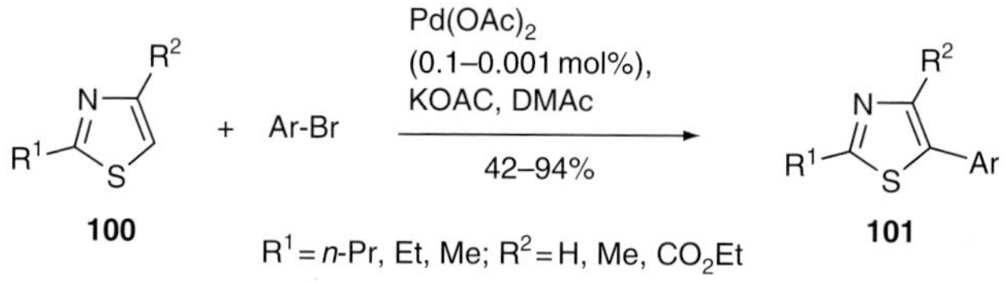

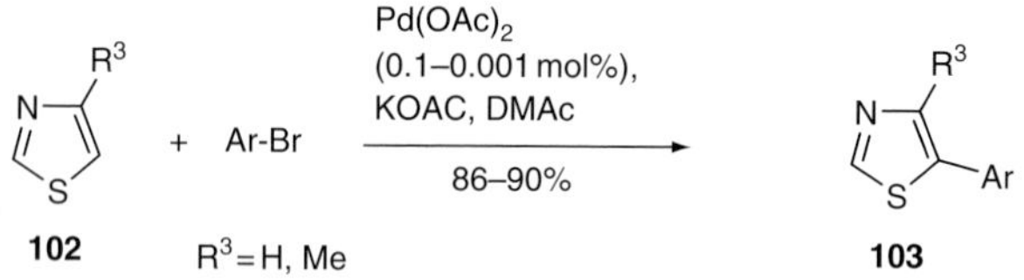

The nickel-catalyzed C-2 arylation of thiazoles with aryl halides/triflates has been described <09OL1733>. Extensive screening studies show that nickel acetate in combination with 2,2′-bipyridyl (bipy) in the presence of lithium *t*-butoxide serves as an efficient catalytic system. With this system, various 2-arylbenzothiazoles are obtained from aryl bromides and iodides in good yields. However, these conditions do not work for aryl triflates and chlorides. These substrates can be coupled with thiazole or benzothiazoles when dppf is used as the ligand. The synthetic utility of this methodology is demonstrated in a short synthesis of febuxostat, a selective, nonpurineinhibitor of xanthine inhibitor. Under the optimized conditions for aryl iodides, thiazole **104** couples with iodide **105** to furnish febuxostat after deprotection of the *tert*-butyl group.

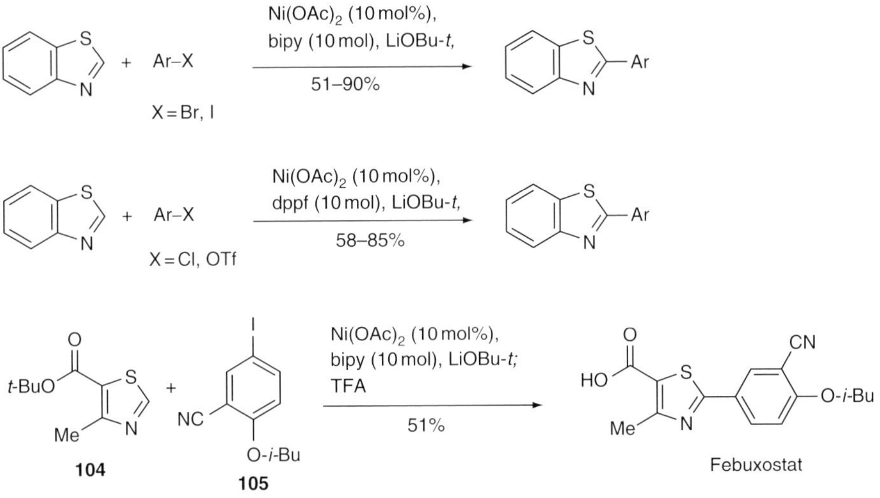

Broadly applicable reaction conditions for the palladium-catalyzed direct C-2 and C-5 arylations of thiazole derivatives have been established <09JOC1826>. These conditions employ a stoichiometric ratio of both thiazoles and aryl halides as well as an substoichiometric quantity of pivalic acid (30 mol%). The arylations are typically preformed in the presence of palladium acetate (catalyst), tricyclohexylphosphine (in its bench-stable phosphonium tetrafluoroborate form) (PCy$_3$•HBF$_4$) (4 mol%) (ligand) and potassium carbonate (base) in DMA at 100 °C. In terms of the regioselectivity, the C-5 position is generally favoured over the C-2 position when available. This method has been modified for the benzylation of thiazoles <09OL4160>. Thus, treatment of thiazoles **110** with benzyl chloride in the presence of Pd(OPiv)$_2$ (2 mol%), 2-Ph$_2$P-2′-(Me$_2$N)biphenyl (4 mol), PivOH (20 mol %), and cesium carbonate (1.5 eq.) in toluene at 110 °C affords 2-benzylated thiazoles **111** in good yields.

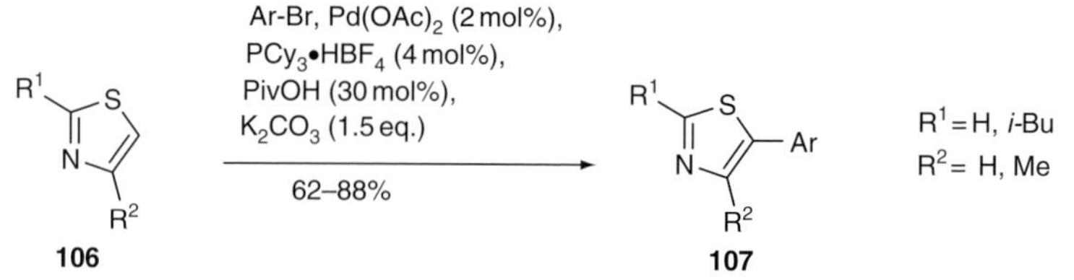

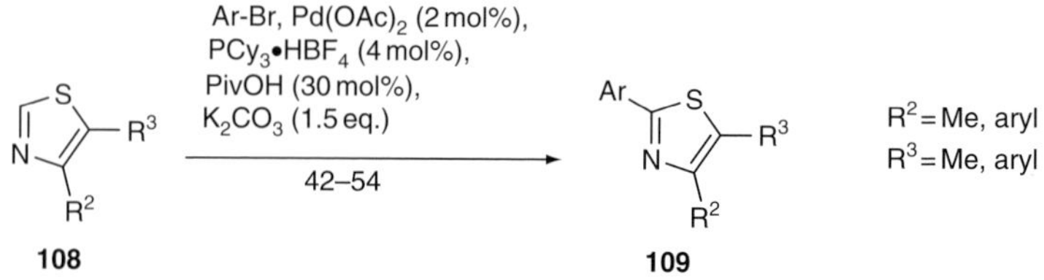

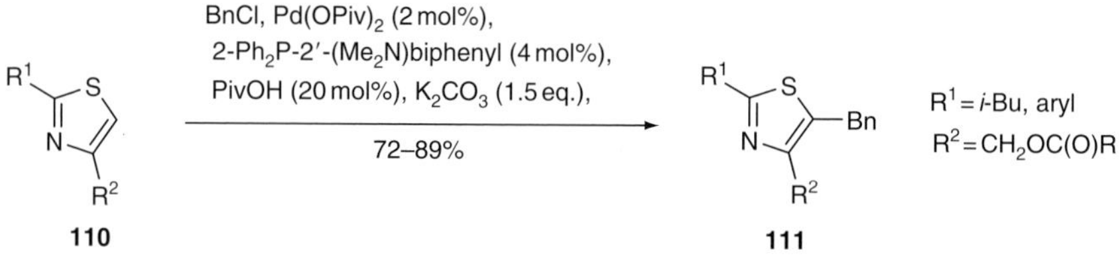

**Ar-Br, Pd(OAc)$_2$ (2 mol%),
PCy$_3$•HBF$_4$ (4 mol%),
PivOH (30 mol%),
K$_2$CO$_3$ (1.5 eq.)**

106 → 107, 62–88%, R^1 = H, i-Bu; R^2 = H, Me

**Ar-Br, Pd(OAc)$_2$ (2 mol%),
PCy$_3$•HBF$_4$ (4 mol%),
PivOH (30 mol%),
K$_2$CO$_3$ (1.5 eq.)**

108 → 109, 42–54, R^2 = Me, aryl; R^3 = Me, aryl

**BnCl, Pd(OPiv)$_2$ (2 mol%),
2-Ph$_2$P-2'-(Me$_2$N)biphenyl (4 mol%),
PivOH (20 mol%), K$_2$CO$_3$ (1.5 eq.),**

110 → 111, 72–89%, R^1 = i-Bu, aryl; R^2 = CH$_2$OC(O)R

Despite many recent advances in the direct arylation of thiazoles, regioselective C2 and C5 arylation can still be problematic, and a high yielding C4 arylation remains challenging. While useful activity has been achieved at C-2 and C-5, direct arylation at C-4, even C4/C5 regioselectivity issues aside, is extremely rare. To this end, a chlorine functional group is introduced at C-5 to not only to block reaction at C-5 but also to enhance activity at C-4 toward direct arylation <09JOC1047>. Under the modified arylation conditions, the C-4 arylation products **113** are obtained in good yields, and subsequent dehydrochlorination of **113** results in functionalized 5-H thiazoles **114**. The 5-chloroarylthiazoles may also be used for further functionalization, a feature that should find application in the synthesis of more complex thiazole compounds.

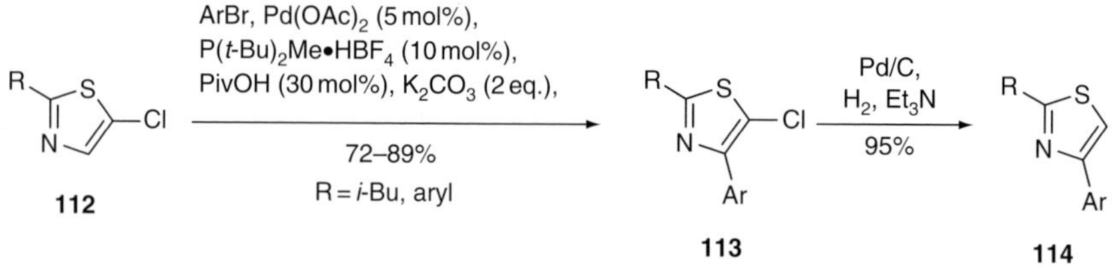

**ArBr, Pd(OAc)$_2$ (5 mol%),
P(t-Bu)$_2$Me•HBF$_4$ (10 mol%),
PivOH (30 mol%), K$_2$CO$_3$ (2 eq.),**
72–89%, R = i-Bu, aryl

112 → 113 → (Pd/C, H$_2$, Et$_3$N, 95%) → 114

The palladium-catalyzed direct arylation of thiazole N-oxides has been described in detail <09JA3291>. The thiazole oxides are readily prepared from thiazoles by treatment with mCPBA or H$_2$O$_2$ in the presence of catalytic MeReO$_3$, and the N-oxide moiety can be easily deoxygenated via zinc powder in aqueous ammonium chloride. The *N*-oxide group not only increases reactivity in direct arylation at all positions of the thiazole ring but also provides a reliable regioselectivity. In general, preferential arylation reaction of thiazole oxides is observed at C2 which occurs under very mild conditions. Subsequent reactions occur at C-5 followed by arylation at C4. Treatment of an aryl halide with thiazole *N*-oxide **115** in the presence of palladium acetate, biphenyl ligand **117**, pivalic acid (PivOH), and potassium carbonate in toluene at room temperature results in C2 arylation to give **116** as the exclusive product. These are rare examples of direct C-2 arylation occurring under such mild conditions. When the C2 position is blocked, the thiazole *N*-oxide **118** can then undergo selective C5 arylation using palladium acetate, *t*-Bu$_3$P•HBF$_4$, and potassium carbonate in toluene at 70 °C. Interestingly, addition of pivalic acid to the C5 arylation reactions is detrimental to the C5/C4 selectivity. The optimal C4 arylation conditions include palladium acetate in the presence of triphenylphosphine and potassium carbonate in toluene at 110 °C. This arylation approach using thiazole *N*-oxides provides a valuable alternative to the use of traditional cross-coupling methodology for the synthesis of arylthiazole compounds.

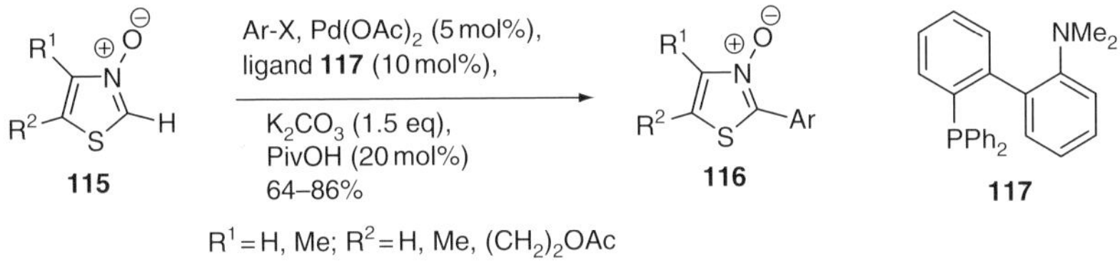

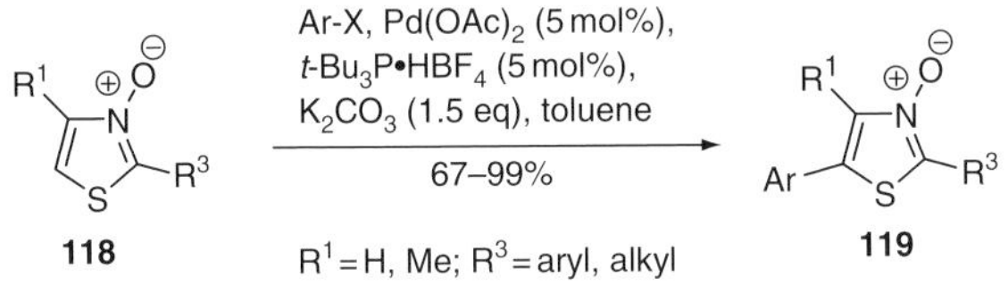

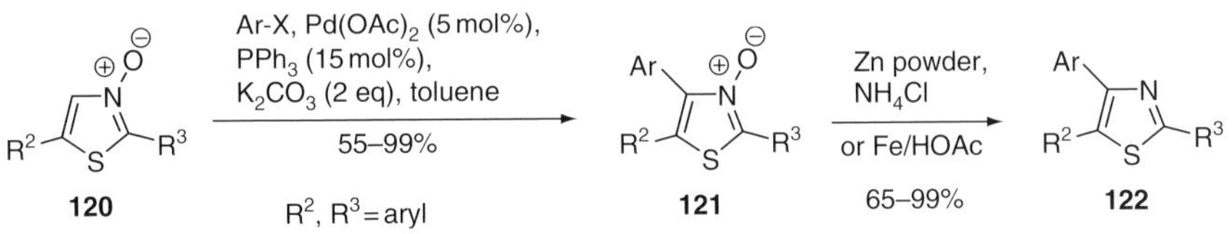

It is well known that addition of thiazoles to dimethyl acetylenedicarboxylate (DMAD) leads to the formation of zwitterions such as **126**. This intermediate can be trapped with chrome-3-carboxaldehyde **123**, according to a recent study <09TL1196>. Reaction of 4,5-dimethylthiazole with DMAD and aldehyde **123** generates the thiazolo[3,2-*a*]pyridinedicarboxylate **124** and chromenothiazolopyridinedicarboxylate **125** in 4% and 45% yields, respectively. According to the proposed mechanism, the zwitterion **126** undergoes conjugate addition with aldehyhe **123**, and the resulting intermediate **127** cyclises to give the minor product **124** (path b). The major pathway (path a) involves the chromone ring opening of the intermediate **127**, formation of a new six-membered ring, deformylation, cleavage of the thiazole ring via 1,5-sigmatropic shift, and finally 5-membered ring closure to give the major product **125**.

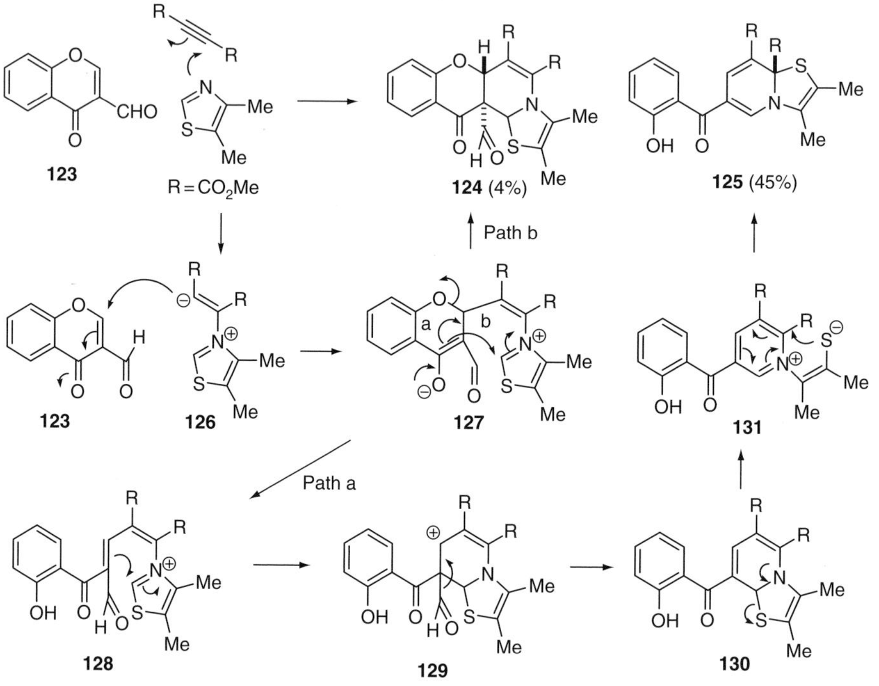

5.5.2.5 Thiazole Intermediates in Synthesis

The utility of thiazolidinethione chiral auxiliaries in asymmetric aldol reactions is demonstrated in a recent enantioselective synthesis of the macrocyclic core of (−)-pironetin <09OL1635>. Addition of aldehyde **135** to the titanium enolate solution of *N*-propionyl thiazolidinethione **132** produces aldol product **136** in a 64% yield with high diastereoselectivity (*dr* 98 : 2). Reductive cleavage of the chiral auxiliary affords aldehyde **137**, which is subjected to an acetate aldol reaction with thiazolidinethione **133** to give alcohol **138** in 88% yield and 95 : 5 *dr*. Protection of alcohol **138** as its triethylsilyl ether followed by reductive removal of the auxiliary furnishes aldehyde **139**. This aldehyde reacts with an excess of the enolate of thione **134** under "Evans" *syn* aldo reaction conditions to generate aldol adduct **140** in >20 : 1 *dr* and a 65% yield. Thus, three iterative aldol reactions involving thiazolidinethione chiral auxiliaries establish five of the chiral centers present in (−)-pironetin. Adduct **140** is converted to (−)-pironetin in several steps.

Thiazolidinethione chiral auxiliariy has also been applied to the stereoselective preparation of C1–C10 and C11–O14 fragments of narbonolide <09S474> and stereoselective synthesis of tertiary methyl esters <09OL2193>.

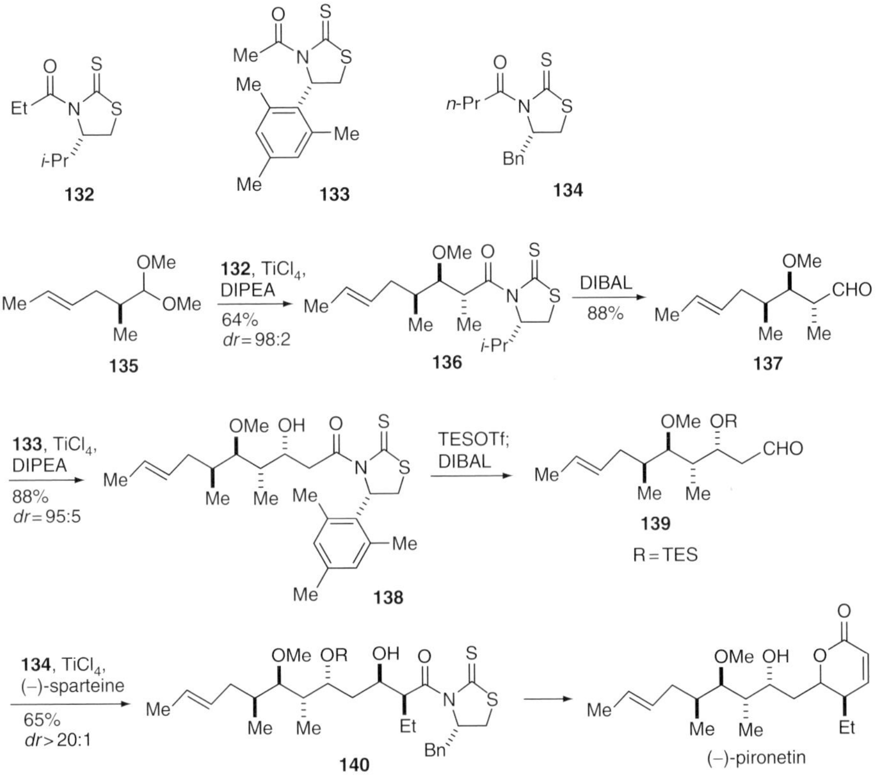

The Julia olefination reaction involving alkylsulfonyl benzothiazoles remains one of the most effective methods for the stereoselective formation of olefins. The power of this reaction is reflected in the synthesis of (+)-myxothiazol <09TA298>. Coupling of sulfone **141** with aldehyde **142** using lithium hexamethyldisilazide (LHMDS) produces a 3.3 : 1 mixture of *E* and *Z* isomers, of which the *E* isomer **143** is isolated in 59% yield. Oxidation of **143** affords sulfone **144**, which is subjected to another Julia olefination with (2*E*)-4-methylpentenal **145** to give (+)-myxothiazol A. Other applications of the Julia olefination reaction involving alkylsulfonyl benzothiazoles include the synthesis of (+)-piericidin A_1 and (−)-piericidin B_1 <09TA1975> and prostaglandin <09OL3104>.

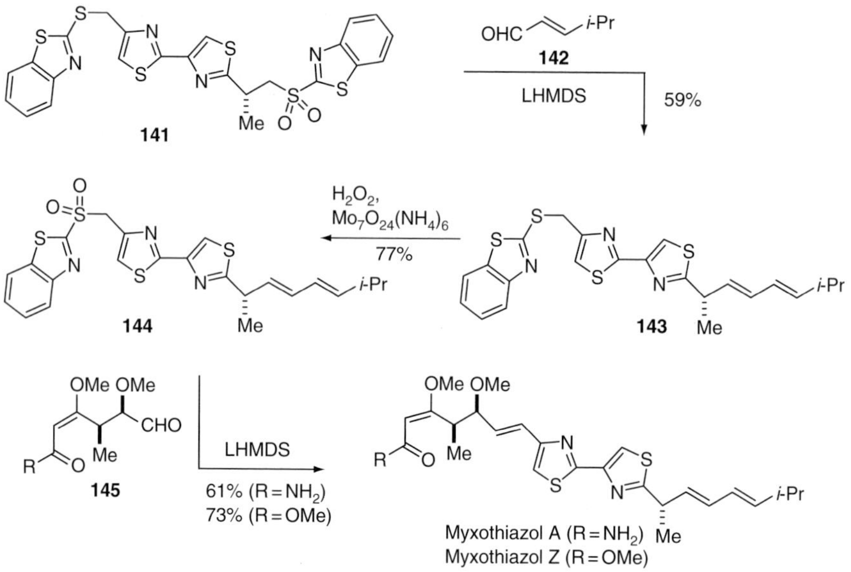

The stereoselectivity of the Julia olefinations has been well documented, and as a rule of thumb, α-lithioalkyl benzothiazolyl sulfones deliver *cis*-olefins and α-lithiobenzyl benzothiazolyl sulfones *trans*-olefins <05SL289>. However, unexpected high *cis*-selectivity is observed in the Julia coupling reaction of benzothiazolyl sulfone **146** with aldehyde **147** <09TL4874>. Deprotonation of sulfone **146** with LiHMDS in the presence of the aldehyde **147** furnishes exclusively the *cis*-olefin **148** in 75% yield. The origin of the high *cis*-selectivity observed remains to be determined.

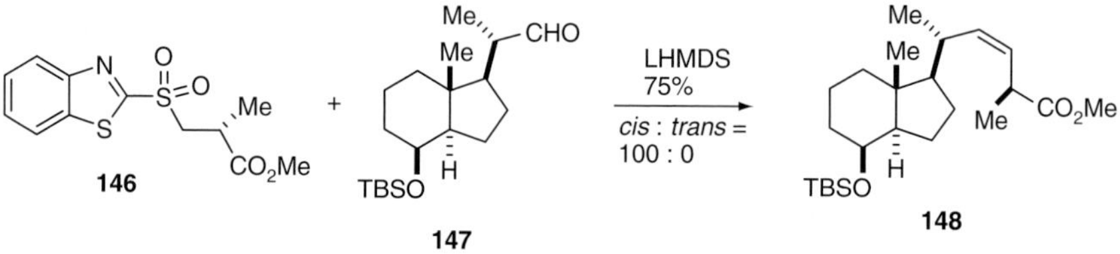

There are several new methodologies based on the Julia olefination reaction. For example, 2-(benzo[*d*]thiazol-2-ylsulfonyl)-*N*-methoxy-*N*-methylacetamide **149**, prepared in two steps from 2-chloro-*N*-methoxy-*N*-methylacetamide, reacts with a variety of aldehydes in the presence of sodium hydride to furnish the α,β-unsaturated Weinreb amides **150** <09JOC3689>. An efficient synthesis of fluorinated olefins **152** features the Julia olefination of aldehydes or ketones with α-fluoro 1,3-benzothiazol-2-yl sulfones **151**, readily available from the corresponding des-fluoro sulfones via electrophilic fluorination <09JOC3689>.

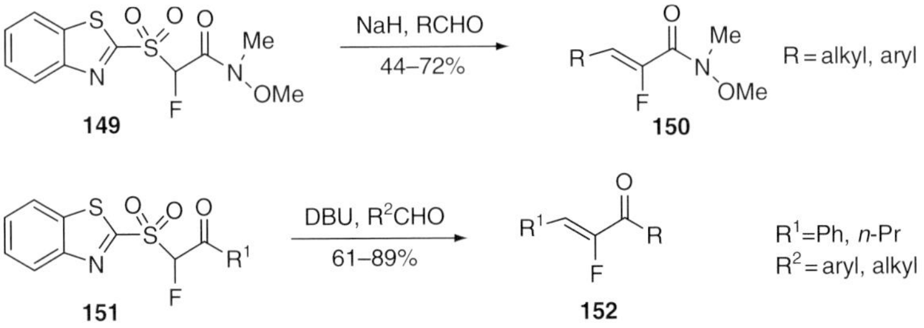

Thiazolyl thioglycosides such as **153** have been developed as glycosyl donors. Glycosylation of **153** with **154** using benzyl bromide as a promoter proceeds stereoselectively to give disaccharide **155** in high yield <09OL799>. In this case, benzyl bromide promotes selective activation of S-thiazolinyl donor **153** over S-benzothiazolinyl acceptor **154**. The S-benzothiazolinyl moiety of **155** is then activated with AgBF$_4$ and glycosidated with acceptor **156** to give trisaccharide **157**. This strategy provides an expeditious assembly of various oligosaccharides.

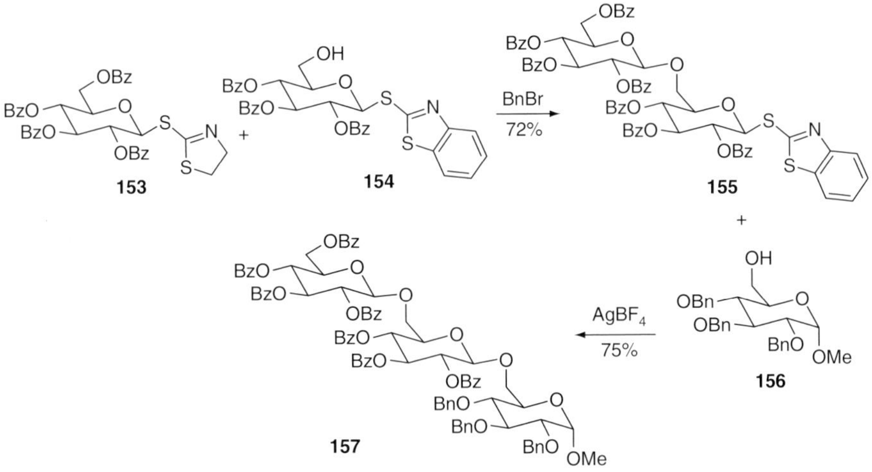

The thiazoline–oxazoline ligand **160** has been evaluated in the asymmetric Friedel–Crafts reaction <09EJOC4833>. Asymmetric alkylation of indole with nitroalkene **158** catalyzed by ligand **160** and zinc triflate provides adduct **159** in good yield; however, the enantioselectivity is moderate.

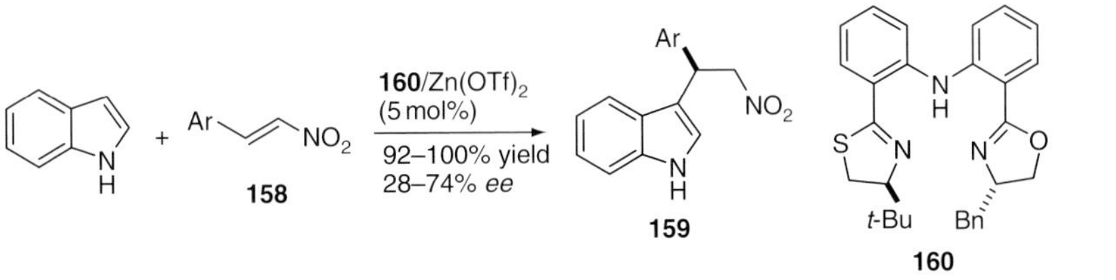

5.5.2.6 Reactions of Thiazolium and Thiazolinium salts

The thiazolium-catalyzed addition of an aldehyde-derived acyl anion with a receptor is a valuable synthetic tool leading to the synthesis of highly funtionalized products. When the acyl anion receptor is a Michael acceptor, the reaction is commonly known as the Stetter reaction <09SL1189>. An intramolecular version of this reaction has been recently applied to the synthesis of *trans*, *syn*, *trans*-fused pyran **165** <09SL233>. Reaction of aldehyde **162** with thiazolium salt **161b** (1.5 eq.) in the presence of DBU gives pyranone **163** in excellent yield. Ketoester **163** is elaborated into aldehyde **164**, which undergoes another intramolecular Stetter reaction to give tricyclic pyranone **165**. This represents the first example of an intramolecular Stetter reaction between an aliphatic aldehyde and a deactivated α,β-unsaturated ester to form a six-membered ring. The intramolecular Stetter reaction is also used in the preparation of chromone **167** <09SL23>.

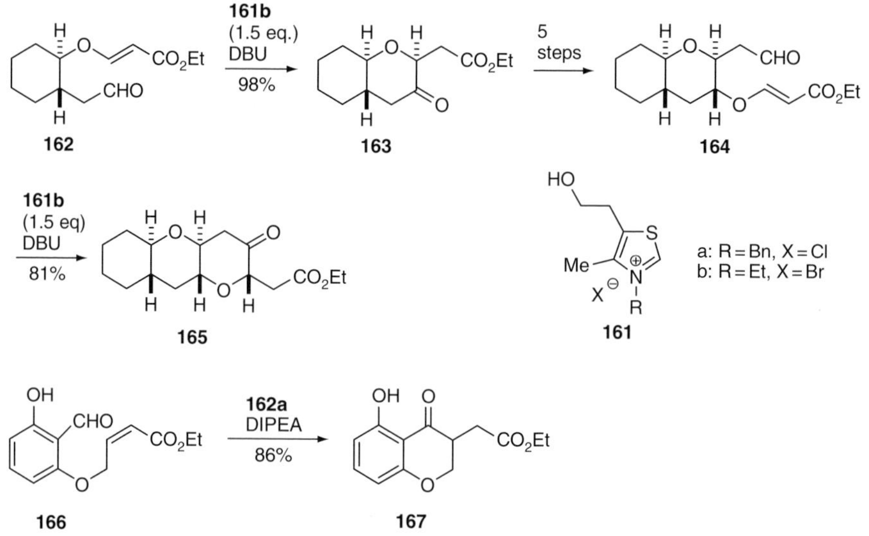

The N-heterocyclic carbenes derived from thiazolium salts are used to catalyze the cross-coupling of aldehydes with arylsulfonyl indoles **168** <09OL3182>. Presumably, the Breslow intermediate **170** reacts with olefin **169** (derived from **168** via elimination of the sulfonyl group) to give the intermediate **171**, which generates the coupling product **172** while releasing the carbene catalyst. This approach provides an easy entry to a series of α-(3-indolyl)ketones **172**.

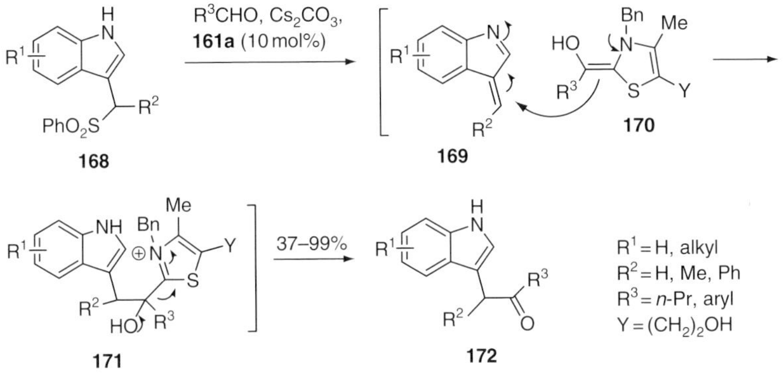

The synthesis of highly functionalized indane derivatives **175** takes advantage of a novel domino Stetter-Michael reaction involving thiazolium-derived *N*-heterocyclic carbenes <09JOC7536>. Reaction of enone **173** with thiazolium salt **161b** (30 mol%) and DBU (27 mol%) provides **175** as a mixture of diastereomers in good yield. This process represents the first example of a domino reaction involving the enolate intermediates such as **174** generated from a Stetter reaction.

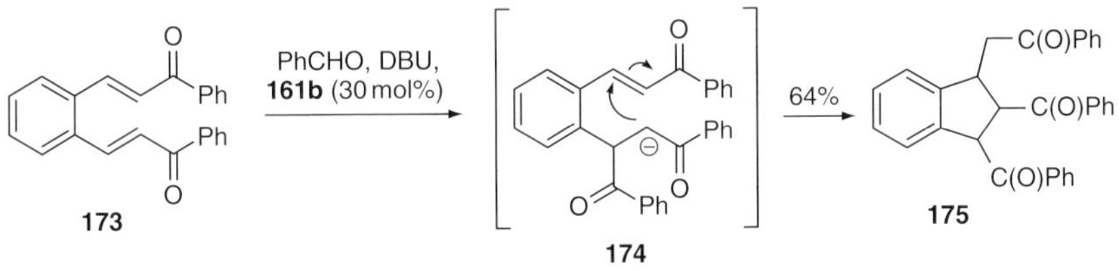

Thiazolium salts has also been utilized in the construction of highly functionalized selenopheno[2,3-*b*]pyrazines <09OBC3264>. For example, thiazole carbene **177**, generated *in situ* from the thiazolium salt **176** with sodium hydride, reacts with aryl isoselenocyanate **178** to produce 2-selenocarbamoylthiazolium inner salt **179**. Exposure of this inner salt to dimethyl acetylenedicarboxylate (DMAD) results in a good yield of the selenopheno[2,3-*b*]pyrazine **184**. The formation of **184** is rationalized by the [3 + 2] cycloaddition of inner salt **179** with DMAD. The resulting cycloadduct **180** undergoes ring cleavage to give zwitterion **181**. Electrocyclization followed by isomerization leads to pyrazine **183**, which transforms into the final product **184** via Michael addition with DMAD.

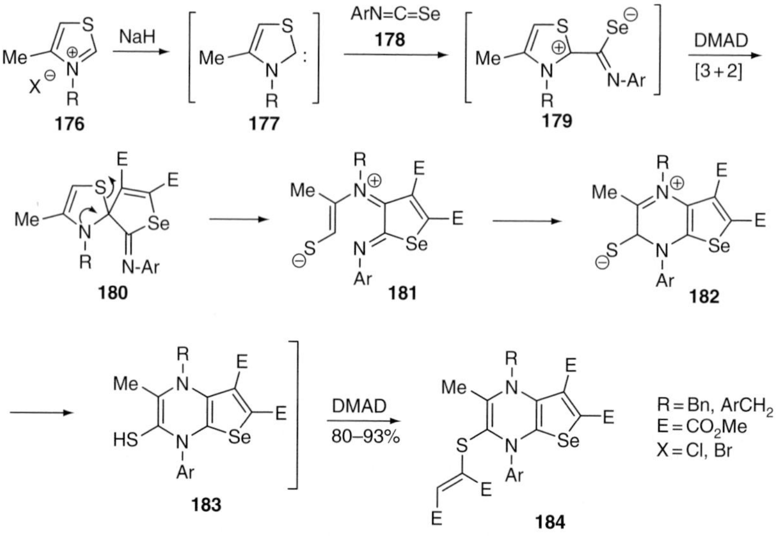

Reaction of nucleophilic thiazole carbenes **187** (derived from thiazolium salts **186**) with ketenimines **185** results in polyfunctionalized thiazole-spiro-pyrrole compounds **188** <09JOC850>. The reaction may proceed via a tandem nucleophilic addition of the carbene **187** to the C=N bond of the ketenimine **185** followed by a stepwise [3 + 2] cycloaddition of the 1,3-dipolar intermediate with the C=C bond of the ketenimine. This methodology also works for benzothiazole carbene substrates.

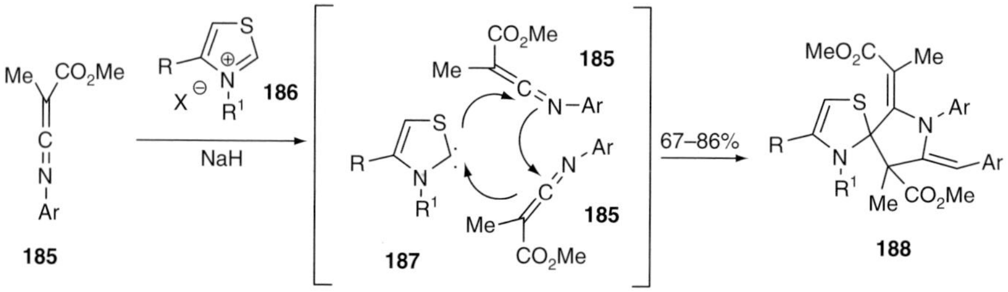

Thiazolinium salts **190**, readily available from thiazoline **189** by alkylation with various alkyl halides, are treated with mesyl chloride and triethylamine to afford thiazolinium mesylates **193** in moderate to good yields <09EJOC4357>. Conceivably, deprotonation of **190** at the α position to the heterocycle by triethylamine gives thiazolidine **191**, which reacts with mesyl chloride to form the α-sulfonylthiazolinium salt **192**. Subsequent deprotonation at the α position to the heterocycle affords sulfonylthiazolidine **193**.

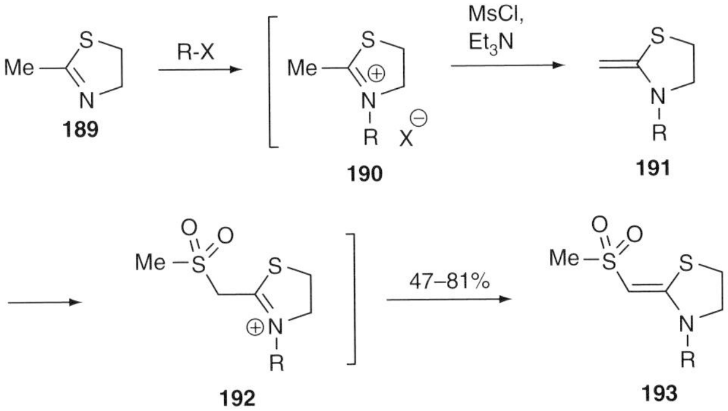

5.5.2.7 New Thiazole-Containing Natural Products

Hoiamide A, a novel cyclic depsipeptide, was isolated from an environmental assemblage of the marine cyanobacteria *Lyngbya majuscula* and *Phormidium gracile* collected in Papua New Guinea <09CB893>. The stereochemically complex metabolite bears two α-methylated thiazolines and one thiazole. It activates sodium influx in mouse neocortical neurons and exhibits modest cytotoxicity to cancer cells.

Nocardithiocin, a novel thiopeptide antibiotic containing six thiazole rings, was produced by the pathogenic *Nocardia pseudobrasiliensis* strain IFM 0757 <09JAN613>. It is highly active against rifampicin susceptible and resistant Mycobacterium tuberculosis strains.

Pretubulysin was identified as a direct biosynthetic precursor of the tubulysins <09AG(E)1>. It exhibits reduced cytotoxicity as compared with tubulysins A and D.

Neobacillamide A, a novel thiazole-containing alkaloid, was isolated from the marine bacterium *Bacillus vallismortis* C89, which is associated with the South China Sea sponge *Dysidea avara* <09HCA607>. Preliminary studies of this compound showed no antitumor activity.

Two thiazole-containing cyclopeptides, sanguinamides A and B, were isolated from the nudibranch *Hexabranchus sanguineus* <09JNP732>. Their biological activities have not yet been disclosed.

Six new thiazolyl peptides, thiazomycins B-D and truncated thiazomycins (lacking an indole residue) E1-E3, were isolated *Amycolatopsis fastidiosa* <09JNP841>. Thiazomycin B exhibits antibacterial activity comparable to thiazomycin, while thiazomycins C and D show intermediate activities. Interestingly, the truncated thiazomycins E1-E3 are essentially inactive. The discovery of the truncated compounds has revealed the minimal structural requirements for antibacterial activity.

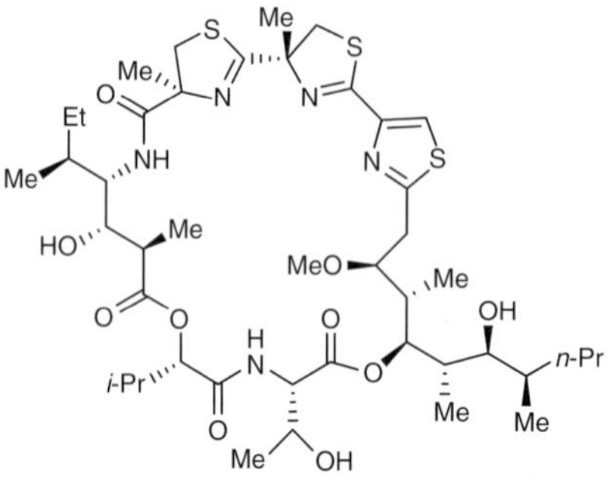

Hoiamide A

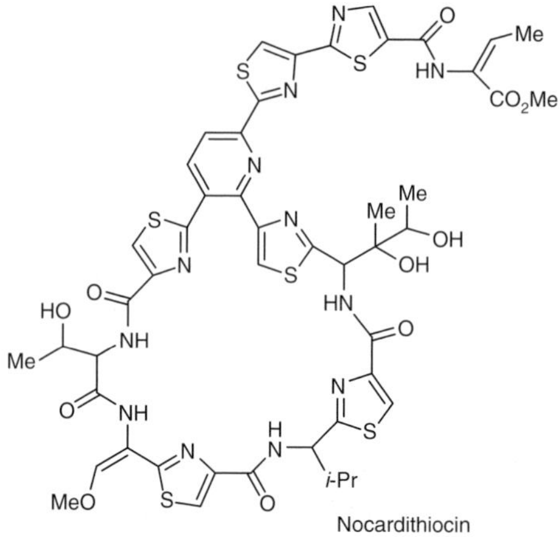

Nocardithiocin

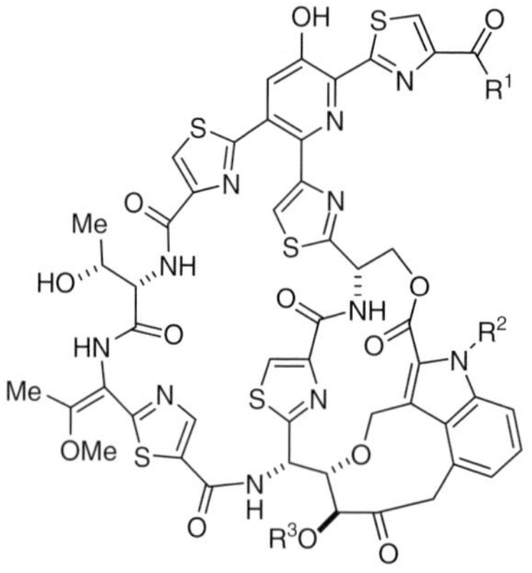

Thiazomycin B
$R^1 = $ (HN-C(=CH$_2$)-C(=O)-NH$_2$)
$R^2 = R^3 = H$

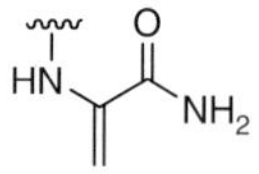

Thiazomycin C
$R^1 = NH_2$ $R^2 = H$
$R^3 = $ (sugar: Me, HO, NMe$_2$, Me)

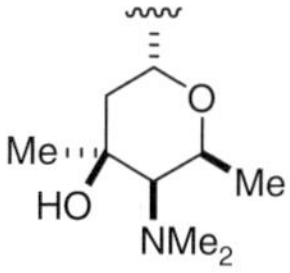

Thiazomycin D
$R^1 = $ (HN-C(=CH$_2$)-C(=O)-NH$_2$)
$R^2 = OH$
$R^3 = $ (sugar: Me, HO, OH, Me)

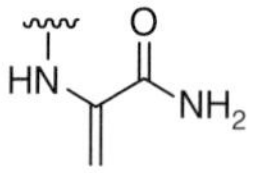

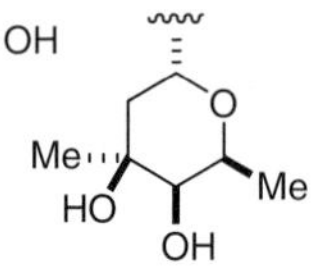

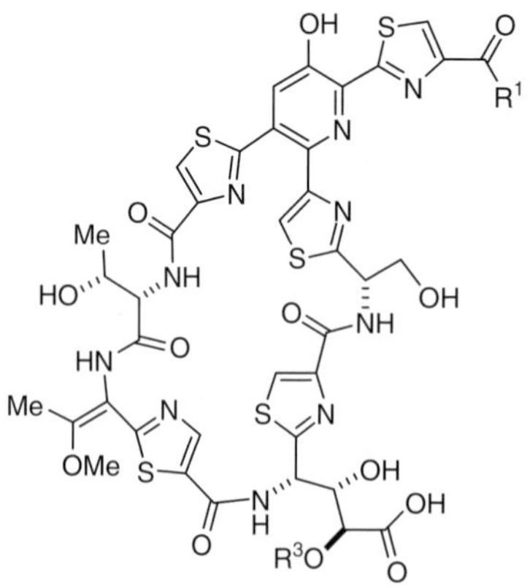

Thiazomycin E$_1$
$R^1 = $ (HN-C(=CH$_2$)-C(=O)-NH$_2$)
$R^3 = $ (sugar: Me, HO, NMe$_2$, Me)

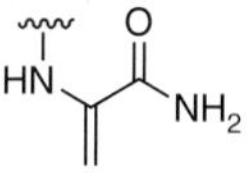

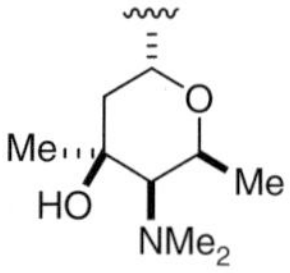

Thiazomycin E$_2$
$R^1 = $ (HN-C(=CH$_2$)-C(=O)-NH$_2$)
$R^3 = H$

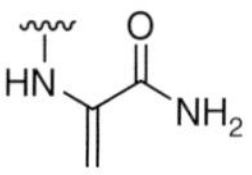

Thiazomycin E$_3$
$R^2 = NH_2$ $R^2 = H$

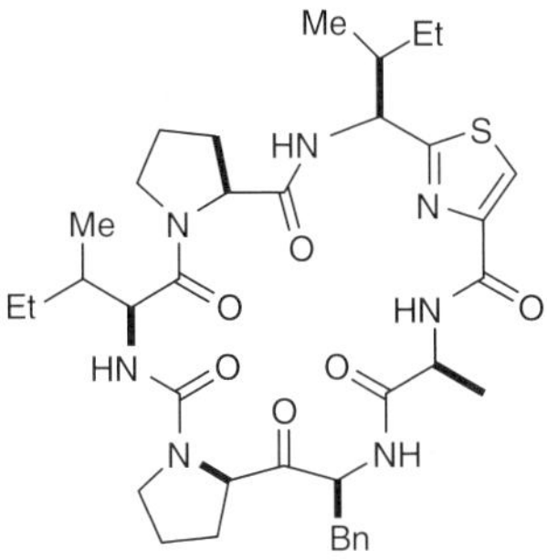

Sanguinamide A

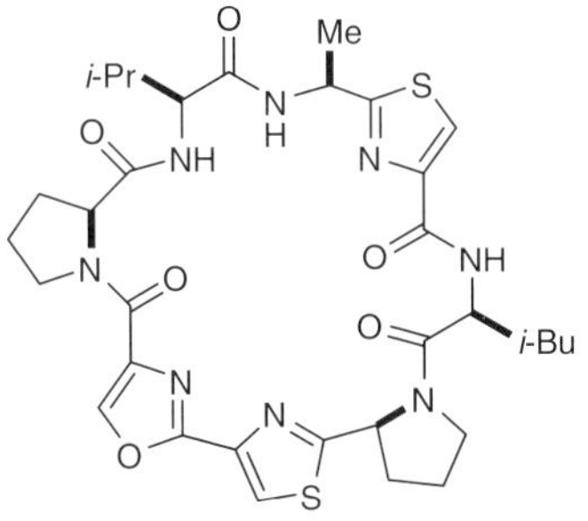

Sanguinamide B

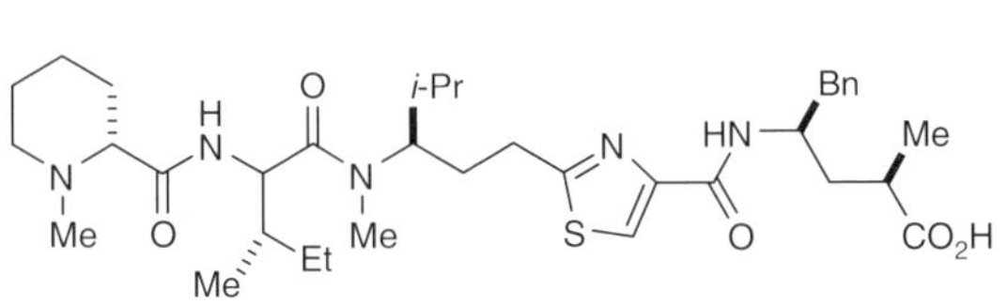

Pretubulysin

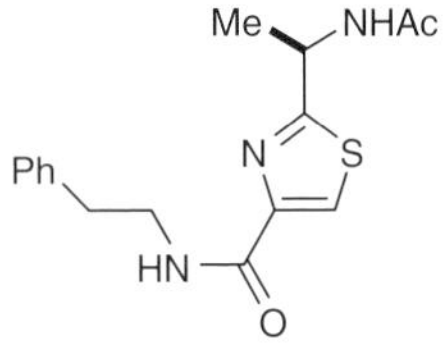

Neobacillamide A

5.5.2.8 Synthesis of Thiazole-Containing Natural Products

The thiopeptide antibiotic micrococcin P_1 (MP_1) is the major component of "micrococcin P", a cytotoxic extract isolated from *Bacillus pumilus*. Even though MP_1 is one of the structurally less complex thiopeptides, its exact structure has remained uncertain for the past five decades. Recently, a total synthesis of MP1 has been completed, thereby establishing its constitution and configuration <09AG(E)4198>. Other total syntheses of thiazole-containing natural products include (−)-apratoxin <09CAJ111> and tubulysins U and V <09JMC238>. A stereoselective synthesis of the C1–C12 fragment of the thuggacins <09TA2027> and an improved synthesis of the pyridine-thiazole cores of thiopeptide antibiotics <09JOC5750> have been described.

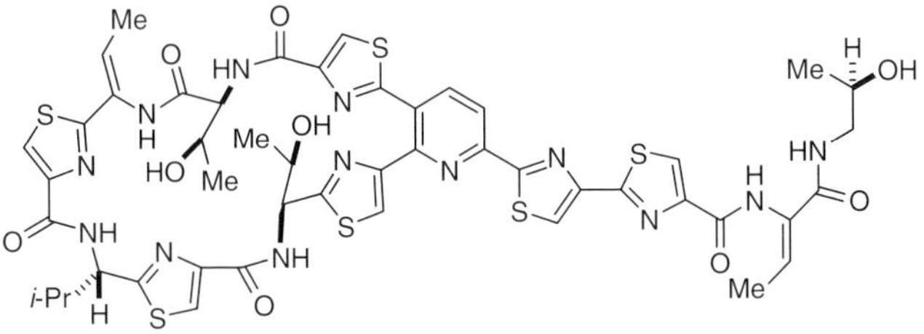

Micrococcin P1

Apratoxin

Thuggacin A

Tubulysin U (R = Ac)
Tubulysin V (R = H)

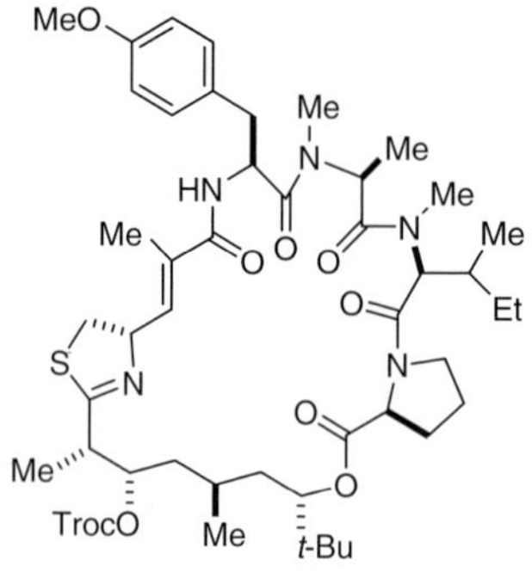

5.5.2.9 Pharmaceutically Important Thiazoles

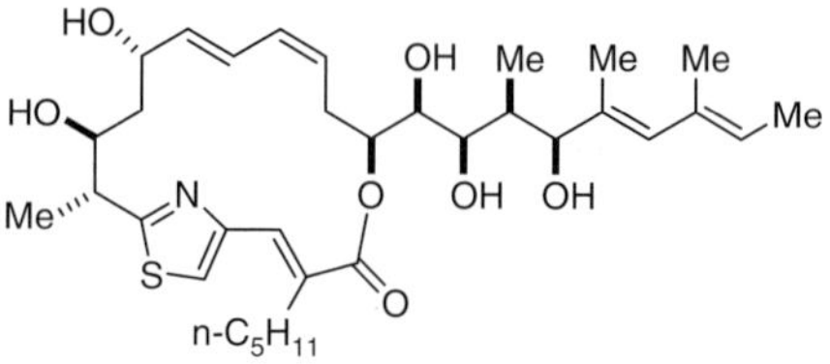

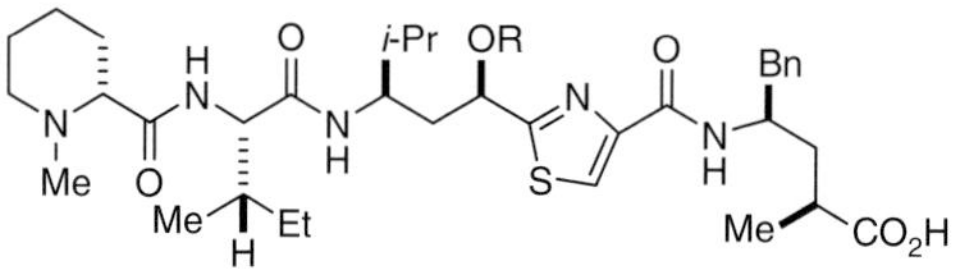

A number of biologically important thiazole analogs have been disclosed. For example, compound **194** is a cyclopentane-containing macrocyclic inhibitor of the hepatitis C virus NS3/4A protease <09JMC7014>. It displays a favourable pharmacokinetic profile in preclinical species and shows sustained liver levels following oral administration in rats. PR-047 has been identified as an orally bioavailable and selective peptide epoxyketone proteasome inhibitor. It exhibits oral antitumor activity in multiple animal models and may have potential for the treatment of malignant diseases <09JMC3028>.

Edoxaban, an orally active Factor Xa (FXa) inhibitor, is in phase III clinical studies as an anticoagulant for the potential treatment of cardiovascular indications, including venous thromboembolism (VTE) and non-valvular atrial fibrillation (NVAF) <09DF861, TPC>.

Voreloxin is a naphthyridine cell cycle inhibitor and apoptosis stimulator that inhibits topoisomerase II and intercalates into DNA, and this agent is undergoing phase II clinical evaluations for the potential iv treatment of a range of cancers, including solid tumours and acute myeloid leukemia (AML) <09DF363, TPC>.

TMC435350 is a cyclopentane-containing macrocyclic inhibitor of the hepatitis C virus NS3/4A protease <08BMCL4853>. This compound is currently being evaluated in phase II clinical trials for the potential oral treatment of HCV infection <09DF545, TPC>.

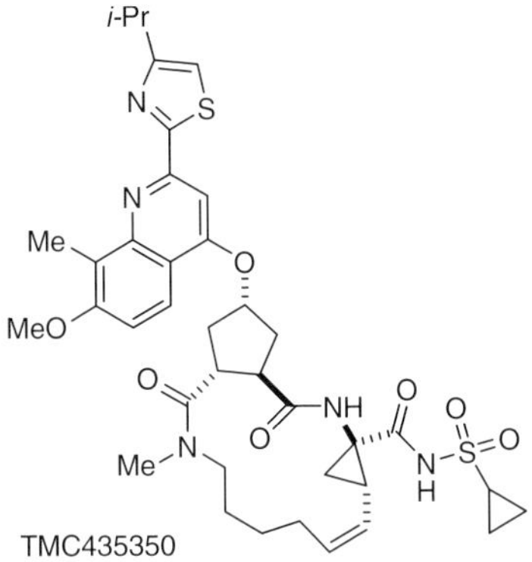

TMC435350
Phase II
Tibotec/medivir

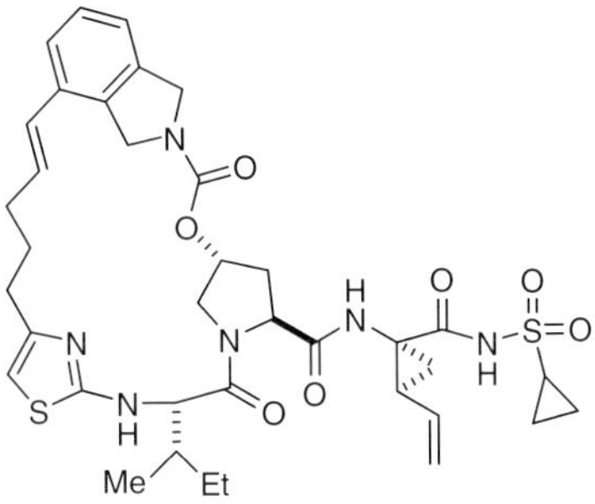

194

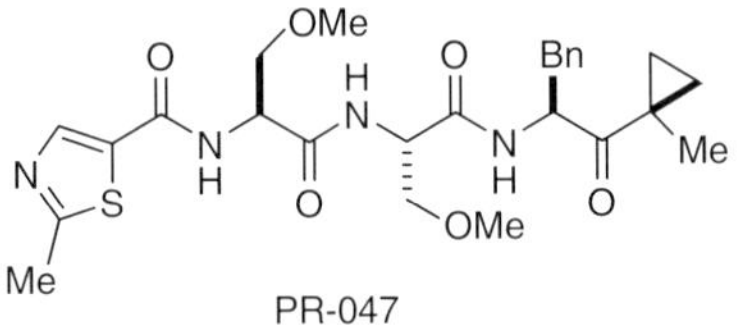

PR-047

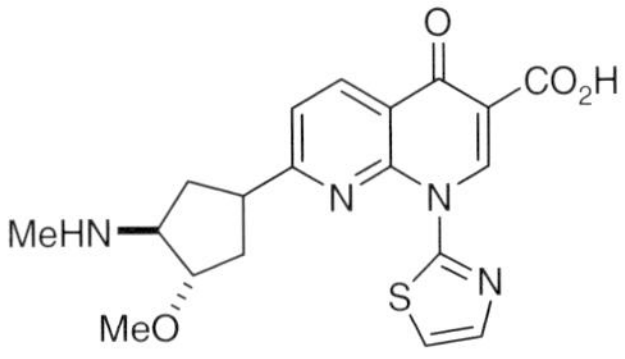

Voreloxin
Phase II
Sunesis/dainippon

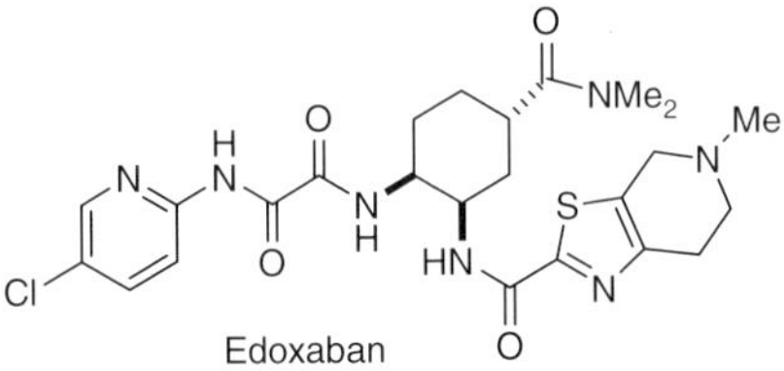

Edoxaban
Phase III
Daiichi sankyo

5.5.3. ISOTHIAZOLES

5.5.3.1 Synthesis of Isothiazoles

An efficient synthesis of 4,5-disubstituted-3-trihalomethylisothiazoles **197** and **198** from trihaloacetonitriles and 2-cyanothioacetamides **195** has been developed <09TL7286>. The reactivity of the necessary trihaloacetonitriles has a significant impact on the observed reaction pathways. Reactions with CF_3CN require an oxidant (H_2O_2 or NIS) to mediate cyclization, while CCl_3CN functions as both the reactant and oxidant.

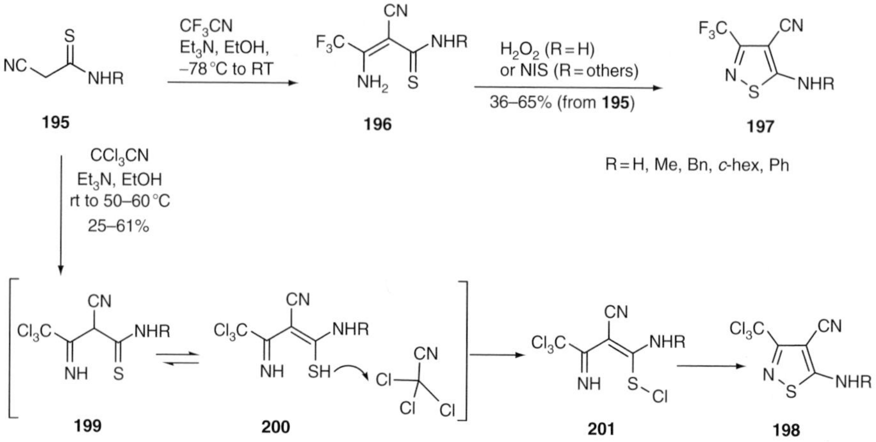

Paul Hanson has reported a diversity-oriented synthesis (DOS) strategy termed "click, click, cyclize" to form a series of skeletally diverse 5-membered sultams based on a central tertiary vinyl sulfonamide linchpin <09OL531>. As exemplified by **202**, the "click, click, cyclize" reaction manifold includes the sequential N-sulfonylation with **203** and N-alkylation reactions to afford the N-(prop-2-ynyl)ethenesulfonamide **204**, followed by a ring-closing enyne metathesis (RCEM) catalyzed by cat-B (Grubbs' catalyst B) leading to the 5-membered sultam **205**. The ring closure step can also be accomplished with oxidation/intramolecular Baylis-Hillman reaction (for preparation of **208**), RCM (ring-closing metathesis, for **209**), and Pauson-Khand (for **211**) reactions.

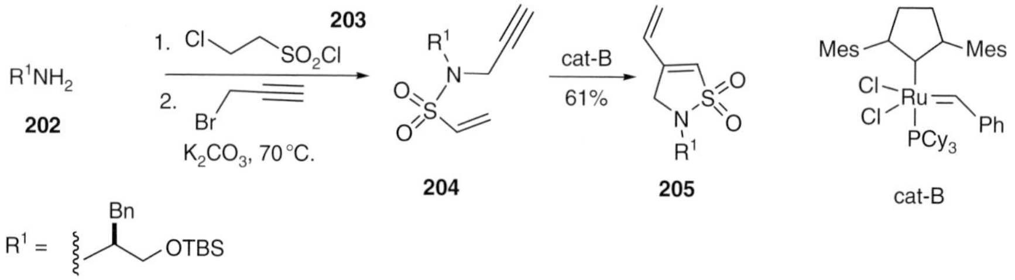

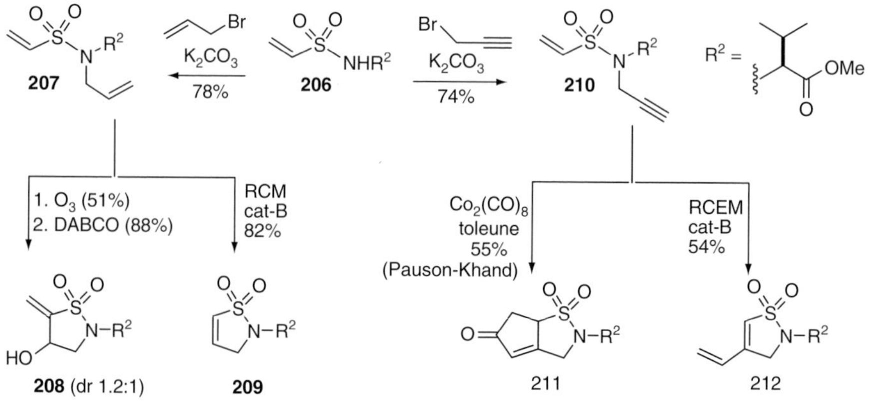

When the tertiary vinyl sulfonamides engage in a ring closure via intramolecular Diels–Alder reaction followed by cascade reactions including ring-opening metathesis, ring-closing metathesis and cross-metathesis (ROM–RCM–CM), a diverse collection of tricyclic sultams is conveniently synthesized <09T4992>. In this study, functionalized sultam scaffolds **216**, **221** and **223**, derived from intramolecular Diels–Alder reactions, undergo metathesis cascades with a variety of CM partners, alkenes **217**, ethylene and **224**, to yield tricyclic sultams **218**, **222**, and **225**, respectively.

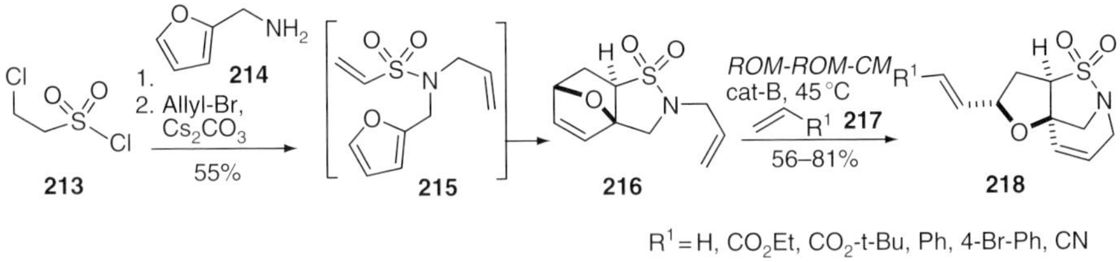

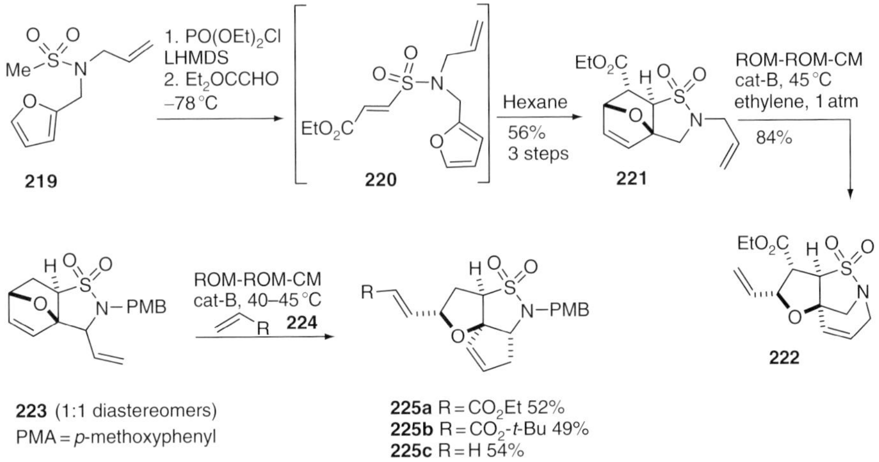

New approaches to five-membered benzosultams based on the α-halobenzene sulfonamides **227** and **230** feature Sonogashira and intramolecular hydroamination reactions either by stepwise or in a one-pot fashion, generating the benzosultams **229** and **231**, respectively <09T3180>.

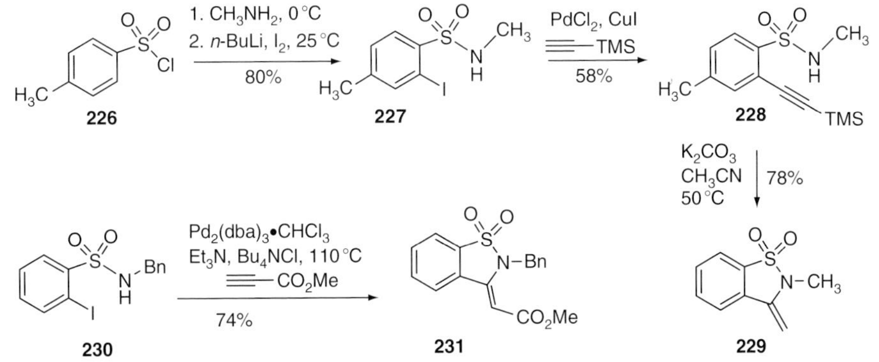

A facile synthesis of bicyclic sultams **234** utilizes the readily available N-(ω-1, ω-dibromoalkyl)(methoxycarbonyl)methanesulfanilide **232** <09EJOC2635>. The key step involves an intramolecular dialkylation of the dibromide **233**, available from bromination of **232**, affording the bicyclic product containing 1-methoxycarbonyl-cyclopropane moiety. The 4-methoxyphenyl N-protecting group can be oxidatively removed.

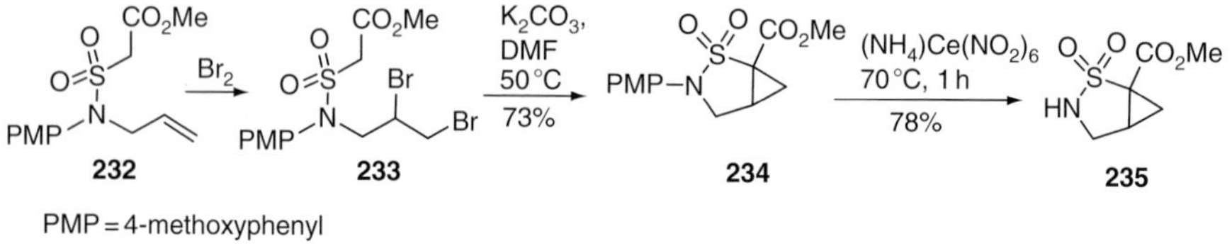

PMP = 4-methoxyphenyl

A variety of substituted spiro(benzoisothiazole-pyrazoles) **241** are prepared by the condensation of dilithiated C(α),N-carboalkoxyhydrazones **237** with aminosulfonylbenzoate **238** followed by the cyclization of intermediates **239** in the presence of acetic anhydride, resulting in the formation of spiro N-acetylated products (when R^1 = Me, Et), or spiro NH products (when R^1 = t-Bu) <09JHC231>. The tetrahydronaphthalene hydrazinecarboxylate **242** affords the corresponding spiro-benzoisothiaole **243** under the same reaction conditions.

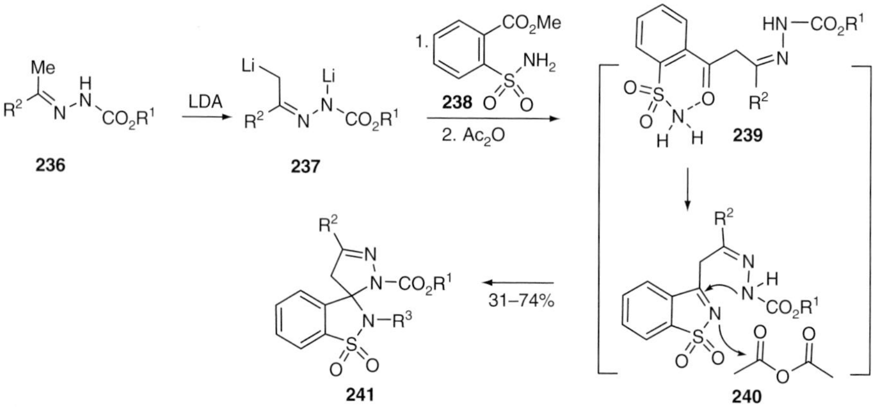

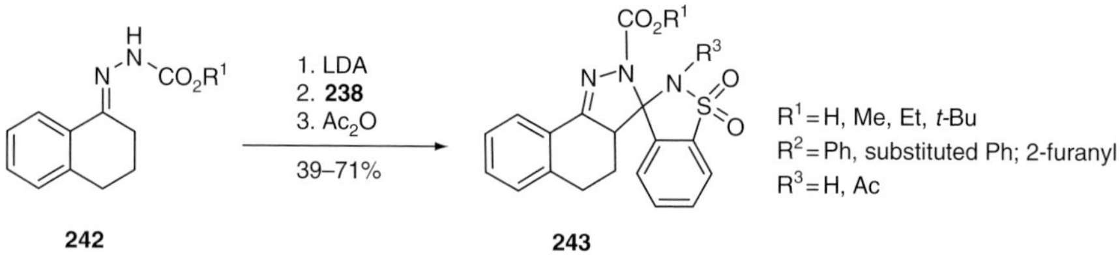

R^1 = H, Me, Et, t-Bu
R^2 = Ph, substituted Ph; 2-furanyl
R^3 = H, Ac

A one-pot microwave-assisted reaction facilitates the synthesis of novel tetracyclic sultams **246** from benzyl amines **244** and 2-cyanobenzenesulfonyl chloride **245** <09SC3607>. The proposed mechanism is likely to involve generation of a sulfonamide **247** followed by intramolecular cyclization with the nitrile, providing **248**. A second ring is formed via reaction of the amidine with the primary amino functionalities and subsequent elimination of ammonia.

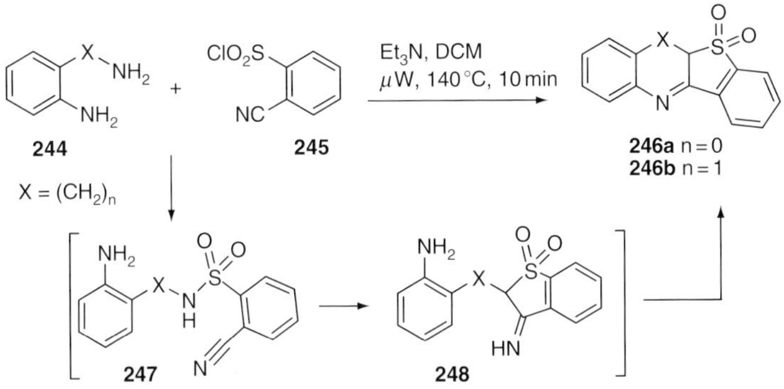

A recent review has discussed the development in the novel synthesis of saccharin related five- and six-membered benzosultams of the past 10 years <09H(78)1387>.

5.5.3.2 Reactions of Isothiazoles

N-Heterocyclic carbene (NHC)-catalyzed homoenolate addition to saccharin derivatized ketimines **253** was reported previously as a simple and convenient solution to the synthesis of stereochemically defined tricyclic γ-lactam **254** <08JA17266>. The use of IMes•HCl precatalyst **250** and DBU proves most effective for the formation of lactams predominantly with the two substituents R^1 and R^2 at *cis* orientation. The substrate scope of these reactions, however, is limited by the difficulties of preparing the starting α,β-unsaturated aldehydes **249**. Recently, α′-hydroxyenones **257**, which can be easily prepared in a single convenient step from aryl aldehydes **255**, have been introduced as efficient surrogates for enals in the annulation reactions <09JA8714>. The facile access to the starting α′-hydroxyenones makes possible a dramatic expansion of the substrate scope and use of an assortment of heterocyclic substituents. NHC-catalyzed annulations of aldehyde **249** are thought to occur via the intermediacy of **252**, which arises via tautomerization of **251**, the initial adduct of IMes•HCl **250** and aldehyde **249**. The identical intermediate **252** may also be reached by a *retro*-benzoin reaction of the α′-hydroxyenone adduct **258**.

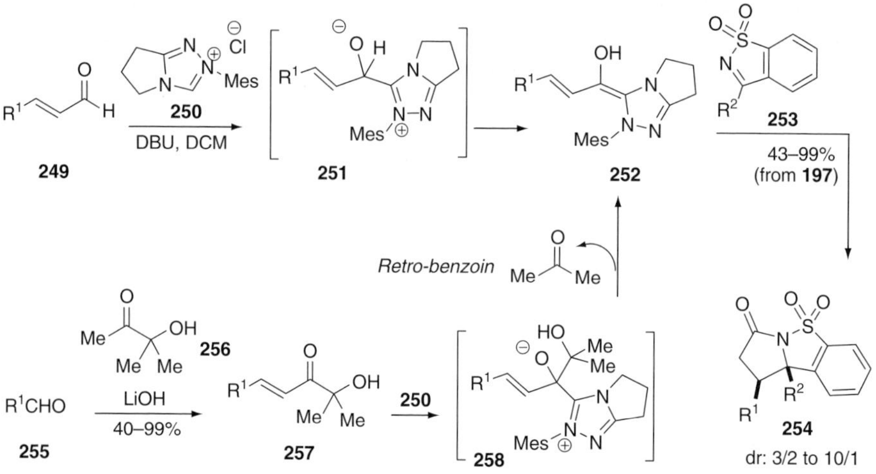

R¹ = Et, Ph, substituted Ph, 2-furyl, 2-py, 4-methylthiazolly, 2-quinolinyl
R² = Ph, substituted Ph, heteroaryl, *i*-Bu

The copper-promoted intramolecular aminooxygenation of alkene substrates has demonstrated its utility in the synthesis of disubstituted pyrrolidines from benzosultam <09OL1915>. Under the reaction condition (Cu(EH)$_2$, Cs$_2$CO$_3$ and 1.5 eq. of TEMPO in hot xylene), benzosultam **259** is converted to the *trans*-pyrrolidine **261**. The reaction pathway involves a primary carbon radical intermediate **264** that is trapped efficiently with TEMPO **260**. Since the *N*-substituent is directly tethered to the α-carbon of the sulfonamide unit in **259**, the reaction occurs via transition state **262**, which places the α-substituent in a pseudoequatorial position, thereby favoring the formation of the 2,5-*trans* pyrrolidine adduct **263**. The *N*-methoxy piperidine moiety can be conveniently converted to an aldehyde by a known procedure employing *m*CPBA.

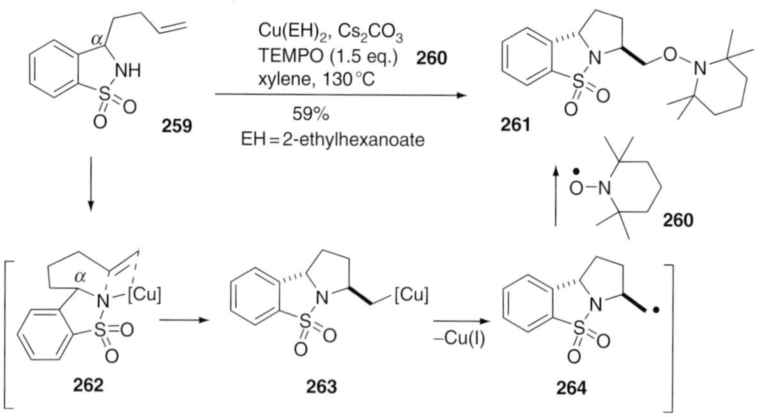

Using $Pd(CF_3CO_2)_2$ /(S,S)-f-binaphane **266** as the catalyst, an efficient enantioselective synthesis of sultams **267** and **269** was developed via asymmetric hydrogenation of the corresponding cyclic imines with high enantioselectivities <09JOC5633>. The hydrogenation products **269** can be easily transformed to chiral homoallylic amines **271** without erosion of the enantioselectivity.

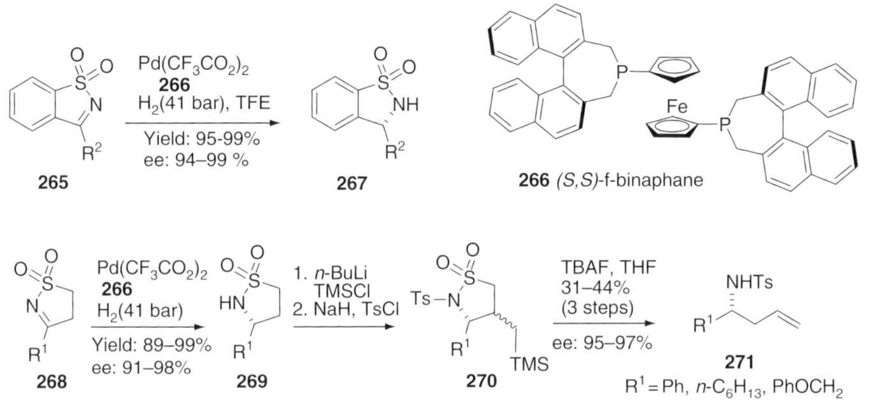

R^1 = alkyl, Ph, substituted Ph, Bn, substituted $PhOCH_2$, 2-naphthyl OCH_2
R^2 = alkyl, Ph, substituted Ph, Bn

The conversion of 3-chloro-4-cyano-isothiazoles **272** into pyrazoles **273** is accomplished upon treatment with hydrazine <09T7023>. The rational mechanism for this transformation is proposed wherein the initial attack of hydrazine occurs at C-5 position to afford the 2,5-dihydroisothiazole **275** that could be in equilibrium with its ring opened form **276**. Then, loss of HCl and sulfur would give the hydrazinyl acrylonitrile **277**. Intramolecular cyclization and subsequent tautomerization afford the 3-aminopyrazole **273**.

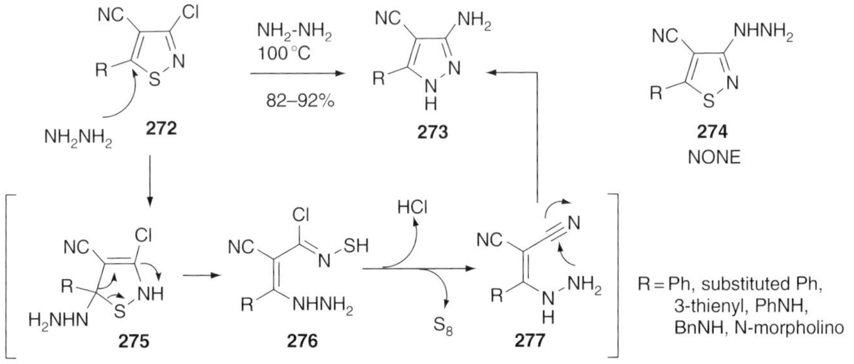

A ring-closing metathesis-based strategy has allowed access to a novel pyrido-sultam **282** <09JOC1982>. The synthetic route takes advantage of the ready availability of *N*-allylsaccharin **278** and proceeds via the incorporation of an allyl group to the sultam ring. Sultam **282** is transformed under direct irradiation at 350 nm into thienopyridine **283** and **284** via [1,3]-sigmatropy involving the S-N bond and heterocyclic ring cleavage, respectively.

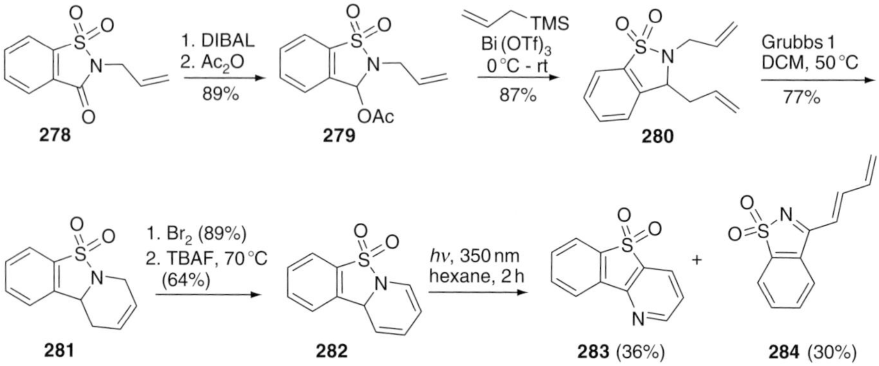

A short multigram process for the preparation of the analgesic compound SCP-123, **289** and its sodium salt **290** has been developed via hydrolysis of the saccharin ring in **288** <09OPRD820>. Saccharin **287** is added to 2-chloroacetamide **286**, prepared from the reaction of 4-aminophenol **285** and 2-chloroacetyl chloride, to afford the saccharin derivative **288**. This process route is deemed to be of great merit due to the low cost of the commercially available starting materials and also because both intermediates and final products **289** and **290** can be obtained in high purity (>95%) by precipitation or recrystallization; no chromatography is required in the entire process.

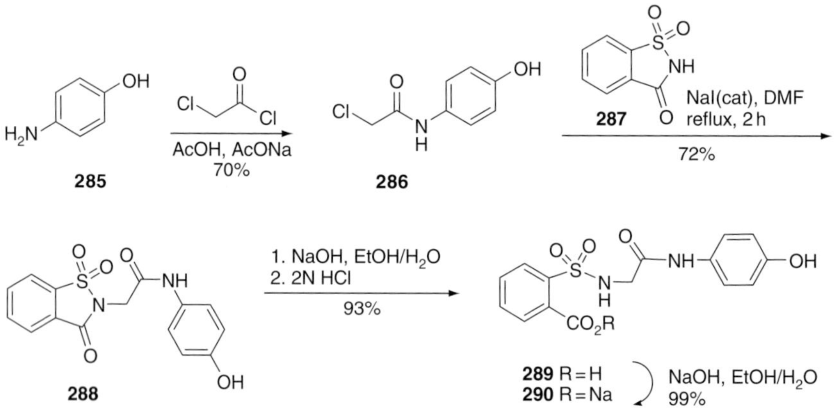

The 1-methyl-4-vinylbenzene **292** undergoes smooth hydroamination with saccharin **291** in the presence of 10 mol % of iodine to furnish benzyl saccharin in 78% yield. The use of inexpensive molecular iodine makes this protocol quite simple and practical <09TL5351>.

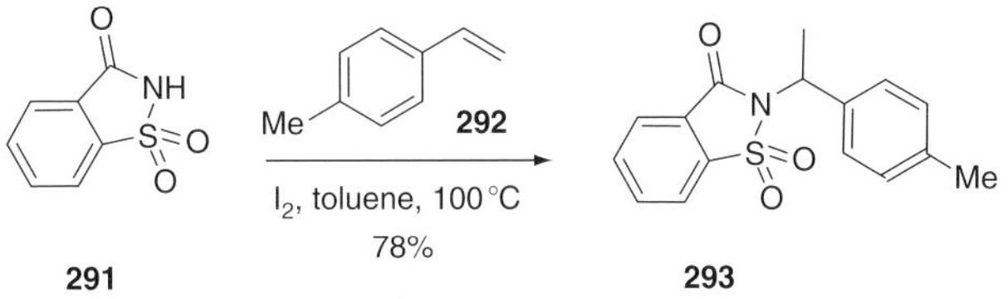

5.5.3.3 Isothiazoles as Auxiliaries and Reagents in Organic Syntheses

The Oppolzer camphorsultam and its derivatives continue demonstrating great versatility in asymmetric synthesis. Utilizing the Oppolzer's glycylsultam **296R** in the asymmetric [C+NC+CC] coupling reactions catalyzed by either Ag(I) <06OL3647> or CuI <07TL3867> offers convenient access to a variety of 4,5-*cis* pyrrolidine *endo*-products **297** and 4,5-*trans* pyrrolidine *exo*-products **298**, respectively. An efficient, scalable, and environmentally friendly synthesis of glycylsultam **296R** has been developed based on the Delépine reaction <09S1261>. A two-step process involves nucleophilic displacement of the activated bromide **301** by HMTA, followed by decomposition of the intermediate quaternary hexamethylenetetramine salt **302** with ethanolic hydrogen chloride. This reaction sequence results in a mixture of ammonium salts and diethoxymethane, from which the desired primary amine, glycylsultam **296R** can be obtained, after neutralization, in good overall yield.

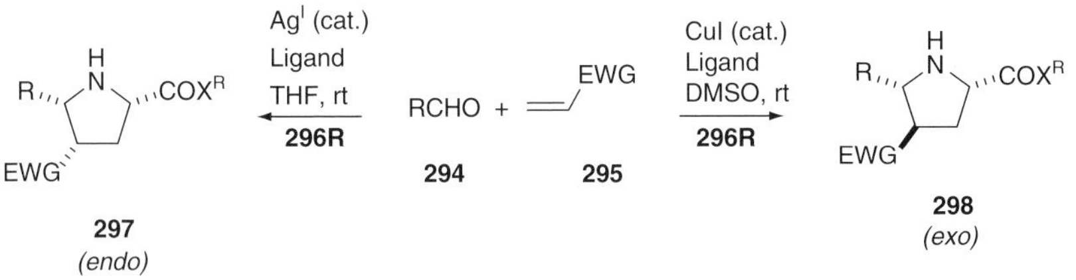

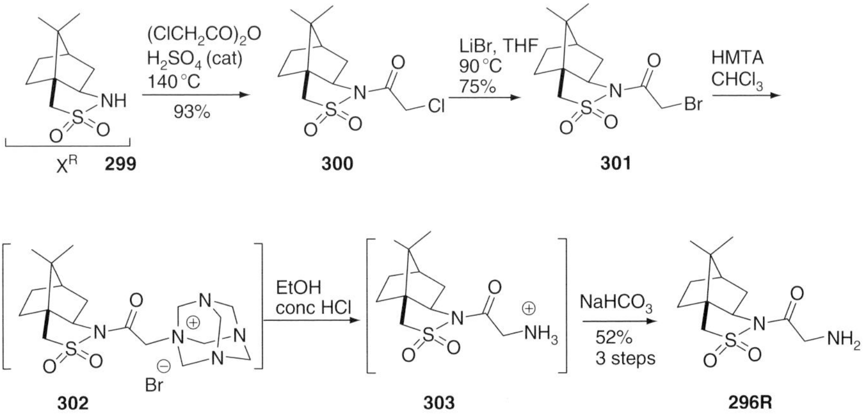

An enantioselective total synthesis of (−)-prestalotiopsin A **309** has engaged a [2+2] cycloaddition of *N*-propioloyl Oppolzer's L-camphorsultam [(2*R*)-bornane-10,2-sultam] **304** and ketene dialkyl acetal **305** in the presence of a catalytic amount of $ZrCl_4$ followed by stereoselective 1,4-hydride addition with lithium tri-secbutyl borohydride (L-Selectride) and protonation of the resulting enolate to provide highly enantioenriched cyclobutane derivative **307** <09JOC6452>. The observed high level of stereoselection in the tandem 1,4-hydride addition/protonation can be reasoned from the transition state of the more favorable conformation **311**, wherein the carbonyl group directs *anti* to the SO_2 group and adopts *S-trans* conformation to the α,β-unsaturated bond to avoid a steric repulsion between the ethoxy and the SO_2 groups in the *S-cis* conformer **310**. Then, the 1,4-hydride addition to the *S-trans* conformer **311** generates the *Z*-enolate intermediate **312**, which is reorganized to the more stable lithium-chelated conformer **313**. A proton approaches predominantly from the side opposite the bulky auxiliary (from the front side), giving **307** almost exclusively. The same synthetic venture has also provided (+)-prestalotiopsin with the employment of D-camphorsultam in the [2+2] cycloaddition reaction.

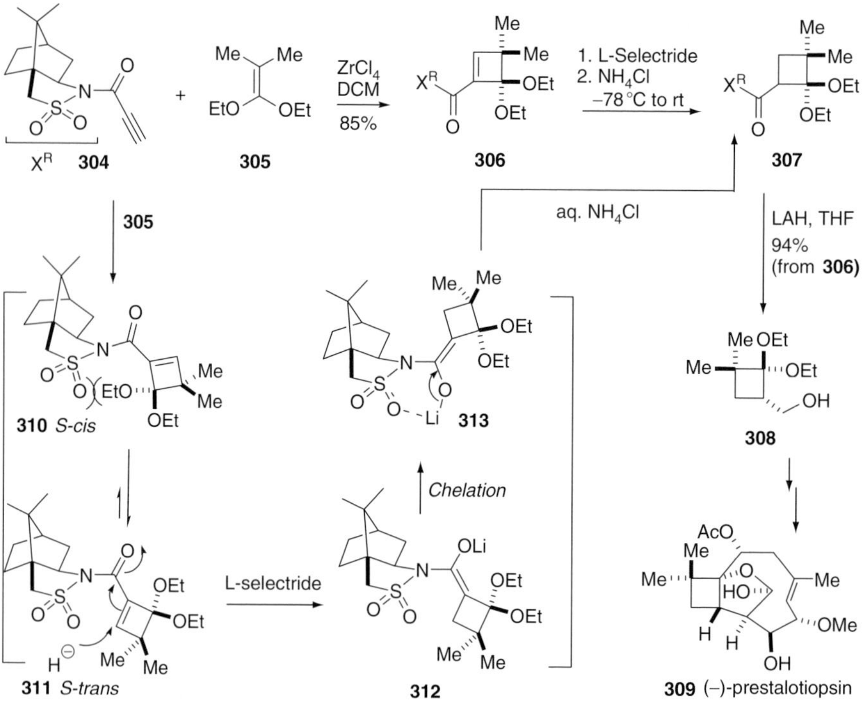

Additional application of Oppolzer's sultams is exemplified with an Oppolzer *anti*-aldol approach for the synthesis of the sex pheromone (3*S*,5*R*,6*S*)-3,5-dimethyl-6-isopropyl-3,4,5,6-tetrahydropyran-2-one <09TA1806>.

A novel oxidation reagent **315**, a benzosultam sulfonyloxaziridine bearing a pendant aldehyde moiety, offers a direct approach to the selective conversion of primary amines to chiral hydroxylamines with retention of stereochemistry <09JA3864>. Using tripeptide **314** as a model substrate, reagent **315** demonstrates its utility to effect the synthesis of solid supported *N*-terminal peptide hydroxylamine **316** in the sequential steps involving nitrone formation and hydrolysis to the hydroxylamine. Further ligation of **315** with a peptide α-ketoacid **317** affords a tetrapeptide **318** after its cleavage from the resin. The oxidation selectivity of reagent **315** arises from the rapid formation of *N*-terminal imine **323** that effects an intramolecular oxygen atom transfer to provide nitrone **322**. The rapid pre-complexation of the peptide **314** and reagent **315** reduces competing, and possibly more rapid, oxidation reactions at other sites of the peptide **314**. As a single reagent, **315** effects both imine formation and oxidation to avoid prolonged intermediacy of the stereochemically labile imine, thereby precluding erosion of the amine stereochemistry. The substrate scope of reagent **315** with respect to the identity of the *N*-terminal amine residue has been extended to a series of constant pentapeptides bearing different *N*-terminal amines, with reaction yields of 18–25%.

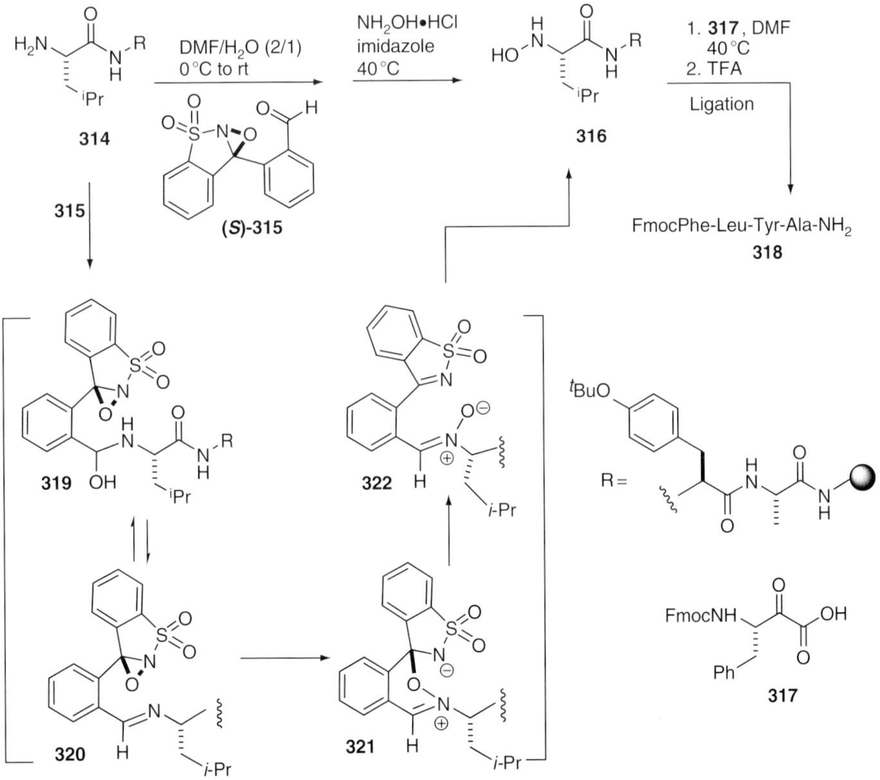

Davis oxaziridine **324**, the classic chiral oxidizing reagent, is a camphorylsulfonyl oxaziridine derivative. A mild and efficient method to synthesize chiral 3-aminosubstituted isothiazole sulfoxides (*S*)-**325** takes advantage of Davis oxaziridine (+)-**324** under microwave irradiation <09TA2247>. The determination of the absolute configuration of the chiral sulfoxide was achieved by theoretical calculation of the CD spectra. The reason for the observed stereoselectivity was enlightened by means of analysis of the data using DFT (density functional theory) calculations. Oxidation reaction using Davis oxaziridine (−)-**324** affords comparable results in terms of yields and ee, obviously in favour of the opposite enantiomer, (*R*)-**325**.

323 Yield: 50–80% ee: 50–65% (*S*)-**325** (major) (*R*)-**325** (minor) (−)-**324**

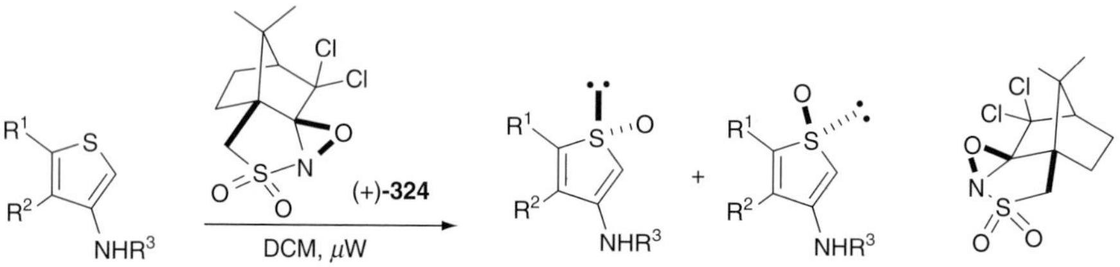

R^1 = H; R^2 = H, Cl, Br; R^3 = Bn, CH(Me)Ph, Me.

A recent application of Davis oxaziridine (+)-**324** in asymmetric oxidation of bis (alkylsulfanyl)-bipyridines **326** has provided a series of optically active mono sulfoxides **327** and bis-sulfoxides **328** <09PSS1247>. Both sulfoxides present possibilities of promoting enantioselective reaction as chiral ligands in the asymmetric addition of diethylzinc to benzaldehyde **329**.

326 **327** **328**

R = Me, Et, i-Pr

327: Yield 36–54% ee: 47–76%

328: Yield 8–13% ee: 42–>99%

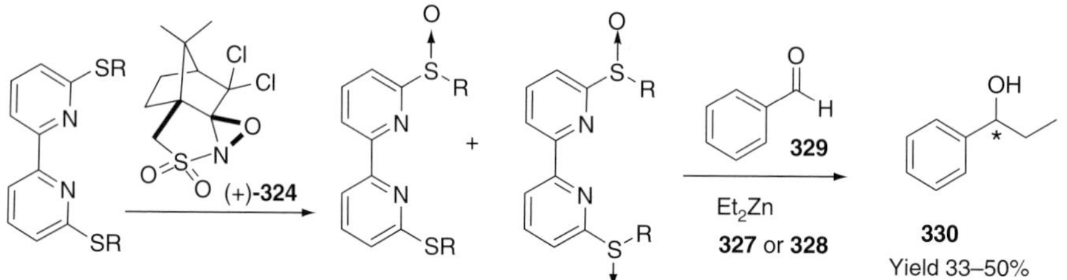

5.5.3.4 Pharmaceutically Interesting Isothiazoles

Several biologically active isothiazoles and their saturated and/or oxygenated analogs were reported in 2009. Sultams have been incorporated into HCV (hepatitis C virus) NS5 protease inhibitor **331** <09BMCL1105>, and HCV polymerase NS5B inhibitor **332** <09BMCL5652>, and isothiazole into inhibitor **333** of VEGF (vascular endothelial growth factor) receptors 1 and 2 <09BMCL1195>.

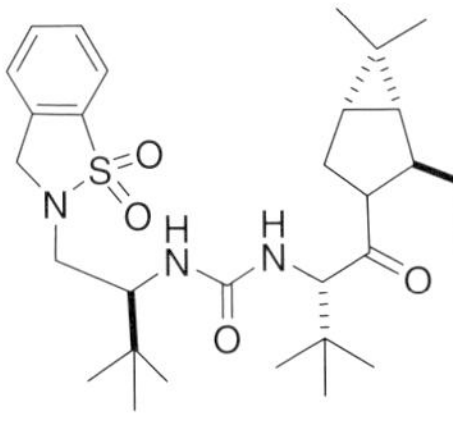

5.5.4. THIADIAZOLES AND SELENODIAZOLES

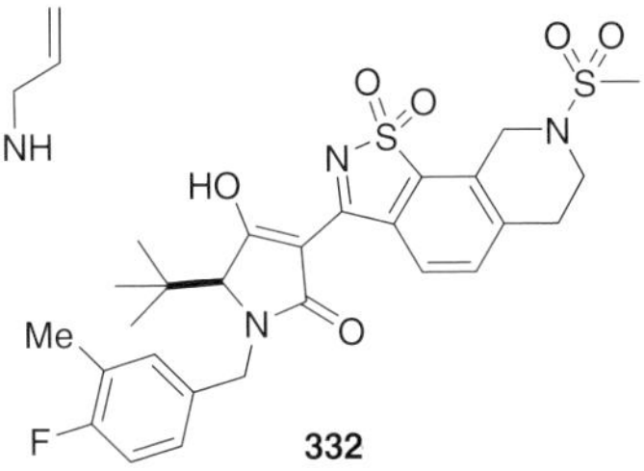

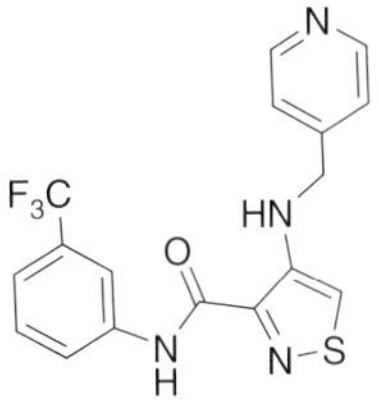

5.5.4.1 Syntheses of Thiadiazoles and Selenodiazoles

5-Amino-substituted 3-phenyl-1,2,4-thiadiazoles **338** can be efficiently synthesized by a concise sequence employing initial cyclization reaction of carboxamidine dithiocarbamate **335** with p-toluenesulfonyl chloride <09S913>. The carboxamidine dithiocarbamate **335** was produced by a three-component nucleophilic substitution reaction with carbon disulfide, benzamidine **334**, and benzyl chloride. The key intermediate, 5-(benzylsulfanyl)-3-phenyl-1,2,4-thiadiazole **336**, is then transformed to the desired 1,2,4-thiadiazoles **338** via oxidation of the sulfide group to form the corresponding sulfone **337** and subsequent substitution reactions with various amines.

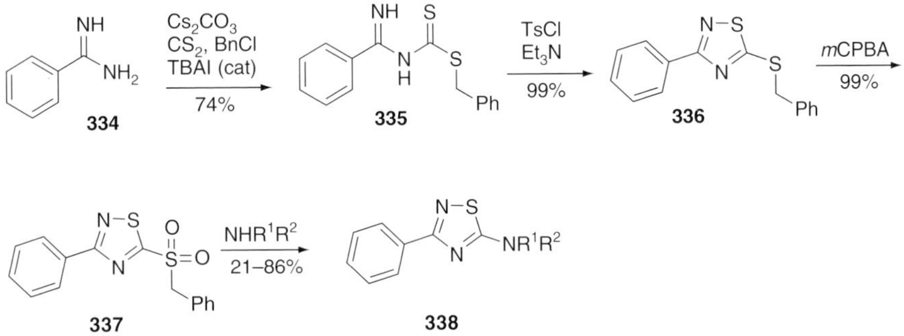

NHR¹R² = pyrrolidine, morpholine, N-alkyl-piperazine, 4-substituted-piperidine, 4-alkyl-piperazine, 4-aryl-piperazine, alkylamine, cycloalkylamine, substituted benzyl amine

The hypervalent iodine (V) reagent, o-iodoxybenzoic acid (IBX) **340** has been used as an effective promoter of oxidative dimerization of thiomides **339** to yield 3,5-disubstituted 1,2,4-thiadiazoles **341** in the presence of TEAB (tetraethylammonium bromide) <09TL5802>.

R = Ph, substituted Ph, Bn, 2-furanyl, 3-pyridyl

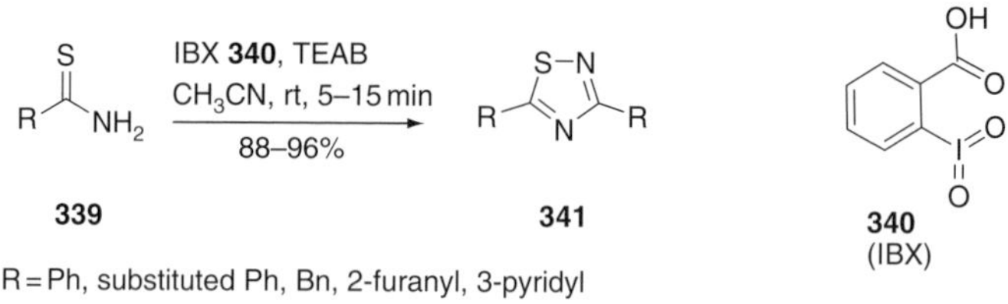

A solid-phase parallel synthesis procedure also provides easy access to 5-amino-substituted 1,2,4-thiadiazoles as well as 5-amido-1,2,4-thiadiazoles, that involves a cyclization reaction of resin-bound carboximidine thiourea **345** promoted by *p*-toluenesulfonyl chloride <09SL999>. The thiourea **345**, produced by addition of arylcarboximidine **344** to an isocyanate terminated resin **343**, serves as key intermediates that undergo cyclization to generate 5-amino-thiadiazole resins **346**. *N*-Alkylation or *N*-acylation reaction of **346** yields various functionalized thiadiazoles, **348** and **349**, after cleavage from the respective resins.

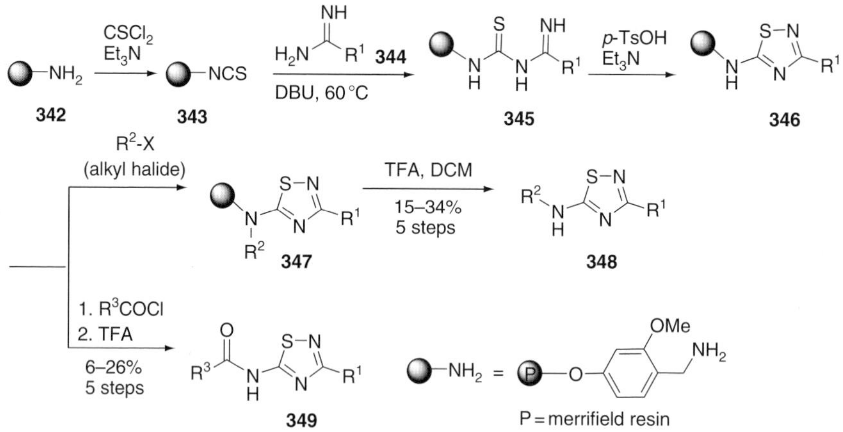

R^1 = Ph, 3-NO_2Ph; R^2 = Bn, substituted Bn, allyl, *i*-Bu, PhCH=CH-; R^3 = Ph, substituted Ph, heteroaryl.

Propylphosphonic anhydride (T3P®) **352** has been demonstrated to be an efficient and mild reagent for the one-pot synthesis of 1,3,4-thiadiazoles **353** from carboxylic acids **350** and acid hydrazide **351** via cyclodehydration reaction <09T9989>. As T3P® produces only water-soluble by-products, in most cases, an aqueous work up was sufficient to obtain pure products.

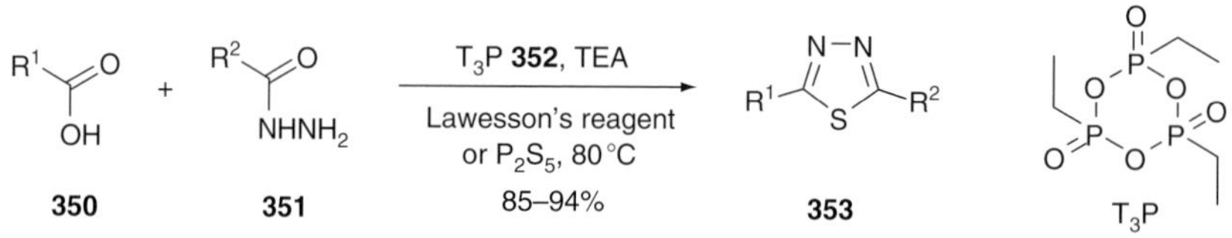

R^1 = alkyl, c-alkyl, substituted Ph, substituted-furan-2-yl, substiuted styryl
R^2 = H, alkyl, substituted Ph, substituted pyridin-3-yl

N-Chlorosuccinimide **356** has been identified as a convenient and safe alternative oxidant for the oxidative condensation of isothiocyanates **354** and isocyanates **355** to afford 1,2,4-thiadiazolidine-3,5-diones **357** <09TL257>. Evidence for the lack of an accelerating effect of radical initiators suggests a plausible classical ionic mechanism, involving an initial attack of the sulfur of **354** on the chlorine of NCS **356** affording **358**, which is immediately converted into **359**. Sulfenyl chloride **359** undergoes an electrophilic attack by the nitrogen of isocyanate **355**, forming the highly electrophilic species **360**, followed by cyclization to **361**. When exposed to air, iminium chloride **361** can be hydrolyzed to yield **357**.

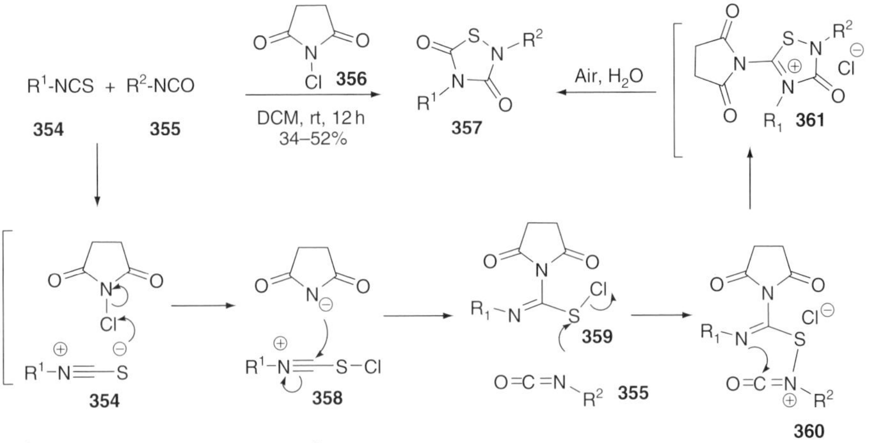

R^1 = Bn, Hex, Ph, c-Hex, *m*-anisyl; R^2 = Et, Bn, allyl, Ph, 2-Cl-Et

[1,2,4]Triazolo[3,4-*b*][1,3,4]thiadiazoles have attracted increased attention because of their interesting biological properties. Most synthetic routes construct the bicyclic system based on one existing ring, imidazole or thiadiazole. Recently an expeditious novel approach renders carboxylic acid **362** and thiocarbohydrazide **363** as starting materials using silica-supported dichlorophosphate as a recoverable cyclodehydrant under microwave irradiation to provide triazolo-thiadiazole **364** <09SC3816>. The protocol has the benefit of green chemistry with easy workup procedure.

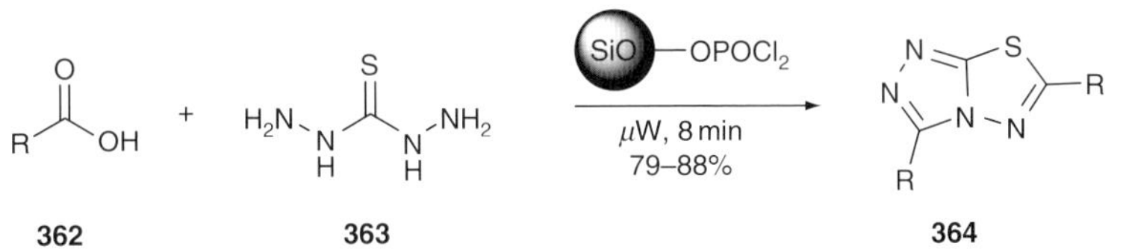

R = Ph, substituted Ph, 2-furanyl

5.5.4.2 Reactions of Thiadiazoles and Selenodiazoles

The sequential selective cross coupling of polyhalogenated heterocycles has emerged as an efficient strategy for the construction of polysubstituted heterocycles. A novel approach for the synthesis of substituted 5-amino- and 3-amino-1,2,4-thiadiazoles highlights the sequential selective palladium-catalyzed Suzuki-Miyaura coupling reactions from 3-bromo-5-chlorothiadiazole **365** <09OL5666>. Interestingly, under the reaction conditions the cross-coupling preferentially takes place at the chloride of 3-bromo-5-chloro-1,2,4-thiadiazole **365**, that enables a convenient protocol for the expeditious preparations of both 3-substituted-5-amino- and 3-amino-5-substituted-1,2,4-thiadiazoles, **369** and **371**, from this common starting material. FMO calculations rationalize the observed chemoselectivity for coupling at chlorine.

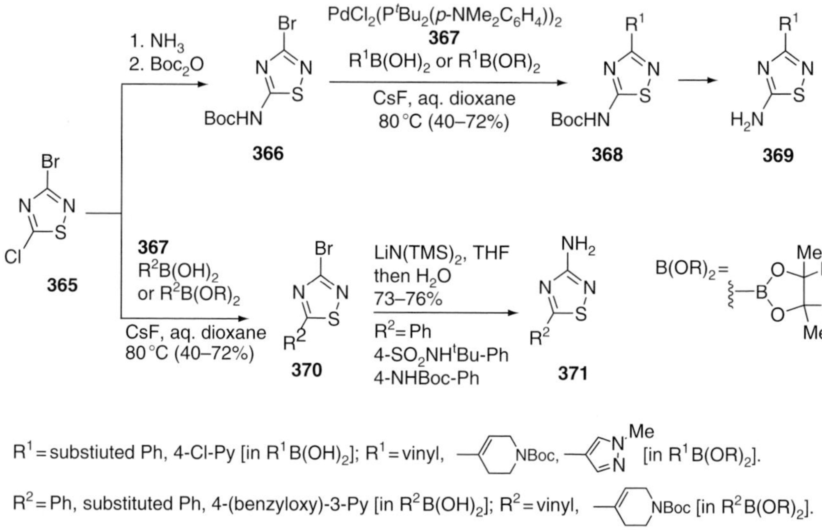

A general method for the synthesis of benzimidazole-4-sulfonamides **375** employs commercially available benzothiadiazole sulfonyl chloride **372** as a benzimidazole equivalent <09TL1219>. Reaction with a variety of amines followed by highly chemoselective reductive desulfurization gives intermediate 1,2-phenylenediamine **374**, which may then react with aryl, heteroaryl and alkyl aldehydes to provide substituted benzimidazole sulfonamides **375**.

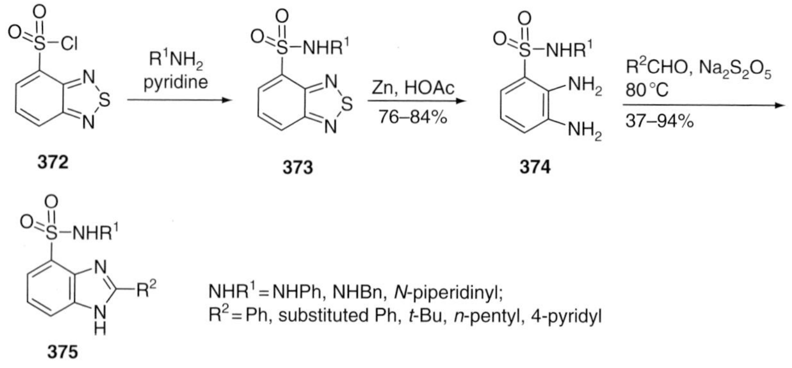

High reactivity of the selenenamide bond in selenazoles toward nucleophiles has been exploited for synthesis of other organoselenium compounds <09SC3141>. Treatment of benzselenadiazole **376** and related selenaheterocycle **380** containing a Se-N moiety with Grignard reagent provides unsymmetrical aryl-aryl and aryl-alkyl selenides **378** and **381**, respectively.

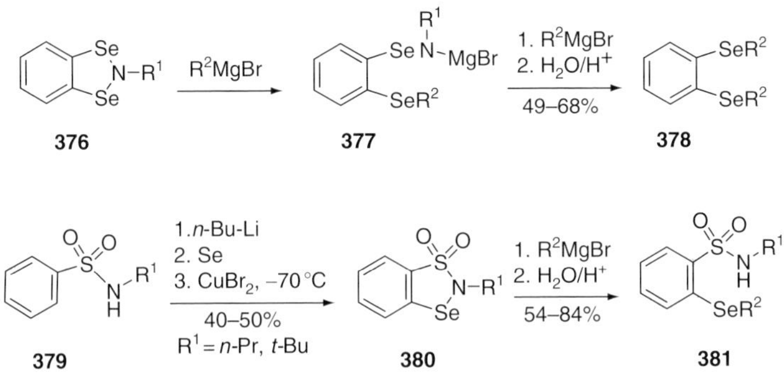

R^1 = Ph, i-Pr; R^2 = i-Pr, Ph, c-Hex, t-Bu, 3-methoxyPh

5.5.4.3 Pharmaceutically Interesting Thiadiazoles

The search for biologically active compounds among thiadiazoles and their saturated and/or oxygenated analogs continues to attract increasing attention in drug discovery. 1,2,5-Thiadiazolidin-1,1-dioxide derivative **382** has been identified as a potent γ-secretase inhibitor <09JMC3441>. 1,3,4-Thiadiazoles are incorporated into the selective EP3 receptor antagonist **383** <09BMCL4292>, c-Jun N-terminal kinase inhibitor **384** <09JMC1943>, cannabinoid CB1 receptor antagonist **386** <09BMCL142>, and the carbonic anhydrases IX and XII inhibitor **385** with *in vivo* tumor-targeting capability, which restricts binding to carbonic anhydrases present on tumor cells <09BMCL4851>. 1,2,5-Thiadiazole is incorporated into CXCR2 (CXC-chemokine receptors 2) antagonists **387** <09BMCL1434>, and 1,2,3-thiadiazole into HIV-1 NNRT (non-nucleoside reverse transcriptase) inhibitor **388** <09BMCL5920>. Interestingly, the selenadiazole analog **389** possesses NNRT inhibitory activity similar to the prototype series of thiadiazole <09BMCL6374>.

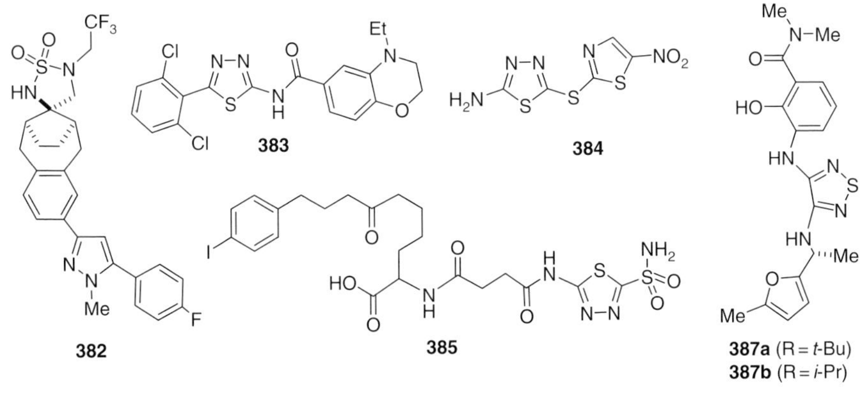

386

388

389

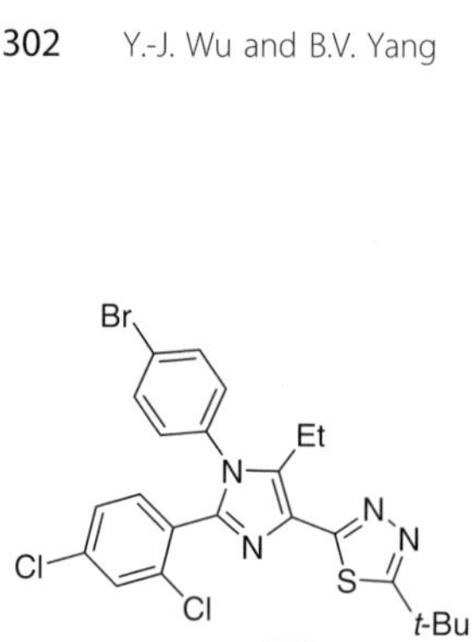

5.5.5. SELENAZOLES, 1,3-SELENADOLIDINES AND TELENAZOLES

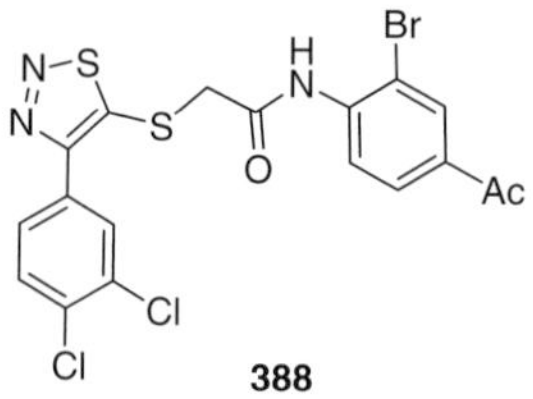

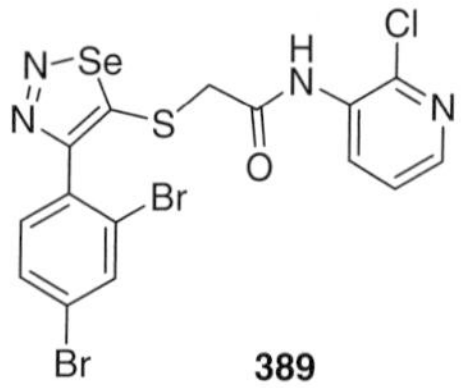

A new cycloaddition reaction of 3-selanylpropargyl alcohols **390** to benzoselenoamide **391** offers an effective means to access selenamides **392** <09OL2952>. α-Selanyl propadienyl cations **396** were easily generated in the catalytic system of scandium triflate in nitromethane-H_2O-Bu_4NHSO_4, forming the multifunctionalized selenazoles **392** with complete regioselectivities. Removal of the phenylselanyl group can be achieved by treatment with MeLi or Bu_3SnH/AIBN then MeLi. The catalytic cycle may involve in situ generation of the R-selanyl propadienyl cation **395** from scandium-catalyzed activation and dehydroxylation of the alcohol **390**. The propargyl cation **395** would isomerize to the propadienyl cation **396** stabilized by the α-selenium atom. The reaction of **396** with thioamides **391** probably gives the intermediate **397**, which easily cyclizes in the 5-*exo* mode, leading to selenazole **392**. Tetrabutylammonium hydrogensulfate acts as a scavenger of the eliminated hydroxyl group.

R = substituted Ph, 2-thienyl

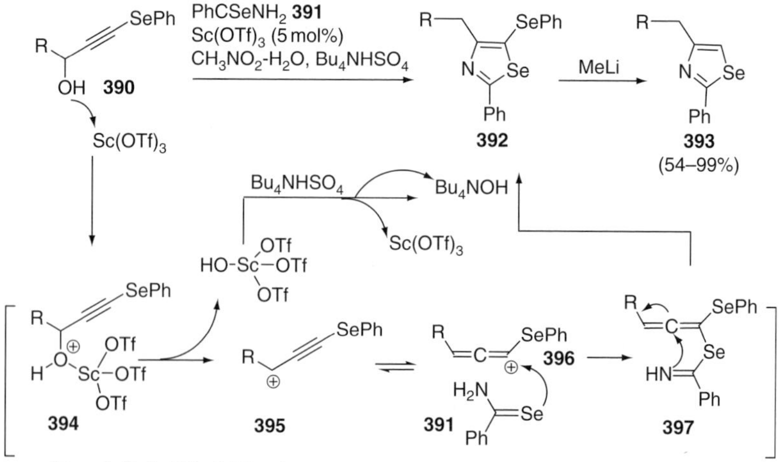

The syntheses of 2,4,5-trisubstituted-1,3-selenazoles **404** have been achieved by an easy one-pot four-step procedure from dimethyl cyanodithioimidocarbonate **398** <09T2982>. Chemical diversity was introduced on the selenazole rings by varying the amines **399** and the activated halides **402**. Diazotization and chlorination of the obtained 2,4-disubstituted 1,3-selenazole-5-carbonitrile **404** (EWR = CN) affords 4-chlorotriazine fused selenazole **405** <09S3472>.

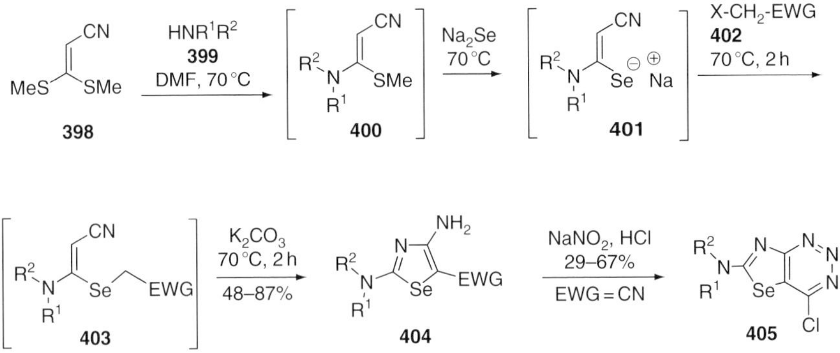

NR^1R^2 = 4-morpholinyl-, 1-pyrolidinyl, 1-piperidinyl; X = Cl, Br; EWG = CN, CO_2Et, Ac, $-NO_2$, Bz, 4-Cl-Ph-CO-

The first example of a sugar-derived 2-amino-1,3-selenazole has been prepared in three steps: coupling of benzoyl isoselenocyanate **407** with O-protected glucosamine **408**, Se-alkylation of the corresponding N-benzoyl selenourea **409** with phenacyl bromide, and acid-promoted intramolecular cyclization of **410** <09T2556>.

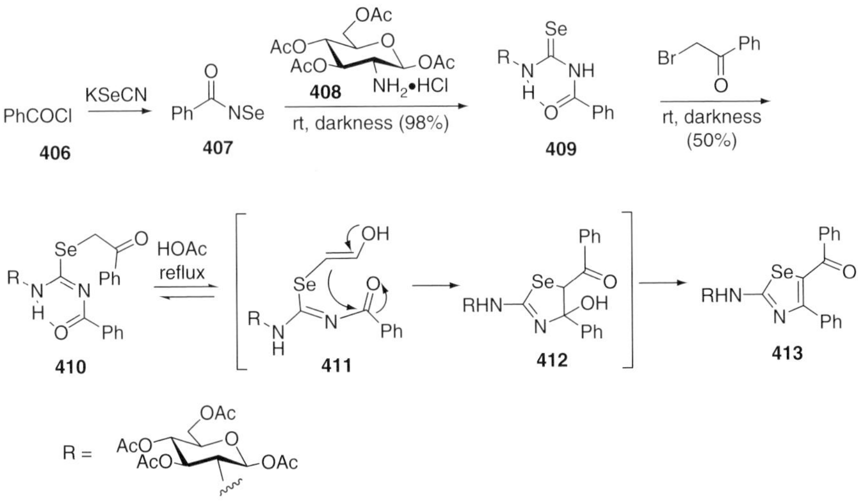

High reactivity of the selenenamide N-Se bond in benzisoselenazol-3(2H)-ones **414** has been exploited for synthesis of a series of organoselenium compounds **415**

with the utility of a Grignard reagent <09SC3141>. This reaction has a synthetic value because it is highly selective and can be realized under mild conditions.

R^1 = Ph, *i*-Pr;
R^2 = *i*-Pr, Ph, *c*-Hex, t-Bu, 3-methoxyPh

The tandem addition-cyclization reactions of 2-pyridinetellurenyl chloride **416** with indene **417** and 3,4-dihydro-2H-pyran **419** in methylene chloride at 20 °C gives condensed dihydro-1,3-tellurazole **418** and **420** <09CHC884>.

REFERENCES

93T5359	G.C. Mulqueen, G. Pattenden and D.A. Whiting, *Tetrahedron*, **49**, 5359 (1993).
95T7321	R.J. Boyce, G.C. Mulqueen and G. Pattenden, *Tetrahedron*, **51**, 7321 (1995).
03AG(E)83	S.L. You, H. Razavi and J.W. Kelly, *Angew. Chem. Int. Ed.*, **42**, 83 (2003).
05SL289	A. Sorg and R. Bruckner, *Synlett*, 289 (2005).
06OL3647	P. Garner, H.U. Kaniskan, J. Hu, J. Wiley and M. Panzner, *Org. Lett.*, **8**, 3647 (2006).
07S3535	E.A. Merritt and M.C. Bagley, *Synthesis*, 3535 (2007).
07SL954	E.A. Merritt and M.C. Bagley, *Synlett*, 954 (2007).
07TL3867	P. Garner, J. Hu, C.G. Parker, W.J. Youngs and D. Medvetz, *Tetrahedron Lett.*, **48**, 3867 (2007).
08BMCL4853	P. Raboisson, H. de Kock, A. Rosenquist, M. Nilsson, L. Salvador-Oden, T. Lin, N. Roue, V. Ivanov, K. Vickstrom, E. Hamelink, M. Edlund, L. Vrang, S. Vendeville, W. Van de Vreken, D. McGowan, A. Tahri, L. Hu, C. Boutton, O. Lenz, F. Delouvroy, G. Pille, D. Surleraux, P. Wigerinck, B. Samuelsson and K. Simmen, *Bioorg. Med. Chem. Lett.*, **18**, 4853 (2008).
08JA17266	T. Rommel, T. Fukuzumi and J.W. Bode, *J. Am. Chem. Soc.*, **130**, 17266 (2008).
09AG(E)1	A. Ullrich, Y. Chai, D. Pistorius, Y.A. Elnakady, J.E. Herrmann, K.J. Weissman, U. Kazmaier and R. Muller, *Angew. Chem. Int. Ed.*, **48**, 1 (2009).
09AG(E)4198	D. Lrfranc and M.A. Ciufolini, *Angew. Chem. Int. Ed.*, **48**, 4198 (2009).
09AG(E)4222	D. Ma, S. Xie, P. Xue, X. Zhang, J. Dong and Y. Jiang, *Angew. Chem. Int. Ed.*, **48**, 4222 (2009).
09BMCL142	J.Y. Kim, H.J. Seo, S.-H. Lee, M.E. Jung, K. Ahn, J. Kim and J. Lee, *Bioorg. Med. Chem. Lett.*, **19**, 142 (2009).
09BMCL1105	K.X. Chen, B. Vibulbhan, W. Yang, L.G. Nair, X. Tong, K.-C. Cheng and F.G. Njoroge, *Bioorg. Med. Chem. Lett.*, **19**, 1105 (2009).
09BMCL1195	K.A. Kiselyov, M. Semenova and V.V. Semenov, *Bioorg. Med. Chem. Lett.*, **19**, 1195 (2009).
09BMCL1434	P. Biju, A.G. Taveras, Y. Yu, J. Zheng, R.W. Hipkin, J. Fossetta, X. Fan, J. Fine and D. Lundell, *Bioorg. Med. Chem. Lett.*, **19**, 1434 (2009).
09BMCL4292	M.A. Hilfiker, N. Wang, X. Hou, Z. Du, M.A. Pullen, M. Nord, R. Nagilla, H.E. Fries, C.W. Wu, A.C. Sulpizio, J.-P. Jaworski, D. Morrow, R.M. Edwards and J. Jin, *Bioorg. Med. Chem. Lett.*, **19**, 4292 (2009).

09BMCL4851	J.K. Ahlskog, C.E. Dumelin, S. Truessel, J. Marlind and D. Neri, *Bioorg. Med. Chem. Lett.*, **19**, 4851 (2009).
09BMCL5652	C. Yee, G. Adjabeng, T.R. Elworthy, J. Li, B. Wang, J.Y. Bamberg, S.F. Harris, A. Wong, V.J.P. Leveque, I. Najera, S. Le Pogam, S. Rajyaguru, G. Ao-Ieong, L. Alexandrova, S. Larrabee, M. Brandl, A. Briggs, S. Sukhtankar and R. Farrell, *Bioorg. Med. Chem. Lett.*, **19**, 5652 (2009).
09BMCL5920	P. Zhan, X. Liu, Z. Li, Z. Fang, Z. Li, D. Wang, C. Pannecouque and E. De Clercq, *Bioorg. Med. Chem. Lett.*, **19**, 5920 (2009).
09BMCL6374	P. Zhan, X. Liu, Z. Fang, C. Pannecouque and E. De Clercq, *Bioorg. Med. Chem. Lett.*, **19**, 6374 (2009).
09CAJ111	Y. Numajiri, T. Takahashi and T. Doi, *Chem. Asian J.*, **4**, 111 (2009).
09CB893	A. Pereira, Z. Cao, T.F. Murray and W.H. Gerwick, *Chem. Biol.*, **16**, 893 (2009).
09CEJ10280	K. Snegaroff, J. L'Helgoual'ch, G. Bentabed-Ababsa, T.T. Nguyen, F. Chevallier, M. Yonehara, M. Uchiyama, A. Derdour and F. Mongin, *Chem. Eur. J.*, **15**, 10280 (2009).
09CHC884	A.V. Borisov and Z.V. Matsulevich, *Chem. Heterocyclic Compounds*, **45**, 884 (2009).
09CR1371	A. Gaumont, M. Gulea and J. Levillain, *Chem. Rev.*, **109**, 1371 (2009).
09DF363	H. Moualla, D.A. Mills, R. Hromas, C.F. Verschraegen, *Drug Fut.*, **34**, 363 (2009).
09DF545	S. Davies, *Drug Fut.*, **34**, 545 (2009).
09DF861	G. Escolar and M. Diaz-Ricart, *Drug Fut.*, **34**, 861 (2009).
09EJOC2635	V.A. Rassadin, A.A. Tomashevskiy, V.V. Sokolov, A. Ringe, J. Magull and A. de Meijere, *Eur. J. Org. Chem.*, 2635 (2009).
09EJOC3228	M. Schnurch, A.F. Khan, M.D. Mihovilovic and P. Stanetty, *Eur. J. Org. Chem.*, 3228 (2009).
09EJOC4357	G. Mercey, J. Lohier, A. Gaumont, J. Levillain and M. Gulea, *Eur. J. Org. Chem.*, 4357 (2009).
09EJOC4833	S.C. Mckeon, H. Muller-Bunz and P.J. Guiry, *Eur. J. Org. Chem.*, 4833 (2009).
09H(78)1387	Z. Liu and Y. Takeuchi, *Heterocycles*, **78**, 1387 (2009).
09HCA607	L. Yu, C. Peng, Z. Li and Y. Guo, *Helv. Chim. Acta*, **92**, 607 (2009).
09JA3291	L. Campeau, D.R. Stuart, J. Leclerc, M. Bertrand-Laperle, E. Villemure, H. Sun, S. Lasserre, N. Guimond, M. Lecavallier and K. Fagnou, *J. Am. Chem. Soc.*, **131**, 3291 (2009).
09JA3864	T. Fukuzumi and J.W. Bode, *J. Am. Chem. Soc.*, **131**, 3864 (2009).
09JA8714	P.-C. Chiang, M. Rommel and J.W. Bode, *J. Am. Chem. Soc.*, **131**, 8714 (2009).
09JA17052	H. Do and O. Daugulis, *J. Am. Chem. Soc.*, **131**, 17052 (2009).
09JAN613	A. Mukai, T. Fukai, Y. Hoshino, K. Yazawa, K. Harada and Y. Mikami, *J. Antibiotics*, **62**, 613 (2009).
09JHC231	A.C. Dawsey, C. Potter, J.D. Knight, Z.C. Kennedy, E.A. Smith, A.M. Acevedo-Jake, A.J. Puciaty, C.R. Metz, C.F. Beam, W.T. Pennington and D.G. Van Derveer, *J. Heterocycl. Chem.*, **46**, 231 (2009).
09JMC238	R. Balasubramanian, B. Raghavan, A. Begaye, D.L. Sackett and R.A. Fecik, *J. Med. Chem.*, **52**, 238 (2009).
09JMC1943	S.K. De, J.L. Stebbins, L.-H. Chen, M. Riel-Mehan, T. Machleidt, R. Dahl, H. Yuan, A. Emdadi, E. Barile, V. Chen, R. Murphy and M. Pellecchia, *J. Med. Chem.*, **52**, 1943 (2009).
09JMC3028	H. Zhou, M.A. Aujay, M.K. Bennett, M. Dajee, S.D. Demo, Y. Fang, M.N. Ho, J. Jiang, C.J. Kirk, G.J. Laidig, E.R. Lewis, Y. Lu, T. Muchamuel, F. Parlati, E. Ring, K.D. Shenk, J. Shields, P.J. Shwonek, T. Stanton, C.M. Sun, C. Sylvain, T.M. Woo and J. Yang, *J. Med. Chem.*, **52**, 3028 (2009).
09JMC3441	L.E. Keown, I. Collins, L.C. Cooper, T. Harrison, A. Madin, J. Mistry, M. Reilly, M. Shaimi, C.J. Welch, E.E. Clarke, H.D. Lewis, J.D.J. Wrigley, J.D. Best, F. Murray and M.S. Shearman, *J. Med. Chem.*, **52**, 3441 (2009).
09JMC7014	M.E. Di Francesco, G. Dessole, E. Nizi, P. Pace, U. Koch, F. Fiore, S. Pesci, J. Di Muzio, E. Monteagudo, M. Rowley and V. Summa, *J. Med. Chem.*, **52**, 7014 (2009).
09JNP732	D.S. Dalisay, E.W. Rogers, A.S. Edison and T.F. Molinski, *J. Nat. Prod.*, **72**, 732 (2009).

09JNP841	C. Zhang, K. Herath, H. Jayasuriya, J.G. Ondeyka, D.L. Zink, J. Occi, G. Birdsall, J. Venugopal, M. Ushio, B. Burgess, P. Masurekar, J.F. Barett and S.B. Singh, *J. Nat. Prod.*, **72**, 841 (2009).
09JOC850	Y. Cheng, Y. Ma, X. Wang and J. Mo, *J. Org. Chem.*, **74**, 850 (2009).
09JOC1047	B. Liegault, I. Petrov, S.I. Gorelsky and K. Fagnou, *J. Org. Chem.*, **74**, 1047 (2009).
09JOC1179	J. Roger, F. Pozgan and H. Doucet, *J. Org. Chem.*, **74**, 1179 (2009).
09JOC1826	B. Liegault, D. Lapointe, L. Caron, A. Vlassova and K. Fagnou, *J. Org. Chem.*, **74**, 1826 (2009).
09JOC1982	L.A. Paquette, R.D. Dura and I. Modolo, *J. Org. Chem.*, **74**, 1982 (2009).
09JOC2578	F.G. Simeon, M.T. WEndahl and V.W. Pike, *J. Org. Chem.*, **74**, 2578 (2009).
09JOC3689	A.K. Ghosh, S. Banerjee, S. Sinha, S. Kang and B. Zajc, *J. Org. Chem.*, **74**, 3689 (2009).
09JOC5633	C.-B. Yu, D.-W. Wang and Y.-G. Zhou, *J. Org. Chem.*, **74**, 5633 (2009).
09JOC5750	V.S. Aulakh and M.A. Ciufolini, *J. Org. Chem.*, **74**, 5750 (2009).
09JOC6452	K.-I. Takao, N. Hayakawa, R. Yamada, T. Yamaguchi, H. Saegusa, M. Uchida, S. Samejima and K.-I. Tadano, *J. Org. Chem.*, **74**, 6452 (2009).
09JOC7084	M.J. Thompson and B. Chen, *J. Org. Chem.*, **74**, 7084 (2009).
09JOC7536	E. Sanchez-Larios and M. Gravel, *J. Org. Chem.*, **74**, 7536 (2009).
09JOC8309	I. Popov, H. Do and O. Daugulis, *J. Org. Chem.*, **74**, 8309 (2009).
09JOC8988	J.F. Sanz-Cervera, R. Blasco, J. Piera, M. Cynamon, I. Ibanez, M. Murguia and S. Fustero, *J. Org. Chem.*, **74**, 8988 (2009).
09JOMC455	F. Derridj, J. Roger, F. Geneste, S. Djebbar and H. Doucet, *J. Organometallic Chem.*, **694**, 455 (2009).
09OBC739	M.J. Peterson, I.D. Jenkins and W.A. Loughlin, *Org. Biomol. Chem.*, **7**, 739 (2009).
09OBC3264	J. Zhang and Y. Cheng, *Org. Biomol. Chem.*, **7**, 3264 (2009).
09OL531	A. Zhou, D. Rayabarapu and P.R. Hanson, *Org. Lett.*, **11**, 531 (2009).
09OL799	S. Kaeothip, P. Pornsuriyasak, N.P. Rath and A.V. Demchenko, *Org. Lett.*, **11**, 799 (2009).
09OL1301	A.A. Bowers, N. West, T.L. Newkirk, A.E. Troutman-Youngman, S.L. Schreiber, O. Wiest, J.E. Bradner and R.M. Williams, *Org. Lett.*, **11**, 1301 (2009).
09OL1607	D. Monguchi, T. Fujiwara, H. Furukawa and A. Mori, *Org. Lett.*, **11**, 1607 (2009).
09OL1635	M.T. Crimmins and A.R. Dechert, *Org. Lett.*, **11**, 1635 (2009).
09OL1733	J. Canivet, J. Yamaguchi, I. Ban and K. Itami, *Org. Lett.*, **11**, 1733 (2009).
09OL1915	M.C. Paderes and S.R. Chemler, *Org. Lett.*, **11**, 1915 (2009).
09OL2193	B. Checa, E. Galvez, R. Parello, M. Sau, P. Romea, F. Urpi, M. Font-Bardia and X. Solans, *Org. Lett.*, **11**, 2193 (2009).
09OL2952	M. Yoshimatsu, T. Yamamoto, A. Sawa, T. Kato, G. Tanabe and O. Muraoka, *Org. Lett.*, **11**, 2952 (2009).
09OL3104	A. Vazquez-Romero, L. Cardenas, E. Blasi, X. Verdaguer and A. Riera, *Org. Lett.*, **11**, 3104 (2009).
09OL3182	Y. Li, F. Shi, Q. He and S. You, *Org. Lett.*, **11**, 3182 (2009).
09OL3690	T. Martin, C. Laguerre, C. Hoarau and F. Marsais, *Org. Lett.*, **11**, 3690 (2009).
09OL4160	D. Lapointe and K. Fagnou, *Org. Lett.*, **11**, 4160 (2009).
09OL4672	P. Ghosh, J.R. Cusick, J. Inghrim and L.J. Williams, *Org. Lett.*, **11**, 4672 (2009).
09OL5666	P.M. Wehn, P.E. Harrington and J.E. Eksterowicz, *Org. Lett.*, **11**, 5666 (2009).
09OPRD820	L. Miao, L. Xu, K.W. Narducy and M.L. Trudell, *Org. Process Res. Dev.*, **13**, 820 (2009).
09PSS1247	J. Lawecka, B. Bujnicki, J. Drabowicz, J. Luczak and A. Rykowski, *Phosphorus, Sulfur, Silicon*, **184**, 1247 (2009).
09S474	C.P. Narasimhulu and P. Das, *Synthesis*, 474 (2009).
09S913	J.Y. Park, I.A. Ryu, J.H. Park, D.C. Ha and Y.-D. Gong, *Synthesis*, 913 (2009).
09S1261	K.-I. Takao, N. Hayakawa, R. Yamada, T. Yamaguchi, H. Saegusa, M. Uchida, S. Samejima and K.-i. Tadano, *J. Org. Chem.*, **74**, 6452 (2009).
09S2249	V.M. Muzalevskiy, V.G. Nenajdenko, A.V. Shastin, E.S. Balenkova and G. Haufe, *Synthesis*, 2249 (2009).
09S3472	E. Perspicace, D. Thomae, G. Hamm, S. Hesse, K. Kirsch and P. Seck, *Synthesis*, 3472 (2009).
09SC3141	R. Lisiak and J. Młochowski, *Synth. Commun.*, **39**, 3141 (2009).
09SC3607	A.G. Cole, S.G. Kultgen and I. Henderson, *Synth. Commun.*, **39**, 3607 (2009).

09SC3816	Z. Li and Y. Zhao, *Synth. Commun.*, **39**, 3816 (2009).
09SL23	J.G.M. Morton, L.D. Kwon, J.D. Freeman and J.T. Njardarson, *Synlett*, 23 (2009).
09SL999	I.A. Ryu, J.Y. Park, H.C. Han and Y.-D. Gong, *Synlett*, 999 (2009).
09SL233	C.S.P. McErlean and A.C. Willis, *Synlett*, 233 (2009).
09SL1189	J.R. de Alaniz and T. Rovis, *Synlett*, 1189 (2009).
09SL1341	S. Shankar, P.M. Sani, G. Terraneo and M. Zanda, *Synlett*, 1341 (2009).
09T2102	A. Sakakura, R. Kondo, S. Umemura and K. Ishihara, *Tetrahedron*, **65**, 2102 (2009).
09T2556	O. Lopez, S. Maza, V. Ulgar, I. Maya and J.G. Fernandez-Bolanos, *Tetrahedron*, **65**, 2556 (2009).
09T2982	D. Thomae, E. Perspicace, Z. Xu, D. Henryon, S. Schneider, S. Hesse, G. Kirsch and P. Seck, *Tetrahedron*, **65**, 2982 (2009).
09T3180	D.K. Rayabarapu, A. Zhou, K.O. Jeon, T. Samarakoon, A. Rolfe, H. Siddiqui and P.R. Hanson, *Tetrahedron*, **65**, 3180 (2009).
09T4992	K.O. Jeon, D. Rayabarapu, A. Rolfe, K. Volp, I. Omar and P.R. Hanson, *Tetrahedron*, **65**, 4992 (2009).
09T5739	N. Primas, A. Bouillon, J. Lancelot, H. El-Kashef and S. Rault, *Tetrahedron*, **65**, 5739 (2009).
09T7023	H.A. Ioannidou and P.A. Koutentis, *Tetrahedron*, **65**, 7023 (2009).
09T9989	J.K. Augustine, V. Vairaperumal, S. Narasimhan, P. Alagarsamy and A. Radhakrishnan, *Tetrahedron*, **65**, 9989 (2009).
09TA298	Y. Iwaki, M. Kaneko and H. Akita, *Tetrahedron: Asymmetry*, **20**, 298 (2009).
09TA1806	P. Prabhakar, S. Rajaram and Y. Venkateswarlu, *Tetrahedron: Asymmetry*, **20**, 1806 (2009).
09TA1975	R. Kikuchi, M. Fujii and H. Akita, *Tetrahedron: Asymmetry*, **20**, 1975 (2009).
09TA2027	S. Tang, Z. Xu and T. Ye, *Tetrahedron: Asymmetry*, **20**, 2027 (2009).
09TA2247	A. Casoni, G. Celentano, F. Clerici, A. Contini, M.L. Gelmi, G. Mazzeo, S. Pellegrino and C. Rosini, *Tetrahedron: Asymmetry*, **20**, 2247 (2009).
09TL257	S. Nasim and P.A. Crooks, *Tetrahedron Lett.*, **50**, 257 (2009).
09TL1196	M.A. Terzidis, J. Stephanidou-Stephanatou and C.A. Tsoleridis, *Tetrahedron Lett.*, **50**, 1196 (2009).
09TL1219	M.D. Rosen, Z.M. Simon, K.T. Tarantino, L.X. Zhao and M.H. Rabinowitz, *Tetrahedron Lett.*, **50**, 1219 (2009).
09TL3161	S. Shen, M. Lei, Y. Wong, M. Tong, P. Teo, S. Chiba and K. Narasaka, *Tetrahedron Lett.*, **50**, 3161 (2009).
09TL4874	Z. Gandara, M. Perez, X. Perez-Garcia, G. Gomez and Y. Fall, *Tetrahedron Lett.*, **50**, 4874 (2009).
09TL5351	J.S. Yadav, B.V. Subba Reddy, T.S. Rao and B.B.M. Krishna, *Tetrahedron Lett.*, **50**, 5351 (2009).
09TL5660	A.A. Thorave, P.N. Prajapati, J.P. pethani, K.C. Kothari, M.R. Jain, P.R. Patel and R.K. Kharul, *Tetrahedron Lett.*, **50**, 5660 (2009).
09TL5802	V. Kikelj, K. Julienne, J.-C. Meslin and D. Deniaud, *Tetrahedron Lett.*, **50**, 5802 (2009).
09TL7286	M.P. Zawistoski, S.M. Decker and D.A. Griffith, *Tetrahedron Lett.*, **50**, 7286 (2009).
TPC	https://www.thomson-pharma.com.

CHAPTER 5.6

Five-Membered Ring Systems: With O & S (Se, Te) Atoms

R. Alan Aitken*, Lynn A. Power**
*School of Chemistry, University of St. Andrews, UK
raa@st-and.ac.uk
**IOTA NanoSolutions, Liverpool, UK

5.6.1. 1,3-DIOXOLES AND DIOXOLANES

Hydroxyacetophenones are conveniently converted into the corresponding 1,3-dioxolanes by treatment with ethanediol and triisopropyl orthoformate in the presence of catalytic cerium triflate <09S1318> and reaction of ketones with $PhICl_2$ and ethanediol gives the α-chloro-1,3-dioxolanes <09S2324>. Perfluoroacetophenone reacts with 2-chloroethanol and potassium carbonate to give the dioxolane **1** which is converted into **2** with chlorine, and **2** can be further reacted with SbF_3, SbF_5 and $TiCl_4$ to give **3** <09RJA2156>. The competition between formation of 4-hydroxymethyl-1,3-dioxolanes and 5-hydroxy-1,3-dioxanes in the reaction of carbonyl compounds with glycerol has been examined using both MoO_3/SiO_2 <09JMOA(310)150> and phosphomolybdic acid <09S557>. MoO_3/SiO_2 is also effective in catalysing reactions of aldehydes and ketones with epichlorohydrin to give 4-chloromethyl-1,3-dioxolanones **4** <09MI1404> while reaction of salicylaldehydes with epichlorohydrin gives the bicyclic products **5** <09T8407>. The dimerisation of styrene epoxides to give dioxolanes **6** catalysed by phosphoric acid has been shown to involve a 1,2-hydride shift <09TL5927>. A synthetic and theoretical study of the addition of the carbonyl ylides formed by thermolysis of epoxides **7** with aldehydes, Ar^2CHO, to give dioxolane products **8** has appeared, including the X-ray structures of eight examples of compounds **8** <09JOC2120>.

1

2

3

4

5

6

7 **8**

9

ISSN 0959-6380, DOI: 10.1016/S0959-6380(11)22010-5

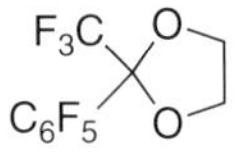

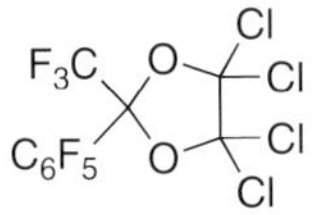

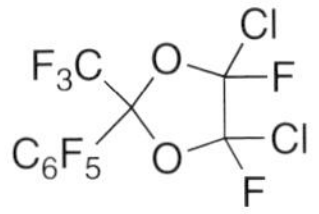

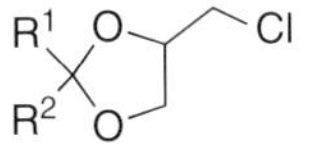

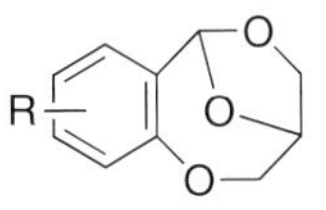

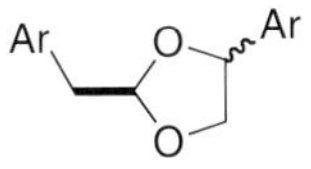

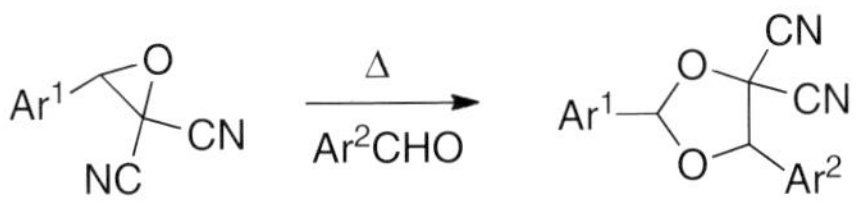

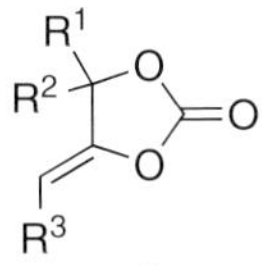

A review on the synthesis of cyclic carbonates from CO_2 includes various routes to 1,3-dioxolan-2-ones <09CC1312>. New effective catalysts for the reactions of epoxides with CO_2 to give 1,3-dioxolan-2-ones include zinc terephthalate/2-aminoterephthalate metal-organic frameworks <09EJI3552>, S-bridged bismuth bis(phenolate) compounds <09CC1136>, an aluminium salen complex bearing quaternary ammonium groups <09CC2577>, and a silica-supported bimetallic aluminium salen complex <09CEJ11454> which is proposed as a means of removing CO_2 from power station waste gases in a continuous flow reactor. A mechanistic study on the use of aluminium salen and Bu_4NBr for this process has revealed an unexpected role of the latter in the catalytic cycle <09AGE2946>. An enantioselective synthesis of (R)-propylene carbonate starting from ethyl (S)-lactate has been reported <09S1403> and preparation of methylenedioxolanones **9** can be achieved either by reacting propargylic alcohols with CO_2 and an N-heterocyclic carbene/CO_2 adduct as catalyst <09AGE4194> or by treatment of t-butyl propargyl carbonates with catalytic $(Ph_3P)AuNTf_2$ <09T1889>.

New theoretical studies have appeared on 2-methylene-1,3-dioxolane <09STC961> and 1,3-dioxol-2-one and 1,3-dioxole-2-thione and their radical cations <09SAA16>. Newly reported X-ray structures of 1,3-dioxolanes include **10** <09AXEo579>, the series of diamides **11**, important intermediates in synthesis of anti-cancer platinum drugs <09AXEo764, 09AXEo841, 09AXEo1960>, the dioxolodioxasilepine **12** which is a chiral dopant for liquid crystals <09AXEo135>, the chiral diphosphine ligand **13** <09AXEo1153>, and the bis(spiro) biphenyl **14** <09T2279>.

Oxidative cleavage of 2-substituted-1,3-dioxolanes to give 2-hydroxyethyl esters occurs upon treatment with 2-iodoxybenzoic acid and Et_4NBr in water <09S929> and a new method for the asymmetric conversion of aldehydes, RCHO into the deuterated alcohols, RCH(D)OH involves formation of chiral dioxolanes **15** using a diol derived from mandelic acid, their reductive cleavage using Et_3SiD, and then further degradation <09TA351>. Photolysis of crystals of the diazo compound **16** has allowed investigation of the reactivity of the corresponding carbene, which includes both abstraction-recombination to give the apparent insertion product **17**, and formation of the oxonium ylide **18** which goes on to give Stevens rearrangement products <09OBC1106>.

The DBU catalysed reaction of anilines with ethylene carbonate to give oxazolidin-2-ones under solvent-free conditions has been reported <09H(78)2093> and reaction of 4-tosyloxymethyl-1,3-dioxolan-2-one with a range of nucleophiles gives either alkyl glycidyl carbonates **19** with hard nucleophiles such as alkoxides or the simple substitution products **20** with softer nucleophiles <09T8571>. Alkylation at C-5 of the chiral 1,3-dioxolan-4-ones **21** of either diastereomeric series forms the key step in the synthesis of trachypsic acid <09TL1566> and squalene synthase inhibitors <09TL3388>.

Reaction of glycosyl 1,3-dioxolane-2-thione **22** with alkyl halides, RX, to give thiocarbonates **23** has been reported <09T8885>. Photochromism of the dioxolane-containing compounds **24** has been investigated and the X-ray structures of five examples are reported <09JST(919)100>. The liquid crystalline dioxolanone **25** forms a one-dimensional ionic conductor with LiOTf <09AM1591>.

5.6.2. 1,3-DITHIOLES AND DITHIOLANES

The reaction of carbonyl compounds with ethanedithiol to give 2-substituted 1,3-dithiolanes can be achieved under solvent-free conditions using either the ionic liquid $Et_3N^+(CH_2)_4SO_3H$ TsO^- <09SL1974> or P_2O_5/Al_2O_3 and microwave irradiation <09JMOA(301)39>. Reaction of the simplest tricarbonyl compound, 2-oxopropane-1,3-dial, with ethanedithiol and $BF_3 \bullet Et_2O$ gives both the bis(dithiolanyl) ketone **26** and the ter(dithiolane) **27** the X-ray structure of which is reported <09EJO1417>. Treatment of S-allyl dithiocarbamates with iodine affords the 2-imino-1,3-dithiolanes **28** <09TL2747> and dithiolane **29** has been used to prepare the anti-tumour amino acid derivatives **30** <09BMC6085>. Reaction of divinyl sulfide with SeX_2 (X = Cl, Br) initially gives the 6-membered ring 1,4-thiaselenane but this rearranges to afford the 1,3-thiaselenolanes **31**, which lose HX at room temperature (X = Br) or on heating (X = Cl) to give the corresponding thiaselenoles **32** <09JOM(694)3369, 09TL306>. Reaction of divinyl selenide with $SeCl_2$ similarly gives **33** which is converted into 1,3-diselenole **34** upon vacuum distillation <09RJC1225>. Treatment of the dibromodiselenolane corresponding to **33** with DBU results in ring opening to regenerate divinyl selenide <09RJC1758>.

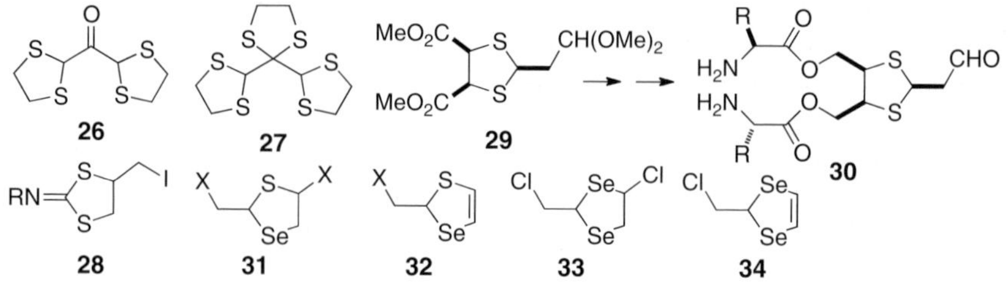

Theoretical studies have appeared on 2-methylene-1,3-dithiolane <09STC961> and 1,3-dithiol-2-one and 1,3-dithiole-2-thione and their radical cations <09SAA16>. Oxidative deprotection of 2-substituted-1,3-dithiolanes to give the corresponding aldehydes or ketones can be achieved using either NBS, 1,3-benzene-bis(N,N-dibromosulfonamide) or a polymeric N-bromosulfonamide <09ARK(ii) 44>, or by treatment with H_2O_2 and iodine in an aqueous miscellar system <09S1393>. Diels–Alder reaction of the o-quinodimethane form of spiro dithiolane 35 with acetylenic dienophiles provides access to highly substituted 1,4-dihydronaphthalene products 36 <09EJO1016>. Sequential treatment of alkynyldithiolanes 37 with butyllithium and an aldehyde R^2CHO, mercuric acetate, and then iodine gives the 3-iodofuran products 38 <09TL3263>. The quinoxaline-containing dithiolanone 39 is an effective fluorescent sensor for Pb^{2+} that works over a wide pH range and in the presence of other metal ions <09AGE3996>. The 2-cinnamoylmethylene-1,3-dithiolanes 40 have proved useful in multi-component synthesis of highly substituted cyclohexanones, as illustrated by reaction with an aldehyde, Ar^2CHO, and an acidic methylene compound, $E^1CH_2E^2$, to afford products 41 <09JOC3116>. The mechanism by which the tris(carboxybenzobis(dithiole))yl radicals 42 act as EPR probes for O_2^{-} has been elucidated for the first time <09CC1416>.

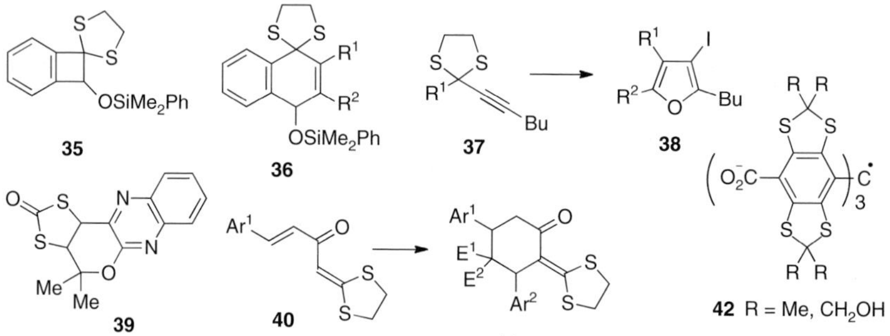

Charge transfer complexes between tetrathiafulvalene (TTF) and both tetracyanopyrazine <09NJC545> and 2,4,5,7-tetranitrofluorenone <09SM45> have been examined and their X-ray structures and electrical properties described. Both TTF and tetraselenafulvalene have been used to form single-molecule nanofabricated mechanically controllable break junctions between gold electrodes <09JA14146>. The synthesis and properties of a variety of new TTF donor–acceptor compounds have

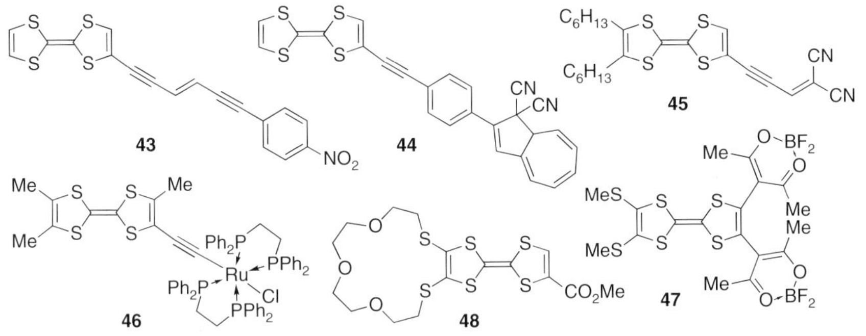

been reported including **43** <09JOC375>, **44** <09EJO1855>, **45** <09OBC3474>, **46** <09CC7200> and **47** <09T6123>. A new (third) polymorph of dibenzoTTF has been reported <09AXEo2083> and organic field-effect transistors based on soluble alkyl-substituted dibenzoTTFs have been described <09CL200>. The preparation and X-ray structure of the TTF crown ether **48** has been reported <09AXEo1057> and other similar crown ether-containing TTFs have been prepared and characterised <09DP(81)40>, and also used as sensors for Pb^{2+} <09NJC813>.

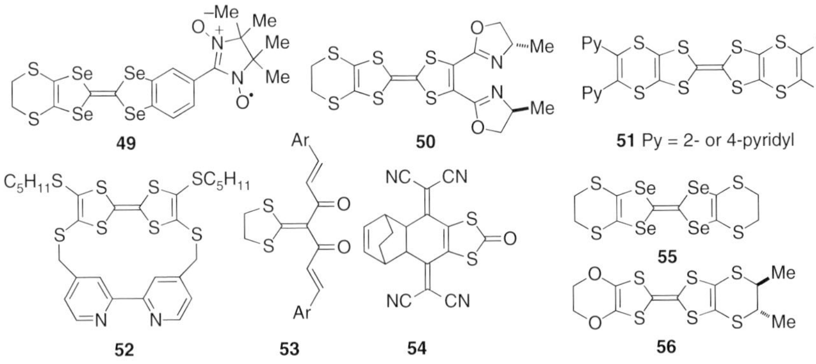

The new organic radical donor **49** has been prepared <09PO1996> and complexes of the chiral bis(oxazolinyl)TTF **50** with $Mo(CO)_4$ and $W(CO)_4$ have been reported <09CC3753>. The pyridyl functionalised TTF donors **51** have been prepared <09EJI3084> and a copper complex of the bipyridyl TTF derivative **52** has been characterised by X-ray and ESR methods <09SM2075>. A series of donor-acceptor-donor compounds **53** have been studied <09SM153> and the properties of donor-acceptor compound **54** have been reported <09EJO6341>.

Charge transfer complexes of the tetraselenafulvalene donor **55** with both $CdBr_4^{2-}$ and $HgBr_4^{2-}$ <09SM1072>, and the Keggin polyoxometallate $SMo_{12}O_{40}^{3- \text{ or } 4-}$ <09IC11314> have been examined, and conductivity studies on the organic metal formed from its all-sulfur analogue with $SF_5CH_2CF_2SO_3^-$ have been reported <09SM1043>. Electrical measurements on **56**•$AuBr_2$ have appeared <09SM2387>.

New X-ray structure determinations in the TTF area include **57** <09AXEo2920>, **58** <09AXEo2716>, **59** <09AXEo1082>, and the simple dithiolodithiolethione **60** <09AXEo1050>. Salts of dithiadiselenafulvalene **61** with I- and Br-containing anions act as organic superconductors <09CMA3521> and radical cation salts of **62** have been studied in detail for the first time <09SM2381>. The conductivity of the complex $(63)_2$ SbF_6 has been examined as a function of temperature and pressure <09SM2394>. Various oligomeric TTF derivatives have been prepared including the "molecular tweezers" **64** <09T10348>, compound **65** which shows electrochromism <09H(77)837> and the pentakis(TTF) **66** <09SM735>. Extended TTF systems incorporating ferrocenes <09S1000> and C_{60} <09AGE815, 09CC5374> have also been reported.

5.6.3. 1,3-OXATHIOLES AND OXATHIOLANES

Reaction of carbonyl compounds with 2-mercaptoethanol to give 2-substituted 1,3-oxathiolanes proceeds efficiently using a silica-supported sulfuric acid catalyst that is also effective in catalysing the reverse reaction <09CCL1457>. A clean synthesis of 1,3-oxathiolanes **67** from the corresponding 1,3-dicarbonyl compounds with CS_2, 2-bromoethanol and potassium carbonate in water has been described <09S824> and reaction of isoselenocyanates with sodium hydride and 2-bromoethanol gives the 2-imino-1,3-oxaselenolanes **68** <09H(78)449>, which, on thermolysis rearrange to the isomeric selenazolidin-2-ones **69**. Reaction of **70** with the diazo ketone **71** proceeds via a thiocarbonyl ylide to give the oxathiole product **72** <09T8191>. The X-ray structure of oxathiolanone **73** related to the anti-viral agent lamivudine has appeared <09AXEo1483> and theoretical studies on 2-methylene-1,3-oxathiolane have been reported <09STC961>.

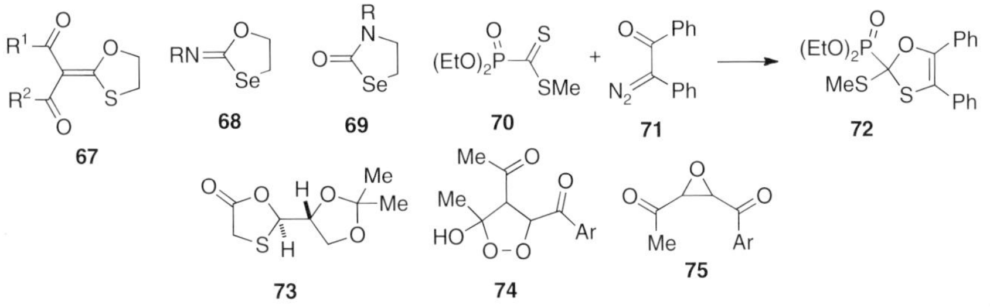

5.6.4. 1,2-DIOXOLANES

In a remarkable process, reaction of alkynes, ArC≡CH, with acetylacetone and Mn $(OAc)_3$ in acetic acid under air gives the 1,2-dioxolanes **74** which lose acetic acid upon treatment with silica to give the diacylepoxides **75** <09T3745>.

5.6.5. 1,2-DITHIOLES AND DITHIOLANES

The formation of 1,2-dithiol-3-ones and 1,2-dithiole-3-thiones from S_2Cl_2 has been reviewed <09MC55> and reaction of 2-methylindoles with S_2Cl_2 / DABCO followed by triethylamine gives products **76** which undergo cycloaddition with DMAD to give **77** <09T2178>. Both *cis* and *trans* isomers of the 1,2-dithiole disulfoxide **78** have been prepared by stepwise oxidation of the corresponding dithiole and their X-ray structures determined. Photolysis of these causes *cis/trans* interconversion, rearrangement to the isomeric sulfide/sulfone, and formation of the sulfoxide/sulfone by an unknown route <09AGE4832>. It is notable that, in the absence of the *t*-butyl groups, no such disulfoxides are formed. Reaction of the 1,2-diselenolane **79** <09OM1039> or the spiro 1,2-dithiolane, diselenolane or ditellurolane **80** <09OM6666> with $Fe_3(CO)_{12}$ leads to cleavage of the rings to form derivatives of type **81** the X-ray structures of which were determined. The "caged" lipoic acid-like 1,2-thiaselenolane **82** has been patented for the prevention of ocular damage <09WOP111633>, and compounds such as **83** act as protein tyrosine phosphatase inhibitors and are in clinical trials as potential anti-diabetic agents <09EJM3147>.

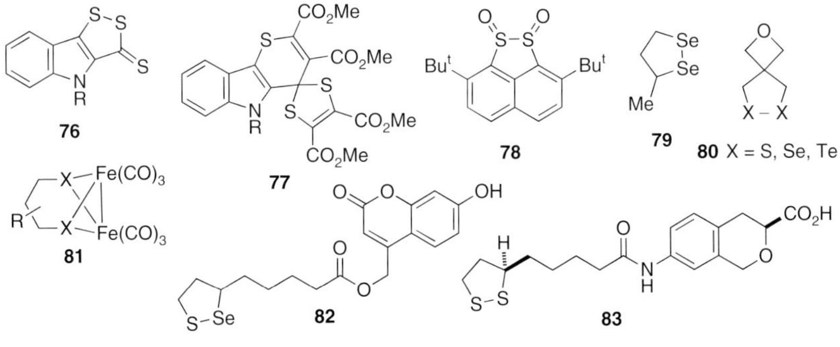

5.6.6. 1,2-OXATHIOLES AND OXATHIOLANES

The 1,2-oxathiole S,S-dioxides **84** have been prepared and are active against a range of viruses including human cytomegalovirus <09JME1582>.

5.6.7. THREE HETERO ATOMS

A series of tetracyclic 1,2,4-trioxolanes **85** have been prepared and evaluated as potential plant growth regulators <09JFA10109>. The X-ray structure was determined for R = Me, and the compounds with R = H and Et showed good activity against weeds. A large number of further spiro 1,2,4-trioxolanes with structures such as **86** and **87** have been prepared and evaluated as anti-malarial agents <09BML4542, 09WOP058859, 09WOP091433>.

Improved methods have been described for synthesis of carbohydrate-derived 1,3,2-dioxathiolane 2-oxides and 2,2-dioxides <09T6341> and both enantiomers of the simple cyclic sulfate **88** have been used in asymmetric synthesis of adrenergic agonists <09TA322>. The cyclic sulfates **89** have been examined as additives for battery electrolytes <09MI714>.

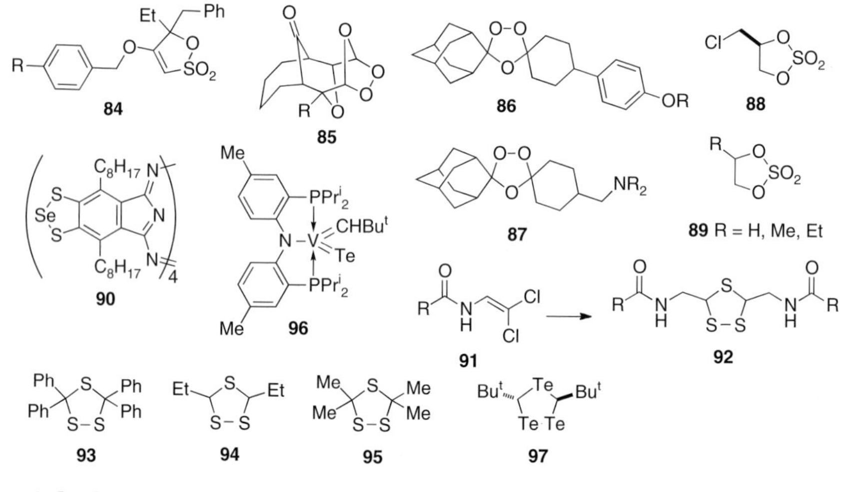

A [2,2]-paracyclophane-fused 1,2,3-trithiole has been reported <09JSU293> and phthalocyanins with fused 1,3,2-dithiaselenole rings **90** have been prepared and their electrochemical and optical properties studied <09H(79)1081>. Reaction of the dichlorovinyl amides **91** with H$_2$S and triethylamine affords the diamido 1,2,4-trithiolanes **92** <09RJC500>. The tetraphenyl-1,2,4-trithiolane **93** reacts with norbornene-platinum (0)-bis(phosphine) complexes by initial cycloreversion to thiobenzophenone and its S-sulfide which are then both complexed to the palladium <09EJI3545>. The diethyltrithiolane **94** occurs in garlic oil <09MI1234> while the tetramethyl compound **95** has been detected in cooked goat meat <09MI1081>. Reaction of the alkylidene vanadium complex **96** with elemental tellurium gives a range of products including the *trans*-1,2,4-tritellurolane **97** the X-ray structure of which was determined <09AGE2394>.

REFERENCES

09AGE815	S.S. Gayathri, M. Wielopolski, E.M. Pérez, G. Fernández, L. Sánchez, R. Viruela, E. Ortí, D.M. Guldi and N. Martín, *Angew. Chem. Int. Ed.*, **48**, 815 (2009).
09AGE2394	U.J. Kilgore, J.A. Karty, M. Pink, X. Gao and D.J. Mindiola, *Angew. Chem. Int. Ed.*, **48**, 2394 (2009).
09AGE2946	M. North and R. Pasquale, *Angew. Chem. Int. Ed.*, **48**, 2946 (2009).
09AGE3996	L. Marbella, B. Serli-Mitasev and P. Basu, *Angew. Chem. Int. Ed.*, **48**, 3996 (2009).
09AGE4194	Y. Kayaki, M. Yamamoto and T. Ikariya, *Angew. Chem. Int. Ed.*, **48**, 4194 (2009).
09AGE4832	R.S. Grainger, B. Patel and B.M. Kariuki, *Angew. Chem. Int. Ed.*, **48**, 4832 (2009).
09AM1591	H. Shimura, M. Yoshio, A. Hamasaki, T. Mukai, H. Ohno and T. Kato, *Adv. Mater. (Weinheim, Ger.)*, **21**, 1591 (2009).
09ARK(ii)44	R. Ghorbani-Vaghei and H. Veisi, *Arkivoc*, (ii), 44 (2009).
09AXEo135	Y.M. Hijji, P.F. Hudrlik, A.M. Hudrlik, R.J. Butcher and J.P. Jasinski, *Acta Crystallogr. Sect. E*, **65**, o135 (2009).
09AXEo579	J.V. Bjerrum, T. Ulven and A.D. Bond, *Acta Crystallogr. Sect. E*, **65**, o579 (2009).
09AXEo764	W. Xu, Z. Yang, X.-H. Li, B.-N. Liu and D.-C. Wang, *Acta Crystallogr. Sect. E*, **65**, o764 (2009).
09AXEo841	H. Liu, D.-C. Wang, W. Xu, Z. Yang and T. Gai, *Acta Crystallogr. Sect. E*, **65**, o841 (2009).
09AXEo1050	M. Tomura and Y. Yamashita, *Acta Crystallogr,. Sect. E*, **65**, o1050 (2009).
09AXEo1057	R. Hou, B. Li, B. Yin and L. Wu, *Acta Crystallogr. Sect. E*, **65**, o1057 (2009).
09AXEo1082	M. Tomura and Y. Yamashita, *Acta Crystallogr. Sect. E*, **65**, o1082 (2009).
09AXEo1153	L.-Y. Jian, X.-J. He, Y.-X. Sun, Q.-H. Jiang and X. Zhu, *Acta Crystallogr. Sect. E*, **65**, o1153 (2009).
09AXEo1483	Q.-P. Wu, D.-X. Shi, H. Wang and Q.-S. Zhang, *Acta Crystallogr. Sect. E*, **65**, o1483 (2009).
09AXEo1960	D.-C. Wang, J. Bai, W. Xu, T. Gai and H.-Q. Liu, *Acta Crystallogr. Sect. E*, **65**, o1960 (2009).
09AXEo2083	M. Mamada and Y. Yamashita, *Acta Crystallogr. Sect. E*, **65**, o2083 (2009).
09AXEo2716	K. Ueda and K. Yoza, *Acta Crystallogr. Sect. E*, **65**, o2716 (2009).
09AXEo2920	K. Ueda and K. Yoza, *Acta Crystallogr. Sect. E*, **65**, o2920 (2009).
09BMC6085	F. Huang, M. Zhao, X. Zhang, C. Wang, K. Qian, R.-Y. Kuo, S. Morris-Natschke, K.-H. Lee and S. Peng, *Bioorg. Med. Chem.*, **17**, 6085 (2009).
09BML4542	X. Wang, D.J. Creek, C.E. Chiaffo, Y. Dong, J. Chollet, C. Scheurer, S. Wittlin, S.A. Charman, P.H. Dussault, J.K. Wood and J.L. Vennerstrom, *Bioorg. Med. Chem. Lett.*, **19**, 4542 (2009).
09CC1136	S.-F. Yin and S. Shimada, *Chem. Commun.*, 1136 (2009).
09CC1312	T. Sakakura and K. Kohno, *Chem. Commun.*, 1312 (2009).
09CC1416	C. Decroos, Y. Li, G. Bertho, Y. Frapart, D. Mansuy and J.-L. Boucher, *Chem. Commun.*, 1416 (2009).
09CC2577	J. Meléndez, M. North and P. Villuendas, *Chem. Commun.*, 2577 (2009).
09CC3753	F. Riobé and N. Avarvari, *Chem. Commun.*, 3753 (2009).
09CC5374	B.M. Illescus, J. Santos, M. Wielopolski, C.M. Atienza, N. Martín and D.M. Guldi, *Chem. Commun.*, 5374 (2009).
09CC7200	A. Vacher, F. Barrière, T. Roisnel and D. Lorcy, *Chem. Commun.*, 7200 (2009).
09CCL1457	F. Shirini, P. Sadeghzadeh and M. Abedini, *Chin. Chem. Lett.*, **20**, 1457 (2009).
09CEJ11454	M. North, P. Villuendas and C. Young, *Chem. Eur. J.*, **15**, 11454 (2009).
09CL200	T. Yoshino, K. Shibata, H. Wada, Y. Bando, K. Ishikawa, H. Takezoe and T. Mori, *Chem. Lett.*, **38**, 200 (2009).
09CMA3521	Y. Bando, T. Kawamoto, T. Mori, T. Kakiuchi, H. Sawa, K. Takimiya and T. Otsubo, *Chem. Mater.*, **21**, 3521 (2009).
09DP(81)40	R. Wang, W. Liu, Y. Chen, J.-L. Zuo and X.-Z. You, *Dyes Pigments*, **81**, 40 (2009).
09EJI3084	A.C. Brooks, P. Day, S.I.G. Dias, S. Rabaça, I.C. Santos, R.T. Henrques, J.D. Wallis and M. Almeida, *Eur. J. Inorg. Chem.*, 3084 (2009).
09EJI3545	T. Weisheit, H. Petzold, H. Görls, G. Mloston and W. Weigand, *Eur. J. Inorg. Chem.*, 3545 (2009).

09EJI3552	W. Kleist, F. Jutz, M. Maciejewski and A. Baiker, *Eur. J. Inorg. Chem.*, 3552 (2009).
09EJM3147	N. Lakshminarayana, Y.R. Prasad, L. Gharat, A. Thomas, P. Ravikumar, S. Narayanan, C.V. Srinivasan and B. Gopalan, *Eur. J. Med. Chem.*, **44**, 3147 (2009).
09EJO1016	K. Benda, W. Regehardt, E. Schaumann and G. Adiwidjaja, *Eur. J. Org. Chem.*, 1016 (2009).
09EJO1417	S. Goswami, A.C. Maity, H.-K. Fun and S. Chantrapromma, *Eur. J. Org. Chem.*, 1417 (2009).
09EJO1855	M.Å. Petersen, A.S. Andersson, K. Kilså and M.B. Mielsen, *Eur. J. Org. Chem.*, 1855 (2009).
09EJO6341	F. Dumur, X. Guégano, N. Gautier, S.-X. Liu, A. Neels, S. Decurtins and P. Hudhomme, *Eur. J. Org. Chem.*, 6341 (2009).
09H(77)837	M. Hasegawa, Y. Kobayashi, K. Hara, H. Enozawa and M. Iyada, *Heterocycles*, **77**, 837 (2009).
09H(78)449	Y. Toyoda, D.R. Garud and M. Koketsu, *Heterocycles*, **78**, 449 (2009).
09H(78)2093	H. Gong and N. Yang, *Heterocycles*, **78**, 2093 (2009).
09H(79)1081	T. Kimura, T. Namauo, A. Yamakawa and Y. Takaguchi, *Heterocycles*, **79**, 1081 (2009).
09IC11314	E. Coronado, S. Curreli, C. Giménez-Saiz, C.J. Gómez-García, A. Alberola and E. Canadell, *Inorg. Chem.*, **48**, 11314 (2009).
09JA14146	M. Taniguchi, M. Tsutsui, K. Shoji, H. Fujiwara and T. Kawai, *J. Am. Chem. Soc.*, **131**, 14146 (2009).
09JFA10109	L.C.A. Barbosa, C.R.A. Maltha, R.C. Cusati, R.R. Teixeira, F.F. Rodrigues, A.A. Silva, M.G.B. Drew and F.M.D. Ismail, *J. Agr. Food. Chem.*, **57**, 10109 (2009).
09JME1582	S. De Castro, C. García-Aparicio, G. Andrei, R. Snoeck, J. Balzarini, M.-J. Camarasa and S. Velázquez, *J. Med. Chem.*, **52**, 1582 (2009).
09JMOA(301)39	A. Zarei, A.R. Hajipourc, L. Khazdoozd, B.F. Mirjalili and S. Zahmatkesh, *J. Mol. Catal. A:. Chem.*, **301**, 39 (2009).
09JMOA(310)150	S.B. Umbarkara, T.V. Kotbagia, A.V. Biradara, R. Pasrichab, J. Chanalea, M.K. Dongarea, A.-S. Mamedec, C. Lancelot and E. Payenc, *J. Mol. Catal. A: Chem.*, **310**, 150 (2009).
09JOC375	A.S. Andersson, L. Kerndrup, A.Ø. Madsen, K. Kilså, M.B. Nielsen, P.R. La Porta and I. Biaggio, *J. Org. Chem.*, **74**, 375 (2009).
09JOC2120	G. Bentabed-Ababsa, A. Derdour, T. Roisnel, J.A. Sáez, P. Pérez, E. Chamorro, L.R. Domingo and F. Mongin, *J. Org. Chem.*, **74**, 2120 (2009).
09JOC3116	Y. Ma, M. Wang, D. Li, B. Bekturhun, J. Liu and Q. Liu, *J. Org. Chem.*, **74**, 3116 (2009).
09JOM(694)3369	S.V. Amosova, M.V. Penzik, A.I. Alabanov and V.A. Potapov, *J. Organomet. Chem.*, **694**, 3369 (2009).
09JST(919)100	S. Pu, M. Li, C. Fan, G. Liu and L. Shen, *J. Mol. Struct.*, **919**, 100 (2009).
09JSU293	S. Ogawa, S. Nakajo, H. Muraoka, Y. Kawai and R. Sato, *J. Sulfur Chem.*, **30**, 293 (2009).
09MC55	L.S. Konstantinova and O.A. Rakitin, *Mendeleev Commun.*, **19**, 55 (2009).
09MI714	A. Sano and S. Maruyama, *J. Power Sources*, **192**, 714 (2009).
09MI1081	M.S. Madruga, J.S. Elmore, A.T. Dodson and D.S. Mottram, *Food. Chem.*, **115**, 1081 (2009).
09MI1234	H.B. Sowbhagya, K.T. Purnima, S.P. Florence, A.G.A. Rao and P. Srinivas, *Food. Chem.*, **113**, 1234 (2009).
09MI1404	A.P. Amrute, S. Sahoo, A. Bordoloi, Y.K. Hwang, J.-S. Hwang and S.B. Halligudi, *Catal. Commun.*, **10**, 1404 (2009).
09NJC545	S.V. Rosokha, J. Lu, B. Han and J.K. Kochi, *New J. Chem.*, **33**, 545 (2009).
09NJC813	Y.-P. Zhao, X.-J. Wang, J.-J. Wang, G. Si, Y. Liu, C.-H. Tung and L.-Z. Wu, *New J. Chem.*, **33**, 813 (2009).
09OBC1106	M.A. Garcia-Garibay and H. Dang, *Org. Biomol. Chem.*, **7**, 1106 (2009).
09OBC3474	A.S. Andersson, F. Diederich and M.B. Nielsen, *Org. Biomol. Chem.*, **7**, 3474 (2009).

09OM1039	M.H. Karb, T. Niksch, J. Winghager, H. Görls, R. Holze, L.T. Lockett, N. Okamura, D.H. Evans, R.S. Glass, D.L. Lichtenberger, M. El-Khateeb and W. Weigand, *Organometallics*, **28**, 1039 (2009).
09OM6666	M.H. Karb, U.-P. Apfel, J. Kübel, H. Görls, G.A.N. Felton, T. Sakamoto, D.H. Evans, R.S. Glass, D.L. Lichtenberger, M. El-Khateeb and W. Weigand, *Organometallics*, **28**, 6666 (2009).
09PO1996	H. Komatsu, R. Mogi, M.M. Matsushita, T. Miyagi, Y. Kawada and T. Sugawara, *Polyhedron*, **28**, 1996 (2009).
09RJA2156	L.M. Popova, *Russ. J. Appl. Chem.*, **82**, 2156 (2009).
09RJC500	B.A. Demidchuk, S.O. Seferov, A.N. Vasilenko, V.S. Brovarets and B.S. Drach, *Russ. J. Gen. Chem.*, **79**, 500 (2009).
09RJC1225	V.A. Potapov, K.A. Volkova, M.V. Penzik and S.V. Amosova, *Russ. J. Gen. Chem.*, **79**, 1225 (2009).
09RJC1758	V.A. Potapov, K.A. Volkova and S.V. Amosova, *Russ. J. Gen. Chem.*, **79**, 1758 (2009).
09S557	N.W. Fadnavis, R. Gowrisankar, G. Ramakrishna, M.K. Mishra and G. Sheelu, *Synthesis*, 557 (2009).
09S824	Y. Li, Q. Zhang, X. Cheng, Q. Liu and X. Xu, *Synthesis*, 824 (2009).
09S929	C. Kuhakarn, W. Panchan, S. Chiampanichayakul, N. Samakkanad, M. Pohmakotr, V. Reurtrakul and T. Jaipetch, *Synthesis*, 929 (2009).
09S1000	A.A.O. Sarhan and C. Bolm, *Synthesis*, 1000 (2009).
09S1318	F. Ono, H. Takenaka, T. Fujikawa, M. Mori and T. Sato, *Synthesis*, 1318 (2009).
09S1393	N.C. Ganguly and S.K. Barik, *Synthesis*, 1393 (2009).
09S1403	J.M. Whitaker and R.C. Ronald, *Synthesis*, 1403 (2009).
09S2324	J. Yu and C. Zhang, *Synthesis*, 2324 (2009).
09SAA16	S. Jaiswal, A. Kushwaha, R. Prasad, R.L. Prasad and R.A. Yadav, *Spectrochim. Acta, Part A*, **74**, 16 (2009).
09SL1974	A.R. Hajipour, A. Azizi and A.E. Ruoho, *Synlett*, 1974 (2009).
09SM45	E.W. Reinheimer, J.R. Galán-Mascarós and K.R. Dunbar, *Synth. Met.*, **159**, 45 (2009).
09SM153	P. Peng, Y. Zhou, K. Li, H. Zhang, H. Xia, Y. Zhao and W. Tian, *Synth. Met.*, **159**, 153 (2009).
09SM735	W. Hong, W. Xu, Q. Wang and D. Zhu, *Synth. Met.*, **159** (2009).
09SM1043	M. Glied, S. Yasin, S. Kaiser, N. Drichko, M. Dressel, J. Wosnitza, J.A. Schlueter and G.L. Gard, *Synth. Met.*, **159**, 1043 (2009).
09SM1072	E.I. Zhilyaeva, O.A. Bogdanova, G.V. Shilov, R.B. Lyubovskii, S.I. Pesotskii, S.M. Aldoshin, A. Kobayashi, H. Kobayashi and R.N. Lyubovskaya, *Synth. Met.*, **159**, 1072 (2009).
09SM2075	L. Kaboub, A.-K. Gouasmia, J.-P. Legros, E. Harte, C. Coulon and J.M. Fabre, *Synth. Met.*, **159**, 2075 (2009).
09SM2381	Y. Nakano, M. Takahashi, M. Sakata, H. Yamochi, G. Saito and K. Tanaka, *Synth. Met.*, **159**, 2381 (2009).
09SM2387	H. Yoshino, G.C. Papavassiliou and K. Murata, *Synth. Met.*, **159**, 2387 (2009).
09SM2394	Y. Weng, H. Yoshino, N. Hiratani, H. Akutsu, J. Yamada, K. Kikuchi and K. Murata, *Synth. Met.*, **159**, 2394 (2009).
09STC961	D.J. Beard, C.R. Pace, C.U. PittmanJr. and S. Saebø, *Struct. Chem.*, **20**, 961 (2009).
09T1889	A.K. Buzas, F.M. Istrate and F. Gagosz, *Tetrahedron*, **65**, 1889 (2009).
09T2178	L.S. Konstantinova, K.A. Lysov, S.A. Amelichev, N.V. Obruchnikova and O.A. Rakitin, *Tetrahedron*, **65**, 2178 (2009).
09T2279	E. Cadoni, E. Perra, C. Fattuoni, G. Bruno, M.G. Cabiddu, S. De Montis and S. Cabiddu, *Tetrahedron*, **65**, 2279 (2009).
09T3745	T. Tsubusaki and H. Nishino, *Tetrahedron*, **65**, 3745 (2009).
09T6123	M. Guerro, T. Roisnel and D. Lorcy, *Tetrahedron*, **65**, 6123 (2009).
09T6341	C. Hardacre, I. Messina, M.E. Migaud, K.A. Ness and S.E. Norman, *Tetrahedron*, **65**, 6341 (2009).
09T8191	G. Mloston, K. Urbaniak, A. Linden and H. Heimgartner, *Tetrahedron*, **65**, 8191 (2009).

09T8407	D. Janeliunas, M. Daskeviciene, T. Malinauskas and V. Getautis, *Tetrahedron*, **65**, 8407 (2009).
09T8571	J. Rousseau, C. Rousseau, B. Lynikaite, A. Sackus, C. de Leon, P. Rollin and A. Tatibouët, *Tetrahedron*, **65**, 8571 (2009).
09T8885	M. Benazza, R. Kanso and G. Demailly, *Tetrahedron*, **65**, 8885 (2009).
09T10348	M. Skibinski, R. Gómez, E. Lork and V.A. Azov, *Tetrahedron*, **65**, 10348 (2009).
09TA322	K. Kulig, A. Boba, A. Bielejewska, M. Gorska and B. Malawska, *Tetrahedron: Asymmetry*, **20**, 322 (2009).
09TA351	C.F. Morelli, P. Cairoli, T. Marigolo, G. Speranza and P. Manitto, *Tetrahedron: Asymmetry*, **20**, 351 (2009).
09TL306	S.V. Amosova, M.V. Penzik, A.I. Albanov and V.A. Potapov, *Tetrahedron Lett.*, **50**, 306 (2009).
09TL1566	F. Calo, J. Richardson, A.J.P. White and A.G.M. Barrett, *Tetrahedron Lett.*, **50**, 1566 (2009).
09TL2747	A.Z. Halimehjani, H. Maleki and M.R. Saidi, *Tetrahedron Lett.*, **50**, 2747 (2009).
09TL3263	C.-W. Chen and T.-Y. Luh, *Tetrahedron Lett.*, **50**, 3263 (2009).
09TL3388	F. Calo, A. Bondke, J. Richardson, A.J.P. White and A.G.M. Barrett, *Tetrahedron Lett.*, **50**, 3388 (2009).
09TL5927	O. Mazimba, R.R. Majinda and I.B. Masesane, *Tetrahedron Lett.*, **50**, 5927 (2009).
09WOP058859	J.L. Vennerstrom, Y. Dong, S.A. Charman, S. Wittlin, J. Chollet, D.J. Creek, X. Wang, K. Sriraghavan, L. Zhou, H. Matile and W.N. Charman, *PCT Int. Appl. WO*, 058859 (2009). [Chem. Abstr.**2009**, 150, 494879].
09WOP091433	J.L. Vennerstrom, Y. Dong, S.A. Charman, S. Wittlin, J. Chollet, X. Wang, K. Sriraghavan, L. Zhou, H. Matile and W.N. Sharman, *PCT Int. Appl. WO*, 091433 (2009). [Chem. Abstr.**2009**, 151, 198428].
09WOP111633	W. Garner, M. Garner and R.D. Blum, *PCT Int. Appl. WO*, 111633 (2009). [Chem. Abstr.**2009**, 151, 328960].

CHAPTER 5.7
Five-Membered Ring Systems with O & N Atoms

Stefano Cicchi, Franca M. Cordero, Donatella Giomi
Università degli Studi di Firenze, Italy
donatella.giomi@unifi.it

5.7.1. ISOXAZOLES

The use of primary nitroalkanes as nitrile oxide precursors in 1,3-dipolar cycloadditions (1,3-DCs) with active methylene compounds is precluded due to competitive interactions between the commonly used dehydrating agents and enolic dipolarophiles. On the other hand, the treatment of activated nitro compounds **1** and dipolarophiles **2** with tertiary amines such as *N*-methylpiperidine (NMP) in the presence of a Cu(II) salt did allow the synthesis of highly functionalized isoxazoles **3** by cycloaddition/condensation reactions <09EJO5971>.

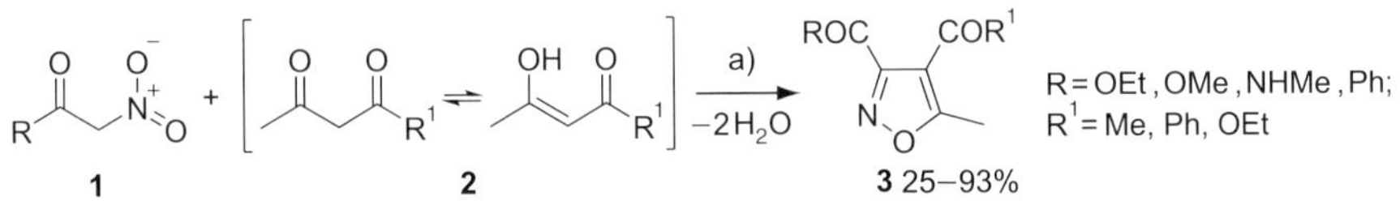

a): NMP (20 mol%), Cu(OAc)$_2$ (10 mol%), CHCl$_3$ or toluene, 60 °C, 38–72 h

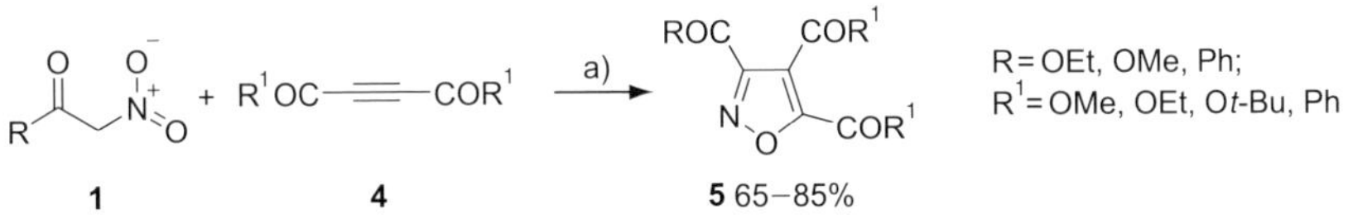

a): pyridine or *N*-methylimidazole (20 mol%), H$_2$O/MeCN (5:1), rt, 6 h

Nitromethane derivatives **1** also reacted with electron-deficient acetylenes **4** in the presence of pyridine or *N*-methylimidazole in H$_2$O/MeCN to give polyfunctionalized isoxazoles **5** in good yields <09T2067>.

Highly functionalized isoxazoles **8** were synthesized in moderate to excellent yields under mild conditions *via* regioselective tandem 1,3-DC of cyclopropene-1,1-diesters **6** with nitrile oxides generated from **7** <09T9146>.

R = Et, Me; R^1 = H, *n*-Bu, Ar; R^2 = *n*-Pr, *i*-Pr, Ar

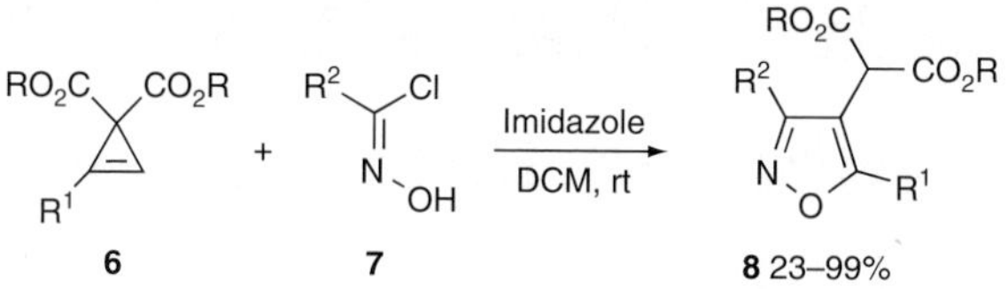

The regioselective 1,3-DC of pentafluorophenyl α-bromovinylsulfonate **9** with nitrile oxides derived from aryl hydroxymoyl chlorides afforded 3,5-disubstituted isoxazoles **10** that were converted directly into the corresponding sulfonamides **11** <09OBC4349>.

R^1, R^2 = H, Me, *t*-Bu, *i*-Pr, Bn, Ar

R = Ar, Het, alkyl, cycloalkyl, sugar
X = Ph, Bn, OAc, CN, CO_2Et, CH_2OTBS, CH_2OTHP, CH_2NHBoc, TMS

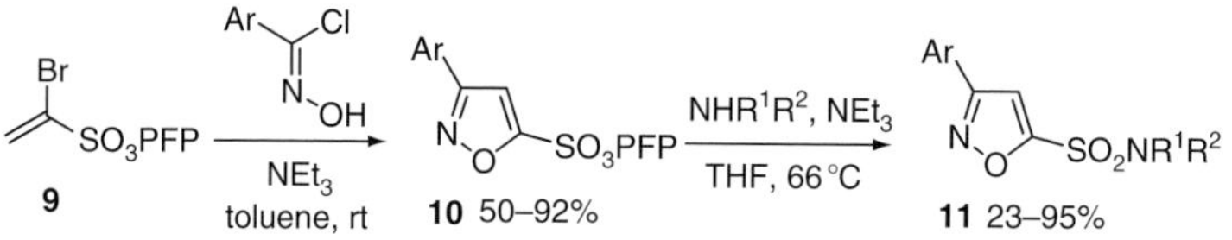

Treatment of aldoximes **12** with Magtrieve™ (CrO_2) in the presence of alkene and alkyne dipolarophiles gave a variety of 3,5-disubstituted isoxazolines and isoxazoles **13**. The methodology has been extended to intramolecular nitrile oxide cycloadditions (INOCs) to access chromane derivatives **14** in 75-80% yields <09TL3948>.

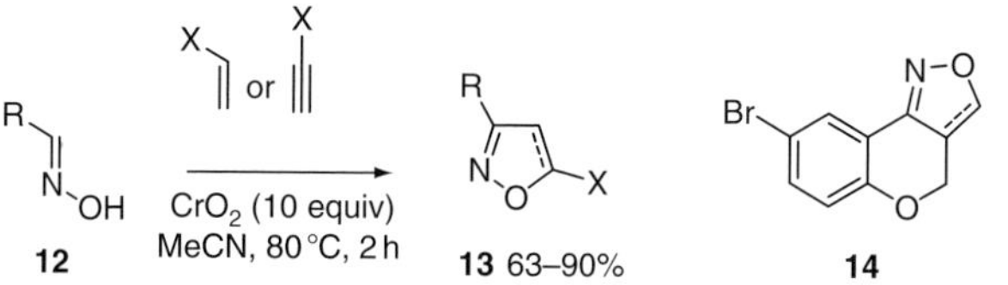

5-Substituted isoxazole- and isoxazoline-3-phosphonates **15** were regioselectively prepared by 1,3-DC of (diethoxyphosphoryl)formonitrile oxide with monosubstituted alkenes and alkynes <09S591>. On the other hand, 3-substituted-4-(diethoxyphosphoryl)isoxazoles **18** were obtained from 3-azidopenta-1,3-dienylphosphonate **16** *via* a chemoselective 1,3-DC on the terminal double bond leading to **17**, followed by cyclization using manganese dioxide <09S3405>.

R = Ph, Bu, CO_2Me, CH_2OH **15** 18–48%

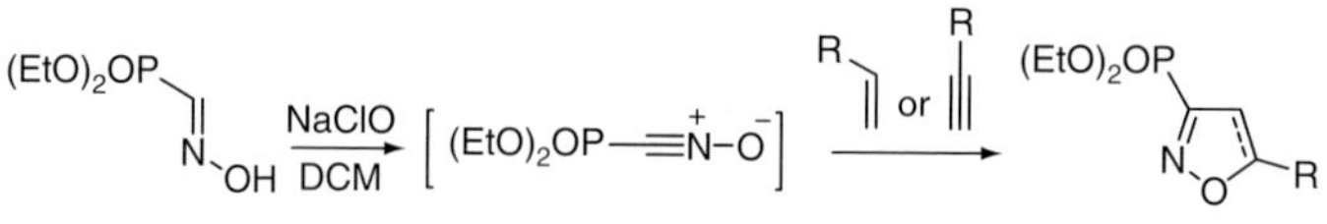

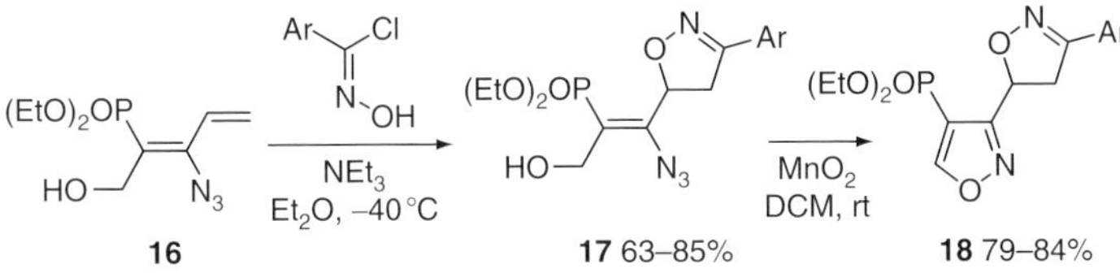

A solid-phase nitrile oxide-alkyne click reaction has been used to form isoxazole conjugated oligonucleotides. Commercially available 500 Å CPG-succinyl nucleoside support was selected and the reactivity of the support-bound alkynes in nitrile oxide click cycloadditions was tested by exposing them to aldoximes and chloramine-T. After deprotection and cleavage, near quantitative conversions to isoxazole-nucleotide conjugates **19** was evidenced *via* HPLC analyses. The procedure is selective, convenient and fast, occurs in atmospheric aqueous conditions, within minutes, and is high yielding and highly regioselective <09CC3276>.

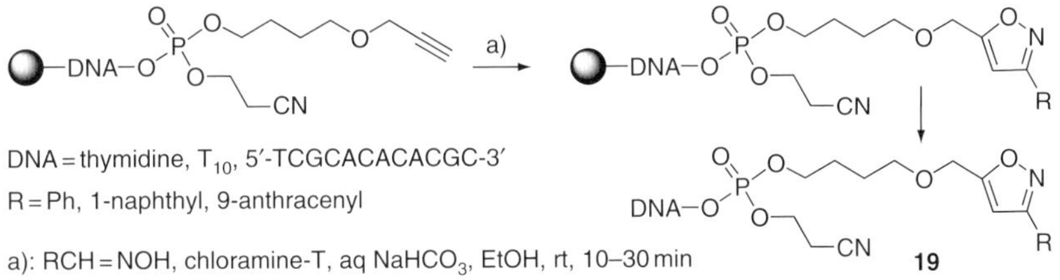

Operating as above, with chloramine-T as the dipole-generating agent, the nitrile oxide-alkyne click chemistry was also exploited as an efficient 'metal-free' tool for functionalizing well-defined biocompatible polymers <09MM5411>. The click polymerization of a homo ditopic nitrile oxide (generated *in situ* from the corresponding hydroxymoyl chloride with molecular sieves 4 Å) and various diynes produced polyisoxazoles **20** in high yields. Selective transformations of the isoxazole moieties into β-aminoenones or β-aminoalcohols afforded reactive polymers <09MM7709>.

A one-pot three-component synthesis involving a consecutive coupling-cycloaddition sequence of ferrocenyl substituted acid chlorides, terminal alkynes, and *in situ* generated nitrile oxides gave ferrocenyl redox active isoxazoles in moderate to good yields <09JOM(694)942>.

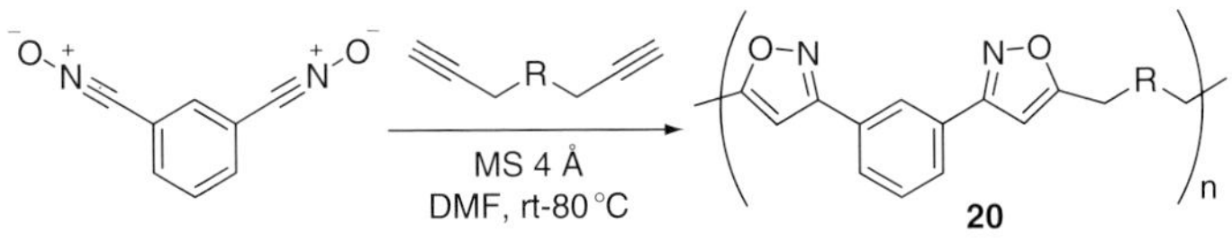

A series of 3-substituted and 3,5-disubstituted isoxazoles **21** has been efficiently synthesized with moderate to excellent yields by the reaction of *N*-hydroxy-4-toluenesulfonamide with α,β-unsaturated aldehydes/ketones. This strategy is associated with readily available starting materials, mild conditions, high regioselectivity, and wide scope <09OL3982>.

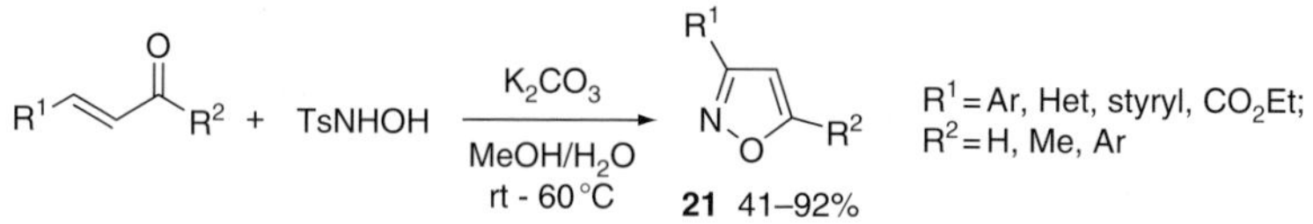

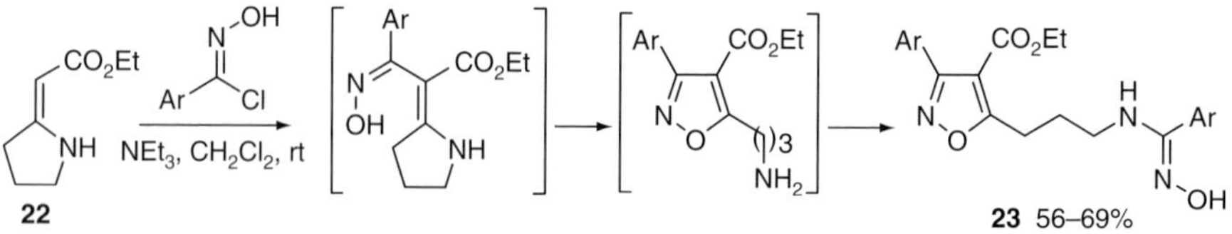

Reactions of alkylidenepyrrolidines **22** with α-chlorooximes gave isoxazoles **23** via an acylation reaction followed by ring isomerisation <09TL7392>.

3-Amino-5-substituted isoxazoles **26** were synthesized through a two-step procedure involving the addition-elimination of amines on 3-bromoisoxazolines **24** in the presence of a base. An iodine-mediated oxidation protocol allowed the conversion of 3-aminoisoxazolines **25** into **26** in satisfactory yields <09OL1159>.

An efficient three-component, two-step 'catch and release' solid-phase synthesis of trisubstituted isoxazoles has been developed. The first step involves a base-promoted condensation of a 2-sulfonylacetonitrile derivative with an isothiocyanate and in situ immobilization of the sulfonyl intermediate on Merrifield resin. Reaction with hydroxylamine, followed by release from the resin and intramolecular cyclization, afforded isoxazoles **27** <09JCO697>. A series of 3,4-diaryl-5-aminoisoxazoles has been synthesized and the heterocycles evaluated for their biological activities <09BMC6279>.

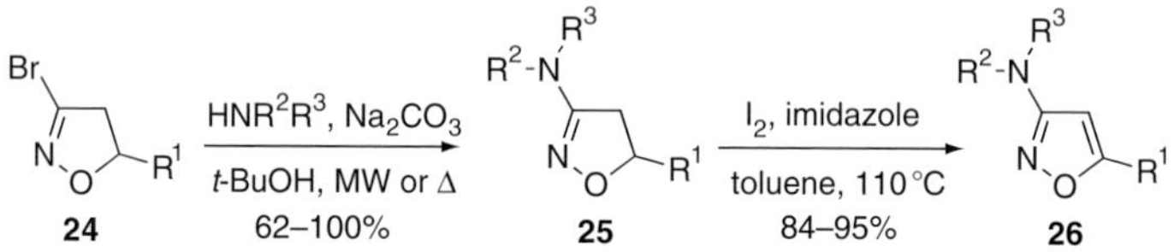

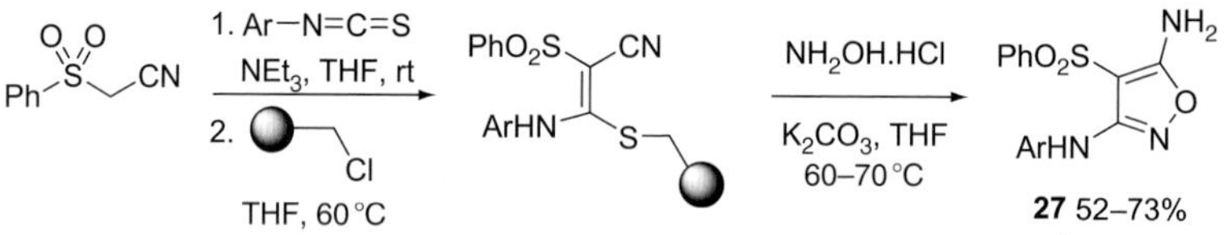

4-Arylisoxazoles were easily prepared by a Pd-catalyzed C-H bond activation/arylation of 3,5-disubstituted isoxazoles using aryl or heteroaryl bromides substituted with various functional groups. The use of 0.1-0.5 mol% of an air-stable $PdCl_2$ complex gave the arylation products in satisfactory yields (44–95%) <09EJO4041>. Isoxazolo[5,4-*b*]pyridines were synthesized by microwave-assisted three-component reactions of 1,3-dicarbonyl compounds, aromatic aldehydes, and 5-amino-3-methylisoxazole in water <09JCO428>. Bromination of 3-methyl-4-nitro-5-styrylisoxazoles followed by treatment with NEt_3 allowed the synthesis of (*E*)-5-(1-bromo-2-arylvinyl)-3-methyl-4-nitroisoxazoles able to react with 1,3-dicarbonyl compounds to give isoxazolyl-cyclopropanes or -dihydrofurans in high yields and diastereoselectivity <09T5402>. A series of 3,5-diarylisoxazoles was synthesized as a novel class of anti-hyperglycemic and lipid lowering agents <09BMC5285>.

5.7.2. ISOXAZOLINES

The Mg(II)-mediated, hydroxyl-directed nitrile oxide cycloaddition to allyl alcohols has been extended to dipolarophiles such as homoallylic alcohols and monoprotected homoallylic diols. The process is *anti*-diastereoselective and can be used to synthesize a variety of isoxazolines as polyketide building blocks. The syntheses of isoxazolines **28** and **29** are representative examples <09CEJ12065>.

R	%yield
CO_2Me	89
$CONMe_2$	100
CN	85

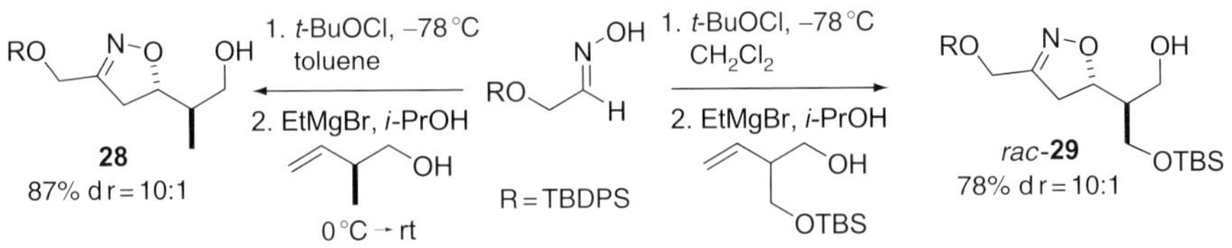

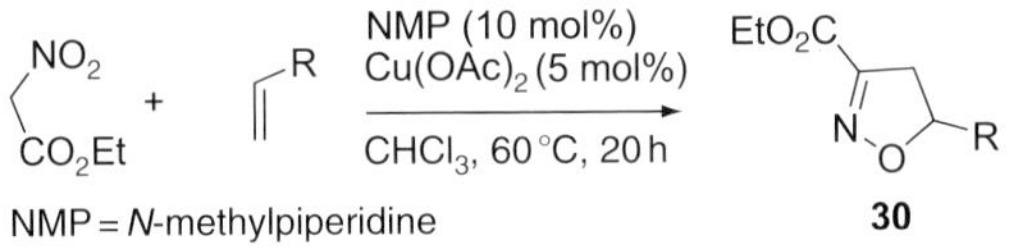

A study on chemoselectivity of cycloaddition condensations over conjugate additions of ethyl nitroacetate with electron-poor olefins has been reported. The proportions of the two competitive processes could be modified using different reaction conditions and catalytic systems. In particular, the formation of Michael adducts is favoured in the presence of a strong base whereas weaker bases such as NMP, DABCO or NMM combined with Cu(II) salts led to a selective formation of isoxazolines **30** <09CEJ7940>.

Isoxazoline *N*-oxides **32** were obtained through a highly stereoselective three-step one-pot sequence from simple starting materials. The first intermediate **31** was formed through an organocatalyzed enantioselective α-bromination of aldehydes in the presence of the diaryl-prolinol **34**. Then, **31** underwent an *in situ* base-promoted face-selective Henry addition of nitroacetates followed by a stereospecific O-alkylation to afford **32** in good yield and with excellent diastereo- and enantioselectivity <09AGE6844>.

R	%yield	%ds	%ee
i-Pr	77	>95	95
c-Pent	86	>95	96
c-Hex	68	>95	96

R	%yield	%ds	%ee
cis-2-pentenyl	85	>95	94
n-Bu	69	77	93
$C_{15}H_{31}$	90	73	94

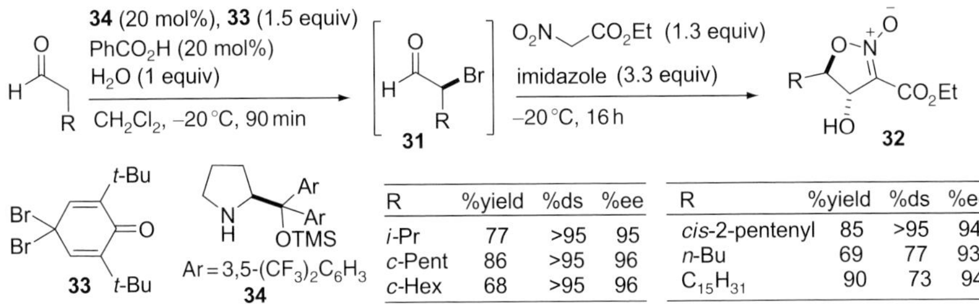

The same organocatalyst **34** was found to promote the enantioselective formation of 2-isoxazolines through one-pot oxime conjugate addition followed by cyclization *via* transoximation. This approach afforded isoxazolines **35** in moderate yields and high enantioselectivities (nine examples were reported) <09CEJ3960>.

4-Silyl 4-isoxazolines such as **36** were obtained by reduction of 4-silylated 4-isoxazolium tetrafluoroborates with $NaBH_4$ or $LiAlH_4$. Analogously, 5-silyl 4-isoxazolines were prepared by the corresponding isoxazolium salts <09T5472>.

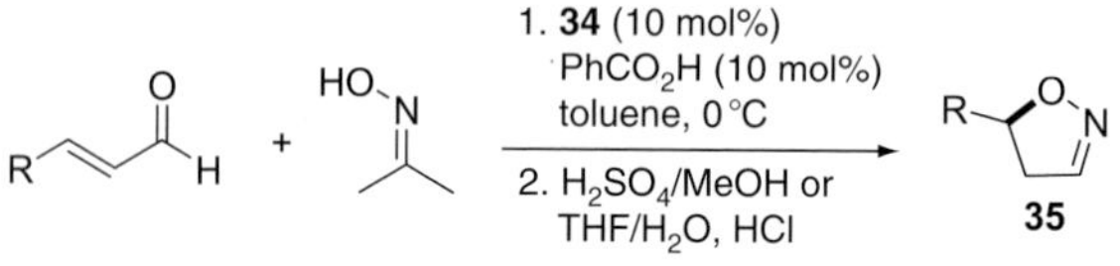

R	%yield	%ee
$BnCH_2$	63	91
c-Hex	45	94
$BnOCH_2$	54	91
$PMBO(CH_2)_2$	52	90

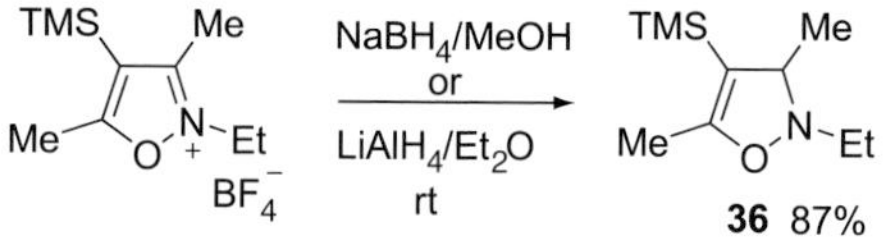

Isoxazolino[4,5-c]quinolines **39** have been synthesized by 1,3-DC of N-methyl nitrones to α,β-ynones **37** followed by cyclocondensation of intermediates **38**. The polycyclic products were obtained as single regioisomers (11 examples, 43-87% yield) <09EJO1027>.

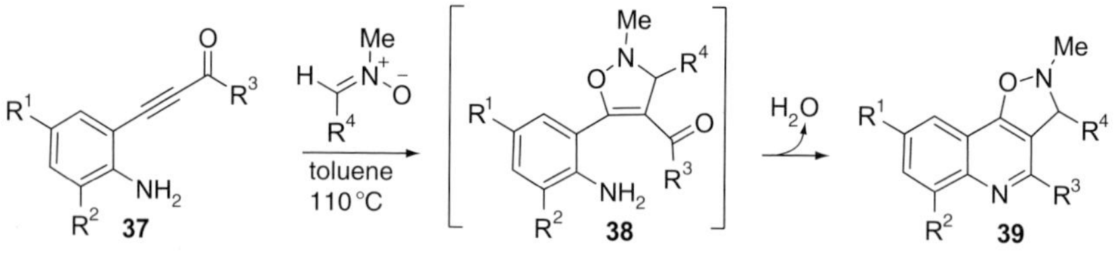

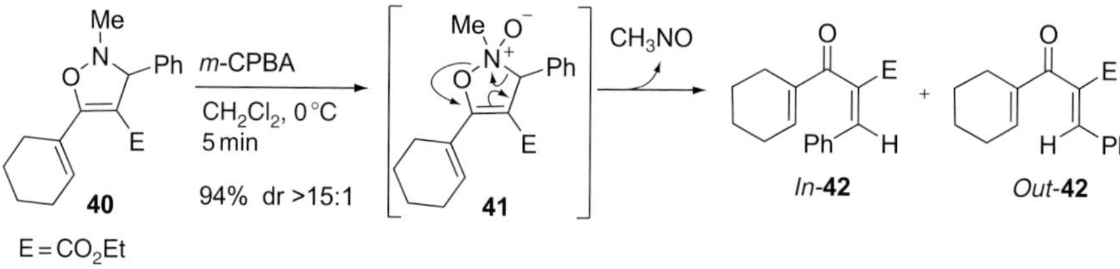

4-Isoxazolines such as **40** were converted into Nazarov cyclization substrates (divinyl ketones) through a stereoselective extrusion of nitrosomethane triggered by oxidation with *m*-CPBA. The torquoselectivity of the extrusion step depends on the substitution pattern, and in the majority of the reported examples, the formation of the *in*-isomer was favoured. Computational evidences suggest that the sequence proceeds through a stepwise pathway, and that stereoelectronic effects in the diradical intermediate influence the direction of rotation during extrusion <09T3165>.

The asymmetric 1,3-DC of nitrile oxides with alkenoyl oxazolidinones and pyrazolidinone catalyzed by chiral binaphthyldiimine-Ni(II) complexes has been reported <09JOC1099>.

The spiro bis(isoxazoline) **43** and its enantiomer have been used as chiral ligands in Pd-catalyzed enantioselective transformations such as intramolecular oxidative aminocarbonylation of alkenylureas to tetrahydropyrrolo[1,2-*c*]pyrimidine-1,3-diones <09JOC9274>, isotactic copolymerization of CO with styrene derivatives <09SL310>, and oxidative cyclization of enynes to bicyclo[3.1.0]hexanes <09JA3452>. The last process is the first example of asymmetric Pd(II)/Pd(IV) catalysis. The development of chiral spiro ligands such as spiro bis(isoxazolines), spiro bis(isoxazoles) and spiro bis(oxazolines) for metal-catalyzed asymmetric reactions has been reviewed <09BCJ285>.

Suitable substituted isoxazolines were used as molecular components in the synthesis of new liquid crystals such as **44**, which exhibits a nematic phase <09EJO889>.

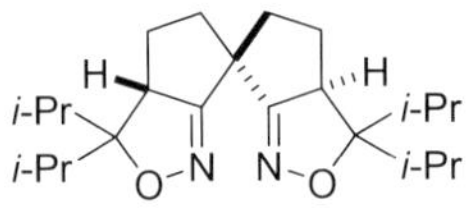

1,3-DC reactions of fulminic acid and the parent *N*-methyl nitrone to curved arenes, fullerenes, and nanotubes has been investigated with density functional theory and ONIOM methods <09CEJ13219>. Developments in microwave-assisted 1,3-DC of nitrile oxides and nitrones and the applications of these cycloaddition reactions to the synthesis of antiviral agents have been reviewed <09EJO5287; 09OBC4567>.

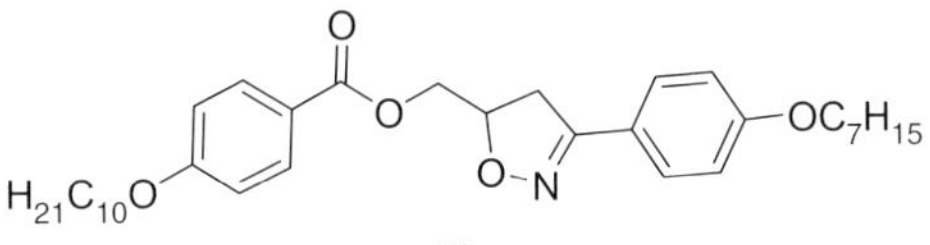

5.7.3. ISOXAZOLIDINES

1,3-DC of nitrones and alkenes followed by cleavage of the isoxazolidine N–O bond is a common approach to 1,3-amino alcohols, which are frequently used as intermediates in synthesizing a large range of structurally and stereochemically diverse azocompounds. During the last year, new applications of this strategy have been published. For example, 4-Amido-N^5-acetyl-4-deoxyneuraminic acid **48** was obtained in four steps starting from isoxazolidine **46**, which was in turn prepared through a highly diastereoselective cycloaddition of nitrone **45** with methyl acrylate <09OL3678>. A similar approach was used to introduce an 1,3-hydroxy amino moiety in the synthesis of polyhydroxylated 8-aminoindolizidinones <09T2322>.

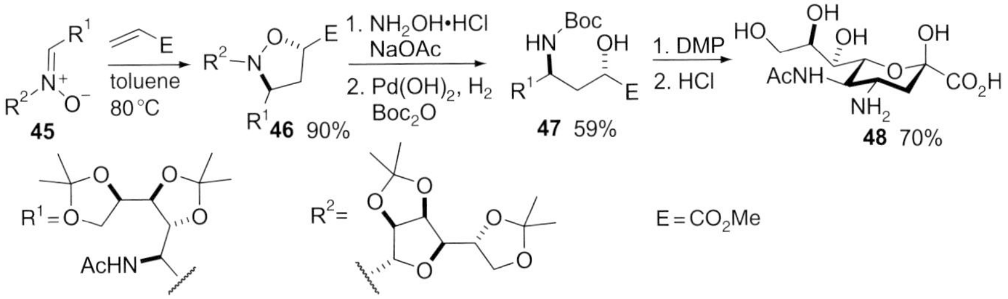

Stereocontrolled 1,3-DC of enantiopure pyrroline N-oxides followed by reduction of the N–O bond and N-cyclization by intramolecular nucleophilic substitution or aminolysis of esters and amides is a general approach to pyrrolizidine and indolizidine alkaloids <09CEJ7808>. Goti et al. described a new total synthesis of naturally occurring casuarine **52**. Nitrone **49** derived from arabinose reacted with the (Z)-alkene **50** with high regio- and stereoselectivity giving isoxazolidine **51** in 79% yield. Suitable elaboration of the adduct **51** afforded **52** in four steps and 55% overall yield <09CEJ1627>. Adducts of the same nitrone **49** with protected enantiopure 3-buten-1,2-diols were converted into different pentahydroxylated pyrrolizidines and indolizidines <09JOC5679>. Analogous strategies were applied to the synthesis of the bicyclic skeleton of casuarine-related derivatives, polyhydroxylated 7-aminopyrrolizidines, and 7- and 8-aminoindolizidines <09CAR167; 09T7056, 09ASC1155>.

n	m	54 % yield	55 % yield
1	1	89	92
1	2	64	98
2	1	47	97

Tricyclic isoxazolidines **54** were prepared in one step from acyclic precursors **53** through a cascade process involving condensation of the aldehyde with hydroxylamine, cyclization to nitrone followed by intramolecular 1,3-DC <09JOC2290>. Reductive cleavage of the N–O bond gave the bicyclic amino alcohols **55** in high yield. The overall process was completely regio- and diastereoselective affording each product as a single isomer. The same methodology was applied to a synthesis of (±)-myrioxazine A <09OL1515>.

A diversity-oriented synthesis of 3- and 2,3-disubstituted piperidines, featuring an intramolecular 1,3-DC of N-alkenyl nitrone enoates was reported <09JOC254>. Isoxazolidines can be prepared by 1,3-DC of alkenes with N-alkylated nitrones generated in situ from aldoximes and electron-poor olefins in organized aqueous media (nanoreactor system) using dodecylbenzenesulfonic acid (DBSA) as surfactant <09GC169>.

The effects of the position of a *trans*-acetonide group and of the relative configuration of the substituents on regio- and stereoselectivity in intramolecular nitrone/alkene cycloaddition (INAC) of nitrones derived from hept-6-enoses has been studied <09CEJ2693>. In the presence of a dioxolane ring on C-3 and C-4 and of an *anti* relationship between C-2 and C-3, such as in **56**, only the bridged adducts were obtained. 2,3-Acetonide derivatives gave a mixture of bridged and fused isoxazolidines arising respectively from an *endo* and *exo* cyclization mode. Finally, compounds possessing a 4,5-acetonide group led exclusively to the fused isoxazolidines.

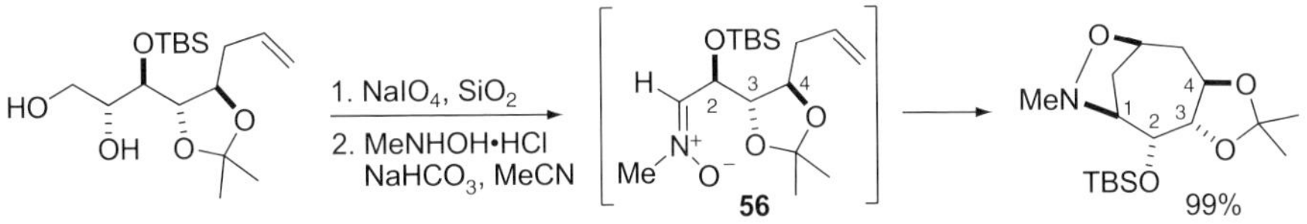

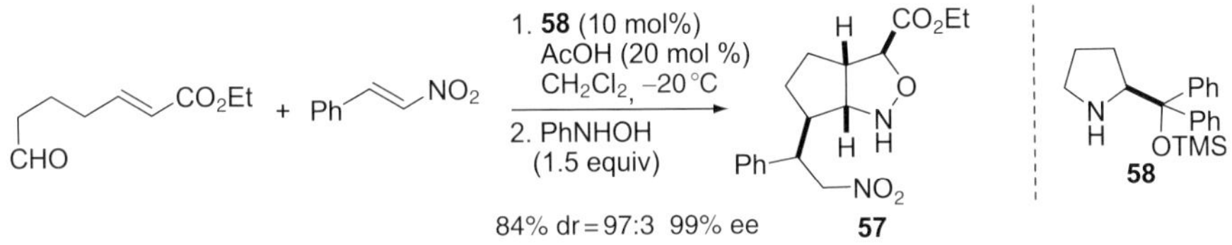

Zhong *et al.* reported an organocatalytic one-pot asymmetric synthesis of bicyclic isoxazolidines by a domino process involving a Michael addition/*in situ* condensation/INAC sequence starting from 7-oxo-2-heptenoate and a nitroalkene <09AGE6089>. Jørgensen's catalyst **58** was showed to be the best catalyst. All 17 examples analyzed afforded bicyclic compounds such as **57** with excellent control of the relative and absolute configuration of the five new formed stereocenters (51-92% yield, 89-97% ds, >98-99% ee).

Diamines derived by 1,1'-binaphthyl-2,2'-diamine and α-amino acids were tested as organocatalysts in the 1,3-DC of aromatic nitrones with (*E*)-crotonaldehyde <09SL2261>.

Optically active N-Boc and N-Cbz-protected isoxazolidines were obtained through an organocatalytic asymmetric formal [3+2] cycloaddition of *in situ* generated N-carbamoyl nitrones <09JA9614>. For example, the N-hydroxy-α-amido sulfone **59** reacted with dimethyl glutaconate in the presence of the quinine-derived ammonium salts Q^+Cl^- to afford isoxazolidine **60** as a sole diastereomer in >99% ee and quantitative yield.

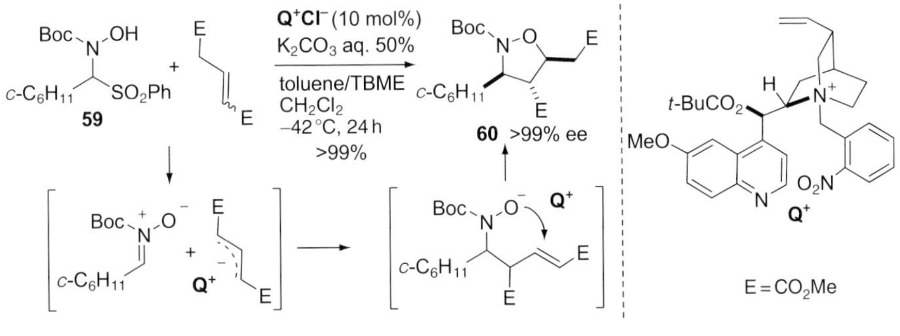

1,3-DC reactions of C-aryl-N-phenyl nitrones with the dimethyl ester **61** of Feist's acid, afforded the corresponding 4-spiro-fused isoxazolidines **62** as single diastereoisomers in moderate yield <09EJO525>.

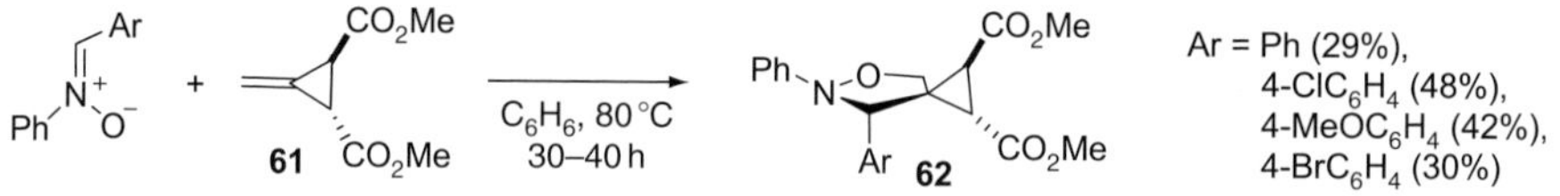

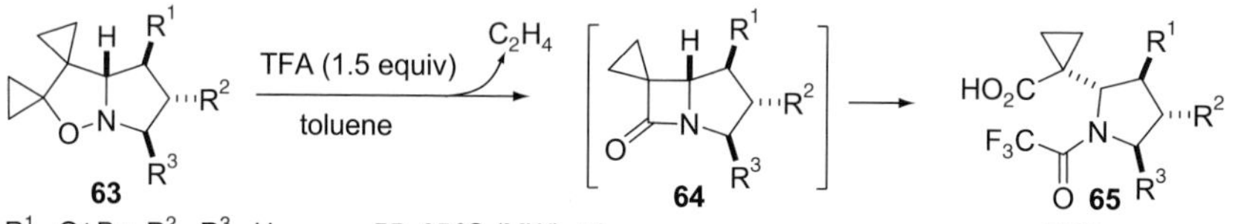

By heating in the presence of trifluoroacetic acid (TFA), enantiopure dispiro-fused pyrrolo[1,2-*b*]isoxazolidines such as **63** directly afforded N-trifluoroacetyl α-cyclopropyl-β-homoprolines **65** in high yield <09JOC4225>. The thermal fragmentative rearrangement of **63** is believed to go through the intermediate formation of spiro-fused carbapenams **64** which undergo trifluoroacetolysis followed by O-N trifluoroacetyl shift under the reaction conditions.

A highly stereoselective synthesis of disubstituted isoxazolidines such as **66** was achieved *via* Pd-catalyzed carboamination reaction of *N*-Boc-*O*-(but-3-enyl) hydroxylamine derivatives <09JOC2533>. The observed diastereoselectivity (dr > 20:1) was superior to those typically obtained in 1,3-DC providing the same kind of products and significantly higher than that of related C–O bond-forming carboetherification reactions of *N*-benzyl-*N*-(but-3-enyl)hydroxylamines.

R^1/R^2 = Me/Ph (77%, dr = 3.8:1); Me/OEt (90%, *trans* only); Me/CO_2Me (72%, dr = 11:1); Et/Ph (77%, dr = 3.5:1); Bn/Ph (71%, dr = 5.5:1); $(CH_2)_2CO_2Me$/Ph (90%, dr = 3:1); Bn/OEt (89%, *trans* only)

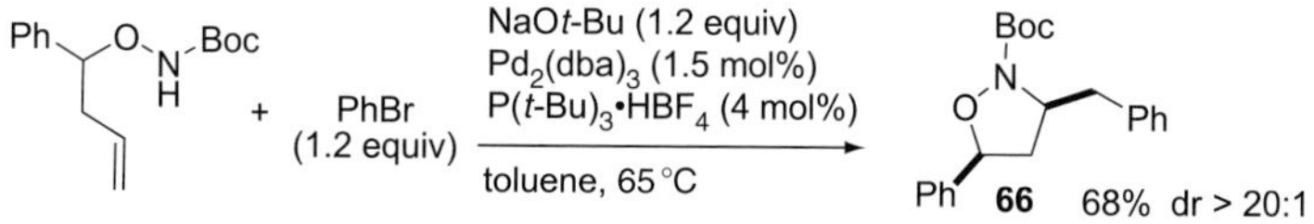

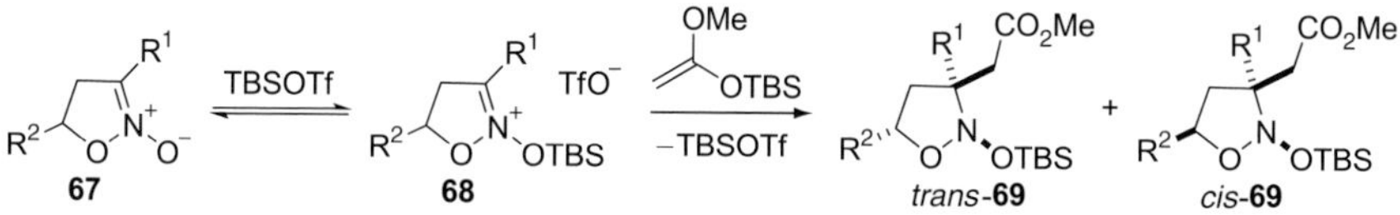

3,3-Disubstituted N-OTBS isoxazolidines **69** were obtained starting from five-membered cyclic nitronates **67** through C–C coupling with silyl ketene acetal catalyzed by TBSOTf. Usually, a mixture of *cis* and *trans* isomers was obtained except in the case of 5-alkoxy substituted nitronates, which were converted with complete *trans*-diastereoselectivity <09EJO3066>.

Isoxazolidines were connected to perylene tetracarboxylic bisimide to form stable dyads that can lead to charge separation upon absorption of visible light <09CEJ12733>.

5.7.4. OXAZOLES

Some new synthetic procedures were described. The Schöllkopf reaction for the synthesis of oxazoles affords 4,5-disubstituted derivatives **70** through the use of strong bases as BuLi. By switching to milder reaction conditions, using 2,6-lutidine as a base, it is possible to obtain 2,5-disubstituted oxazoles **71** in a remarkable example of a base-induced chemoselective process in isocyanide chemistry <09CC3907>.

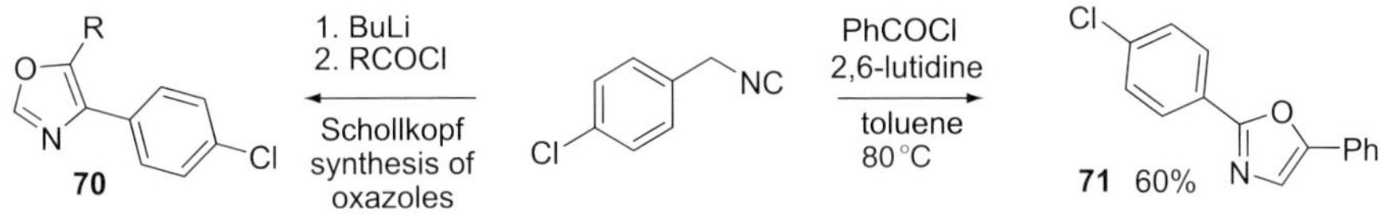

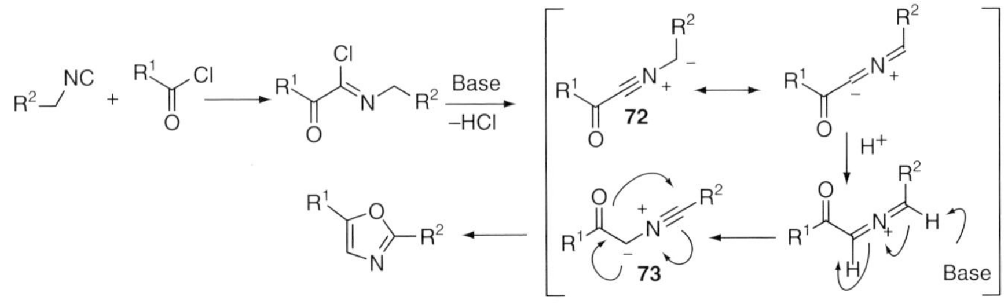

The proposed mechanism suggests that the nitrilium ylide **72** is converted into intermediate **73** by a protonation deprotonation sequence. The final cyclization of the new nitrilium ylides **73** affords the oxazole. The isocyanide substituent (R^2 in the previous scheme) is only aromatic or heteroaromatic while the acyloyl chloride can be aromatic as well as aliphatic and the yield ranges from 40% to 81%.

A new one-pot procedure for the synthesis of oxazoles was developed using a dirhodium(II) catalyzed reaction of α-diazo-β-ketoesters (carboxylic or phosphonic) with carboxamides. Under the 'conventional' two-step method using dirhodium tetraacetate as catalyst, the intermediate 1,4-dicarbonyl compounds **74** were readily isolated and subsequently dehydrated to oxazole-4-carboxylates **75** in modest overall yield (27–54% over two steps). In this one-pot route to oxazoles **75**, the first step was carried out using the same catalyst–solvent combination but under microwave irradiation at 80 °C for 5 minutes, followed by addition of phosphorus oxychloride (2 equiv.) and heating to 110 °C for 30 minutes again using microwave irradiation <09CC3291>.

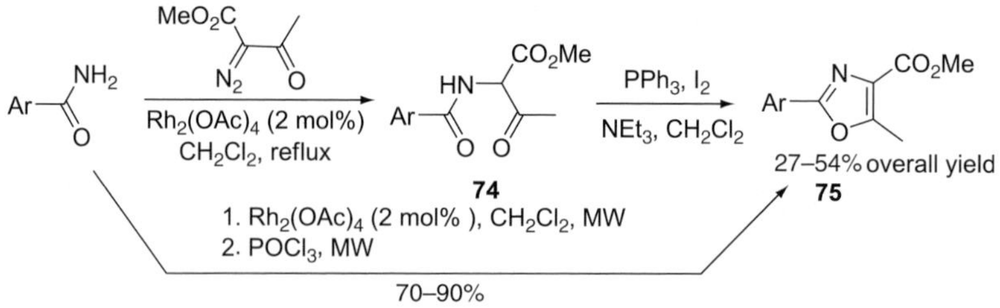

In a three component reaction alkoxyallenes **76**, nitriles and carboxylic acids afforded alkoxy-substituted enamines **77** which, upon treatment with TFA were transformed into 2,4-disubstituted 5-acetyloxazoles **78** <09CEJ5432>.

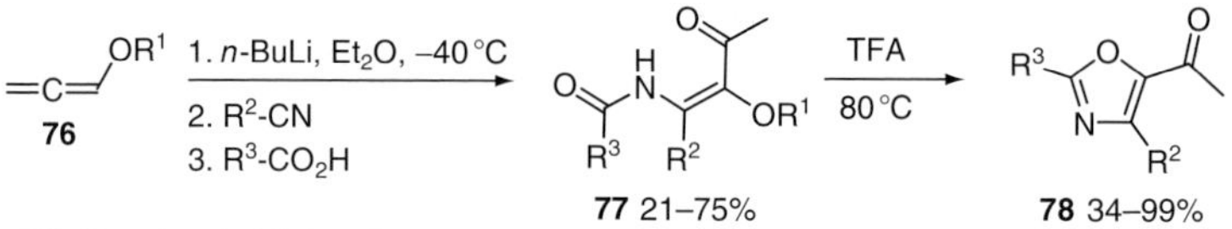

R^1 = Bn, trimethylsilylethyl
R^2 = Ph, n-C_9H_{19}, c-Pr; R^3 = Ph, 2-pyridyl, 2-thienyl, CF_3, acetyl, c-Pr, C≡CH

Oxazoles with a 5-acetyl group are not easily available by other methods. Only a few methods deal with the synthesis of related oxazol-5-yl ketones. This functional group enables the preparation of a variety of other oxazole derivatives.

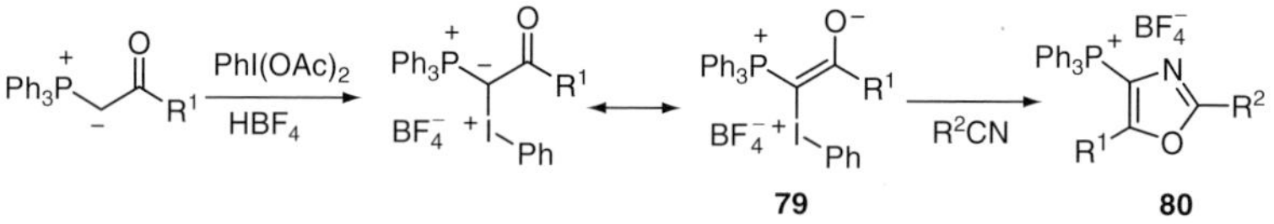

R^1 = OEt, OMe, Ph; R^2 = Me, Et, Ph, NEt_2, 4-pyridyl

Mixed phosphonium ylides **79** can be prepared, through an already known procedure, by treatment of carbonyl-stabilized phosphonium ylides with λ^3-iodanes such as $Ar(I)Hal_2$, ArIO, or $ArI(OAc)_2$ in the presence of HBF_4. These mixed ylides react with nitriles, in a process that is greatly accelerated by UV irradiation, to afford phosphonium substituted oxazoles **80** <09EJO2323>.

A new one-pot process was described for the synthesis of substituted oxazoles starting from amides and propargyl alcohols **81** with the bifunctional catalysis of p-toluensulfonic acid (PTSA) <09JOC3148>.

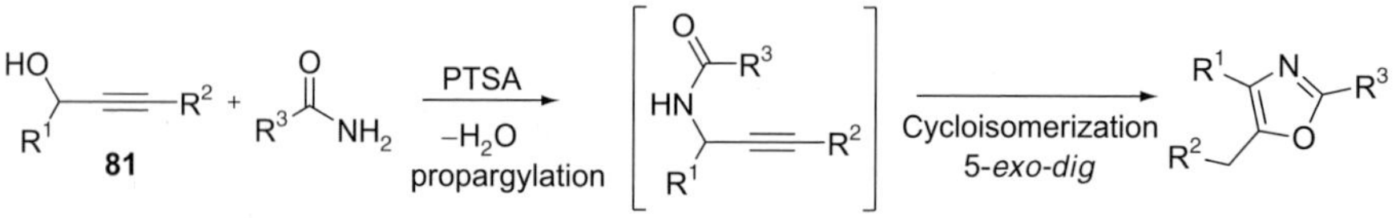

The reaction is very versatile and has been applied to a wide range of substrates (always aromatic amides) with good yields.

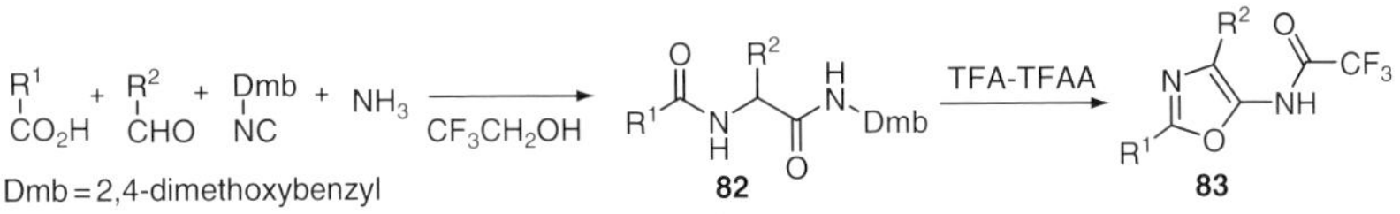

A series of Ugi reactions between carboxylic acids, aldehydes and isonitriles has been successfully performed using ammonia as the amine component, and 2,2,2-trifluoroethanol as a non-nucleophilic solvent in order to suppress alternative side reactions. Utilizing concentrated aqueous ammonia as a convenient source, this approach offered a simple, one-step assembly of Ugi adducts **82** suitable for elaboration into a variety of 5-aminooxazoles **83** through post-condensation modifications <09JOC7084>.

It is known that the carbene tautomer, 2,3-dihydrooxazol-2-ylidene **84**, corresponding to oxazole, is not stable. It is, however, possible to obtain coordination complexes containing carbene ligands. Applying a methodology previously developed for imidazole derivatives, oxazole and thiazole molecules coordinated to manganese(I), are easily transformed to obtain oxazolin-2-ylidene and thiazolin-2-ylidene complexes **85**, respectively <09CC2741>.

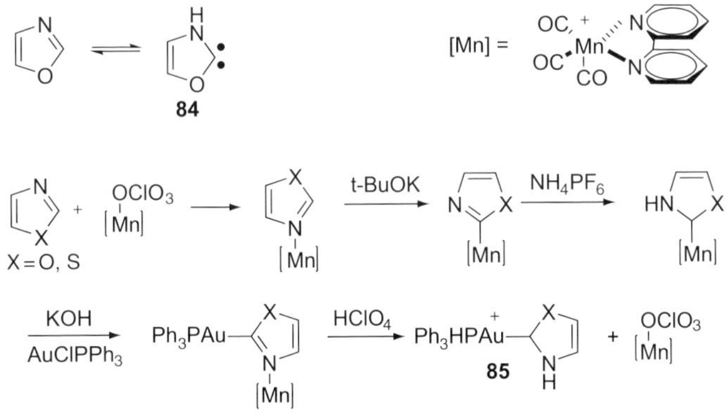

A highly versatile route to 5-acylaminooxazoles **89** was described: the synthesis proceeds through the conversion of readily accessible 5-trifluoroacetylaminooxazoles **86** into the Boc-protected 5-aminooxazoles **87**. These compounds are amenable to parallel amide synthesis utilizing a reliable, one-pot, acylation-deprotection procedure. Through a proper choice of the reaction conditions it was possible to minimize the formation of the side product N-(oxazol-5-yl)-2,2,2-trifluoroacetimidates **88** <09JOC3856>. This work was optimized for a diversity oriented synthesis and a large number of different compounds were obtained in good yields.

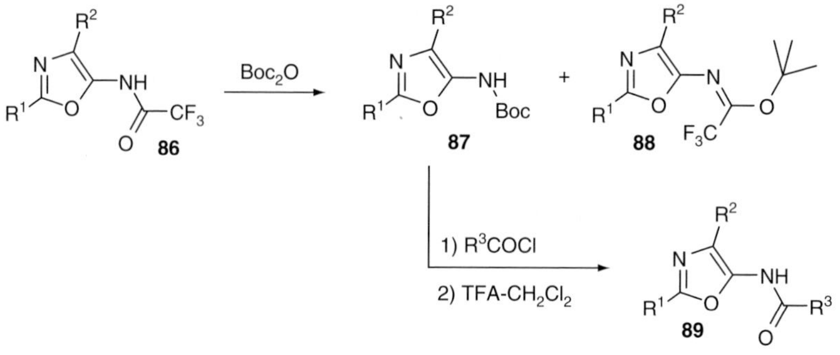

Cuprous iodide, together with triphenylphosphine and Na_2CO_3, was revealed as an efficient catalyst for the direct arylation of 5-arylisoxazoles **90** with aryl iodides. The reaction afforded, in high yields, a number of 2,5-diaryl oxazoles **91** <09TL3273>.

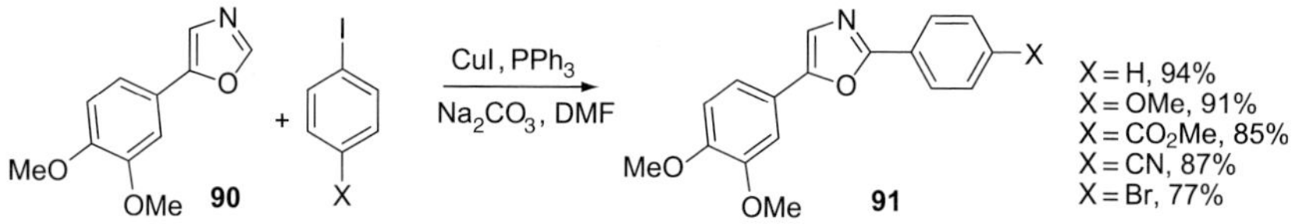

The ethyl oxazole-4-carboxylate **92** was directly and regioselectively alkenylated, benzylated and alkylated with alkenyl, benzyl, allyl and alkyl halides in the presence of catalytic amounts of palladium acetate with cesium carbonate using Buchwald's JohnPhos ligand [P(biphen-2-yl)Cy_2]. As an example the alkenylination reaction is illustrated <09OBC647>.

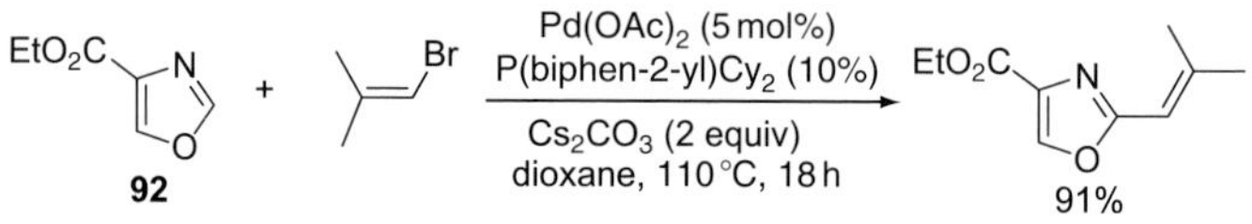

2,5-Bis(methylthio)-oxazoles **93** (available *via* 5-lithiation of 2-methylthiooxazole) can be converted into 2,5-disubstituted oxazoles **94** *via* successive nickel-catalyzed cross-coupling reactions with various organozinc reagents <09OL1457>.

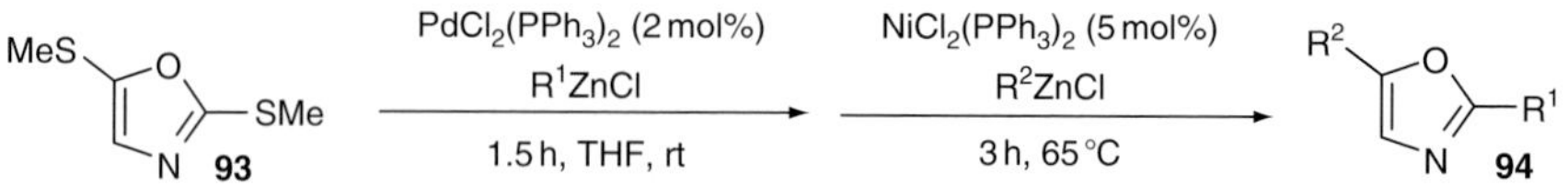

Hexabranchus sanguineus, a nudibranch from the Indo-Pacific region yielded several oxazole containing macrolides in submicromolar amounts. The structures of these cyclic compounds were determined by analyses of 1D and 2D NMR spectra recorded with a state-of-the-art 1 mm ^{1}H NMR high-temperature superconducting microcryoprobe, together with mass spectra <09JNP732>. 4-Ethoxycarbonyl-5-

methyl-2-(phenylsulfonyl)methyloxazole **95** was revealed as a key intermediate for the total synthesis of siphonazoles <09JOC9140, 09OL2389, 09OBC3908>.

Originally proposed structure of neopeltolide   Correct reassigned structure of neopeltolide

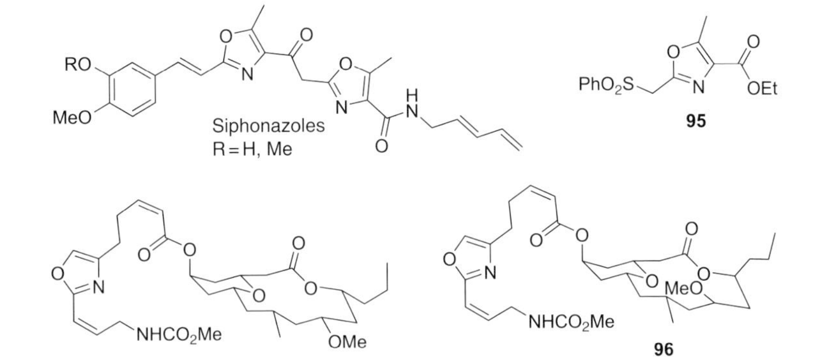

The previously assigned structure of neopeltolide, a marine alkaloid containing the oxazole ring, was revised <09JA12406>.

5.7.5. OXAZOLINES

Treatment of N-allylacetamides (**97a**, Z = Ac) with aryl halides in the presence of sodium *t*-butoxide and a palladium catalyst led to intramolecular arylative cyclization providing the corresponding benzyl-substituted oxazolines **98** in high yields, while N-allylanilines (**97b**, Z = Ph) afforded the corresponding aziridines **99** <09CC5754>.

Variously substituted 2-oxazolines **100** were easily synthesized from N-(2-hydroxyethyl)-amides using a PPh_3-DDQ system <09TL6838>.

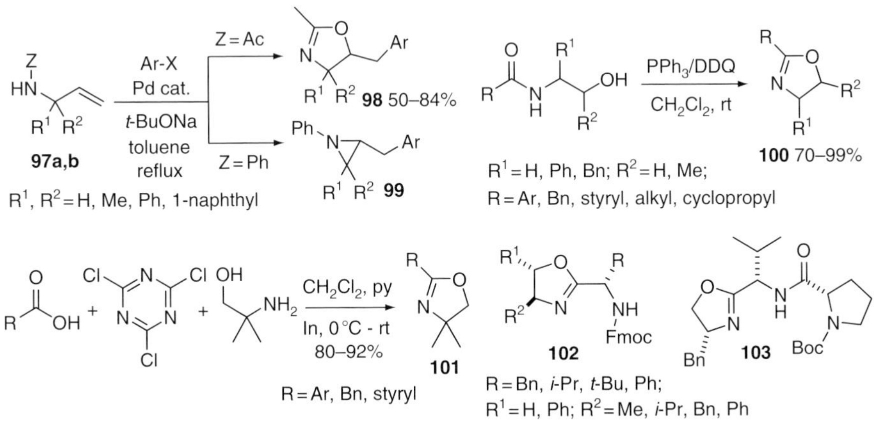

A novel and direct synthesis of oxazolines **101**, as well as 1,3,4-oxadiazoles, from carboxylic acids was achieved under very mild conditions with cyanuric chloride/indium

<09TL5332>. Analogous microwave-assisted condensations of carboxylic acids with 2-amino-2-methyl-1-propanol at 170 °C were described using an open vessel technique to allow the dehydration process <09TL5780>. Amine functionalized oxazolines **102** were synthesized in satisfactory yields from amino acids and amino alcohols, as well as modular oxazoline ligands of type **103** prepared on large scale (> 5g) and purified by simple recrystallization <09T3110>.

The synthesis of 2-substituted-2-oxazolines by reaction of various nitriles with 2-aminoethanol under Lewis acid catalysis was performed using an automated synthesizer. On the basis of reproducibility tests and optimization steps, zinc acetate as catalyst and chlorobenzene as solvent were found to represent the optimal conditions <09JCO274>. Sugar oxazolines **104** were directly synthesized from N-acetyl-2-amino sugars in water using a chloroformamidinium-type dehydrating reagent. The method was applied to various monosaccharides and oligosaccharides <09JOC2210>. A practical and efficient solution-phase parallel synthesis of spirooxazolinoisoxazolines **105** from serine methyl ester has been developed. A 100-membered library was synthesized in high yield, high purity, and excellent regioselectivity <09JCO281>.

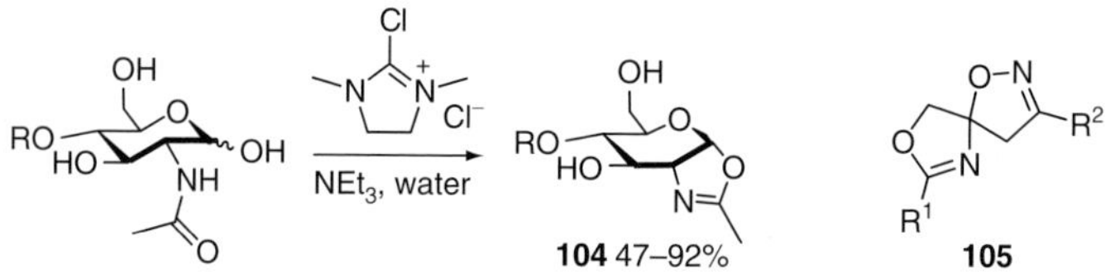

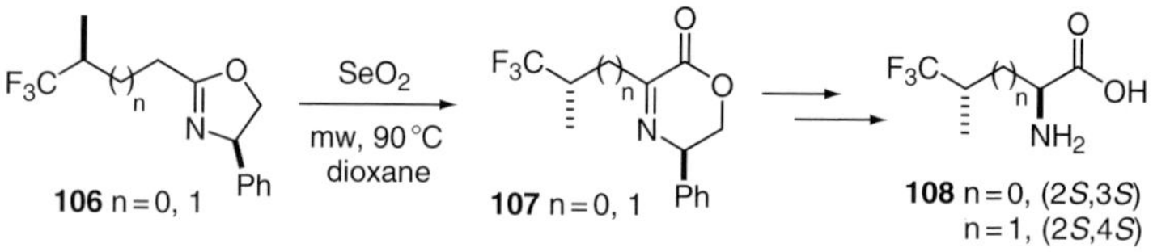

Oxazoline-oxazinone rearrangement was exploited for the stereoselective syntheses of fluorinated amino acids (2S,3S)-4,4,4-trifluorovaline and (2S,4S)-5,5,5-trifluoroleucine **108**. SeO$_2$-Promoted oxidative rearrangement of chiral oxazolines **106** led to dihydro-2H-oxazinones **107**, and face-selective hydrogenation of the C=N double bond, followed by hydrogenolysis-hydrolysis gave the final amino acids <09JOC5510>.

4-Substituted oxazolines **109**, which are readily synthesized from naturally occurring α-amino acids, are converted efficiently and with high stereoselectivity to β-aminoaldehydes **110** using hydrogen, carbon monoxide, and the inexpensive precatalyst dicobalt octacarbonyl <09CC5704>.

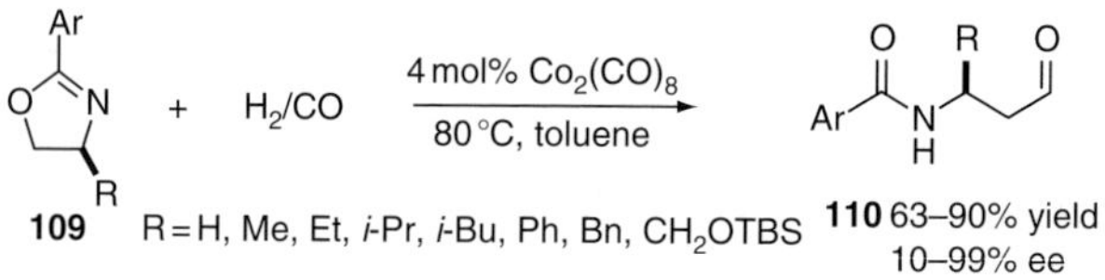

A solid-phase synthesis of α-alkylserines *via* phase-transfer catalytic alkylation of polymer-supported 2-phenyl-2-oxazoline-4-carboxylate has been developed <09T8839>.

A new asymmetric synthesis of (–)-swainsonine has been reported starting from an enantiopure *trans*-oxazoline precursor **111** <09JOC3962>.

Terminal oxazolinyloxiranes **112** were synthesized, even in enantioenriched form, and applied in reactions with amines and in regioselective β-lithiation <09T8745>.

The first enantiopure 2-(*o*-iodoxyphenyl)-oxazolines **113** have been synthesized and applied to the enantioselective oxidation of phenols <09OL1221>.

The living cationic ring-opening polymerization of 2-oxazolines is a versatile method which allows copolymerization of a variety of 2-oxazoline monomers to give a range of tunable polymer properties that enable an easy access to hydrophilic, hydrophobic, and fluorophilic poly(2-oxazoline)s. This class of polymers exhibits numerous potential applications as biomaterials and thermoresponsive materials, as well as the easy access to amphiphilic structures for self-assembly <09AGE7978>.

A general asymmetric synthesis of inherently chiral calix[4]arenes with high enantiomeric excess has been described using a chiral oxazoline derived from L-valine and an *ortho*-lithiation strategy <09OL4986>.

According with the great interest associated with the use of oxazoline derivatives in asymmetric synthesis, recent applications of mono(oxazoline) and bis(oxazoline) ligands in asymmetric catalysis have been reviewed <09CRV2505>. Moreover, being fundamental to the understanding of the relationship between catalyst structure and enantioselectivity to plan the synthesis of new catalysts, the use of Taft steric parameters to correlate the substituent size of a ligand with the enantiomeric ratio (*er*) of a reaction has been studied. Linear free energy relationships have been constructed by plotting log(*er*) versus the steric parameters reported by Taft and modified by Charton. Successful correlations were found for different enantioselective reactions and in particular for chromium-catalyzed aldehyde and ketone allylation using modular oxazoline ligands of type **103** <09JOC7633>.

A new class of chiral phosphine-oxazoline ligands (SpinPHOX, **114**) based on the spiro[4,4]-1,6-nonadiene backbone has been developed from racemic spiro[4,4]nonane-1,6-dione. The cationic iridium complexes showed high efficiency in the hydrogenation of ketimines <09AGE5345>. New HetPHOX ligands **115** were synthesized and applied to the intermolecular asymmetric Heck reaction of 2,3-dihydrofuran with aryl and cyclohexenyl triflates <09EJO1889>. Thiazoline-oxazoline ligands **116** were also prepared in a four-step, high yielding process and applied in the asymmetric Friedel–Crafts alkylation of indole <09EJO4833>.

114 (*R,S*) or (*S,S*)
R = Ph, Bn, *i*-Pr, *i*-Bu, *t*-Bu

115 R = Ph, *i*-Pr, *t*-Bu

116 R^1, R^2 = Ph, Bn, *i*-Pr, *t*-Bu

The development of modular, polymer-supported enantiopure phosphinooxazoline ligands containing a sterically tunable alkoxymethyl group has been optimized for palladium-catalyzed asymmetric allylic amination under batch and continuous flow conditions <09ASC1539>. A library of modular pyranoside phosphite-oxazoline ligands has been screened for Pd-catalyzed asymmetric allylic substitution reactions <09ASC3217>. Phosphite-oxazoline ligands were successfully applied for the first time in the Ir-catalyzed asymmetric hydrogenation of a broad range of 1,1-disubstituted terminal alkenes (up to > 99% ee) <09JA12344>. Chromium catalysts derived from chiral oxazolinyl sulfonamides, having three distinct sites for structural modifications, effect the couplings of aldehydes with vinyl, allyl, and alkyl halides <09JA15387>.

A rational design of a new class of bis(oxazoline) ligands **117**, bearing 4,4′-sulfonamidomethyl groups, for asymmetric Diels–Alder (DA) reactions of various dienes with acryloyloxazolidinone has been reported. Their Cu(NTf$_2$)$_2$ complexes gave high reactivities and enantioselectivities (up to 99%) <09JA17762>. New fluorous-tagged azabis(oxazoline) ligands of type **118** were prepared using the copper-catalyzed azide-alkyne 1,3-DC as ligation method. The ligands were tested in copper-catalyzed asymmetric benzoylations (up to 99% *ee*), nitroaldol (up to 90% *ee*), and Michael reactions (up to 82% *ee*). Depending on the number of fluorous tags and triazole moieties, as well as the reaction solvent, the Cu(II) complexes can be easily recovered and recycled <09ASC1961>. A facile synthesis of the chiral urea bridged box ligand **119** has been described. This ligand readily forms a bis-copper complex with unusual structural features <09CC7309>. New carbohydrate-based bis(oxazoline) ligands **120** were prepared from inexpensive D-glucosamine and applied in cyclopropanation reactions revealing a strong dependence of enantioselectivity on the nature of the 3-O-residue <09EJO997>. Other C$_2$-symmetric chiral bis(oxazolines) containing an EDT-TTF (ethylenedithio-tetrathiafulvalene) backbone **121** were synthesized as suitable precursors for chiral multifunctional molecular materials and coordination metal complexes <09CC3753>. A series of heteroarylidenemalonate-derived box ligands **122** were synthesized and their Cu(II) complexes were applied to the catalytic Friedel–Crafts alkylation of indole (up to > 99% *ee*) <09ASC3113>.

117 R = H, Me; R[1] = Me, CF_3

118 $R^F = CH_2CH_2C_8F_{17}$

119

120 R = Me, Bn, TES, Ac, Bz, Piv

121 (*S*,*S*) and (*R*,*R*)

X = S, O

122 R = *i*-Pr, *i*-Bu, Bn, Ph

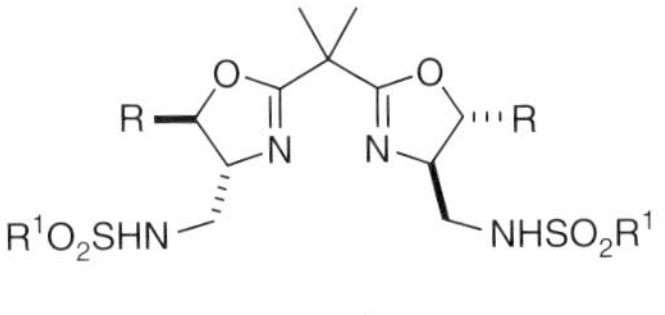

Bis- and tris(oxazolines) containing an alkynyl unit in the ligand backbone were covalently attached to carbosilane dendrimers to give polyfunctional ligands for recyclable Cu(II) Lewis acid catalysts that were immobilized in a membrane bag. The general catalytic potential of these dendritic systems was tested in the α-hydrazination of a β-keto ester and in the Henry reaction of 2-nitrobenzaldehyde with nitromethane <09CEJ5450>.

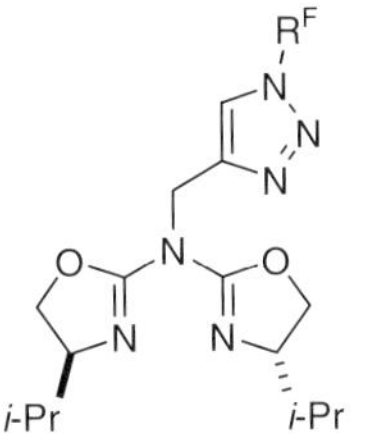

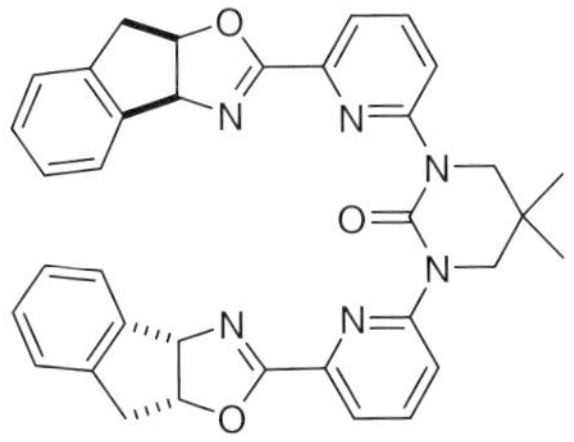

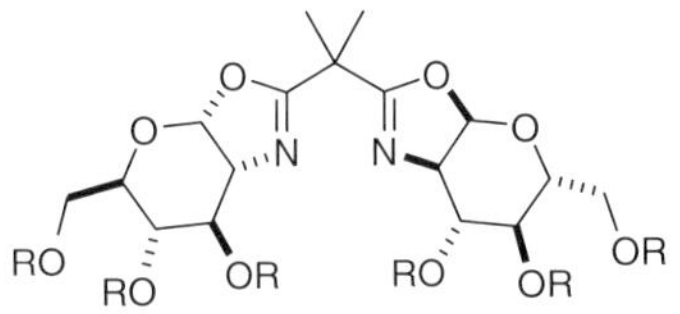

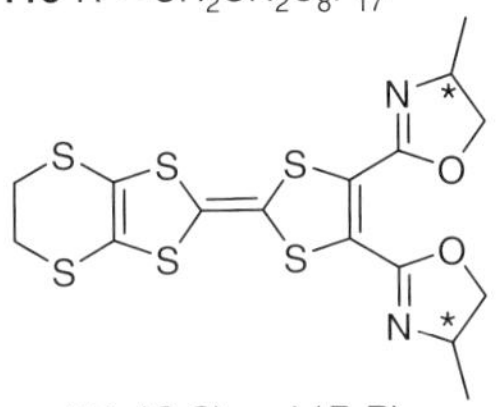

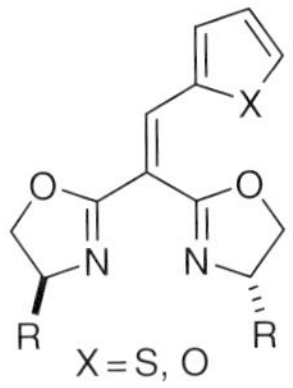

Palladium(II) bis(oxazoline) complexes were exploited in the intermolecular methoxycarbonylation of terminal alkynes. The reaction conditions are tolerant of various functionalities <09AG(E)3326>.

5.7.6. OXAZOLIDINES

An improved process for the synthesis of oxazolidines was developed using $Sc(OTf)_3$ as catalyst in a reaction between 2-methyl-*N*-tosylaziridine **123** and aldehydes or ketones <09CC3928>. With respect to previous procedures, this one appears wide in scope allowing the use of other 2-alkylaziridines. The reaction afforded a mixture of two regioisomers in a ratio ranging from 1:1 to 5:1 (in favour of the 5-alkyl regioisomer, **124**) for aldehydes and higher than 7:1 for ketones, respectively.

127 98%

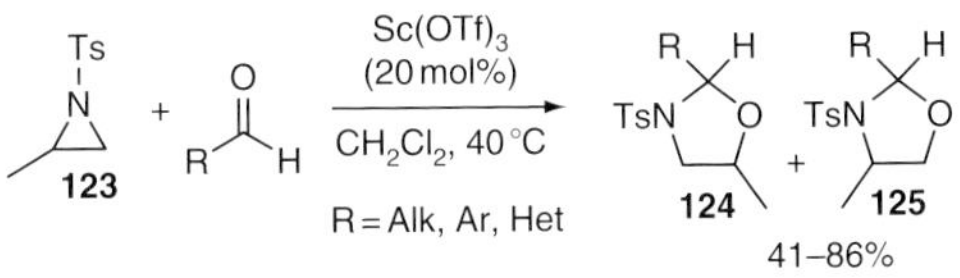

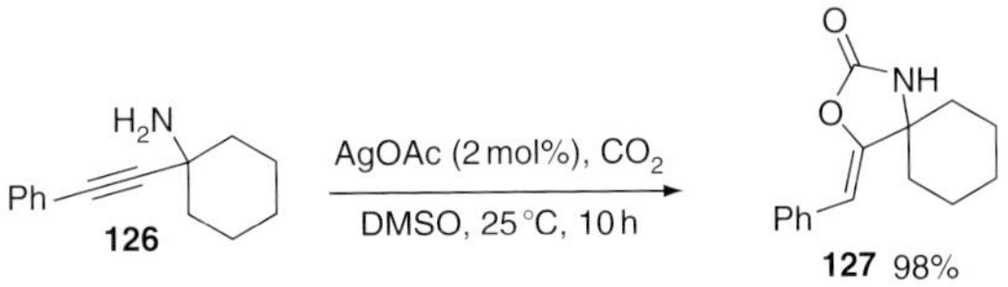

2-Aryl-N-alkyl aziridines can also react with CO_2 in the presence of zirconyl chloride to afford oxazolidinones in high yield <09T6204>.

Propargylamine **126** was transformed into oxazolidinone derivative **127** by reaction with CO_2 and with the catalysis of silver salts <09CL786>. A wide number of examples demonstrated the scope of the reaction affording always almost quantitative yields.

In a similar procedure α-allenylamine **128** was transformed in the corresponding substituted oxazolidinone through a palladium-catalyzed reaction with CO_2 <09TL6491>.

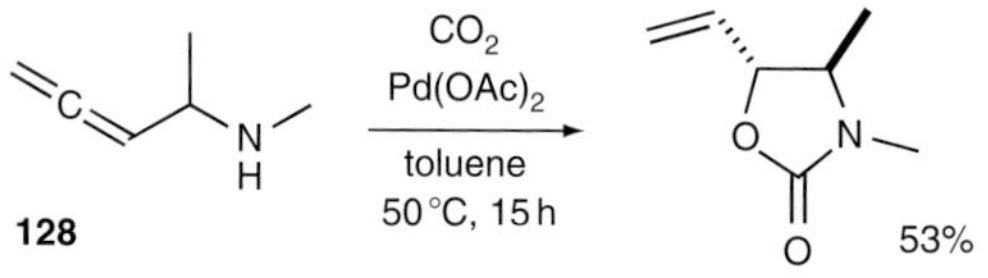

Enantiopure biciclic oxazolidines **129** were obtained in a palladium-catalyzed cyclization of β-aminoalcohols bearing a terminal double bond <09OL3746>. The reaction is highly selective and affords acceptable yields (32-84%).

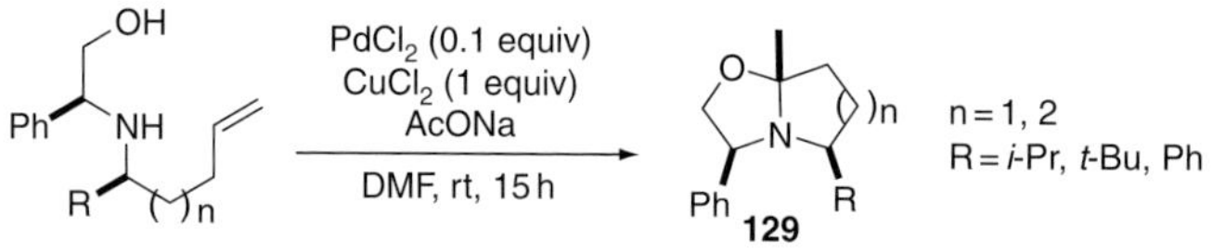

Two distinct procedures for the synthesis of 4-methylene-1,3-oxazolidine derivatives starting from propargyl alcohols were described <09TL60, 09TL4700>.

Concerning the transformation and application of the oxazolidine nucleus, besides the countless examples of use as chiral auxiliaries, some special reactions deserve some consideration.

3-Aminopropionitrile **131** was easily obtained, in good yield, by treating 2-oxazolidinone **130** with KCN in the presence of 18-crown-6 with no solvent <09TL4857>. The nucleophilic attack of the cyanide ion onto the 5-position of the 2-oxazolidinone induces the ring opening and subsequent decarboxylation.

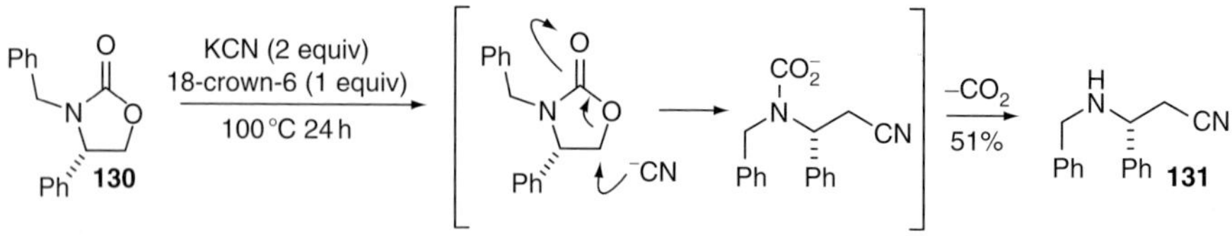

The hydroaminomethylation of N-olefinic oxazolidinones **132**, catalyzed by Rh complexes, followed by hydrolysis, afforded polyaminoalcohols <09ASC2113>. This procedure was used for the production of simple dendrimeric polyamines useful as dendritic cores or as ligand precursors.

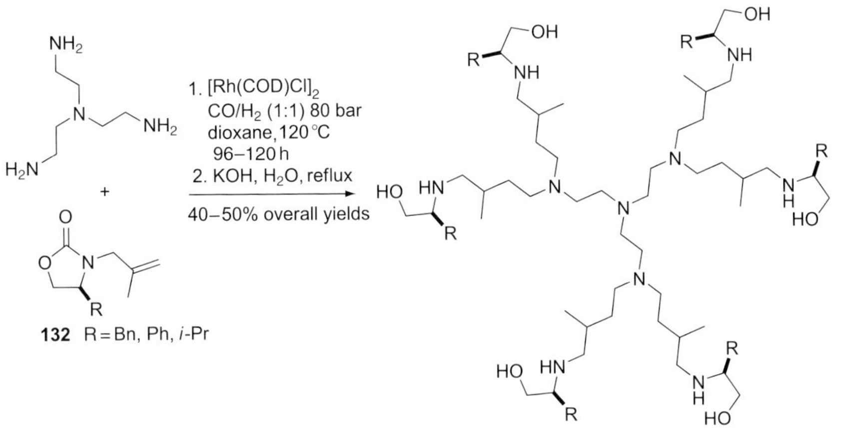

Short oligopeptides containing oxazolidinone moieties were synthesized and their foldamers analyzed to understand their secondary structure <09CEJ8037>.

A new and general model to rationalize the results obtained in D-A reactions using the Evans auxiliary was proposed <09CEJ7665>. The model is based on NMR spectroscopic data and molecular modelling. The observed selectivities are explained by a chirality-transfer concept, in which an achiral Lewis acid works as a bridge for the transfer of chirality between a chiral auxiliary and a prochiral reactive center. The D-A cycloaddition reaction was also the subject of another investigation in which the experimental data of cycloaddition of silyloxydienes with α-β-unsaturated N-acyloxazolidinones were matched with computational analysis results <09JA1947>.

Chiral N-(diethoxymethyl)oxazolidinones **133**, prepared from the corresponding oxazolidinones by heating in triethyl orthoformate, can be used as organozinc carbenoid precursors for the enantioselective amidocyclopropanation of alkenes. Many examples were described with a variety of substituents both on the oxazolidinone moiety and the olefin <09EJO1532>. In most cases the *trans/exo* product is obtained.

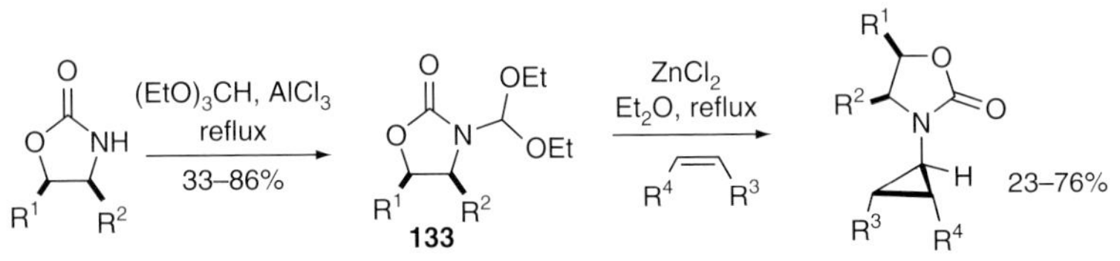

A non-cross-linked polystyrene-supported oxazolidinone **134** was used as a chiral auxiliary in a stereoselective Michael addition of an organocopper reagent onto a crotonamide. The stereoselectivity was extremely high and allowed the synthesis of a pheromone with a 98% *ee* <09EJO1078>.

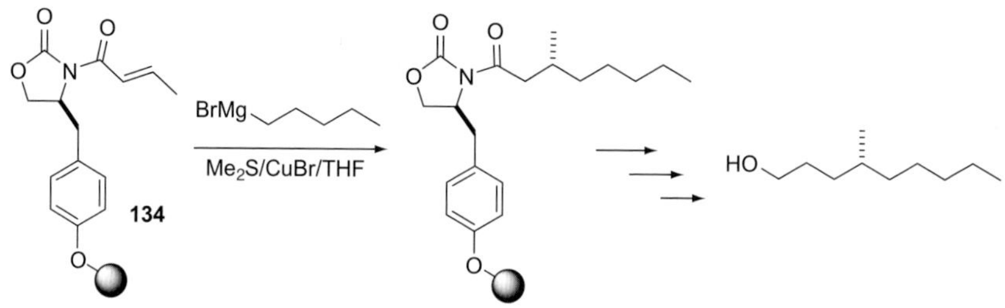

Chiral enamide **136** prepared from cyclohexanone and oxazolidinone **135** was used in hetero-Diels–Alder reactions using Eu(fod)$_3$ as catalyst <09OL3060>. The reaction proceeded with high stereoselectivity and yield, regardless of the nature of the cyclic ketone and the substituents on the dienophile.

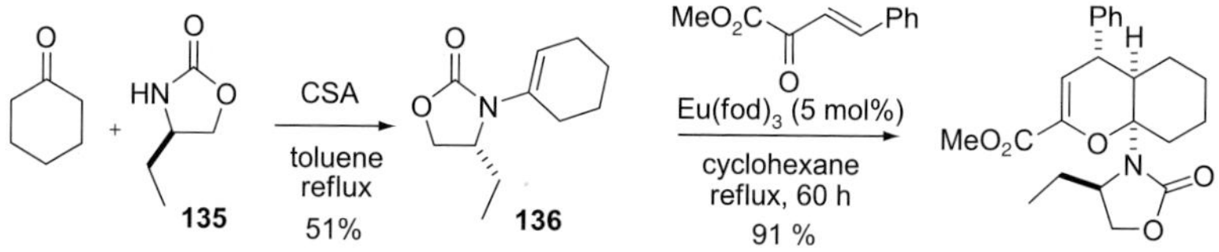

A detailed mechanistic study was made to understand the factors that influence the post-electron transfer steps in the SmI$_2$-promoted carbon-carbon bond forming reaction between N-acyl oxazolidinones and acrylamides <09JA10523>. The most important result is that the reactivity of simple N-acyloxazolidinones correlates with the activation barrier for C–N bond rotation.

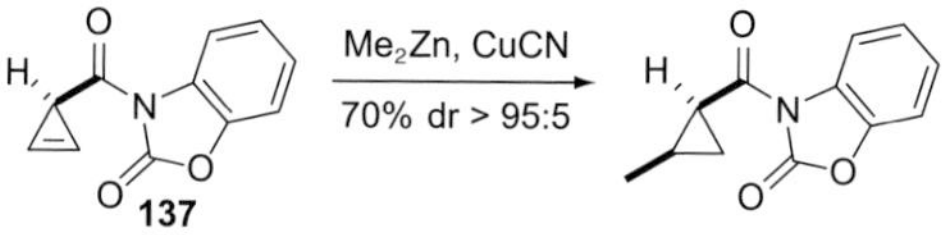

Finally, acyloxazolidinone auxiliaries can direct carbozincation reactions of cyclopropenes, as in the case of compound **137**. The reaction proceeded with high distereoselectivity <09JA5382>.

5.7.7. OXADIAZOLES

2-Aryl- and 2-alkyl-1,3,5-oxadiazoles undergo direct alkenylation at the 5-position upon treatment with alkynes in the presence of a Ni catalyst. The reaction with styrenes in place of alkynes affords the branched isomers such as **138** with complete regioselectivity <09JOC6410>. A copper-mediated direct arylation at the same position is also possible using aryl iodides in the presence of 1,10-phenanthroline (phen) as a ligand and Cs$_2$CO$_3$ as a base <09OL3072>.

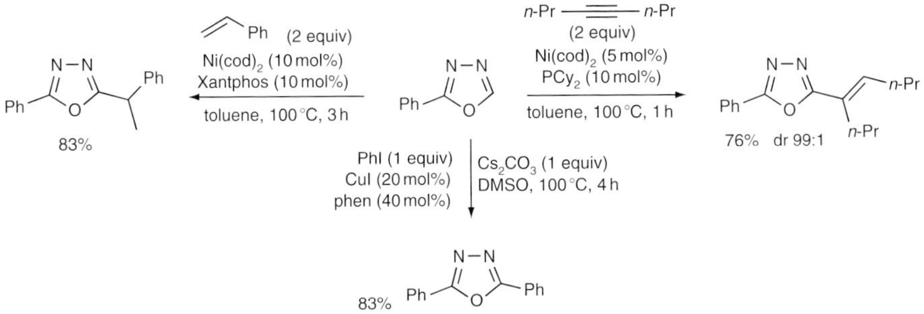

The thermal rearrangement of N-1,2,4-oxadiazol-3-yl hydrazones **139** under solvent-free conditions led to 1,2,4-triazole derivatives **140**. This process represents the first example of a three-atom side-chain rearrangement involving an NNC sequence linked at C-3 of the oxadiazole. The precursor accessibility, the reaction conditions, the easy work-up, and the good product yields make this methodology significant for the synthesis of functionalized triazolamines <09OL4018>.

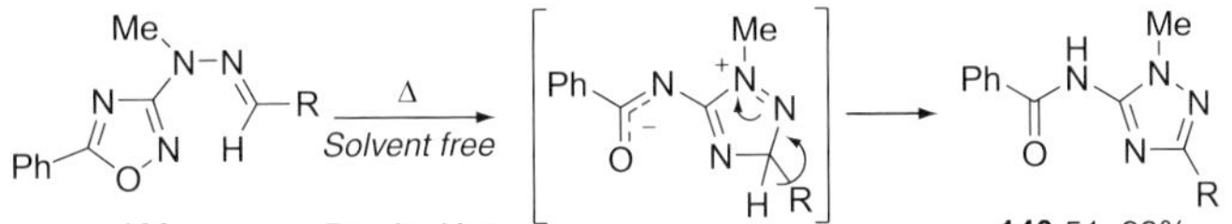

A new series of heterocyclic oligomers based on the 1,3,4-oxadiazole ring were synthesized and the physical properties, thin film morphologies, and field-effect transistor characteristics were evaluated. Compound **141** is the first example of an oxadiazole-containing organic semiconductor oligomer in an n-channel organic field-effect transistor and shows moderate mobilities <09JA1692>.

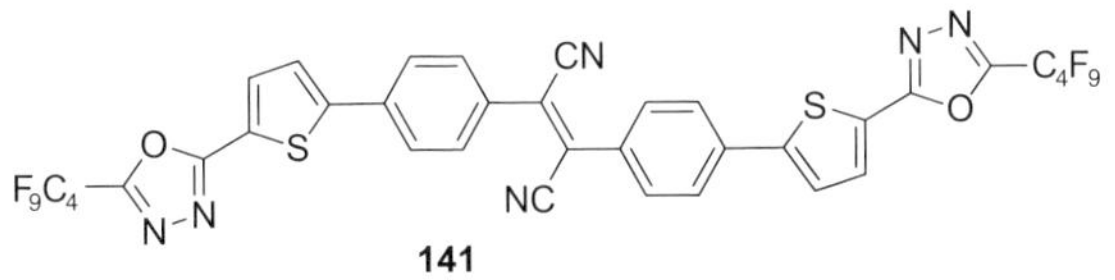

REFERENCES

09AG(E)3326	K. Kato, S. Motodate, T. Mochida, T. Kobayashi and H. Akita, *Angew. Chem. Int. Ed.*, **48**, 3326 (2009).
09AGE5345	Z. Han, Z. Wang, X. Zhang and K. Ding, *Angew. Chem. Int. Ed.*, **48**, 5345 (2009).
09AGE6089	D. Zhu, M. Lu, L. Dai and G. Zhong, *Angew. Chem. Int. Ed.*, **48**, 6089 (2009).
09AGE6844	H. Jiang, P. Elsner, K.L. Jensen, A. Falcicchio, V. Marcos and K.A. Jørgensen, *Angew. Chem. Int. Ed.*, **48**, 6844 (2009).
09AGE7978	R. Hoogenboom, *Angew. Chem. Int. Ed.*, **48**, 7978 (2009).
09ASC1155	F.M. Cordero, P. Bonanno, S. Neudeck, C. Vurchio and A. Brandi, *Adv. Synth. Catal,* **351**, 1155 (2009).
09ASC1539	D. Popa, R. Marcos, S. Sayalero, A. Vidal-Ferran and M.A. Pericàs, *Adv. Synth. Catal,* **351**, 1539 (2009).

09ASC2113	M.A. Subhani, K.-S. Muller and P. Eilbracht, *Adv. Synth. Catal.*, **351**, 2113 (2009).
09ASC3217	Y. Mata, O. Pàmies and M. Diéguez, *Adv. Synth. Catal*, **351**, 3217 (2009).
09ASC1961	R. Rasappan, T. Olbrich and O. Reiser, *Adv. Synth. Catal.*, **351**, 1961 (2009).
09ASC3113	Y.-J. Sun, N. Li, Z.-B. Zheng, L. Liu, Y.-B. Yu and Z.-H. Qin, *Adv. Synth. Catal.*, **351**, 3113 (2009).
09BCJ285	G.B. Bajracharya, M.A. Arai, P.S. Koranne, T. Suzuki, S. Takizawa and H. Sasai, *Bull. Chem. Soc. Jpn.*, **82**, 285 (2009).
09BMC5285	A. Kumar, R.A. Maurya, S. Sharma, P. Ahmad, A.B. Singh, A.K. Tamrakar and A.K. Srivastava, *Biorg. Med. Chem*, **17**, 5285 (2009).
09BMC6279	T. Liu, X. Dong, N. Xue, R. Wu, Q. He, B. Yang and Y. Hu, *Biorg. Med. Chem*, **17**, 6279 (2009).
09CAR167	S. Stecko, J. Solecka and M. Chmielewski, *Carbohydr. Res.*, **344**, 167 (2009).
09CC2741	J. Ruiz and B.F. Perandones, *Chem. Commun.*, 2741–2743 (2009).
09CC3276	I. Singh, J.S. Vyle and F. Heaney, *Chem. Commun.*, 3276 (2009).
09CC3291	B. Shi, A.J. Blake, I.B. Campbell, B.D. Judkins and C.J. Moody, *Chem. Commun.*, 3291 (2009).
09CC3753	F. Riobé and N. Avarvari, *Chem. Commun.*, 3753 (2009).
09CC3907	A. dos Santos, L. El Kaım, L. Grimaud and C. Ronsseray, *Chem. Commun.*, 3907 (2009).
09CC3928	B. Kang, A.W. Miller, S. Goyal and S.T. Nguyen, *Chem. Commun.*, 3928 (2009).
09CC5704	D.S. Laitar, J.W. Kramer, B.T. Whiting and E.B. Lobkovsky, *Chem. Commun.*, 5704 (2009).
09CC5754	D. Fujino, S. Hayashi, H. Yorimitsu and K. Oshima, *Chem. Commun.*, 5754 (2009).
09CC7309	R. Mal, N. Mittal, T.J. Emge and D. Seidel, *Chem. Commun.*, 7309 (2009).
09CEJ1627	F. Cardona, C. Parmeggiani, E. Faggi, C. Bonaccini, P. Gratteri, L. Sim, T.M. Gloster, S. Roberts, G.J. Davies, D.R. Rose and A. Goti, *Chem. Eur. J.*, **15**, 1627 (2009).
09CEJ2693	T.K.M. Shing, W.F. Wong, T. Ikeno and T. Yamada, *Chem. Eur. J.*, **15**, 2693 (2009).
09CEJ3960	A. Pohjakallio and P.M. Pihko, *Chem. Eur. J.*, **15**, 3960 (2009).
09CEJ5432	T. Lechel, D. Lentz and H.-U. Reissig, *Chem. Eur. J.*, **15**, 5432 (2009).
09CEJ5450	M. Gaab, S. Bellemin-Laponnaz and L.H. Gade, *Chem. Eur. J.*, **15**, 5450 (2009).
09CEJ7665	S.M. Bakalova, F.J.S. Duarte, M.K. Georgieva, E.J. Cabrita and A.G. Santos, *Chem. Eur. J.*, **15**, 7665 (2009).
09CEJ7808	A. Brandi, F. Cardona, S. Cicchi, F.M. Cordero and A. Goti, *Chem. Eur. J.*, **15**, 7808 (2009).
09CEJ7940	E. Trogu, F. De Sarlo and F. Machetti, *Chem. Eur. J.*, **15**, 7940 (2009).
09CEJ8037	G. Angelici, G. Falini, H.-J. Hofmann, D. Huster, M. Molinari and C. Tomasini, *Chem. Eur. J.*, **15**, 8037 (2009).
09CEJ12065	N. Lohse-Fraefel and E.M. Carreira, *Chem. Eur. J.*, **15**, 12065 (2009).
09CEJ12733	H. Langhals, A. Obermeier, Y. Floredo, A. Zanelli and L. Flamigni, *Chem. Eur. J.*, **15**, 13219 (2009).
09CEJ13219	S. Osuna and K.N. Houk, *Chem. Eur. J.*, **15**, 13219 (2009).
09CL786	S. Yoshida, K. Fukui, S. Kikuchi and T. Yamada, *Chem. Lett.*, **38**, 786 (2009).
09CRV2505	G.C. Hargaden and P.J. Guiry, *Chem. Rev.*, **109**, 2505 (2009).
09EJO525	V.V. Diev, T.Q. Tung and A.P. Molchanov, *Eur. J. Org. Chem.*, 525 (2009).
09EJO889	A. Tavares, P.H. Schneider and A.A. Merlo, *Eur. J. Org. Chem.*, 889 (2009).
09EJO1027	A. Abbiati, Arcadi, F. Marinelli, E. Rossi and M. Verdecchia, *Eur. J. Org. Chem.*, 1027 (2009).
09EJO1078	C. Lu, D. Li, Q. Wang, G. Yang and Z. Chen, *Eur. J. Org. Chem.*, 1078 (2009).
09EJO1532	G. Begis, D.E. Cladingboel, L. Jerome, W.B. Motherwell and T.D. Sheppard, *Eur. J. Org. Chem.*, 1532 (2009).
09EJO1889	M.O. Fitzpatrick, H. -Bunz and P.J. Guiry, *Eur. J. Org. Chem.*, 1889 (2009).
09EJO2323	E.D. Matveeva, T.A. Podrugina, A.S. Pavlova, A.V. Mironov, R. Gleiter and N.S. Zefirov, *Eur. J. Org. Chem.*, 2323 (2009).
09EJO3066	V.O. Smirnov, A.S. Sidorenkov, Y.A. Khomutova, S.L. Ioffe and V.A. Tartakovsky, *Eur. J. Org. Chem.*, 3066 (2009).

09EJO4041	Y. Fall, C. Reynaud, H. Doucet and M. Santelli, *Eur. J. Org. Chem.*, 4041 (2009).
09EJO4833	S.C. McKeon, H. Müller-Bunz and P.J. Guiry, *Eur. J. Org. Chem.*, 4833 (2009).
09EJO5287	M. Pineiro and T.M.V.D. Pinho e Melo, *Eur. J. Org. Chem.*, 5287 (2009).
09EJO5971	E. Trogu, L. Cecchi, F. De Sarlo, L. Guideri, F. Ponticelli and F. Machetti, *Eur. J. Org. Chem.*, 5971 (2009).
09EJO997	T. Minuth, M. Irmak, A. Groschner, T. Lehnert and M.M.K. Boysen, *Eur. J. Org. Chem.*, 997 (2009).
09GC169	S.K. Hota, A. Chatterjee, P.K. Bhattacharya and P. Chattopadhyay, *Green Chem.*, **11**, 169 (2009).
09JA1692	T. Lee, C.A. Landis, B.M. Dhar, B.J. Jung, J. Sun, A. Sarjeant, H.-J. Lee and H.E. Katz, *J. Am. Chem. Soc.*, **131**, 1692 (2009).
09JA17762	A. Sakakura, R. Kondo, Y. Matsumura, M. Akakura and K. Ishihara, *J. Am. Chem. Soc.*, **131**, 17762 (2009).
09JA1947	Y. Lam, P.H.-Y. Cheong, J.M. Blasco Mata, S.J. Stanway, V. Gouverneur and K.N. Houk, *J. Am. Chem. Soc.*, **131**, 1947 (2009).
09JA3452	T. Tsujihara, K. Takenaka, K. Onitsuka, M. Hatanaka and H. Sasai, *J. Am. Chem. Soc.*, **131**, 3452 (2009).
09JA5382	V. Tarwade, X. Liu, N. Yan and J.M. Fox, *J. Am. Chem. Soc.*, **131**, 5382 (2009).
09JA9614	C. Gioia, F. Fini, A. Mazzanti, L. Bernardi and A. Ricci, *J. Am. Chem. Soc.*, **131**, 9614 (2009).
09JA10523	R.H. Taaning, K.B. Lindsay, B. Schiott, K. Daasbjerg and T. Skrydstrup, *J. Am. Chem. Soc.*, **131**, 10253 (2009).
09JA12344	J. Mazuela, J.J. Verendel, M. Coll, B. Schäffner, A. Börner, P.G. Andersson, O. Pàmies and M. Diéguez, *J. Am. Chem. Soc.*, **131**, 12344 (2009).
09JA12406	D.W. Custar, T.P. Zabawa, J. Hines, C.M. Crews and K.A. Scheidt, *J. Am. Chem. Soc.*, **131**, 12406 (2009).
09JA15387	H. Guo, C.-G. Dong, D.-S. Kim, D. Urabe, J. Wang, J.T. Kim, X. Liu, T. Sasaki and Y. Kishi, *J. Am. Chem. Soc.*, **131**, 15387 (2009).
09JCO274	K. Kempe, M. Lobert, R. Hoogenboom and U.S. Schubert, *J. Comb. Chem.*, **11**, 274 (2009).
09JCO281	H.-W. Shih, C.-W. Guo, K.-H. Lo, M.-Y. Huang and W.-C. Cheng, *J. Comb. Chem.*, **11**, 281 (2009).
09JCO428	S.-J. Tu, X.-H. Zhang, Z.-G. Han, X.-D. Cao, S.-S. Wu, S. Yan, W.-J. Hao, G. Zhang and N. Ma, *J. Comb. Chem.*, **11**, 428 (2009).
09JCO697	W. Ma, B. Peterson, A. Kelson and E. Laborde, *J. Comb. Chem.*, **11**, 697 (2009).
09JNP732	D.S. Dalisay, E.W. Rogers, A.S. Edison and T.F. Molinski, *J. Nat. Prod.*, **72**, 732 (2009).
09JOC254	B.E. Stephens and F. Liu, *J. Org. Chem.*, **74**, 254 (2009).
09JOC1099	H. Suga, Y. Adachi, K. Fujimoto, Y. Furihata, T. Tsuchida, A. Kakehi and T. Baba, *J. Org. Chem.*, **74**, 1099 (2009).
09JOC2210	M. Noguchi, T. Tanaka, H. Gyakushi, A. Kobayashi and S. Shoda, *J. Org. Chem.*, **74**, 2210 (2009).
09JOC2290	A.J.M. Burrell, I. Coldham, L. Watson, N. Oram, C.D. Pilgram and N.G. Martin, *J. Org. Chem.*, **74**, 2290 (2009).
09JOC2533	G.S. Lemen, N.C. Giampietro, M.B. Hay and J.P. Wolfe, *J. Org. Chem.*, **74**, 2533 (2009).
09JOC3148	Y.-M. Pan, F.-J. Zheng, H.-X. Lin and Z.-P. Zhan, *J. Org. Chem.*, **74**, 3148 (2009).
09JOC3856	M.J. Thompson, H. Adams and B. Chen, *J. Org. Chem.*, **74**, 3856 (2009).
09JOC3962	Y.-S. Tian, J.-E. Joo, B.-S. Kong, V.-T. Pham, K.-Y. Lee and W.-H. Ham, *J. Org. Chem.*, **74**, 3962 (2009).
09JOC4225	F.M. Cordero, M. Salvati, C. Vurchio, A. de Meijere and A. Brandi, *J. Org. Chem.*, **74**, 4225 (2009).
09JOC5510	J.A. Pigza, T. Quach and T.F. Molinski, *J. Org. Chem.*, **74**, 5510 (2009).
09JOC5679	J.A. Tamayo, F. Franco, D. Lo Re and F. Sánchez-Cantalejo, *J. Org. Chem.*, **74**, 5679 (2009).
09JOC6410	T. Mukai, K. Hirano, T. Satoh and M. Miura, *J. Org. Chem.*, **74**, 6410 (2009).
09JOC7084	M.J. Thompson and B. Chen, *J. Org. Chem.*, **74**, 7084 (2009).
09JOC7633	M.S. Sigman and J.J. Miller, *J. Org. Chem.*, **74**, 7633 (2009).

09JOC9140	J. Zhang, E.A. Polishchuk, J. Chen and M.A. Ciufolini, *J. Org. Chem.*, **74**, 9140 (2009).
09JOC9274	T. Tsujihara, T. Shinohara, K. Takenaka, S. Takizawa, K. Onitsuka, M. Hatanaka and H. Sasai, *J. Org. Chem.*, **74**, 9274 (2009).
09JOM(694)942	B. Willy, W. Frank, F. Rominger and T.J.J. Müller, *J. Organomet. Chem.*, **694**, 942 (2008).
09MM5411	I. Singh, Z. Zarafshani, J.-F. Lutz and F. Heaney, *Macromolecules*, **42**, 5411 (2009).
09MM7709	Y.-G. Lee, Y. Koyama, M. Yonekawa and T. Takata, *Macromolecules*, **42**, 7709 (2009).
09OBC647	C. Verrier, C. Hoarau and F. Marsais, *Org. Biomol. Chem.*, **7**, 647 (2009).
09OBC3908	J. Linder, A.J. Blake and C.J. Moody, *Org. Biomol. Chem.*, **6**, 3908 (2008).
09OBC4349	C.C. Lee, R.J. Fitzmaurice and S. Caddick, *Org. Biomol. Chem.*, **7**, 4349 (2009).
09OBC4567	C. Nájera and J.M. Sansano, *Org. Biomol. Chem.*, **7**, 4567 (2009).
09OL1159	M. Girardin, P.G. Alsabeh, S. Lauzon, S.J. Dolman, S.G. Ouellet and G. Hughes, *Org. Lett.*, **11**, 1159 (2009).
09OL1221	J.K. Boppisetti and V.B. Birman, *Org. Lett.*, **11**, 1221 (2009).
09OL1457	K. Lee, C.M. Counceller and J.P. Stambuli, *Org. Lett.*, **11**, 1457 (2009).
09OL1515	A.J.M. Burrell, I. Coldham and N. Oram, *Org. Lett.*, **11**, 1515 (2009).
09OL2389	J. Zhang and M.A. Ciufolini, *Org. Lett.*, **11**, 2389 (2009).
09OL3060	F. Gallier, H. Hussain, A. Martel, A. Kirschning and G. Dujardin, *Org. Lett.*, **11**, 3060 (2009).
09OL3072	T. Kawano, T. Yoshizumi, K. Hirano, T. Satoh and M. Miura, *Org. Lett.*, **11**, 3072 (2009).
09OL3678	Z.-X. Gao, M. Wang, S. Wang and Z.-J. Yao, *Org. Lett.*, **11**, 3678 (2009).
09OL3746	J. Alladoum, E. Vrancken, P. Mangeney, S. Roland and C. Kadouri-Puchot, *Org. Lett.*, **16**, 3746 (2009).
09OL3982	S. Tang, J. He, Y. Sun, L. He and X. She, *Org. Lett.*, **11**, 3982 (2009).
09OL4018	A. Palumbo Piccionello, A. Pace, S. Buscemi and N. Vivona, *Org. Lett.*, **11**, 4018 (2009).
09OL4986	S.A. Herbert and G.E. Arnott, *Org. Lett.*, **11**, 4986 (2009).
09S591	P. Conti, A. Pinto, L. Tamborini, P. Dunkel, V. Gambaro, G.L. Visconti and C. De Micheli, *Synthesis*, 591 (2009).
09S3405	V.K. Brel, *Synthesis*, 3405 (2009).
09SL310	G.B. Bajracharya, P.S. Koranne, T. Tsujihara, S. Takizawa, K. Onitsuka and H. Sasai, *Synlett*, 310 (2009).
09SL2261	Ł. Weselińskia, P. Stępniaka and J. Jurczak, *Synlett*, 2261 (2009).
09T2067	I. Yavari, M. Piltan and L. Moradi, *Tetrahedron*, **65**, 2067 (2009).
09T2322	X. Li, Z. Zhu, K. Duan, H. Chen, Z. Li, Z. Li and P. Zhang, *Tetrahedron*, **65**, 2322 (2009).
09T3110	J.J. Miller, S. Rajaram, C. Pfaffenroth and M.S. Sigman, *Tetrahedron*, **65**, 3110 (2009).
09T3165	D.P. Canterbury, I.R. Herrick, J. Um, K.N. Houk and A.J. Frontier, *Tetrahedron*, **65**, 3165 (2009).
09T5402	M.F.A. Adamo, S. Suresh and L. Piras, *Tetrahedron*, **65**, 5402 (2009).
09T5472	A.M. González-Nogal and M. Calle, *Tetrahedron*, **65**, 5472 (2009).
09T6204	Y. Wu, L.N. He, Y. Du, J.Q. Wang, C.-X. Miao and W. Li, *Tetrahedron*, **65**, 6204 (2009).
09T7056	S. Stecko, M. Jurczak, O. Staszewska-Krajewska, J. Solecka and M. Chmielewski, *Tetrahedron*, **65**, 7056 (2009).
09T8745	L. Degennaro, V. Capriati, C. Carlucci, S. Florio, R. Luisi, I. Nuzzo and C. Cuocci, *Tetrahedron*, **65**, 8745 (2009).
09T8839	J. Lee, M.W. Ha, T.-S. Kim, M.-J. Kim, J.-M. Ku, S. Jew, H. Park and B.-S. Jeong, *Tetrahedron*, **65**, 8839 (2009).
09T9146	S. Chen, J. Ren and Z. Wang, *Tetrahedron*, **65**, 9146 (2009).
09TL60	H.-F. Jiang and J.-W. Zhao, *Tetrahedron Lett.*, **50**, 60 (2009).
09TL3273	T. Yoshizumi, T. Satoha, K. Hirano, D. Matsuo, A. Orita, J. Otera and M. Miura, *Tetrahedron Lett.*, **50**, 3273 (2009).
09TL3948	S. Bhosale, S. Kurhade, U.V. Prasad, V.P. Palle and D. Bhuniya, *Tetrahedron Lett.*, **50**, 3948 (2009).

09TL4700	N. Fleury-Bregeot, A. Voituruez, P. Retailleau and A. Marinetti, *Tetrahedron Lett.*, **50**, 4700 (2009).
09TL4857	T. Taniguchi, N. Goto and H. Ishibashi, *Tetrahedron Lett.*, **50**, 4857 (2008).
09TL5332	C.O. Kangani and B.W. Day, *Tetrahedron Lett.*, **50**, 5332 (2009).
09TL5780	R. Sharma, S.K. Vadivel, R.I. Duclos, Jr. and A. Makriyannis, *Tetrahedron Lett.*, **50**, 5780 (2009).
09TL6491	Y. Kayaki, N. Mori and T. Ikariya, *Tetrahedron Lett.*, **50**, 6491 (2009).
09TL6838	Q. Xu and Z. Li, *Tetrahedron Lett.*, **50**, 6838 (2009).
09TL7392	C. Altug, Y. Dürüst and M.C. Elliott, *Tetrahedron Lett.*, **50**, 7392 (2009).

CHAPTER 6.1

Six-Membered Ring Systems: Pyridines and Benzo Derivatives

Philip E. Alford
Dartmouth College, Hanover, NH 03755
philip.alford@dartmouth.edu

6.1.1. INTRODUCTION

Pyridine and pyridine-derived structures are privileged pharmacophores in medicinal chemistry and an essential functionality for organic chemists. As the prototypical π-deficient heterocycle, pyridine illustrates distinctive chemistry as both substrate and reagent. The basicity and metallophilic high donor number of these π-deficient systems has long favored them as ligands in metal catalysis. The last decade saw pyridine assume a stronger role as functional group for directed C–H oxidation/activation. Pyridine derivatives lend themselves to many roles in the spirited field of supramolecular chemistry – whether as the ligand backbone of metal-organic polymers or presiding over the key electronic stations of nanodevices. In biochemistry, pyridine-containing cofactors are necessary nutrients on which our lives depend. The blockbuster proprietary drugs Nexium®, Singulair®, Takepron®, and Actos® accounted for over $20 billion in sales last year – each containing the auspices of a pyridine moiety. Yet, as industry proves its ingenuity towards the synthesis of novel azine containing compounds, nature continues to inspire us with new structures.

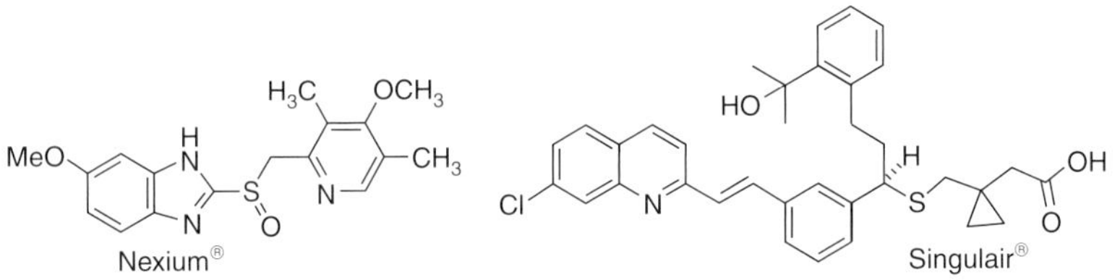

Scheme 1

The plumieride-family of anticancer alkaloids gained a new representative from the flowers of *Plumeria rubra* L. cv. *Acutifolia*. The dashing structure of plumericidine has been supported by X-ray crystallography <09HCA2790>. Moss-like evergreens from the genus *Lycopodium* produce impressive alkaloids with complex frameworks; these compounds have been the focus of several synthetic efforts as well as a review <09H(77)679; 09H(79)791; 09EJO643; 09T6584>.

From murky origins in 1948, the thiopeptide micrococcin P had evaded identification for more than half a century until elucidation by total synthesis this past year

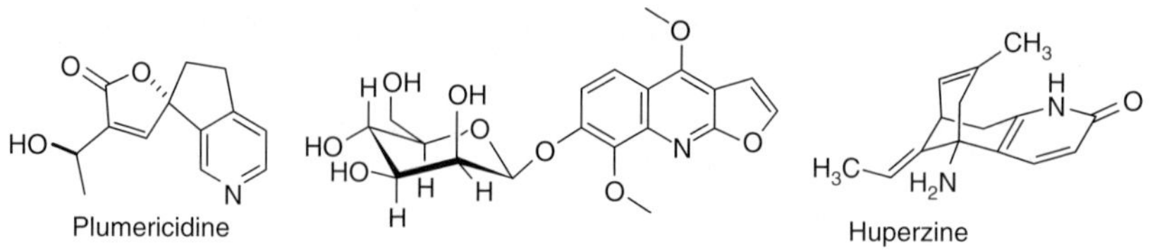

Scheme 2

<09AG(I)4198>. The perennial weed Goldenrod, *Solidago canadensis*, has provided novel fused[2,3-*b*]quinoline alkaloids, 8-methoxydictamnine-7-ß-D-mannopyranoside as well as the demethoxy derivative <09HCA928>. Pyrinadone A (Scheme 3) is a dimeric marine alkaloid containing an exotic azoxy moiety; synthesis of this natural product was achieved by careful oxidative coupling of the hydroxylamine monomer units <09T5834>. In honor of John Daly, chemical and historical reviews of the frog alkaloid epibatidine have been compiled <09H(79)207; 09H(79)99>.

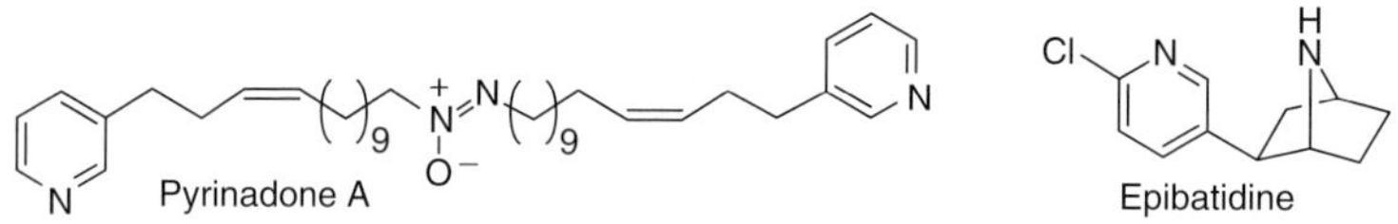

Scheme 3

A myriad of synthetic methods accomplish the synthesis of these systems. Classic methods continue to be improved upon and modern synthetic advances innovate new strategies. The unique reactivity of pyridine remains relevant to all fields that touch upon heterocyclic chemistry. This chapter covers selected advances from the past year made towards the novel preparation and transformation of pyridines, quinolines, and isoquinolines.

6.1.2. PYRIDINES

6.1.2.1 Preparation of Pyridines

Despite the awesome power of modern ring-closing metathesis (RCM), the method has made few appearances for the construction of pyridine rings. Illustrated by two recent reviews, RCM methods have generally been oriented towards the construction of 2-pyridinones and benzo-fused derivatives <09CR3743; 09H(78)2735>. Innovative access to 3-hydroxypyridines was realized by RCM of readily available allylated amino acid derivatives (Scheme 4). The method was also extended to 3-aminopyridines by addition of $NH_2OH \cdot HCl$ to 1,6-dihydro-2*H*-pyridin-3-one before aromatization <09OL515>.

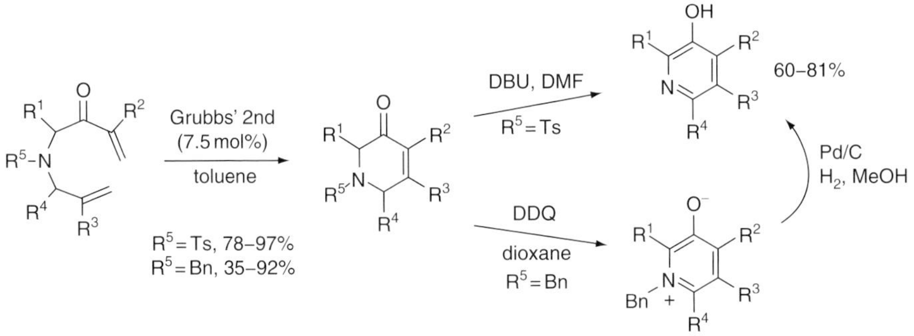

Scheme 4

Comparatively, metal-mediated [2+2+2] cyclotrimerization is a much more common approach for synthesis of pyridines. Many methods involve tethered alkynes to produce fused polycyclic pyridine ring systems <09OL341>. One novel

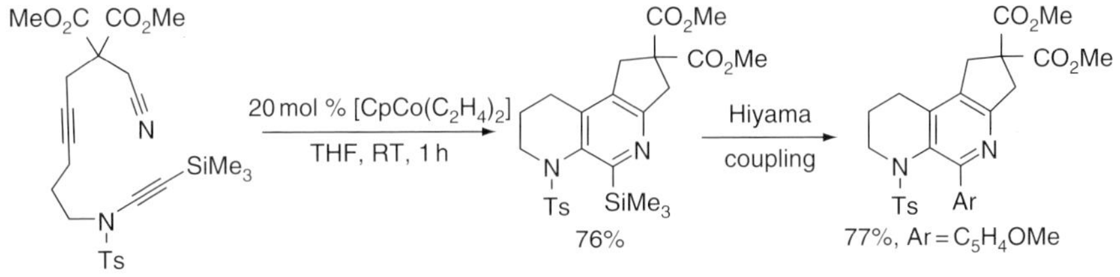

Scheme 5

cobalt-catalyzed [2+2+2] cycloaddition between a tethered ynamide, nitrile, and alkyne produces tricyclic fused 3-amino-2-silylpyridines (Scheme 5). The consequent silyl pyridine was also shown to be an excellent substrate for Hiyama cross-couplings <09CEJ2129>.

A rhodium cyclotrimerization of phenol-linked 1,6-diynes with activated nitriles demonstrates strong regioselectivity and exclusively forms the 3-azadibenzofuran product <09OL2361>. The method has also been used to generate planar-chiral pyridinophanes in high ee <09OL3906>. Gandon *et al.* have designed a new air-stable cobalt catalyst for these reactions and have included optimized conditions for the synthesis of bicyclic pyridines <09AG(I)1810>. Rhodium catalyzes formal [4+2] processes as well, also through oxidative cyclization <09S1400>. Beauchemin *et al.* have reported a novel synthesis of pyridines by metal-free intramolecular hydroamination of acyclic alkynyl oximes (Scheme 6) <09AG(I)8325>.

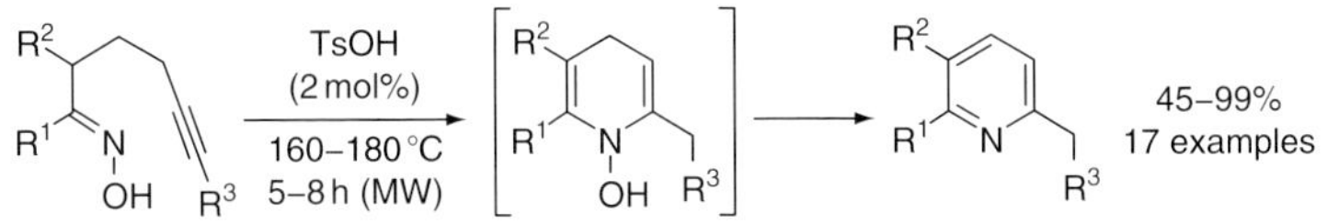

Scheme 6

The aza-Diels–Alder reaction is a well-studied approach to the pyridine nucleus, generating tetrahydropyridines from a variety of azadienes and dienophiles <09S113>. 1,2,4-Triazines are commonly used as conformationally rigid masked 2-azadienes for inverse-electron-demand Diels–Alder reactions. An intramolecular example of this strategy was recently used to generate 2,3-dihydrofuro[2,3-*b*]pyridines from alkyne tethered 1,2,4-triazines <09SL92>. Moody et al. report a superb new sequence for the synthesis of pyridines via 1,2,4-triazines <09OL3686>.

Readily available hydrazides (Scheme 7) undergo copper-mediated insertion of a carbene intermediate generated from α-diazo-β-ketoesters; subsequent condensation

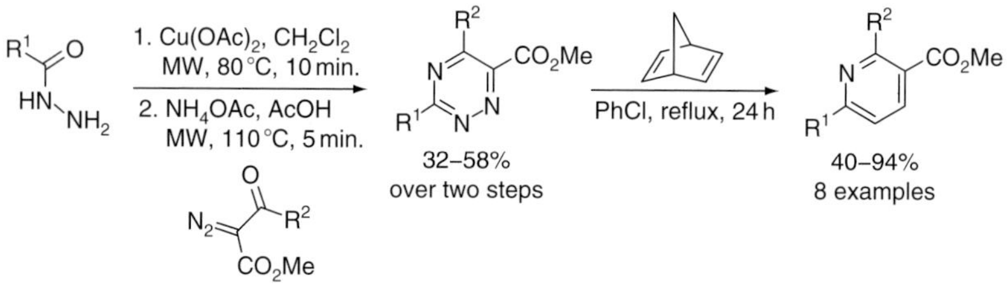

Scheme 7

with ammonia produces the triazine. In extension of previous 1,2,4-triazine work, Stanforth et al. have described a one-pot approach to 2,2′-bipyridines from α-acetoxy-α-chloro-β-ketoesters, amidrazones, and 2,5-norbornadiene <09T975>.

Methods designed around electrocyclic ring closure and sigmatropic rearrangements typically find less use than their other pericyclic counterparts. One captivating example, reported by Yudin et al., produces pyridones from δ-lactams through a tandem ring-expanding [3,3]-sigmatropic rearrangement and 6π-electrocyclization (Scheme 8) <09OL1281>.

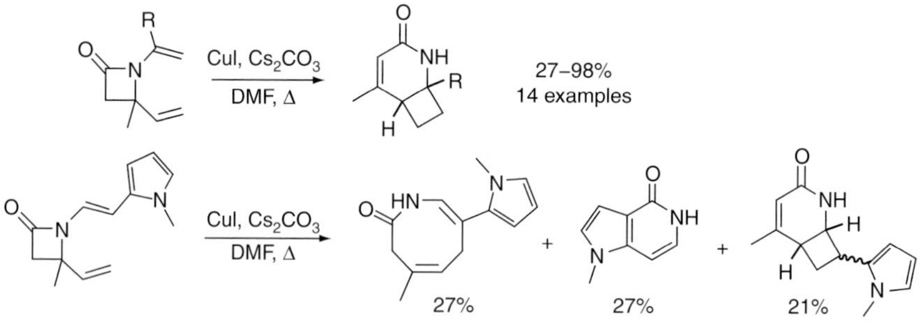

Scheme 8

Other electrocyclizations exploited multicomponent couplings to produce the aza-triene equivalent necessary for ring closure. A spectacular one-pot cascade involves the orchestration of a Staudinger–Meyer reaction, a Wolff rearrangement, and an aza-Wittig reaction – all before the final 6π-electrocyclization (Scheme 9) <09JOC903>.

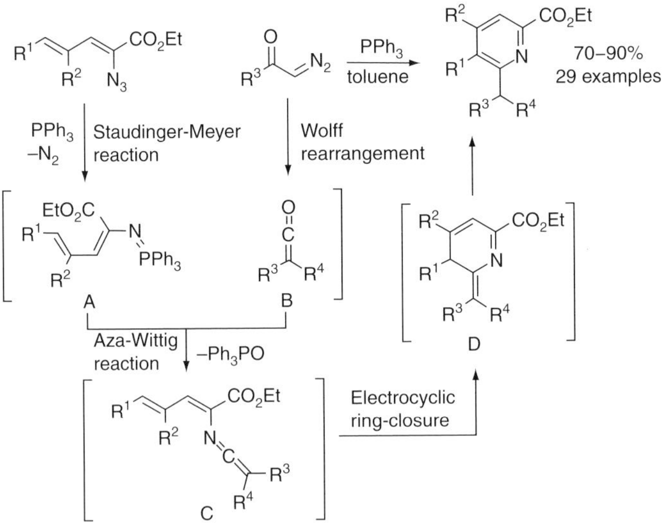

Scheme 9

While multicomponent reactions of all types gained favor in recent years, cyclocondensations in particular have historically dominated the literature. The Hantzsch synthesis, perhaps the most widely known (dihydro)pyridine synthesis, is representative of these reactions <09JHC336>. Conversion of the dihydropyridine to pyridine is facile in most cases and many conditions yield a fully aromatized product <09JHC931>. Cyclic ß-keto-esters and ß-keto-enamines can be used to generate bicyclic products <09JHC465>.

In evidence of the utility of this reaction, Ciufolini et al. enlisted the Hantzsch synthesis for construction of the central trithiazolylpyridine core of micrococcin P1 (Scheme 10) <09AG(I)4198>. An improved assembly, reported later in the year, employed the Bagley variation of the related Bohlmann–Rahtz cyclocondensation reaction to synthesize the key trithiazolylpyridine core <09JOC5750>. While exploring these syntheses, Ciufolini et al. noted that a one-pot Bagley protocol produced higher yields than the more sedulous Eiden–Herdeis method (Scheme 11).

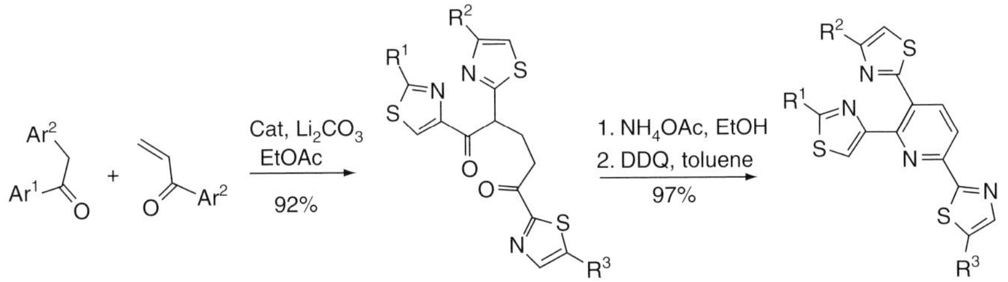

Scheme 10

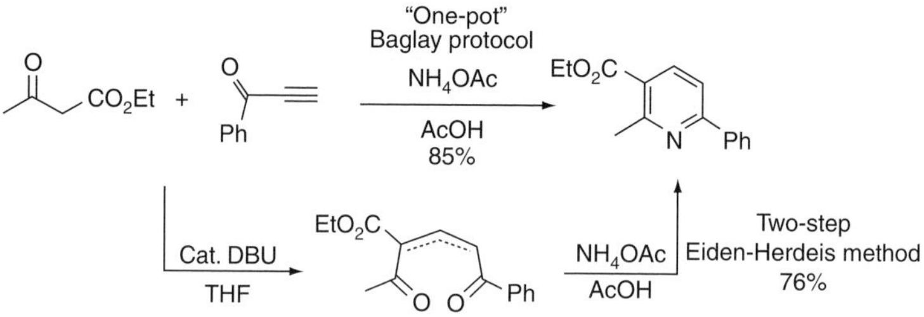

Scheme 11

A textbook example of the closely related Guareschi–Thorpe pyridine synthesis illustrates both the utility of cyanoesters as well as the resulting 3-cyanopyridin-2-ones <09H(78)2067>. Lithium nitride has been reported as a convenient source of ammonia in these particular reactions <09H(78)977>. Malononitrile based cyclocondensations have become very popular, especially when constituent to multicomponent reactions with thiols <09JHC54>. These condensations rapidly generate unsymmetrical 2-hydroxy- or 2-amino-3,5-disubstituted-6-thiopyridines under acidic <09TL3897> or basic conditions <09JHC69; 09JHC1208>. Chen et al. compared the effects of ionic and amine bases in these reactions, uncovering different mechanisms for the penultimate oxidation in each case <09JOC6999>.

Caerulomycins A and E were synthesized through a new cyclocondensation sequence (Scheme 12). The method also highlights a useful trick: using nonaflates to convert highly polar heterocycles to less polar, more easily isolable compounds <09CEJ6811>.

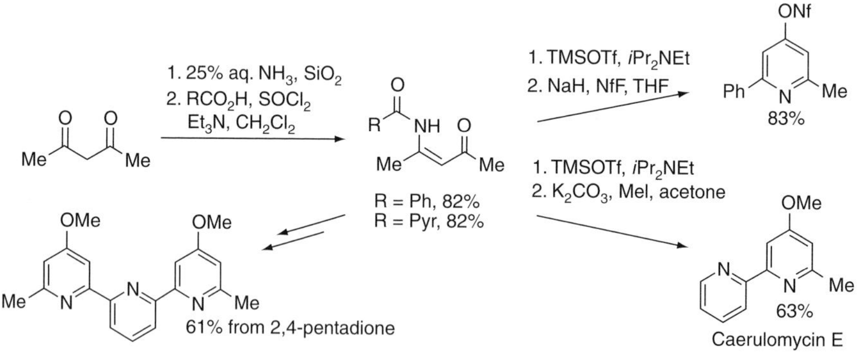

Scheme 12

Showcasing an asymmetric three-component cascade, an alkoxyallenyllithium-based MCR incorporated various amino acids with no detectable loss of enantiopurity (Scheme 13) <09ASC1162>.

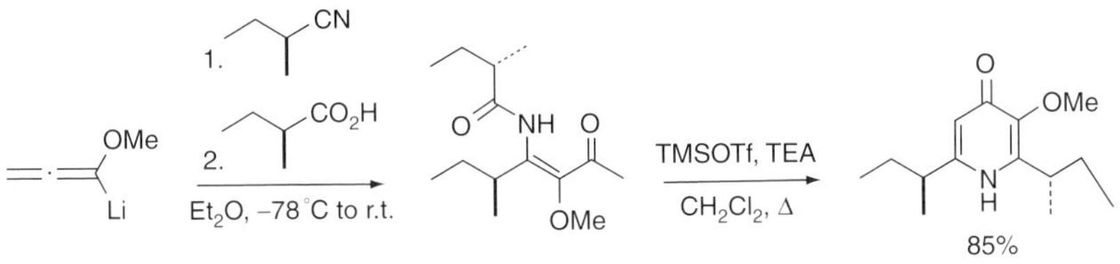

Scheme 13

β-Oxo-amides undergo self-condensation in the presence of TEA and a Vilsmeier-like N,N,N′,N′-tetramethylchloroformamidinium chloride reagent to produce α-pyridinones <09ASC2217>. Baylis–Hillman acetates are another useful precursor to α-pyridinones; one particularly simple cyclocondensation procedure has recently been described <09TL4229>. For the synthesis of nitropyridinones, β-formyl-β-nitro-enamines provide a versatile C3 synthon which integrate a *meta*-nitro-group into the final product <09OBC325>.

As the basis for a total synthesis of lyconadin A, Castle *et al.* have reported a Reformatsky-type condensation protocol for the synthesis of 5-alkyl-6-carbomethoxy-2-pyridones (Scheme 14). The authors discuss and compare several other pyridinone syntheses before presenting their new method <09T6584>.

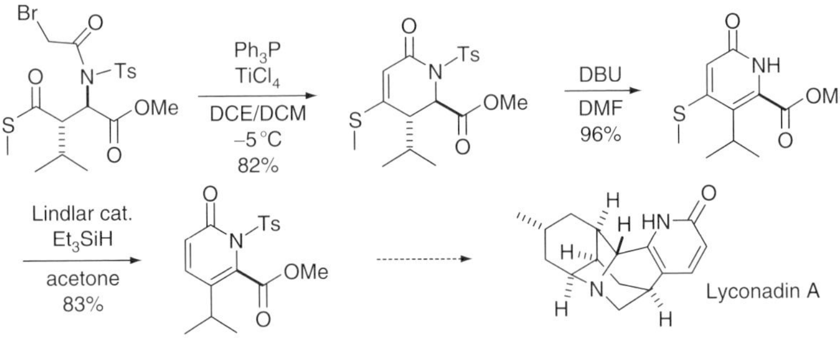

Scheme 14

Two innovative new syntheses couple the chemistry of azides with the reactivity of strained rings. Wang *et al.* have taken advantage of transition metal catalyzed β-carbon elimination to effect ring cleavage of a cyclopentenol (Scheme 15). Tandem elimination of molecular nitrogen affords an iminyl palladium species that simultaneously templates, activates, and attacks the vestigial aldehyde to form a six-membered nitrogen heterocycle. Examination of both *cis* and *trans* azidoalcohols revealed that the less strained *cis*-metallopentacycle produced significantly lower yields <09JA12886>.

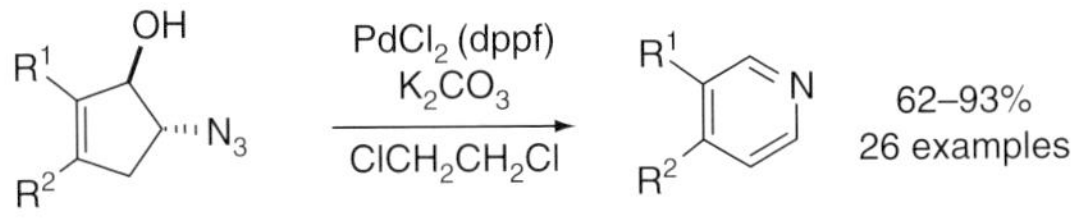

Scheme 15

Another excellent use of azides takes a radical approach to effect formation of the pyridine ring (Scheme 16). Mn(III) catalyzed single-electron oxidation of phenylcyclopropanol produces a β-keto radical which readily adds to a vinyl azide with evolution of nitrogen gas. The resulting iminyl radical undergoes intramolecular cyclization at the carbonyl to produce a tetrahydropyridine. After reductive elimination of manganese, pyridine is generated by loss of water and oxidative aromatization <09JA12570>.

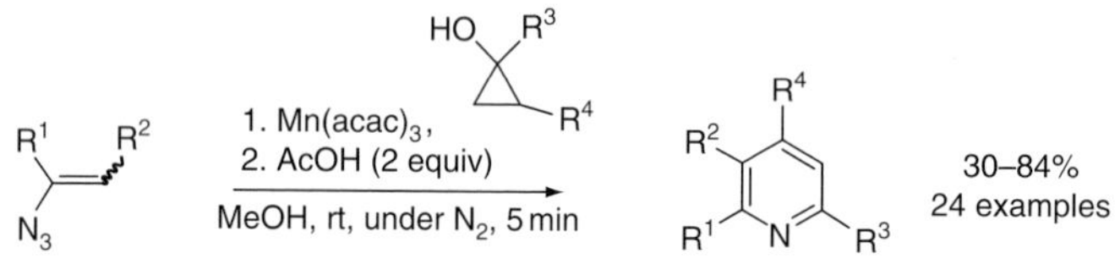

Scheme 16

Lastly, Yamamoto et al. exercised an unusual strategy for the incorporation of pyridine in their total synthesis of 8-oxoerymelanthine (Scheme 17). Erymelanthine and the oxo-derivative are unique among the *Erythrina* alkaloids due to the presence of a pyridine ring. The co-isolation of deaza-analogues suggests a biosynthetic pathway involving aromatic cleavage and ammonium insertion. Carrying a trimethoxyphenyl moiety through most of the synthesis, Yamamoto et al. were able to effect selective oxidative cleavage of the electron-rich aromatic ring by ozonolysis; a subsequent Dieckmann condensation and aminolysis generates the pyridinone ring. While quite common in nature, this type of sequence is rare in the laboratory due to difficulty of selective oxidative cleavage <09JOC6010>.

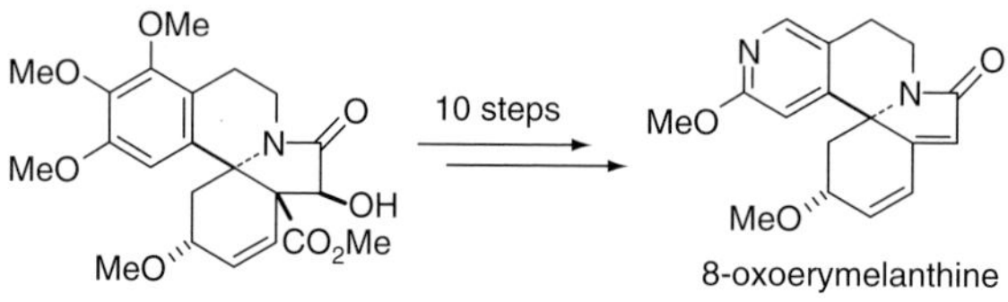

Scheme 17

6.1.2.2 Reactions of Pyridines

The reactivity of pyridine extends far beyond the confines of its structure. While functionalization of this system is exceptionally important, the pyridine moiety is also a crucial Lewis base, functional group, and synthon. Demonstrably versatile, pyridine is synthetically equivalent to a host of frameworks that are not immediately apparent.

For a new total synthesis of Tamiflu®, Fukuyama et al. eschewed shikimic acid in favor of unfunctionalized pyridine as a starting substrate (Scheme 18) <09T3239>. The total synthesis of G. B. 13 boasts an unusual late-stage reduction of pyridine to piperidine <09JA13244>. A novel route to the antibacterial monobactam family has been achieved by a photochemical cyclization/ring-opening metathesis sequence <09T3580> (Scheme 19).

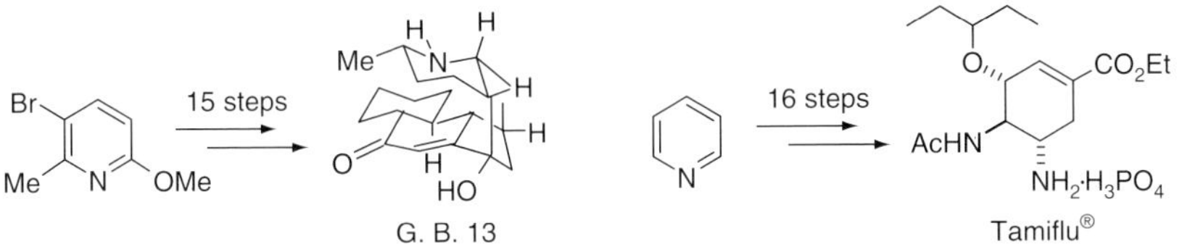

Scheme 18

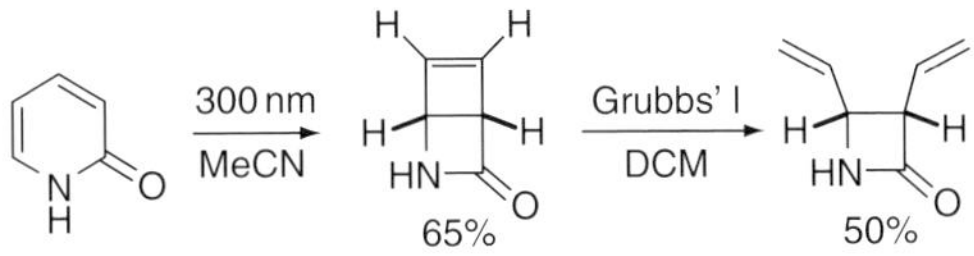

Scheme 19

Privileged with over a century of inquiry, the classic portrayal of pyridine describes a resistance to electrophilic substitution and vulnerability to nucleophilic attack. The π-deficient nature of the system, responsible for this hallmark disposition, also alters the reactivity of side-chains. For instance, vinyl side-chains act as Michael acceptors – provided the group is conjugated to the pyridine nitrogen (Scheme 20) <09TL1928>.

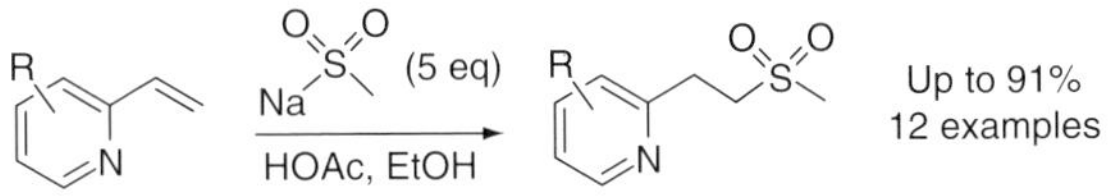

Scheme 20

While this tactic is well established, new methods have used these systems for the exploration of asymmetric conjugate addition (Scheme 21). In metal-catalyzed examples, coordination between the pyridine nitrogen and catalyst is an important intricacy <09JA10386>.

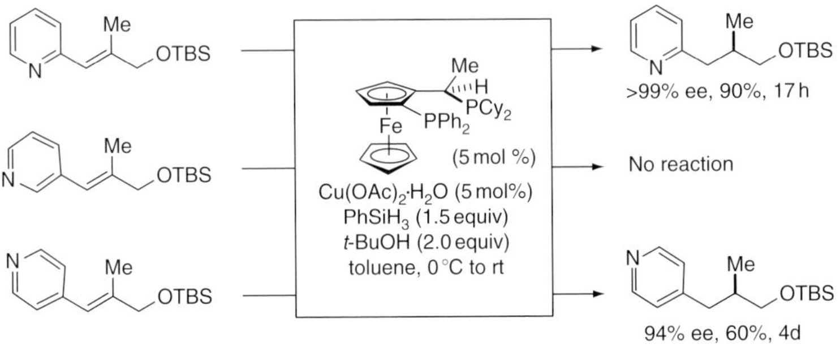

Scheme 21

The alkyl side-chains of pyridines exhibit higher acidity at the α- and γ-positions due to resonance stabilization of the resulting anion. Lateral lithiation of lutidines and other 2-picoline derivatives has been used as a key strategy in total synthesis. Towards the synthesis of cananodine, these pyridylmethyllithium species were used to open hindered trisubstituted epoxides (Scheme 22) <09JOC1374>. Lipshutz et al. used lateral lithiation for the selective chlorination of a highly functionalized pyridine to produce coupling partners for piericidin A1 (Scheme 23) <09JA1396>.

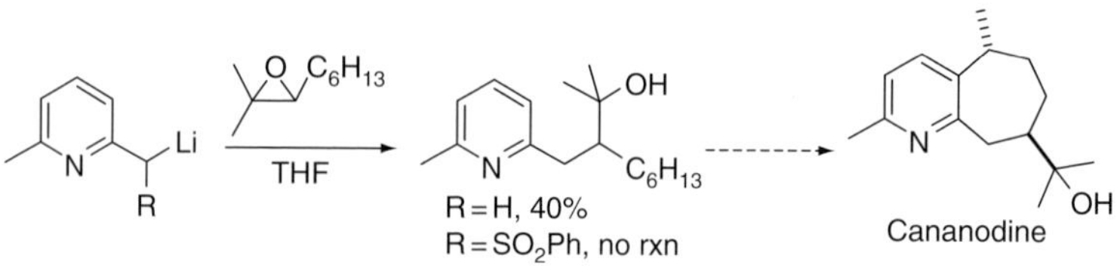

Scheme 22

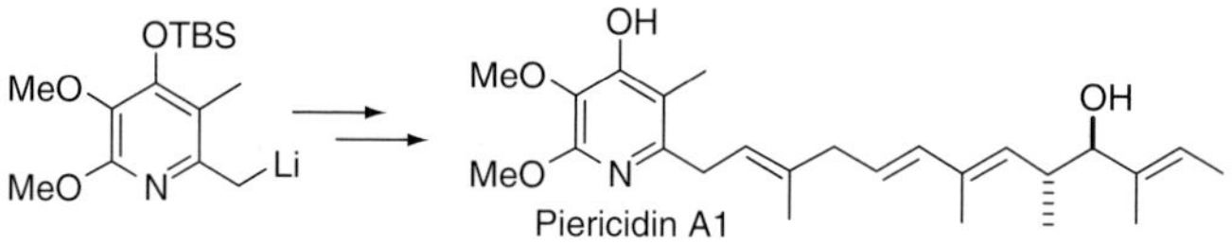

Scheme 23

While performing a side-chain lithiation of 2-benzyloxypyridines, Dudley et al. discovered an unexpected rearrangement <09JOC7998>. Rather than generating a stabilized benzyllithium salt, [1,2]-migration occurs via *ipso*-attack with preferential elimination of alcohol to produce aryl pyridyl carbinols (Scheme 24). Though formally evocative of the [1,2]-Wittig rearrangement, this new pyridine-mediated migration proceeds through an anionic process rather than the radical mechanism of the Wittig. *ipso*-Substitution of pyridine side-chains has accounted for other rearrangements as well, such as the Smiles <09AJC176>.

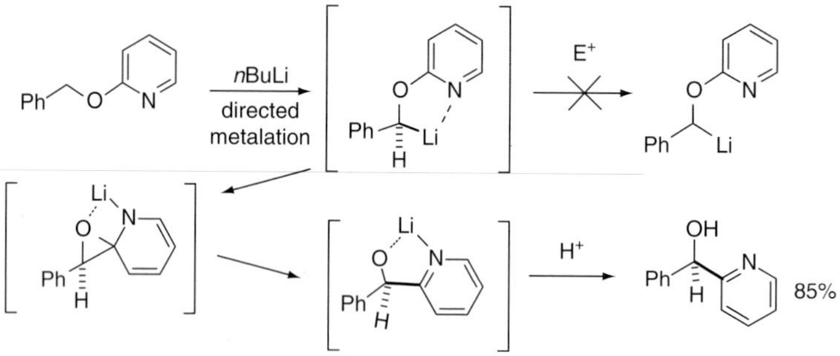

Scheme 24

Direct lithiation of the pyridine ring provides important access to halopyridines. *ortho*-Lithiation is commonly employed to produce substituted halopyridines for use as transition-metal coupling partners. A concise synthesis of louisianins C and D was reported which demonstrates this strategy to produce high overall yields (>20%) of each of these compounds <09T748>.

One recent study of *ortho*-lithiation by Hoarau *et al.* investigates the nature of carboxamide directing metalating groups on halopyridines (Scheme 25). In describing a regiocontrolled lithiation protocol, Hoarau *et al.* relate an excellent comparison of the steric effects between *N*-alkoxyamides and oxazoline when lithiated with LDA or LTMP <09TL1768>.

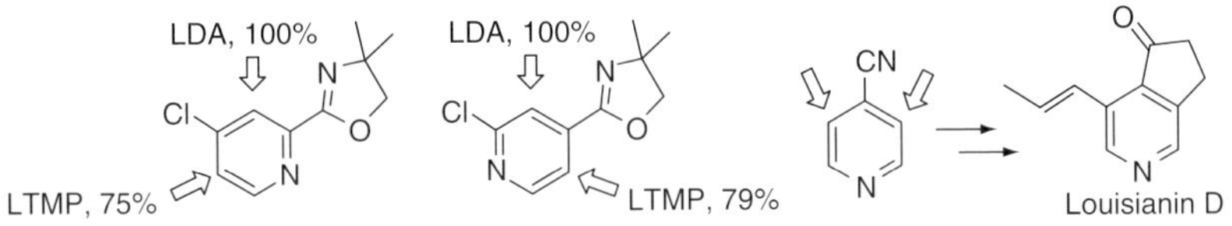

Scheme 25

Due to the π-deficient nature of pyridine, direct lithiation with *n*BuLi often produces side products from nucleophilic addition – for this reason, sterically hindered bases are generally required. The combination of alkyllithium with lithium aminoalkoxides produces reagents with inhibited nucleophilicity but increased basicity. The most popular preparation, *n*BuLi/LiDMAE, results in direct lithiation with indifference to the influence of any directing groups. A review of the use of these reagents specifically for the metallation of pyridine was published this past year <09EJO4199>.

Knochel *et al.* continue to expand the breadth of their lithium-magnesium-amide based metalation methodologies <09AG(I)7256>. The group gives much appreciated attention to heterocyclic compounds in their papers; one recent report encourages the use of mixed Li/Mg and Li/Mg/Zn amides for the metalation of pyridines <09EJO1781>. For related magnesation work, Mayr *et al.* reported a convenient comparison of the relative reactivities of various bromoheteroarenes towards bromine-magnesium exchange (Scheme 26) <09OL3502>.

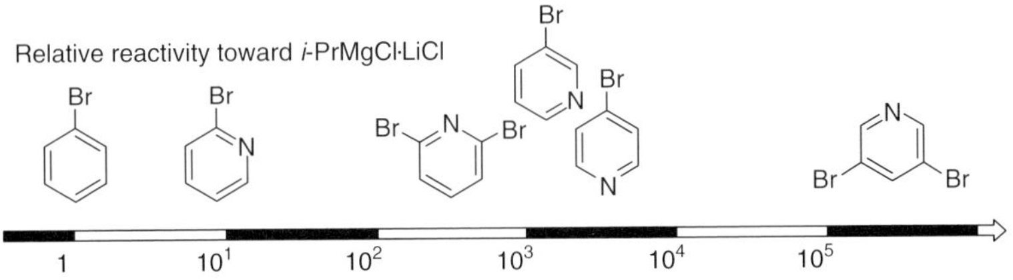

Scheme 26

Transition metals allow the direct C–H arylation of pyridine. Gold, Cy_3PAuCl, has been shown to catalyze (5 mole %) arylation at C–2 though progress remains to be made in terms of yield and selectivity (Scheme 27) <09TL1478>. Nickel mediates the direct coupling of pyridine with arylbromides, though this process also lacks regioselectivity and produces a mixture of 2-, 3-, and 4-arylpyridines <09OL2679>. While these examples are important for exploratory purposes, the maturation of palladium-catalyzed methods has proven supremely effective for the arylation of heteroaromatics.

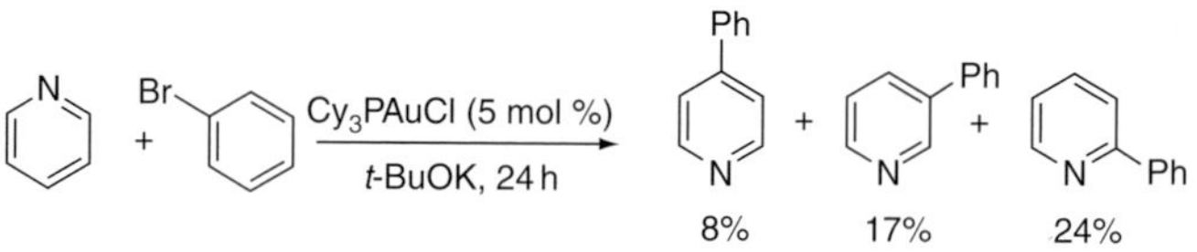

Scheme 27

The Suzuki–Miyaura reaction is quickly becoming the defacto method for palladium-catalyzed cross-couplings. The low toxicity and accessibility of borates has made this reaction especially valuable to the pharmaceutical industry. While phenyl boronates and boronic acids are generally noted for their stability and ease of use, pyridyl borates are notoriously challenging systems. Due to the exceptional importance of the pyridine system in medicinal chemistry, tremendous effort has been put into reconciling this issue.

Until a 2008 report by Buchwald et al., the use of 2-pyridyl nucleophiles in the Suzuki–Miyaura reaction was virtually unknown; the few precedent conditions involved aryl iodides and found little success in general use <08AG(I)4695>. The electron deficient nature of the pyridine ring causes a slow rate of transmetalation that is not always competitive with protodeboronation. Buchwald's solution involved formation of highly nucleophilic lithium borate salts. Unfortunately, the air-sensitive preparation and low stability of these species limited both their functional group tolerance and popularity.

This past year, a critical advancement was reported by chemists at Merck (Scheme 28). Readily available 2-pyridylpinacol boronates have been coupled with aryl bromides in the presence of copper to produce 2-arylpyridines in exceptional yields under general and mild conditions <09OL345>. The low cost and availability

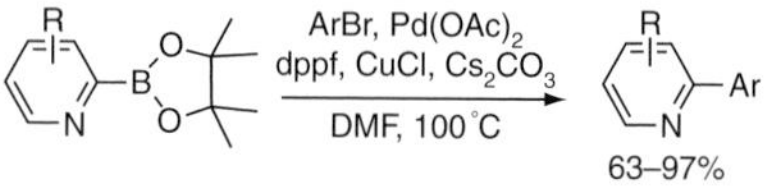

Scheme 28

of copper chloride make this method especially appealing. Other satisfactory solutions for this difficult coupling were described as well <09JA6961>. Ackermann et al. describes efficient coupling with TADDOL based secondary phosphine oxide <09SL2852> and a report from CombiPhos suggests use of their PXPd ligand – commercially available for $250/g <09OL381>.

3-Pyridylboronic acids pose fewer problems; with proper selection of ligands, coupling at this position proceeds smoothly <09OBC2155>. Chen et al. offer a comparison of diphosphane-based ligands for their couplings at the 3-position <09EJO2051>. When possible, methods using Suzuki–Miyaura chemistry employ halopyridines rather than pyridylborate derivatives – particularly at the 2-position. For the synthesis of 2,3′-bipyridyls, such as noranabasamine, 3-pyridylboronic acid is coupled with a 2-halopyridine (Scheme 29) <09OL1579>.

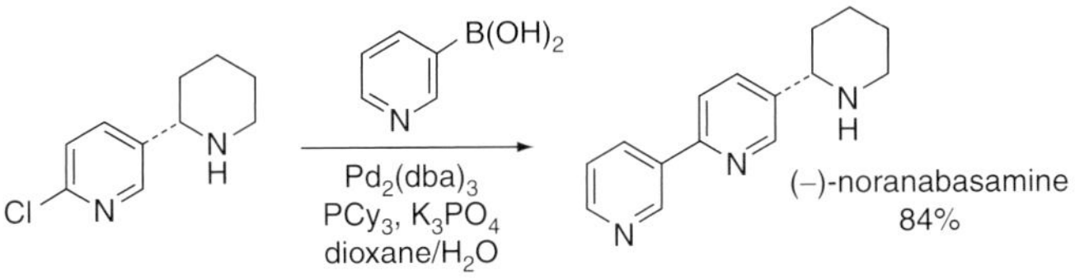

Scheme 29

The regioselectivity of the Suzuki-Miyaura with dihalopyridines is demonstrated by preferential formation (76%) of 3-bromo-2-phenylpyridine from 2,3-dibromopyridine and phenylboronic acid <09SL1081>. This selectivity is determined by the C–X bond distortion and dissociation energies as described in a recent computational study <09JA6632>. A one-pot double cross-coupling produces good yields of unsymmetrical 2,6-diarylpyridine from dibromopyridine products (Scheme 30). The authors report that even minor differences in reactivity between two simultaneously present arylboronic acids produces unsymmetrical products <09OL1801>.

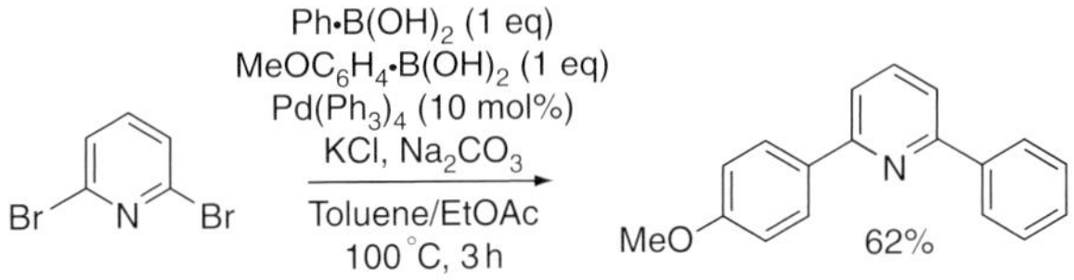

Scheme 30

The compatibility of the Suzuki–Miyaura with a vast array of coupling partners has firmly established the reaction as a method of choice for SAR studies <09BMCL894; 09BMCL1807>. The d-series of thiopeptide antibiotics offer a rigorous proving ground for palladium-catalyzed heterocyclic chemistry. A new approach to the micrococcinate pyridine core relies on Suzuki chemistry <09OL3690> as have strategies to other natural products <09JOC4547>. Excellent variations on the Suzuki reaction have seen development <09JA6961> and the Molander group's work with organotrifluoroborate reagents continues to expand in scope <09JOC973; 09OL4330>.

In addition to halogens, 2-tosyl- and 2-thiomethyl-substituted pyridines undergo palladium cross-coupling (Scheme 31). Both the Knochel and Fukuyama groups reported new Pd-catalyzed C–C bond formation between organozinc reagents and thiomethyl-substituted heterocycles <09OL4228; 09H(77)233>. Pyridyl zinc reagents react with a variety of electrophiles and undergo Negishi couplings <09TL5329; 09TL6985>. The Mizoroki–Heck reaction can be performed on 2-

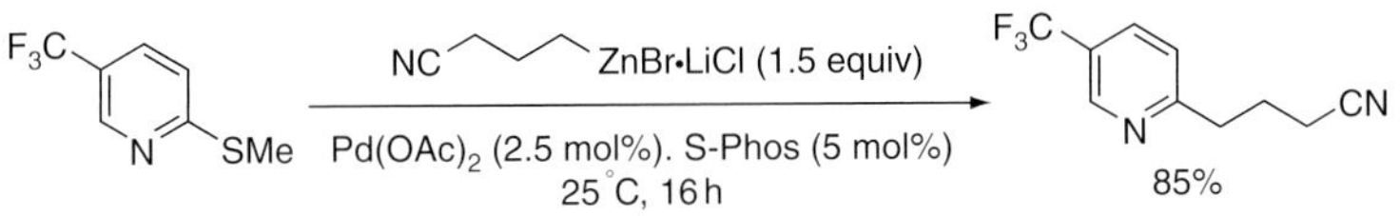

Scheme 31

tosylpyridines <09CEJ5950>. While investigating palladium-catalyzed crossing-coupling of aryl bromides and triflates with amides, Pujol et al. observed that nicotinamide acted as an efficient ammonium surrogate (Scheme 32). The group then employed this novel source of ammonia for a direct synthesis of primary aryl amines <09T1951>.

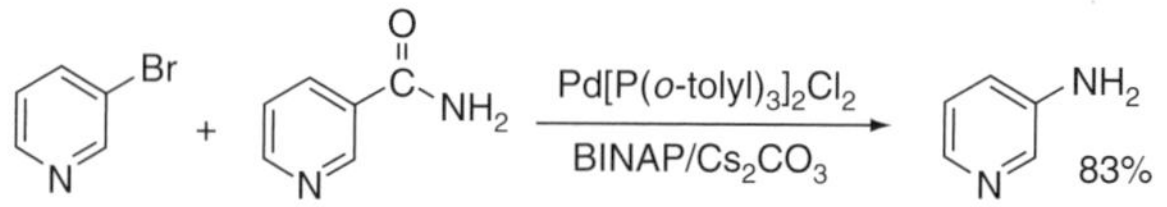

Scheme 32

Amination of pyridine has come a long way since the Chichibabin reaction was first reported almost a century ago. Modern methods are more mild and either involve the S_NAr displacement of a halogen <09S115; 09T1180; 09TL2481> or transition metal-catalyzed coupling <09EJO4586>. Alkylation of these aminopyridines can be difficult, particularly in the case of 2- and 4-aminopyridines where the amine is in direct conjugation with the pyridine nitrogen. While highly reactive electrophiles and microwave heating can be effective <09OBC1410>, other methods enlist transition-metal catalysts to effect alkylation. Iridium catalysis, $[IrCl(cod)]_2$, provides a way of circumventing the low nucleophilicity of 2-aminopyridines by catalyzing the highly selective monoalkylation of amines with alcohols (Scheme 33) <09CEJ3790>.

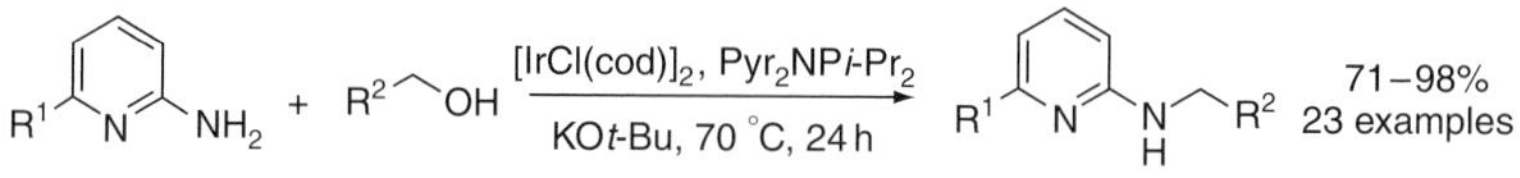

Scheme 33

For the synthesis of N-arylated 2,4-diaminopyridines, a protocol has been developed for the site selective monoarylation of these compounds <09T8950>. Palladium-catalyzed aminocarbonylation offers another route; with recent use of $Mo(CO)_6$ as a CO source <09EJO2820>. Transition metal-catalyzed amination can be coupled with other transformations; α-carbolines <09JOC3152> and dipyrrolopyridines <09JOC4246> are two fused systems recently accessed by domino sequences (Scheme 34).

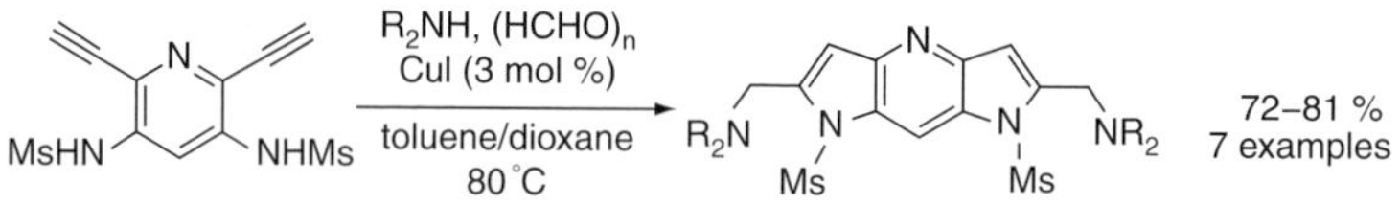

Scheme 34

Aminopyridines are versatile synthons for the formation of fused heteroaromatic ring systems – many of which are valuable drug scaffolds. Of the many fused systems accessible from 2- and 3-aminopyridines, special attention has been paid to imidazo[1,2-a]pyridines. Beyond the success of zolpidem, this pharmacophore has shown promise as a proton-pump inhibitor <09BMCL3602> as well as an analgesic and anti-inflammatory agent <09BMC74>.

Imidazo[1,2-*a*]pyridines are easily produced by the Ugi reaction of 2-aminopyridine, acetaldehyde, and *t*-BuNC; one new protocol favors the use of $ZrCl_4$ (10%) in PEG-400 (Scheme 35) <09SL628>. Alternatively, 2-aminopyridines readily con-

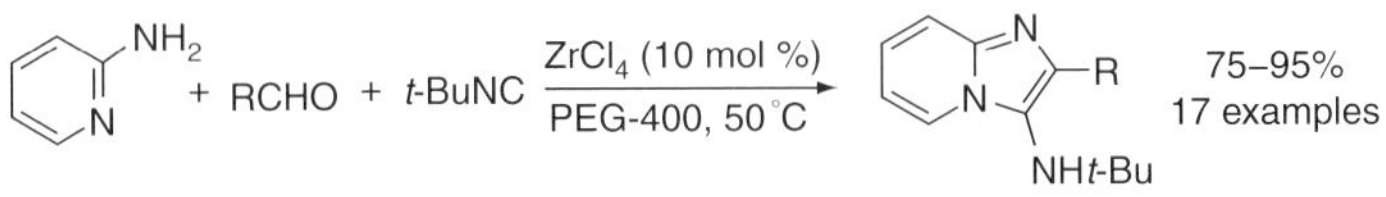

Scheme 35

dense with cyanohydrins and cyclize to imidazo[1,2-*a*]pyridines under silica-sulfuric acid catalysis <09TL4389>. Imidazo[1,5-*a*]pyridines can be formed via oxidative condensation/cyclization of 2-pyridylmethylamines, aromatic aldehydes, and mildly oxidative elemental sulfur <09JOC3566>. A one-pot approach to the imidazo [1,5-*a*]pyridine system has been reported as well <09TL4916>. Pd-catalyzed amidation of 2-chloro-3-nitropyridine produces *N*-(3-nitropyridyl)amides, which were subsequently used for the synthesis of fused imidazo[4,5-*b*]pyridines <09TL3798>. If a green process is a desired, imidazo[4,5-*b*]pyridines can simply be generated from 2,3-diaminopyridine and benzaldehyde in boiling water <09TL1780>.

The development of unnatural nucleoside mimics is one important application of fused heterocyclic systems (Scheme 36) <09JA5488>. These compounds have found

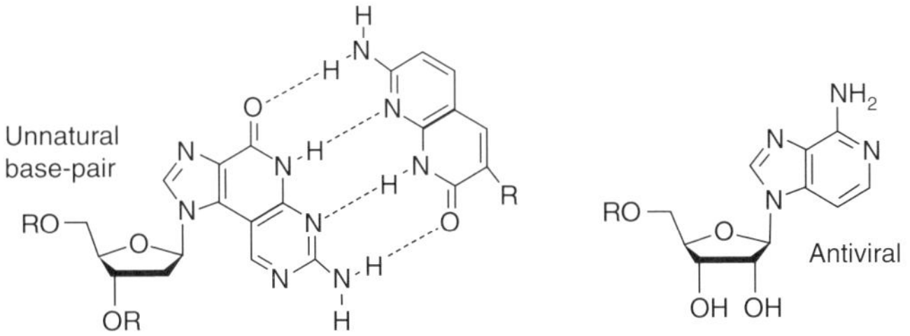

Scheme 36

medicinal use as inhibitors of reverse transcriptase. A recent review discusses the synthesis and biological application of *C*-linked nucleosides <09CR6729>. Such systems have commonly been designed by replacing the pyrimidine or imidazole portion of a purine nucleoside with a different heteroaromatic system. A series of related papers from the Ukraine have described pyridine-containing nucleosides fused to pyrroles, pyrazoles, indoles, pyridines, pyrimidines, and imidazoles <09S1375; 09S1851; 09S1858; 09S1865; 09S2393>. Another preparation of unnatural purines involves the insertion of an aromatic ring at the ring juncture <09S1271>; these compounds have been used as unnatural base pairs <09JA1644>.

Pyridyl nitrenes are useful intermediates for the synthesis of fused pyridine ring systems. Pyridyl azides are convenient precursors to these reactive intermediates, though the requisite thermolysis and inherent reactivity of nitrenes brings many side products (Scheme 37). Identification of some of these compounds has revealed a ring cleavage mechanism <09T3668>.

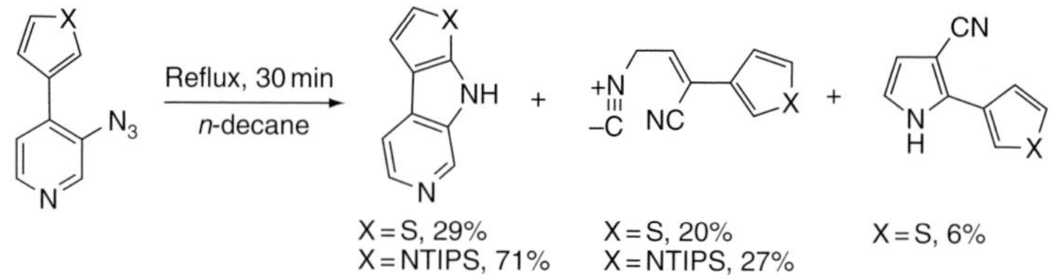

Scheme 37

X = S, 29%
X = NTIPS, 71%

X = S, 20%
X = NTIPS, 27%

X = S, 6%

In spite of the many recent advances in metal-catalyzed functionalization, S_NAr remains a valuable tool for the preparation of substituted pyridines <09T757>. For the synthesis of iodopyridines, a convenient transformation of 2-hydroxypyridines to 2-iodopyridine utilizes Tf_2O in toluene to generate the reactive triflate <09JOC5111>. High yields were shown to be general, (Scheme 38) though it is worth noting that alternate conditions employing mesylates, tosylates, or TFA resulted in no or low conversion. Banwell et al. reported a series of trans-halogenations to generate iodoazines <09JOC4893>. Conversely, two new methods have been reported for the reductive dehalogenation of azines <09H(78)2735; 09H(77)1163>. Activation

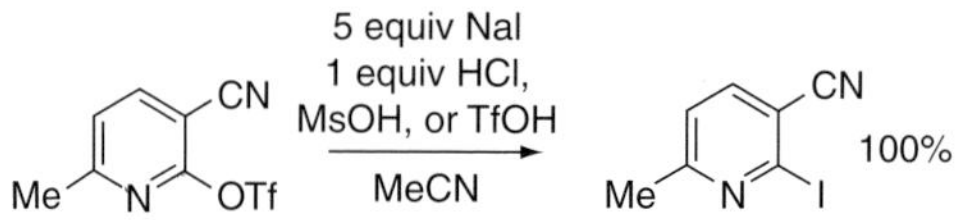

Scheme 38

with a powerful Lewis acid permits nucleophilic attack to yield dearomatized dihydropyridines. Clayden's group showed that tethered π-deficient and π-excessive heterocycles can be activated to effect an intramolecular cross-coupling in the presence of Tf_2O and 2,6-lutidine (Scheme 39) <09CC1964>.

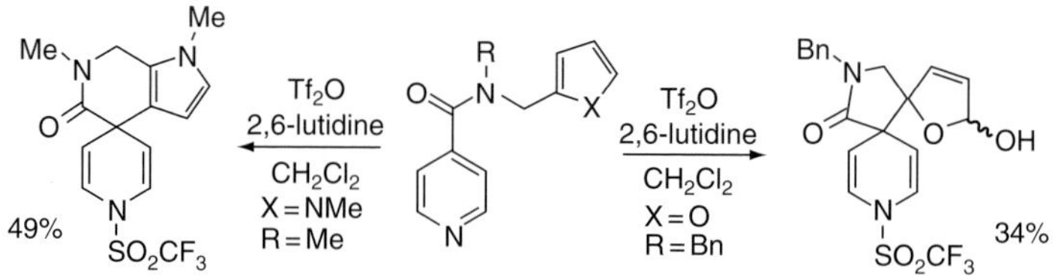

Scheme 39

Several useful transformations of carbonyl side-chains have been disclosed. During the development of novel EP1 antagonists, chemists at GlaxoSmithKline have performed a thorough investigation of the phenolic aldol reaction of 2-formylpyridines <09SL1609>. Also from formylpyridines, an attractive transformation to the corresponding esters is achieved via oxidative benzoin-like condensation in the presence of thiamine and alcohol <09CL484>. For the decarboxylation of pyridine carboxylic acids, the Larrosa group reported a highly efficient Ag-catalyzed method (Scheme 40) <09OL5710>.

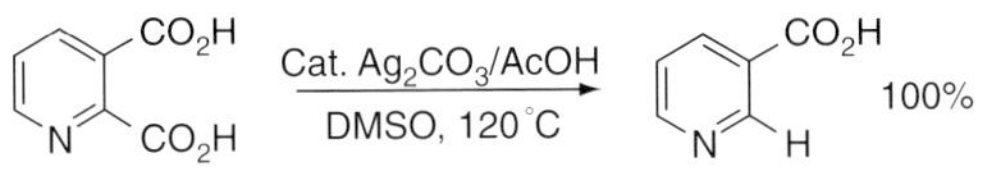

Scheme 40

Acylpyridines are efficiently synthesized by addition of alkyl zinc reagents to any of the pyridine carboxylic acid chlorides <09SL1091>. Alternatively, the reaction of 2-pyridyl-magnesium chlorides with N,N'-dialkylarylamides confers high yield with no sign of tertiary alcohol side product <09SC1835>.

As a powerful directing group, pyridine has held a central role in the advent of modern C–H activation chemistry, particularly as part of ligand-containing substrates for *ortho*-functionalization. This important topic has earned a special topics issue of *Chemical Reviews* due early next year <10CR575>. A variety of pyridine-containing groups allows *ortho*-functionalization as illustrated by a recent *ortho*-acylation using benzaldehyde (Scheme 41) <09OL3120>. Dong *et al.* have demonstrated a novel C–H functionalization with arylsulfonyl chloride producing either C–S or C–Cl bonds depending on conditions (Scheme 42) <09JA3466>.

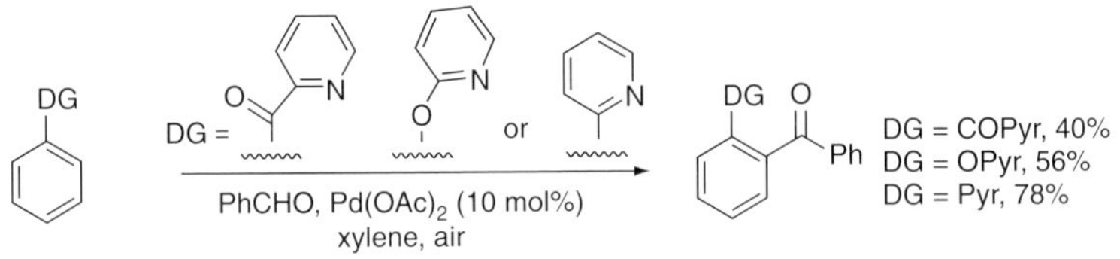

Scheme 41

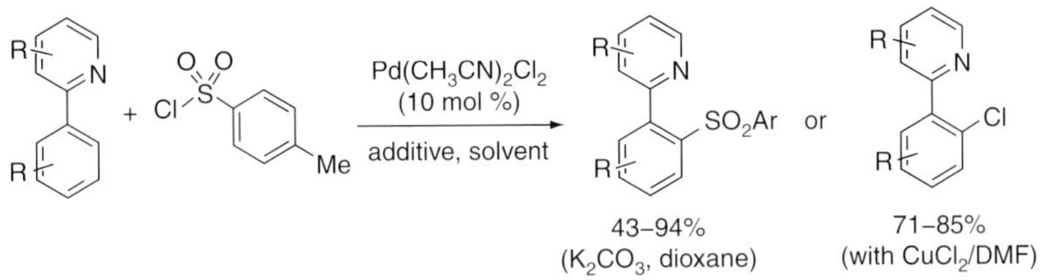

Scheme 42

Addressing the practicality of palladium C–H activation, Yu *et al.* have reviewed progress to date and discuss future challenges for this methodology <09AG(I)5094>. In addition to the numerous *ortho*-functionalization possible with palladium catalysts <09OL3174; 09AG(I)6511>, a host of other late transition metals are used as well. Following up on an unexpected result from 2006, <06JOC6790> members of the Yu group detailed their Pd-free Cu(II) mediated oxidative-dimerization of 2-phenylpyridine <09T3085>. Rhodium permits decarbonylative *ortho*-arylation <09OL1317>, benzoxylation <09OL3974>, *ortho*-vinylation <09CL118; 09JOC7094>, and alkylation <09AG(I)6045>. Rhenium systems have also been described <09OL2711>.

Though many new reports of C–H functionalization are being produced, elucidation of the mechanistic underpinnings represents the most vital work being done

<09JA9651; 09JA11234>. Sanford et al. have worked extensively to illuminate the mechanism of C–H acetoxylation and arylation – drawing on crossover studies, Hammett plots, Eyring analyses, solvent effects, and more <09JA10974>. Acetoxylation progesses by initial dissociation of an acetoxylate, the resulting cationic 5-coordinate intermediate fosters C–O bond formation (Scheme 43). For formation of

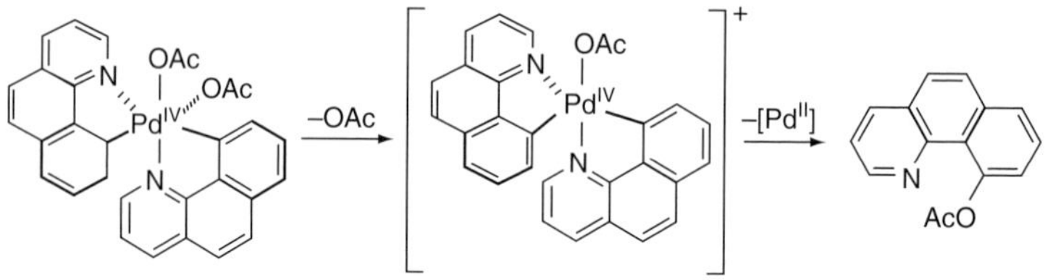

Scheme 43

C–C bonds, the initial octahedral Pd^{IV} complex directly undergoes reductive elimination. A comparison between the mechanisms for acetoxylation and chlorination has been presented as well <09OL4584>. Douglas et al. investigated a system in which C–C and C–H activation processes were in competition, reporting a novel olefin carboacylation in the process <09AG(I)6121>.

The use of ligand-containing substrates has enabled other strategies in addition to C–H bond cleavage <09JOC5656; 09ASC107; 09TL4217>. Miller et al. used an adjacent pyridine to effect the regio- and stereoselective cleavage of nitrosocycloadducts <09JOC7990; 09OL449>. Proximal 2-pyridylsulfonyl groups have also been used to engage chiral phosphoric acids to effect an enantioselective aza-Friedel–Crafts reaction. It is interesting to note that the corresponding 2-quinolylsulfonyl group provides higher ee (99%) but lower yield <09SL1639>. The presence of a pyridinyl or quinolylsulfonyl group on a 1-azadiene permits the metal-catalysis of an otherwise obstinate aza-Diels–Alder reaction (Scheme 44) <09S113>.

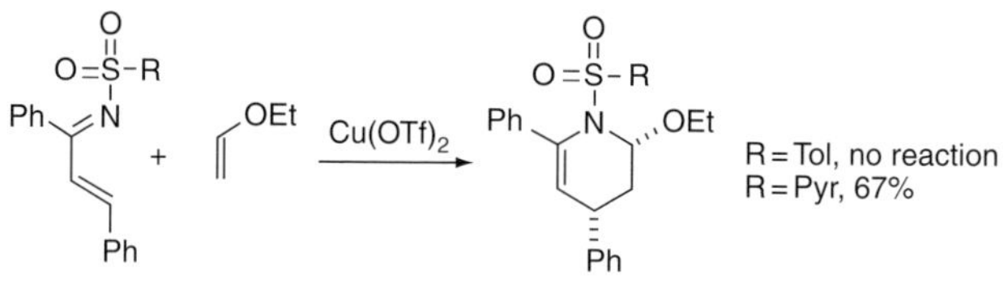

Scheme 44

The picolinate group has recently been discovered to be a powerful leaving group, particularly for allylic substitution reactions <08OL1719; 09TL3547>. Due to chelation between the carbonyl oxygen and pyridyl nitrogen, $MgBr_2$ activates picolinate as both a leaving group and an electron-withdrawing group. Most compelling, allyl picolinates afford the anti S_N2' product following the addition of organocopper-derived Grignard reagents (Scheme 45) <09JOC1939>. Lithium copper acetylides generated in the presence of $MgBr_2$ also produced excellent (Scheme 46) results <09JOC7489>.

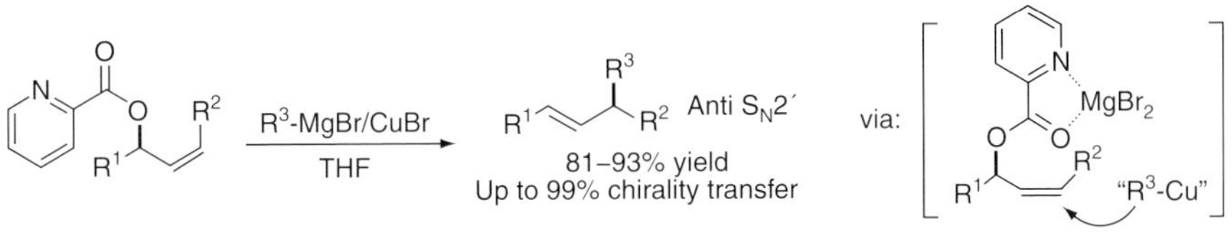

Scheme 45

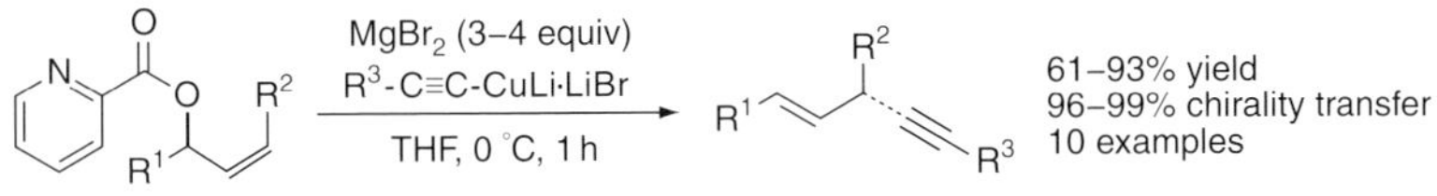

Scheme 46

In the past year, Kobayashi's group has also reported several total syntheses that take advantage of this methodology: (S)-(+)-imperanene <09JOC5920>, (−)-sesquichamaenol <09JOC1939>, and the active form of loxoprofen <09OL1103>.

6.1.2.3 Pyridine *N*-Oxides and Pyridinium Salts

The ambiphilic nature of pyridine *N*-oxide results in a remarkably versatile species capable of both electrophilic and nucleophilic substitution. The push–pull presence of the *N*-oxide is commonly employed to alter the functionality of side-chains or activate the ring to substitution reactions. For example, the synthesis of 4-chlorodipicolinic acid chloride usually requires a lengthy multistep procedure. Parquette *et al.* harnessed the increased reactivity of pyridine *N*-oxides to produce an improved two-step protocol <09S713>. Dehydrative amide couplings with 2-aminopyridines have historically suffered from low yields due to the weak nucleophilicity of 2-aminopyridine (Scheme 47). To accomplish these difficult amidations, chemists at Pfizer have taken advantage of the increased nucleophilicity of 2-aminopyridine-*N*-oxides <09TL1986>.

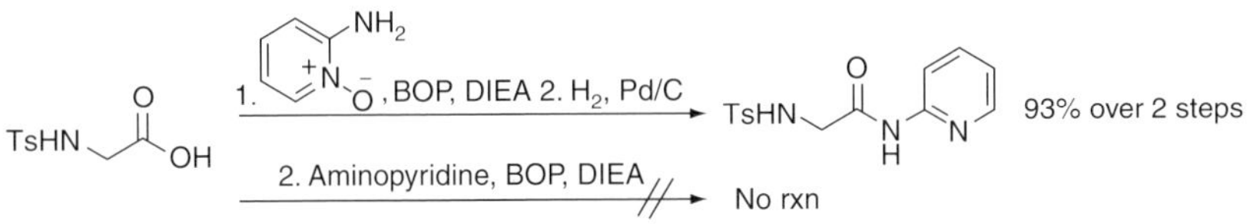

Scheme 47

Movassaghi's group recently devised an excellent new method for the syntheses of these compounds by applying Tf$_2$O activated amides to *N*-oxides; N-pyridyl-amides were quickly generated in high yields through a Katada-like rearrangement (Scheme 48) <09JOC1341>.

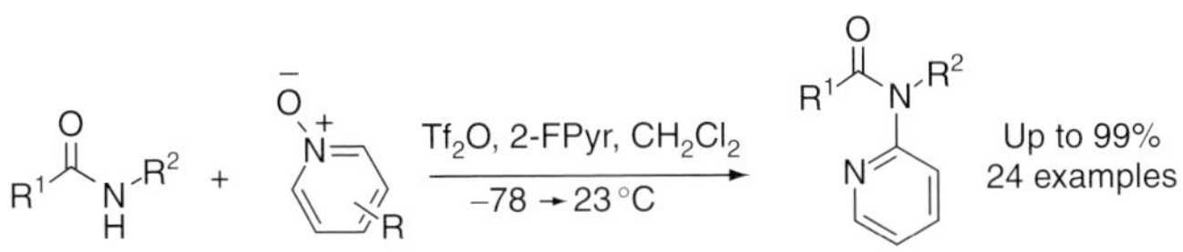

Scheme 48

Nucleophilic addition to pyridine *N*-oxides is an important practice for generating 2-substituted pyridines and dihydropyridines. The addition of alkynes to 3-substituted pyridinium salts reveals a notable regioselectivity that is not obvious from the electronic nature of the starting substrates (Scheme 49). In this case, regiochemistry

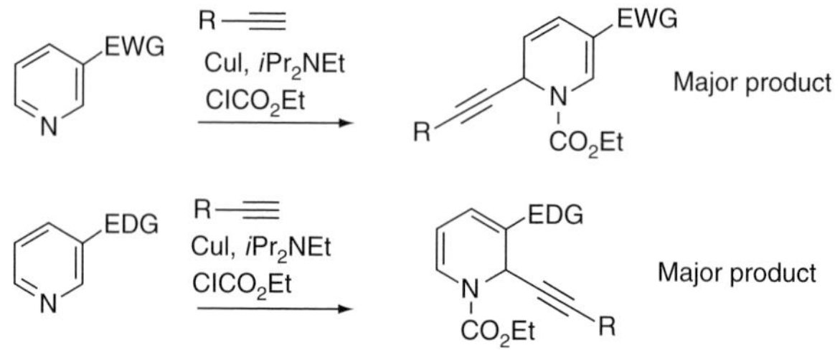

Scheme 49

was explained by considering product stabilities <09T2329>. The addition of lithium acetylides can also directly produce 2-alkynylpyridines via auto-aromatization, though this method is plagued by ring-opening decomposition products <09TL1444>. Though the addition of Grignard reagents has historically had similar problems, a recent report from Almqvist *et al.* describes high yields of *trans*-2,3-dihydropyridines when addition is followed with an electrophilic quench at low temperature <09AG(I)3288>.

With the higher C-2 acidity of pyridine *N*-oxides, generation of the aryl anion is possible with relatively weak bases such as *t*-BuOLi. Daugulis *et al.* report mild non-cryogenic conditions for the metalation/halogenation of pyridine *N*-oxides <09OL421>. The magnesation of bromo- and iodopyridine *N*-oxide was studied and used to synthesize caerulomycins A and E <09JOC939>. Appealing to transition metal catalysis, copper(I) iodide produced direct arylation with bromotoluene <09AG(I)3296>. The use of palladium with pyridine *N*-oxides grew greatly in utility this past year – largely due to the excellent work of Keith Fagnou.

Fagnou *et al.* noted that the use of pyridine in palladium chemistry commonly required the use of *N*-oxides to generate the halopyridine coupling substrates (Scheme 50). The development of coupling methodology that harnesses pyridine *N*-oxides directly would eliminate the halogenation step and improve efficiency. Presence of the *N*-oxide may promote C–2 metalation by blocking the counter

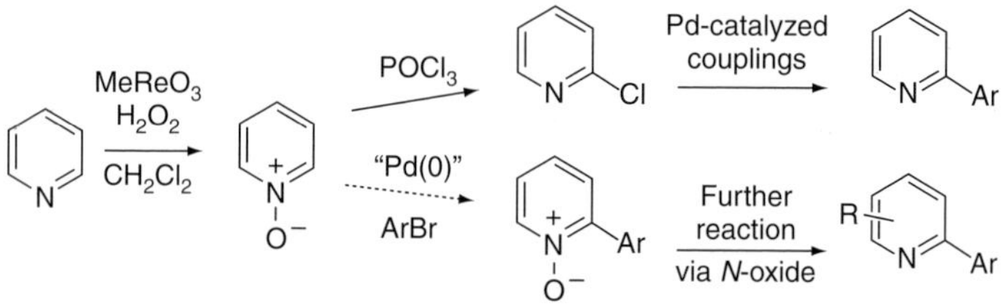

Scheme 50

productive nitrogen lone pair and by increasing the acidity of ring protons. Further, the N-oxide remains after arylation, allowing subsequent functionalization (Scheme 50). A full investigation has produced optimized conditions for these reactions as well as a rationale for regiochemical and substituent effects (Scheme 51) <09JA3291>.

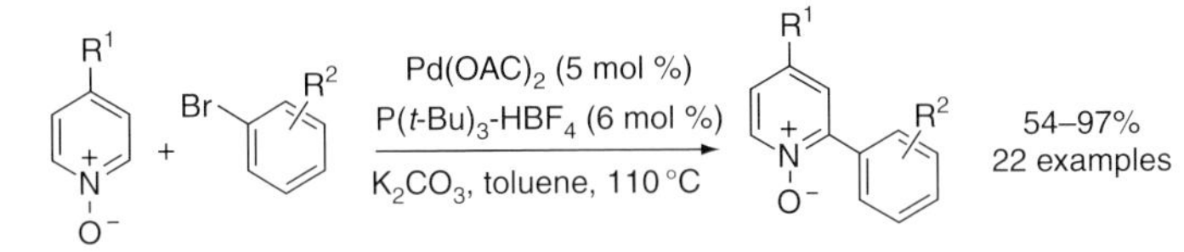

Scheme 51

A clever comparison of the reactivities of pyridine N-oxides and azoles is illustrated by the arylation of azaindoles. Normally, azaindoles will undergo palladation at C–2 on the pyrrole ring – but use of an N-oxide allows preferential arylation of the azine <09OL1357>. A protocol for completely site selective (sp^2 or sp^3) arylation of 2-picoline N-oxides was achieved by careful selection of base <09T3155>. This method was used for an efficient synthesis of crykonisine and papaverine (Scheme 52). A related investigation of quinoline-N-oxides has shown that these substrates can act as both substrate and oxidant <09JA13888>.

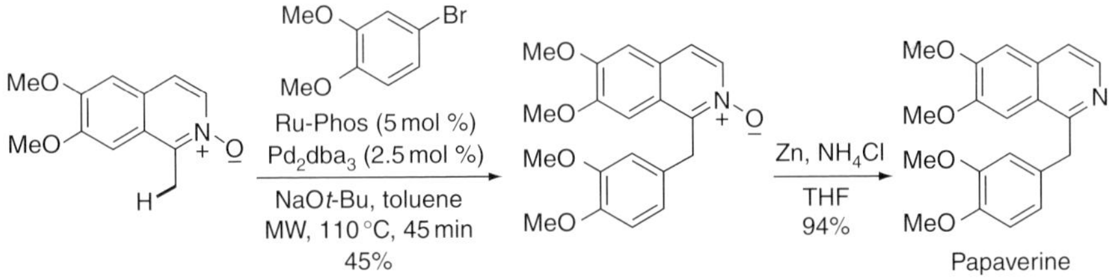

Scheme 52

In the presence of *tert*-butylperoxide, a metal-free direct C-H coupling is observed between pyridine and cyclohexane at trace amounts. An investigation of this reaction revealed that *tert*-butylperoxide catalyzes the C-H alkylation of pyridine N-oxides <09CEJ333>. To return to pyridine, MoO_2Cl_2 has been shown to be an excellent catalyst for deoxygenation <09TL949>.

Several novel syntheses of indolizines and quinolizines were reported in 2009 <09OL3398>. Pyridinium salts provide crucial routes to the formation of these compounds, commonly accessed via multicomponent coupling of an α-haloacetate,

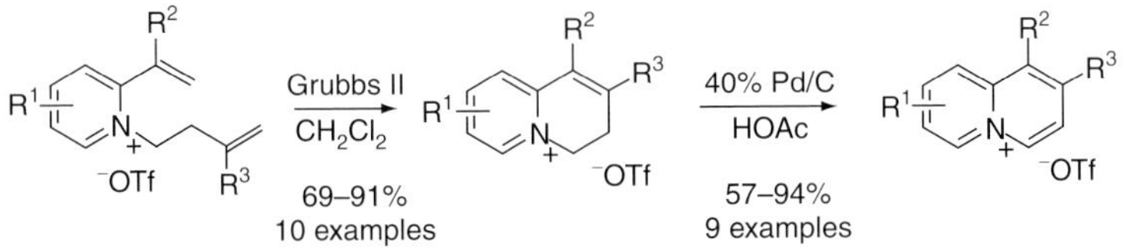

Scheme 53

DMAD, and azine <09JHC1203>. A new RCM method has been reported for the synthesis of quinolizinium triflates from 1-butenyl-2-vinylpyrdinium salts (Scheme 53) <09JOC4166>.

The availability of Baylis–Hillman bromides has prompted their use for the synthesis of indolizines and pyrrolo[1,2-*a*]isoquinolines via 1,5-rearrangement of an allyl pyridinium salt (Scheme 54) <09SL411>. Pyridinium salts, known to accumulate in

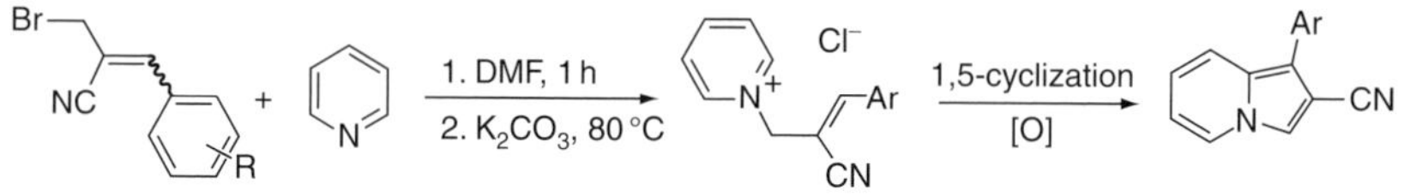

Scheme 54

mitochondria, have been synthesized for use as radical scavengers containing nitrones to act as cell specific antioxidants. The Zincke reaction used to generate these compounds was shown to be compatible with the nitrone functionality for the first time <09T5284>. Lastly, a report from the Wu group finds excellent correlation between Hammett substituent constants and the pyridinium *N*-methyl NMR chemical shift <09TL5018>.

6.1.3. QUINOLINES
6.1.3.1 Preparation of Quinolines

The Friedländer synthesis is one of the most popular preparations of quinolines. This past year benefited from a review celebrating the last 30 years of Friedländer chemistry <09CR2652>. While this review covered part way through 2009, several noteworthy advances have since been reported <09H(78)487>. Excellent new conditions for the Friedländer synthesis were found using cerium(IV) ammonium nitrate (CAN) at catalytic amounts in ethanol (Scheme 55). This method is quite general

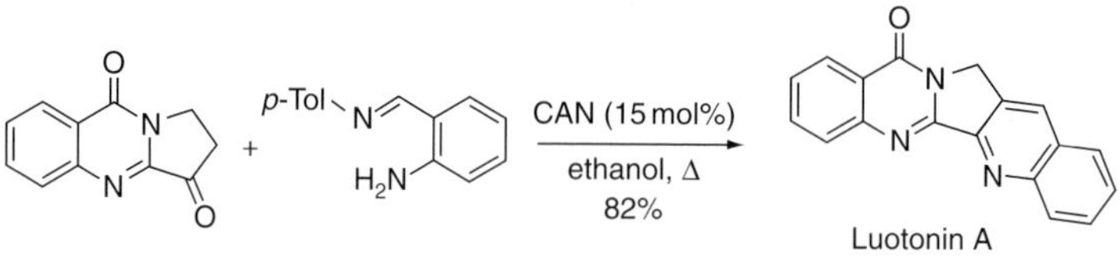

Scheme 55

and was used for the synthesis of luotonin A under both Friedländer and Friedländer–Borsche conditions <09JOC5715>. Green conditions <09CR4140> continue to be of interest and several common surface catalysts have been compared during the design of an improved, solvent-free version of this classic transformation (Scheme 56) <09CJC1122>.

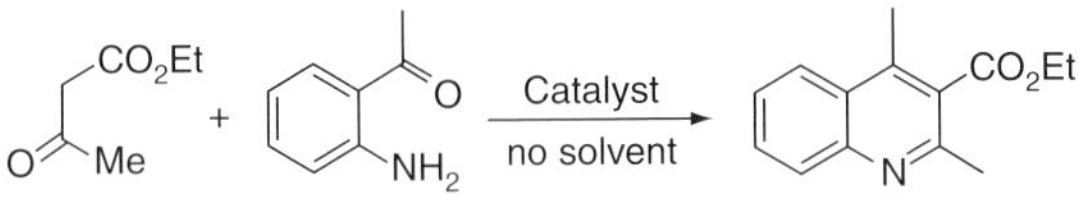

Scheme 56

Surface	Time	Yield
MgO	24 h	40%
TiO$_2$	8 h	90%
CaO	24 h	73%
ZnO	11 h	90%
Al$_2$O$_3$	6 h	82%
Al$_2$O$_3$	9 h	80%
ZrO$_2$	4 h	98%

Several classic quinoline syntheses, such as the Conrad–Limpach, Doebner–Miller, and Skraup reactions, begin with unsubstituted anilines but each require harsh reaction conditions. The originally reported yields of these ancient reactions are not always easy to reproduce by the modern chemist. In pursuit of large quantities of 4-hydroxyquinoline, chemists at Microbiotix reevaluated the solvent of choice for the classic Conrad–Limpach reaction, ultimately suggesting the use of 2,6-di-*tert*-butylphenol <09SC1563>. Rebranded as a multi-component reaction, the Beyer modification of the Doebner–Miller reaction has been popular for the synthesis of fused polycycles <09JHC1222; 09JHC1229>.

The relatively low nucleophilicity of aniline is one of several complicating factors in many of these syntheses. It has been shown, however, that the nucleophilicity of aniline is highly solvent dependent – even approaching that of ethylamine when in water. The 1,4-addition of anilines to Michael acceptors, essential to many quinoline syntheses, has been evaluated in several highly polar solvents. Hexafluoroisopropyl alcohol promotes the mild addition-cyclization of aniline and methyl vinyl ketone in a Doebner–Miller fashion at modest temperatures (Scheme 57) <09JOC6260>. The addition/elimination mechanism of the Gould–Jacobs reaction addresses issues of nucleophilicity, but the ultimate cyclization still requires harsh conditions <09AJC150>.

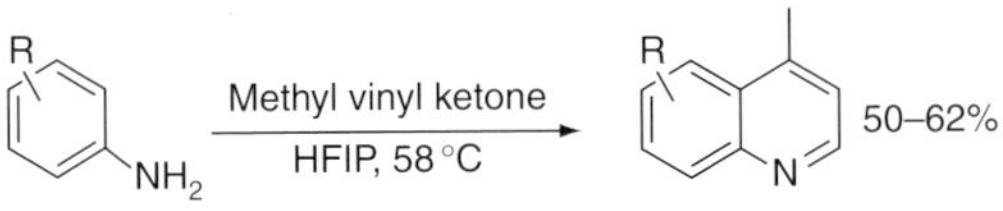

Scheme 57

Strategies styled after the Meth-Cohn quinoline synthesis are especially useful for the construction of 2-quinolones. Dong *et al.* report modified conditions for this classic reaction – replacing POCl$_3$ with PBr$_3$ in DMF <09EJO4165>. A new cyclocondensation approach has been reported for the synthesis of 3-hydroxyquinolin-2-ones along with a preparation of the fungal metabolite viridicatin <09OL1603>. This two-step protocol starts with a one-pot tandem Knoevenagel condensation and epoxidation. Acid catalyzed opening of the oxirane is followed by arene cyclization and loss of cyanide to produce 3-hydroxy-4-substituted-quinolinones (Scheme 58).

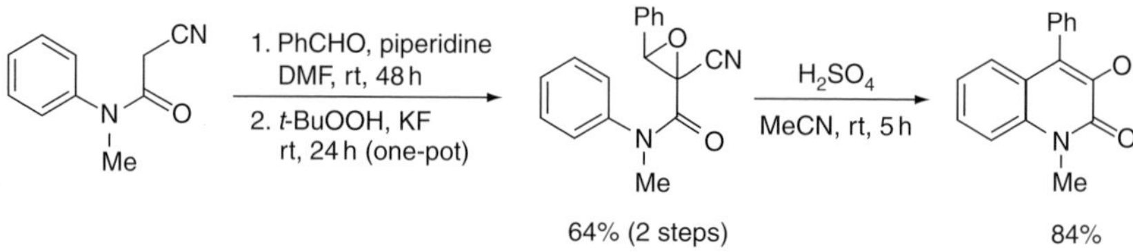

Scheme 58

Cycloadditions account for many important syntheses of quinolines. The Povarov reaction is a powerful, commonly multicomponent, synthesis of dihydro- and tetrahydroquinolines <09JHC796>. The first multicomponent catalytic enantioselective example of this reaction was achieved by use of a chiral Brønsted acid <09JA4598>. An intramolecular version of this imino-Diels–Alder reaction was used as the key step towards preparing the quinoline moiety of uncialamycin (Scheme 59). A comparison between alkynes and alkene dienophiles is made as well as a survey of oxidation conditions <09JOC6728>.

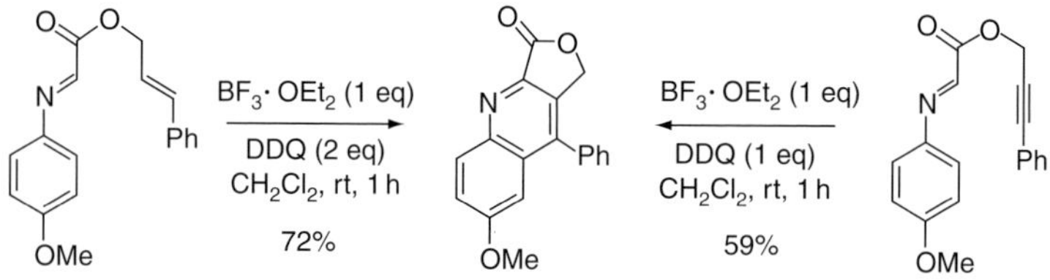

Scheme 59

Quinolines may also form through a hetero-Diels–Alder reaction via aza-quinodimethane intermediates <09H(77)1341>. A creative use of 1,3-dipolar cycloaddition between a nitrone and ß-(2-aminophenyl)-α,ß-ynone induces a conformation ripe for cyclocondensation (Scheme 60) <09EJO1027>.

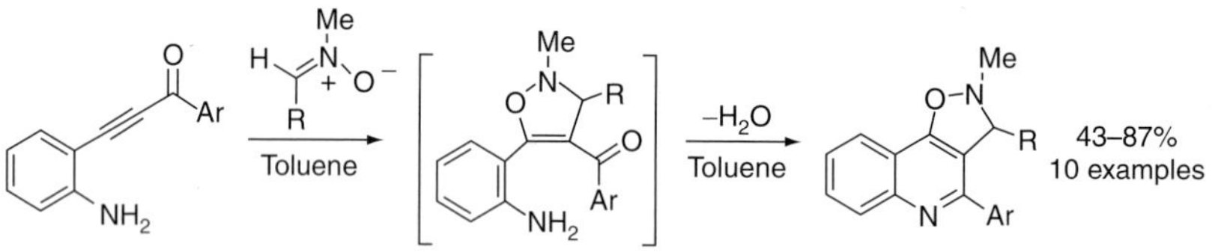

Scheme 60

In an attempt to identify alkaloid isolates from woad (*Isatis tinctoria*), quinolo[3,4-*b*]quinolines were synthesized by following up the classic but under used Diels–Reese reaction with a Friedel–Crafts arylation (Scheme 61) <09HCA668>.

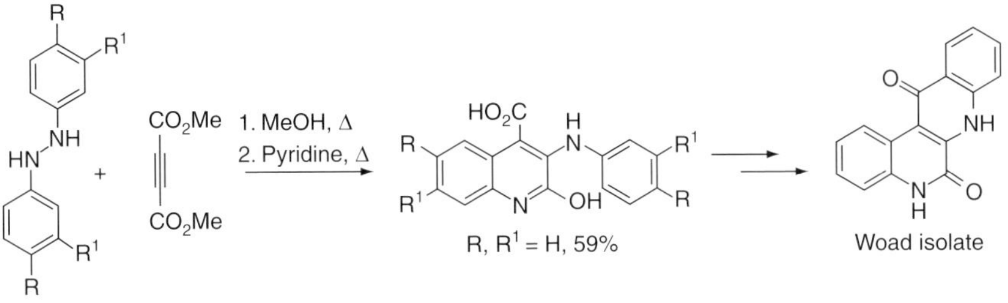

Scheme 61

Another thermal cyclization produces [1,8]naphthyridines, generating two quinoline rings in one step (Scheme 62) <09CPB393>. An impressive one-pot formation of both an indole and quinoline ring produces the indolo[2,3-*b*]quinoline system of cryptotackieine through a palladium-catalyzed oxidative cyclization <09CL772>.

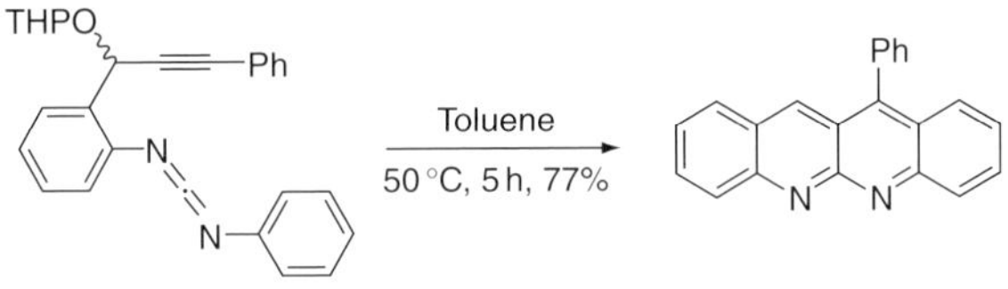

Scheme 62

Some electrocyclization substrates may be better suited for iodocyclization <09OBC85>. As reported by Likhar *et al.*, the advantage of an electrophilic process is apparent in the mild conditions and high yields – excepting an instance where a *para*-methoxy-group preferentially produces the *ipso*-substituted spirocycle (Scheme 63).

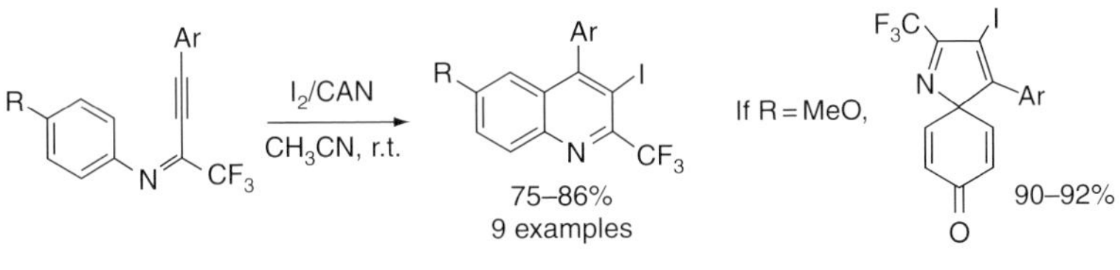

Scheme 63

Following a similar ring-closure, a new method for synthesizing quinoline-2-carboxylate derivatives involves a copper(II) triflate catalyzed addition of terminal alkynes to Schiff bases followed by electrophilic cyclization <09JOC5476>. Lastly, an unusual radical cyclization was reported that involves the use of phenyltellanyl radicals. The resulting telluro groups can be cleaved by lithium-tellurium exchange to produce unsubstituted quinolines (Scheme 64) <09OL3422>.

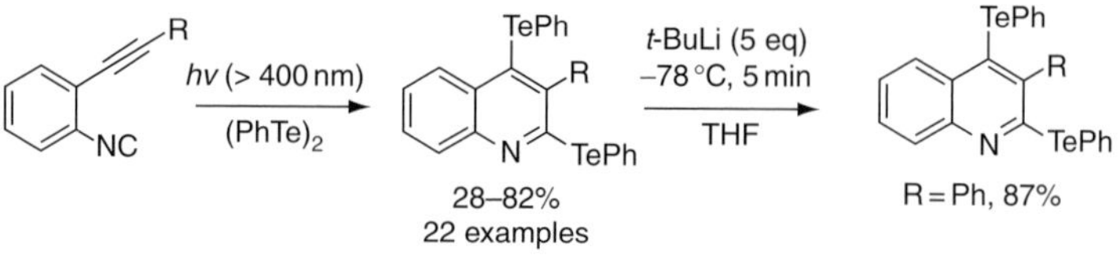

Scheme 64

6.1.3.2 Reactions of quinolines

The dominant aromaticity of the benzene portion of quinoline results in decreased aromaticity at the azine ring. Compared to pyridine, this leaves quinoline more susceptible to substitution and nucleophilic addition. Demonstrating the great reactivity of the pyridine half of quinoline, Heck reactions performed on 4-bromo-8-tosyloxy-quinolines are regioselective to the 4-position <09T4422>. Lewis acids and quinolinium salts provide a means of further activating the system. Activated quinolines are especially well suited for 1,2-addition and this has been the basis for many methodologies <09EOC4158>.

A recent 1,2-addition/oxidation strategy for a Ni-catalyzed direct C–H arylation of azines demonstrates the enhanced reactivity of quinoline (Scheme 65) <09JA12070>. Another direct C–H alkylation of quinoline was reported by Li et al. who imagined using a Lewis acid to evoke a complex with an electronic nature similar to N-oxides <09OL1171>.

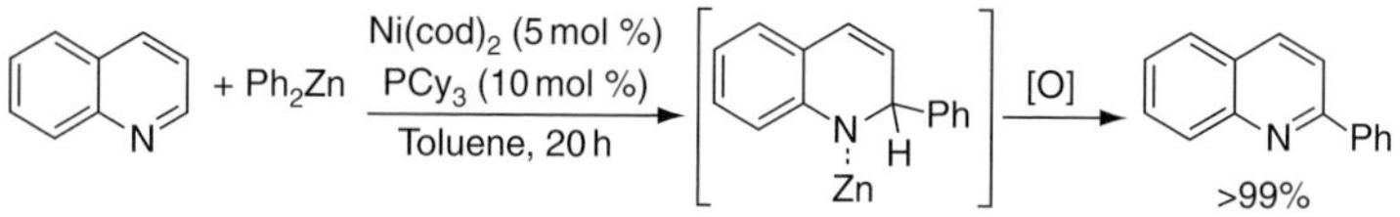

Scheme 65

Traditional Minisci reaction conditions were used to incorporate oxetane and azetidine groups into anti-malarial quinolines and other azines in decent yields (Scheme 66) <09JOC6354>. Other applications of the Minisci reaction have been more problematic; competing cyclizations proved routes to 4-azaindoles and 1,5-naphthyridines to be nonviable <09TL6772>.

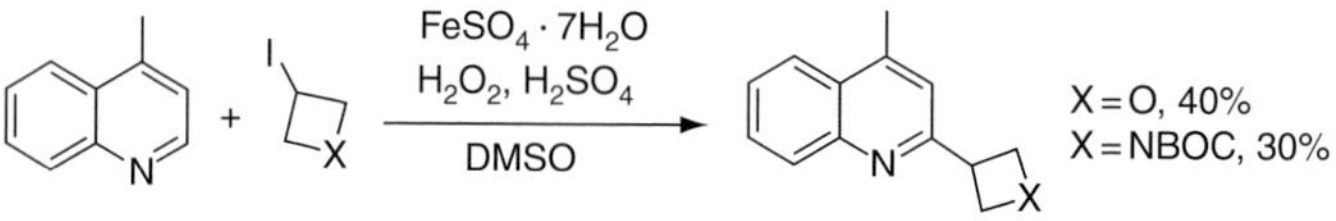

Scheme 66

Multicomponent reactions frequently exploit the reactivity of the 2-position and the formal 1,3-dipolar cycloaddition of quinoline ylides remains one of the most useful approaches to pyrrolo[1,2-*a*]quinolines (Scheme 67) <09SL1795; 09ARK(12) 243; 09H(78)177; 09JHC1203>.

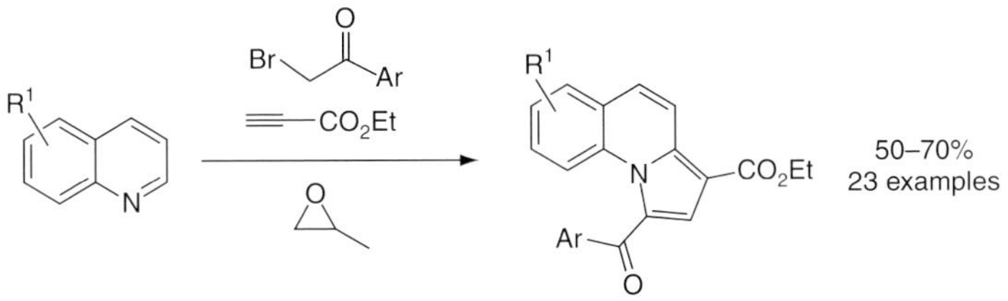

Scheme 67

For a highly enantioselective reduction of quinoline, the [Ir(COD)Cl]$_2$/bisphosphine/I$_2$ system furnishes tetrahydroquinolines in up to 96% ee <09JOC2780>. One interesting set of conditions allows either hydrogenation or dehydrogenation by simply switching between a hydrogen or argon atmosphere (Scheme 68) <09JA8410>. For benzylic oxidation of the 4-position of tetrahydroquinoline, a mixture of Cr(CO)$_6$ and *t*-BuOOH was favored over other common conditions <09H(78)3011>.

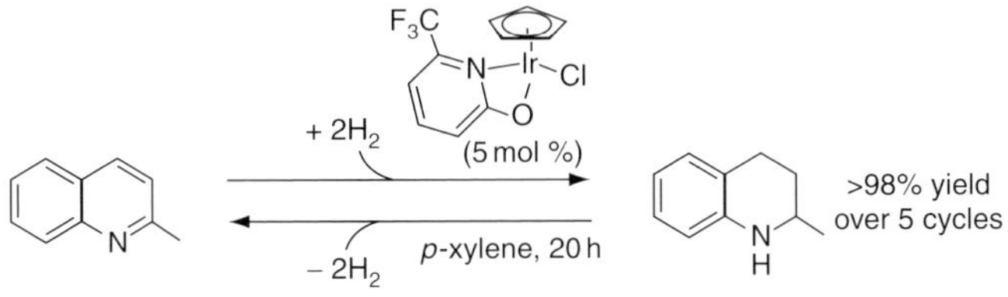

Scheme 68

2-Chloro-3-formylquinolines, readily available via the Meth–Cohn synthesis, are versatile annulation substrates. Morita–Baylis–Hillman reaction of these substrates provides access to pyridoquinolines <09S2333>. A novel entrance to the pyrano[4,3-*b*] quinoline system involves a tandem Sonogashira coupling-annulation (Scheme 69) <09JOC5664>. Fused pyranoquinolines show a wide range of biological activities: anti-inflammatory, antiallergenic, psychotropic, and estrogenic <09CR2703; 09CPB557>.

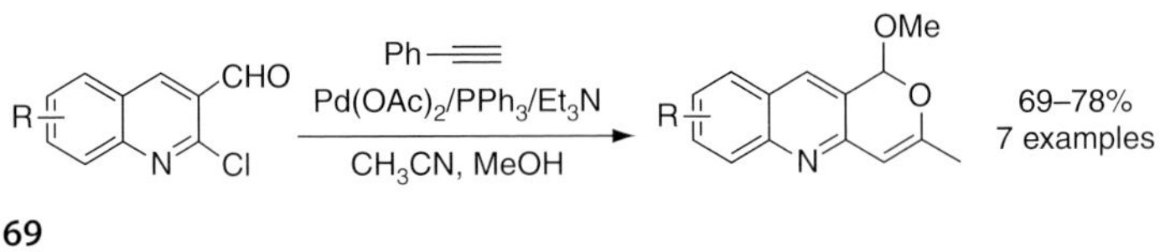

Scheme 69

The uncommon quisylate leaving group received some attention during an inquiry into the stereochemical outcome of nucleophilic displacement reactions of arylsulfonate-based leaving groups. Quisylates were reported to be substantially more reactive than the corresponding tosylates under Finkelstein conditions <09JOC6042>. 4- and 7-Quinolinesulfonyl chlorides have been used to generate the corresponding sulfonamides <09H(78)93>. Quinolinesulfonyl groups have also been used as DMGs <09SL1639; 09S113>.

From Woodward to Perkin, the history of organic chemistry has been woven with Cinchona alkaloids. The landmark synthesis of quinine helped shaped the modern era of organic chemistry; even today, quinine derivatives are being designed for their antimalarial properties <09CEJ7637>. Cinchona alkaloids have been celebrated for their application in organocatalytic methods – their use continues to produce excellent results <09OL4010; 09JOC4650; 09OL437; 09OL2205>. These compounds may react selectively at either the quinoline or quinuclidine nitrogen depending on the nature of the electrophile. An indispensible new work by Mayr et al. asks the question "which nitrogen is more nucleophilic?" As for the answer, readers are urged to consult the primary reference – though it may suffice to say that large, bulky electrophiles prefer attack from the quinoline ring while smaller nucleophiles favor the quinuclidine nitrogen (Scheme 70) <09JOC7157>.

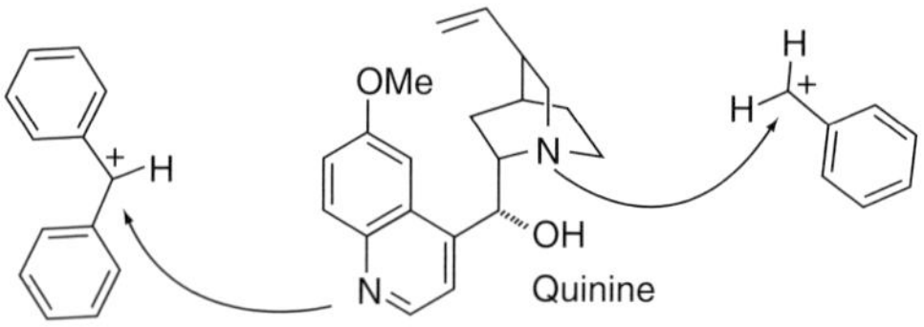

Scheme 70

Cinchona alkaloids have catalyzed a wide variety of operations; one inventive new method uses these compounds to effect a centrifugal continuous asymmetric extraction <09CEJ2111>. As an example of organocatalysis, these compounds have been used for the synthesis of ß-sultams from electron deficient imines with alkylsulfonyl chlorides. To extend this method to a wider scope of imines (Scheme 71), N-(2-pyridylsulfonyl)imines were used in conjunction with Yb(OTf)$_3$ to generate a bidentate coordinated activated complex <09CEJ8204>.

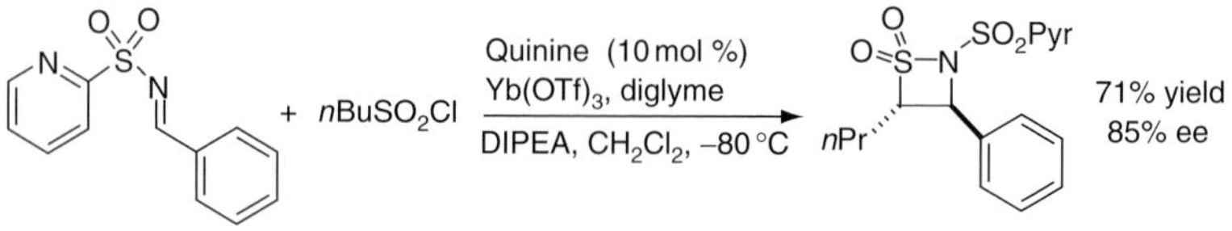

Scheme 71

6.1.4. ISOQUINOLINES
6.1.4.1 Preparation of Isoquinolines

The Larock isoquinoline synthesis has quickly proven itself as one of the most versatile isoquinoline syntheses available to the heterocyclic chemist (Scheme 72) <09JCC1061>. Still, new modifications and related syntheses seek to improve this method.

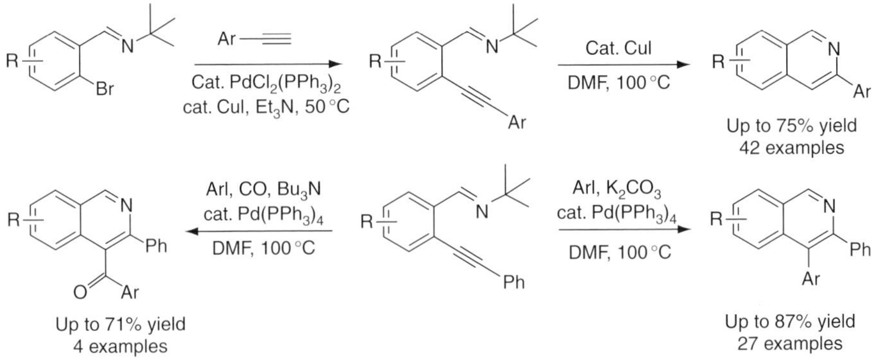

Scheme 72

Fagnou et al. sought to develop an isoquinoline synthesis along the same disconnections as the Larock synthesis but without the requirement of an *ortho*-halogen. This goal was realized by using rhodium to induce C–H bond cleavage under mild conditions (Scheme 73) <09JA12050>. A very similar transformation was reported by Miura et al. – rhodium-catalyzed oxidative coupling of *N*-phenyl aromatic imines with alkynes to produce quinoline derivatives <09CC5141>. Another [4+2] disconnection produces isoquinolines by use of palladium-mediated benzyne chemistry <09AG(I)572>.

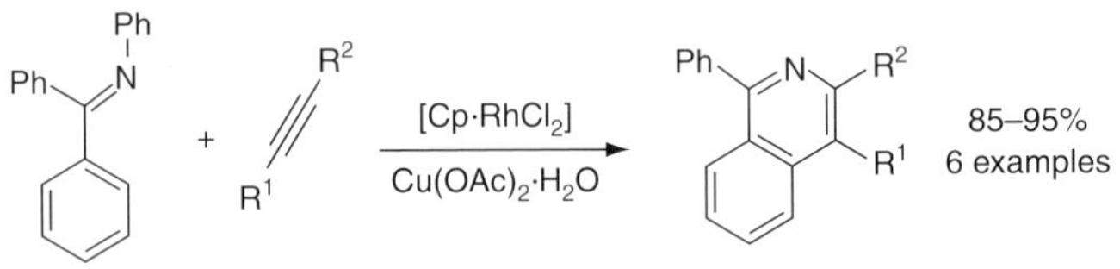

Scheme 73

Several reports illustrated the metal-catalyzed cyclization of *ortho*-alkynylbenzaldoximes. One such synthesis exploits the easily oxidized benzyloxy moiety to generate isoquinolines in high yield from 2-alkynylbenzaldoximes <09TL2305>. Zhang et al. observed an interesting Katada-like rearrangement when these reactions are performed with AgOTf on O-acetyl oximes (Scheme 74) <09ASC85>. As many of these cyclizations utilize silver, it is appropriate to note a recent review on silver catalyzed cycloisomerizations <09EJO6075>.

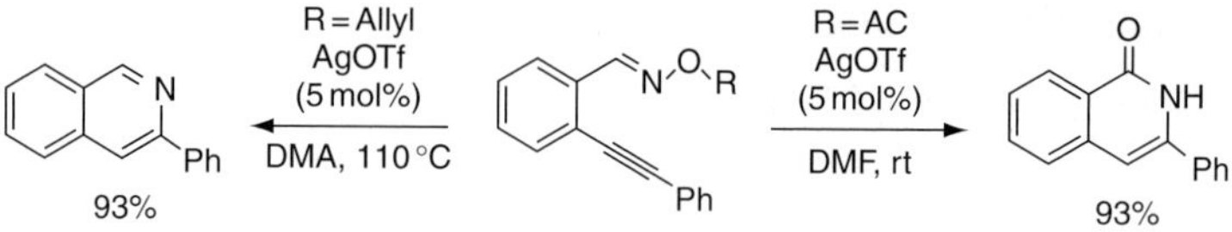

Scheme 74

A novel four-component example generates fused polycyclic isoquinolines from an *ortho*-alkynylbenzaldehyde, diamine, and Mannich reagent (Scheme 75) <09JOC6299>. Under microwave irradiation, similar tandem imination/cyclization reactions have also been performed <09EJO2852>.

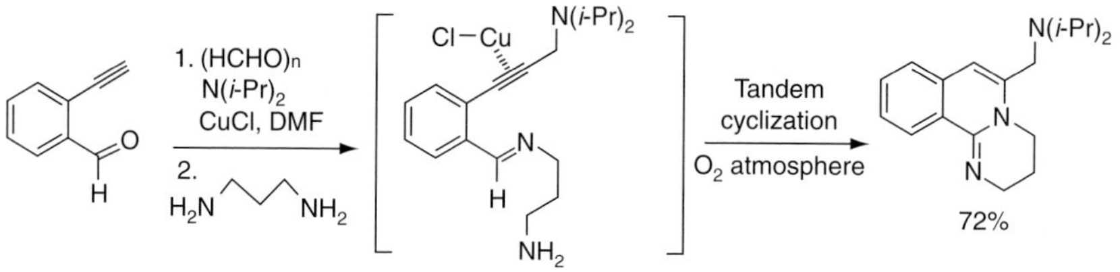

Scheme 75

A recent approach to isoquinolines involves the cyclization of *o*-alkynyl-benzyl-azides (Scheme 76). Liang *et al.* report a silver-catalyzed cyclization to generate 1,3-substituted isoquinolines <09JOC2893>. *o*-Vinyl benzyl azides form isoquinolines as well, but through a substantially different mechanism (Scheme 77). In this case,

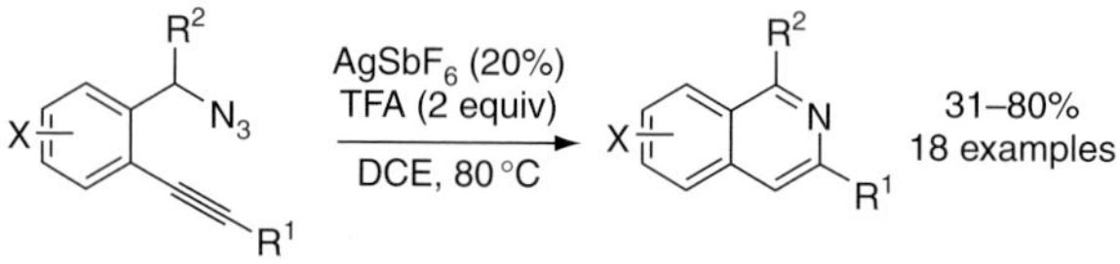

Scheme 76

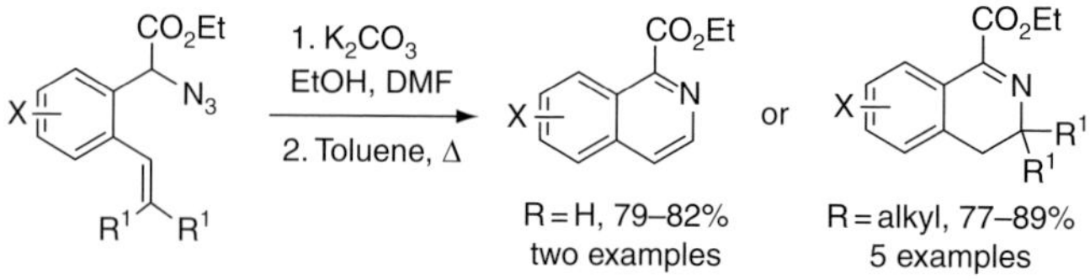

Scheme 77

base-catalyzed loss of molecular nitrogen engenders an azatriazene capable of 6π-electrocyclization to an easily oxidizable dihydropyridine <09OL729>.

Baudoin et al. orchestrated a tandem ring-opening/6π-electrocyclization to generate the dihydroisoquinoline core (Scheme 78). This methodology was applied to a synthesis of (±)-coralydine <09AG(I)179>. Styryl isocyanates are another azatriene capable of cyclizing to a quinoline; generated from styryl acylazides or styryl carbamates <09H(78)2979>.

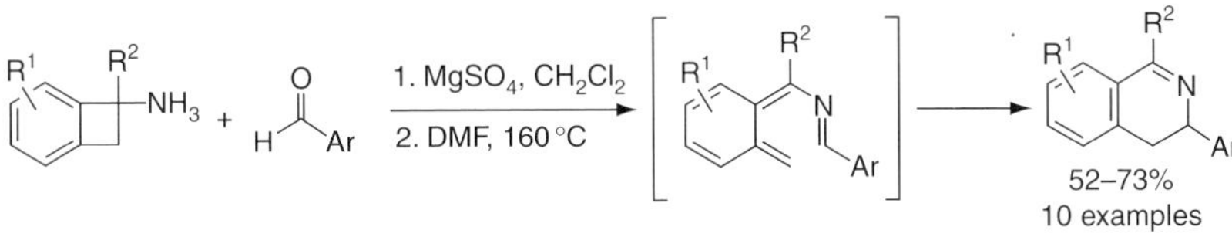

Scheme 78

Diels–Alder reactions can be performed to generate the benzene portion of the isoquinoline ring, one recent example takes advantage of Guareschi–Thorpe pyridinones to generate thieno[3,4-c]pyridinones which are used as reactive dienes <09H(78)2067>. An Aza-Diels–Alder reaction allows the formation of isoquinolines by multicomponent reaction of arynes, isocyanides, and alkynes <09AG(I)3458>.

Classic isoquinoline syntheses continue to be used to generate dihydro- and tetrahydro-isoquinolines; in particular, the Pictet–Spengler (Scheme 79) <09EJO292>

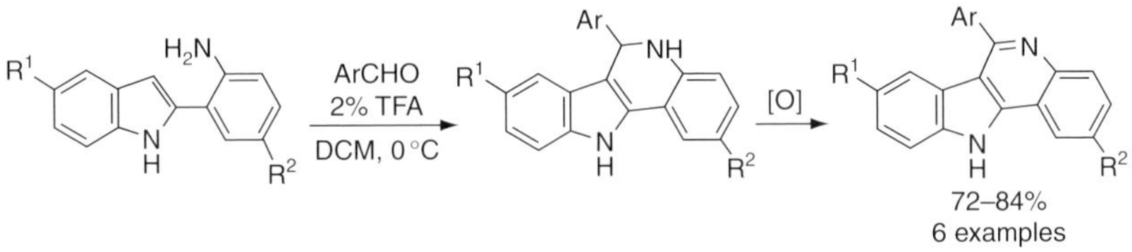

Scheme 79

and Bischler–Napieralski <09T8412> reactions. Crispine B was synthesized for the first time by using a Pictet–Gams variation of the Bischler–Napieralski reaction to generate the fully aromatized isoquinoline core (Scheme 80) <09H(77)1397>.

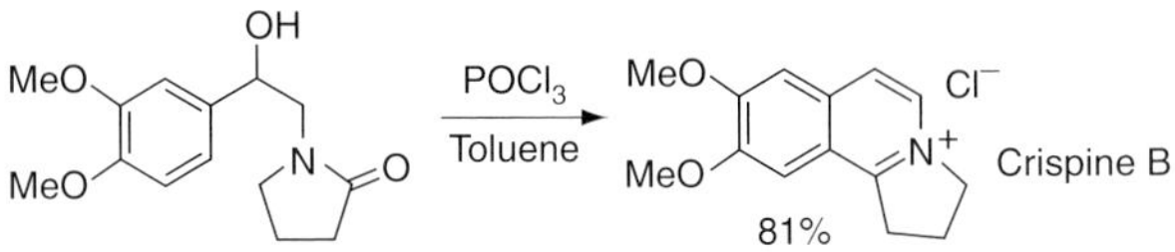

Scheme 80

6.1.4.2 Reactions of isoquinoline

Most traditional methods of generating isoquinoline do so via oxidative aromatization of dihydro- or tetrahydro-isoquinolines and each year many new such conditions are reported – usually offering some environmental or economic advantage

<09EJO530>. Though the reactivity of isoquinoline generally mirrors or transposes that of quinoline and pyridine, several reports during the past year demonstrate unique differences between these heterocycles.

Lewis acid activation and nucleophilic attack can produce multiple regioisomers in quinolines and pyridines; isoquinolines generally only undergo attack at the 1-position. This effect is also observed in many direct C–H alkylation methods as these reactions often operate via an addition/oxidation mechanism (Scheme 81) <09OL1171>.

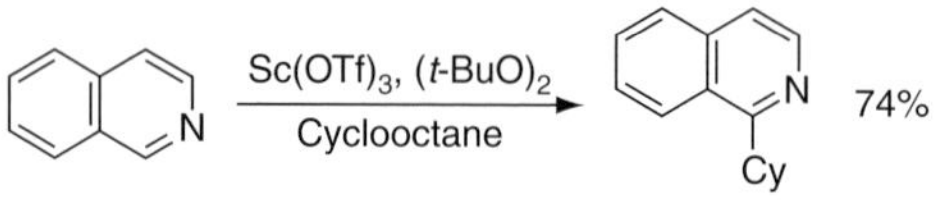

Scheme 81

Zwitterionic species derived from isoquinoline sometimes react in a manner divergent from the other pyridine species. The 1,4-dipolar DMAD-isoquinoline complex readily engages in three-component reactions (Scheme 82) with activated alkenes, imines, and carbonyl compounds – in contrast, the pyridine-DMAD can induce molecular rearrangements <09SL229>.

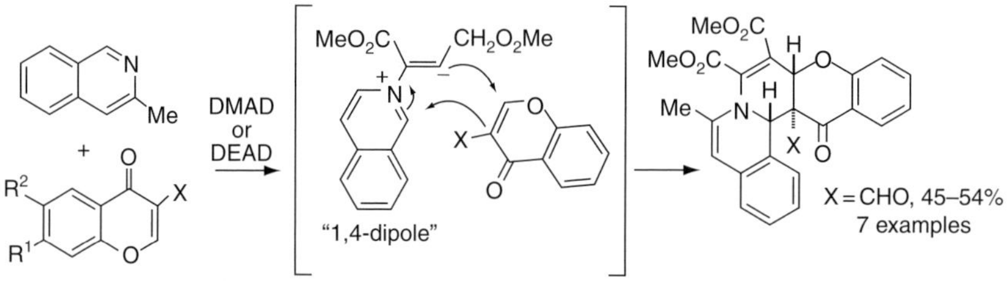

Scheme 82

1,3-Dipolar complexes produced from isoquinoline are also useful intermediates. While these zwitterionic species are also available to their pyridine and quinoline counterparts, several new examples of stabilized N-ylides have been unique to isoquinoline. An MCR between isoquinoline, isocyanides, and trifluoroacetic anhydride results in an interesting mesoionic structure (Scheme 83). When exposed to identical conditions, quinoline produces no reaction <09EJO617>. The Jie Wu

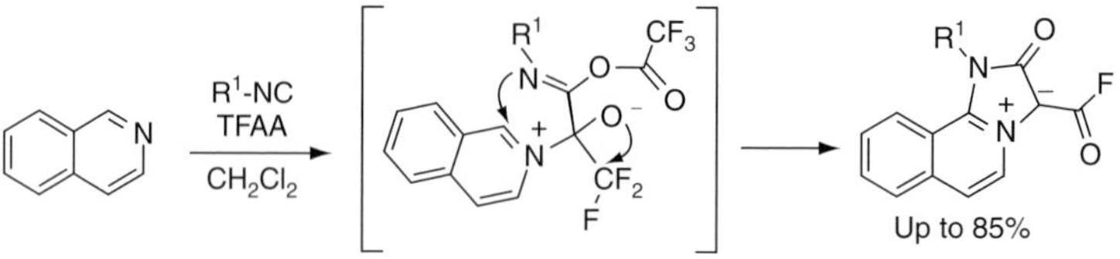

Scheme 83

group has emphasized the integration of this type of chemistry into domino sequences – encompassing, in several cases, formation of the isoquinoline *in situ* (Scheme 84) <09ASC2702>.

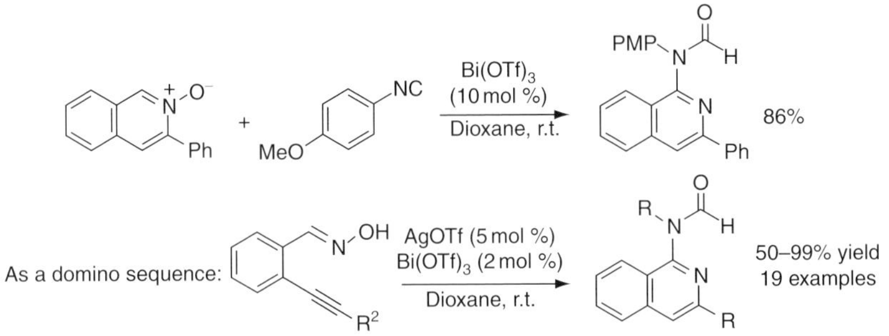

Scheme 84

Donor-acceptor cyclopropanes have been successfully applied as dienophiles in a domino cyclization-[3+3] cycloaddition (Scheme 85). As in the previous method, isoquinoline *N*-oxide is generated by silver-catalyzed cyclization <09TL198>.

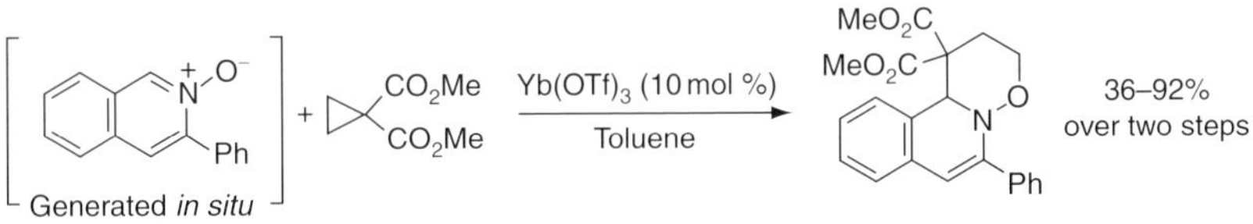

Scheme 85

Ventures to incorporate these methods into tandem multicomponent reactions have produced some unexpected results. Rather than the expected fused isoxazole, one MCR involving DMAD produced rearranged products (Scheme 86). The unusual azomethine ylides were remarkably stable and consented to Suzuki–Miyaura couplings <09JOC921>.

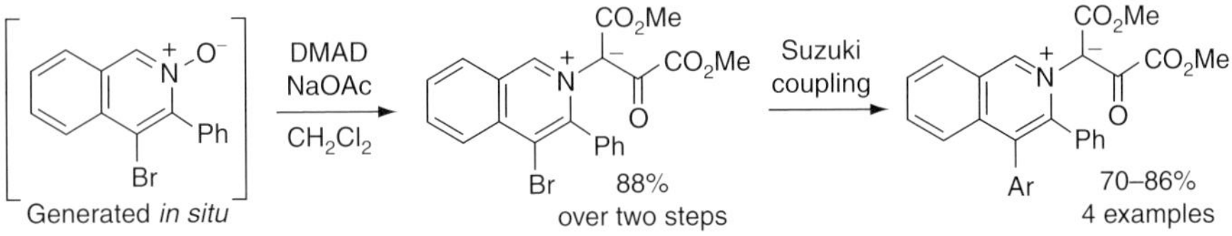

Scheme 86

Wu *et al.* also found similar results when starting from hydrazone derivatives – though conditions were ultimately identified which generate the originally intended tricyclic pyrrazolo[3,2-*a*]isoquinoline (Scheme 87) <09ASC1692; 09OBC4526; 09TL340>.

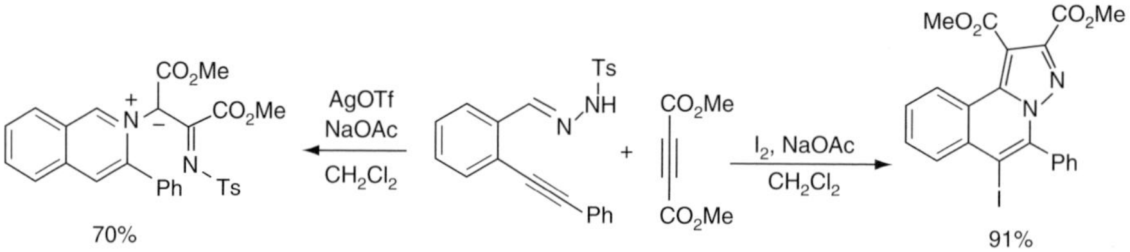

Scheme 87

An investigation of isoquinolyl nitrenes has shed light on the complex chemistry of pyridyl nitrenes. Generation of ring-expanded diazacycloheptatriene and other rearrangment products requires the invocation of transient fused azirene. These intermediates have been long proposed, but had yet to be observed – until this past year. Wentrup et al. reported the first direct observation of a fused azirene from pyridyl nitrenes (Scheme 88). This compound is stable to temperatures above 110 K and undergoes automerization as revealed by an ^{15}N-labeling study <09JOC1171>.

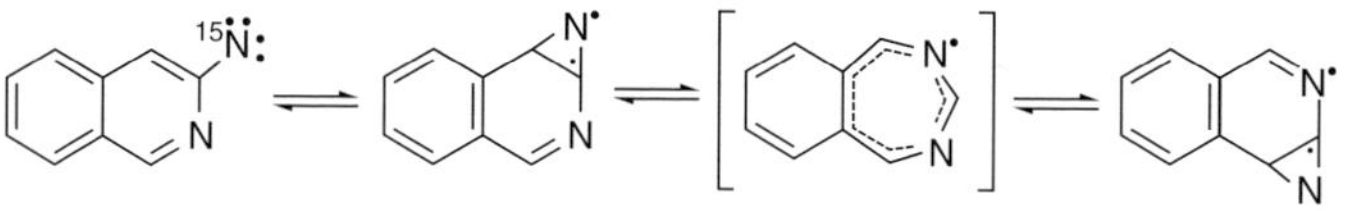

Scheme 88

6.1.5. SPECIAL TOPIC: SUPRAMOLECULAR CHEMISTRY

It is in part due to supramolecular chemistry that heterocycles have been privileged with a central role in biological processes. The supramolecular interactions that govern enzymatics have long favored the involvement of heterocycles at active sites; synthetic mimics of these systems have taken a similar path <09EJO4515>. Synthetic substrate-selective recognition receptors are constructed by the careful arrangement of H–bond donor and acceptor units – a role which pyridines and even C–2 acidic pyridiniums are well suited (Scheme 89). Growing interest in abiotic foldamers as protein models has prompted the synthesis of pyridine oligomers, several theoretical surveys, and a review <09OBC2534; 09T1679; 09CC692; 09EJO5699>. Quinoline-based foldamers have been used as dynamic scaffolds to study photo-induced charge transfer <09JA4819>.

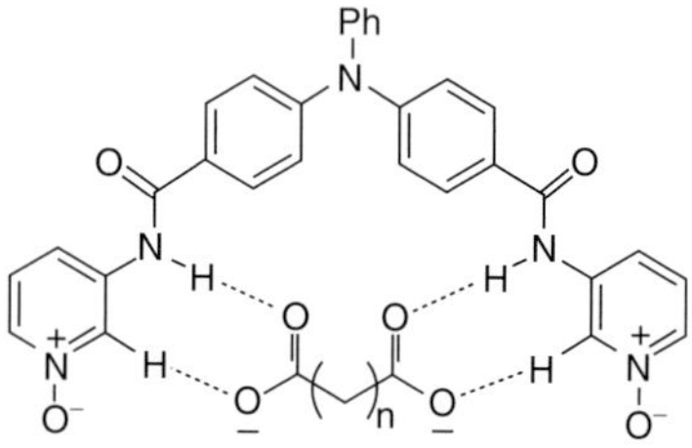

Scheme 89

Scores of new metal-specific ion sensors have been reported, many of which gain their specificity from dipicolylamine groups (Scheme 90) <09OL795; 09OL1655; 09OL3454; 09JOC2992>. These compounds also often include quinoline moieties as fluorescent reporting groups < 09OL1269; 09OL4426>. The Lewis basicity of pyridine has played a vital role in pH probes <09JA3016> and acid controlled switches <09JA18269>. By utilizing pyridine functionality, Aprahamian *et al.* produced a novel chemically controlled rotary switch (Scheme 91).

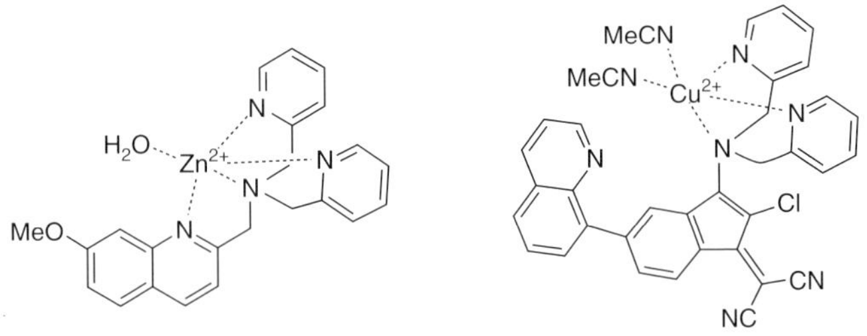

Scheme 90

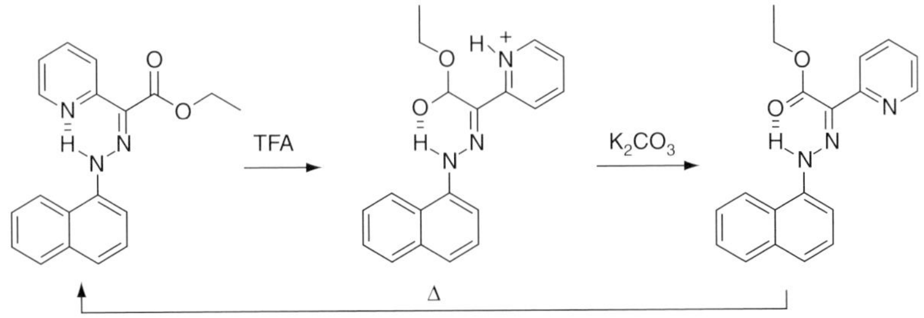

Scheme 91

Nanomolecular devices incorporate pyridiniums as stable cationic moieties <09OL613; 09OL385>. Rotaxane switches and memory devices use cyclobisparaquot for their macrocyclic ring due to the many strong supramolecular interactions these groups induce – pyridinium and viologen base-stations are effective for the same reasons (Scheme 92) <09OBC142>. Many of these devices rely on a redox

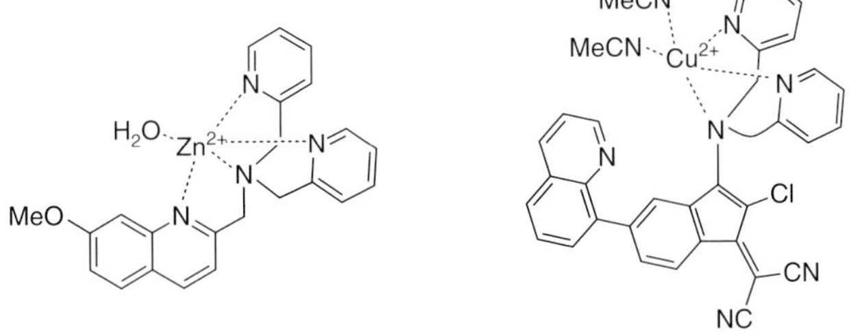

Scheme 92

cycle at a pyridinium center; suitably, the reductive potential of viologen and similar compounds is currently of interest <09OBC1445; 09CC1876; 09OL685>.

Functional structures have been designed with supramolecular chemistry in mind. New calixarenes containing pyridines – calixpyridines – have been synthesized as novel cavitands (Scheme 93) <09CC2899; 09JOC5361; 09JOC8595>. Calix[4]

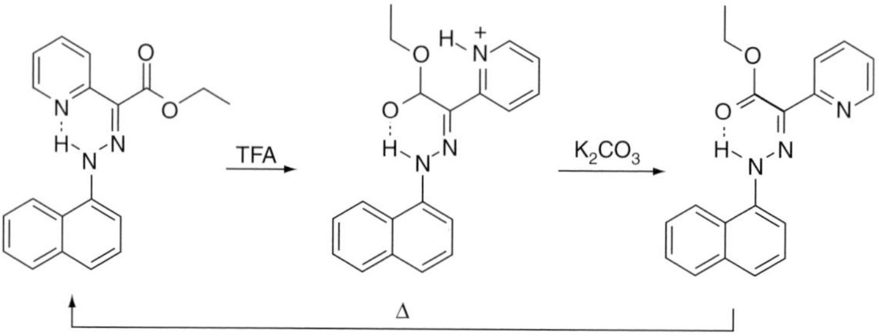

Scheme 93

pyrroles show molecular recognition of pyridine N-oxide in water <09JA3178>. Metal-organic frameworks and supramolecular polymers continue to use pyridine-containing units as linkers and ligands at the metal-organic interface (Scheme 94) <09OL3562>.

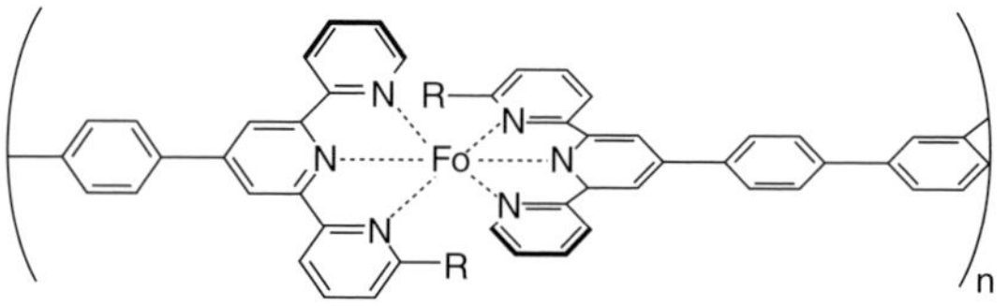

Scheme 94

Offering application in these systems, methods toward terpyridines and other oligopyridines have benefited from renewed attention <09CEJ6811; 09T5413; 09EJO801>. While these topics may not be of interest to all heterocyclic chemists, it is our chemistry upon which many of these structures depend. The involvement of heterocycles in supramolecular structures is both a solicitation and opportunity for new methods in heterocyclic chemistry.

REFERENCES

06JOC6790	N. Hara, S. Nakamura, N. Shibata and T. Toru, *J. Org. Chem.*, **71**, 6790 (2006).
08AG(I)4695	K.L. Billingsley and S.L. Buchwald, *Angew. Chem. Int. Ed.*, **47**, 4695 (2008).
08OL1719	Y. Kiyotsuka, H.P. Acharya, Y. Katayama, T. Hyodo and Y. Kobayashi, *Org. Lett.*, **10**, 1719 (2008).
09AG(I)179	M. Chaumontet, R. Piccardi and O. Baudoin, *Angew. Chem. Int. Ed.*, **48**, 179 (2009).
09AG(I)572	T. Gerfaud, L. Neuville and J. Zhu, *Angew. Chem. Int. Ed.*, **48**, 572 (2009).

09AG(I)1810	A. Geny, N. Agenet, L. Iannazzo, M. Malacria, C. Aubert and V. Gandon, *Angew. Chem. Int. Ed.*, **48**, 1810 (2009).
09AG(I)3288	H. Andersson, M. Gustafsson, D. Boström, R. Olsson and F. Almqvist, *Angew. Chem. Int. Ed.*, **48**, 3288 (2009).
09AG(I)3296	D. Zhao, W. Wang, F. Yang, J. Lan, L. Yang, G. Gao and J. You, *Angew. Chem. Int. Ed.*, **48**, 3296 (2009).
09AG(I)3458	F. Sha and X. Huang, *Angew. Chem. Int. Ed.*, **48**, 3458 (2009).
09AG(I)4198	D. Lefranc and M.A. Ciufolini, *Angew. Chem. Int. Ed.*, **48**, 4198 (2009).
09AG(I)5094	X. Chen, K.M. Engle, D.-H. Wang and J.-Q. Yu, *Angew. Chem. Int. Ed.*, **48**, 5094 (2009).
09AG(I)6045	L. Ackermann, P. Novák, R. Vicente and N. Hofmann, *Angew. Chem. Int. Ed.*, **48**, 6045 (2009).
09AG(I)6121	M.T. Wentzel, V.J. Reddy, T.K. Hyster and C.J. Douglas, *Angew. Chem. Int. Ed.*, **48**, 6121 (2009).
09AG(I)6511	A. García-Rubia, R.G. Arrayás and J.C. Carretero, *Angew. Chem. Int. Ed.*, **48**, 6511 (2009).
09AG(I)7256	S.H. Wunderlich, M. Kienle and P. Knochel, *Angew. Chem. Int. Ed.*, **48**, 7256 (2009).
09AG(I)8325	T. Rizk, E.J.-F. Bilodeau and A.M. Beauchemin, *Angew. Chem. Int. Ed.*, **48**, 8325 (2009).
09ASC85	H. Gaoa and J. Zhang, *Adv. Synth. Catal.*, **351**, 85 (2009).
09ASC107	S. Barroso, G. Blay, M.C. Muñoz and J.R. Pedro, *Adv. Synth. Catal.*, **351**, 107 (2009).
09ASC1162	C. Eidamshausa and H.U. Reissiga, *Adv. Synth. Catal.*, **351**, 1162 (2009).
09ASC1692	Z. Chen, Q. Ding, X. Yu and J. Wu, *Adv. Synth. Catal.*, **351**, 1692 (2009).
09ASC2217	Y. Wang, X. Xin, Y. Liang, Y. Lin, H. Duan and D. Donga, *Adv. Synth. Catal.*, **351**, 2217 (2009).
09ASC2702	Z. Chen, X. Yu, M. Su, X. Yang and J. Wu, *Adv. Synth. Catal.*, **351**, 2702 (2009).
09AJC150	A. Hanna-Elias, D.T. Manallack, I. Berque-Bestel, H.R. Irving, I.M. Coupar and M.N. Iskander, *Aust. J. Chem.*, **62**, 150 (2009).
09AJC176	J. Li and L. Wang, *Aust. J. Chem.*, **62**, 176 (2009).
09BMC74	R.B. Lacerda, C.K.F. de Lima, L.L. da Silva, N.C. Romeiro, A.L.P. Miranda, E.J. Barreiro and C.A.M. Fraga, *Bioorg. Med. Chem.*, **17**, 74 (2009).
09BMCL894	S.P. East, C.B. White, O. Barker, S. Barker, J. Bennett, D. Brown, E.A. Boyd, C. Brennan, C. Chowdhury, I. Collins, E. Convers-Reignier, B.W. Dymock, R. Fletcher, D.J. Haydon, M. Gardiner, S. Hatcher, P. Ingram, P. Lancett, P. Mortenson, K. Papadopoulos, C. Smee, H.B. Thomaides-Brears, H. Tye, J. Workman and L.G. Czaplewski, *Bioorg. Med. Chem. Lett.*, **19**, 894 (2009).
09BMCL1807	B.A. Johns, J.G. Weatherhead, S.H. Allen, J.B. Thompson, E.P. Garvey, S.A. Foster, J.L. Jeffrey and W.H. Miller, *Bioorg. Med. Chem. Lett.*, **19**, 1807 (2009).
09BMCL3602	N. Bailey, M.J. Bamford, D. Brissy, J. Brookfield, E. Demont, R. Elliott, N. Garton, I. Farre-Gutierrez, T. Hayhow, G. Hutley, A. Naylor, T.A. Panchal, H.-X. Seow, D. Spalding and A.K. Takle, *Bioorg. Med. Chem. Lett.*, **19**, 3602 (2009).
09CC692	F. Chevallier, M. Charlot, C. Katan, F. Mongin and M. Blanchard-Desce, *Chem. Commun.*, 692 (2009).
09CC1876	D. Wang, W.E. Crowe, R.M. Stronginb and M. Sibrian-Vazquez, *Chem. Commun.*, 1876 (2009).
09CC1964	H. Brice and J. Clayden, *Chem. Commun.*, 1964 (2009).
09CC2899	B. Yao, D.-X. Wang, Z.-T. Huang and M.-X. Wang, *Chem. Commun.*, 2899 (2009).
09CC5141	T. Fukutani, N. Umeda, K. Hirano, T. Satoh and M. Miura, *Chem. Commun.*, 5141 (2009).
09CR2652	J. Marco-Contelles, E. Pérez-Mayoral, A. Samadi, M.D.C. Carreiras and E. Soriano, *Chem. Rev.*, **109**, 2652 (2009).
09CR2703	N.R. Candeias, L.C. Branco, P.M.P. Gois, C.A.M. Afonso and A.F. Trindade, *Chem. Rev.*, **109**, 2703 (2009).
09CR3743	W.A.L. van Otterlo and C.B. de Koning, *Chem. Rev.*, **109**, 3743 (2009).

09CR4140	M.A.P. Martins, C.P. Frizzo, D.N. Moreira, L. Buriol and P. Machado, *Chem. Rev.*, **109**, 4140 (2009).
09CR6729	J. Stambasky, M. Hocek and P. Kocovsky, *Chem. Rev.*, **109**, 6729 (2009).
09CEJ333	G. Deng, K. Ueda, S. Yanagisawa, K. Itami and C.-J. Li, *Chem. Eur. J.*, **15**, 333 (2009).
09CEJ2111	A.J. Hallett, G.J. Kwant and J.G. de Vries, *Chem. Eur. J.*, **15**, 2111 (2009).
09CEJ2129	P. Garcia, S. Moulin, Y. Miclo, D. Leboeuf, V. Gandon, C. Aubert and M. Malacria, *Chem. Eur. J.*, **15**, 2129 (2009).
09CEJ3790	B. Blank, S. Michlik and R. Kempe, *Chem. Eur. J.*, **15**, 3790 (2009).
09CEJ5950	T.M. Gøgsig, A.T. Lindhardt, M. Dekhane, J. Grouleff and T. Skrydstrup, *Chem. Eur. J.*, **15**, 5950 (2009).
09CEJ6811	J. Dash and H.-U. Reissig, *Chem. Eur. J.*, **15**, 6811 (2009).
09CEJ7637	C. Bucher, C. Sparr, W.B. Schweizer and R. Gilmour, *Chem. Eur. J.*, **15**, 7637 (2009).
09CEJ8204	M. Zajac and R. Peters, *Chem. Eur. J.*, **15**, 8204 (2009).
09CJC1122	M. Hosseini-Sarvari, *Can. J. Chem.*, **87**, 1122 (2009).
09CL118	T. Katagiri, T. Mukai, T. Satoh, K. Hirano and M. Miura, *Chem. Lett.*, **38**, 118 (2009).
09CL484	S. Goswami and A. Hazra, *Chem. Lett.*, **38**, 484 (2009).
09CL772	H. Takeda, T. Ishida and Y. Takemoto, *Chem. Lett.*, **38**, 772 (2009).
09CPB393	H. Kimura, K. Torikai and I. Ueda, *Chem. Pharm. Bull.*, **57**, 393 (2009).
09EJO292	P.K. Agarwal, D. Sawant, S. Sharma and B. Kundu, *Eur. J. Org. Chem.*, 292 (2009).
09EJO530	X. Wu and A.E.V. Gorden, *Eur. J. Org. Chem.*, 530 (2009).
09EJO617	M.J. Arévalo, N. Kielland, C. Masdeu, M. Miguel, N. Isambert and R. Lavilla, *Eur. J. Org. Chem.*, 617 (2009).
09EJO643	S. Vanlaer, A. Voet, C. Gielens, M.D. Maeyer and F. Compernolle, *Eur. J. Org. Chem.*, 643 (2009).
09EJO801	A. Winter, C. Friebe, M.D. Hager and U.S. Schubert, *Eur. J. Org. Chem.*, 801 (2009).
09EJO1027	G. Abbiati, A. Arcadi, F. Marinelli, E. Rossi and M. Verdecchia, *Eur. J. Org. Chem.*, 1027 (2009).
09EJO1781	C.J. Rohbogner, S.H. Wunderlich, G.C. Clososki and P. Knochel, *Eur. J. Org. Chem.*, 1781 (2009).
09EJO2051	X.-L. Fu, L.-L. Wu, H.-Y. Fu, H. Chen and R.-X. Li, *Eur. J. Org. Chem.*, 2051 (2009).
09EJO2820	A. Begouin and M.-J.R.P. Queiroz, *Eur. J. Org. Chem.*, 2820 (2009).
09EJO2852	M. Alfonsi, M. Dell'Acqua, D. Facoetti, A. Arcadi, G. Abbiati and E. Rossi, *Eur. J. Org. Chem.*, 2852 (2009).
09EJO4165	Y. Wang, X. Xin, Y. Liang, Y. Lin, R. Zhang and D. Dong, *Eur. J. Org. Chem.*, 4165 (2009).
09EJO4199	P.C. Gros and Y. Fort, *Eur. J. Org. Chem.*, 4199 (2009).
09EJO4515	K. Ghosh, G. Masanta and A.P. Chattopadhyay, *Eur. J. Org. Chem.*, 4515 (2009).
09EJO4586	G. Toma, K.-I. Fujita and R. Yamaguchi, *Eur. J. Org. Chem.*, 4586 (2009).
09EJO5699	G. Maayan, *Eur. J. Org. Chem.*, 5699 (2009).
09EJO6075	P. Belmont and E. Parker, *Eur. J. Org. Chem.*, 6075 (2009).
09H(77)233	T. Koshiba, T. Miyazaki, H. Tokuyama and T. Fukuyama, *Heterocycles*, **77**, 233 (2009).
09H(77)679	Y. Hirasawa, J. Kobayashi and H. Moritaa, *Heterocycles*, **77**, 679 (2009).
09H(77)1163	E. Fukuda, Y. Takahashi, N. Hirasawa, O. Sugimoto and K.-I. Tanji, *Heterocycles*, **77**, 1163 (2009).
09H(77)1341	G. Cremonesi, P.D. Croce, F. Fontana and C.L. Rosa, *Heterocycles*, **77**, 1341 (2009).
09H(77)1397	T. Yasuhara, N. Zaima, S. Hashimoto, M. Yamazaki and O. Muraoka, *Heterocycles*, **77**, 1397 (2009).
09H(78)93	K. Marciniec and A. Maślankiewicz, *Heterocycles*, **78**, 93 (2009).
09H(78)177	N.A. Kheder, E.S. Darwish and K.M. Dawood, *Heterocycles*, **78**, 177 (2009).
09H(78)487	H.-M. Wang, R.-S. Hou, H.-T. Cheng and L.-C. Chen, *Heterocycles*, **78**, 487 (2009).
09H(78)977	L. Wu, C. Yang, L. Yang and L. Yang, *Heterocycles*, **78**, 9777 (2009).
09H(78)2067	K.D. Khalil, H.M. Al-Matar and M.H. Elnagdi, *Heterocycles*, **78**, 2067 (2009).

09H(78)2735	A. Sato, O. Sugimoto and K.I. Tanji, *Heterocycles*, **78**, 2067 (2009).
09H(78)2979	C.-C. Chen, Li.-Y. Chen, R.-Y. Lin, C.-Y. Chu and S.A. Dai, *Heterocycles*, **78**, 2979 (2009).
09H(78)3011	R.C. Higgins, N.O. Townsend and Y.A. Jackson, *Heterocycles*, **78**, 3011 (2009).
09H(79)99	F.I. Carroll, *Heterocycles*, **79**, 99 (2009).
09H(79)207	H.M. Garraffo, T.F. Spande and M. Williams, *Heterocycles*, **79**, 207 (2009).
09H(79)791	J. Ward and V. Caprio, *Heterocycles*, **79**, 791 (2009).
09HCA668	H.M. Riepl and M. Kellermann, *Helv. Chim. Acta*, **92**, 668 (2009).
09HCA928	Y.-K. Lia, Q.-J. Zhaob, J. Huc, Z. Zoua, X.-Y. Hea, H.-B. Yuana and X.-Y. Shi, *Helv. Chim. Acta*, **92**, 928 (2009).
09HCA2790	G. Yea, Z.-X. Lia, G.-X. Xiaa, H. Pengb, Z.-L. Suna and C.-G. Huang, *Helv. Chim. Acta*, **92**, 2790 (2009).
09JA1396	B.H. Lipshutz and B. Amorelli, *J. Am. Chem. Soc.*, **131**, 1396 (2009).
09JA1644	N. Minakawa, S. Ogata, M. Takahashi and A. Matsuda, *J. Am. Chem. Soc.*, **131**, 1644 (2009).
09JA3016	B. Tang, F. Yu, P. Li, L. Tong, X. Duan, T. Xie and X. Wang, *J. Am. Chem. Soc.*, **131**, 3016 (2009).
09JA3178	B. Verdejo, G. Gil-Ramírez and P. Ballester, *J. Am. Chem. Soc.*, **131**, 1644 (2009).
09JA3291	L.-C. Campeau, D.R. Stuart, J.-P. Leclerc, M. Bertrand-Laperle, E. Villemure, H.-Y. Sun, S. Lasserre, N. Guimond, M. Lecavallier and K. Fagnou, *J. Am. Chem. Soc.*, **131**, 3291 (2009).
09JA3466	X. Zhao, E. Dimitrijevic and V.M. Dong, *J. Am. Chem. Soc.*, **131**, 3466 (2009).
09JA4598	H. Liu, G. Dagousset, G. Masson, P. Retailleau and J. Zhu, *J. Am. Chem. Soc.*, **131**, 4598 (2009).
09JA4819	M. Wolffs, N. Delsuc, D. Veldman, N.V. Anh, R.M. Williams, S.C.J. Meskers, R.A.J. Janssen, I. Huc and A.P.H.J. Schenning, *J. Am. Chem. Soc.*, **131**, 4819 (2009).
09JA5488	S.K. Jarchow-Choy, E. Sjuvarsson, H.O. Sintim, S. Eriksson and E.T. Kool, *J. Am. Chem. Soc.*, **131**, 5488 (2009).
09JA6632	Y. Garcia, F. Schoenebeck, C.Y. Legault, C.A. Merlic and K.N. Houk, *J. Am. Chem. Soc.*, **131**, 6632 (2009).
09JA6961	D.M. Knapp, E.P. Gillis and M.D. Burke, *J. Am. Chem. Soc.*, **131**, 6961 (2009).
09JA8410	R. Yamaguchi, C. Ikeda, Y. Takahashi and K.-I. Fujita, *J. Am. Chem. Soc.*, **131**, 8410 (2009).
09JA9651	K.L. Hull and M.S. Sanford, *J. Am. Chem. Soc.*, **131**, 9651 (2009).
09JA10386	L. Rupnicki, A. Saxena and H.W. Lam, *J. Am. Chem. Soc.*, **131**, 10386 (2009).
09JA10974	J.M. Racowski, A.R. Dick and M.S. Sanford, *J. Am. Chem. Soc.*, **131**, 10974 (2009).
09JA11234	N.R. Deprez and M.S. Sanford, *J. Am. Chem. Soc.*, **131**, 11234 (2009).
09JA12050	N. Guimond and K. Fagnou, *J. Am. Chem. Soc.*, **131**, 12050 (2009).
09JA12070	M. Tobisu, I. Hyodo and N. Chatani, *J. Am. Chem. Soc.*, **131**, 12070 (2009).
09JA12570	Y.-F. Wang and S. Chiba, *J. Am. Chem. Soc.*, **131**, 12570 (2009).
09JA12886	S. Chiba, Y.-J. Xu and Yi.-F. Wang, *J. Am. Chem. Soc.*, **131**, 12886 (2009).
09JA13244	K.K. Larson and R. Sarpong, *J. Am. Chem. Soc.*, **131**, 13244 (2009).
09JA13888	J. Wu, X. Cui, L. Chen, G. Jiang and Y. Wu, *J. Am. Chem. Soc.*, **131**, 13888 (2009).
09JA18269	S.M. Landge and I. Aprahamian, *J. Am. Chem. Soc.*, **131**, 13888 (2009).
09JCC1061	S. Roy, S. Roy, B. Neuenswander, D. Hill and R.C. Larock, *J. Comb. Chem.*, **11**, 1061 (2009).
09JHC54	S. Tu, D. Zhou, L. Cao, C. Li and Q. Shao, *J. Heterocyclic Chem.*, **46**, 54 (2009).
09JHC69	R. Mamgain, R. Singh and D.S. Rawat, *J. Heterocyclic Chem.*, **46**, 69 (2009).
09JHC336	B.R.P. Kumar, P. Masih, C.R. Lukose, N. Abraham, D. Priya, R.M. Xavier, K. Saji and L. Adhikaryb, *J. Heterocyclic Chem.*, **46**, 336 (2009).
09JHC465	L. Rong, H. Han, H. Jiang and S. Tu, *J. Heterocyclic Chem.*, **46**, 465 (2009).
09JHC796	M.-G. Shen, C. Cai and W.-B. Yi, *J. Heterocyclic Chem.*, **46**, 796 (2009).
09JHC931	M. Nasr-Esfahani, B. Karami and M. Behzadi, *J. Heterocyclic Chem.*, **46**, 931 (2009).
09JHC1203	E. Kianmehr, H. Estiri and A. Bahreman, *J. Heterocyclic Chem.*, **46**, 1203 (2009).
09JHC1208	B. Das, B. Ravikanth, A.S. Kumar and B.S. Kanth, *J. Heterocyclic Chem.*, **46**, 1208 (2009).

09JHC1222	X.-S. Wang, Q. Li, J. Zhou and S.-J. Tua, *J. Heterocyclic Chem.*, **46**, 1222 (2009).
09JHC1229	X.-S. Wang, Q. Li, J.-R. Wu, C.-S. Yao and S.-J. Tua, *J. Heterocyclic Chem.*, **46**, 1229 (2009).
09JOC903	Z.-B. Chen, D. Hong and Y.-G. Wang, *J. Org. Chem.*, **74**, 903 (2009).
09JOC921	Q. Ding, Z. Wang and J. Wu, *J. Org. Chem.*, **74**, 921 (2009).
09JOC973	G.A. Molander, B. Canturk and L.E. Kennedy, *J. Org. Chem.*, **74**, 973 (2009).
09JOC939	X.-F. Duan, Z.-Q. Ma, F. Zhang and Z.-B. Zhang, *J. Org. Chem.*, **74**, 939 (2009).
09JOC1171	M. Vosswinkel, H. Lüerssen, D. Kvaskoff and C. Wentrup, *J. Org. Chem.*, **74**, 1171 (2009).
09JOC1341	J.W. Medley and M. Movassaghi, *J. Org. Chem.*, **74**, 1341 (2009).
09JOC1374	J.R. Vyvyan, R.C. Brown and B.P. Woods, *J. Org. Chem.*, **74**, 1374 (2009).
09JOC1939	Y. Kiyotsuka, Y. Katayama, H.P. Acharya, T. Hyodo and Y. Kobayashi, *J. Org. Chem.*, **74**, 1939 (2009).
09JOC2780	D.-W. Wang, X.-B. Wang, D.-S. Wang, S.-M. Lu, Y.-G. Zhou and Y.-X. Li, *J. Org. Chem.*, **74**, 2780 (2009).
09JOC2893	Y.-N. Niu, Z.-Y. Yan, G.-L. Gao, H.-L. Wang, X.-Z. Shu, K.-G. Ji and Y.-M. Liang, *J. Org. Chem.*, **74**, 2893 (2009).
09JOC2992	J.V. Carolan, S.J. Butler and K.A. Jolliffe, *J. Org. Chem.*, **74**, 2992 (2009).
09JOC3152	J.K. Laha, P. Petrou and G.D. Cuny, *J. Org. Chem.*, **74**, 3152 (2009).
09JOC3566	F. Shibahara, R. Sugiura, E. Yamaguchi, A. Kitagawa and T. Murai, *J. Org. Chem.*, **74**, 3566 (2009).
09JOC4166	A. Nuñez, B. Abarca, A.M. Cuadro, J. Alvarez-Builla and J.J. Vaquero, *J. Org. Chem.*, **74**, 4166 (2009).
09JOC4246	Y. Suzuki, Y. Ohta, S. Oishi, N. Fujii and H. Ohno, *J. Org. Chem.*, **74**, 4246 (2009).
09JOC4547	P.D. O'Shea, D. Gauvreau, F. Gosselin, G. Hughes, C. Nadeau, A. Roy and C.S. Shultz, *J. Org. Chem.*, **74**, 4547 (2009).
09JOC4650	L. Cheng, L. Liu, H. Jia, D. Wang and Y.-J. Chen, *J. Org. Chem.*, **74**, 4650 (2009).
09JOC4893	A.C. Bissember and M.G. Banwell, *J. Org. Chem.*, **74**, 4893 (2009).
09JOC5111	K.M. Maloney, E. Nwakpuda, J.T. Kuethe and J. Yin, *J. Org. Chem.*, **74**, 5111 (2009).
09JOC5361	B. Yao, D.-X. Wang, H.-Y. Gong, Z.-T. Huang and M.-X. Wang, *J. Org. Chem.*, **74**, 5361 (2009).
09JOC5476	H. Huang, H. Jiang, K. Chen and H. Liu, *J. Org. Chem.*, **74**, 5476 (2009).
09JOC5656	X. Ding, D. Ye, F. Liu, G. Deng, G. Liu, X. Luo, H. Jiang and H. Liu, *J. Org. Chem.*, **74**, 5656 (2009).
09JOC5664	A. Chandra, B. Singh, R.S. Khanna and R.M. Singh, *J. Org. Chem.*, **74**, 5664 (2009).
09JOC5715	V. Sridharan, P. Ribelles, M.T. Ramos and J.C. Menéndez, *J. Org. Chem.*, **74**, 5715 (2009).
09JOC5750	V.S. Aulakh and M.A. Ciufolini, *J. Org. Chem.*, **74**, 5750 (2009).
09JOC5920	Y. Takashima and Y. Kobayashi, *J. Org. Chem.*, **74**, 5920 (2009).
09JOC6010	Y. Yoshida, K. Mohri, K. Isobe, T. Itoh and K. Yamamoto, *J. Org. Chem.*, **74**, 6010 (2009).
09JOC6042	D.C. Braddock, R.H. Pouwer, J.W. Burton and P. Broadwith, *J. Org. Chem.*, **74**, 6042 (2009).
09JOC6260	K. De, J. Legros, B. Crousse and D. Bonnet-Delpon, *J. Org. Chem.*, **74**, 6260 (2009).
09JOC6299	Y. Ohta, Y. Kubota, T. Watabe, H. Chiba, S. Oishi, N. Fujii and H. Ohno, *J. Org. Chem.*, **74**, 6299 (2009).
09JOC6354	M.A.J. Duncton and M.A. Estiarte, *J. Org. Chem.*, **74**, 6354 (2009).
09JOC6728	S. Desrat and P.V. de Weghe, *J. Org. Chem.*, **74**, 6728 (2009).
09JOC6999	K. Guo, M.J. Thompson and B. Chen, *J. Org. Chem.*, **74**, 6999 (2009).
09JOC7094	N. Umeda, K. Hirano, T. Satoh and M. Miura, *J. Org. Chem.*, **74**, 7094 (2009).
09JOC7157	M. Baidya, M. Horn, H. Zipse and H. Mayr, *J. Org. Chem.*, **74**, 7157 (2009).
09JOC7489	Y. Kiyotsuka and Y. Kobayashi, *J. Org. Chem.*, **74**, 7489 (2009).
09JOC7990	B. Yang and M.J. Miller, *J. Org. Chem.*, **74**, 7990 (2009).
09JOC7998	J. Yang and G.B. Dudley, *J. Org. Chem.*, **74**, 7998 (2009).

09JOC8595	E.-X. Zhang, D.-X. Wang, Z.-T. Huang and M.-X. Wang, *J. Org. Chem.*, **74**, 8595 (2009).
09OBC85	P.R. Likhar, M.S. Subhas, S. Roy, M.L. Kantam, B. Sridhar, R.K. Seth and S. Biswas, *Org. Biomol. Chem.*, **7**, 85 (2009).
09OBC142	W. Abraham, A. Wlosnewski, K. Buck and S. Jacob, *Org. Biomol. Chem.*, **7**, 142 (2009).
09OBC325	Y. Nakaike, D. Hayashi, N. Nishiwaki, Y. Tobea and M. Arigab, *Org. Biomol. Chem.*, **7**, 325 (2009).
09OBC1410	W.-J. Hao, B. Jiang, S.-J. Tu, X.-D. Cao, S.-S. Wu, S. Yan, X.-H. Zhang, Z.-G. Hana and F. Shia, *Org. Biomol. Chem.*, **7**, 1410 (2009).
09OBC1445	M. Albrecht, O. Schneiderb and A. Schmidt, *Org. Biomol. Chem.*, **7**, 1445 (2009).
09OBC2155	K.M. Clapham, A.S. Batsanov, M.R. Brycea and B. Tarbitb, *Org. Biomol. Chem.*, **7**, 2155 (2009).
09OBC2534	H.-Y. Hu, J.-F. Xianga and C.-F. Chen, *Org. Biomol. Chem.*, **7**, 2534 (2009).
09OBC4526	X. Yu, X. Yang and J. Wu, *Org. Biomol. Chem.*, **7**, 4526 (2009).
09OL341	N. Nicolaus, S. Strauss, J.-M. Neudoörfl, Aram Prokop and H.-G. Schmalz, *Org. Lett.*, **11**, 341 (2009).
09OL345	J.Z. Deng, D.V. Paone, A.T. Ginnetti, H. Kurihara, S.D. Dreher, S.A. Weissman, S.R. Stauffer and C.S. Burgey, *Org. Lett.*, **11**, 345 (2009).
09OL381	D.X. Yang, S.L. Colletti, K. Wu, M. Song, G.Y. Li and H.C. Shen, *Org. Lett.*, **11**, 381 (2009).
09OL385	C.-J. Chuang, W.-S. Li, C.-C. Lai, Y.-H. Liu, S.-M. Peng, I. Chao and S.-H. Chiu, *Org. Lett.*, **11**, 385 (2009).
09OL421	H.-Q. Do and O. Daugulis, *Org. Lett.*, **11**, 421 (2009).
09OL437	J. Luo, L.-W. Xu, R.A.S. Hay and Y. Lu, *Org. Lett.*, **11**, 437 (2009).
09OL449	W. Lin, A. Gupta, K.H. Kim, D. Mendel and M.J. Miller, *Org. Lett.*, **11**, 449 (2009).
09OL515	K. Yoshida, F. Kawagoe, K. Hayashi, S. Horiuchi, T. Imamoto and A. Yanagisawa, *Org. Lett.*, **11**, 515 (2009).
09OL613	T.-C. Lin, C.-C. Lai and S.-H. Chiu, *Org. Lett.*, **11**, 613 (2009).
09OL685	E.L. Clennan and A.K.S. Warrier, *Org. Lett.*, **11**, 685 (2009).
09OL729	B.W.-Q. Hui and S. Chiba, *Org. Lett.*, **11**, 729 (2009).
09OL795	Z. Liu, C. Zhang, Y. Li, Z. Wu, F. Qian, X. Yang, W. He, X. Gao and Z. Guo, *Org. Lett.*, **11**, 795 (2009).
09OL1103	T. Hyodo, Y. Kiyotsuka and Y. Kobayashi, *Org. Lett.*, **11**, 1103 (2009).
09OL1171	G. Deng and C.-J. Li, *Org. Lett.*, **11**, 1171 (2009).
09OL1269	E. Ballesteros, D. Moreno, T. Gómez, T. Rodríguez, J. Rojo, M. García-Valverde and T. Torroba, *Org. Lett.*, **11**, 1269 (2009).
09OL1281	L.L.W. Cheung and A.K. Yudin, *Org. Lett.*, **11**, 1281 (2009).
09OL1317	W. Jin, Z. Yu, W. He, W. Ye and W.-J. Xiao, *Org. Lett.*, **11**, 1317 (2009).
09OL1357	M.P. Huestis and K. Fagnou, *Org. Lett.*, **11**, 1357 (2009).
09OL1579	L. Miao, S.C. DiMaggio, H. Shu and M.L. Trudell, *Org. Lett.*, **11**, 1579 (2009).
09OL1603	Y. Kobayashi and T. Harayama, *Org. Lett.*, **11**, 1603 (2009).
09OL1655	L. Xue, C. Liu and H. Jiang, *Org. Lett.*, **11**, 1655 (2009).
09OL1801	F. Beaumard, P. Dauban and R.H. Dodd, *Org. Lett.*, **11**, 1801 (2009).
09OL2205	M.A. Calter and J. Wang, *Org. Lett.*, **11**, 2205 (2009).
09OL2361	Y. Komine, A. Kamisawa and K. Tanaka, *Org. Lett.*, **11**, 2361 (2009).
09OL2679	O. Kobayashi, D. Uraguchi and T. Yamakawa, *Org. Lett.*, **11**, 2679 (2009).
09OL2711	Y. Kuninobu, Y. Fujii, T. Matsuki, Y. Nishina and K. Takai, *Org. Lett.*, **11**, 2711 (2009).
09OL3120	X. Jia, S. Zhang, W. Wang, F. Luo and J. Cheng, *Org. Lett.*, **11**, 3120 (2009).
09OL3174	W.-Y. Yu, W.N. Sit, Z. Zhou and A.S.C. Chan, *Org. Lett.*, **11**, 3174 (2009).
09OL3398	G. Barbe, G. Pelletier and A.B. Charette, *Org. Lett.*, **11**, 3398 (2009).
09OL3422	T. Mitamura, K. Iwata and A. Ogawa, *Org. Lett.*, **11**, 3422 (2009).
09OL3454	L. Xue, Q. Liu and H. Jiang, *Org. Lett.*, **11**, 3454 (2009).
09OL3502	L. Shi, Y. Chu, P. Knochel and H. Mayr, *Org. Lett.*, **11**, 3502 (2009).
09OL3562	R.R. Pal, M. Higuchi and D.G. Kurth, *Org. Lett.*, **11**, 3562 (2009).
09OL3686	B. Shi, W. Lewis, I.B. Campbell and C.J. Moody, *Org. Lett.*, **11**, 3686 (2009).
09OL3690	T. Martin, C. Laguerre, C. Hoarau and F. Marsais, *Org. Lett.*, **11**, 3690 (2009).

09OL3906	T. Shibata, T. Uchiyama and K. Endo, *Org. Lett.*, **11**, 3906 (2009).
09OL3974	Z. Ye, W. Wang, F. Luo, S. Zhang and J. Cheng, *Org. Lett.*, **11**, 3974 (2009).
09OL4010	A.E. Nibbs, A.-L. Baize, R.M. Herter and K.A. Scheidt, *Org. Lett.*, **11**, 4010 (2009).
09OL4228	A. Metzger, L. Melzig, C. Despotopoulou and P. Knochel, *Org. Lett.*, **11**, 4228 (2009).
09OL4330	Y.A. Cho, D.-S. Kim, H.R. Ahn, B. Canturk, G.A. Molander and J. Ham, *Org. Lett.*, **11**, 4330 (2009).
09OL4426	X.-Y. Chen, J. Shi, Y.-M. Li, F.-L. Wang, X. Wu, Q.-X. Guo and L. Liu, *Org. Lett.*, **11**, 4426 (2009).
09OL4584	K.J. Stowers and M.S. Sanford, *Org. Lett.*, **11**, 4584 (2009).
09OL5710	P. Lu, C. Sanchez, J. Cornella and I. Larrosa, *Org. Lett.*, **11**, 5710 (2009).
09S113	J. Esquivias, I. Alonso, R.G. Arrayás and J.C. Carretero, *Synthesis*, **14**, 113 (2009).
09S115	T. Yoshizumi, A. Ohno, T. Tsujita, H. Takahashi, O. Okamoto, I. Hayakawa and H. Kigoshi, *Synthesis*, **14**, 115 (2009).
09S713	K. Mitsui and J.R. Parquette, *Synthesis*, **14**, 713 (2009).
09S1271	J.A. Crawford, W. Fraser and C.A. Ramsden, *Synthesis*, **14**, 1271 (2009).
09S1375	E.A. Muravyova, S.V. Shishkina, V.I. Musatov, I.V. Knyazeva, O.V. Shishkin, S.M. Desenko and V.A. Chebanov, *Synthesis*, **14**, 375 (2009).
09S1400	K. Parthasarathy and C.-H. Cheng, *Synthesis*, **14**, 1400 (2009).
09S1851	V.O. Iaroshenko, Y. Wang, D.V. Sevenard and D.M. Volochnyuk, *Synthesis*, **14**, 1851 (2009).
09S1858	A.P. Mityuk, D.M. Volochnyuk, S.V. Ryabukhin, A.S. Plaskon, A. Shivanyuk and A.A. Tolmachev, *Synthesis*, **14**, 1858 (2009).
09S1865	V.O. Iaroshenko, D.V. Sevenard, D.M. Volochnyuk, Y. Wang, A. Martiloga and A.O. Tolmachev, *Synthesis*, **14**, 1865 (2009).
09S2333	W. Zhong, F. Lin, R. Chen and W. Su, *Synthesis*, **14**, 2333 (2009).
09S2393	V.O. Iaroshenko, Y. Wang, B. Zhang, D. Volochnyuk and V.Y. Sosnovskikh, *Synthesis*, **14**, 2393 (2009).
09SC1563	J.-C. Brouet, S. Gu, N.P. Peet and J.D. Williams, *Synth. Commun.*, **39**, 1563 (2009).
09SC1835	G.V. Rao, B.N. Swamy, P.H. Kumar and G.C. Reddy, *Synth. Commun.*, **39**, 1835 (2009).
09SL92	Y. Hajbi, F. Suzenet, M. Khouili, S. Lazar and G. Guillaumet, *Synlett*, **10**, 92 (2009).
09SL229	M.A. Terzidis, C.A. Tsoleridis and J. Stephanidou-Stephanatou, *Synlett*, **10**, 229 (2009).
09SL411	D. Basavaiah, B. Devendar, D.V. Lenin and T. Satyanarayana, *Synlett*, **10**, 411 (2009).
09SL628	S.K. Guchhait and C. Madaan, *Synlett*, **10**, 628 (2009).
09SL1081	C.-G. Dong, T.-P. Liu and Q.-S. Hu, *Synlett*, **10**, 1081 (2009).
09SL1091	T. Iwai, T. Nakai, M. Mihara, T. Ito, T. Mizuno and T. Ohno, *Synlett*, **10**, 1091 (2009).
09SL1609	M. Whiting, M.C. Wilkinson and K. Harwood, *Synlett*, **10**, 1609 (2009).
09SL1639	S. Nakamura, Y. Sakurai, H. Nakashima, N. Shibata and T. Toru, *Synlett*, **10**, 1639 (2009).
09SL1795	E. Georgescu, M.R. Caira, F. Georgescu, B. Draghici, M.M. Popa and F. Dumitrascu, *Synlett*, **10**, 1795 (2009).
09SL2852	L. Ackermann and H.K. Potukuchi, *Synlett*, **10**, 2852 (2009).
09T748	C.-Y. Chang, H.-M. Liu and R.-T. Hsu, *Tetrahedron*, **65**, 748 (2009).
09T757	R. Morgentin, F. Jung, M. Lamorlette, M. Maudet, M. Ménard, P. Plé, G. Pasquet and F. Renaud, *Tetrahedron*, **65**, 757 (2009).
09T975	M. Altuna-Urquijo, A. Gehre, S.P. Stanforth and B. Tarbit, *Tetrahedron*, **65**, 975 (2009).
09T1180	J.L. Bolliger and C.M. Frech, *Tetrahedron*, **65**, 1180 (2009).
09T1679	C.-K. Liang, P.-S. Wang and M.-K. Leung, *Tetrahedron*, **65**, 1679 (2009).
09T1951	M. Romero, Y. Harrak, J. Basset, J.A. Oru' e and M.D. Pujol, *Tetrahedron*, **65**, 1951 (2009).
09T2329	S. Yamada, A. Toshimitsu and Y. Takahashi, *Tetrahedron*, **65**, 2329 (2009).

09T3085	X. Chen, G. Dobereiner, X.-S. Hao, R. Giri, N. Maugel and J.-Q. Yu, *Tetrahedron*, **65**, 3085 (2009).
09T3155	D.J. Schipper, L.-C. Campeau and K. Fagnou, *Tetrahedron*, **65**, 3155 (2009).
09T3239	N. Satoh, T. Akiba, S. Yokoshima and T. Fukuyama, *Tetrahedron*, **65**, 3239 (2009).
09T3580	M.F.A. Adamo, P. Disetti and L. Piras, *Tetrahedron*, **65**, 3580 (2009).
09T3668	V. Stockmann, J.M. Bakke, P. Bruheim and A. Fiksdahl, *Tetrahedron*, **65**, 3668 (2009).
09T4422	W.A.E. Omar and O.E.O. Hormi, *Tetrahedron*, **65**, 4422 (2009).
09T5284	L. Robertson and R.C. Hartley, *Tetrahedron*, **65**, 5284 (2009).
09T5413	G. Burzicki, A.S. Voisin-Chiret, J.S.O. Santos and S. Rault, *Tetrahedron*, **65**, 5413 (2009).
09T5834	M. Anwar and V. Lee, *Tetrahedron*, **65**, 5834 (2009).
09T6584	Y. Zhang, B.M. Loertscher and S.L. Castle, *Tetrahedron*, **65**, 6584 (2009).
09T8412	G. Szántó, L. Hegedüs, L. Mattyasovszky, A. Simon, Á. Simon, I. Bitter, G. Tóth, L. Töke and I. Kádas, *Tetrahedron*, **65**, 8412 (2009).
09T8950	M.M. Bio, E. Cleator, A.J. Davies, S.E. Hamilton, A. Lawrence, F.J. Sheen, G.W. Stewart and R.D. Wilson, *Tetrahedron*, **65**, 8950 (2009).
09TL198	Q. Ding, Z. Wang and J. Wu, *Tetrahedron Lett.*, **65**, 198 (2009).
09TL340	Q. Ding, Z. Chen, X. Yu, Y. Peng and J. Wu, *Tetrahedron Lett.*, **65**, 340 (2009).
09TL949	P.M. Reis and B. Royo, *Tetrahedron Lett.*, **65**, 949 (2009).
09TL1444	A.M. Prokhorov, M. Makosza and O.N. Chupakhin, *Tetrahedron Lett.*, **65**, 1444 (2009).
09TL1478	M. Li and R. Hua, *Tetrahedron Lett.*, **65**, 1478 (2009).
09TL1768	N. Robert, T. Martin, J. Grisel, J. Lazaar, C. Hoarau and F. Marsais, *Tetrahedron Lett.*, **65**, 1768 (2009).
09TL1780	R.P. Kale, M.U. Shaikh a, G.R. Jadhav and C.H. Gill, *Tetrahedron Lett.*, **65**, 1780 (2009).
09TL1928	G.M. Schaaf, S. Mukherjee and A.G. Waterson, *Tetrahedron Lett.*, **65**, 1928 (2009).
09TL1986	A.T. Londregan, G. Storer, C. Wooten, X. Yang and J. Warmus, *Tetrahedron Lett.*, **65**, 1986 (2009).
09TL2305	S. Hwang, Y. Lee, P.H. Lee and S. Shin, *Tetrahedron Lett.*, **65**, 2305 (2009).
09TL2481	C.E. Quevedo, V. Bavetsias and E. McDonald, *Tetrahedron Lett.*, **65**, 2481 (2009).
09TL3547	T. Hyodo, Y. Katayama and Y. Kobayashi, *Tetrahedron Lett.*, **65**, 3547 (2009).
09TL3798	C. Salomé, M. Schmitt and J.-J. Bourguignon, *Tetrahedron Lett.*, **65**, 3798 (2009).
09TL3897	M. Sridhar, B.C. Ramanaiah, C. Narsaiah, B. Mahesh, M. Kumaraswamy, K.K.R. Mallu, V.M. Ankathi and P.S. Rao, *Tetrahedron Lett.*, **65**, 3897 (2009).
09TL4217	P. Saisaha, C. Nerungsi, S. Iamsaard and T. Thongpanchang, *Tetrahedron Lett.*, **65**, 4217 (2009).
09TL4229	M. Ravinder, P.S. Sadhu and V.J. Rao, *Tetrahedron Lett.*, **65**, 4229 (2009).
09TL4389	A.I. Polyakov, V.A. Eryomina, L.A. Medvedeva, N.I. Tihonova, A.V. Listratova and L.G. Voskressensky, *Tetrahedron Lett.*, **65**, 4389 (2009).
09TL4916	J.M. Crawforth and M. Paoletti, *Tetrahedron Lett.*, **65**, 4916 (2009).
09TL5018	S. Huang, J.C.S. Wong, A.K.C. Leung, Y.M. Chan, L. Wong, M.R. Fernendez, A.K. Miller and W. Wu, *Tetrahedron Lett.*, **65**, 5018 (2009).
09TL5329	S.-H. Kim and R.D. Rieke, *Tetrahedron Lett.*, **65**, 5329 (2009).
09TL6772	R.N. Burgin, S. Jones and B. Tarbit, *Tetrahedron Lett.*, **65**, 6772 (2009).
09TL6985	S.-H. Kim and R.D. Rieke, *Tetrahedron Lett.*, **65**, 6985 (2009).
10CR575	Various authors, *Chem. Rev.*, **110**, 575 (2010).

CHAPTER 6.2

Six-Membered Ring Systems: Diazines and Benzo Derivatives

Michael M. Miller, Albert J. DelMonte
Bristol-Myers Squibb Company, Princeton, NJ, USA
michael.miller@bms.com

6.2.1. INTRODUCTION

This review provides a glimpse into the numerous publications involving diazine chemistry reported in 2009 with the idea that a reader may be intrigued by a brief description, example transformation, or structure, and then decide to read the primary article. Due to the large volume of disclosures published throughout the year and the space limitations of this chapter, the authors regret that not all contributions to the field can be highlighted herein.

Diazines are an important class of compounds and can be found in many areas of chemistry. There have been numerous reports on the synthesis, reactions, and applications of diazines within 2009. In addition, several reviews covering this chemical motif and their use within the industry have appeared in the literature <09JHC1420; 09OBC2841; 09AHC1; 09OPP479; 09JJC1; 09CRV2275; 09CRV2880>. Throughout this review, we will be exploring recent advances in diazine chemistry and their applications to the chemical enterprise. While there is a wealth of rich applications that can be found in the patent literature; unfortunately, these examples are considered outside the scope of this document. For the purposes of this summary, the three diazine systems (i.e., pyridazine, pyrimidines, and pyrazine) and their benzo analogs (i.e., phthalazine, cinnoline, quinazoline, pteridine, quinoxaline, and phenazine) will be described.

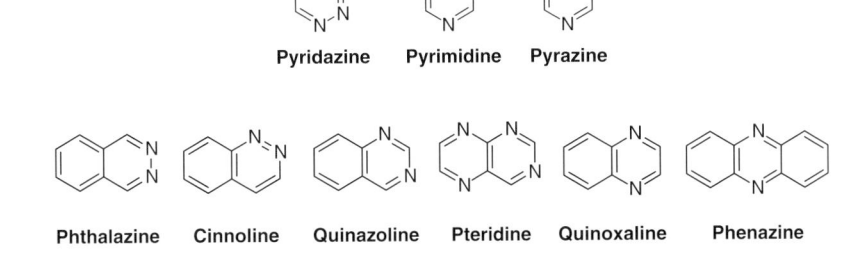

Pyridazine   Pyrimidine   Pyrazine

Phthalazine   Cinnoline   Quinazoline   Pteridine   Quinoxaline   Phenazine

6.2.2. PYRIDAZINES AND BENZO DERIVATIVES
6.2.2.1 Syntheses

Reactions with hydrazines were found to be useful in the construction of pyridazines and their corresponding adducts. It was reported that the *in situ* ruthenium-catalyzed isomerization of 1,4-alkyne diols to their respective 1,4-diketones in the presence of hydrazine can afford substituted pyridazines (**1**) <09T8981>. Treatment of (*Z*)-4-bromo-1,3-diphenylbut-2-en-1-one (**2**) with arylhydrazine **3** afforded pyridazine derivatives, as well as pyrroles depending on the substituents and reaction conditions <09CHE815>. The conversion of anthenes to tetracyclic cinnolinocinnolines via a domino reaction was reported and a mechanism was proposed <09OBC1171>. The reaction of hydroxy naphthyl derivative **4** with hydrazine provides tetraazapyrene (**5**) <09RJOC848>. Phthalazine **7** was produced from the corresponding benzalphthalide **6** <09T1574>. Interestingly, under microwave conditions, the addition of only one hydrazine was observed.

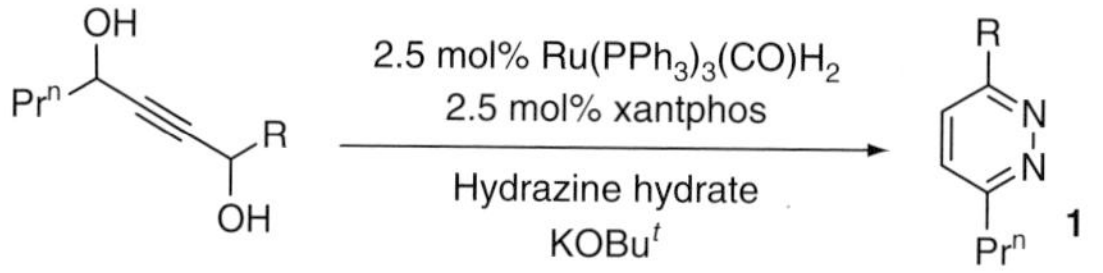

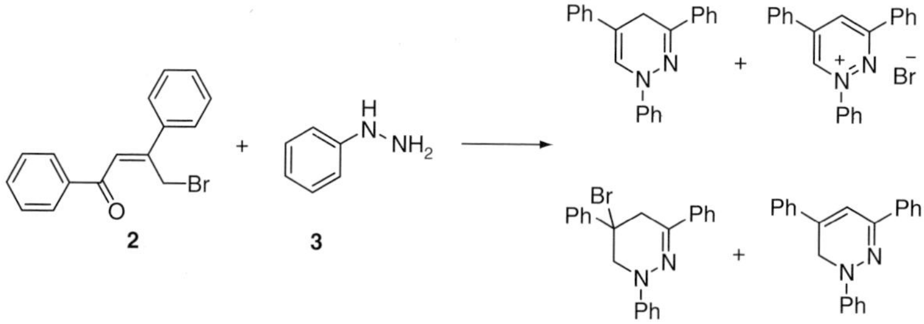

R	Yield (%)
Bn	76
Styrl	73
n-Butyl	79
i-Butyl	68

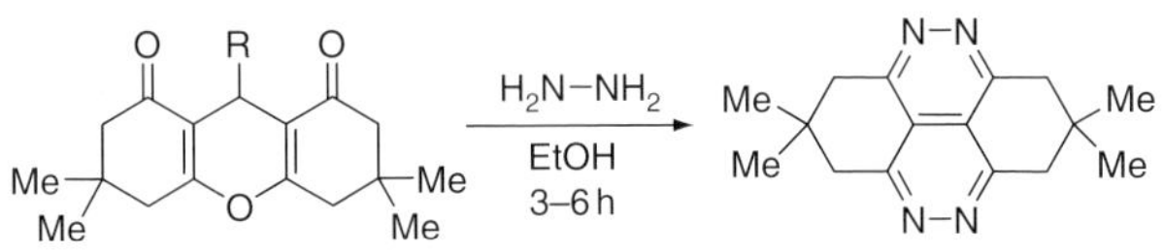

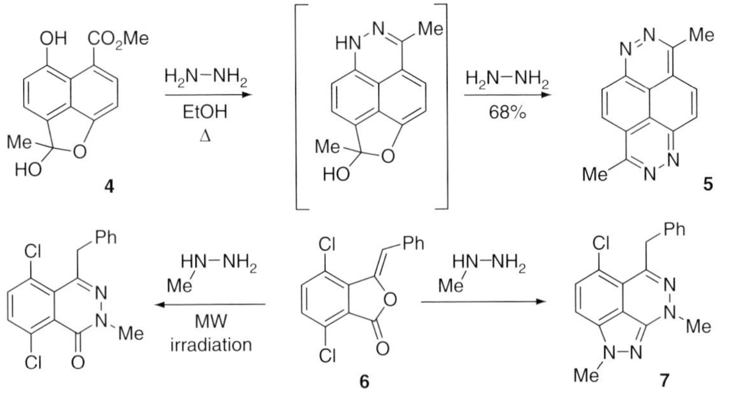

A series of pyridazines and quinazolines was synthesized by the reactions of enaminones with aromatic amines <09SC4088>.

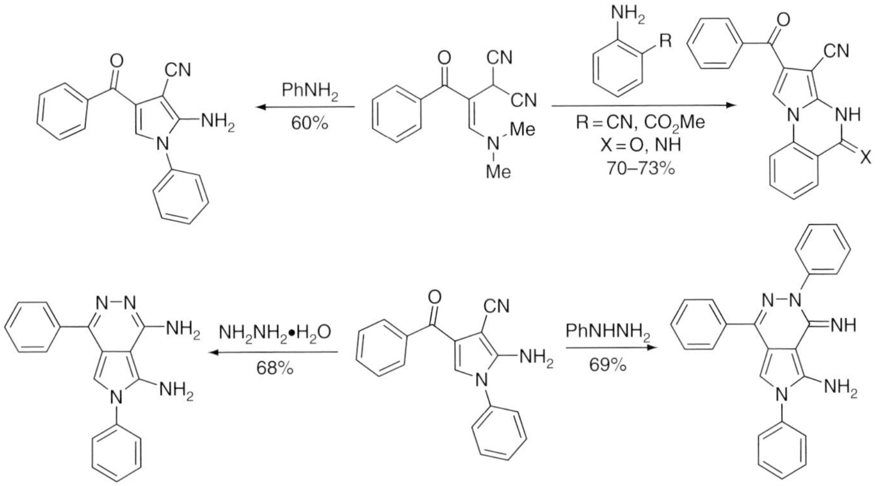

A variety of arylhydrazones were converted to their pyridazines and fused pyridazine analogs via an amino polysaccharide (chitosan) catalyzed reaction with α,β-unsaturated nitriles (**8**) <09ARK302>. When similar chemistry was attempted with 2-nitro substitution, a new product was afforded which, after $ZnCl_2$-mediated cyclization, produced phthalazine **9** <09TL6411>. Interestingly, spiro pyridazine derivative **10** was produced via a synthesis which utilized a chitosan-catalyzed reaction to condense malononitrile with the corresponding isatin as the first step in the synthesis <09SL625>. Subsequently, a DBU-catalyzed approach to a similar structural motif was reported <09T10069>.

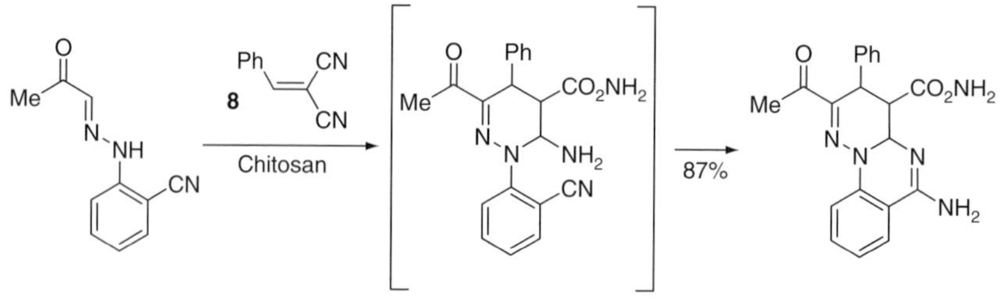

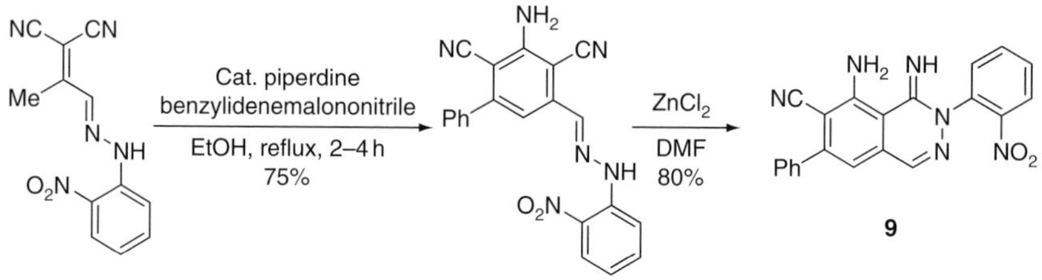

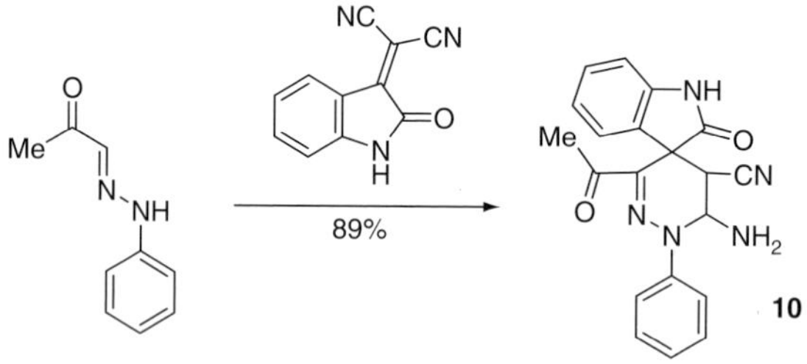

Several papers described the construction of pyridazines via the application of a Diels–Alder process. Pyridazine **13** was synthesized via a [4+2] cycloaddition between ketenimine **11** with PTAD (**12**, 4-phenyl-1,2,4-triazoline-3,5-dione) <09JOC3558>. A microwave facilitated hetero Diels–Alder reaction was utilized to produce pyridazine **14** <09S1876>. DFT calculations suggest a concerted but highly asynchronous transition state.

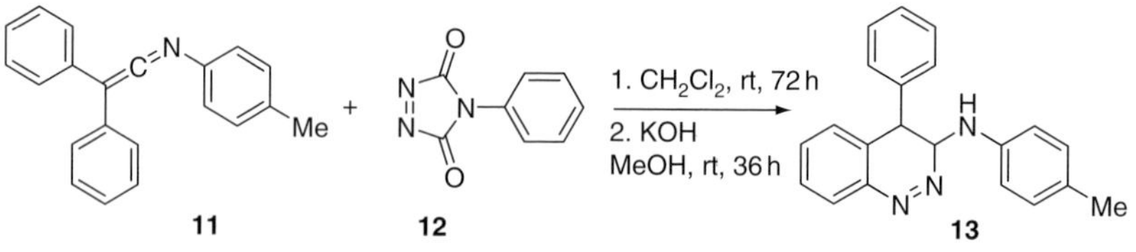

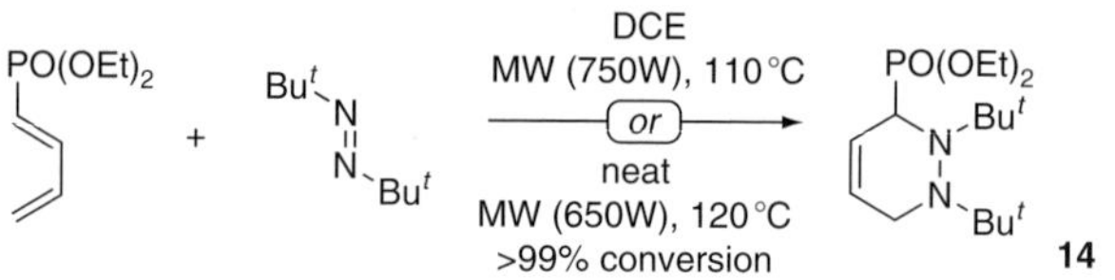

Alternate cycloaddition chemistry was the theme of a number of pyridazine-based manuscripts in 2009. A silver carbonate mediated 1,3-dipolar cycloaddition of nitrile imines with fulvenes was utilized to synthesize a series of cyclopentapyridazines <09TL6698>. A series of pyridazine derivatives was synthesized via a gold(III) catalyzed [3+3] cycloaddition <09JA11654>.

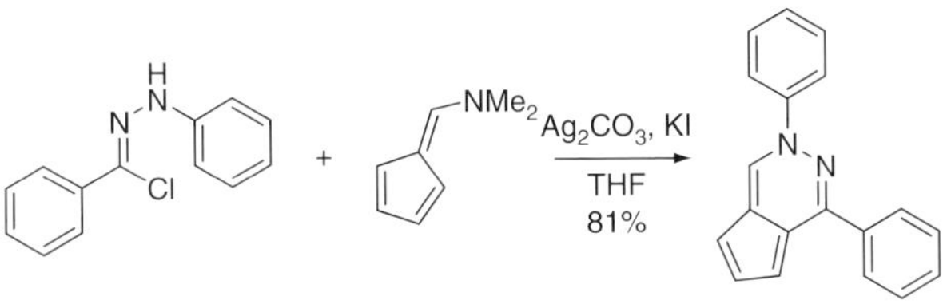

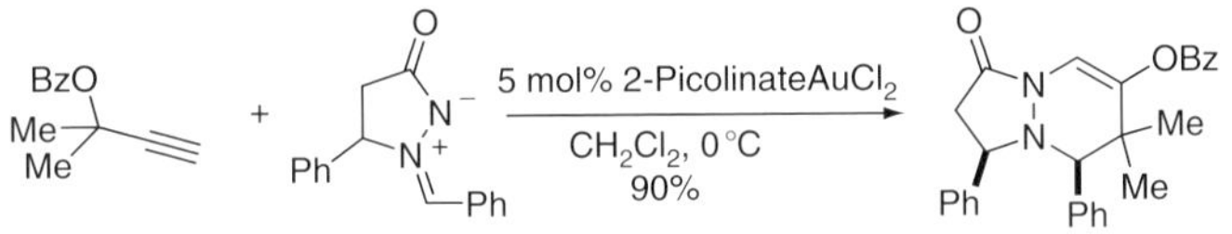

The synthesis of pyrazolocinnolines was reported via diazotization of 2-methyl-4-(3,4-dimethyoxyphenyl)-1H-5-aminopyrazole and subsequent intramolecular azo coupling <09RJOC211>.

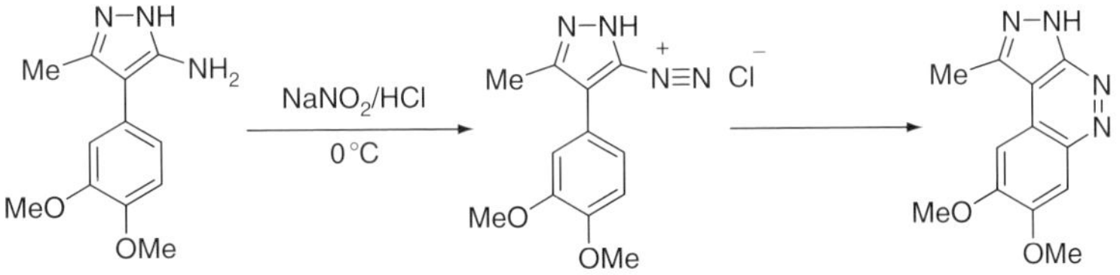

The synthesis of alkynyl-substituted cinnolines, which can be further functionalized, was achieved via HCl or HBr mediated Richter-type cyclization of the corresponding triazines <09TL6358>.

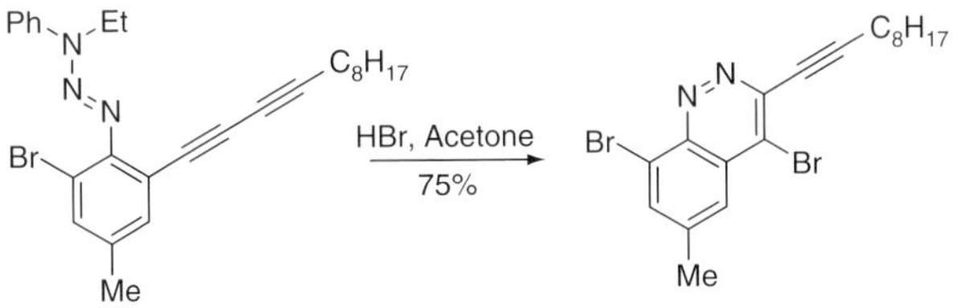

A series of imido-phthalazines was synthesized via direct arylation of N-aminoimidazoles with bromo-benzaldehydes <09S1715>.

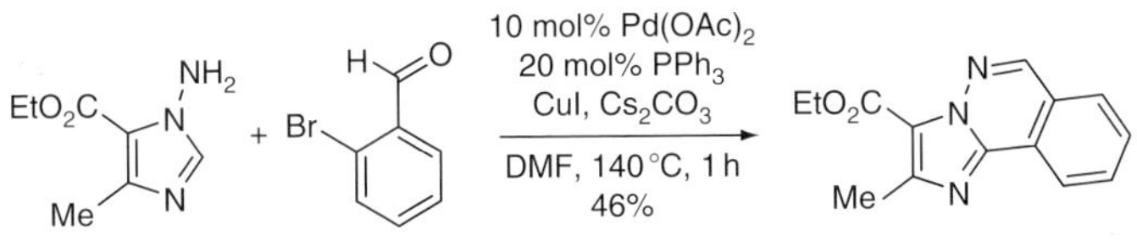

In 2009, there were two reports in which pyridazine derivatives were generated via ring-closing metathesis and subsequent elimination of a sulfinate <09CC3008; 09T8969>.

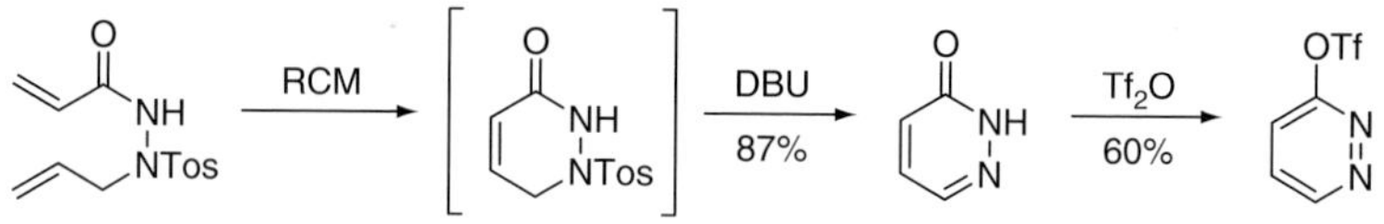

6.2.2.2 Reactions

The reactions of pyridazinium ylides were investigated under non-classical conditions including microwave conditions in the liquid phase and under phase-transfer conditions. A variety of new pyridazine derivatives were produced, characterized, and investigated for antibacterial and antifungal activity <09BMC2823>.

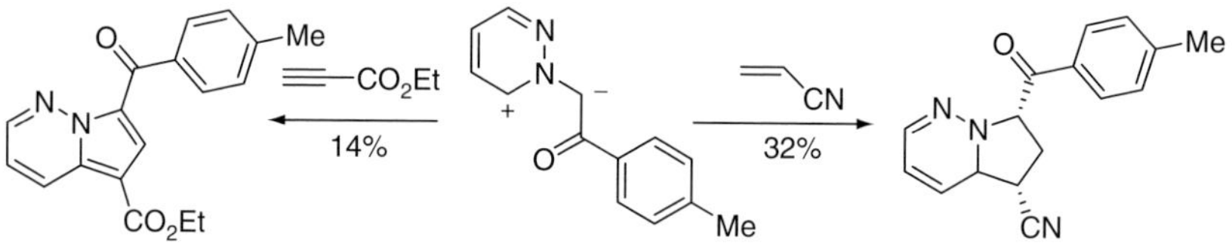

A novel method for the preparation of spirodiazine derivatives in moderate yields (56-63%) was disclosed <09TL7205>.

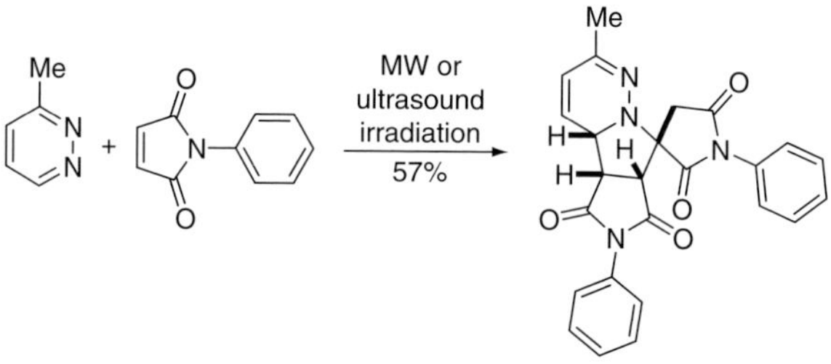

A pyridazine containing phthalocyanine compound **15** was synthesized and treated with C_{60} to afford a unique C_{60} derivative **16**. Characterization, photophysical, and electrochemical data were generated and DFT calculations were performed <09CAJ1678>.

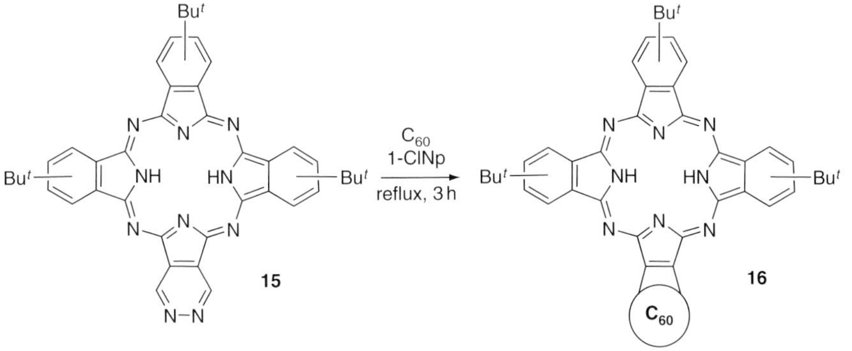

A series of pyridazine derivatives was synthesized for eventual evaluation as cardiotonic agents <09PCJ87>.

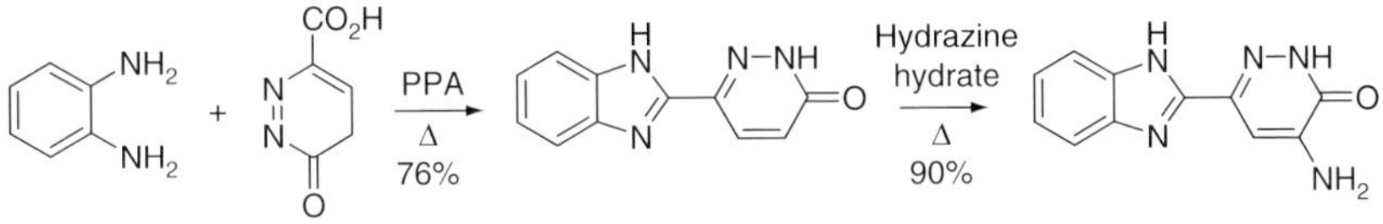

6.2.2.3 Applications

Several papers in 2009 explored various physical organic studies of pyridazines and their benzo analogs. A variety of symmetric quadrupolar and asymmetric dipolar bipyridazines were synthesized and their solvatochromic fluorescence was investigated <09L9405>. An investigation of the isomerization and cyclization of azobenzenes to benzocinnolines was reported in an attempt to consolidate previously published data, as well as newly generated computational information <09IECR10120>. A series of compounds (**17**), including a pyridazine group and its N-oxide derivative, was evaluated as macroscopic compasses <09JOC8554>. Polyimides incorporating pyridazines in their main chains were appraised for stability and their optical properties <09JPS4886>. A helical foldamer (**18**) was synthesized and its local and molecular conformational preferences were investigated. <09CEJ10030>.

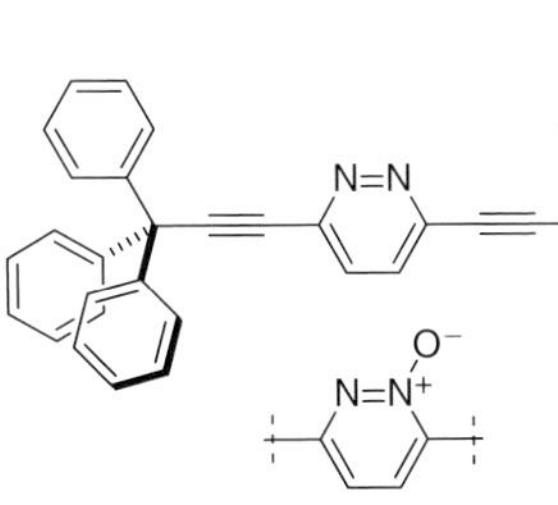

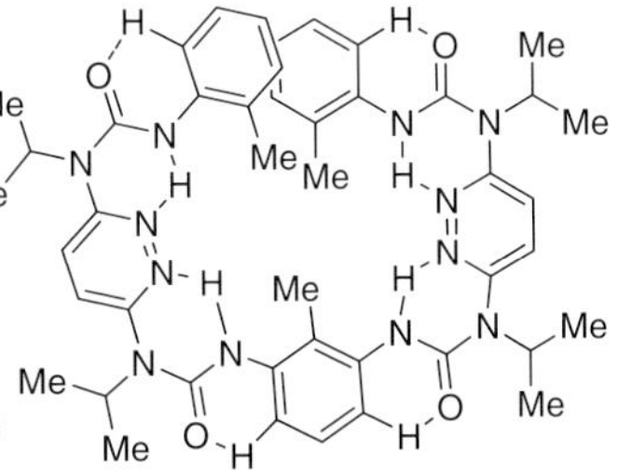

Pyridazine scaffolds were incorporated into ligands of metal complexes. The synthesis, characterization, and ability of pyridazine derivative **19** to form a dinuclear palladium complex was reported <09OM3256>. In another paper, a pyridazine-based ligand complexed with two cobalt atoms (**20**) was synthesized and evaluated <09CC6729>. These systems can reversibly support five oxidation states and can be used as an active proton reduction electrocatalyst. Lastly, the structural, thermal, spectroscopic and magnetic properties of a trinuclear nickel(II) complex, [Ni_3 $(NCS)_6$(pyridazine)$_6$], were reported <09ZAAC2459>.

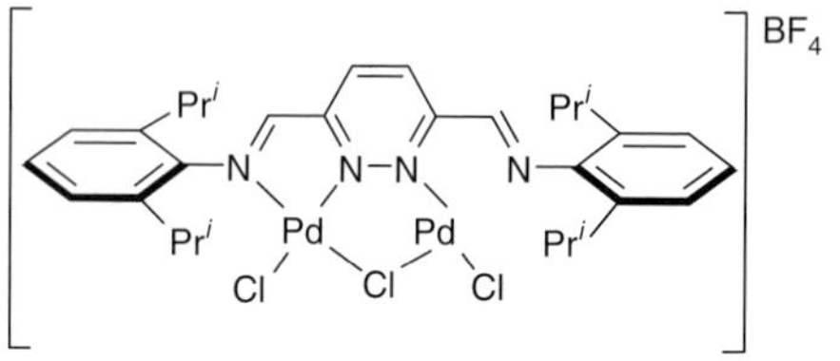

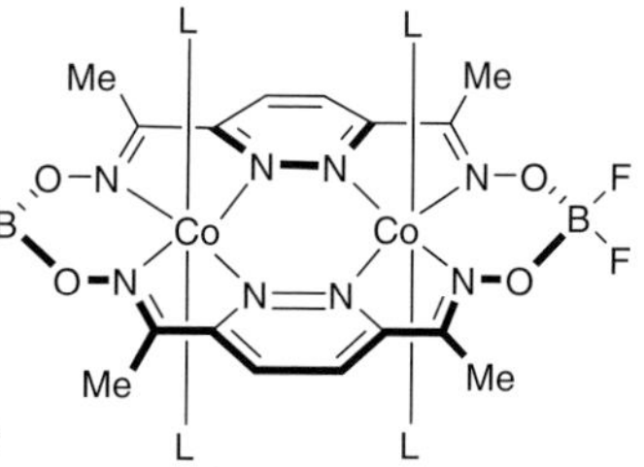

A large number of biologically active compounds have been described that contained pyridazine-based skeletons. Pyrazolopyridazine **21** was synthesized and evaluated for herbicidal activity <09JHC584>, while new phthalhydrathiazole derivatives were found to have antiproliferative activity against hepatocellular carcinoma <09JHC674>. Bicyclic triazolo[4,3-*b*]pyridazine **22** was synthesized and evaluated as a PIM-1 kinase inhibitor for the treatment of cancer <09BMCL3019>. Diazine derivatives of minozac and minaprine were evaluated by CYP2D6 enzyme kinetic analysis to determine the effect of SAR on substrate status <09DMD2204>. Racemic and enantiomerically enriched dihydrophthalazine derivatives (**23a-c**) <09AMAC3620>, pyrrolopyridazine adducts <09BMC2823>, as well as some new heterocycles based on phthalazine <09EJM4448> were investigated for antibacterial activity. While a series of cinnolines was prepared and evaluated as agonists of liver X receptors (LXRβ) <09BMC3519>, a process development and kilogram scale synthesis of a phthalazine based p38 MAP Kinase inhibitor **24** was described <09OPRD230; 09JOC795>.

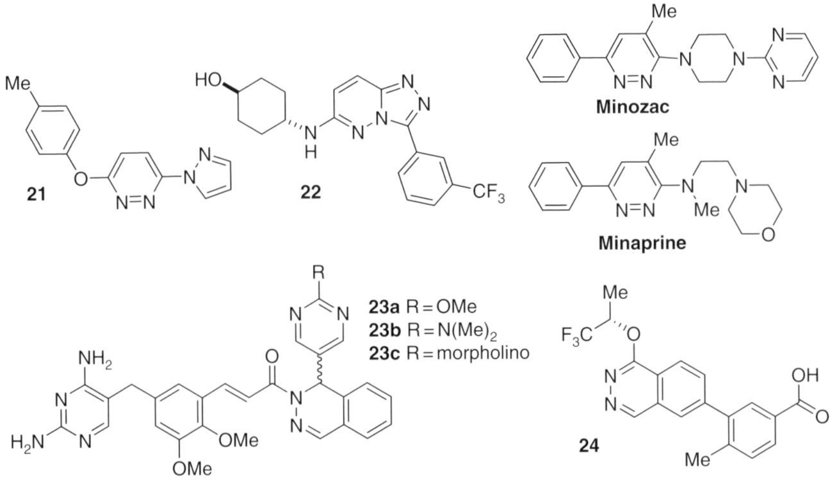

6.2.3. PYRIMIDINES AND BENZO DERIVATIVES

6.2.3.1 Syntheses

A versatile synthesis of substituted pyrimidines (**25**) and quinazolines was described in which R^1, R^2, and R^3 were alkyl or aryl groups. Other examples where R^1 included $N(PMB)_2$, $N(Me)OMe$, or OPh and R^4 included SMe, NPhth, OPh, Br, N$(CH_2CH_2)_2O$, and ^{c}Hx were also reported <09JOC8460>.

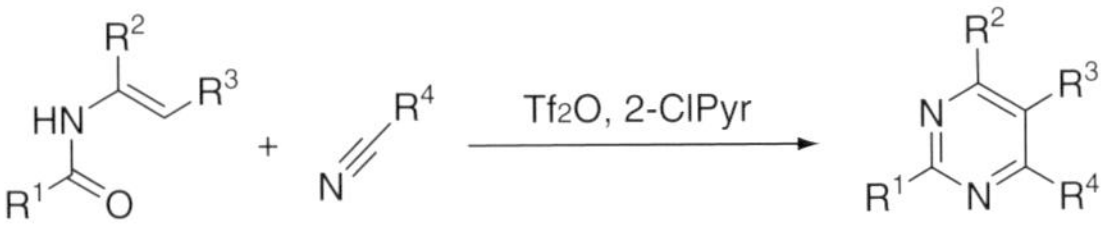

Pyrimidine **27** was prepared from alkyne **26** upon treatment with guanidinium hydrochloride <09CEJ5006>. Compound **26** and related heterocycles had been generated via a decarbonylative Sonogashira reaction on indoles, azaindoles, and pyrroles. In addition, a similar method was employed to prepare optically active fluorinated pyrimidines from (*S*)-butynol <09JOC4646>. Pyrimidine analogs were also synthesized via a solvent-free microwave-assisted cyclocondensation reaction of *o*-aminocarbonyl substrates with nitriles <09JHC178>.

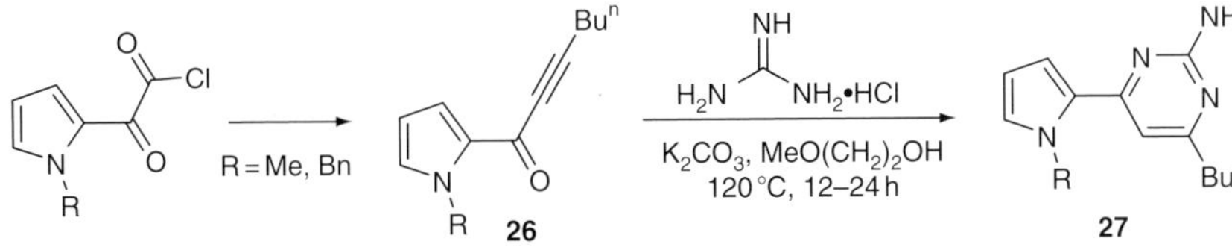

Multi-component preparation of pyrimidines remained an active area of research in 2009. In one study, a three-component ZnCl$_2$-catalyzed reaction for the construction of pyrimidines **28** was reported <09OL2161>. A multi-component synthesis of pyrrolo-pyrimidine derivatives was reported <09JHC708>. The absorption and fluorescence properties of the resulting compounds (**29**, **30**) were studied. Solvent-free three-component one-pot reactions were employed to furnish a series of pyrimidines and quinazolines derivatives such as **31** and **32** <09JHC152>. In an effort to identify a replacement for volatile and toxic solvents, polyethylene glycol (PEG) was used as a recyclable reaction medium for the generation of substituted benzo[4,5]imidazo[1,2-*a*]pyrimidine **33** via a microwave-assisted multi-component sequence <09JHC664>.

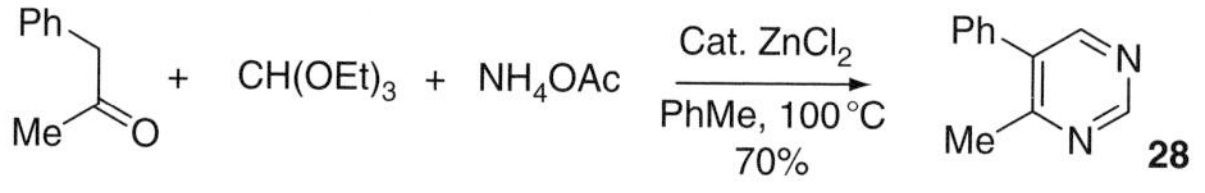

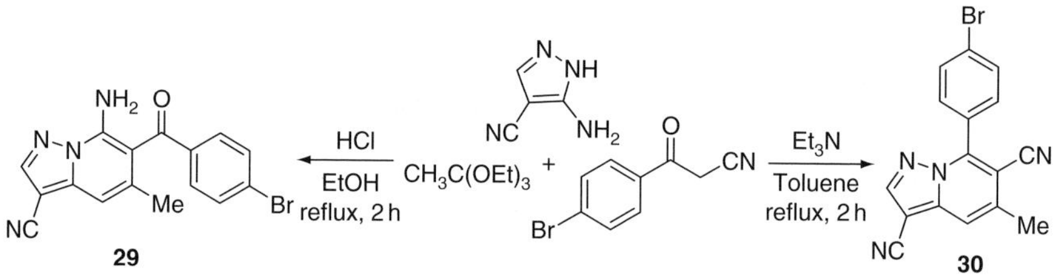

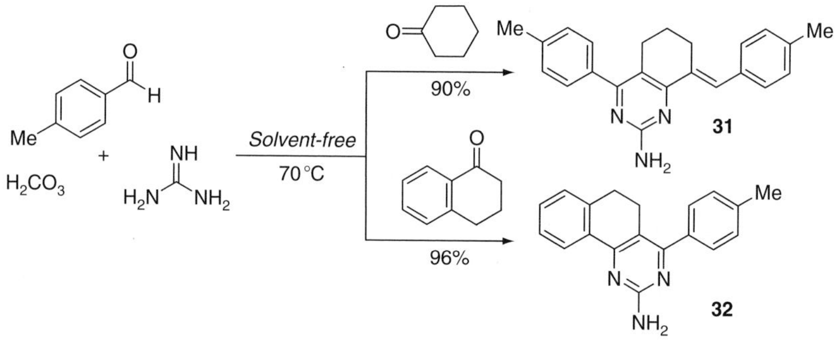

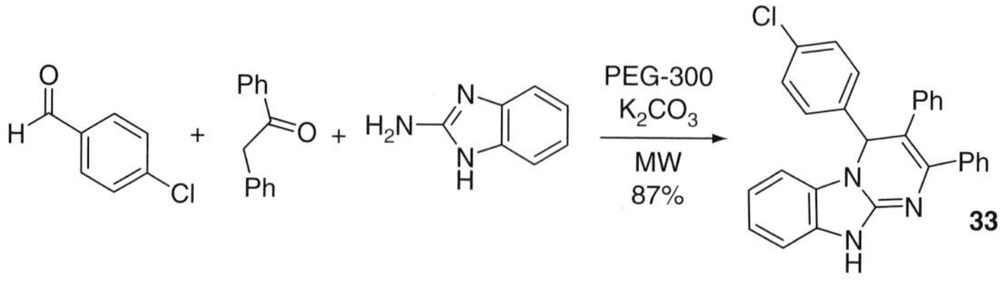

Several papers described the synthesis of compounds with spiro-motifs. A KAl(SO$_4$)$_2$•12H$_2$O catalyzed three-component approach was utilized to synthesize spiro-isoindoline-quinazoline derivative **34** <09T3804>, and a reaction between isatin **35**, barbituric acid **36**, and pyrazole **37** was utilized to build a library of spiro [indoline-pyrazolo[4′,3′:5,6]pyrido[2,3-*d*][pyrimidine]triones <09JC0393>.

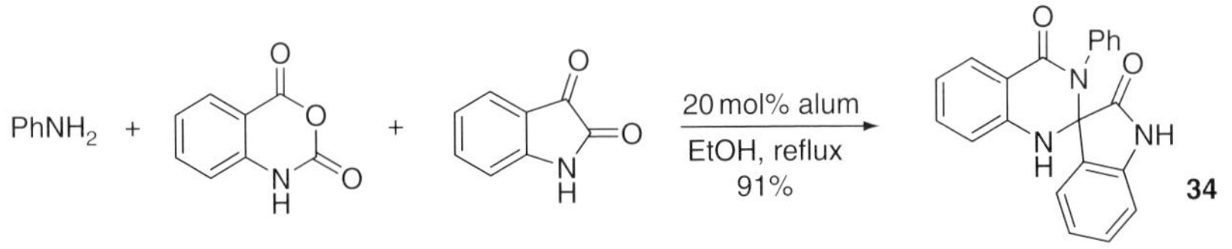

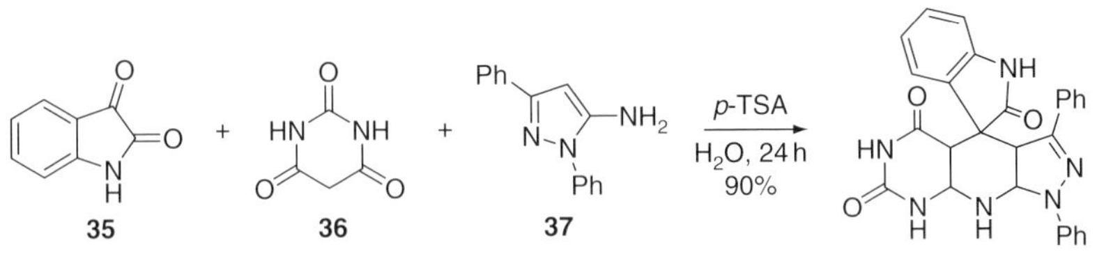

A series of quinazoline derivatives was synthesized via a four-component domino reaction under microwave conditions in which four stereocenters, including one quaternary carbon, are generated <09JA11660>.

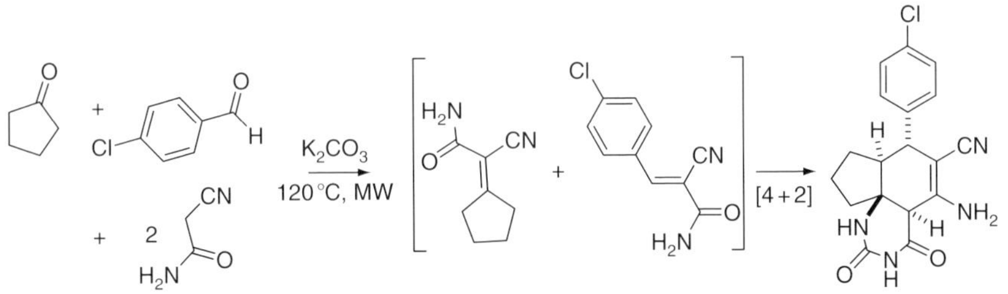

An inverse electron-demand Diels–Alder reaction of electron-deficient 1,3,5-triazine **38** and 2-aminoindole **39** was found to effectively provide the desired pyrimido[4,5-*b*]indole **40** in good yield <09SL3206; 09S3967>. Interestingly, due to the reactivity of **39**, these reactions proceeded nicely at much lower temperatures than the typical inverse electron-demand Diels–Alder processes reported to date.

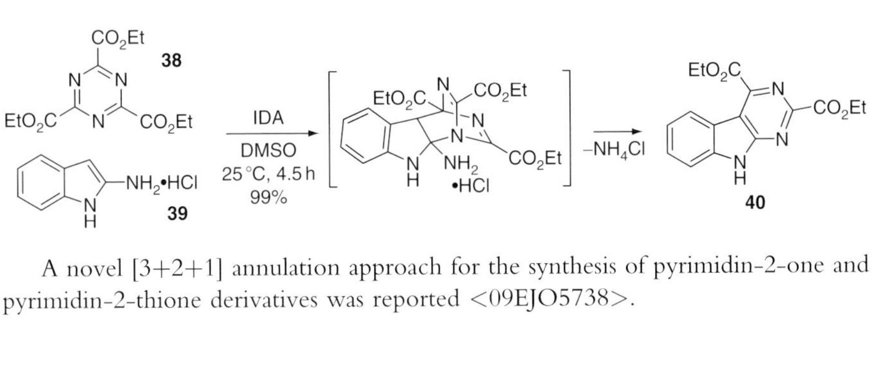

A novel [3+2+1] annulation approach for the synthesis of pyrimidin-2-one and pyrimidin-2-thione derivatives was reported <09EJO5738>.

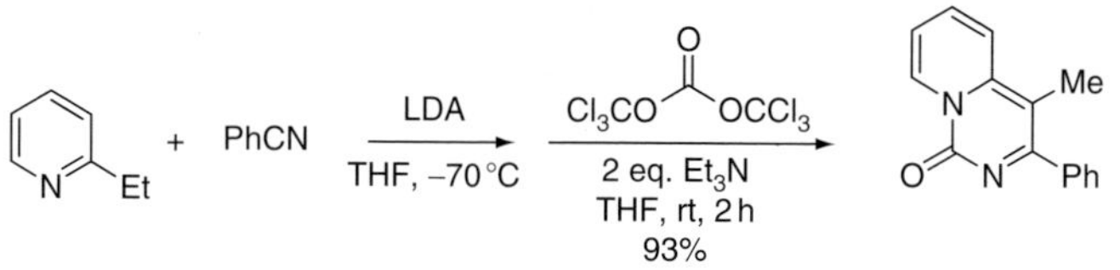

A variety of dihydroazolopyrimidines were synthesized from 3-amino-1,2,4-triazole with arylidene-5-acetyl barbituric acids <09JHC285>.

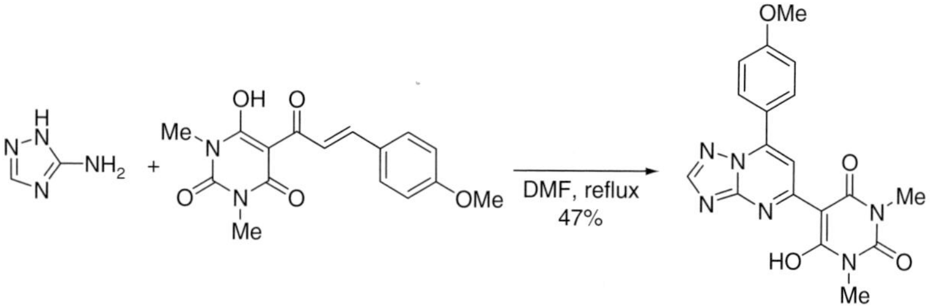

The preparation of quinazoline derivatives was an area of focus and numerous approaches were employed to access these systems. Keggin-type heteropolyacids catalyze reactions between 2-aminobenzamide **41**, aniline **42**, and orthoester **43** to afford 4-arylaminoquinazoline **44** <09TL943>. Structurally constrained quinazoline derivative **45** was synthesized from *N*-substituted 2-aminobenzamide <09T8582>. A series of quinazoline derivatives was synthesized via a free-radical reaction of *O*-phenyl oximes with aldehydes. Dihydroquinazolines were produced, and when $ZnCl_2$ was utilized during the reaction, fully aromatic quinazolines (**46**) were obtained <09JOC4934>. A one-step synthesis of quinazoline derivatives under microwave conditions was reported to yield motifs such as **47** and **48** <09TL6048>. Similar pyrimidine-fused heterocycles were furnished under thermal conditions using Vilsmeier–Haack formylation product **49** and pyrazole **50** <09JHC327>.

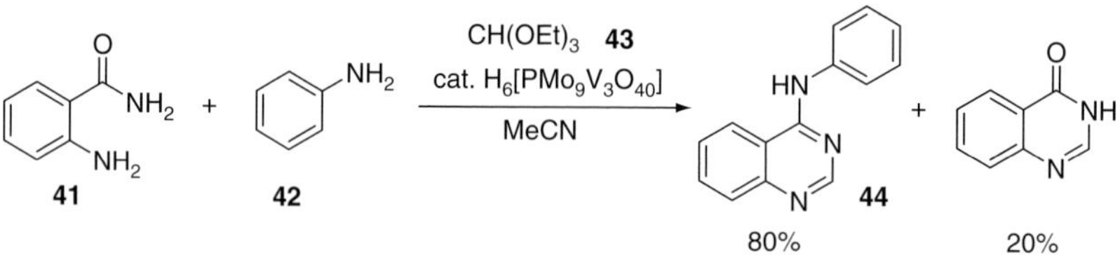

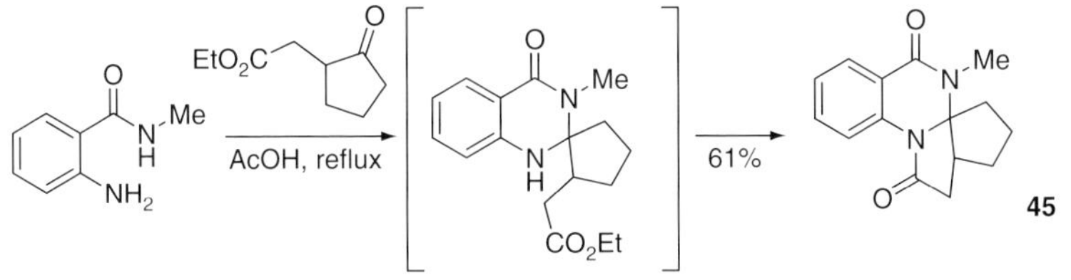

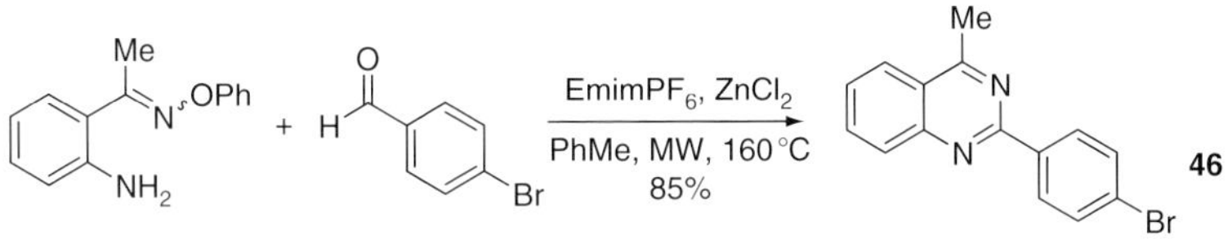

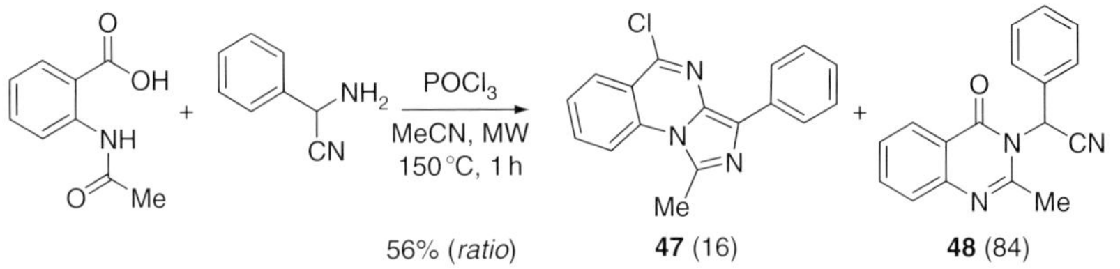

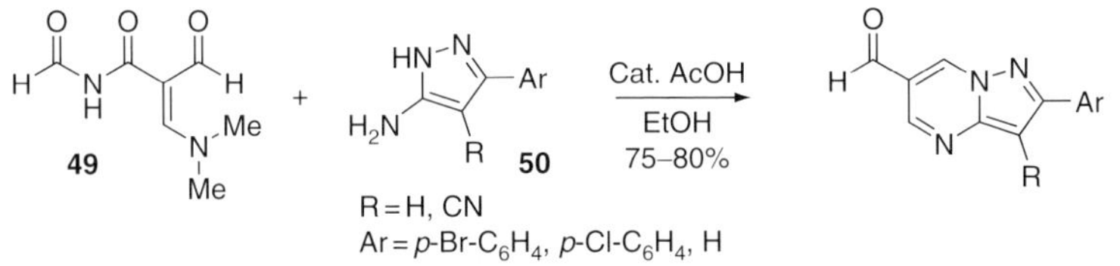

Several monofluorinated quinazoline derivatives (**51** and **55**) were synthesized via cyclocondensations of benzoyl chloride **53** with 2-aminopyridine **52** or S-alkylisothiourea **54** <09RJOC904>.

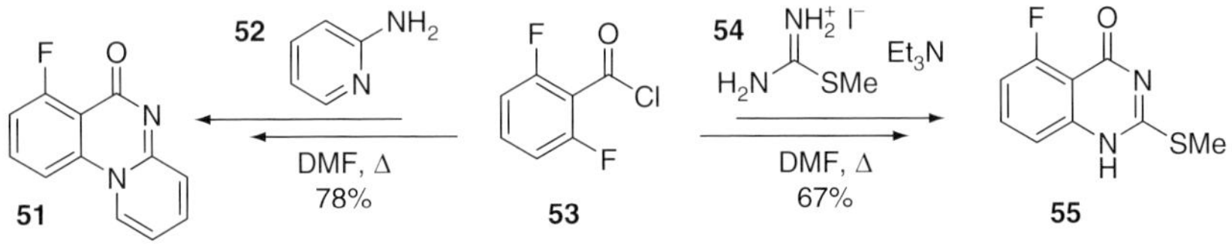

The synthesis of quinazolinediones and quinazolinones was reported in multiple publications during 2009. Quinazolinedione derivative **56** was synthesized from the MgO/ZrO$_2$-catalyzed reaction of CO$_2$ and 4,5-dimethoxy-2-aminobenzonitrile <09CAL201>. Compound **59** was prepared from 2-aminomethylbenzoate **57** and isocyanate **58** <09RJOC1691>. *ortho*-Lithiophenyl isocyanide was generated from **60** and treated with isocyanate **61** to afford 3*H*-quinazolin-4-one **62** <09OL389>.

56

57 58 59

60 61 62

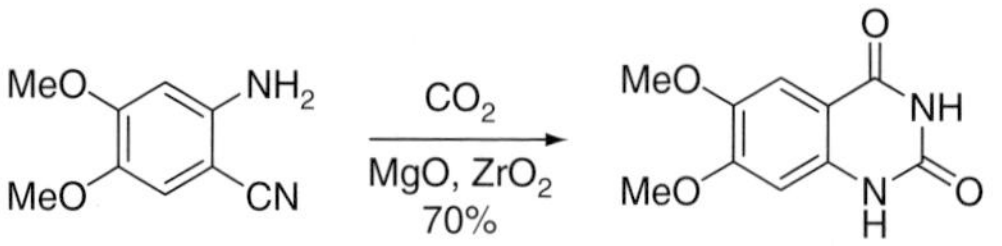

The Aza–Wittig reaction was leveraged in the synthesis of quinazolinone **63** and oxazolo[3,2-*c*]quinazoline **64** <09EJO2490; 09SL611>.

63

64

X = S, NPh

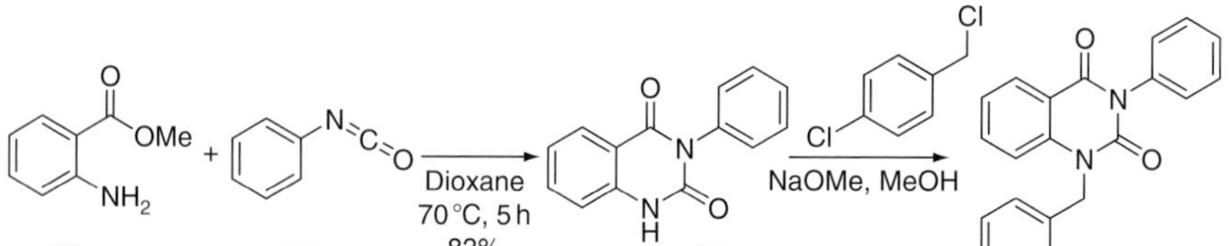

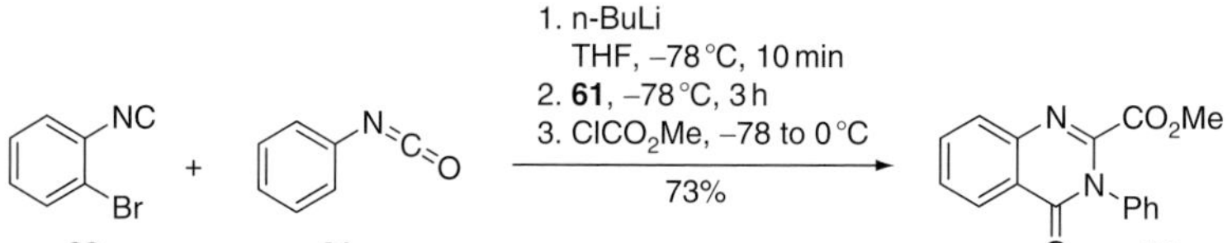

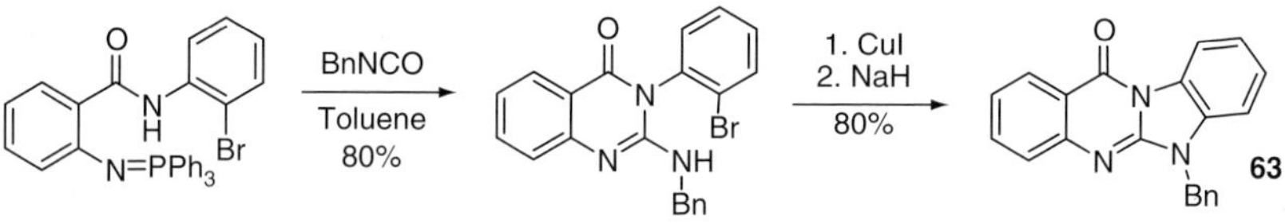

6.2.3.2 Reactions

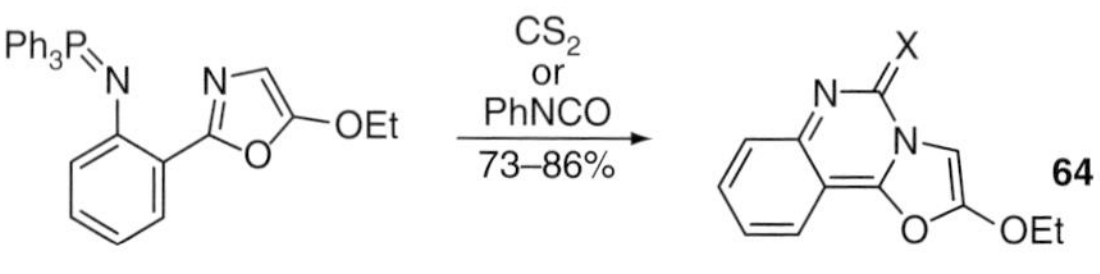

Several reports explored the functionalization of pyrimidines via cross coupling reactions. The Suzuki—Miyaura reaction was exploited in order to decorate densely substituted arylpyrimidines <09JHC960>. An investigation of 4,6-dichloropyrimidine **65** containing reactive groups in position 2 and/or 5 of the skeleton was reported and yielded the corresponding arylpyrimidine **66** and/or **67**. In another study, an unexpected-CO bond was formed when using non-degassed solvent systems, thereby generating **68** instead of **69** <09JHC459>. 2-Substituted pyrido[3,2-*d*] pyrimidine **71** was furnished via Suzuki and Stille palladium-catalyzed reactions of 2-chloropyrido[3,2-*d*]pyrimidine **70** <09S2379>. Furthermore, in the same study

70 was elaborated via aminations (S_NAr) and amidations, thereby demonstrating the versatility of the moiety.

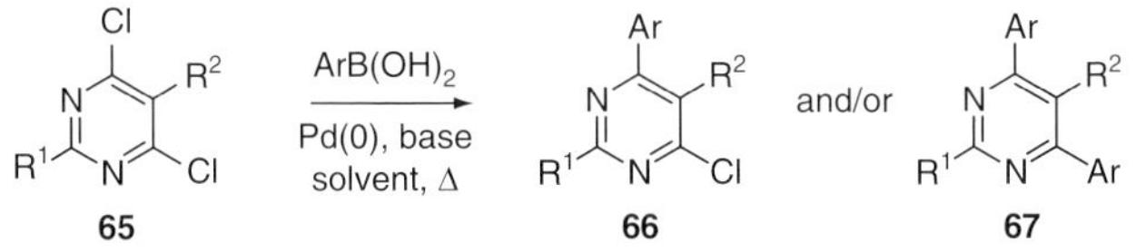

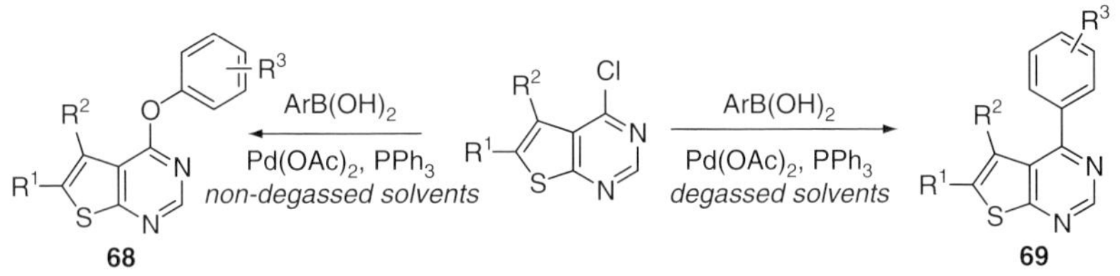

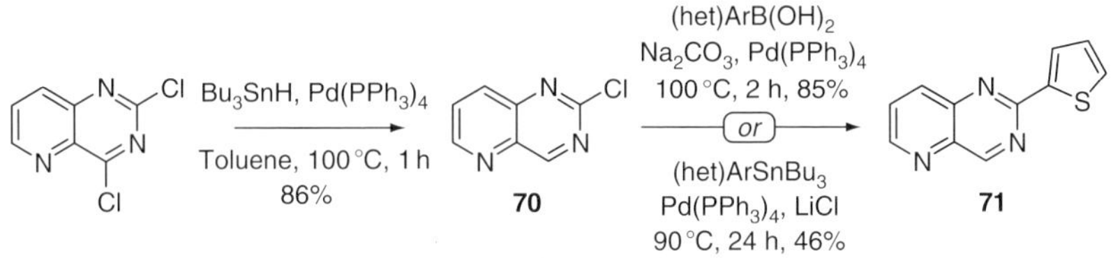

An investigation into the regioselective and chemoselective metalation of chloropyrimidine was reported <09CEJ1468>. The paper detailed the large scope of applicability which allows modification at all positions of the pyrimidine moiety.

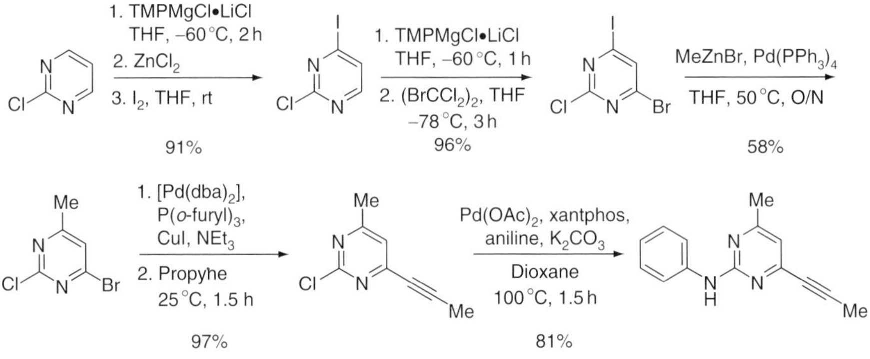

A diastereoselective domino reaction with aryl aldehydes and barbituric acids afforded spirosubstituted pyridopyrimidines <09JCO612>.

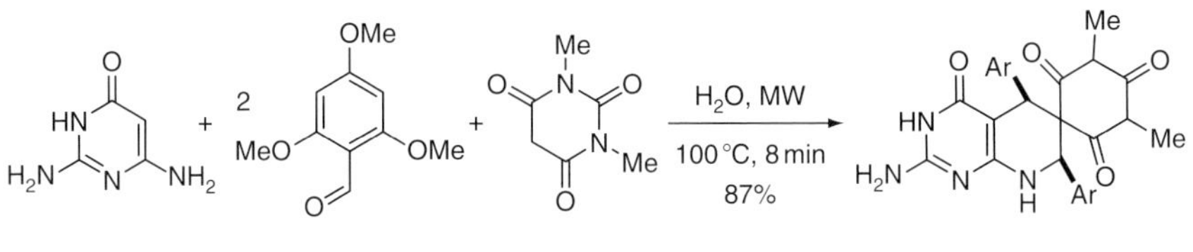

The reaction of 2-chloropyrimidine with aminoquinuclidine afforded the rearranged pyrrolopyrimidine product **72** <09CHIR681>. A mechanism was postulated for the unanticipated formation of *cis*- and *trans*-octahydropyrrolo[2,3]pyrimidine derivatives.

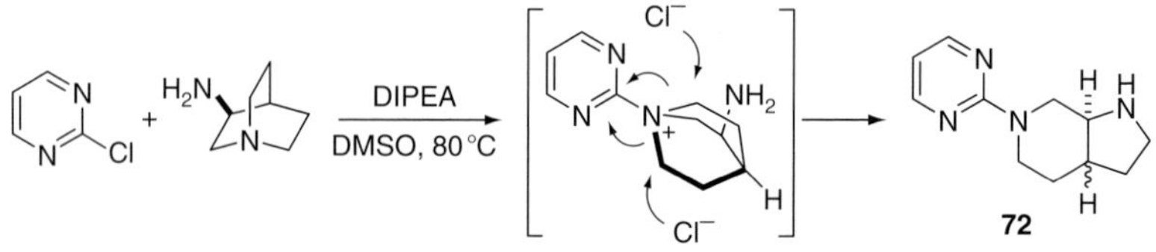

Numerous fused heterocycles were furnished from pyrimidine derivatives. An efficient synthesis of hexahydropyridopyrimidine **73** from arylmethylidenepyruvic acids was reported <09HCA932>. The synthesis of pyrimido[5′,4′:4,5]pyrrolo[2,1-*c*][1,4]oxazine **74** was reported <09CHE1285>. The addition of hydrazine to dihydrotriazolopyrimidine **75** was investigated <09JHC1413>. During the development of potential methionine synthase inhibitors, 2-amino-4-hydroxyl-6-hydroxymethyl-5,6,7,8-tetrahydropyrido[3,2-d]pyrimidine **77** was required, and thus the synthesis of **76** from 2-aminouracil was reported <09JHC1151>. 6-Cyanopurine **78** was treated with benzylhydroxylamine to afford an intermediate which, once isolated as its HCl salt, could easily be converted to the corresponding *N*-oxide **79** <09EJO4867>.

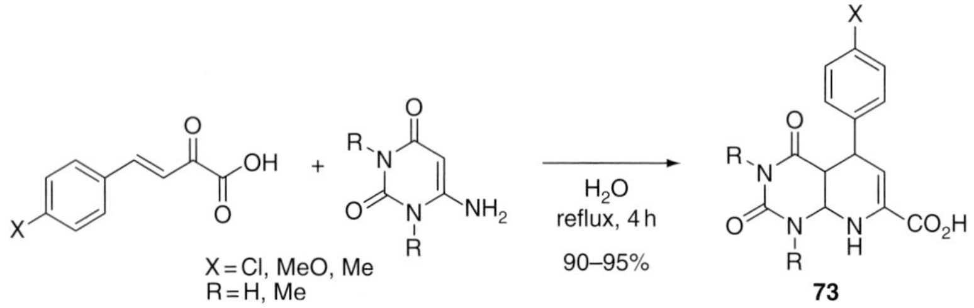

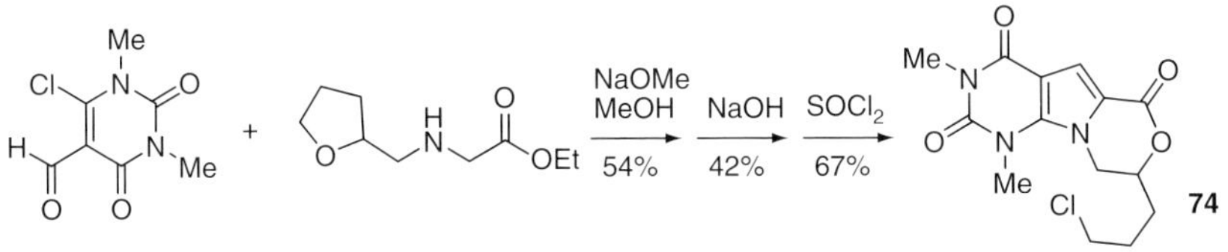

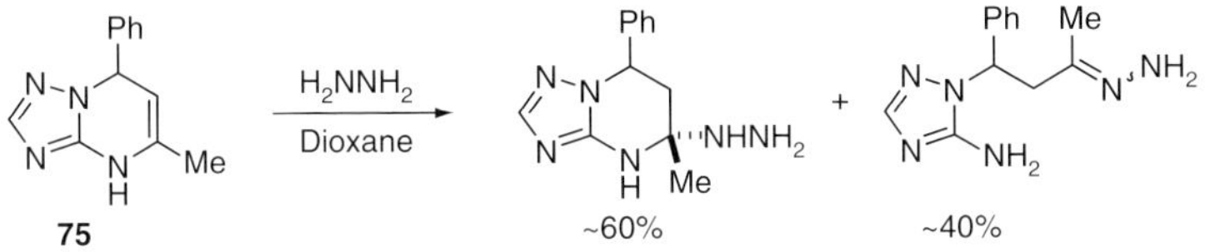

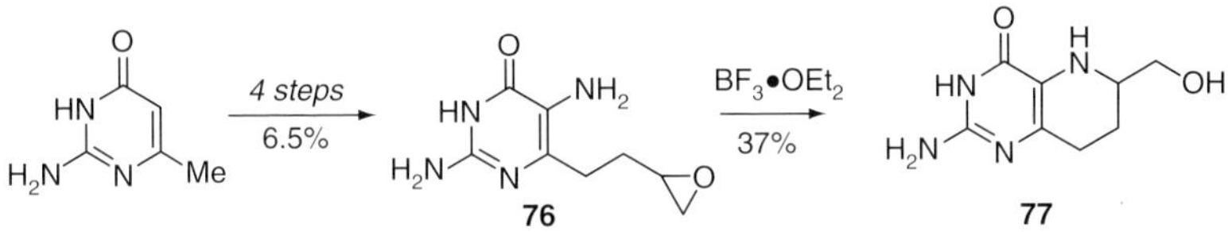

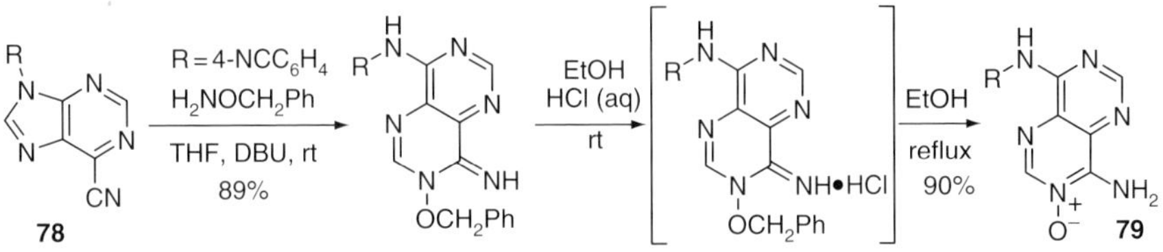

Three-component reactions were also employed in the elaboration of pyrimidine scaffolds. For example, the efficient generation of spiro[indoline-3,5′-pyrano[2,3-d]pyridine]-6′-carbonitrile **80** was generated upon cyclocondensation of malononitrile, 2-methylpyrimidine-4,6-diol and isatin in the presence of acid <09JHC1266>.

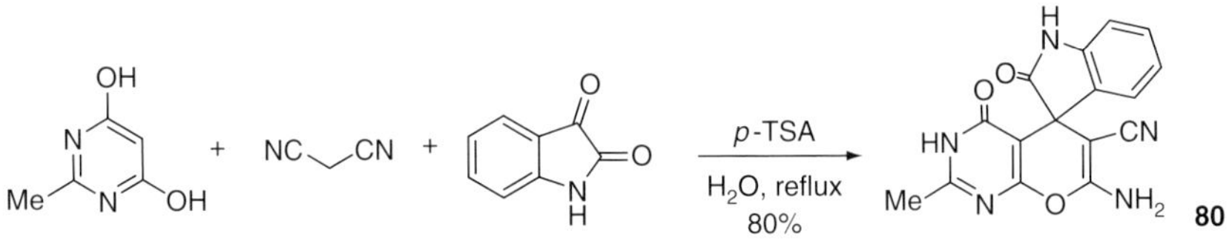

The reactivity of quinazoline moieties was investigated throughout 2009 due to their prevalence in alkaloids and importance in biologically active molecules. In one paper, the reactions of 4-hydroxyquinazoline, 4-chloroquinazoline, and quinazoline with a variety of C-nucleophiles were investigated <09TL2899>. The formation of **81** and **82** from unsubstituted quinazoline upon treatment with C-nucleophiles was explored <09CHE115>. The reactions of 1,3-dialkyl-6-chlorosulfonylquinazoline-2,4-dione **83** with nucleophiles were investigated <09CHE1508>. A sequence of reactions was reported which enabled control over substitution at the C-4, C-6, and C-7 positions of a gefinitib quinazoline core **84** <09TL1600>. And lastly, a

1,3,4-triazapyrene ring system was generated from the reaction of 1-methylbenzo[*f*] quinazolines with 1,3,5-triazine <09CHE119>.

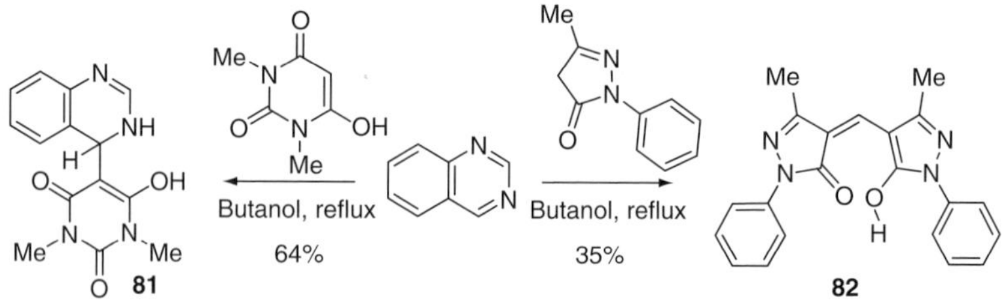

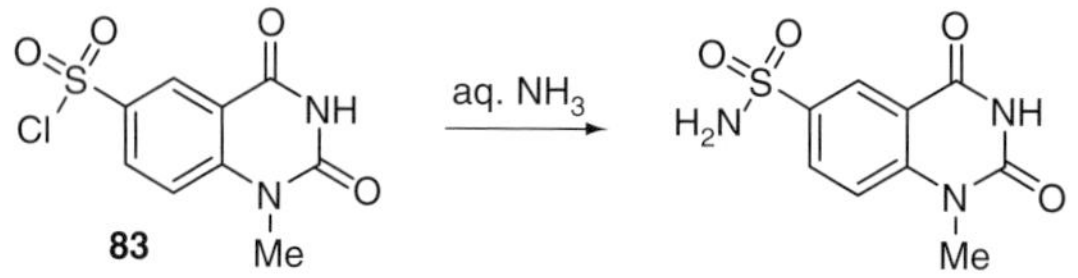

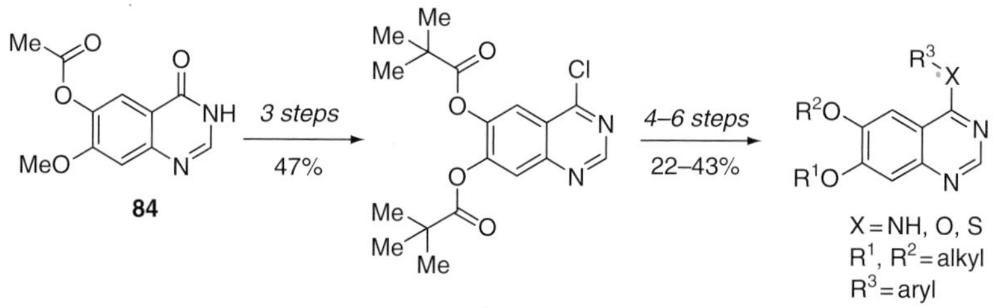

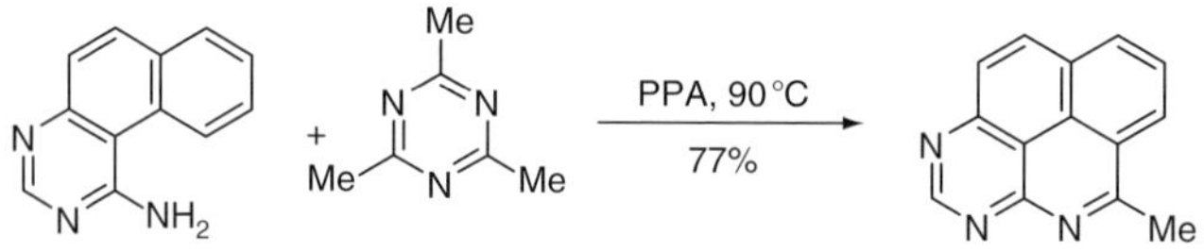

6.2.3.3 Applications

As in the past <08CHECIII117>, pyrimidines have played a prominent role in medicinal compounds which possess biological activity. Functionalized pyrimidines have been considered to be privileged structures and have been incorporated into a wide variety of pharmaceutical agents. Based on this diazine, efficacious compounds have been identified with antifungal (**85**) <09JHC895>, antimycobacterial (**86**, **87**) <09APCL94; 09EJM3330>, antiinflammatory (**88**) <09EJM609>, antiparasitic (**89**, **90**) <09BMCL5474; 09BMCL2542; 09BMCL3031; 09BMC4313>, antidepressant <09APCL671>, and broad spectrum anticancer (**91**) <09APCL238> activities. Several papers evaluated compounds for multiple activities. In particular, **92** was highlighted as a dual antimicrobial/anticancer agent <09APCL299>, while others were reported as potential antiinflammatory/analgesic compounds <09EJM4249>.

85

86

87

88, Ar = 4-methoxyphenyl

89

90

91

92

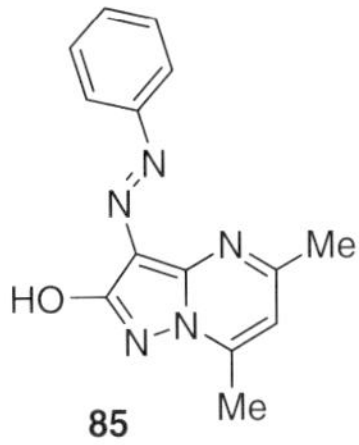

Pyrimidine-based molecules were designed, synthesized, and investigated for cox-2 (**93**) <09APCL321>, ribonuclease A <09JMC932>, histone lysine methyltransferase G9a (**94**) <09JMC7950>, protein kinase C (**95**) <09JMC6193>, cytosolic phospholipase $A_2\alpha$ (**96**) <09BMC4383>, and dipeptidyl peptidase IV (**97**) <09BMCL1991> inhibition. Pyrazolo[1,5-*c*]quinazoline **98** was studied for its ability to bind the glycine co-agonist site of the NMDA receptor for the treatment of neurological diseases <09CPB826>, while SAR was performed around **99** to identify a potent activator of glucokinase for potential treatment of diabetes <09BMCL5531> and RS-1154 was identified from a library screen as an effective CCR4 antagonist as demonstrated by *in vitro* and *in vivo* data <09EJP38>. In addition, a series of nucleoside analogs, including compounds containing pyrimidine derivatives such as **100**, was synthesized and used to systematically investigate the role of a cytosine nucleobase in the active site of the HDV ribosome <09JOC8021>. This diazine substructure is also found in the marketed HBV drug Baraclude® (**101**) and a radiolabelled synthesis of [$^{13}C_4$]Baraclude® (entecavir) was described <09JLCR485>.

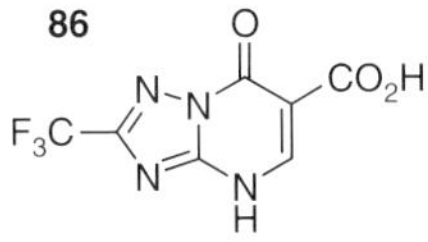

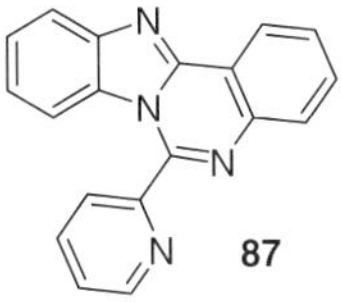

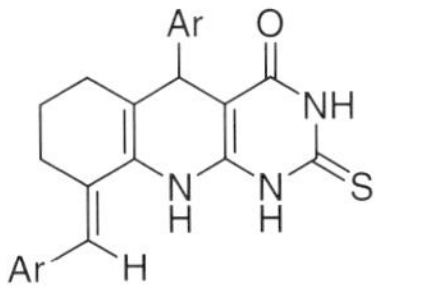

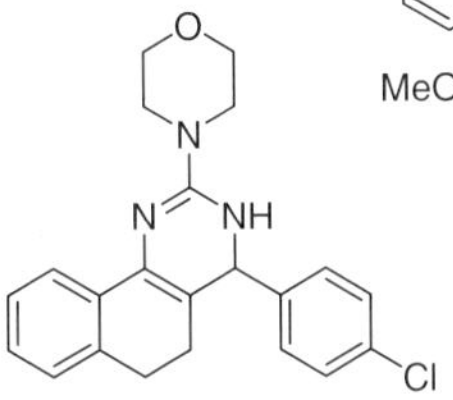

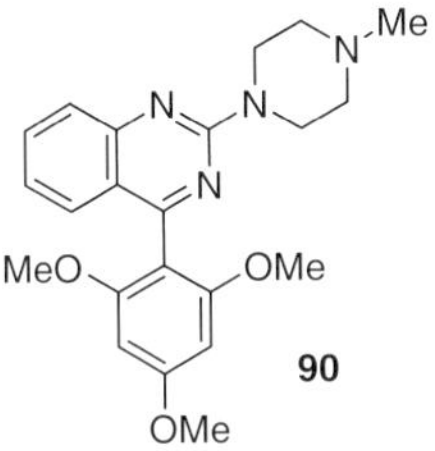

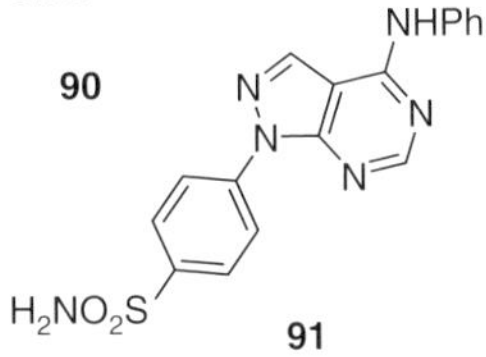

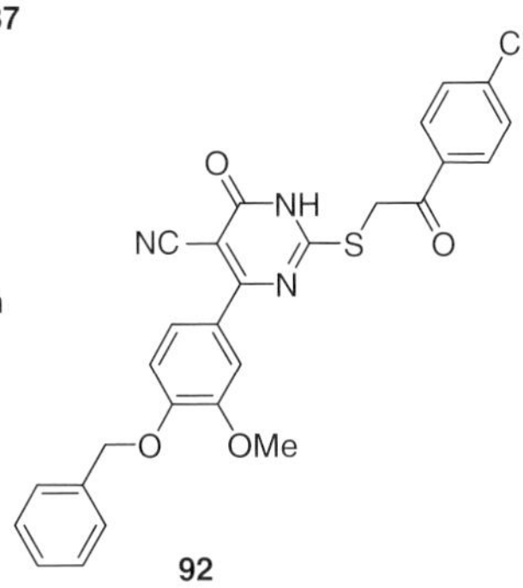

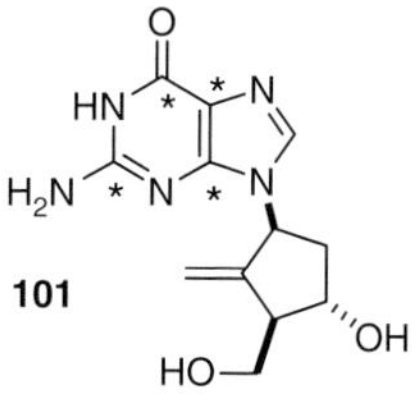

Pyrimidine-based heterocycles have been architectural motifs in a wide variety of compounds pursued for the treatment of cancer. A review of purine and pyrimidine antimetabolites used as anticancer drugs was recently published <09CRV2880>. A series of 2-aminothienol pyrimidines (**102**) was synthesized and tested for Hsp90 molecular chaperone inhibition <09JMC4794>. Pyrrolo[3,4-*h*]quinazoline **103** was prepared from a tetrahydroisoindole-4-one derivative and showed antitumor activity against 59 cell lines <09TL5389>. Researchers reported that boron-conjugated quinazoline **104** was an inhibitor of EGFR tyrosine kinase and used theoretical mechanical docking calculations to suggest that the boronic acid moiety can react with Asp800, thus accounting for the prolonged inhibition observed <09OBC4415>. A number of other papers also described structurally unique EGFR kinase inhibitors <09NAT1070; 09MCT2546> and/or quinazoline-based complexes for EGFR-TK imaging (**105**) <09EJM4021; 09JBIC261>. Substituted quinazolines were evaluated for the potential treatment of colorectal (**106**) <09BMCL4980> and prostate cancers <09BMC3152>. And lastly, quinazoline derivative **107** was identified as an orally available cyclin dependent kinase inhibitor which has been advanced into Phase II clinical trials <09JMC5152>.

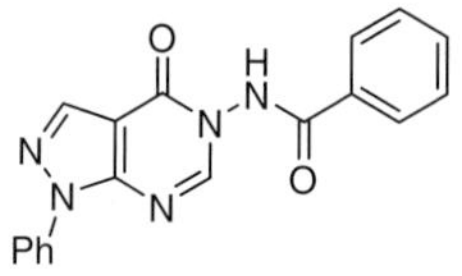

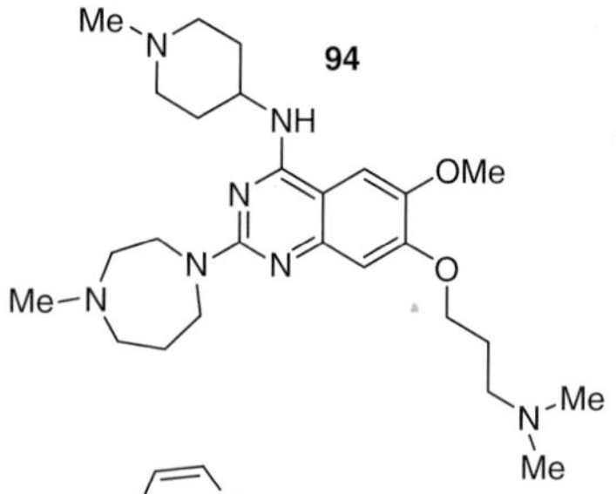

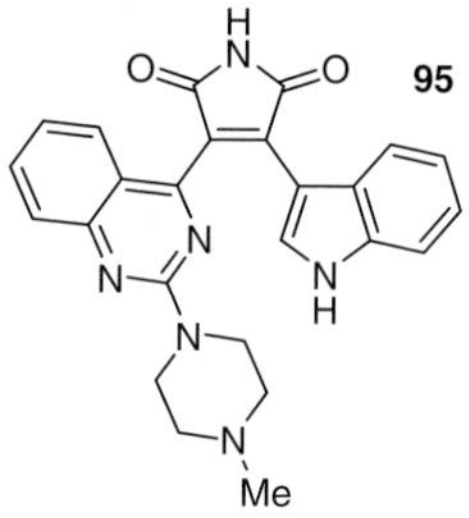

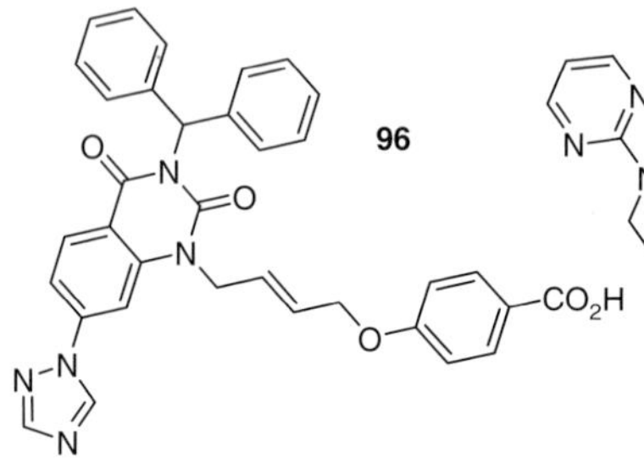

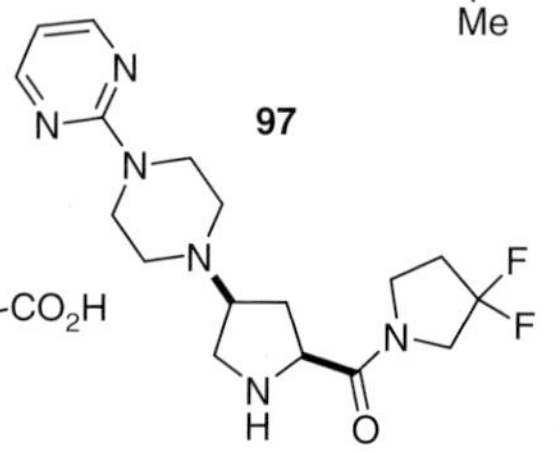

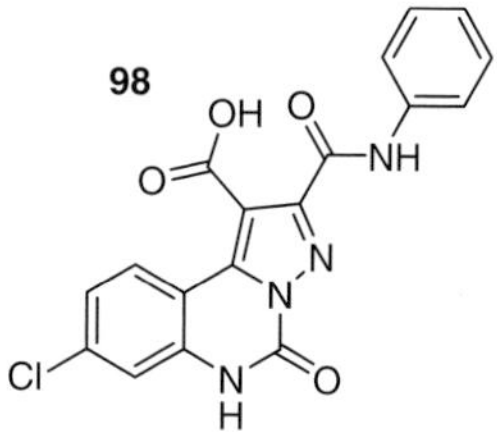

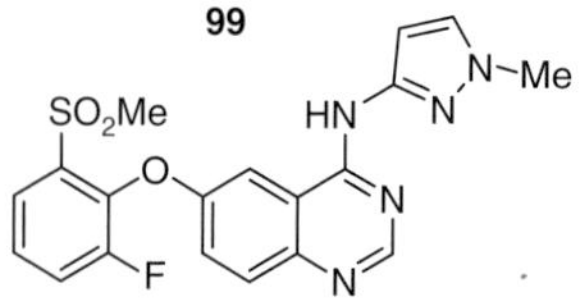

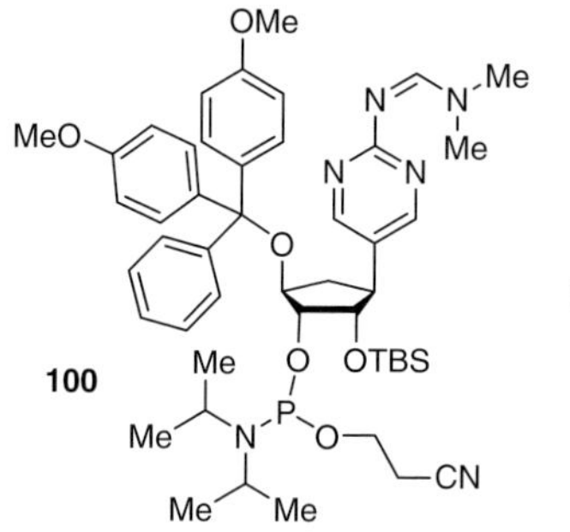

102 103 104

105a, M = Re, X = Br

105b, M = ^{99m}Tc, X = Cl

106 107

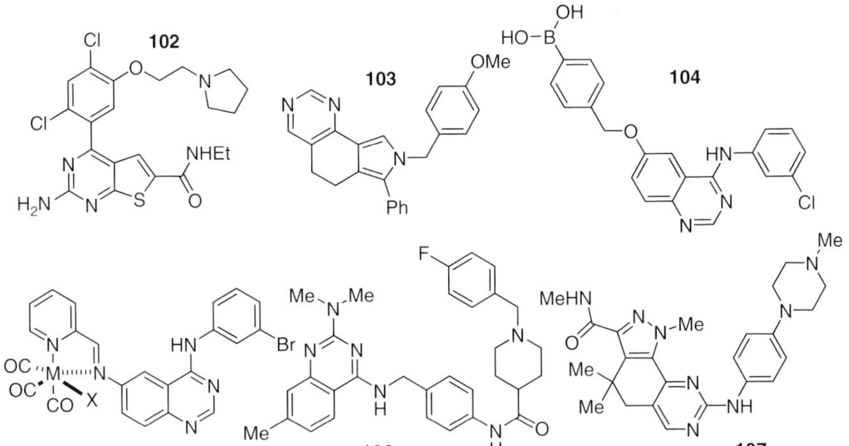

In 2009, pyrimidines played a role in a number of non-pharmaceutical applications. The optical absorption and emission properties, as well as use as pH indicators were investigated of a pyrimidine-based oligomer <09JOC3711>. Pyrimidines were used with gold surfaces <09L11486> and as part of a series of carbene derived palladium ligand systems found to be effective in the CH activation of methane as well as the Mizoroki–Heck reaction <09OM2142>.

6.2.4. PYRAZINES AND BENZO DERIVATIVES

6.2.4.1 Syntheses

In 2009, the preliminary substrate scope for the formation of substituted pyrazines via an intramolecular hydroamination of oxime precursor **108** was reported <09AG(I) 8325>.

108

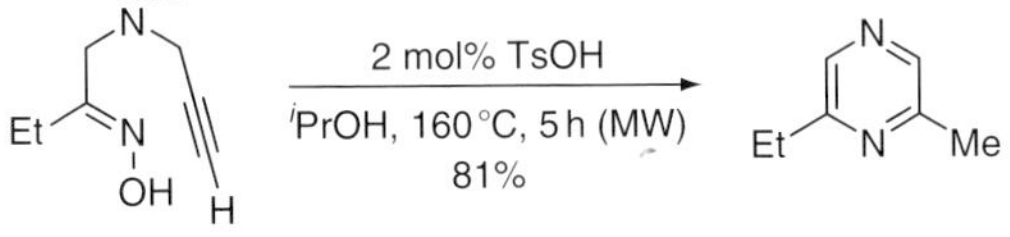

tert-Butyl amide **113** produced via an Ugi multi-component reaction of **109**, **110**, **111**, and **112** was converted to dihydropyrazolopyrazine **114** when irradiated with microwaves <09SL260>.

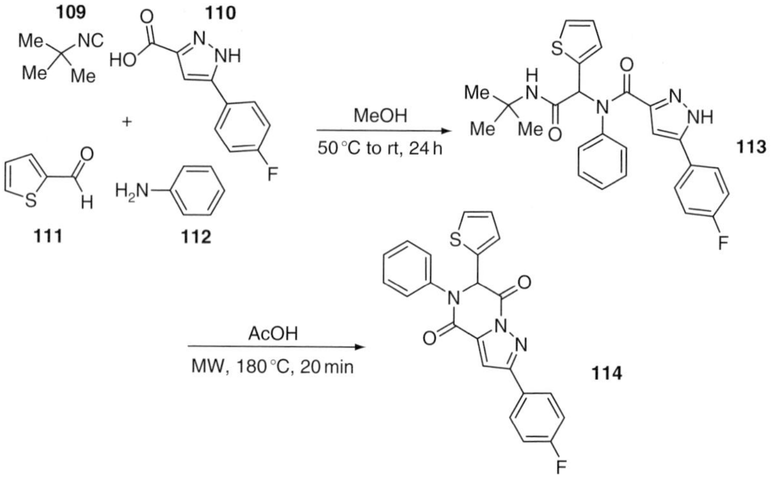

A productive cascade event was developed to rapidly construct 8-trifluoromethyltetrahydro-6H-pyrido[1,2-a]pyrazine-6-one (**115**), a key intermediate in the synthesis of potential chemokine (CCR2) <09TL4050>. Due to the envisioned utility, this methodology was further explored and access to several substituted fused heterocycles was demonstrated.

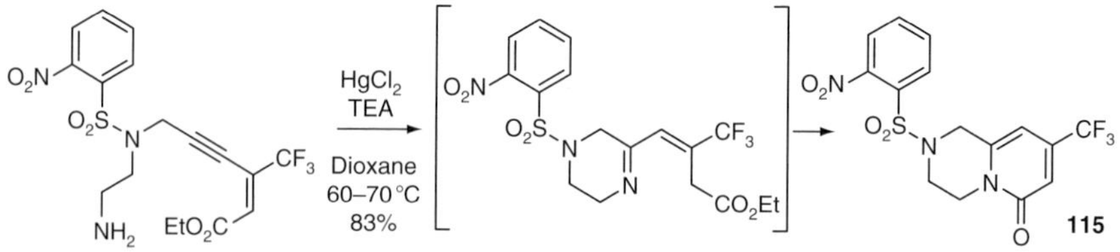

1,3-Diarylpyrrolo[1,2-a]quinoxalines **116** or **118** were found to be easily furnished from 2-[2,4-phenyl-1H-pyrrol-1-yl]phenamine **117** either upon formylation followed by intramolecular condensation or acylation with acetic acid followed by cyclization in the presence of POCl$_3$, respectively <09CHE1396>.

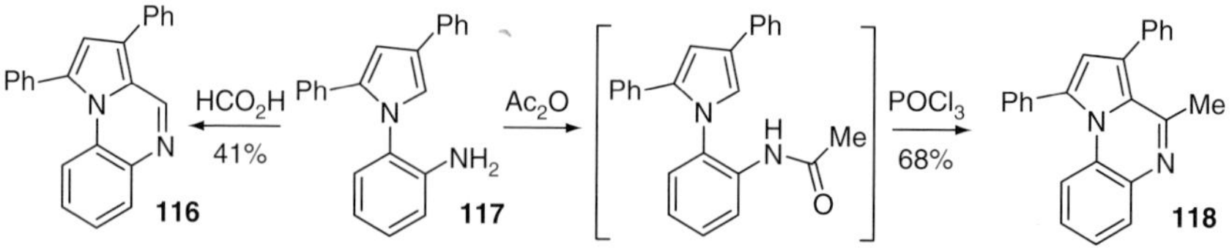

An oxidation-annulation approach was disclosed to afford 4H-imidazo[4,5-b]quinoxaline **120** from 4H-imidazole **119** upon treatment with CAN in the presence of potassium carbonate <09S4049>. A postulated mechanism and DFT-calculations were presented in support of a radical process.

R¹	R²	Yield (%)
2-FC$_6$H$_4$	8-F	52
Ph	H	84
4-F$_3$CC$_6$H$_4$	6-F$_3$C	78

Interestingly, fluororubine derivative **122** was observed as a byproduct upon extended sunlight irradiation of 1,2-diazetine **121** <09CEJ12799>. Upon isolation, these compounds were fully characterized and their photophysical and chemical properties were investigated.

6.2.4.2 Reactions

Pyrazines were the focus of several computational modeling studies to explore regioselectivity and reactivity. Aminations of 2-substitued pyrazines by O-mesitylenesulfonylhydroxylamine were performed and DFT calculations were conducted to understand the observed regioselectivity with results supporting an S$_N$2 mechanism <09TL6779>. A theoretical study on the titanium-mediated trimerization of pyrazine was conducted using DFT calculations as well as MP2 and CCSD(T) calculations <09JCTC2044>. These studies indicated that the driving force for the trimerization arose from the six resulting Ti-N bonds formed.

In 2009, a series of papers employed pyrazines in metal-catalyzed coupling reactions to provide useful reagents and products. Pyrazine derivative **124** which incorporated a hexatriene chain tethered to either a donating or accepting group was synthesized from the commercially available chloropyrazine **123** via a Negishi cross coupling reaction followed by a Wittig olefination <09T4190>. The light emitting properties of such "push-pull" molecules were investigated. Numerous heteroaryl chlorides, including pyrazine **123**, were successfully coupled with alkynes using Pd-catalysis <09T7146>. The functionalization of imidazo[1,5-*a*]pyrazine **125** was demonstrated <09OL5118>. Substituted imidazopyrazine **126** was synthesized via a regioselective metalation on the parent heterocycle **125**. Furthermore, an application of the directed remote metalation concept provided access to the tricyclic ring system, triazadibenzo[*cd,f*]azulen-7(6*H*)-one **127**. Alternately, the successful AlCl$_3$-mediated coupling of arenes and heteroarenes with chloropyrazine was reported <09TL1618>. This strategy avoids expensive organometallic reagents and provided a means to access **128** from readily available materials under mild conditions and in good yield. In addition, a gold-catalyzed C-H arylation of pyrazine **129** with arylbromide **130** was reported <09TL1478>.

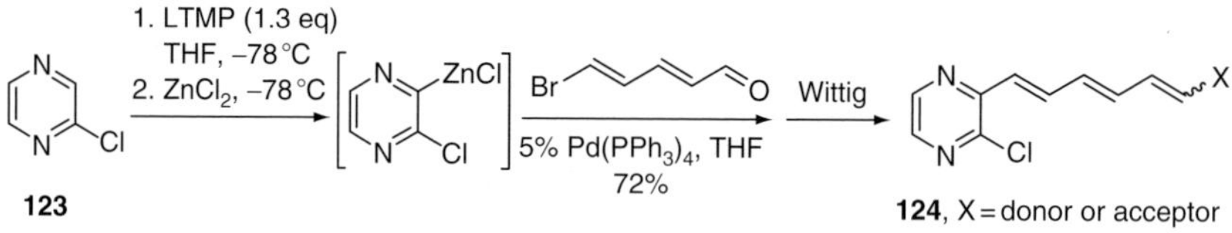

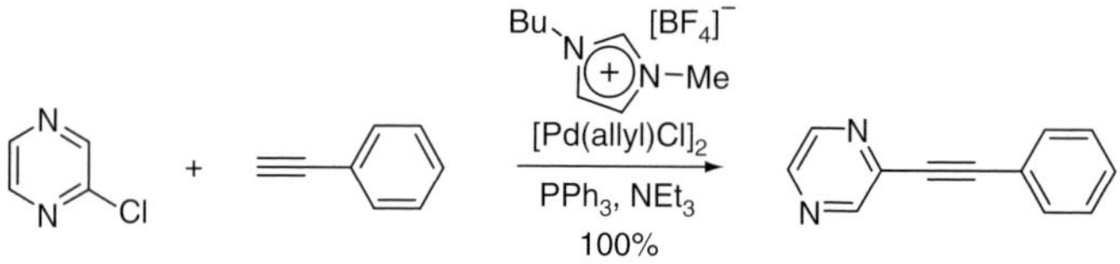

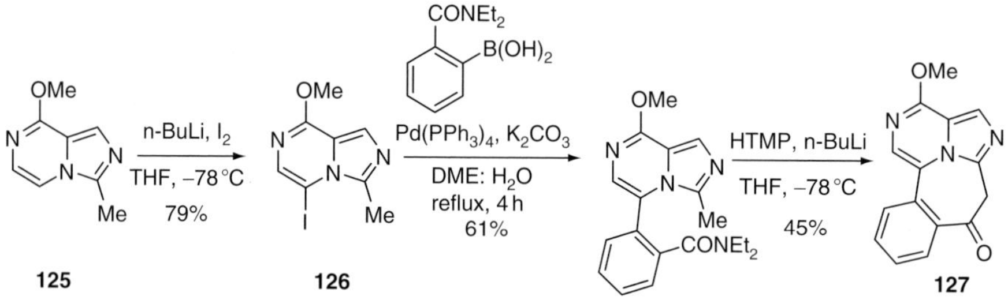

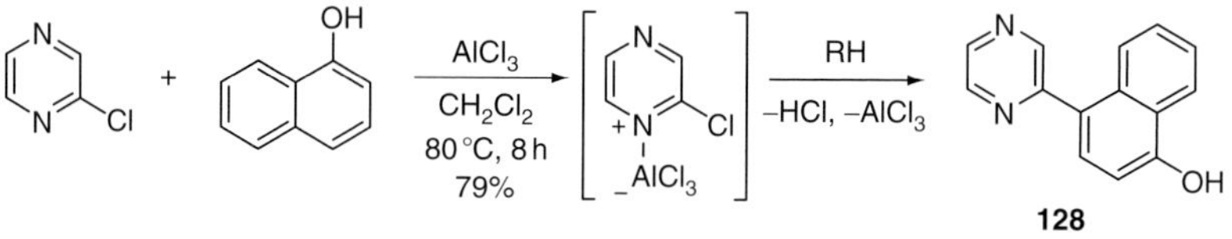

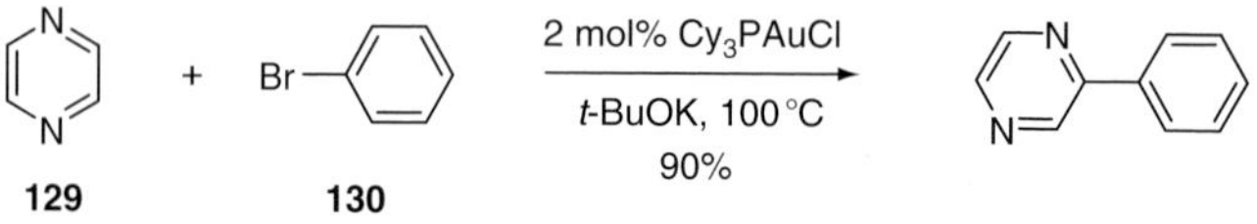

An interesting paper demonstrated that aromatic compounds, including methoxypyrazine **131**, can be deprotonated by a cadmium-lithium base followed by trapping with I_2 <09CEJ10280>. Separately, the lithium cadamates obtained from **132** were subjected to Pd-catalyzed cross couplings or quenched with acid chlorides.

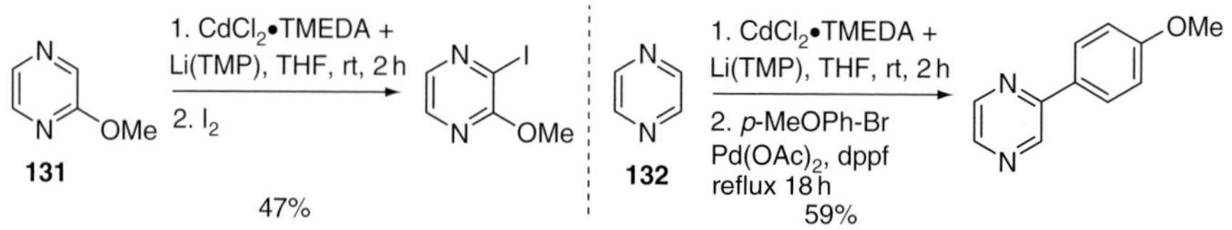

Two reports describing the asymmetric hydrogenation of quinoxaline **133** under iridium-catalyzed reaction conditions were disclosed <09AG(I)9135; 09ASC2549>.

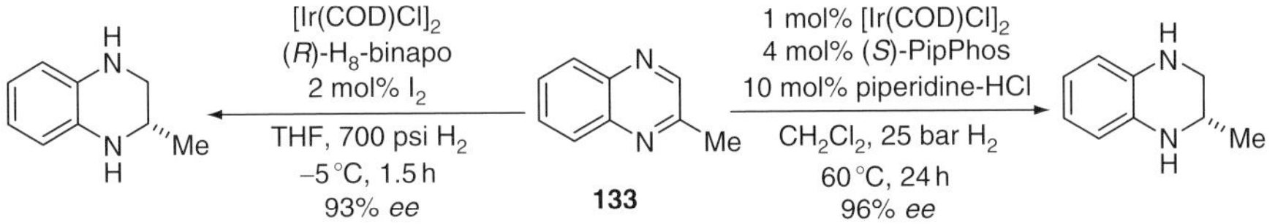

Pyrazines were found to be useful in cycloaddition reactions. The 1,3-dipolar cycloaddition between asymmetric benzodiazines and nitrilimines was investigated <09TL7333>. The cycloadditions of dihydropyrazine **134** with ketenes was explored experimentally as well as with DFT calculations <09CPB846>.

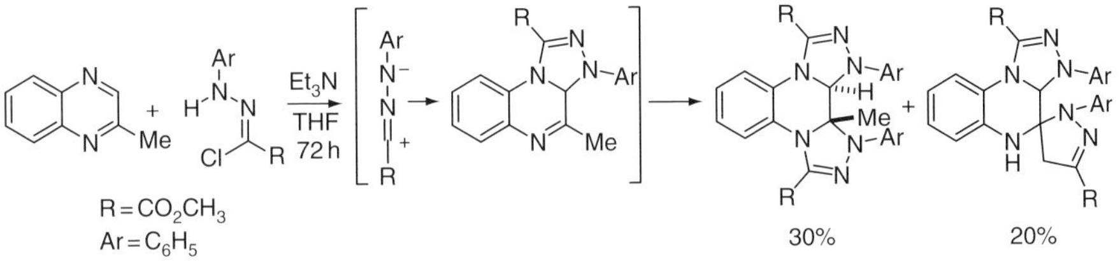

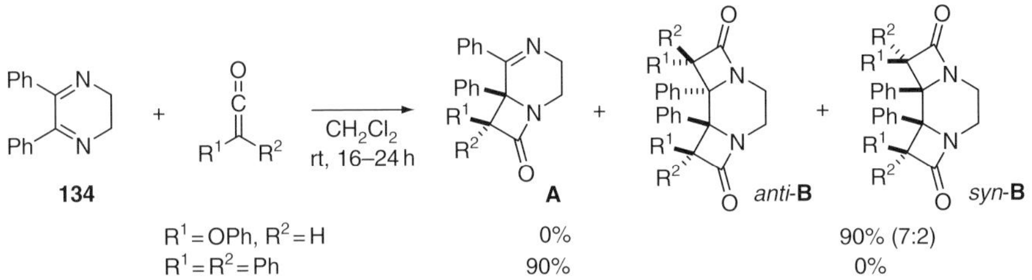

Quinoxaline derivative **135** was subjected to oxalyl chloride resulting in a one-pot synthesis of two different quinoxaline derivatives, **136** and **137** <09RJOC1730>.

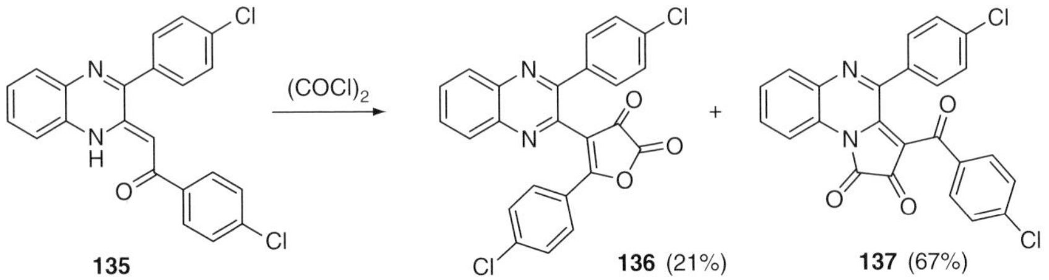

6.2.4.3 Applications

Pyrazines have found utility as part of metal complexes. A phenazine-based ligand, dipyrido[3,2-a:2′,3′-c]phenazine (dppz), was utilized as part of a dirhodium complex (**138**) <09JA11353>, as well as similar copper <09IC9120> and ferrocene-conjugated copper <09OM1495> complexes. These complexes were characterized and explored for their ability to interact with DNA, among other uses. The coordination modes of phenazine to molybdenum were investigated <09JA7828> and the reaction of the molybdenum complex **139** with H_2 was described. The formation of pyrazine bridged diimidazolium salts was reported and their palladium complexes (**140**) were shown to be adequate catalysts for Heck-type coupling reactions <09T909>. Pyrazines were also the central feature of paramagnetic radical anionic triple-decker metallo-organic diazine derivatives <09OM6194>. A single crystal of a pyrazyl-containing triple decker (**141**) was successfully isolated.

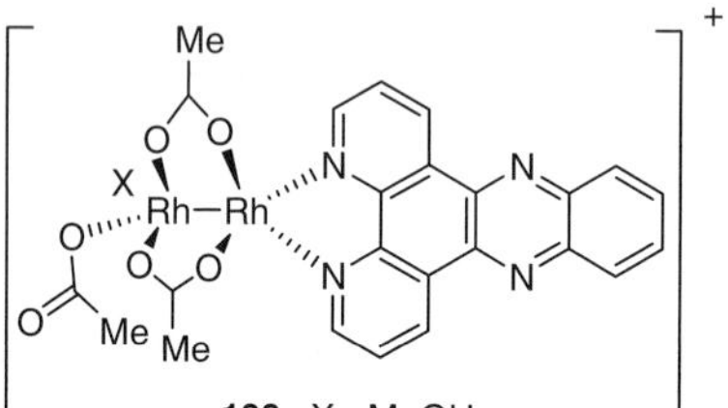

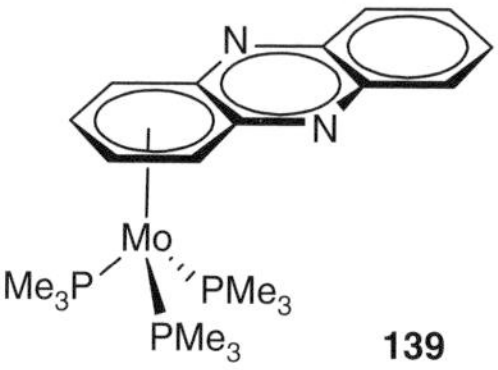

Substituted pyrazines have been found as subunits of multiple synthetically constructed therapeutic agents, as well as several natural products. In 2009, the reported biological activity which these pyrazine-based scaffolds possessed spanned the gamut. Specifically, compounds were furnished as inhibitors of type II FMS (**142**) <09BMCL1206> and EphB4 (**143**) <09BMCL6991> tyrosine receptor kinases, as well as human aldose reductase (**144**) <09T3019>, and agonists were evaluated for the 5-HT$_3$ receptor (**145**) <09JMC6946>. In addition, pyrazine motifs have been reported as antitubercular agents <09CHE1058>, as antagonists of gonadotropin releasing hormone (GnRH) receptor (**146**) <09JMC2148>, and as antimicrobial compounds <09CHE1370>. A study was disclosed to better understand the mechanism, and time frame, in which tetramethylpyrazine treats neural ischemia/reperfusion injury in rats <09TR727>, while another investigation was conducted on benzenesulfonamide **147** to evaluate its ability to affect leishmanicidal activity <09BMC7449>.

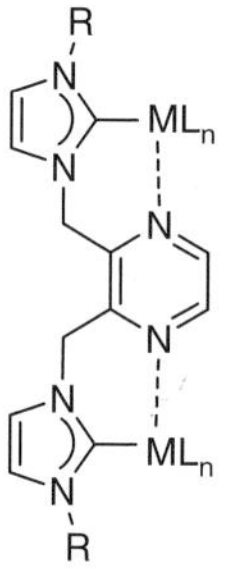

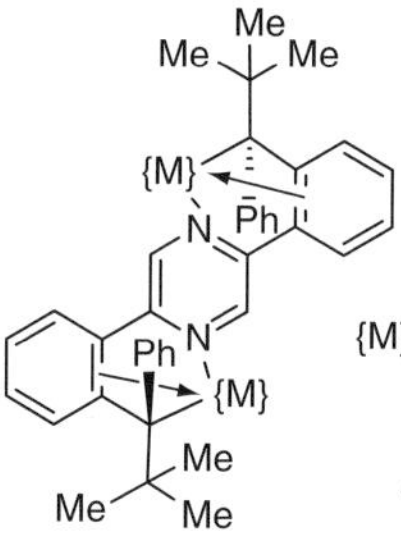

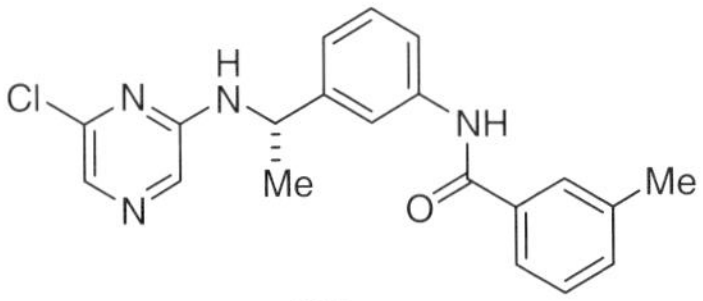

Pyrazines and pyrazine derivatives were also reported to be useful in a host of other miscellaneous applications. These include uses to pre-organize porphyrin monomers <09BB635>, to bind group I, II and III cations <09RCB89>, and to identify protein targets which bind to aloisine A via affinity chromatography <09BMC5572>. In addition, quinoxaline derivatives were evaluated as bulk-heterojunction solar cells <09OL4898>, and characterized as functionalized π-conjugated dendrimers with electronic applications <09OL4500>.

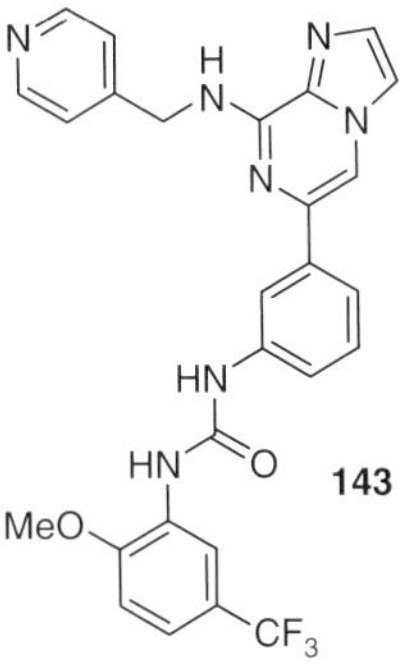

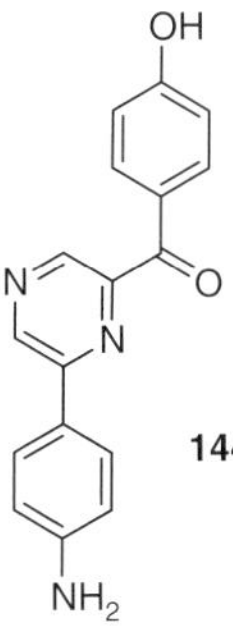

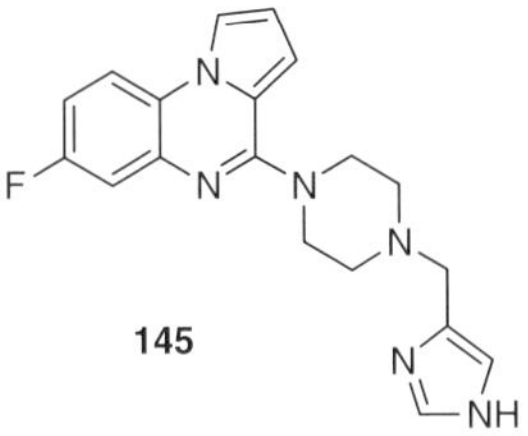

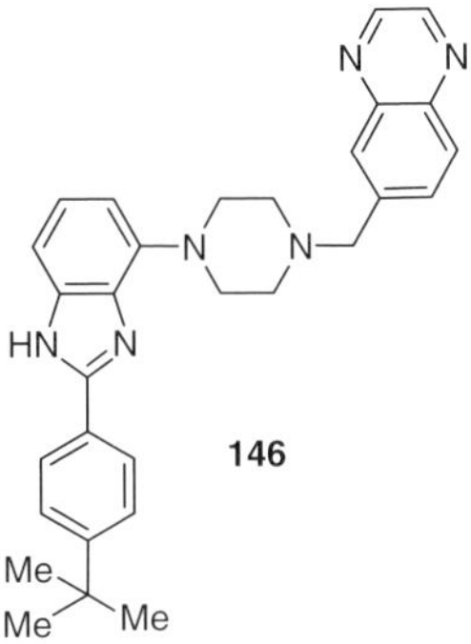

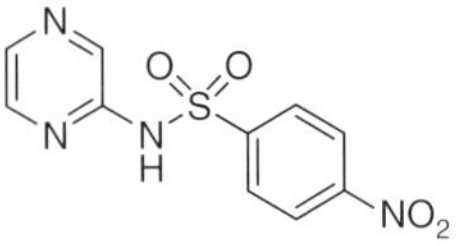

REFERENCES

08CHECIII117	G.W. Rewcastle, Pyrimidines and their Benzo Derivatives. (A.R. Katritzky, C.A. Ramsden, E.F.V. Scriven, and R.J.K. Taylor, eds.), **8**, p. 117. Elsevier, Oxford (2008).
09AG(I)8325	T. Rizk, E.J.-F. Bilodeau and A.M. Beauchemin, *Angew. Chem. Int. Ed.*, **48**, 8325 (2009).
09AG(I)9135	W. Tang, L. Xu, Q.-H. Fan, J. Wang, B. Fan, Z. Zhou, K.-H. Lam and A.S.C. Chan, *Angew. Chem. Int. Ed.*, **48**, 9135 (2009).
09AHC1	M.H. Elnagdi, N.A. Al-Awadi and I.A. Abdelhamid, *Adv. Heterocycl. Chem.*, **97**, 1 (2009).
09AMAC3620	P. Caspers, L. Bury, B. Gaucher, J. Heim, S. Shapiro, S. Siegrist, A. Schmitt-Hoffmann, L. Thenoz and H. Urwyler, *Antimicrob. Agents Ch.*, **53**, 3620 (2009).
09APCL94	H.M. Abdel-Rahman, N.A. El-Koussi and H.Y. Hassan, *Arch. Pharm. Chem. Life. Sci.*, **342**, 94 (2009).
09APCL238	H.M.A. Ashour and A.E.A. Wahab, *Arch. Pharm. Chem. Life Sci.*, **342**, 238 (2009).
09APCL299	S.A.F. Rostom, H.M.A. Ashour and H.A.A. El Razik, *Arch. Pharm. Chem. Life Sci.*, **342**, 299 (2009).

09APCL321	D. Raffa, B. Maggio, F. Plescia, S. Cascioferro, S.V. Raimondi, S. Plescia and M.G. Cusimano, *Arch. Pharm. Chem. Life. Sci.*, **342**, 321 (2009).
09APCL671	H.-J. Wang, C.-X. Wei, X.-Q. Deng, F.-L. Li and Z.-S. Quan, *Arch. Pharm. Chem. Life Sci.*, **342**, 671 (2009).
09ARK302	S.A.S. Ghozlan, M.H. Mohamed, A.M. Abdelmoniem and I.A. Abdelhamid, *ARKIVOC*, **10**, 302 (2009).
09ASC2549	N. Mršić, T. Jerphagnon, A.J. Minnaard, B.L. Feringa and J.G. De Vries, *Adv. Synth. Catal.*, **351**, 2549 (2009).
09BB635	J. Matsuia, T. Sodeyamab, Y. Saiki, T. Miyazawab, T. Yamadab, K. Tamakib and T. Murashima, *Biosens. Bioelectron.*, **25**, 635 (2009).
09BMC2823	R.M. Butnariu and I.I. Mangalagiu, *Biorg. Med. Chem.*, **17**, 2823 (2009).
09BMC3152	K. Lee, J. Kim, K.-W. Jeong, K.-W. Lee, Y. Lee, J.Y. Song, M.S. Kim, G.S. Lee and Y.B. Kim, *Biorg. Med. Chem.*, **17**, 3152 (2009).
09BMC3519	B. Hua, R. Unwalla, M. Collini, E. Quinet, I. Feingold, A. Goos-Nilsson, A. Wihelmsson, P. Nambi and J. Wrobel, *Biorg. Med. Chem.*, **17**, 3519 (2009).
09BMC4313	P. Verhaeghe, N. Azas, S. Hutter, C. Castera-Ducros, M. Laget, A. Dumètre, M. Gasquet, J.-P. Reboul, S. Rault, P. Rathelot and P. Vanelle, *Biorg. Med. Chem.*, **17**, 4313 (2009).
09BMC4383	S.J. Kirincich, J. Xiang, N. Green, S. Tam, H.Y. Yang, J. Shim, M.W.H. Shen, J.D.C. Clark and J.C. McKew, *Biorg. Med. Chem.*, **17**, 4383 (2009).
09BMC5572	C. Corbel, R. Haddoub, D. Guiffant, O. Lozach, D. Gueyrard, J. Lemoine, M. Ratin, L. Meijer, S. Bach and P. Goekjian, *Biorg. Med. Chem.*, **17**, 5572 (2009).
09BMC7449	M.A. Dea-Ayuela, E. Castillo, M. Gonzalez-Alvarez, C. Vega, M. Rolón, F. Bolás-Fernández, J. Borrás and M.E. González-Rosende, *Biorg. Med. Chem.*, **17**, 7449 (2009).
09BMCL1206	C.J. Burns, M.F. Harte, X. Bu, E. Fantino, M. Giarrusso, M. Joffe, M. Kurek, F.S. Legge, P. Razzino, S. Su, H. Treutlein, S.S. Wan, J. Zeng and A.F. Wilks, *Biorg. Med. Chem. Lett.*, **19**, 1206 (2009).
09BMCL1991	M.J. Ammirati, K.M. Andrews, D.D. Boyer, A.M. Brodeur, D.E. Danley, S.D. Doran, B. Hulin, S. Liu, R.K. McPherson, S.J. Orena, J.C. Parker, J. Polivkova, X. Qiu, C.B. Soglia, J.L. Treadway, M.A. VanVolkenburg, D.C. Wilder and D.W. Piotrowski, *Biorg. Med. Chem. Lett.*, **19**, 1991 (2009).
09BMCL2542	S. Kumar, N. Shakya, S. Gupta, J. Sarkar and D.P. Sahu, *Biorg. Med. Chem. Lett.*, **19**, 2542 (2009).
09BMCL3019	R. Grey, A.C. Pierce, G.W. Bemis, M.D. Jacobs, C.S. Moody, R. Jajoo, N. Mohal and J. Green, *Biorg. Med. Chem. Lett.*, **19**, 3019 (2009).
09BMCL3031	A. Cavalli, F. Lizzi, S. Bongarzone, R. Brun, R.L. Krauth-Siegel and M.L. Bolognesi, *Biorg. Med. Chem. Lett.*, **19**, 3031 (2009).
09BMCL4980	Z. Chen, A.M. Venkatesan, C.M. Dehnhardt, O.D. Santos, E.D. Santos, S. Ayral-Kaloustian, L. Chen, Y. Geng, K.T. Arndt, J. Lucas, I. Chaudhary and T.S. Mansour, *Biorg. Med. Chem. Lett.*, **19**, 4980 (2009).
09BMCL5474	K.C. Agarwal, V. Sharma, N. Shakya and S. Gupta, *Biorg. Med. Chem. Lett.*, **19**, 5474 (2009).
09BMCL5531	T. Iino, T. Sasaki, M. Bamba, M. Mitsuya, A. Ohno, K. Kamata, H. Hosaka, H. Maruki, M. Futamura, R. Yoshimoto, S. Ohyama, K. Sasaki, M. Chiba, N. Ohtake, Y. Nagata, J.-I. Eiki and T. Nishimura, *Biorg. Med. Chem. Lett.*, **19**, 5531 (2009).
09BMCL6991	S.A. Mitchell, M.D. Danca, P.A. Blomgren, J.W. Darrow, K.S. Currie, J.E. Kropf, S.H. Lee, S.L. Gallion, J.-M. Xiong, D.A. Pippin, R.W. DeSimone, D.R. Brittelli, D.C. Eustice, A. Bourret, M. Hill-Drzewi, P.M. Maciejewski and L.L. Elkin Biorg, *Med. Chem. Lett.*, **19**, 6991 (2009).
09CAJ1678	T. Fukuda, N. Hashimoto, Y. Araki, M.E. El-Khouly, O. Ito and N. Kobayashi, *Chem. Asian J.*, **4**, 1678 (2009).
09CAL201	Y.P. Patil, P.J. Tambade, K.D. Parghi, R.V. Jayaram and B.M. Bhanage, *Catal. Lett.*, **133**, 201 (2009).
09CC3008	T.J. Donohoe, L.P. Fishlock, J.A. Basutto, J.F. Bower, P.A. Procopiou and A.L. Thompsonz, *Chem. Commun.*, 3008 (2009).
09CC6729	N.K. Szymczak, L.A. Berben and J.C. Peters, *Chem. Commun.*, 6729 (2009).

09CEJ1468	M. Mosrin and P. Knochel, *Chem. Eur. J.*, **15**, 1468 (2009).
09CEJ5006	E. Merkul, T. Oeser and T.J.J. Müller, *Chem. Eur. J.*, **15**, 5006 (2009).
09CEJ10030	J.J. Mousseau, L. Xing, N. Tang and L.A. Cuccia, *Chem. Eur. J.*, **15**, 10030 (2009).
09CEJ10280	K. Snégaroff, J.M. L'Helgoual'ch, G. Bentabed-Ababsa, T.T. Nguyen, F. Chevallier, M. Yonehara, M. Uchiyama, A. Derdour and F. Mongin, *Chem. Eur. J.*, **15**, 10280 (2009).
09CEJ12799	J. Fleischhauer, R. Beckert, Y. Jüttke, D. Hornig, W. Günther, E. Birckner, U.W. Grummt and H. Gçrls, *Chem. Eur. J.*, **15**, 12799 (2009).
09CHE115	Y.A. Azev and S.V. Shorshnev, *Chem. Heterocycl. Compd.*, **45**, 115 (2009).
09CHE119	A.V. Aksenov, I.V. Aksenova and A.S. Lyakhovnenko, *Chem. Heterocycl. Compd.*, **45**, 119 (2009).
09CHE815	L.M. Potikha1, V.A. Kovtunenko1 and A.V. Turov1, *Chem. Heterocycl. Compd.*, **45**, 815 (2009).
09CHE1058	I.V. Ukrainets, L.A. Grinevich, A.A. Tkach, O.V. Bevz and S.V. Slobodzian, *Chem. Heterocycl. Compd.*, **45**, 1058 (2009).
09CHE1285	O.B. Smolii, L.V. Muzychka1 and E.V. Verves, *Chem. Heterocycl. Compd.*, **45**, 1285 (2009).
09CHE1370	O.O. Ajani, C.A. Obafemi, C.O. Ikpo, K.O. Ogunniran and O.C. Nwinyi, *Chem. Heterocycl. Compd.*, **45**, 1370 (2009).
09CHE1396	L.M. Potikha and V.A. Kovtunenko, *Chem. Heterocycl. Compd.*, **45**, 1396 (2009).
09CHE1508	R.S. Kuryazov, N.S. Mukhamedov and K.M. Shakhidoyatov, *Chem. Heterocycl. Compd.*, **45**, 1508 (2009).
09CHIR681	G. Goljer, A. Molinari, Y. He, L. Nogle, W. Sun, B. Campbell and O. McConnell, *Chirality*, **21**, 681 (2009).
09CPB826	F. Varano, D. Catarzi, V. Colotta, D. Poli, G. Filacchioni, A. Galli and C. Costagli, *Chem. Pharm. Bull.*, **57**, 826 (2009).
09CPB846	K. Nakahara, K. Yamaguchi, Y. Yoshitake, T. Yamaguchi and K. Harazo, *Chem. Pharm. Bull.*, **57**, 846 (2009).
09CRV2275	S. Lee, T.G. LaCour and P.L. Fuchs, *Chem. Rev.*, **109**, 2275 (2009).
09CRV2880	W.B. Parker, *Chem. Rev.*, **109**, 2880 (2009).
09DMD2204	L.K. Chico, H.A. Behanna, W. Hu, G. Zhong, S.M. Roy and D.M. Watterson, *Drug Metab. Dispos.*, **37**, 2204 (2009).
09EJM609	A.B.A. El-Gazzar, M.M. Youssef, A.M.S. Youssef, A.A. Abu-Hashem and F.A. Badria, *Eur. J. Med. Chem.*, **44**, 609 (2009).
09EJM3330	R. Rohini, K. Shanker, P.M. Reddy, Y.-P. Ho and V. Ravinder, *Eur. J. Med. Chem.*, **44**, 3330 (2009).
09EJM4021	A. Bourkoula, M. Paravatou-Petsotas, A. Papadopoulos, I. Santos, H.-J. Pietzsch, E. Livaniou, M. Pelecanou, M. Papadopoulos and I. Pirmettis, *Eur. J. Med. Chem.*, **44**, 4021 (2009).
09EJM4249	A.-R.B.A. El-Gazzar, H.N. Hafez and H.-A.S. Abbas, *Eur. J. Med. Chem.*, **44**, 4249 (2009).
09EJM4448	A.M. Khalil, M.A. Berghot and M.A. Gouda, *Eur. J. Med. Chem.*, **44**, 4448 (2009).
09EJO2490	J.A. Bleda, P.M. Fresneda, R. Orenes and P. Molina, *Eur. J. Org. Chem.*, **15**, 2490 (2009).
09EJO4867	A. Ribeiro, M.A. Carvalho and M.F. Proença, *Eur. J. Org. Chem.*, **15**, 4867 (2009).
09EJO5738	T. Sasada, M. Moriuchi, N. Sakai and T. Konakahara, *Eur. J. Org. Chem.*, **15**, 5738 (2009).
09EJP38	Y. Nakagami, K. Kawashima, K. Yonekubo, M. Etori, J. Jojima, S. Miyazaki, R. Sawamura, K. Hirahara, F. Nara and M. Yamashita, *Eur. J. Pharm.*, **624**, 38 (2009).
09HCA932	S. Balalaie, S. Abdolmohammadib and B. Soleimanifarda, *Helv. Chim. Acta*, **92**, 932 (2009).
09IC9120	A. Barve, A. Kumbhar, M. Bhat, B. Joshi, R. Butcher, U. Sonawane and R. Joshi, *Inorg. Chem.*, **48**, 9120 (2009).
09IECR10120	H.K. Cammenga, V.N. Emel'yanenko and S.P. Verevkin, *Ind. Eng. Chem. Res.*, **48**, 10120 (2009).
09JA7828	A. Sattler, G. Zhu and G. Parkin, *J. Am. Chem. Soc.*, **131**, 7828 (2009).

09JA11353	J.D. Aguirre, A.M. Angeles-Boza, A. Chouai, J.-P. Pellois, C. Turro and K.R. Dunbar, *J. Am. Chem. Soc.*, **131**, 11353 (2009).
09JA11654	N.D. Shapiro, Y. Shi and F.D. Toste, *J. Am. Chem. Soc.*, **131**, 11654 (2009).
09JA11660	B. Jiang, S.-J. Tu, P. Kaur, W. Wever and G. Li, *J. Am. Chem. Soc.*, **131**, 11660 (2009).
09JBIC261	R. Garcia, P. Fousková, L. Gano, A. Paulo, P. Campello, É. Tóth and I. Santos, *J. Biol. Inorg. Chem.*, **14**, 261 (2009).
09JC0393	R. Ghahremanzadeh, M. Sayyafi, S. Ahadi and A.J. Bazgir, *Comb. Chem.*, **11**, 393 (2009).
09JC0612	B. Jiang, L.-J. Cao, S.-J. Tu, W.-R. Zheng and H.-Z. Yu, *J. Comb. Chem.*, **11**, 612 (2009).
09JCTC2044	T. Jung, R. Beckhaus, T. Klüner, S. Höfener and W. Klopper, *J. Chem. Theory Comput.*, **5**, 2044 (2009).
09JHC152	L. Rong, H. Han, H. Wang, H. Jiang, S. Tu and D. Shi, *J. Heterocycl. Chem.*, **46**, 152 (2009).
09JHC178	K.S. Jain, J.B. Bariwal, M.S. Phoujdar, M.A. Nagras, R.D. Amrutkar, M.K. Munde, R.S. Tamboli, S.A. Khedkar, R.H. Khiste, N.C. Vidyasagar, V.V. Dabholkar and M.K. Kathiravan, *J. Heterocycl. Chem.*, **46**, 178 (2009).
09JHC285	R.V. Rudenko, S.A. Komykhov, V.I. Musatov and S.M. Desenko, *J. Heterocycl. Chem.*, **46**, 285 (2009).
09JHC327	M.G. Ghagare, D.R. Birari, D.P. Shelar, R.B. Toche and M.N. Jachak, *J. Heterocycl. Chem.*, **46**, 327 (2009).
09JHC459	E. Perspicace, S. Hesse, G. Kirsch, M. Yemloul and C. Lecomtec, *J. Heterocycl. Chem.*, **46**, 459 (2009).
09JHC584	F.Z. Hu, G.F. Zhang, B. Liu, X.M. Zou, Y.Q. Zhu and H.Z. Yang, *J. Heterocycl. Chem.*, **46**, 584 (2009).
09JHC664	S.-L. Wang, W.-J. Hao, S.-J. Tu, X.-H. Zhang, X.-D. Cao, S. Yan, S.-S. Wu, Z.-G. Han and F. Shi, *J. Heterocycl. Chem.*, **46**, 664 (2009).
09JHC674	M.C. Cardia, S. Distinto, E. Maccioni, A. Plumitallo, L. Sanna, M.L. Sanna and S. Vigo, *J. Heterocycl. Chem.*, **46**, 674 (2009).
09JHC708	B.K. Ghotekar, M.N. Jachak and R.B. Toche, *J. Heterocycl. Chem.*, **46**, 708 (2009).
09JHC895	A.A. Aly, *J. Heterocycl. Chem.*, **46**, 895 (2009).
09JHC960	S. Tumkevicius, J. Dodonova, I. Baskirova and A. Voitechovicius, *J. Heterocycl. Chem.*, **46**, 960 (2009).
09JHC1151	Z. Zhang, J. Liu, W. Liu, X. Wang, Z. Cheng and J. Liu, *J. Heterocycl. Chem.*, **46**, 1151 (2009).
09JHC1266	R. Ghahremanzadeh, T. Amanpour and A. Bazgir, *J. Heterocycl. Chem.*, **46**, 1266 (2009).
09JHC1413	S.A. Komykhov, K.S. Ostras, K.M. Kobzar, V.I. Musatov and S.M. Desenko, *J. Heterocycl. Chem.*, **46**, 1413 (2009).
09JHC1420	S. Ferro, S. Agnello, M.L. Barreca, L. De Luca, F. Christ and R. Gitto, *J. Heterocycl. Chem.*, **46**, 1420 (2009).
09JJC1	S. Ostrowski, *Jordan J. Chem.*, **1**, 1 (2009).
09JLCR485	S.B. Tran, I.V. Ekhato and J.K. Rinehart, *J. Label Compd. Rad.*, **52**, 485 (2009).
09JMC932	A. Samanta, D.D. Leonidas, S. Dasgupta, T. Pathak, S.E. Zographos and N.G. Oikonomakos, *J. Med. Chem.*, **52**, 932 (2009).
09JMC2148	J.C. Pelletier, M.V. Chengalvala, J.E. Cottom, I.B. Feingold, D.M. Green, D.B. Hauze, C.A. Huselton, J.W. Jetter, G.S. Kopf, J.T. Lundquist, IV, R.L. Magolda, C.W. Mann, J.F. Mehlmann, J.F. Rogers, L.K. Shanno, W.R. Adams, C.O. Tio and J.E. Wrobel, *J. Med. Chem.*, **52**, 2148 (2009).
09JMC4794	P.A. Brough, X. Barril, J. Borgognoni, P. Chene, N.G.M. Davies, B. Davis, M.J. Drysdale, B. Dymock, S.A. Eccles, C. Garcia-Echeverria, C. Fromont, A. Hayes, R.E. Hubbard, A.M. Jordan, M.R. Jensen, A. Massey, A. Merrett, A. Padfield, R. Parsons, T. Radimerski, F.I. Raynaud, A. Robertson, S.D. Roughley, J. Schoepfer, H. Simmonite, S.Y. Sharp, A. Surgenor, M. Valenti, S. Walls, P. Webb, M. Wood, P. Workman and L. Wright, *J. Med. Chem.*, **52**, 4794 (2009).
09JMC5152	M.G. Brasca, N. Amboldi, D. Ballinari, A. Cameron, E. Casale, G. Cervi, M. Colombo, F. Colotta, V. Croci, R. D'Alessio, F. Fiorentini, A. Isacchi, C. Mercurio, W. Moretti, A. Panzeri, W. Pastori, P. Pevarello, F. Quartieri, F. Roletto, G. Traquandi, P. Vianello, A. Vulpetti and M. Ciomei, *J. Med. Chem.*, **52**, 5152 (2009).

09JMC6193	J. Wagner, P. Von Matt, R. Sedrani, R. Albert, N. Cooke, C. Ehrhardt, M. Geiser, G. Rummel, W. Stark, A. Strauss, S.W. Cowan-Jacob, C. Beerli, G. Weckbecker, J.-P. Evenou, G. Zenke and S. Cottens, *J. Med. Chem.*, **52**, 6193 (2009).
09JMC6946	S. Butini, R. Budriesi, M. Hamon, E. Morelli, S. Gemma, M. Brindisi, G. Borrelli, E. Novellino, I. Fiorini, P. Ioan, A. Chiarini, A. Cagnotto, T. Mennini, C. Fracasso, S. Caccia and G. Campiani, *J. Med. Chem.*, **52**, 6946 (2009).
09JMC7950	A.M. Quinn, G.A. Wasney, A. Dong, D. Barsyte, I. Kozieradzki, G. Senisterra, I. Chau, A. Siarheyeva, D.B. Kireev, A. Jadhav, J.M. Herold, S.V. Frye, C.H. Arrowsmith, P.J. Brown, A. Simeonov, M. Vedadi and J. Jin, *J. Med. Chem.*, **52**, 7950 (2009).
09JOC795	M. Achmatowicz, O.R. Thiel, P. Wheeler, C. Bernard, J. Huang, R.D. Larsen and M.M. Faul, *J. Org. Chem.*, **74**, 795 (2009).
09JOC3558	M. Alajarin, B. Bonillo, B. Marin-Luna, A. Vidal and R.A. Orenes, *J. Org. Chem.*, **74**, 3558 (2009).
09JOC3711	S. Achelle, I. Nouira, B. Pfaffinger, Y. Ramondenc, N. Plé and J. Rodríguez-López, *J. Org. Chem.*, **74**, 3711 (2009).
09JOC4646	P. Bannwarth, A. Valleix, D. Gre'e and R. Grée, *J. Org. Chem.*, **74**, 4646 (2009).
09JOC4934	F. Portela-Cubillo, J.S. Scott and J.C. Walton, *J. Org. Chem.*, **74**, 4934 (2009).
09JOC8021	J. Lu, N.-S. Li, S.C. Koo and J.A. Piccirilli, *J. Org. Chem.*, **74**, 8021 (2009).
09JOC8460	O.K. Ahmad, M.D. Hill and M. Movassaghi, *J. Org. Chem.*, **74**, 8460 (2009).
09JOC8554	B. Rodriguez-Molina, M.E. Ochoa, N. Farfán, R. Santillan and M.A. García-Garibay, *J. Org. Chem.*, **74**, 8554 (2009).
09JPS4886	N.-H. You, Y. Nakamura, Y. Suzuki, T. Higashihara, S. Ando and M. Ueda, *J. Polym. Sci. Pol. Chem.*, **47**, 4886 (2009).
09L9405	J. Do, J. Huh and E. Kim, *Langmuir*, **25**, 9405 (2009).
09L11486	A. Patra, J. Ralston, R. Sedev and J. Zhou, *Langmuir*, **25**, 11486 (2009).
09MCT2546	M.P. Morelli, A.M. Brown, T.M. Pitts, J.J. Tentler, F. Ciardiello, A. Ryan, J.M. Jürgensmeier and S.G. Eckhardt, *Mol. Canc. Ther.*, **8**, 2546 (2009).
09NAT1070	W. Zhou, D. Ercan, L. Chen, C.-H. Yun, D. Li, M. Capelletti, A.B. Cortot, L. Chirieac, R.E. Iacob, R. Padera, J.R. Engen, K.K. Wong, M.J. Eck, N.S. Gray and P.A. Jänne, *Nature*, **462**, 1070 (2009).
09OBC1171	B. Jiang, W.-J. Hao, J.P. Zhang, S.-J. Tu and F. Shia, *Org. Biomol. Chem.*, **7**, 1171 (2009).
09OBC2841	M. Radi, S. Schenone and M. Botta, *Org. Biomol. Chem.*, **7**, 2841 (2009).
09OBC4415	H.S. Ban, T. Usui, W. Nabeyama, H. Morita, K. Fukuzawac and H. Nakamura, *Org. Biomol. Chem.*, **7**, 4415 (2009).
09OL389	A.V. Lygin and A. De Meijere, *Org. Lett.*, **11**, 389 (2009).
09OL2161	T. Sasada, F. Kobayashi, N. Sakai and T. Konakahara, *Org. Lett.*, **11**, 2161 (2009).
09OL4500	M. Mastalerz, V. Fischer, C.-Q. Ma, R.A.J. Janssen and P. Bäuerle, *Org. Lett.*, **11**, 4500 (2009).
09OL4898	M. Velusamy, J.-H. Huang, Y.-C. Hsu, H.-H. Chou, K.-C. Ho, P.-L. Wu, W.-H. Chang, J.T. Lin and C.-W. Chu, *Org. Lett.*, **11**, 4898 (2009).
09OL5118	J. Board, J.X. Wang, A.P. Crew, M. Jin, K. Foreman, M.J. Mulvihill and V. Snieckus, *Org. Lett.*, **11**, 5118 (2009).
09OM1495	B. Maity, M. Roy, S. Saha and A.R. Chakravarty, *Organometallics*, **28**, 1495 (2009).
09OM2142	D. Meyer, M.A. Taige, A. Zeller, K. Hohlfeld, S. Ahrens and T. Strassner, *Organometallics*, **28**, 2142 (2009).
09OM3256	K. Ohno, K. Arima, S. Tanaka, T. Yamagata, H. Tsurugi and K. Mashima, *Organometallics*, **28**, 3256 (2009).
09OM6194	S. Choua, J.P. Djukic, J. Dalléry, R. Welter and P. Turek, *Organometallics*, **28**, 6194 (2009).
09OPP479	J.I. Bardag'ı and R.A. Rossi, *Org. Prep. Proc. Inter.*, **41**, 479 (2009).
09OPRD230	O.R. Thiel, M. Achmatowicz, C. Bernard, P. Wheeler, C. Savarin, T.L. Correll, A. Kasparian, A. Allgeier, M.D. Bartberger, H. Tan and R.D. Larsen, *Org. Process. Res. Dev.*, **13**, 230 (2009).
09PCJ87	N.N. Smolyar, Y.M. Yutilov and S.V. Gres'ko, *Pharm. Chem. J.*, **43**, 87 (2009).
09RCB89	V.V. Yanilkin, N.V. Nastapova, A.S. Stepanov, A.A. Kalinin and V.A. Mamedov, *Russ. Chem. Bull.*, **58**, 89 (2009).

09RJOC211	V.V. Didenko, V.A. Voronkova and K.S. Shikhaliev, *Russ. J. Org. Chem.*, **45**, 211 (2009).
09RJOC848	R.V. Tyurin, O.V. Kosygina and V.V. Mezheritskii, *Russ. J. Org. Chem.*, **45**, 848 (2009).
09RJOC904	E.V. Nosova, A.A. Laeva, T.V. Trashakhova, A.V. Golovchenko, G.N. Lipunova, P.A. Slepukhin and V.N. Charushin, *Russ. J. Org. Chem.*, **45**, 904 (2009).
09RJOC1691	A.S. Shestakov, O.E. Sidorenko, I.S. Bushmarinov, K.S. Shikhaliev and M.Y. Antipin, *Russ. J. Org. Chem.*, **45**, 1691 (2009).
09RJOC1730	P.S. Silaichev, M.A. Kryuchkova and A.N. Maslivets *Russ, J. Org. Chem.*, **45**, 1730 (2009).
09S1715	A. Heim-Riether and K.R. Gipson, *Synthesis*, **10**, 1715 (2009).
09S1876	J.-C. Monbaliu and J. Marchand-Brynaert, *Synthesis*, **11**, 1876 (2009).
09S2379	A. Tikad, S. Routier, M. Akssira, J.M. Léger, C. Jarry and G. Guillaumeta, *Synthesis*, **14**, 2379 (2009).
09S3967	V.O. Iaroshenko, *Synthesis*, **23**, 3967 (2009).
09S4049	S. Herzog, G. Buehrdel, R. Beckert, S. Klimas, E.U. Würthwein, S. Grimme and H. Görls, *Synthesis*, **23**, 4049 (2009).
09SC4088	F.M. Abdelrazek and N.H. Metwally, *Synth. Commun.*, **39**, 4088 (2009).
09SL260	M. Nikulnikov, S. Tsirulnikov, V. Kysil, A. Ivachtchenko and M. Krasavin, *Synlett*, 260 (2009).
09SL611	N.-Y. Huang, Y.-B. Nie and M.-W. Ding, *Synlett*, 611 (2009).
09SL625	I.A. Abdelhamid, *Synlett*, 625 (2009).
09SL3206	G. Xu, L. Zheng, S. Wang, Q. Dang and X. Bai, *Synlett*, 3206 (2009).
09T909	M.C. Jahnke, M. Hussain, F. Hupka, Y. Pape, S. Ali, F.E. Hahn and K.J. Cavell, *Tetrahedron*, **65**, 909 (2009).
09T1574	D. Viña, E. Del Olmo, J.L. Lopez-Pérez and A. San Feliciano, *Tetrahedron*, **65**, 1574 (2009).
09T3019	R. Saito, M. Tokita, K. Uda, C. Ishikawa and M. Satoh, *Tetrahedron*, **65**, 3019 (2009).
09T3804	A.A. Mohammadi, M. Dabiri and H. Qaraat, *Tetrahedron*, **65**, 3804 (2009).
09T4190	N. Hebbar, Y. Ramondenc, G. Plé, G. Dupas and N. Plé, *Tetrahedron*, **65**, 4190 (2009).
09T7146	S. Saleh, M. Picquet, M.P. Meunier and J.-C. Hierso, *Tetrahedron*, **65**, 7146 (2009).
09T8582	R.T. Iminov, A.V. Tverdokhlebov, A.A. Tolmachev, Y.M. Volovenko, S.V. Shishkina and O.V. Shishkin, *Tetrahedron*, **65**, 8582 (2009).
09T8969	T.J. Donohoe, J.F. Bower, J.A. Basutto, L.P. Fishlock, P.A. Procopiou and C.K.A. Callens, *Tetrahedron*, **65**, 8969 (2009).
09T8981	S.J. Pridmore, P.A. Slatford, J.E. Taylor, M.K. Whittlesey and J.M.J. Williams, *Tetrahedron*, **65**, 8981 (2009).
09T10069	I.A. Abdelhamid, M.H. Mohamed, A.M. Abdelmoniem and S.A.S. Ghozlan, *Tetrahedron*, **65**, 10069 (2009).
09TL943	M.M. Heravi, S.S. Sadjadi, N.M. Haj, H.A. Oskooie, R.H. Shoar and F.F. Bamoharram, *Tetrahedron Lett.*, **50**, 943 (2009).
09TL1478	M. Li and R. Hua, *Tetrahedron Lett.*, **50**, 1478 (2009).
09TL1600	C.S. Harris, L.F. Hennequin and O. Willerval, *Tetrahedron Lett.*, **50**, 1600 (2009).
09TL1618	A. Kodimuthali, B.C. Chary, P.L. Prasunamba and M. Pal, *Tetrahedron Lett.*, **50**, 1618 (2009).
09TL2899	Y.A. Azev, S.V. Shorshnev and B.V. Golomolzin, *Tetrahedron Lett.*, **50**, 2899 (2009).
09TL4050	S. Kothandaraman, D. Guiadeen, G. Butora, G. Doss, S.G. Mills, M. MacCoss and L. Yang, *Tetrahedron Lett.*, **50**, 4050 (2009).
09TL5389	P. Barraja, V. Spanò, P. Diana, A. Carbone and G. Cirrincione, *Tetrahedron Lett.*, **50**, 5389 (2009).
09TL6048	G. Li, R. Kakarla, S.W. Gerritz, A. Pendri and B. Ma, *Tetrahedron Lett.*, **50**, 6048 (2009).
09TL6358	O.V. Vinogradova, V.N. Sorokoumov and I.A. Balova, *Tetrahedron Lett.*, **50**, 6358 (2009).

09TL6411	S.M. Al-Mousawi, M.S. Moustafa and M.H. Elnagdi, *Tetrahedron Lett.*, **50**, 6411 (2009).
09TL6698	K.J. Lee, J.K. Choi, E.K. Yumb and S.Y. Cho, *Tetrahedron Lett.*, **50**, 6698 (2009).
09TL6779	G.I. Borodkin, A.Y. Vorobév, M.M. Shakirov and V.G. Shubin, *Tetrahedron Lett.*, **50**, 6779 (2009).
09TL7205	C.C. Moldoveanu, P.G. Jones and I.I. Mangalagiu, *Tetrahedron Lett.*, **50**, 7205 (2009).
09TL7333	A. Lauria, A. Guarcello, G. Macaluso, G. Dattolo and A.M. Almerico, *Tetrahedron Lett.*, **50**, 7333 (2009).
09TR727	J. Jia, X. Zhang, Y.-S. Hu, Y. Wua, Q.-Z. Wang, N.N. Li, C.-Q. Wu, H.-X. Yu and Q.-C. Guo, *Thromb. Res.*, **123**, 727 (2009).
09ZAAC2459	M. Wriedt and C. Näther, *Z. Anorg. Allg. Chem.*, 2459 (2009).

CHAPTER 6.3

Triazines, Tetrazines and Fused Ring Polyaza Systems

Dmitry N. Kozhevnikov*, Anton M. Prokhorov**

*Institute of Organic Synthesis, Ural Division of RAS, S. Kovalevskoy 20, Ekaterinburg, 620990, Russia
dnk@ios.uran.ru
**Ural Federal University, Mira 19, Ekaterinburg, 620002, Russia

6.3.1. TRIAZINES

6.3.1.1 1,2,3-Triazines

Complexes of 1,2,3-triazine, 1,3,5-triazine stabilized by X−Li· · ·N bonds with F−Li, H−Li, and CH_3Li as the Lewis acids have been studied by *ab initio* calculations <09PCA10327>.

The reactions of 1-amino-2,3-diferrocenylcyclopropenylium tetrafluoroborate **1** with sodium azide afforded 5-amino-4,6-diferrocenyl-1,2,3-triazines **2** <09JHC477>.

1 **2**

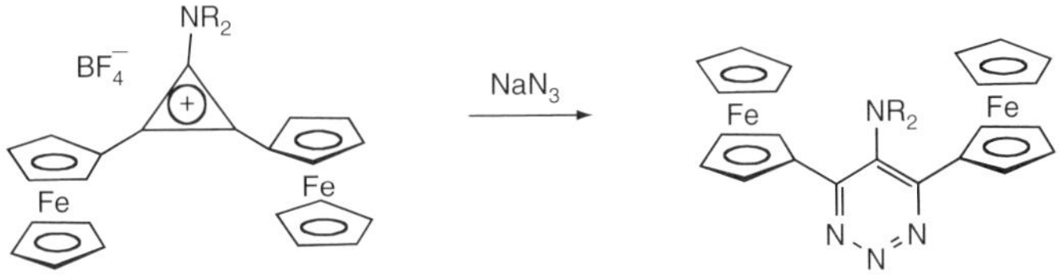

6.3.1.2 1,2,4-Triazines

One of the most useful and widely used properties of 1,2,4-triazines is their application as dienes in pyridine synthesis through (*i*) an inverse electron demand Diels–Alder (D–A) reaction with an appropriate electron-rich dienophile, *e.g.* an enamine (*ii*) nitrogen molecule extrusion by a retro-D–A reaction, and (*iii*) rearomatisation of the intermediate dihydropyridine, *e.g.* by elimination of an amine.

All four members of the louisianin family, simple pyridine and 2-pyridone alkaloids that display both antibacterial and anticancer activity, were synthesized *via* a common cyclopentenopyridine **3**. The latter was obtained from a conveniently prepared 1,2,4-triazine **4** by reaction with morpholinocyclopentene <09JOC8343>.

Progress in Heterocyclic Chemistry, Volume 22
ISSN 0959-6380, DOI: 10.1016/S0959-6380(11)22014-2

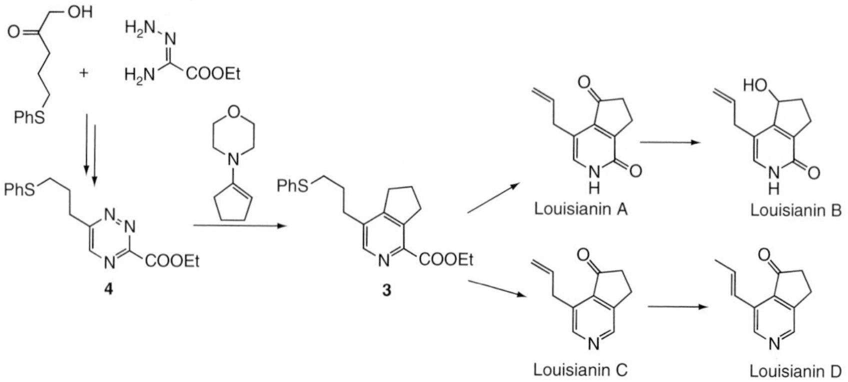

A simple method for the synthesis of pyridines **5** from hydrazides has been reported in which the key steps were reaction of the hydrazide **6** with a carbene intermediate derived from α-diazo-β-keto-esters **7**, cyclisation with ammonium acetate giving 1,2,4-triazines **8**, followed by D–A reaction with norbornadiene <09OL3686>.

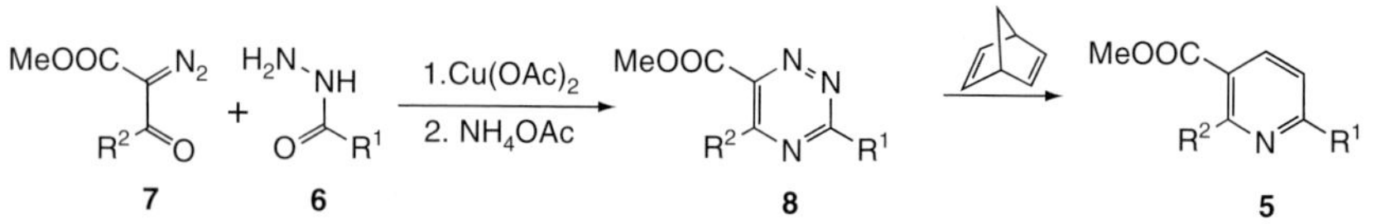

α-Chloro-α-acetoxy-β-keto-esters **9**, prepared from β-keto-esters **10**, reacted with amidrazones **11** yielding 1,2,4-triazines **12** in good yields. Reaction of triazines **12** with dienophiles, e.g. norbornadiene, gave derivatives **13** of nicotinic acid. A 'one-pot' reaction of the α,β-diketo-esters with amidrazones **11** in the presence of norbornadiene in boiling ethanol yielded the pyridines **13** directly without the need to isolate the corresponding triazines **12** <09T975>.

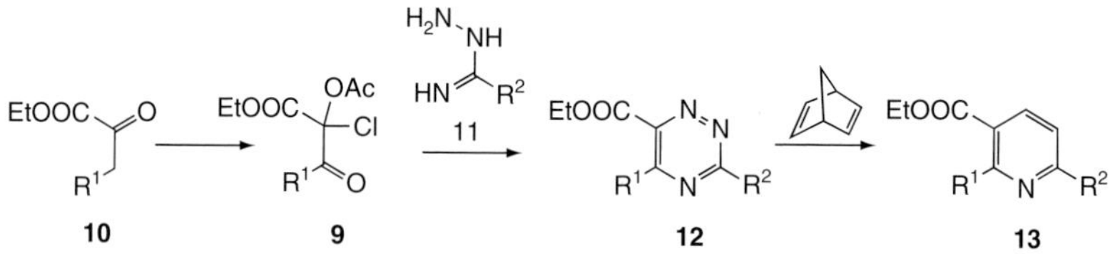

A similar approach was used for the synthesis of 2,2′:6′,2″-terpyridine-5,5″-dicarboxylates **14** from α-acetoxy-α-chloro-β-keto-esters **9** with pyridine-2,6-bis-amidrazone **15** and 2,5-norbornadiene via 2,6-bis(1,2,4-triazin-3-yl)pyridines **16** <09T1115>.

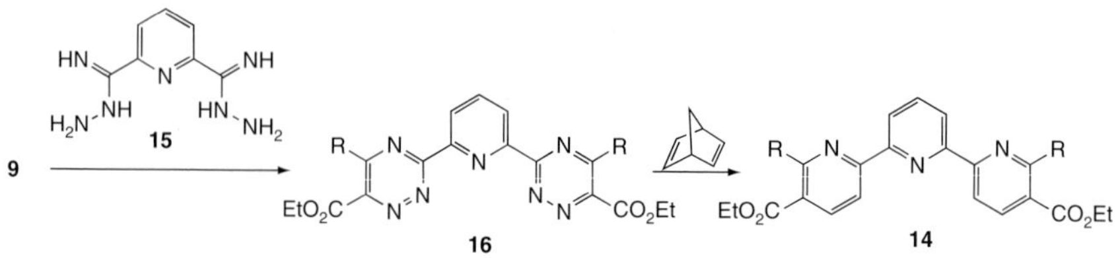

A series of new 5-aryl-2-thienylpyridines **17**, **18** and their phosphorescent cyclometallated platinum(II) complexes **19**, **20** were obtained *via* 3-(thien-2-yl)-1,2,4-triazines **21** and their reactions with norbornadiene or an enamine. Triazines **21** are readily available due to their easy synthesis from hydrazide **22** and bromoacetophenones **23** <09IC4179>.

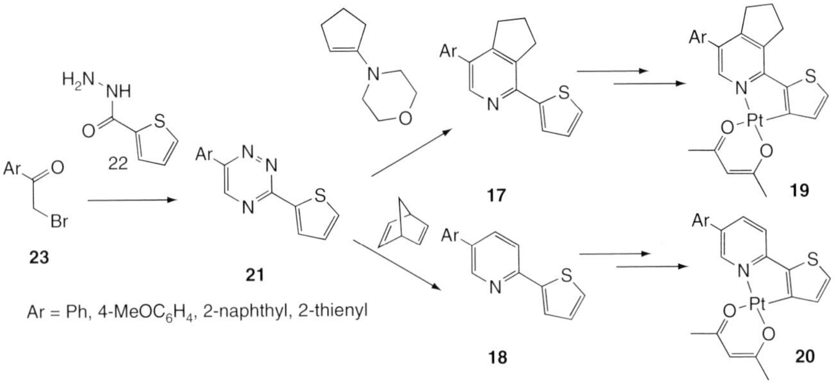

In the same manner, liquid crystalline cyclometallated platinum(II) complexes [PtLacac] **24**, where H*L* = 2,5-bis(4-alkoxyphenyl)pyridines **25** were obtained from ready available 2,6-bis(4-methoxyphenyl)-1,2,4-triazine **26** through D–A reaction with norbornadiene and a demethylation-alkylation sequence <09CM3831>.

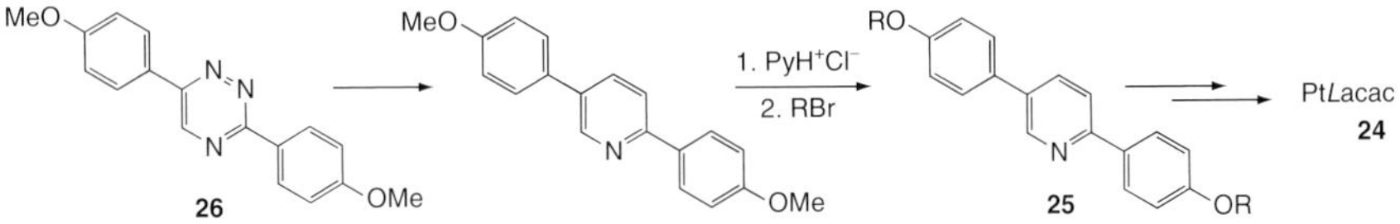

The efficient palladium-catalyzed amination of 1,2,4-triazines has been reported. Starting from 3-methylthio-1,2,4-triazine **27**, Buchwald–Hartwig-type reactions with a wide range of amines gave 3-amino-1,2,4-triazines **28** in good to excellent yields <09SL2137>.

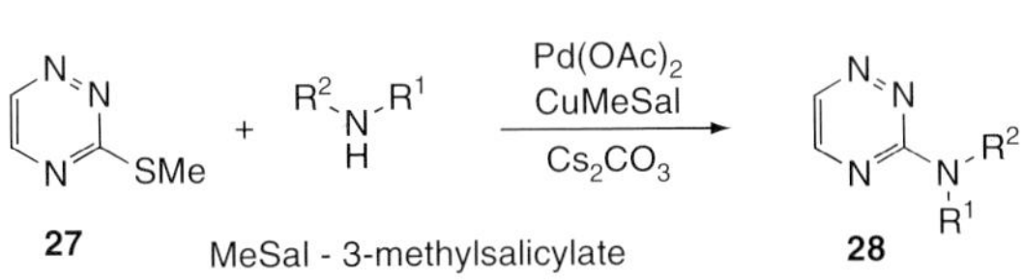

In general, 1,2,4-triazines have a high reactivity towards nucleophiles which allows direct introduction of various substituents at C-5 of the triazine ring. For example, reaction of 1,2,4-triazine 4-oxide **29** with lithium trimethylsilylacetylide followed by addition of acetyl chloride gave ethynyltriazine **30**. The reaction proceeded *via* formation of an intermediate adduct **31**, which could be isolated by carefully treating the reaction mixture with acetic acid <09TL1444>.

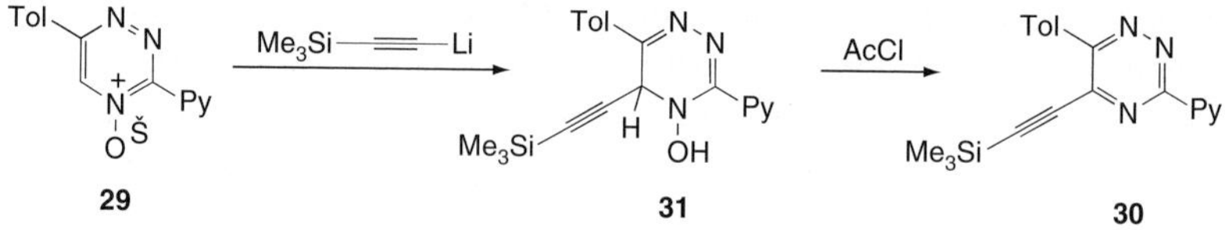

Other organolithium compounds can be used as nucleophiles in this reaction. Triazine 4-oxide **32** reacted with lithium ferrocene **33** yielding 5-ferrocenyl-1,2,4-triazine **34**, after treatment of the adduct **35** with acetyl chloride, or 5-ferrocenyl-1,2,4-triazine 4-oxide **36**, if oxidation with dichlorodicyanoquinone (DDQ) was used instead of acylation <09MI208>.

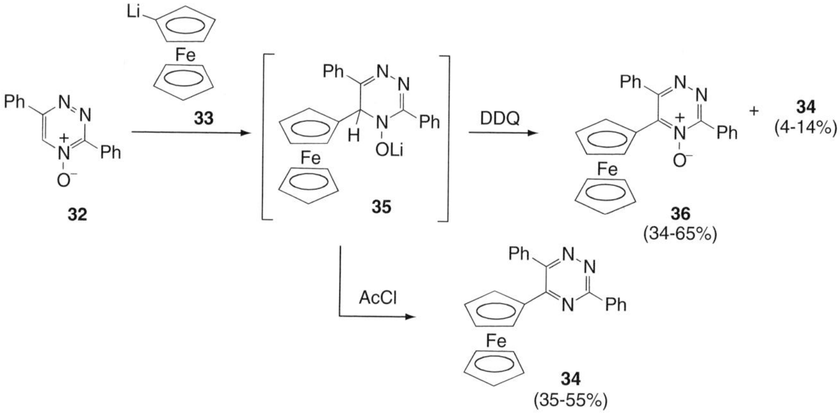

Reaction of triazine 4-oxide **32** with lithium derivative of nitronyl nitroxide radical **37** gave triazinyl nitronyl nitroxide **38** with retention of the *N*-oxide group in the triazine ring <09JOC2870>.

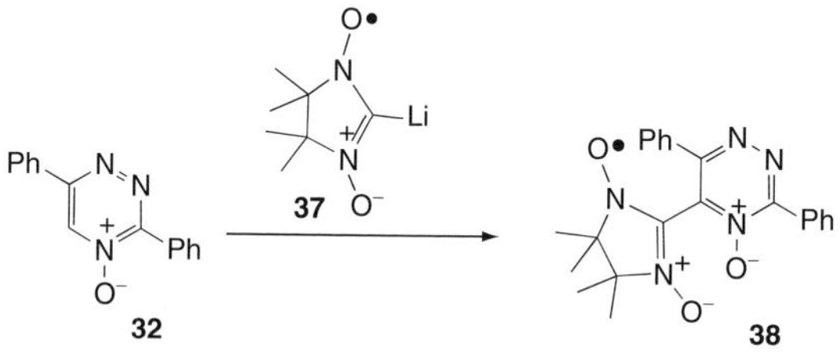

Triphenylphosphine-mediated reactions of diazoimides **39** in water resulted in formation of several 1,2,4-triazine derivatives **40** in high yields <09TL1331>.

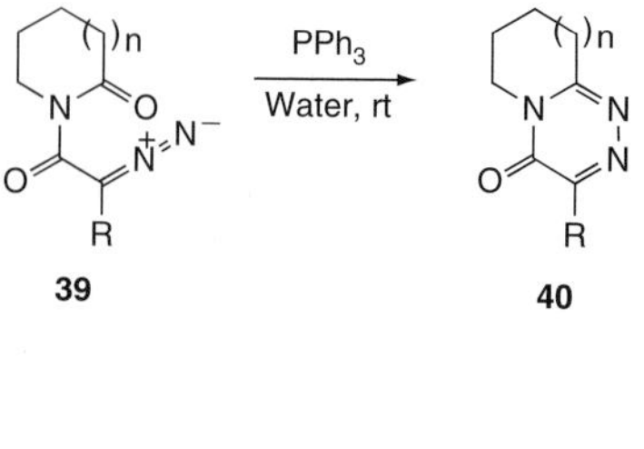

A method for the synthesis of 3,3'-dichloro-5,5'-bi-1,2,4-triazine **41** and its application to the synthesis of 3,3'-diamino-5,5'-bi-1,2,4-triazines **42** by nucleophilic aromatic substitution have been described <09H(78)623>.

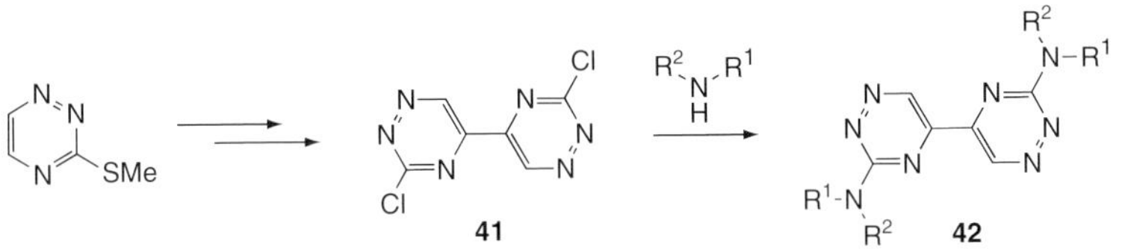

1,2,4-Triazines were used as ligands for transition metals, *e.g.* new copper(II) <09JCC1207>, zinc(II) and cadmium(II) <09JCC2155> complexes of 1,2,4-triazines have been described.

6.3.1.3 1,3,5-Triazines

Ruthenium(II) complexes of 6-phenyl-2,4-dipyridyl-1,3,5-triazine and terpyridine were obtained and their intense deep-red light emission at room temperature was described. Solid-state electroluminescent devices were prepared using one of the complexes. These devices emit deep-red light at low voltages and exhibit extraordinary stabilities <09IC3907>.

2,4,6-Tris(2-pyridyl)-1,3,5-triazine (tpt) was widely used as a ligand for transition metal complexes, e.g. complexes with copper(II) <09IC6630>, nickel(II) <09DT2510>, lead (II) <09JCC1972>, ruthenium(II) <09DT4012> have been described.

Metal complexes with more complex 1,3,5-triazine-based ligands, 2,4-bis-(2-diphenylmethylene)-hydrazinyl-6-piperidin-1-yl-1,3,5-triazine <09JCC1902>, 2,4,6-tris (N-7-azaindolyl)-1,3,5-triazine or 2,4,6-tris(2,2'-dipyridylamino)-1,3,5-triazine <09DT1776> have been demonstrated. A trinuclear star-shaped organometallic host based on 1,3,5-triazine with a π-acidic interior cavity was shown to have a strong anion–π interaction with PF_6^- <09CC1511>.

Data on the recyclization reactions of 1,3,5-triazines have been reviewed. Classification of the reactions was based on the fragments of the 1,3,5-triazine molecules used in the construction of new heterocycles: from a single C-atom to five CNCNC-atoms <09CHE130>.

Several dienophiles were used to extend the scope of 1,3,5-triazine inverse electron-demand Diels–Alder (D–A) reactions. The reactions of various 1,3,5-triazines with latent dienophiles 2-amino-3-thiophenecarboxylic acids **43** and 3-amino-2-thiophenecarboxylic acid have been described in the one-step synthesis of both thieno[2,3-*d*]pyrimidines **44** and thieno[3,2-*d*]pyrimidines <09TL2874>.

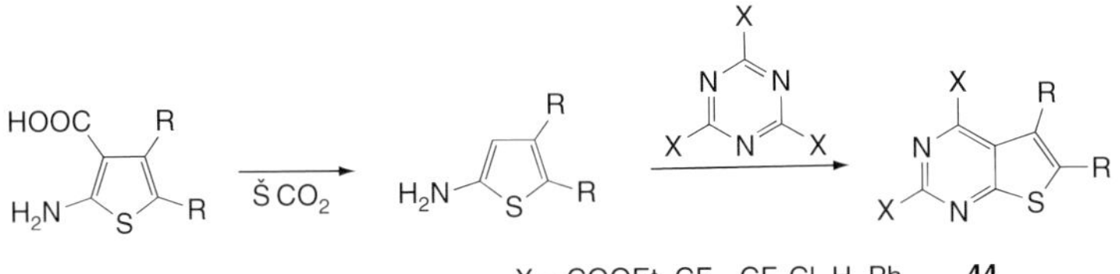

X = COOEt, CF_3, CF_2Cl, H, Ph

2-Aminoindoles were used as dienophiles in D–A reactions with 1,3,5-triazines to give various 3-aza-α-carbolines in excellent yields <09SL3206>. Furo[2,3-*d*]pyrimidines **45** were readily prepared *via* D–A reaction 2-aminofurans **46** with 1,3,5-triazines <09TL6758>.

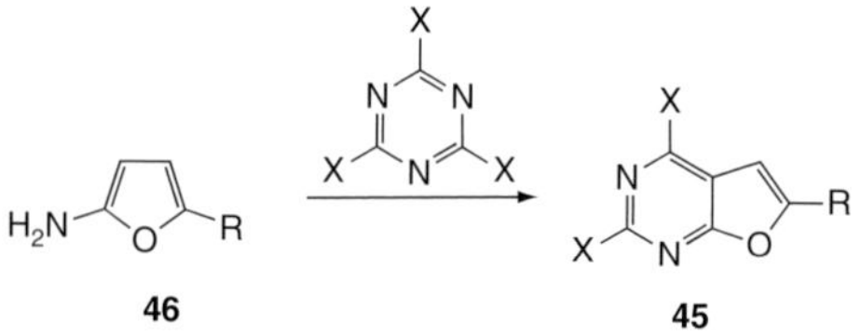

D–A reactions of 2,4,6-tris(polyfluoroalkyl)-1,3,5-triazines **47** with enamines **48**, amino heterocycles **49** (pyrazole, imidazole, thiazole, furan, thiophene, pyrrole, indole) and anilines resulted in the synthesis of polyfluoroalkyl pyrimidines **50** and heteroannulated pyrimidines **51** <09S3967>.

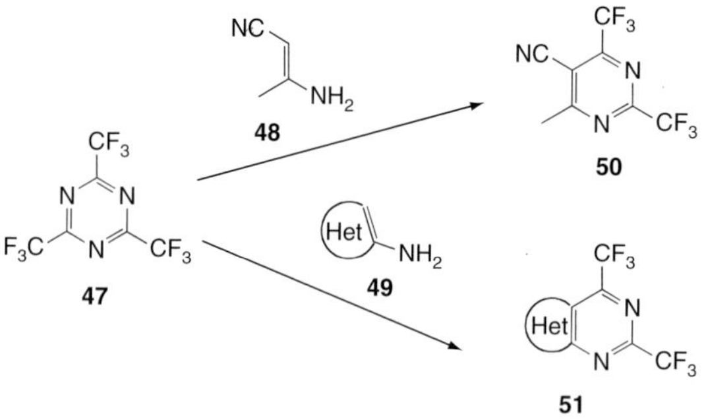

The mechanism of the D–A reaction of 1-*tert*-butyl-2-aminopyrrole with 2,4,6-tris(trifluoromethyl)-1,3,5-triazine to give a pyrrolo[2,3-*d*]pyrimidine was studied by NMR spectroscopy <09JOC319>.

Significant attention has been paid to C_3-symmetric star-shaped molecules based on a 1,3,5-triazine central core. The synthesis and properties of a series of new star-shaped donor-acceptor molecules containing the 2,4,6-tris(thiophen-2-yl)-1,3,5-triazine unit have been published in two independent works. In both cases, cross-couplings (Stille, Suzuki or Sonogashira) were used to extend conjugated side-chains using 2,4,6-tris(5-bromothien-2-yl)-1,3,5-triazine **52** as starting material. Reaction of **52** with tributylstannylthiophenes **53** resulted in triazines **54** bearing short oligothiophene side-chains <09TL5673>. Cross-couplings of **52** with arylboronic acid esters **55** or ethynylarenes **56** gave a series of star-shaped molecules **54**, **57** with various conjugated side-chains. All novel compounds exhibited two-photon absorption activity in the range of 720–880 nm <09EJO5587>.

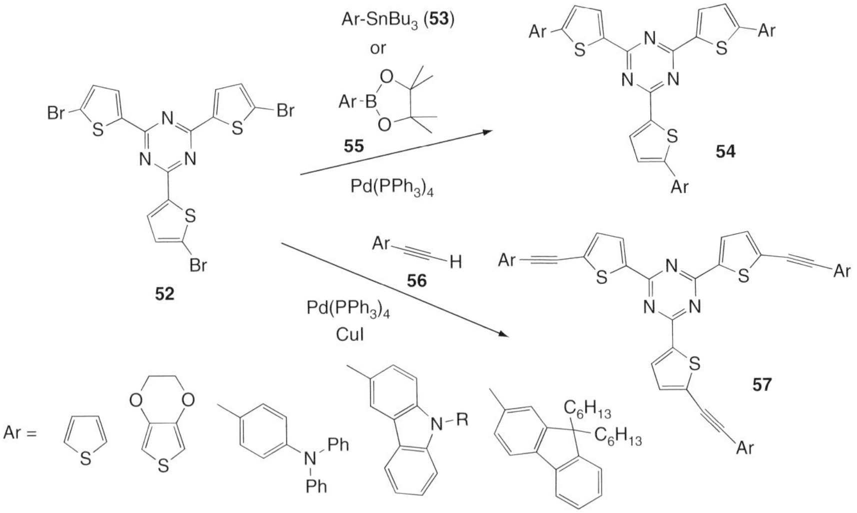

Suzuki couplings of 2,4,6-tris(3-bromophenyl)-1,3,5-triazine **58** (obtained by cyclisation of 3-bromobenzonitrile **59**) with arylboronic acids **60** gave oligophenyl star-shaped 1,3,5-triazine derivatives **61** as new electron transport host materials for green phosphorescent organic light-emitting devices. The morphological, thermal, and photophysical properties and the electron mobilities of these host materials are influenced by the nature of the aryl substituents attached to the triazine core <09JMC8112>.

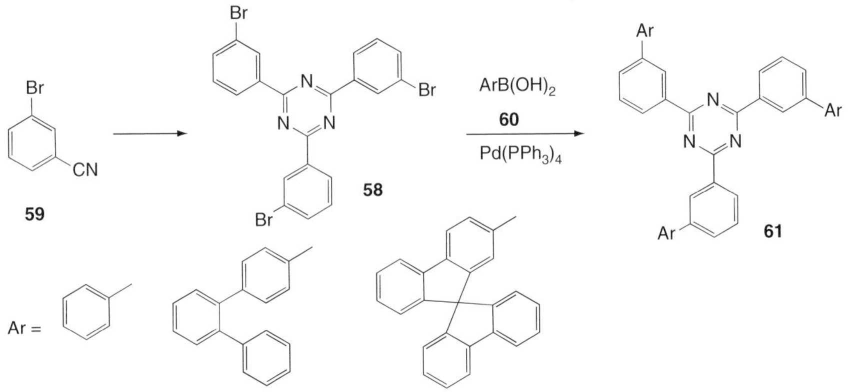

A new 1,3,5-triazine-based C_3 building block, bearing three phosphonate groups, has been prepared in a simple two-step synthesis starting from 4-bromomethylbenzonitrile; the new building block easily underwent further olefination reactions to afford in a straightforward manner 2,4,6-tris(tetrathiafulvalene)- and 2,4,6-tris(ferrocene)-1,3,5-triazines <09OL5398>.

1,3,5-Triazines **62** bearing a diphenylphosphanyl group were selectively synthesized from cyanuric chloride **63** by the one-pot step-by-step addition of silylphosphane and other nucleophiles. The Pd complexes of new phosphanyltriazines were isolated, and the crystal structures were determined. Mizorogi-Heck reaction and the Suzuki-Miyaura coupling reaction using the phosphanyltriazine-Pd catalysts have been described <09EJO4956>.

$$\text{63} \xrightarrow[\text{2. Nu}^1\text{H}]{\text{1. Ph}_2\text{PSiMe}_3} \text{62}$$

NuH = alkylamines or alcohols

Reaction of cyanuric chloride **63** or 4-alkoxy-2,6-dichloro-1,3,5-triazine with phenylenediamines resulted in formation of calixtriazines. These macrocycles tend to form flattened conformations, leading to a stable π-conjugated system, presumably due to the electronic features of the alkoxy-substituent on the triazine rings <09TL3923, 09S542>.

A variety of functionalized aryl bromides or benzyl chlorides were coupled with 2-chloro-4,6-dimethoxy-1,3,5-triazine in good yields by a one-step procedure via cobalt catalysis <09SL3192>.

A traceless solid-phase synthesis of 6-amino- and 6-hydroxyimino-1,3,5-triazine-2,4-diones and 1,3,5-triazine-2,4,6-triones has been developed. The strategy included linking a preformed N-carbamothioylcarbamate to bromomethyl resin to give an S-linked isothiourea, which then underwent cyclization with isocyanates to yield the resin-bound 1,3,5-triazine-2,4-diones. Subsequent cleavage was accomplished either by substitution with an amine or by oxidative activation of the thioether functionality followed by nucleophilic substitution <09JCO1050>.

5-Aryl-1,3,5-triazaspiro[5.5]undeca-1,3-diene-2,4-diamines prepared by cyclisation of anilines with cyclohexanone and dicyandiamide were found to have the best neuronal sodium binding activity among the four groups of triazines evaluated <09BML5644>.

A new organic fluorescent dye, 2,4-dichloro-6-[p-(N,N-diethylamino)biphenylyl]-1,3,5-triazine (DBQ) has been synthesized. DBQ exhibits high fluorescence quantum yields (0.96 in hexane and 0.71 in THF), high extinction coefficients, and an excitation window extending up to ^{TM}480 nm <09PCA5066>.

2-Methylthio-4,6-dimethoxy-1,3,5-triazine underwent smooth Pd-catalyzed cross-coupling reactions with functionalized aryl-, heteroaryl-, benzylic-, and alkylzinc reagents using Pd(OAc)$_2$/S-Phos as the catalytic system. No copper salt was required to perform these reactions <09OL4228>.

A family of 1,3,5-triazine dendrimers differing in their core flexibility was prepared and evaluated for their ability to accomplish gene transfection <09BCC1799>.

Bifunctional triazine-based ligands **64** containing both a TEMPO and a bipyridine moiety have been synthesized. These bpy/TEMPO-based molecules **64** have been used as catalyst precursors for the copper-catalyzed aerobic oxidation of alcohols to aldehydes and ketones <09DT3559>.

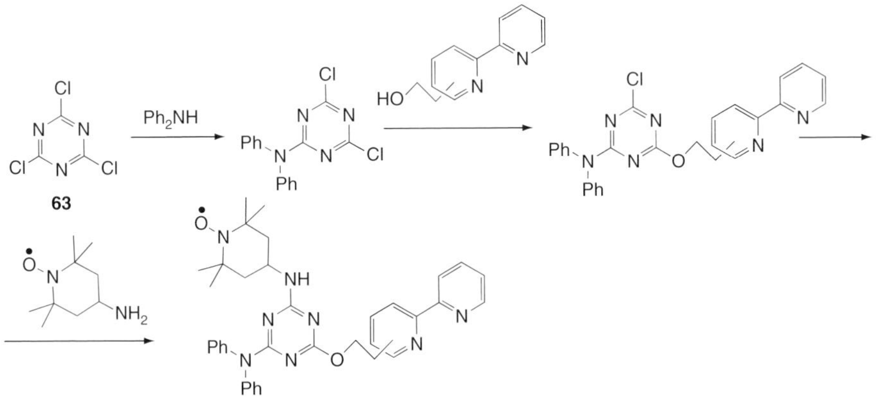

An original oxone-induced ring contraction of 4-methylthio-1,3,5-triazine-2 (1H)-thiones to 3-methylthio-5-amino-1,2,4-thiadiazoles has been described <09TL5802>.

An efficient method for the Lossen rearrangement that uses cyanuric chloride as a promoter was reported. This procedure allowed the preparation of various carbamates, thiocarbamates, and ureas in good yields directly from the corresponding hydroxamic acids <09TL6800>.

Monodisperse Fe_3O_4 nanoparticles originally synthesized with a hydrophobic oleylamine capping ligand were made water soluble and conjugated to the anticancer drug methotrexate using cyanuric chloride as linker <09JMC6400>.

A series of poly(methyl methacrylate)-based copolymers have been obtained through free radical copolymerizations of methyl methacrylate in the presence of the either 2-vinyl-4,6-diamino-1,3,5-triazine or vinylbenzylthymine. Hydrogen-bonding interactions within blends of the two copolymers poly(2-vinyl-4,6-diamino-1,3,5-triazine-co-methyl methacrylate) and poly(vinylbenzylthymine-co-methyl methacrylate) have been studied <09MM4701>.

Translational spectroscopy coupled with coincidence detection techniques has been used to investigate the two-body dissociation of 1,3,5-triazine to HCN + $(HCN)_2$ upon electronic excitation from charge exchange between the 1,3,5-triazine cation and cesium <09PCA8834>.

Hybrid organic–inorganic supramolecular aggregates displaying liquid crystal properties at room temperature have been synthesized by linking equimolar amounts of 2,4,6-triarylamino-1,3,5-triazine and metallo-acids $[Fe(CO)_4(CNC_6H_4CO_2H)]$ and $[M(CO)_5(CNC_6H_4CO_2H)]$ (M = Cr, Mo, W) through hydrogen bond formation <09CM3282>.

The π-stacking between phenylacetylene and 1,3,5-triazine was studied by IR–UV double resonance spectroscopy, IR-spectroscopy and DFT and MP2 calculations. π-Stacked heterodimer is formed between phenylacetylene and 1,3,5-triazine <09PCP11207>.

The surface-confined coupling reaction between 1,3,5-triazine-2,4,6-triamine and 1,4-phenylene diisocyanate has been investigated on Au(111) by scanning tunneling microscopy <09JA16706>.

The role played by hydrogen bonding in formation of stable glassy phases under ambient conditions in diaminotriazine derivatives with 3,5-dimethylphenyl groups has been studied <09JMC2747>.

Equimolar mixtures of melamine with the semiperfluorinated benzoic acids formed discrete hydrogen-bonded heterodimers with an elongated central core with liquid crystalline properties <09CM491>.

A method for the detection of melamine utilising the color change induced by triple hydrogen-bonding recognition between melamine and a cyanuric acid derivative grafted on the surface of gold nanoparticles has been suggested <09JA9496>.

6.3.2. TETRAZINES

The synthesis of the first vinyltetrazine derivative was described. 3,6-Divinyl-1,2,4,5-tetrazine **65** was obtained by methodology involving cyclization from an imidate **66** and use of 2-phenylsulfonylethyl groups as precursors for vinyl entities. The compound **65** can be seen as a tetraaza analogue of the widely used 1,4-divinylbenzene <09SL731>

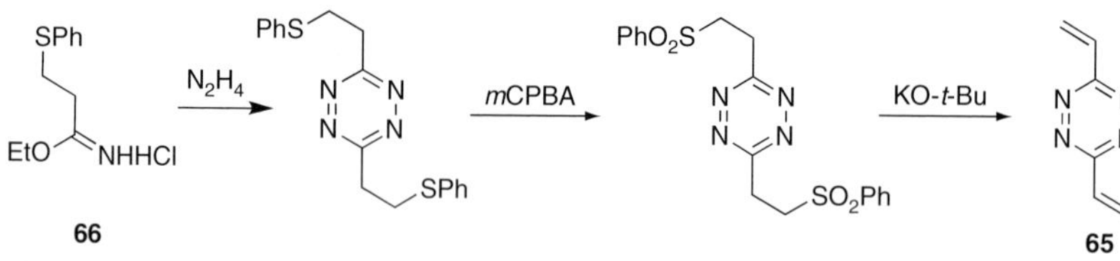

The efficient palladium-catalyzed amination of methyl sulfur derivatives of 1,2,4,5-tetrazines *via* methyl sulfur release was reported <09SL2137>.

The inverse electron demand Diels–Alder (D–A) reaction remains an important method for the synthesis of highly substituted pyridazines and related systems. A series of substituted 8,9-diazafluoranthenes **67** was synthesized from disubstituted tetrazines **68** and acenaphthylene **69** <09OBC2082>.

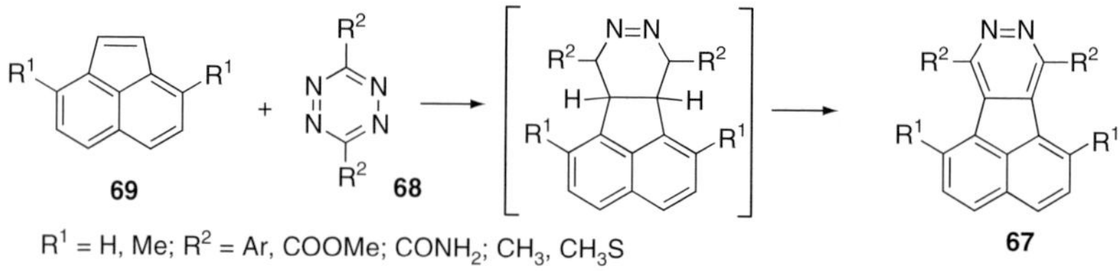

R^1 = H, Me; R^2 = Ar, COOMe; CONH$_2$; CH$_3$, CH$_3$S

The direct synthesis of pyridazine difluoroboranes **70** from *N*-heterocycle substituted tetrazines **71** by *N*-directed cycloaddition reaction has been reported <2010OL160>.

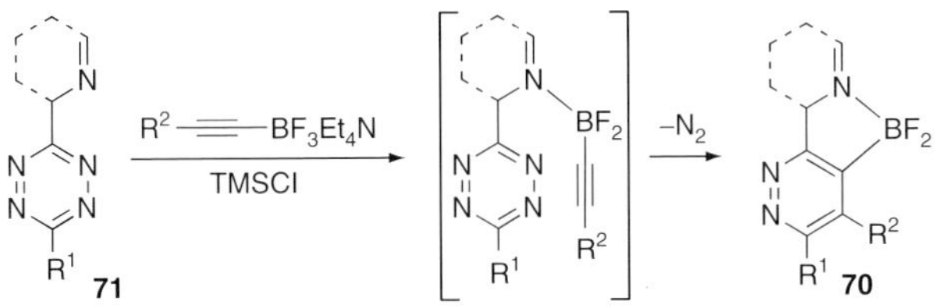

Devaraj et al. developed a highly sensitive technique for the covalent labelling of live cancer cells on the basis of the cycloaddition of a tetrazine to a highly strained *trans*-cyclooctene <09AGE7013>.

The D–A reactions of terminal alkyne units of SiO_2-supported [2]pseudorotaxanes with 1,2,4,5-tetrazine derivatives proceeded efficiently through solid-to-solid contact to provide both asymmetric and symmetric [2]pseudorotaxanes, where a pyridazine unit acts as a stopper for locking the macrocycle within a [2]rotaxane <09T2824>.

Formation of furopyridazines in reaction of symmetric 3,6-disubstituted 1,2,4,5-tetrazines with but-3-yn-1-ol was reported as an example of D–A reaction leading to new heterocyclic systems <09RJO1102>.

A computational approach was used to study the diazocinone and pyridazine forming cascades that result from the D–A reaction of tetrazines with cyclic enolates <09JOC4804>.

The regioselective hetero-D–A reaction of asymmetric tetrazines with electron-rich ethylenes has been studied theoretically by DFT methods <09JOC2726>.

A series of new heteroatom or aromatic substituted 1,2,4,5-tetrazines **72**, **73** and **74** was prepared *via* two routes: **A** (by nucleophilic substitutions of dichlorotetrazine **75**) and **B** (by cyclisation of nitriles **76**), and their electrochemistry and fluorescence efficiency evaluated. The occurrence of fluorescence, as well as its wavelength, were found to be strongly dependent on the substituents, which have to be electronegative heteroatoms <09EJO6121>.

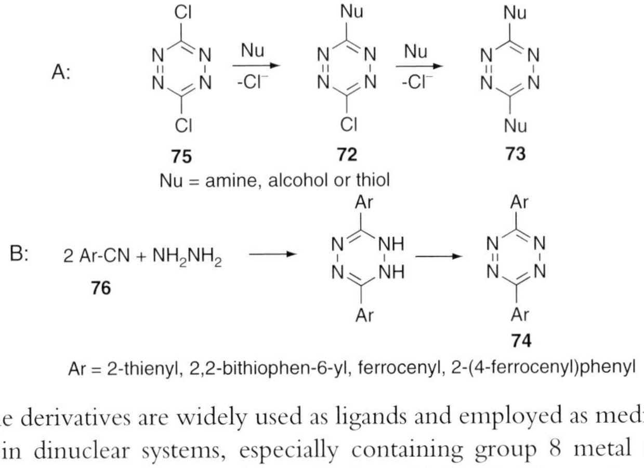

Tetrazine derivatives are widely used as ligands and employed as mediating bridging ligands in dinuclear systems, especially containing group 8 metal compounds. New diruthenium complexes with 3,6-bis(2-pyridyl)-1,2,4,5-tetrazine (bptz) as a bridging ligand were synthesised. It was shown that the bptz ligand induces strong electronic communication between the two metal centres <10ICA163>. In addition, a Re(CO)$_3$Cl(me$_2$bptz) complex was evaluated from the viewpoints of locations of electronic and structural changes in the excited state of the complex <09JA11656>.

Cu(I) pseudorotaxanes self-assembled with the binucleating ligand 3,6-bis(5-methylpyridin-2-yl)-1,2,4,5-tetrazine (me$_2$bptz) were reported. The electrochemical reduction of the tetrazine-based redox active ligand and the high lability of Cu(I) provide the means to create a reversible supramolecular switch that exchanges between two- and three-component pseudorotaxanes <09JA1305>.

A series of pyridyltetrazine ligand-substituted oxo-centred triruthenium derivatives with low valence were obtained by displacing one of six bridging acetates as well as one or two axial ligands in the parent triruthenium cluster. Ligand substitution of one of the six bridging acetates by a neutral pyridinyltetrazine exerts a dramatic influence on the electronic and redox characteristics of triruthenium cluster derivatives <09DT8696>.

A series of framework coordination polymers based on silver(I) and copper(I) complexes of 1,2,4,5-tetrazines have been discussed. All four nitrogen atoms were functional as lone pair donors leading to an unprecedented μ$_4$-coordination of the ligands <09DT2856>.

3,6-Diamino-1,2,4,5-tetrazine was used for the first application for incorporating 1,2,4,5-tetrazine directly into the backbone of a linear segmented elastomer with high mechanical properties and strong metal-complexation capability <09JAP3915>.

Tetrazine-based electrofluorochromic windows were prepared using electroactive fluorescent tetrazines. The fluorescence of the tetrazine fluorophore can be switched on and off reversibly, according to its redox state <09JEC201>.

Another widely studied application of tetrazines is development of high energy materials. Synthesis of 3,6-bis(1H-tetrazol-5-ylamino)-1,2,4,5-tetrazine **77** and 3-(1H-tetrazol-5-ylamino)-6-(3,5-dimethyl-pyrazol-1-yl)-1,2,4,5-tetrazine monohydrate **78** by nucleophilic substitution reaction of dipyrazolyltetrazines **79** with aminotetrazole **80**, their characterization and thermolysis studies have been reported. <09JHM306>.

Also, the synthesis and detonation properties of high energy density salts with ethylene- and propylene bis(nitroiminotetrazolate) as the anions were reported <09CEJ3198>.

Interactions of disubstituted tetrazines with single walled carbon nanotubes via Diels–Alder reaction resulted in debundling of the nanotubes and their chemical functionalisation <09MI253>.

6.3.3. FUSED [6]+[5] POLYAZA SYSTEMS
6.3.3.1 Nonpurine [6+5] fused systems

3,6-Diaryl[1,2,4]triazolo[4,3-b]pyridazine was obtained from 6-phenylpyridazin-1-one by initial treatment with $POCl_3$, followed by reaction with hydrazine hydrate, then condensation with benzaldehyde, and finally oxidative cyclization of intermediate hydrazone with diethyl azodicarboxylate <09BML4963>. An aryl substituent can be introduced in the pyridazine ring by Suzuki <09BML3686> or Stille <09BML6307> cross-coupling of 6-chloropyrazolopyridazine with an arylboronic acid or a tributylstannyl(hetero)arene.

Functionalized 3,6-disubstituted-[1,2,4]triazolo[4,3-b]pyridazines **81** were obtained with high yields from 3,6-dichloropyridazines **82** through cyclisation with hydrazides followed by nucleophilic aromatic substitution with amines <09TL212>. A slightly different pathway included synthesis of chlorotriazolopyridazines by reaction of hydrazinochloropyridazine **83** with benzoylchlorides <09BML3019>. New triazolopyridazines proved to be selective inhibitors of Pim-1 kinase <09BML3019> and M1 antagonists <09TL212>.

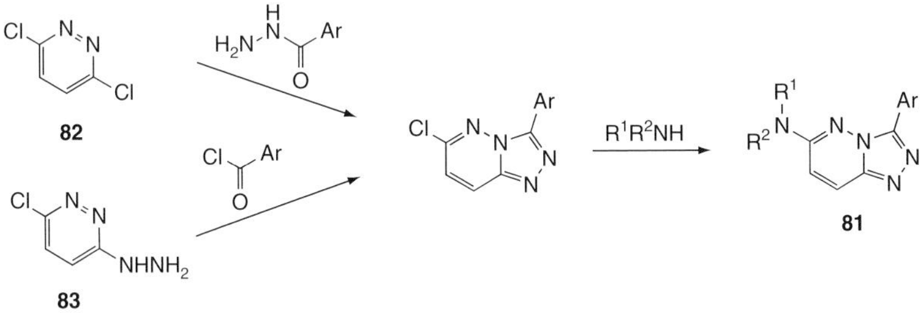

Metal-catalysed direct (hetero)arylations at the azole ring of imidazo[2,3-b][1,2,4]triazines have been described <09T10269>.

New pyrazolo[1,5-a]pyrimidines and triazolo[3,4-b]pyrimidines, containing a benzofuran moiety, have been reported <09JHC680>. Synthesis of several pyrazolo[1,5-c]pyrimidines, pyrazolo[1,5-a]pyrimidines and pyrazolo[1,5-a][1,3,5]triazines with potent activity against *Herpes simplex* viruses has been described <09BML5689>. Cyclisation of ethyl 2-ethoxymethylidene-3-polyfluoroalkyl-3-oxo-propionates with 3-amino-1,2,4-triazole resulted in formation of 7-polyfluoroalkyltriazolo[1,5-a]pyrimidines <09H(78)435>. Coupling of E-1-(1-methylbenzimidazol-2-yl)-3-(N,N-dimethylamino)prop-2-enone with the diazonium salts prepared from aminopyrazole and aminotriazole gave pyrazolo[5,1-c][l,2,4]triazine and [1,2,4]-triazolo[5,1-c][1,2,4]triazine <09H(78)699>.

3-Aroylpyrazolo[1,5-*a*]pyrimidines were obtained by condensation of 4-aroyl-pyrazole-3-amines with enaminones <09T9421>.

Bromination of *tert*-butoxy-β-(trifluoromethyl)styrenes **84** proceeded with formation of aryl(bromo)methyl trifluoromethyl ketones **85**. The latter compounds were found to be useful starting materials for the synthesis of different condensed heterocycles, *e.g.* 3-aryl-2-(trifluoromethyl)imidazo[1,2-*a*]pyrimidines **86** <09S2249>.

Synthesis, cytotoxic activity and inhibition of tubulin polymerization of pyrazolo[1,5-*a*][1,3,5]triazine myoseverin derivatives have been described <09BMC3471, 09JME655>. Synthesis of potential antidepressant 3-aryl-7-aminopyrazolo[1,5-*a*][1,3,5]triazines has been reported <09JME3073>.

[1,2,4]triazolo[1,5-*a*]pyrimidines are widely used as ligands to obtain complexes with copper(II), zinc(II), cobalt(II) <09POL3143>, iron(II), nickel(II) <09ICA861>, rhenium(II) <09POL2571>, platinum(II) <09DT10736>.

6.3.3.2 Purines

Purine nucleosides and related structures are not included in this review.

Alkyldiamine derivative of 2,6-diaminopurine, N-9-aminoethylaminopropyl-2,6-diaminopurine, reacted with Pd(II) as a terdentate ligand to give an N3-coordinated complex <09IC11085>. The effect of the 2-amino group on metal ion binding at the N-3-position of a purine base has been investigated using chelate-tethered derivatives <09IC10295>.

A series of trisubstituted purinones was synthesized from 2,6-dichloro-5-nitropyridine by a sequence of nucleophilic substitutions of halogen atoms with amines, then reduction of the nitro group followed by cyclisation *via* reaction with carbonyl diimidazole <09BML1399>.

Two structural isomers (9-aryl-6-cyanopurines and imidazole-4,5-dicarbonitriles) were isolated from the reaction of (Z)-N1-aryl-N2-(2-amino-1,2-dicyanovinyl)formamidines with triethyl orthoacetate or propionate. However, 9-aryl-6-cyanopurines were the only product, when triethyl orthoformate was used <09H(78)2245>.

The reaction of benzylhydroxylamine with 6-cyanopurines **87** resulted in formation of 7-benzyloxy-8-imino-7,8-dihydropyrimido[5,4-*d*]pyrimidines **88**. Refluxing the hydrochlorides of these compounds in ethanol gave the Dimroth-rearranged products, 8-benzyloxyimino-7,8-dihydropyrimido[5,4-*d*]pyrimidines **89** <09EJO4867>.

A method for the synthesis of 6,7,8-trisubstituted purines **90** *via* copper-catalyzed amidation from easily accessible 4-iodo-5-methylamino-6-benzylthiopyrimidine **91** has been reported. Furthermore, oxidation of the resulting 6-benzylthiopurine derivatives **90** followed by nucleophilic substitution gave 6,7,8-trisubstituted purines **92** for biological screening purposes <09JOC463>.

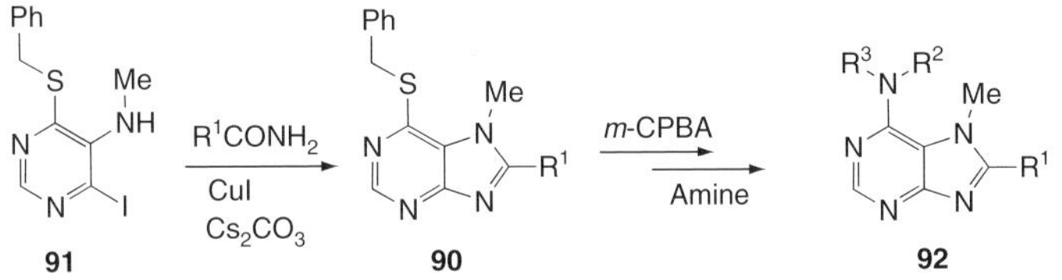

New purine derivatives **93** were prepared by condensation of 6-chloropurine **94** with 3,4-dihydro-2*H*-pyran or 2,3-dihydrofuran. Reaction of these intermediates **93** with the corresponding benzylamines gave novel 6,9-disubstituted purines **95** <09BMC1938>.

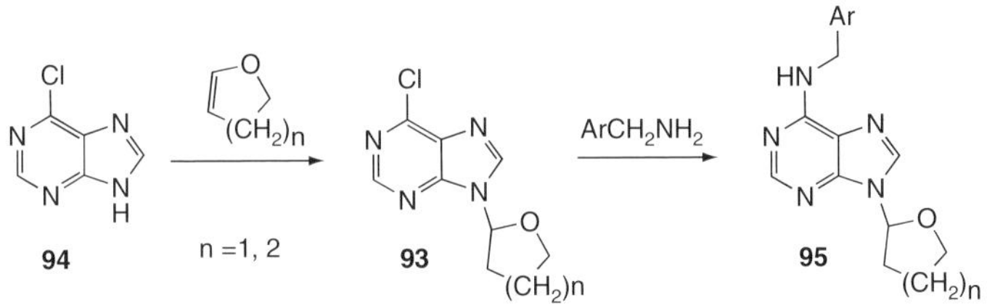

The scope and limitations of the use of the palladium-catalyzed cross-coupling reactions of diverse alkyl- and aryltrifluoroborates with halopurines have been studied. While aryl- and hetaryltrifluoroborates reacted readily with both 6-chloropurines and 8-bromoadenines to give the corresponding 6- or 8-aryl derivatives in high yields, the alkyltrifluoroborates were much less reactive <09S1309>.

Sonogashira protocol in the reaction of 9-benzyl-2,6-dichloropurines **96** with 1-ethynyl-2-hydroxymethylbenzene **97** resulted in formation of 6-[1(3*H*)-isobenzofuranylidene-methyl]purines **98** displaying profound antimycobacterial activity <09BMC6512>.

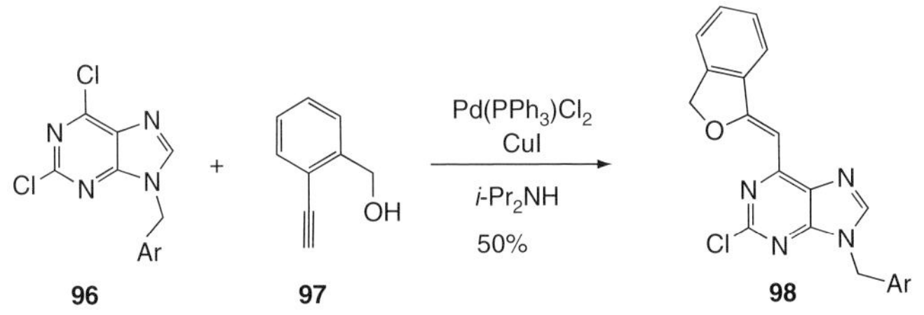

A series of N-3 and N-9 aryladenines **99** and **100** was prepared by arylation of 8-bromoadenine **101** with arylboronic acid **102** in the presence of Cu(II), followed by reductive debromination with hydrogen on palladium. The selectivity of the reaction varied with temperature and the ligand used <09H(78)1205>.

The presence of the bromine atom at the 8-position of 9-substituted adenines promotes in general interaction with the adenosine receptors, in particular at the A_{2A} subtype <09BMC2812>.

It was found that in the nucleophilic aromatic substitution of the nitro group of 2-nitro-6-aryl-9-benzylpurines, the nucleophile initially attacks C-8 with opening of the imidazole ring. Open-chain formylaminopyrimidines were found to be in equilibrium with cyclic nitropurines. However the final substitution of the nitro group with hydroxyl shifts the equilibrium totally towards cyclic purinone <09BML3297>.

Highly regio- and enantioselective iridium-catalyzed N-allylations of purines have been developed. N-Allylated purines were obtained in high yields (up to 91%) with high N-9/N-7 selectivity (up to 96:4), high branched-to-linear selectivity (98:2), and high enantioselectivity (up to 98% ee) by reaction of purines with unsymmetrical allylic carbonates in the presence of single component, ethylene-bound, metallacyclic iridium catalysts <09JA8971>.

6-Biarylamino purine derivatives of roscovitine inhibit cyclin dependent kinases and demonstrate potent antiproliferative activity <09BML6613>. 2-Cyanopurines represented a novel antimalarial scaffold, and a potential starting point for the development of new inhibitors <09BML3546>.

The tautomeric equilibria of purine and some purine derivatives in methanol and N,N-dimethylformamide solutions were investigated by low-temperature ^{1}H and ^{13}C NMR spectroscopy. The N-7–H and N-9–H tautomeric forms were quantified by integrating the individual ^{1}H NMR signals at low temperatures <09EJO1377>.

2-Phenyl-9-benzyl-8-azapurines, bearing at the 6-position an amido group interposed between the 8-azapurine moiety and an alkyl or a substituted phenyl group, have been synthesised and assayed as ligands for adenosine receptors. No benzyl cleavage during reduction of the nitro group was found <09BMC1817>.

Under different reaction conditions, 6-halopurine derivatives reacted with ethyl acetoacetate efficiently to yield 2-(purin-6-yl)acetoacetic acid ethyl esters (purin-6-yl)acetates and 6-methylpurines respectively. No metal catalyst and ligand were required <09OL1745>.

A series of 8,9-disubstituted adenines, 6-substituted aminopurines and 9-(p-fluorobenzyl/cyclopentyl)-6-substituted aminopurines with antimicrobial activities have been reported <09BMC1693>.

The use of microwave-assisted chemistry for the successful one-pot synthesis of 8-arylmethyl-9H-purin-6-amines has been reported to produce a large chemical diversity in the 8-arylmethyl-9H-purin-6-amine series <09BML415>.

"Push–pull" purines have been synthesized by the introduction of electron-accepting functional groups ($\mathbf{A}$ = CN, CO_2Me, and CONHR) to the heterocyclic C-8 position to complement typical electron-donating substituents at C-2 ($\mathbf{D^1}$) and

C-6 (D^2). The donor–acceptor purines show significantly altered, and overall improved photophysical properties relative to their acceptor-free precursors (**A** = H) <09JA623>.

6.3.4. FUSED [6]+[6] POLYAZA SYSTEMS

The synthesis of a series of protected as well as free heterocyclic aldehydes (pterins **103**, **104** and pyrido[2,3-*b*]pyrazines **105**, **106**) from diaminocytosine **107** and diaminopyridine **108** by the use of the appropriate tricarbonyl compounds was reported <09EJO1417>.

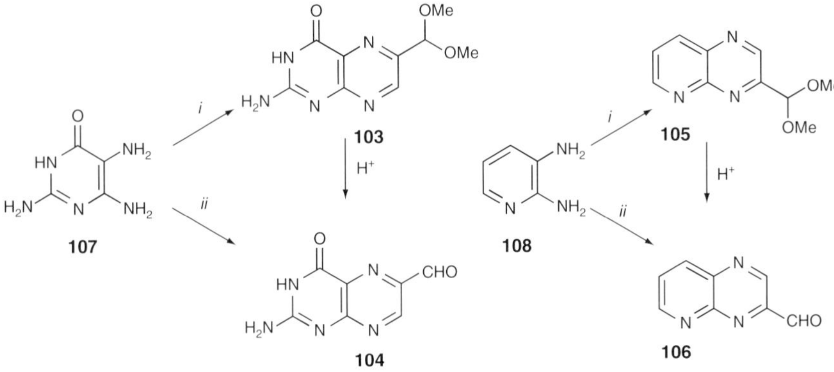

i: 1,1-dihydroxy-3,3-dimethoxy-2-propanone;   *ii*: 2,2-dihydroxy-1,3-propanedial

A novel method for the synthesis of a library of cyano *N*-heterocycles (including pterin **109** and 1,8-naphthyridine **110**) using triselenium dicyanide (TSD) under microwave conditions was developed <09SC407>.

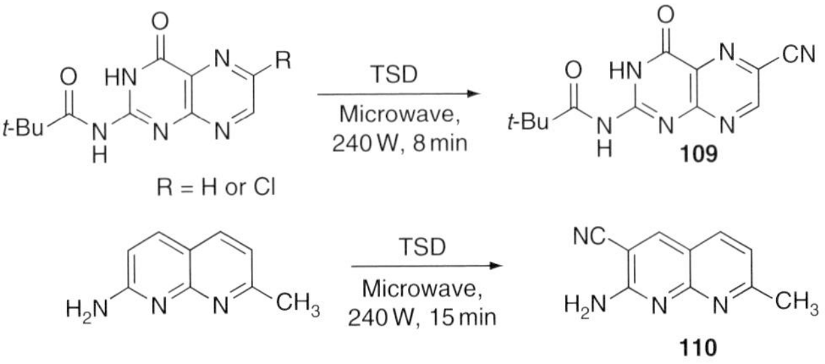

Studies of the photophysical properties of aromatic unconjugated pterins revealed that their fluorescence in aqueous solutions can be quenched by nucleotides and that this process depends on the pH conditions and on the chemical structure of the quencher <09PCA1794>.

Antimicrobial and fungicidal activity of 6-thioxo[1,2,4,5]tetrazino[4,3-*a*]quinazolin-8-ones have been reported <09EJM1188>.

REFERENCES

<09AGE7013>	N.K. Devaraj, R. Upadhyay, J.B. Haun, S.A. Hilderbrand and R. Weissleder, *Angew. Chem. Int. Ed.*, **48**, 7013 (2009).
<09BCC1799>	O.M. Merkel, M.A. Mintzer, J. Sitterberg, U. Bakowsky, E.E. Simanek and T. Kissel, *Bioconjugate Chem.*, **20**, 1799 (2009).
<09BMC1693>	M. Tunçbilek, Z. Ateş-Alagöz, N.A. Altanlar, S. Karayel and Özbey, *Bioorg. Med. Chem.*, **17**, 1693 (2009).
<09BMC1817>	I. Giorgi, M. Leonardi, D. Pietra, G. Biagi, A. Borghini, I. Massarelli, O. Ciampi and A.M. Bianucci, *Bioorg. Med. Chem.*, **17**, 1817 (2009).
<09BMC1938>	L. Szüčová, L. Spíchal, K. Doležal, M. Zatloukal, J. Greplová, P. Galuszka, V. Kryštof, J. Voller, I. Popa, F.J. Massino, J.E. Jørgensen and M. Strnad, *Bioorg. Med. Chem.*, **17**, 1938 (2009).
<09BMC2812>	C. Lambertucci, I. Antonini, M. Buccioni, D.D. Ben, D.D. Kachare, R. Volpini, K.N. Klotz and G. Cristalli, *Bioorg. Med. Chem.*, **17**, 2812 (2009).
<09BMC3471>	F. Popowycz, C. Schneider, S. DeBonis, D.A. Skoufias, F. Kozielski, C.M. Galmarini and B. Joseph, *Bioorg. Med. Chem.*, **17**, 3471 (2009).
<09BMC6512>	M. Brændvang, V. Bakken and L. Gundersen, *Bioorg. Med. Chem.*, **17**, 6512 (2009).
<09BML415>	H. Tao, Y. Kang, T. Taldone and G. Chiosis, *Bioorg. Med. Chem. Lett.*, **19**, 415 (2009).
<09BML1399>	Y. Shao, A.G. Cole, M.R. Brescia, L.Y. Qin, J. Duo, T.M. Stauffer, L.L. Rokosz, B.F. McGuinness and I. Henderson, *Bioorg. Med. Chem. Lett.*, **19**, 1399 (2009).
<09BML3019>	R. Grey, A.C. Pierce, G.W. Bemis, M.D. Jacobs, C.S. Moody, R. Jajoo, N. Mohal and J. Green, *Bioorg. Med. Chem. Lett.*, **19**, 3019 (2009).
<09BML3297>	M. Brændvang, C. Charnock and L. Gundersen, *Bioorg. Med. Chem. Lett.*, **19**, 3297 (2009).
<09BML3546>	J.P. Mallari, W.A. Guiguemde and R.K. Guy, *Bioorg. Med. Chem. Lett.*, **19**, 3546 (2009).
<09BML3686>	A.P. Skoumbourdis, C.A. LeClair, E. Stefan, A.G. Turjanski, W. Maguire, S.A. Titus, R. Huang, D.S. Auld, J. Inglese, C.P. Austin, S.W. Michnick, M. Xia and C.J. Thomas, *Bioorg. Med. Chem. Lett.*, **19**, 3686 (2009).
<09BML4963>	A.E. Kümmerle, M.M. Vieira, M. Schmitt, A.L.P. Miranda, C.A.M. Fraga, J.J. Bourguignon and E.J. Barreiro, *Bioorg. Med. Chem. Lett.*, **19**, 4963 (2009).
<09BML5644>	X. Ma, T. Poon, P.T.H. Wong and W. Chui, *Bioorg. Med. Chem. Lett.*, **19**, 5644 (2009).
<09BML5689>	K.S. Gudmundsson, B.A. Johns and J. Weatherhead, *Bioorg. Med. Chem. Lett.*, **19**, 5689 (2009).
<09BML6307>	A.A. Boezio, L. Berry, B.K. Albrecht, D. Bauer, S.F. Bellon, C. Bode, A. Chen, D. Choquette, I. Dussault, S. Hirai, P. Kaplan-Lefko, J.F. Larrow, M.H.J. Lin, J. Lohman, M.H. Potashman, K. Rex, M. Santostefano, K. Shah, R. Shimanovich, S.K. Springer, Y. Teffera, Y. Yang, Y. Zhang and J.C. Harmange, *Bioorg. Med. Chem. Lett.*, **19**, 6307 (2009).
<09BML6613>	M.P. Trova, K.D. Barnes, L. Alicea, T. Benanti, M. Bielaska, J. Bilotta, B. Bliss, T.N. Duong, S. Haydar, R.J. Herr, Y. Hui, M. Johnson, J.M. Lehman, D. Peace, M. Rainka, P. Snider, S. Salamone, S. Tregay, X. Zheng and T.D. Friedrich, *Bioorg. Med. Chem. Lett.*, **19**, 6613 (2009).
<09CC1511>	C.Y. Hung, A.S. Singh, C.W. Chen, Y. Wen and S.S. Sun, *Chem. Commun.*, 1511 (2009).
<09CEJ3198>	Y.H. Joo and J.M. Shreeve, *Chem. Eur. J.*, **15**, 3198 (2009).
<09CHE130>	A.V. Aksenov and I.V. Aksenova, *Chem. Heterocycl. Comp.*, **45**, 130 (2009).
<09CM3282>	S. Coco, C. Cordovilla, C. Dominguez, B. Donnio, P. Espinet and D. Guillon, *Chem. Mater.*, **21**, 3282 (2009).
<09CM3831>	A. Santoro, A.C. Whitwood, J.A.G. Williams, V.N. Kozhevnikov and D.W. Bruce, *Chem. Mater.*, **21**, 3871 (2009).
<09CM491>	A. Kohlmeier, A. Nordsieck and D. Janietz, *Chem. Mater.*, **21**, 491 (2009).
<09DT1776>	E. Wong, J. Li, C. Seward and S. Wang, *Dalton Trans.*, 1776 (2009).

<09DT2510> M.C. Aragoni, M. Arca, M. Crespo, F.A. Devillanova, M.B. Hursthouse, S.L. Huth, F. Isaia, V. Lippolis and G. Verani, *Dalton Trans.*, 2510 (2009).
<09DT2856> I.A. Gural'skiy, D. Escudero, A. Frontera, P.V. Solntsev, E.B. Rusanov, A.N. Chernega, H. Krautscheid and K.V. Domasevitch, *Dalton Trans.*, 2856 (2009).
<09DT3559> Z. Lu, T. Ladrak, O. Roubeau, J. Van Der Toorn, S.J. Teat, C. Massera, P. Gamez and J. Reedijk, *Dalton Trans.*, 3559 (2009).
<09DT4012> M. Schwalbe, M. Karnahl, H. Görls, D. Chartrand, F. Laverdiere, G.S. Hanan, S. Tschierlei, B. Dietzek, M. Schmitt, J. Popp, J.G. Vos and S. Rau, *Dalton Trans.*, 4012 (2009).
<09DT8696> F.R. Dai, H.Y. Ye, B. Li, L.Y. Zhanga and Z.N. Chen, *Dalton Trans.*, 8696 (2009).
<09DT10736> I. Łakomska, H. Kooijman, A.L. Spek, W. Shen and J. Reedijk, *Dalton Trans.*, 10736 (2009).
<09EJM1188> S.K. Pandeya, A. Singha, A. Singha and Nizamuddin, *Eur. J. Med. Chem.*, **44**, 1188 (2009).
<09EJO1377> T. Bartl, Z. Zacharová, P. Sečkářová, E. Kolehmainen and R. Marek, *Eur. J. Org. Chem.*, 1377 (2009).
<09EJO1417> S. Goswami, A.C. Maity, H.K. Fun and S. Chantrapromma, *Eur. J. Org. Chem.*, 1417 (2009).
<09EJO4867> A. Ribeiro, M.A. Carvalho and M.F. Proenç, *Eur. J. Org. Chem.*, 4867 (2009).
<09EJO4956> M. Hayashi, T. Yamasaki, Y. Kobayashi, Y. Imai and Y. Watanabe, *Eur. J. Org. Chem.*, 4956 (2009).
<09EJO5587> L. Zou, Z. Liu, X. Yan, Y. Liu, Y. Fu, J. Liu, Z. Huang, X. Chen and J. Qin, *Eur. J. Org. Chem.*, 5587 (2009).
<09EJO6121> Y.H. Gong, F. Miomandre, R. Méallet-Renault, S. Badré, L. Galmiche, J. Tang, P. Audebert and G. Clavier, *Eur. J. Org. Chem.*, 6121 (2009).
<09H(78)1205> J. Krouželka and I. Linhart, *Heterocycles*, **78**, 1205 (2009).
<09H(78)2245> A. Al-Azmi and K. Anita Kumari, *Heterocycles*, **78**, 2245 (2009).
<09H(78)435> M.V. Goryaeva, Y.V. Burgart, V.I. Saloutin, E.V. Sadchikova and E.N. Ulomskii, *Heterocycles*, **78**, 435 (2009).
<09H(78)623> E. Wolińska, *Heterocycles*, **78**, 623 (2009).
<09H(78)699> M.R. Shaaban, T.S. Saleh and A.M. Farag, *Heterocycles*, **78**, 699 (2009).
<09IC3907> H.J. Bolink, E. Coronado, R.D. Costa, P. Gaviña, E. Orti and S. Tatay, *Inorg. Chem.*, **48**, 3907 (2009).
<09IC4179> D.N. Kozhevnikov, V.N. Kozhevnikov, M.M. Ustinova, A. Santoro, D.W. Bruce, B. Koenig, R. Czerwieniec, T. Fischer, M. Zabel and H. Yersin, *Inorg. Chem.*, **48**, 4179 (2009).
<09IC6630> C. Yuste, L. Cañadillas-Delgado, A. Labrador, F.S. Delgado, C. Ruiz-Pérez, F. Lloret and M. Julve, *Inorg. Chem.*, **48**, 6630 (2009).
<09IC10295> M.A. Galindo, D. Amantia, A.M. Martinez, W. Clegg, R.W. Harrington, V.M. Martinez and A. Houlton, *Inorg. Chem.*, **48**, 10295 (2009).
<09IC11085> M.A. Galindo, D. Amantia, A.M. Martinez, W. Clegg, R.W. Harrington, V.M. Martinez and A. Houlton, *Inorg. Chem.*, **48**, 11085 (2009).
<09ICA861> J.M. Balkaran, S.C.P. van Bezouw, J. van Bruchem, J. Verasdonck, P.C. Verkerk, A.G. Volbeda, I. Mutikainen, U. Turpeinen, G.A. van Albada, P. Gamez, J.G. Haasnoot and J. Reedijk, *Inorg. Chim. Acta*, **362**, 861 (2009).
<09JA623> R.S. Butler, P. Cohn, P. Tenzel, K.A. Abboud and R.K. Castellano, *J. Am. Chem. Soc.*, **131**, 623 (2009).
<l09JA1305> K.A. McNitt, K. Parimal, A.I. Share, A.C. Fahrenbach, E.H. Witlicki, M. Pink, D. Kwabena Bediako, C.L. Plaisier, N. Le, L.P. Heeringa, D.A. Vander Griend and A.H. Flood, *J. Am. Chem. Soc.*, **131**, 1305 (2009).
<09JA8971> L.M. Stanley and J.F. Hartwig, *J. Am. Chem. Soc.*, **131**, 8971 (2009).
<09JA9496> K. Ai, Y. Liu and L. Lu, *J. Am. Chem. Soc.*, **131**, 9496 (2009).
<09JA11656> G. Li, K. Parimal, S. Vyas, C.M. Hadad, A.H. Flood and K.D. Glusac, *J. Am. Chem. Soc.*, **131**, 11656 (2009).
<09JA16706> S. Jensen, H. Früchtl and C.J. Baddeley, *J. Am. Chem. Soc.*, **131**, 16706 (2009).
<09JAP3915> A.R. Sayed and J.S. Wiggins, *J. App. Polym. Sci.*, **114**, 3915 (2009).
<09JCO1050> K.H. Kong, C.K. Tan and Y. Lam, *J. Comb. Chem.*, **11**, 1050 (2009).

<09JCC1207>   L. Li, C.P. Landee, M.M. Turnbull and B.M. Foxmanx, *J. Coord. Chem.*, **62**, 1207 (2009).
<09JCC1902>   T. Peppel and M. Köckerling, *J. Coord. Chem.*, **62**, 1902 (2009).
<09JCC1972>   F. Marandi and H. Fun, *J. Coord. Chem.*, **62**, 1972 (2009).
<09JCC2155>   F. Marandi, H.K. Fun and S. Chantrapromma, *J. Coord. Chem.*, **62**, 2155 (2009).
<09JEC201>   Y. Kim, J. Do, E. Kim, G. Clavier, L. Galmiche and P. Audebert, *J. Electroanal. Chem.*, **632**, 201 (2009).
<09JHC477>   E.I. Klimova, T. Klimova, M.F. Álamo, D.M. Iturbide and M.M. García, *J. Heterocycl. Chem.*, **46**, 477 (2009).
<09JHC680>   A.O. Abdelhamid, *J. Heterocycl. Chem.*, **46**, 680 (2009).
<09JHM306>   A. Saikia, R. Sivabalan, B.G. Polke, G.M. Gore, A. Singh, A.S. Rao and A.K. Sikder, *J. Hazard. Mat.*, **170**, 306 (2009).
<09JMC2747>   R. Wang, C. Pellerin and O. Lebel, *J. Mater. Chem.*, **19**, 2747 (2009).
<09JMC6400>   K.L. Young, C. Xu, J. Xie and S. Sun, *J. Mater. Chem.*, **19**, 6400 (2009).
<09JMC8112>   H.F. Chen, S.J. Yang, Z.H. Tsai, W.Y. Hung, T.C. Wang and K.T. Wong, *J. Mater. Chem.*, **19**, 8112 (2009).
<09JME655>   F. Popowycz, G. Fournet, C. Schneider, K. Bettayeb, Y. Ferandin, C. Lamigeon, O.M. Tirado, S. Mateo-Lozano, V. Notario, P. Colas, P. Bernard, L. Meijer and B. Joseph, *J. Med. Chem.*, **52**, 655 (2009).
<09JME3073>   P.J. Gilligan, L. He, T. Clarke, P. Tivitmahaisoon, S. Lelas, Y.W. Li, K. Heman, L. Fitzgerald, K. Miller, G. Zhang, A. Marshall, C. Krause, J. McElroy, K. Ward, H. Shen, H. Wong, S. Grossman, G. Nemeth, R. Zaczek, S.P. Arneric, P. Hartig, D.W. Robertson and G. Trainor, *J. Med. Chem.*, **52**, 3073 (2009).
<09JOC319>   M. De Rosa and D. Arnold, *J. Org. Chem.*, **74**, 319 (2009).
<09JOC463>   N. Ibrahim and M. Legraverend, *J. Org. Chem.*, **74**, 463 (2009).
<09JOC2726>   L.R. Domingo, M.T. Picher and J.A. Saez, *J. Org. Chem.*, **74**, 2726 (2009).
<09JOC2870>   O.N. Chupakhin, I.A. Utepova, M.V. Varaksin, E.V. Tretyakov, G.V. Romanenko, D.V. Stass and V.I. Ovcharenko, *J. Org. Chem.*, **74**, 2870 (2009).
<09JOC4804>   M.W. Lodewyk, M.J. Kurth and D.J. Tantillo, *J. Org. Chem.*, **74**, 4804 (2009).
<09JOC8343>   N. Catozzi, M.G. Edwards, S.A. Raw, P. Wasnaire and R.J.K. Taylor, *J. Org. Chem.*, **74**, 8343 (2009).
<09MI208>   O.N. Chupakhin, M.V. Varaksin, I.A. Utepova and V.L. Rusinov, *Arkivoc*, (vi), 208 (2009).
<09MI253>   H. Hayden, Y.K. Gun'ko, T. Perova, S. Grudinkin, A. Moore and E.D. Obraztsova, *Plast. Rubber Compos.*, **38**, 253 (2009).
<09MM4701>   S.W. Kuo and H.T. Tsai, *Macromolecules*, **42**, 4701 (2009).
<09OBC2082>   N. Rahanyan, A. Linden, K.K. Baldridge and J.S. Siegel, *Org. Biomol. Chem.*, **7**, 2082 (2009).
<2010OL160>   J.F. Vivat, H. Adams and J.P.A. Harrity, *Org. Lett.*, **12**, 160 (2010).
<09OL1745>   G.R. Qu, Z.J. Mao, H.Y. Niu, D.C. Wang, C. Xia and H.M. Guo, *Org. Lett.*, **11**, 1745 (2009).
<09OL3686>   B. Shi, W. Lewis, I.B. Campbell and C.J. Moody, *Org. Lett.*, **11**, 3686 (2009).
<09OL4228>   A. Metzger, L. Melzig, C. Despotopoulou and P. Knochel, *Org. Lett.*, **11**, 4228 (2009).
<09OL5398>   A. García, B. Insuasty, M.A. Herranz, R. Martínez-Álvarez and N. Martín, *Org. Lett.*, **11**, 5398 (2009).
<09PCA10327> J.E. Del Bene, I. Alkorta and J. Elguero, *J. Phys. Chem. A*, **113**, 10327 (2009).
<09PCA1794>   G. Petroselli, M. Laura Dantola, F.M. Cabrerizo, C. Lorente, A.M. Braun, E. Oliveros and A.H. Thomas, *J. Phys. Chem. A*, **113**, 1794 (2009).
<09PCA5066>   Y. Ma, R. Hao, G. Shao and Y. Wang, *J. Phys. Chem. A*, **113**, 5066 (2009).
<09PCA8834>   J.D. Savee, J.E. Mann and R.E. Continetti, *J. Phys. Chem. A*, **113**, 8834 (2009).
<09PCP11207> M. Guin, G.N. Patwari, S. Karthikeyan and K.S. Kim, *Phys. Chem. Chem. Phys.*, **11**, 11207 (2009).
<09POL2571>   B. Machura, M. Jaworska, P. Lodowski, J. Kusz, R. Kruszynski and Z. Mazurak, *Polyhedron*, **28**, 2571 (2009).

<09POL3143>	J.H. Adriaanse, S.H.C. Askes, Y. van Bree, S. van Oudheusden, E.D. van den Bos, E. Günay, I. Mutikainen, U. Turpeinen, G.A. van Albada, J.G. Haasnoot and J. Reedijk, *Polyhedron,* **28**, 3143 (2009).
<09RJO1102>	R.I. Ishmetova, N.I. Latosh, I.N. Ganebnykh, N.K. Ignatenko, S.G. Tolshchina and G.L. Rusinov, *Rus. J. Org. Chem.,* **45**, 1102 (2009).
<09S542>	K. Hioki, K. Ohshima, Y. Sota, M. Tanaka and M. Kunishima, *Synthesis,* 542 (2009).
<09S1309>	Z. Hasník, R. Pohl and M. Hocek, *Synthesis,* 1309 (2009).
<09S2249>	V.M. Muzalevskiy, V.G. Nenajdenko, A.V. Shastin, E.S. Balenkova and G. Haufe, *Synthesis,* 2249 (2009).
<09S3967>	V.O. Iaroshenko, *Synthesis,* 3967 (2009).
<09SC407>	S. Goswami, A.C. Maity, N.K. Das, D. Sen and S. Maity, *Synth. Commun.,* **39**, 407 (2009).
<09SL731>	S. Pican, V. Lapinte, J.F. Pilard, E. Pasquinet, L. Beller, L. Fontaine and D. Poullain, *Synlett,* 731 (2009).
<09SL2137>	L. Pellegatti, E. Vedrenne, J.-M. Leger, C. Jarry and S. Routier, *Synlett,* 2137 (2009).
<09SL3192>	J.M. Bégouin, S. Claudel and C. Gosmini, *Synlett,* 3192 (2009).
<09SL3206>	G. Xu, L. Zheng, S. Wang, Q. Dang and X. Bai, *Synlett,* 3206 (2009).
<09T975>	M. Altuna-Urquijo, A. Gehre, S.P. Stanforth and B. Tarbit, *Tetrahedron,* **65**, 975 (2009).
<09T1115>	A. Gehre, S.P. Stanforth and B. Tarbit, *Tetrahedron,* **65**, 1115 (2009).
<09T2824>	C.C. Hsu, C.C. Lai and S.H. Chiu, *Tetrahedron,* **65**, 2824 (2009).
<09T9421>	K.D. Khalil, H.M. Al-Matar, D.M. Al-Dorri and M.H. Elnagdi, *Tetrahedron,* **65**, 9421 (2009).
<09T10269>	F. Bellina and R. Rossi, *Tetrahedron,* **65**, 10269 (2009).
<09TL212>	L.N. Aldrich, E.P. Lebois, L.M. Lewis, N.T. Nalywajko, C.M. Niswender, C.D. Weaver, P.J. Conn and C.W. Lindsley, *Tetrahedron Lett.,* **50**, 212 (2009).
<09TL1331>	S. Muthusamy and P. Srinivasan, *Tetrahedron Lett.,* **50**, 1331 (2009).
<09TL1444>	A.M. Prokhorov, M. Makosza and O.N. Chupakhin, *Tetrahedron Lett.,* **50**, 1444 (2009).
<09TL2874>	Q. Dang, E. Carruli, F. Tian, F.W. Dang, T. Gibson, W. Li, H. Bai, M. Chung and S.J. Hecker, *Tetrahedron Lett.,* **50**, 2874 (2009).
<09TL3923>	J. Clayden, S.J.M. Rowbottom, M.G. Hutchings and W.J. Ebenezer, *Tetrahedron Lett.,* **50**, 3923 (2009).
<09TL5673>	P. Leriche, F. Piron, E. Ripaud, P. Frère, M. Allain and J. Roncali, *Tetrahedron Lett.,* **50**, 5673 (2009).
<09TL5802>	V. Kikelj, K. Julienne, J. Meslin and D. Deniaud, *Tetrahedron Lett.,* **50**, 5802 (2009).
<09TL6758>	Q. Dang and Y. Liu, *Tetrahedron Lett.,* **50**, 6758 (2009).
<09TL6800>	F. Hamon, G. Prié, F. Lecornué and S. Papot, *Tetrahedron Lett.,* **50**, 6800 (2009).
<10ICA163>	M.M. Vergara, M.E. Garcia Posse, F. Fagalde, N.E. Katz, J. Fiedler, B. Sarkar, M. Sieger and W. Kaim, *Inorg. Chim. Acta,* **363**, 163 (2010).

CHAPTER 6.4

Six-Membered Ring Systems: With O and/or S Atoms

John D. Hepworth*, B. Mark Heron**
*University of Central Lancashire, Preston, UK
j.d.hepworth@tinyworld.co.uk
**Department of Colour Science, School of Chemistry, University of Leeds, Leeds, UK
b.m.heron@leeds.ac.uk

6.4.1. INTRODUCTION

As usual, much work has appeared on the isolation, synthesis, structure determination and reactions of naturally occurring oxygen heterocycles. Reviews of transannulation reactions in the synthesis of natural products <09S691>, the chemistry and biology of mycotoxins <09CR3903>, marine natural products <09NPR170> and a dictionary of marine natural products <08B1> have been published.

A review of the halichondrins has appeared <09CR3044> and other work in this area includes a total synthesis of norhalichondrin B <09AGE2346> and approaches to halichondrin C-14–C-38 building blocks <09JA15636, 09JA15642>. Of the ladder polyethers, mention can be made of the progress towards the synthesis of the gambieric acids <09CL866, 09JOC4024, 09OL113> and a total synthesis of gambierol <09OL4382>. A total synthesis of ciguatoxin <09AGE2941> and a convergent synthesis of the A–E ring segment of ciguatoxin CTX3C <09T7784> have appeared. A total synthesis of brevetoxin A is based on the coupling of two major sub-units <09CEJ9223, 09CEJ9235, 09OL489>.

Amongst the macrolides, total syntheses have been reported for spongistatins 1 and 2 <09T6470, 09T6489>, spirastrellolide F methyl ester <09AGE9940, 09AGE9946>, various members of the laulimalide family <09CEJ5979> and of (+)-neopeltolide <09CEJ12807>. Interest continues in the extremophile derived berkelic acid <09AGE1283, 09JA11350, 09JOC6245>.

The asymmetric hetero Diels–Alder (hDA) reactions of carbonyl compounds have been reviewed <09T2839> and the scope and limitations of the hDA approach to spiroacetals have been discussed <09OBC1053>. Reviews on the synthesis of five- and six-membered heterocycles involving metathesis reactions <09H(78) 1109> and by metal-catalysed heterocyclisations <09H(78)2661> contain material relevant to this chapter.

O-Heterocycles feature in various applications as reviews on the biological activity of complexes of coumarins, coumarins and flavones <09CCR2588>, fluorescent labelling of biomolecules with organic probes <09CR190> and molecular design and synthesis of organic dyes for dye-sensitized solar cells <09EJO2903> testify.

6.4.2. HETEROCYCLES CONTAINING ONE OXYGEN ATOM
6.4.2.1 Pyrans

The reaction of cyclic 1,3-diketones with α,β-unsaturated aldehydes is catalysed by the phosphoric acid **1** and yields annulated 2*H*-pyrans under mild conditions (Scheme 1). Application to 4-hydroxycoumarin and 6-methyl-4-hydroxypyran-2-one leads to the pyrano[3,2-*c*]chromenone and pyrano[4,3-*b*]pyranone systems respectively <09JOC8963>.

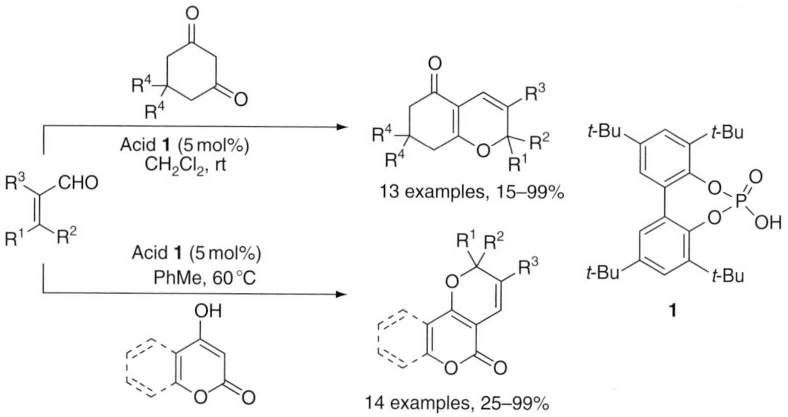

Scheme 1

Cross-conjugated enaminones yield highly substituted 2*H*-pyrans on reaction with DMAD. It is proposed that an initial [4+2] cycloaddition is followed by a [1,3]-H shift. In like manner, enaminothiones generated from the enaminones by treatment with Lawesson's reagent are trapped by acrylates to give 2*H*-thiopyrans. Reaction of the enaminones with ketenes affords 3,6-disubstituted pyran-2-ones (Scheme 2) <09T8478>.

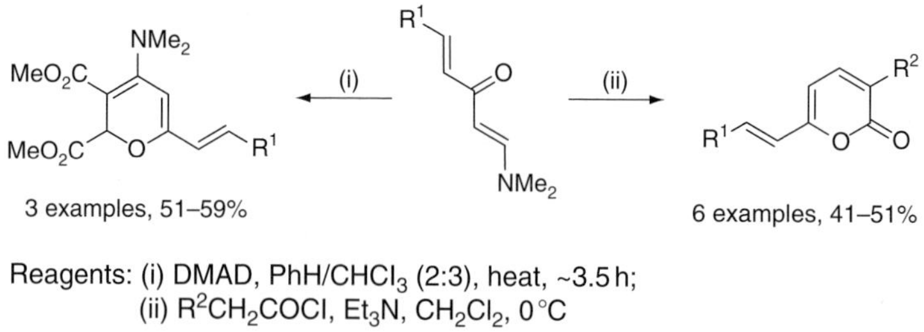

Reagents: (i) DMAD, PhH/CHCl$_3$ (2:3), heat, ~3.5 h;
(ii) R^{2}CH$_2$COCl, Et$_3$N, CH$_2$Cl$_2$, 0 °C

Scheme 2

The [4+2] cycloaddition of enones and alkynes is efficiently catalysed by Ni (cod)$_2$ and gives highly substituted 4*H*-pyrans (Scheme 3). An intramolecular example leads to a cyclohexa[*c*]pyran <09JA1350>. A number of 2,4-diaryl cycloalka[*b*]

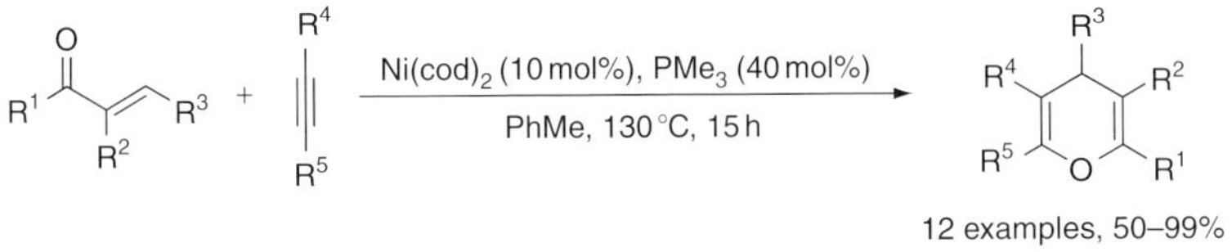

Scheme 3

pyrans has been obtained by the In-catalysed reaction between aryl propargyl alcohols and cyclic 1,3-diketones; I_2 is also a suitable catalyst (Scheme 4) <09TL3963>.

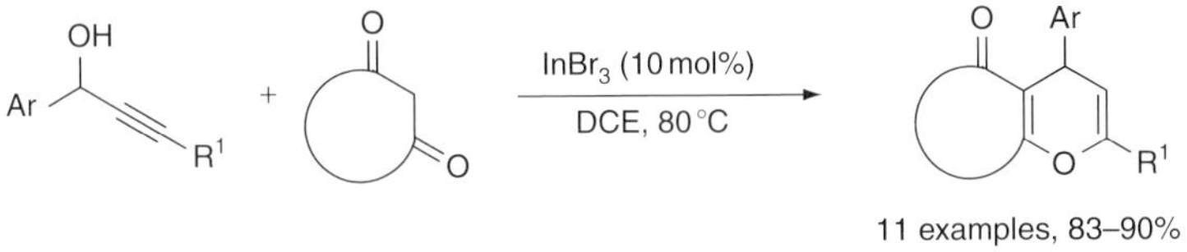

Scheme 4

A one-pot three-component reaction carried out in a temperature-dependent biphasic system of toluene and a PEG-based ionic liquid and involving aromatic aldehydes, malononitrile and 5,5-dimethylcyclohexa-1,3-dione yields 2-aminocyclohexa[b]pyrans **2** <09CC2878>.

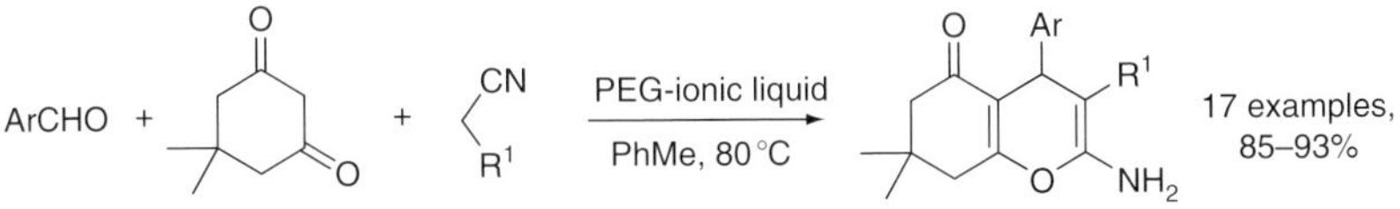

The Au-catalysed cycloisomerisation of 3,3-disubstituted alk-4-yn-1-ones gives 4H-pyrans through a 6-*endo-dig* cyclisation. In the absence of one of the 3-substituents, a 5-*exo-dig* process supervenes and furans result (Scheme 5) <09OBC1221>.

The choice of catalyst decides the course of the tandem Michael addition–cyclisation reaction of 2-(1-alkynyl)alk-2-en-1-ones with 1,3-dicarbonyl

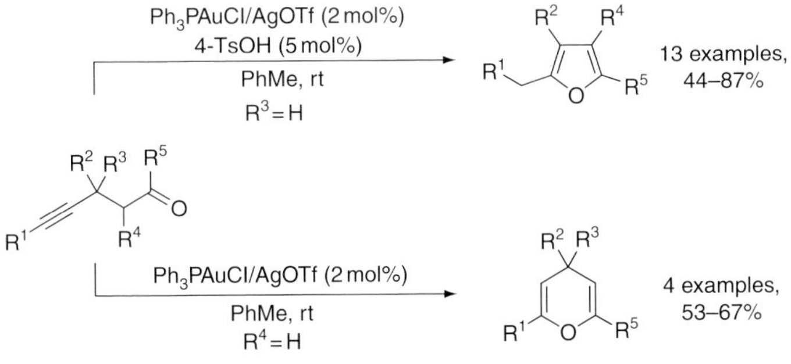

Scheme 5

compounds. 4H-Pyrans are formed in high yield using DBU as a basic catalyst whereas a cationic Pd(II) catalyst leads to furans (Scheme 6) <09CC3594>.

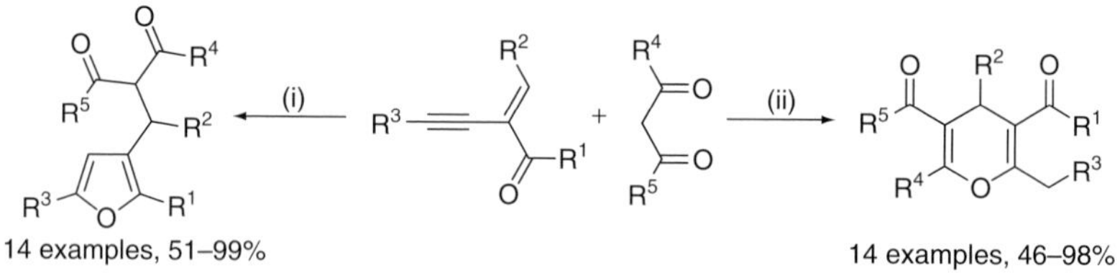

Reagents: (i) [Pd(dppp)(H$_2$O)$_2$](OTf)$_2$, (5 mol%), DCE, rt; (ii) DBU (5 mol%), DMF, 100 °C

Scheme 6

An analogous regiodivergence is observed in the intramolecular hydroalkoxylation of γ-allenols. Thus, 6-*exo-dig* selectivity is effected by Sn or Zn triflates and leads to a pyran derivative, but furans result from a 5-*exo-trig* cyclisation under the influence of AgOTf. The cyclisation of 2,2-diphenylhexa-4,5-dien-1-ol **3** is illustrative <09CC7125>.

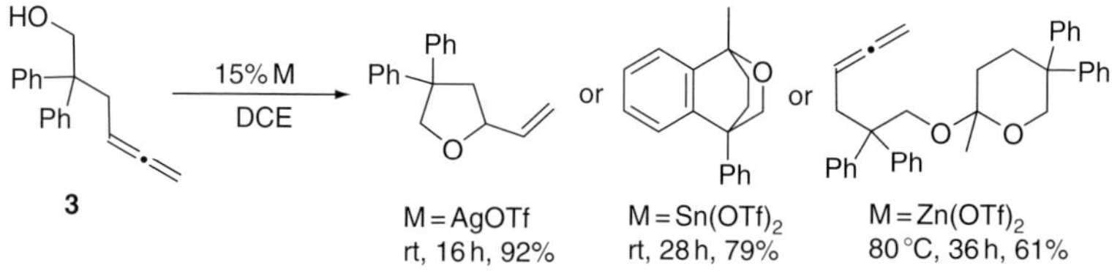

In a 3-component, one-pot diastereoselective reaction catalysed by BiBr$_3$, an initial Mukaiyama aldol reaction between ketene silyl acetals or silyl enol ethers and a β,γ-unsaturated aldehyde is followed by the addition of a second aldehyde and an intramolecular silyl-Prins reaction. The product is a 2,6-*cis*-disubstituted 3,6-dihydropyran (Scheme 7) <09T6834>.

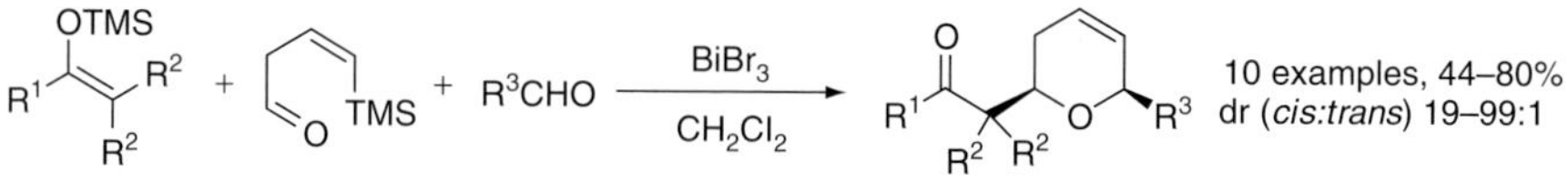

Scheme 7

Similar products arise from a tandem carbonylallylation–silyl-Prins cyclisation of aldehydes with γ-trimethylsilylallyltri-*n*-butylstannane, either in solution or as the polymer-supported dibutyl derivative (Scheme 8) <09T3953>.

2,6-*trans*-Dihydropyrans are available from ester-containing allenic alcohols and aldehydes through an In-catalysed Prins cyclisation. It is considered that the ester moiety suppresses the alternative oxonia-Cope rearrangement in addition to stabilising the oxo-carbenium intermediate (Scheme 9) <09OL1741>. 2-Substituted 4-halogendihydropyrans are obtained in good yields from a Fe-catalysed Prins

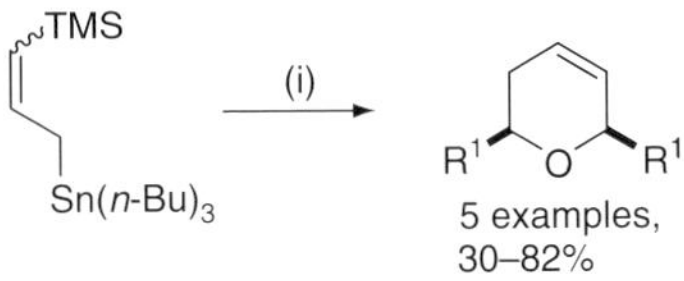

Scheme 8

Reagents: (i) 2 eq. R^1CHO, $InCl_3$, MeCN

Scheme 9

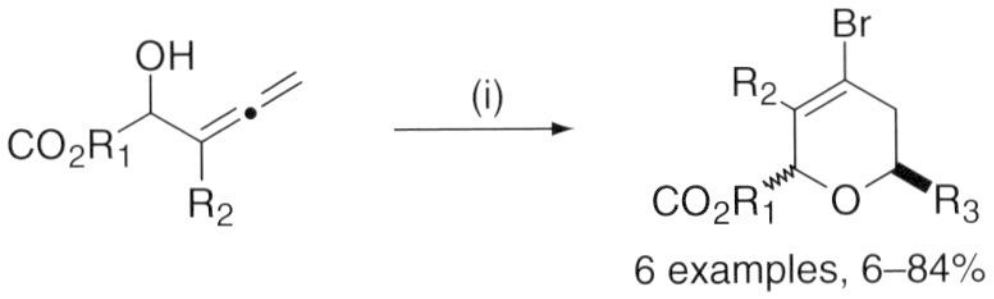

Reagents: (i) R^3CHO, $In(OTf)_3$ (10 mol%), TMSBr, CH_2Cl_2, 0 °C

reaction between homopropargylic alcohols and aldehydes. When this approach is applied to homoallylic alcohols, the corresponding tetrahydropyrans result <09OL357>.

Under the influence of a chiral phosphoric acid **4** the hDA reaction between ethyl glyoxylate and an electron-rich diene exhibits enantio- and *anti*-diastereoselectivity (Scheme 10) <09JA12882>.

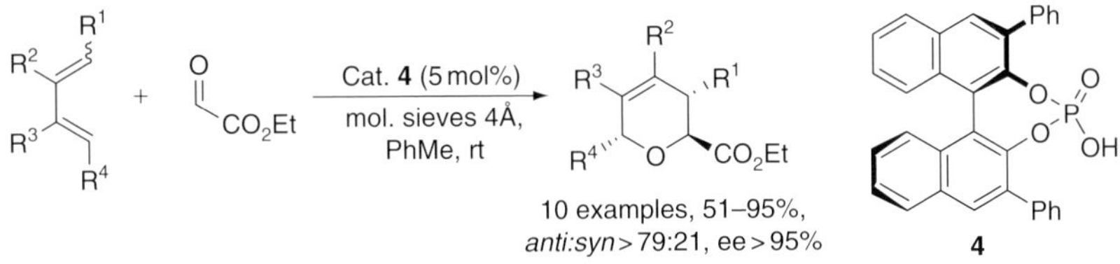

Scheme 10

A three-component microwave assisted reaction between an alkyne, ethyl vinyl ether and ethyl glyoxylate also leads predominantly to the *trans*-2,6-disubstituted dihydropyran **5**. Here, an initial cross-metathesis produces a diene and an hDA reaction follows, the *exo* attack being a consequence of a favourable anomeric effect and 1,3-diaxial interactions <09TL1526>. A RCM features in a stereodivergent synthesis of the stereoisomers of centrolobine from protected 3-(4-hydroxyphenyl)propanal <09CEJ11948>.

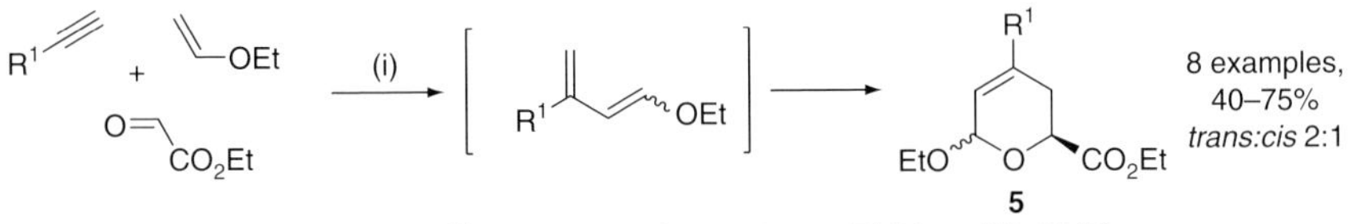

Reagents: (i) 2nd generation Grubbs' cat., PhMe, μW, 80 °C

Alkynyl benzyl ethers undergo a Rh-catalysed cyclisation to 3,6-dihydropyrans. Substitution adjacent to the ether O atom results in diastereoselective ring closure to the *cis*-2,6-disubstituted pyran but substitution at the propargyl position leads to the *trans*-2,5 product (Scheme 11) <09JA3166>.

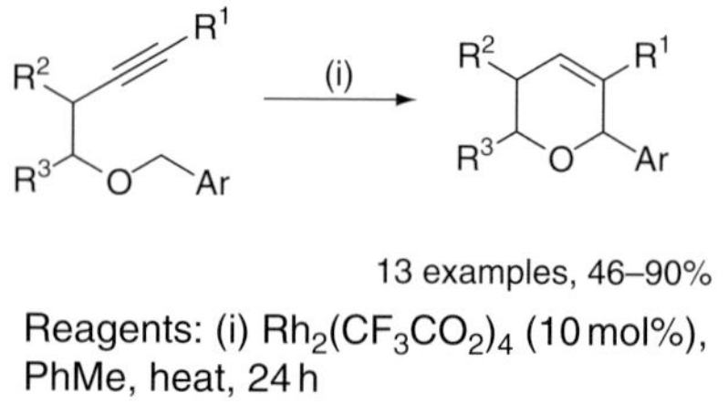

13 examples, 46–90%
Reagents: (i) $Rh_2(CF_3CO_2)_4$ (10 mol%), PhMe, heat, 24 h

Scheme 11

1,7-Dioxaspiro[5.5]undec-4-enes are readily prepared by the Au-catalysed cyclisation of the propargylic triols **6**; an allene intermediate is considered likely (Scheme 12) <09OL121>.

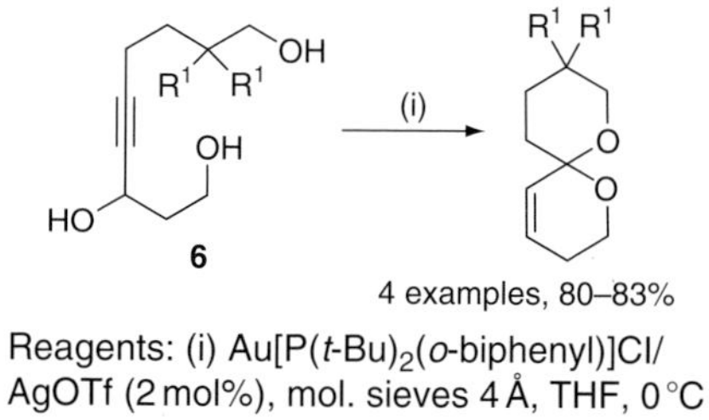

4 examples, 80–83%
Reagents: (i) Au[P(*t*-Bu)$_2$(*o*-biphenyl)]Cl/ AgOTf (2 mol%), mol. sieves 4 Å, THF, 0 °C

Scheme 12

Borylation of the pyranyl triflate **7** results in a formal asymmetric isomerisation and a subsequent one-pot reaction with aldehydes allows the stereoselective synthesis of α-hydroxyalkyldihydropyrans (Scheme 13) <09JA9612>.

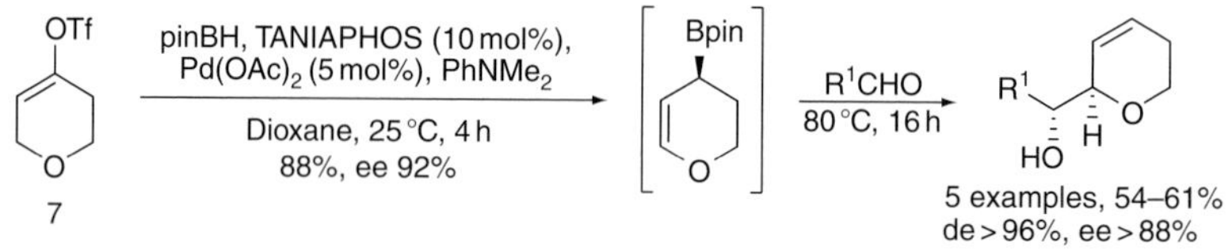

pinBH, TANIAPHOS (10 mol%), Pd(OAc)$_2$ (5 mol%), PhNMe$_2$
Dioxane, 25 °C, 4 h
88%, ee 92%

R^1CHO
80 °C, 16 h

5 examples, 54–61%
de > 96%, ee > 88%

Scheme 13

There have been several developments in the application of the Prins reaction to tetrahydropyran synthesis. The one-electron oxidation of a benzylic or allylic sp^3 C–H bond is the precursor to a Lewis acid-induced intramolecular attack by an unactivated alkene, which leads to *cis*-2,4,6-trisubstituted tetrahydropyrans (Scheme 14) <09OL3442>. Unactivated hex-5-en-1-ols are efficiently cyclised by lanthanide triflates in ionic liquids <09OL1523>. In a different context, Friedel–Crafts alkylation

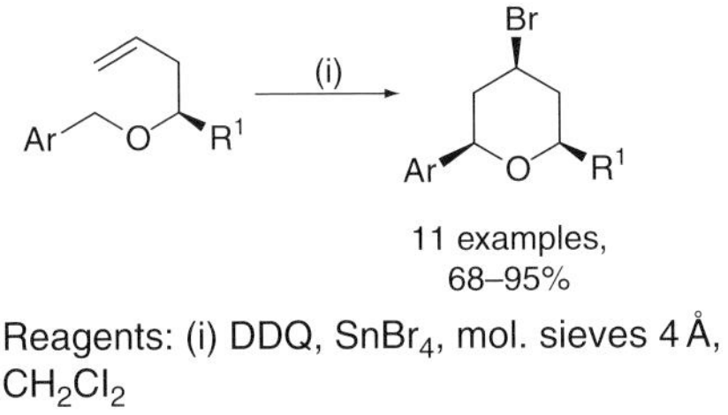
Br
(i)
Ar
O
R1
Ar
O
R1
11 examples,
68–95%
Reagents: (i) DDQ, SnBr4, mol. sieves 4 Å,
CH2Cl2

Scheme 14

of a C–H bond α to a primary ether has been achieved by generation of an oxocarbenium ion from an alkenic acetal which prompts a hydride ion transfer and subsequent cyclisation (Scheme 15) <09JA402>.

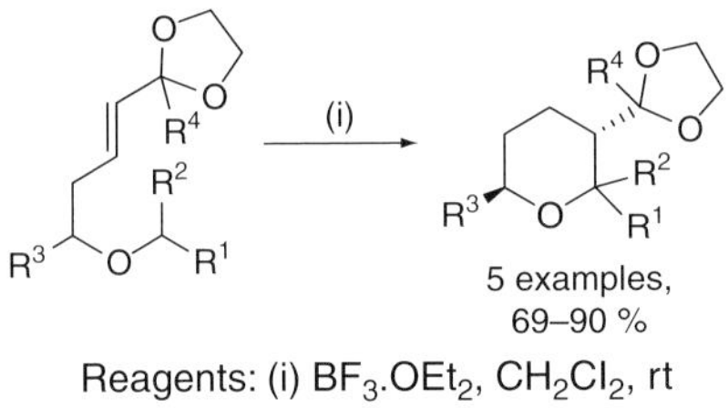
O
O
R4
R2
R3
O
R1
(i)
R4
O
O
R2
R3
O
R1
5 examples,
69–90 %
Reagents: (i) BF3.OEt2, CH2Cl2, rt

Scheme 15

A diastereoselective synthesis of 2,6-disubstituted 4-aryltetrahydropyrans involves a one-pot, three-component reaction between allyltrimethylsilane, aldehydes and arenes. The initial formation of a homoallylic alcohol is followed by a Prins cyclisation and a non-regioselective Friedel–Crafts attack on the arene (Scheme 16) <09EJO1625>. In a similar manner, homoallylic alcohols react directly with carbonyl compounds <09JOC2605>.

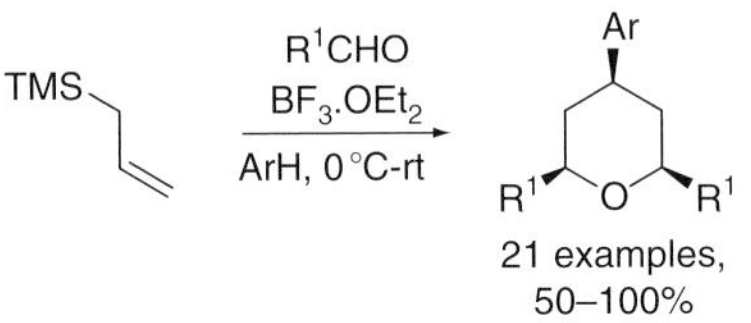
TMS
R1CHO
BF3.OEt2
ArH, 0 °C-rt
Ar
R1
O
R1
21 examples,
50–100%

Scheme 16

When the reaction between homoallylic alcohols and aldehydes is carried out in ionic liquid HF salts, 4-fluorinated tetrahydropyrans are obtained with high stereoselectivity. This approach is also successful with homoallylic thiols, leading to *cis*-substituted tetrahydrothiopyrans (Scheme 17) <09EJO103> and has been applied to the formation of 4-fluorinated tetrahydropyran-based polymers <09CC2932>.

The Pt or Au-catalysed tandem hydroalkoxylation and Prins cyclisation of diallyl-substituted alkynols **8** offers a facile route to [3.3.1] bicyclic systems in which structural diversity and enantioselectivity are readily achieved (Scheme 18)

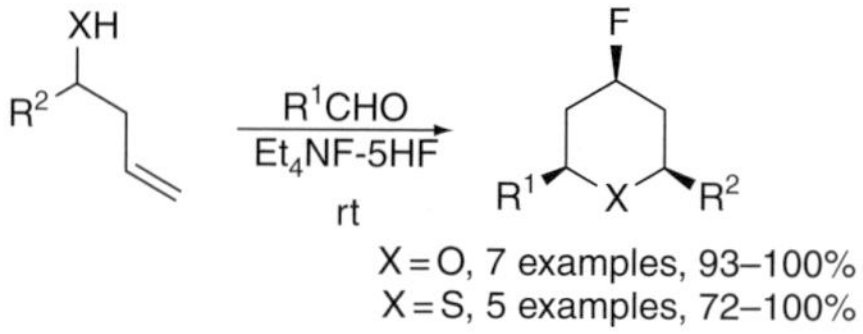

Scheme 17

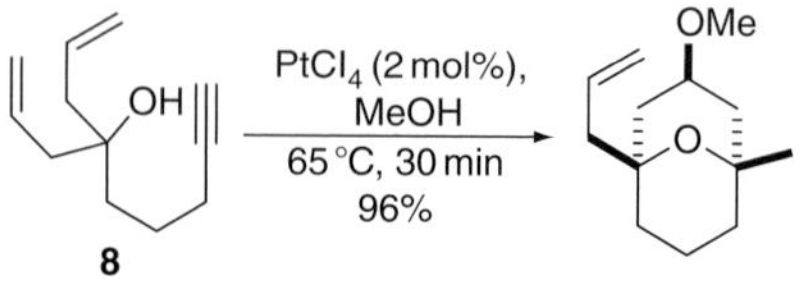

X = O, 7 examples, 93–100%
X = S, 5 examples, 72–100%

Scheme 18

<09CEJ11660>. An intramolecular Prins reaction in which hex-3-en-1,6-diols are coupled to aldehydes provides hexahydro-2*H*-furo[3,2-*c*]pyrans with high diastereoselectivity <09TL5998>.

Cyclobutanediesters undergo a Sc-catalysed formal [4+2] cycloaddition with aldehydes to give *cis*-2,6-disubstituted tetrahydropyrans. Improved yields result when a one-pot procedure is adopted in which sequential [2+2] and [4+2] reactions occur (Scheme 19) <09JA14202>.

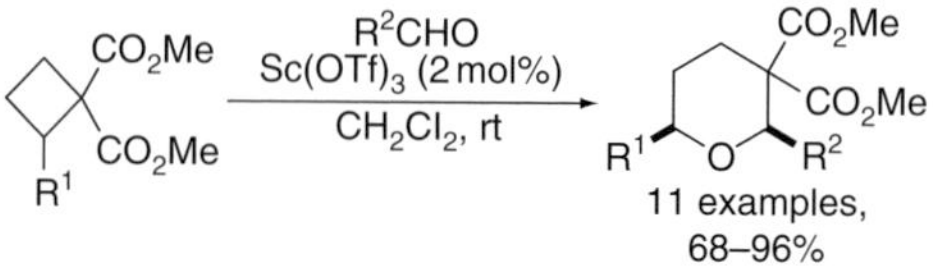

Scheme 19

1,1-Cyclopropanediesters **9** ring open and react with propargyl alcohols to give 3-methylene-tetrahydropyrans in a one-pot process catalysed by $In(OTf)_3$ / $ZnBr_2$ and with added base to effect the final cyclisation step. In some instances the intermediate propargyl ether can be isolated <09JOC8414>.

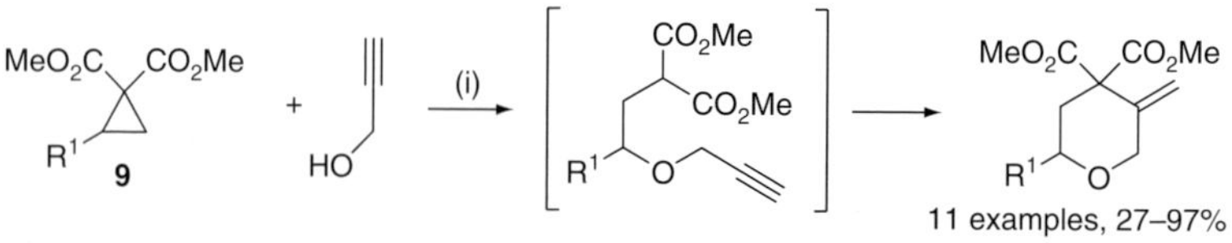

Reagents: (i) $In(OTf)_3$ (20 mol%), PhMe, rt then Et_3N, $ZnBr_2$, heat

A chiral Mo complex brings about an enantioselective ring-opening of oxabicycles **10** and cross metathesis with aryl alkenes which delivers *Z*-substituted tetrahydropyrans <09JA3844>.

In the presence of a nucleophile, 2-methylenetetrahydropyrans react with activated carbonyl compounds to give tetrahydropyranyl ketides without any competing isomerisation of the double bond (Scheme 20) <09CC6457>.

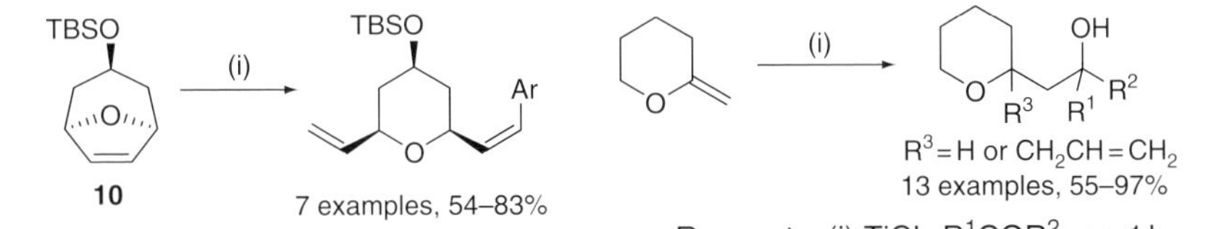

Scheme 20

Reagents: (i) chiral Mo complex, ArCH=CH$_2$, neat, 22 °C, ~1 h

Reagents: (i) TiCl$_4$, R^{1}COR^2, ca. 1 h then either Et$_3$SiH or TMSCH$_2$CH=CH$_2$

2-Bromotetradec-1-en-7,13-diynes undergo a Pd-catalysed tricyclisation and the bis-ether, the 4,10-dioxa derivative, affords the cyclopenta[1,2-c;3,4-c']dipyran **11** <09HCA1729>. A series of diamondoid ketones has been converted into the respective oxadiamondoids, of which **12** is illustrative, through reaction with MeMgI and subsequent oxidation with trifluoroperacetic acid <09OL3068>.

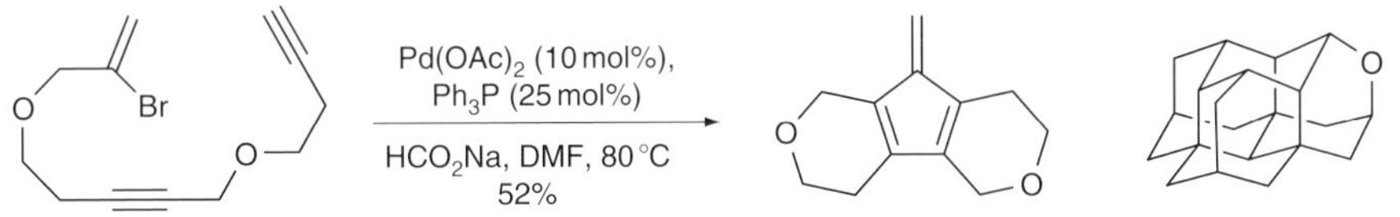

The kinetic spirocyclisation of glycal epoxides provides benzannulated 5,6- and 6,6-spiroketals with either retention or inversion at C-1 according to the conditions used <09OL3670>. A one-pot approach to diastereomerically pure chroman spiroacetals involves a Pd-catalysed cascade between a salicylaldehyde, pent-4-yn-1-ols and either an amine or an orthoester (Scheme 21) <09AGE1644>. A further approach to 5,6- and 6,6-spiroketals is based on the intramolecular radical cyclisation of γ-benzopyranylalkanols **13**, assembled using cross-metathesis methodology (Scheme 22) <09SL793>.

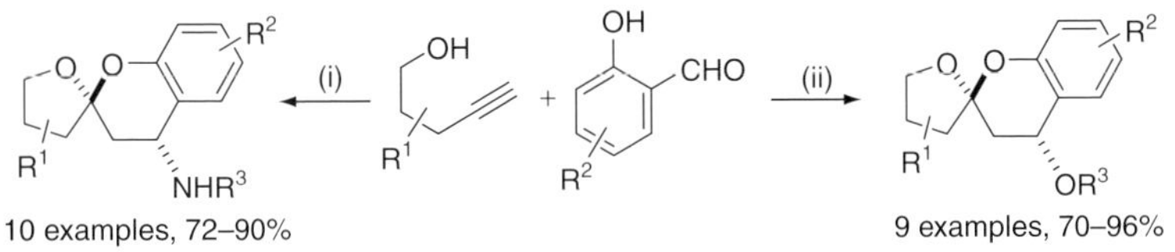

Reagents: (i) R^{3}NH$_2$, [Pd(MeCN)$_4$](BF$_4$)$_2$ (5 mol%), MeCN, rt then Mg(ClO$_4$)$_2$, HClO$_4$, CH$_2$Cl$_2$, MeCN, rt; (ii) HC(OR^3)$_3$, [Pd(MeCN)$_4$](BF$_4$)$_2$ (5 mol%), MeCN, rt

Scheme 21

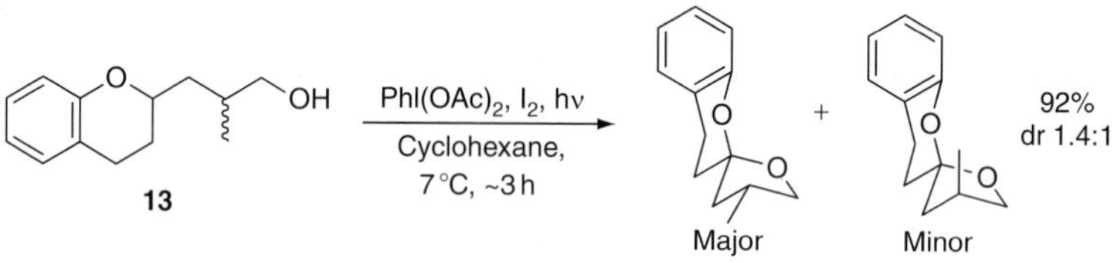

Scheme 22

Reagents: PhI(OAc)$_2$, I$_2$, hv, Cyclohexane, 7 °C, ~3 h; 92%, dr 1.4:1 (Major + Minor from 13)

6.4.2.2 [1]Benzopyrans and Dihydro[1]benzopyrans (Chromenes and Chromans)

Routes to naphtho[2,1-*b*]pyrans and their 2,3-dihydro analogues <09JHC1098> and the asymmetric synthesis of chromans <09T3931> have been reviewed.

Developments in the aryl propargyl ether route to chromenes include the fast direct synthesis of 3-bromo derivatives through Pd-catalysed cyclisation in the presence of stoichiometric amounts of CuBr$_2$ and added LiBr in acetic acid (Scheme 23)

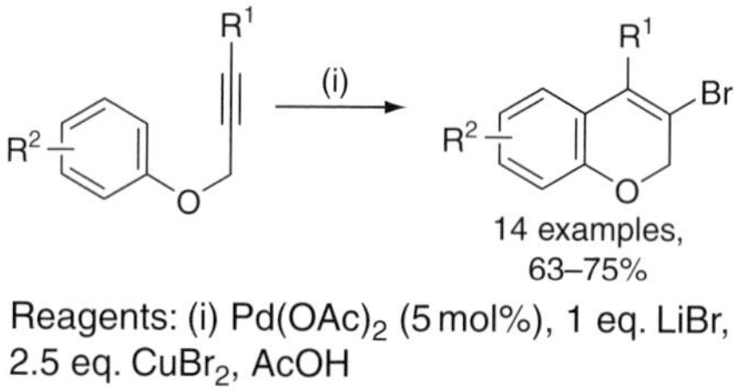

Scheme 23

14 examples, 63–75%

Reagents: (i) Pd(OAc)$_2$ (5 mol%), 1 eq. LiBr, 2.5 eq. CuBr$_2$, AcOH

<09SL2079>. Introduction of a S or Se substituent into the terminal position of the alkyne followed by base-catalysed electrophilic cyclisation using I$_2$ or ICl leads to 4-chalcogenyl-3-iodobenzopyrans; further manipulation through the halogen function is possible (Scheme 24) <09JOC3469>. Both aryl propargyl ethers and aryl propargylates undergo a facile intramolecular hydroarylation reaction in the presence of an Au catalyst leading to chromenes and coumarins, respectively <09JOC8901>. A solid-supported Hg triflate very efficiently cyclises an aryl propargyl ether to the chromene at room temperature <09AGE1244>.

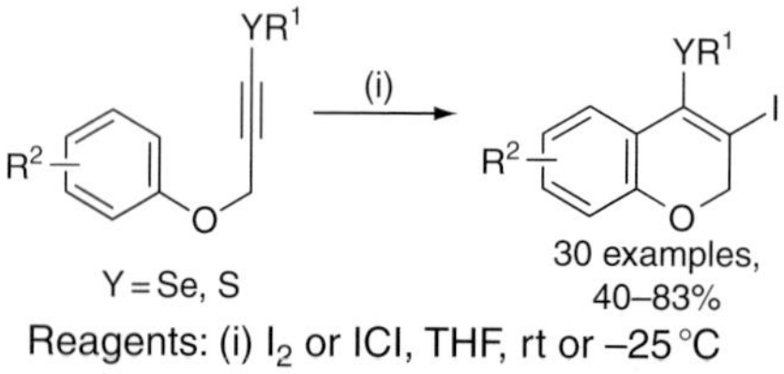

Scheme 24

Y = Se, S

30 examples, 40–83%

Reagents: (i) I$_2$ or ICl, THF, rt or –25 °C

Good yields of 1,3-disubstituted naphthopyrans are obtained in a Ga-catalysed one-pot reaction involving a naphthol, an alkyne and a benzaldehyde (Scheme 25) <09TL5798>.

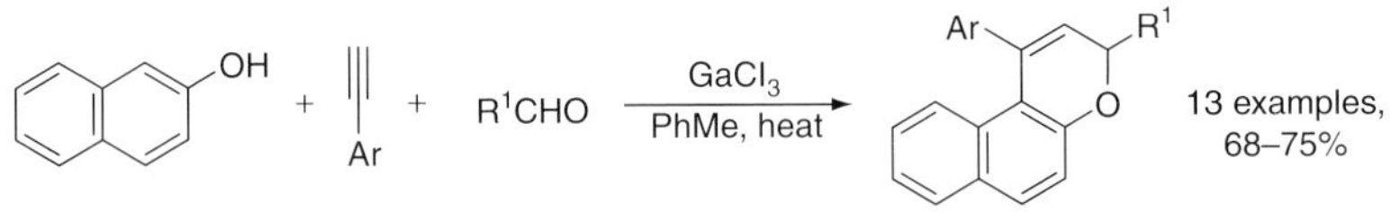

Scheme 25

High enantioselectivity and yields of 2-substituted benzopyran-3-carboxaldehydes result from chiral amine – chiral acid catalysed tandem oxa-Michael – aldol reactions between cinnamaldehydes and salicylaldehydes (Scheme 26) <09OBC4539>. Activated phenols react directly with 3-methylbut-2-enal under

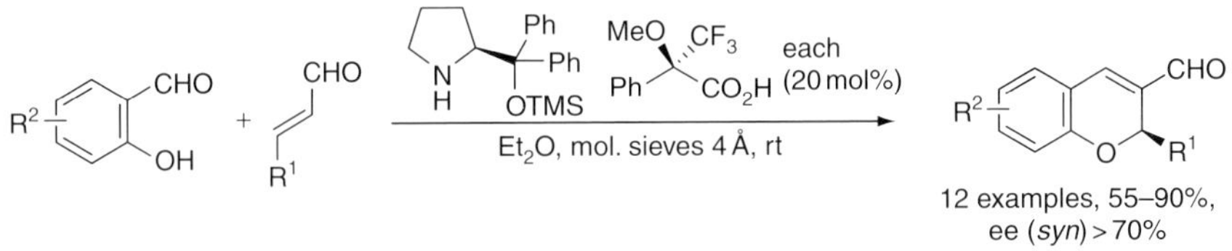

Scheme 26

microwave irradiation in $CHCl_3$ to give 2,2-dimethylchromenes, including octandrenolone and the precocenes, albeit in quite variable yields <09TL5075>. Using ethylenediamine diacetate (EDDA) as catalyst, pinosylvin and α,β-unsaturated aldehydes afford naturally occurring pyranostilbenes <09S2146>. rac-α-Tocopherol acetate is readily accessible from trimethylhydroquinone through the pyridine-catalysed condensation with α,β-unsaturated aldehyde acetals and subsequent hydrogenation of C3=C4 in the chromene <09BCJ843>.

Salicylaldehydes react with allenylphosphonates in DBU/DMSO to give phosphonochromenes **14** in which E/Z isomerisation is observed at room temperature <09JOC5395> and salicylaldehydes and vinyl boronic acids undergo a Petasis borono-Mannich reaction in water leading to 2-substituted chromenes <09EJO1859>.

An Au(I)-catalysed asymmetric carboalkoxylation of propargyl esters affords 4,4-disubstituted 4H-chromenes; a rearrangement of an allylic oxonium intermediate is considered to be the likely pathway (Scheme 27) <09JA3464>.

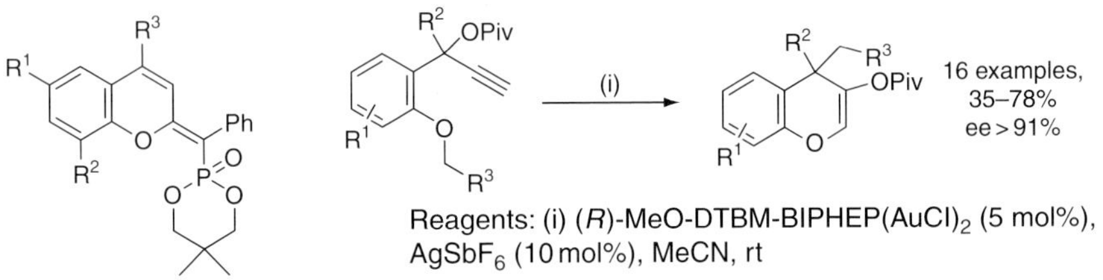

Reagents: (i) (R)-MeO-DTBM-BIPHEP(AuCl)$_2$ (5 mol%), AgSbF$_6$ (10 mol%), MeCN, rt

14

Scheme 27

Aryl propargyl ethers bearing an o-aminoacetonitrile function **15** undergo an anionic cyclisation to a mixture of a 3-methyl-4H-chromene-4-nitrile **16** and a 3-methylenechroman-4-nitrile under phase transfer conditions. The former rearrange to benzofuranones on elution from alumina <09S2029>. 2-Amino-4H-chromen-4-ylphosphonic acid esters are formed in a one-pot reaction between salicylaldehydes, malononitrile and triethyl phosphite catalysed by InCl$_3$ <09SL917>.

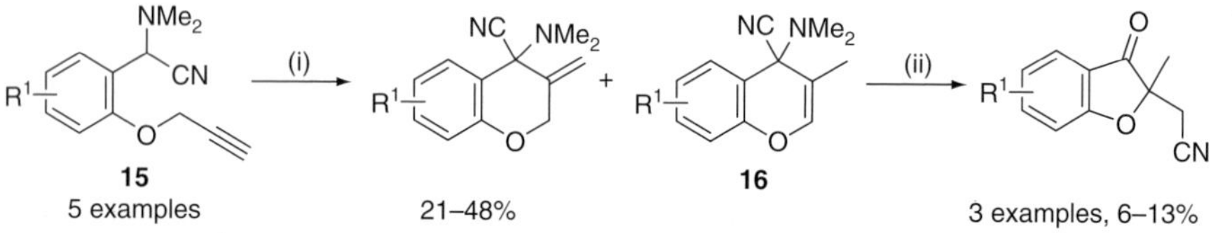

Reagents: (i) NaOH, DMSO, TEBAC, 35 °C, 2 h; (ii) Aluminium oxide (basic), EtOAc, hexane

Sequential benzylation, cyclisation and dehydration of 1,3-dicarbonyl compounds with 2-[hydroxy(phenyl)methyl]phenol in a recyclable Brønsted acid ionic liquid produces 2,3,4-trisubstituted 4H-chromenes (Scheme 28) <09T7457>.

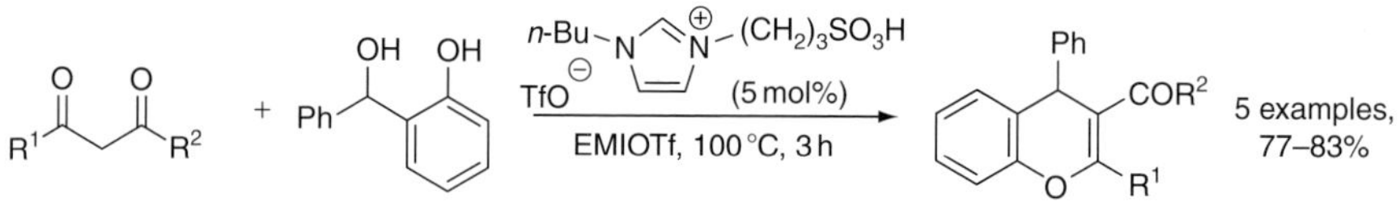

Scheme 28

New illustrations of the value of 3-nitrochromenes in synthesis include their conversion into 1,4-dihydrobenzopyrano[4,3-d]triazoles through the facile 1,3-dipolar cycloaddition of azide ion <09T5799> and the diastereoselective 3-alkylation of indoles by 2-aryl-3-nitrochromenes under aqueous conditions <09EJO4503>. Microwave irradiation assists the Rh-catalysed decarbonylation <09SL1383> and Wittig olefination <09T1300> of 3-formylchromenes. The direct arylation of 4-acetamidochromenes can be achieved through the Pd-catalysed Hiyama coupling with trialkoxy aryl silanes <09AGE5355> and 2-substituents have been introduced into isoflav-3-enes through initial conversion to the isoflavylium salt and subsequent nucleophilic addition <09SL306>. 4-Acylbenzopyrans and their thio analogues undergo a diastereoselective Nazarov cyclisation which affords annulated cyclopenta[c](thio)chromans <09OBC1858>. The Heck arylation of 4H-chromenes produces 2-aryl-2H-chromenes <09TL1222>.

When applied to the propargylic alcohol **17**, the Ru-catalysed cycloisomerisation of the enyne unit results in the construction of the tetracycle **19** through the simultaneous formation of both pyran and cyclohexadiene rings. The same ring system is formed by cyclisation of 1-propynyl-2-ethenylnaphtho[2,1-b]pyran; chroman- and thiochroman-based enynes similarly afford dibenzo[b,d](thio)pyrans <09AGE2534>.

The *6-endo-dig* intramolecular phenoxycyclisation of 1,5-enynes **20** leads to fused chromans (hexahydroxanthenes) under the influence of a cationic gold complex; extension to the synthesis of a reduced naphtho[2,1-*b*]chroman from a 1,5,9-dienyne has been accomplished <09OL2888>. Gold catalysis also features in the synthesis of the furan- and pyrrole-bridged chroman derivatives **21** from complex propargyl ethers <09AGE5848>.

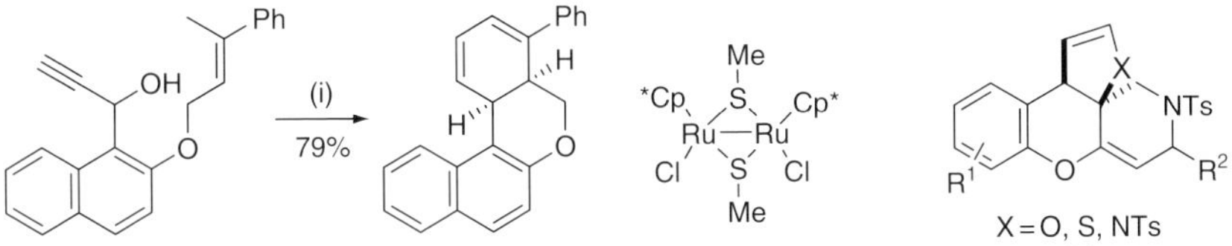

17 **19** Catalyst **18** X = O, S, NTs **21**

Reagents: (i) cat. **18** (10 mol%), NH_4BF_4 (20 mol%), DCE, 80 °C, 20 h

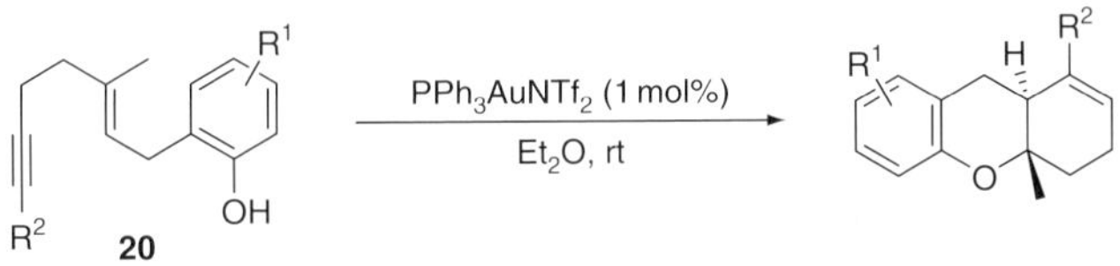

20

8 examples, 76–96%

Carbonickelation of 2-iodophenyl pentynyl ether **22** results in the generation of nucleophilic vinylnickel species which can be trapped by electrophiles to give 4-alkenylchromans (Scheme 29) <09CC4753>. In(0) in the presence of Boc-protected glycine brings about the cyclisation of chiral hydrazones derived from aryl allyl ethers **23** to *cis*-4-amino-3-vinylchromans (Scheme 30) <09JOC7183>.

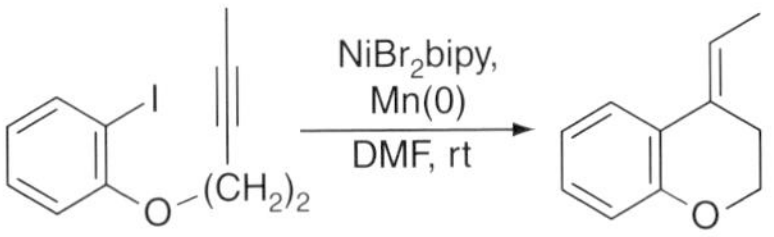

22 62%

Scheme 29

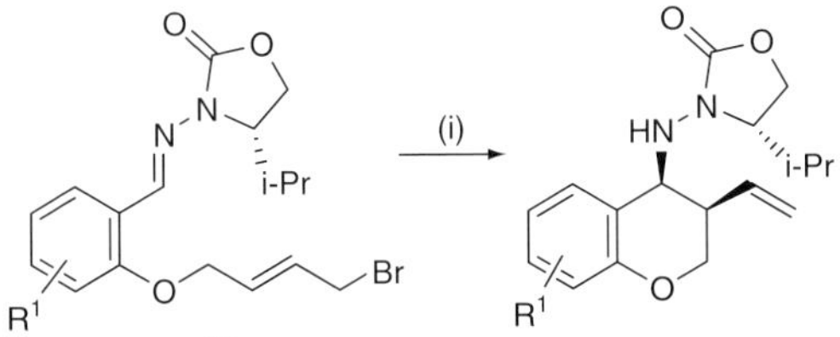

23

10 examples, 68–91%
dr 99:1

Reagents: (i) In(0), Boc-Gly-OH, THF, rt

Scheme 30

The Pd-catalysed reaction between propargylic carbonates and 2-(2-hydroxyphenyl)acetates proceeds with high stereo- and enantioselectivity to afford *trans*-3,4-disubstituted chromans bearing a (Z)-alkenyl moiety at the 2-position (Scheme 31) <09OL4752>.

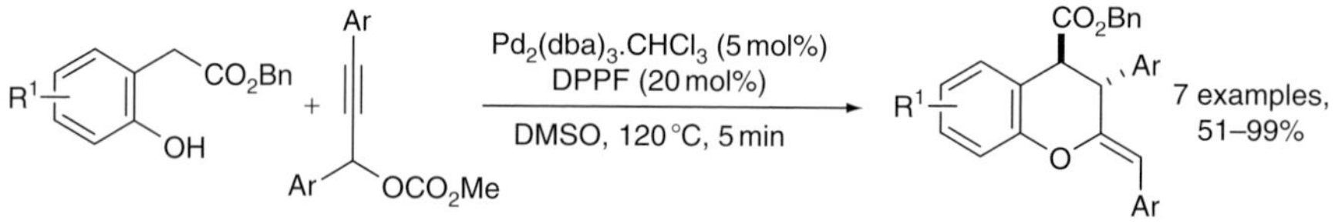

Scheme 31

A chiral proline catalyst effects the enantioselective synthesis of highly substituted chromans from cinnamaldehydes and *trans*-β-nitrostyrenes which proceeds through an oxa-Michael – Michael cascade sequence. A hemiaminal intermediate **24** behaves as a nucleophile in the catalytic sequence during which three stereogenic centres are created (Scheme 32) <09OL1627>. A similar catalyst promotes a sequence of three Michael reactions and an aldol condensation when 2-[(*E*)-2-nitrovinyl]phenol reacts

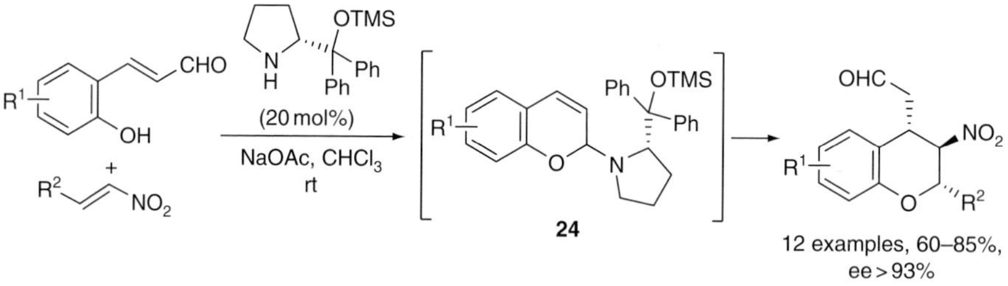

Scheme 32

with two equivalents of α,β-unsaturated aldehydes that leads to tetrahydro-5*H*-benzo[*c*]chromenes **25** as single enantiomers; the synthesis is successful when two different enals are used (Scheme 33) <09TL704>.

A domino aldol reaction – hDA reaction between *O*-allyl salicylaldehydes and resorcinols catalysed by EDDA/TEA leads to benzopyrano[3,4-*c*]benzopyrans. An intermediate *o*-quinone methide is postulated, the *endo* form of which cyclises to generate the *cis*-fused tetracycle (Scheme 34) <09T101>. The use of 1-methylindoline-2-thione in related Knoevenagel – hDA sequences leads to indole-annulated

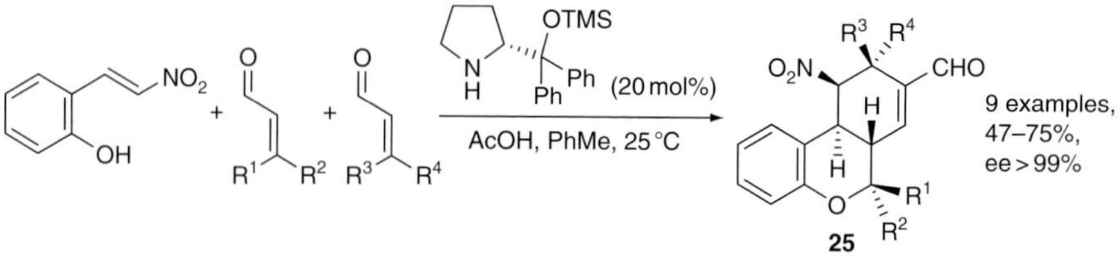

Scheme 33

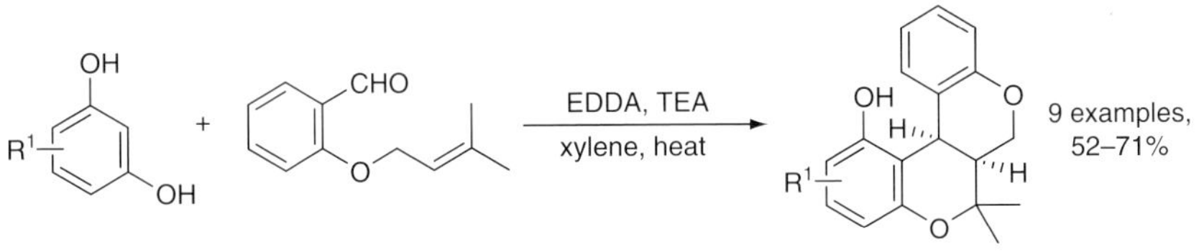

Scheme 34

thiopyrano[3,4-*c*]benzopyrans <09TL3889, 09TL6723>. The Knoevenagel product from 2,4-dihydro-3*H*-pyrazol-3-ones and *O*-propargyl salicylaldehydes undergoes an *in situ* Cu(I)-catalysed hDA reaction to produce benzopyrano[3,4-*c*]pyrano[2,3-*c*]pyrazoles (Scheme 35) <09SL55>. Quinone methides feature in a synthesis of 2-arylchromans through reaction with styrenes <09JOC4009>. Quinone methides formed from phenanthrols through a proline-catalysed condensation with formalde-

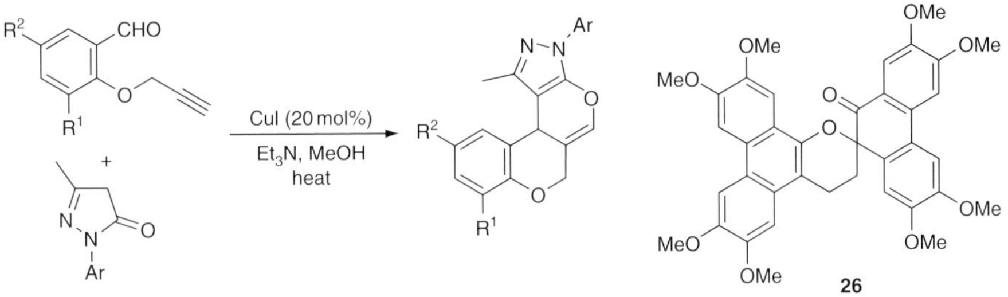

Scheme 35

hyde dimerise to produce spirophenanthrones *e.g.* **26** <09SL1501>. The synthesis of amino-substituted naphthopyrans through the microwave-assisted reaction between a naphthol, 3-hydroxy-2,2-dialkylpropanal and a secondary amine is considered to involve a quinone methide derived from a Mannich base <09TL51>. Oxidative homocoupling of electron-rich styrenylphenols using a Cu(II)/(-)-sparteine catalyst followed by a spontaneous inverse electron demand DA offers a route to the polemannones, benzoxanthenone lignans <09TL3084>.

A proline-catalysed enantioselective synthesis of naphthopyrans involves an asymmetric Friedel–Crafts alkylation – cyclisation sequence between 1-naphthols and α,β-unsaturated aldehydes <09JOC6881>. Aryl butenyl ethers undergo a Friedel–Crafts cyclisation when treated with PhSeBr and a Ag salt at low temperature; a seleniranium ion is implicated <09OL2924>. The Friedel–Crafts-like allylation of phenols is efficiently conducted under microwave irradiation using a Mo complex catalyst with chloranil as oxidant. Good yields of 2-substituted chromans result through a formal [3+3] cyclocoupling (Scheme 36) <09OL717>. Cyclopenta[*c*]

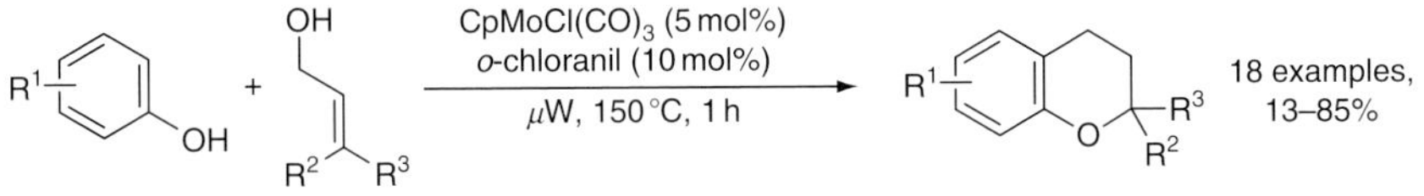

Scheme 36

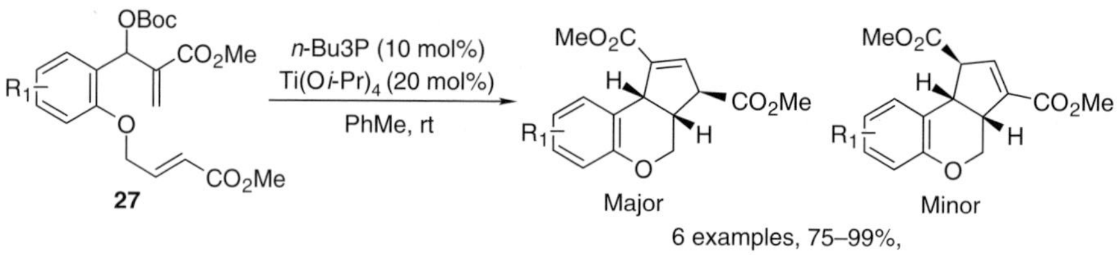

Scheme 37

chromans are formed with high diastereoselectivity by the phosphine-catalysed intramolecular [3+2] cycloaddition of aryl ethers **27** (Scheme 37) <09JOC3394> and Ph$_3$P catalyses the reaction between salicyl N-thiophosphinylimines and ethyl penta-2,3-dienoate in which the γ-methyl group undergoes cyclisation to the 2,4-disubstituted chromans (Scheme 38) <09OL991>. 2-(Phenylthio)chroman is obtained through the allylation of phenol with 3-chloro-1-(phenylthio)propene

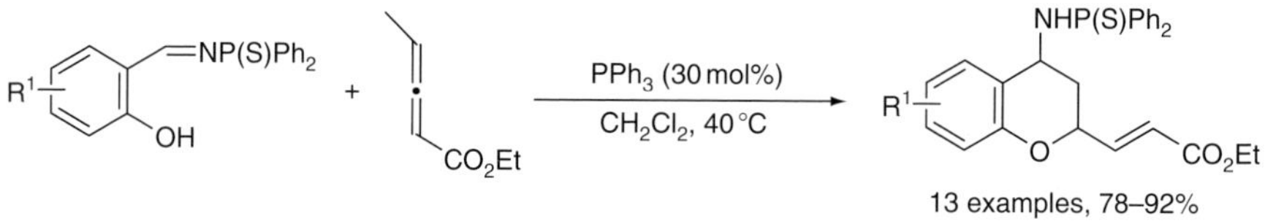

Scheme 38

and subsequent cyclisation; the propene functions as a three-carbon annulating species <09OL4576>.

Chromans are formed when o-vinylaryl alkyl ethers, readily available in two steps from salicylaldehydes, are treated with scandium triflate, although yields are much influenced by the overall structure of the ether. Initial hydride transfer promotes cyclisation involving an sp^3. C–H bond and an activated alkene (Scheme 39) <09OL2972>. In a related manner, 2-aroylchromans result from a base-catalysed intramolecular cyclisation of 1-aryl-2-(2-vinylphenoxy)ethanones also available

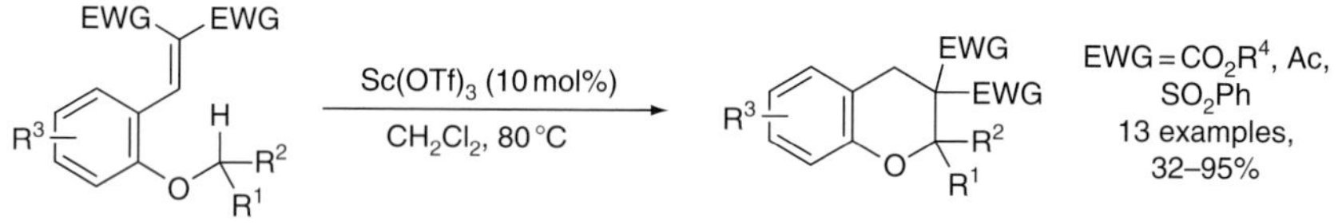

Scheme 39

from salicylaldehydes by a three-step one-pot reaction (Scheme 40) <09T8702, 09TL5748>.

Sharpless asymmetric epoxidation of the allyl alcohol **28** followed by intramolecular epoxide ring opening by phenoxide ion offers an attractive route to enantiomerically pure 2-hydroxymethylchromans <09S1886>. 2,3-trans-Flavan-3-ols are

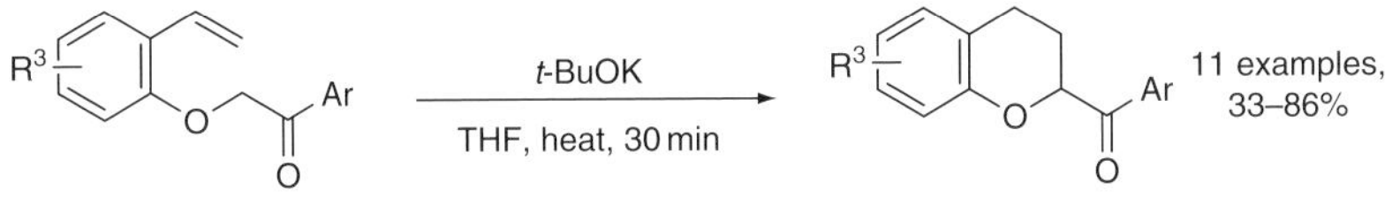

Scheme 40

accessible through a Mitsunobu cyclisation of diols **29** <09S779>. Both (R)- and (S)-2-hydroxymethyl-2,5,7,8-tetramethylchroman-6-ol are available from 2,3,5,6-tetramethylbenzoquinone *via* reaction with (R)- or (S)-2-benzyloxymethyl-2-methyloxirane <09EJO833>.

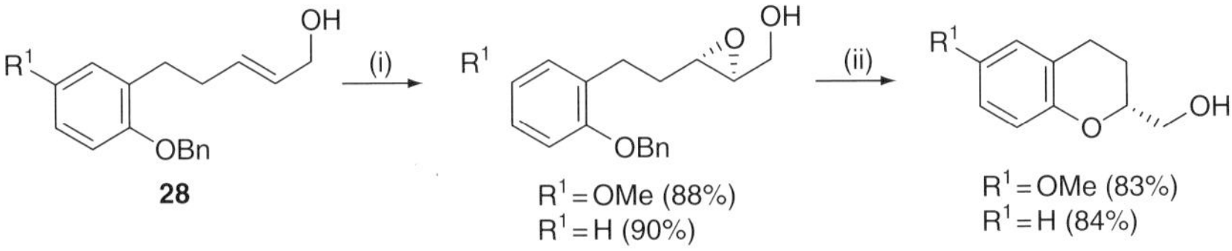

Reagents: (i) D-(−)-DIPT, Ti(O-*i*-Pr)$_4$, TBHP, CH$_2$Cl$_2$, −25 °C, 18 h; (ii) 10% Pd/C, EtOAc, H$_2$, 2 h then aq. NaOH saturated with NaCl, 0 °C, 3 h

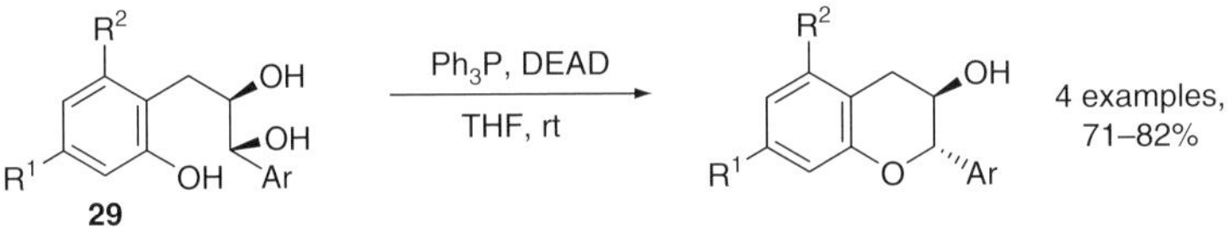

6.4.2.3 [2]Benzopyrans and Dihydro[2]benzopyrans (Isochromenes and Isochromans)

2-Ethynylbenzyl alcohols undergo a Ru-catalysed cycloisomerisation to 1*H*-[2]benzopyrans in the presence of an amine which is essential for the catalytic cycle (Scheme 41) <09OL5350>. Incorporation of a phosphonate unit at the benzylic carbon atom affords 1-phosphonylated isochromenes through a Pd-catalysed 6-*endo-dig* cyclisation <09OBC2848>.

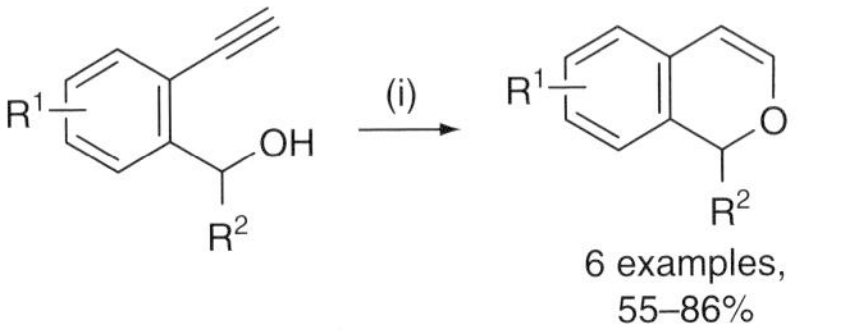

Reagents: (i) CpRuCl(PPh$_3$)$_2$, *n*-BuNH$_2$, 90 °C

Scheme 41

The rearrangement of a chiral 4-naphthyldioxolane into a naphtho[1,2-*c*]pyran effected by TiCl$_4$ is considered to involve two aromatic substitutions; initial loss of a tosyloxy group is followed by electrophilic chlorination involving a titanium species (Scheme 42) <09TL6361>.

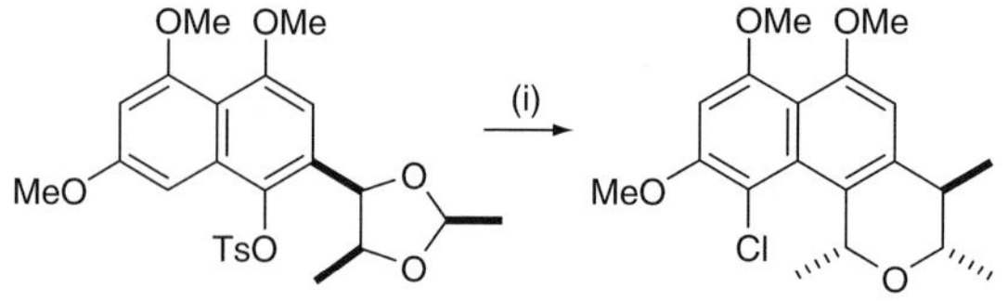

Reagents: (i) TiCl$_4$, −65 °C (42%)

Scheme 42

Activation of isochromans by *N*-hydroxyphthalimide (NHPI) enables oxidative coupling at C-1 to be achieved with malonates and ketones in the presence of oxygen and Cu and In salts (Scheme 43) <09SL138>.

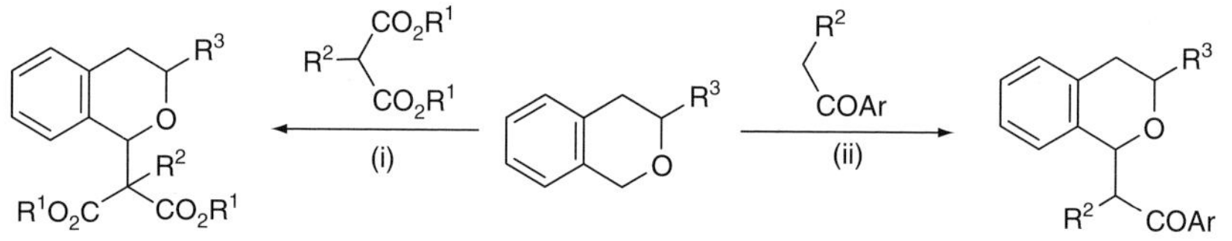

7 examples, 42–83% 8 examples, 1–75%

Reagents: (i) Cu(OTf)$_2$ (5 mol%), InCl$_3$ (5 mol%), NHPI (20 mol%), O$_2$, 55 °C; (ii) Cu(OTf)$_2$ (5 mol%), InCl$_3$ (5 mol%), NHPI (20 mol%), O$_2$, 75 °C

Scheme 43

6.4.2.4 Pyrylium Salts

Both oxygen and sulfur have been incorporated into large cationic polycyclic aromatic hydrocarbons. 14-Phenyl-14*H*-dibenzo[*a,j*]xanthenes and the thio analogues were oxidised to the (thio)xanthenylium salts whereupon photocyclisation gave the purple benzo[5,6]-naphthaceno[1,12,11,10-*jklmna*](thio)xanthylium salts **30** <09OL5686>.

The biscamphorpyrylium **31**, prepared from 3-benzoylcamphor, gives the chiral biscamphorphosphabenzene on reaction with P(TMS)$_3$ <09T9368>.

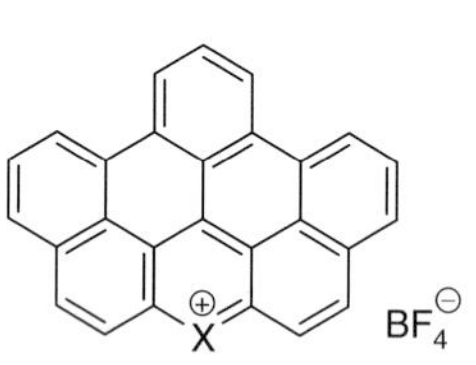

X = O, S
30

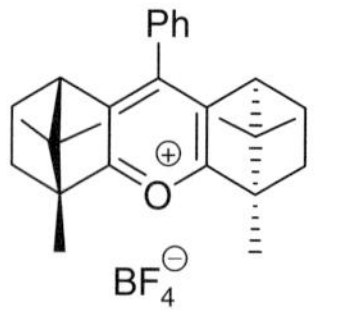

31

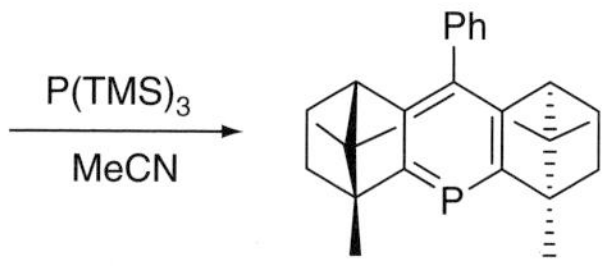

82%

A range of bridged and multi-ring structures have been derived from the reaction of isochromenylium salts with alkenes <09JOC8787>. A thermochromic spiropyran results from the treatment of the flavylium salt **32** with NEt$_3$ (Scheme 44) <09OL1769>.

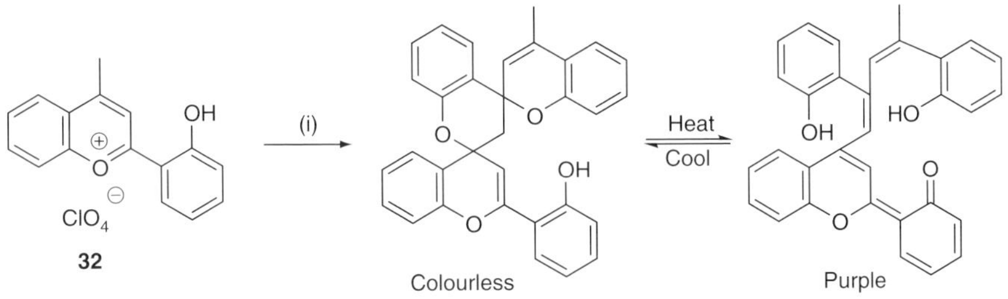

Scheme 44

Reagents: (i) Et₃N, CH₂Cl₂, 0 °C, 30 min (65%)

6.4.2.5 Pyranones

The value of pyran-2-ones in synthesis through ring transformations brought about by nucleophilic reagents has been discussed <09H(77)657>.

The Rh-catalysed oxidative coupling of acrylic acids with alkynes affords pyran-2-ones (Scheme 45) <09JOC6295> and the synthesis of 4-methylsulfanylpyran-2-ones from ketene dithioacetals and ketones has been discussed <09H(78)555>.

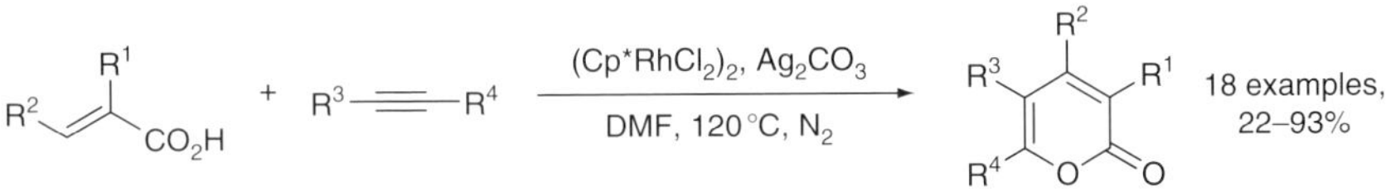

Scheme 45

Mention can be made of the synthesis of benzo[*b*]pyrano[2,3-*d*]oxepines by this method <09SL2992> and of the conversion of the sulfanylpyranones into blue-emitting fluorenes <09OL1289>.

A synthesis of substituted pyran-2-ones is based on the hydroalkylation of electron-deficient alkynes with activated methylene compounds, with cyclisation following an initial nucleophilic addition. In the presence of added formaldehyde, 5,6-dihydropyranones are produced (Scheme 46) <09T2110>.

The bicyclo[3.2.2]nonadiene ring system is accessible through a PPh₃-catalysed [4+3] annulation of 5-carboxymethylpyran-2-one with allylic carbonates <09OL3978> and functionalised octadienes result from an intramolecular [4+4] photocycloaddition of ketals **33** derived from complex pyran-2-ones <09TL1188>.

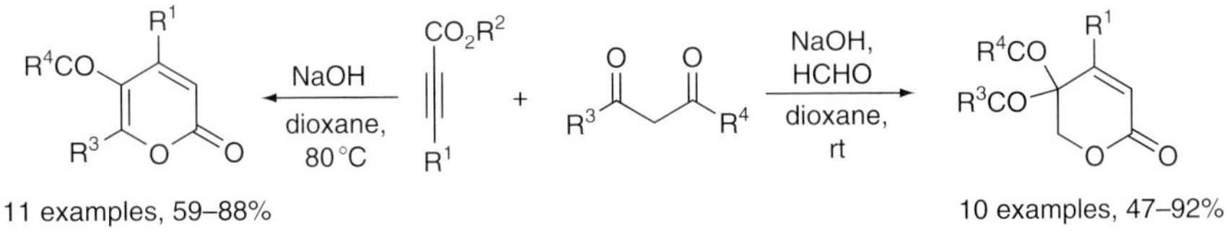

11 examples, 59–88%   10 examples, 47–92%

Scheme 46

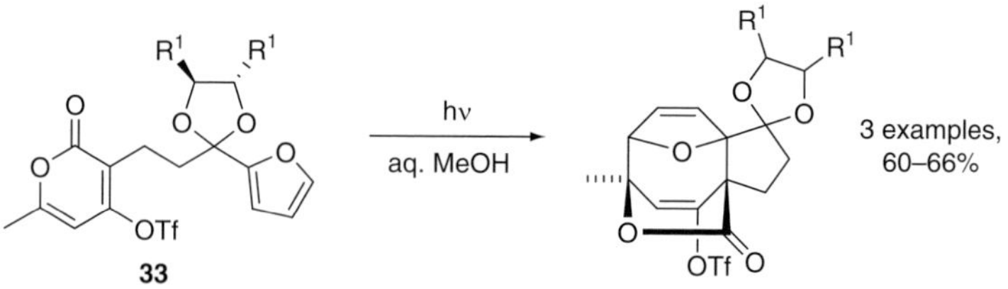

Scheme 47

The course of the reaction between 1,3-bis(trimethylsilyloxy)-1,3-butadienes and 1,1-dimethoxy-4,4-dichlorobut-1-en-3-one is influenced by choice of Lewis acid catalyst. Thus, although $TiCl_4$ leads to dichloromethyl-substituted salicylates, TMS triflate affords 6-substituted 2-(dichloromethyl)pyran-4-ones (Scheme 47)

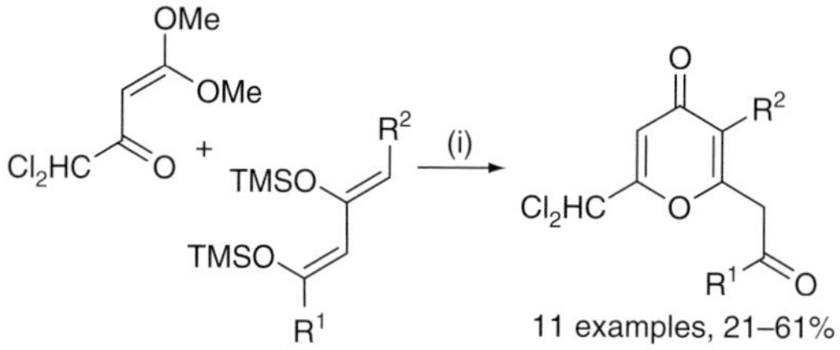

<09T9271>. An enantioselective hDA reaction ensues between Rawal's diene **34** and aldehydes under the influence of a Rh catalyst leading to 2,3-dihydropyran-4-ones (Scheme 48) <09CC7294>.

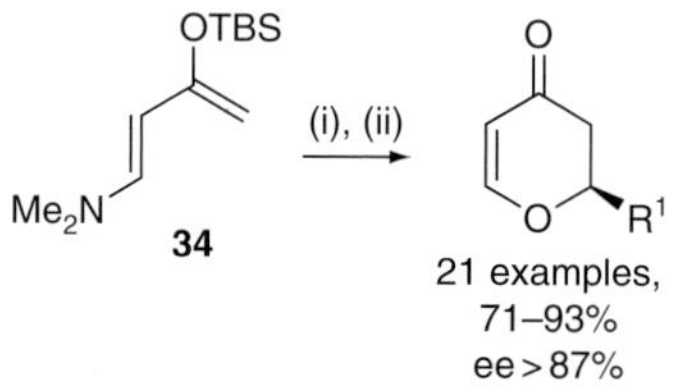

Scheme 48

Imidazolium salts are efficient catalysts for the formation of fused 3,4-dihydropyran-2-ones from α,β-unsaturated enol esters. The initial generation of an α,β-unsaturated acyl imidazolium and an enolate prompts a conjugate addition and an intramolecular acylation completes the sequence. Similar products can also be prepared from α,β-unsaturated acyl fluorides and TMS enol ethers (Scheme 49) <09JA14176>. N-Heterocyclic carbenes (NHCs) also effect the ring expansion of 2-acyl-1-formylcyclopropanes to 3,4-dihydropyran-2-ones <09OL1623>.

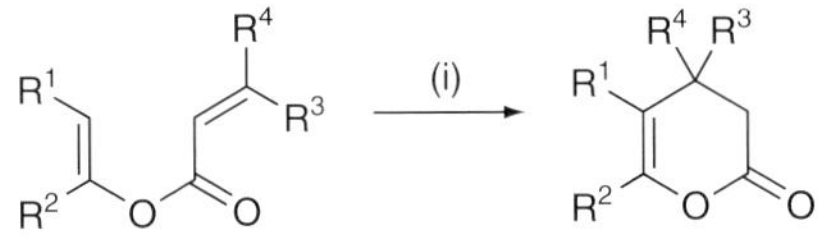

Reagents: (i) Imidazolium salt (10 mol%), KO*t*-Bu (20 mol%), PhMe, heat, 16 h

12 examples, 32–87%

Scheme 49

Tetrahydropyran-2-ones are similarly formed through the NHC-catalysed ring expansion of tetrahydrofuran-2-aldehydes <09OL891>.

A one-pot hydrosilylation–RCM–protodesilylation protocol allows the synthesis of 5,6-dihydropyran-2-ones bearing an ω-alkenyl chain at the 6-position from the alkyne **35** (Scheme 50). This approach has been used in total syntheses of (+)-gonothalamin and (−)-pironetin <09SL565>.

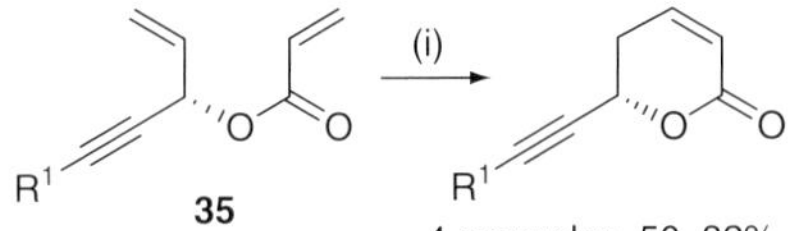

Reagents: (i) [Cp*Ru(MeCN)$_3$]PF$_6$, HSi(OEt)$_3$, then Grubbs' II then AgF, MeOH, H$_2$O, THF

4 examples, 50–82%

Scheme 50

Under basic conditions, but with an acidic work-up, 1,3-diketones react with aromatic aldehydes to give 6-aryl-2,3-dihydropyran-4-ones <09TL3020> and a related product arises from a thiazolium salt-catalysed Mukaiyama aldolisation between benzaldehyde and Danishefsky's diene <09TL7239>. The synthesis of 2,3-dihydropyran-4-ones by the acid-catalysed cyclisation of α,β-unsaturated 1,3-diketones has been optimised <09JOC6973>.

The *cis*-fused lactone **37** is formed with high diastereoselectivity when the intramolecular DA reaction of ester-tethered decatrienes **36** is carried out in an ionic liquid <09OBC3657>.

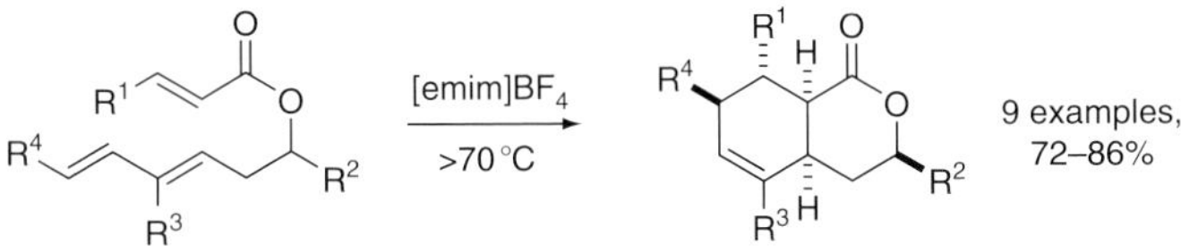

9 examples, 72–86%

The diastereoselective 1,4-bromolactonisation of enynoic acids **38** is catalysed by DABCO and leads to 6-(bromoallenyl)tetrahydropyran-2-ones. When an O-heteroatom is present in the alkanoic acid chain, the corresponding 1,4-dioxanes are produced <09JA3832>.

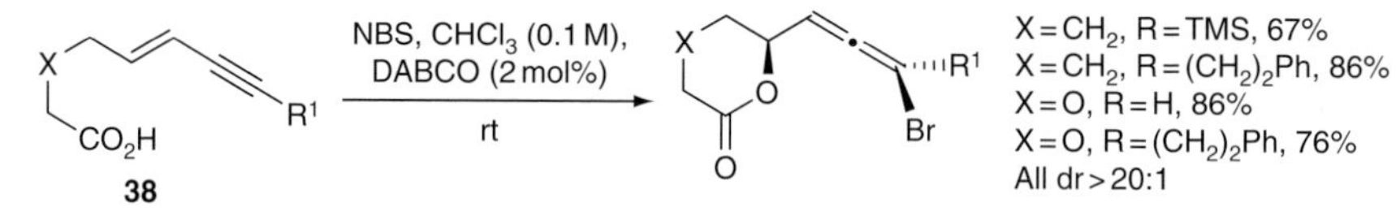

An intramolecular Stetter reaction involving an aliphatic aldehyde and an acrylate moiety leads diastereoselectively to a *trans,syn*-fused cyclohexa[*b*]pyran-3-one **39**. Further manipulation of the product and a second Stetter reaction produces a *trans, syn,trans* fused polyether array (Scheme 51) <09SL233>.

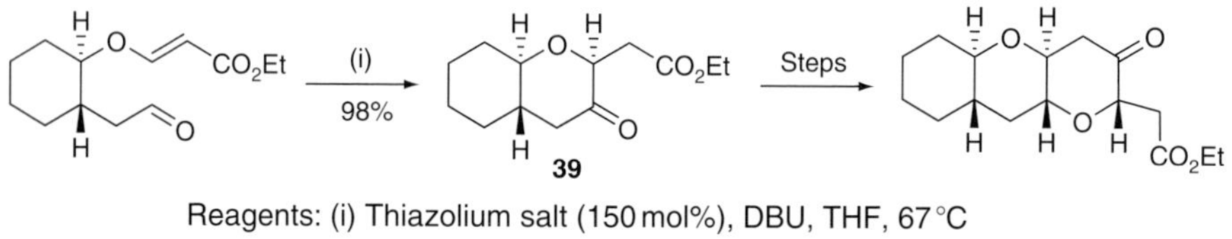

Reagents: (i) Thiazolium salt (150 mol%), DBU, THF, 67 °C

Scheme 51

6.4.2.6 Coumarins

Several new approaches to coumarins use salicylaldehydes as the starting point, as for example in two one-pot processes. Reaction of the derived oxyanion with the ylide triphenyl(α-carboxymethylene)phosphorane imidazolide is followed by an intramolecular Wittig reaction (Scheme 52) <09TL236> and reaction with 2-phenyl-1,3-

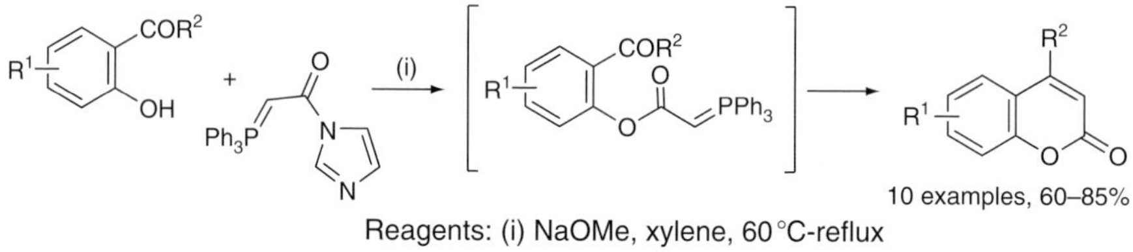

Reagents: (i) NaOMe, xylene, 60 °C-reflux

Scheme 52

oxazolan-5-one in an ionic liquid proceeds through a Knoevenagel reaction and yields 3-benzamidocoumarins (Scheme 53) <09TL2208>.

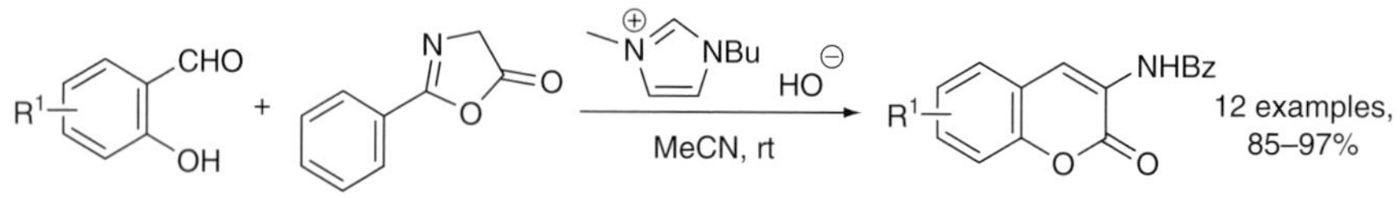

Scheme 53

High yields of 3-aroylbenzopyran-2-thiones result from reaction with β-oxo-dithioesters and urea in a $SnCl_2$-catalysed Biginelli cyclocondensation <09JOC3141>. 3-Cyanochromenes, available from salicylaldehydes, afford

3-alkoxymethyl-2-phenyliminochromenes on treatment with alkoxide and aniline. Subsequent hydrolysis yields 3-alkoxymethylcoumarins <09JOC8798>.

Naphtho[2,1-b]pyran-2-ones have been synthesised from tetralones in which cis-hydrogenation of an enynal, electrocyclisation and oxidation of a 2H-pyran moiety are the key steps. The route has been extended to furo-, thieno- and pyrido-[f]fused coumarins (Scheme 54) <09CC5618>.

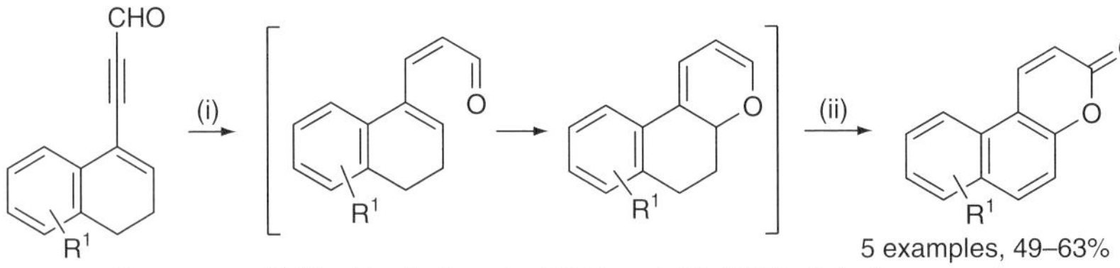

Reagents: (i) H$_2$, Lindlar's cat., EtOAc, rt; (ii) DDQ, 1,4-dioxane, rt

Scheme 54

A good range of hetero-fused coumarins have been obtained from aryl O-carbamoyl ortho-boronic acids through coupling with heteroaromatic iodides and subsequent LDA-induced carbamoyl migration (Scheme 55) <09JOC4094>. A similar sequence of directed ortho metalation, cross-coupling and directed remote meta-

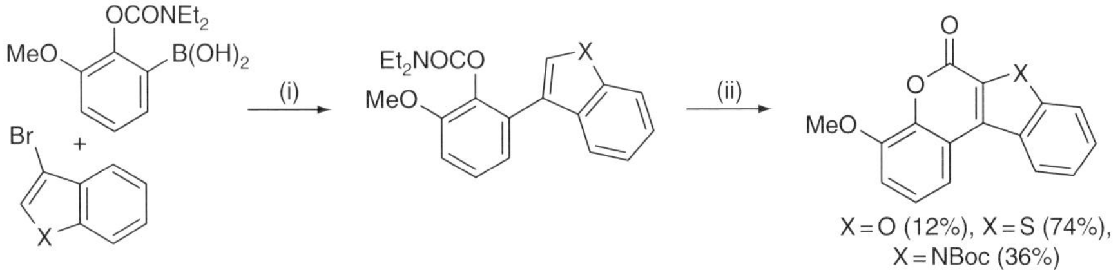

Reagents: (i) Pd(PPh$_3$)$_4$, PhMe, reflux; (ii) LDA, THF, 0 °C then AcOH reflux

Scheme 55

lation has been used in syntheses of aglycones of the gilvocarcins and arnottin I, bioactive naphtho[b,d]benzopyran-6-ones <09JOC4080>.

4-Hydroxycoumarin is a useful synthon as indicated by the Michael addition to α,β-unsaturated ketones which yields enantiopure warfarin analogues <09EJO5192> and by the synthesis of various 3-substituted derivatives through addition to Baylis-Hillman acetate adducts <09S399>. C-Alkylation with alcohols is efficiently catalysed by Ir complexes <09T7468> and by I$_2$ <09T9233>, while a Pd-catalysed cross-coupling with alkynes <09T6810> and arylboronic acids <09TL2103> affords 4-alkynylcoumarins and 4-arylcoumarins, respectively. Direct sulfanylation has been achieved with thiols under aqueous conditions <09TL2405> and the derived triflates react with amides and similar compounds to give 4-aminocoumarins <09S3689>.

4-Cyanocoumarins undergo a DA reaction with oxygenated dienes to give mainly the *endo* adduct which are subsequently aromatised by way of a base-catalysed elimination to dibenzo[*b,d*]pyranones (Scheme 56) <09OL757>. Under high pressure, 3-cyanocoumarins cycloadd to butadienes to give 6a-cyanobenzo[*c*]coumarins in good yield <09JOC4311>.

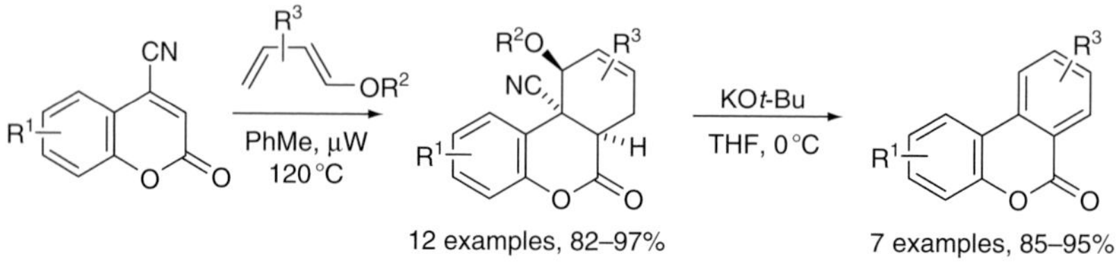

Scheme 56

The microwave-assisted zincation of coumarin offers a route to 3-substituted coumarins <09CC5615> and 3-allylcoumarins are available from allyl esters of coumarin-3-carboxylic acids through a Pd-catalysed decarboxylative sp^2–sp^3 coupling (Scheme 57) <09OL3434>. The asymmetric reduction of 4-alkyl/arylcoumarins, achieved in high yield and with high ee using a Cu hydride catalyst, has been applied in the synthesis of various natural products and drugs <09OL5374>. The synthesis of dihydrocoumarins by the reduction of coumarins has been reviewed <09S3533>.

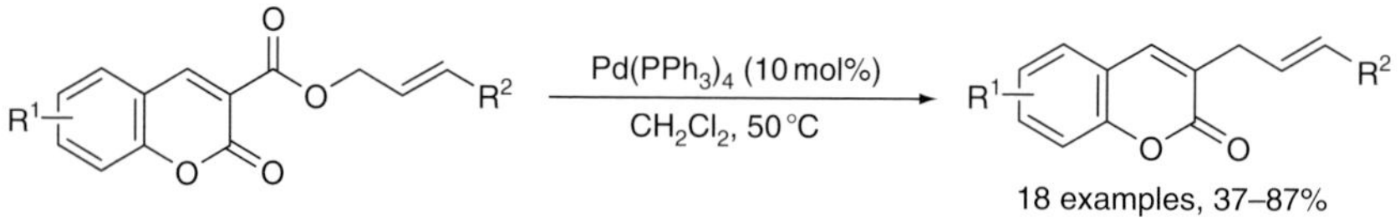

Scheme 57

Triazolium salts serve as NHCs and catalyse the conversion of 2-hydroxycinnamaldehydes into 3,4-dihydrocoumarins (Scheme 58) <09JOC1759>.

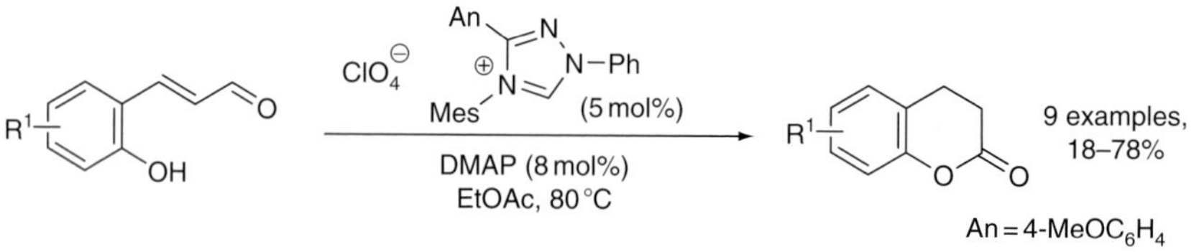

Scheme 58

The TiCl$_4$-catalysed diastereoselective hydroarylation of benzylidene malonates by phenols leads to 3,4-disubstituted dihydrocoumarins <09JOC4612>.

The Pd-catalysed *ortho*-alkylation of benzoic acids with 1,2-dichloroethane can be achieved without the need for I^--scavenging Ag^+ salts and 3,4-dihydroisocoumarins result through a spontaneous lactonisation (Scheme 59) <09AGE6097>.

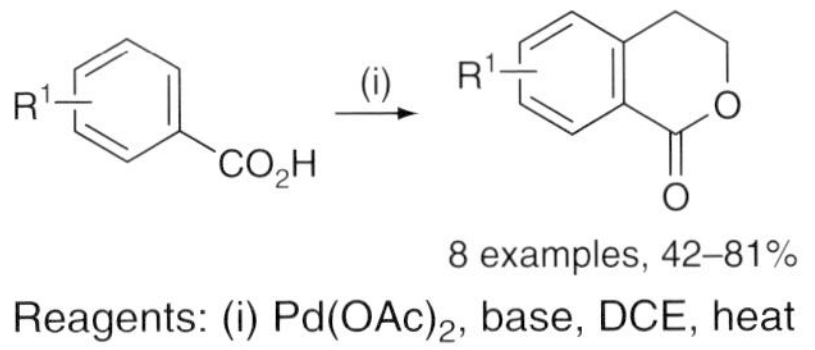
R1
CO2H
(i)
R1
O
O
8 examples, 42–81%
Reagents: (i) Pd(OAc)2, base, DCE, heat

Scheme 59

A solid-phase synthesis of isocoumarins commences with the attachment of 2-bromobenzoic acid to a Wang resin and a subsequent Sonogashira cross-coupling with terminal alkynes. An electrocyclisation occurs efficiently on treatment with ICl which both releases the 3-substituted 4-iodoisocoumarin and allows recycling of the resin (Scheme 60) <09JOC4158>.

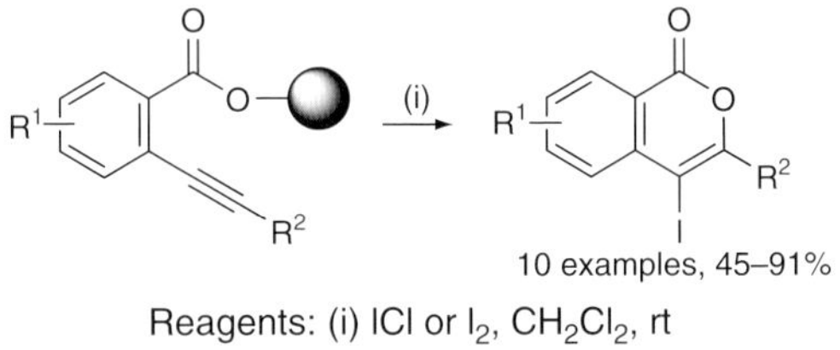
O
O
R1
R2
(i)
O
O
R1
R2
I
10 examples, 45–91%
Reagents: (i) ICl or I2, CH2Cl2, rt

Scheme 60

The enolates derived from α-(2-haloaryl)ketones undergo a Pd-catalysed carbonylation and a subsequent intramolecular acylation leads to a variety of 3,4-disubstituted and 3,4-fused isocoumarins (Scheme 61). The synthesis can be adapted to a one-pot, two-step procedure starting from a 1,2-dihalogenobenzene and a simple ketone <09CC6744>.

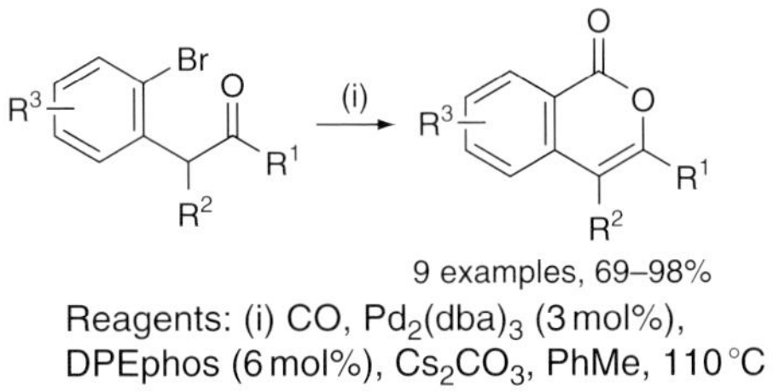
Br
O
R3
R1
R2
(i)
O
O
R3
R1
R2
9 examples, 69–98%
Reagents: (i) CO, Pd2(dba)3 (3 mol%),
DPEphos (6 mol%), Cs2CO3, PhMe, 110 °C

Scheme 61

Internal alkynes couple with various 2-substituted benzoic acids in the presence of air and a Ru/Cu catalyst to give 8-substituted isocoumarins. The procedure has been extended to heteroaryl carboxylic acids and aromatic diacids when hetero-[*c*] fused pyran-2-ones and benzo[1,2-*c*:4,5-*c*′]dipyrandiones are produced respectively (Scheme 62) <09JOC3478>.

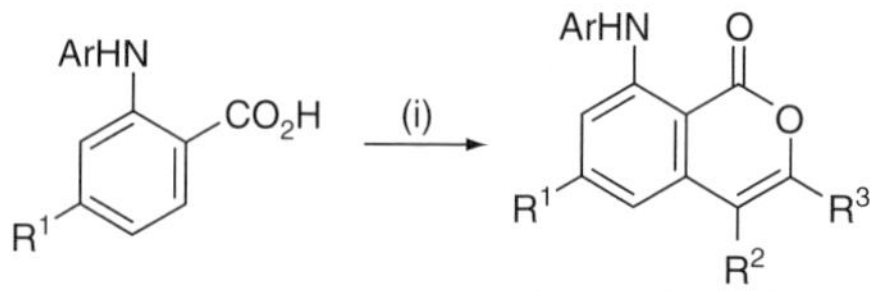

Scheme 62

Reagents: (i) (Cp*RhCl$_2$)$_2$, Cu(OAc)$_2$, air, R^2———R^3, xylene, 120 °C

2-(Alkoxycarbonyl)benzenediazonium bromides are converted into dihydroisocoumarins through CuBr-catalysed reaction with various acrylates (Scheme 63) <09TL6112> and in the presence of a base benzenediazonium-2-carboxylates lose N$_2$ to generate a betaine, trapping of which with C$_{60}$ gives the C$_{60}$-fused isocoumarin **40** <09CC1769>.

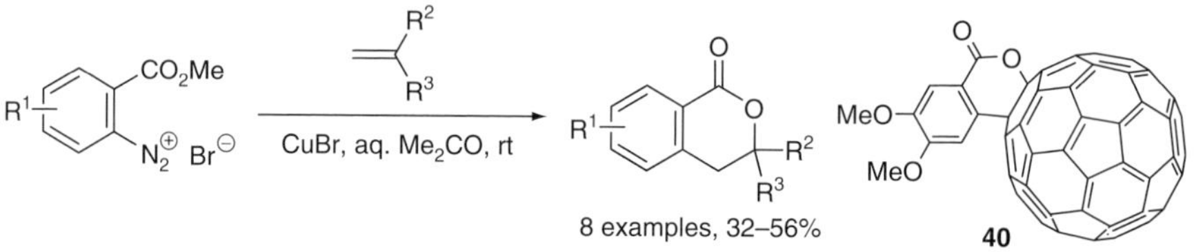

Scheme 63

6.4.2.7 Chromones

A facile, microwave-mediated one-pot synthesis of flavones combines a Sonogashira reaction of aryl iodides and a terminal alkyne with a carbonylative annulation with 2-iodophenols (Scheme 64) <09OL3210>. Isoflavones are available through the Pd-catalysed oxidative cyclisation of α-methylene deoxybenzoins <09TL1542> and through cross-coupling of 3-iodochromones with triarylbismuths <09SL2597>.

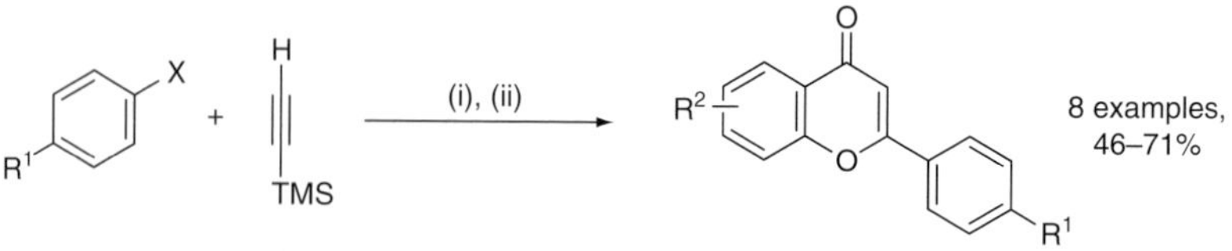

Reagents: (i) Pd$_2$(dba)$_3$ (1.5 mol%), PA-Ph (3 mol%), DMF, DBU, μW, heat;
(ii) 2-iodophenol, Pd$_2$(dba)$_3$ (1.5 mol%), PA-Ph (3 mol%), CO, TBAF, DMF, DBU, μW, heat

Scheme 64

Flavones are efficiently reduced to *cis*-flavan-4-ols by $NaBH_4$ in the presence of Co (II) phthalocyanine <09CC6397>.

3-Acrylates of chromones result from the PPh_3-catalysed reaction of 1-(2-hydroxyaryl)-3-alkyl-1,3-diones with ethyl propiolate (Scheme 65) <09CC6089> and Ni

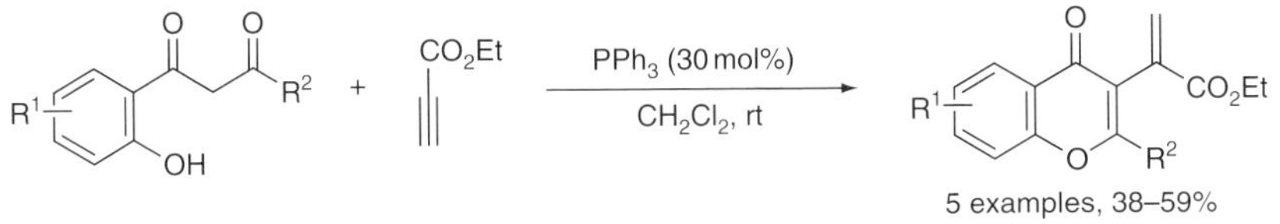

Scheme 65

(cod)$_2$ catalyses the reaction of salicylic acid ketals with internal alkynes which yields 2,3-disubstituted chromones <09JA13194>.

6,6-Bisbenzannulated spiroketals **41** which incorporate a chromanone unit have been obtained by the application of a double intramolecular hetero-Michael addition to protected dihydroxy ynones derived from salicylaldehydes and 4-arylbutynes <09TL3245>.

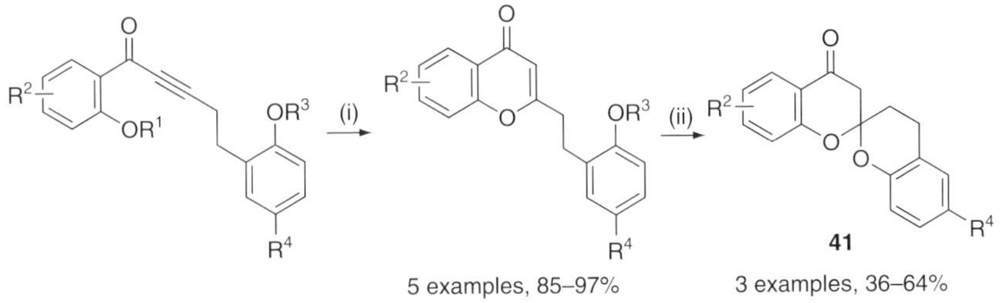

Reagents: (i) Et_2NH, CH_2Cl_2, reflux; (ii) CBr_4, *i*-PrOH

It appears that only 2-(alkynyl)phenylacetic acids in which an alkyl substituent is present on the alkyne moiety undergo a 6-*exo-dig* iodolactonisation to the isochroman-3-ones. With aryl-substituted alkynes, a 7-*endo-dig* pathway dominates and benzo[*d*]oxepinones are produced. The two heterocyclic systems are distinguishable through their ^{1}H and ^{13}C NMR spectra (Scheme 66) <09TL1385>.

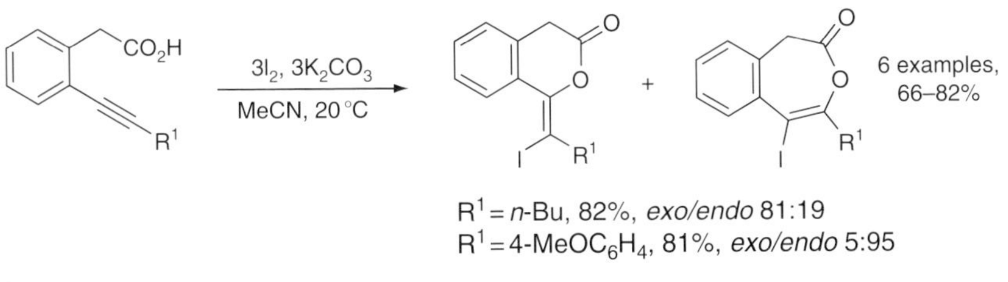

R^1 = *n*-Bu, 82%, *exo/endo* 81:19
R^1 = 4-MeOC$_6$H$_4$, 81%, *exo/endo* 5:95

Scheme 66

UV-irradiation of chromone affords a 1:1 mixture of the *cis-cis-anti* and *cis-trans* head-to-tail dimers <09CC2379>.

The asymmetric intramolecular Stetter reaction catalysed by NHCs has been reviewed <09SL1189>. O-Allylated salicylaldehydes are cyclised to chroman-4-ones on heating in the presence of an NHC, providing the first example of hydroacylation of unactivated alkenes <09JA14190> and high yields of chroman-4-ones are reported when 4-(2-formylphenoxy)butenoates are heated under microwave irradiation with an alkylthiazolium-based ionic liquid and NEt_3 <09SL1915>. Interestingly, the use of ionic liquids favours the synthesis of bis-spirochromanones by the Kabbe reaction <09TL2643> and microwave irradiation assists in the formation of 2-substituted chromanones from 2′-hydroxyacetophenones and aliphatic aldehydes <09JOC2755>.

Chiral fluorinated flavanones result from the alkaloid-catalysed tandem intramolecular oxa-Michael addition and electrophilic fluorination of the activated α,β-unsaturated ketone **42** (Scheme 67) <09CEJ13299, 09JOC1400> and *trans*-3-

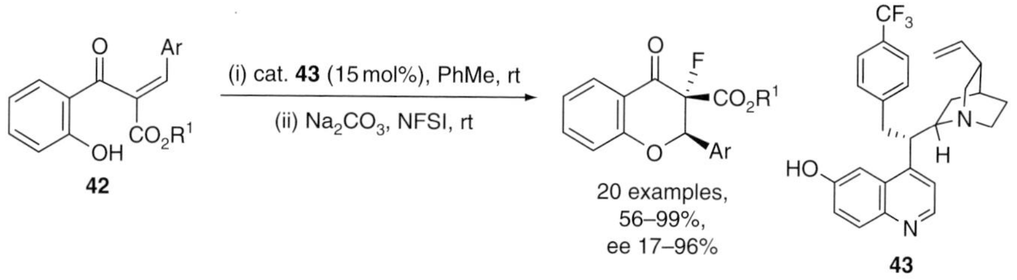

Scheme 67

hydroxyflavanones have been obtained by the cycloisomerisation of epoxyalkenes *e.g.* **44** catalysed by $Yb(OTf)_3$. Related epoxyalkynes afford 4-substituted 3-methylenechromans <09EJO3129>.

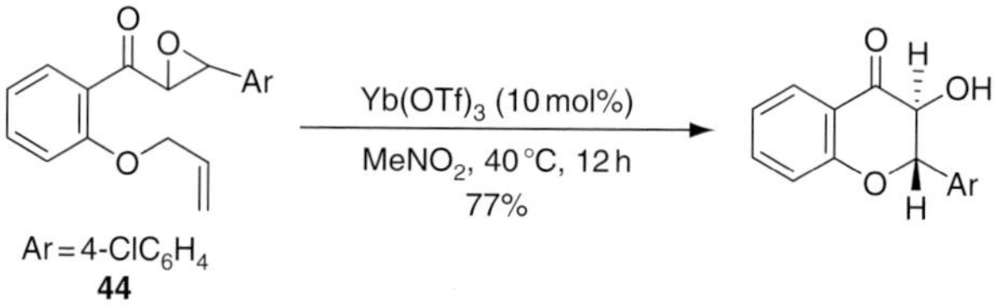

Reaction of the alkyne-tethered aryl iodides **45** with boronic acids leads to 4-arylalkylidene isochroman-3-ones (Scheme 68) <09JOC2234>.

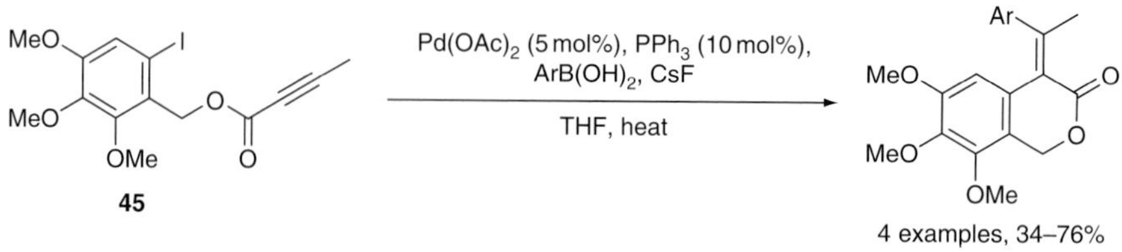

Scheme 68

6.4.2.8 Xanthones and Xanthenes

In the presence of CsF and under basic conditions, benzyne derived from *o*-trimethylsilyl-phenyl triflate reacts with salicylaldehydes to give 9-hydroxyxanthene. In the absence of base, the xanthenol disproportionates and a mixture of xanthenes and xanthones results (Scheme 69). Since the reaction fails or is only low yielding with alternative sources of the aryne, Cs^+ plays a critical role in the procedure <09OL169>.

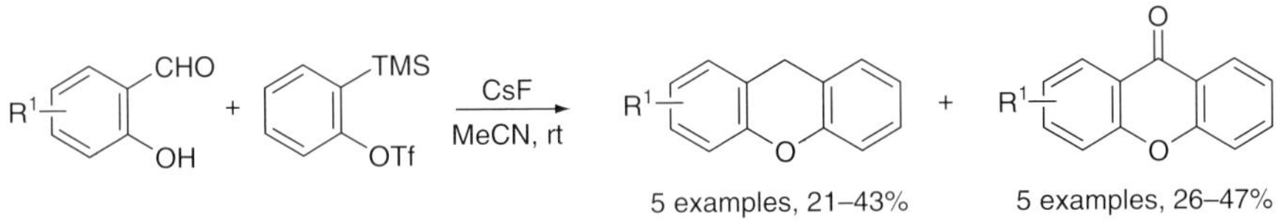
CHO
OH
TMS
OTf
CsF
MeCN, rt
5 examples, 21–43%
5 examples, 26–47%

Scheme 69

Acceptable yields of xanthones result from the Pd-catalysed reaction between salicylaldehydes and 1,2-dibromoarenes <09CC6469>.

Under aqueous basic conditions 2-halogeno-2′-hydroxybenzophenones cyclise through an S_NAr process to give high yields of xanthones <09T5729> and 3-(1-alkynyl)chromones undergo a base-promoted Michael addition of 1,3-dicarbonyl compounds which initiates a sequence of reactions terminating in the formation of functionalised xanthones (Scheme 70) <09AGE6520>.

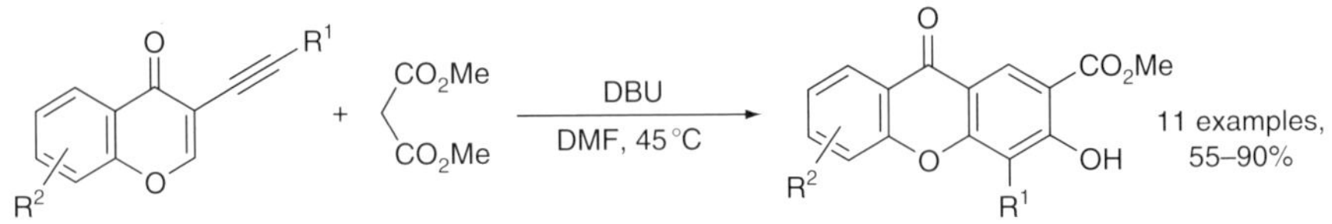
DBU
DMF, 45 °C
11 examples, 55–90%

Scheme 70

Both 9-arylxanthenes and thioxanthenes are accessible from 2-fluorobenzaldehydes by a three-step procedure which features a $FeCl_3$-catalysed intramolecular diarylmethylation of electron-rich arenes (Scheme 71) <09EJO4757>. In a variation of

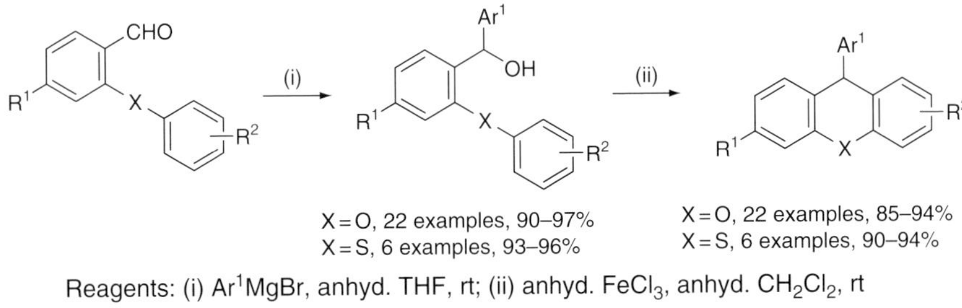
(i)
(ii)
X = O, 22 examples, 90–97%
X = S, 6 examples, 93–96%
X = O, 22 examples, 85–94%
X = S, 6 examples, 90–94%
Reagents: (i) Ar¹MgBr, anhyd. THF, rt; (ii) anhyd. FeCl₃, anhyd. CH₂Cl₂, rt

Scheme 71

this approach, the 2-aryloxybenzaldehydes react directly with the arene to give the xanthene <09JOC6797>.

2-(9-Xanthenyl)malonates undergo a Mn(OAc)$_3$-promoted 1,2-aryl radical rearrangement which leads to dibenzo[b,f]oxepincarboxylates (Scheme 72) <09JOC3978>.

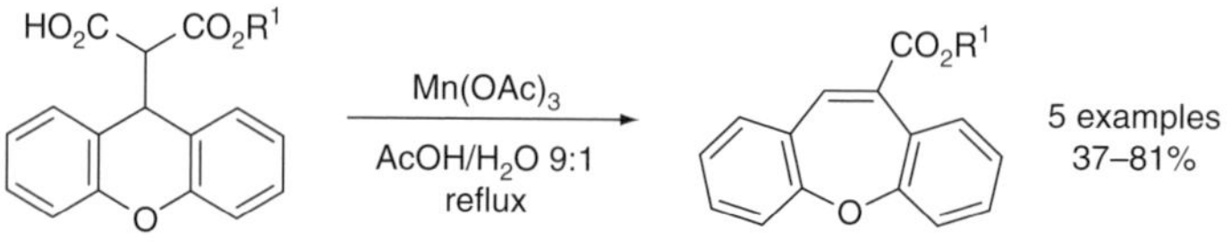

Scheme 72

6.4.3. HETEROCYCLES CONTAINING ONE SULFUR ATOM

6.4.3.1 Thiopyrans and analogues

Nitrobutadienes derived from 3-nitrothiophene undergo a base-catalysed intramolecular Michael addition which yields 4-nitro-3,6-dihydrothiopyran 1,1-dioxides (Scheme 73) <09T336>.

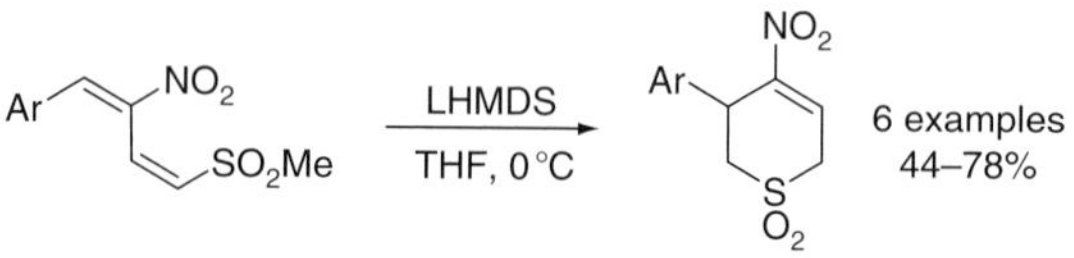

Scheme 73

The course of the cyclisation of alkylsulfonyl diazoacetates through C–H insertion is dependent on the choice of catalyst and also on substitution both at the insertion site and adjacent to the sulfone group. The latter favours formation of the thiophene at the expense of the thiopyran 1,1-dioxide (Scheme 74) <09TL1954>.

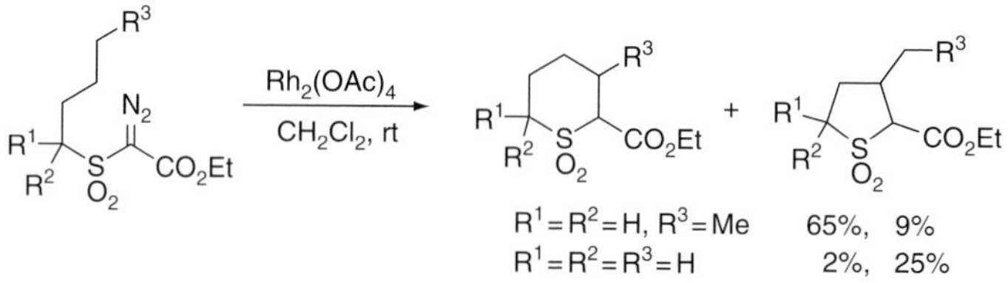

Scheme 74

By careful choice of the hydroxyl protecting group and of the enolate used, aldehyde **46** and ketone **47**, both of which show high diastereofacial selectivity, undergo aldol reactions which give enantio enriched adducts through kinetic resolution (Scheme 75) <09JOC4447>.

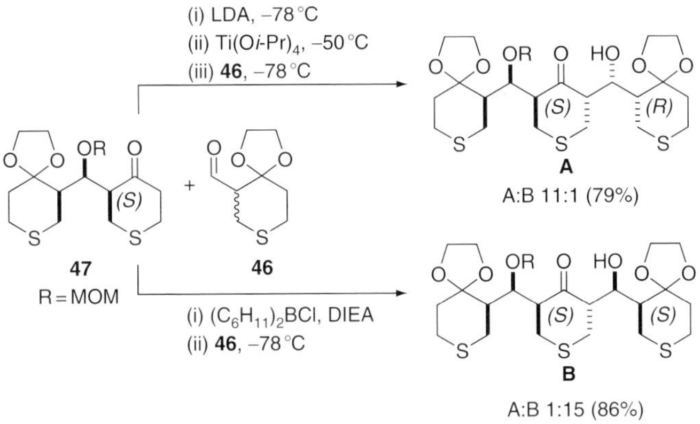

Scheme 75

The final step in a synthesis of 2-methylthiaadamantane and its CD_3 analogue from Meerwein's ester, is a Wolff–Kishner reduction of the diketone **48** <09CC595>.

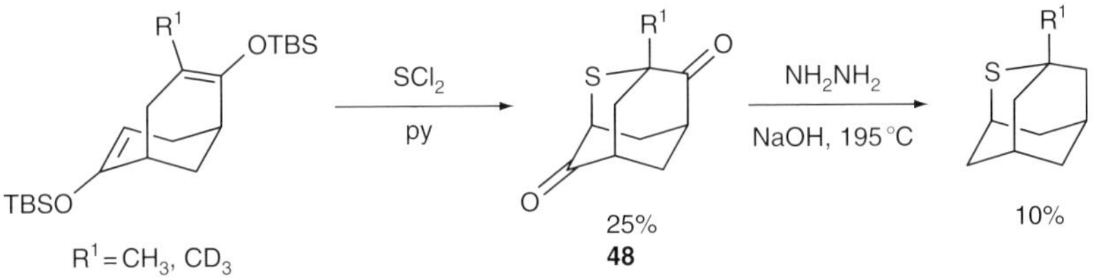

A one-pot three-component Pd/Cu catalysed reaction between 2-halogenoaroyl chlorides, alkynes and Na_2S nonahydrate proceeds through a coupling – Michael addition – cyclisation sequence which yields thiochromones (Scheme 76) <09SL1255>.

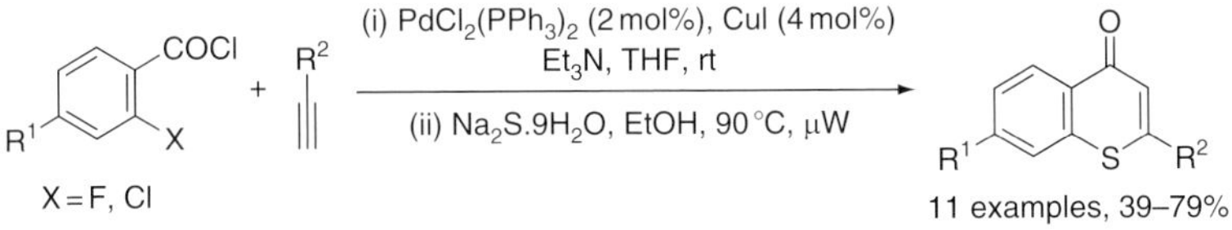

Scheme 76

The sequential treatment of 2-bromostyrenes with n-BuLi, CS_2 and I_2 leads to a separable mixture of the 5-*exo* and 6-*endo* cyclisation products. However, the benzothiophene can often be readily converted into the isothiochromene-1-thione (Scheme 77) <09H(78)169> while addition of NH_4Cl to the Li 2-(vinyl)dithiobenzoate intermediates affords isothiochroman-1-thiones <09H(78)2077>.

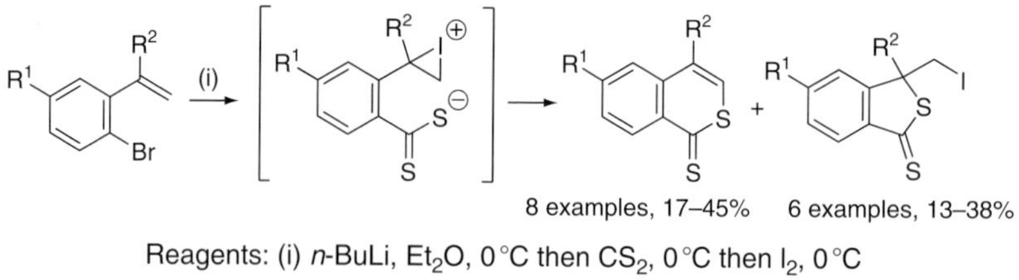

Scheme 77

Reagents: (i) *n*-BuLi, Et$_2$O, 0 °C then CS$_2$, 0 °C then I$_2$, 0 °C

8 examples, 17–45%   6 examples, 13–38%

Chalcones derived from 4-acetyl-1,3-benzoxathiol-2-ones can be cyclised to thioflavanones under basic conditions (Scheme 78) <09S1811>.

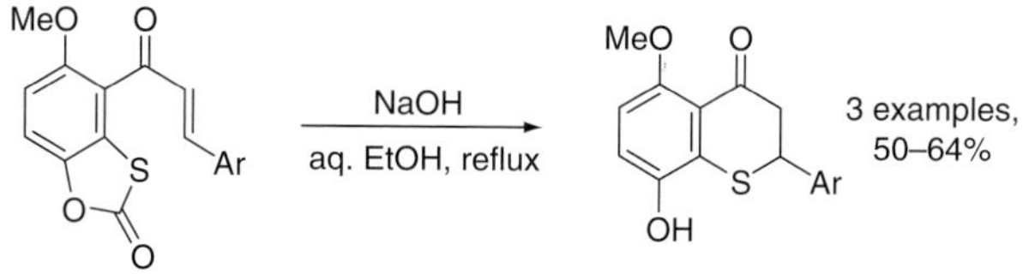

Scheme 78

3 examples, 50–64%

4-Dicyanovinylthiochromans undergo an enantio- and diastereoselective vinylogous Mannich reaction with *N*-sulfonyl alkylimines using a combined BINOL–cinchona chiral catalyst (Scheme 79) <09CC6994>.

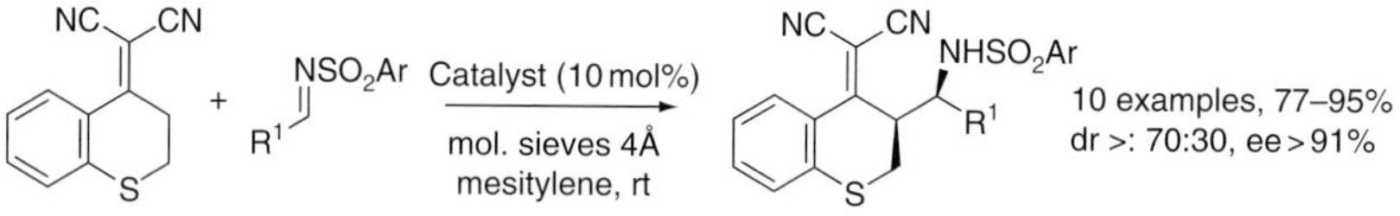

Scheme 79

10 examples, 77–95%
dr >: 70:30, ee > 91%

6.4.4. HETEROCYCLES CONTAINING TWO OR MORE OXYGEN ATOMS

6.4.4.1 Dioxins and Dioxanes

The peroxycyclisation involving the intramolecular addition of hydroperoxides to a pendant alkene is best achieved using Pd(OAc)$_2$ as catalyst with benzoquinone acting as a sacrificial oxidant. Although diastereoselectivity is only moderate, functional groups which allow further manipulation are tolerant of the conditions (Scheme 80) <09OL3290>.

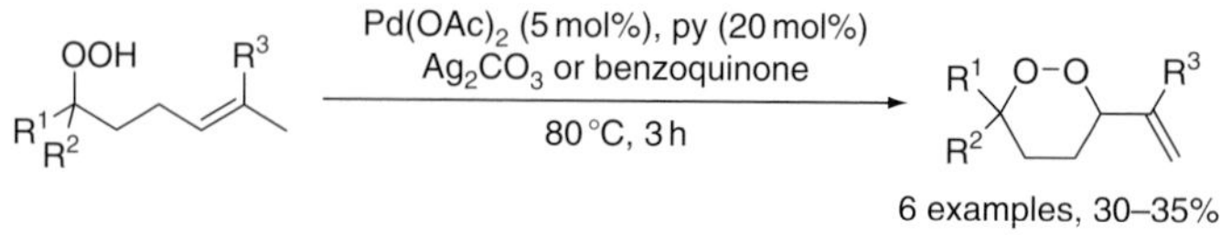

Pd(OAc)$_2$ (5 mol%), py (20 mol%)
Ag$_2$CO$_3$ or benzoquinone
80 °C, 3 h

6 examples, 30–35%

Scheme 80

An asymmetric synthesis of the 1,2-dioxane propionate core of the marine natural products the peroxyplakoric acids features a stereospecific intramolecular alkylation of a hydroperoxyacetal. The latter are available from an asymmetric aldol reaction between ester **49** and aldehyde **50** (Scheme 81) <09T9680>.

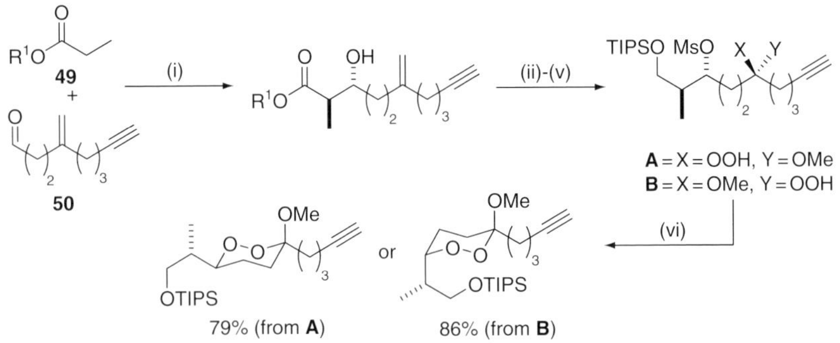

Reagents: (i) Cy$_2$BOTf, Et$_3$N (86%); (ii) LiAlH$_4$ (73%), (iii) TIPSCl (85%); (iv) MsCl, Et$_3$N (81%); (v) O$_3$, MeOH, CH$_2$Cl$_2$ (88%); (vi) KO*t*-Bu, 18-c-6, PhMe.

Scheme 81

The urea–H$_2$O$_2$ complex is the source of the peroxy unit in a synthesis of some 1,2,7-trioxaspiro[5.5]undecanes, analogues of the peroxyplakoric acids which exhibit antimalarial activity, from ethyl 10-hydroxy-6-oxodec-2-enoates <09T6972>.

Initial 6-*endo-dig* cyclisation of the enynols **51**, derived from a one-pot, three component reaction between allylic epoxides, benzynes and terminal alkynes, is followed by dimerisation to the 1,4-dioxane **52** <09AGE391>.

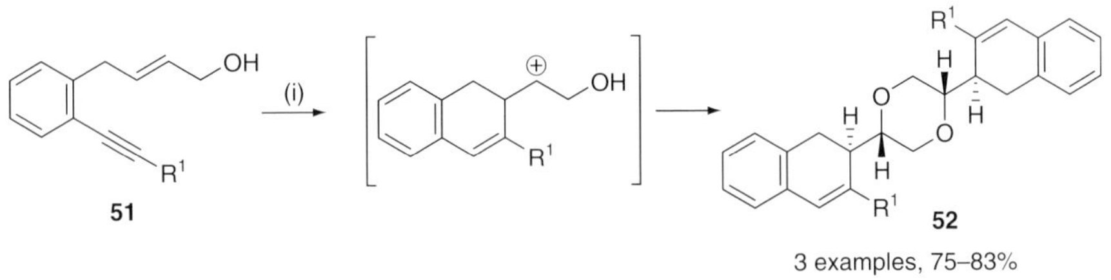

Reagents: (i) [AuCl(PPh$_3$)], AgSbF$_6$, CH$_2$Cl$_2$, rt, 6 h

1,2-Bis(allyloxy)benzenes and their mono- and dithio analogues afford 1,4-benzo- dioxins, dithiins and oxathiins by way of a combined isomerisation and RCM route <09T10650>.

Alcohols can be protected as their 1,4-dioxanyl ethers by reaction with O,Se-acetals which undergo a selective Cu(II)-catalysed cleavage of the C–Se bond; full recovery of the Se is normal <09SL2429>.

The diastereoselectivity of the DA reaction of the chiral dienes **53** is influenced by the size of the substituents on the dienophile which can hinder an *exo* approach <09TL7144>.

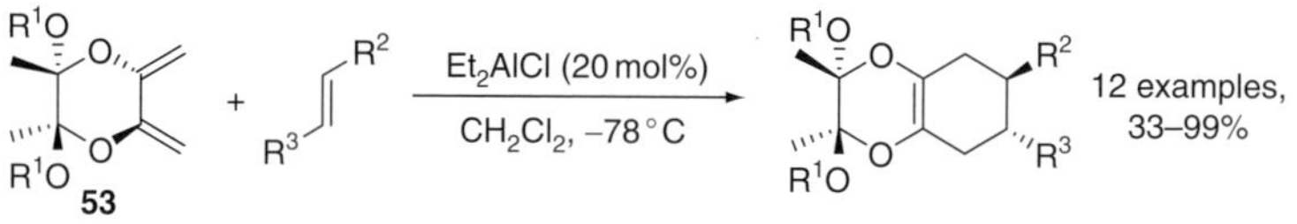

[Scheme: compound 53 + alkene → bicyclic product, Et$_2$AlCl (20 mol%), CH$_2$Cl$_2$, −78 °C, 12 examples, 33–99%]

6.4.4.2 Trioxanes and Tetraoxanes

Bicyclic 1,2,4-trioxanes are formed from γ,δ-unsaturated ketones through a sequence of addition of H$_2$O$_2$, epoxidation and rearrangement (Scheme 82) <09OL507>. Chiral β-hydroperoxy alcohols react with cyclic ketones to give enantio enriched 1,2,4-trioxanes <09T8531>.

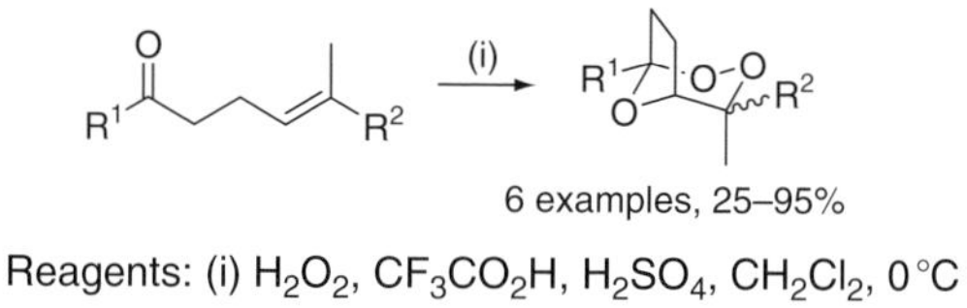

6 examples, 25–95%

Reagents: (i) H$_2$O$_2$, CF$_3$CO$_2$H, H$_2$SO$_4$, CH$_2$Cl$_2$, 0 °C

Scheme 82

Ketones and 1,1-dihydroperoxides undergo a Re(VII)-catalysed condensation which yields 1,2,4,5-tetraoxanes. In a one-pot variant, a carbonyl compound is converted into the hydroperoxide whereupon a second ketone is added (Scheme 83) <09OL213>.

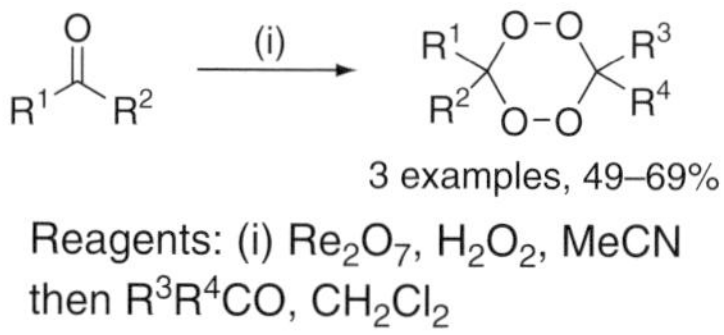

3 examples, 49–69%

Reagents: (i) Re$_2$O$_7$, H$_2$O$_2$, MeCN then R^{3}R^{4}CO, CH$_2$Cl$_2$

Scheme 83

6.4.5. HETEROCYCLES CONTAINING TWO OR MORE SULFUR ATOMS

6.4.5.1 Dithianes and Trithianes

The bis-thioacetate derived from the corresponding 1,2-bis(bromomethyl)benzene affords 1,4-dihydrobenzo[*d*][1,2]dithiines on mild hydrolysis with concomitant air oxidation <09TL3023> and cyclisation of the optically pure 1,4-diphenylbutan-1,4-dimesylate with Na$_2$S nonahydrate yields (−)-(3*S*,6*S*)-3,6-diphenyl-1,2-dithiane <09S1739>.

Vicinal tri- and tetracarbonyl compounds form stable 1,3-dithianes on reaction with propan-1,3-dithiol <09EJO1417>. 2-Methylene-1,3-dithiane 1-oxide functions as a ketene equivalent and undergoes a radical addition of alkyl halides <09CL248> and it yields the 2-arylmethylene derivatives through a Mizoroki–Heck arylation <09CL624>. Replacement of a carbonyl oxygen atom by the difluoromethyl group can be achieved through the intermediacy of the ketene dithioacetal (Scheme 84) <09T1361>. Stable eight-membered cyclic allenes result from the Au-catalysed rearrangement of propargylic 1,3-dithianes (Scheme 85) <09CC2535>.

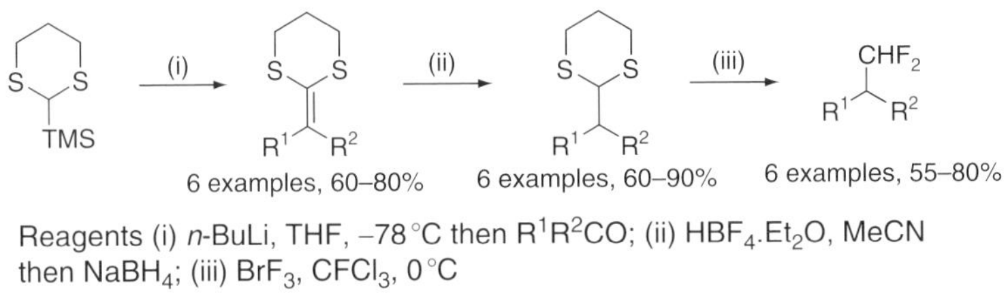

Reagents (i) *n*-BuLi, THF, −78 °C then R^{1}R^{2}CO; (ii) HBF$_4$.Et$_2$O, MeCN then NaBH$_4$; (iii) BrF$_3$, CFCl$_3$, 0 °C

Scheme 84

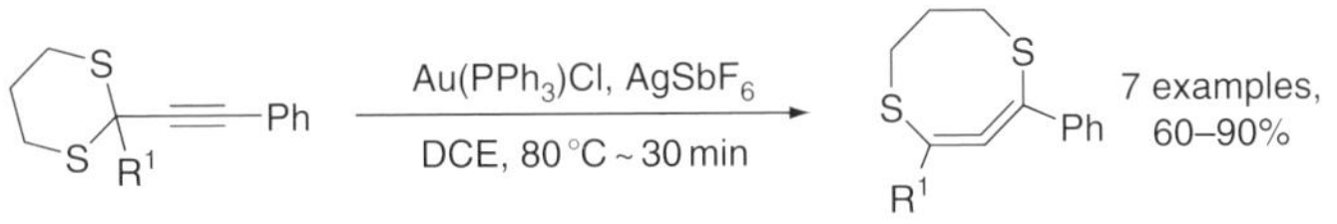

Scheme 85

1,2-Diarylsulfides yield thianthrenes through an oxidative coupling induced by MoCl$_5$ either alone or with TiCl$_4$ <09CEJ13313> and a Mo complex catalyses the [4+2] cycloadditions of bis-*o*-phenylene tetrasulfide with alkenes which provides a new route to dihydrobenzodithiins (Scheme 86) <09CC7572>.

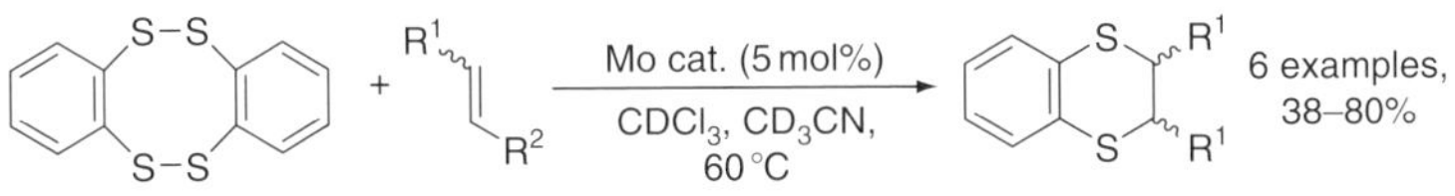

Scheme 86

Benzaldehydes are converted into trimers, 1,3,5-trithianes, using SiCl$_4$/Na$_2$S in MeCN as the thionating agent <09TL5933>.

6.4.6. HETEROCYCLES CONTAINING BOTH OXYGEN AND SULFUR IN THE SAME RING

6.4.6.1 Oxathianes

Polycyclic sultones e.g. **54** are formed through the Pd-catalysed intramolecular coupling of aromatic sulfonates <09TL4781>.

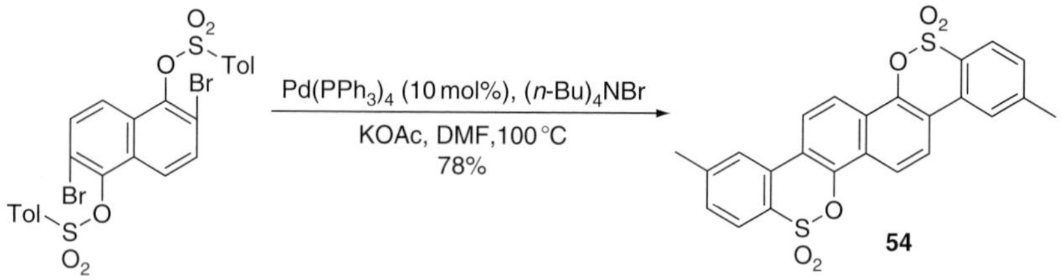

1,3-Benzoxathianes are accessible from 2-thio-substituted phenols through sequential cyclisation and β-elimination initiated by treatment with diiodomethane in basic conditions (Scheme 87) <09S3848>.

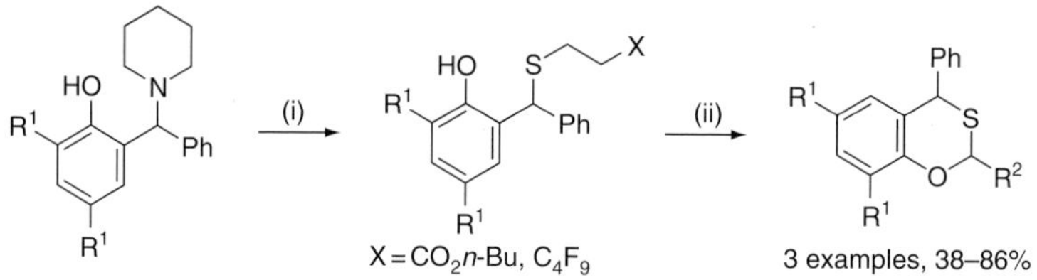

Reagents: (i) HS(CH$_2$)$_2$X, PhMe, 140 °C; (ii) CH$_2$I$_2$ or R^{2}CHCl$_2$, NaOH, sulfolane, rt

Scheme 87

Phenoxathiine is selectively aroylated adjacent to the oxygen heteroatom by directed alumination followed by transmetalation and acylation <09AGE1501>.

REFERENCES

08B1	J.W. Blunt, and M.H.G. Munro (eds.). Dictionary of Marine Natural Products Chapman Hall / CRC (Boca Raton) (2008).
09AGE391	M. Jeganmohan, S. Bhuvaneswari and C.-H. Cheng, *Angew. Chem. Int. Ed.*, **48**, 391 (2009).
09AGE1244	H. Yamamoto, I. Sasaki, Y. Hirai, K. Namba, H. Imagawa and M. Nishizawa, *Angew. Chem. Int. Ed.*, **48**, 1244 (2009).
09AGE1283	X. Wu, J. Zhou and B.B. Snider, *Angew. Chem. Int. Ed.*, **48**, 1283 (2009).
09AGE1501	S.H. Wunderlich and P. Knochel, *Angew. Chem. Int. Ed.*, **48**, 1501 (2009).
09AGE1644	J. Barluenga, A. Mendoza, F. Rodríguez and F.J. Fañanás, *Angew. Chem. Int. Ed.*, **48**, 1644 (2009).
09AGE2346	K.L. Jackson, J.A. Henderson, H. Motoyoshi and A.J. Phillips, *Angew. Chem. Int. Ed.*, **48**, 2346 (2009).
09AGE2534	K. Fukamizu, Y. Miyake and Y. Nishibayashi, *Angew. Chem. Int. Ed.*, **48**, 2534 (2009).
09AGE2941	A. Hamajima and M. Isobe, *Angew. Chem. Int. Ed.*, **48**, 2941 (2009).
09AGE5355	H. Zhou, Y.-H. Xu, W.-J. Chung and T.-P. Loh, *Angew. Chem. Int. Ed.*, **48**, 5355 (2009).
09AGE5848	A.S.K. Hshmi, M. Rudolph, J. Huck, W. Frey, J.W. Bats and M. Hamzić, *Angew. Chem. Int. Ed.*, **48**, 5848 (2009).
09AGE6097	Y.-H. Zhang, B.-F. Shi and J.-Q. Yu, *Angew. Chem. Int. Ed.*, **48**, 6097 (2009).
09AGE6520	L. Zhao, F. Xie, G. Cheng and Y. Hu, *Angew. Chem. Int. Ed.*, **48**, 6520 (2009).
09AGE9940	G.W. O'Neil, J. Ceccon, S. Benson, M.-P. Collin, B. Fasching and A. Fürstner, *Angew. Chem. Int. Ed.*, **48**, 9940 (2009).

09AGE9946	S. Benson, M.-P. Collin, G.W. O'Neil, J. Ceccon, B. Fasching, M.D.B. Fenster, C. Godbout, K. Radkowski, R. Goddard and A. Fürstner, *Angew. Chem. Int. Ed.*, **48**, 9946 (2009).
09BCJ843	V. Gembus, N. Sala-Jung and D. Uguen, *Bull. Chem. Soc. Jpn.*, **82**, 843 (2009).
09CC595	V. Russo, J. Allen and Z.T. Ball, *Chem. Commun.*, 595 (2009).
09CC1769	G.-W. Wang and B. Zhu, *Chem. Commun.*, 1769 (2009).
09CC2379	M. Sakamoto, M. Kanehiro, T. Mino and T. Fujita, *Chem. Commun.*, 2379 (2009).
09CC2535	X. Zhao, Z. Zhong, L. Peng, W. Zhang and J. Wang, *Chem. Commun.*, 2535 (2009).
09CC2878	H. Zhi, C. Lü, Q. Zhang and J. Luo, *Chem. Commun.*, 2878 (2009).
09CC2932	S. Inagi, Y. Doi, Y. Kishi and T. Fuchigami, *Chem. Commun.*, 2932 (2009).
09CC3594	Y. Xiao and J. Zhang, *Chem. Commun.*, 3594 (2009).
09CC4753	M. Durandetti, L. Hardou, M. Clément and J. Maddaluno, *Chem. Commun.*, 4753 (2009).
09CC5615	M. Mosrin, G. Monzon, T. Bresser and P. Knochel, *Chem. Commun.*, 5615 (2009).
09CC5618	Y.-S. Hon, T.-W. Tseng and C.-Y. Cheng, *Chem. Commun.*, 5618 (2009).
09CC6089	L.-G. Meng, B. Hu, Q.-P. Wu, M. Liang and S. Xue, *Chem. Commun.*, 6089 (2009).
09CC6397	P. Kumari, S.M.S. Poonam and Chauhan, *Chem. Commun.*, 6397 (2009).
09CC6457	G. Liang, L.J. Bateman and N.I. Totah, *Chem. Commun.*, 6457 (2009).
09CC6469	S. Wang, K. Xie, Z. Tan, X. An, X. Zhou, C.-C. Guo and Z. Peng, *Chem. Commun.*, 6469 (2009).
09CC6744	A.C. Tadd, M.R. Fielding and M.C. Willis, *Chem. Commun.*, 6744 (2009).
09CC6994	X.-F. Xiong, Z.-J. Jia, W. Du, K. Jiang, T.-Y. Liu and Y.-C. Cheng, *Chem. Commun.*, 6994 (2009).
09CC7125	J.L. Arbour, H.S. Rzepa, A.J.P. White and K.K. Hii, *Chem. Commun.*, 7125 (2009).
09CC7294	Y. Watanabe, T. Washio, N. Shimada, M. Anada and S. Hashimoto, *Chem. Commun.*, 7125 (2009).
09CC7572	D.J. Harrison and U. Fekl, *Chem. Commun.*, 7572 (2009).
09CCR2588	M. Grazul and E. Budzisz, *Coord. Chem. Rev.*, **253**, 2588 (2009).
09CEJ5979	A. Gollner, K.-H. Altmann, J. Gertsch and J. Mulzer, *Chem. Eur. J.*, **15**, 5979 (2009).
09CEJ9223	M.T. Crimmins, J.M. Ellis, K.A. Emmitte, P.A. Haile, P.J. McDougall, J.D. Parrish and J.L. Zuccarello, *Chem. Eur. J.*, **15**, 9223 (2009).
09CEJ9235	M.T. Crimmins, J.L. Zuccarello, P.J. McDougall and J.M. Ellis, *Chem. Eur. J.*, **15**, 9235 (2009).
09CEJ11660	J. Barluenga, A. Fernández, A. Diéguez, F. Rodríguez and F.J. Fañanás, *Chem. Eur. J.*, **15**, 11660 (2009).
09CEJ12807	H. Fuwa, A. Saito, S. Naito, K. Konoki, M. Yotsu-Yamashita and M. Sasaki, *Chem. Eur. J.*, **15**, 12807 (2009).
09CEJ13299	H.-F. Wang, H.-F. Cui, Z. Chai, P. Li, C.-W. Zheng, Y.-Q. Yang and G. Zhao, *Chem. Eur. J.*, **15**, 13299 (2009).
09CEJ13313	A. Spurg, G. Schnakenburg and S.R. Waldvogel, *Chem. Eur. J.*, **15**, 13313 (2009).
09CEJ11948	B. Schmidt and F. Hölter, *Chem. Eur. J.*, **15**, 11948 (2009).
09CL248	S. Yoshida, H. Yorimitsu and K. Oshima, *Chem. Lett.*, **38**, 248 (2009).
09CL624	E. Morita, M. Iwasaki, S. Yoshida, H. Yorimitsu and K. Oshima, *Chem. Lett.*, **38**, 624 (2009).
09CL866	H. Fuwa, N. Noji and M. Sasaki, *Chem. Lett.*, **38**, 866 (2009).
09CR190	M.S.T. Gonçalves, *Chem. Rev.*, **109**, 190 (2009).
09CR3044	K.L. Jackson, J.A. Henderson and A.J. Phillips, *Chem. Rev.*, **109**, 3044 (2009).
09CR3903	S. Bräse, A. Encinas, J. Keck and C.F. Nising, *Chem. Rev.*, **109**, 3903 (2009).
09EJO103	Y. Kishi, S. Inagi and T. Fuchigami, *Eur. J. Org. Chem.*, 103 (2009).
09EJO833	M. Fuchs, Y. Simeo, B.T. Ueberbacher, B. Mautner, T. Netscher and K. Faber, *Eur. J. Org. Chem.*, 833 (2009).
09EJO1417	S. Goswami, A.C. Maity, H.-K. Fun and S. Chantrapromma, *Eur. J. Org. Chem.*, 1417 (2009).
09EJO1625	U.C. Reddy, S. Bondalapati and A.K. Saikia, *Eur. J. Org. Chem.*, 1625 (2009).
09EJO1859	N.R. Candeias, L.F. Veiros, C.A.M. Afonso and P.M.P. Gois, *Eur. J. Org. Chem.*, 1859 (2009).
09EJO2903	Y. Ooyama and Y. Harima, *Eur. J. Org. Chem.*, 2903 (2009).
09EJO3129	L.-Z. Dai and M. Shi, *Eur. J. Org. Chem.*, 3129 (2009).

09EJO4503	P.M. Habib, V. Kavala, B.R. Raju, C.-W. Kuo, W.-C. Huang and C.-F. Yao, *Eur. J. Org. Chem.*, 4503 (2009).
09EJO4757	S.K. Das, R. Singh and G. Panda, *Eur. J. Org. Chem.*, 4757 (2009).
09EJO5192	Z. Dhong, L. Wang, X. Chen, X. Liu, L. Lin and X. Feng, *Eur. J. Org. Chem.*, 5192 (2009).
09H(77)657	F. Požgan and M. Kočevar, *Heterocycles*, **77**, 657 (2009).
09H(78)169	S. Fukamachi, H. Konishi and K. Kobayashi, *Heterocycles*, **78**, 169 (2009).
09H(78)555	M. Hagimori, N. Mizuyama, Y. Shigemitsu, B.-C. Wang and Y. Tominaga, *Heterocycles*, **78**, 555 (2009).
09H(78)1109	K.C. Majumdar, S. Muhuri, R.U. Islam and B. Chattopadhyay, *Heterocycles*, **78**, 1109 (2009).
09H(78)2077	S. Fukamachi, M. Tanmatsu, H. Konishi and K. Kobayashi, *Heterocycles*, **78**, 2077 (2009).
09H(78)2661	K.C. Majumdar, P. Debnath and B. Roy, *Heterocycles*, **78**, 2661 (2009).
09HCA1729	W.M. Tokan, S. Schweizer, C. Thies, F.E. Meyer, P.J. Parsons and A. de Meijere, *Helv. Chim. Acta*, **92**, 1729 (2009).
09JA402	K.M. McQuaid and D. Sames, *J. Am. Chem. Soc.*, **131**, 402 (2009).
09JA1350	I. Koyama, T. Kurahashi and S. Matsubara, *J. Am. Chem. Soc.*, **131**, 1350 (2009).
09JA3166	D. Shikanai, H. Murase, T. Hata and H. Urabe, *J. Am. Chem. Soc.*, **131**, 3166 (2009).
09JA3464	M. Uemura, I.D.G. Watson, M. Katsukawa and F.D. Toste, *J. Am. Chem. Soc.*, **131**, 3464 (2009).
09JA3832	W. Zhang, H. Xu, H. Xu and W. Tang, *J. Am. Chem. Soc.*, **131**, 3832 (2009).
09JA3844	I. Ibrahem, M. Yu, R.R. Schrock and A.H. Hoveyda, *J. Am. Chem. Soc.*, **131**, 3844 (2009).
09JA9612	S. Lessard, F. Peng and D.G. Hall, *J. Am. Chem. Soc.*, **131**, 9612 (2009).
09JA11350	C.F. Bender, F.K. Yoshimoto, C.L. Paradise and J.K. De Brabander, *J. Am. Chem. Soc.*, **131**, 11350 (2009).
09JA12882	N. Momiyama, H. Tabuse and M. Terada, *J. Am. Chem. Soc.*, **131**, 12882 (2009).
09JA13194	A. Ooguri, K. Nakai, T. Kurahashi and S. Matsubara, *J. Am. Chem. Soc.*, **131**, 13194 (2009).
09JA14176	S.J. Ryan, L. Candish and D.W. Lupton, *J. Am. Chem. Soc.*, **131**, 14176 (2009).
09JA14190	K. Hirano, A.T. Biju, I. Piel and F. Glorius, *J. Am. Chem. Soc.*, **131**, 14190 (2009).
09JA14202	A.T. Parsons and J.S. Johnson, *J. Am. Chem. Soc.*, **131**, 14202 (2009).
09JA15636	D.-S. Kim, C.-G. Dong, J.T. Kim, H. Guo, J. Huang, P.S. Tiseni and Y. Kishi, *J. Am. Chem. Soc.*, **131**, 15636 (2009).
09JA15642	C.-G. Dong, J.A. Henderson, Y. Kaburagi, T. Sasaki, D.-S. Kim, J.T. Kim, D. Urabe, H. Guo and Y. Kishi, *J. Am. Chem. Soc.*, **131**, 15642 (2009).
09JHC1098	A. Jha and P.-J.J. Huang, *J. Heterocycl. Chem.*, **46**, 1098 (2009).
09JOC1400	H. Cui, P. Li, Z. Chai, C. Zheng, G. Zhao and S. Zhu, *J. Org. Chem.*, **74**, 1400 (2009).
09JOC1759	K. Zeitler and C.A. Rose, *J. Org. Chem.*, **74**, 1759 (2009).
09JOC2234	M. Arthuis, R. Pontikis and J.-C. Florent, *J. Org. Chem.*, **74**, 2234 (2009).
09JOC2605	U.C. Reddy, S. Bondalapati and A.K. Saikia, *J. Org. Chem.*, **74**, 2605 (2009).
09JOC2755	M. Fridén-Saxin, N. Pemberton, K. da Silva Andersson, C. Dyrager, A. Friberg, M. Grøtli and K. Luthman, *J. Org. Chem.*, **74**, 2755 (2009).
09JOC3141	O.M. Singh and N.S. Devi, *J. Org. Chem.*, **74**, 3141 (2009).
09JOC3394	X. Han, L.-W. Ye, X.-L. Sun and Y. Tang, *J. Org. Chem.*, **74**, 3394 (2009).
09JOC3469	B. Godoi, A. Sperança, D.F. Back, R. Brandão, C.W. Nogueira and G. Zeni, *J. Org. Chem.*, **74**, 3469 (2009).
09JOC3478	M. Shimizu, K. Hirano, T. Satoh and M. Miura, *J. Org. Chem.*, **74**, 3478 (2009).
09JOC3978	Z. Cong, T. Miki, O. Urakawa and H. Nishino, *J. Org. Chem.*, **74**, 3978 (2009).
09JOC4009	P. Batsomboon, W. Phakhodee, S. Ruchirawat and P. Ploypradith, *J. Org. Chem.*, **74**, 4009 (2009).
09JOC4024	H. Fuwa, K. Ishigai, T. Goto, A. Suzuki and M. Sasaki, *J. Org. Chem.*, **74**, 4024 (2009).
09JOC4080	C.A. James and V. Snieckus, *J. Org. Chem.*, **74**, 4080 (2009).
09JOC4094	C.A. James, A.L. Coelho, M. Gevaert, P. Forgione and V. Snieckus, *J. Org. Chem.*, **74**, 4094 (2009).

09JOC4158	M. Peuchmaur, V. Lisowski, C. Gandreuil, L.T. Maillard, J. Martinez and J.-F. Hernandez, *J. Org. Chem.*, **74**, 4158 (2009).
09JOC4311	E. Ballerini, L. Minuti, O. Piermatti and F. Pizzo, *J. Org. Chem.*, **74**, 4311 (2009).
09JOC4447	D.E. Ward, F. Becerril-Jimenez and M.M. Zahedi, *J. Org. Chem.*, **74**, 4447 (2009).
09JOC4612	S. Duan, R. Jana and J.A. Tunge, *J. Org. Chem.*, **74**, 4612 (2009).
09JOC5395	N.N.B. Kumar, M.N. Reddy and K.C.K. Swamy, *J. Org. Chem.*, **74**, 5395 (2009).
09JOC6245	X. Wu, J. Zhou and B.B. Snider, *J. Org. Chem.*, **74**, 6245 (2009).
09JOC6295	S. Mochida, K. Hirano, T. Satoh and M. Miura, *J. Org. Chem.*, **74**, 6295 (2009).
09JOC6797	H. Li, J. Yang, Y. Liu and Y. Li, *J. Org. Chem.*, **74**, 6797 (2009).
09JOC6881	L. Hong, L. Wang, W. Sun, K. Wong and R. Wang, *J. Org. Chem.*, **74**, 6881 (2009).
09JOC6973	F.K. MacDonald and D.J. Burnell, *J. Org. Chem.*, **74**, 6973 (2009).
09JOC7183	D. Samanta, R.B. Kargbo and G.R. Cook, *J. Org. Chem.*, **74**, 7183 (2009).
09JOC8414	A.B. Leduc, T.P. Lebold and M.A. Kerr, *J. Org. Chem.*, **74**, 8414 (2009).
09JOC8787	Z.-L. Hu, W.-J. Qian, S. Wang, S. Wang and Z.-J. Yao, *J. Org. Chem.*, **74**, 8787 (2009).
09JOC8798	J.-C. Tsai, S.-R. Li, M.Y. Chiang, L.-Y. Chen, P.-Y. Chen, Y.-F. Lo, C.-H. Wang, C.-N. Lin and E.-C. Wang, *J. Org. Chem.*, **74**, 8798 (2009).
09JOC8901	R.S. Menon, A.D. Findlay, A.C. Bissember and M.G. Banwell, *J. Org. Chem.*, **74**, 8901 (2009).
09JOC8963	J. Moreau, C. Hubert, J. Batany, L. Toupet, T. Roisnel, J.-P. Hurvois and J.-L. Renaud, *J. Org. Chem.*, **74**, 8963 (2009).
09NPR170	J.W. Blunt, B.R. Copp, W-P. Hu, M.H.G. Munro, P.T. Northcote and M.R. Prinsep, *Nat. Prod. Rep.*, **26**, 170 (2009).
09OBC1053	M.A. Rizzacasa and A. Pollex, *Org. Biomol. Chem.*, **7**, 1053 (2009).
09OBC1221	V. Belting and N. Krause, *Org. Biomol. Chem.*, **7**, 1221 (2009).
09OBC1858	R. Singh, M.K. Parai and G. Panda, *Org. Biomol. Chem.*, **7**, 1858 (2009).
09OBC2848	F. Wang, Z. Miao and R. Chen, *Org. Biomol. Chem.*, **7**, 2848 (2009).
09OBC3657	H. Yanai, H. Ogura and T. Taguchi, *Org. Biomol. Chem.*, **7**, 3657 (2009).
09OBC4539	S.-P. Luo, Z.-B. Li, L.-P. Wang, Y. Guo, A.-B. Xia and D.-Q. Xu, *Org. Biomol. Chem.*, **7**, 4539 (2009).
09OL113	T. Saito and T. Nakata, *Org. Lett.*, **11**, 113 (2009).
09OL121	A. Aponick, C.-Y. Li and J.A. Palmes, *Org. Lett.*, **11**, 121 (2009).
09OL169	K. Okuma, A. Nojima, N. Matsunaga and K. Shioji, *Org. Lett.*, **11**, 169 (2009).
09OL213	P. Ghorai and P.H. Dussault, *Org. Lett.*, **11**, 213 (2009).
09OL357	P.O. Miranda, R.M. Carballo, V.S. Martín and J.I. Padrón, *Org. Lett.*, **11**, 357 (2009).
09OL489	M.T. Crimmins, J.L. Zuccarello, J.M. Ellis, P.J. McDougall, P.A. Haile, J.D. Parrish and K.A. Emmitte, *Org. Lett.*, **11**, 489 (2009).
09OL507	A.P. Ramirez, A.M. Thomas and K.A. Woerpel, *Org. Lett.*, **11**, 507 (2009).
09OL717	Y. Yamamoto and K. Itonaga, *Org. Lett.*, **11**, 717 (2009).
09OL757	M.E. Jung and D.A. Allen, *Org. Lett.*, **11**, 757 (2009).
09OL891	L. Wang, K. Thai and M. Gravel, *Org. Lett.*, **11**, 891 (2009).
09OL991	X. Meng, Y. Huang, H. Zhao, P. Xie, J. Ma and R. Chen, *Org. Lett.*, **11**, 991 (2009).
09OL1289	A. Goel, S. Chaurasia, M. Dixit, V. Kumar, S. Prakash, B. Jena, J.K. Verma, M. Jain, R.S. Anand and S.S. Manoharan, *Org. Lett.*, **11**, 1289 (2009).
09OL1523	A. Dzudza and T.J. Marks, *Org. Lett.*, **11**, 1523 (2009).
09OL1623	G.-Q. Li, L.-X. Din and S.-L. You, *Org. Lett.*, **11**, 1623 (2009).
09OL1627	L. Zu, S. Zhang, H. Xie and W. Wang, *Org. Lett.*, **11**, 1627 (2009).
09OL1741	X.-H. Hu, F. Liu and T.-P. Loh, *Org. Lett.*, **11**, 1741 (2009).
09OL1769	J.-R. Chen and D.-Y. Yang, *Org. Lett.*, **11**, 1769 (2009).
09OL2888	P.V. Toullec, T. Blarre and V. Michelet, *Org. Lett.*, **11**, 2888 (2009).
09OL2924	H.J. Lim and T.V. RajanBabu, *Org. Lett.*, **11**, 2924 (2009).
09OL2972	K.M. McQuaid, J.Z. Long and D. Sames, *Org. Lett.*, **11**, 2972 (2009).
09OL3068	A.A. Fokin, T.S. Zhuk, A.E. Pashenko, P.O. Dral, P.A. Gunchenko, J.E.P. Dahl, R.M.K. Carlson, T.V. Koso, M.A. Serafin and P.R. Schreiner, *Org. Lett.*, **11**, 3068 (2009).
09OL3210	E. Awuah and A. Capretta, *Org. Lett.*, **11**, 3210 (2009).
09OL3290	J.R. Harris, S.R. Waetzig and K.A. Woerpel, *Org. Lett.*, **11**, 3290 (2009).

09OL3434	R. Jana, R. Trivedi and J.A. Tunge, *Org. Lett.*, **11**, 3434 (2009).
09OL3442	B. Yu, T. Jiang, J. Li, Y. Su, X. Pan and X. She, *Org. Lett.*, **11**, 3442 (2009).
09OL3670	G. Liu, J.M. Wurst and D.S. Tan, *Org. Lett.*, **11**, 3670 (2009).
09OL3978	S. Zheng and X. Lu, *Org. Lett.*, **11**, 3978 (2009).
09OL4382	H. Furuta, Y. Hasegawa and Y. Mori, *Org. Lett.*, **11**, 4382 (2009).
09OL4576	T. Liu, X. Zhao, L. Lu and T. Cohen, *Org. Lett.*, **11**, 4576 (2009).
09OL4752	M. Yoshida, M. Higuchi and K. Shishido, *Org. Lett.*, **11**, 4752 (2009).
09OL5350	A. Varela-Fernández, C. González-Rodríguez, J.A. Varela, L. Castedo and C. Sáa, *Org. Lett.*, **11**, 5350 (2009).
09OL5374	B.D. Gallagher, B.R. Taft and B.H. Lipshutz, *Org. Lett.*, **11**, 5374 (2009).
09OL5686	D. Wu, W. Pisula, M.C. Haberecht, X. Feng and K. Müllen, *Org. Lett.*, **11**, 5686 (2009).
09S399	C.R. Reddy, N. Kiranmai, K. Johny, M. Pendke and P. Naresh, *Synthesis*, 399 (2009).
09S691	P.A. Clarke, A.T. Reeder and J. Winn, *Synthesis*, 691 (2009).
09S779	K. Krohn, I. Ahmed and M. John, *Synthesis*, 779 (2009).
09S1739	M. Periasamy, G. Ramani and G.P. Muthukumaragopal, *Synthesis*, 1739 (2009).
09S1811	W. Konieczny and M. Konieczny, *Synthesis*, 1811 (2009).
09S1886	S.K. Dinda, S.K. Das and G. Panda, *Synthesis*, 1886 (2009).
09S2029	T. Zdrojewski, J. Musielak and A. Jończyk, *Synthesis*, 2029 (2009).
09S2146	B.H. Park, Y.R. Lee and W.S. Lyoo, *Synthesis*, 2146 (2009).
09S3848	M. Gerster and M. Mihalic, *Synthesis*, 3848 (2009).
09S3533	V. Semeniuchenko, U. Groth and V. Khilya, *Synthesis*, 3533 (2009).
09S3689	O.G. Ganina, A.Y. Fedorov and I.P. Beletskaya, *Synthesis*, 3689 (2009).
09SL55	M.J. Khoshkholgh, S. Balalaie, H.R. Bijanzadeh and J.H. Gross, *Synlett*, 55 (2009).
09SL138	W.-J. Yoo, C.A. Correia, Y. Zhang and C.-J. Li, *Synlett*, 138 (2009).
09SL233	C.S.P. McErlean, A.C. Willis and C.-J. Li, *Synlett*, 233 (2009).
09SL306	J. Faragalla, A. Heaton, R. Griffith and J.B. Bremner, *Synlett*, 306 (2009).
09SL565	C. Bressy, F. Bargiggia, M. Guyonnet, S. Arseniyadis and J. Cossy, *Synlett*, 565 (2009).
09SL793	Y.-C. (W.) Liu, J. Sperry, D.C.K. Rathwell and M.A. Brimble, *Synlett*, 793 (2009).
09SL917	P. Jayashree, G. Shanthi and P.T. Perumal, *Synlett*, 917 (2009).
09SL1189	J.R. de Alaniz and T. Rovis, *Synlett*, 1189 (2009).
09SL1255	B. Willy and T.J.J. Müller, *Synlett*, 1255 (2009).
09SL1383	M.C. Bröhmer, N. Volz and S. Bräse, *Synlett*, 1383 (2009).
09SL1501	X. Wu, S. Lin, M. Li and T. You, *Synlett*, 1501 (2009).
09SL1915	A. Aupoix and G. Vo-Thanh, *Synlett*, 1915 (2009).
09SL2079	G. Savitha, K. Felix and P.T. Perumal, *Synlett*, 2079 (2009).
09SL2429	A. Temperini, R. Terlizzi, L. Testaferri and M. Tiecco, *Synlett*, 2429 (2009).
09SL2597	M.L.N. Rao, V. Venkatesh and D.N. Jadhav, *Synlett*, 2597 (2009).
09SL2992	V.K. Tandon, H.K. Maurya, B. Kumar, B. Kumar and V.J. Ram, *Synlett*, 2992 (2009).
09T101	Y.R. Lee, Y.M. Kim and S.H. Kim, *Tetrahedron*, **65**, 101 (2009).
09T336	L. Bianchi, M. Maccagno, G. Petrillo, E. Rizzato, F. Sancassan, E. Severi, D. Spinelli, M. Stenta, A. Galatini and C. Tavani, *Tetrahedron*, **65**, 336 (2009).
09T1300	R. Bera, G. Dhananjaya, S.N. Singh, R. Kumar, K. Mukkanti and M. Pal, *Tetrahedron*, **65**, 1300 (2009).
09T1361	O. Cohen, Y. Hagooly and S. Rozen, *Tetrahedron*, **65**, 1361 (2009).
09T2110	W.-B. Liu, H.-F. Jiang and C.-L. Qiao, *Tetrahedron*, **65**, 2110 (2009).
09T2839	H. Pellissier, *Tetrahedron*, **65**, 2839 (2009).
09T3931	H.C. Shen, *Tetrahedron*, **65**, 3931 (2009).
09T3953	G. Fraboulet, V. Fargeas, M. Paris, J.-P. Quintard and F. Zammattio, *Tetrahedron*, **65**, 3953 (2009).
09T5729	N. Barbero, R. SanMartin and E Domínguez, *Tetrahedron*, **65**, 5729 (2009).
09T5799	P.M. Habib, B.R. Raju, V. Kavala, C.-W. Kuo and C.-F. Yao, *Tetrahedron*, **65**, 5799 (2009).
09T6470	A.B. Smith, III, Q. Lin, V.A. Doughty, L. Zhuang, M.D. McBriar, J.K. Kerns, A.M. Boldi, N. Murase, W.H. Moser, C.S. Brook, C.S. Bennett, K. Nakayama, M. Sobukawa and R.E.L. Trout, *Tetrahedron*, **65**, 6470 (2009).

09T6489	A.B. Smith, III, C. Sfouggatakis, C.A. Risatti, J.B. Sperry, W. Zhu, V.A. Doughty, T. Tomioka, D.B. Gotchev, C.S. Bennett, S. Sakamoto, O. Atasoylu, S. Shirakami, D. Bauer, M. Takeuchi, J. Koyanagi and Y. Sakamoto, *Tetrahedron*, **65**, 6489 (2009).
09T6810	Y. Luo and J. Wu, *Tetrahedron*, **65**, 6810 (2009).
09T6834	R.J. Hinkle, Y. Lian, L.C. Speight, H.E. Stevenson, M.M. Sprachman, L.A. Katkish and M.C. Mattern, *Tetrahedron*, **65**, 6834 (2009).
09T6972	Y. Li, Q. Zhang, S. Wittlin, H.-X. Jin and Y. Wu, *Tetrahedron*, **65**, 6972 (2009).
09T7457	K. Funabiki, T. Komeda, Y. Kubota and M. Matsui, *Tetrahedron*, **65**, 7457 (2009).
09T7468	R. Grigg, S. Whitney, V. Sridharan, A. Keep and A. Derrick, *Tetrahedron*, **65**, 7468 (2009).
09T7784	I. Kadota, T. Abe, M. Uni, H. Takamura and Y. Yamamoto, *Tetrahedron*, **65**, 7784 (2009).
09T8478	P. Singh, P. Sharma, K. Bisetty and M.P. Mahajan, *Tetrahedron*, **65**, 8478 (2009).
09T8531	S. Sabbani, L. La Pensée, J. Bacsa, E. Hedenström and P.M. O'Neill, *Tetrahedron*, **65**, 8531 (2009).
09T8702	S.-R. Li, C.-J. Shu, L.-Y. Chen, H.-M. Chen, P.-Y. Chen and E.-C. Wang, *Tetrahedron*, **65**, 8702 (2009).
09T9233	X. Lin, X. Dai, Z. Mao and Y. Wang, *Tetrahedron*, **65**, 9233 (2009).
09T9271	V. Karapetyan, S. Mkrtchyan, G. Ghazaryan, A. Villinger, C. Fischer and P. Langer, *Tetrahedron*, **65**, 9271 (2009).
09T9368	J.R. Bell, A. Franken and C.M. Garner, *Tetrahedron*, **65**, 9368 (2009).
09T9680	C. Xu, C. Schwartz, J. Raible and P.H. Dussault, *Tetrahedron*, **65**, 9680 (2009).
09T10650	G.L. Morgans, E.L. Ngidi, L.G. Madeley, S.D. Khanye, J.P. Michael, C.B. de Koning and W.A.L. van Otterlo, *Tetrahedron*, **65**, 10650 (2009).
09TL51	P.-J.J. Huang, T.S. Cameron and A. Jha, *Tetrahedron Lett.*, **50**, 51 (2009).
09TL236	P.K. Upadhyay and P. Kumar, *Tetrahedron Lett.*, **50**, 236 (2009).
09TL704	P. Kotame, B.-C. Hong and J.-H. Liao, *Tetrahedron Lett.*, **50**, 704 (2009).
09TL1188	L. Li, J.A. Bender and F.G. West, *Tetrahedron Lett.*, **50**, 1188 (2009).
09TL1222	A.H.L. Machado, M.A. de Sousa, D.C.S. Patto, L.F.S. Azevedo, F.I. Bombonato and C.R.D. Correia, *Tetrahedron Lett.*, **50**, 1222 (2009).
09TL1385	M.G.A. Badry, B. Kariuki, D.W. Knight and F.K. Mohammed, *Tetrahedron Lett.*, **50**, 1385 (2009).
09TL1526	D. Castagnolo, L. Botta and M. Botta, *Tetrahedron Lett.*, **50**, 1526 (2009).
09TL1542	E.H. Granados-Covarrubias and L.A. Maldonado, *Tetrahedron Lett.*, **50**, 1542 (2009).
09TL1954	C.S. Jungong, J.P. John and A.V. Novikov, *Tetrahedron Lett.*, **50**, 1954 (2009).
09TL2103	Y. Luo and J. Wu, *Tetrahedron Lett.*, **50**, 2103 (2009).
09TL2208	L.D.S. Yadav, S. Singh and V.K. Rai, *Tetrahedron Lett.*, **50**, 2208 (2009).
09TL2405	Y.-Y. Peng, Y. Wen, X. Mao and Q. Qiu, *Tetrahedron Lett.*, **50**, 2405 (2009).
09TL2643	M. Muthukrishnan, U.M.V. Basavanag and V.G. Puranik, *Tetrahedron Lett.*, **50**, 2643 (2009).
09TL3020	R. Ahmad, R.A. Khera, A. Villinger and P. Langer, *Tetrahedron Lett.*, **50**, 3020 (2009).
09TL3023	S. Espinosa, M. Solivan and C.P. Vlaar, *Tetrahedron Lett.*, **50**, 3023 (2009).
09TL3084	O.O. Fadeyi, R.N. Daniels, S.M. DeGuire and C.W. Lindsley, *Tetrahedron Lett.*, **50**, 3084 (2009).
09TL3245	P.J. Choi, D.C.K. Rathwell and M.A. Brimble, *Tetrahedron Lett.*, **50**, 3245 (2009).
09TL3889	K.C. Majumdar, A. Taher and K. Ray, *Tetrahedron Lett.*, **50**, 3889 (2009).
09TL3963	J.S. Yadav, B.V.S. Reddy, K.V.R. Rao and R. Narender, *Tetrahedron Lett.*, **50**, 3963 (2009).
09TL4781	K.C. Majumdar, S. Mondal and D. Ghosh, *Tetrahedron Lett.*, **50**, 4781 (2009).
09TL5075	M.J. Adler and S.W. Baldwin, *Tetrahedron Lett.*, **50**, 5075 (2009).
09TL5748	L.-Y. Chen, S.-R. Li, P.-Y. Chen, I.-L. Tsai, C.-L. Hsu, H.-P. Lin, T.P. Wang and E.-C. Wang, *Tetrahedron Lett.*, **50**, 5748 (2009).
09TL5798	J.S. Yadav, B.V.S. Reddy, S.K. Biswas and S. Sengupta, *Tetrahedron Lett.*, **50**, 5798 (2009).
09TL5933	T.A. Salama, A.-A.S. El-Ahl, S.S. Elmorsy, A.-G.M. Khalil and M.A. Ismail, *Tetrahedron Lett.*, **50**, 5933 (2009).

09TL5998	J.S. Yadav, P.P. Chakravarthy, P. Borkar, B.V.S. Reddy and A.V.S. Sarma, *Tetrahedron Lett.*, **50**, 5998 (2009).
09TL6112	M.D. Obushak, V.S. Matiychuk and V.V. Turytsya, *Tetrahedron Lett.*, **50**, 6112 (2009).
09TL6361	R.G.F. Giles and J.D. McManus, *Tetrahedron Lett.*, **50**, 6361 (2009).
09TL6723	M. Kiamehr and F.M. Moghaddam, *Tetrahedron Lett.*, **50**, 6723 (2009).
09TL7144	B. Linclau, P.J. Clarke and M.E. Light, *Tetrahedron Lett.*, **50**, 7144 (2009).
09TL7239	G. Mercey, D. Brégeon, C. Baudequin, F. Guillen, J. Levillain, M. Gulea, J.-C. Plaquevent and A.-C. Gaumont, *Tetrahedron Lett.*, **50**, 7239 (2009).

CHAPTER 7

Seven-Membered Rings

Jason A. Smith*, Peter P. Molesworth**, Christopher J.T. Hyland[†],
John H. Ryan[‡]

*School of Chemistry, University of Tasmania, Hobart, Tasmania, 7001, Australia
Jason.Smith@utas.edu.au
**School of Chemistry, University of Tasmania, Hobart, Tasmania, 7001, Australia
Peter.Molesworth@utas.edu.au
[†]School of Chemistry, University of Tasmania, Hobart, Tasmania, 7001, Australia and Department of Chemistry and Biochemistry, California State University Fullerton, California, 92831, United States
Chris.Hyland@utas.edu.au
[‡]CSIRO Division of Molecular and Health Technologies, Clayton, Victoria, 3168, Australia
Jack.Ryan@csiro.au

7.1. INTRODUCTION

This chapter summarises the chemistry of seven-membered heterocycles published in 2009 and puts particular emphasis on research focussing on the construction and reactions of these heterocyclic systems. In recent years there appears to have been a trend towards more complex molecules with more than one heteroatom. While benzodiazepines and benzothiazepines continue to be important heterocyclic motifs in the synthesis of biologically active molecules there has been a resurgence in the synthesis of azepines and their fused derivatives.

Reviews published during 2009 included a summary of the strategies towards the synthesis of the stemona alkaloids that contain a pyrroloazepine system <09EJO2421> and the synthesis of 3-methylbenzoxepines and 3-methyl-1-benzoxepines <09H(79)243>.

7.2. SEVEN-MEMBERED SYSTEMS CONTAINING ONE HETEROATOM

7.2.1 Azepines and Derivatives

Palladium-catalysed intramolecular oxidative allylic C-H amination favoured the formation of azepines, e.g. **2** over the alternative pyrrolidines, e.g. **3**, which were formed as the minor products of the reactions. The method was shown to be very versatile with numerous examples being reported <09OL2707>.

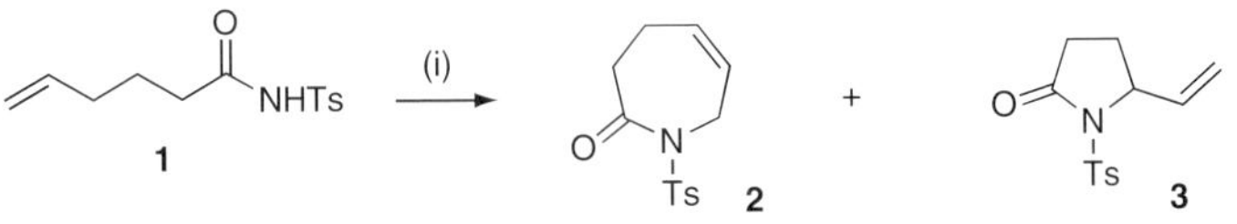

Reagents: (i) $Pd(OAc)_2$, NaOBz, 4Å MS, maleic anhydride, O_2, DMA, 91% (ratio of **2:3**, 86:14)

Intramolecular hydroamination with a zirconium complex gave cyclic amines including azepine **5** and benzazepine derivatives in good yields <09JA18246>. The asymmetric synthesis of spirocyclic caprolactones **6** was reported *via* diastereoselective Birch reductive alkylation with a dihaloalkane. Substitution of the remaining halide with azide followed by reduction and cyclisation results in the seven-membered spirocyclic lactams <09OL963>.

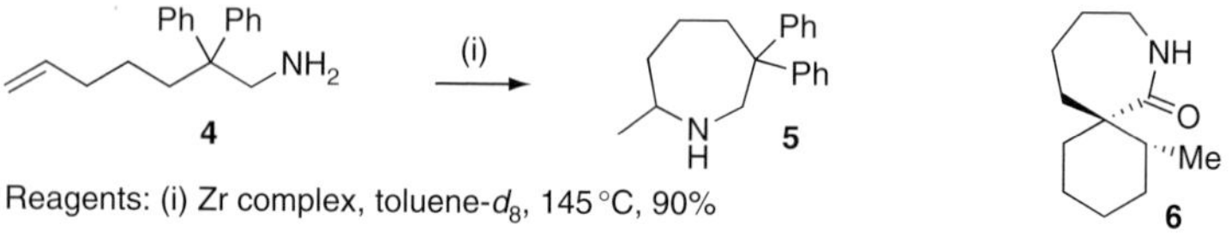

Reagents: (i) Zr complex, toluene-d_8, 145 °C, 90%

Ring closing metathesis (RCM) has been used for the large-scale synthesis of the azepine core **7** of a cathepsin K inhibitor <09T6291> as well as azepine derivatives **8** of sedum alkaloids <09T10192>. The azepine core **9** of analogues of microtubule stabilising marine alkaloid ceratamine A **10** were also prepared by RCM with the resultant alkene of **9** being functionalised to introduce a fused imidazole <09JOC995>.

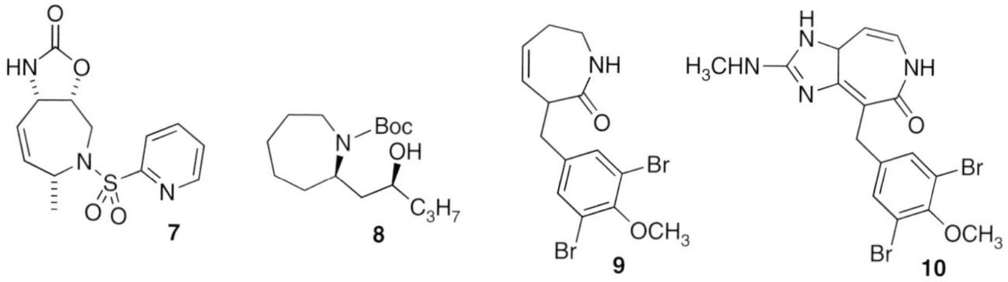

A range of azepine azasugars such as **11** were synthesised from D-arabinose <09BMC5598> while a range of simple substituted azepine sulfonamides **12** was prepared by the ring expansion of 4-oxopiperidine <09BML4563> followed by reaction with aryllithiums.

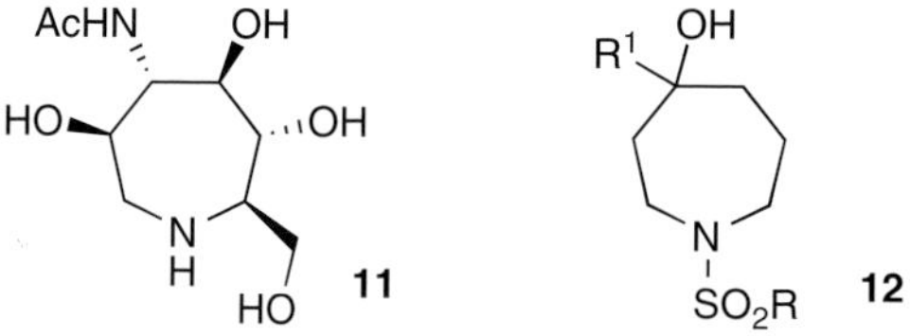

7.2.2 Fused Azepines and Derivatives

The gold-catalysed cyclisation at C-2 of indole onto a tethered alkyne **13** occured preferentially to give the 8-*endo-dig* product except when bulky phosphine ligands were used, in which case the 7-*exo-dig* mode giving **14** predominated <09T9015>. The palladium-mediated cyclisation of bis-allenols **15** gave the dihydrofuran fused azepines **16** in moderate to good yields <09OL1205>.

Reagents: (i) Au catalyst, CH_2Cl_2, rt,

Reagents: (i) $PdCl_2$, DMF, 80 °C, 1 h (70%)

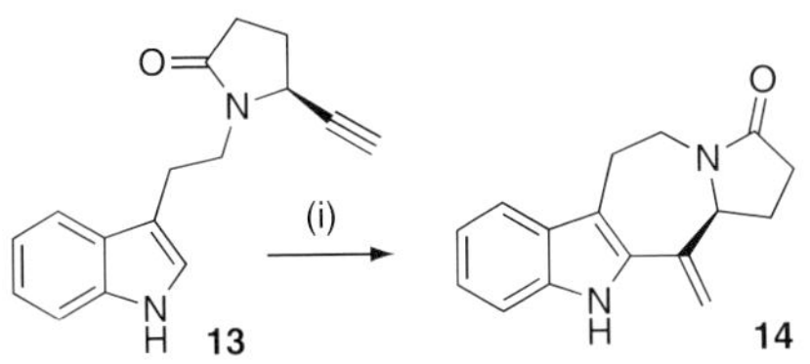

A combination of aldol and Schmidt reactions has been used for the synthesis of pyrroloazepines but a mixture of all four diastereomers was formed <09JOC7618>. However, in combination with an initial Sakurai reaction, greater selectivity was observed and using bulky aldehyde **18** one example resulted in a 42% yield of a single diastereomer **19**. The photochemical [5+2] cycloaddition from direct irradiation of maleimide-tethered alkenes such as 20 gave pyrrolo-fused azepines 21 in good to excellent yields <09AGE8716>. With direct irradiation a singlet state is generated and resulted in cleavage of the imide to give the [5+2] cyclisation, however, if a triplet sensitiser is added than [2+2] cycloaddition resulted.

Reagents: (i) $TiCl_4$, allyl-TMS, CH_2Cl_2, −45 °C, 1 h then aldehyde **18**, −45 - 0 °C, 24 h

Reagents: (i) hν, 125 W Hg lamp, CH_3CN, 1 h (69%)

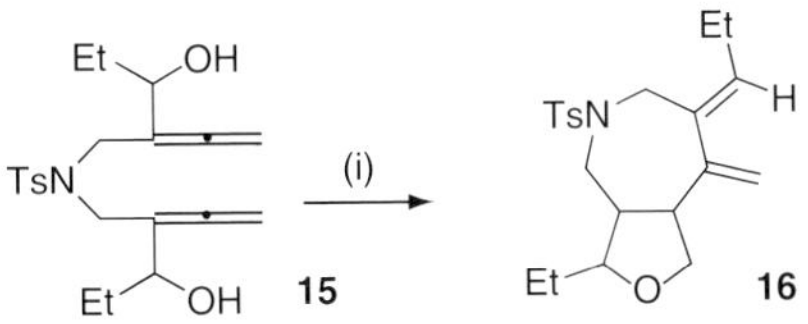

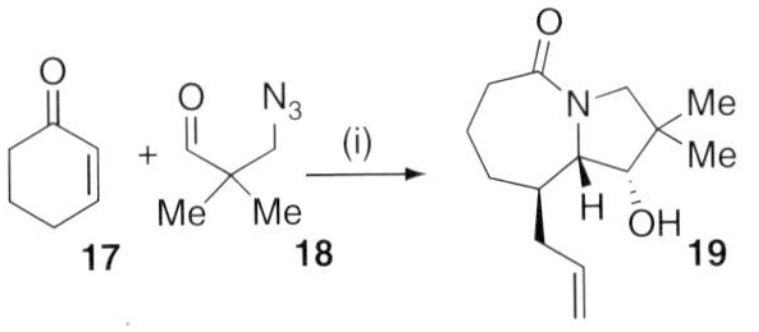

Ring closing metathesis (RCM) was used for the synthesis of scaffolds for the synthesis of hydroxylated pyranoazepines <09TL3657> and bicyclic azepine aza sugars <09TA1217> as potential glucosidase inhibitors.

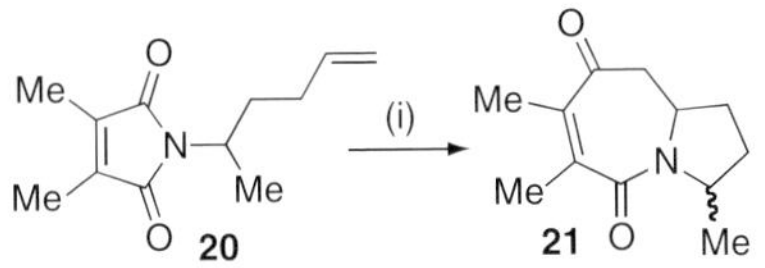

The isolation and characterisation of two new stemona alkaloids (**22** and the *N*-oxide of **22**) that contain a pyrroloazepine ring system have been reported from the root extracts of *Stemono aphylla* <09JNP848>. The total syntheses of the azepine-containing natural products (±)- aurantioclavine **23** <09EJO5752>, (±)-hinckdentine A **24** <09OL197>, (±)-stemoamide **27** <09SL1979> and (±)-13-epieostenine **25** <09T5716> were reported, as was the formal synthesis of (±)-stemonamine **26** <09JOC3211>. An analogue of stemoamide, (+)-9a-*epi*-stemoamide 27 was also reported <09SL2188>.

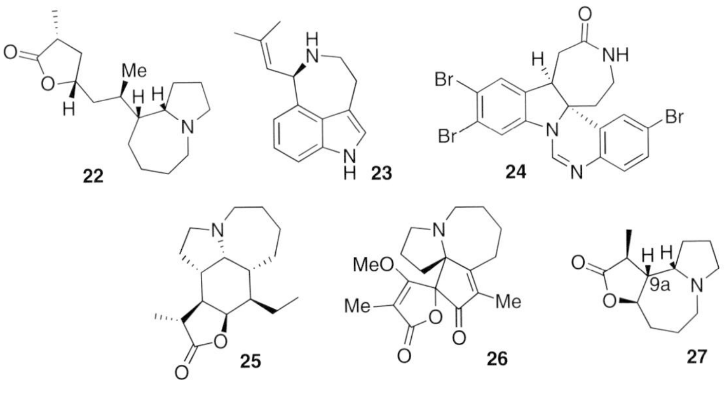

Other syntheses of fused azepines included the palladium-mediated cyclisation of tethered brominated aminonapthoquinones onto C-2 of indole <09BML5753>, the acid-mediated cyclsation of a tethered carboxylic acid onto a thienopyrrole <09BML841>, Beckmann rearrangement of a pyrroloisoquinoline <09BML1849>, a Pictet-Spengler cyclisation onto pyrrole <09BML5289> and the reductive ring expansion of the corresponding oximes with diisobutylaluminium hydride to yield heterocyclic fused derivatives <09H(78)1183>. The Pictet-Spengler cyclisation of N_β-benzylserotonin **28** gave the indole fused azepines **29** in moderate to good yields while in contrast, tryptamine derivatives reacted through C-2 to form carbolines <09H(77)825>.

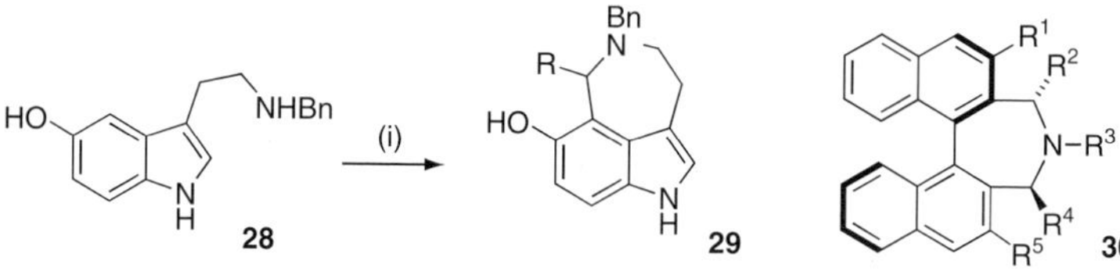

Reagents: (i) RCHO, DABCO, MeOH, reflux, 3 h, (55–86%)

Chiral fused benzazepines continue to be applied as organocatalysts for asymmetric synthesis with reports for aldol <09CEJ6678> <09AGE4363>, Mannich <09CEJ6678> <09AGE1838> <09JOC5734>, Henry (nitroaldol) <09BML3895> conjugate addition <09TL4674> <09BKC1441>, aminoxylation <09S1557>, [3+2] cycloaddition of azomethine ylides <09S1670>, phase transfer catalysts <09SL675> and epoxidation reactions <09EJO3413>. The general structure of these organocatalysts is represented by **30**.

7.2.3 Benzoazepines and Derivatives

The number of reports of the synthesis of benzoazepines has increased during 2009 mostly due to the reported biological activity of compounds containing this structural moiety.

The synthesis of both enantiomers of 2-methyl-3-benzazepines reported in 2008 was extended to the synthesis of a series of 2-substituted derivatives **32** and **33** in good yield. The use of (R)-phenylglycinol as a chiral auxiliary in the first step of the synthesis resulted in ee's of the final product of >99% <09JOC2788>. The substrates with large substituents such as phenyl and t-butyl showed high affinity for the NMDA receptor.

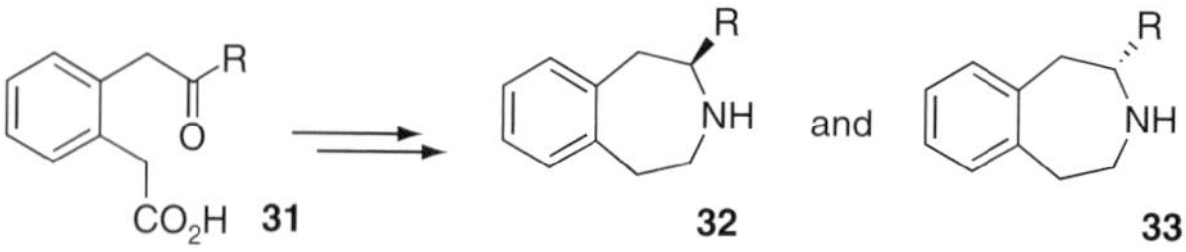

Annulation onto a phenyl ring is a common route to benzazepines and a method using tethered alkynes has been reported. For terminal alkynes a gold catalyst was required but for electron-deficient alkynes, gold was not required <09OL1225>.

The oxygen of the *N*-oxide **34**, generated *in situ*, is transferred to the alkene generating either a carbene or metal carbenoid, depending upon conditions, with subsequent cyclisation onto the aromatic ring giving compounds **35**.

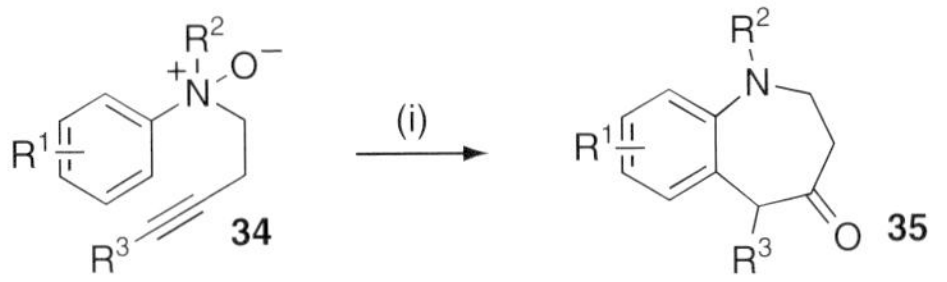

Reagents: (i) R^2=H, Ph$_3$PAuNTf$_2$, CH$_2$Cl$_2$, –20 °C; R^2=EWG, no catalyst required

Ring closing metathesis played a key role in the formation of benzazepines such as **36** <09EJO3741>, **37** <09TL1911> and **38** <09T6454> by formation from either a vinyl or allyl substituted benzene derivative. The more complex indole fused benzoazepine **39** was formed as a scaffold for the synthesis of hepatitis C polymerase inhibitors <09SL1395>.

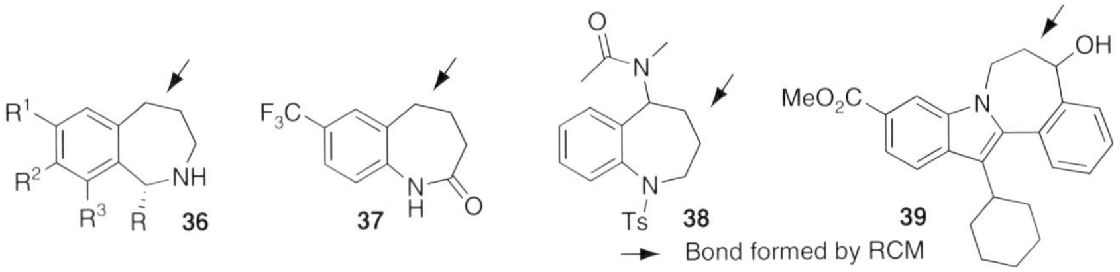

→ Bond formed by RCM

A mechanistic study into the formation of the seven-membered ring by a 7-*endo-trig* radical cyclisation was reported <09TL1727>. The method was used to generate a small library of compounds for evaluation in a wound-healing assay <09BML3193>. Radical cyclisation of aromatic radicals onto N-(methoxycarbonyl)acrylamides **40** was also reported but 7-*exo-trig* cyclisation to give the benzoazepine **41** was the minor product with the 8-*endo-trig* product being the preferred mode of cyclisation <09H(77)575>.

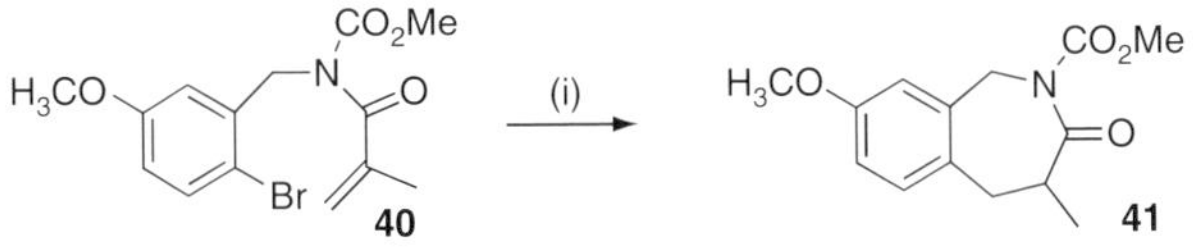

Reagents: (i) Bu$_3$SnH, AIBN, benzene, reflux (49%)

Heck cyclisation onto an annulated allylamine <09EJO1934> was used to form benzoazepine **42** in good yield but also more complex indole fused systems. Heck cyclisation onto a propargylamide was reported to give **43** as an intermediate in a synthetic approach towards the alkaloid apanorhin **44** <09EJO793>.

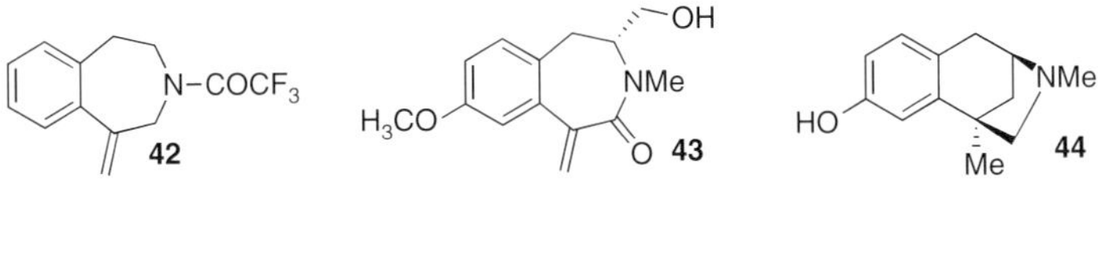

The ring expansion of tetrahydroisoquinolines by intramolecular reaction of a copper carbene on the nitrogen and ring opening yielded benzoazepine derivatives <09TA1154>. The reverse of this was also described with benzoazepine **45** undergoing ring contraction with FeCl$_3$ in nitrobenzene at 120 °C to give tetrahydroisoquinolines **46** in good to high yields <09CEJ11119>. The mechanism was proposed to involve oxidation of the benzylic position. A ring contraction promoted by aerobic oxidation with cobalt(II) acetate yielding indolo[3,2-c]quinolines **47** from benzoazepines was reported to occur in moderate to reasonable yields (27–60%) <09S1185>.

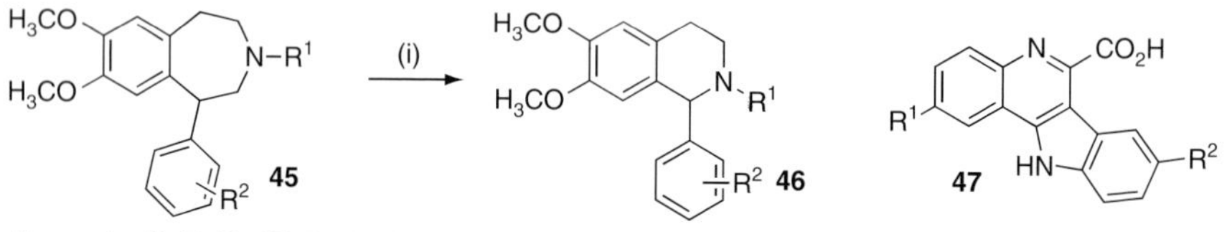

Reagents: (i) FeCl$_3$.6H$_2$O, nitrobenzene, 120 °C (53–87%)

Samarium diiodide was reported to promote the ring expansion of 1-aroyltetrahydroisoquinolines **48** <09OL1857>. The method was employed for the synthesis of natural product bulgaramine **50**.

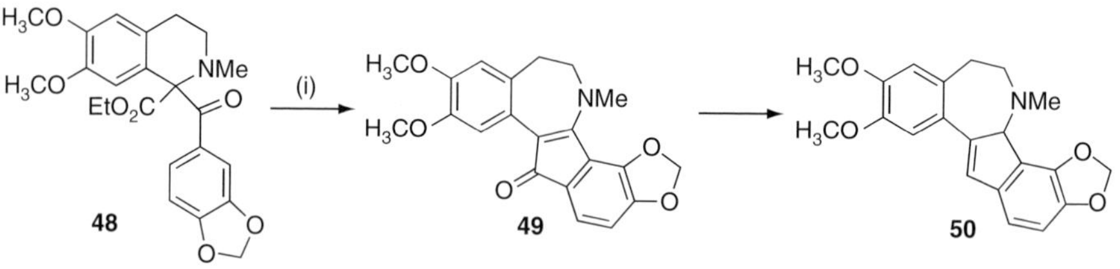

Reagents: (i) SmI$_2$ (4 equiv), MeOH (3 equiv), THF, rt then p-TsOH (0.1 equiv) toluene, reflux (68%)

The rearrangement of N-aryl-2-vinylaziridine **51**, formed *in situ* by the reaction of an arylazide, a diene and a ruthenium catalyst, was promoted thermally or with silica gel to give benzazepine **52** in moderate yield <09CEJ1241>. The mechanism was proposed to be an aza-[3,3]-Claisen rearrangement which was supported by a theoretical investigation. A novel synthesis of benzazepine **54** was reported *via* a Bergman cyclisation of an 11-membered enediyne **53**. The cyclisation required elevated temperatures compared to lower homologues with little or no reaction at 80 °C but increasing the temperature to 97 °C gave the benzazepine in 31% yield <09OBC695>. The benzofused enediyne was even more stable but gave the corresponding naphthoazepine in 60% yield when reacted at 150 °C over nine days, showing the relative stability of the larger ring systems.

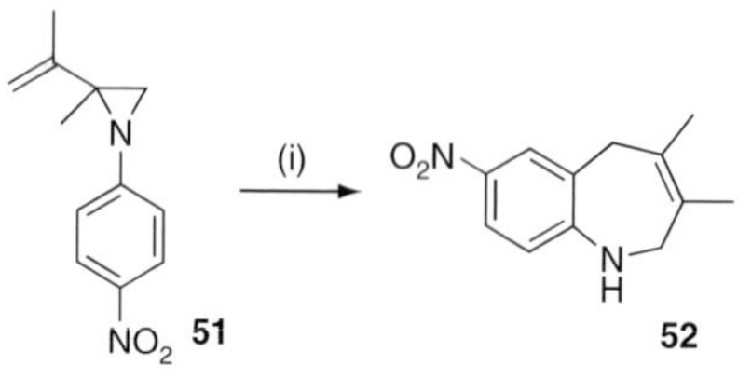

Reagents: (i) benzene, silica gel (65%)

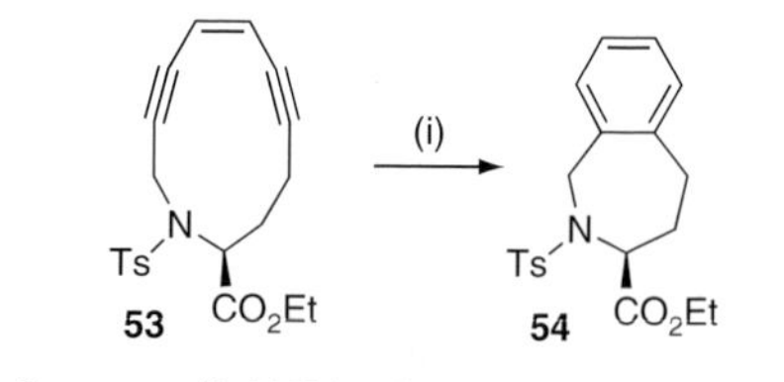

Reagents: (i) EtCN, reflux, N$_2$, 6 days (31%)

Cyclisation reactions for the formation of benzazepines were also reported using intramolecular cycloaddition of a nitrone with subsequent reduction of the N–O bond to give hydroxy substituted benzazepines **55** that were evaluated for anti-parasitic activity <09BML2360>. Electrocyclisation of the anion formed from N-allyl or N-benzyl imines **56** gave benzazepines **57** after trapping with an acid chloride, in moderate to good yields <09EJO2342>.

Reagents: (i) LDA, THF, −78 °C - 0 °C then RCOCl (40–79%)

Annulation of a tethered amine to generate benzazepines was also reported using Pummerer-type <09OBC589> and Pictet–Spengler cyclisations <09EJO1309>. A tribenz-[b,d,f]-azepine was formed by photostimulated $S_{RN}1$ substitution of a diarylamine in good yield when an *ortho*-biphenyl group was one of the substituents rather than formation of a carbazole which was the normal trend <09JOC4490>. The reaction of ω-amino-1-chlorosulfoxides with Grignard reagents gave benzoazepine **59** and also azepines by formation of a magnesium carbenoid <09TA169>.

Reagents: (i) *i*-PrMgCl, toluene, −40 °C then EtOCOCl (68%)

Cyclomethylation of a tethered hydrazine with chloromethyl methyl ether gave an annulated seven-membered ring in moderate yield and cleavage of the N–N bond gave benzazepines <09TA1903>. Intramolecular alkylation was reported from a tethered acetal <09BML1871>, benzylic alcohol <09BML104>, while direct intramolecular arylation gave heterocyclic fused benzoazepines in good yields <09JME5916>. Intramolecular Dieckmann condensation was exploited as the key step in formation of a heterocyclic fused benzazepine that was used as a scaffold for the generation of a library of potential GABA agonists <09BML5746>. Cyclisation of N-arylethyl–acylpyrrolidinium ions, formed *in situ* by triflic acid mediated cyclisation of a tethered acetal onto a 3-phenylpropionamide, gave benz-[c]-azepines in poor to excellent yields (9–90%) depending upon the substituent on the aromatic ring <09OBC3561>. A palladium-catalysed domino *ortho*-alkylation gave benzoazepines **61** in reasonable yield <09JOC1791>. The cyclisation of an alkene onto a tethered N-acyliminium ion, formed using $FeCl_3$ on a hydroxy lactam, was exploited for the synthesis of the tetracyclic core of the antibiotic tetrapetalone <09OL4036>. With simple alkenes, a mixture of diastereomeric alkyl chlorides was formed but the with the tethered benzyl ether **62**, subsequent cyclisation occurred to give a tetracyclic product **63** as a 36:47 ratio of diastereomers.

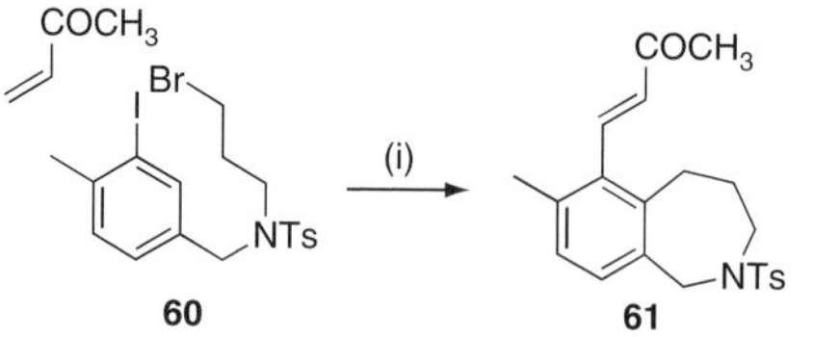

Reagents: (i) Pd(OAc)₂, PPh₃, Cs₂CO₃, norbornene, DME, 80 °C, 16 h, (59%)

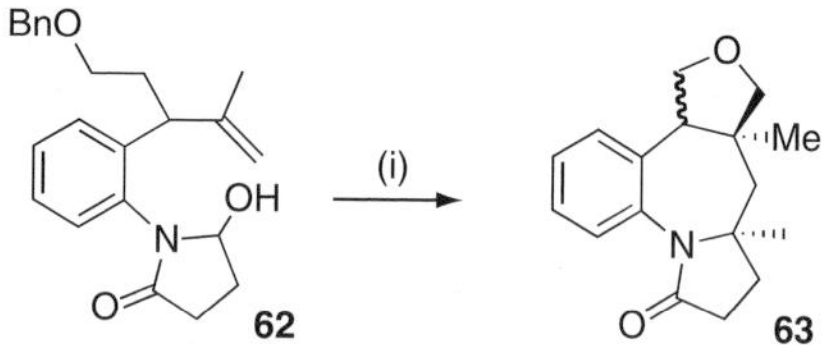

Reagents: (i) FeCl₃, TMSCl, CH₂Cl₂ (83%)

Indole-fused benzoazepines were formed in good yields by a sequence of carbopalladation, *o*-alkylation and C-H functionalisation. When 1-(2-iodobenzyl)-indole **64** was reacted with the alkynyl bromide **65** the pentacyclic products **66** were formed in 70-90% yields <09SL1004>. The same authors also reported an *o*-alkylation reaction to prepare a tetracyclic derivative containing a furan ring system in high yield <09JOC289>. The benzoazepinoquinoline **67**, a mutagen formed from tryptophan and glucose during cooking, was synthesised by lactamisation of the appropriate aryl-quinoline <09SL1781>.

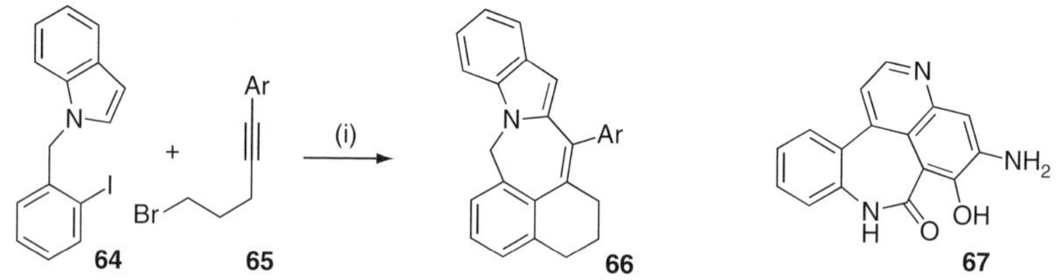

Reagents: (i) PdCl₂, TFP, norbornene, Cs₂CO₃, MeCN, 90 °C, 24 h (70–90%)

The bisbenz-[*b*,*d*]-azepine **69** was formed by intramolecular [4+2] cycloaddition of a tethered alkene onto a furan **68** followed by ring opening and dehydration of the cycloadduct <09TL3145>.

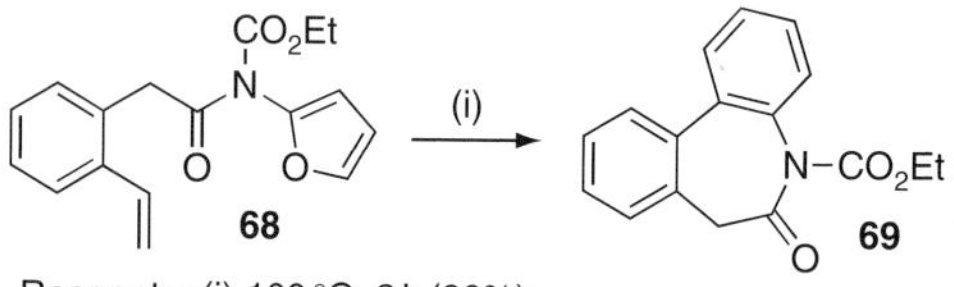

Reagents: (i) 100 °C, 2 h (90%)

Irradiation of N-(alkyladamantyl)phthalimides resulted in the formation of various benzoazepines such as **70** and **71**. If the tether between the adamantane and the phthalimide was short then cyclisation occurred onto the adamantane system but with longer tethers cyclisation occurred onto the alkyl linker <09JOC8219>.

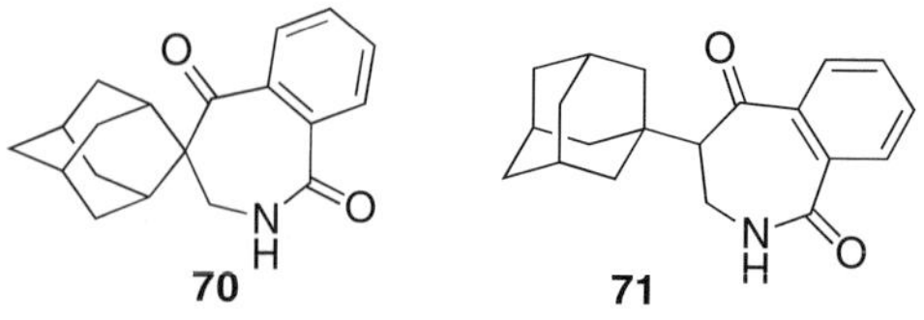

7.2.4 Oxepines and Fused Derivatives

Numerous marine polycyclic ethers contain fused oxepine rings and the synthesis of fragments of ciguatoxin <09T7784> and gambieric acid <09OL113> were reported. The total synthesis of brevanal <09OL2531>, ciguatoxin <09AGE2941> and gambierol <09OL4382> were reported exploiting ring closing metathesis and intramolecular alkylation while the first synthesis of *ent*-dioxepandehydrothyrsiferol <09JA12084> was reported using a bromonium-initiated epoxide-opening cascade of the triepoxide **72** for the synthesis of the tricyclic fragment of the natural product.

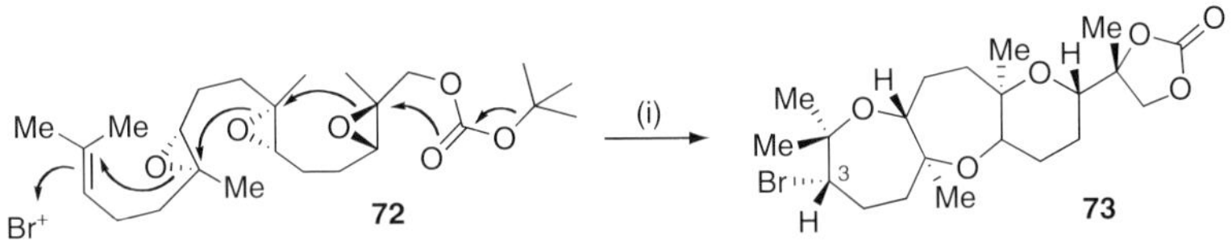

Reagents: (i) NBS, HFIP, 4Å MS, 0 °C, 15 min, rt (36% + C-3 epimer 36%)

Ring closing metathesis was also used for the synthesis of oxepine sugar derivatives <09JOC6486> synthesised from glucose, and benzoxepines as transition state analogues of β-secretases <09BML264>. Two other oxepine containing sugar derivatives were reported <09OL4482> <09CAR448>.

Numerous new natural products containing a benzoxepine were reported including atroindonesiabib E1 **74** from the wood of an Indonesion tree <09APR191>, xylarinol A **75** and B **76** from fungi <09JAN163> <09PM1104> <09JAN533>, benzophomposin A **77** <09JAN533>, 7-hydroxyjanthinone **78** from a mangrove endophytic fungi <09NPC1481>, microsphaeropsones A-C **79** from endophytic fungi <09CEJ12121>, trigonostemone A **80** from a small chinese shrub <09TL2917> and graphisins B **81** isolated from lichen from Thailand <09AJC389>.

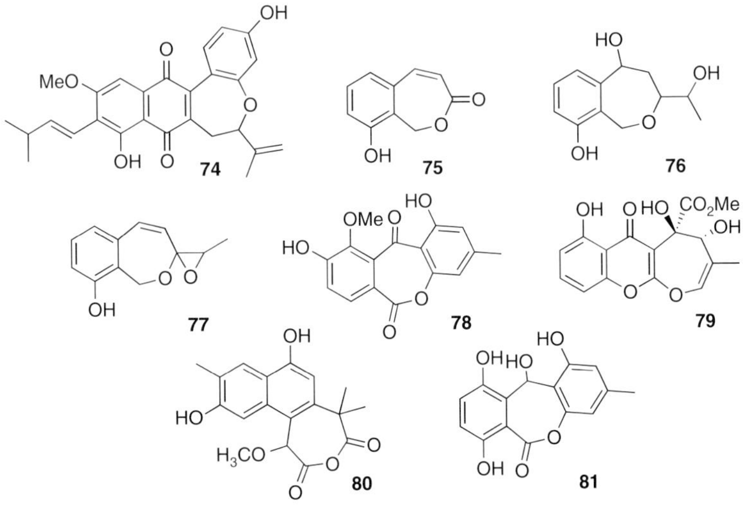

Cyclodehydration using catalytic *p*-TsOH has been used prepare the oxepine ring in a synthesis of novel pseudosteroids **83** containing a fused oxepine ring system, in excellent yields <09BML3977>. Hydrofluoric acid in acetonitrile continues to be used for concomitant deprotection and cyclisation to form dibenzoxepine rings and was used in the preparation of oxepine analogues **83** of *N*-acetylcolchinol **84** <09JOC4329>.

82 **83** **84**

Reagents: (i) TsOH (2 mol %), CH_2Cl_2

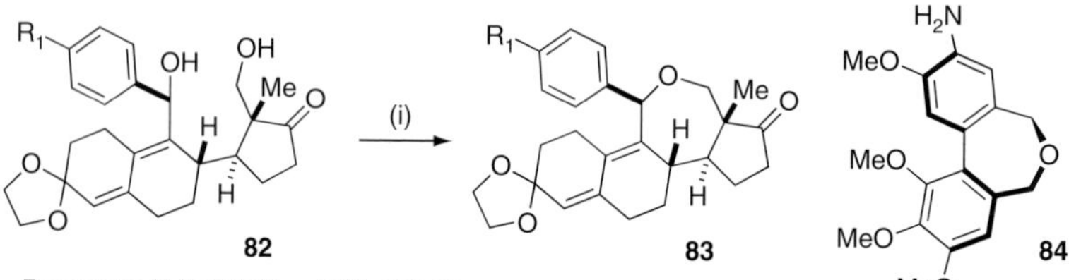

Lactonisation was used as the final step to prepare (±)-velloziolide **86**. Deprotection of the methylenedioxy protecting group was achieved using boron trichloride in dichloromethane, however wet silver nitrate was required to avoid formation of the unwanted chloromethylated product **87** <09JOC7411>. Electron-rich aromatic substituents were shown to be key in controlling formation of benzoxepines **89** *via* a 7-*endo* route in the iodolactonisation of alkynyl phenylacetic acids **88** <09TL1385>.

85 **86** R = H **87** R = CH_2Cl **88** **89**

Reagents: (i) BCl_3, CH_2Cl_2, –78 °C, (ii) $AgNO_3$, THF:H_2O Reagents: (i) I_2, K_2CO_3, MeCN, 20 °C

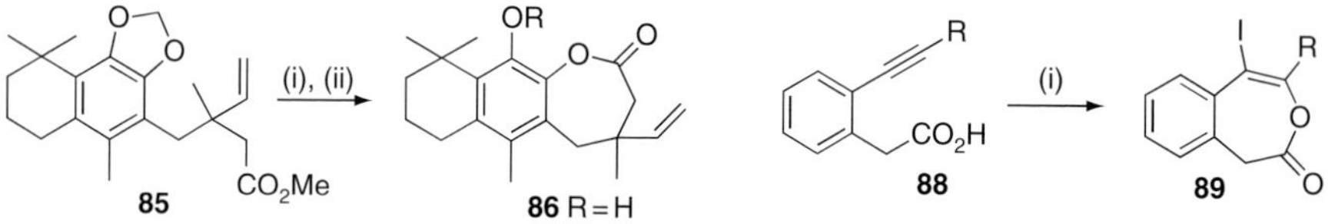

A [2+2+2] cycloaddition approach (alkyne cyclotrimerisation) was used to prepare a series of 6-oxa-allocolchicinoids **91** in moderate to excellent yields <09OL341>.

90 **91**

Reagents: (i) $CpCo(CO)_2$, PPh_3, chlorobenzene, μw (300 W), 30 min, 150 °C, sealed tube

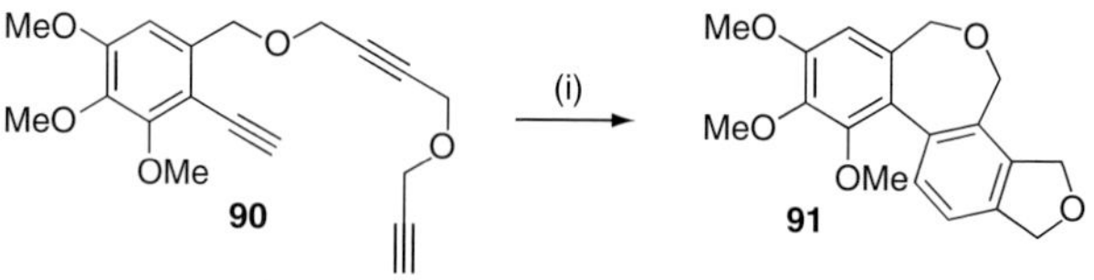

Knoevenagel condensation followed by intramolecular nucleophilic displacement of an aromatic nitro group gave a series of fused indolo-benzoxepines **95** in good yields after base catalysed rearrangement of the alkene **94** <09T6868>. Spirooxepine-oxindole derivatives were prepared *via* Prins type annulation using TMSOTf and chiral alcohols <09OL3366> and <09OL3362>.

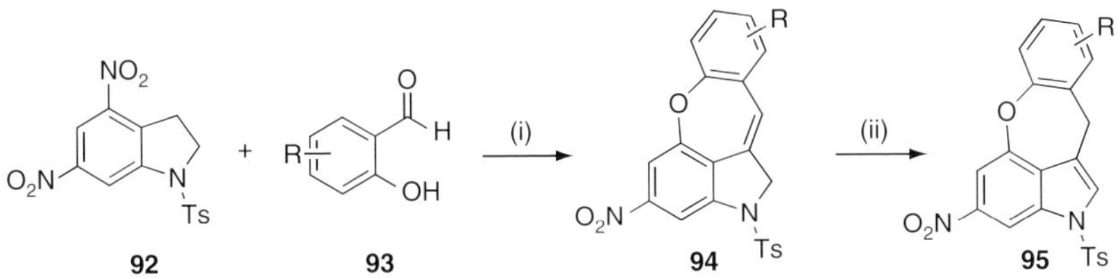

Reagents: (i) C$_6$H$_6$, piperidine, reflux (75–81%); (ii) DBU, C$_6$H$_6$, reflux (82–91%)

Bauhinoxepin J **97** was prepared by decarboxylative radical cyclisation onto the quinone unit of **96** to form the oxepine ring.

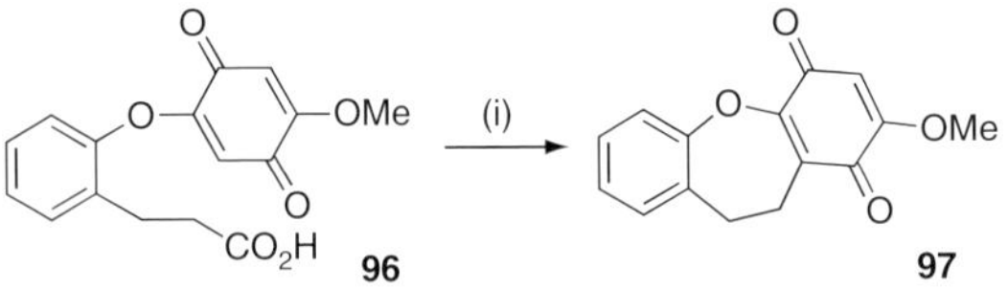

Reagents: (i) (NH$_4$)$_2$S$_2$O$_8$, AgNO$_3$, MeCN, H$_2$O (40%)

Radical cyclisation was used to prepare the oxepine ring of a series of benz[*b*]oxepines and 12 oxobenzo[*c*]phenanthridinones which were tested as topoisomerase inhibitors <09BML2444> while intramolecular photocyclisation of an alcohol onto a terminal alkyne was used to prepare cyclic carbohydrates <09T7921>, <09OL851>.

Sulfanyl radical addition was used achieve 7-*endo-trig* radical cyclisation and thus prepare a series of thioalkenyl oxepines **99** in excellent yields <09TL228>.

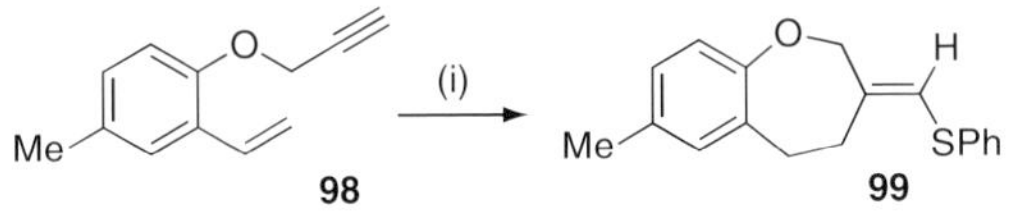

Reagents: (i) PhSH, AIBN, C$_6$H$_6$, (88%)

Ring expansion to form the oxepine ring was commonly used. A series of dibenzoxepines **101** were prepared *via* oxidative 1,2 radical ring expansion, in good yields <09JOC3978> while ring expansion of cyclopropanated carbohydrates was achieved using TMSOTf to give the corresponding heptanose sugars <09JOC7627>. 2,7-Dialkyloxepines were prepared by ring expansion of the corresponding *cis*-tetrahydropyran with either trimethylsilyldiazomethane <09TL5285> or an α-substituted-diazoacetate <09JA6614> with good diastereoselectivity.

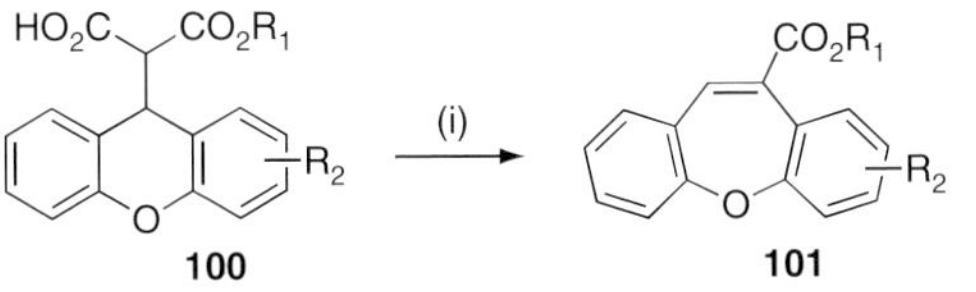

Reagents: (i) Mn(OAc)$_2$, AcOH/H$_2$O, reflux, 5 min (37–81%)

Metal-catalysed (Au(III), Pd(II) and La(III)) fuctionalisation of γ-allenols **102** to yield both oxepines <09CEJ9127> and fused oxepines **103** <09CEJ1901> was extensively studied both experimentally and theoretically <09CEJ1909>.

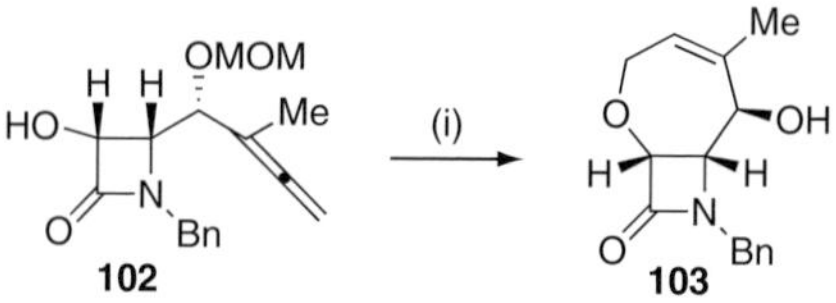

Reagents: (i) La[N(SiMe$_3$)$_2$]$_3$ (5 mol%), toluene, reflux, (58%)

Simple benzoxepines such as **105** could be formed readily prepared *via* a phenoxide ion mediated 7-*endo-tet* carbocylisation of a cyclic sulfate in good yield <09EJO204>.

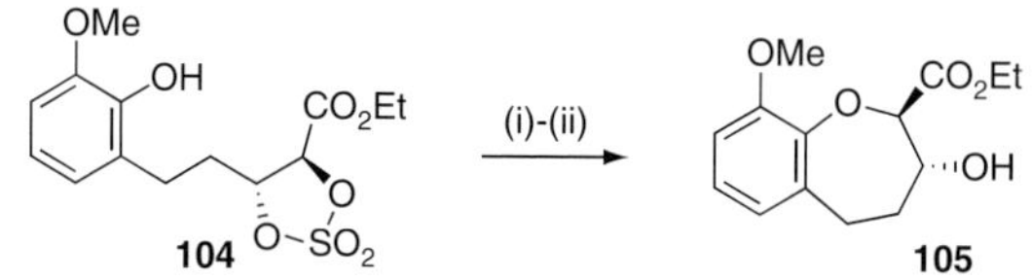

Reagents: (i) K$_2$CO$_3$, acetone, rt, 8 h; (ii) 20% H$_2$SO$_4$, THF, rt (72%)

Metal-catalysed cyclisation was studied extensively: rhodium-catalysed oxygen assisted hydroacylation <09JA6932>, iron-catalysed arene-aldehyde addition/cyclisation <09JOC6797> and palladium-catalysed intramolecular carboesterification <09AGE9690> were all used to produce benzoxepines. Tricyclic fused oxepines **107** and **108** were prepared using an intramolecular tandem Nicholas / Pauson–Khand reaction after activation of an alkyne **106** with cobalt, in 80% overall yield.

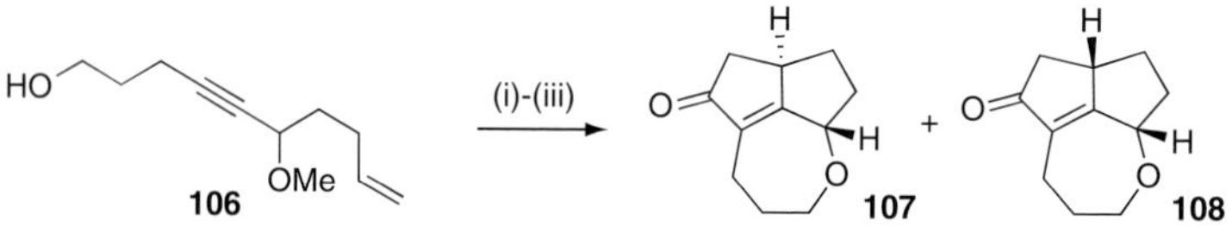

Reagents: (i) Co$_2$(Co)$_8$, CH$_2$Cl$_2$, 30 min, (ii) BF$_3$.OEt$_2$, CH$_2$Cl$_2$; (iii) Cyclohexylamine (80%)

7.2.5 Thiepines and Fused Derivatives

Proline was used as an organocatalyst for the ring opening of cyclopropylcarbaldehydes **110** with thiosalicylaldehydes **109** and subsequent ring closure afforded benzothiepines **111** in moderate yields <09SL1830>.

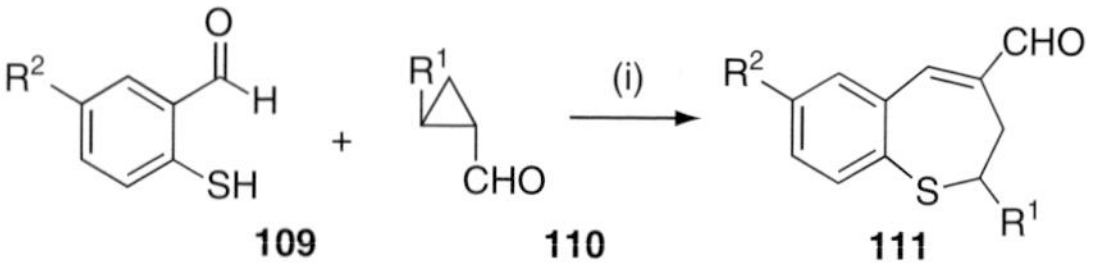

Reagents: (i) (*S*)-proline, 4Å MS, THF, rt, 3 d (34–56%)

A Dieckmann reaction was used to form indolizidinyl-fused thiepines **112** in excellent yields by alkylation of a thiolate anion with ethyl 4-bromobutenoate and formation of a vinylogous anion <09H(78)319 >. The bisbenzothiepine **113** and the selenium analogues were synthesised *via* a McMurray coupling to form the alkene between the two aromatic rings <09T8350>.

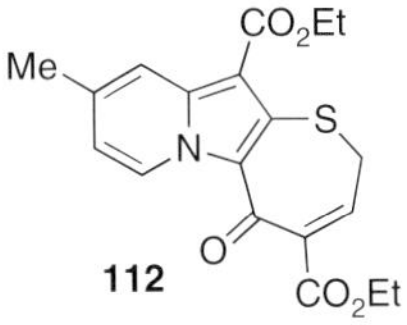

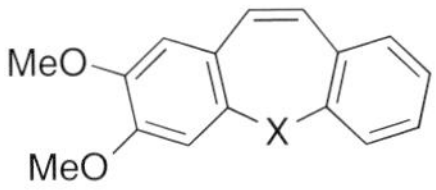

7.3. SEVEN-MEMBERED SYSTEMS CONTAINING TWO HETEROATOMS

7.3.1 Diazepines and Fused Derivatives

This section of the review focuses on literature reporting new synthetic methods for the preparation of diazepines. It should be noted that there continues to be a high level of interest in diazepines as biologically active compounds and full coverage of this work for 2009 is beyond the scope of this review.

Cyclisation of the methyl ester of 2-acetyl-3-indolylacetic acid **114** and 3-acetylamino-1-methyl-indolo[2,3-*c*]pyrylium perchlorate **116** in the presence of hydrazine hydrate have been shown to provide 2-oxo-2,3-dihydro-1*H*-[1,2]diazepino[4,5-*b*]indoles **115** <09CHE726> in moderate yields.

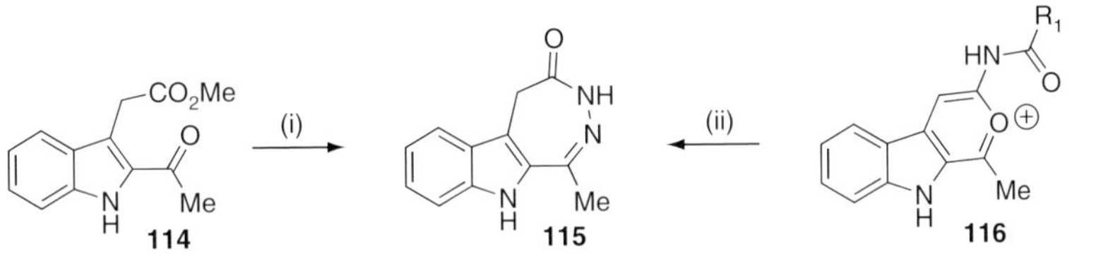

Reagents: (i) H$_2$NNH$_2$, 2-propanol, reflux, 30 min, 52%; (ii) H$_2$NNH$_2$, 2-propanol, reflux, 2 h, 47%

Trifluoromethyl-substituted 1-amino-1-azapenta-1,4-dien-3-ones **117** have been shown to produce iminium ion **118** in the presence of excess trifluoromethanesulfonic acid. This iminium ion can then undergo an 8π, 1,7-electrocyclisation to form dihydroindenodiazepines **119** or 6π, 1,5-electrocyclisation to form dihydrospiroindenepyrazoles **120**. Unfortunately, the reaction displays poor selectivity for all substrates tested <09JOC4584>.

117 118 119 120

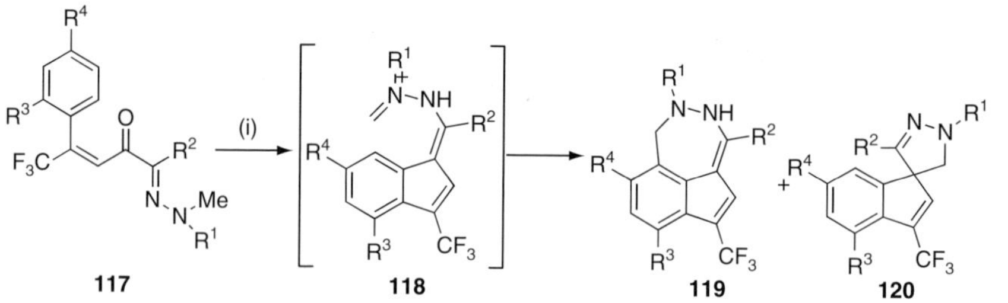

Reagents: (i) CF_3SO_3H, Ac_2O, CH_2Cl_2 (16–66%)

Building on the lack of investigation into the photochemistry of the C=N bond, a synthesis of 1,3-diazepines **123** from maleimides **121** has been developed <09AGE2514>. Based on previous work the authors propose biradical **122** forms *via* a singlet mechanism with n→π★ excitation and C–N bond cleavage. Subsequent [5+2] cycloaddition occurs with a variety of C=N systems to provide fused polycyclic 1,3-diazepines in good to excellent yields.

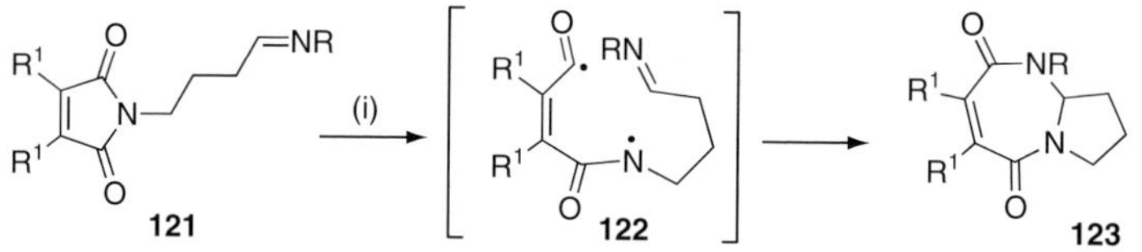

121 122 123

Reagents: (i) hν, MeCN, 1.5 h (40–90%)

Base-induced cyclisations of the hydroxyguanidine *O*-sulfonic acid **124** have been shown to yield 6-amino-5*H*-dibenzo[*d*,*f*][1,3]diazepine **126** along with 2-aminobenzimidazole **127**. It is proposed that **125** is an intermediate in the formation of **126** and may be produced *via* a base-induced oxidative coupling. It is postulated that **125** is likely to result from a Cope rearrangement of an iminodiaziridine <09EJO3940>.

124 125 126 127

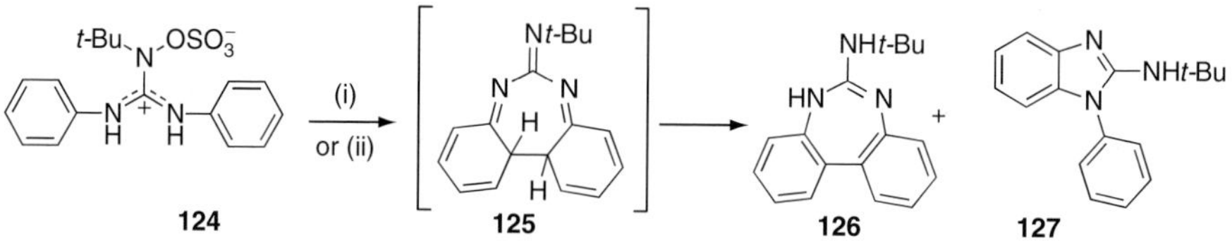

Reagents: (i) KOH (40%), MeOH/benzene (91%, **126**:**127**, 15:85)
(ii) Aqueous NaOH (2 M) (98%, **126**:**127**, 75:25)

In the burgeoning field of *N*-heterocyclic carbene ligands a synthesis of seven-membered ring amidinium salts based on the methods of Bertrand, Grubbs and Carvell has been reported. Amidinium salts **130** were prepared in moderate to excellent yields <09JOR2454>.

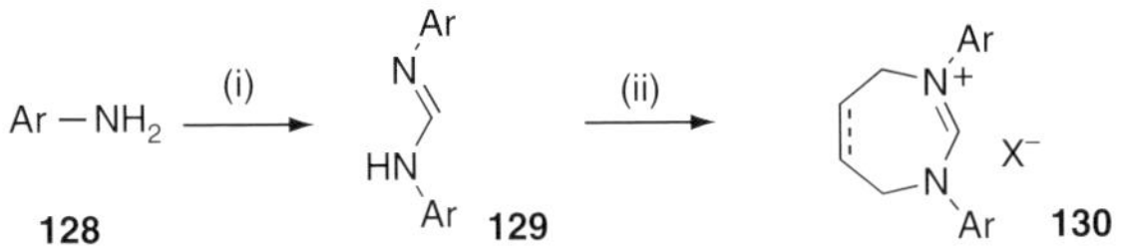

Reagents: (i) CH(OEt)$_3$; (ii) *n*-BuLi, DIPEA or K$_2$CO$_3$ and X-(CH$_2$)$_4$-X or X-CH$_2$CH=CHCH$_2$-X, (46–92%)

The 1,3-diazepan-2-iminium natural product monanchorin **132** has recently been the subject of a synthetic study by Snider and coworkers. The protected guanidino alcohol **131**, which was prepared from an enantiomerically pure epoxy acetal, underwent efficient cyclisation in CDCl$_3$/TFA to yield monanchorin **132** in good yield <09OL1031>.

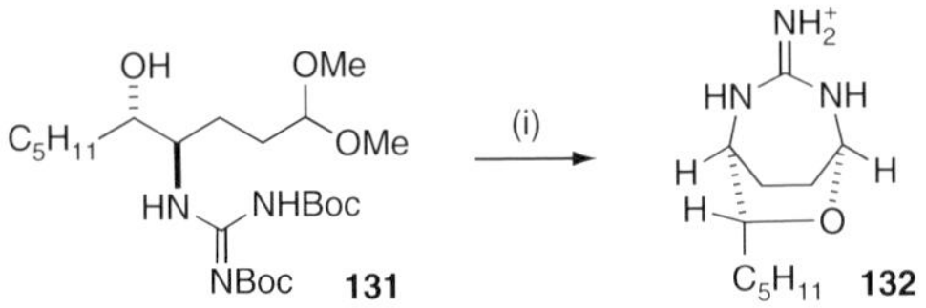

Reagents: (i) CDCl$_3$, TFA, 25 °C (73%)

Nucleophile-initiated ring expansion of 4-mesyloxymethyl- and 4-tosyloxymethyl-5-tosyl-1,2,3,4-tetrahydropyrimidin-2-ones **134** allowed for reliable preparation of 6-tosyl-2,3,4,5-tetrahydro-1*H*-1,3-diazepin-2-ones. The reaction is compatible with a range of nucleophiles and the starting pyrimidin-2-ones were readily synthesised in five steps from simple commercially available materials <09T2344>.

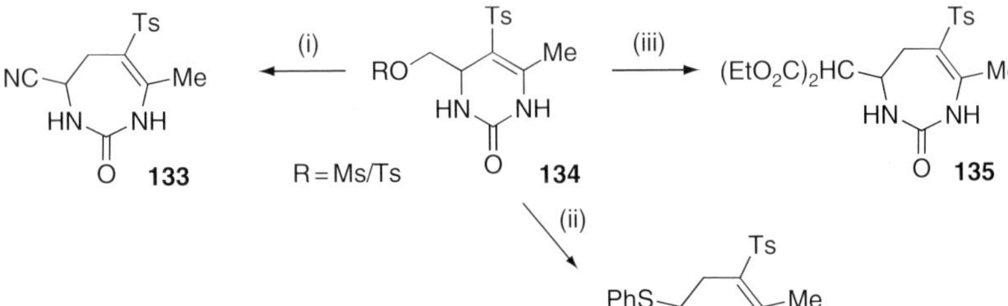

Reagents: (i) NaCN, DMF, rt (95%)
(ii) PhSH, NaH, rt (97%)
(iii) CH$_2$(CO$_2$Et)$_2$, NaH, rt (95%)

Aza 3-(piperidin-4-yl)-4,5-dihydro-1*H*-benzo[*d*][1,3]diazepin-2(3*H*)-ones **138** have been prepared simply by reaction of diamines such as **137** to form the cyclic urea with carbonyl diimidazole (CDI) <09TL386>.

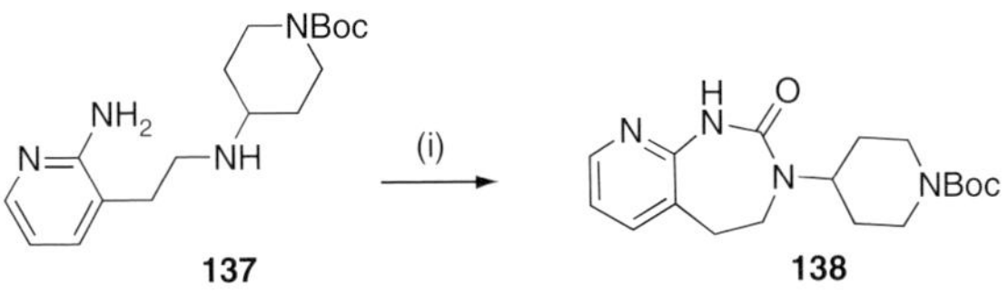

Reagents: (i) CDI, Et$_3$N, MeCN, rt, (90%)

Under microwave heating, 4-arylthio-3-oxazolin-5-ones **139** were converted into pyrrolo[1,3]diazepines **144**. It is proposed that the reaction proceeds *via* loss of CO_2 to form a nitrile ylide **140**, which can then undergo [3+2] cycloaddition to acetylenic dipolarophiles, followed by a *retro*-Mannich reaction and ring closure by iminium ion attack **143**. While the powerful dipolarophile dimethyl acetylenedicarboxylate provided a pyrrolo[1,3]diazepine in good yield, other acetylenes delivered only poor yields <09TL6810>.

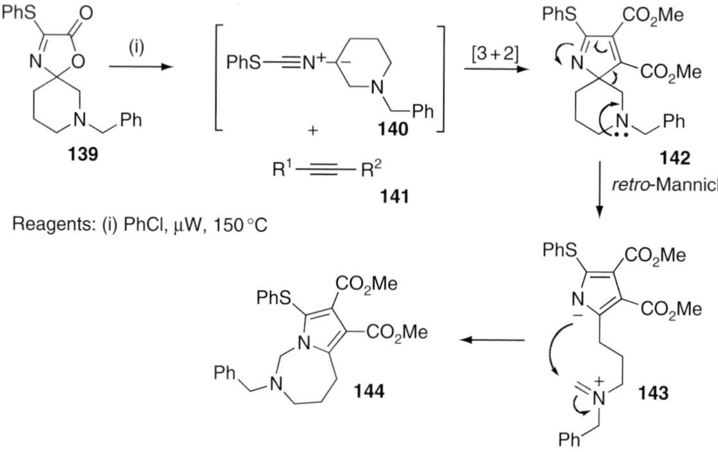

A range of enantiopure tetracyclic triazole-fused benzodiazepines **147** was prepared by coupling aromatic azido-acids **146** with the proline-derived alkyne **145** followed by *in situ* 1,3-dipolar cycloaddition. The compounds prepared were evaluated as enzymatic protease inhibitors and showed weak to good activity <09BML5241>.

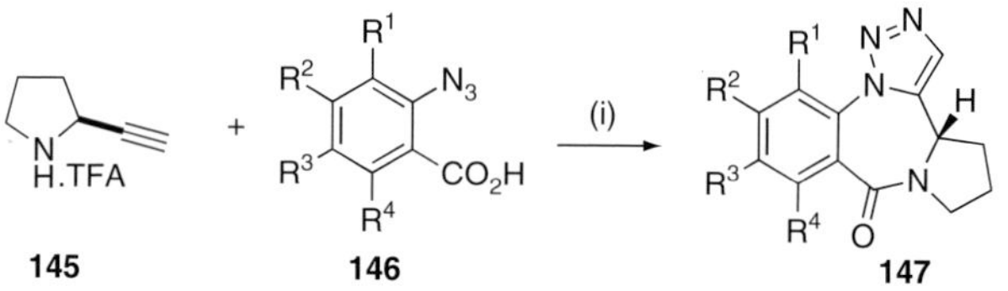

Similarly, a [3+2] cycloaddition of azido-alkynes derived from protected β-amino esters has given *trans*-disubstituted triazolodiazepines **149** <09JOC2004>. Another useful multi-component sequence involving an intramolecular [3+2] cycloaddition led to triazole-fused benzodiazepine **148** <09T6454>. The [3+2] cycloaddition pathway for azides does not always occur and it has been shown that they can form aziridino-fused systems **150** *via* nitrene addition to alkenes <09SL3043>.

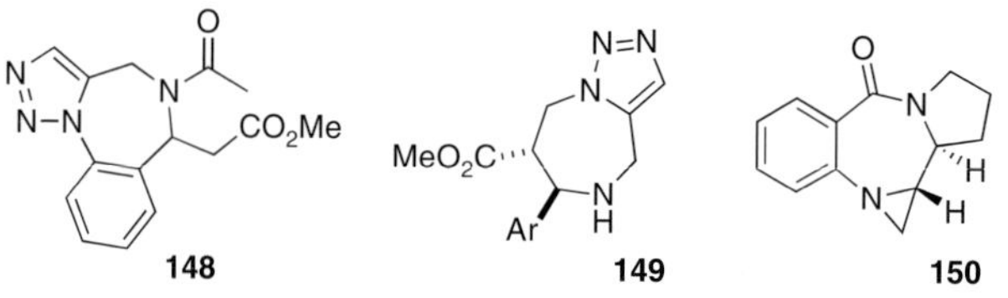

An intermolecular azide 1,3-dipolar cycloaddition has been used to assemble the precursor **151** which reacted with ammonia to give the thieno[2,3-*f*][1,2,3]triazolo [1,5-*a*]-[1,4]diazepine derivative **152** <09H(78)2837>. Thienodiazepinones **155** have also been prepared from amino acids and 3-thiaisatoic anhydride **153** under aqueous conditions <09S389>.

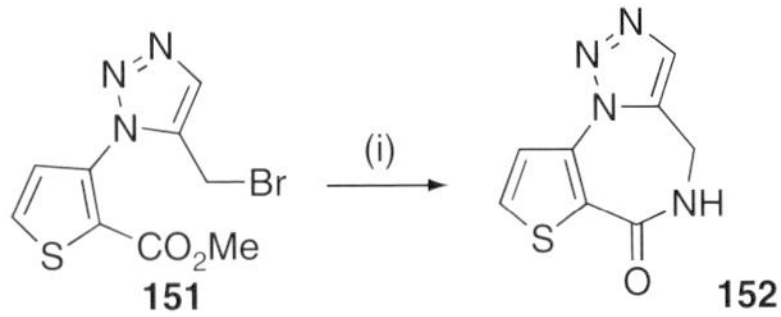

151 152

Reagents: (i) NH_3, EtOH, NH_4OH (67%)

153 154 155

Reagents: (i) H_2O, reflux (35–81%)

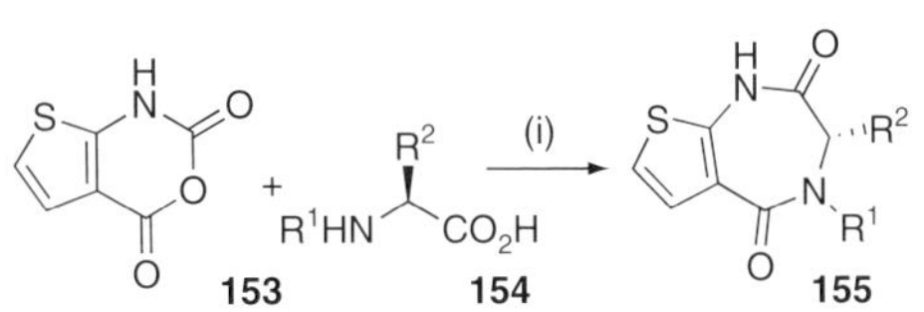

A four-component stereoselective synthesis of dihydro-1,5-benzodiazepines containing a phosphonate **160** or phosphite ylide **161** has been demonstrated. The reaction proceeds *via* benzodiazepine formation from the reaction of *o*-phenylenediamine **156** with diketene **157**. The benzodiazepine intermediate then reacts with trialkyl phosphite–dialkyl acetylenedicarboxylate zwitterions, formed *in situ*, to give **160** in wet dichloromethane and **161** under anhydrous conditions <09T2684>.

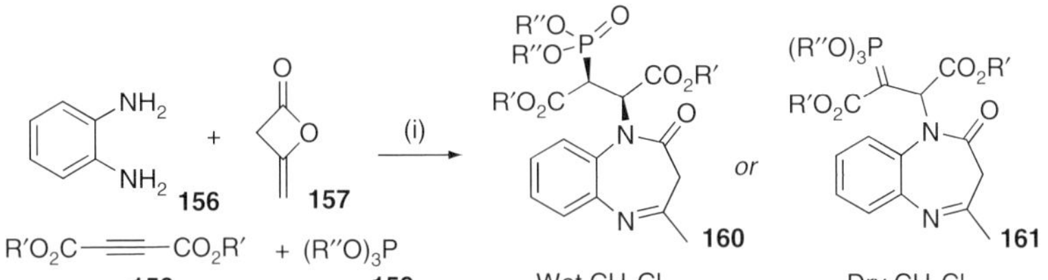

156 157 158 159 160 161

Reagents: (i) CH_2Cl_2

Quinazolinones fused with a benzodiazepinedione **163** feature in several natural products, including the asperlicins C and D. These structures have been prepared by a Sn(OTf)$_2$-catalysed double-cyclisation of the tri-peptide derivative **162** <09CC445>.

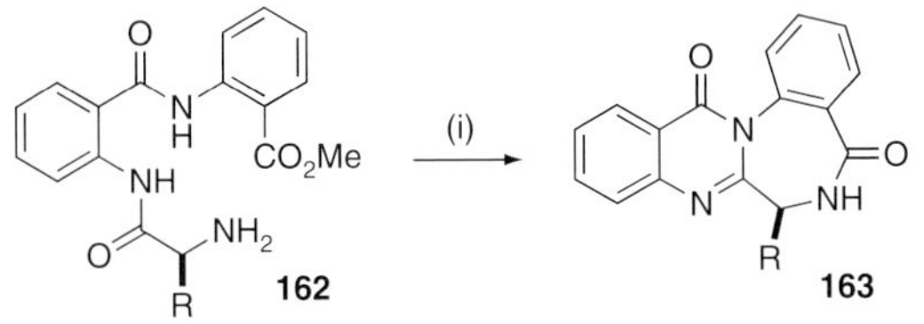

162 163

Reagents: (i) Sn(OTf)$_2$, DMF, 5–15 min, μw, 140 °C (34–85%)

Spirocyclic 1,4-diazepin-2-ones **165** have been formed by cyclisation of a pyrrole ring onto a pyridine ring activated by trifluoromethanesulfonic anhydride <09CC1964>.

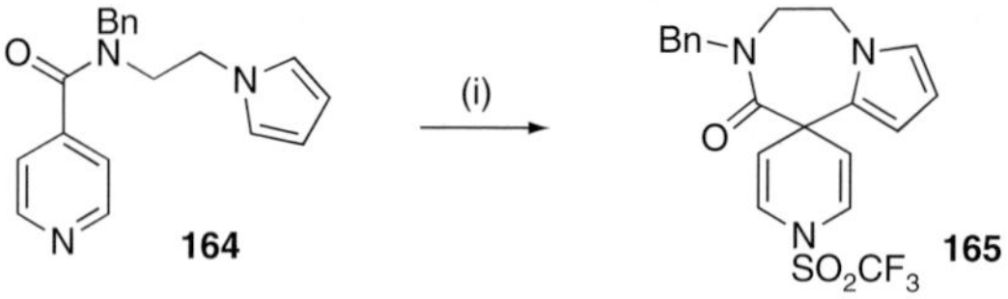

164 → **165**

Reagents: (i) (CF$_3$SO$_2$)$_2$O, 2,6-lutidine, CH$_2$Cl$_2$, 0–20 °C (41%)

Cyclisation of amines with aldehydes or esters continues to be a popular route to 1,4-diazepines and 1,4-diazepinones respectively. Chemoselective azide reduction with aluminium and gadolinium triflates catalysts in conjunction with sodium iodide results in cyclisation *in situ* to provide pyrrolo[2,1-c][1,4]benzodiazepines **167** <09CEJ7215>. On related substrates the same authors demonstrated that a nickel boride catalyst could be used for the synthesis of pyrrolo[2,1-c]-[1,4]benzodiazepine-5,11-diones <09S2163>.

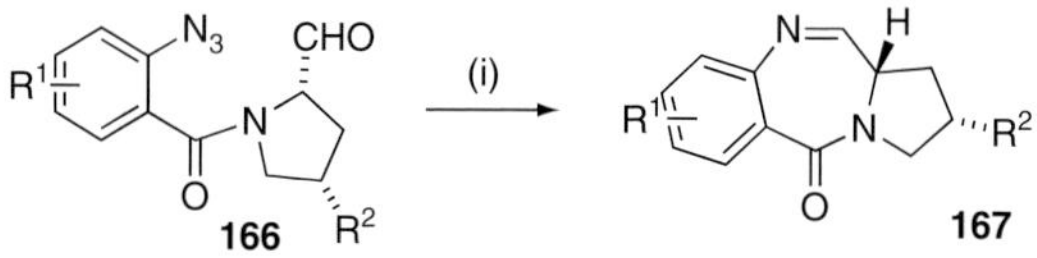

166 → **167**

Reagents: (i) NaI, Al/Gd(OTf)$_3$ (20 mol%), CH$_3$CN, rt

Baylis–Hillman adducts **168** of 2-nitrobenzaldehydes and acrylonitrile have been converted into polycyclic diazepino-quinolines **169** *via* reductive cyclisation <09EJO3454>.

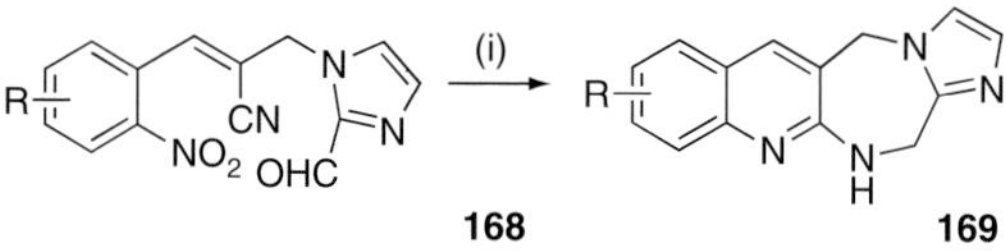

168 → **169**

Reagents: (i) Fe, AcOH, N$_2$, 100 °C, (47–48%)

Enantiopure polyfunctionalised diazepanone scaffolds **172** have been synthesised in an efficient two-step cyclisation from azido *t*-butyl ester **170**, which was prepared from D-isoascorbic acid and L-serine derivatives <09TA2320>.

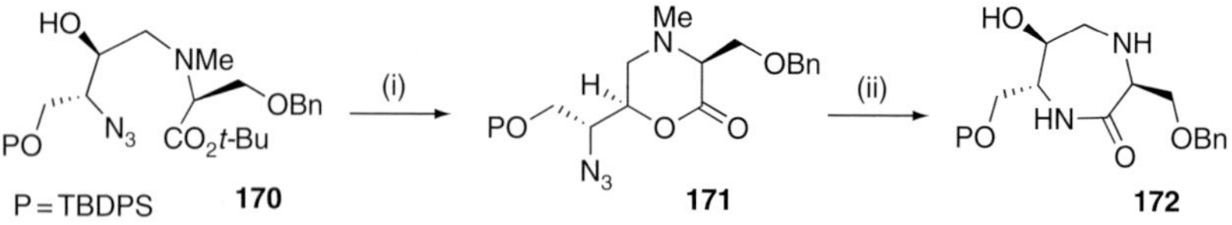

170 → **171** → **172**

P = TBDPS

Reagents: (i) TFA, CH$_2$Cl$_2$ (ii) HCO$_2$NH$_4$, Pd/C (84%)

A convergent approach to indole-derived 1,4-diazepan-2-ones **177** has been reported and proceeds *via* an Ugi reaction and a subsequent Pictet–Spengler reaction. Unfortunately, while convergent and requiring only simple starting materials, the yields for these reactions were generally low <09JOC6895>.

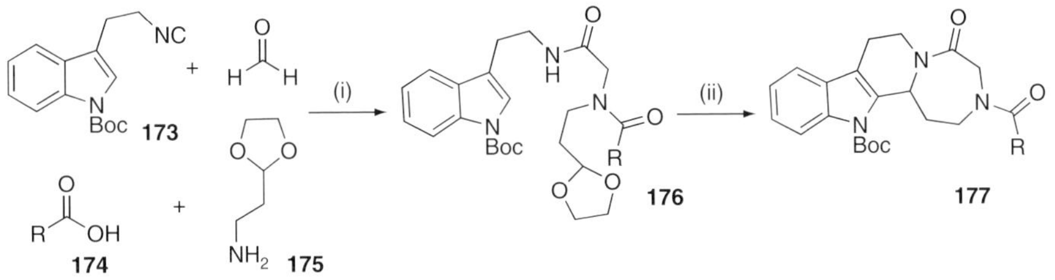

Reagents: (i) MeOH, rt; (ii) HCO$_2$H, rt–60 °C

In another related approach, the Ugi reaction was used to assemble benzodiazepine precursors **182**, which are then cyclized *via* a Staudinger/aza-Wittig reaction to give the benzodiazepine **183** <09JOC2189>.

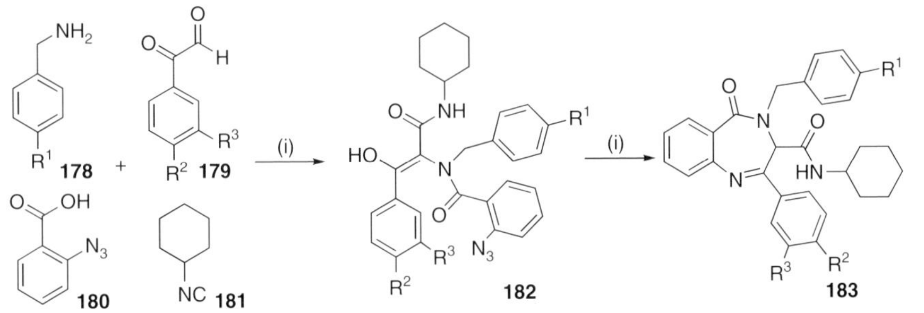

Reagents: (i) MeOH (ii) PPh$_3$, toluene

A multi-component reaction of 1,3-dicarbonyl compounds with aldehydes and 1,2-diamines has been reported as a convergent, one-pot route to 1,4-diazepines **185**. The reaction was extended to a range of aldehydes, diamines and 1,3-dicarbonyl compounds <09OBC1911>.

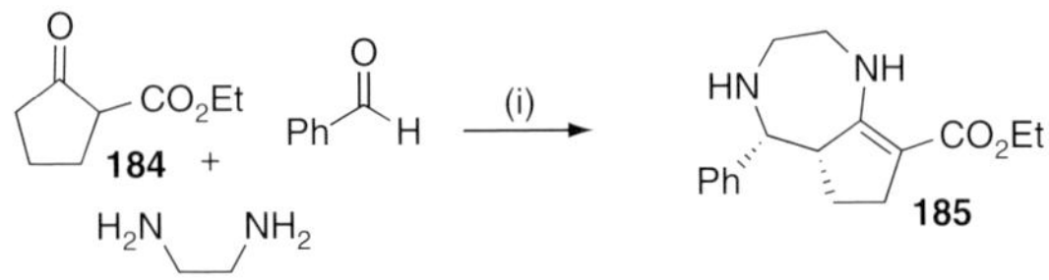

Reagents: (i) 4Å MS, toluene, reflux (21–77%)

Synthesis of benzodiazepines from 1,2-phenylenediamines continues to be a common route. For example, a multi-component reaction of 1,2-phenylenediamines **186**, isonitriles and Meldrum's acid yields benzodiazepines **189** <09OL3342>. A mixture of benzodiazepinothiophenones and benzodiazepines was obtained from the condensation of 2-mercaptopropionic acid, phorone and 1,2-phenylenediamines <09T7741>. Direct synthesis of benzodiazepines from ketones and nitro-aromatics, which were reduced *in situ* with a selective multi-site solid phase catalyst to 1,2-phenylenediamine, has been reported <09CEJ8834>. Difulvene dialdehydes react with 1,2-phenylenediamines in the presence of a CeCl$_3$ catalyst to yield interesting bis-indene-fused benzodiazepines <09T9935>. New catalysts for the condensation of 1,2-phenylenediamines continue to be reported and examples include $MgBr_2$ <09CCL32>, boric acid <09CCL905>, $Ga(OTf)_3$ <09FMC909> and a tetranitrile-silver complex <09LOC17>.

186 187 189

$R^3\text{-}N^+\equiv C^-$ 188

Reagents: (i) CH_2Cl_2, rt (70–90%)

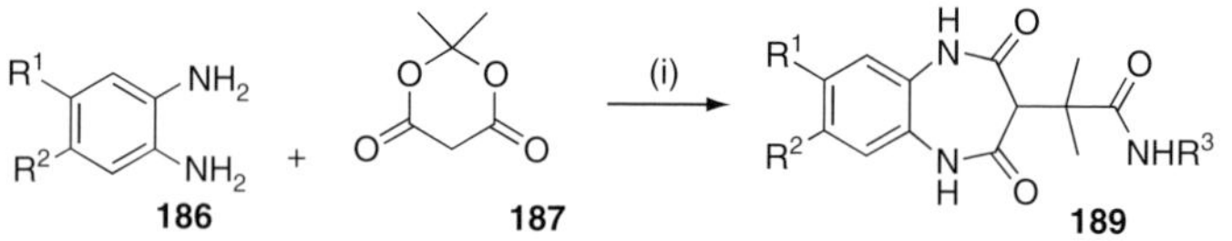

Dynamic kinetic transfer hydrogenation has been carried out on **190** using a chiral phosphoric acid **194** as a catalyst and the Hantzsch ester **191** as the hydride source. The products were obtained as a mixture of diastereoisomers **192** and **193** with up to 8:1 diastereoselectivity and 86% and 94% ee for the major and minor diastereoisomers, respectively <09BML3729>.

190 191 192 193

Reagents: (i) 10 mol% **194**, $CHCl_3$, −10 °C, Na_2SO_4, 5.5 days

194

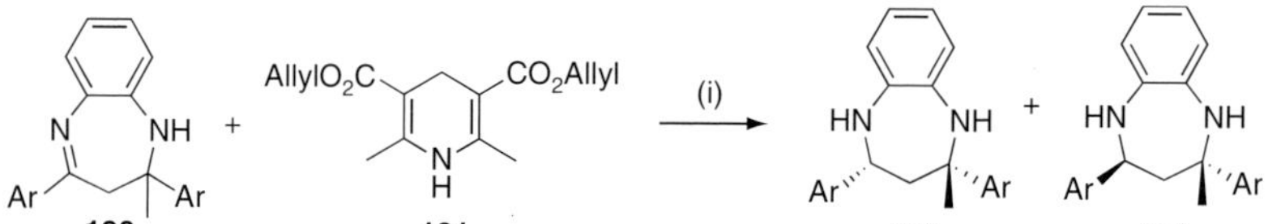

Addition of organometallics to N-(α-haloacyl)-*o*-aminobenzonitrile **195** afforded 2,3-dihydro-1*H*-1,4-benzodiazepin-3-one **196**. When one of the methyl groups on **195** was replaced with a hydrogen, intramolecular hydride transfer to form a 4,5-dihydro-1*H*-1,4-benzodiazepin-3-one was observed and it was found possible to add a second equivalent of the organometallic to the imine of this compound <09OBC1184>.

195 196

Reagents: (i) PhMgBr, THF, reflux (83%)

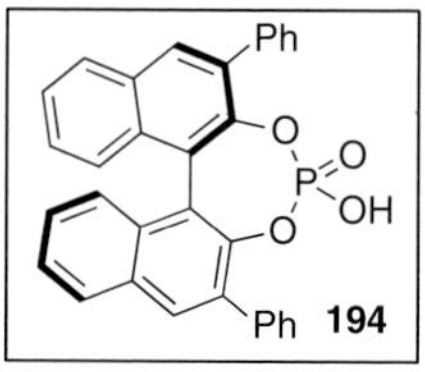

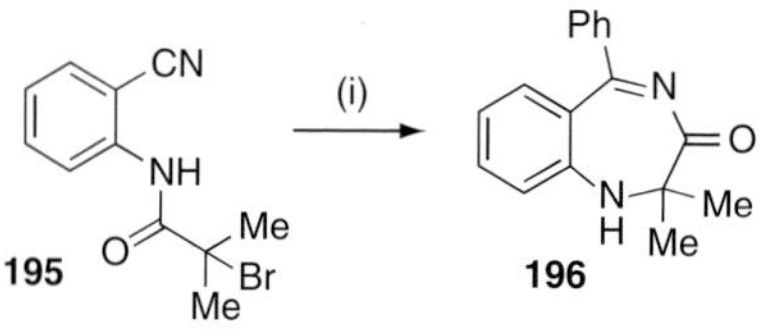

An Ugi reaction has been utilised to prepare precursors **198** which then undergo two tandem ring closing transformations to yield triazadibenzoazulenones **199** <09TL1939>. A sequential benzimidazole benzodiazepine ring forming sequence was proposed.

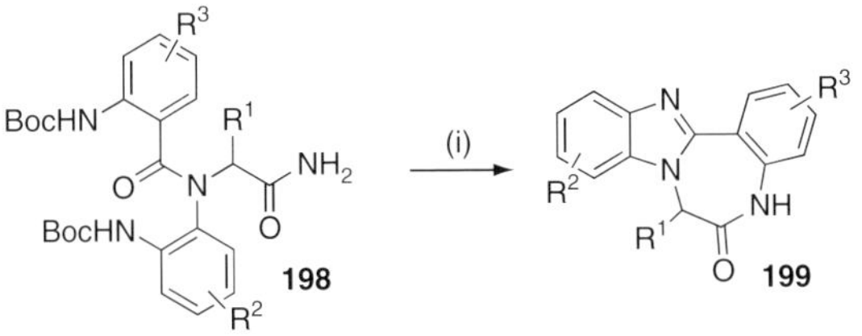

Reagents: (i) μw, TFA, DCE, 130 °C (22–70%)

An isocyanide-based multicomponent reaction has also been applied to the synthesis of 1,4-diazepin-2-amines **201** <09TL2854>.

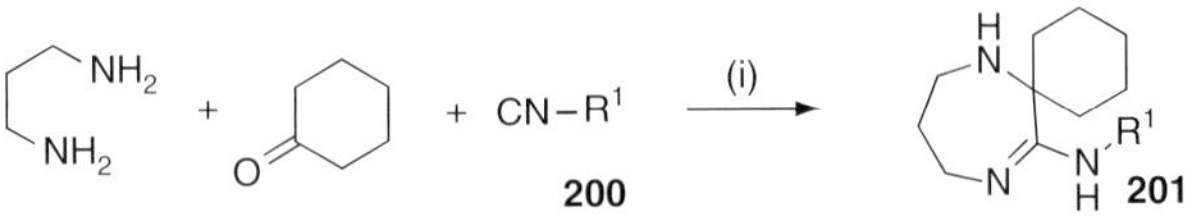

Reagents: (i) MeOH, 45 °C then TMSCl, MeCN (59–91%)

When 1,2-di(pyrrol-1-yl)ethane **202** was treated with oxalyl chloride an unexpected electrophilic substitution–decarbonylation process ensued to yield dipyrrolo[d,g][1,4]diazepine **204** in low yield <09NJC1703>.

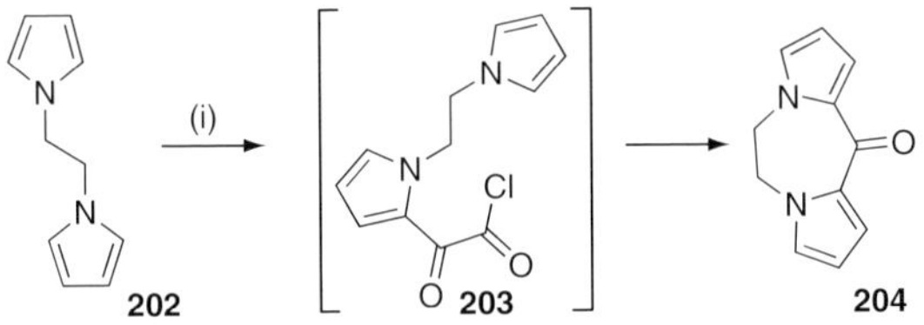

Reagents: (i) Oxalyl chloride, CH$_2$Cl$_2$ (19%)

The versatility of ethenetricarboxylic acid diester **205** was employed in the synthesis of a variety of heterocycles, including a single example of a diazepinone **207** <09OBC655>. The reaction proceeds *via* amide formation and subsequent Michael addition to form the diazepinone ring.

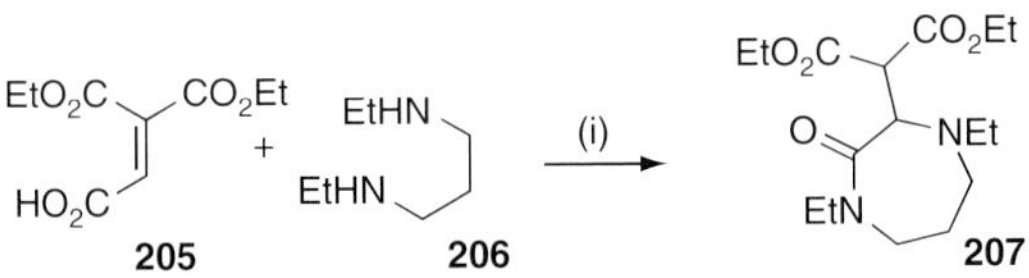

Reagents: (i) EDCI, HOBT, THF (67%)

Palladium-catalysed amination has been shown to be a productive method for the synthesis of diazepine derivatives. Diazepines **209** were synthesised by intramolecular alkene carboamination using *N*-fluorobenzenesulfonamide as an oxidant to produce a Pd(IV) intermediate that is intercepted even by weakly nucleophilic arenes <09JA9488>.

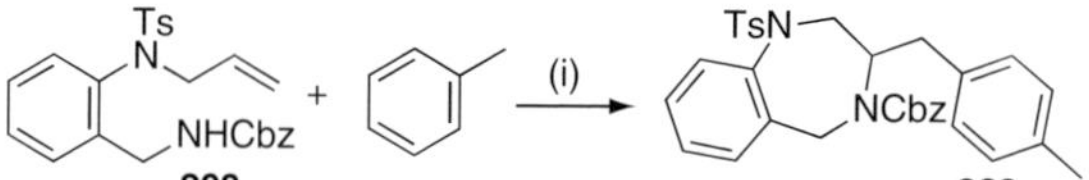

Reagents: (i) Pd(TFA)$_2$, *N*-fluorobenzenesulfonimide, BHT, 3Å MS (53%)

A palladium-catalyzed intramolecular aerobic allylic C–H amidation of alkenes gave diazepine **211**. It was found that the base added had a significant effect on the regioselectivity of this reaction, with NaOBz favoring formation of the seven-membered ring over the five <09OL2707>. Kuok and coworkers described a ligand study on the aryl amidation reaction for the formation of 1-diazepinones and identified the Pd(OAc)$_2$/P(*t*-Bu)$_3$ ligand system as a more accessible alternative to the previously reported MOP ligand system <09T525>.

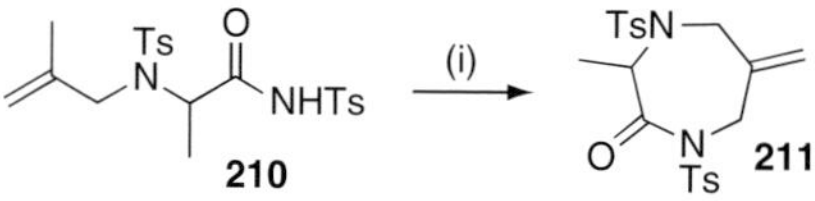

Reagents: (i) Pd(OAc)$_2$, NaOBz, maleic anhydride, DMA, O$_2$ (1atm) (68%)

A fascinating intramolecular Pd(0)-mediated ring-opening of **212** yielded the putative Pd-π-allyl complex **213**, which was then intercepted by the sulfonamide nitrogen to yield benzodiazepine **214** <09OL1575>.

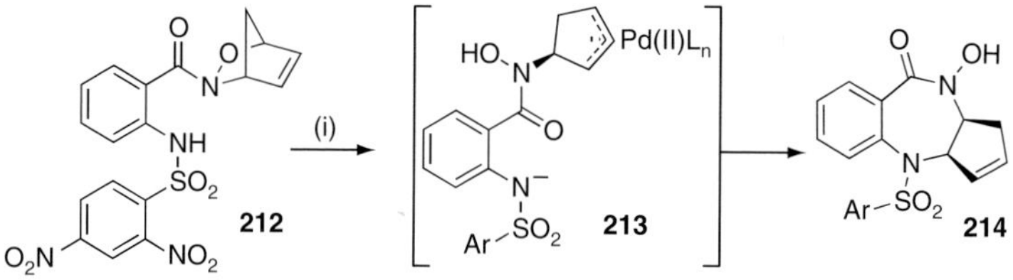

Reagents: (i) PS-PPh$_3$Pd, THF (75%)

Copper-catalysed amination was used for the synthesis of amino acid-derived diazepinones **216** <09JOC6077> as well as well as the azetidine-fused 1,4-benzodiazepin-2-one **218** <09T8965>.

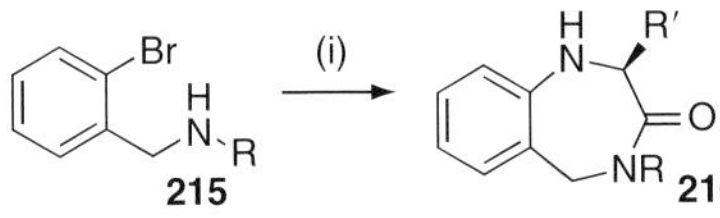

215 (i) → **216**

Reagents:
(i) α-amino acid, CuI, Cs$_2$CO$_3$,
DMF, 90 °C then DPPA 0–5 °C (35–60%)

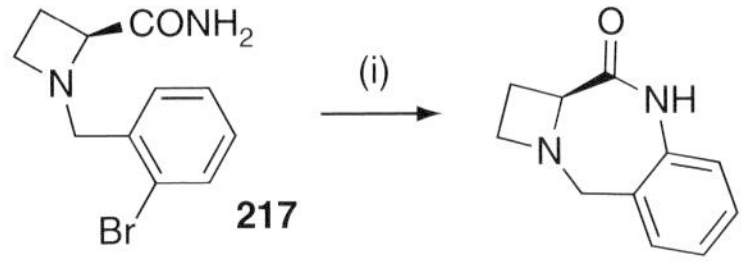

217 (i) → **218**

Reagents:
(i) CuI, *N,N*-dimethylglycine hydrochloride,
Cs$_2$CO$_3$, dioxane, reflux (84%)

The Mitsunobu reaction continues to be a popular ring-closure method in the synthesis of diazepines, with aziridine-derived amino alcohols <09JOC5652> and sulfamidate-derived amino alcohols <09OL5494> being reported as substrates for this reaction. Enaminone **219**, which is readily available from ethyl acetoacetate, was shown to react with chlorocarbonylphenyl ketone **220** to yield diazepinone derivative **221**. Only a single example was reported, but the reaction was efficient and rapid, being complete in two minutes <09JHC96>.

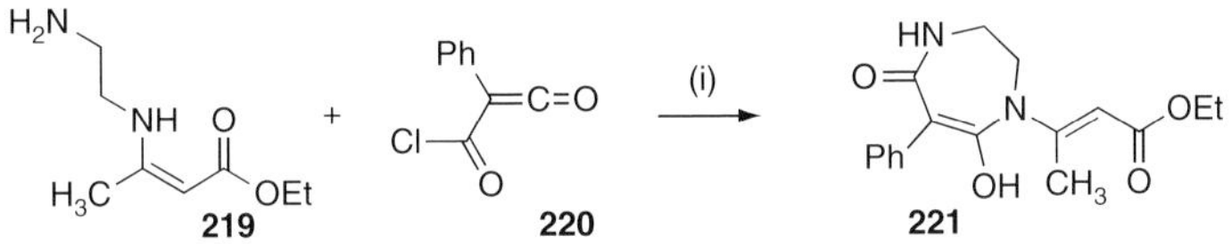

219 + **220** (i) → **221**

Reagents: (i) Toluene, reflux (81%)

7.3.2 Dioxepines, Dithiepines and Fused Derivatives

Gold-catalysed isomerisation of α,β-epoxy ketones that contain a tethered alkyne **222** yields benzodioxepines **223** in moderate yields <09EJO3133>. The first step involves a Lewis acid mediated rearrangement to a 1,3-diketone before addition of the gold catalyst.

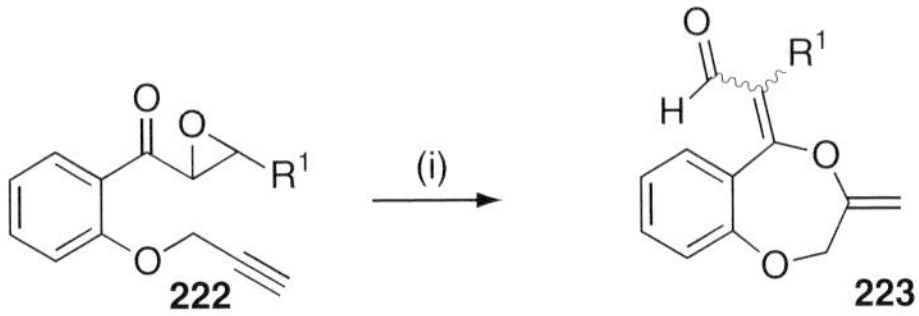

222 (i) → **223**

Reagents: (i) Yb(OTf)$_3$, CH$_3$NO$_2$, 40 °C,
30 min then Au(PPh$_3$)Cl/AgOTf, rt (40–61%)

A mechanistic insight into rhodium-catalysed intramolecular hydroacylation of ketones to yield seven-membered lactones **225** was reported <09JA1077> thus the ketone insertion was suggested as the rate limiting step in these processes. The [2+2] photocycloaddition of alkenyl ether tethered to a cyclopentane **226** occurs rapidly to give the fused dioxepine **227** as a single product <09EJO5953>. The acetal linkage was hydrolysed to yield a bicyclic diol.

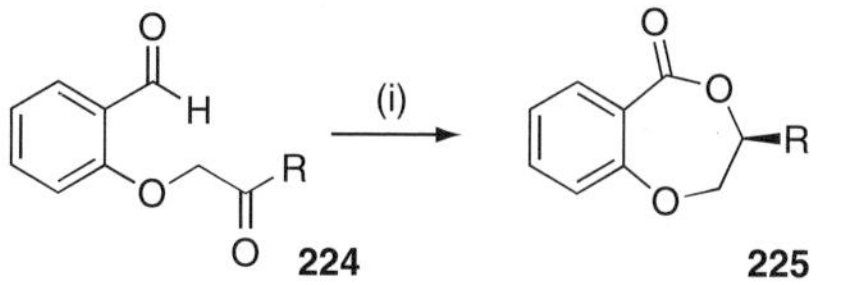

Reagents: (i) [Rh(R)-DTM-SEGPHOS]BF$_4$ (5 mol%), CH$_2$Cl$_2$, rt (up to 99% yield and >99% ee)

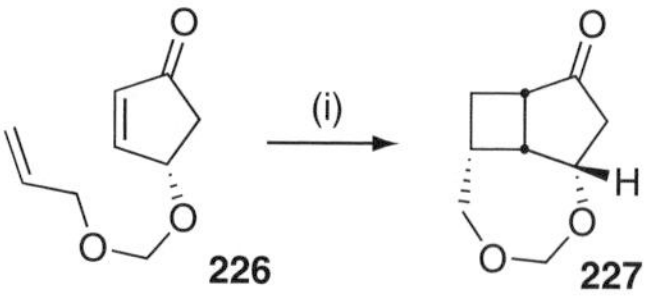

Reagents: (i) hv, acetone, 15 min (60%)

7.3.3 Miscellaneous Derivatives with Two Heteroatoms

Heterocycles with two different heteroatoms continue to be of interest as templates for the synthesis of compound libraries for drug discovery.

Gem-dichloraziridines **228** were reacted with a Lewis acid to generate an imidyl chloride which reacted with a nucleophilic phenoxy group to give a bisbenzoxazepine **229** <09OL979>. These derivatives reacted further to yield the aziridine fused derivative **230** by reduction of the imine which on heating forms a reactive azomethine ylide that undergoes [3+2] cycloaddition chemistry.

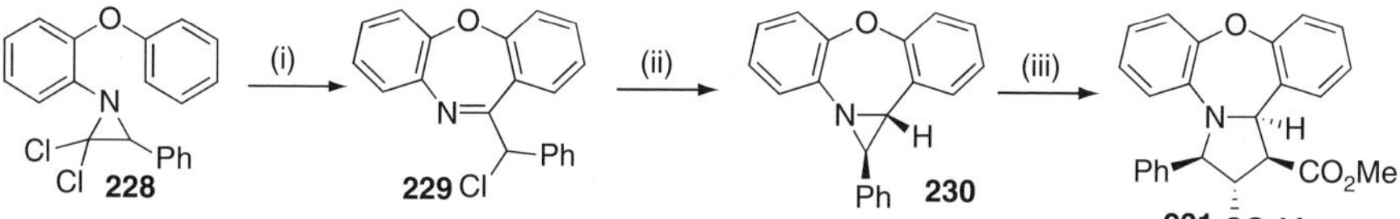

Reagents: (i) AlCl$_3$, MeNO$_2$ (59%); (ii) LiAlH$_4$, Et$_2$O (70%); (iii) Dimethyl fumarate, toluene, 90 °C, 7 h (94%).

Ring opening of epoxides, aziridines and azetidines has been used as a practical method for the synthesis of oxazepines. The N-tethered allenyl epoxide **232** undergoes tandem cyclisation to form the pyran fused oxazepine **233** under gold catalysis <09OL3490>. Copper-mediated ring opening of an N-tosyl azetidine or aziridine with bromo-alcohols and subsequent cyclisation gives N-tosyloxazepines <09JOC7013>. The reaction of fused aziridines **234** with propargyl alcohols was very general and provided numerous examples of fused oxazepines such as **235** <09JOC8814>.

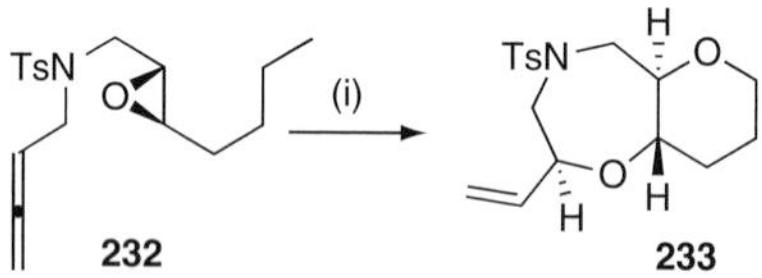

Reagents: (i) (PhO)$_3$PAuCl, AgOTf, CH$_2$Cl$_2$, (59%)

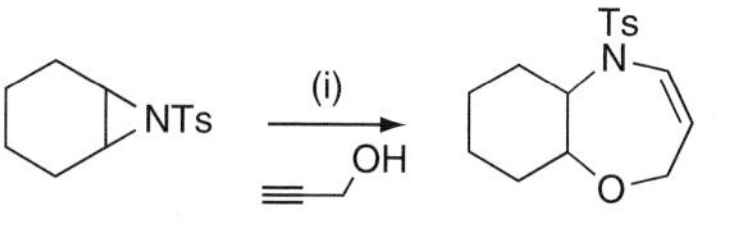

Reagents: (i) [Ag(COD)$_2$]PF$_6$, CH$_2$Cl$_2$, CsCO$_3$ (72%)

A range of fused oxazepines were synthesised from N-allyl-2-allyloxypiperidines by ring closing metathesis (RCM) <09JOC9365>. The yields were generally very high (70-94%) and the method was also applicable to the enyne metathesis. A range of syntheses of oxazepines were reported which were essentially alkylations using Mitsunobu chemistry on a hydroxy-pyrazole <09JME2652>, phenols on chloroacetamides <09ARK185>, the cyclisation of a carboxylate anion onto a tethered acrylonitrile <09TL1423> and the S_NAr substitution of a fluorinated aromatic <09H(78)2209>. The alkylation using bromoethylsulfonium salts **237** with phenolic and alcoholic sulfonamides **238** gave derivatives **236** <09OL257> while the ring opening of cyclic sulfamidates **239** with 2-hydroxymethylphenol, and the thiophenol and aniline equivalents, and subsequent Mitsunobu cyclisation, yielded derivatives such as **240** <09OL5494>.

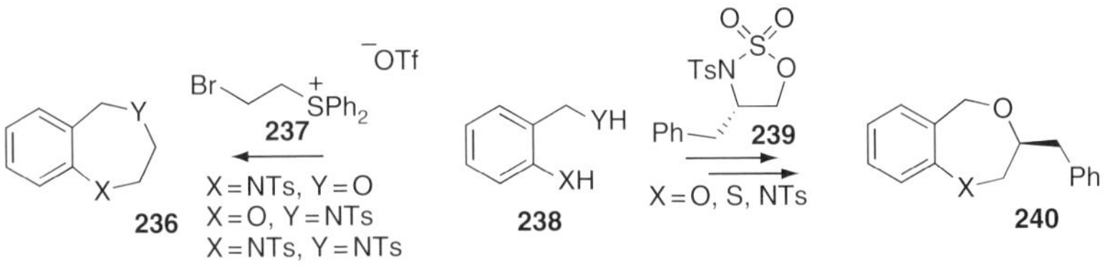

While iodoamination of tethered allyloxy amine **241** usually occurs via 6-*exo-trig* cyclisation hydrozirconation/iodination yielded the 7-*endo* cyclised product **242** <09EJO3726>. The novel cyclopropane-fused oxazepines **244** were formed by alkylation of the alkoxides **243** with an electrophilic cyclopropene that was formed *in situ* by elimination of a bromide <09JA6906>.

Reagents: (i) CpZr(H)Cl, CH_2Cl_2, rt; (ii) I_2, NEt_3 (66%)

Reagents: (i)THF, 18-crown-6, 80 °C (71%)

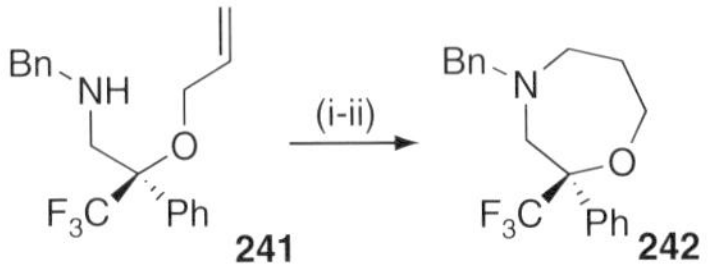

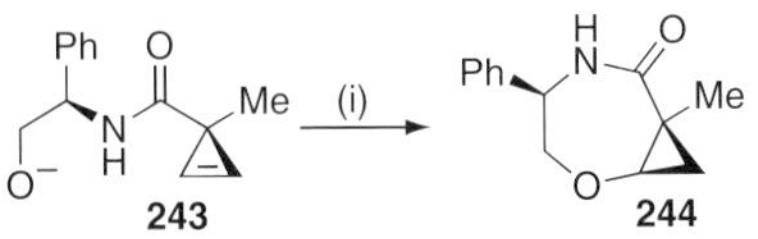

Fused oxazepines **245** were formed by the ring opening of cyclopentane-fused maleimides by imines, such as dihydroisoquinolines, and subsequent cyclisation of the carboxylate anion onto the resultant iminium ion <09OL3802>. The reaction was unique to the cyclopentane fused maleimides as cyclohexyl or monocyclic derivatives gave alternative cyclisation products. A copper- or gold-catalysed tandem cycloisomerisation/formal [4+3] cycloaddition was reported for the synthesis of a furan-fused 1,2-oxazepine **248** from a nitrone **247** and alkynylcyclopropyl ketone **246** <09CEJ8975>; numerous examples were reported, with good yields achieved, except for bulky nitrones.

Reagents: (i) $Cu(OTf)_2$, DCE, 50 °C (83%)

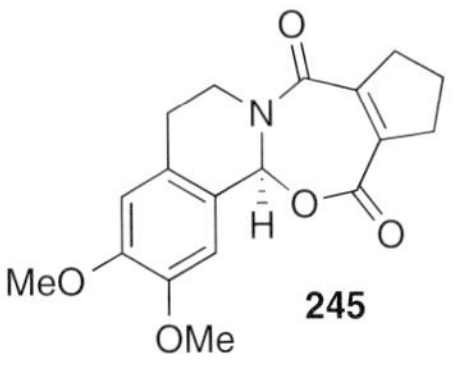

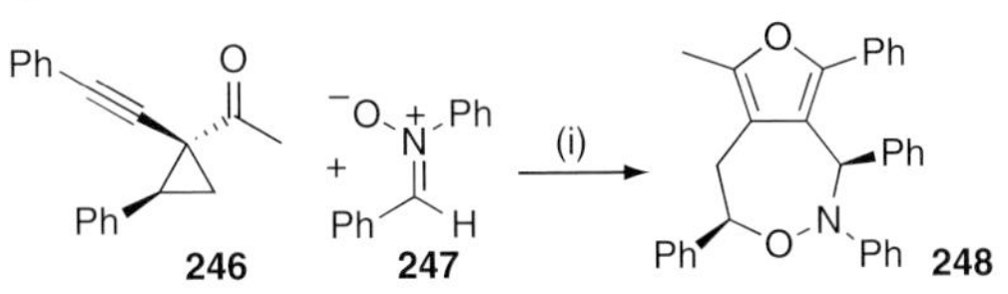

The Ugi four-component coupling was exploited to generate bis-benzoxazepine **252** from S$_N$Ar substitution of the initial Ugi product under the reaction conditions. Intramolecular hydroamination with a palladium catalyst generated a benzoazepine ring **253** yielding complex products in two steps <09SL1162>.

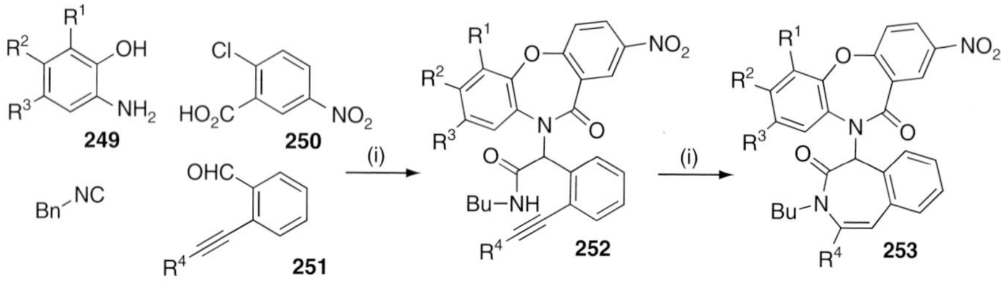

Reagents: (i) MeOH, MW, 80 °C, 20 min (57–81%);
(ii) Pd(PhCN)$_2$Cl$_2$, THF, 60 °C, 24 h (61–74%)

Dioxepinones, oxazepinones and oxathiapinones were formed by the oxidation of tethered diols. Some regioselectivity was observed for the oxidative esterification with oxonium salts with the best results obtained when 2,4,6-collidine was used as the base <09LOC478>.

Benzothiazapines were reported to be synthesised by reaction of 2-aminothiophenol with chalcones <09IJH291>, by the alkylation of triazole-thione with 1,2-bis(chloromethyl)benzene <09IJH91>, by a multi-component coupling of aniline, benzaldehyde and mercaptoacetic acid under microwave conditions <09OBC557>, by alkylation of a sulfonamide carbanion <09EJO2635> and by copper-mediated cyclisation of a thiol onto an aryl iodide <09OL2788>. The ring opening of 2-aminobenzothiazolones **254** with propiolates and triphenylphosphine gave 4-amino-5-oxobenzo[1,4]thiazepine carboxylates **255** in moderate yields. The mechanism proposed involves attack of the phosphorous ylide from triphenylphosphine and propiolate onto the nitrogen of the heterocycle breaking the N–S bond <09T7487>. Bis-benzothiazepines **257** were formed in good yield by the palladium-catalysed cyclisation of the sulfonamides **256** formed from the reaction of 2-halobenzylamines with arylsulfonylchlorides <09SL3127>.

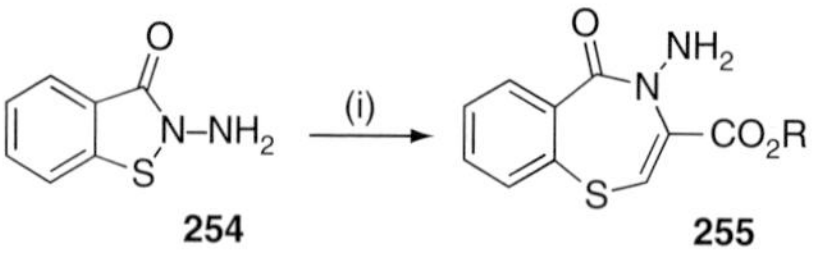

Reagents: (i) PPh$_3$, HC≡CCO$_2$R, toluene, rt (44–46%)

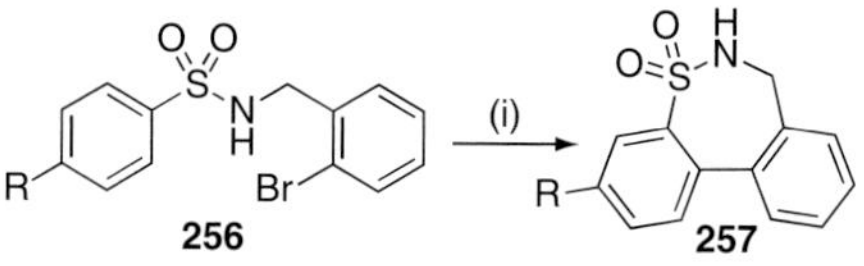

Reagents: (i) PPh$_3$, toluene, rt (44–46%)

The bicyclothioglycolate lactam **258** was exploited for the stereoselective synthesis of the *E*-enolate **259** by reductive enolisation with lithium di-*tert*-butylbiphenylide (LiDBB) <09OL1725>.

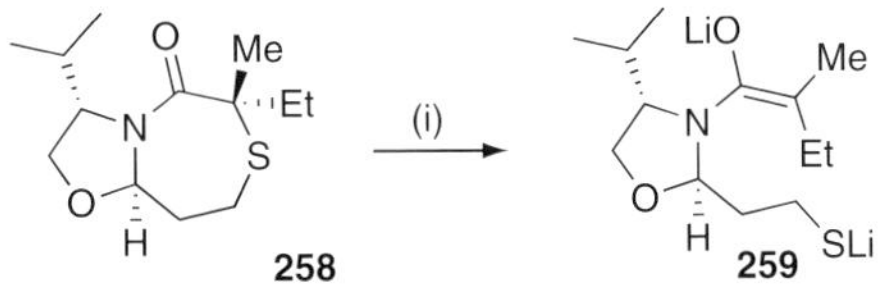

Reagents: (i) LiDBB, THF, −78 °C

Aromatic homolytic substitution occurs when a quinolinyl or pyridinyl radical is generated from sulfinates such as **260** to yield the cyclic sulfinate **261** which contrasts to a phenyl radical, which results in formation of a five-membered ring by homolytic substitution on the sulfur atom <09CEJ10225>.

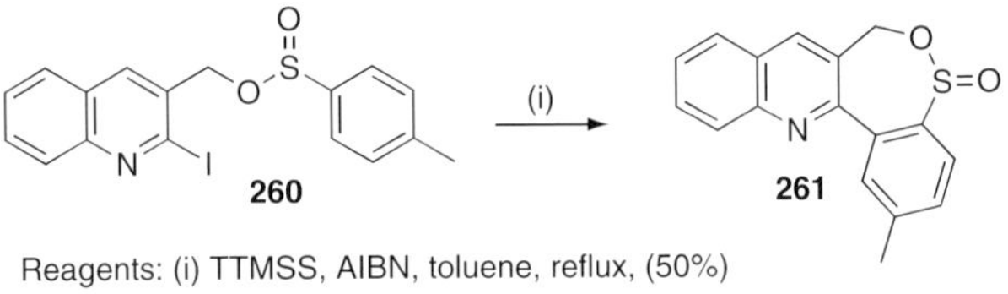

Reagents: (i) TTMSS, AIBN, toluene, reflux, (50%)

7.4. SEVEN-MEMBERED SYSTEMS CONTAINING THREE OR MORE HETEROATOMS

7.4.1 Systems with N, S and/or O

Seven-membered heterocyclic sultams containing a third heteroatom have been formed *via* conjugate addition of an alcohol **264** or *via* peptide methodology **265** <09OL531>. Sulfamate esters tethered to an alkyne **262**, cyclise and form rhodium nitrene intermediates on treatment with a rhodium catalyst. The nitrene can be intercepted with numerous nucleophiles including aromatic systems to give complex heterocyclic systems **263** <09JA2434>.

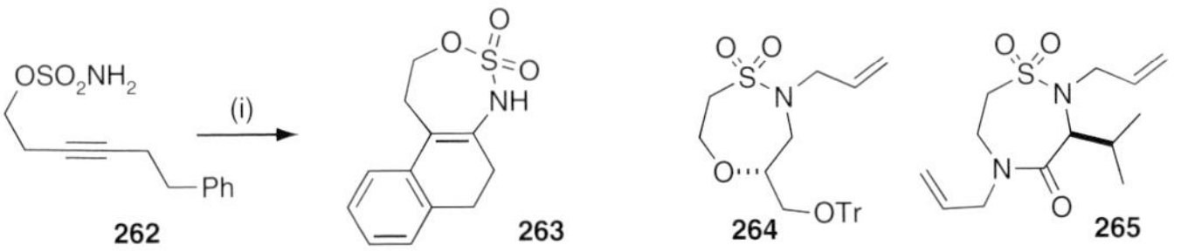

Reagents: (i) $Rh_2(esp)_2$ (2 mol%), $PhI(OAc)_2$, CH_2Cl_2, 40 °C (71%)

Nitrene insertion into an alkene tethered to a sulfonamide **266** yielded the aziridine-fused benzothiadiazepine **267** in 44% yield <09SL3043>. An indole-fused thiadiazepine **268** was synthesised from a 7-aminoindole by formation of a vinyl sulfonamide and intramolecular conjugate addition of the nitrogen of the indole <09BML3669>.

Reagents: (i) MeCN, reflux, 20 h (44%)

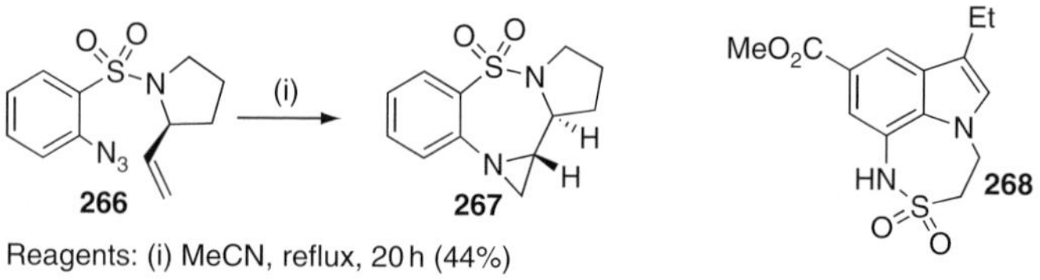

In the burgeoning field of asymmetric catalysis there continue to be multiple reports on the synthesis and use of enantiopure seven-membered heterocycles, both as ligands and as Brønsted acid catalysts. For example, arylaminophosphonium barfates have been reported as a new class of Brønsted acid for the enantioselective activation of nonionic Lewis bases <09JA7242>. A catalytic and enantioselective vinylogous Mukaiyama aldol reaction between 2-siloxypyrrole donors and aromatic/heteroaromatic aldehyde acceptors. This reaction employs an enantiopure bisphosphoramide catalyst in combination with $SiCl_4$ <09TL3428>. A selection of enantiopure monodentate phosphorus triamides based on 1,1′-binaphthyl-2,2′-diamine have been prepared and deployed in the copper-catalysed conjugate addition of diethylzinc to cyclohex-2-enone as well as the nickel-catalysed hydrovinylation of styrene <09EJO6198>.

An enantioselective synthesis of P-stereogenic phosphinates including **270** by means of a molybdenum-catalysed asymmetric ring-closing methathesis has been reported.

Reagents: (i) **271** (10 mol%) CH_2Cl_2, 96% ee (79%)

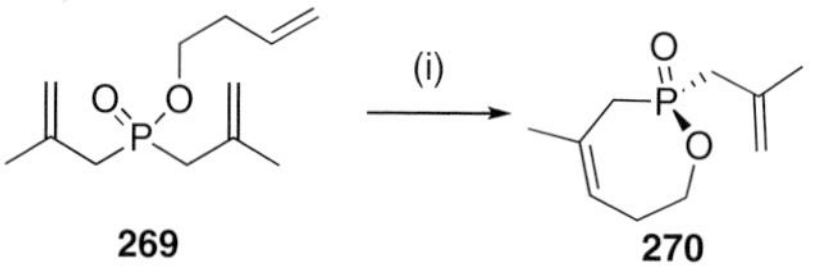

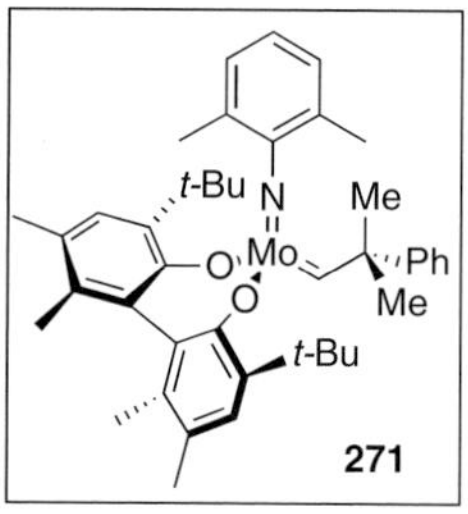

7.5. SEVEN-MEMBERED SYSTEMS OF PHARMACOLOGICAL SIGNIFICANCE

There continues to be very strong interest in pharmacologically active compounds incorporating seven-membered heterocyclic components and several reviews have been published. Seven-membered heterocycles feature prominently in therapeutic approaches for Alzheimer's disease <09AGE3030> while the discovery of batrachotoxin, an extremely toxic, oxazepane-containing steroidal alkaloid, found in the skins of the dart poison frogs of the genus *Phyllobates* from Columbia, was reviewed <09H(79)195>. The preparation of hymenialdisine and analogues as kinase inhibitors <09CME3122> and the imidazolobenzodiazepine conivaptan **272**, the first vasopressin receptor antagonist approved by the FDA for treatment of heart failure

<09EOP2161>; the pyrrolobenzodiazepine lixivaptan **273**, a vasopressin receptor antagonist with high V2 receptor affinity, is in phase III clinical trials for treatment of a range of diseases associated with hyponatremia <09EOD657>. Doxepin **274**, a dibenzooxepine derivative is under investigation for treatment of insomnia <09EOP1649>. Telcagepant **275**, an azepin-2-one derivative, is a CGRP receptor antagonist in phase III clinical trials as a new oral treatment for migraine <09EOP1523>. Diazepinomicin **276**, a farnesylated dibenzodiazepinone isolated from a marine *Micromonospora* sp., is in phase II clinical trials for the treatment of glioblastoma multiforme <09DOF349>.

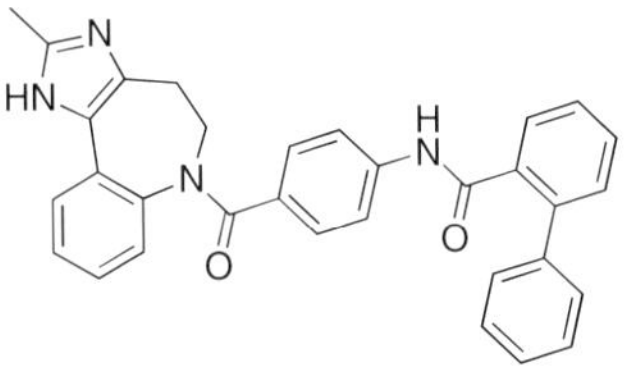

272 Conivaptan

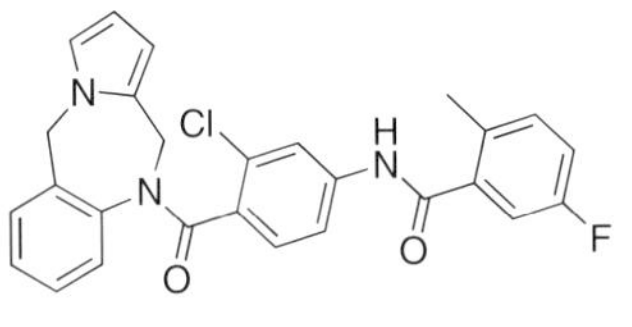

273 Lixivaptan

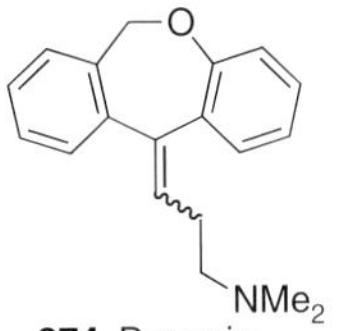

274 Doxepin

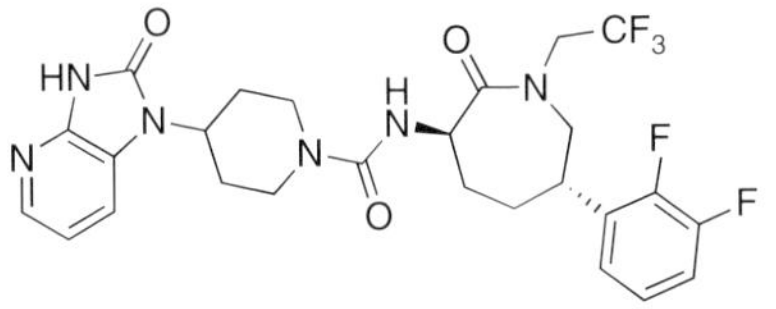

275 Telcagepant (MK-0974)

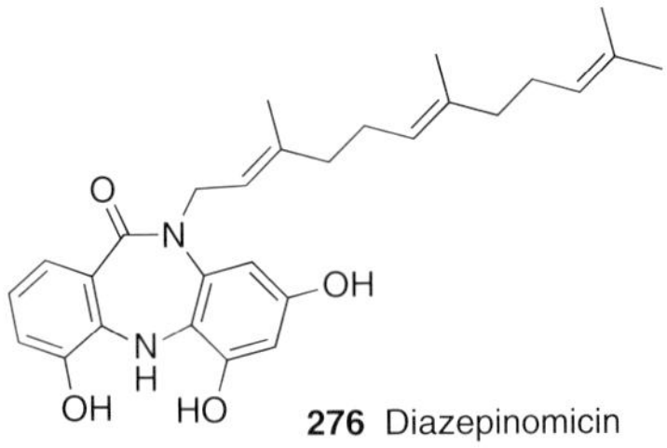

276 Diazepinomicin

7.5.1 Seven Membered Heterocycles in Human Health, Animal Health and Crop Protection

In 2009, a number of seven-membered heterocyclic derivatives were approved by the FDA (http://www.fda.gov). The dibenzooxepinopyrrole, asenapine **277**, a dopamine and serotonin receptor antagonist, was approved for the acute treatment of schizophrenia and manic or mixed episodes associated with bipolar I disorder in adults <09NRD843>. Besifloxacin **278**, an azepine-substituted fourth generation fluoroquinolone antibiotic was approved for treatment of ocular infections and has broad spectrum *in vitro* activity against aerobic and anaerobic bacteria <09AAC3552>. Tolvaptan **279**, a selective vasopressin V2 receptor antagonist, was approved for the treatment of clinically significant hypervolemic and euvolemic hyponatremia <09NRD611>. Eslicarbazepine acetate **280**, a prodrug of eslicarbazepine was approved by the FDA for filing as a new antiepileptic drug <09EOD221>. The azepane-containing azelastine **281** was approved as a first and only once daily nasal antihistamine. Quetiapine **282** was approved as adjunctive treatment to antidepressants in adults with major depressive disorder. Olanzapine **283** is an option for the treatment of schizophrenia and manic or mixed episodes associated with bipolar I disorder in adolescents and in combination with fluoxetine as a first-line medication for treatment-resistant depression. In animal health, the (−)-(6R,7R)-enantiomer of zilpaterol **284** accounts for essentially all the β₂-adrenergic activity in the racemic form which is administered as a production enhancer in cattle <09JME1773>. In crop protection, the discovery of the 1,4,5-oxadiazepane pinoxaden **285** as a novel cereal herbicide was documented <09BMC4241>.

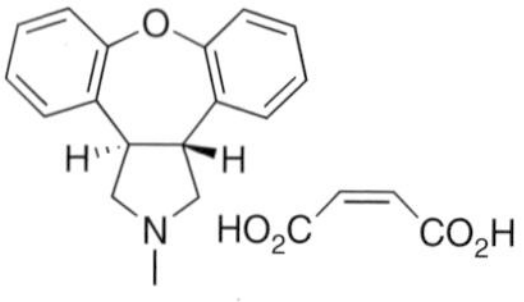

277 Asenapine
278 Besifloxacin 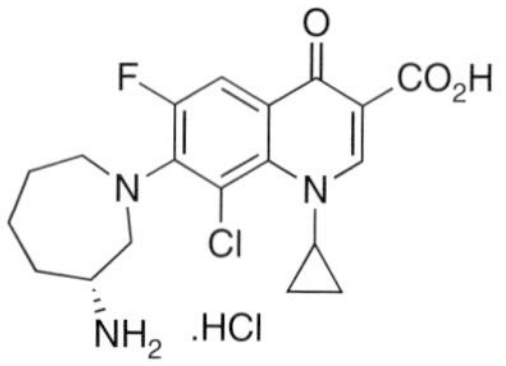
279 Tolvaptan 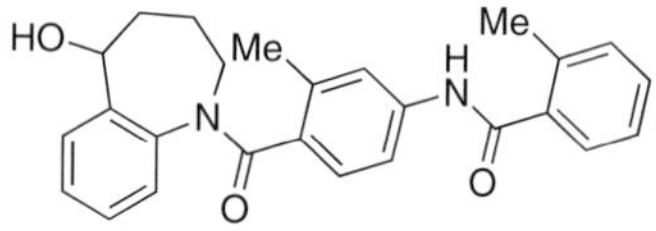
280 Eslicarbazepine acetate 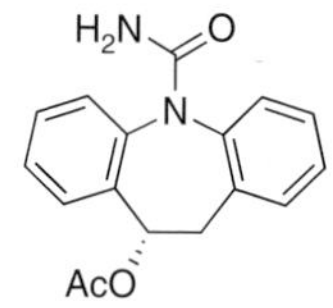
281 Azelastine 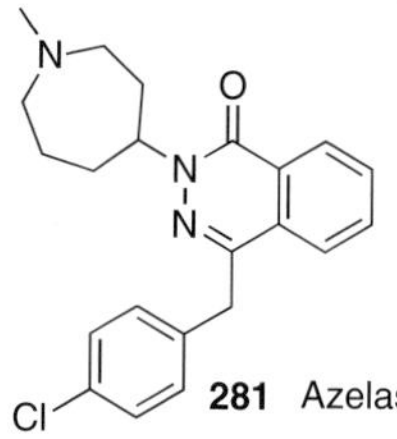
282 Quetiapine 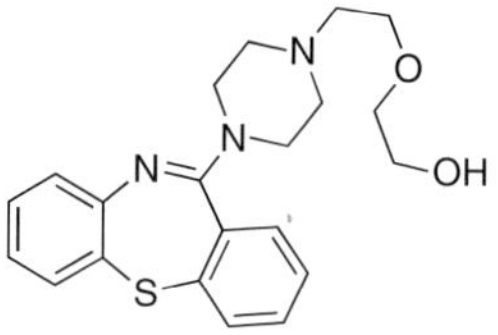
283 Olanzapine 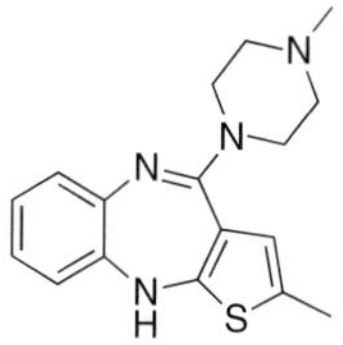
284 (−)-zilpaterol 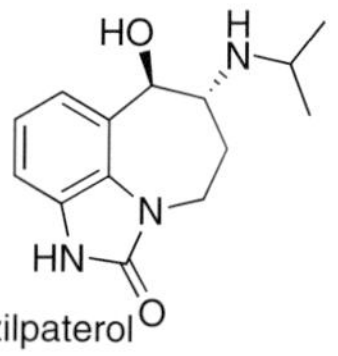
285 Pinoxaden 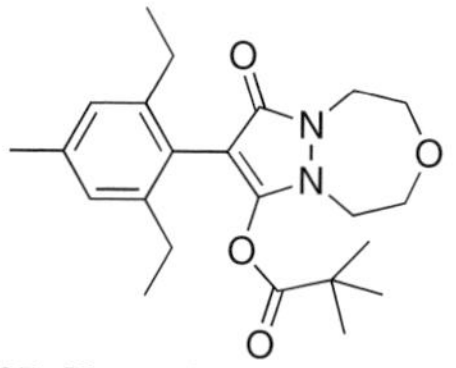

7.5.1.1 CNS – Neurodegeneration and Cognition

The benzoazepinone peptidomimetic semagacestat (LY450139) **286**, a functional γ-secretase inhibitor, is in phase III clinical trials for the treatment of Alzheimer's disease <09EOP1657>. New BACE-1 (= β-sectretase = β-site amyloid precursor protein cleaving enzyme) inhibitors include [1,2,5]thiadiazepinoindole 2,2-dioxide and [1,2] thiazepinoindole 2,2-dioxide derivatives <09BML3669; 09BML3674> and dibenzo-diazepinone derivatives <09JME6484>. A range of galanthamine derivatives was studied as acetylcholinesterase inhibitors <09EJM772>. Imidazotriazolobenzodiazepines were revealed as potent and highly selective GABA_A (γ-aminobutyric acid A) α5 inverse agonists <09BML5746> with potential for the treatment of cognitive dysfunction displaying activity in *in vivo* models for cognitive improvement <09BML5958>. Certain examples of this class (e.g. **287**) have been selected as clinical candidates for further development <09BML5940>. Benzotriazolo-diazepines were discovered with selective efficacy against GABA_A receptor sub-types <09JME1795>.

Dopamine D1 receptor antagonists are of interest as potential therapeutics for Parkinson's disease, psychotic behaviour, substance abuse and obesity. Benzo[d][1,3]diazepines were discovered as novel D1 receptor antagonists <09BML5218>. Linking a tetrahydro-1H-benzo[d]azepine D1 receptor agonist with a 5-HT$_{1A}$ receptor pharmacophore using click chemistry produced D2 receptor activity <09BMC4873>. Tetrahydro-1H-benzo[c]azepin-1-one derivatives were discovered with high affinities and selectivities for D3 receptors <09BML1773>. Diether derivatives of homopiperidine were designed as non-imidazole histamine H3 receptor ligands with potent nm affinities of interest for sleep disorders, obesity and cognition disorders <09BMC3037>. 1-Hydroxy-1-phenyl-tetrahydro-1H-3-benzoazepines showed moderate norepinephrine potentiating activity of interest for development of new treatments of ADHD <09CPB443>. Hexahydro-1H-dibenzo[b,e][1,4]diazepin-1-ones were reported to have similar antipsychotic effects as clozapine <09BJC1445>. Angiotensin IV plays a role in cognition, memory, seizures as well as vascular and renal function. Replacement of His(4) in angiotensin IV by the conformationally constrained residue 4-amino-tetrahydroindolo[2,3-c]azepin-3-one provided highly potent and selective analogues <09JME5612>.

In search of new treatments of insomnia, N,N-disubstituted-1,4-diazepanes were discovered as dual (OX$_1$R/OX$_2$R) orexin receptor antagonists for example **288** was found to be a potent and brain-penetrating compound which was shown to induce REM and non-REM sleep in rats <09CMH1069>. Synthesis of a macrocyclic analogue **289**, suggests that the bioactive conformation has an intramolecular π-stacking interaction and a twist-boat ring conformation <BML2997>. Hexahydro-1H-dibenzo[b,e][1,4]diazepin-1-one derivatives were shown to be potent, low toxicity inhibitors of a Neuromedin B receptor, which is of interest for a range of central and peripheral nervous system processes as well as cellular proliferation <09BML4264>. Enantiopure disubstituted tetrahydro-1H-benzo[d]azepines showed high affinity towards σ1 and σ2 receptors and 2-substituted tetrahydro-3-benzoazepines showed high selectivity for σ1 receptors <09TA1383; 09JOC2788>. 1,4-Diazepanes derived from (S)-serine were found to have improved σ1 receptor affinity with selectivity over σ2 receptors <09EJM519>. The incorporation of conformationally constrained phenylalanine analogs 4-amino-tetrahydro-indolo[2,3-c]azepin-3-one and 4-amino-tetrahydro-2-benzoazepin-3-one scaffolds led to the discovery of novel potent μ-selective agonists as well as potent and selective δ-opioid receptor antagonists <09BML433>. Substituted amino-tetrahydro-2-benzoazepin-3-ones were explored for β-turn mimicry and resulted in potent opioid analogs <09T2266>.

286 Semagacestat (LY450139)

287

288

289

7.5.1.2 Pain – Migraine

Spiropiperidine-substituted azepinones, e.g. **290** are calcitonin gene-related peptide (CGRP) receptor antagonists with similar potency to telcagepant **275** (in phase III clinical trials as an antimigraine agent) and with reduced potential for metabolism <09BML6368>. Synthetic ciguatoxins, containing numerous oxepine units, selectively activate $Na_V1.8$-derived sodium channels expressed on HEK293 cells, of interest for understanding the role of these ion channels in pain signalling <09JBC7597>. Dibenzoimidazodiazepinone derivatives were discovered as non-peptidergic Mas-related G-protein coupled (Mrg)X1 and MrgX2 agonists of interest in nociception (pain reception) <09BML1729>.

7.5.1.3 Inflammatory Diseases – Arthritis, Chronic Inflammation, Neuropathic Pain

Tri- and tetrahydroxyazepines were shown to be potent and selective β-N-acetylhexosaminidase inhibitors of interest in development of treatments for osteoarthritis <09BMC5598>. In the search for new inflammatory agents, 2-azepinyl-5-arylpyridines were discovered as selective cannabinoid CB_2 agonists <09BML6578>. Tetrahydropyrazolo-diazepinones were reported as human purinergic subtype $P2X_7$ receptor antagonists <09BML6053>, tetrahydronaphthoazepinoindolediones as inhibitors of pro-inflammatory cytokines <09BML5753>, 3-acylaminoazepin-2-ones, e.g. **291**, were revealed as highly potent, orally available, anti-inflammatory inhibitors of chemokine-induced chemotaxis that result in the reduction of TNF-α <09JME3591> and dibenzo[b,e]oxepinones were discovered as potent p38 mitogen-activated protein (MAP) kinase inhibitors <09JME1778>. Dihydrobenzothienodiazepinones and 3-(pyridin-3-yl)-tetrahydrodiazepinothienoquinolin-8-ones were discovered as potent MAP-activated protein kinase 2 (MAPK2) inhibitors with selectivity over cyclin-dependent kinase 2 (CDK2) of interest for potential therapies for arthritis <09BML4878; 09BML4882> while diazepinyl-substituted piperazines were discovered as potent proviral insertion site in Moloney murine leukemia virus (PIM) kinase inhibitors of interest in inflammation and immunology <09JME1814>.

7.5.1.4 Cardiovascular and Metabolics: Coagulation, High Cholesterol, Cardioprotection

3-Cyanoguanidinyl and 3-(aroylguanidinyl)-azepin-2-ones were discovered as novel orally bioavailable factor Xa inhibitors, with one found to be a selective orally efficacious activated blood coagulation factor X (FXa) inhibitors in animal models <09BML4034; 09BML6882>. 2,3-Dihydrobenzothiazepinones were designed as factor VIIa/tissue factor inhibitors <09BML1386>. The azepino[4,5-b]indole WAY-362450 **292** was identified as a potent and selective farnesoid X receptor FXR agonist that advanced to clinical trials. While it showed significant reduction of cholesterol and triglycerides in knockout mice it displayed poor aqueous solubility <09JME904>. Analogous pyrrolo[2,3-d]azepino-compounds were prepared with improved solubility but with some loss of FXR agonist activity <09BML5289>. For the development of cardioprotective agents, dihydrobenzo[b][1,4]oxathiepin-3-amines, e.g. **293** were discovered as selective and potent inhibitors of the late current mediated by the cardiac isoform of the sodium channel (Na$_V$1.5) <09JME4149>.

290, **291**, **292**, **293**

7.5.1.5 Metabolic Disorders – Diabetes, Obesity, Hypertension

Azepine sulfonamides have been discovered as potent 11β-hydroxysteroid dehydrogenase type 1 (11β-HSD1) inhibitors of interest for various metabolic disorders such as diabetes, obesity and hypertension <09BML4563>. Dihydropyrazolo[1,4]oxazepinone **294** (PF-514273) is a cannabinoid-1 receptor antagonist that has been advanced to human clinical trials for treatment of obesity <09JME2652>. Dibenzo[1,4]thiazepines are selective cannabinoid-1 receptor (CB$_1$) inverse agonists, of interest for type II diabetes, obesity, drug addiction and smoking cessation, leading to potential preclinical candidate **295**, which promoted weight loss in rodents <09JME1975>. 4,4-Difluoro-1,2,3,4-tetrahydro-5*H*-1-benzoazepine-5-ylidene derivatives were reported as potent and selective agonists, both *in vitro* and *in vivo* for the arginine vasopressin V$_2$ receptor <09BMC3130> <09BMC8161>. 1,2,5-Triazepine derivatives were reported as dipeptidyl peptidase IV (DPP-IV) inhibitors of interest for development of drugs for type II diabetes mellitus <09JFC1001>. Dihydrobenzo[*e*][1,4]diazepinones were designed as β-turn peptidomimetics with agonist activity at the melanocortin receptor <09BCH90>.

7.5.1.6 Urology, Gastrointestinal, Immunosuppression, Wound Healing

Selective 5-HT$_{2C}$ agonists are of interest for new treatments of obesity, schizophrenia, sexual dysfunction and urinary incontinence. Triazolopyrimidoazepines were reported as potent and selective 5-HT$_{2C}$ agonists metabolically stable *in vitro* and efficacious in an *in vivo* model of stress urinary incontinence <09BML4999>. A series of tetrahydrobenzo[*d*]azepine-sulfonamides are potent and selective 5-HT$_{2C}$ agonists <09BML1871>. Tetrahydrobenzo[*d*]azepinesulfonamides and sulfones are potent agonists of the motilin receptor of interest for development of gastroprokinetic agents for a range of gastrointestinal disorders involving delayed gastric emptying <09BML6452>. Dibenzazepinones have been discovered as K$_V$1.3 and IK-1 ion channel blockers and display inhibitory effects immunosuppressant activity within an animal model <09BML2299>. N-Alkyl-2-benzazepine derivatives promote *in vitro* wound healing with human skin epithelial cells <09BML3193>.

7.5.1.7 Women's Health

Oxepine-containing pseudo-steroids are potent and selective progesterone receptor antagonists of similar potency to mifepristone (RU-486) <09BML3977>. In the search for new remedies for postmenopausal symptoms, benzopyranobenzoxapanes were discovered as selective estrogen receptor modulators (SERMS) <09JME7544>. The azepine-containing bazedoxifene **296** is a selective estrogen receptor modulator (SERM) that has undergone phase II clinical trials and is being developed for the prevention and treatment of post-menopausal osteoporosis <09EOP1377>.

7.5.1.8 Oncology – Cancer Cell Proliferation

Structural classes that show antitumour activity include benzo[*d*]azepinediones that are active against colon cancer cell lines <09BML1534>, dibenzo[*b,e*]azepines that inhibit tumour cell proliferation in the G0-G1 phase transition <09BML104>, azepinyl-substituted pyrrolopyrimidinone oxides that inhibit the growth of human solid tumour cell lines, arresting the cell cycle in the G2/M phase <09BMC4955>, 2-azepinone analogues of androstanolone that show anti-tumour activity against Ehrlich ascites carcinoma <09EJM3936> and azepinopyrroloisoqunolinones with activity against U-251 central nervous system, PC-3 prostate and K-562 leukemia cell lines <09BMC1849>. Benzopyridooxathiazepine derivatives inhibited tubulin polymerisation with activity against a range of tumour cell lines <09BMC1132>, 6-oxaallocolchicinoids (containing a dihydrodibenzo-oxepine ring system) induced apoptosis in Burkitt-like Lymphoma (BJAB) tumor cell lines <09OL341>, tetrahydrobenzocyclopenta[1,4]diazepinones showed activity against MCF-7 and PC-3 tumour cell lines <09OL1575> and platinum(II) complexes of azepinyl-substituted thiosemicarbazones showed antitumour activity on leukemia P388-bearing animal models <09EJM1296>.

7.5.1.9 Oncology – Pyrrolobenzodiazepines

The pyrrolo[2,1-*c*][1,4]benzodiazepine (PDB) dimer SJG-136 **297** was about to enter phase II clinical trials as a sequence selective DNA-interactive agent and was found to form sequence-dependent intrastrand DNA cross-links as well as mono-alkylated adducts <09JA13756>. A (PBD) dimer prodrug was developed with improved water solubility and reduced DNA reaction rate <09BML6463>. The covalent reaction between PBDs and double stranded or hairpin DNA is greatly accelerated by microwave irradiation <09CC2875>. Fluorination of PDBs leads to enhanced DNA-binding ability and anticancer activity across a range of cancer cell lines <09BMC1557>.

7.5.1.10 Oncology – Kinase Inhibitors, DNA Regulation
Tetrahydrothienopyrroloazepinone analogues of hymenialdisine were reported to be potent inhibitors of the serine/threonine checkpoint kinase CHK1, towards agents that sensitise tumours towards DNA damaging agents <09BML841>. Tetrahydropyridoazepinoindoles showed selective inhibition of anaplastic lymphoma kinase (ALK) <09BMC3308>. Azapinyl-substituted pyrroloisoquinolinones and pyrazoloquinazolinones are potent poly(ADP-ribase) polymerase-1 (PARP-1) inhibitors, PARP-1 has an important role in regulation of DNA integrity, and is an attractive target for cancer therapy <09BML4042; 09BML4196>.

7.5.1.11 Oncology – Peptidomimetics, New Targets, Prodrugs
The signal transducers and activators of transcription 3 (STAT3) oncogene is a promising target for the design of a new class of anticancer drugs and peptidomimetics containing 3-aminoazepin-2-one and tetrahydroazepinoindolone scaffolds are lead compounds for novel STAT3 inhibitors <09BML1733; 09JME2429>. Somatostatin(sst) mimetics containing a 4-aminoindolo[2,3-*c*]azepin-3-one scaffold, e.g., **298**, were reported as highly potent sst receptor activities and with a range of sst subtype selectivities. A sst5-selective analogue with subnanomolar binding affinity is the most potent agonist reported to date. Selective sst ligands are of interest for a range of diseases involving cell proliferation <09JME95>. Potent second mitochondria-derived activator of caspases (Smac) mimetics/X-inhibitor of apoptosis proteins (XIAPs) inhibitors based on the 6-amino-5-oxooctahydropyrroloazepine-3-carboxylic acid scaffold were designed as proapoptotic agents of interest for cancer therapy <09BMC5834>. A 4-diazepinyl-substituted quinazoline was reported to be a potent and selective inhibitor of histone lysine methyltransferase G9a, for studying chromatin remodelling and its role in various diseases such as cancer <09JME7950>. Hexahydrobenzothienopyrimidoazepinone derivatives are potent and selective 17β-hydroxysteroid dehydrogenase type 1 (17β-HSD1) inhibitors, for the development of new treatments of breast tumours <09JME6660>. Diazepinyl analogues of JS-K, a nitric oxide prodrug anti-cancer lead compound, displayed similar antiproliferative activity against leukemia cells <09BML2760>.

7.5.1.12 Antiviral – Hepatitis C, HIV, Glucosidase and Serine Protease Inhibitors
For Hepatitus C, hexahydrodibenzodiazepinones were reported as potent and selective NS5B RNA polymerase inhibitors <09BML2492> and further optimisation resulted in 6-hydroxy-hexahydrobenzothiopyranodiazepine 1,1-dioxide **299** which had 20-fold greater affinity than the carbonyl analog <09JME4099>. Hexahydroindolopyrrolobenzoazepines displayed good activity in the cell-based NS5B RNA replicon assay in the presence of serum proteins <09BML633>. For HIV,

benzopyrrolooxazepine and pyridopyrrolooxazepine derivatives were potent broad spectrum HIV-1 non-nucleoside reverse transcriptase inhibitors <09JME1224>. Derivatives of the dipyridodiazepinone nevirapine were shown to inhibit HIV-1 reverse transcriptase in wild-type and mutant type enzymes <09BJO36>. Polyhydroxylated 3- and 5-acetamidoazepanes were used to study the molecular basis for inhibition of GH84 glycosidase hydrolases <09JA5390> and dihydro-benzopyrrolotriazolodiazepinones were moderate inhibitors of serine protease <09BML5241>.

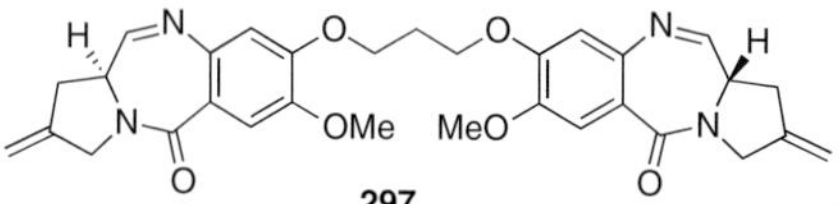

7.5.1.13 Antibiotics, Antitubercular and Antiamoebia

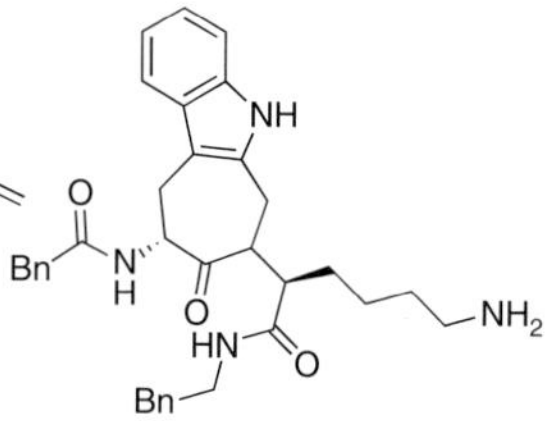

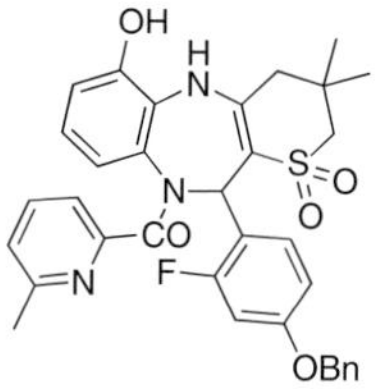

Homomorpholine (1,4-oxazepane) oxazolidinone linezolid analogues were studied as antibacterial agents and showed similar levels of antibacterial activity as linezolid <09BML550>. Tetrahydropyridodiazepines are inhibitors of the bacterial enoyl ACP reductase, FabI (*S. aureus* and *E. Coli*) and efficacious in a mouse infection model <09BML5359>. Dihydrobenzothiazepines were discovered with significant activity against *C. albicans*, *S. epidermis* and *S. aureus* <EJM2815>. Pyrido[2,3-e][1,4]diazepine derivatives were discovered as potent and selective inhibitors of *Heliobactor pylori* glutamate racemase (MurI) of interest due to the realisation that Heliobactor infection can cause chronic peptic ulceration and increase the risk of gastric carcinoma <09BML930>. N,N-Difunctionalised diazepanes (homopiperazine) analogues possessed *in vitro* activity against drug sensitive and drug resistance *M. tuberculosis* <09BML6074>. Tetrahydronaphthoazepine derivatives were found to be active against live forms of *Trypanosoma cruzi* with low mammalian toxicity <09BML2360>. Dibenzooxathiepines were discovered through virtual screening and inhibited trypanothione reductase and showed activity against *T. cruzi* and *T. brucei* <09JME1670>. N-Azepinylthiocarbamoylpyrazolines show antiamoebic activities with better inhibition of *E. histolytica* compared with the control metronidazole <09EJM417>.

7.6. FUTURE DIRECTIONS

Seven membered heterocycles continue to be of interest due to the vast and potent biological activity of these substrates and no doubt this field of research will continue to expand in coming years.

REFERENCES

09AAC3552	W. Haas, C.M. Pillar, G.E. Zurenko, J.C. Lee, L.S. Brunner and T.W. Morris, *Antimicrob. Agents Chemother.*, **53**, 3552 (2009).
09AGE762	J.S. Harvey, S.J. Malcolmson, K.S. Dunne, S.J. Meel, A.L. Thompson, R.R. Schrock, A.H. Hoveyda and A. Gouverneur, *Angew. Chem. Int. Ed.*, **48**, 762 (2009).
09AGE1838	T. Kano, Y. Yamaguchi and K. Maruoka, *Angew. Chem. Int. Ed.*, **48**, 1838 (2009).
09AGE2514	K.L. Cubbage, A.J. Orr-Ewing and K.I. Booker-Milburn, *Angew. Chem. Int. Ed.*, **48**, 2514 (2009).
09AGE2941	A. Hamajima and M. Isobe, *Angew. Chem. Int. Ed.*, **48**, 2941 (2009).
09AGE3030	R. Jakob-Roetne and H. Jasobsen, *Angew. Chem. Int. Ed.*, **48**, 3030 (2009).
09AGE4363	P. Garcia-Garcia, F. Lay, P. Garcia-Garcia, C. Rabalakos and B. List, *Angew. Chem. Int. Ed.*, **48**, 4363 (2009).
09AGE8716	C. Roscini, K.L. Cubbage, M. Berry, A.J. Orr-Ewing and K.I. Booker-Milburn, *Angew. Chem. Int. Ed.*, **48**, 8716 (2009).
09AGE9690	Y. Li, K.J. Jardine, R. Tan, D. Song and V.M. Dong, *Angew. Chem. Int. Ed.*, **48**, 9690 (2009).
09AJC389	P. Pittayakhajonwut, V. Sri-indrasutdhi, A. Dramae, S. Lapanun, R. Suvannakad and M. Tantichareon, *Aust. J. Chem.*, **62**, 389 (2009).
09APR191	I. Musthapa, L.D. Juliawaty, Y.M. Syah, E.H. Hakim, J. Latip and E.L. Ghisalberti, *Arch. Pharm. Rev.*, **32**, 191 (2009).
09ARK185	S. Bilgic, O. Bilgic, M.O. Bilgic, M. Gunduz and N. Karakoc, *Arkivoc*, **xiii**, 185 (2009).
09BCH90	J.Y. Lee, I. Im, T.R. Webb, D. McGrath, M.-R. Song and Y.C. Kim, *Bioorg. Chem.*, **37**, 90 (2009).
09BJO36	N. Khunnawutmanotham, N. Chimnoi1, A. Thitithanyanont, P. Saparpakorn, K. Choowongkomon, P. Pungpo, S. Hannongbua and S. Techasakul, *Beil. J. Org. Chem.*, **5**, 36 (2009).
09BKC1441	B.-K. Kwon and D.Y. Kim, *Bull. Korean Chem. Soc.*, **30**, 1441 (2009).
09BKC1445	O.I. El-Sabbagh and S.M. El-Nabtity, *Bull. Korean Chem. Soc.*, **30**, 1445 (2009).
09BMC1132	S. Gallet, N. Flouquet, P. Carato, B. Pfeiffer, P. Renard, S. Léonce, A. Pierré, P. Berthelot and N. Lebegue, *Bioorg. Med. Chem.*, **17**, 1132 (2009).
09BMC1557	A. Kamal, D.R. Rajender, M.K. Reddy, G. Reddy, T.B. Balakishan, M. Shaik, G.N. Chourasia and Sastry, *Bioorg. Med. Chem.*, **17**, 1557 (2009).
09BMC1849	R. Martinez, M.M. Arzate and M.T. Ramirez-Apan, *Bioorg. Med. Chem.*, **17**, 1849 (2009).
09BMC3037	D. Łazewska, K. Kuder, X. Ligneau, J.-C. Camelin, W. Schunack, H. Stark and K. Kieć-Kononowicz, *Bioorg. Med. Chem.*, **17**, 3037 (2009).
09BMC3130	I. Tsukamoto, H. Koshio, T. Kuramochi, C. Saitoh, H. Yanai-Inamura, C. Kitada-Nozawa, E. Yamamoto, T. Yatsu, Y. Shimada, S. Sakamoto and S.-i. Tsukamoto, *Bioorg. Med. Chem.*, **17**, 3130 (2009).
09BMC3308	P.J. Slavish, Q. Jiang, X. Cui, S.W. Morris and T.R. Webb, *Bioorg. Med. Chem.*, **17**, 3308 (2009).
09BMC4241	M. Muehlebach, M. Boeger, F. Cederbaum, D. Cornes, A.A. Friedmann, J. Glock, T. Niderman, A. Stoller and T. Wagner, *Bioorg. Med. Chem.*, **17**, 4241 (2009).
09BMC4873	J. Zhang, H. Zhang, W. Cai, L. Yu, X. Zhen and A. Zhang, *Bioorg. Med. Chem.*, **17**, 4873 (2009).
09BMC4955	E. Pudziuvelyte, C. Ríos-Luci, L.G. León, I. Cikotiene and J.M. Padrón, *Bioorg. Med. Chem.*, **17**, 4955 (2009).
09BMC5598	H. Li, F. Marcelo, C. Bello, P. Vogel, T.D. Butters, A.P. Rauter, Y. Zhang, M. Sollogoub and Y. Blériot, *Bioorg. Med. Chem.*, **17**, 5598 (2009).
09BMC5834	P. Seneci, A. Bianchi, C. Battaglia, L. Belvisi, M. Bolognesi, A. Caprini, F. Cossu, E. de Franco, M. de Matteo, D. Delia, C. Drago, A. Khaled, D. Lecis, L. Manzoni, M. Marizzoni, E. Mastrangelo, M. Milani, I. Motto, E. Moroni, D. Potenza, V. Rizzo, F. Servida, E. Turlizzi, M. Varrone, F. Vasile and C. Scolastico, *Bioorg. Med. Chem.*, **17**, 5834 (2009).

09BMC8161	I. Tsukamoto, H. Koshio, M. Orita, C. Saitoh, H. Yanai-Inamura, C. Kitada-Nozawa, E. Yamamoto, T. Yatsu, S. Sakamoto and S.-i. Tsukamoto, *Bioorg. Med. Chem.*, **17**, 8161 (2009).
09BML104	R. Al-Qawasmeh, Y. Lee, M.-Y. Cao, X. Gu, S. Viau, J. Lightfoot, J.A. Wright and A.H. Young, *Bioorg. Med. Chem. Lett.*, **19**, 104 (2009).
09BML264	S.R. Chirapu, B. Pachaiyappan, H.F. Nural, X. Cheng, H. Yuan, D.C. Lankin, S.O. Abdul-Hay, G.R.J. Thatcher, Y. Shen, A.P. Kozikowski and P.A. Petukhov, *Bioorg. Med. Chem. Lett.*, **19**, 264 (2009).
09BML433	S. Ballet, D. Feytens, R. De Wachter, M. De Vlaeminck, E.D. Marczak, S. Salvadori, C. de Graaf, D. Rognan, L. Negri, R. Lattanzi, L.H. Lazarus, D. Tourwé and G. Balboni, *Bioorg. Med. Chem. Lett.*, **19**, 433 (2009).
09BML550	J.-Y. Kim, F.E. Boyer, A.L. Choy, M.D. Huband, P.J. Pagano and J.V.N.V. Prasad, *Bioorg. Med. Chem. Lett.*, **19**, 550 (2009).
09BML633	J. Habermann, E. Capitò, M.d.R.R. Ferreira, U. Koch and F. Narjes, *Bioorg. Med. Chem. Lett.*, **19**, 633 (2009).
09BML841	J.-G. Parmentier, B. Portevin, R.M. Golsteyn, A. Pierré, J. Hickman, P. Gloanec and G. De Nanteuil, *Bioorg. Med. Chem. Lett.*, **19**, 841 (2009).
09BML930	B. Geng, G. Basarab, J. Comita-Prevoir, M. Gowravaram, P. Hill, A. Kiely, J. Loch, L. MacPherson, M. Morningstar, G. Mullen, E. Osimboni, A. Satz, C. Eyermann and T. Lundqvist, *Bioorg. Med. Chem. Lett.*, **19**, 930 (2009).
09BML1386	E. Ayral, P. Gloanec, G. Bergé, G. de Nanteuil, P. Mennecier, A. Rupin, T.J. Verbeuren, P. Fulcrand, J. Martinez and J.-F. Hernandez, *Bioorg. Med. Chem. Lett.*, **19**, 1386 (2009).
09BML1534	S.M. Sondhi, R. Rani, P. Roy, S.K. Agrawal and A.K. Saxena, *Bioorg. Med. Chem. Lett.*, **19**, 1534 (2009).
09BML1729	L. Malik, N.M. Kelly, J.-N. Mab, E.A. Currier, E.S. Burstein and R. Olsson, *Bioorg. Med. Chem. Lett.*, **19**, 1729 (2009).
09BML1733	C. Gomez, L. Bai, J. Zhang, Z. Nikolovska-Coleska, J. Chen, H. Yi and S. Wang, *Bioorg. Med. Chem. Lett.*, **19**, 1733 (2009).
09BML1773	R. Ortega, E. Raviña, C.F. Masaguer, F. Areias, J. Brea, M.I. Loza, L. López, J. Selent, M. Pastor and F. Sanz, *Bioorg. Med. Chem. Lett.*, **19**, 1773 (2009).
09BML1871	P.V. Fish, A.D. Brown, E. Evrard and L.R. Roberts, *Bioorg. Med. Chem. Lett.*, **19**, 1871 (2009).
09BML2299	S. Pegoraro, M. Lang, T. Dreker, J. Kraus, S. Hamma, C. Meere, J. Feurle, S. Tasler, S. Prütting, Z. Kuras, V. Visan and S. Grissmer, *Bioorg. Med. Chem. Lett.*, **19**, 2299 (2009).
09BML2360	A. Palma, A.F. Yepes, S.M. Leal, C.A. Coronado and P. Escobar, *Bioorg. Med. Chem. Lett.*, **19**, 2360 (2009).
09BML2444	S.-H. Lee, T.M.V. Hue, S.H. Yang, K.-T. Lee, Y. Kwon and W.-J. Cho, *Bioorg. Med. Chem. Lett.*, **19**, 2444 (2009).
09BML2492	D. McGowan, O. Nyanguile, M.D. Cummings, S. Vendeville, K. Vandyck, W. Van den Broeck, C.W. Boutton, H. De Bondt, L. Quirynen, K. Amssoms, J.-F. Bonfanti, S. Last, K. Rombauts, A. Tahri, L. Hu, F. Delouvroy, K. Vermeiren, G. Vandercruyssen, L. Van der Helm, E. Cleiren, W. Mostmans, P. Lory, G. Pille, K. Van Emelen, G. Fanning, F. Pauwels, T.-I. Lin, K. Simmen and P. Raboisson, *Bioorg. Med. Chem. Lett.*, **19**, 2492 (2009).
09BML2760	R.S. Nandurdikar, A.E. Maciag, M.L. Citro, P.J. Shami, L.K. Keefer, J.E. Saavedra and H. Chakrapani, *Bioorg. Med. Chem. Lett.*, **19**, 2760 (2009).
09BML2997	C.D. Cox, G.B. McGaughey, M.J. Bogusky, D.B. Whitman, R.G. Ball, C.J. Winrow, J.J. Renger and P.J. Coleman, *Bioorg. Med. Chem. Lett.*, **19**, 2997 (2009).
09BML3193	A. Kamimura, M. So, T. Kuratani, K. Matsuura and M. Inui, *Bioorg. Med. Chem. Lett.*, **19**, 3193 (2009).
09BML3669	N. Charrier, B. Clarke, E. Demont, C. Dingwall, R. Dunsdon, J. Hawkins, J. Hubbard, I. Hussain, G. Maile, R. Matico, J. Mosley, A. Naylor, A. O'Brien, S. Redshaw, P. Rowland, V. Soleil, K.J. Smith, S. Sweitzer, P. Theobald, D. Vesey, D.S. Walter and G. Wayne, *Bioorg. Med. Chem. Lett.*, **19**, 3669 (2009).

09BML3674	N. Charrier, B. Clarke, L. Cutler, E. Demont, C. Dingwall, R. Dunsdon, J. Hawkins, C. Howes, J. Hubbard, I. Hussain, G. Maile, R. Matico, J. Mosley, A. Naylor, A. O'Brien, S. Redshaw, P. Rowland, V. Soleil, K.J. Smith, S. Sweitzer, P. Theobald, D. Vesey, D.S. Walter and G. Wayne, *Bioorg. Med. Chem. Lett.*, **19**, 3674 (2009).
09BML3729	Z.Y. Han, H. Xiao and L.Z. Gong, *Bioorg. Med. Chem. Lett.*, **19**, 3729 (2009).
09BML3895	H. Ube and M. Terada, *Bioorg. Med. Chem. Lett.*, **19**, 3895 (2009).
09BML3977	N. Jain, G. Allan, O. Linton, P. Tannenbaum, X. Chen, J. Xu, P. Zhu, J. Gunnet, K. Demarest, S. Lundeen, W. Murray and Z. Sui, *Bioorg. Med. Chem. Lett.*, **19**, 3977 (2009).
09BML4034	Y. Shi, J. Zhang, M. Shi, S.P. O'Connor, S.N. Bisaha, C. Li, D. Sitkoff, A.T. Pudzianowski, S. Chong, H.E. Klei, K. Kish, J. Yanchunas, E.C.-K. Liu, K.S. Hartl, S.M. Seiler, T.E. Steinbacher, W.A. Schumacher, K.S. Atwal and P.D. Stein, *Bioorg. Med. Chem. Lett.*, **19**, 4034 (2009).
09BML4042	D. Branca, M. Cerretani, P. Jones, U. Koch, F. Orvieto, M.C. Palumbi, M. Rowley, C. Toniatti and E. Muraglia, *Bioorg. Med. Chem. Lett.*, **19**, 4042 (2009).
09BML4196	F. Orvieto, D. Branca, C. Giomini, P. Jones, U. Koch, J.M. Ontoria, M.C. Palumbi, M. Rowley, C. Toniatti and E. Muraglia, *Bioorg. Med. Chem. Lett.*, **19**, 4196 (2009).
09BML4264	J. Fu, S.J. Shuttleworth, R.V. Connors, A. Chai and P. Coward, *Bioorg. Med. Chem. Lett.*, **19**, 4264 (2009).
09BML4563	S.F. Neelamkavil, C.D. Boyle, S. Chackalamannil, W.J. Greenlee, L. Zhang and G. Terracina, *Bioorg. Med. Chem. Lett.*, **19**, 4563 (2009).
09BML4878	D.R. Anderson, M.J. Meyers, R.G. Kurumbail, N. Caspers, G.I. Poda, S.A. Long, B.S. Pierce, M.W. Mahoney and R.J. Mourey, *Bioorg. Med. Chem. Lett.*, **19**, 4878 (2009).
09BML4882	D.R. Anderson, M.J. Meyers, R.G. Kurumbail, N. Caspers, G.I. Poda, S.A. Long, B.S. Pierce, M.W. Mahoney, R.J. Mourey and M.D. Parikh, *Bioorg. Med. Chem. Lett.*, **19**, 4882 (2009).
09BML4999	P.E. Brennan, G.A. Whitlock, D.K.H. Hoa, K. Conlon and G. McMurray, *Bioorg. Med. Chem. Lett.*, **19**, 4999 (2009).
09BML5218	Z. Zhu, Z.-Y. Sun, Y. Ye, B. McKittrick, W. Greenlee, M. Czarniecki, A. Fawzi, H. Zhang and J.E. Lachowicz, *Bioorg. Med. Chem. Lett.*, **19**, 5218 (2009).
09BML5241	D.K. Mohapatra, P.K. Maity, M. Shabab and M.I. Khan, *Bioorg. Med. Chem. Lett.*, **19**, 5241 (2009).
09BML5289	J.F. Mehlmann, M.L. Crawley, J.T. Lundquist, R.J. Unwalla, D. Harnish, M.J. Evans, C.Y. Kim, J.E. Wrobel and P.E. Mahaney, *Bioorg. Med. Chem. Lett.*, **19**, 5289 (2009).
09BML5359	J. Ramnauth, M.D. Surman, P.B. Sampson, B. Forrest, J. Wilson, E. Freeman, D.D. Manning, F. Martin, A. Toro, M. Domagala, D.E.E. Awrey, N. Bardouniotis, J. Kaplan and H.W. Berman Pauls, *Bioorg. Med. Chem. Lett.*, **19**, 5359 (2009).
09BML5746	G. Achermann, T.M. Ballard, F. Blasco, P.E. Broutin, B. Buettelmann, H. Fischer, M. Graf, M.-C. Hernandez, P. Hilty, F. Knoflach, A. Koblet, H. Knust, A. Kurt, J.R. Martin, R. Masciadri, R.H.P. Porter, H. Stadler, A.W. Thomas, G. Trube and J. Wichmann, *Bioorg. Med. Chem. Lett.*, **19**, 5746 (2009).
09BML5753	W.S. Phutdhawong, W. Ruensamran, W. Phutdhawong and T. Taechowisan, *Bioorg. Med. Chem. Lett.*, **19**, 5753 (2009).
09BML5940	H. Knust, G. Achermann, T. Ballard, B. Buettelmann, R. Gasser, H. Fischer, M.-C. Hernandez, F. Knoflach, A. Koblet, H. Stadler, A.W. Thomas, G. Trube and P. Waldmeier, *Bioorg. Med. Chem. Lett.*, **19**, 5940 (2009).
09BML5958	B. Buettelmann, T.M. Ballard, R. Gasser, H. Fischer, M.-C. Hernandez, F. Knoflach, H. Knust, H. Stadler, A.W. Thomas and G. Trube, *Bioorg. Med. Chem. Lett.*, **19**, 5958 (2009).
09BML6053	J.-Y. Lee, J. Yu, W.J. Cho, H. Ko and Y.-C. Kim, *Bioorg. Med. Chem. Lett.*, **19**, 6053 (2009).
09BML6074	X. Zhang, Y. Hu, S. Chen, R. Luo, J. Yue, Y. Zhang, W. Duan and H. Wang, *Bioorg. Med. Chem. Lett.*, **19**, 6074 (2009).

09BML6368	C.S. Burgey, C.M. Potteiger, J.Z. Deng, S.D. Mosser, C.A. Salvatore, S. Yu, S. Roller, S.A. Kane, J.P. Vacca and T.M. Williams, *Bioorg. Med. Chem. Lett.*, **19**, 6368 (2009).
09BML6452	J.M. Bailey, J.S. Scott, J.B. Basilla, V.J. Bolton, I. Boyfield, D.G. Evans, E. Fleury, T.D. Heightman, E.M. Jarvie, K. Lawless, K.L. Matthews, F. McKay, H. Mokc, A. Muir, B.S. Orlek, G.J. Sanger, G. Stemp, A.J. Stevens, M. Thompson, J. Ward, K. Vaidya and S.M. Westaway, *Bioorg. Med. Chem. Lett.*, **19**, 6452 (2009).
09BML6463	P.W. Howard, Z. Chen, S.J. Gregson, L.A. Masterson, A.C. Tiberghien, N. Cooper, M. Fang, M.J. Coffils, S. Klee, J.A. Hartley and D.E. Thurston, *Bioorg. Med. Chem. Lett.*, **19**, 6578 (2009).
09BML6578	R.J. Gleave, P.J. Beswick, A.J. Brown, G.M.P. Giblin, C.P. Haslam, D. Livermore, A. Moses, N.H. Nicholson, L.W. Page, B. Slingsby and M.E. Swarbrick, *Bioorg. Med. Chem. Lett.*, **19**, 6578 (2009).
09BML6882	Y. Shi, C. Li, S.P. O'Connor, J. Zhang, M. Shi, S.N. Bisaha, Y. Wang, D. Sitkoff, A.T. Pudzianowski, C. Huang, H.E. Klei, K. Kish, J. Yanchunas, E.C.-K. Liu, K.S. Hartl, S.M. Seiler, T.E. Steinbacher, W.A. Schumacher, K.S. Atwal and P.D. Stein, *Bioorg. Med. Chem. Lett.*, **19**, 6882 (2009).
09CAR448	G. Sizun, D. Dukhan, J.-F. Griffon, L. Griffe, J.-C. Meillon, F. Leroy, R. Storer, J.-P. Sommadossi and G. Gosselin, *Carbohydr. Res.*, **344**, 448 (2009).
09CC445	M.C. Tseng, C.Y. Lai, Y.W. Chu and Y.H. Chu, *Chem. Commun.*, 445 (2009).
09CC1964	H. Brice and J. Clayden, *Chem. Commun.*, 1964 (2009).
09CC2875	K.M. Rahman and D.E. Thurston, *Chem. Commun.*, 2875 (2009).
09CCL32	S.S. Pawar, M.S. Shingare and S.N. Thore, *Chin. Chem. Lett.*, **20**, 32 (2009).
09CCL905	X. Zhou, M.Y. Zhang, S.T. Gao, J.J. Ma, C. Wang and C. Liu, *Chin. Chem. Lett.*, **20**, 905 (2009).
09CEJ1241	S. Fantauzzi, E. Gallo, A. Caselli, C. Piangiolino, F. Ragaini, N. Re and S. Cenini, *Chem. Eur. J.*, **15**, 1241 (2009).
09CEJ1901	B. Alcaide, P. Almendros, T. Martinez del Campo, E. Soriano and J.L. Marco-Contelles, *Chem. Eur. J.*, **15**, 1901 (2009).
09CEJ1909	B. Alcaide, P. Almendros, T. Martinez del Campo, E. Soriano and J.L. Marco-Contelles, *Chem. Eur. J.*, **15**, 1909 (2009).
09CEJ6678	T. Kano, Y. Yamaguchi and K. Maruoka, *Chem. Eur. J.*, **15**, 6678 (2009).
09CEJ7215	A. Kamal, N. Markandeya, Shankaraiah, C.R. Reddy, S. Prabhakar, C.S. Reddy, M.N. Eberlin and L.S. Santos, *Chem. Eur. J.*, **15**, 7215 (2009).
09CEJ8834	M.J. Climent, A. Corma, S. Iborra and L.L. Santos, *Chem. Eur. J.*, **15**, 8834 (2009).
09CEJ8975	Y. Bai, J. Fang, J. Ren and Z. Wang, *Chem. Eur. J.*, **15**, 8975 (2009).
09CEJ9127	B. Alcaide, P. Almendros, T. Martinez del Campo, E. Soriano and J.L. Marco-Contelles, *Chem. Eur. J.*, **15**, 9127 (2009).
09CEJ10225	J. Coulomb, V. Certal, M.-H. Larraufie, C. Ollivier, J.-P. Corbet, G. Mignani, L. Fensterbank, E. Lacote and M. Malacria, *Chem. Eur. J.*, **15**, 10225 (2009).
09CEJ11119	J. Zhang and A. Zhang, *Chem. Eur. J.*, **15**, 11119 (2009).
09CEJ12121	K. Krohn, S.F. Kouam, G.M. Kuigoua, H. Hussain, S. Cludius-Brandt, U. Floerke, T. Kurtan, G. Pescitelli, L. Di Bari, S. Draeger and B. Schulz, *Chem. Eur. J.*, **15**, 12121 (2009).
09CHE726	V.S. Tolkunov, A.B. Eresko, A.I. Khizhan, O.V. Shishkin, G.V. Palamarchuk and S.V. Tolkunov, *Chem. Heterocycl. Compd.*, **45**, 726 (2009).
09CME3122	T.N.T. Nguyen and J.J. Tepe, *Curr. Med. Chem.*, **16**, 3122 (2009).
09CMH1069	D.B. Whitman, C.D. Cox, M.J. Breslin, K.M. Brashear, J.D. Schreier, M.J. Bogusky, R.A. Bednar, W. Lemaire, J.G. Bruno, G.D. Hartman, D.R. Reiss, C.M. Harrell, R.L. Kraus, Y. Li, S.L. Garson, S.M. Doran, T. Prueksaritanont, C. Li, C.J. Winrow, K.S. Koblan, J.J. Renger and P.J. Coleman, *ChemMedChem*, **4**, 1069 (2009).
09CPB443	M. Ikeuchi, M. Ikuta, M. Hariki, M. Ikeuchi, S. Maruyama, M. Nakase, K. Sakamoto, Y. Yoshioka, A. Yamauchi and M. Kihara, *Chem. Pharm. Bull.*, **57**, 443 (2009).
09DOF349	C. Campàs, *Drugs of the Future*, **24**, 349 (2009).
09EJM129	D. Kovala-Demertzi, A. Papageorgiou, L. Papathanasis, A. Alexandratos, P. Dalezis, J.R. Miller and M.A. Demertzis, *Eur. J. Med. Chem.*, **44**, 129 (2009).

09EJM417	M. Abid, A.R. Bhat, F. Athar and A. Azam, *Eur. J. Med. Chem.*, **44**, 417 (2009).
09EJM519	S. Bedürftig and B. Wünsch, *Eur. J. Med. Chem.*, **44**, 519 (2009).
09EJM772	P. Jia, R. Sheng, J. Zhang, L. Fang, Q. He, B. Yang and Y. Hu, *Eur. J. Med. Chem.*, **44**, 772 (2009).
09EJM2815	L. Wang, P. Zhang, X. Zhang, Y. Zhang, Y. Li and Y. Wang, *Eur. J. Med. Chem.*, **44**, 2815 (2009).
09EJM3936	M. El-Far, G.A. Elmegeed, E.F. Eskander, H.M. Rady and M.A. Tantawy, *Eur. J. Med. Chem.*, **44**, 3936 (2009).
09EJO204	S.K. Das, S.K. Dinda and G. Panda, *Eur. J. Org. Chem.*, 204 (2009).
09EJO793	P.A. Donets, J.L. Goeman, J. Van der Eycken, K. Robeyns, L. Van Meervelt and E.V. Van der Eycken, *Eur. J. Org. Chem.*, 793 (2009).
09EJO1309	S.K. Sharma, S. Sharma, P.K. Agarwal and B. Kundu, *Eur. J. Org. Chem.*, 1309 (2009).
09EJO1934	S. Stewart, C.H. Heath and E.L. Ghisalberti, *Eur. J. Org. Chem.*, 1934 (2009).
09EJO2342	M. Sajitz, R. Froehlich and E.-U. Wuerthwein, *Eur. J. Org. Chem.*, 2342 (2009).
09EJO2421	R. Alibes and M. Figueredo, *Eur. J. Org. Chem.*, 2421 (2009).
09EJO2635	V.-A. Rassadin, A.A. Tomashevskiy, V.V. Sokolov, A. Ringe, J. Magull and A. de Meijere, *Eur. J. Org. Chem.*, 2635 (2009).
09EJO3129	L.Z. Dai and M. Shi, *Eur. J. Org. Chem.*, 3129 (2009).
09EJO3413	P.C.B. Page, B.R. Buckley, M.M. Farah and A.J. Blacker, *Eur. J. Org. Chem.*, 3413 (2009).
09EJO3454	V. Singh, S. Hutait and S. Batra, *Eur. J. Org. Chem.*, 3454 (2009).
09EJO3726	J. Nonnenmacher, F. Grellepois and C. Portella, *Eur. J. Org. Chem.*, 3726 (2009).
09EJO3741	D. Dumoulin, S. Lebrun, E. Deniau, A. Couture and P. Grandclaudon, *Eur. J. Org. Chem.*, 3741 (2009).
09EJO3940	H. Quast, K.H. Ross, G. Philipp, M. Hagedorn, H. Hahn and K. Banert, *Eur. J. Org. Chem.*, 3940 (2009).
09EJO5752	K. Yamada, Y. Namerikawa, T. Haruyama, Y. Miwa, R. Yanada and M. Ishikura, *Eur. J. Org. Chem.*, 5752 (2009).
09EJO5953	M. Le Liepvre, J. Ollivier and D.J. Aitken, *Eur. J. Org. Chem.*, 5953 (2009).
09EJO6198	K. Barta, M. Eggenstein, M. Hölscher, G. Franciò, G. Leitner, *Eur. J. Org. Chem.* 6198 (2009).
09EOD221	J. Ferreira and T. Mestre, *Expert Opin. Investig. Drugs*, **18**, 221 (2009).
09EOD657	E. Ku, N. Nobakht and V.M. Campese, *Expert Opin. Investig. Drugs*, **18**, 657 (2009).
09EOP1377	A.W.C. Kung, E.Y.W. Chu and L. Xu, *Expert Opin. Pharmacother.*, **10**, 1377 (2009).
09EOP1523	S.A. Doggrell, *Expert Opin. Pharmacother.*, **10**, 1523 (2009).
09EOP1649	H.W. Goforth, *Expert Opin. Pharmacother.*, **10**, 1649 (2009).
09EOP1657	D.B. Henley, P.C. May, R.A. Dean and E.R. Siemers, *Expert Opin. Pharmacother.*, **10**, 1657 (2009).
09EOP2161	M.Z. Hoque, R. Arumugham, N. Huda, N. Verma, M. Afiniwala and D.H. Karia, *Expert Opin. Pharmacotherapy*, **10**, 2161 (2009).
09FMC909	G.K. Surya Prakash, A. Vaghoo, A. Venkat, C. Panja, S. Chacko, T. Mathew and G.A. Olah, *F. Med. Chem.*, **1**, 909 (2009).
09H(77)575	T. Taniguchi, H. Zaimoku and I. Ishibashi, *Heterocycles*, **77**, 575 (2009).
09H(77)825	K. Yamada, Y. Namerikawa, T. Abe and M. Ishikura, *Heterocycles*, **77**, 825 (2009).
09H(78)1183	H. Cho, Y. Iwama, K. Sugimoto, E. Kwon and H. Tokuyama, *Heterocycles*, **78**, 1183 (2009).
09H(78)2209	A. Aoyama, H. Aoyama, M. Makishima, Y. Hashimoto and H. Miyachi, *Heterocycles*, **78**, 2209 (2009).
09H(78)319	H. Isawa, H. Suga and A. Kakehi, *Heterocycles*, **78**, 319 (2009).
09H(78)635	D. Muroni, M. Mucedda and A. Saba, *Heterocycles*, **78**, 635 (2009).
09H(78)2837	G. Molteni and P. Del Buttero, *Heterocycles*, **78**, 2837 (2009).
09H(79)195	H.M. Garraffo and T.F. Spande, *Heterocycles*, **79**, 195 (2009).
09H(79)243	S. Yamaguchi, *Heterocycles*, **79**, 243 (2009).
09IJH91	A. Davoodnia, M. Roshani, A. Monfared and N. Tavakoli-Hoseini, *Indian J. Heterocycl. Chem.*, **19**, 91 (2009).
09IJH291	S.G. Jagdhani, S.K. Narwade, S.B. Kale and B.K. Karale, *Indian J. Heterocycl. Chem.*, **18**, 291 (2009).

09JA1077	Z. Shen, P.K. Dornan, H.A. Khan, T.K. Woo and V.M. Dong, *J. Am. Chem. Soc.*, **131**, 1077 (2009).
09JA2434	A.R. Thornton, V.I. Martin and S.B. Blakey, *J. Am. Chem. Soc.*, **131**, 2434 (2009).
09JA5390	F. Marcelo, Y. He, S.A. Yuzwa, L. Nieto, J. Jiménez-Barbero, M. Sollogoub, D.J. Vocadlo, G.D. Davies and Y. Blériot, *J. Am. Chem. Soc.*, **131**, 5390 (2008).
09JA6614	T. Hashimoto, Y. Naganawa and K. Maruoka, *J. Am. Chem. Soc.*, **131**, 6614 (2009).
09JA6906	B.K. Alnasleh, W.M. Sherrill, M. Rubina, J. Banning and M. Rubin, *J. Am. Chem. Soc.*, **131**, 6906 (2009).
09JA6932	M.M. Coulter, P.K. Dornan and V.M. Dong, *J. Am. Chem. Soc.*, **131**, 6932 (2009).
09JA7242	D. Uraguchi, D. Nakashima and T. Ooi, *J. Am. Chem. Soc.*, **131**, 7242 (2009).
09JA9488	C.F. Rosewall, P.A. Sibbald, D.V. Liskin and F.E. Michael, *J. Am. Chem. Soc.*, **131**, 9488 (2009).
09JA12084	J. Tanuwidjaja, S.-S. Ng and T.F. Jamison, *J. Am. Chem. Soc.*, **131**, 12084 (2009).
09JA13756	K.M. Rahman, A.S. Thompson, C.H. James, M. Narayanaswamy and D.E. Thurston, *J. Am. Chem. Soc.*, **131**, 13756 (2009).
09JA18246	D.C. Leitch, P.R. Payne, C.R. Dunbar and L.L. Schafer, *J. Am. Chem. Soc.*, **131**, 18246 (2009).
09JAN163	I.-K. Lee, Y.-W. Jang, Y.-S. Kim, S.-H. Yu, K.J. Lee, S.-M. Park, B.-T. Oh, J.C. Chae and B.S. Yun, *J. Antibiot.*, **62**, 163 (2009).
09JAN533	Y. Shiono, A. Nitto, K. Shimanuki, T. Koseki, T. Murayama, T. Miyakawa, J. Yoshida and K. Kimura, *J. Antibiot.*, **62**, 533 (2009).
09JBC7597	K. Yamaoka, M. Inoue, K. Miyazaki, M. Hirama, C. Kondo, E. Kinoshita, H. Miyoshi and I. Seyama, *J. Biol. Chem.*, **284**, 7597 (2009).
09JFC1001	W.S. Park, M.A. Jun, M.S. Shin, S.W. Kwon, S.K. Kang, K.Y. Kim, S.D. Rhee, M.A. Bae, B. Narsaiah, D.H. Lee, H.G. Cheon, J.H. Ahn and S.S. Kim, *J. Fluorine Chem.*, **130**, 1001 (2009).
09JHC96	M. Abaszadeh, H. Sheibani and K. Saidi, *J. Heterocycl. Chem.*, **46**, 96 (2009).
09JME95	D. Feytens, M. De Vlaeminck, R. Cescato, D. Tourwe and J.C. Reubi, *J. Med. Chem.*, **52**, 95 (2009).
09JME904	B. Flatt, R. Martin, T.L. Wang, P.E. Mahaney, B. Murphy, X.-H. Gu, P. Foster, J. Li, P. Pircher, M. Petrowski, I. Schulman, S. Westin, J. Wrobel, G. Yan, E. Bischoff, C. Daige and R.J. Mohan, *J. Med. Chem.*, **52**, 904 (2009).
09JME1224	S. Butini, M. Brindisi, S. Cosconati, L. Marinelli, G. Borrelli, S. Sanna Coccone, A. Ramunno, G. Campiani, E. Novellino, S. Zanoli, A. Samuele, G. Giorgi, A. Bergamini, M. Di Mattia, S. Lalli, B. Galletti, S. Gemma and G. Maga, *J. Med. Chem.*, **52**, 1224 (2009).
09JME1670	R. Perez-Pineiro, A. Burgos, D.C. Jones, L.C. Andrew, H. Rodriguez, M. Suarez, A.H. Fairlamb and D.S. Wishart, *J. Med. Chem.*, **52**, 1670 (2009).
09JME1773	C. Kern, T. Meyer, S. Droux, D. Schollmeyer and C. Miculka, *J. Med. Chem.*, **52**, 1773 (2009).
09JME1778	S.C. Karcher and S.A. Laufer, *J. Med. Chem.*, **52**, 1778 (2009).
09JME1795	F.M. Rivas, J.P. Stables, L. Murphree, R.V. Edwankar, C.R. Edwankar, S. Huang, H.D. Jain, H. Zhou, S. Majumder, S. Sankar, B.L. Roth, J. Ramerstorfer, R. Furtmüller, W. Sieghart and J.M. Cook, *J. Med. Chem.*, **52**, 1795 (2009).
09JME1814	K. Qian, L. Wang, C.L. Cywin, B.T. FarmerII, E. Hickey, C. Homon, S. Jakes, M.A. Kashem, G.L.S. Leonard, J. Li, R. Magboo, W. Mao, E. Pack, C. Peng, A. ProkopowiczIII, M. Welzel, J. Wolak and T. Morwick, *J. Med. Chem.*, **52**, 1814 (2009).
09JME1975	H. Pettersson, A. Bülow, F. Ek, J. Jensen, L.K. Ottesen, A. Fejzic, J.-N. Ma, A.L. Del Tredici, E.A. Currier, L.R. Gardell, A. Tabatabaei, D. Craig, K. McFarland, T.R. Ott, F. Piu, E.S. Burstein and R. Olsson, *J. Med. Chem.*, **52**, 1975 (2009).
09JME2429	P.K. Mandal, D. Limbrick, D.R. ColemanIV, G.A. Dyer, Z. Ren, J.S. Birtwistle, C. Xiong, X. Chen, J.M. Briggs and J.S. McMurray, *J. Med. Chem.*, **52**, 2429 (2009).
09JME2652	R.L. Dow, P.A. Carpino, J.R. Hadcock, S.C. Black, P.A. Iredale, P. Da Silva-Jardine, S.R. Schneider, E.S. Paight, D.A. Griffith, D.O. Scott, R.E. O'Connor and C.I. Nduaka, *J. Med. Chem.*, **52**, 2652 (2009).

09JME3591	D.J. Fox, J. Reckless, H. Lingard, S. Warren and D.J. Grainger, *J. Med. Chem.*, **52**, 3591 (2009).
09JME4099	K. Vandyck, M.D. Cummings, O. Nyanguile, C.W. Boutton, S. Vendeville, D. McGowan, B. Devogelaere, K. Amssoms, S. Last, K. Rombauts, A. Tahri, P. Lory, L. Hu, D.A. Beauchamp, K. Simmen and P. Raboisson, *J. Med. Chem.*, **52**, 4099 (2009).
09JME4149	B. Le Grand, C. Pignier, R. Letienne, F. Colpaert, F. Cuisiat, F. Rolland, A. Mas, M. Borras and B. Vacher, *J. Med. Chem.*, **52**, 4149 (2009).
09JME5612	A. Lukaszuk, H. Demaegdt, D. Feytens, P. Vanderheyden, G. Vauquelin and D. Tourwe, *J. Med. Chem.*, **52**, 5612 (2009).
09JME5916	A. Putey, F. Popowycz, Q.-T. Do, P. Bernard, S.K. Talapatra, F. Kozielski, C.M. Galmarini and B. Joseph, *J. Med. Chem.*, **52**, 5916 (2009).
09JME6484	T.H. Al-Tel, R.A. Al-Qawasmeh, M.F. Schmidt, A. Al-Aboudi, S.N. Rao, S.S. Sabri and W. Voelter, *J. Med. Chem.*, **52**, 6484 (2009).
09JME6660	A. Lilienkampf, S. Karkola, S. Alho-Richmond, P. Koskimies, N. Johansson, K. Huhtinen, K. Vihko and K. Wähälä, *J. Med. Chem.*, **52**, 6660 (2009).
09JME7544	N. Jain, J. Xu, R.M. Kanojia, F. Du, G. Jian-Zhong, E. Pacia, M.-T. Lai, A. Musto, G. Allan, M. Reuman, X. Li, D. Hahn, M. Cousineau, S. Peng, D. Ritchie, R. Russell, S. Lundeen and Z. Sui, *J. Med. Chem.*, **52**, 7544 (2009).
09JME7950	F. Liu, X. Chen, A. Allali-Hassani, A.M. Quinn, G.A. Wasney, A. Dong, A. Barsyte, I. Kozieradzki, G. Senisterra, I. Chau, A. Siarheyeva, D.B. Kireev, A. Jadhav, J.M. Herold, S.V. Frye, C.H. Arrowsmith, P.J. Brown, A. Simeonov, M. Vedadi and J. Jin, *J. Med. Chem.*, **52**, 7950 (2009).
09JNP848	P. Mungkornasawakul, S. Chaiyong, T. Sastraruji, A. Jatisatienr, C. Jatisatienr, S.G. Pyne, A.T. Ung, J. Korth and W. Lie, *J. Nat. Prod.*, **72**, 848 (2009).
09JOC289	A. Rudolph, N. Rackelmann, M.-O. Turcotte-Savard and M. Lautens, *J. Org. Chem.*, **74**, 289 (2009).
09JOC995	M. Nodwell, A. Pereira, J.L. Riffell, C. Zimmerman, B.O. Patrick, M. Roberge and R.J. Andersen, *J. Org. Chem.*, **74**, 995 (2009).
09JOC1791	P. Thansandote, C. Gouliaras, M.O. Turcotte-Savard and M. Lautens, *J. Org. Chem.*, **74**, 1791 (2009).
09JOC2004	V. Declerck, L. Toupet, J. Martinez and F. Lamaty, *J. Org. Chem.*, **74**, 2004 (2009).
09JOC2189	M. Sanudo, M. Garcia-Valverde, S. Marcaccini, J.J. Delgado, J. Rojo and T. Torroba, *J. Org. Chem.*, **74**, 2189 (2009).
09JOC2788	S.M. Husain, R. Fröhlich, D. Schepmann and B. Wünsch, *J. Org. Chem.*, **74**, 2788 (2009).
09JOC3211	Y.-M. Zhao, P. Gu, H.-J. Zhang, Q.-W. Zhang, C.-A. Fan, Y.-Q. Tu and F.-M. Zhang, *J. Org. Chem.*, **74**, 3211 (2009).
09JOC3680	K.D. Closser, M.M. Quintal and K.M. Shea, *J. Org. Chem.*, **74**, 3680 (2009).
09JOC3978	Z. Cong, T. Miki, O. Urakawa and H. Nishino, *J. Org. Chem.*, **74**, 3978 (2009).
09JOC4329	V. Colombel and O. Baudoin, *J. Org. Chem.*, **74**, 4329 (2009).
09JOC4490	M.E. Buden, V.A. Vaillard, S.E. Martin and R.A. Rossi, *J. Org. Chem.*, **74**, 4490 (2009).
09JOC4584	N. Ghavtadze, R. Frohlich and E.U. Wurthwein, *J. Org. Chem.*, **74**, 4584 (2009).
09JOC5652	F. Crestey, M. Witt, J.W. Jaroszewski and H. Franzyk, *J. Org. Chem.*, **74**, 5652 (2009).
09JOC5734	Y.K. Kang and D.Y. Kim, *J. Org. Chem.*, **74**, 5734 (2009).
09JOC6077	D.H. Leng, D.X. Wang, J. Pan, Z.T. Huang and M.X. Wang, *J. Org. Chem.*, **74**, 6077 (2009).
09JOC6486	V.H. Jadhav, O.P. Bande, R.V. Pinjari, S.P. Gejji, V.G. Puranik and D.D. Dhavale, *J. Org. Chem.*, **74**, 6486 (2009).
09JOC6797	H. Li, J. Yang, Y. Liu and Y. Li, *J. Org. Chem.*, **74**, 6797 (2009).
09JOC6895	H.X. Liu and A. Domling, *J. Org. Chem.*, **74**, 6895 (2009).
09JOC7013	M.K. Ghorai, D. Shukla and K. Das, *J. Org. Chem.*, **74**, 7013 (2009).
09JOC7411	J.R. Green and A.A. Tjeng, *J. Org. Chem.*, **74**, 7411 (2009).
09JOC7618	C.W. Huh, G.K. Somal, C.E. Katz, H. Pei, Y. Zeng, J.T. Douglas and J. Aube, *J. Org. Chem.*, **74**, 7618 (2009).
09JOC7627	R. Batchelor, J.E. Harvey, P.T. Northcote, P. Teesdale-Spittle and J.O. Hoberg, *J. Org. Chem.*, **74**, 7627 (2009).

09JOC8219	M. Horvat, H. Gorner, K.-D. Warzecha, J. Neudorfl, A.G. Griesbeck, K. Mlinaric-Majerski and N. Basaric, *J. Org. Chem.*, **74**, 8219 (2009).
09JOC8814	M. Bera and S. Roy, *J. Org. Chem.*, **74**, 8814 (2009).
09JOC9365	V. Sridharan, S. Maiti and J.C. Menendez, *J. Org. Chem.*, **74**, 9365 (2009).
09JOR2454	E.L. Kolychev, I.A. Portnyagin, V.V. Shuntikov, V.N. Khrustalev, M.S. Nechaev, *J. Organomet. Chem.*, **694**, 2454 (2009).
09LOC17	G.R. Krishnan, R. Sreerekha and K. Sreekumar, *Lett. Org. Chem.*, **6**, 17 (2009).
09LOC478	A. Hassannia, G. Piercy and N. Merbouh, *Lett. Org. Chem.*, **6**, 478 (2009).
09NJC1703	K.A. Johnston and H. McNab, *New J. Chem.*, **33**, 1703 (2009).
09NPC1481	Z. Guo, F. Cheng, K. Zou, J. Wang, Z. She and Y. Lin, *Nat. Prod. Commun.*, **4**, 1481 (2009).
09NRD611	J.K. Ghali, B. Hamad, U. Yasothan and P. Kirkpatrick, *Nat. Rev. Drug Disc.*, **8**, 611 (2009).
09NRD843	H.Y. Meltzer, A. Dritselis, U. Yasothan and P. Kirkpatrick, *Nat. Rev. Drug Disc.*, **8**, 843 (2009).
09OBC557	S.-J. Tu, X.-D. Cao, W.-J. Hao, X.-H. Zhang, S. Yan, S.-S. Wu, Z.-G. Han and F. Shi, *Org. Biomol. Chem.*, **7**, 557 (2009).
09OBC1184	B. Pettersson, A. Rydbeck and J. Bergman, *Org. Biomol. Chem.*, **7**, 1184 (2009).
09OBC1911	E. Sotoca, C. Allais, T. Constantieux and J. Rodriguez, *Org. Biomol. Chem.*, **7**, 1911 (2009).
09OBC3561	F.D. King, A.E. Aliev, S. Caddick and R.C.B. Copley, *Org. Biomol. Chem.*, **7**, 3561 (2009).
09OBC589	M. Miller, J.C. Vogel, W. Tsang, A. Merrit and D.J. Procter, *Org. Biomol. Chem.*, **7**, 589 (2009).
09OBC655	S. Yamazaki, Y. Iwata and Y. Fukushima, *Org. Biomol. Chem.*, **7**, 655 (2009).
09OBC695	J. Kaiser, B.C.J. van Esseveldt, M.J.A. Segers, F.L. van Delft, J.M.M. Smits, S. Butterworth and F.P.J.T. Rutjes, *Org. Biomol. Chem.*, **7**, 695 (2009).
09OL113	T. Saito and T. Nakata, *Org. Lett.*, **11**, 113 (2009).
09OL197	K. Higuchi, Y. Sato, M. Tsuchimochi, K. Sugiura, M. Hatori and T. Kawasaki, *Org. Lett.*, **11**, 197 (2009).
09OL257	M. Yar, E.M. McGarrigle and V.K. Aggarwal, *Org. Lett.*, **11**, 257 (2009).
09OL341	N. Nicolaus, S. Strauss, J.-M. Neudörfl, A. Prokop and H.-G. Schmalz, *Org. Lett.*, **11**, 341 (2009).
09OL531	A. Zhou, D. Rayabarapu and P.R. Hanson, *Org. Lett.*, **11**, 531 (2009).
09OL851	M.A. Boone, F.E. McDonald, J. Lichter, S. Lutz, R. Cao and K.I. Hardcastle, *Org. Lett.*, **11**, 851 (2009).
09OL963	S.M. Gueret, P.D. O'Connor and M.A. Brimble, *Org. Lett.*, **11**, 963 (2009).
09OL979	A.F. Khlebnikov, M.S. Novikov, P.P. Petrovskii, J. Magull and A. Ringe, *Org. Lett.*, **11**, 979 (2009).
09OL1031	M. Yu and B.B. Snider, *Org. Lett.*, **11**, 1031 (2009).
09OL1205	Y. Deng, Y. Shi and S. Ma, *Org. Lett.*, **11**, 1205 (2009).
09OL1225	L. Cui, G. Zhang, Y. Peng and L. Zhang, *Org. Lett.*, **11**, 1225 (2009).
09OL1575	L.P. Tardibono and J.M. Miller, *Org. Lett.*, **11**, 1575 (2009).
09OL1725	E.A. Tiong and J.L. Gleason, *Org. Lett.*, **11**, 1725 (2009).
09OL1857	T. Honda, E. Aranishi and K. Kaneda, *Org. Lett.*, **11**, 1857 (2009).
09OL2531	H. Takamura, S. Kikuchi, Y. Nakamura, Y. Yamagami, T. Kishi, I. Kadota and Y. Yamamoto, *Org. Lett.*, **11**, 2531 (2009).
09OL2707	L. Wu, S.F. Qiu and G.S. Liu, *Org. Lett.*, **11**, 2707 (2009).
09OL2788	J. Gan and D. Ma, *Org. Lett.*, **11**, 2788 (2009).
09OL3342	A. Shaabani, A.H. Rezayan, S. Keshipour, A. Sarvary and S.W. Ng, *Org. Lett.*, **11**, 3342 (2009).
09OL3362	M.P. Castaldi, D.M. Troast and J.A. Porco, *Org. Lett.*, **11**, 3362 (2009).
09OL3366	Y. Zhang and J.S. Panek, *Org. Lett.*, **11**, 3366 (2009).
09OL3490	M.A. Tarselli, J.L. Zuccarello, S.J. Lee and M.R. Gagne, *Org. Lett.*, **11**, 3490 (2009).
09OL3802	Y. Tang, J.C. Fettinger and J.T. Shaw, *Org. Lett.*, **11**, 3802 (2009).
09OL4036	C. Li, X. Li and R. Hong, *Org. Lett.*, **11**, 4036 (2009).
09OL4382	H. Furuta, Y. Hasegawa and Y. Mori, *Org. Lett.*, **11**, 4382 (2009).
09OL4482	J. Saha and M.W. Peczuh, *Org. Lett.*, **11**, 4482 (2009).
09OL5494	J. Rujirawanich and T. Gallagher, *Org. Lett.*, **11**, 5494 (2009).

09PM1104	J.R. Kesting, D. Staerk, M.V. Tejesvi. K.R. Kini, H.S. Prakash and J.W. Jaroszewski, *Planta Med.*, **75**, 1104 (2009).
09S389	Y. Brouillette, G. Sujol, J. Martinez and V. Lisowski, *Synthesis*, 389 (2009).
09S1185	A. Becker, S. Kohfeld, T. Pies, K. Wieking, L. Preu and C. Kunick, *Synthesis*, 1185 (2009).
09S1557	T. Kano, A. Yamamoto, F. Shirozu and K. Maruoka, *Synthesis*, 1557 (2009).
09S2163	N. Shankaraiah, N. Markandeya, M. Espinoza-Moraga, C. Arancibia, A. Kamal and L.S. Santos, *Synthesis*, 2163 (2009).
09SL675	B. Lygo, B. Allbutt, D.J. Beaumont, U. Butt and J.A.R. Gilks, *Synlett*, 675 (2009).
09SL1004	V. Aureggi, M. Davoust, K.M. Gericke and M. Lautens, *Synlett*, 1004 (2009).
09SL1162	J. Wu, Y. Jiang and W.-M. Dai, *Synlett*, 1162 (2009).
09SL1395	S. Ponzi, J. Habermann, M.R. Ferreira and F. Narjes, *Synlett*, 1395 (2009).
09SL1670	M. Nakano and M. Terada, *Synlett*, 1670 (2009).
09SL1781	M. Ozeki, A. Muroyama, T. Kajimoto, T. Watanabe, K. Wakabayashi and M. Node, *Synlett*, 1781 (2009).
09SL1830	L. Li, Z. Li and Q. Wang, *Synlett*, 1830 (2009).
09SL1979	R.W. Bates and S. Sridhar, *Synlett*, 1979 (2009).
09SL2188	P. Gao, Z. Tong, H. Hu, P.-F. Xu, W. Liu, C. Sun and H. Zhai, *Synlett*, 2188 (2009).
09SL3043	N. Patel, C.S. Chambers and K. Hemming, *Synlett*, 3043 (2009).
09SL3127	K.C. Majumdar, S. Chakravorty, T. Ghosh and B. Sridhar, *Synlett*, 3127 (2009).
09T525	E.L. Cropper, A.P. Yuen, A. Ford, A.J.P. White and K.K. Hii, *Tetrahedron*, **65**, 525 (2009).
09T2266	R. De Wachter, L. Brans, S. Ballet, I. Van den Eynde, D. Feytens, A. Keresztes, G. Toth, Z. Urbanczyk-Lipkowska and D. Tourwe, *Tetrahedron*, **65**, 2266 (2009).
09T2344	A.A. Fesenko, M.L. Tullberg and A.D. Shutalev, *Tetrahedron*, **65**, 2344 (2009).
09T2684	A. Alizadeh, N. Zohreh and L.G. Zhu, *Tetrahedron*, **65**, 2684 (2009).
09T5716	M. Tang, C.-A. Fan, F.-M. Zhang and Y.-Q. Tu, *Tetrahedron*, **65**, 5716 (2009).
09T6291	H. Wang, H. Matsuhashi, B.D. Doan, S.N. Goodman, X. Ouyang and W.M. Clark, *Tetrahedron*, **65**, 6291 (2009).
09T6454	J.D. Sunderhaus, C. Dockendorff and S.F. Martin, *Tetrahedron*, **65**, 6454 (2009).
09T6868	A.V. Samet, A.N. Yamskov, Y.A. Strelenko and V.V. Semenov, *Tetrahedron*, **65**, 6868 (2009).
09T7487	M. Incerti, D. Acquotti, P. Sandor and P. Vicini, *Tetrahedron*, **65**, 7487 (2009).
09T7741	M. Pozarentzi, J. Stephanidou-Stephanatou, C.A. Tsoleridis, C. Zika and V. Demopoulos, *Tetrahedron*, **65**, 7741 (2009).
09T7784	I. Kadota, T. Abe, M. Uni, H. Takamura and Y. Yamamoto, *Tetrahedron*, **65**, 7784 (2009).
09T7921	S. Castro, C.S. Johnson, B. Surana and M.W. Peczuh, *Tetrahedron*, **65**, 7921 (2009).
09T8350	H. Shirani, J. Bergman and T. Janosik, *Tetrahedron*, **65**, 8350 (2009).
09T8965	H. Wang, Y. Jiang, K. Gao and D. Ma, *Tetrahedron*, **65**, 8956 (2009).
09T9015	C. Ferrer, A. Escribano-Cuesta and A.M. Echavarren, *Tetrahedron*, **65**, 9015 (2009).
09T9935	R.N. Davis and T.D. Lash, *Tetrahedron*, **65**, 9935 (2009).
09T10192	S.G. Davies, A.M. Fletcher, P.M. Roberts and A.D. Smith, *Tetrahedron*, **65**, 10192 (2009).
09TA1154	D. Muroni, M. Mucedda and A. Saba, *Tetrahedron: Asymmetry*, **20**, 1154 (2009).
09TA1217	B. Chandrasekhar, B.V. Rao, K.V.M. Rao and B. Jagadeesh, *Tetrahedron: Asymmetry*, **20**, 1217 (2009).
09TA1383	S.M. Husain, M.T. Heim, D. Schepmann and B. Wünsch, *Tetrahedron: Asymmetry*, **20**, 1383 (2009).
09TA1697	S. Mitsunaga, T. Ohbayashi, S. Sugiyama, T. Saitou, M. Tadokoro and T. Satoh, *Tetrahedron: Asymmetry*, **20**, 1697 (2009).
09TA1903	D. Dumoulin, S. Lebrun, A. Couture, E. Deniau and P. Grandclaudon, *Tetrahedron: Asymmetry*, **20**, 1903 (2009).
09TA2320	O. Monasson, M. Ginisty, J. Mravljak, G. Bertho, C. Gravier-Pelletier and Y. Le Merrer, *Tetrahedron: Asymmetry*, **20**, 2320 (2009).
09TL228	K.C. Majumdar and A. Taher, *Tetrahedron Lett.*, **50**, 228 (2009).
09TL386	X.J. Han, R.L. Civiello, S.E. Mercer, J.E. Macor and G.M. Dubowchik, *Tetrahedron Lett.*, **50**, 386 (2009).

09TL1385	M.G. Ali Badry, B. Kariuki, D.W. Knight and F.K. Mohammed, *Tetrahedron Lett.*, **50**, 1385 (2009).
09TL1423	L.D.S. Yadav, V.P. Srivastava and R. Patel, *Tetrahedron Lett.*, **50**, 1423 (2009).
09TL1727	A. Kamimura, Y. Ishihara, M. So and T. Hayashi, *Tetrahedron Lett.*, **50**, 1727 (2009).
09TL1911	S.B. Hoyt, C. London and M. Park, *Tetrahedron Lett.*, **50**, 1911 (2009).
09TL1939	C. Hulme, S. Chappeta and J. Dietrich, *Tetrahedron Lett.*, **50**, 4054 (2009).
09TL2854	V. Kysil, A. Khvat, S. Tsirulnikov, S. Tkachenko and A. Ivachtchenko, *Tetrahedron Lett.*, **50**, 2854 (2009).
09TL2917	X.-J. Hu, Y.-H. Wang, L.-Y. Kong, H.-P. He, S. Gao, H.-Y. Liu, J. Ding, H. Xie, Y.-T. Di and X.-J. Hao, *Tetrahedron Lett.*, **50**, 2917 (2009).
09TL3145	D.R. Bobeck, S. France, C.A. Leverett, F. Sanchez-Cantalejo and A. Padwa, *Tetrahedron Lett.*, **50**, 3145 (2009).
09TL3428	C. Curt, A. Sartori, L. Battistini, G. Rassu, F. Zanardi and G. Casiraghi, *Tetrahedron Lett.*, **50**, 3428 (2009).
09TL3657	D.L. Laventine, P.M. Cullis, M.D. Garcia and P.R. Jenkins, *Tetrahedron Lett.*, **50**, 3657 (2009).
09TL4674	Y. Oh, S.M. Kim and D.Y. Kim, *Tetrahedron Lett.*, **50**, 4674 (2009).
09TL5285	G. Pazos, M. Perez, Z. Gandara, G. Gomez and Y. Fall, *Tetrahedron Lett.*, **50**, 5285 (2009).
09TL5303	G.A. Kraus, A. Thite and F. Liu, *Tetrahedron Lett.*, **50**, 5303 (2009).
09TL6810	M. Liang, C. Saiz, C. Pizzo and P. Wipf, *Tetrahedron Lett.*, **50**, 6810 (2009).

CHAPTER 8

Eight-Membered and Larger Rings

George R. Newkome
The University of Akron, Akron, Ohio USA
newkome@uakron.edu

8.1. INTRODUCTION

Numerous reviews as well as perspectives, feature articles, tutorials, and mini-reviews have appeared throughout 2009 that are of particular interest to the macroheterocyclic enthusiast and those delving into supramolecular chemistry at the macromolecular level, as well as those studying nanoconstructs: multiple multicomponent macrocyclizations; <09CR796> 1,4,7,10-tetraazacyclododecane-N,N',N'',N'''-tetraacetic acid <09CCR1906>; cyclodextrin-based supramolecular polymers <09CSR875>; tetrathiafulvalenes <09CC2245>; thiacalixarenes <09JIPMC1>; calixcrowns <09JIPMC189>; cyclic and multicyclic polymeric crown ethers <09JPSA1971>; cyclic porphyrin arrays, as solar collectors <09ACR1922>; metal complexes of tetrapyrrolic ligands <09CSR2716>; supramolecular assemblies of acyclic oligopyrroles <09JIPMC193>; porphyrinoids <09ACIE4284>; phosphole-containing calixpyrroles, calixphyrins, and porphyrins <09ACR1193>; aza-deficient porphyrins <09CCR2036>; merging porphyrins with organometallics <08ACIE7396>; phthalocyanines <08TCR75>; porphyrin-containing catanenes and rotaxanes <09CSR422>; photophysical properties of expanded porphyrins <09CC261>; melding porphyrin chemistry with metal-catalyzed reactions <09CC1011>; supramolecular polymerizations <09CR5687>; supramolecular complexes <09ACIE3924>; supramolecular coordination chemistry <08ACIE8794>; supramolecular chemistry of metalloporphyrins <09CR1659>; self-organized porphyrins <09CR1630>; photoactive corrole-based arrays <09CSR1635>; catanenes <09CR6024, 09CSR1674, 09CSR1562, 09CSR1530, 09TCR136>; rotaxanes <09CC6329, 09TCR136, 09CSR1542, 09CSR1530>; polymeric rotaxanes <09CR5974>; rotaxanes and catenane structures for sensing charged guests <09OBC415>; cyclodextrin polyrotaxanes <09JPSA6333>; proton coordination <09CSR1663>; recognition and sensing of the fluoride anion <09CC2809>; chromogenic and fluorogenic hybrid chemosensors <09CSR1904>; metallosupramolecular systems <08TCR240>; electrides, ionic solids with cavity-trapped electrons <09ACR1564>; coordination chemistry of macrocyclic hexamine-dithiophenolate ligands <09CCR2244>; functional oligothiophenes <09CR1141>; phosphines in coordination-based self-assembly <09CSR1744>; sulfate anion templation of macrocyclic structures

<08JOC3336>; anion receptors <09CC513, 09CSR506>; halide anions with macrocyclic receptors <08COC1231>; supramolecular metallacycles and metallacages <08T11495>; metal-organic macrocycles, polyhedral, and frameworks <09CC3326>; metal-organic polyhedral, as supermolecular building blocks <09CSR1400>; macrocyclic approach to transition metal and uranyl Pacman complexes <09CC3154>; tellurium-containing macrocycles <09CCR1947>; giant macrocycles with thiophene subunits 09CRC395>; selena-macrocycles <09CCR1056>; polydentate phosphines <09JOMC3982>; chemistry within discrete, self-assembled host structures <09ACIE3418>; trinuclear metallacycles <09CR4979>; metallosupramolecular architectures, formation and scanning tunneling microscopy <09ACR249>; artificial polymerases and molecular chaperones <09JPSA4469>; the chemistry of the mechanical bond <09CSR1802>; and solvolysis to self-assembly (self-assembly of finite supramolecular ensembles) <09JOC2>.

Most synthetic chemist's lifes are an on-going (ad)venture in twists and turns – some planned – some by luck – but no one more so than Professor Jean-Pierre Sauvage, who just turned 65. As Fraser Stoddart aptly titled the history of Jean-Pierre: "The master of chemical topology" <09CSR1521> or Nazario Martín, who overviews Sauvage's birthday symposium: in the "Lord of the Rings" <10EJOC1407>, whose research is the perfect example of molecular twists and turns. Jean-Pierre's ingenious chemistry is a perfect blending of the arts and science as reflected by his leadership approach to metal template-directed procedures to [2]catenane, then [3] catenanes and on to the first molecular trefoil knot. His chemistry has made the difficult seems so simple. Congratulations to a molecular topology genius – Jean-Pierre Sauvage, who has truly earned the golden ring.

As always, because of space limitations, only meso- and macrocycles possessing heteroatoms and/or subheterocyclic rings have been reviewed; in general, lactones, lactams, and cyclic imides have been excluded. In view of the delayed availability of some articles appearing in previous years, several have been incorporated, where appropriate. I apologize in advance since it is impossible to do justice to this topic and the numerous researchers that have elegantly contributed to the field in the space allocated.

8.2. CARBON–OXYGEN RINGS
8.2.1 Crown Ethers

Since the advent of crown ethers in the mid-sixties, the attachment of substituents and their fusion into ring systems continues to expand. The macrocyclization of disilyl ether **1** with **2** *via* a Nicholas ether-exchange reaction with TMSOTf in CH_2Cl_2 gave an interesting dicobalt hexacarbonyl complex **3** appended to the 20-membered benzocrown ethereal frame; treatment of this cobalt complex with cerium ammonium nitrate (CAN) in MeOH generated (85%) the free cyclic alkyne, which was transformed to the corresponding vinyl iodide in 80% yield <09OL1313>. The aminomethyl-18-crown-6, diaminodibenzo-18-crown-6, and carboxybenzo-18-crown-6 ethers have been used to create a series of immobilized crown ethers that were used in a continuous flow reactor <09T1618>. The attachment *via* an alkyne linker of a tetrathiafulvalene (TTF) to a benzo-18-crown ether has given rise to a sensor for Pb(II) over other cations <09NJC813>.

The 4,6-dicarboxyl- and 4,6-*bis*(4-pyridinylethynyl)-*m*-phenylene-*m'*-phenylene-32-crown-10 were prepared *via* a macrocyclization procedure, then the later was treated with an appropriate *bis*-Pt(II) reagent to form a tetrametallo-rhomboid structure <09JOC3905>. Synthesis of *bis*[5-(phenylureidomethylene)-*m*-phenylene]-32-crown-10 has been reported and shown to complex divalent salts of paraquat by introducing ion-pair recognition sites <09JOC1322>. The introduction of a rigid, aromatic, H-shaped scaffold, pentiptycene into the dibenzocrown ether gave rise to a new type of molecular tweezers (**4**) <09OL4446> or **5** possessing a *bis*(crown ether) <09CC1987>. An improved three-step procedure for the macrocyclic 1,5-dinaphtho[38]-crown-10 has appeared <10TL983>. Also a practical and regioselective route to *syn*-substituted, specifically nitro, formyl, and carbomethoxy groups, dibenzo-3-crown-10 ethers has been reported <09T2285>.

Numerous rotaxanes, molecular machines, have been reported in which a large crown ether, generally a dibenzo[3*n*]crown-*n*, is a component combined with an encapsulated dumbbell-shaped rod, which can be composed of: a mannoside azide containing a pyridinium amide moiety and an alkyne ammonium moiety coupled *via* a click reaction <09CEJ5186>; the known [(C$_5$H$_4$NCH$_2$C$_6$H$_4$CH$_2$Br)$_2$(PF$_6$)$_2$] derivative with *tris*(4-methoxyphenyl)phosphane <09EJOC1053>; the known extended [C$_6$H$_4$(CH$_2$NC$_5$H$_4$C$_5$H$_4$NCH$_2$C$_6$H$_4$CH$_2$Br)$_2$(PF$_6$)$_4$] derivative with *tris*(4-methoxyphenyl)phosphane <09EJOC6128>; *bis*-pyridine-terminated naphthalenediimide thread capped with either Ru(II) carbonyl or Rh(III) iodide porphyrin <09CSR1701>; and a guanidium salt thread capped with 3,5-di-*tert*-butylbenzyl bromide <09OL613>. The use of molecular cage **6** has different threading modes thus opening the door to mimicking muscle-like behavior; the thread {[(CH$_2$)$_4$NH$_2$(CH$_2$)$_6$NC$_5$H$_4$C$_6$H$_3$(CMe$_3$)$_2$]$_2$ (PF$_6$)$_4$} was capable of contraction in the presence of Cu(II) and relaxation in the presence of tetrabutylammonium fluoride

(TBAF) in MeCN <09OL385>. This same molecular cage **6** was also capable of hosting *bis*diazonium, *bis*pyridinium, anthraquinone guests <09OL4604>. The use of cyclotriveratrylene caps gave rise to a more globular container (**7**) that was shown to encapsulate dimethyldiazapyrenium and 4,4′-biphenyl*bis*diazonium ions <09CC5814>. A versatile route to a double threaded, bistable [*c*2]daisy chain has been described in which propargyl and pentenyl moieties have been grafted onto the [*c*2]daisy chains using a template-directed procedure; these functionalized daisy-chains as well as the polymeric structures "undergo quantitative, efficient and fully reversible switching processes in solution." <09JA7126> A family of rigid-strut-containing crown ethers and [2]catenanes have been constructed for the incorporation into metal-organic structures <09CEJ13356>. Based on the dibenzo-24-crown-8-terminated four-arm star, poly(ε-caprolactone) (PCL) and dibenzyl-ammonium-terminated two-arm PCL were prepared by a combination of ring-opening polymerization and click-chemistry, responsive supramolecular gels generated and subsequently shown to undergo thermal and pH-induced reversible sol-gel transitions <09ACIE1798>.

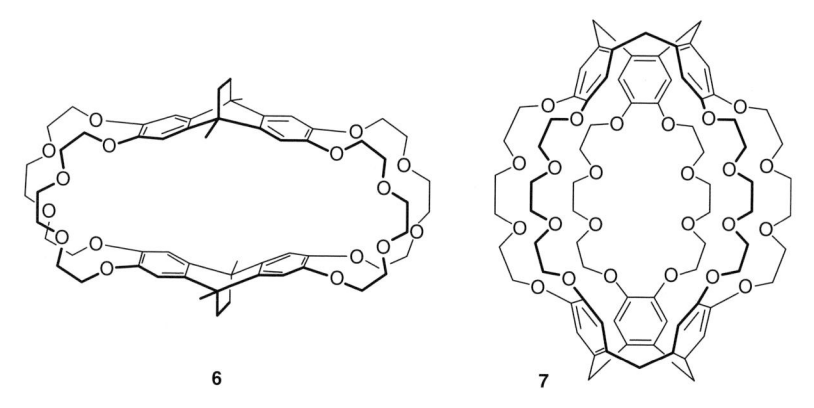

The use of *bis*(*m*-phenylene)-32-crown-10-based cryptand (**8**) was treated with two different *bis*-paraquat rings to generate 1:1 catenane (**9**) with the smaller molecular ring as well as a 2:1 catenane (**10**) when a larger *bis*-paraquat was utilized <09OBC1288>.

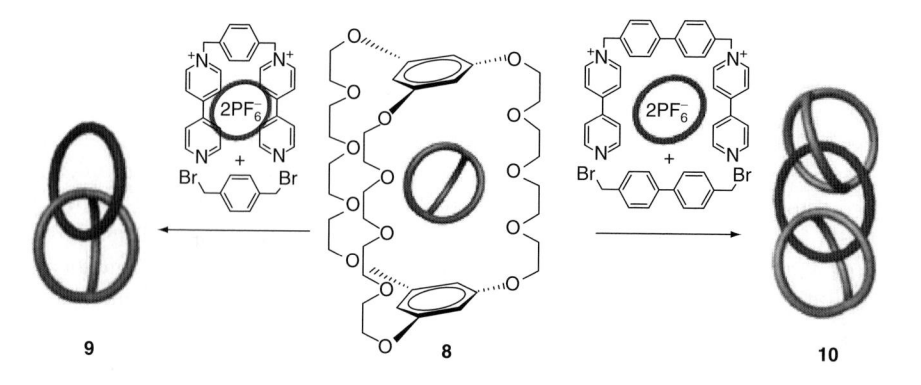

8.2.2 Calixarenes

A series of calix[4]arene-1,3-crown-4 possessing ionizable groups, such as oxyacetic acid and N-(X)sulfonyl oxyacetamide, where X = -Me, -C$_6$H$_5$, 4-C$_6$H$_4$NO$_2$, and -CF$_3$ in order to tune the acidity <09T5893>. The synthesis of 7,15,23-tri-*tert*-butyl-25,26,27-triadamantylketone-2,3,10,11,18,19-hexahomo-3,11,19-trioxacalix-[3]arene was prepared (22%) from *p-tert*-butylhexahomotrioxacalix[3]arene and 1-adamantyl bromomethyl ketone; this triketone is a weak extractant but shows a strong selectivity for Na(I) and some preference for Ag(I) <09T496>. A copolymer possessing both calix[4]pyrrole and benzo[15]crown-5 has been prepared as well as the calixpyrrole crown ether pseudodimer **11** <08ACIE9648>.

11

8.2.3 Cyclophanes

A double Sonogashira coupling of a diiodoparacyclophane with alkyne reagents generated planar, chiral 2,5-dialkynylparacyclophanes using a chiral Pd catalyst prepared *in situ* from PdCl$_2$(MeCN)$_2$ and Taniaphos; the asymmetric coupling gave up to *ca.* 80% ee <09CC1870>.

8.3. CARBON–NITROGEN RINGS

8.3.1 Azamacrocycles

Due to its expense, a convenient and inexpensive method for the synthesis of 1-(acetic acid)-4,7,10-*tris*(*tert*-butoxycarbonylmethyl)-1,4,7,10-tetraazacyclododecane [DOTA-*tris-t*-bu ester] has been reported <09TL2929>. A chiral tetraazamacrocycle with four pendant arms was prepared by repeated ring-openings of an *N*-nosylaziridine and a secondary benzylamine, both of which were derived from tyrosine, followed by an intramolecular alkylation <09OL2289>. A series of new 20- and 25-membered polyacetylenic azamacrocycles have been prepared and subsequently subjected to a catalytic [2+2+2]cyclotrimerization using Wilkinson's catalyst; the 25-membered ring led to the expected product in contrast to the smaller 20-membered ring <09CEJ5289>. The addition of two equivalents of HCl to *N*,*N'*-diisopropyl-1,8-diazacyclotetradeca-3,5,10,12-tetrayne generated a tricyclic species possessing a central cyclooctatetraene ring <09EJOC3006>. Macrocycle **12** was prepared from nitro-functionalized *m*-terphenyl moieties *via* reductive dimerization <09EJOC2562>. A new and efficient procedure for the preparation of unsymmetrical difunctionalized cyclen and its closely related derivatives using a modified

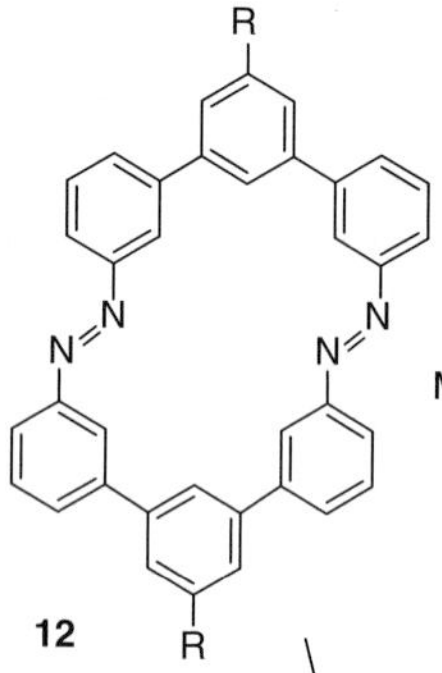

Ugi reaction has appeared <09OL417>. The Pd-catalyzed aryl amination of 1,3,5-tris[di(4-chlorophenyl)amino]benzene with 1,3-bis[(4-methoxyphenyl)amino]benzene gave (ca. 10%) the trimacrocycle **13** in a one-pot reaction <08CC6573>. Treatment of equimolar amounts of N-methyl-2,2′-diaminodiethylamine and terephthaldehyde under high-dilution conditions afforded a macrocyclic Schiff base, which was reduced to give the desired hexaaza macrocycle; in its tetraprotonated form, it coordinated two bromides <09TL6537>. Tetranitroazacalix[4]arenes have been synthesized by the aromatic nucleophilic substitution of 1,5-difluoro-2,4-dinitrobenzene with 1,3-diaminobenzene <09TL620>. A polyaza cryptand (**14**) with pyridinyl spacers was synthesized (54% overall) by a [2+3]-Schiff-base condensation of 1,3,5-tris(aminomethyl)-2,4,6-triethylbenzene and 2,6-diformylpyridine, followed by NaBH$_4$ reduction <09JOC8638>.

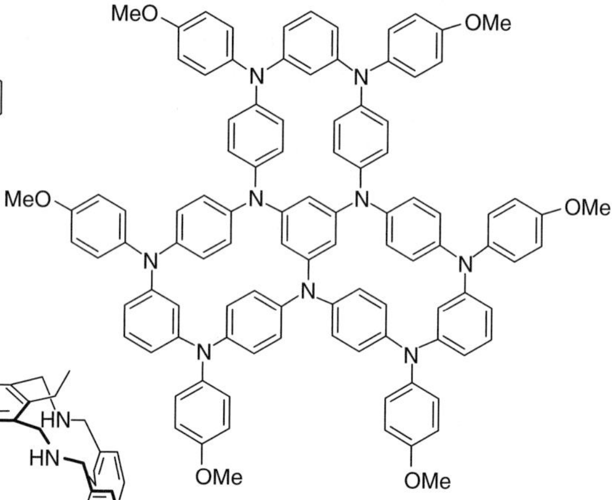

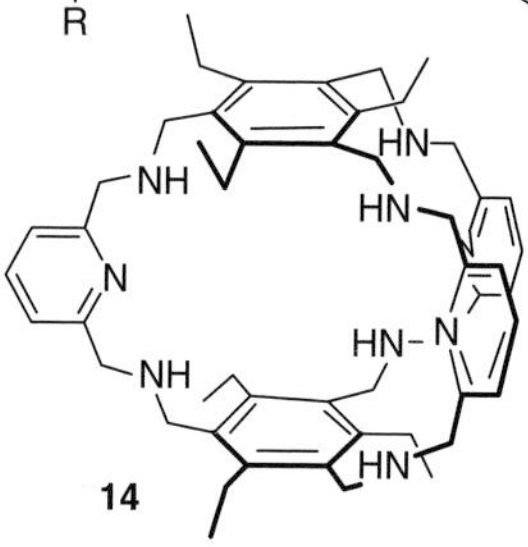

New calix[4]pyrroles strapped with the chromogenic dipyrrolylquinoxaline <09JOC1065> or 1,2,3-triazole <09CC3017> were prepared and shown to act as a selective colorimetric response agent to fluoride, dihydrogen phosphate and acetate ions, or a high affinity for chloride and lipid bilayer chloride transport properties, respectively. The fusion of a tetrathiafulvalene to calix[4]pyrroles has appeared and was used to electrochemically sense chloride ions <09CEJ8128>. Reaction of 1,1′-[1,4-phenylenebis(methylene)]bis-4,4′-pyridinylpyridium hexafluorophosphate with 3,5-bis(chloromethyl)-1H-1,2,4-triazole gave the desired cyclophanes, 2,11,22,31-tetraazonia[1.0.1.1.0]paracyclo[1](3,5)triazolophane; the closely related 2,11,21,30-tetraazonia<[1.0]paracyclo[1](3,5)triazolophane>^{2} possessing two 1,2,4-triazole moieties was also prepared <09NJC300>. Similarly, treatment of 1,1′-[1,4-phenylene-bis(methylene)]-bis-4,4′-pyridinylpyridinium hexafluorophosphate with 2,5-di(bromomethyl)benzyl pent-4-ynoate generated the desired cyclophanes possessing a functionalized side arm capable of undergoing click chemistry with a polymer possessing azido groups <09T400>. The thermodynamic forecasting of mechanically interlocked molecular switches to be incorporated in either molecular electronic devices or nanoelectromechanical systems has appeared and should prove to be a great starting point for interest in this field <09OBC4391>.

8.3.2 Pyridine-Containing Macrocycles

Two new macrocyclic chelators possessing either a pyridine or 2,2′-bipyridine subunit as well as triethylenetetraaminetetraacetic acid core were synthesized from 2,6-*bis*(bromomethyl)pyridine or 6,6′-*bis*(bromomethyl)-2,2′-bipyridine, respectively <09TL6522>. Use of the related 5,5′-*bis*(bromomethyl)-2,2′-bipyridine with 5-*N*-*tert*-butyltriazinan-2-one gave a triazinanone-protected macrocycle, which was deprotected with HN(CH$_2$CH$_2$OH)$_2$ affording the *bis*-urea-bridged macrocycle **15** <09JA17620>. The tetra-aza-calix[1]arene[3]pyridine (**16**) was synthesized using 2,6-di(methylamino)pyridine and 1,3-di(6-bromopyridinyl-*N*-methylamino)benzene in the presence of Cu(II) forming the aryl-Cu(III) complex, which with different nucleophiles gave 2-aryl substitution in high yield <09CC2899, 09JOC5361>. A series of unusual chiral (2,5)pyrido[7$_4$]allenoacetylenic cyclophanes (**17**) has been reported and their reaction with [Re(CO)$_5$Br] led to an uncommon tetracarbonyl rhenium complex [Re(CO)$_4$L] <09CEJ6495>. A new efficient route to 4-carbomethoxy-6,6′-dimethyl-2,2′-bipyridine has appeared and has opened an interesting avenue to [N$_2$(bpy)$_3$(CO$_2$Me)]cryptands that can easily be attached or structurally tuned <09T7673>. Treatment of 3,6,9,15-tetraazabicyclo-[9.3.1]pentadeca-1(15),11,13-triene with either BrCH$_2$CONHCH$_2$CO$_2$CMe$_3$ or ClCH$_2$CONHCH$_2$PO(OBu)$_2$ generating the 3,6,9-tri-*N*-alkylated products, which readily form the desired lanthanide(III) complexes <09CEJ13188>.

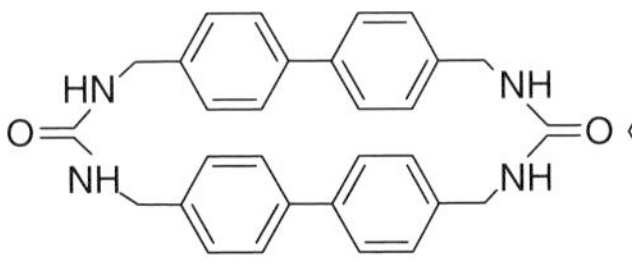

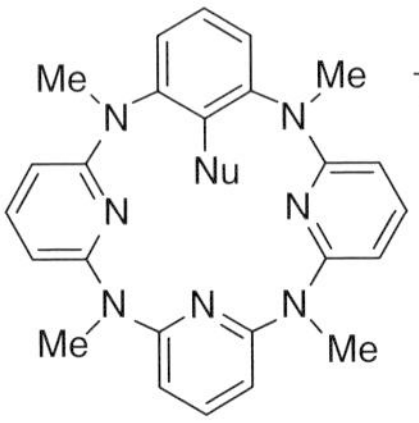

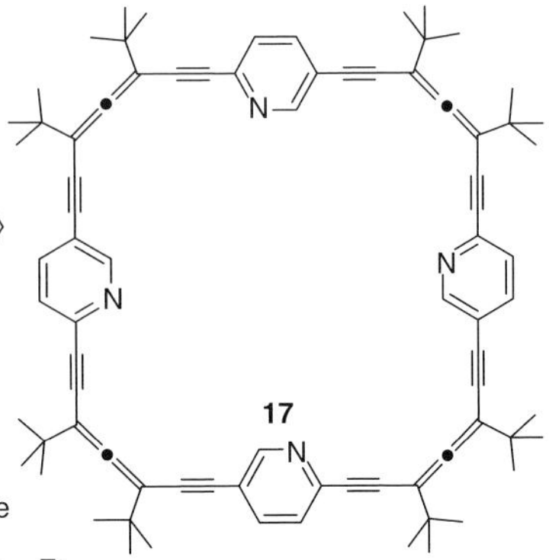

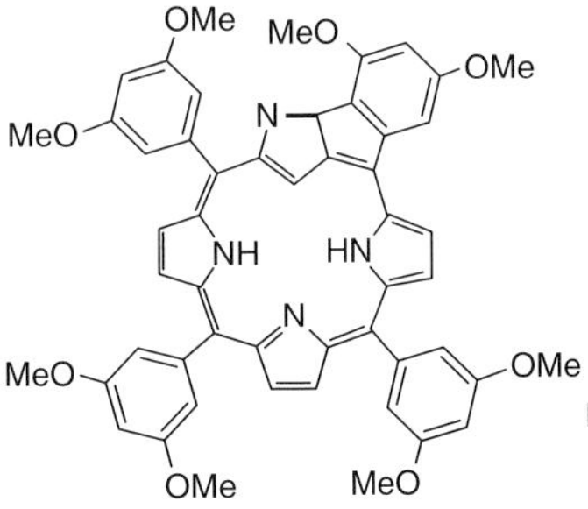

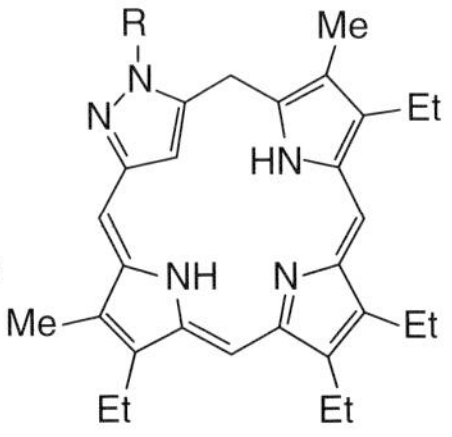

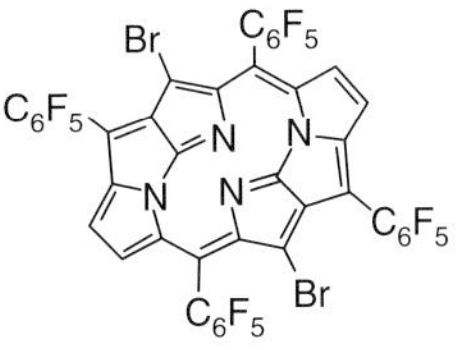

8.3.3 Porphyrins and Confused Porphyrins

Acid-catalyzed exocyclic ring formation in *meso*-tetra*kis*(3′,5′-dimethoxyphenyl)-2-aza-21-carbaporphyrin regioselectively occurred to give (70%) **18**; whereas, refluxing in toluene in air gave the dimer in variable yields <09EJOC3930>. The direct amination at the inner carbon of carbaporpholactone has been shown; the analogous reactivity of *N*-confused porphyrin has also been considered <09JOC8547>. The *N*-substituted pyrazole dialdehydes reacted with tripyrrane under typical [3+1]-conditions to give aza-analogues of the *N*-confused porphyrins **19** <08CC6309>. Construction of a porphyrin-*N*-confused porphyrin dyad by the acid-catalyzed [2+2] condensation of a porphyrin-attached dipyrromethane with *N*-confused dipyrromethane dicarbinol has been reported; an enhanced emission quantum yield as well as acceleration of excitation energy transfer was induced by anion binding <09OBC3027>. The Pt(II) complex of *meso*-tetra*kis*(pentafluorophenyl) *N*-confused porphyrin efficiently bound Cl$^-$, Br$^-$, and I$^-$ anions at the peripheral NH$^-$; whereas, deprotonation occurred with F$^-$, H$_2$PO$_4^-$, and OH$^-$ in CH$_2$Cl$_2$ <09DT6151>. In a study to prepare A,D-di-*p*-benzi[28]hexaphyrin(1.1.1.1.1.1), when hindered aryl substituents, *e.g.*, 2,4,6-triethylphenyl, were utilized in the 1,4-di(pyrrolylarylmethyl)benzene intermediate, its acid-catalyzed reaction with benzaldehyde, followed by DDQ oxidation gave a doubly confused benzihexaphyrin, namely 8,23-dioxo-7,22-diaza-33,37-dicarba-A,D-di-*p*-benzi[28]hexaphyrins-(1.1.1.1.1.1) <09OL3930>. Using a "fusion approach", the doubly *N*-fused porphyrin **20** was prepared from an *N*-confused, *N*-fused porphyrin, then bromination with 1,3-dibromo-2,2-dimethylhydantoin (to be noted: NBS did not work), followed by treatment with Hünig's base <08ACIE8913>. *N*-Substitution of 20-π-electron β-tetra*kis*(trifluoromethyl)-*meso*-tetraphenylporphyrin with α,ω-dibromoalkanes in DMF at ambient temperature in the presence of base gave various *N*-substituted isophlorins <09S3860>.

An interesting approach to subphthalocyanines (**21**) has been reported, in which 4-iodophthalonitrile was reacted with BCl$_3$ in a one-pot cyclotrimerization procedure, followed by chlorine exchange with phenol; the iodo moieties were subjected to a Sonogashira catalytic cross coupling reaction to functionalize the periphery <09TL2041>. Although the boron is locked within the core of the subphthalocyanines, the axial group (Br < OH ≤ OPh ~ Cl) is usually thermally the first moiety to be lost <09JOMC1617>. In an attempt to react a pyrrole *bis*acrylaldehyde with a polyfunctionalized dipyrrylmethane to generate the [18]triphyrin(3.1.3), an expanded subporphyrin, the reaction course led to a [26]pentaphyrin(3.1.0.1.3) (**22**), an expanded sapphyrin ring system and trace (6%) of the vinylogous hexaphyrin <09OL1249>. The [26]hexaphyrin (**23**) was prepared (20%) by condensation of the appropriate dipyrromethane-dicarbinol and 3,4-diphenylpyrrole under acidic conditions, followed by DDQ oxidation; its reduction gave the [28]hexaphyrin in quantitative yield <09CC6047>. Treatment of 3,5-dibenzoylporphyrins with ammonium acetate gave an unusual oxypyridochlorins (**24**), which is the first example of a pyridine-fused porphyrinoid <09CC1028>. The syntheses and dynamic figure-eight loop-structures of series of polyalkylated *meso*-tetraaryl[32]-

octaphyrins-(1.0.1.0.1.0.1.0) have appeared <09JOC3579>. *Bis*-picket-fence corrole ligands were prepared from (2,6-dichlorophenyl)dipyrromethane in two steps <09JOMC1011>.

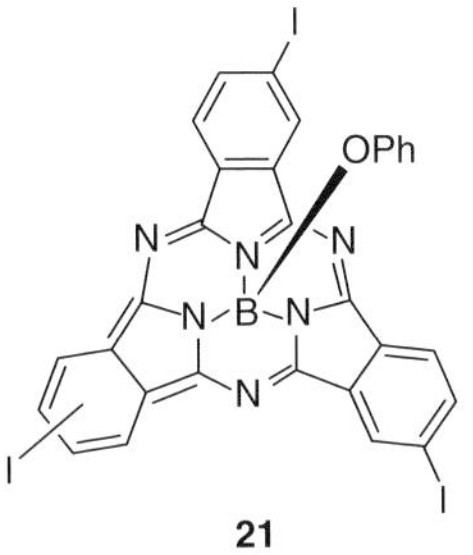

8.4. CARBON–SULFUR RINGS

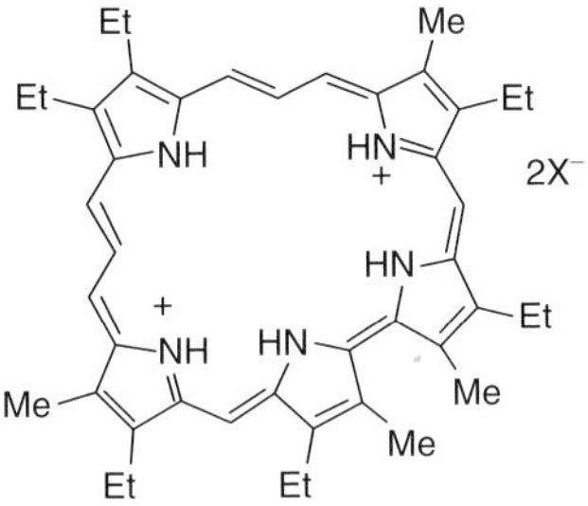

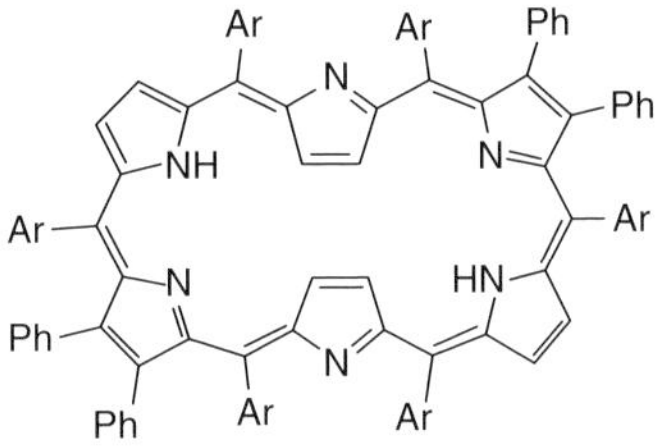

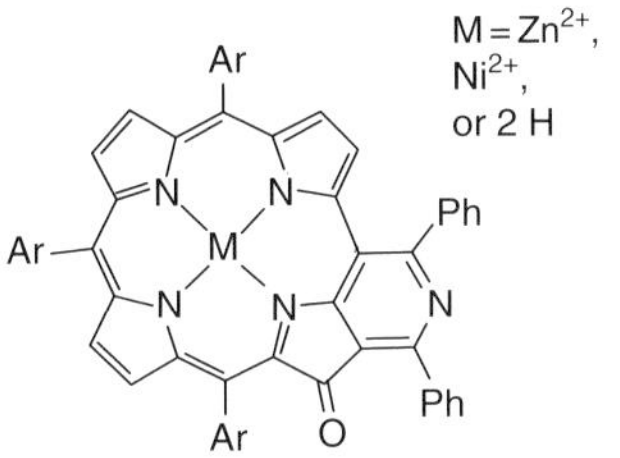

The rearrangement of diphenylpropargylic 1,3-dithiane catalyzed by Au(I) gave (67%) the stable 8-membered dithio-substituted cyclic allene **25**, which remained unchanged after 24 hours at 100 °C under nitrogen; diverse catalysts were utilized but AuCl and Au(PPh$_3$)Cl/AgSbF$_6$ gave the best yields and aryl substituents had minor effects on the conversion (60-90%) <09CC2535>. Although *p-tert*-butylthiacalix[4]arenes have been previously prepared, their functionalization has been utilized in an attempt to create selective chemosensors <09EJOC4534, 09T7109>. A communication concerning the microwave assisted synthesis of thiacalix[*n*]arenes has appeared <09JIPMC379>. Treatment of either the *cis*- or *trans*-hex-3-ene-1,6-dithiol with CsF/Celite and air in MeCN gave 3,4,7,8-tetrahydro[1,2]dithiocine (65%) or 1,2,9,10-tetrathiacyclohexadeca-5,13-diene (21%), respectively <09T1257>. The formation of giant cyclic polythiophenes, *via* a cheletropic expulsion from a Pt(II) macrocyclic intermediate, has been reported <09ACIE6632>.

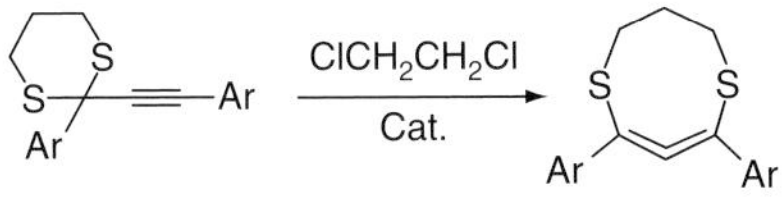

8.5. CARBON–SELENIUM RINGS

A THF solution of 1,3-*bis*(bromomethyl)-5-*tert*-butyl-2-methoxybenzene was treated with a molar equivalent of NaSeH in EtOH over 15 minutes resulting 86% in a kinetic mixture of homoselenacalix[*n*]arenes (**26**; *n* = 0-5) with a 37:20:14:8:7 product distribution, respectively. The tetraselena[3.3.3.3]calixarene (**26**, *n*=1) was prepared by the initial conversion of a *bis*-bromomethyl reagent with KSeCN to create the *bis*-selenocyanate, which with one equivalent of the same *bis*-bromomethyl reagent gave the desired **26** (*n*=1) in 67% yield along with **26** (*n*=3) in 18% yield <09OL3040>. Two macrocyclic polyselenaferrocenophanes, 1,5,9-triselena[9]-ferrocenophane and 1,5,9,21,25,29-hexaselena[9.9]ferrocenophane, have been prepared from 1,1′-*bis*(3-bromopropylseleno)ferrocene and Na$_2$Se <09ICC846>. The simultaneous addition of Se[(CH$_2$)$_3$OTs]$_2$ and *o*-C$_6$H$_4$(CH$_2$SeCN)$_2$ or NCSe(CH$_2$)$_3$SeCN in THF/EtOH to a suspension of NaBH$_4$ in THF/EtOH gave high yields of the desired 13- and 12-membered triselenomacrocycles; a related 11-membered ring (**27**) was similarly prepared <09DT4569>.

8.6. CARBON–NITROGEN–OXYGEN RINGS

The aza-locus on azacrown ethers has been the ideal site to append diverse functional groups giving rise to molecular devices with monoazacrowns: chemiluminescence with benzylideneacridans <09JOC1014>, fluorogenic chemosensor with 9-methylanthracene <09CC2399>, di- and tripeptides of luminescent benzocrown ether aminocarboxylic acids <09T690>, aza-24-crown-8 attached to the 2,6-pyridinylene ethynylene polymer for saccharide recognition <09CC2121>, *N*-alkyl-aza-15-crown-5 possessing a fused β-D-glycopyranoside moiety for asymmetric synthesis of substituted α-amino phosphonates <09SL1429>, and for the diazacrowns: 1- and 2D-coordination polymers with *N*,*N*′-*bis*(4-pyridinylmethyl)diaza-18-crown-6 <09CC5579>, Cu(II) complex from *N*,*N*′-[1,2,3]triazolo ligands derived from *N*,*N*′-dipropargyl-4,13-diaza-18-crown-6 <09ICC382>, a modular approach *via* click chemistry afforded a series of functionalized and expanded crown ethers <09ACIE6654>; membrane-length amphiphiles with ion channel activity has been shown for *bis*- and *tris*-diazacrowns <09CEJ10543>. The use of the known <07HCA1439> 39-membered (**28**) or 41-membered (**29**) macrocycles, both possessing the 8,8′-diphenyl-3,3′-biquinoline subunit created a nonsterically

                                28   29

unhindered binding locus permitting the introduction of a metal with two bipyridine ligands; the introduction of substituted 8,8′-diphenyl-3,3′-biquinoline subunits that can be capped with hindered tetraphenylmethane stoppers, using Click chemistry, gave entre to double threaded rotaxanes in good yield <09JA6794>. Different bibrachial cyclo*bis*intercaland-type C,N,O-macrocycles with 2,6-naphthalene subunits and pyrene, ferrocene or primary amino moieties were synthesized by a [2+2]-cyclocondensation or the related functionalized diethylenetriamine derivatives with 2,6-diformylnaphthalene, followed by reduction of imine intermediate <09T1349>.

The Ag(I)-mediated self-assembly of discrete molecular cage has been demonstrated for oxacalix[2](1,2)benzene[2]pyrazine; however, an infinite chain of coordination cages was obtained from the isomeric oxacalix[2](1,3)benzene[2]pyrazine <09JA8338>. The successful reversible controlling of the helical pitch length and the handiness of the induced helix by photochemical and thermal isomerizations of the chiral benzenophane 30 in an induced cholesteric liquid crystal phase has been shown <09CC3609>. A fluorescence macrocyclic receptor 31 has been prepared and developed for the molecular recognition of amino acids in water/DMF solution; there was selectivity for *L*-asparate and *L*-cysteine <09ICC815>. The fluorescent *bis*anthracene macrocycle 32 has been prepared <09CEJ1314>; its fluorescence could be modulated upon bonding with mismatched base pairs in DNA. A series of substituted oxacalix[3]arene[3]pyrimidines as well as the unsymmetrical oxacalix [4]- and oxacalix[8]-arenes were prepared *via* nucleophilic aromatic substitution <09OL1681>. A series of "clicked" triazole-bridging calix[4]crowns possessing both soft and hard binding loci, has been prepared <09NJC725>. Click chemistry has also been applied to the simple procedure in which (HC≡CCH$_2$)$_3$N is reacted with 3,5-(OHC)C$_6$H$_3$OCH$_2$CH$_2$N$_3$, subsequent treatment with ethylenediamine gave (99%) the hexaimine molecular cage 33; chirality can be incorporated by the use of (1*R*,2*R*)-1,2-diaminocyclohexane <09CC343>. The chiral *N*-benzyl-5,11,17,25-tetra-*tert*-butyl-25,27-dihydroxy (and methoxy)-26,28-(4′*R*,8′*R*-diphenyl-6′-aza-3′,9′-dioxaundecane)dioxycalix[4]arenes have been prepared and used in the chiral recognition of the enantiomers of phenylalanine and alanine methyl ester hydrochlorides <09T3014>.

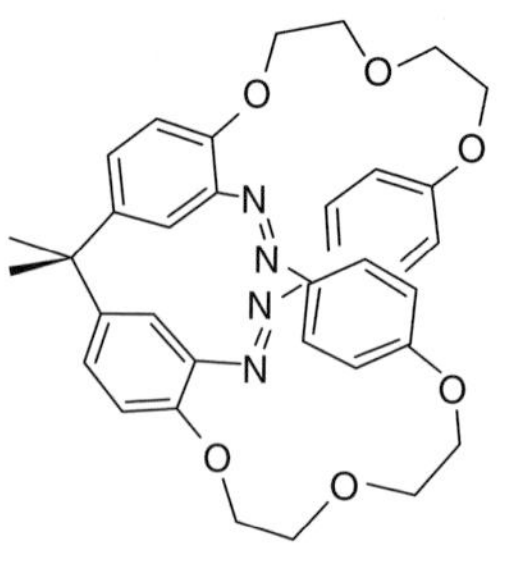

30 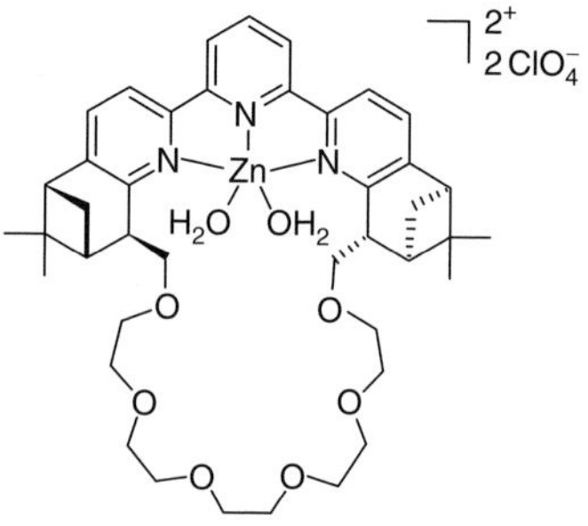 **31**

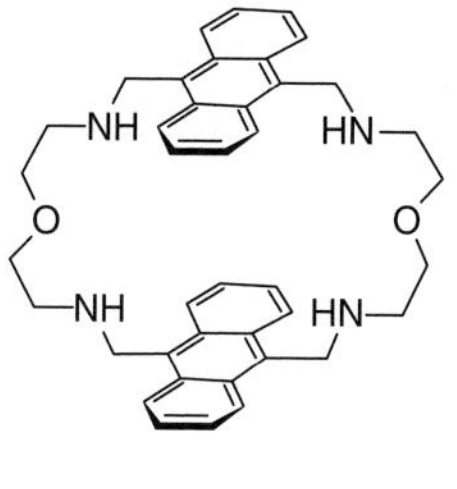

32 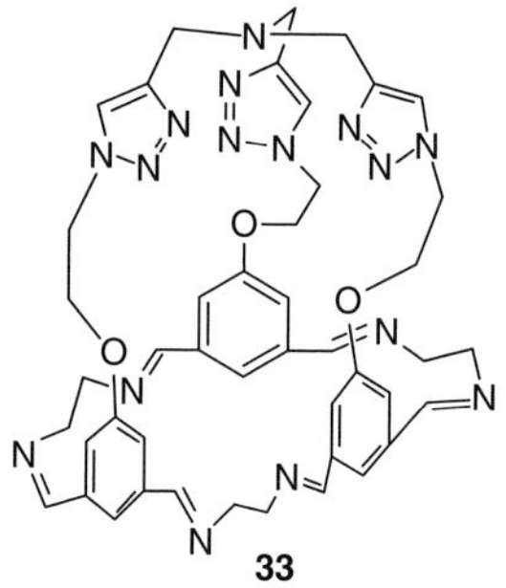 **33**

The reaction of the *N,N'*-bis(2-aminoethyl) derivative of the dibenzofuran[22]N_2O_3 macrocycle with *m*-xylyldicarbonyl dichloride gave the desired macrobicycle, which is a receptor for anion recognition <09JOC4819>. The interesting aza-oxa cryptand **34** possessing three different fluorophores, specifically 7-nitrobenz-2-oxa-1,3-diazole, anthracene, and quinoline, was demonstrated when a transition metal is added to this dye; it enters the cavity and occupies the N_4 locus thus assisting signal transduction <09CC4982>.

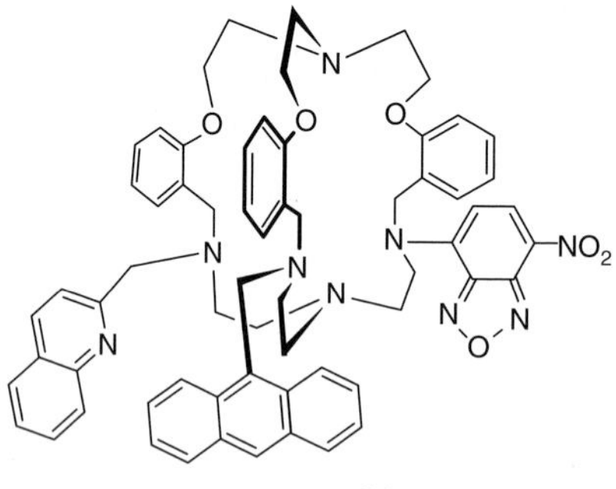

34 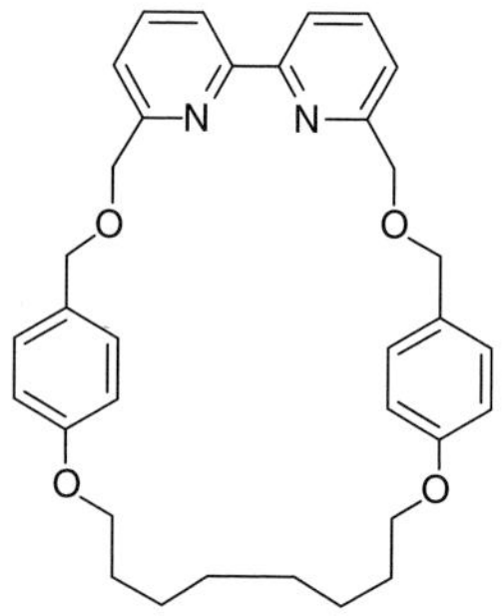 **38**

The rotaxanes and catenanes are important components to the family of artificial motors using transition metal complexes and open potential avenues to logic machines. The C,N,O-macrocycle **35** with a bipyridine rod readily rotates by a simple redox process; Cu(I) favors a bipyridine-phenanthroline (bpy-phen) complex,

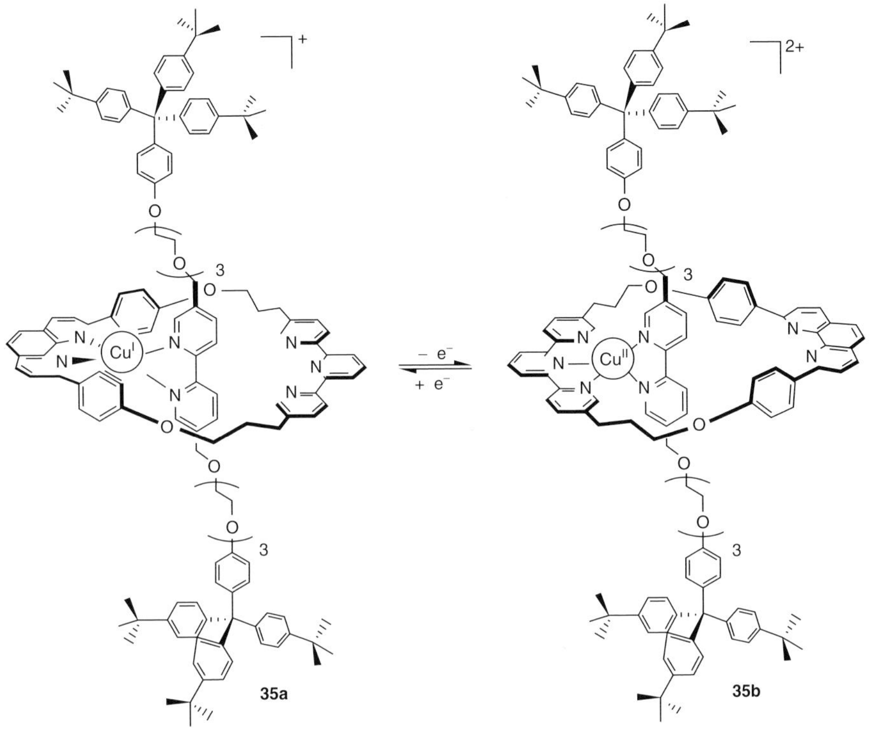

35a **35b**

whereas, Cu(II) favors the bpy-terpyridine (tpy) combination <09CEJ1310, 09JA5609, 09CEJ4124>. A Cu(I)-driven quadruple threaded [4]pseudorotaxane was quantitatively formed when the rigid *bis*-bipyridine rod **36** was treated with the rigid azacrown ether **37** <09CC1706, 09JA5609>; also see <09NJC2148> for related [3](pseudo)rotaxanes. The use of Cu(I) in the templation process is nicely presented in the creation of interlocked catenane *via* the use of **38** in which the Cu(I)

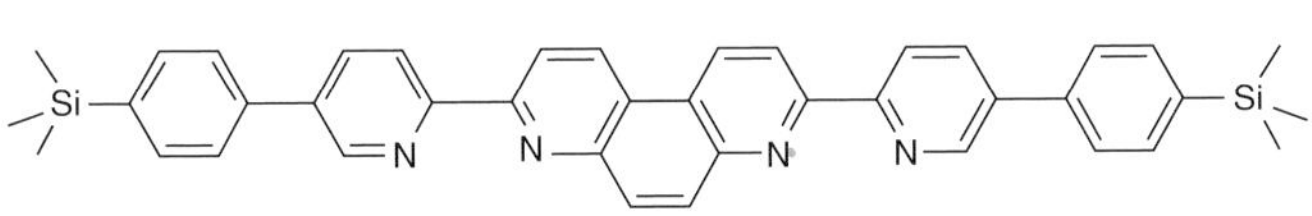

36

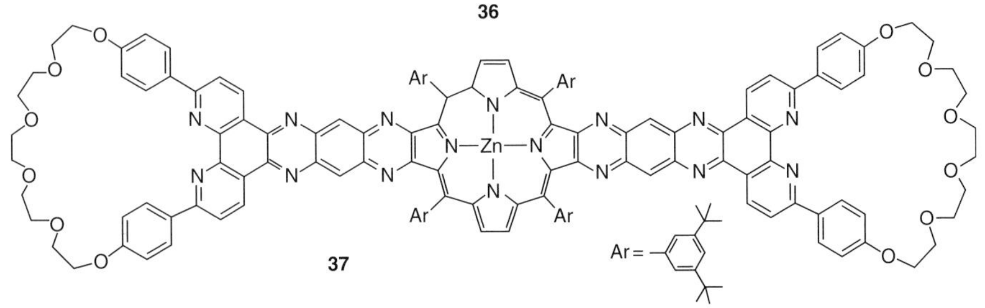

37

is initially complexed with a *bis*-alkyne and subsequently transformed to the *bis*-alkyne by a modified Cadiot-Chodkiewicz reaction <09JA15924>; Co(III)-templation has also been reported to give rotaxanes and catenanes <09JA3762>. Other examples of this oxidative intramolecular diyne coupling mediated by macrocyclic Cu(I) complexes have appeared <09ACIE504>. The formation of functionalized catenanes *via* the initial creation of different rotaxanes terminated with either azide or alkyne moieties permitted the use of Click chemistry to generate the desired interlocked structures <09CEJ5444>.

8.7. CARBON–NITROGEN–SULFUR RINGS

Macrocyclization of a benzylthiol-terminated *tris*(pyrazolyl)methane <04OL747> with either 1,3,5-*tris*(bromomethyl)benzene or 1,3,5-*tris*(bromomethyl)-2,4,6-trimethylbenzene in DMF at 60 °C under high-dilution conditions gave the macrobicycles **39** in *ca.* 30%; the Cu(I) complexes were investigated <09NJC327>. A *meso*-alkylidenylthia(*p*-benzi)porphyrin (**40**) was synthesized (24%) using the traditional [3+1] strategy starting with an appropriate tripyrrane analogue with 2,5-*bis*(hydroxyarylmethyl)thiophene; the related dimer was also isolated in 7% yield <09CC5877>.

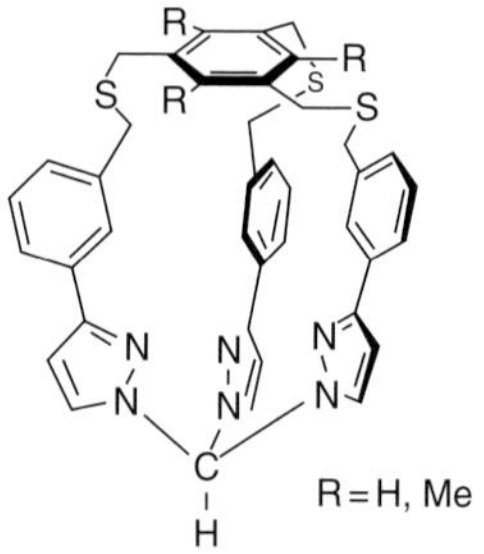

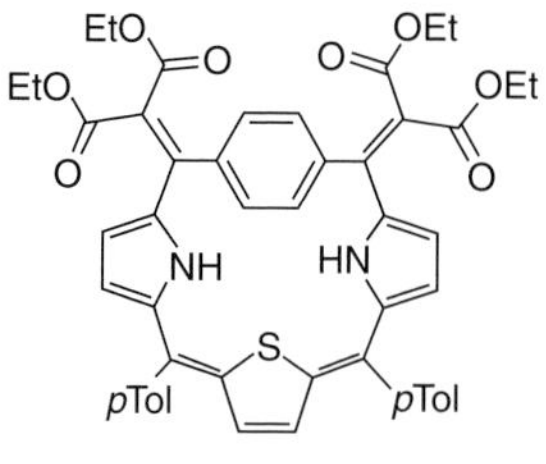

8.8. CARBON–SULFUR–OXYGEN RINGS

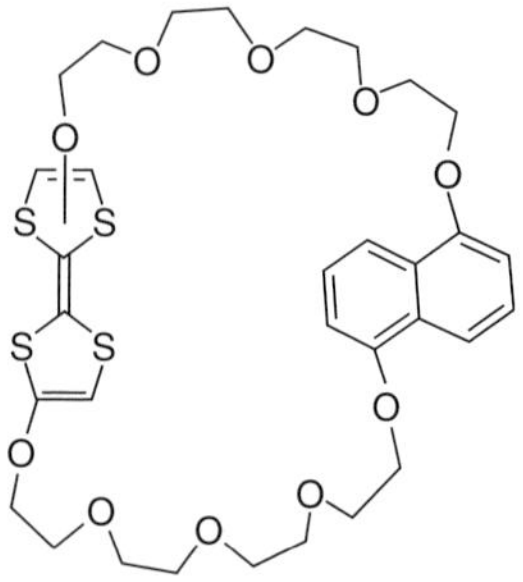

Treatment of 1,3-alternate 26,28-*bis*(5′-chloro-3′-oxapentyloxy)calix[4]arene with *cis*-1,2-dicyano-1,2-ethylenedithiolate gave the 1,3-alternate 26,28-(35,36-dicyano-34,37-dithia-29,32,40,43-tetraoxa-35-en)calix[4]arene-crown-5, which was attached to a magnesium porphyrazine core <09ICC304>. Other magnesium and metal-free porphyrazines possessing four tetrathiacrown ether-linked tetrathiafulvene moieties have also appeared and have been shown to possess two reductive and three oxidative processes as well as formed a charge-transfer complex <09ICC739>. A tetrathiafulvalene-based catenane, derived from **41**, has been easily synthesized by a "threading-followed-by-clipping" protocol using a mild as well as efficient Cu(II)-catalyzed Eglinton coupling <09JA11571>; its inclusion in poly[2]catenanes has appeared and shown to generate novel nanosuperstructures <09ACIE1792> as well adapted to form a bistable pretzelane <09CC4844>.

8.9. CARBON–OXYGEN–SILICON RINGS

Ring-contraction of nine-membered triperoxides, e.g., 1,2,4,5,7,8-hexaoxa-3-silonanes (**42**), upon treatment with reducing agents, such as PPh$_3$, P(C$_8$H$_{17}$)$_3$ or H$_2$NC(S)NH$_2$, gave (67-91%) the related previously unknown class of seven-membered rings namely the 1,3,5,6-tetraoxa-2-silapanes <09JOC1917>.

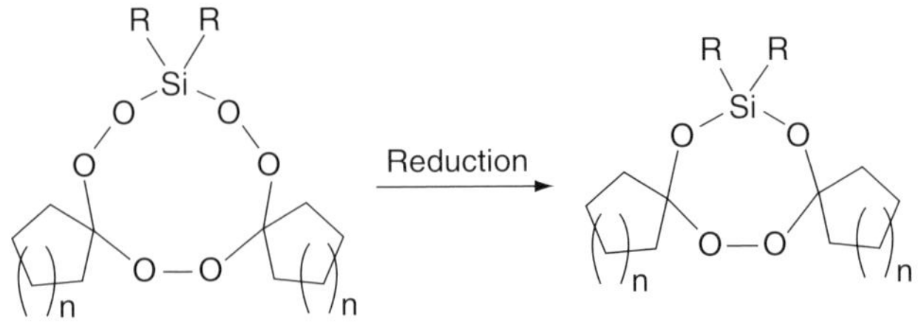

8.10. CARBON–OXYGEN–PHOSPHORUS RINGS

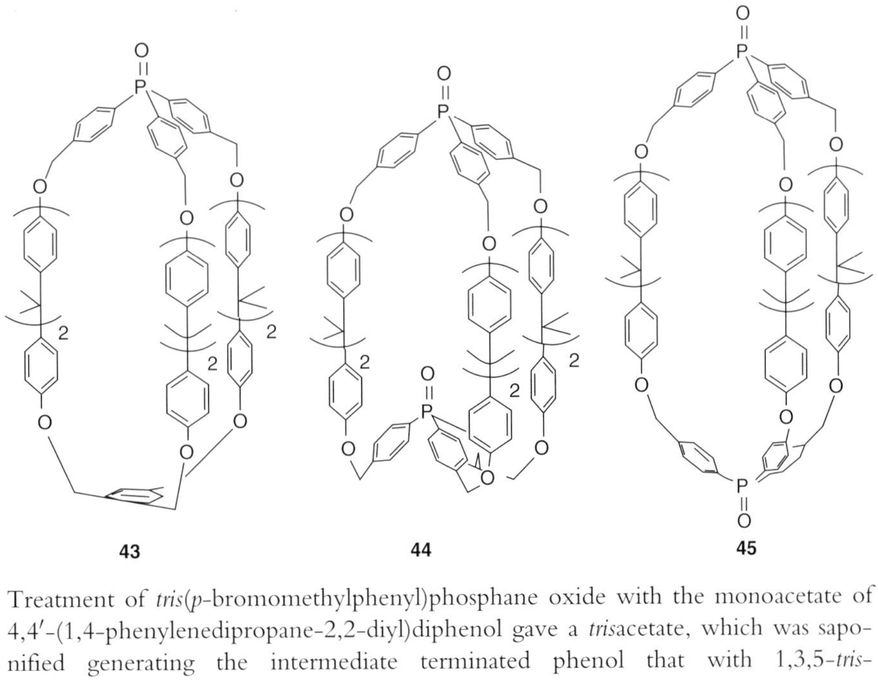

Treatment of *tris*(p-bromomethylphenyl)phosphane oxide with the monoacetate of 4,4′-(1,4-phenylenedipropane-2,2-diyl)diphenol gave a *tris*acetate, which was saponified generating the intermediate terminated phenol that with 1,3,5-*tris*-

(bromomethyl)benzene in the presence of K_2CO_3 and KI in dry DMF gave the desired macrobicyclic phosphane oxide **43** in 52% yield. The crystal structure of **43** showed an *out*-position for the P=O; the related *in*-isomer was not detected <09EJOC5571>. The above intermediate terminated phenol can also be treated with *tris*(p-bromomethylphenyl)phosphane oxide under similar conditions to generate a 2:1 mixture of the *in,out*- (**44**) and *out,out*- (**45**) isomers; each was reduced to the corresponding phosphanes <09T2995>. The first practical procedure for the synthesis of optically pure P-stereogenic 18-diphosphacrown-6 ethers containing a chiral *bis*phosphine moiety has appeared <09OL2241>.

8.11. CARBON–NITROGEN–PHOSPHORUS RINGS

The self-assembly of either 2,4,6-*tris*(isopropyl)phenyl- or mesityl-phosphine with formaldehyde and 1,4-*bis*[α-(4′-aminophenyl)isopropyl]benzene gave a new type of heterophane, specifically, $1^3,1^7,7^3,7^7$-tetra*kis*aryl-3,3,5,5,9,9,11,11-octamethyl-1,7(1,5)-*bis*(1,5-diaza-3,7-di-phosphacyclooctana)-2,4,6,8,10,12(1,4)-hexabenzenacyclododecaphane **46** <09DT490>.

8.12. CARBON–NITROGEN–SELENIUM RINGS

The simultaneous addition of 2,6-*bis*(bromomethyl)pyridine in THF/EtOH and either o-$C_6H_4(CH_2SeCN)_2$ or $NCSe(CH_2)_3SeCN$ in THF/EtOH gave a quantitative yield of either **47** or **48**; the crystal structures of both **47** as well as its Pt complex were ascertained <09DT4569>.

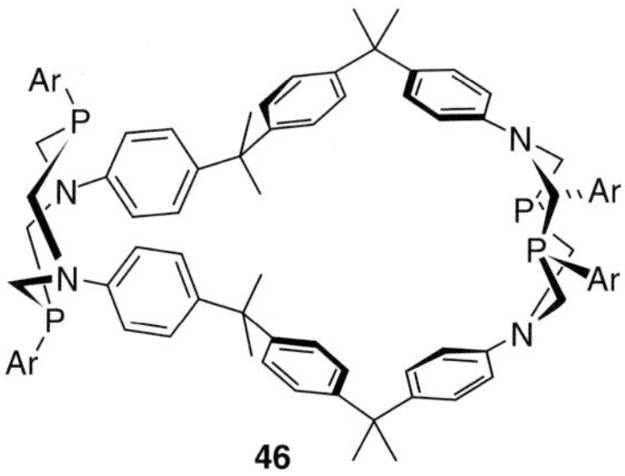

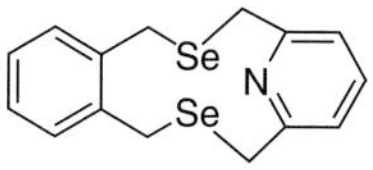

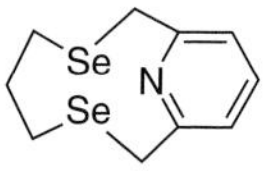

8.13. CARBON–NITROGEN–SULFUR–OXYGEN RINGS

The azo-coupled macrocyclic chromoionophores (**49**) incorporating either a benzene (X=CH) or pyridine (X=N) subunit have been synthesized by treatment of the appropriate *bis*(chloromethyl) derivative with $C_6H_5N(CH_2CH_2SH)_2$ with Cs_2CO_3 under high-dilution conditions, followed by reaction of the diazonium salt of p-nitroaniline <09OL1393>. Although the 1,4-dioxa-7,13-dithia-10-azacyclopentadecane is known, attachment with a 9-aminoacridizinium fluorophore (**50**) generated (47%) a fluorescent probe allowing the selective detection of Hg(II) in an aqueous environment <09CC3175>. The cyclocondensation of aromatic amines with 1,1'-bis(2,4-dinitrophenyl)-4,4'-bipyridinium salts gave a family of doubly or quadruply charged, macrocyclic N,N'-diarylbipyridinium cations <09OL5238>.

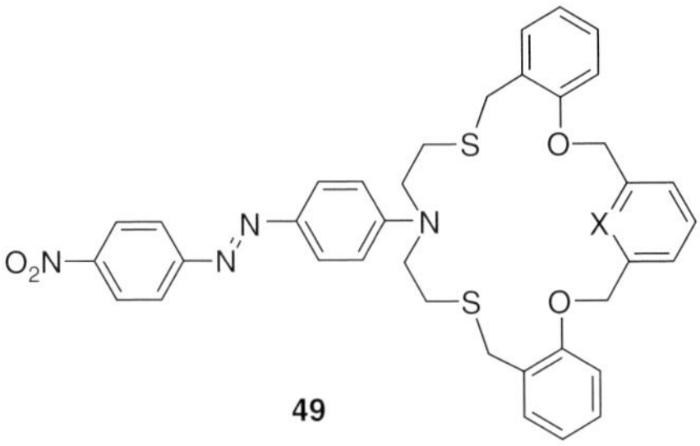

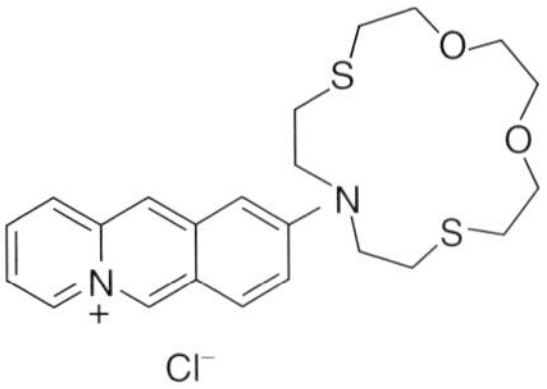

Meso-(4-Borylphenyl)thiaporphyrins with N_3S and N_2S_2 porphyrin moieties have been created and used for the preparation of meso-meso- and β-1,4-phenylene-bridged unsymmetrical porphyrin dyads <10SL67>.

8.14. CARBON–NITROGEN–PHOSPHORUS–SULFUR RINGS

Recently, the phosphole-containing P,S,N_2-hybrid calixphyrins (**51**, **52**) were reported <08JA990, 09JA14123> and these structures are interconvertible by redox processes; thus in order to evaluate the electronic effects, new P,S,N_2-hybrid calixarenes possessing *para*-substituted aryl moieties were prepared. The related Pd(II) and Rh(I)-P,S,N_2-complexes were also formed and evaluated <09O6213>. A theoretical study of the related P,S-containing hybrid calixarenes **53**, **54** has also appeared <09JA10955>.

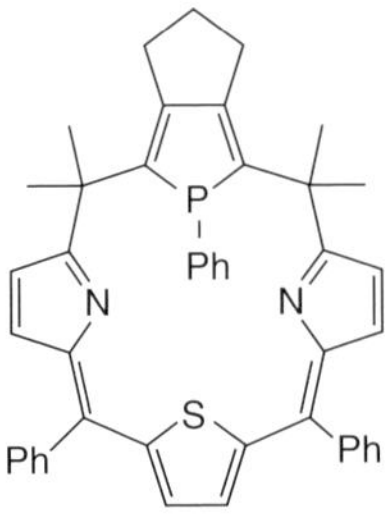

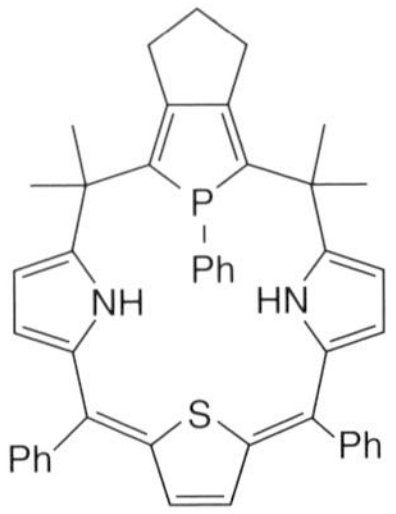

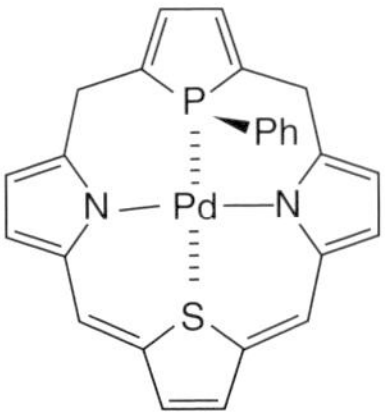

8.15. CARBON–NITROGEN–METAL RINGS

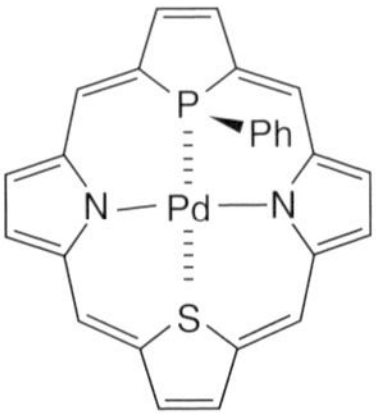

Treatment of stoichiometric amounts of the racemic *bis*(nitrile) ligand based on Tröger's base with (dppp)Pd(OTf)$_2$ gave rise to the dinuclear metallosupramolecular rhomboid **55** in quantitative yield <09CC2320>. Self-assembly of ligands, based on *trans*-1,2-*bis*(4-pyridinyl)ethylene, with either Pd(II) or Pt(II) complexes gave quadrangular metallacycles; 1,1'-methylene-*bis*[4-[(*E*)-2-(pyridin-4-yl)vinyl]pyridinium gave a square metallocycle **56** <09JOC6577>. The assembly of 2,4,6-*tris*(4-pyridinyl)triazine (the panels) with pyrazine (the columns) and (en)Pt(II) (corners) gave the desired three-dimensional molecular prism cage (**57**), which can encapsulate tetraazaporphine that can readily rotate within this D_{3h}-symmetric host <09JA12526>. A closely related prism can be constructed utilizing verdazyl panels <09CC4245>. The above triazine panel was treated with [HLPtX]$_2$(4,4′-bipy)], where LH$_2$ = 2,6-diphenylpyridine, X = (+)-camphor-10-sulfonate, to generate prism **58**, which disassembled to give [LPt)$_2$(4,4′-bipy)] possessing two Pt-phenyl bonds <09JA16398>.

The construction of three *endo*-functionalized two-dimensional supramolecular metallocycles such as two [2+2] rhomboids and a [3+3]hexagon, based on methyl 4-(nitrobenzyloxy)-3,5-di(4′-pyridinylethynyl)benzoate, has been reported <09JOC8516>. The self-assembly of a stabilized M$_{12}$L$_{24}$ sphere **59** was comprised

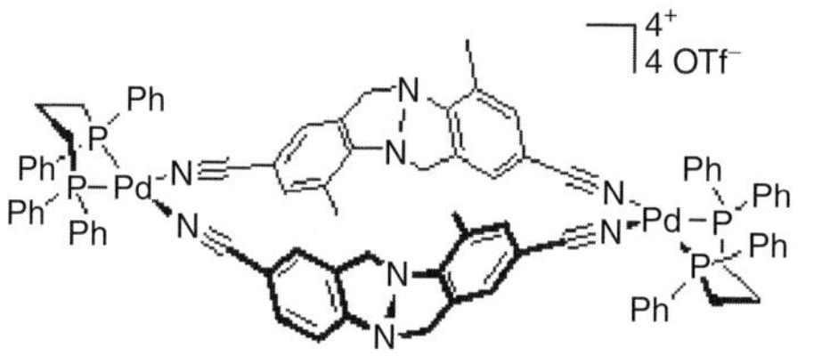

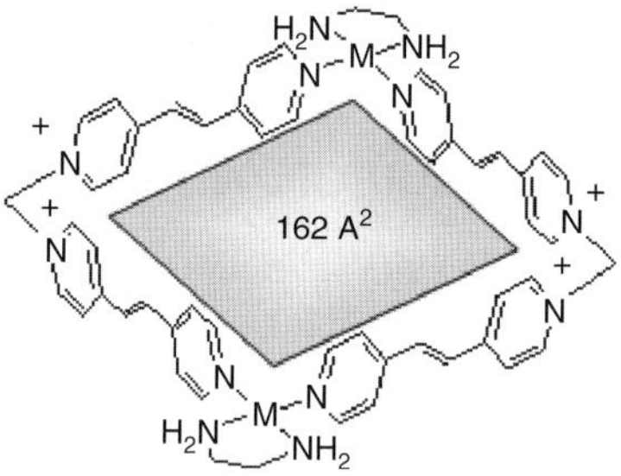

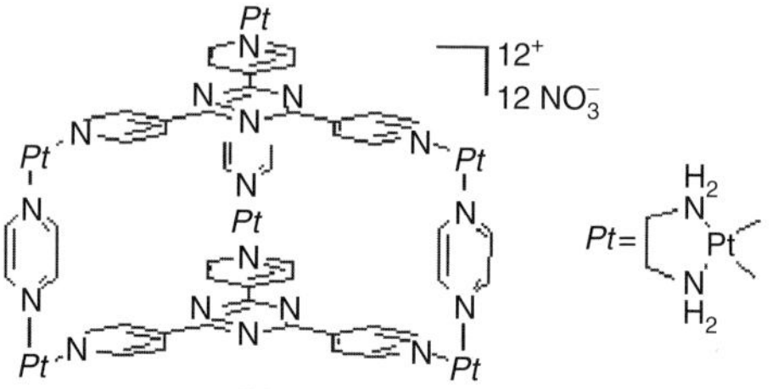

of 48 Pd(II)-pyridine connections utilizing 1,3-(4′-pyridinylethynyl)-2-alkoxybenzene, as the bridging ligand <09JA6064>. Photo-induced self-assembly of 1,4-di(4′-pyridinyl)-2,3,5,6-tetramethylbenzene or 1,3,5-*tris*(4′-pyridinyl-2,3,5,6-tetramethylphenyl)benzene with (en)Pt(II) led to two- and three-dimensional structures *via* the photolability of the Pt(II)-py bond <09NJC264>.

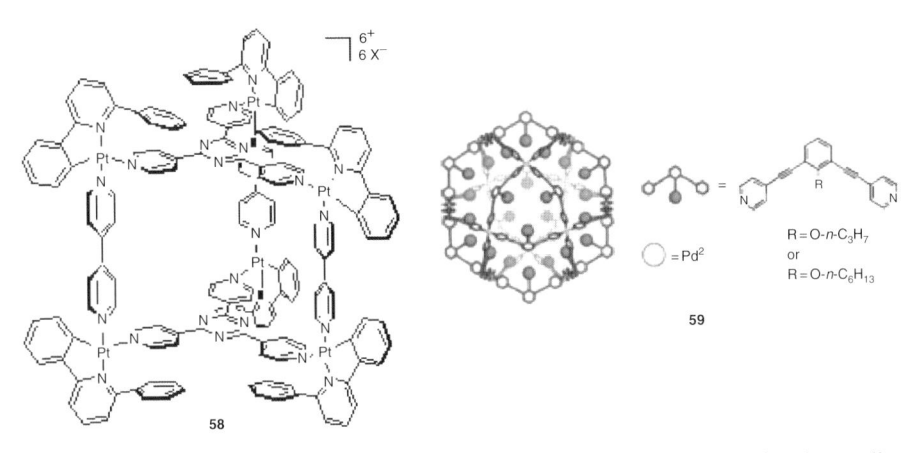

21-Telluraporphyrin **60** with PCl$_3$ in Et$_3$N gave (55%) the dihydro-*N*-fused 21-telluraporphyrin **61**, the removal of the phosphorus(V) center was not accomplished <09CEJ10924>.

8.16. CARBON–NITROGEN–OXYGEN–METAL RINGS

A remarkably stable, hollow molecular cube **62** possessing 3 nm sides was self-assembled in quantitative yield by the reaction of 2,8-di(4′-pyridinylethynyl)dibenzofuran with Pd(NO$_3$)$_2$ in DMSO; the host-guest chemistry should offer interesting new insight to the occupation of void volumes <09CC1638>.

62

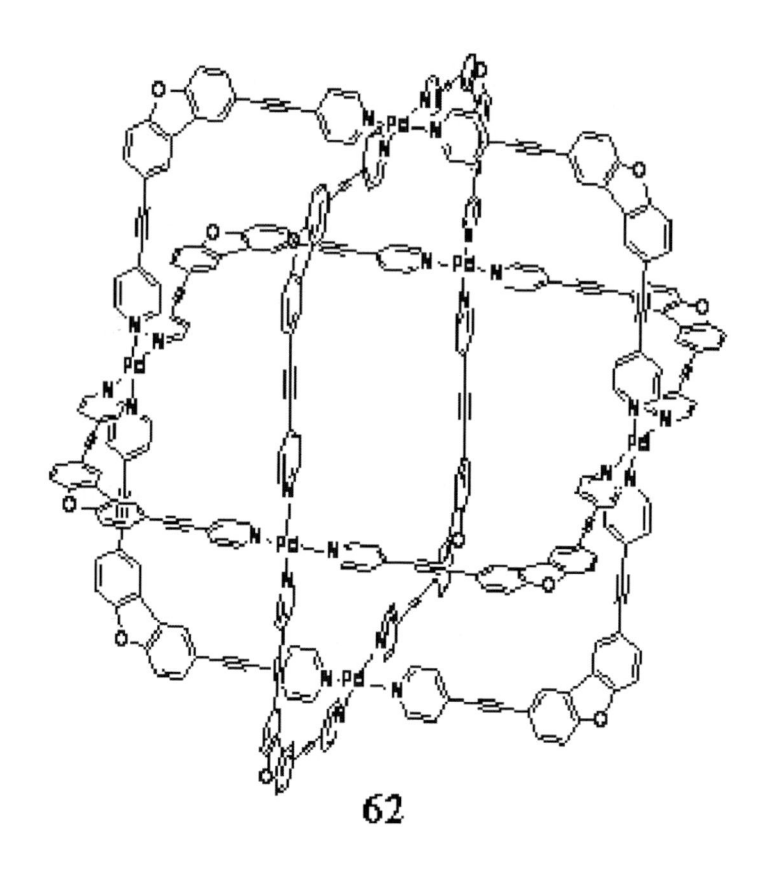

8.17. CARBON–NITROGEN–SULFUR/PHOSPHORUS–METAL RINGS

The complexation of $Ph_2P(CH_2CH_2O)_3CH_2CH_2PPh_2$ with $PdCl_2$ in MeCN/THF gave the *cis*-<$PdCl_2[Ph_2P(CH_2CH_2O)_3CH_2CH_2PPh_2$-$P,P'$]>; *cis-trans* isomerization and oligomerization equilibria were followed by $^{31}P[^{1}H]$ NMR <09EJOC4710>.

REFERENCES

04OL747	L. Wang and J.-C. Chambron, *Org. Lett.*, **6**, 747 (2004).
07HCA1439	F. Durola, J.-P. Sauvage and O.S. Wenger, *Helv. Chim. Acta*, **90**, 1439 (2007).
08ACIE7396	B.M.J.M. Suijkerbuijk and R.J.M.K. Gebbink, *Angew. Chem. Int. Ed.*, **47**, 7396 (2008).
08ACIE8794	R.W. Saalfrank, H. Maid and A. Scheurer, *Angew. Chem. Int. Ed.*, **47**, 8794 (2008).
08ACIE8913	M. Toganoh, T. Kimura, H. Uno and H. Furuta, *Angew. Chem. Int. Ed.*, **47**, 8913 (2008).
08ACIE9648	A. Aydogan, D.J. Coady, S.K. Kim, A. Akar, C.W. Bielawski, M. Marquez and J.L. Sessler, *Angew. Chem. Int. Ed.*, **47**, 9648 (2008).
08CC6309	T.D. Lash, A.M. Young, A.L. Von Ruden and G.M. Ferrence, *Chem. Commun.*, 6309 (2008).

08CC6573	A. Ito, Y. Yamagishi, K. Fukui, S. Inoue, Y. Hirao, K. Furukawa, T. Kato and K. Tanaka, *Chem. Commun.*, 6573 (2008).
08COC1231	M.A. Hossain, *Curr. Org. Chem.*, **12**, 1231 (2008).
08JOC3336	K.M. Mullen and M.J. Gunter, *J. Org. Chem.*, **73**, 3336 (2008).
08T11495	B.H. Northrop, D. Chercka and P.J. Stang, *Tetrahedron*, **64**, 11495 (2008).
08TCR75	C.G. Claessens, U. Hahn and T. Torres, *Chem. Rec.*, **8**, 75 (2008).
08TCR240	T. Nabeshima and S. Akine, *Chem. Rec.*, **8**, 240 (2008).
09ACIE504	Y. Sato, R. Yamasaki and S. Saito, *Angew. Chem. Int. Ed.*, **48**, 504 (2009).
09ACIE1792	M.A. Olsen, A.B. Braunschweig, L. Fang, T. Ikeda, R. Klajn, A. Trabolsi, P.J. Wesson, D. Benitez, C.A. Mirkin, B.A. Grzybowski and J.F. Stoddart, *Angew. Chem. Int. Ed.*, **48**, 1792 (2009).
09ACIE1798	Z. Ge, J. Hu, F. Huang and S. Liu, *Angew. Chem. Int. Ed.*, **48**, 1798 (2009).
09ACIE3418	M. Yoshizawa, J.K. Klosterman and M. Fujita, *Angew. Chem. Int. Ed.*, **48**, 3418 (2009).
09ACIE3924	H.-J. Schneider, *Angew. Chem. Int. Ed.*, **48**, 3924 (2009).
09ACIE4284	N. Jux, *Angew. Chem. Int. Ed.*, **48**, 4284 (2009).
09ACIE6632	F. Zhang, G. Götz, H.D.F. Winkler, C.A. Schalley and P. Bäuerle, *Angew. Chem. Int.Ed.*, **48**, 6632 (2009).
09ACIE6654	S. Binauld, C.J. Hawker, E. Fleury and E. Drockenmuller, *Angew. Chem. Int. Ed.*, **48**, 6654 (2009).
09ACR249	S.-S. Li, B.H. Northrop, Q.-H. Yuan, L.J. Wan and P.J. Stang, *Acc. Chem. Res.*, **42**, 249 (2009).
09ACR1193	Y. Matano and H. Imahori, *Acc. Chem. Res.*, **42**, 1193 (2009).
09ACR1564	J.L. Dye, *Acc. Chem. Res.*, **42**, 1564 (2009).
09ACR1922	N. Aratani, D. Kim and A. Osuka, *Acc. Chem. Res.*, **42**, 1922 (2009).
09CC261	J.M. Lim, Z.S. Yoon, J.-Y. Shin, K.S. Kim, M.-C. Yoon and D. Kim, *Chem. Commun.*, 261 (2009).
09CC343	V. Steinmetz, F. Couty and O.R.P. David, *Chem. Commun.*, 343 (2009).
09CC513	V. Amendola and L. Fabbrizzi, *Chem. Commun.*, 513 (2009).
09CC1011	H. Shinokubo and A. Osuka, *Chem. Commun.*, 1011 (2009).
09CC1028	S. Tokuji, Y. Takahashi, H. Shinmori, H. Shinokubo and A. Osuka, *Chem. Commun.*, 1028 (2009).
09CC1638	K. Suzuki, M. Tominaga, M. Kawano and M. Fujita, *Chem. Commun.*, 1638 (2009).
09CC1706	J.-P. Collin, F. Durola, J. Frey, V. Heirz, J.-P. Sauvage, C. Tock and Y. Trolez, *Chem. Commun.*, 1706 (2009).
09CC1870	K. Kanda, T. Koike, K. Endo and T. Shibata, *Chem. Commun.*, 1870 (2009).
09CC1987	J. Cao, Y. Jiang, J.-M. Zhao and C.-F. Chen, *Chem. Commun.*, 1987 (2009).
09CC2121	H. Abe, S. Takashima, T. Yamamoto and M. Inouye, *Chem. Commun.*, 2121 (2009).
09CC2245	D. Canevet, M. Sallé, G. Zhang, D. Zhang and D. Zhu, *Chem. Commun.*, 2245 (2009).
09CC2320	T. Weilandt, U. Kiehne, G. Schnakenburg and A. Lützen, *Chem. Commun.*, 2320 (2009).
09CC2399	F.A. Khan, K. Parasuraman and K.K. Sadhu, *Chem. Commun.*, 2399 (2009).
09CC2535	X. Zhao, Z. Zhong, L. Peng, W. Zhang and J. Wang, *Chem. Commun.*, 2535 (2009).
09CC2809	M. Cametti and K. Rissanen, *Chem. Commun.*, 2809 (2009).
09CC2899	B. Yao, D.-X. Wang, Z.-T. Huang and M.-X. Wang, *Chem. Commun.*, 2899 (2009).
09CC3017	M.G. Fisher, P.A. Gale, J.R. Hiscock, M.B. Hursthouse, M.E. Light, F.P. Schmidtchen and C.C. Tong, *Chem. Commun.*, 3017 (2009).
09CC3154	J.B. Love, *Chem. Commun.*, 3154 (2009).
09CC3175	M. Tian and H. Ihmels, *Chem. Commun.*, 3175 (2009).
09CC3326	M.J. Prakash and M.S. Lah, *Chem. Commun.*, 3326 (2009).
09CC3609	M. Mathews and N. Tamaoki, *Chem. Commun.*, 3609 (2009).
09CC4245	Y. Ozaki, M. Kawano and M. Fujita, *Chem. Commun.*, 4245 (2009).
09CC4844	Y.-L. Zhao, A. Trabolsi and J.F. Stoddart, *Chem. Commun.*, 4844 (2009).
09CC4982	K.K. Sadhu, S. Banerjee, A. Datta and P.K. Bharadwag, *Chem. Commun.*, 4982 (2009).

09CC5579	M.B. Duriska, S.M. Neville and S.R. Batten, *Chem. Commun.*, 5579 (2009).
09CC5814	M.-J. Li, C.-C. Lai, Y.-H. Liu, S.-M. Peng and S.-H. Chiu, *Chem. Commun.*, 5814 (2009).
09CC5877	S.-D. Jeong, K.J. Park, H.-J. Kim and C.-H. Lee, *Chem. Commun.*, 5877 (2009).
09CC6047	T. Koide, K. Youfu, S. Saito and A. Osuka, *Chem. Commun.*, 6047 (2009).
09CC6329	J.J. Gassensmith, J.M. Baumes and B.D. Smith, *Chem. Commun.*, 6329 (2009).
09CSR506	J.W. Steed, *Chem. Soc. Rev.*, **38**, 506 (2009).
09CCR1056	A. Panda, *Coord. Chem. Rev.*, **253**, 1056 (2009).
09CCR1906	N. Viola-Villegas and R.P. Doyle, *Coord. Chem. Rev.*, **253**, 1906 (2009).
09CCR1947	A. Panda, *Coord. Chem. Rev.*, **253**, 1947 (2009).
09CCR2036	E. Pacholska-Dudziak and L. Latos-Grazynski, *Coord. Chem. Rev.*, **253**, 2036 (2009).
09CCR2244	V. Lozan, C. Loose, J. Kortus and B. Kersting, *Coord. Chem. Rev.*, **253**, 2244 (2009).
09CEJ1310	G. Periyasamy, J.-P. Collin, J.-P. Sauvage, R.D. Levine and F. Remacle, *Chem. Eur. J.*, **15**, 1310 (2009).
09CEJ1314	A. Granzhan and M.-P. Teulade-Fichou, *Chem. Eur. J.*, **15**, 1314 (2009).
09CEJ4124	F. Durola, J. Lux and J.-P. Sauvage, *Chem. Eur. J.*, **15**, 4124 (2009).
09CEJ5186	F. Coutrot and E. Busseron, *Chem. Eur. J.*, **15**, 5186 (2009).
09CEJ5289	A. Dachs, A. Torrent, A. Roglans, T. Parella, S. Osuna and M. Solà, *Chem. Eur. J.*, **15**, 5289 (2009).
09CEJ5444	J.D. Megiatto, Jr. and D.I. Schuster, *Chem. Eur. J.*, **15**, 5444 (2009).
09CEJ6495	J.L. Alonso-Gómez, A. Navarro-Vázquez and M.M. Cid, *Chem. Eur. J.*, **15**, 6495 (2009).
09CEJ8128	L.G. Jensen, K.A. Nielsen, T. Breton, J.L. Sessler, J.O. Jeppesen, E. Levillain and L. Sanguinet, *Chem. Eur. J.*, **15**, 8128 (2009).
09CEJ10543	W. Wang, R. Li and G.W. Gokel, *Chem. Eur. J.*, **15**, 10543 (2009).
09CEJ10924	E. Pacholska-Dudziak, F. Ulatowski, Z. Ciunik and L. Latos-Grazynski, *Chem. Eur. J.*, **15**, 10924 (2009).
09CEJ13188	F.A. Rojas-Quijano, E.T. Benyó, G. Tiresó, F.K. Kálmán, Z. Baranyai, S. Aime, A.D. Sherry and Z. Kovács, *Chem. Eur. J.*, **15**, 13199 (2009).
09CEJ13356	Y.-L. Zhao, L. Liu, W. Zhang, C.-H. Sue, Q. Li, O.S. Miljanic, O.M. Yaghi and J.F. Stoddard, *Chem. Eur. J.*, **15**, 13356 (2009).
09CR796	L.A. Wessjohann, D.G. Rivera and O.E. Vercillo, *Chem. Rev.*, **109**, 796 (2009).
09CR1141	A. Mishra, C.-Q. Ma and P. Bäuerle, *Chem. Rev.*, **109**, 1141 (2009).
09CR1630	C.M. Drain, A. Varotto and I. Radivojevic, *Chem. Rev.*, **109**, 1630 (2009).
09CR1659	I. Beletskaya, V.S. Tyurin, A.Y. Tsivade, R. Guilard and C. Stern, *Chem. Rev.*, **109**, 1659 (2009).
09CR4979	E. Zangrando, M. Casanova and E. Alessio, *Chem. Rev.*, **108**, 4979 (2009).
09CR5687	T.F.A. De Greef, M.M.J. Smulders, M. Wolffs, A.P.H.J. Schenning, R.P. Sijbesma and E.W. Meijer, *Chem. Rev.*, **109**, 5687 (2009).
09CR5974	A. Harada, A. Hashidzume, H. Yamaguchi and Y. Takashima, *Chem. Rev.*, **109**, 5974 (2009).
09CR6024	Z. Niu and H.W. Gibson, *Chem. Rev.*, **109**, 6024 (2009).
09CRC395	M. Iyoda, *C. R. Chim.*, **12**, 395 (2009).
09CSR422	J.A. Faiz, V. Heitz and J.-P. Sauvage, *Chem. Soc. Rev.*, **38**, 422 (2009).
09CSR875	A. Harada, Y. Takashima and H. Yamaguchi, *Chem. Soc. Rev.*, **38**, 875 (2009).
09CSR1400	J.J. Perry, IV, J.A. Perman and M.J. Zaworotko, *Chem. Soc. Rev.*, **38**, 1400 (2009).
09CSR1521	J.F. Stoddart, *Chem. Soc. Rev.*, **38**, 1521 (2009).
09CSR1530	J.D. Crowley, S.M. Goldup, A.-L. Lee, D.A. Leigh and R.T. McBurney, *Chem. Soc. Rev.*, **38**, 1530 (2009).
09CSR1542	V. Balzani, A. Credi and M. Venturi, *Chem. Soc. Rev.*, **38**, 1542 (2009).
09CSR1562	D.B. Amabilino and L. Pérez-García, *Chem. Soc. Rev.*, **38**, 1562 (2009).
09CSR1635	L. Flamigni and D.T. Gryko, *Chem. Soc. Rev.*, **38**, 1635 (2009).
09CSR1663	J.-C. Chambron and M. Meyer, *Chem. Soc. Rev.*, **38**, 1663 (2009).
09CSR1674	E. Coronado, P. Gaviña and S. Tatay, *Chem. Soc. Rev.*, **38**, 1674 (2009).
09CSR1701	K.M. Mullen and P.D. Beer, *Chem. Soc. Rev.*, **38**, 1701 (2009).
09CSR1744	S.L. James, *Chem. Soc. Rev.*, **38**, 1744 (2009).
09CSR1802	J.F. Stoddart, *Chem. Soc. Rev.*, **38**, 1802 (2009).

09CSR1904	W.S. Han, H.Y. Lee, S.H. Jung, S.J. Lee and J.H. Jung, *Chem. Soc. Rev.*, **38**, 1904 (2009).
09CSR2716	L. Cuesta and J.L. Sessler, *Chem. Soc. Rev.*, **38**, 2716 (2009).
09DT490	A.A. Karasik, D.V. Kulikov, A.S. Balueva, S.N. Ignat'eva, O.N. Kataeva, P. Lönnecke, A.V. Kozlov, S.K. Latypov, E. Hey-Hawkins and O.G. Sinyashin, *Dalton Trans.*, 490 (2009).
09DT4569	W. Levason, J.M. Manning, G. Reid, M. Tuggey and M. Webster, *Dalton Trans.*, 4569 (2009).
09DT6151	D.-H. Won, M. Toganoh, H. Uno and H. Furuta, *Dalton Trans.*, 6151 (2009).
09EJOC1053	S. Li, K. Zhu, B. Zhang, X. Wen, N. Li and F. Huang, *Eur. J. Org. Chem.*, 1053 (2009).
09EJOC2562	M. Müri, K.C. Schuermann, L. De Cola and M. Mayor, *Eur. J. Org. Chem.*, 2562 (2009).
09EJOC3006	R. Gleiter, K. Hovermann, B. Esser and A. Bandyopadhyay, *Eur. J. Org. Chem.*, 3006 (2009).
09EJOC3930	P.J. Chmielewski, J. Maciolek and L. Szterenberg, *Eur. J. Org. Chem.*, 3930 (2009).
09EJOC4534	M. Kumar, A. Dhir and V. Bhalla, *Eur. J. Org. Chem.*, 4534 (2009).
09EJOC4710	S.B. Owans, Jr., D.C. Smith, Jr., C.H. Lake and G.M. Gray, *Eur. J. Inorg. Chem.*, 4710 (2009).
09EJOC5571	F. Dädritz, A. Jäger and I. Bauer, *Eur. J. Org. Chem.*, 5571 (2009).
09EJOC6128	S. Li, M. Liu, J. Zhang, B. Zheng, X. Wen, N. Li and F. Huang, *Eur. J. Org. Chem.*, 6128 (2009).
09ICC304	N. Kabay, S. Söyleyici and Y. Gök, *Inorg. Chem. Commun.*, **12**, 304 (2009).
09ICC382	J.-P. Joly, M. Beley, K. Selmeczi and E. Wenger, *Inorg. Chem. Commun.*, **12**, 382 (2009).
09ICC739	R. Hou, L. Jin and B. Yin, *Inorg. Chem. Commun.*, **12**, 739 (2009).
09ICC815	H.-L. Kwong, W.-L. Wong, C.-S. Lee, C.-T. Yeung and P.-F. Teng, *Inorg. Chem. Commun.*, **12**, 815 (2009).
09ICC846	S. Jing, C.Y. Gu, W. Ji and B. Yang, *Inorg. Chem. Commun.*, **12**, 846 (2009).
08JA990	Y. Matano, T. Miyajima, N. Ochi, T. Nakabuchi, M. Shiro, Y. Nakao, S. Sakaki and H. Ihahori, *J. Am. Chem. Soc.*, **130**, 990 (2008).
09JA3762	D.A. Leigh, P.J. Lusby, R.T. McBurney, A. Morelli, A.M.Z. Slawin, A.R. Thomson and D.B. Walker, *J. Am. Chem. Soc.*, **131**, 3762 (2009).
09JA5609	J.-P. Collin, J. Frey, V. Heitz, J.-P. Sauvage, C. Tock and L. Allouche, *J. Am. Chem. Soc.*, **131**, 5609 (2009).
09JA6064	S. Sato, Y. Ishido and M. Fujita, *J. Am. Chem. Soc.*, **131**, 6064 (2009).
09JA6794	A.I. Prikhod'ko and J.-P. Sauvage, *J. Am. Chem. Soc.*, **131**, 6794 (2009).
09JA7126	L. Fang, M. Hmadeh, J. Wu, M.A. Olson, J.M. Spruell, A. Trabolsi, Y.-W. Yang, M. Elhabiri, A.-M. Albrecht-Gary and J.F. Stoddart, *J. Am. Chem. Soc.*, **131**, 7126 (2009).
09JA8338	M.-L. Ma, X.-Y. Li and K. Wen, *J. Am. Chem. Soc.*, **131**, 8338 (2009).
09JA10955	N. Ochi, Y. Nakao, H. Sato, Y. Matano, H. Imahori and S. Sakaki, *J. Am. Chem. Soc.*, **131**, 10955 (2009).
09JA11571	J.M. Spruell, W.F. Paxton, J.-C. Olsen, D. Benitez, E. Tkatchouk, C.L. Stern, A. Trabolsi, D.C. Friedman, W.A. Goddard, III and J.F. Stoddart, *J. Am. Chem. Soc.*, **131**, 11571 (2009).
09JA12526	K. Ono, J.K. Klosterman, M. Yoshizawa, K. Sekiguchi, T. Tahara and M. Fujita, *J. Am. Chem. Soc.*, **131**, 12526 (2009).
09JA14123	Y. Matano, T. Miyajima, N. Ochi, T. Nakabuchi, M. Shiro, Y. Nakao, S. Sakaki and H. Ihahori, *J. Am. Chem. Soc.*, **131**, 14123 (2009).
09JA15924	S.M. Goldup, D.A. Leigh, T. Long, P.R. McGonigal, M.D. Symes and J. Wu, *J. Am. Chem. Soc.*, **131**, 15924 (2009).
09JA16398	P.J. Lusby, P. Mukker, S.J. Pike and A.M.Z. Slawin, *J. Am. Chem. Soc.*, **131**, 16398 (2009).
09JA17620	L. Tian, C. Wang, S. Dawn, M.D. Smith, J.A. Krause and L.S. Shimizu, *J. Am. Chem. Soc.*, **131**, 17620 (2009).
09JIPMC1	N. Iki, *J. Incl. Phenom. Macrocyc. Chem.*, **64**, 1 (2009).
09JIPMC189	J.S. Kim and J. Vicens, *J. Incl. Phenom. Macrocyc. Chem.*, **64**, 189 (2009).
09JIPMC193	H. Maeda, *J. Incl. Phenom. Macrocyc. Chem.*, **64**, 193 (2009).

09JIPMC379	M.H. Patel and P.S. Shrivastav, *J. Incl. Phenom. Macrocyc. Chem.*, **64**, 379 (2009).
09JOC2	P.J. Stang, *J. Org. Chem.*, **74**, 2 (2009).
09JOC1014	J. Motoyoshiya, T. Tanaka, M. Kuroe and Y. Nishii, *J. Org. Chem.*, **74**, 1014 (2009).
09JOC1065	J. Yoo, M.-S. Kim, S.-J. Hong, J.L. Sessler and C.-H. Lee, *J. Org. Chem.*, **74**, 1065 (2009).
09JOC1322	K. Zhu, S. Li, F. Wang and F. Huang, *J. Org. Chem.*, **74**, 1322 (2009).
09JOC1917	A.O. Terent'ev, M.M. Platonov, A.I. Tursina, V.V. Chernyshev and G.I. Nikishin, *J. Org. Chem.*, **74**, 1917 (2009).
09JOC3579	M. Mori, T. Okawa, N. Iizuna, K. Nakayama, J.H. Lintuluoto and J. Setsune, *J. Org. Chem.*, **74**, 3579 (2009).
09JOC3905	K. Zhu, J. He, S. Li, M. Liu, F. Wang, M. Zhang, Z. Abliz, H.-B. Yang, N. Li and F. Huang, *J. Org. Chem.*, **74**, 3905 (2009).
09JOC4819	N. Bernier, S. Carvalho, F. Li, R. Delgado and V. Félix, *J. Org. Chem.*, **74**, 4819 (2009).
09JOC5361	B. Yao, D.X. Wang, H.-Y. Gong, Z.-T. Huang and M.-X. Wang, *J. Org. Chem.*, **74**, 5361 (2009).
09JOC6577	V. Blanco, A. Gutiérrez, C. Platas-Iglesias, C. Peinador and J.M. Quintela, *J. Org. Chem.*, **74**, 6577 (2009).
09JOC8516	L. Zhao, K. Ghosh, Y.-R. Zheng and P.J. Stang, *J. Org. Chem.*, **74**, 8516 (2009).
09JOC8547	N. Grzegorzek, M. Pawlicki and L. Latos-Grazynski, *J. Org. Chem.*, **74**, 8547 (2009).
09JOC8638	P. Mateus, R. Delgado, P. Brandão and V. Félix, *J. Org. Chem.*, **74**, 8638 (2009).
09JOMC1011	M. Bröring, C. Milsmann, S. Ruck and S. Köhler, *J. Organomet. Chem.*, **694**, 1011 (2009).
09JOMC1617	D. González-Rodríguez, T. Tomas, E.L.G. Denardin, D. Samios, V. Stafani and D.S. Corrêa, *J. Organomet. Chem.*, **694**, 1617 (2009).
09JOMC3982	A. Pascariu, S. Iliescu, A. Popa and G. Ilia, *J. Organomet. Chem.*, **694**, 3982 (2009).
09JPSA1971	H.R. Kricheldorf, *J. Polym. Sci. A Polym. Chem.*, **47**, 1971 (2009).
09JPSA4469	A. Harada, *J. Polym. Sci. A Polym. Chem.*, **47**, 4469 (2009).
09JPSA6333	G. Wenz, *J. Polym. Sci. A Polym. Chem.*, **47**, 6333 (2009).
09NJC264	K. Yamashita, K. Sato, M. Kawano and M. Fujita, *New J. Chem.*, **33**, 264 (2009).
09NJC300	S. Ramor, E. Alcalde, J.F. Stoddart, A.J.P. White, D.J. Williams and L. Pérez-García, *New J. Chem.*, **33**, 300 (2009).
09NJC327	L. Wang, J.-C. Chambron and E. Espinosa, *New J. Chem.*, **33**, 327 (2009).
09NJC725	J. Zahn, D. Tian and H. Li, *New J. Chem.*, **33**, 725 (2009).
09NJC813	Y.-P. Zhao, X.-J. Wang, J.-J. Wang, G. Si, Y. Liu, C.-H. Tung and L.-Z. Wu, *New J. Chem.*, **33**, 813 (2009).
09NJC2148	J.-P. Collin, J.-P. Sauvage, Y. Trolez and K. Rissanen, *New J. Chem.*, **33**, 2148 (2009).
09O6213	Y. Matano, M. Fujita, T. Miyajima and H. Imahori, *Organometallics*, **28**, 6213 (2009).
09OBC415	M.J. Chmielewski, J.J. Davis and P.D. Beer, *Org. Biomol. Chem.*, **7**, 415 (2009).
09OBC1288	M. Liu, S. Li, M. Zhang, Q. Zhou, F. Wang, M. Hu, F.R. Fronczek, N. Li and F. Huang, *Org. Biomol. Chem.*, **7**, 1288 (2009).
09OBC3027	M. Toganoh, H. Miyachi, H. Akimaru, F. Ito, T. Nagamura and H. Furuta, *Org. Biomol. Chem.*, **7**, 3027 (2009).
09OBC4391	M.A. Olsen, A.B. Braunschweig, T. Ikeda, L. Fang, A. Trabolsi, A.M.Z. Slawin, S.I. Khan and J.F. Stoddart, *Org. Biomol. Chem.*, **7**, 4391 (2009).
09OL385	C.-J. Chuang, W.-S. Li, C.-C. Lai, Y.-H. Liu, S.-M. Peng, I. Chao and S.-H. Chiu, *Org. Lett.*, **11**, 385 (2009).
09OL417	G. Piersanti, F. Remi, V. Fusi, M. Formica, L. Giorgi and G. Zappia, *Org. Lett.*, **11**, 417 (2009).
09OL613	T.-C. Lin, C.-C. Lai and S.-H. Chiu, *Org. Lett.*, **11**, 613 (2009).
09OL1249	Z. Zhang, G.M. Ferrence and T.D. Lash, *Org. Lett.*, **11**, 1249 (2009).
09OL1313	N. Kihara and K. Kidoba, *Org. Lett.*, **11**, 1313 (2009).
09OL1393	H. Lee and S.S. Lee, *Org. Lett.*, **11**, 1393 (2009).
09OL1681	W.V. Rossom, M. Ovaere, L.V. Merrvelt, W. Dehaen and W. Maes, *Org. Lett.*, **11**, 1681 (2009).

09OL2241	Y. Morsaki, H. Imoto, K. Tsurui and Y. Chujo, *Org. Lett.*, **11**, 2241 (2009).
09OL2289	S. Kamioka, T. Takahashi, S. Kawaychi, H. Adachi, Y. Mori, K. Fujii, H. Uekusa and T. Doi, *Org. Lett.*, **11**, 2289 (2009).
09OL3040	J. Thomas, W. Maes, K. Robeyns, M. Ovaere, L.V. Meervelt, M. Smet and W. Dehaen, *Org. Lett.*, **11**, 3040 (2009).
09OL3930	M. Stepien, B. Szyszko and L. Latos-Grazynski, *Org. Lett.*, **11**, 3930 (2009).
09OL4446	J. Cao, H.-Y. Lu, X.-J. You, Q.-Y. Zhang and C.-F. Chen, *Org. Lett.*, **11**, 4446 (2009).
09OL4604	M.-L. Yen, N.-C. Chen, C.-C. Lai, Y.-H. Liu, S.-M. Peng and S.-H. Chiu, *Org. Lett.*, **11**, 4604 (2009).
09OL5238	H.M. Colquhoun, B.W. Greenland, Z. Zhu, J.S. Shaw, C.J. Cardin, S. Burattini, J.M. Elliott, S. Basu, C.J. Gasa and J.F. Stoddart, *Org. Lett.*, **11**, 5238 (2009).
09S3860	X.-G. Chen, C. Liu, D.-M. Shen and Q.-Y. Chen, *Synthesis*, 3860 (2009).
09SL1429	Z. Jászay, T.S. Pham, G. Németh, P. Bakó, I. Petneházy and L. Töke, *Synlett*, 1429 (2009).
09T400	M. Bria, J. Bigot, G. Cooke, J. Lyskawa, G. Rabani, V.M. Rotello and P. Woisel, *Tetrahedron*, **65**, 400 (2009).
09T496	P.M. Marcos, J.R. Ascenso, M.A.P. Segurado, R.J. Bernardino and P.J. Cragg, *Tetrahedron*, **65**, 496 (2009).
09T690	A. Späth and B. König, *Tetrahedron*, **65**, 690 (2009).
09T1257	E.L. Ruggles, P.B. Deker and R.J. Hondal, *Tetrahedron*, **65**, 1257 (2009).
09T1349	A. Granzhan and M.-P. Teulade-Fichou, *Tetrahedron*, **65**, 1349 (2009).
09T1618	G.P. Wild, C. Wiles, P. Watts and S.J. Haswell, *Tetrahedron*, **65**, 1618 (2009).
09T2285	P. Piatek and A. Litwin, *Tetrahedron*, **65**, 2285 (2009).
09T2995	F. Däbritz, G. Theumer, M. Gruner and I. Bauer, *Tetrahedron*, **65**, 2995 (2009).
09T3014	H.N. Demirtas, S. Bozkurt, M. Durmaz, M. Yilmaz and A. Sirit, *Tetrahedron*, **65**, 3014 (2009).
09T5893	X. Liu, K. Surowiec and R.A. Bartsch, *Tetrahedron*, **65**, 5893 (2009).
09T7109	I.I. Stoikov, E.A. Yushkova, I. Zharov, I.S. Antipin and A.I. Konovalov, *Tetrahedron*, **65**, 7109 (2009).
09T7673	F. Havas, N. Leygue, M. Danel, B. Mestre, C. Galaup and C. Picard, *Tetrahedron*, **65**, 7673 (2009).
09TCR136	P.C. Haussmann and J.F. Stoddart, *Chem. Rec.*, **9**, 136 (2009).
09TL620	H. Konishi, S. Hashimoto, T. Sakakibara, S. Matsubara, Y. Yasukawa, O. Morikawa and K. Kobayashi, *Tetrahedron Lett.*, **50**, 620 (2009).
09TL2041	L. Lapok, C.G. Claessens, D. Wöhrle and T. Torres, *Tetrahedron Lett.*, **50**, 2041 (2009).
09TL2929	C. Li, P. Winnard, Jr. and Z.M. Bhujwalla, *Tetrahedron Lett.*, **50**, 2929 (2009).
09TL6522	G. Bechara, N. Leygue, C. Galaup, B. Mestre and C. Picard, *Tetrahedron Lett.*, **50**, 6522 (2009).
09TL6537	D. Gibson, K.R. Dey, F.R. Fronczek and M.A. Hossain, *Tetrahedron Lett.*, **50**, 6537 (2009).
10EJOC1407	N. Martin, *Eur. J. Org. Chem.*, 1407 (2010).
10SL67	M. Yedukondalu and M. Ravikanth, *Synlett*, 67 (2010).
10TL983	C.J. Bruns, S. Basu and J.F. Stoddart, *Tetrahedron Lett.*, **50**, 983 (2010).

INDEX

A

Acceptor-donor-acceptor (A-D-A) sensor, 41
Achmatowicz oxidation, 185
Alkylidenecyclopropylcarboxaldehydes, 148
4-Alkynyl-3-aminocoumarins, 144
Alkynyldithiolanes, 312
3-Alkynyl selenophene (ASP), 136
γ-Aminoalkynes, 144
3-(Aminomethyl)-2-(bromomethyl)-1,
 4-naphthoquinones, 145
3-Aminopropionitrile, 340
1-Aroyltetrahydroisoquinolines, 494
3-Aryl-3-azetidinyl acetic acid esters, 86
2-Aryl-2,3-dihydropyrroles, 144
Aza–Diels–Alder reaction, 352
Azaindoles, 172–173
Azamacrocycles, 541–543
Azetidines and azetines
 L-azetidine-2-carboxylic acid, 86
 β-chloro N-oxides, 88
 DFT calculations, 86
 enantioselective epoxidation, 85
 3-fluoroazetidine-3-carboxylic acid,
  86–87
 N-allyl azetidiniums, 87
 N-heterocyclic carbenes, 88
 N-tosylaldimines, 86
 oxazocines preparation, 88
 Sonogashira coupling reaction, 85
 tricyclic 3-aminopyridines, 87
Aziridines
 alkenes, 70
 α-alkylation reactions, 78
 aminoglycosides, 71
 aza-Darzens approach, 73
 aza-MIRC reaction, 72
 chiral allylic alcohols, 71
 cycloaddition reaction, 75
 Diels–Alder cycloaddition, 78
 2,3-dihydropyrroles, 77
 dimeric yttrium-salen, 74
 Grignard reagents, 76
 imines, 73
 nucleophilic ring-opening reactions, 74
 palladium-catalyzed reactions, 72
 propargylic, 77
 ring-opening reactions, 75–76
 sulfonyloxaziridine, 78
 tandem aziridine ring-opening/closing
  cascade reactions, 75
 zirconyl chloride, 75

B

1,3-Benzenebis(N,N-dibromosulfonamide),
 312
Benzo[b]furans
 C–H bond, 201
 electrophilic cyclization, 203
 furo[2,3-b]pyridines, 201
 intramolecular Heck reaction, 203
 Pd-catalyzed Suzuki–Miyaura reaction, 201
 phosphine-catalyzed domino reaction, 202
 rhodium-catalyzed C–C bond,
  formation 206–207
 three-component reaction, 204
Benzo[b]pyrrole. See Indoles
Benzo[c]furans, 207–208
Benzodifurans (BDFs), 22
Benzo[4,5]imidazo[1,2-a]pyrimidine, 402
Bischler–Napieralski reaction, 379
Bischler reaction, 163
1,3-Bis(2,4,6-trimethylphenyl)imidazol-
 2-ylidene, 10
Bis(phenylvinyl)quinoxaline, 47–48
Bohlmann–Rahtz cyclocondensation
 reaction, 353
Boron dipyrromethene (BODIPY) dyes,
 38–39
3-Bromoisoxazolines, 324
3-Bromopyridine, 3
Buchwald–Hartwig-type reactions, 429

C

Carbazoles, 171–172
Carboline analogs, 172–173
Carbon–nitrogen–oxygen rings
 Ag(I)-mediated self-assembly, molecular
  cage, 547
 aza-oxa cryptand, 548
 chiral benzenophane, 547–548

563

Carbon–nitrogen–oxygen rings (*Continued*)
  diazacrowns, 546
  fluorescence macrocyclic receptor, 547–548
  monoazacrowns, 546
  N,N'-bis(2-aminoethyl) derivative, 548
  rotaxanes and catenanes, 548–549
  templation process, 549–550
Cerium(IV) ammonium nitrate (CAN), 368
Cetyltrimethylammonium bromide (CTAB), 41
Chiral *N*-(diethoxymethyl)oxazolidinones, 341
4-Chloromethyl-1,3-dioxolanones, 309
β-Chloro *N*-oxides, 88
Chromenes and chromans
  4-acetamidochromenes, 460
  arylation, 460
  aryl propargyl ethers, 458, 460
  asymmetric epoxidation, 464
  carbonickelation, 461
  C—H bond, 464
  chiral proline catalyst, 462
  1,3-disubstituted naphthopyrans, 459
  domino aldol reaction, 462–463
  enantioselective synthesis, naphthopyrans, 463
  6-*endo-dig* intramolecular phenoxycyclisation, 461
  Pd-catalysed reaction, 462
  quinone methides, 463
  Ru-catalysed cycloisomerisation, 460
  salicylaldehydes, 459
  three-step one-pot reaction, 464–465
Chromones
  3-acrylates, 475
  alkyne-tethered aryl iodides, 476
  asymmetric intramolecular Stetter reaction, 476
  6,6-bisbenzannulated spiroketals, 475
  chiral fluorinated flavanones, 476
  flavones, 474–475
  isoflavones, 474
  UV-irradiation, 475
2-Cinnamoylmethylene-1,3-dithiolanes, 312
Clauson-Kaas reaction, 148
Coumarins
  2-(alkoxycarbonyl)benzenediazonium bromides, 474
  3-aroylbenzopyran-2-thiones, 470–471
  3-benzamidocoumarins, 470
  4-cyanocoumarins, 472
  3,4-dihydrocoumarins, 472
  4-hydroxycoumarin, 471
  microwave-assisted zincation, 472
  naphtho[2,1-*b*]pyran-2-ones, 471
  Pd-catalysed *ortho*-alkylation, 473
  salicylaldehydes, 470
  solid-phase synthesis, isocoumarins, 473
Crown ethers
  bis(crown ether), 539
  1:1 and 2:1 catenane, 540
  dibenzo[3*n*]crown-*n*, 539
  dicobalt hexacarbonyl complex, 538–539
  molecular cage, 539–540
  molecular tweezers, 539
Cyclocondensation reactions, 147

D

2,5-Dialkoxytetrahydrofurans, 147
Di- and tetrahydrofurans
  reactions, 186–188
  synthesis
    2,3- and 2,5-dihydrofuran derivatives, 195
    exo-methylene group, 197
    gold-catalyzed cycloisomerization, 198
    Pauson–Khand reaction, 197
    O-tethered enynes, 198
    tetrahydrofuran derivatives, 199–200
3,4-Diaryl-5-aminoisoxazoles, 324
Diaryl-prolinol, 325
3,6-Diaryl[1,2,4]triazolo[4,3-*b*]pyridazine, 439
Diazines and benzo derivatives
  pyrazines (*see* Pyrazines and benzo derivatives)
  pyridazines (*see* Pyridazines and benzo derivatives)
  pyrimidines (*see* Pyrimidines and benzo derivatives)
α-Diazomethanesulfonate, 185
Dichlorodicyanoquinone (DDQ), 430
Diels—Alder (D—A) reaction, 157
  1,3-dithioles and dithiolanes, 312
  furans, 191
  isoquinolines, 379
  oxazolidines, 341

oxazolines, 338
pyridazines and benzo derivatives, 396
pyridines preparation, 352
pyridinoids, 46
pyrimidines and benzo derivatives, 403
quinolines, 372
tetrazines, 436–437
triazines, 427
Dienyl sulfonamides, 186
Dihydroindenodiazepines, 503–504
Diketopyrrolopyrroles (DPPs), 39–40
Dimethylaminophenyl derivative, 47–48
2,5-Dimethyl-3,4-di-*tert*-butylfuran, 188
2,5-Dimethyl-*N*-tosylpyrroles, 147
2,7-Dioxabicyclo[2.2.1]heptane, 200
trans-Dioxasilacyclooctene, 69
Dioxetanes. See β-Lactones
Dioxetanones. See β-Lactones
1,3-Dioxoles and dioxolanes
 aluminium salen complex, 310
 DBU catalysed reaction, 311
 enantioselective synthesis, (*R*)-propylene carbonate, 310
 hydroxyacetophenones, 309
 X-ray structures, 310–311
Dipyrido[3,2-*a*:2′,3′-*c*]phenazine (dppz), 417–418
3,5-Disubstituted isoxazoles, 323
3,4-Di-*tert*-butylhex-3-ene-2,5-dione, 188
Dithienylethenes (DTEs), 50–51
1,3-Dithioles and dithiolanes
 carbonyl compounds reaction, 311
 charge transfer complex, 312, 314
 D–A reaction, 312
 dithiadiselenafulvalene, 314
 S-allyl dithiocarbamates, 311
 TTF, 312–314
 X-ray structure, 314
Dithioloids, 51–52
Diversity-oriented synthesis (DOS), 286
Domino reaction, 148

E

Edoxaban, 284
Eiden–Herdeis method, 353
Eight-membered and larger rings
 carbon–nitrogen–metal rings, 553–555
 carbon–nitrogen–oxygen–metal rings, 555–556
carbon–nitrogen–oxygen rings
 Ag(I)-mediated self-assembly, molecular cage, 547
 aza-oxa cryptand, 548
 chiral benzenophane, 547–548
 diazacrowns, 546
 fluorescence macrocyclic receptor, 547–548
 monoazacrowns, 546
 N,N'-bis(2-aminoethyl) derivative, 548
 rotaxanes and catenanes, 548–549
 templation process, 549–550
carbon–nitrogen–phosphorus rings, 552
carbon–nitrogen–phosphorus–sulfur rings, 553
carbon–nitrogen rings
 azamacrocycles, 541–543
 porphyrins, 544–545
 pyridine-containing macrocycles, 543
carbon–nitrogen–selenium rings, 552
carbon–nitrogen–sulfur–oxygen rings, 552–553
carbon–nitrogen–sulfur/phosphorus–metal rings, 556
carbon–nitrogen–sulfur rings, 550
carbon–oxygen–phosphorus rings, 551–552
carbon–oxygen rings
 calixarenes, 541
 crown ethers, 538–540
 cyclophanes, 541
carbon–oxygen–silicon rings, 551
carbon–selenium rings, 546
carbon–sulfur–oxygen rings, 550–551
carbon–sulfur rings, 545
Electron-withdrawing groups (EWG), 72
Electrophilic
 Friedel–Crafts acylation, 153–154
 regioselective bromination, 152
 Vilsmeier–Haack reaction, 153
Epoxides
 aldehyde, 63
 alkynyl, 70
 allylic, 67
 amino derivatives, 66
 anion derivatives, 68
 asymmetric allylic oxidation, 64
 t-Bu-perbenzoate, 64
 carbanions, 67–68

Epoxides (*Continued*)
  carbonyls, 65
  chalcone and chalcone derivatives, 60
  chiral dioxiranes, 61
  cyclic and tertiary allylic alcohols, 61
  cyclization reaction, 65
  diazocarbonyl derivative, 63
  Friedel–Crafts reaction, 70
  gem-dihydroperoxide, 62
  heterocyclic ring, 59
  intramolecular cyclization reaction, 64–65, 68–69
  Julia–Colonna-type, 62
  Mo-catalyst, 59–60
  radical reactions, 67
  RCM catalyst, 59
  Rh-catalyzed reaction, 70
  rhodium complexes, 70
  ring-opening reactions, 64–65
  silacyclopropane, 69
  Sonagashira coupling, 69
  sulfur ylide, 63
  tertiary allylic alcohol, 61
  thiol/selenol, 65
  titanium and vanadium, 60
  use of enzymatic, 62
4-Ethoxycarbonyl-5-methyl-2-(phenylsulfonyl)methyloxazole, 334–335
Ethylenedithio-tetrathiafulvalene (EDT-TTF), 338
Ethylidene-3-tosylmethyl-2,3-dihydrofuran, 190
2-Ethynylbenzyl alcohols, 465
Excited-state intramolecular proton-transfer (ESIPT), 46–47
3-Exo-methylenetetrahydrofuran, 198

F
Fischer indole synthesis, 164
Five-membered ring systems
  azaindoles (*see* Azaindoles)
  carbazoles (*see* Carbazoles)
  carboline analogs (*see* Carboline analogs)
  di- and tetrahydrofurans (*see* Di- and tetrahydrofurans)
  1,2-dioxolanes, 315
  1,3-dioxoles and dioxolanes (*see* 1,3-Dioxoles and dioxolanes)
  1,2-dithioles and dithiolanes, 315
  1,3-dithioles and dithiolanes (*see* 1,3-Dithioles and dithiolanes)
  dithioloids (*see* Dithioloids)
  furanoids (*see* Furanoids)
  furans (*see* Furans)
  imidazoles (*see* Imidazoles)
  imidazoloids (*see* Imidazoloids)
  indoles (*see* indoles)
  isothiazoles (*see* Isothiazoles)
  isoxazoles (*see* Isoxazoles)
  isoxazolidines (*see* Isoxazolidines)
  isoxazolines, 325–327
  oxadiazoles, 342–343
  1,2-oxathioles and oxathiolanes, 316
  1,3-oxathioles and oxathiolanes, 314–315
  oxazoles (*see* Oxazoles)
  oxazolidines (*see* Oxazolidines)
  oxazolines (*see* Oxazolines)
  oxindoles (*see* Oxindoles)
  pyrazoles (*see* Pyrazoles)
  pyrroles (*see* Pyrroles)
  pyrroloids (*see* Pyrroloids)
  selenazoles, 1,3-selenadolidines and telenazoles, 302–304
  spirooxindoles (*see* Spirooxindoles)
  synthesis
    benzo[*b*]furans, 201–207
    benzo[*c*]furans, 207–208
  tetrazoles (*see* Tetrazoles)
  thiadiazoles and selenodiazoles (*see* Thiadiazoles and selenodiazoles)
  thiazoles (*see* Thiazoles)
  thiophene (*see* Thiophene)
  thiophenoids (*see* Thiophenoids)
  three hetero atoms, 316
  1,2,3-triazoles (*see* 1,2,3-Triazoles)
  1,2,4-triazoles (*see* 1,2,4-Triazoles)
Fluorescence resonance energy transfer (FRET), 26, 31
β-Fluoropyrroles, 144
Four-membered ring systems
  azetidines and azetines (*see* Azetidines and azetines)
  β-lactams (*see* β-Lactams)
  β-lactones (*see* β-Lactones)
  silicon and phosphorus heterocycles (*see* Silicon and phosphorus heterocycles)
  thietanes and β-sultams, 98–99

Friedel—Crafts reaction, 70
Furanoids
  aminobenzodifurans, 23
  benzodifuranones, 22
  dyes, 23
  p-hydroxymandelic acid, 22
  tetracyclic naphthofuranones, 24
Furans. *See also* Di- and tetrahydrofurans
  reactions
    Achmatowicz oxidation, 185
    [4+2] cycloaddition, 182, 184
    Diels—Alder cycloaddition, 183—184
    enantiomeric excess, 184
    gold-catalyzed reaction, 184
    2-lithio-5-methoxyfuran, 185
    Michael addition, 185
    $PtCl_2$ catalysis, 183
    Rh-/Cu-catalyzed reaction, 185
    silyloxyfurans 186
    γ-spiroketal, 186
    triethylsilyl group, 184
  synthesis
    Ag-catalyst, 189
    Au-catalyzed cycloisomerization, 192
    3-bromo-2,5-disubstituted furans, 191
    copper-catalyzed domino reaction, 194
    Cu-catalyzed reaction, 191
    dehydrogenative coupling reaction, 192—193
    D—A reaction, 191
    electrophiles, 190
    furo[3,4-*c*]furanones, 188
    3-furylmethyl ether, 190
    Garcia-Gonzalez reaction, 188
    hydroarylation, 190
    Lewis acid-catalyzed reaction, 188
    Michael addition, 193
    Pd-catalyzed reaction, 195
    scandium triflate-catalyzed, β-ketoesters, 191
    tetra-*tert*-butylethylene, 188
Fused [6]+[5] polyaza systems
  nonpurine [6+5] fused systems, 439—440
  purines
    alkyldiamine derivative, 440
    benzylhydroxylamine, 440
    6-biarylamino purine, 442
    6-halopurine derivatives, 442
    6-[1(3*H*)-isobenzofuranylidene-methyl] purines, 441
    N-3 and N-9 aryladenines, 441—442
    one-pot synthesis, 442
    palladium-catalyzed cross-coupling reaction, 441
    "push—pull" purines, 442—443
    6,7,8-trisubstituted purines, 441
Fused [6]+[6] polyaza systems, 443

G
Gewald domino protocol, 131—132
Gould—Jacobs reaction, 371

H
trans-3,4-2*H*-1,4-benzoxazines, 75
Heck cyclisation, 495
Heck-type coupling reactions, 418
Hemetsberger—Knittel indole synthesis, 161
2-Heteroarylation, 156
Heterocyclic dyes
  applications, 21
  five-membered rings
    dithioloids, sulfur, 51—52
    furanoids, oxygen, 22—24
    imidazoloids, 40—42
    pyrroloids, nitrogen (*see* Pyrroloids)
    thiophenoids, sulfur, 49—51
  functional dyes, 21
  six-membered rings
    pyranoids (*see* Pyranoids)
    pyrazinoids, 47—48
    pyridinoids, 42—47
2,5-Hexanedione, 148
Hinsberg approach, 149
Huisgen cycloaddition, 236—237
Human histone deacetylase (HDAC), 89
3-Hydroxydihydroindole, 13
β-Hydroxy phosphonates, 65
3-Hydroxypyrroles, 145

I
Imidazoles
  2-aminoimidazoles, 227
  benzimidazoles, 227—228, 230
  benzoin-type condensation reaction, 231
  catalyzed/norbornene-mediated one-step synthesis, 234

Imidazoles (*Continued*)
  chemoselective reductive desulfurization, 227
  combinatorial synthesis, 235
  copper-catalyzed methods, 228–229
  cyclocondensation, 226
  direct C—H amination and arylation, 230
  6-*endo-dig*/5-*exo-dig* cyclization, 232
  imidazolopyridines, 233
  Kumada and Suzuki cross-coupling reactions, 230
  Kumada–Tamao–Corriu cross-coupling, 230
  microwave-accelerated tandem process, 234
  microwave-assisted hydrazinolysis, 227
  microwave-assisted solvent-free synthesis, 226
  N-arylation, 231
  N-heterocyclic carbene-catalysed oxidative carboxylation, 232
  nucleophilic substitution reactions, 229
  one-pot, four-component synthesis, 226
  one-pot synthesis, 233
  oxidative condensation-cyclization, 233
  palladium-catalyzed Heck-type coupling reactions, 231
  regio-and enantioselective iridium-catalyzed *N*-allylations, 229
  regioselective nucleophilic substitution, 233
  S-alkylation, 229
  solid-phase synthesis, 234
  Staudinger reaction, 232
  2-substituted-benzimidazoles, 228–229
  three-component condensation, 233
  tosylation, 231
  Ugi-type multicomponent reaction, 233
Imidazoloids
  A-D-A sensor, 41
  CTAB micellar system, 41
  cyanuric chloride-based reactive UV-absorber, 40
  Debus synthesis, 41
  photochromic molecules, 41
  photochromism, 42
  sulfonated benzimidazole component, 40
Imino-Diels–Alder reaction, 372
Indoles
  copper-catalysed couplings, 14–16
  cross-coupling reactions
    Negishi coupling, 4
    Stille coupling, 2–3
    Suzuki coupling, 3–4
    transmetallation and metallation, 4
  intermolecular approaches, 162–165
  intramolecular approaches, 160–162
  natural products, 173–174
  oxidative coupling reactions, 17–19
  palladium-catalysed couplings
    annulated indoles, 9
    azaindoles, 12
    base-free direct arylation, 7–8
    C-2 and C-3 arylation, 6–7
    cationic Pd complex, 8–9
    (pseudo)haloarene, coupling partners, 13
    N-acetyl-tryptophan methyl ester, 12
    NHC ligands 6–7
    N-methylindole, C-2 arylation, 5, 8, 10, 14
    phosphine-free system, 7
    room temperature C–H arylation, 10–11
    TEMPO, 13
  reactions
    benzene ring functionalization, 168
    C3/C2 substitution, 166–167
    nitrogen substitutions, 167–168
    pericyclic transformations, 165–166
    side-chains, 168–170
  rhodium-catalysed couplings, 16–17
Intramolecular nitrile oxide cycloadditions (INOCs), 322
Intramolecular Wittig-like reaction, 146
Isochromenes and isochromans, 465–466
Isoquinolines
  donor-acceptor cyclopropanes, 381
  crispine B, 379
  cyclization, 378–379
  Diels–Alder reactions, 379
  fused polycyclic isoquinoline generation, 378
  Katada-like rearrangement, 377–378
  Larock synthesis, 377
  pyridyl nitrenes, 382
  tricyclic pyrazolo[3,2-*a*]isoquinoline, 381–382
  zwitterionic species, 380–381
Isothiazoles

synthesis
  Davis oxaziridine, 295–296
  Oppolzer camphorsultam, 293–294
  oxidation reagent, 295
pharmaceutical applications, 296–297
reactions
  chiral homoallylic amines, 291
  copper-promoted intramolecular aminooxygenation, 290
  hydroamination, 292–293
  intramolecular cyclization and subsequent tautomerization, 291
  NHC-catalyzed homoenolate addition, 289
  ring-closing metathesis-based strategy, 291–292
  short multigram process, 292
synthesis
  benzosultams generation, 287
  intramolecular cyclization, 289
  intramolecular dialkylation, 288
  one-pot microwave-assisted reaction, 288–289
  RCM, 286–287
  tricyclic sultams, 287
Isoxazoles
  acylation reaction, 324
  aldoximes, 322
  3-amino-5-substituted isoxazoles, 324
  4-arylisoxazoles, 324–325
  biocompatible polymers, 323
  cycloaddition/condensation reactions, 321
  N-hydroxy-4-toluenesulfonamide, 323
  N-methylimidazole, 321
  solid-phase nitrile oxide-alkyne click reaction, 323
  sulfonamides, 322
Isoxazolidines
  casuarine synthesis, 328
  diamines, 329
  diastereoisomers, 330
  nitrones and alkenes, 327
  N—O bond, bicyclic amino alcohols, 328
  stereoselective synthesis, 330
Isoxazolines
  cycloaddition condensations, 325
  1,3-DC reactions, 327
  isoxazolino[4,5-c]quinolines, 326
  N-oxides, 325

J
Julia olefination, 276

K
4-Keto-4,5,6,7-tetrahydrobenzofurans, 203
Kinetic isotope effect (KIE), 6, 8
Knorr pyrrole synthesis, 146

L
β-Lactams
  fused and spirocyclic
    6-aminopenicillanic acid, 92
    Kinugasa reaction, 94
    non-traditional approach, 93
  monocyclic 2-azetidinones
    α-alkylidene-β-lactams, 90
    antiaromaticity, 91
    carnosine derivatives, 92
    human histone deacetylase isoforms, 89
    isoxazoline-fused cispentacins, 91
    Kinugasa reaction, 90
    one-pot synthesis, 89
    statin therapy, 88–89
    Staudinger ketene–imine cycloaddition reaction, 91
β-Lactones
  decomposition pathways, 96
  hydrolytic kinetic resolution, 95
  Lewis acid–Lewis base bifunctional catalyst, 97
  Paternò–Büchi reaction, 96
  quantum-chemical calculations, 95
  α-stabilized phosphorus ylides, 98
  Wacker-type reaction, 97
2-Lithio-5-methoxyfuran, 185

M
Metal-mediated cross-coupling reactions, 155
Methylenecyclopropane fragmentation, 148
Michael initiated ring closure (MIRC) reaction, 72
Mizoroki–Heck reaction, 361–362
Morita–Baylis–Hillman reaction, 375

N
N-acetylcolchinol, 500
N-acetyl-tryptophan methyl ester, 12

1,4,5,8-Naphthalenetetracarboxylic dianhydride (NTCDA), 43
1,8-Naphthyridine, 443
N-Bus-phenylaziridine, 77
N-(2-chloroethylidene)-*tert*-butylsulfinamide, 73
N-chlorosuccinimide (NCS), 8
Negishi cross coupling reaction, 4, 415–416
Nexium®, 349
N-heterocyclic carbene (NHC), 6–7
N-(2-hydroxyethyl)-amides, 335
N-hydroxyphthalimide (NHPI), 466
N-hydroxypyrroles, 147
N-methyl-7-azaindole, 12
N-methyl-2-(4-nitrophenyl)indole, 8
N-methylpiperidine (NMP), 321
N-(oxazol-5-yl)–2,2,2-trifluoroacetimidates, 333
N-protected indol-2-ylboronic acid, 3
N-sulfonyl-1,2,3-triazoles, 151
N-triisopropylsilylpyrrole, 158

O

One heteroatom
  azepines and derivatives, 491–492
  benzoazepines and derivatives
    annulation, amine, 497
    aromatic ring, 493, 495
    benzoazepinoquinoline, 498
    Bergman cyclisation, 496
    bisbenz-[b,d]-azepine, 498
    electrocyclisation, 497
    formation, cyclisation reactions, 497
    Heck cyclisation, 495
    intramolecular reaction, 494
    N-(alkyladamantyl)phthalimides irradiation, 498
    bulgaramine, 496
    phenyl ring, 494
    radical cyclisation, 493
    RCM, 495
    tethered hydrazine, cyclomethylation, 497
    tetrahydroisoquinolines ring expansion, 496
  fused azepines and derivatives
    chiral fused benzazepines, 494
    [5+2] cyclisation, 493
    gold-catalysed cyclisation, 492–493
    palladium-mediated cyclisation, 492, 494
    Pictet–Spengler cyclisation, 492
    RCM, 493
    Sakurai reaction, 493
  oxepines and fused derivatives
    bauhinoxepin, 501
    ciguatoxin synthesis, 499
    [2+2+2] cycloaddition approach, 500
    cyclodehydration, 500
    dibenzoxepines, 500–501
    gambieric acid synthesis, 499
    Knoevenagel condensation, 500–501
    lactonisation, 500
    metal-catalysed cyclisation, 502
    radical cyclisation, 501
    tricyclic fused oxepine, 502
  thiepines and fused derivatives, 502–503
Oppolzer camphorsultam, 293–294
Oxadiazoles, 342–343
Oxazoles
  5-acetyl group, 332
  5-acylaminooxazoles, 333–334
  5-arylisoxazoles, 334
  macrolides, 334–335
  one-pot procedure, 331
  Schöllkopf reaction, 331
Oxazolidines
  3-aminopropionitrile, 340
  chiral enamide, 342
  D–A reactions, 341
  enantiopure biciclic oxazolidines, 340
  regioisomer mixtures, 339
Oxazolines
  cationic ring-opening polymerization, 337
  D–A reactions, 338
  Henry reaction, 339
  N-allylacetamides treatment, 335
  2-substituted-2-oxazolines, 336
  Taft steric parameters, 337
Oxetanes. *See* β-Lactones
2-Oxetanones. *See* β-Lactones
Oxindoles, 170–171

P

Paal–Knorr reaction, 147
2,4-Pentanedione, 26
Pharmacological compounds

antibiotics, antitubercular and antiamoebia, 526
antiviral, 525–526
asenapine, 519–520
azepane-containing azelastine, 519–520
besifloxacin, 519–520
cancer cell proliferation, 524
cardiovascular and metabolics, 522–523
inflammatory diseases, 522
kinase inhibitors, DNA regulation, 525
metabolic disorders, 523
migraine, 522
neurodegeneration and cognition, 520–521
olanzapine, 519–520
1,4,5-oxadiazepane pinoxaden, 520
peptidomimetics, new targets, prodrugs, 525
pyrrolobenzodiazepines, 524–525
quetiapine, 519–520
tolvaptan, 519–520
urology, gastrointestinal, immunosuppression, wound healing, 523
women's health, 524
zilpaterol, 520
(N-tosylimino)phenyliodinane, 70
Pictet–Spengler reaction, 157, 379
Piloty–Robinson pyrrole synthesis, 146
Poly(ε-caprolactone) (PCL), 540
Prins cyclisation, 454–456
Pyranoids
alkylation-cyclization protocol, 25
iminocoumarin dye, 28
merocyanine-type dyes, 29
RBH, 26
Rhodamine B, 24, 28
"trimethyl lock" approach, 27
V-shaped nonlinear optical chromophore, 30
xanthene system, 24–25
Pyranones
bicyclo[3.2.2]nonadiene ring system, 467
cis-fused lactone, 469
2,3-dihydropyran-4-ones, 468
5,6-dihydropyranones, 467
5,6-dihydropyran-2-ones, 469
enantioselective hDA reaction, 468
imidazolium salts, 468–469
Rh-catalysed oxidative coupling, 467
Stetter reaction, 470
$TiCl_4$, rearrangement 468
Pyrans
alkynyl benzyl ethers, 454
Au-catalysed cycloisomerisation, 451
cross-conjugated enaminones, 450
cross-metathesis methodology, 457–458
[4+2] cycloaddition, enones, 450–451
cyclobutane diesters, 456
diastereoselective synthesis, 455
6-exo-dig selectivity, 452
Friedel−Crafts alkylation, 454–455
furans, 451–452
α-hydroxyalkyldihydropyrans, 454
intramolecular Prins reaction, 456
one-pot three-component reaction, 451
spirocyclisation, glycal epoxides, 457
three-component microwave assisted reaction, 453
2,6-trans-dihydropyrans, 452
Pyrazines and benzo derivatives
applications, 417–419
reactions
cycloaddition reactions, 417
lithium cadamates, 416
metal-catalyzed coupling reactions, 415
molybdenum complex, 418
Negishi cross coupling reaction, 415–416
quinoxaline derivative, 417
synthesis
1,3-diarylpyrrolo[1,2-a]quinoxalines, 414
dihydropyrazolopyrazine, 413–414
fluororubine derivative, 415
oxidation-annulation approach, 414–415
oxime precursor, 413
Pyrazinoids, 47–48
Pyrazoles
pyrazole-fused ring systems
larger fused pyrazoles, 225
pyranopyrazoles, 223
pyrazolopyridines, 223–224
pyrazolopyrimidines, 224
tricyclic fused pyrazole ring systems, 225
reactions and synthesis
C−H arylation, 221

Pyrazoles (*Continued*)
   cross-coupling reactions, 221
   domino reaction sequence, 218
   microwave irradiation, 220
   N-alkylation, 221–222
   N-arylation, 222
   one-pot synthesis, 218–219
   preparation, hydrazine, 217
   reagents, synthetic transformations, 223
   regioselective metalation, 222
   regioselective *O*-alkylation,
      indazolinones, 222
   regioselective synthesis, 217–218, 224
   solvent-free coupling reactions, 222
   Suzuki cross coupling approach, 219
   Suzuki–Miyaura cross-coupling/
      hydroxydeboronation reaction, 221
Pyridazines and benzo derivatives
   applications, 399–401
   reactions
      cardiotonic agents, 399
      C_{60} derivative, 398–399
      microwave conditions, 398
      spirodiazine preparation, 398
   synthesis
      cycloaddition chemistry, 397
      Diels–Alder reaction, 396
      enaminones reaction, 395
      hydrazine reaction, 394
      phthalazine, 394–396
      pyrazolocinnolines, 397
      tetraazapyrene, 394–395
Pyridines
   N-oxides and pyridinium salts
      2-aminopyridine, 367
      bromo- and iodopyridine magnesation, 368
      crykonisine and papaverine
         synthesis, 369
      halopyridine coupling substrates, 368
      Katada-like rearrangement, 367
      nucleophilic addition, 368
      quinolizinium triflates, 369–370
   preparation
      allylated amino acid derivatives,
         350–351
      aza-Diels–Alder reaction, 352
      azides use, 355–356
      cyclocondensation sequence, 354–355
      cyclopentenol, 355
      cyclotrimerization, 351
      Eiden–Herdeis method, 353–354
      Hantzsch synthesis, 353
      hydrazides, 352
      lyconadin A synthesis, 355
      multicomponent couplings, 352–353
      8-oxoerymelanthine, 356
      RCM methods, 350
   reactions
      acylpyridines, 365
      alkyl side-chains, 357–358
      allyl picolinates, 366–367
      aminopyridines, 362
      bromine-magnesium exchange, 359
      C–H activation, 365
      C–H arylation, 359–360
      C–O bond formation, 366
      copper chloride, 360
      2,6-diarylpyridine, 361
      2-formylpyridines, 364
      imidazo[1,2-*a*]pyridines, 363
      iodopyridines synthesis, 364
      Mizoroki–Heck reaction, 361–362
      nucleoside mimics, 363
      ortho-lithiation, 358–359
      piericidin A1, 358
      3-pyridylboronic acid, 360
      pyridyl nitrenes and azides, 363–364
      Suzuki–Miyaura reaction, 360–361
      Tamiflu®, 356–357
      vinyl side-chains, 357
Pyridinoids
   angular substitution motif, 45
   blue-emitting napthalimide moiety, 42
   cyclic voltammetry, 44
   Diels–Alder reaction, 46
   ESIPT, 46–47
   fluorescent dye, 46
   naphthalene bisimide dyes, 43
   perylenes, 44
   tetrasubstituted anthrazoline, 46
Pyrimidines and benzo derivatives
   applications
      cancer treatment, 412–413
      medicinal compounds, 410–411
      molecules, design and synthesis,
         411–412
      non-pharmaceutical applications, 413

reaction
  chloropyrimidine, 406–407
  C-nucleophiles, 409–410
  hexahydropyridopyrimidine, 408
  N-oxide, 408–409
  pyrrolopyrimidine, 408
  Suzuki–Miyaura reaction, 406–407
synthesis
  aza-Wittig reaction, 406
  Diels–Alder reaction, 403
  dihydroazolopyrimidines, 404
  monofluorinated quinazoline derivatives, 405
  multi-component preparation, 402
  quinazolinediones and quinazolinones, 405–406
  quinazolines, 401, 403–404
  spiro-isoindoline-quinazoline, 402–403
Pyrinadone A, 350
Pyrroles
  carbon, substitution reactions
    C–H activation, 155–156
    electrophilic, 152–154
    miscellaneous, 157
    organometallics, 154–155
    ring annelation, 156–157
    side-chain functionalization of, 157–158
    stereoselective, electrophilic, 154
  intermolecular approaches
    multi-component reactions, 147–150
  intramolecular approaches, 144–145
  nitrogen, 151–152
  transformations, 150–151, 158
Pyrrolo[1,2-a]quinolines, 149
Pyrroloids
  BODIPY dyes, 38–39
  carbazoles, 36–37
  "catch and release" protocol, 32
  diiodopyridine, 37
  dipyrroloazo dye, 40
  3,6-disubstitution motif, 37
  DPPs, 39–40
  FRET, 31
  indocyanine dye synthesis, 31
  indolocarbazole, 38
  indolospirobenzofurans, 34
  Lawesson's reagent, 32–33
  leuco-merocyanine equilibrium, 35
  malonaldehyde dianilide, 32
  N-substituents, 34
  pyrimidine-fused benzoindolene, 33
  squarylium dyes, 34
  symmetrical cyanines, 33
  synthetic strategy, 33
  tartaric acid, 37
Pyrylium salts, 466–467

Q

Quinolines
  preparation
    alkaloid isolation, 372–373
    aniline, 371
    electrophilic process, 373
    Friedländer synthesis, 370–371
    hetero-Diels–Alder reaction, 372
    4-hydroxyquinoline, 371
    3-hydroxy-4-substituted-quinolinones, 371–372
    multi-component reaction, 371
    telluro groups, 373–374
  reaction
    cinchona alkaloids, 376
    Heck reaction, 374
    hydrogenation/dehydrogenation, 375
    Meth-Cohn synthesis, 375
    Minisci reaction, 374
    multicomponent reactions, 375
    quinine synthesis, 376
Quinoxaline, 47–48

R

Rainbow perylene monoimides, 46
Rhodamine B hydrazide (RBH), 26
Rhodium-catalyzed coupling, 172
Ring-closing enyne metathesis (RCEM), 286
Ring closing metathesis (RCM), 145, 286, 350
  benzazepines and derivatives, 495
  epoxides, 59
  fused azepines and derivatives, 493
  pyridines, 350

S

Selenazoles, 1,3-selenadolidines and telenazoles
  chlorination and diazotization, 303
  cycloaddition reaction, 302

Selenazoles, 1,3-selenadolidines and telenazoles (*Continued*)
  one-pot four-step procedure, 303
  sugar-derived 2-amino-1,3-selenazole, 303–304
  tandem addition-cyclization reactions, 304
Selenophenes/tellurophenes (Se/Te)
  ASP, 136
  multifunctional dihydroselenophenes, 134
  2-selenophen-2-pyrroles, 135
  semiconducting copolymers, 135
  Stille coupling, 134
Seven-membered rings
  N, S and O systems, 517–518
  one heteroatom (*see* One heteroatom)
  pharmacology (*see* Pharmacological compounds)
  two heteroatoms (*see* Two heteroatoms)
Silicon and phosphorus heterocycles
  triphosphacyclobutadiene, 100
  vinylidene ruthenium complexes, 101
  Woollins' reagent, 100
Singulair®, 349
Six-membered ring systems
  benzo derivatives
    pyrazines (*see* Pyrazines and benzo derivatives)
    pyridazines (*see* Pyridazines and benzo derivatives)
    pyrimidines (*see* Pyrimidines and benzo derivatives)
    [1]benzopyrans and dihydro[1]benzopyrans (*see* Chromenes and chromans)
    [2]benzopyrans and dihydro[2]benzopyrans (*see* Isochromenes and isochromans)
    chromones, 474–476
    coumarins, 470–474
    dioxins and dioxanes, 480–482
    dithianes and trithianes, 482–483
    isoquinolines (*see* Isoquinolines)
    oxathianes, 483–484
    pyranoids (*see* Pyranoids)
    pyranones, 467–470
    pyrans (*see* Pyrans)
    pyrazinoids (*see* Pyrazinoids)
    pyridines (*see* Pyridines)
    pyridinoids (*see* Pyridinoids)
  pyrylium salts, 466–467
  quinolines (*see* Quinolines)
  supramolecular chemistry, 382–384
  thiopyrans and analogues, 478–480
  trioxanes and tetraoxanes, 482
  xanthones and xanthenes, 477–478
Sonogashira coupling reactions, 85, 161
Spirooxazolinoisoxazolines, 336
Spirooxindoles, 171–172
Stille coupling, 2–3
β-Sultams, 98–99
Supramolecular chemistry
  calixpyridines, 384
  chemically controlled rotary switch, 383
  heterocycles, 382
  metal-specific ion sensors, 383
  nanomolecular devices, 383–384

T

Tetrabutylammonium fluoride (TBAF), 539–540
Tetracyclic naphthofuranones, 24
2,2,6,6-Tetramethylpiperidine N-oxyl radical (TEMPO), 13
Tetrathiafulvalene (TTF), 52, 312, 539
Tetrazines
  cycloaddition reaction, 436–437
  D–A reaction, 436–437
  3,6-diamino-1,2,4,5-tetrazine, 438
  8,9-diazafluoranthenes, 436
  furopyridazines formation, 437
  heteroatom/aromatic substituted 1,2,4,5-tetrazines, 437
  high energy materials, 438
  ligand, 437–438
  pyridazine difluoroboranes, 436–437
  triruthenium derivatives, 438
  vinyltetrazine derivative, 436
Tetrazoles
  antimony trioxide and cadmium chloride, 246
  azido and nitro-containing tetrazoles, 247
  natrolite zeolite, 246
  nitriles treatment, 245–246
  reagent, 247
  three-component condensation, 246
Thiadiazoles and selenodiazoles
  pharmaceutical applications, 301–302
  reactions

aryl-aryl and aryl-alkyl selenides, 301
chemoselective reductive
  desulfurization, 300
sequential selective cross coupling,
  299–300
syntheses
  5-amino-substituted 3-phenyl-
    1,2,4-thiadiazoles, 297
  cyclization reaction, 297
  microwave irradiation, 299
  one-pot synthesis, 298
  oxidative condensation, 299
  oxidative dimerization, 297–298
  solid-phase parallel synthesis procedure,
    298
Thiazoles
  natural products
    apratoxin, 283–284
    hoiamide A, 281–282
    micrococcin P_1 (MP_1), 283
    neobacillamide A, 281, 283
    nocardithiocin, 281–282
    pretubulysin, 281, 283
    sanguinamides A and B, 281, 283
    thiazomycins B-D, 281–282
  pharmaceutical applications, 284–285
  reactions
    deprotonation, 268
    direct amination, 269
    Glaser–Hay reaction, 269
    halogen dance reaction, 267–268
    2-iodobenzothiazole, 268–269
    nickel-catalyzed C-2 arylation, 271
    palladium-catalyzed direct arylation,
      271–273
    palladium-catalyzed direct C-5
      heteroarylation, 270
    regioselective halogenation, 268
    zwitterions formation, 273–274
  synthesis
    aldol reactions, 274–275
    cycloaddition reaction, 262–263
    1,4-dicarbonyl compounds, 261
    1,4-dioxaspiro[2.2]pentanes, 260
    Friedel–Crafts reaction, 277
    fused thiazoles, 263
    glycosylation, 277
    Hantzsch reaction, 259–260
    intramolecular nucleophilic
      substitution, 262
    Julia olefination reaction, 275–276
    nucleophilic substitution, 261
    stereoselective synthesis, 275
  thiazolines
    cyclocondensation, 266–267
    intramolecular Michael addition, 265
    Kelly's method, 264
    molybdenum(VI) oxides-catalyzed
      dehydrative cyclization, 265–266
    Pattenden's approach, 266
  thiazolium and thiazolinium salts reaction
    alkylation, 280
    coupling product, 278–279
    [3 + 2] cycloaddition, 279–280
    deprotonation, 280–281
    Stetter reaction, 278, 279
Thietanes and β-sultams, 98–99
β-Thiolactones, 99
Thiophene
  and benzothiophenes elaboration
    bromothiophenes, 121
    carbene, 121
    condensation, 121–122
    cross-coupling reactions, 117
    diastereoselective domino reactions, 126
    facile synthesis, 118
    Friedel–Crafts acylations, 126
    functionalization, 123–124
    gold and lead, 123
    ligand-less palladium-catalyzed
      method, 117
    metal-free oxidation coupling, 117–118
    MIDA boronic acid, 119
    stereochemical and regioselective
      syntheses, 124–125
    Suzuki–Miyaura cross-coupling, 118–120
    thiophenecarboxaldehyde, 123
    TMPZnCl·LiCl, 120
  reactions, 109–110
  derivatives, medicinal chemistry, 131–134
  fused thiophene
    flash vacuum pyrolysis, 116
    synthons synthesis, 116
    TMS, 115
  material science
    fusion, diarylamino groups, 128

Thiophene (*Continued*)
   heteroacene, 127–128
   monomers synthesis, 129–130
   oligomers, 130–131
   semiconducting polymers, 130
   zirconium-mediated coupling reaction, 129
  synthesis
   condensation, fluorobenzonitrile, 114–115
   Gewald synthesis, 112–113
   iodine-mediated cyclization reaction, 114
   microwave-assisted thiophene ring synthesis, 113
   one-pot four-step synthesis, 111–112
   one-pot procedure, 110
   one-pot three-step synthesis, 112
   Pummerer reaction, 111
   regioselective synthesis, 112
   solution-phase parallel synthetic methodology, 114
   sulfanylation-acylation, 111
Thiophenoids
  DTEs, 50–51
  merocyanine dye, 49
  photovoltaic devices, 49
  thiophene ring, 50
  triazinyl ring, 49–50
Thiopyrans
  chalcones, 480
  4-dicyanovinylthiochromans, 480
  isothiochromene-1-thione, 479–480
  2-methylthiaadamantane, 479
  nitrobutadienes, 478
  one-pot three-component Pd/Cu, 479
Three-component reaction, 150
Three-membered ring systems
  aziridines (*See* Aziridines)
  epoxides (*See* Epoxides)
2,3,5-Triarylpyrroles, 147
Triazines
  1,2,3-triazines, 427
  1,2,4-triazines
   5-aryl-2-thienylpyridines, 429
   derivatives, diazoimides reaction, 430
   D–A reaction, 427
   louisianin family, 427–428
   nucleophilic aromatic substitution, 431
   one-pot reaction, 428
   organolithium compounds, 430
   palladium-catalyzed amination, 429
   pyridines synthesis, 428
   triazinyl nitronyl nitroxide, 430
  1,3,5-triazines
   bifunctional triazine-based ligands, 434–435
   C_3 building block, 433
   cross-couplings, 432–433
   dendrimers, 434
   dienophiles, D–A reactions, 431–432
   diphenylphosphanyl group, 434
   hydrogen bonding, 435–436
   recyclization reactions, 431
   ruthenium(II) complexes, 431
   Suzuki couplings, 433
   traceless solid-phase synthesis, 434
1,2,3-Triazoles, 150
  benzotriazoles, 238, 240
  click chemistry, 235, 240–241
  click reaction, 235–237
  copper-catalyzed 1,3-dipolar cycloaddition reaction, 235–236
  domino Michael-ketalization method, 240
  fused systems, 241
  in situ formation, 237
  microwave-assisted [3+2] cycloadditions, 238
  one-pot microwave-assisted solvent free synthesis, 235
  organic azides, 237
  palladium-catalyzed direct arylation, 239
  primary amines, one-pot procedure, 237
  reagents, 240
  Suzuki cross-coupling reaction, 239
1,2,4-Triazoles, 245
  copper-mediated direct arylation, 243
  cyclocondensation, 241–242
  iodogen, 242
  microwave-assisted organic synthesis, 244
  microwave-assisted reaction, 242
3-Trifluoroacetyl-4,5-dihydrofuran, 186
3-Trifluoroacetylpyrrole, 151
Trifluoroacylation, 153
Two heteroatoms
  diazepines and fused derivatives